PULP AND PAPER MANUFACTURE
THIRD EDITION

VOLUME 7

PAPER MACHINE OPERATIONS

Technical Editor

Benjamin A. Thorp
Vice President Engineering
James River Corporation
Richmond, VA, USA

Series Editor

Dr. Michael J. Kocurek, *Director*
Herty Foundation Research & Development Center
Savannah, GA, USA

Contributing Authors

B.W. Attwood, G.A. Baum, J.A. Charles, W.G. Corboy, Jr., K. Cutshall, M.M. Cutts, R.P. Derrick, C.B. Edgar, K.G. Frye, K.C. Hill, V.K. Holland, M.C. Johnson, C.P. Klass, D.W. Manson, M.D. Moskal, D.P. Newcombe, H. Paulapuro, J.D. Pfeiffer, R.A. Reese, R.H. Reeves, R.H. Root, N. Ryti, W.E. Scott, B.A. Stowe, B.A. Thorp, D.A. Wensley, R.H. Van Eperen, R.C. Vonderau, J.A. Zwinak

Published by the
Technical Section
Canadian Pulp and Paper Association,
1155 Metcalfe Street, Suite 1900
Montreal, QC, Canada H3B 4T6
Tel. (514) 866-6621
Fax (514) 866-3035
e-mail: techpubs@cppa.ca

PULP AND PAPER MANUFACTURE

A series of volumes detailing the principles and practice of pulping and papermaking for mill staff and university students by contributors from the manufacturing, academic and supply sectors of the industry.

THIRD EDITION

Volume 1. Properties of Fibrous Raw Materials and their Preparation for Pulping
Volume 2. Mechanical Pulping
Volume 3. Secondary Fibers and Non-Wood Pulping
Volume 4. Sulfite Science & Technology
Volume 5. Alkaline Pulping
Volume 6. Stock Preparation
Volume 7. Paper Machine Operations
Volume 8. Coating, Converting, and Specialty Processes
Volume 9. Mill Control & Control Systems
Volume 10. Mill-Wide Process Control & Information Systems

Prepared under the direction of
THE JOINT TEXTBOOK COMMITTEE OF THE PAPER INDUSTRY

PUBLISHING HISTORY OF THE SERIES

Initiated in 1918 at a joint meeting of the Eduction committees of the Technical Section, CPPA and of TAPPI.
Published under the Joint Textbook Committee in the following editions:

1921	The Manufacture of Pulp and Paper	1st ed., 5 vol., Ed. J.N. Stephenson	McGraw-Hill
1925	The Manufacture of Pulp and Paper	2nd ed., 5 vol., Ed. J.N. Stephenson	McGraw-Hill
1930	La fabrication des pâtes et papiers	1st ed., 5 vol., Trans. L. Vidal	McGraw-Hill
1938	The Manufacture of Pulp and Paper	3rd ed., 5 vol., Ed. J.N. Stephenson	McGraw-Hill
1950	Pulp and Paper Manufacture	1st ed., 4 vol., Ed. J.N. Stephenson	McGraw-Hill
1969	Pulp and Paper Manufacture	2nd ed., 3 vol., Eds. R.G. Macdonald, J.N. Franklin	McGraw-Hill

PULP AND PAPER MANUFACTURE VOL. 7
PAPER MACHINE OPERATIONS

With bibliographic references and index.

REPRINTED 1998

Canadian Cataloguing in Publication Data
Main entry under title:
Pulp and paper manufacture

3rd ed.
Includes bibliographical references and indexes.
Partial contents: v. 7. Paper machine operations.
ISBN 0-919893-04-X (set) — ISBN 1-895288-13-4 (v. 7).

1. Paper industry. 2. Wood-pulp
I. Kocurek, M.J. II. Joint Textbook Committee of the Paper Industry of the United States and Canada.

TS1105.J65 1983 676 C83-090155-8

CONTENTS

PREFACE vii

I. PAPER PROPERTIES AND TESTING PROCEDURES **1**
A. Introduction 1
B. Structural properties 2
C. Strength properties 7
D. Liquid absorption properties 13
E. Optical properties 15
F. Electrical properties 19
G. Relative humidity and properties 20
H. Standardization and calibration 22
I. Future trends in testing 24
REFERENCES **25**
BIBLIOGRAPHY **25**

II. FIBROUS RAW MATERIALS FOR PAPERMAKING **26**
A. Introduction 26
B. Sources of papermaking fibers 26
C. Linking pulp to paper properties 27
D. Role of pulping process 32
E. Pulp suitability for papermaking 35
F. Pulp/fiber sourcing 41
G. Basic fiber features 44
H. Pulpwood characteristics 48
I. Summary 51
REFERENCES **52**

III. SHEET STRUCTURE CONSIDERATIONS — PAPER AS AN ENGINEERED MATERIAL **54**
A. Introduction 54
B. Consolidation of the web 56
C. Characterization of the sheet structure 60
D. Physical properties 64
E. Refining 66
F. Stock delivery 68
G. Wet pressing 69
H. Draws 70
I. Drying 73
J. Convertibility 75
K. End use performance 78
L. Closing comments 82
M. Acknowledgements 83
REFERENCES **83**

IV. STOCK AND WHITE WATER SYSTEMS **85**
A. Introduction 86
B. Short circulation 86
C. Long circulation design principles 89
D. Closing the white water system 92
E. Stock and water tanks 94
F. Process management 95
REFERENCES **96**

V. ROLL HEADBOXES AND THEIR APPROACH FLOW **97**
A. Introduction 97
B. Flow approach system 97
C. Manifold design 100
D. Headbox design 100
E. Slice evolution 104
F. Distributor rolls 110
G. Liquid level control 111
H. General 114
REFERENCES **116**

VI. HYDRAULIC HEADBOXES **117**
A. Introduction 117
B. Headbox design 119
C. Evaluation of headboxes 129
D. Headbox operation 130
E. Headbox cleanliness — troubleshooting 134
F. Jet exit angle and point of impact 134
G. Lip opening 135
H. Skimmer operations 135
I. Head control 136
J. Header control 137
REFERENCES **139**

VII.	**PAPERMAKING CHEMISTRY**	**140**
A.	Introduction	140
B.	Colloid chemistry concepts important to papermaking chemistry	141
C.	Applications of colloid chemistry concepts to papermaking chemistry	146
D.	Colloidal aspects of sheet formation	154
E.	Internal sizing	160
F.	Dry strength agents	164
G.	Coloring materials	170
H.	Interfering substances ("cationic demand" or "anionic trash")	174
I.	Association colloids	174
J.	Deposit control in papermaking	177
K.	Alkaline papermaking	178
L.	Basic papermaking chemistry laboratory measurements	179
M.	First-pass retention performance of a paper machine	182
N.	Laboratory investigation of first-pass retention	184
O.	Papermaking chemistry process control	185
REFERENCES		**187**
VIII.	**FOURDRINIER PAPERMAKING**	**192**
A.	History	192
B.	The modern fourdrinier machine	194
C.	Wire flow and drainage	202
D.	Formation	207
E.	Acknowledgements	214
REFERENCES		**215**
IX.	**PRINCIPLES OF TWIN-WIRE AND MULTIPLE-WIRE FORMATION AND DRAINAGE**	**216**
A.	Introduction	216
B.	Definitions	217
C.	Selected sheet properties	218
D.	Setting the jet	221
E.	Gap roll former	222
F.	Tissue formers	223
G.	Gap blade forming	226
H.	Hybrid or preformers	228
I.	Gap roll/blade forming	233
J.	Two-wire and/or multiple formers	234
REFERENCES		**236**
X.	**MULTI-PLY FORMING**	**237**
A.	History	237
B.	Fundamentals of multi-ply forming	239
C.	The structured web	239
D.	Multi-ply forming equipment	239
E.	Stock delivery	239
F.	Forming equipment	240
G.	Multi-ply forming methods	240
H.	High-consistency forming	254
I.	Dry forming	255
J.	Multi-ply paperboard and paper products	255
K.	Multi-ply paper grades	258
REFERENCES		**259**
GENERAL READING		**259**
XI.	**PAPERMAKING OPERATIONS**	**260**
A.	Introduction	260
B.	Press mechanical aspects	263
C.	Types of presses	265
D.	Sheet transfer from wire to press section	270
E.	Types of press arrangements	272
F.	Press configuration considerations	277
G.	Press clothing	278
H.	Press section variables	278
REFERENCES		**281**
BIBLIOGRAPHY		**281**
XII.	**PAPER DRYING**	**282**
A.	Introduction	282
B.	Conventional cylinder drying	282
C.	Basic drying theory	283
D.	Evaporation	285
E.	Steam and condensate system	287
F.	Dryer ventilation system	293
G.	Hoods and hood exhaust	297
H.	Pocket ventilation	298
I.	Dryer section runnability	301
REFERENCES		**305**
XIII.	**SURFACE SIZING**	**306**
A.	Introduction	306
B.	Starch as a surface size	307
C.	Starch preparation	307
D.	Film formers other than starch	311
E.	Wax emulsions	311
F.	Other specialty surface sizes	312
G.	Conventional size press configurations	312
H.	Size press design improvements	315
I.	Modified size press designs	316
J.	Alternatives to the size press	318
K.	Effects on sheet properties	318
L.	Effects on drying and machine speed	319
M.	Web handling and drying of the surface-sized sheet	320
N.	Calender sizing of paperboard	320
REFERENCES		**322**

XIV. YANKEE DRYERS 323
A. Introduction 323
B. The tissuemaking process 324
C. Design and construction 325
D. Steam systems and condensate removal equipment 326
E. Bearings and lubrication 327
F. Design stresses 328
G. Dryer safety 330
H. Shell metallurgy 331
I. Deformation analysis 332
J. Surface wear model 334
K. Yankee drying 335
L. The creping process 338
M. Summary 340
N. Nomenclature 340
REFERENCES 341

XV. WINDING 343
A. Winding definition 343
B. Winding theory 344
C. Equipment definition 372
D. Winding processes 376
E. Winder structure control types 378
F. Structure control function by winder type 379
G. Wet control function 390
H. Slitting function 406
I. Unwinding functions 423
J. Productivity and manpower 426
GLOSSARY 431
REFERENCES 435

XVI. CALENDERING 436
A. Introduction 436
B. Calender roll construction 441
C. Heating systems for metal rolls 441
D. Heated oil for temperature and anti-deflection control 442
E. Induction heating 443
F. Radiant heating 444
G. Rolls with non-metallic coverings 444
H. The operating variables for calender stacks 445
I. Sheet variables 446
J. Calender machinery variables 447
K. Reel building 455
L. Supercalenders: special features 456
M. Gloss calenders, soft calenders, temperature gradient calenders, and substrata thermal molding 457
REFERENCES 461

XVII. BROKE SYSTEMS 462
A. Introduction 462
B. Broke system design 462
C. Broke system control 463
D. Dry broke 466
E. Broke storage 468
F. Wet broke 469
G. Finishing broke 470
H. Coated broke 470
I. Colored broke 471
J. Wet strength 471
REFERENCES 471

XVIII. CROSS-DIRECTION CONTROL 472
A. Introduction 472
B. Benefits of improved CD uniformity 474
C. Defining the problem (opportunity) 481
D. CD basis weight control 487
E. CD moisture control 490
F. CD caliper control 500
G. CD coat weight control 501
H. CD smoothness/gloss control 502
I. Acknowledgement 502
REFERENCES 503
BIBLIOGRAPHY 504

XIX. PAPER MACHINE SHOWERS AND DOCTORS 507
A. Introduction 507
B. Showers 507
C. Process water systems 512
D. Shower applications 513
E. Doctors 518
F. Doctor applications 519
G. Doctor installation 521
H. Doctor maintenance 524
I. Summary 524
REFERENCES 524

XX. PAPER MACHINE AND WINDER DRIVES 525
A. Types of drives 525
B. Types of regulators 527
C. Operator's control functions 528
D. Reliability and maintenance 529
E. Power requirements 530
F. Press sections 538
G. Dryer section 541
H. Calender section 543
I. Reel section 543
J. Winder drives 544
K. Summary 546
USEFUL REFERENCES 547

XXI. PROPER DESIGN OF PAPER MACHINE VACUUM SYSTEMS **548**
A. Introduction 548
B. Basic fundamentals of vacuum systems 548
C. Vacuum generating equipment 550
D. System requirements 557
E. Piping design 561
REFERENCES **562**

XXII. PAPER MACHINE CLOTHING **563**
A. Forming fabrics 564
B. Press fabrics 571
C. Dryer fabrics 585
REFERENCES **593**
BIBLIOGRAPHY **593**

XXIII. PAPER MACHINE AND SUCTION ROLL CORROSION **594**
A. Paper machine corrosion 595
B. Suction roll corrosion 604
REFERENCES **613**

XXIV. PAPER MACHINE AND ROLL VIBRATION **615**
A. Introduction 615
B. Fundamentals of vibration theory 616
C. Vibration measurement techniques and equipment 622
D. Examples of paper machine vibration 624
E. Control of vibration and its impact 631
F. On-line vibration monitoring systems 636
G. Summary 638
REFERENCES **638**

XXV. COVERED ROLL CARE AND GRINDING **640**
A. Introduction 640
B. Roll handling 641
C. In-service operation 642
D. General grinding instructions 648
REFERENCES **649**

XXVI. PAPER MACHINE SAFETY **650**
A. Introduction 650
B. Paper machine safety 650
C. General practices for operation and maintenance 651
D. Noise safety 656
E. Paper machine lighting 658
F. Recommended safety reminders 658
G. A closer look at the paper machine 659
REFERENCES **679**
BIBLIOGRAPHY **679**

INDEX **681**

METRIC CONVERSION CHART **693**

PREFACE

The previous edition of the Pulp and Paper Manufacture Series was valued and used for nearly 20 years. The enduring usefulness of this unique reference is a tribute to the authors. Their presentation of the fundamental principles and practices of technology has been a major contribution to the industry.

The longevity of previous editions of this series indicated that this volume should stress fundamentals, as these change slowly with time. The "what do we do", and "why we do it" received considerable emphasis. One goal was to develop material that would be meaningful to a spectrum of readers from the young process engineer to the experienced manager who, perhaps, has just moved into the paper mill from another part of the facility or from a different industry. A second goal was to present expert opinion and fact. That led to the process of selecting a recognized authority in each unit process with both theoretical and practical knowledge of the field wherever that additional knowledge was advantageous. The belief is that authors with this rounded background have a unique insight to be shared with the reader. An attempt was made to use worldwide expertise, and the reader will find several nationalities represented. A third goal was to develop material that would be of "immediate practical value" to the reader. Much of the technical detail was left to the references, which can be used to learn much more about a specific area of interest.

The plot in this volume is a relatively straightforward walk through the "unit processes" of papermaking, with the exception of the last chapter – Safety. Safety must receive more emphasis in our industry. Safety must begin as a fundamental value for all of us, if we are, or hope to be, role models in our industry. Professors can kindle this process of awareness by making the safety material required reading.

I am indebted to many. First, to the Joint Textbook Committee for giving me the opportunity of putting this volume together. Second, to T. C. Kennedy of BE&K and to R. C. Williams of James River who supported my effort. I am indebted to many in the industry who have been mentors or role models. In addition to Ted and Bob, I am indebted to the late Irv Peters of Huyck Felt, Gordon Chalmers of Valmet-Enerdry and Jim McNamee the "godfather" of the respected TAPPI Wet End Operations Short Course. There is a large debt to the TAPPI members and staff who have helped me learn to serve.

We owe sincere thanks and appreciation to the authors who selflessly distilled their hard earned years of experience and expertise into succinct contributions for the benefit of all. Sincere thanks and appreciation go to the expert peer reviewers who enriched the material with their profound insights and challenges. Finally, my thanks and appreciation go to Michael Kouris whose encouragement and facilitating skills helped pull it all together.

B.A. Thorp
Volume Editor

Series Editor:

MICHAEL KOCUREK

Dr. Michael Kocurek is Director of the Herty Research & Development Center in Savannah, GA and formerly Professor and Chairman of the Paper Science & Engineering Department at the University of Wisconsin — Stevens Point. He has been a central figure in the education of the industry's engineers and scientists for 17 years. Dr. Kocurek received his B.S., M.S. and Ph.D in Paper Science and Engineering from the SUNY College of Forestry at Syracuse University. He has served as a consultant/instructor to over 30 different pulp and paper companies. He has been, and still is, very active in TAPPI activities, particularly in the area of continuing education and professional development. He is a member of the Joint Textbook Committee and has served on the Board of Directors of TAPPI. He was elected a TAPPI Fellow in 1985. Dr. Kocurek is the author of numerous papers.

Technical Editor:

BENJAMIN A. THORP

Benjamin Thorp received a B.S. in Physics from the University of Maryland and also did graduate studies in Mechanical Engineering at the University of Bridgeport, in Management Engineering at Rensselaer Polytechnic Institute, and in Marketing at the University of Tennessee. In 1964, he joined the Huyck Corporation, and worked for them in various locations, starting as a Development Engineer, and working up to the position of Vice President & Director of Research, responsible for worldwide research and development of paper machine equipment and clothing. Between 1981 and 1984, Mr. Thorp was President of Poyry-Bek Inc. (now Jakko Poyry Inc.), and in 1984 he became the Vice President, Engineering, of its parent company, BE&K Engineering, in Birmingham, AL. Since 1985, he has been the Vice President, Engineering, at James River Corporation in Richmond, VA, where he is responsible for corporate engineering, manufacturing technology and environmental affairs. Mr. Thorp is a holder of a number of U.S. Patents. He has been an active member of TAPPI and was elected a TAPPI Fellow in 1986.

I

PAPER PROPERTIES AND TESTING PROCEDURES

R.H. VAN EPEREN

Roger Van Eperen is a Technical Specialist in the Research Services Division of the Institute of Paper Science and Technology in Atlanta, GA. He has more than 40 years experience in measurement technology for paper and paper products. He is a lecturer at CE courses on property measurement, has presented papers at seminars, and is author or coauthor of numerous technical articles and publications. He is an active member of the Process and Product Quality Division of TAPPI, serving as division chairman in 1986.

REVIEWED BY

William E. Scott
Professor & Chairman
Paper Science and Engineering Dept.
Miami University
Oxford, OH

A. INTRODUCTION

Instruments for measuring paper properties, along with procedures for making tests, have been in existence for well over 100 years. Prior to the introduction of test instruments, decisions on paper purchases were based on subjective handling tests. Since that time hundreds of paper properties have been defined, and most of these can now be measured by instrumental means.

It is not the intent of this chapter to provide exact procedures for measuring these properties as such information is already available in TAPPI test methods and from other sources. Rather, it is hoped the information herein will be helpful in understanding the significance of selected properties which are common throughout the paper industry. In pursuing this objective, the attempt is made to point out areas where our ability to define certain properties by instrumental measurements may need improvement. An example is the common practice of defining the surface of paper for printing purposes by air-leak smoothness tests. The reader may disagree with some of the opinions expressed and, in some cases, that is desirable. Such disagreements may stimulate debate leading ultimately to improved methods of paper testing.

1. Why test paper?

Apart from the obvious need for a means to define the product, paper tests are made for a variety of reasons. Among these are:

- Quality control of the manufactured product.
- Meeting customer specifications.
- Responding to customer complaints.
- Measuring progress in product development and research.

In each of these cases, different testing methodology may be used to meet specific needs.

Testing for quality control purposes, for example, often requires that the test data be obtained in a relatively short period of time, perhaps before a newly manufactured roll can be shipped to the customer. In these cases the required conditioning time may be shortened, or the test procedure otherwise modified, to speed up the generation of test data. The accuracy of the test results may then suffer, but the advantages of speedy testing usually outweigh the disadvantages of lost accuracy.

Testing to answer customer complaints requires the laboratory to be particularly innovative. The first step, and usually the simplest (although it often will not resolve the problem) is to retest the paper to ensure that specified properties are within required tolerances. Careful observation of the sample during this testing will often yield more useful information than the actual test results. Next, the complaint needs to be reviewed both internally and with the customer to verify its validity. Perhaps the customer has modified a converting process and the particular grade of paper being used is no longer suitable. Evaluation of the use requirements may suggest important tests and properties which have been ignored in the specifications. Required changes should then be discussed with the customer and, if necessary, new specifications developed. It is almost always helpful in trying to resolve complaints to obtain a sample which the customer has found satisfactory. Test results for the "good" and "bad" sample can then be compared.

Testing to measure progress in product development and in research usually requires greater accuracy than control testing. Additionally, time constraints are not usually as rigid. In these cases, testing procedures should be rigidly enforced. As in the case of customer complaints, information on unusual behavior of the sample during testing can be extremely helpful. Such information should be passed on to the product development engineer or researcher along with the test data.

2. Classification of paper properties

Paper properties can be classified in a number of different ways. A particularly useful classification is to separate fundamental properties from those which are arbitrarily defined. Fundamental properties are those which give basic information about the product. Such properties should be independent of the method of measurement, should be mathematically related to other fundamental properties, and should correlate quite well with papermaking and converting operations. Arbitrarily defined properties, on the other hand, are usually complex combinations of several fundamental properties. These properties are dependent on carefully defined systems of measurement, and often may not correlate very well with machine operations. Proponents argue that a thorough understanding of fundamental properties should make it possible to predict the performance of paper in any converting process or end use.

Another method of classifying paper properties is to define them in terms of the kind of property being measured, and this method has been used in this chapter. Structural properties are those dealing primarily with dimensions and weight. Included are basis weight or grammage, thickness, smoothness, air permeability, density, formation, moisture content, and hygroexpansivity. Strength properties are those in which the paper is subjected to stresses and/or strains. These include load-elongation, zero-span (fiber strength), flexural stiffness, tear resistance, folding endurance, bursting strength, edgewise compression, and surface strength. Optical properties are those related to appearance of the sheet and include brightness, color, gloss, opacity, light scattering, and light absorption. Electrical properties relate to the use of paper as a dielectric. Hence, such properties as resistivity, dielectric constant, loss, and breakdown strength are included in this classification. Liquid absorption properties relate to the interaction between paper and liquids with which it may come in contact such as inks, adhesives, semi-liquid products, etc.

3. Moisture content effects

Almost all properties of paper are affected by the moisture content of the paper at the time of measurement. Because of these effects, most paper tests are made in controlled humidity environments. Basic information on these effects, and on the measurement and control of test environments are included in this chapter.

B. STRUCTURAL PROPERTIES

Structural properties are generally those related to weight, dimensions, and composition of paper. Hence, this category includes basis weight (grammage), thickness, apparent density, air resistance, moisture content, formation, and surface smoothness. Other properties that might be included in this category but which are not discussed are fiber type and type and amount of additives and fillers.

1. Basis weight (grammage)

Most paper is bought and sold in terms of its weight per unit area. Therefore basis weight has great significance both to the consumer and the producer in defining prices. Basis weight is also important because of its effect on other sheet properties. It is almost impossible to alter the basis weight of paper such that only the weight, and no other property, is changed (see Sec. B6).

The weight per unit area of paper can, of course, be determined accurately and unambiguously. A specimen of known dimensions is first preconditioned at a low relative humidity (RH), and then conditioned in a standard testing atmosphere of 50% RH, to equilibrate its moisture content. The preconditioning is necessary because of moisture hysteresis discussed later in the section on moisture content.

Table 1. Common ream sizes for basis weight.

Grade of Paper	Basis Weight Specifications Dimension, in.	No. of sheets	Conversion Factor, Basis Wt. to Grammage
Writing/Printing	17 × 22	500	3.760
Paperboard	1000 sq. ft.		4.882
Wrapping Tissue	24 × 36	480	1.695
Book	25 × 38	500	1.480
News/Tissue	24 × 36	500	1.627
Bag Paper	3000 sq. ft.		1.627

The conditioned sample is then weighed on either an analytical balance or basis weight scale. This weight will include both the weight of the solid materials in the sheet and the moisture absorbed from the atmosphere. The weight is usually expressed for a given area of paper.

Historically, in English-speaking countries, basis or ream weight is used to describe the weight in pounds of a prescribed number of sheets having specified dimensions. The number of sheets is usually 500, but sometimes 480 or 1000. The sheet dimensions, or ream size, are usually standardized for each paper grade, depending on use dimensions. Examples of some common ream sizes are 17 x 22 inch for writing and printing papers; 24 x 36 inch for newsprint, wrapping, tissue; and 25 x 38 inch for book paper. The basis weight of other grades, linerboard and sack paper for example, is expressed in pounds of a specified square footage of paper.

In most other countries, and increasingly in English-speaking countries, the weight per unit area of paper is expressed in units of grams per square meter. Through the efforts of the ISO TC6 committee on paper, the weight so expressed is universally called grammage. Some common ream sizes, and factors for converting ream weight to grammage, are shown in Table 1. A more complete listing is included in the TAPPI method for measuring grammage and in the Dictionary of Paper [1].

2. Effect of weight on properties

From a practical standpoint, the assumption of a linear proportional relationship between basis weight and many strength properties (e.g., tensile and burst) is a good approximation [2]. Within the narrow range of weight at which specific grades are manufactured, any effect of non-linearity is only of minor significance. Over the entire range of weights at which paper is manufactured, it is usually found that the strength to weight ratio is less for very low and very high weights than it is for intermediate weights.

A notable exception to the above is tearing resistance, which almost always increases more rapidly than weight. Studies have shown that the effect of basis weight on tear is non-linear, can vary with the type of pulp and process variables, and is complicated.

In the ideal case, an increase in basis weight would bring about an identical percentage increase in thickness. This is not usually the case. Laboratory studies have shown that thickness increases are almost always smaller than weight increases. Consequently, properties related to thickness (density, air resistance, bending stiffness, etc.) will not always respond to weight changes to the expected degree.

Most surface properties are unaffected by basis weight. Among these are smoothness, friction, surface strength, surface conductivity, and surface absorption of liquids [3]. Other properties which usually are not affected by weight are moisture content (when expressed as a percentage), dirt, permanence, and hygroexpansivity.

The Kubelka-Munk equation describes the effect of basis weight on light scattering and opacity. This relationship is discussed later under Optical Properties (see Sec. E1).

3. Thickness, apparent density, and bulk

Thickness (or caliper) of paper is useful in defining an end-use product. More importantly, it is needed to determine the apparent density of paper. Density is particularly useful because of the many important relationships that exist between it and mechanical properties of the sheet. Bulk is the volume occupied by a given weight of paper. It is the inverse of apparent density.

Thickness of paper is, at best, a poorly defined property. The standardized procedure for measuring thickness (sometimes called TAPPI thickness) is to hold the paper between two steel platens with a prescribed pressure. The separation of the platens with the paper inserted is taken as a measure of thickness.

There are two obvious errors in this type of measurement. First, paper is usually not uniformly thick due to the roughness of the sheet surface. Second, the pressure exerted by the platen will cause a compression of the sheet structure. Since papers differ both in surface roughness and in compressibility, considerable errors can result when thickness comparisons are made.

Because of the inherent error in this type of thickness measurement, the results are not very useful for predicting other properties. For example, when papers are stacked, as in a book, the high points on the surface of one sheet may fill a void created by low points on an adjacent sheet. Hence, the thickness of a book is always less than would be predicted by single-sheet TAPPI thickness measurements. Similarly, the density of paper calculated from TAPPI thickness will always be too low.

Other methods are available which minimize, or eliminate, the errors associated with the measurement of TAPPI thickness. These include the measurement of multi-sheet thickness, soft-platen thickness gauges [4], calculation of thickness from pycnometric density measurements, and the simultaneous solution of equations defining bending and extensional stiffness. In the latter method, it can be shown that sheet thickness is proportional to the square root of the ratio of the two stiffnesses [5].

4. Smoothness

Many subjective definitions of paper "smoothness" have evolved over the years [6]. If all the special meanings which depend on merely minor differences of subjective modes of judgement are removed, it is found that only two important types of smoothness exist: smoothness of the free surface; and smoothness of the surface under pressure.

Undoubtedly, numerous special cases can be cited where the smoothness of the free surface is of interest. A number of techniques have been used to measure this property. Most of these are tedious, with the attendant difficulty of defining a single number to describe the property of interest.

Probably the greatest interest in smoothness lies in printing of paper where smoothness is importantly related to the ink transfer process, and thus to the visual appearance of the printed surface. In these cases, some pressure is usually applied to the sheet during the printing process and smoothness of the second type is of greatest interest.

The best way to determine whether a paper has adequate smoothness for printing purposes is to test print it under conditions which exactly duplicate the intended printing process. Such conditions are difficult to achieve reproducibly in the laboratory. A number of laboratory techniques for the indirect measure of paper smoothness as it relates to printing performance have been proposed. The most widely used of these are the "air leak" tests described below and illustrated in Fig. 1.

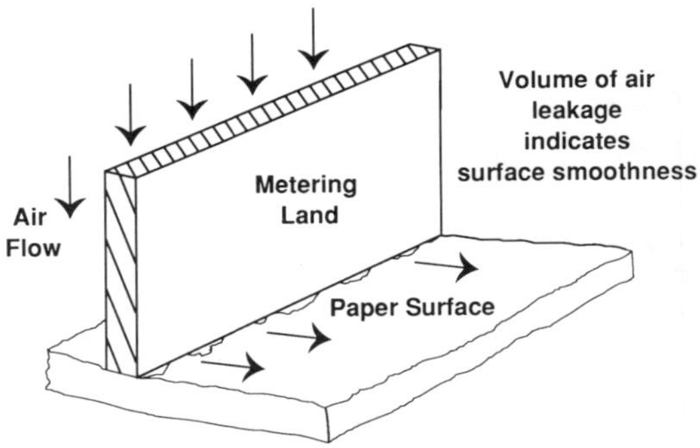

Fig. 1. Schematic of common air leak smoothness testers.

The four most widely used "air leak" smoothness testers are the Bekk, Sheffield, Bendtsen, and Print-Surf. The Bekk instrument is the oldest of these types of instruments. Its use is slowly diminishing as specifications for printing papers are rewritten. The Sheffield instrument has had wide acceptance in the United States and is probably the most widely used. The Bendtsen instrument is used extensively in some European markets and, to a lesser extent, in the United States. The Print-Surf instrument is claimed to more closely measure paper surface properties important to printing, and its use is rapidly increasing.

The principle of operation of all air leak testers is the same. An annular pressure member is impressed on the paper surface. A prescribed air pressure difference is applied across the annulus, and the air escaping through void spaces between the paper and annulus surfaces is measured. In the Bekk system, the time required for a given volume of air to escape is measured. Both the Sheffield and Bendtsen measure the rate of air flow, the latter in mL/min and the former in "Sheffield Units". The Print-Surf instrument uses the rate of air flow to calculate the "average" distance between the paper surface and the annulus surface in micrometers.

The important differences between the four types of air leak testers is in the width of the annulus (ranging from 13.5 mm for the Bekk to 0.051 mm for the Print-Surf), the pressure exerted on the sheet surface, and in the rigidity of the surface backing the specimen. Because of these differences, correlations between the different instruments is not very good.

5. Air resistance

The porous structure of paper plays an important role in the effectiveness of paper in many applications. Included are its ability to protect packaged products; its ability to absorb liquids such as ink, oil, and water; and its filtering efficiency. This property is difficult to measure, and more difficult to describe in simple terms. A property which is easy to measure, and which can usually be correlated with the porous structure of the sheet, is air resistance [7]. (This property is often incorrectly referred to as porosity. Porosity is the ratio of the void volume to the total volume of the paper. There is often an inverse correlation between porosity and air resistance.)

The most common instrument for measuring air resistance is the Gurley "Densometer". (The manufacturer's choice of "densometer" for this instrument is a poor one, since the instrument does not measure density.) In the test, a pressure difference is developed across the sheet and the time required for a specified volume of air to flow through the sheet is measured. Both the test area and pressure difference are controlled.

For very dense materials, a single test may take as long as 30 min. Other instruments are now available which measure the rate of air flow and, hence, give an instantaneous test result. These include the Sheffield, Bendtsen, and Print-Surf. For these, the same flow rate measurement system is used as for smoothness, but with a different measurement head. These instruments differ in the level of pressure difference and in the sample test area. If these differences are taken into account, and if the measurement system is properly calibrated, then good agreement should be found between the different instruments. The Gurley results will, of course, be inversely correlated with the others because of the quantity measured. For the same reason, the results obtained with the flow rate instruments should be referred to as air permeability rather than air resistance.

6. Moisture content

The amount of moisture in paper depends on the temperature and relative humidity of the air surrounding the paper. When paper is exposed to a given atmosphere it will either absorb or desorb moisture until an equilibrium condition is reached. The weight of the moisture in the paper will be higher if this condition is reached by desorption than if it is reached by absorption. This difference, known as moisture hysteresis, is usually about 0.5 to 1.5% of the fiber weight, the actual value depending on properties such as fiber type, refining level, etc.

Moisture in paper is usually measured and controlled with on-line sensors. Depending on the grade of paper, and its ultimate use, this reel moisture can vary from only a few percent to more than 10%. For purposes of standardization, moisture content is often defined as that existing at 50% RH and 23°C with the paper reaching its equilibrium moisture by absorption. Moisture content is then determined from the weight lost when the sheet is dried at 105°C to constant weight. It is reported as a percentage of the weight of the sheet (including moisture) at 50% RH. For some purposes, moisture is calculated as a percentage of the oven-dry weight of the solid material in the sheet. The former is referred to as moisture content, air-dry (a.d.) basis, and the latter as moisture content, oven-dry (o.d.) or bone-dry (b.d.) basis.

The air in ovens controlled at 105°C will still hold some moisture, the amount depending on the temperature and relative humidity of the surrounding atmosphere. The resulting error in determination of moisture content of paper is small (usually less than 0.5% in moisture content) and can usually be ignored for routine purposes. However, when precise measurements are needed, then this error should be determined.

7. Hygroexpansivity

Hygroexpansivity refers to the dimensional changes that occur when paper is exposed to different relative humidities. Over the range of relative humidities, these changes are typically a few tenths of a percent in the machine direction of the paper, and 1% or more in the cross direction [8]. The property is usually measured in

Fig. 2. Instrument for measuring hygroexpansivity of paper.

a hygroexpansimeter. The instrument (see Fig. 2) is equipped so that as many as 20 specimens can be tested simultaneously. It is designed to generate a number of different relative humidities between about 11 and 93%. The test specimens need to be exposed to each humidity long enough to achieve equilibrium dimensions. This time can be as short as a few hours or as long as several days, depending primarily on the moisture sorption rate of the material being tested.

Hygroexpansivity of paper can be divided into a recoverable and a non-recoverable component. The recoverable component is that related to the usual shrinking and swelling of fibers with changes in moisture content. The non-recoverable component is that related to the relaxation of dried-in strains in the paper. This is discussed later in the section on Relative Humidity and Properties (see Sec. G).

The non-recoverable component of hygroexpansivity is effected by the sequence of relative humidities to which the paper is exposed. Hence a standard procedure for the measurement of hygroexpansivity is not practical. Rather, a sequence of relative humidity exposures needs to be custom selected, depending on the ultimate use of the paper.

8. Formation

The term "formation" is often used to describe the variations in many properties of paper [9]. Probably the most frequent use, and the subject of this discussion, is in describing small-scale variations in macro- or micro-basis weight. A paper with poor formation can cause many problems in converting and end use. The average strength, for example, will be less than that for a comparable sheet with good formation. The quality of printing on a poorly formed sheet is usually inferior to that for a well-formed sheet. Coating consumption will usually increase as formation becomes worse.

Formation is sometimes determined by visually comparing the test sheet to a series of "standards" that have been assigned specific values. More frequently, instruments are used to rapidly scan the test sheet and measure the variation in light transmission through the sheet. This variation, or sometimes an index calculated from it, is taken as a measure of formation. Test results obtained with these techniques should not be regarded as a valid measure of weight differences, since variation in density and in optical properties can influence the result.

The most straightforward, and least used, method for evaluating weight variation is to cut very small specimens and weigh them. The difficulty in this technique is in the ability to cut the specimens accurately, and in the time needed to cut and weigh enough specimens to obtain a valid distribution function.

A procedure that is gaining popularity both for on-line and laboratory measurements is the use of beta rays. Unlike light waves which are scattered within the sheet structure, beta waves have relatively long wavelengths and are not scattered. Hence, the absorption and/or transmission of beta rays depend only on the mass of the viewed area, and the variation of beta ray transmission is more closely related to basis weight variation than light transmission [10].

9. Structural non-uniformity

A common difficulty in describing the structure and properties of paper is that it is both non-homogeneous and anisotropic. The former refers to the non-uniformity in structure and composition, usually in reference to the thickness dimension of the sheet. The latter refers to the effect of the angular orientation (e.g., MD, CD) of the test specimen on strength and other properties of the paper.

One source of anisotropy is the preferential alignment of fibers in the machine direction of the paper, with fewer fibers aligned across the machine direction, and almost none in the thickness direction. Another source is straining of the web during drying. Because of these effects, strength properties are usually highest in the machine direction, lower across the machine direction, and lowest in the thickness direction. A direct measurement of fiber orientation is tedious. Procedures are available to give an estimate of fiber orientation. Among these are microwave attenuation and ultrasound wave propogation.

A principal reason for non-homogeneity in paper relates to the manufacturing process. Stock in the headbox is composed of fibers, fibrous debris (fines), filler material, and other additives. Fines and filler particles are usually small compared to the interstitial spaces of the forming fabric on the paper machine. On the wire side of the sheet, many of these small particles are lost through the forming fabric with the white water. Nearer the top side, the built-up mat acts as a filter and retains more of these small particles. Hence, many properties of the sheet will be different, depending on which side of the sheet the measurement is made.

In many cases, paper is designed to increase non-homogeneity and/or anisotropy. For example, one or both sides of the sheet may be surface sized. The MD/CD fiber orientation ratio may be altered to develop either a square sheet if equal strength is needed in all directions, or a high ratio if strength is important in only one direction. Multiple headboxes permit structuring a sheet with lower cost raw materials in the center part of the sheet structure.

Regardless of the cause, all of these factors need to be considered in the testing laboratory. Data on surface properties must include a statement identifying the side of the sheet tested. The results of strength tests must be related to sample orientation, e.g., machine direction, cross direction, or thickness direction. In some cases, test procedures need to include a description of sample orientation during the test. Bursting strength and tear resistance results, for example, may be as much as 20% different for some papers, depending on whether the wire side of the sample is oriented toward or away from the instrument during the test.

C. STRENGTH PROPERTIES

Many converting and use applications apply stresses and strains to the paper. The response of the paper to these stresses and strains will govern, in a large way, how well it performs in sheet-fed operations, whether web breaks occur in printing presses, the customer's acceptance of the paper in handling, and the ability of the paper to protect packaged goods.

Most laboratory tests designed to measure strength properties of paper apply a controlled orientation force or strain to a test strip and then measure one or more of several responses. The usual responses are force, elongation, or energy. Notable exceptions are fatigue-type tests, such as folding endurance or bag drop, which measure the number of loading cycles the sample can sustain before failure occurs.

The strength of paper depends on the strength and arrangement of fibers in the sheet, the level and extent of molecular bonding between fibers, and the contribution of additives designed to enhance strength. The first two, and especially the second, are strongly affected by process variables. Chemical processes such as bleaching, for example, can reduce fiber strength. Refining and wet pressing will usually enhance bonding between fibers, while wet straining can reduce bonding.

Before discussing strength measurements, it is desirable to define what is meant by paper strength. The term in itself has little meaning unless the method of test, or the use to which the paper is to be put, is carefully described. This is so because, first, the measured strength is dependent on the orientation of the test specimen with respect to the applied force. For example, machine direction tensile is usually higher than cross-machine direction tensile, and both are much higher than thickness direction tensile (ZDT) [11].

Second, the capability of paper to absorb an applied force is dependent on the time period over which the force is applied. This capability is attributable to the viscous component of the visco-elastic response of paper to applied force. For example, a paper having a certain strength when tested in the few seconds required for the usual laboratory tensile test would have a much lower measured strength if the force were applied over a longer period of time. It is this phenomenon which causes boxes and bags to fail when stacking forces are applied over long periods and which permit a weaker paper with high elongation to respond better to converting shocks than a stronger paper with low elongation.

Hence, the strength of paper must always be defined in terms of the method of measurement. Fortunately, for most common strength properties, this is accomplished by using standardized testing procedures. For example, the term "tensile strength" is usually accepted as the failure strength of a 180 mm long, 25 mm wide strip tested at a speed of 25 mm/min as described in TAPPI Test Method T 494. It should be recognized that the above applies not only to strength properties, but to many other properties of paper as well.

1. Load elongation

The load–elongation properties of paper are often regarded as fundamental rather than use-oriented properties of paper. In this regard, attempts are often made to use the results of these tests to predict other strength properties of paper. A good example of this is the bursting strength test which has been shown to be related to the average tensile strength for the machine direction and cross direction of the paper and the stretch in the machine direction.

The structural factors in paper which can influence load–elongation properties include fiber strength, the bonded area between fibers in the sheet, the strength of these bonds, and fiber dimensions. Most process variables will also play a role in determining the ultimate tensile strength of paper as well as its elongation properties. Included in these are the distribution of fiber orientation, the level of wet pressing, drying tensions, and pulping variables. Several of these, particularly fiber orientation and drying tensions, cause differences in the load-elongation response for the machine and cross-machine directions. The machine direction of paper is almost always found to have higher strength and less elongation than the cross-machine direction.

A factor often overlooked in the use of tensile results to predict performance is the effect of formation, or the uniformity of weight distribution. Paper cannot be manufactured so that its weight, and consequently its strength, is the same at all points. Hence, the measured tensile strength of a strip of paper reflects the strength of the weakest point, rather than the average strength, of the tested specimen. Because of this effect, a paper having poor formation will have a lower tensile strength than one of better formation, even when all other structural aspects of the sheets are equal. Similarly, the measured tensile strength of paper will decrease as the test length is increased, because the probability of including a still weaker point in a test specimen increases with length.

The measurement of the load-elongation properties of paper is relatively straightforward. A specimen of specified width and length is subjected to increasing strain until the specimen breaks. Many instruments provide a plot of load versus elongation. This plot, as shown in Fig. 3, reflects the visco-elastic nature of paper. At low strain levels, usually only a few tenths of a percent, paper extends elastically. In this region, the resultant force on the sample increases linearly with strain and, if the force were removed, the sample would return to its original dimensions. When the sample is strained beyond this elastic region, non-reversible changes occur in the sheet structure and the force on the sample increases more slowly with further increases in strain. Finally, of course, the force becomes greater than can be sustained by the weakest point and the specimen breaks.

Four different properties of the sheet can be determined from the load-elongation curve. These are tensile strength, stretch, tensile energy absorption, and tensile stiffness.

Tensile strength in the paper industry is defined as the maximum force sustained by the specimen divided by the specimen width. In most other materials testing (metals, plastics, wood, etc.), tensile strength is defined as the maximum force divided by the cross-sectional area of the specimen. This is usually not done in paper testing because paper is not uniformly thick, and the thickness of the specimen at the point where failure initiates is not usually known.

The elongation of the sample per unit length is referred to as strain and, when this is determined as a percentage of the length, it is called stretch.

The area between the load–elongation plot

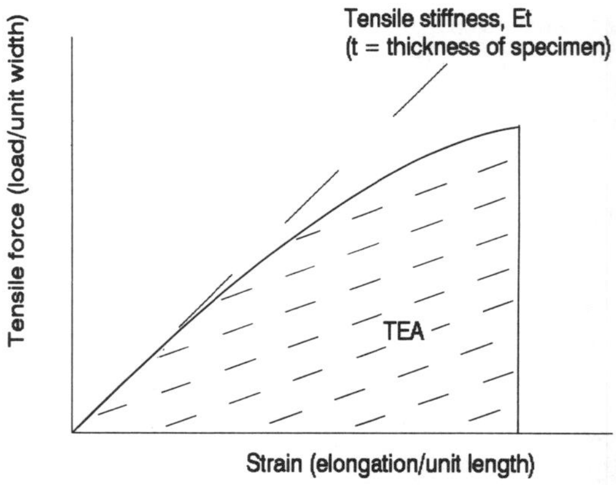

Fig. 3. Typical load-elongation relationship for paper. TEA — Tensile Energy Absorption.

and the elongation axis is a measure of energy absorbed by the specimen during the test. When the measured energy is divided by the specimen dimensions (length x width) the result is called tensile energy absorption, often shortened to TEA.

The modulus of elasticity, E, of the sample can be determined from the ratio of tensile force to elongation in the elastic region of the load–elongation curve and the specimen dimensions. This modulus is equal to stress divided by strain. If the sample thickness is not included in the calculation, the property then determined is tensile stiffness, Et. Tensile stiffness is often calculated because it has been found to be related to the performance of paper in many converting processes. Hence, it yields a useful elastic property without the attendent difficulty of accurately defining sample thickness.

The determination of the above properties from a recorded load–elongation plot is a tedious manual procedure. Most modern instruments are equipped with micro-processors which can provide a hard copy print of all of the above data along with statistical properties.

The relatively simple definition of the tensile strength test should result in good agreement between test data from different laboratories. Collaborative testing data, however, often do not show the expected agreement. For this reason, it is worthwhile to consider the factors that can lead to inaccurate data. The calibration of the load and extension-measuring scales of the instrument should involve standards traceable to the Bureau of Standards. These are usually available through instrument makers. The moisture content of the sample should be controlled through proper conditioning. Sample cutting is critical. A dull or knicked knife can damage the specimen edges causing premature failure during the tensile test. A common practice of cutting multiple specimens on cutters designed for single-specimen cutting usually results in changing dimensions for sheets in the pad, and often non-parallel edges. The clamps used to grip the specimen must be adequate to prevent slippage and, at the same time, not grip the sample so tightly that damage occurs. Finally, because paper is visco-elastic, the test result will be affected by the rate of straining. Hence, the testing speed needs to be specified for comparative testing.

2. Zero-span tensile

The zero-span tensile test is designed to yield a measure related to fiber strength. In the usual tensile strength test, where the test length is greater than fiber length, the measured strength of a specimen is dependent on the strength of the fibers and on the extent and strength of the bonds between fibers. In the zero-span test a special pair of clamps, shown in Fig. 4, is used to reduce the test length to "zero" so that all fibers involved in the test are gripped by both clamps. Hence, in theory, the measured strength is dependent only on fiber strength. As a practical matter, the test length is never truly zero. Some fibers will thus terminate in the small distance between the clamps and bonding will play a role in determining the final zero-span result.

Quite obviously, the limiting factor in paper strength is fiber strength. Hence, the zero-span test can be useful in predicting the limiting strength of paper. As a practical matter, paper tensile strength will rarely exceed 75% of this limiting value, and will often be much less. This is so for two reasons. First, paper is usually made so that its bonding strength is less than the fiber strength. Second, non-uniformity of the sheet structure causes a reduction in measured strength as the test length is increased. For paper with "bad formation" this reduction can be greater than 50% of the measured zero-span strength.

The zero-span instrument has also been used to obtain information on fiber bonding and fiber length. For this purpose tests are made over a series of small, increasing test lengths, shorter than fiber length. These tests are made for both air-dry (bonded) and wet (unbonded) sheets. For the air-dry tests, the rate of decrease in strength with increasing test length will depend both on fiber length and bonding. For the wet sheets, the rate of decrease depends only on fiber length. Hence, as shown in Fig. 5, indices related to fiber strength, fiber length, and bonding can be extracted from the data. The Pulmac instrument is specially designed for testing at various test lengths.

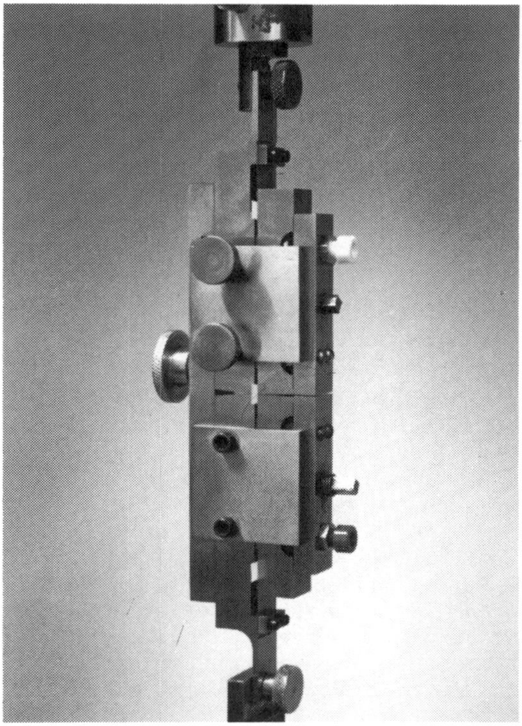

Fig. 4. Zero-span instrument — Institute of Paper Science and Technology.

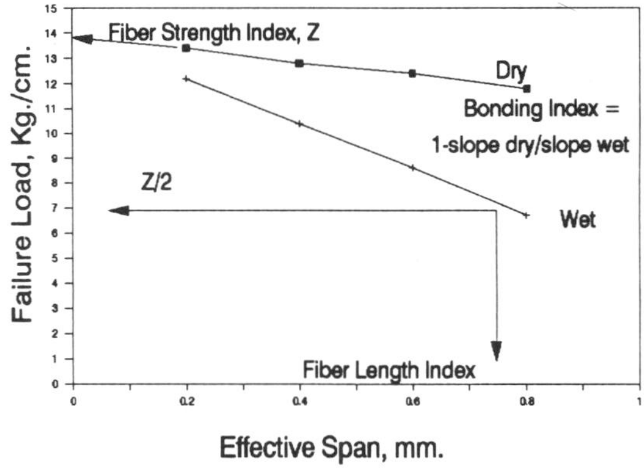

Fig. 5. Pulp properties determined from short-span tensile measurements.

If, during a zero-span test, all of the fibers in the sheet were aligned with their long dimension parallel to the testing direction, then the measured zero-span stress should be equal to the average fiber stress. Paper is not made with this kind of fiber orientation. Pulp handsheets, for example, will have completely random orientation of the fibers. Quite obviously, fibers aligned at some angle to the testing direction will not make the same contribution to the failure load as fibers aligned parallel to the testing direction. It can be shown that the final failure load for these handsheets will only be 3/8 of that for a hypothetical sheet in which all of the fibers are aligned in the testing direction [12]. More complicated theories have been developed for machine-made sheets in which the distribution of fiber orientation is usually not random.

3. Bending stiffness

An external force applied to the surface of a strip of paper will cause the strip to deflect. If the force is applied perpendicular to the plane of the strip, and if the resultant deflection does not exceed the elastic limits of the material, then the relationship between force and deflection is a measure of the bending stiffness of the strip.

This relationship will be governed by two other properties of the strip, an elastic property, E, and a dimensional quantity, I. Here E is the modulus of elasticity (see Sec. C1) of the material I is the moment of inertia of the cross-section of the strip and is dependent on its cross-sectional area. For a rectangular cross-section:

$$I = wt^3/12$$

where w is the strip width and t is the sample thickness. The fundamental measure of the stiffness of a material is its flexural rigidity calculated as the product EI [13].

Instruments for measuring stiffness fall in one of two groups. In one group are those instruments which measure the modulus of elasticity, E. The modulus is then multiplied by the moment of inertia, I, to obtain the flexural stiffness. Instruments in the second group measure the force required to bend the strip a controlled distance.

The sample is not usually bent out of its plane by instruments in the first group. For example, in the tensile strength test the test specimen is strained in its plane, and the resulting force is measured. Within the elastic region of the load-elongation relationship, the modulus of elasticity can be determined by dividing tensile stress by tensile strain.

Another method in this group is the resonance method. A frequency generator is used to vibrate the test strip. Then, depending on the method used, either the frequency or the strip length are adjusted until resonance vibrations occur. The modulus of elasticity can then be calculated from the resonance frequency and the sample dimensions.

A third method involves measuring the velocity of ultrasound waves in the plane of the sheet. For this method, the modulus of elasticity is calculated from the measured velocity and the density of the sheet. A significant advantage of this procedure is its amenability to on-line testing [14].

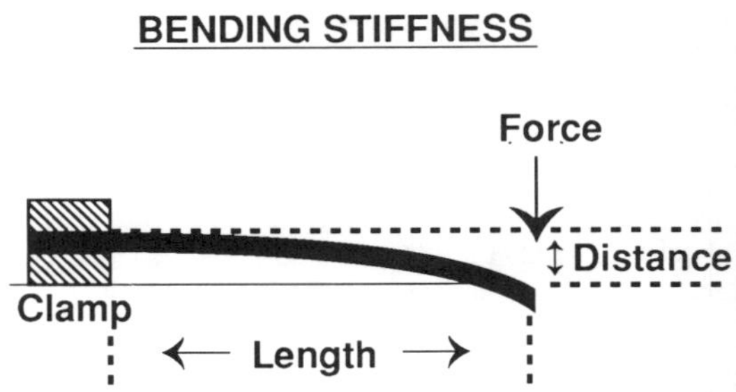

Fig. 6. Schematic of cantilever beam stiffness testers.

Most of the methods used for measuring bending properties in the laboratory are of the second type. The more commonly used of these (Gurley, Taber, L&W) use cantilever beam geometry (illustrated in Fig. 6). For this purpose, the test strip is gripped at one end and a bending force is applied to the opposite end. The force required to deflect the strip a given distance is then indicated by the scale of the instrument. The results obtained with these different instruments should be proportional to each other. They will not, however, be equal. This non-equality results from different sample dimensions being used in the various tests. If this were accounted for by calculating the flexural rigidity for each test, then the results would be equal.

There is some risk in calculating flexural properties from in-plane modulus measurements. Beam theory, for this purpose, requires that the modulus be constant throughout the sheet thickness. This is not always the case. In multi-ply sheets, for example, the boundary layer between adjacent plies may have a lower modulus than the individual plies, or the moduli of the individual plies may vary. Even in single-ply sheets, modulus variations may occur due to variations in fines retention, drying rate, fiber orientation, etc., through the sheet thickness. In each of the above cases, flexural rigidity calculated from in-plane modulus measurements will be higher than that determined from bending tests if the high-modulus zones are near the sheet surfaces, and lower if they are more centrally located.

The effect of process variables on flexural rigidity needs to be considered separately from the effect on other strength properties. Many of these – refining, wet pressing, calendering –

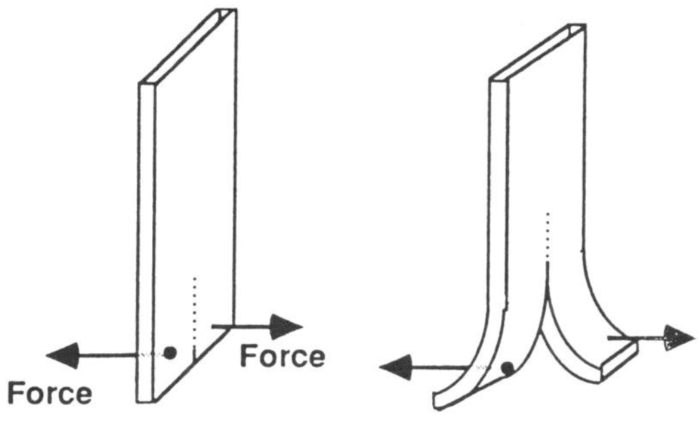

Fig. 7. Measurement principle of Elmendorf-type tear test.

may increase the modulus of the sheet but will also reduce sheet thickness. Since flexural rigidity is dependent on both these properties, the hoped-for result is not always obtained.

4. Tearing strength

Tearing strength, as defined by the paper industry (illustrated in Fig. 7), is the minimum force required to continue an initiated tear in the paper, with the anti-parallel tearing forces exerted perpendicular to the plane of the paper. Attempts to measure this force directly show that the force varies considerably from point to point along the path of tear. Hence, it is exceedingly difficult to define an average force. Elmendorf solved this problem rather elegantly by proposing the use of an energy-measuring pendulum to determine the total energy required to tear a controlled length of paper. This energy is simply divided by the tear length to yield the average force. Although the Elmendorf instrument measures energy, the scale is graduated in force units, usually grams or milli-newtons (mN).

The most commonly used instrument in the U.S. has a capacity of 1600 g. The scale is designed to give the average force for a single sheet when 16 sheets are torn together and when the tear length per sheet is 43 mm. Common practice is to tear the number of sheets together that will give a reading near the middle of the scale, divide the result by the number of sheets, and then multiply by 16. This practice permits the use of a single instrument for papers differing greatly in tear strength. The number of sheets required can be as low as 1 for paperboard, whereas as many as 32 sheets may be required for very weak papers.

Two segments of the industry where tear strength is particularly important are in packaging papers and in groundwood printing papers, primarily newsprint. Bags are expected to withstand handling shocks, and newsprint to withstand web tension shocks in the printing press, without tearing. The ability of paper to withstand these shocks is called toughness, and the assumption is that paper tear strength correlates with toughness.

It is important to our understanding of tear test results to note that any work-consuming mechanisms involved in the test will contribute to that result. For example, the sample is bent through 90° at the start of the test, and through 180° at the end of the test. The work involved in bending the specimen becomes part of the total measured work. When tearing more than one sheet at a time, the individual sheets get in each others' way, thus contributing to the measured work. This inter-ply effect is especially significant for strong, long-fibered pulps for which the effect is very large at low densities and diminishes as density increases.

In addition to the procedural problems discussed in the preceding paragraph, the tear test has been criticized on two theoretical considerations. First, in most converting processes the forces which could start a tear usually occur in the plane of the paper, rather than perpendicular to it as in the tear test. Second, the initiating cut in the tear test is not similar to the weak spots occurring normally in paper. For these reasons, researchers have been examining alternative methods for evaluating this important property of paper. This research includes the measurement of tear by an in-place mode, and the measurement of fracture resistance.

Studies have shown that these new measurements are better predictors of the shock absorption capability of paper than the present tear test. Industry, however, is slow to accept these new methods. Presumably, problems associated with this type of failure are judged to be too infrequent to justify change.

5. Bursting strength

Bursting strength is probably the most frequently used empirical test for assessing the strength of paper. The test has been in use for more than 100 years. Its popularity probably relates to the ease of performing the test, the relatively short time required to make a test, and the single-number assessment of average sheet strength.

In the test, as shown in Fig. 8, hydraulic pressure is applied to a circular area on one surface of the sheet through a rubber diaphragm. The paper is securely clamped around the periphery of this circular area. The increasing hydraulic pressure causes the rubber diaphragm

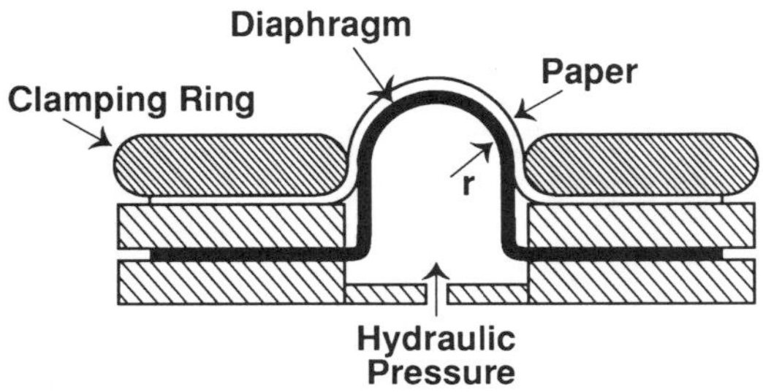

Fig. 8. Schematic of bursting strength test.

and paper to expand into the area encompasssed by the clamp. When this expansion exceeds the maximum extensibility of the paper, rupture occurs. The pressure at the moment of rupture is indicated by a maximum reading Bourdon-type pressure gauge or, in newer instruments, by electronic digital pressure indicators.

It should be obvious that the burst test is synonymous with simultaneous tensile tests for all orientations of the paper. In the burst test, of course, rupture will occur when the strain exceeds that for the least extensible direction of the paper, usually the machine direction. The force contribution for other orientations at the moment of rupture will then depend on the load-elongation relationship for each orientation. It has been found that the average of the machine and cross-machine direction tensile provides a reasonable approximation of the sum of the force contributions for all orientations. Bursting strength is then found to be proportional to the product of this average tensile strength and the stretch of the paper in the least extensible direction.

It is often thought that changes in bursting strength will parallel those in tensile when process variables are changed. This is often the case, but the exceptions are too frequent to accept this as a general rule. For example, increased wet straining in the machine direction will usually increase tensile strength but could cause a decrease in burst if the stretch is reduced sufficiently.

Two models of bursting strength testers are used in the industry, a Model A for board grades and corrugated board, and a Model C for fine paper grades and pulp testing handsheets. The major difference between the two is in the rubber diaphragm. For the Model C tester, the diaphragm is thin and flat, and its normal distension pressure is 30 kPa. The diaphragm for the Model A tester is molded so that its thickness is greater near its center, and its distension pressure is 159–207 kPa. These pressures are those required to distend the diaphragm 9.5 mm. The actual distension of the diaphragm during testing is almost always less than 9.5 mm, and depends on the extensibility of the sample being tested. Consequently, the correlation between the Model C and A testers is not always reliable.

The bursting strength tester requires considerable care in calibration and maintenance if reliable results are to be achieved. An extensive review of this subject was published by Whitsitt [15].

6. Surface strength

The difficulties resulting from defective bonding between the various plies of paperboard, within the sheet structure of linerboard, and between the surface and body stock of printing papers, are well known. The strength between layers or plies within the sheet structure is often referred to as “internal bonding strength”, whereas that at or near the surface is referred to as “surface strength”.

A common printing trouble is the picking of paper, resulting in small areas becoming detached from its surface during the printing operation. This can leave unprinted specks in the printed areas, and particles of the paper or coating adhering to the press rollers and/or contaminating the ink.

Terms often used to describe the properties of paper associated with these problems are:

1. Pick resistance – that property which resists the lifting or removal of fibers, small fiber bundles, coating particles or fragments from the sheet surface.
2. Rupture resistance – that property which resists complete removal of the surface layers, or delamination of inner layers, within the sheet structure.
3. Linting and dusting – not related to sheet strength but refers to the removal of loose debris from the sheet surface during the printing process.
4. Piling – is the accumulation of materials from picking, linting, or dusting on printing plates or blankets of sufficient quantity to affect print quality.

The strength property of interest is generally measured through the application of a force perpendicular to the plane of the sheet. Depending on the procedure used for testing, three types of failure can be discerned: pick, blister, and rupture. Blister failures, which occur at a force level intermediate between pick and rupture failure, are the delamination of small areas within the sheet structure and often appear as small raised areas on the sheet surface.

Some test procedures are designed to measure the force needed to create only one of these types of failure, while others can evaluate all three. Rupture-type failures usually occur at well-defined force levels. Pick and blister failures, on the other hand, usually begin at isolated areas of the sheet and gradually become more continuous as the applied force is increased. Thus, the pick and/or blister strength of paper is less well defined and its measurement depends on subjective discretion.

Picking strength is commonly measured with the wax pick test or with the IGT (Instituut voor Grafische Techniek) instrument. The latter can also provide a measure of blister and rupture strength. The most common instruments for measuring rupture strength are the ZDT and Scott ply-bond testers.

In the wax pick test a series of hard resin waxes are used. These are formulated for increasing adhesiveness to the paper surface. The ends of the wax sticks are softened to the melting point over a low flame and immediately attached to the sheet surface. The wax is allowed to cool and then pulled from the sheet. The highest wax number in the series which does not visibly disrupt the sheet surface is reported as the critical wax number. Several problems have been encountered with this procedure. First is the subjective nature of the end point. Experience has shown that different operators can disagree by as many as 3 wax numbers. Second, it has been found that paper can be surface sized sufficiently to pass the wax pick test, but the paper will still fail internally during conversion. Third, the use of thermoplastic resins in some coating formulations can increase the cohesion between the wax stick and the coating. The critical wax number is thus reduced, even though the coating strength has increased.

In the IGT test, a viscous film of controlled thickness is doctored onto the periphery of a wheel. The wheel is then rolled over the paper surface with increasing velocity. The shear forces involved in splitting the viscous film between the wheel and paper surface increases with velocity, thus exerting an increasing force to the paper surface. If the film viscosity and wheel velocity are selected properly, the sample will be torn apart. The result is reported as the product of viscosity and velocity (VVP). Common test films used are tack-graded inks and polybutene, the latter selected because of its newtonian properties. Often, in a single test, all three types of failure (pick, blister, and rupture) can be determined, with each type occurring at different velocities. This test is also known to suffer because of differences of opinion between operators in selecting an end point, particularly that associated with pick.

7. Edgewise compression strength

An important function of rigid-side containers (corrugated or solid fiber) is to protect the product contained therein. The filled containers are frequently stacked in warehouses or for transportation. The bottom container in the stack may thus be subjected to considerable compressive force. The most important factor governing the compressive strength of the container is the edgewise compression strength of the components making up the container walls.

A common difficulty in measuring the edgewise compressive strength of thin flat materials is preventing the sample from bowing or buckling during the test. The test result for samples which buckle or bow may be influenced more by the flexural stiffness of the material than by its true compression strength. In the common ring crush test, a 152.4 mm long by 12.7 mm wide sample is formed into a ring and then crushed in the 12.7 mm dimension. The ring structure is intended to minimize buckling. Although this test has been the standard in the industry for many years, it is well known that the ring configuration is inadequate to prevent buckling for most lighter weight grades of linerboard and corrugating medium.

A newer test which is gaining wide acceptance in the industry is the short-span compression stength test (often called STFI test). The sample is gripped by opposing pairs of clamps with a short distance of only 0.7 mm between them. Taking into account the sample thickness, column formulae will show that buckling should not occur at this test length.

Regression relationships between ring crush and short-span compression strength show considerable curvature for lighter weight grades. This is believed to be due to buckling of the thinner samples in the ring crush test. The short-span compression strength test should improve the ability of board mills to predict container performance, based on component testing.

D. LIQUID ABSORPTION PROPERTIES

Almost all paper products are designed to respond in a controlled way to contact with liquids [16]. Many packaging materials, for example, must protect the product being packaged and thus have a high level of resistance to penetration of liquids. Printing papers are expected to absorb ink vehicles at an optimum rate for good printing. Personal products like tissue and towelling should absorb liquid readily.

In all of these cases, steps are taken during manufacture to impart the desired liquid absorption or resistance properties to the paper. Tissue and towelling are made from lightly refined stock to retain a high level of absorptivity. Most other papers are treated, either internally or on the surface, with a low surface energy sizing material to increase their resistance to liquid penetration. For some papers, surface coatings and/or films may be used so that liquids do not directly contact the fibrous sheet structure.

1. Tests for absorbance

Hundreds of tests have been proposed for measuring the liquid permeability of paper. A commonality among them is that the test liquid (water, oil, grease, ink, etc.) of interest is brought into contact with the sheet. The liquid used for the test is usually consistent with the intended use of the paper. Where the tests differ is in the method of defining the reaction to be measured. As in the case of liquid selection, the reaction should also relate to the intended paper use. Because of the many different uses of paper, it is not possible to design a single test suitable for all grades of paper.

Most liquid penetration tests meaure either the time for the paper to absorb a specified volume of liquid, the volume of liquid absorbed in a specified period of time, or the rate of penetration of the liquid into the sheet surface.

In many converting processes, the liquid is in contact with the sheet surface for a relatively short period of time, typically a few milliseconds. The interactions which occur during this short time period are complex, and involve properties of both the paper and the liquid. Important among these are the ability of the liquid to wet the paper surface, and the rate of penetration of the liquid into the structure of the sheet once wetting occurs. Both of these are independent phenomena depending on different properties of the paper and liquid, except that wetting must occur before penetration can begin. The penetration phenomenon is further complicated if aqueous liquids are used. These liquids cause swelling of the fibers as they are absorbed into the fiber wall, undoubtedly resulting in changes in the volume and pore structure of the sheet. These changes can complicate equations usually used for predicting the penetration rate of liquids into porous structures.

Tests which measure the rate of penetration into the sheet surface at times comparable to those encountered in converting processes are most appropriate for predicting reactions in those processes. The Bristow wheel is particularly attractive for this purpose. In this test, the liquid is spread over the sheet surface in a controlled way, and the total area covered by a known volume of liquid determined. Tests are made at different speeds so that, for each speed, the time available for penetration and consequently the spread area are different for each test. A typical plot of the area-time relationship is shown in Fig. 9. This plot is used to determine three parameters of the test sample: a wetting delay time, the rate of penetration of the liquid after the wetting delay, and a measure related to surface roughness. The latter can be determined since part of the liquid is transported through the metering system by filling the low areas of the sheet surface. A relatively long period of time is required to complete the number of tests needed to evaluate a single sample. Hence, the use of the test is usually restricted to research laboratories.

Another common group of tests are those for which printing inks are the penetrating medium. These include the IGT printability tester, K & N ink smears, and laboratory printing presses. These tests can be quite useful for evaluating the suitability of paper for printing purposes. They do not, however, provide enough information to separate liquid penetration affects from other printing variables.

Most laboratories rely on tests in which the liquid is in contact with the sample for a relatively long period of time. The most common among these are water drop tests and the Cobb size test. In the former, a controlled volume drop of water is placed on the sheet surface. The time required for the water to either reach the opposing surface or to be completely absorbed into the sheet structure is measured. In some cases the contact angle is measured at near zero time and then at one or more longer time intervals. The near zero measurement supposedly indicates the wettability of the sheet surface, and the rate of change in contact angle the penetration rate of the liquid. In the Cobb test, the weight of water that can be absorbed into the sheet in a specified period of time (usually 2 min but sometimes less) is measured. The result is reported per unit area of paper surface. Another common test which uses ink as the penetrating medium is the Hercules Size Test (HST). In this test, ink is poured on one surface of the sheet, and the opposite surface is illuminated from a controlled light source. As the ink penetrates into the sheet the reflection from the opposite side will decrease. The test is terminated when the reflection decreases by a preselected amount. All of these tests are particularly attractive in a product control laboratory because the results

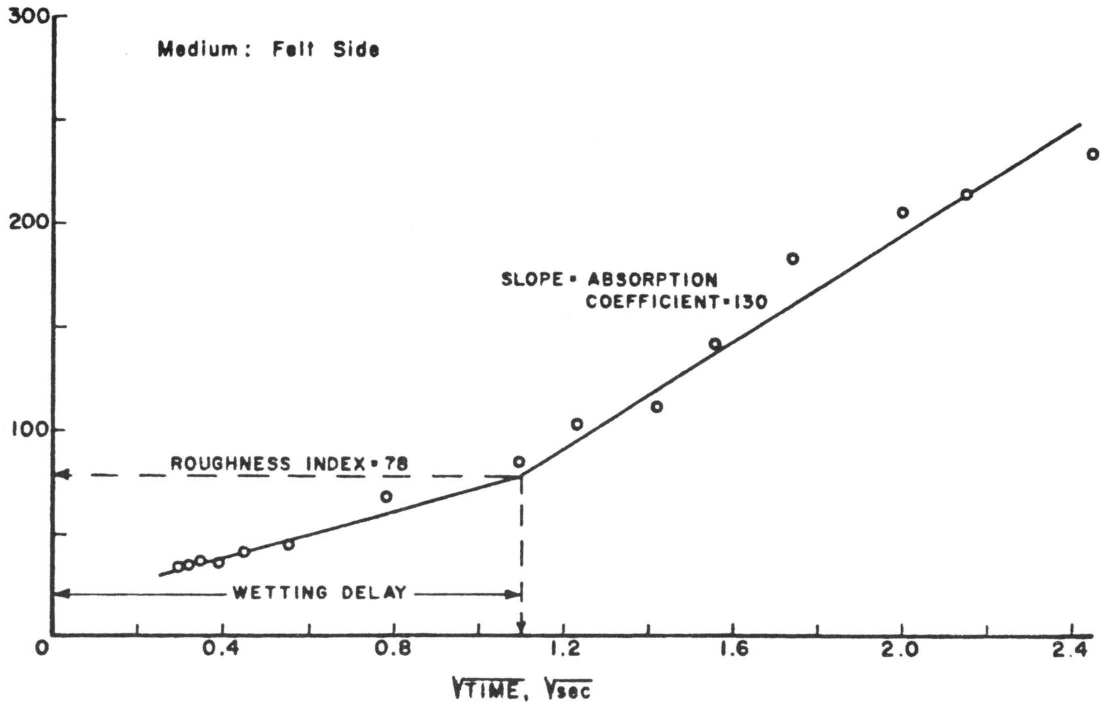

Fig. 9. Liquid absorption data measured with the Bristow tester.

can be obtained in a relatively short period of time. They are good indicators of the water resistance, or sizing, of the test sample. They cannot, however, provide reliable information on liquid-paper interactions in dynamic situations.

The adage that the only reliable test for printing papers is the press run itself seems still to have some validity today. Similar statements might well apply to other applications of paper involving interactions with liquids. Laboratory tests can provide useful information on liquid penetration behavior. However, the test data need to be interpreted only with regard to the specifics of the actual test made. Extreme caution should be exercised in predicting penetration behavior for other liquids, exposure times, etc., from those test results.

E. OPTICAL PROPERTIES

The appearance or optical properties of paper are of considerable importance to the papermaker, the converter, and the end user. Books, for example, carry the writers' thoughts to the reader, and the function of the paper is as a medium to carry the printing. If the paper has visually unpleasant attributes, the reader will be distracted and will lose some of the thought the writer is trying to convey.

The appearance of paper should, therefore, be regarded as a subjective property, and its evaluation should include an entire array of physical and psychological factors which can affect visual assessment. The intent of optical properties is to define objectively attributes which can be measured accurately, repeatably, and, most importantly, which relate to human observation. Examples of such properties which are commonly measured are brightness, gloss, opacity, color, light scattering, and light absorption.

All of these properties involve the interaction of paper with light. They differ in the method of illuminating the sample and in the component of the source light which exits the sample and is detected by the instrument sensors. It is, therefore, useful to consider the behavior of light when it strikes paper.

Light incident on a sheet of paper will follow one or more of three paths: reflection, scattering, or transmission; or it will be absorbed. Reflection is that part of the incident light returned from the sheet surface. The portion of the reflected light which is returned at an angle equal to that of the incident light is referred to as "specular reflection". That portion of the incident light which is not reflected from the sheet surface, and thus enters the sheet, will encourter numerous fiber-air, fiber-fiber, filler-fiber, and filler-air interfaces within the sheet structure. At each interface the light can again be reflected, transmitted or absorbed. Unless the fiber is totally transparent, the transmitted component will be bent, or refracted, as it passes through the fiber. Part of the light entering the sheet structure, after numerous reflections and refractions, will exit the sheet on the viewer side and is referred to as scattered light. Lignin and other impurities in the sheet will selectively absorb certain wavelengths of light and convert it to heat. Light energy lost in this manner is referred to as "absorbed light". Finally, part of the incident light will pass completely through the sheet and this component is referred to as "transmitted light".

1. Opacity

If an illuminated sheet of paper is backed by a reflecting surface, some of the light energy lost by transmission will be reflected back into the sheet and will add to the total light exiting the viewed side of the sheet. The amount of light reflected back into the sheet will depend on the reflectance of the backing material. A sheet which is relatively opaque will reflect about the same amount of light when backed by either a high or a low reflectance material. On the other hand, a more transparent sheet will reflect different amounts of light for these two backing materials. Opacity measurements are based on the reflectances for three standardized backing materials.

The first of these is a black cavity which absorbs all or most of the light transmitted through the sheet. The reflectance of the sheet when backed by this black cavity is called R0. The second reflectance, called R0.89, is for the sheet backed by a layer of magnesium oxide (MgO) under a cover glass. The requirement is that the cover glass-MgO combination have an effective reflection equal to 89% of that for pure MgO alone. The third reflectance is for the sheet backed by a pad of sheets of the same material. The pad is made thick enough so that further increases in its thickness will not change the reflectance of the test sheet. This third reflectance is referred to as $R\infty$.

Opacity is calculated as the ratio of reflectance R0 to one of the other two reflectances. If the second reflectance is R0.89, the ratio is commonly called TAPPI Opacity. When the $R\infty$ reflectance is used, the ratio is called Printing Opacity. TAPPI Opacity is measured more frequently than Printing Opacity. However, many people believe that the latter agrees more closely with the eye's perception when viewing the page of a book.

The spectral aspects of opacity instruments are selected to approximate the sensitivity of the human eye. Hence opacity is heavily weighted near the center, or green-yellow, region of the visible spectrum, with reduced response near the extremes, or red and blue, regions.

Most opacimeters today use one of two standard geometries. The geometry of the TAPPI Method, which dates back to 1930, is 15° illumination and diffuse viewing. With diffuse viewing all of the light exiting the sheet on the viewed side is measured, regardless of the angle of exit. The standard geometry for the ISO method is diffuse illumination and normal viewing. Diffuse illumination causes the incident light to strike the sheet at all angles. In normal viewing, only that portion of light exiting the sheet at right angles is measured. Because opacity, by definition, is a ratio of two reflectances, the effect of different geometries will be small compared to other optical measurements.

Almost any change in the papermaking process will result in a change in opacity and can, therefore, be used in its control. Opacity will increase with basis weight, thickness, percent of filler, coating weight and groundwood content. It will decrease with refining, bleaching, and calendering. Of course, the effect of these changes on other properties needs to be considered.

2. Brightness

One of the more important visual attributes of bleached papers used for printing and, undoubtedly, other purposes, is its color. Paper which is white or near-white will make a greater contrast with printing and thus enhance readability.

The most precise method for measuring this attribute would be to measure the reflection of light at all wavelengths across the visible spectrum, or perhaps at three specified wavelengths as discussed later under color measurements (See Sec. E6). These measurements, when plotted, produce a spectrophotometric curve which can then be interpreted to describe the color of the sheet (see also Chap. VI in Vol. 9 of this series).

Typical spectrophotometric curves for an unbleached kraft pulp show that the pulp reflects about half of long-wavelength (red and yellow) light, but absorbs most of the short-wavelength (blue and green) light. The selective absorption of blue light gives the pulp its brownish appearance. The curve for a bleached pulp shows increased reflection across the entire spectrum and, more importantly, a flattening of the curve which makes the sample appear white.

The change in the spectrophotometric curve resulting from bleaching suggests that a reflectance measurement at a single wavelength could be a useful, and more easily understood, indicator of the effect of bleaching on color. This observation led to the development of papermakers brightness. Quite obviously, this single-point measurement could be made at any wavelength. A short wavelength was selected because the greatest change occurs at this end of the spectrum. This was a fortuitous selection since the yellowing of paper with age has been found to reduce reflection at short wavelengths, while little, if any, change occurs at longer wavelengths. Hence, brightness is a useful property to measure in permanence studies on paper.

The selection of a specific wavelength for brightness measurements was purely arbitrary. However, it is extremely important that the selected wavelength be precisely defined so that all brightness testing instruments will yield the same result. The amount of light at various wavelengths which excites the detection system is known as the spectral response of the instrument. The spectral response of all brightness testers today has an effective wavelength of 457 nm, spectrally distributed in a specified way over the range of about 400 to 500 nm.

Except for a perfect mirror, the light exiting the illuminated side of a material will be reflected over a wide range of angles. Light energy exiting a sheet at an angle different from the incident angle is referred to as diffuse reflectance. Standard brightness of pulp and paper is defined as the ratio of the diffuse reflectance of an opaque pad of test sheets, divided by the reflectance of a pure magnesium oxide surface. This description, as well as the spectral response described above, is the standard in both TAPPI and ISO methods, and in the methods of most countries of the world.

The earliest method for measuring brightness specified that the sample be illuminated at an angle of 45° over a circular area about 12 mm in diameter. Reflectance measurements were made normal to the plane of the sheet. This brightness scale has not changed since its inception and is still widely used today. It is commonly referred to as "45/0" or "directional" brightness.

Another geometry commonly used today involves diffuse illumination of the sample over a circular area 34 mm in diameter. As in the case of directional geometry, reflectance measurements are made normal to the plane of the sheet. This measurement is commonly called "diffuse/0" or simply "diffuse" brightness. Because of the diffuse illumination, a small part of the light entering the viewing aperture is specularly reflected light. Because the definition of "brightness" requires that only diffusely reflected light be measured, most instruments of this geometry are now equipped with a "gloss trap" which blocks the specular component.

Attempts have been made to establish correlations between these two brightness scales. None of these has been totally successful. Thus it is important that all brightness data include a description of the geometry used in their measurement.

The expression "TAPPI Brightness" is sometimes used for measurements made with a directional geometry instrument. The use of this expression should be discouraged, since TAPPI methods now exist describing both geometries.

3. Fluorescent component of brightness

Optical brightening agents are sometimes added to paper which increase the blue reflectance and cause the paper to appear whiter [16]. These agents absorb ultraviolet energy in the incident beam and re-emit it at longer wavelengths in the blue region of the spectrum, a phenomenon known as fluorescence.

Some brightness testers are equipped with a green filter which can be inserted in the incident light before it strikes the sample. The function of the filter is to remove all ultraviolet energy from the incident light. Optical brighteners then become ineffective since they require ultraviolet energy for exitation. The difference in brightness with and without the green filter is a measure of the fluorescent component.

4. Gloss

Light rays striking a mirror-like surface will reflect all of the light in one direction, the angle of reflection being equal to the angle of illumination (specular reflection). A very dull matte-like surface, on the other hand, will scatter the light and reflect it with equal intensity in all directions (diffuse reflection). Most surfaces, including those on paper, will have both a specular and a diffuse component of reflection. The glossiness of the surface is then the ratio of specular to diffuse reflection.

A complete characterization of the glossiness of the surface of a sheet of paper would require measurements at all angles of illumination. Goniophotometers are available which can accommodate these measurements. However, the test is time consuming and seldom used except for research purposes.

The TAPPI test method describes a simplified procedure in which gloss is calculated as the ratio of the specular reflection of the paper to that of a black mirror surface. In the method, the angle for measurement is standardized at 75° (15° from the plane of the sample). Measurements are made with the light rays both parallel and perpendicular to the machine direction, and then averaged to obtain a single number.

Most printing papers have a specification for gloss level, but usually ignore non-uniformity of gloss. This non-uniformity can sometimes detract from the perceived quality of the printed page more than the gloss level. Papermakers need to give more consideration to this important attribute.

5. Light scattering and absorption

As discussed earlier in this section, scattering occurs when light is either reflected or refracted at the numerous air-fiber boundaries within the sheet structure. Scattering can also occur at other boundaries such as those provided when fillers are added to paper. The numerous boundaries created with fine fillers such as titanium dioxide account for the significant increase in opacity when these are added.

Absorption occurs when the source light strikes a colored substance in the sheet. The colored substance selectively absorbs light energy in specific wavelengths. This absorbed energy is transformed to heat. The amount of light absorbed is a good indication of the amount of colored material in the sheet. Unbleached pulp, for example, will absorb a significant part of a source light, while a bleached pulp will absorb only a little.

The measurements of the amount of light scattered and absorbed by a given pulp is very useful for predicting the effect of basis weight, additives, pressing, etc. on other optical properties.

Scattering and absorption coefficents, first defined by Kubelka and Munk, can be calculated based on absolute reflectance measurements of R_0 and $R\infty$. TAPPI methods include charts and equations showing how these data can then be used to predict changes in other optical properties.

6. Color

To be useful, a color measurement system must agree with the color assessment of the "average" human eye. Monochromatic light of different wavelengths could be used, and the reflectance at each wavelength plotted, to develop a spectrophotometric curve. Each point on the curve would then need to be properly weighted to account for the effect of wavelength on eye sensitivity. This would be a time-consuming procedure, and the resultant information would be difficult to describe in a few words or numbers.

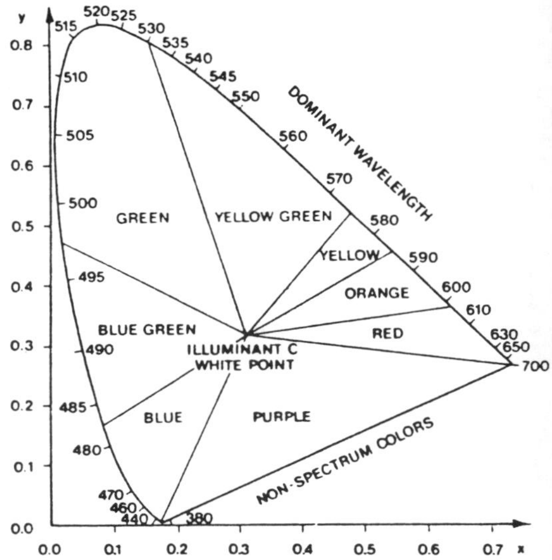

Fig. 10. Dominant wavelength theory for color measurement.

Most color systems today are based on reflectance measurements for three standardized light sources: red, green, and blue. Numerous experiments have shown that the proper mixture of these three will duplicate most other colors. These light sources, each having a precisely defined spectral distribution, are referred to as the CIE[1] tristimulus functions X(red), Y(green), and Z(blue).

Three different systems, all based on the CIE functions, are in common use in the paper industry today. These are the dominant wavelength, purity, and luminosity system; the Hunter Lab system; and the CIE STARLAB system. The three systems differ in the method of defining color, once the CIE functions have been measured.

In the dominant wavelength system, the measured tristimulus values (X Y Z) are used to calculate chromaticity coordinates (x y z) such that $x = X/(X+Y+Z)$, $y = Y/(X+Y+Z)$, and $z = Z/(X+Y+Z)$. Since the sum of x, y and z must equal 1, only two of the coordinates are needed to describe a point in color space. Hence, z is usually not calculated. Color space can then be described by a plot of x and y within the visible spectrum (Fig. 10). The center of this plot, where both x and y equal 0.33, is the white point. The direction and distance of a sample point from the white point describes its dominant

1. Commission Internationale de l'Eclairage

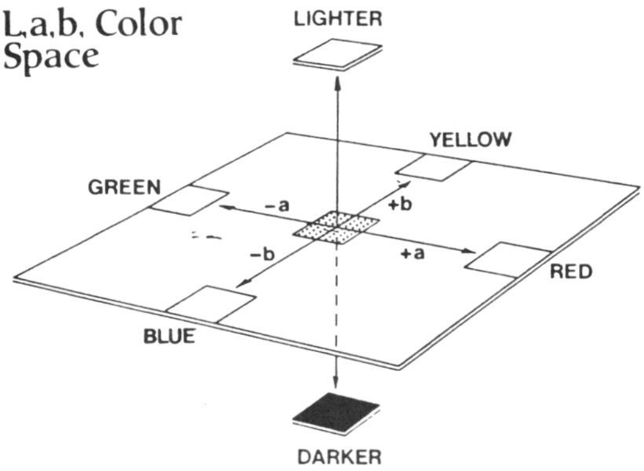

Fig. 11. Hunter L, a, b theory for color measurement.

wavelength and purity, respectively. The luminosity, or lightness, of the sample is the Y function measured by the colorimeter. Similarly, the coordinates can be plotted for two different samples and the dominant wavelength of one with respect to the other determined.

In the Hunter Lab system, illustrated in Fig. 11, the quantities L, a and b are calculated from the measured CIE tristimulus values. The equations for making these calculations are given in TAPPI test methods. In these equations, L is the measure of lightness and varies from 0 to 100. A positive a value indicates redness, and a negative value greenness. A positive b value indicates yellowness, and a negative value blueness. Hence, a blue-green sample would be negative in both a and b.

The CIE STARLAB system is similar to the Hunter Lab system except that the equations for calculating L, a and b are somewhat different. The quantities calculated are referred to as L*, a*, and b*. The claim is that the CIE System gives an improved definition of color over the Hunter system.

Both the Hunter Lab and STARLAB systems also provide for calculation of a color difference (Δ E) value between two samples. The intent is that a given Δ E value will represent the same visual perception of color difference anywhere in color space. Δ E is calculated as the square root of the sum of the squares of Δ L, Δ a, and Δ b; or L*, a*, and b* in the STARLAB system.

F. ELECTRICAL PROPERTIES

Paper is particularly suitable for electrical insulating purposes. It has good dielectric properties, mechanical and structural properties amenable to the required converting processes, and is low in cost. The most frequent uses in this area are as insulating material for electrical wiring and as dielectrics in capacitors. For certain applications, where high dielectric strength is required, the paper may be impregnated with oils, waxes, or resins. Some specialty papers used in certain photographic, recording, and printing processes will require more conductivity than the fiber/air structure of the sheet can provide. In these cases the paper may be treated to increase its conductivity.

Papers intended for use as insulating material have important requirements in electrical, mechanical, and chemical properties. The important electrical properties are resistivity (volume and/or surface), dielectric constant, dielectric loss, and dielectric breakdown strength. Mechanical properties include high strength (usually MD tensile and CD tear), and controlled levels of weight, thickness, density, and surface smoothness. Important chemical properties include moisture content, acidity, ash, and the number of conducting particles per unit area of paper.

Some of the important electrical properties, particularly resistivity and dielectric constant, depend on the orientation of the paper. For these properties, different results can be expected for tests made in the machine direction, the cross-machine direction, and the thickness direction of the paper. For papers intended to be used as an insulating material, the thickness direction property, or volume conductivity, is of primary interest. In other applications measurements made in the plane of the sheet, or surface conductivity, may be of concern. In the latter case, measurements may be in the machine direction, across the machine direction or, more commonly, the conductivity for all directions in the plane of the sheet are averaged together to obtain a single value.

1. Conductivity/resistivity

Conductivity of paper is usually determined by applying a controlled voltage through two electrodes which contact the sheet surface(s), and measuring the resultant current flow. Measurements may be made with either AC or DC equipment, and the reported results should indicate which was used.

The resistance of the material is then calculated as the ratio of the applied voltage to the resultant current. Conductivity or its reciprocal, resistivity, can then be calculated from the measured resistance and the geometry of the measurement system.

Commonly, but not always, planar measurements are calculated as surface resistivity and thickness direction measurements as volume conductivity.

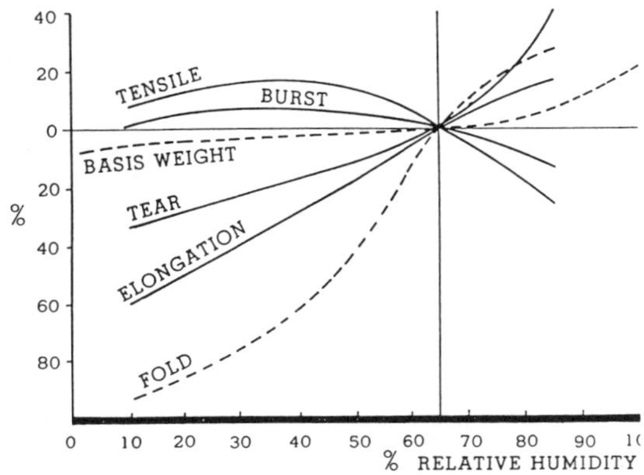

Fig. 12. Effect of relative humidity on paper properties.

2. Dielectric constant/loss

Dielectric properties are usually measured by placing the paper to be tested between the plates of a parallel plate capacitor. The measured capacitance will be greater when the space between the plates is filled with paper then when it is under vacuum. The dielectric constant is measured as the ratio of the two capacitances.

Since paper is not a perfect insulator, energy losses will occur when the capacitor, with paper between the plates, is charged. The energy loss must be dissipated as heat, a problem often encountered in capacitors. The amount of energy lost for a given voltage gradient is referred to as either dielectric loss, dissipation factor, or loss tangent.

3. Dielectric strength

Dielectric strength is a measure of the maximum voltage stress that the paper can withstand without irreversible changes in the resistivity of the paper. It is measured by placing the paper between electrodes, and then increasing the applied voltage until a measurable change in conductivity occurs. At this point the paper becomes highly conductive and is no longer suitable as an insulating material. The voltage at which this occurs is called the breakdown voltage and dielectric strength is usually expressed as the ratio of this breakdown voltage to the paper thickness.

G. RELATIVE HUMIDITY AND PROPERTIES

Almost all properties of paper are affected by the amount of moisture in the paper [18]. The most significant effect is on physical strength properties. Typical efffects are illustrated in Fig. 12. Tensile strength, for example, will change 50% or more over the range of relative humidities, with the measured strength decreasing as humidity increases. Tensile elongation, on the other hand, may increase more than 80%. Changes in compressive strength are similar to those for tensile. Tear resistance will increase about 50% and fold endurance more than 200%. Since increasing moisture causes tensile to decrease and elongation to increase, the effect of humidity on most other strength properties will depend on the relative contribution of force and elongation to their measurement. Bursting strength and tensile energy absorption, for example, increase with both increasing tensile and elongation. Hence, the effect of humidity on these properties is relatively small, except at very high humidities where significant decreases are usually observed.

Structural properties are usually affected to a lesser degree by changes in moisture content than strength properties. Thickness increases of 10% are common. Basis weight, of course, also increases. Hence the calculated apparent density (basis weight/thickness) is only mildly affected by changes in moisture, even though the absolute density decreases with increasing moisture. Air resistance of paper usually decreases, perhaps 10 to 20%. Hence, results for the Gurley instrument, which measures air resistance, decrease with increasing humidity, while the results for instruments which measure air permeability, such as the Sheffield and Bendtsen, increase.

The direction of change (increase or decrease) of most properties with changes in moisture content will generally be the same for all paper products. However, the magnitude of change can differ significantly. Hence, it is risky to make assumptions on the affect of moisture, and actual measurements should be made when the affect of moisture is critical to converting processes or use applications of the product.

Properties of paper measured in different laboratories, or at different times in the same laboratory, will not agree unless care is taken to ensure that the moisture content of the sample tested is controlled. For this purpose a specified testing environment of 23°C and 50% RH is used in most paper-testing laboratories. The moisture content also depends upon whether the sample is brought into this testing environment from a higher or lower relative humidity. Depending on the composition of the sample, its measured moisture at 50% RH will be about 0.5 to 1.5% lower in moisture content if moisture is absorbed from the air than if it is desorbed (see Fig. 13). This phenomenon is known as moisture hysteresis. This effect is controlled by specifying that all samples for testing be first "preconditioned"

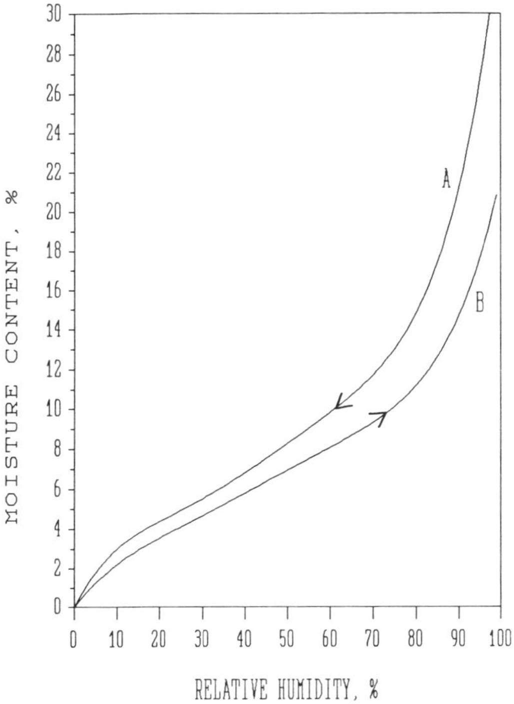

Fig. 13. Typical relationship between relative humidity and moisture content for paper.

at a low relative humidity before being exposed to the 50% RH test atmosphere.

1. Testing environments

The essential elements of any system for maintaining temperature and relative humidity within the required tolerances are the measurement system and the control system. The measurement system consists of a temperature sensor and a humidity sensor. The control system includes provisions for heating, cooling, humidification, and dehumidification. Humidification is usually accomplished by passing the air through a water or steam spray. For purposes of dehumidification, the air passes over chilled coils, operated at or below the required dew point temperature, so that condensation occurs. This dehumidification, of course, cools the air and reheating will be required. The condensor coils of the refrigeration system are often used to provide most of the required reheat, although steam or hot water coils can also be used. The heating stage is usually controlled with steam coils, hot water coils, or electrical heating elements.

In most modern humidity control systems, all four functions—heating, cooling, humidification, and dehumidification—operate continuously. The output of the in-room and in-system sensors modulate the level of control for each function. This system provides better control than on-off systems in which each function is controlled independently and turned on or off, depending on room demands.

The control limits specified in TAPPI and other test methods allow a tolerance of ±1°C for temperature and ±2% for relative humidity. It should be noted that temperature changes of 1° will not significantly effect paper properties. However, temperature shifts of that magnitude will make it difficult for any system to control the relative humidity within the required tolerance. Good temperature control is relatively simple and it is recommended that short-term fluctuations be controlled to better than ±0.3°.

The most common methods for measuring relative humidity are hygrometers of the mechanical, gravimetric, or electrical resistance type. These, however, require a reference standard for calibration. The basic simplicity and fundamental calibration of the psychrometric method make it particularly suitable for this purpose. The elemental form of the psychrometer comprises two thermometers. The bulb of one thermometer is covered with a moistened cotton or linen wick and is called the wet bulb, whereas the bulb of the other thermometer is left bare and is referred to as the dry bulb. The evaporation of water from the moistened wick of the wet bulb produces a lowering in temperature. The relative humidity can be calculated from the wet and dry bulb temperatures and the barometric pressure. A common practice is to use hygrometers for control purposes, and a psychrometer for monitoring.

2. Strain relaxation

Thus far we have only considered reversible effects associated with changes to moisture content. However, exposure of paper to high relative humidities can, and usually does, cause irreversible changes in the structure and mechanical behavior of the paper. The assumption is that the drying restraints imposed on the paper during manufacture are partially relaxed when the moisture content of the sheet is increased [19]. Hence, the sheet "shrinks" when the moisture is again reduced, by an amount greater than can be accounted for by normal hygroexpansivity. This net "shrinkage" results in significant increases in tensile elongation and in properties affected by elongation. There is some evidence that strain relaxation occurs at all moisture levels and is time dependent. At high moisture levels, measurable strain relaxation can occur in minutes or even seconds, while at lower levels months or years may be required. Strain relaxation has also been observed to occur upon mechanical working of the paper, and upon exposure of the paper to elevated temperatures.

H. STANDARDIZATION AND CALIBRATION

Probably the greatest challenge in paper-testing laboratories is that associated with the calibration of test instruments, and standardization of testing procedures. In this respect, tests for measuring paper properties can usually be placed in one of three categories.

First are those tests which can be easily checked in terms of recognized standards of force, mass, and distance. A good example of tests in this classification is basis weight or grammage. The important requirements are accurate measurements of the dimensions and weight of the test sample. Quite obviously the test sample must be a good statistical sampling of the test lot it represents, and the moisture in the sample must be controlled within specified tolerances. Both of these sources of variance are, however, independent of the accuracy of the test method itself.

A second classification are tests designed to closely simulate converting or end use processes. Tests of this type are often made on pilot-scale equipment differing from commercial equipment only in size. These types of tests do not often yield a test result in terms that can be checked against any standard value. Instead, correlations are often developed between performance on the pilot and commercial equipment.

The largest group of tests are those purported to measure properties related to converting processes and/or end use performance. These tests are usually simple to perform with inexpensive equipment, relative to tests made with pilot equipment. The proper selection of a combination of these tests will usually suffice in defining a satisfactory paper product for a given use. Examples of tests in this classification are bursting strength, tear resistance, and water drop. Test instruments in this classification often require custom-developed calibration and standardization procedures.

The essential elements of any instrument standardization system should include:

– Instrument calibration.
– Preventive and corrective maintenance.
– Testing procedure.
– Reference samples.

No single system is suitable for all instruments in the test laboratory, or even for the same type of instrument in different laboratories. The magnitude of the standardization system for a given instrument will largely govern its cost. This cost must then be balanced against the importance of the test to the product being manufactured. Regardless of the magnitude of the system finally selected, the custom-developed system for each instrument should include each of the above four factors to some degree.

1. Instrument calibration

In most cases, calibration of a measurement system involves independent checks of the components of the system. Some of these checks are required only once during the life of an instrument, others should be made on a periodic schedule, while still others should be made each time the instrument is used. For example, with a thickness gauge, the platen diameters, lowering speed of the upper platen, and linearity of the scale need to be checked only when a new or refurbished instrument is first placed in service. These should not then change during the life of the instrument. The frequency of other checks would depend on the amount of use of the instrument. The pressure applied to the sample might be checked quarterly, the platen parallelism monthly, and the scale calibration weekly. The zero setting of the scale and the cleanliness of the platen surfaces should be checked each time the instrument is used.

TAPPI test methods include calibration procedures for many test instruments. These have been developed based on years of experience, and following them exactly will usually yield satisfactory results. Instruction manuals provided by instrument manufacturers can also be quite helpful. In some cases the laboratory will need to be innovative, depending on available resources. With the thickness gauge, for example, the required pressure on the sample is specified, but the laboratory will need to develop the auxiliary equipment needed to check it.

Some laboratories prefer to contract their instrument calibration to instrument manufacturers and other outside sources. In some cases this may be necessary because of the cost of calibration equipment or because of customer requirements. In most cases the development of in-house expertise is preferred.

2. Preventive and corrective maintenance

Good preventive maintenance practices will help ensure satisfactory performance and long life for expensive test equipment. Included in these practices are cleaning, lubrication of moving parts, frequent inspection of critical components, and necessary steps to prevent damage to the instrument and parts. With regard to the latter, instruments and their supporting tables should be located to avoid bumping by laboratory traffic.

The major consideration in corrective maintenance is to restore the instrument to proper operating condition. Minor repairs can usually be handled in-house, whereas instrument manufacturers are often better equipped to handle major repairs. In-house repairs should not be undertaken unless the person doing the repair work has a good appreciation for the relationship between the quality of his work and the ultimate performance of the repaired instrument. Regardless of where the corrective action is taken, it should be followed by a complete calibration check of all affected components before the instrument is placed back in service.

3. Testing procedures

Most paper properties are affected by a number of variables, many of which are separate from the instrument measurement system itself. Among these are sample orientation and dimensions, testing environment, testing speed, and definition of the final test result. In many cases there may be no technical justification for selecting any particular value for these. Numerous examples can be cited: the length of the specimen for tensile testing; the exact spectral distribution of light for brightness tests; the relative humidity at which tests are made; the hydraulic pumping speed of the burst test – to name a few. However, different laboratories cannot be expected to obtain comparable test results unless the same procedures are used in each laboratory.

Hence, a major function of test methods is to assign specific values for all pertinent variables in a given test. The most widely recognized methods in the United States are those developed by the Technical Association of the Pulp and Paper Industry (TAPPI). These methods have been agreed to by representatives of all segments of the industry. Other countries of the world have developed their own methods which sometimes differ from the TAPPI methods. The International Standards Organization (ISO) has been instrumental in developing compromise methods for use in world trade.

Many laboratories have developed their own testing procedures which may differ somewhat from the TAPPI methods. In many cases changes are made as an economy measure to increase the productivity of the laboratory. Doubling the testing speed for the tensile test, for example, could save $18,000 per year if tests are run continuously. In other cases, procedural changes may be required by customer specifications, by product size available for testing, or by time available to complete the testing. In any case, the laboratory should determine the effect of procedural changes on test results for the product being tested before the changes are made.

4. Reference samples

A complete standardization system for any test instrument should include periodic testing of a sample material for which a reference value has been assigned. Depending on the importance of the test, this might include in-house samples, purchased reference materials, round robin testing with laboratories having a common interest such as customer and supplier, participation in industry-wide collaborative testing programs, or any combination of these.

The use of in-house control samples is the preferred method for long-term monitoring of the performance of a single instrument. The sample selected for a given instrument should be one for which the property of interest is not likely to change over a period of at least several years. In this regard, a sample which is at least several months old is preferred over a newly manufactured material. Preferential storage is in a dry atmosphere well below 50% RH. Steps should be taken to ensure that each periodic sample is randomly chosen from the entire lot. The procedure for conditioning, cutting if necessary, and testing of the control sample should be carefully prescribed. The periodic test results should then be evaluated using control chart procedures. These procedures can be found in many statistical handbooks. It should be remembered that in-house control samples provide a reliable monitoring of the precision, but not necessarily the accuracy, of the test to which it is applied.

Purchased reference samples provide the best opportunity for monitoring the accuracy of a test method. These samples will have an actual value assigned to the sample when it is received by the participating laboratory. The assigned value will have been determined in a central laboratory having a high degree of technical competence in the property of interest. An example of this type of program is the brightness and opacity standarization conducted by The Institute of Paper Science and Technology (formerly The Institute of Paper Chemistry).

Round-robin testing is an effective way of monitoring the relative performance of two or more instruments in several laboratories. Frequently a third laboratory, usually a disinterested laboratory with a high level of recognized competence, will be invited to participate in the role of referee. Randomized samples are distributed to each participating laboratory along with detailed procedures for performing the test. The sample used is usually similar to the one in which the laboratories have a common interest. Care should be exercised in preparing the samples for shipment so that the property of interest is not changed due to handling or exposure of the sample to harsh environments. As a minimum, the sample should be protected between sheets of heavy board and wrapped in a moisture-proof package. Following testing, the results from the participating laboratories are compared using statistical techniques.

Collaborative reference testing permits a laboratory to compare their test results to a value representing the average results for a large number of participating laboratories. A central laboratory distributes randomized samples to all participating laboratories along with instructions for testing the samples. After all the test results are collected, the central laboratory tabulates the data and reports the results to the participants. The most significant programs in the U.S. are those administered by Collaborative Testing Services in Herndon, Virginia. Two programs are presently available of interest to the paper industry. The first is co-sponsored by the Process and Product Quality Division of TAPPI. It includes a wide variety of tests commonly found in paper-testing laboratories. Each laboratory can select the particular tests in which it is interested. Samples for this program are sent out six times per year.

The second program is co-sponsored by the Containerboard and Kraft Paper Group of the American Paper Institute. Tests included in this program are those of special interest to manufacturers of linerboard and corrugating medium. Samples for this program are tested weekly.

I. FUTURE TRENDS IN TESTING

Converting processes of the future will continue to become more complex, and they will be designed to operate at continuously increasing speeds. This trend will translate to greater demand for paper of uniform properties. The requirements for this paper will be custom designed to meet the needs of the converting process.

Many of the present-day tests which are not fundamental in character will gradually disappear. These will be replaced by more sophisticated tests which provide more useful information related to specific converting processes or end use performance. More of these will be performed on-line, during the manufacturing process.

Laboratory tests will become increasingly automated to provide data closer to real time. The demands for improved quality in product will be matched by similar demands on testing data. Managers of testing facilities will need to apply all of their skills to meet this demand.

In the longer term, it will be possible to determine a set of process conditions which will yield the desired fundamental properties. These process conditions will be measured and controlled manually (e.g., open loop) or controlled automatically (e.g., closed loop). Those desiring more information on this mechanization should consult material on Statistical Process Control (SPC) or Statistical Quality Control (SQC).

REFERENCES

1. The Dictionary of Paper, 4th ed., The American Paper Institute, Inc. (1980).
2. Nissan, A.H., "Lectures on Fiber Science in Paper", in Pulp and Paper Technology Series No. 4, Joint Textbook Committee of the Pulp and Paper Industry, CPPA and TAPPI (1977).
3. Bradway, K.E., "Factors Which Affect the Printing Smoothness of Handsheets", Tappi 56(8) (Aug. 1973).
4. Wink, W.A. and Baum, G.A., "A Rubber Platen Caliper Gage - A New Concept in Measuring Paper Thickness", Tappi 66(9) (Sept. 1983).
5. Setterholm, V.C., "A New Concept in Paper Thickness Measurement", Tappi 57(3) (Mar. 1974).
6. Fabbri, I. and Laurenzi, A., "Methods to Evaluate Paper Smoothness", Penrose Ann., Vol. 68 (1975).
7. Parker, J.R., "Development and Applications of a Novel Roughness Tester", Tappi 54(6) (June 1971).
8. Gallay, W., "Stability of Dimensions and Form of Paper", Tappi 56(11) (Nov. 1973).
9. Britt, K.W., "Physical and Chemical Relationships in Paper Sheet Formation", Tappi 63(5) (May 1980).
10. Crawford, E.A. and Strain, M., "The Beta-Ray Basis Weight Gage", Tappi 33(4) (Apr. 1950).
11. Silvy, J., "Effects of Drying on Web Characteristics; Part 1 — Web Structure and Physical Characteristics of the Sheet", Paper Technology 12(5) (Oct. 1971).
12. Van den Akker, J.A. et al., "Importance of Fiber Strength to Sheet Strength", Tappi 41(8) (Aug. 1958).
13. Setterholm, V.C., "Factors that Affect the Stiffness of Paper", from Properties of Paper Related to its Uses, Vol 1., Bolan F., ed., p. 253 (1976).
14. Baum, G.A., et al., "Orthotropic Elastic Constants of Paper", Tappi 64(8) (Aug. 1981).
15. Whitsitt, W.A., "Theory, Calibration, and Use of Bursting Strength Testers" Parts 1, 2, and 3, Boxboard Containers 80(1) (Aug. 1972), 80(2) (Sept. 1972), 80(3) (Oct. 1972).
16. Davison, R.W., "The Sizing of Paper", Tappi 58(3) (Mar. 1975).
17. Crouse, B.W., and Snow, G.H., "Fluorescent Whitening Agents in the Paper Industry; Their Chemistry and Measurement", Tappi 64(7) (July 1981).
18. Nissan, A.H. "Effects of Water on Youngs Modulus of Paper", Tappi 60(10) (Oct. 1977).
19. Schultz, O.H., "The Effect of Straining During Drying on the Mechanical and Viscoelastic Behaviour of Paper", Tappi 44(10) (Oct. 1961).

BIBLIOGRAPHY

1. Handbook of Physical and Mechanical Testing of Paper and Paperboard, Vols. 1 & 2, R.E. Mark, ed., Marcel Dekker, Inc. (1983).
2. Handbook of Paper Science, Vol. 2, H.F. Rance, ed., Elsevier Scientific Publishing Company (1982).
3. Waterhouse, J.F. and Van Eperen, R.H., Relative Humidity and Paper. Part 1. The Influence of Environmental Conditions on the Physical Properties of Paper and Board; Part 2. The Production and Management of Testing Environments, TAPPI Notes: Management Techniques for Optimizing Laboratory Operations Seminar (1985).
4. Sara, H., Factors Affecting the Practical Measurement of Formation, EUCEPA Conf. (London) 19, Paper no. 25 (May 21-24, 1979).
5. Seth, R.S. and Page, D.H., The Stress-Strain Curve of Paper, Role of Fundamental Research in Papermaking, BPBIF Symposium held at Cambridge (Sept. 1981).
6. Swanson, J.W., Mechanism of Paper Wetting: Dynamic Wettability, Proc. TAPPI Sizing Short Course, Atlanta, GA (1980).
7. Mendel, J.W. and Davis, C.V., Physical Test Lab Automation VIA Dedicated Minicomputer, Tappi 58(3):107-109 (Mar. 1975).
8. Gallay, W., Stability of Dimensions and Form of Paper. Parts 1 and 2, Tappi 56(11) (Nov. 1973), 56(12) (Dec. 1973).
9. Subject Index and Listing of American Paper Institute Instrumentation Reports and Series, Institute of Paper Science and Technology (revised annually).
10. Lyne, M.B. et al., The Fundamental Properties of Paper Related to Its Uses, BPBIF (London), Vol. 1 p. 269 (1976).
11. Van den Akker, J.A. et al., Evaluation of Absolute Reflectance for Standardization Purposes. J. Optical Soc. Am. 46 (1956) and 56 (1966).
12. Bureau, W.H., Opacity Gets a Bigger Role, Graphic Arts Monthly, p. 68-70 (Mar. 1977), p. 88-89 (Apr. 1977).
13. Popson, S.J., Diffuse vs. Directional Brightness Measurement, Technidyne Corp., Tech. Bull. 101 (1982).

II
FIBROUS RAW MATERIALS FOR PAPERMAKING

R.H. REEVES

R. Heath Reeves has been involved with pulp and fiber use or processing through most of his business and teaching career. He has a B.S. degree in Chemical Engineering from the University of California, Berkeley followed by M.S. and Ph.D. degrees from The Institute of Paper Chemistry. Today, Dr. Reeves is a James River Fellow at the Corporate R&D Center in Neenah, WI providing fiber technology support to various business units for both product and process improvements. He has been an active member of TAPPI since 1965 and has authored several papers.

REVIEWED BY

John A. Heitmann, Jr.
Associate Professor
Dept. of Wood and Paper Science
North Carolina State University
Raleigh, NC

Hasan Jameel
Associate Professor
Dept. of Wood and Paper Science
North Carolina State University
Raleigh, NC

A. INTRODUCTION

Paper and paperboard products have specific use requirements which depend on an empirical mix of properties, market needs, price, and the ability to do its intended job. An unusual feature is the variable composition or furnish from which paper products are made. Raw materials range from inexpensive recycled fibers and mechanical pulps to higher valued, fully bleached pulps and specialty fibers. Equally important to papermaking are the many additives used to modify specific properties. The important role of the non-fibrous components, such as wet and dry strength resins, fillers, pigments, dyes and the like, are discussed in Vol. 6 of this series [1]. As new uses for paper and board products are found, the non-fibrous materials play a major role in the developments.

Most relationships between fiber characteristics and paper properties are defined broadly in terms of the fiber origin (hardwood, softwood, or non-wood) or the nature of the processes used to liberate them (kraft, sulfite, or mechanical pulping). Recent studies have substantiated some of the relationships between fiber morphology and product properties. Among the most notable would be the extensive radiata pine studies reported by Kibblewhite et al. [2-5] and the equally valuable contributions on various eucalypts by Claudio-da-Silva et al. [6-8].

Fiber features considered to be important were reviewed by Dinwoodie [9, 10], who also pointed out apparent anomalies in the relationships between fiber or wood characteristics and the properties of paper made from them. A detailed discussion of each wood or fiber characteristic or every paper and board property is clearly impractical. In this chapter, attention is directed toward the major types of fibers (pulps) we use and how their properties impact some products made from them.

B. SOURCES OF PAPERMAKING FIBERS

Almost any vegetable fiber can be used to make some kind of paper, but only a few fiber sources are commercially important. Figure 14 shows that wood fibers account for about 94% of the papermaking pulps used worldwide while the 6% of non-wood fibers comprise mainly flax, cotton, bagasse, bamboo, and various straw pulps. To assess the papermaking potential of fibers, one should consider their basic morphological properties, plant source and processing conditions, along with climatic or other factors which affected the plant growth. (Secondary fibers or recycled waste papers are derived

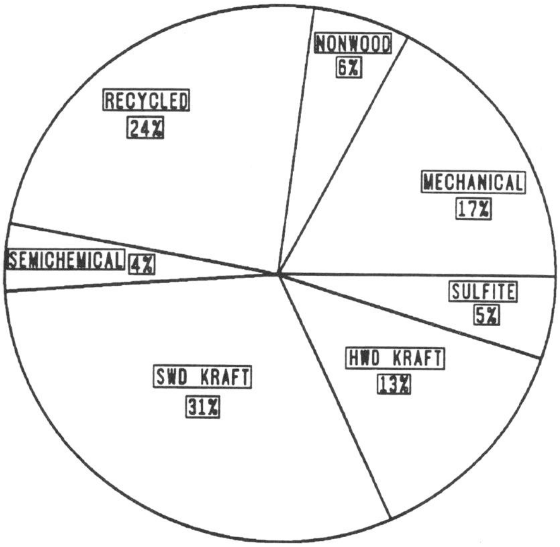

Fig. 14. World use of papermaking fibers by type.

Fig. 15. Mixed northern hardwood and softwood kraft pulps (1:1).

mainly from woodpulps so their properties are essentially those of the initial pulps used except as modified by being dried or refined one or more times.)

Good correlations are reported for wood and fiber features and the paper properties that result. Considering all the intervening processes (i.e., chipping, cooking, screening, cleaning, washing, bleaching, and refining), such relationships reinforce the large influence of inherent wood and fiber properties on the final pulp quality [11], whether from one species or many. Despite rigorous pulping and other treatments, individual fiber properties are still recognizable in the final product and contribute to the usefulness of the finished paper.

Differences seen in pulp, paper, or board properties usually originate with the intrinsic structural characteristics of the fibers from which they are made and many paper/board functional properties can be inferred broadly from the known gross differences between wood species. Thus, fiber length and width, dimensional ratios, basic fiber strength, wood density, fiber coarseness, cell wall thickness, early- and latewood fractions, number of fibers per gram, and other wood/fiber properties have been reported to account for the performance variability of products made from the fibers.

With a typical sheet of paper comprising millions of fibrous elements, it is not easy to picture the importance of an individual fiber's structure or other characteristic on product properties. Paper furnishes are commonly blends of long and short fibers as seen in Fig. 15 which is a 1:1 mixture of northern hardwood and softwood kraft pulps. The heterogeneity of the fibrous elements is typical for pulp blends with the long, flattened fibers coming from softwood species and most of the remaining elements originating with the hardwood pulp. The short, narrow, relatively opaque elements in the figure are tracheids and fibers from hardwoods as are the vessel elements and small parenchyma cells found in almost all broadleaf species. Clearly, the softwood fibers are wider and much longer and are expected to contribute materially different properties to the product. Due to natural size differences, there may be 3-7 times more short-fibered hardwood elements than softwood fibers (at equal weight). Some of the fragments in Fig. 15 constitute the fines which are a natural artifact of the processing stages and are an important part of most papermaking furnishes.

Chemical properties, such as the cellulose, hemicellulose, or lignin content, also influence papermaking. These, however, seem to be less reliable indicators of quality so that knowledge of the chemical composition is of more limited value in selecting the best pulp, or combination of pulps, to make paper products with existing mill equipment. This does not suggest that such differences in chemical composition can be ignored as much of the chemical variation between hardwood, softwood, and non-wood fibers remains after pulping and bleaching. These operations clearly can be used to control some aspects of the papermaking potential of the pulp.

C. LINKING PULP TO PAPER PROPERTIES

Pulp properties are related to both the chemical and physical nature of the fiber elements. Fiber structure determines the ease and manner in which fibrillation and fiber breakage occur so that intrinsic fiber strength is important along with the plasticity and length of the fibers. In this regard, the distribution curve for the properties can often be more critical than just average values. Much has been done already gathering specific knowledge concerning interrelationships between wood structure, chemical composition, and strength properties of various pulps as will be seen here.

To evaluate a new pulp, it is common practice to compare its properties with another on the basis of laboratory beater curves and handsheet tests. Improved use of papermaking pulps, however, requires more knowledge of the basic factors that influence pulp behavior, i.e., how do wood and fiber characteristics affect final pulp strength and the properties of the products made from pulps.

1. Evaluating pulps

No particular pulp is universally suitable for all papermaking but each must be evaluated for use as a component in specific products. Analyses should involve comparison with realistic alternatives, i.e., pulps of known performance. The ultimate objective of the pulp testing then is to obtain wood or fiber quality criteria that can be used to predict the properties of papers made from them. The data on relationships between wood and pulping variables mainly involve the classical tests for tensile, burst, and tear strength along with some optical and appearance properties. Very often, much effort is expended in testing pulp samples but more information is available than just an indication of pulp quality or how it compares to another sample. Fundamental causes for changes in the balance of physical properties, large enough to be important, can be established by closely examining the results of the normal pulp tests.

2. Bonding

Virtually all mechanical strength properties of paper depend on inter-fiber bonding. In addition to bonding, the mechanical strength properties are influenced by other factors such as fiber length, strength, stretch properties and others. The extent of fiber bonding is related to the inter-fiber contact area and to the bond strength per unit area of inter-fiber contact. As the contact area between fibers is related to the number of collapsed fibers or fiber conformability and sheet density, then changes in specific bond strength should lead to changes in the sheet strength. The influence of fiber strength on this relationship becomes evident as inter-fiber bonding increases. In well-bonded sheets, failure occurs as a result of fiber failure rather than bonds breaking but, for damaged fibers, the chance for rupture is greater. In this case, the strength of a well-bonded sheet is lower than it would be if fiber bonding was the sole controlling factor [12].

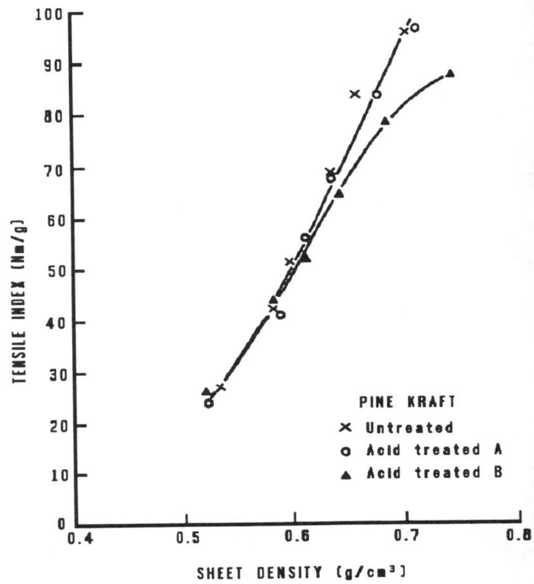

Fig. 16. Effect of fiber damage as seen in the tensile index and sheet density relationship.

Tensile strength is higher, at a given sheet density, when bond strength is better, and the slope of the line relating the two parameters is also greater in most cases. As the tensile strength of the sheet increases, the intrinsic fiber strength exercises a greater influence on the paper strength and tensile is lower for damaged fibers. The greater the fiber damage, the lower the value at which the tensile/density curve deviates from that of undamaged fibers as seen in Fig. 16 where fibers were treated with dilute hydrochloric acid. A 14% reduction in the intrinsic fiber strength had little effect; however, a further reduction in intrinsic fiber strength to only 75% of the untreated value begins to show an early loss in paper tensile strength.

Nordman [13, 14] observed that tensile was directly proportional to relative bonded area for many pulps within certain limits of refining and wet pressing and Van den Akker [15] established that both fiber-to-fiber bonds and fibers are broken when a sheet fails. Theoretical considerations [16, 17] indicate that sheet structure, intrinsic fiber strength, and inter-fiber bond strength are all contributing to the strength of paper.

Structural differences are a factor also in pulp strength and other properties. Fibers from southern and western species minimize, relatively, the inter-fiber contacts so that bonding potential is lower than with northern pulps. Such fibers do not form as well, give a rougher surface and, owing to the solidity of such fibers, they are also more difficult to refine than most northern fibers. Changes in sheet structure due to refining can be monitored by observing the collapse of fiber lumens. As expected, increased refining or wet pressing causes a higher proportion of collapsed fibers.

3. Chemical properties

The chemical composition of wood pulps is of indirect value as a measure of pulp quality but suggests expected physical behavior when viewed together with data on the origin and processing history of the fibers in question. Limited work has been done regarding the influence of small changes in composition on pulp strength properties. The chemical composition and ultrastructure of cellulosic fibers is presented in detail in Vol. 1 [18] of this series.

The main component of papermaking pulps is cellulose and in high-yield pulps much of the other wood constituents are present also. Differences in the lignin content, alpha cellulose, or percent extractives all contribute to differences in pulp yields and properties. For chemical pulps, quality is largely decided by the fiber character, although the chemical constituents are also important to sheet strength. Hemicellulose and lignin, for example, influence strength development favorably and unfavorably, depending on the content of these constituents.

Viscosity testing of pulps is an attempt to measure the chain length of the cellulose in the fibers. Pulp strength tends to rise with increasing viscosity, however, different properties are influenced to a varying extent. In any case, beyond a limiting viscosity, the various strength properties all tend to level off.

4. Mechanical properties

In poorly bonded sheets, intrinsic fiber strength may be of little significance but, in well-bonded sheets, this feature may be the limiting factor [19]. Paper with poor formation (or fiber mass distribution) is also weaker than a corresponding sheet with better formation. Leech [20] observed large increases in burst and tensile strength with improved formation, while tear resistance was less sensitive to the fiber distribution.

For the mechanical strength properties of paper, fiber length influences the stress distribution; fiber bonding is needed to hold the structure itself together, while intrinsic fiber strength is vital to prevent failure in the fibers themselves. Chemical pulps increase in tensile with increased fiber length while mechanical and high-yield pulps show an opposite trend. Normally, sulfite pulps behave like kraft pulps in that the whole pulp is stronger than any of the fractions. This may be why contradictions are found in the literature [21] concerning the role of long fibers, short fibers, and fines in predicting bonding and tensile strength properties for chemical pulps.

For tear strength, chemical pulps show a marked reduction as fiber length is reduced. That is, the tear strength for chemical pulps is strongly influenced by longer fibers which distribute the stresses over a wider area. For groundwood pulps, there is a smaller tear dependence on fiber length [22], but the newer high-yield pulps show a better fiber length/tear relationship. Papers that have relatively strong bonds spaced far apart from each other give high tearing strength compared to a more densely bonded network. The tearing forces are distributed over a large area around the tear path in more loosely bonded structures and this is especially true for long and strong fibers. In dense papers, the same tearing force is applied at the fracture point and this stress concentration ruptures the fibers along the progressing failure line. Thus, the ability to distribute the applied stress greatly influences the tearing resistance of papers.

Sheet density and zero-span tensile account for most of the burst variance with some fiber length influence [15]. High burst strength correlates with thin-walled, long, collapsible, and flexible latewood fibers. For breaking length, fiber properties similar to those for high burst are necessary. Fiber length variations with tree age and species have been studied by Jayme [22].

To a first approximation, the mechanical strength of paper depends on the strength of the individual building units and the forces which hold the fibers together in the structure. The intrinsic fiber strength is important for maximum tensile and tear levels, especially as refining increases. Equations for predicting paper properties have been developed that include an expression for intrinsic fiber strength [16, 23-25].

Sheet apparent density and zero-span tensile correlate with most of the variance in breaking length. Paper tensile strength reaches a limit that is always well below the experimentally determined intrinsic fiber strength. This suggests that fiber-to-fiber bonding (and not intrinsic fiber strength) is still the most important factor determining paper strength, especially when there is a mixture of fibers in the furnish.

Fiber strength and sheet zero-span tensile are related and the relationship improves when sheet apparent density is taken into account. Both fiber strength and sheet apparent density account for much of the total variation in pulp/sheet tensile and burst. These factors, together with fiber length, are also responsible for most of the variation shown in tear. For some hardwood kraft pulps, fiber length correlates well with such sheet properties as zero-span tensile, breaking length, burst, and tear factor [26, 27]. Fiber rigidity and sheet bending stiffness are also significantly correlated while apparent density and fiber bending stiffness accounted for the modulus of elasticity of the handsheets.

Krkoska et al. [28], in a study of nineteen hardwoods and three softwoods, found that morphological features did not correlate that well with strength properties and that refining reduced the influence of the species morphology on the papermaking properties of the pulps.

5. Optical and surface properties

The free surface area in a sheet is reduced by fiber bonding which diminishes the light scattering ability of the pulp which, in turn, depends on both the fiber surface area and the size and distribution of voids in the structure. The scattering coefficient is quite uniform for chemical pulps prepared from dense woods but falls off linearly with bulk for pulps from intermediate- or low-density woods. Pulps from high-density wood species, as part of the furnish, offer the advantage that developing paper strength by beating will not necessarily reduce the opacity and brightness proportionally [29].

Many workers report that fine, less coarse pulps made smoother papers for letterpress printing. For southern softwoods and hardwoods, the smoothest sheets result when final bulk comes from calendering action rather than compaction resulting from refining or wet pressing. Bradway [30] developed smoothness constants which were related to coarseness as seen in Fig. 17. The good correlation indicates that smoothness is governed largely by this property. This is not surprising as one expects coarse fibers positioned at the surface to make for a rougher finish. For many species, coarseness is related to fiber length.

The three-dimensional texture of paper is influenced by the shape and dimensions of the fibers so that fiber flexibility, the filling of voids by fine fibers, fiber length and the degree of bonding all affect bulk. Wood density correlates well with the strength, optical, and bulk properties of paper. Basic wood density also correlates with coarseness and wall thickness which affect smoothness and bulk. Clearly, for fiber crossings at or near the sheet surface, collapsed fibers protrude less than uncollapsed fibers. Thus, cell wall thickness is a controlling factor in roughness, as the cell double wall represents the limiting thickness of the collapsed fiber.

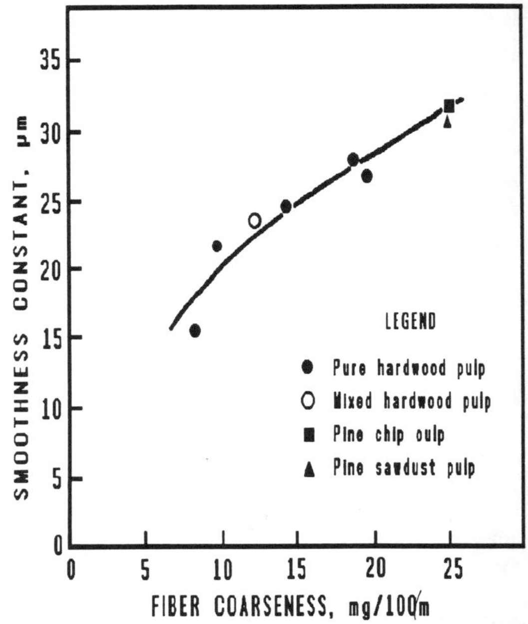

Fig. 17. Smoothness as a function of fiber coarseness.

Thin-walled fibers, which already have sufficient flexibility and lateral conformability to provide a smooth surface, benefit little from refining, while thick-walled fibers become more conformable with concurrent smoothness improvement. The relative changes in tensile, burst, tear, and bulk as a result of refining are smaller for pulps from low-density wood than for pulps from intermediate- or high-density wood species.

6. End-use properties

Common strength measurements are often emphasized with less attention given to properties such as linting, fiber puffing, water absorption, dimensional stability, bulk loss during wet pressing or calendering, and so on. In papermaking we identify critical pairs of properties where

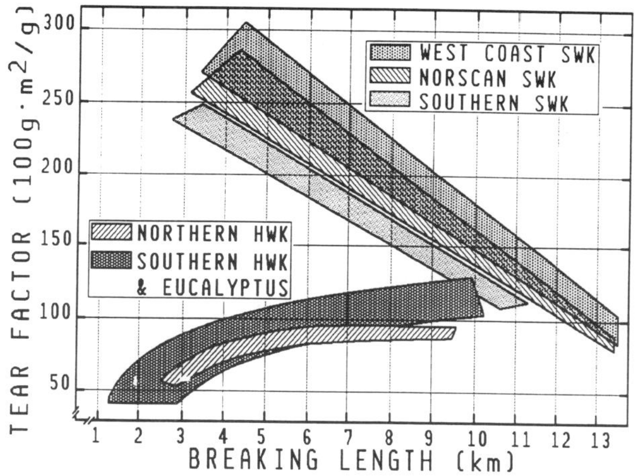

Fig. 18. Tear-tensile relationship for various market pulps.

raising the value of one is always accompanied by a decrease in the other. Tear versus tensile or opacity versus density are examples of such critical pairs. Strength relative to drainage properties is also important as many paper machines are limited by their ability to remove water. For chemical pulps, increased refining for strength development slows the stock. High-yield CTMP has a higher drainage rate than TMP or SGW at given strength levels and tear remains high and almost constant over a wide tensile range. However, the opacity of CTMP is lower than TMP or SGW so that more CTMP may be needed to meet important optical properties. High-yield pulps are less flexible when wet than chemical fibers and in some cases exhibit a greater tendency toward flocculation.

Figure 18 shows the important inter-relationship between tear and tensile, a critical pair of properties often used to assess pulp quality. Wood pulps from various growth regions cover a range of tear-tensile values but with hardwood and softwood pulps occupying clearly different regimes. Results from standard pulp evaluations show West Coast and Norscan softwoods with considerable overlap in their tear-tensile values while Southern softwoods exhibit a lower balance of properties. This is a typical output from standard pulp evaluations but does not directly relate to the results one expects in papermaking where Southern fibers find wide use in very strong products such as shipping containers and multi-wall sacks. In a similar fashion, hardwood pulps show an overlap in properties but in a lower range. Short-fibered, hardwood pulps bond well and develop good tensile strength but virtually never reach the high tear values of the long fibered, softwood pulps.

7. Interdependence of properties

Good correlations are often seen between some fiber and paper properties which does not mean that the fiber parameter concerned is the only causative factor. The inter-relationships observed between features within a species, i.e., fiber length and fibril angle or wall thickness, are not as clearly demonstrated between species. Therefore, it is not surprising to find disagreements over the significance of various fiber characteristics and some mutual dependence between many sheet properties is commonly accepted. For example, a high tensile and burst strength are usually seen together, especially if the stress-strain curve shows high elongation. High stretch correlates with good fold endurance while high burst and tensile are usually obtainable only at the expense of tear. Size press treatments can raise burst and fold materially but invariably with a sacrifice in tear. For most papers, increased tear, fold, and stretch are observed at higher moisture contents. In general, the papermaking properties of pulps are intimately dependent on fiber characteristics [31, 32].

Much effort was directed at measuring the surface area of fibers as it relates to fiber bonding and paper strength. The work of Mason [33] and Thode [31] suggests that the external specific surface area of fibers does not correlate very well with strength properties. Increases in specific surface area are reflected in the Water Retention Value (WRV) which measures both inter- and intra-fiber water in a pad. Thode [31] and others reported good correlations for WRV and many paper properties. Luner [34] showed that changes in WRV depend upon how the fibers were treated and identified two mechanisms for water retention. First, well-fibrillated fibers have a high specific surface area which increases WRV and ultimately sheet density. Secondly, cut fibers show an increase in WRV related to changes in fiber length and ultimately in the pad structure.

High-intensity refining caused changes in fiber length which influenced the water-holding capacity of the pad and the WRV increased while wet fiber flexibility was unchanged. The shorter fibers formed a more uniform pad but the pad density remained low since the fibers had not been fibrillated or delaminated. In contrast, low-intensity refining increased specific surface area leading to greater wet flexibility and WRV due to fibrillation and delamination of the fibers. It is these changes in fiber structure that are responsible for the improved paper properties observed for fibers with higher WRV values. Alone, an increase in either specific surface area or water retention value does not ensure good correlation with sheet strength.

D. ROLE OF PULPING PROCESS

The fibrous raw materials that we use are broadly classified as either mechanical or chemical pulps. The major pulping processes are thoroughly presented in Vols. 2, 4 and 5 [35–37] of this series. Mechanical and high-yield pulps are those where fiber separation results mainly from mechanical treatment so that most of the original wood substance is preserved. In contrast, chemical pulp is recovered from wood by removing much of the original material by chemical treatments. There are also many intermediate pulp types that combine various chemical and mechanical treatments.

The properties obtained from pulps of the same wood origin by different processes reflect both chemical and physical changes to the fibers. Fiber dimensional changes are of greatest concern for the mechanical processes where fiber bundles are torn from the wood chip or bolt and subsequently broken down. The chemical processes liberate the fibers more or less intact. Kraft pulping does less damage to the cell wall structure than sulfite pulping and makes the strongest pulps; mechanical pulps give the lowest strength. There are, however, other fiber requirements besides strength which means we must maintain a proper balance between preserving fiber length and the generation of good quality fines in any of the basic processes.

When focusing on the influence of wood and fiber structure on the properties of pulps or papers derived from them, one might overlook the influence of wood or fiber morphology on the pulping process itself. For production throughput, pulping operations benefit from higher density woods of the same species. However, quality considerations suggest that pulp properties may change in an undesirable direction as they appear to be inherent functions of wood properties affected by density. In chemical pulping, for example, differences between earlywood and latewoood create some variability as the lower density, porous earlywood lets cooking liquors penetrate readily, while dense latewood will require a longer treatment time. Also, earlywood tends to have a higher lignin and extractives content [38].

1. Mechanical pulps

The principal mechanical and high-yield pulps are groundwood (SGW/PGW), chemi(thermo)mechanical (CMP/CTMP), and refiner mechanical (RMP/TMP) along with several variants or subgrades. Process descriptions for these pulps are covered in earlier volumes of this series. Stone groundwood (SGW) has very high yields as only fine particles and water-soluble materials are lost during the process. These pulps have low strength but possess the good bulk, ink absorption, opacity and stiffness desirable for many products. When reinforcing, long fibers are needed to provide strength, as for newsprint, the furnish may still comprise 80-90% groundwood pulp. Pressurized groundwood (PGW) yields pulps with superior tensile and tear at a given freeness over SGW with little change in either the optical properties or energy consumption.

Refiner mechanical pulps (RMP/TMP) exhibit a shift to longer fibers and better strength properties than groundwood. RMP may replace or be mixed with groundwood in many papers. With TMP, from pressurized refiners, the wood fibers are separated more completely and with less damage giving pulps with strength properties beyond the normal capability of other mechanical processes. The strength increases, however, are usually accompanied by a lower scattering coefficient.

Chemimechanical pulping combines chemical and mechanical treatment of wood chips in disk refiners that may be pressurized (CTMP) or atmospheric (CMP) depending on the process needs. These systems offer improved process control, pulp uniformity, production flexibility, and reduced dependence on chemical pulps in many grades without jeopardizing paper quality or runnability. In terms of fiber length distributions, CTMP is more like a chemical pulp than a mechanical pulp; however, in all other respects, CTMP behaves very much as expected for a high-yield or mechanical pulp.

Other developments include extrusion mechanical pulp (BIVIS), explosion pulps (STAKE), ultra-high-yield sulfite pulps and the like. These modifications are not simply upgrades of coarse groundwood through post-refining systems but represent major changes to high-yield, mechanical pulp processing. The diverse chip refining systems also permit the use of a greater variety of wood species or wood wastes for improved resource utilization.

The major factors which determine high-yield pulp quality appear to be the nature and composition of the fiber surfaces, the surface chemistry and morphology of the fines, and the fiber collapse behavior [39]. High-yield and mechanical pulps are characterized as having low brightness and strength with greater opacity and increased tendency toward brightness reversion than chemical pulps. Therefore, they are used mainly where opacity and bulk are beneficial but the lower brightness and greater reversion can be tolerated. In general, high-yield or mechanical pulps do not hydrate as readily during processing so may provide good drainage in papermaking.

The quality of mechanical or high-yield pulps depends largely on the fiber length distribution, mainly the relative proportions of long fibers and fine particles. The good printing response of groundwood is due to its high bulk, resiliency, good smoothness and ink absorption. These pulps require freshly harvested wood to achieve the highest unbleached and bleached brightness levels and to avoid potential discoloration from micro-organisms. Fresh wood also ensures that the stock to the machines will be free of impurities, shives, pitch, and other undesirables.

Norscan spruce species set the standards for high quality in mechanical and high-yield pulps while pine species are less desirable as both the strength and printing characteristics are clearly poorer. Hemlock is sometimes used but tends toward lower brightness levels. Southern pines give high-yield pulps that are lower in tear and tensile along with poorer light scattering coefficient and brightness; however, TMP and CTMP processing improves the potential use of southern pines. Problems can arise with high-yield pulps from mature southern pine due to the high latewood percentage and resin content. Earlywood fibers from southern pine approach spruce in papermaking ability and, thus, present no particular problems. The latewood fibers, however, are thick walled and nonconforming when made into paper so that mechanical pulp from southern pine generally gives rough, porous, and lower density products. Multi-stage refining has been used to reduce the southern fibers to usable form, but the resulting freeness levels are very low.

The quality of mechanical pulps from low-density hardwoods (i.e., *Populus* species) is characterized as low in strength but having excellent optical and printing properties. Aspen CTMP is becoming important as its strength approaches softwood TMP but has higher brightness and opacity. In fact, the use of aspen and similar species is one of the most important developments of the Eighties. Most other hardwoods are unsuitable for pure mechanical pulping, however, they can be processed as CTMP offering fair strength and relatively low opacity. Eucalyptus CTMP becomes interesting as it has both good strength and opacity; however, the brightness and bleachability may not be as good as aspen.

Mechanical pulps typically consist of fiber bundles, shives, long fibers, fiber fragments and various amounts and shapes of fine particles which can interfere with runnability and coating quality. Shives are bad for print quality in both offset and gravure printing and fiber fragments give rise to linting. Stiff mechanical fibers give problems in gravure where a very smooth surface is needed. Poorly bonded long fibers cause poor print quality and, at any level of surface roughness, weakly bonded fibers are detrimental to printability. Fortunately, modern screening and cleaning methods eliminate most of the shives, fiber bundles and other undesirable fragments so that mechanical pulps are really not limited in their use. Good rejects processing and control of the main flow refining minimize problems with oversized or very small fibrous elements in high-yield or mechanical pulps.

Differences noted between high-yield and chemical pulps are understood best in terms of their overall strength and bonding properties. The longest fibers in mechanical pulps are expected to contribute most to the whole pulp strength, but do not do so due to their generally poor flexibility and bonding. Also, the highly lignified fibers are brittle, weak, and inelastic. Fibers that are not well bonded are easily pulled from the sheet under load, and this appears to be the case, as shown by the low tear, density, and breaking length for mechanical pulps [21]. Usually, the intrinsic strength or zero-span tensile of fibers decreases with lower yields.

Mechanical pulps usually need reinforcement pulps and in most cases they are not necessarily the strongest pulp available. It is more important that the reinforcing pulp be selected so the blend provides the desired balance of properties. For papers containing high-yield pulps, runnability becomes critical in converting and printing so that cleanliness and high strength are essential for meeting rising runnability demands. As four-color printing grows, the need for dimensional stability increases and, again, high-yield pulps can play a major role along with their contribution to high opacity and low print-through.

2. Chemical pulps

Much of the non-cellulosic material is removed in chemical pulping which allows for fiber separation with essentially no mechanical treatment. The fiber liberation point is reached at about 55% yield where only 15-20% of the original hemicellulose and lignin remain giving fibers with the greatest strength and brightness potential. The disadvantage of chemical pulping is the substantial yield loss in spite of the complex technology used and the high capital and operational costs.

Differences in chemical composition give hardwoods a distinct advantage over softwoods in chemical pulping as they contain both less lignin and a less condensed lignin which is easier to remove and bleach. Also, hardwood pulps retain more hemicellulose after chemical pulping so they are more easily refined, but they need to be beaten to lower freenesses to obtain comparable tensile and burst strengths as softwoods.

Chemical pulps are characterized by high strength potential, high bleached brightness and brightness stability, and general utility in almost all grades of paper. Kraft pulps are preferred for board and packaging products where strength is the main requirement while the easy beating sulfites are desirable for dense papers such as glassine or greaseproof. High-yield or mechanical pulps, which are weaker still, yield resilient and absorbent fibers with good printability and thus have proved useful for newsprint.

In general, long-fibered softwood species are used to provide strength and a skeletal framework for the paper or paperboard structure. These are usually preferred in the order:

Kraft > Sulfite > High Yield > Mechanical

Norscan > West Coast > Southern

CTMP > TMP > PGW/RMP > SGW

Hardwood species contribute little to strength, especially when used in their unrefined condition, but offer good bulk and opacity. Hardwoods are also desired for their contribution to smoothness and improved formation with preference in the general order:

Kraft > Sulfite/Soda > Mechanical

Eucalypts ≥ Northern > Southern

Kraft pulps have good yields and better strength properties than sulfite along with reduced resin and pitch troubles. Also, virtually any tree species or other fiber source can be pulped by the kraft process and this is especially significant for some resinous pines. Unbleached sulfite pulps are light in color and used in many grades without bleaching; however, they may be less pure than kraft as they retain more resins and hemicelluloses. Sulfite pulps form well, and are plainly weaker than kraft pulps. Today, semichemical and kraft pulps have replaced sulfite grades in many uses as few new sulfite mills have been built.

Sheet density is a basic paper property and those things that affect sheet density have a great influence on most other sheet properties. For high-yield pulps, sheet density rises rapidly as the level of short fibers or fines increases. Stiff, inelastic, long fibers yield bulky sheets having many voids; thus low sheet density and relatively low strength. In contrast, the near constant sheet density observed for chemical pulp fractions suggests equal flexibility of the different length fibers, i.e., equal conformability to packing. Fiber length, also, exhibits little relation to sheet bulk for chemical pulps.

3. Secondary fibers

Pulps recovered from paper stock (waste papers) are the second largest source of fibers for the paper industry and processing of these materials is described in Vol. 3 of this series [40]. Waste paper is always a mixture of long and short fibers and many grades of paper stock are mixtures of chemical and mechanical pulp fibers, so sorting at the source is critical to their use. Secondary fibers account for 20-25% of all the fibrous material used throughout the world (Fig. 14), however, many countries consume recycled fibers at rates exceeding 40-50% of virgin pulp production. Many of these countries rely on imported secondary fiber to achieve these rates.

Cellulosic fibers are not degraded much in most papermaking operations and, in principle, may be reused several times. The principal problem areas associated with recycling fibers are efficient sorting, fiber debris, economical collection, and contamination. Due to difficulties in removing all contaminants, most secondary fiber is reused in products of similar or lower quality. Good, functional material collected mainly from print shops or offices is deinked and bleached to produce "chemically cleaned", bright and attractive products. However, old corrugated cartons from supermarkets, and newsprint overissues are generally reused in similar products.

While originally derived from hardwood and softwood pulps, data on the differences between recycled fibers and the virgin pulps from which they were made are more limited. Fibers from repulped paper or broke appear flattened with many of the external fibrils lost. The fiber structure has been altered and the surfaces are more difficult to fibrillate. Fibers that have been dried one or more times are more resilient and resultant sheets tend to be bulkier. Some shortening of the original fibers has usually occurred and, if again refined, they may be subjected to further shortening.

Fibers dried under restraint (as in making paper) have a degree of curl that is heat-set into the fiber. Although few consider this important, it may explain many phenomena observed with dried fibers from broke or market pulps. Never-dried pulps produce denser webs than once-dried fibers and the tear strength is usually lower than if the fibers were previously dried. (This is true over an extended range of refining.) Thus, differences seen in once-dried fibers are probably related to the closing of voids within the cell wall and much of this effect appears to be irreversible.

When fibers are over-dried or dried so rapidly that the surface dries while some moisture still remains within the fiber wall, the fibers become somewhat case-hardened. This results because the fiber surface shrinks more than the center, putting the surface in tension and the interior portion of the fiber under compression. Concurrently, the fines may also bond very strongly to other fibers during such over-drying. In repulping, it may be difficult to relieve these stresses in the fibers and regenerate the fines.

In practice, rewetting after drying does not fully restore the properties of the undried pulp. Upon drying, moisture gradients across the fiber wall may cause resins to migrate to the surface and coat the surface thus decreasing the subsequent wettability of the fiber. The ease of beating of a pulp is adversely, and in some cases substantially, altered upon drying. Often a reduction in the fiber shrinkage is seen when beaten stock is once-dried and then reformed into paper products.

The trend toward using more post-consumer and household waste presents new difficulties, especially in keeping a stable supply-price-demand condition. Mixed papers from these sources require extensive presorting and precleaning before use even in lower grade products. The cleaning operations themselves produce waste streams that are expensive and difficult to dispose of in an environmentally safe manner. Larger secondary fiber operations might cope with waste disposal or processing problems and remain competitive, but small mills find these obstacles insurmountable.

Household and other post-consumer wastes are expected to be relatively weaker due to contamination of the fiber surfaces which may inhibit bonding. Refining can help recover some of the strength but will also increase debris levels and reduce the drainage and web drying rates. While some debris is removed with efficient screening, reject disposal remains a problem. Still, waste paper use will continue its growth due to the tight and expensive supply of virgin raw materials, especially in Europe and Asia.

E. PULP SUITABILITY FOR PAPERMAKING

The principal uses for pulp are in the manufacture of paper and board products; however, our fiber knowledge of what pulps will or will not do for making certain products is not complete. Typically, we use fiber blends either to achieve combinations of paper properties not easily attained with single components, or to reach a desired quality requirement with the greatest economy. Pulp selection for a specific product will depend on the end-use properties needed and on the papermaking process itself. Changes in technology, such as higher machine speeds or the increased use of fillers, will make pulps with good strength properties more important even when used as a minor component. Changes in grade mix or product requirements, i.e., growth in sheet fed printing, may also influence pulp selection [41].

Long, slender fibers are usually preferred for papers needing strength, smoothness, and good folding qualities with short, fine fibers desired for book, most printing papers, and grades where formation is critical. Softwood chemical pulps from low-density wood species have relatively thin cell walls and are the type of fiber often selected. The situation changes for high-yield pulps where thin-walled fibers may be fractured into short pieces in manufacture; however, Norscan softwoods are still the best source of groundwood for newsprint and many other uses.

The converting response and property combinations attainable in paper or board manufacture are determined by the inherent qualities of the base pulps used and on the refining, forming, sizing, and drying conditions. The wood species, pulping method, and refining technique used define what qualities can be built into pulps. The number and variety of end-use requirements, as well as the great range of property combinations needed to meet specific purposes, make it quite impossible for the pulp producer to design one base pulp that answers all papermaking demands.

While specific paper properties are largely controlled by the papermaker, they still depend on the pulp quality from which the product is made. In turn, pulp properties depend on the native fiber source and methods used to process the raw materials. In practice, we use trees with continuously variable properties both radially and longitudinally in the stem. Most of this variation remains in the pulps we use. Many relationships between fiber morphology, pulp and sheet characteristics can be cited, but the numerous structural variables and cross correlations between them confound our understanding. For example, if coarseness is used to describe the contribution of the fiber cell wall, we should also consider a shape factor, as coarseness reflects the weight per unit length rather than the actual cross-sectional dimensions of the fiber [42].

While there is much common technology applied throughout paper and board making, specific product categories can have important concerns that require specific techniques. Even the simplest of discussions regarding available paper products is beyond the scope here so that only a few of the more common broad categories of paper and board will be addressed.

1. Newsprint

The most widely used and lowest cost medium for printing is newsprint and related papers such as catalog, rotogravure, or advertising supplements where absolute strength and permanence are of secondary importance. They are lightweight, nondurable grades made with mechanical pulps and account for about one third of all paper production in North America. Recycled newspapers are also used as a substitute for groundwood.

The critical product attributes for newsprint grades are press runnability with good print and optical properties. For press runnability we want that combination of properties that permits uninterrupted press operation. Appropriate sheet strength, roll quality, and the absence of shives, slivers, holes, edge defects, etc. are important in this regard. News grades are made from a high percentage of groundwood, which discolors on aging, so are not recommended for permanent or semi-permanent records.

Printability depends on the printing method used along with a complex mix of sheet properties such as smoothness, absorbency, formation, moisture content, bulk and compressibility. These properties will determine the ability of the web to accept and preserve the ink pattern with a minimum of show-through or rub-off. The important optical and appearance properties of opacity, brightness, and cleanliness add then to the print contrast and general quality of the printed product.

The mechanical pulp in the furnish contributes many desirable sheet properties to newsprint, all related to printability. The strength of mechanical pulps, however, is poor and other pulps are frequently added as reinforcement fiber. Because unbleached sulfite or semi-bleached kraft pulps adversely affect printing, the usualy strategy is to use only enough of these more costly fibers to provide the minimum strength required. Newsprint today may be made with 100% TMP and no chemical pulp requirement.

The trend toward lower basis weights and/or use of fillers in newsprint requires some bleached softwoods (reinforcing pulp) to sustain runnability. This also requires high-quality mechanical pulps that simultaneously can deliver more opacity and strength. The printability and runnability requirements for newsprint are not static so that improvements in high-yield pulp processing are needed to provide superior pulps from less select raw materials.

2. Printing and writing grades

Printing and writing papers include those used for magazines, brochures, books, writing, drawing, advertising, etc. which need to run at high speed through printing presses as continuous webs or sheets. The major printing and converting papers are coated or uncoated book, and uncoated groundwood while writing papers include such diverse products as bond, text, reprographic and bible paper or heavier grades such as bleached bristols used for post cards and index cards. A need for all printing and writings is that they receive and carry print without excessive show-through. Also important are high quality, handle or feel, smooth printing surfaces, and generally good appearance; however, some of these characteristics often yield to price.

Intuitively, printability is a key property and this involves print uniformity, show-through, and contrast. In papermaking terms, printing and writing papers need good compressibility, smoothness, and control of ink penetration which determines the show-through, rub-off, and set-off. They require low linting tendency, good dimensional stability and, optically, they should have a certain brightness level, opacity and cleanliness.

Printing and writing papers usually need only enough strength and stiffness to run well on the paper machines and in converting equipment. Since strength is not the governing factor, sulfite, hardwood or sawdust pulps are widely used. Also, there should be as little two-sidedness as possible, which means a uniform distribution of filler and fines through the thickness of the sheet. Fine paper furnishes receive light to moderate refining so as not to adversely influence the optical and appearance properties.

Printing paper quality is defined by the individual printing process to be used. Offset printing papers need good surface strength, dimensional stability, and low water absorptivity. For rotogravure, compressiblity and smoothness are the important quality parameters. As the industry moves toward more offset and gravure or to higher printing speeds, the demands for surface strength and smoothness increase. Printers often are willing to pay more for papers having the best balance of properties such as rigidity, dimensional stability, water and oil absorption and runnability. In this sense, runnability is identified with the "non-break" property of both web and roll stock.

Minimum opacity and brightness levels are normally achieved with bleached kraft pulps (SWD and HWD) with varying amounts and kinds of fillers. Filler materials (up to 30%) are added for improved opacity and smoothness, but at the higher levels these mineral components also reduce strength. The trend in printing and writing papers toward lower grammages requires increased use of mechanical pulps and fillers to maintain the expected quality levels. Mechanical pulps produce papers with good opacity at lower basis weights and some new high-yield processes permit many papers to be made without the need for reinforcing pulps.

Commercially, a great variety of furnishes are used to meet optical and strength requirements at minimum costs. For optimum quality, coated papers provide the best opacity and smoothness, control of ink penetration and, thus, the best printability. The coating is the key to this superior performance of coated paper; generally, the thicker the coating, the better the printing. The premium coated papers are usually free of any mechanical pulp.

Pigment-coated papers are classified as coated freesheets and coated groundwood as a convenient means of denoting the expected quality and the high-yield pulp content ranges from 10-90%. Such a distinction is no longer adequate since papers with low-level content of mechanical pulp are virtually indistinguishable from freesheets in performance. Chemimechanical pulps also confound the issue as their properties fall between those of traditional chemical and mechanical pulps.

Paper quality is always a compromise. Of all quality-related tests for printing papers, smoothness is considered the most important, but obtaining this property results in a loss in bulk and stiffness. Smoothness is determined by the fiber furnish, filler, fines, fines distribution, formation, wire mark, and calendering. High surface strength and low moisture content add to the cost of making suitable heat-set web offset papers.

Ink penetration depends on the surface structure, pore size distribution, and wettability of the fibers. Surface strength is controlled by the furnish, surface sizing, and sheet density with linting and surface strength being correlated. For groundwood sheets, freedom from linting requires well-fibrillated pulps from fresh wood (chips), proper surface sizing, and control of the fines content. A perplexing problem with wood-containing papers has been fiber rise or heat roughening where the ends of several fibers lift from the paper surface. This phenomenon is observed in surface sizing, coating, offset litho and flexo printing where contact with water results in irreversible changes to the surface structure of the paper. Stress relaxation, fiber swelling, bond breakage or combinations of these effects are induced by the rewetting such that fibers or fiber ends rise above the paper surface. Fibers that are partially detached are then more easily pulled from the sheet in subsequent operations. For papers containing mechanical or other high-yield pulps, correlations have been reported between linting and the amount of fiber rise; thus, well-beaten fibers with good bonding potential are usually required to minimize these problems [43].

Printing papers include many diverse grades and print quality becomes important only if the paper can meet the runnability demands in the making and on the presses. Runnability and print-through (or in terms of paper tests: strength and opacity) are a critical pair of properties, both essential and contradictory. The nominal strength and wide range of quality demands for printed products permits the use of almost any available pulp in the making. The best quality papers are woodfree; usually made from eucalyptus or northern hardwood kraft fibers reinforced with a minimum of strong softwood pulps. Where density, formation, good strength and smoothness are desired, the Norscan softwood pulps are the most satisfactory as they can be developed to high strength without impairing other quality factors. Good quality secondary fibers are also used extensively in many of these products.

Hardwood pulps offer highly opacity and smoother surfaces than softwoods so they are ideal for good printability. A drawback, especially from ring porous hardwoods, is that surface picking of vessel elements may occur during offset printing. The upper limit for hardwood use is determined by the product needs where the wet web or sheet surface strengths constitute the determining factors. Usually, hardwood pulps have a positive influence on the performance of printing papers.

Fiber dimensions affect printing smoothness and both fiber length and wall thickness are important in this regard. With decreasing fiber length, smoothness and print quality often improve. Short fibers fill the unevenness between larger fibers to yield a smoother or more closely packed surface with improved formation [44]. With only long fibers available, refining must provide some cutting as fibrillation alone does not provide as good formation. When beating is increased for surface strength, important properties such as bulk and opacity decrease. Hardwood pulps are better than softwoods for smoothness and sulfite pulps are sometimes preferred to kraft for the same species.

The number of optically refracting surfaces and light-scattering particles in the structure determines the opacity. According to Giertz [45, 46], the shape and dimensions of the fibers and the nature of their surfaces are the main factors influencing this property. Short-fibered material has higher opacity due to a greater number of particles and thick-walled fibers are more opaque than thin-walled fibers. At a given grammage there is the same amount of cell wall material, but thin cell walls collapse so that the lumen cavity is not available for opacity. Many fiber treatments during pulping and papermaking also influence opacity. The opacity of softwood kraft pulps is higher than for sulfite at equal yields but this does not necessarily hold for hardwood pulps. Also there is no possibility of opacity being improved through the agency of bleaching.

Supercalendered (SC), light-weight coated (LWC), and related grades are very demanding for pulp quality. For these products, Norscan spruce and fir species are preferred but aspen CTMP is also attractive. Conventional woodfree printing grades are a growing outlet for high-yield pulps which replace some bleached chemical pulps. Bleached CTMP should find increased use in fine papers and multilayer structures in the future. For woodfree printing and writing papers in North America, pricing and the manner of defining grade quality by fiber type (versus function or performance) has retarded the use of CTMP in these products. The incentives for using high-yield pulps in printing papers are good opacity (versus chemical pulps) and the potential for basis weight reductions with increased bulk. The CTMP must be chosen carefully to give the best compromise between porosity, formation, bulk, and surface strength. Multilayering techniques should greatly improve our ability to use more of these pulps.

Neutral sizing sytems allow the manufacture of fine papers of acceptable brightness with CTMP, but little progress has been made on the problem of CTMP brightness reversion. CTMP improvements have reduced the drainage problems and dry strength and retention aids are available that can be used in either acid or neutral sizing systems.

3. Wrapping, packaging and industrial papers

This classification includes a range of strong, durable and specialty papers which serve other manufacturers rather than the communications or cultural markets. Many of these are essential intermediates to other industries and do not reach the individual consumer. Most packaging, wrapping, and industrial papers are simply coarse papers that were adapted to various uses, but some are designed for very specific end-uses and require specialized treatments in manufacture.

Packaging papers have two principal and different functions: containing and protecting the contents of the package. A third important role, especially for consumer goods, is to display the product so that the printability and message-carrying ability of paper is a great asset. Packaging grades serve industrial rather than commercial markets and almost always require high strength. An important characteristic for the pulp raw material is the ease of refining or development of the desired product properties. Virtually all long-fibered kraft pulp is used in packaging and wrapping papers and the prime requirement is then to understand the character and potential of the fiber being used.

In the production of coarse papers where high strength, good porosity, and resistance to tear are desirable, kraft pulps from the southern and western growth regions are preferred. Specialty packaging grades, i.e., greaseproof and glassine, require high sheet density and extensive refining so these use Norscan pulps that refine easily and use the least energy to develop the needed properties. Most industrial grades are heavier and stronger than printing papers and unbleached kraft is also widely employed along with smaller amounts of bleached kraft or other pulps.

Wrapping, bag, and sack papers require good tensile, tear, and burst strengths along with high tensile energy absorption (TEA) (a combination of tensile and stretch). Often, to meet strength requirements, these grades are made from well-refined, unbleached softwood kraft pulps. Examining the stress/strain relationships and TEA helps identify those pulps most likely to give superior products because of better stretch characteristics at any level of inter-fiber bonding. High-consistency refining is also used to provide the increased stretch needed in some of these grades.

A substantial gain in many strength properties comes with increased dryness out of the press section. These benefits vary with different pulps and are lower for secondary fibers than for virgin fibers; the improvements also decrease with high yields. A large increase in stretch (to enhance TEA) is obtained if the sheet is dried with a minimum of restraint. While this is not easy to accomplish on conventional machines, essentially free shrinkage can occur in air impingement dryers.

No segment of the paper industry involves so many converted forms and treatments as the packaging, wrapping, and industrial papers, but this sector of the industry has also been penetrated more by various plastics and films. The industrial papers group is characterized today by combinations of plastics and film with paper which have encroached on other markets with various specialty, protective, and decorative products.

4. Tissue products

Tissues are light weight ($<30\ g/m^2$), fairly transparent products for hygienic and household use and include towels, napkins and facial or toilet tissues. Virtually all types of pulps are used to some extent in tissue manufacture including significant amounts of secondary fiber. Cellulose wadding, a related grade, is a loose mat of fibers (fluff pulp) used as an absorbent material in hospitals or for the manufacture of personal hygiene grades so that bleached chemical pulps are most often used for these products. The lower density and cost of high-yield pulps makes them attractive in fluff applications with CTMP being the most interesting of these mechanical pulps. Fluff grades make many of the same demands on high-yield and mechanical pulps as tissue products, except that softness, shive content, and strength are less critical.

For tissue grades, the important properties are absorbency, softness, a certain network strength, and acceptable brightness. Absorbency is a function of bulk so one looks for relatively stiff, resilient fibers with a minimum of fines; of course, a precondition is good fiber wettability. Bulk or tactile softness is controlled mainly by tensile strength, degree of creping, calendering, and/or embossing, with creping being very critical in determining tissue and napkin softness. Fibers that protrude from the surface, if soft and not stiff, contribute to tactile softness. Higher caliper, associated with less bonding, is beneficial to bulk softness and is controlled by the furnish, pressing operation, and destructuring due to creping. Brightness is important for high-quality tissue and towel grades and the sheet should have low brightness reversion and be free from dirt.

With its light weight, tissue needs high intrinsic strength for machine runnability, especially in converting, but strength takes second place to softness. As towel grades are heavier and wet strength treated, physical strength is less critical. Basis weight uniformity and moisture control are as important for the tissue grades as for other products due to their influence on quality, machine speed and efficiency. Formation and the MD/CD orientation affect strength-related properties but are controlled largely by headbox and operating conditions.

Stratified forming permits selecting fiber types to put next to the Yankee for creping efficiency and product softness. Two plies layered together with the long or coarse fiber sides in and the short or fine fiber sides out makes tissues that feel soft and have good ply-adhesion. CTMPs provide bulk as an inner layer for some products; however, these coarse pulps always contribute to a harsh feel if exposed on the surface.

Hardwood kraft pulps are more slender than softwood sulfites which makes them well suited for tissue. This is largely an effect of fiber dimensions with some influence of the pulping process. The highest quality tissues use fine fibers comprising the Norscan softwoods and hardwoods with considerable eucalyptus being used today. Integrated operations rely on local species so that many Southern and West Coast pulps are also used in tissue products. However, the fibers from these latter regions tend to be coarser so that achieving comparable softness levels is more difficult. However, stratified forming has improved our ability to use some of these coarser fibers.

High-yield and mechanical pulps are not used as much in tissue and towel as in the past; however, CTMP is finding new uses in toweling and fluff pulp. High-yield pulps are low in brightness, which may restrict the amount used. Secondary fibers and pulp substitutes are employed in significant quantity even in some higher quality products. Recycled fibers offer cost advantages but only when the mills are equipped to handle the contaminants that accompany these sources.

5. Board products

Paperboard represents the second large category of industry products. These grades have many requirements similar to paper but are differentiated primarily by their greater basis weight and thickness, although some overlap exists. The main raw materials for paperboard production are chemical or semichemical wood pulps with recycled fibers being an important alternative fiber source. In the U.S., 70% or more of all the recycled fibers used are for making folding boxboard and containerboard.

Most board production is for packaging purposes requiring a combination of good edge compression, bending stiffness, and bulk with good printability. In the U.S., paperboard is classified as either combination or solid board depending on the structure and fiber furnish used. Solid board uses one furnish throughout while combination or multi-ply boards may have different blends in the various layers. Folding boxboard and solid board are often coated for optimum printability while maintaining good internal bond, creasability, and dimensional stability.

The major use properties required of containers are adequate resistance to bulging due to pressure of contents and resistance to breakdown in stacking and handling. The important mechanical properties are bending stiffness and crease quality where both MD and CD values are of equal and separate significance. Along with puncture resistance for combined board, bending stiffness and rigidity are necessary for compression strength.

The strength requirements are met by the pulps used in board manufacture. Raw material selection is the key for these grades, however, papermaking conditions are also involved in achieving the best combination of stiffness and bulk with smooth printing surfaces. Smoothness depends on good formation and the action of the calenders while bulk is controlled mainly through pressing, calendering, and raw material selection.

Large-volume commodity grades, such as corrugating medium or liner, use coarse, strong, unbleached pulps and are most often produced in integrated operations. These products are converted into shipping containers and account for about 65% of the total board production. Containers have obvious strength requirements so high-yield kraft pulp is used in the bottom liner with lower yield, well-refined pulp used for the top liner. The fluted sheet or corrugating medium mainly uses relatively inexpensive semichemical pulps (mostly hardwood) although secondary fibers are also used. Although weak with respect to common strength measures, unbleached semichemical pulps are stiff and crush resistant due to their high residual lignin and hemicellulose content which makes them ideally suited for this purpose.

For liner and medium, CD edge compression strength is the key quality parameter defining its utility; burst is no longer the single parameter used to define liner quality. Also, puncture resistance has importance and properties such as printability, surface strength, and glueability are secondary to the above.

Foodboard is a generic term applied to a variety of paperboard grades used in packaging food. The furnish for these products is bleached or semibleached softwood and hardwood blends to meet the combined requirement for smoothness and carton rigidity. The use of fully bleached pulps is due largely to cleanliness and sanitary perceptions rather than any real added product protection.

Solid bleached board in addition to good formation has similar requirements as for duplex board, but here CTMP has provided new grades with the stiffness of duplex board and overall strength of traditional solid board. Solid bleached board is used mainly for packaging frozen foods and liquids and, in the U.S., they are made from bleached kraft pulps; in some cases 100% hardwood kraft. In Europe, multi-ply forming has allowed partial substitution of bleached kraft with other fibers. Multi-ply technology and raw material substitution will probably be applied much more in the future.

Folding boxboard requires good printing surfaces, strength enough to permit folding without rupture, and is used to make a variety of carton shapes. The list of applications is endless as almost every important item produced for general consumption is now packaged, to some degree, in folding cartons. These grades protect the contents but also enhance the marketability. The traditional furnish for folding boxboard is recycled pulp which accounts for about 80% of the fiber used. Since the waste papers are usually

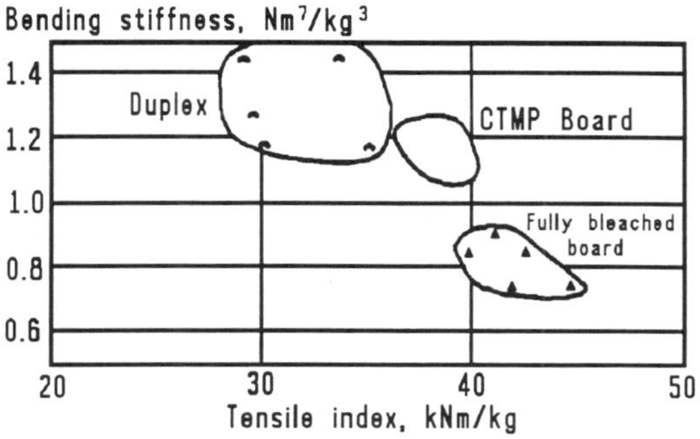

Fig. 19. Bending stiffness and tensile for various board grades.

not deinked, folding cartons are recognized by their gray color as in cereal or shoe boxes. Cost is the over-riding factor and, to improve the appearance of the board, coated or bleached surface layers are common. Set-up boxes are similar in many respects to folding boxboard except they are shipped in a three-dimensional shape ready for use.

The bending stress and the modulus of rupture in boards result from a complex interaction between fiber length, width, cell wall thickness, and lumen diameter. Bending stress and modulus both improve with short slender fibers with thin cell walls while the modulus of elasticity increased with decreasing fiber length for all levels of fiber width and cell wall thickness sampled. Thus, narrow, thin-walled fibers are more desirable than thick-walled fibers of large outside diameter, as they are more comformable when stressed in bending. The short fibers are also reported to have more fiber crossings per unit weight than do long-fibered pulps [47, 48].

Essential paperboard properties are low grammage, good bending stiffness, ply bond, and printability. For bending stiffness, we need good caliper and modulus of elasticity as the stiffness varies with the thickness cubed at the same modulus. This makes thickness or bulk increases relatively more important than an equivalent increase in modulus. CTMP has both high bulk and a higher modulus of elasticity than other mechanical pulps which makes it use in duplex board important (Fig. 19). At present, CTMP is more widely used in Europe than in the U.S. in such products. For duplex board, the ideal structure is a symmetrical 3-ply sheet with a middle layer of greatest possible bulk and outer layers of high modulus of elasticity. Chemical pulps lose bulk when wet pressed while mechanical pulps can be pressed to high solids with little loss of caliper or thickness. CTMP lies somewhere in between but, due to modulus differences, the bending stiffness at a given bulk is higher for CTMP than for TMP and many other pulps. While this is one explanation for the rapid increase in use of CTMP in duplex board, CTMP must be used at higher freenesses to compensate for the slightly greater density after pressing.

Most board products need a smooth printing surface with the dilemma that this is only gained through processes which result in a stiffness loss. At the pulp level, important factors for smoothness are the forming ability of the different fiber layers and the shives content of the high-yield pulps.

Construction board includes heavy board grades for roofing insulation, sheathing, book binding, asphalt impregnation, shoes liners, etc. Many of these products are saturated with binders so that an open and porous structure is an important requirement. For many purposes, coarse, unbleached fibers such as chemi- or semi-chemical pulps are entirely suitable. In a few specialty products, bleached pulps are used with no refining or high-consistency refining to preserve the porous structure.

6. Miscellaneous specialties

Generalizations serve little useful purpose in this category as most of the important properties identified with the larger product categories are needed in these grades. These less common products usually require an unusual combination of properties that must be met through use of some specialty and non-wood pulps.

F. PULP/FIBER SOURCING

Among the plant species used by the industry, different fiber sources show broad ranges for the properties considered to be important. Much of the data reported here was compiled largely from the recent literature and tests on commercial pulps. Thus, the average and range values will differ from those gathered on individually selected, intact fibers as often reported in older literature. Also, a degree of inconsistency in the data can be expected as each plant species is itself so highly variable.

1. Wood species

Worldwide, wood fibers are the predominant raw material used to make paper and we readily recognize differences between hard- and softwood species, that some trees grow faster than others, and so on. We should not assume, however, that all hardwoods or softwoods can be indiscriminately substituted for the same end uses as they do not all perform alike in papermaking.

The climatic conditions of the growth site, different plant species common to an area, soil characteristics, plant spacing, and other external factors can all influence fiber dimensions and structure. Within tree or within species variations often exceed 20% for some features. It is not surprising then, that some of the property ranges reported may be found in two trees from the same stand or even within a single growth ring.

Many important fiber sources are categorized to emphasize the major differences between the growth regions and between hardwood, softwood, or non-wood fiber types; at least with respect to the properties listed. Following acid or alkaline pulping, with or without anthraquinone, chemical differences that remain may be relatively minor. This leveling effect through full chemical pulping occurs with both hardwood and softwood species. Therefore, differences that are observed in papermaking properties and response to refining or other treatments, for similar pulps, must be due largely to differences in fiber morphology.

Pulpwood selection depends on the wood species available in the mill draw area. An important observation in comparing pulps made from different sources has been the influence of the basic fiber dimensions [5, 49]. Wood fibers vary in length, width, and cell wall thickness and these differ widely within and between wood species. Hardwood fibers are shorter than softwoods, make lower strength pulps, and usually exhibit a narrower distribution for their important dimensions. Softwood tracheids or hardwood fibers are both papermaking fibers and the size differences are inherent features. Studies on the influence of fiber length on paper properties (using wood pulps of varying average fiber length) show inconsistent relationships because so many other factors may be changing at the same time. See also Vol. 1 [18] of this series.

(a) Softwoods

Fibers from northern softwood species produce a different balance of product properties than those from Douglas fir and southern pines. Northern fibers exhibit better fiber bonding, higher tensile and burst, and a lower tear strength than kraft pulps of Douglas fir or southern pine. While studies show that the chemical composition does not vary significantly between these fiber types, the fiber dimensions are very different. This implies that quality differences for pulps from different growth regions are largely due to species variation.

In North America we identify four main forest and papermaking regions: Northeast, Midwest, South, and Pacific Coast. For simplification, the Northeast and Midwest growth regions (which can include Scandinavia) are considered together as many of the wood species and their fiber properties tend to be similar. This grouping is identified in Table 2 as the Norscan growth region.

The inherent species/fiber properties from each growth region are mainly responsible for the variations observed in the pulps we use. Norscan spruce and pine species have fibers of medium length, 3–3.5 mm and widths of 25–30 μm (0.025–0.030 mm). The fiber dimensions for slow-grown northern species contribute to good formation and smoothness. The forming attributes and relatively high pentosan content offer good interfiber bonding and easy development of high burst and tensile strengths with refining. These characteristics largely reflect the ease of collapse for these slender softwood fibers.

Hemlock and Douglas fir are the main species found in west coast pulps and they have large fibers that may average 3.5 – 4 mm in length and 35 – 45 μm wide (up to 55 μm for Douglas fir). These larger dimensions contribute to the less uniform formation, rough surface finish, and high tearing resistance of papers from that region. West coast woods, and pulps made from them, have lower hemicellulose content which accounts for some of the slower strength development when compared to Norscan pulps. For Douglas fir, these effects are magnified by the thick-walled, latewood fibers which reduce tensile and finish but provide higher tear strength than virtually any other wood pulp. In contrast to Douglas fir and hemlock, western red cedar has fiber dimensions more like Norscan species and performs in a manner similar to the northern pulps.

Pulps from the broad class of southern pines are very coarse with dimensions and papermaking properties comparable to coastal Douglas fir. Also, southern pines have a higher weight percentage of latewood fibers than most other species which, in practice, makes them even more resistant to strength development. Good formation can be more difficult to achieve with southern softwood pulps; burst and tensile strengths develop more slowly; and the tear resistance is high. Again, smooth surfaces are difficult to achieve with the coarse southern pulps.

(b) Hardwoods

Hardwood fibers also display differences associated with the region in which they are grown. These fibers are typically about 0.8 – 1.5

Table 2. Morphological data for selected softwood species.

SPECIES	Fiber Length (mm) Avg.	Range	Fiber Width (μm) Avg.	Range	Wall Thickness (μm) Avg.	Range	Coarseness (dg) Avg.	Range	Wood Density Avg.	(g/cm^3)
NORCAN SOFTWOODS										
White Spruce	3.2	1.0-4.3		25-35		3-4				0.37
Black Spruce	3.3	1.3-4.9		25-35		3-4				0.40
Jack Pine	3.0	1.5-5.7	30	25-40	4.9		18.0			0.39
Balsam Fir	3.5	1.9-5.6		30-40						0.34
Eastern Larch	3.5	1.7-5.6		25-35			22.3			0.49
Norway Spruce	2.7	1.7-3.7	36	24-50	3.3	2-8				0.37
Scotch Pine	2.5	1.8-4.4	32		4.4	2-10				0.41
White Pine	3.3	1.6-5.0		25-35			19.8			0.34
Silver Fir	3.4			30-40						0.38
Alpine Fir	3.0		29		2.0					0.37
Norway Pine	3.4	1.2-5.2		30-40			21.4			0.44
Eastern Hemlock	3.3			28-40						0.38
SOUTHERN SOFTWOODS										
Loblolly Pine	3.6	1.2-5.9	4.3	35-45	7.0		26.0	22-32		0.47
Shortleaf Pine	3.5	2-7		35-45						0.46
Longleaf Pine	3.5	2.7		35-45						0.54
Slash Pine	3.4		40		3.8					0.56
Parana Pine	4.6		40							
Maritime Pine	3.1		40		3.7					
WEST COAST SOFTWOODS										
Douglas Fir	3.5	1.7-7.0	43	35-45	6.0	2-7		25-32	0.40	0.45
Western Hemlock	3.1	1.8-6.0	41	30-45	4.8	1.7	19.0			0.38
Redwood	6.0	2.9-9.3	52	50-65	3.6	3.7	26.0			0.38
Western Red Cedar	3.5	1.4-5.9		30-40			16.0			0.31
Ponderosa Pine	3.5			35-45			26.0			0.39
Western Larch	3.3			40-50			32.0			0.48
Sitka Spruce	5.0	3.6-7.3	38	35-45	3.5					0.37
Englemann Spruce	3.5		30	25-35	2.9					0.31
Lodgepole Pine	3.1		40	35-45	3.0		23.0			0.38

mm in length and 16–25 μm (0.016–0.025 mm) at their widest point. The fiber length, width, and cell wall thickness are greater for southern hardwoods compared to hardwoods from other regions as seen in Table 3. Southern hardwoods and the plantation-grown eucalypts are stronger than northern hardwood pulps but, due to their smaller dimensions, all hardwood pulps are weaker than softwoods in almost every strength property.

Commercially, hardwood pulps are used in paper and board grades intended for printing where the highest quality with respect to surface properties and appearance is needed. The fine hardwood fibers promote good formation, bulk, smoothness, and opacity compared with longer and coarser softwood fibers [50].

Other characteristics that influence hardwood pulp utility include: non-fibrous elements (vessels) found only in hardwood species tend to reduce pulp uniformity and contribute to poor surface strength and linting (dusting) problems in printing; the increased level of ray and parenchymal cells may reduce some strength properties; and the relatively short, thick-walled fibers give sheets of lower strength [51]. Many of these smaller, non-fibrous elements have a large effect on sheet properties, even though they represent only a small amount of the dry matter.

2. Non-wood fibers

The importance of non-wood fibers for pulping and papermaking is obscured by wood pulp usage and the high production rates in countries which have large wood supplies. While non-wood fibers represent a small portion of the global fiber supply, in some countries they may be the only material available [52]. (Nearly 70% of the pulping capacity in China is based on non-wood fibers.) Non-wood fibers are even more diverse than fibers from wood with lengths which average less than 1 mm or exceed 120 mm and fiber widths cover the whole range, 10–50 μm (0.010–0.050 mm), exhibited by wood fibers.

Table 3. Morphological data for selected softwood species.

SPECIES	Fiber Length (mm) Avg.	Range	Fiber Width (μm) Avg.	Range	Wall Thickness (μm) Avg.	Range	Coarseness (dg) Avg.	Range	Wood Density Avg. (g/cm^3)
NORTHERN HARDWOODS									
Bigtooth Aspen	1.1		22		3.0		8.6		0.35
Quaking Aspen	1.0	0.4-1.9	21	10-27	3.0	1.3-5.3			0.35
Balsam Poplar	1.2			23-40					0.30
Black Cottonwood	1.4		39	23-40					0.32
Yellow Birch	1.5	0.8-2.7	18	20-36	3.7	2.4-7.2			0.55
Paper Birch	1.3			20-25	3.8		13.1		0.48
Red Alder	1.2			16-40					0.37
Beech	1.2	0.6-1.9		16-22	5.2	2.5-9.0			0.56
European Oak	1.1	0.4-1.9	21	10-35	6.0	2.5-9.0			0.59
Sugar Maple	0.8	0.3-1.3	18	16-30	2.5			8-12	0.56
Red Maple	0.7	0.3-1.1	20	16-30				7-12	0.49
Ash	1.2	0.5-1.6	22	10-29	3.8	2.4-6.0			0.45
SOUTHERN HARDWOODS									
Black Willow	1.1	0.5-2.3	20	10-32	2.4	1.2-3.9			0.34
Southern Cottonwood	1.2			25-40					0.37
White Oak	1.4			14-22	5.0		14.1		0.59
Hickory	3.4	1.3-1.4							0.65
Yellow Poplar	1.8	0.8-2.7		24-40					0.38
Red Gum	1.7		29	20-40	3.5		20.0		0.44
Sycamore	1.7			20-36					0.46
Tupelo	1.8	0.8-2.7		20-35					0.40
EUCALYPTUS									
E. globulus	0.9	0.6-1.4	13	10-15	1.6				0.60
E. tereticornis	0.9		14						
E. camaldulensis	0.8	0.5-1.4	11		3.6				0.68
E. saligna	1.0		13		1.4				0.51

Processing information for non-wood fibers is discussed in Vol. 3 [40] of this series.

G. BASIC FIBER FEATURES

It is not always clear how wood and fiber properties affect the final pulps and papers made from them as all the steps from wood preparation through papermaking influence the final product properties. Even in combination, the process effects tend to be less effective than species selection or species blending in the manufacture of specific grades of pulp or paper. This is because each fiber type makes its own contribution to the product and the fiber characteristics influence or principally account for particular paper properties. A better understanding of the relationships between fiber and paper properties requires the study of many structural factors, i.e., the length and width of fibers, the thickness of the cell wall, their cross-sectional area, and the ratio of early- to latewood.

Individual fiber properties from commercial pulps have wide distributions and some studies were based on too small a sample set to reach statistically valid conclusions. Early studies estimated the most evident effects leading to conclusions that were not supported by later work which increased our awareness of the role that fiber morphology has in governing sheet properties. Statistical methods help to account for much of the variability in pulp properties from diverse wood samples and the influence of wood and fiber morphology on pulp or sheet properties has been clearly demonstrated.

Various reviews identify the wood properties considered to be important in pulp and papermaking:

- Specific gravity for its relationship to yield, strength and pulp behavior;
- Fiber length which correlates well with several sheet properties; and
- Latewood percent as it influences both the specific gravity and sheet characteristics.

The fibril angle, i.e., the angle the fibrils make with the axis of the fiber, is another structural parameter important in pulp and papermaking [53-55]. A low angle, for example, is associated with high density. While the composition and distribution of chemicals within the fiber wall are important, it is the anatomical features of fibers that serve as orienting criteria for the papermaking value of a given wood species or non-wood fiber source [56].

1. Fiber length

By itself, fiber length may not be very useful as a primary pulp quality parameter as it is the length relative to the cell wall thickness and fiber diameter that is important [57]. This results because fiber dimensions such as length, width, cell wall thickness, and coarseness are interrelated and the combination of properties influences much of what we do in papermaking.

Since pulps are blended on a weight basis, it is the weight-weighted fiber length that is needed. This is emphasized by Clark [58] as some reports are not always clear as to which fiber length is involved. A definite correlation exists between the weight-weighted length of fibers longer than 0.1 mm and the number of fibers per gram which itself relates to the surface area of the pulp on a unit weight basis.

Determining the weighted average length using a microscope was a tiresome task and common practice was to count 40–400 intact, separate fibers. Today, the Kajaani Fiber Size Analyzer [59] electronically makes a sequential count and length measurement of several thousand fibers in a few minutes. It then automatically prints the number-average, length-weighted average, and weight-weighted average fiber lengths. An advanced model determines coarseness so that these tests are quite routine in pulp evaluation reports for many mills.

Fiber length correlates with many strength properties, but a point is reached where increased length disrupts the uniformity of formation and strength levels decline. Also, long fibers tend to be wider and thicker leading to increased stiffness. Coarse long fibers, i.e., southern pine latewood, do not collapse easily giving bulkier, more open structures with less bonded contact area and lower burst and tensile strengths. The slenderness and length of fibers are attributes that are often interchangeable so that short, finer fibers from hardwoods may be more desirable for many papers. This suggests that fibers, however long, should not be coarse.

The essential importance of fiber length is obvious as it contributes strongly to the sheet structure. Also, fiber length distribution needs to be considered if only to assess cutting and to separately identify the effect of fines. Fiber length, along with flexibility, slenderness, and sheet density, controls the fiber contacts which arise per single fiber. Therefore, fiber length has a certain, if not leading, influence on the papermaking properties of pulp. Processing affects fiber length by altering flexibility and the capacity to form inter-fiber bonds. Changes in fiber length and length distribution are correlated with other pulp properties but such parallel phenomena should not be confused with actual interactions between the variables.

Other things being substantially equal, longer fibers provide higher strength levels. By raising the average fiber length of a given pulp, strength tests show improved maximums even though one needs to increase beating to reach those higher levels. Einspahr [60], working with *Populus* species, found that fiber length and growth were associated so that selection for rapid growth could result in both improved fiber length and strength.

Fiber classifiers separate pulps into different length classes but also sort pulps according to fiber coarseness. With so many fiber characteristics uncontrolled, it is not surprising to find disagreement among workers who used various pulp fractions to study the influence of fiber length on paper strength properties.

Long fibers may adversely affect sheet formation yet provide greater bonding area and better distribution of applied stresses. Therefore, fiber length is often accepted as the dimension that provides the most useful information on pulp and sheet strength properties. Softwood fibers are 2–4 times longer and wider than hardwood fibers and, on average, have thicker cell walls. These natural differences are an advantage for strength development but may be contrary to other desired properties. Large differences exist in the way fibers are processed and used which supports the importance of fiber morphology in determining the papermaking potential of the pulp. It is always the combined structural features that prove the value of any fibers in practice.

2. Fiber width

For wood species and non-wood fibers, fiber widths vary over a wide range. Northern softwood fibers are narrower than southern or west coast species, and are considered fine fibered. Hardwood fiber widths are about one half the width of softwoods but some overlapping occurs. Non-wood fibers are typically narrower than wood fibers of comparable length and this slenderness accounts, in part, for some special properties. Fiber width influences formation, bonding, and stiffness so that small-diameter thin-walled fibers indirectly offer a better balance of desirable paper properties than comparable large-diameter fibers. Overall, fiber width shows inconsistent correlation with sheet strength compared with other properties so is of little value in ranking trees for wood quality. Again, the complex interaction between all the structural dimensions of the fibers should be recognized.

3. Cell wall thickness

Fibers with thick cell walls are more rigid which limits their response to consolidation forces in papermaking. This results in more open structures with higher light scattering coefficient, bulk and porosity. Similar trends follow variations in beating which indicates these properties are strongly dependent on fiber morphology [7]. Cell wall thickness usually tracks fiber length and is one of the more important properties in determining fiber quality.

The basic concepts of pulping processes help explain some observed fiber property/paper strength relationships or account for reported discrepancies. Generally, thick-walled fibers of small diameter do not collapse in pulping, are stiff, and retain their rounded shape during sheet formation. They resist beating and do not bond readily but may improve tearing strength in the resultant sheet. Such bulky sheets have low density, are more opaque, have rougher surfaces, and are lower in burst and tensile strengths. In contrast, wide, thin-walled fibers associated with low-density wood produce sheets much weaker in tear than sheets from thick-walled fibers of equivalent length. These fibres collapse in the conversion steps, and form compact, dense sheets which are low in opacity, uniform in formation, and smooth. Furthermore, such fibers provide effective fiber-to-fiber contact which results in strong bonds, as revealed by high tensile and bursting strengths.

Cell wall thickness provides much useful information about fiber quality and correlates well with pulp and fiber strength properties. For both hardwoods and softwoods, the relationship between wall thickness and wood density is quite definitive within one species. Breeding for low density may be desirable for certain reinforcement pulps or for many strength properties but high density would increase yield and tearing strength. Cell wall thickness is lower in juvenile wood which supports our intuition regarding fiber strength and juvenile wood relationships. Lumen width does not consistently correlate with handsheet or fiber strength properties and has been of limited use in ranking wood species for fiber quality [60].

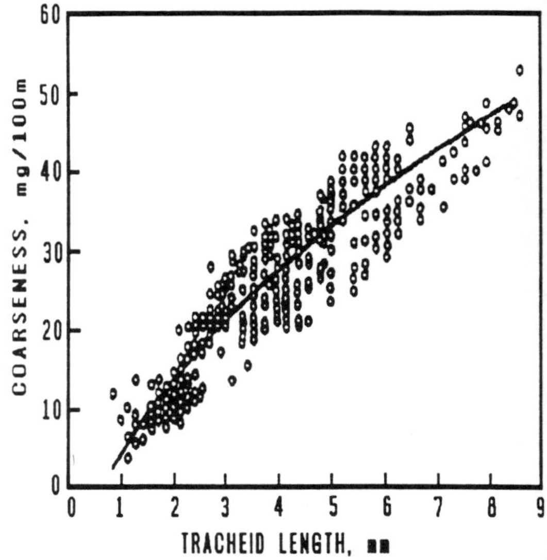

Fig. 20. The relationship between fiber length and the weight per unit length.

4. Coarseness

Fiber coarseness is the weight/unit length reported as the dry weight in milligrams/hundred meters. While coarseness may be as important as fiber length in papermaking, it received less attention due to difficulties in measurement. Figure 20, which shows the coarseness and fiber length relationship, suggests that some conclusions regarding the influence of fiber length, may have been due to differences in coarseness. The coarsest fibers for papermaking have values in excess of 30 mg/100 m while very fine fibers will be less than 10 mg/100 m.

Coarseness strongly influences most strength properties, air and liquid permeability and surface smoothness of paper. There is an inverse relationship between coarseness and the number of fibers per gram which can affect many strength characteristics of the product [61]. Sastry [62, 63] found a near linear relationship between fiber weight and length showing that longer wood fibers tend to be coarser than shorter fibers. Coarseness is important in establishing pulp quality, has a strong effect on paper properties, and it is not usually affected by refining.

Determining coarseness with a microscope was difficult and tedious, but the Kajaani Fiber Size Analyzer [59] measures this property routinely. Multiplying the fiber analysis factors by twenty estimates coarseness for different wood species but this approximation is not an exact measure for any specific sample.

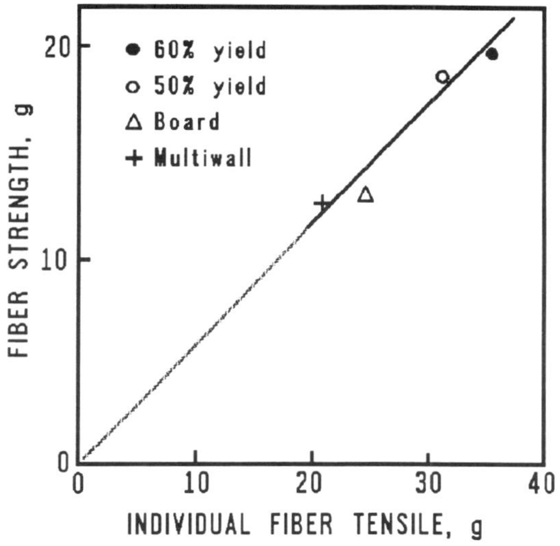

Fig. 21. Single fiber tensile strength versus fiber strength (calculated).

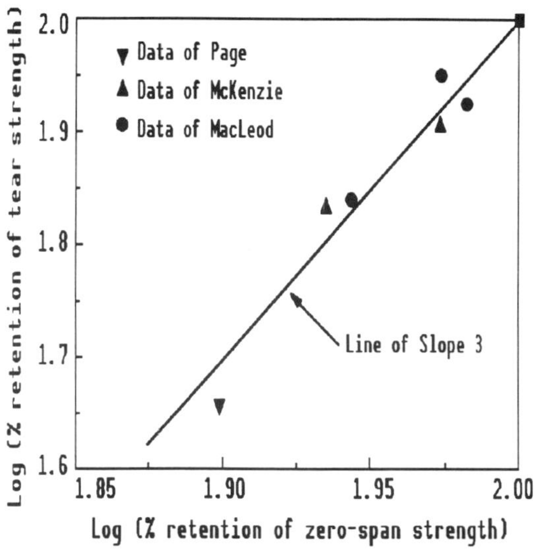

Fig. 22. Log-log relationship between tear and intrinsic fiber strength.

5. Intrinsic fiber strength

The individual fiber strength is essential in determining the strength of a sheet of paper, but the exact manner in which it influences sheet strenght is not resolved [64]. For handsheet properties, i.e., burst, tensile, and tear, no simple correlation with individual fiber strength has been found. Sheet strength depends on several factors, including inter-fiber bond strength, the length of the fibers, and the inherent strength of the fibers themselves.

Van den Akker [15] showed that in tear or tensile tests many fibers fail in rupture even in weakly bonded papers thus focusing attention on the individual fiber strength. Sheet strength was already credited to the influence of inter-fiber bonds inferring that, for strong papers, tensile strength was limited by fiber strength rather than failure of inter-fiber bonds. Figure 21 shows that the fiber strength predicted from the z-span tensile theory correlates well with individual fiber strength [24].

Data support the use of zero-span tensile tests as an indirect measure of fiber strength [65]. For both tensile and burst, fiber strength becomes increasingly important as sheet densities are increased. For tearing resistance, the influence increases up to a sheet density of about 0.60 g/cm^3 but, beyond that level, fiber strength has less influence on tear. It was recently reported that tear strength has a larger dependence on intrinsic fiber strength than previously recognized. Tear at a given tensile strength was approximately proportional to the cube of the fiber strength as seen in Fig. 22 [66] in contrast to the square of the fiber strength as reported earlier [19]. The relationship between fiber bonding and intrinsic fiber strength has been studied by Wink [67], Page [68], Stone [69], Cowan [70], and their co-workers and Matolcsy [71], and the reader is referred to their papers.

Paper strength depends on the strength of the bonds between the structural elements, the strength of the elements themselves, and the distribution of weak spots, i.e., formation. The greater the strength of the fibers, the more they can contribute to paper strength through inter-fiber bonding. Zero-span tensile tests are used to monitor fiber degradation in cooking or bleaching and may identify the limiting potential of a pulp. Beyond this, the intrinsic fiber strength has less of an influence than some other properties, unless of course it is very low. We observe reduced fiber strength and stiffness in pulps from cooks of increasing severity confirming that fiber degradation occurs with processing to higher levels of purity.

6. Wet compactability

Wet compactability or conformability relates to the ease with which fibers conform when pressed against other wet fibers and remain together when dried [58]. This property describes how two fiber surfaces mutually adapt to provide an intimate contact area between them in the network structure. The wet compactability depends on fiber wall thickness and affects many paper properties including the beating response and drainage characteristics of the pulp.

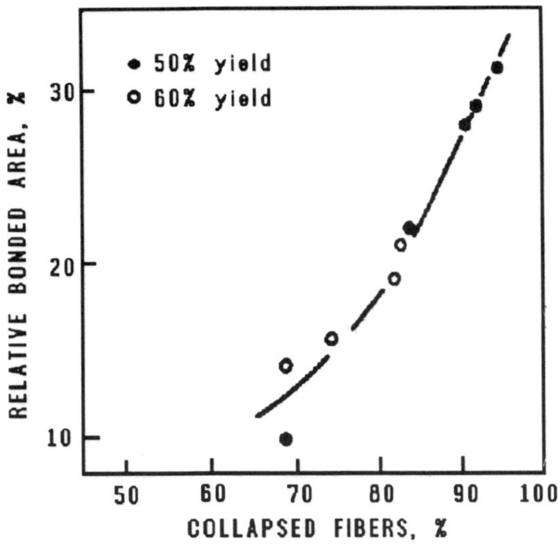

Fig. 23. Effect of fiber collapse on relative bonded area.

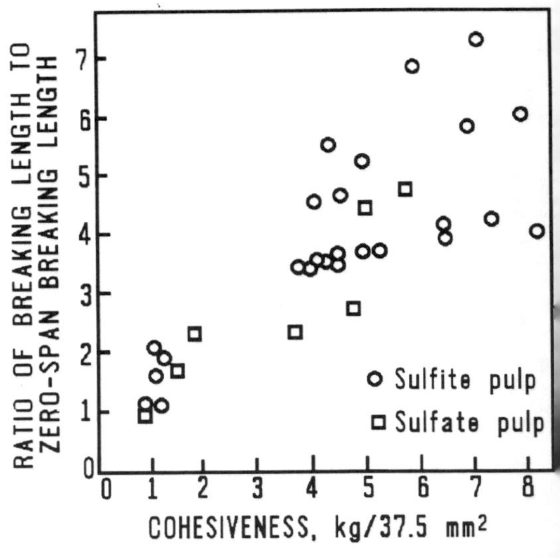

Fig. 24. Ratio of tensile to zero-span tensile as a measure of cohesiveness.

Wet flexibility determines sheet density [72] and fiber collapse can be a measure of fiber conformability. Collapse is an all or nothing phenomenon for any individual fiber often controlling the relative bonded area between adjacent fibers. When one fiber collapses against another, the apparent contact area between the fibers increases. As wet conformability increases or more fibers are readily collapsed, the mat is more easily compacted, both by drainage forces and wet pressing, which leads to increased sheet density and relative bonded area, along with reduced effective pore size [12]. It is inter-fiber bonding which gives paper its tensile and most other mechanical strength properties so that collapsed or flexible fibers increase the area of mutual contact which increases the potential strength of the overlapping joints. Figure 23 shows that pulps with high fiber collapse have relatively greater bonded area which will effect higher levels of sheet tensile strength [24].

7. Cohesiveness

Fiber cohesiveness is a measure of the bonding potential between fibers and may be easier to understand if we accept a distinction between adhesion and cohesion. Adhesion is the molecular attraction between surfaces of different materials while cohesion is molecular attraction by which particles of the same material are united within the structure. Surface tension forces orient fibrillar elements and shrinkage during drying causes the finer elements to be pulled in directions parallel to the plane of the sheet. Cohesiveness then arises from the orientation by liquid water of the fibrillar elements on the surface of the fibers and their subsequent hydrogen bonding into coherent structure.

Attempts made to determine this property are complicated and the results questionable. The best test appears to be a ratio of normal to zero-span tensile, but the results are quite variable. This ratio of dry tensile to dry zero-span strength is interpreted as an index of cohesive strength. Figure 24 shows the suggested ratio is only a rough indicator of fiber cohesiveness as clearly other factors are involved [73].

H. PULPWOOD CHARACTERISTICS

The high correlations found for certain wood characteristics and sheet properties of refined kraft pulps confirms the large influence on quality of natural wood and fiber differences within and between species, regardless of the intermediate processing steps [74]. These properties may be discussed separately but a complete account should describe the correlations seen between wood and pulp and the important relationships that exist between one wood or fiber feature and another. Single wood or fiber properties often show a strong correlation with more than one pulp or sheet property, and frequently the correlations are positive with one but negative with another.

For example, fiber length correlates positively with tear but negatively with formation while cell wall thickness correlates directly to fiber strength but inversely to flexibility. In both cases, an increase in cell wall thickness or fiber length is desirable up to a point, beyond which

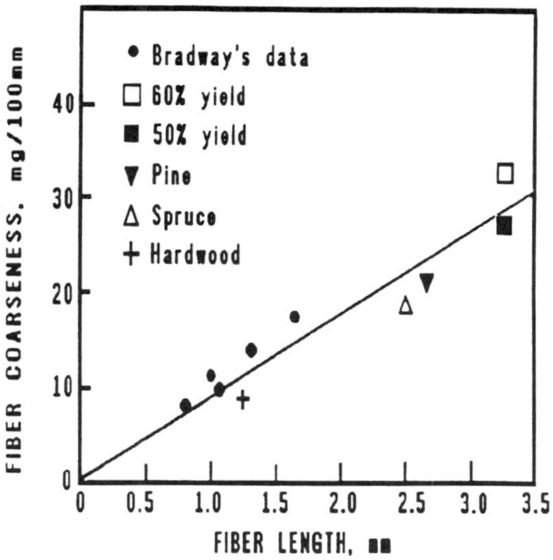

Fig. 25. Correlation between fiber length and coarseness.

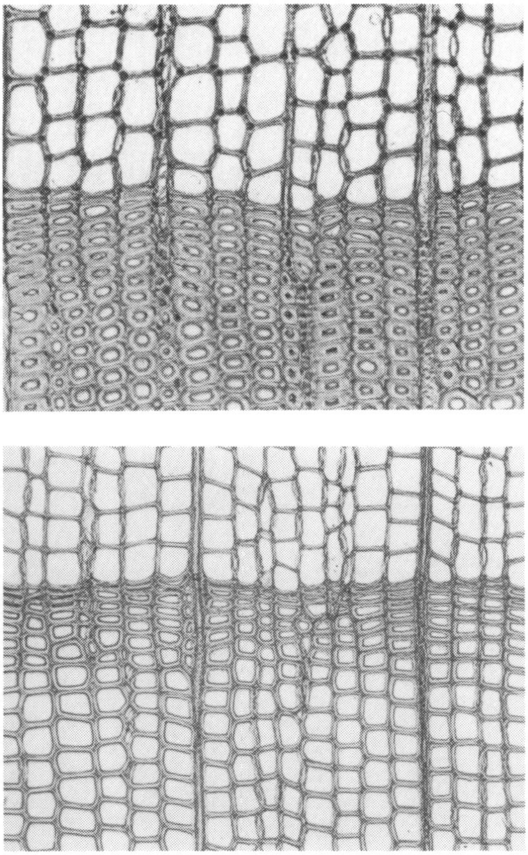

Fig. 26. Cross-sectional views of a southern (top) and northern (bottom) softwoods (X250).

further increases result in another important property being adversely affected. The main morphological features, i.e., length, width, and wall thickness, are all interrelated in a complicated and little-understood manner. Figure 25 shows that long fibers are coarser than short ones [68] so the increase in coarseness with length may be due to an increase in cell width, cell wall thickness, or both.

1. Wood density

Dense wood species have fibers with thick cell walls and, often, smaller lumens. Compared to less dense wood of the same or similar species, pulp from higher density wood makes bulkier products with high tearing strength, but lower burst and tensile strength. Wood density is a complex parameter influenced as much by the chemical composition as the cell wall thickness and, in turn, by the early- to latewood ratio.

On average, hardwoods have higher densities than softwoods and southern softwoods are more dense than those from other growth regions. Figure 26 shows cross-sectional views of southern and northern softwoods where the contrast in fiber size and cell wall thickness is seen as well as differences between the early- and latewood for each species. The combined effect of individual fiber properties is averaged in wood density, which is one of the main hereditary characteristics [7]. The X150 corner-cut views of southern and northern softwood handsheets (Fig. 27) also show the greater number of fibers in the northern pulp handsheet and the greater degree of fiber collapse. Clearly, there are fewer voids both in the surface layers and in the interior structure of the sheet.

Figure 28 shows the range of values obtained with changes in southern pine wood density for total yield and some paper properties [11]. In the case of total yield, a 0.01 density increase gives a 4% increase in absolute yield while the same density change correlates with a 1 km loss in breaking length and a 5 unit increase in tear. The pulp yield per unit volume charged to the digester depends on wood density and is responsible for wood species dominance in contrast to the fibers from non-wood sources.

Trends in tear with wood density are dependent on the amount of refining. At lower levels of refining, tear is lower for denser woods but, after extensive mechanical treatment, no correlation is apparent. Also, the porosity of handsheets, which is related to structure, shows an exponential increase with wood density [7].

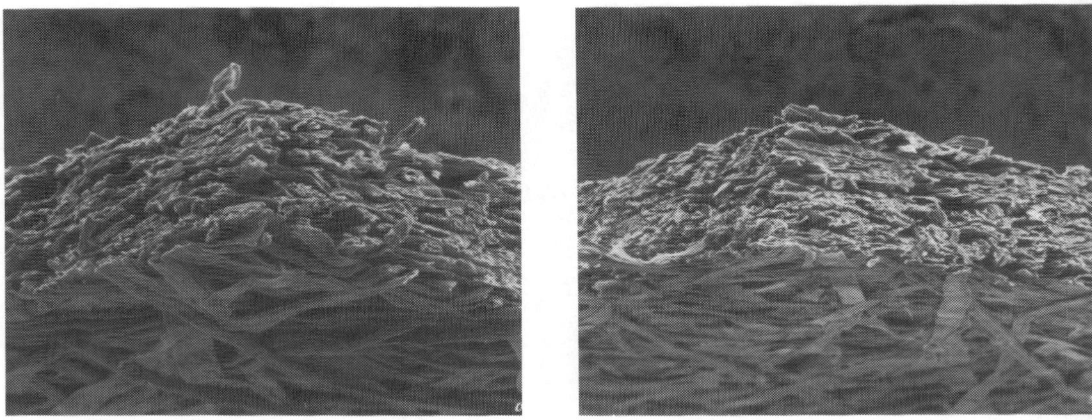

Fig. 27. Corner section view of handsheets from southern (left) and northern (right) softwoods (X150).

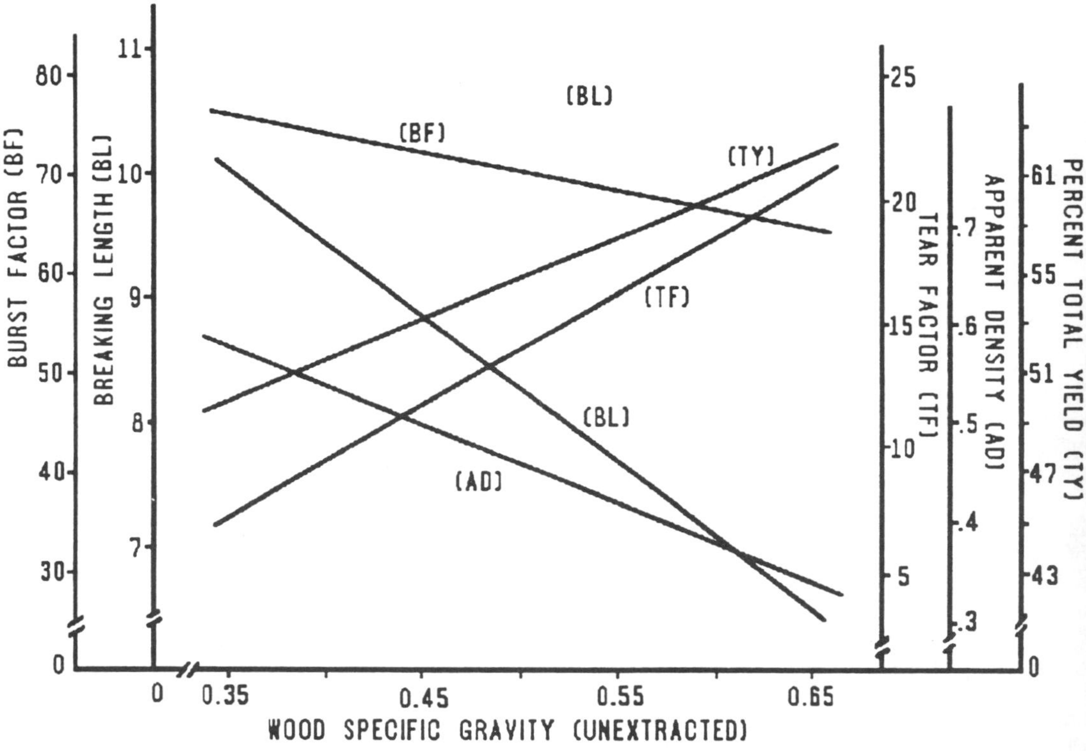

Fig. 28. The effect of wood density on pulp properties at 500 CSF. TY — total yield; BP — burst factor; BL — breaking length; TF — tear factor; AD — apparent density.

2. Influence of age and early-/latewood ratio

Tree age influences anatomical structure, cell dimensions, and the chemical composition of various species [75]. Hardwoods and softwoods are affected similarly by growth rates, but the effect is greater for softwoods. Growth rate shows some correlation with wood density but most of the density variation is associated with the age of the wood (Fig. 29).

Compared to latewood, earlywood fibers are wider, have thinner cell walls, and readily collapse in papermaking without refining. Earlywood fibers form smooth, dense sheets with good burst and tensile; however, they are less resistant to the cutting and bruising action of refiners if

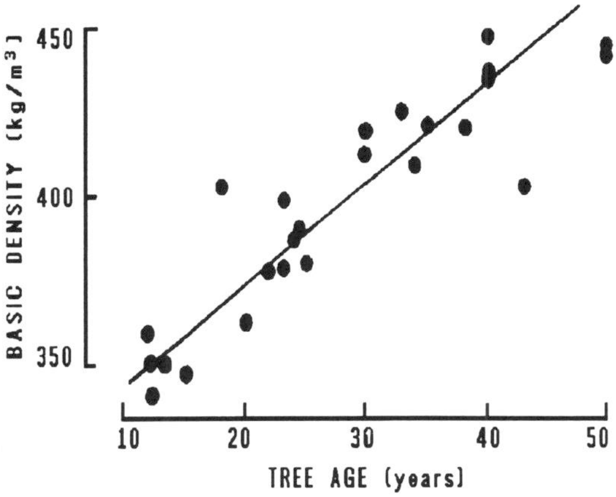

Fig. 29. Change in wood density with tree age.

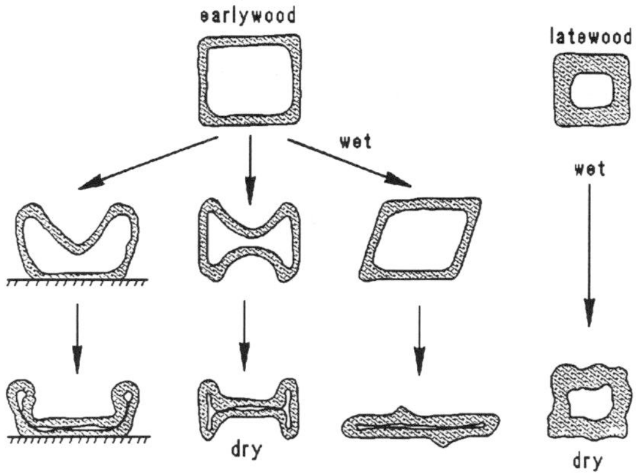

Fig. 30. Collapse behavior of wood fibers on drying.

high specific edge loads are used. Summerwood fibers make bulkier structures with low burst and high tear and these fibers are very resistant to refining. Often, extensive fibrillation of earlywood fibers may occur with little effect on latewood fibers being observed.

The latewood percent is important as it relates to properties easily controlled in tree improvement programs. Wood density is mainly a function of latewood content while fiber length is the only wood or fiber parameter not well correlated with latewood content. The earlywood-to-latewood ratio is also dependent on the latitude and/or altitude of the growth site. Where the two fiber types were studied separately, more completely bonded and denser papers are made from earlywood fibers. But, when sheets failed in tension, more earlywood than latewood fibers were ruptured.

The original tubelike fiber is altered in processing and the remaining structure collapses to a flatter, ribbonlike form upon drying. This influences strength development, i.e., tensile and tear, as it controls the number of bonds-per-fiber formed with neighboring fibers and the average bonded area between fibers. All fibers do not actually flatten out like a ribbon, but assume many shapes contributing to the interfiber bonding in the web (Fig. 30). Thick-walled fibers, such as softwood latewood or some hardwood fibers, often retain much of their tubular shape.

The differences between early- and latewood fibers are more pronounced for softwoods than for hardwoods. Hardwood fibers have relatively thick walls and narrow lumens; however, compared to softwoods, hardwood dimensions are smaller. With their large solid wall fraction, hardwood fibers preserve their original shape more through refining and forming. With less bonding than thin-walled fibers and fewer bonds than long fibers, the strength properties of hardwood pulps are almost always less than softwoods.

Fractionation of coarse southern pine or some West Coast pulps may permit the optimization of cell wall thickness differences so the fractions can be refined separately for strength development and then reblended to produce better paper. It appears that the latewood fibers form smaller bonded areas and thus weaker fiber-to-fiber bonds requiring a larger bonded area to take advantage of their strength [76]. The influence of refining on bonding is greater for these summerwood fibers which, initially, are much stiffer than springwood fibers. The bonded area of summerwood fibers increases with refining which also decreases their bending stiffness but contributes to a more homogeneous sheet structure.

I. SUMMARY

Many studies have shown the importance of wood and fiber characteristics in the manufacture of paper and board products. The anatomical features of most importance are: the kind and amount of cellular elements in hardwoods and softwoods; the fiber length, width, shape, and cell wall thickness; and the ratio of earlywood to latewood. Also important are the sub-microscopic structure of the fiber surface, the nature and distribution of chemical components in the cell wall, and the changes to the fiber surface and structure in response to refining.

We see that the papermaking potential of pulps depends on the chemical, physical and morphological character of the basic raw material and on the changes induced by various pulping processes and stock preparation. Of all the fiber features presented, the most essential appear to be fiber length, width, and cell wall thickness along with the residual lignin and hemicellulose content. The superiority of kraft pulping is clearly accepted, but modified sulfite processes also make very desirable pulps. The nature of the fiber surface and lower pentosan level make it more difficult to obtain transparent, low-porosity papers from kraft pulps than from sulfite pulps. But, the outstanding role of kraft pulps as the main raw material for papermaking has clearly been borne out.

The most important pulp requirement for any application is uniformity of supply. This means getting as close to the same material and properties day in and day out. There is little chance to optimize fiber use or product quality if the principal raw material varies widely in its potential for developing the desired balance of properties in response to refining.

REFERENCES
(Each reference here can lead the interested reader to many more sources of related information.)

1. Pulp and Paper Manufacture, Third Edition, Vol. 6, Stock Preparation, Kocurek, M.J., ed., The Joint Textbook Committee of the Paper Industry, TAPPI/CPPA (in prep.).
2. Duffy, G.G. and Kibblewhite, R.P., Appita 42(3):209 (1989).
3. Corson, S.R. and Kibblewhite, R.P., Appita 39(5):379 (1986).
4. Kibblewhite, R.P., Papermaking Raw Materials (Punton, ed.), Vol. 1:93, Mechanical Eng. Publns. Ltd., London (c1985).
5. Kibblewhite, R.P. and Hamilton, K.A., N.Z. J. Forestry Sci. 14(3):391 (1984).
6. Carpim, M.A., Barrichelo, L.E.G., Claudio-da-Silva, E. and Vasconcellos Dias, R.L. de, ABCP Congr. Anual 20th (Sao Paulo) 183 (1987).
7. Vasconcellos, R.L. de and Claudio-da-Silva, E., Papermaking Raw Materials (Punton, ed.), Vol. 1:7, Mechanical Eng. Publns. Ltd., London (c1985).
8. Vasconcellos, R.L. de and Claudio-da-Silva, D., ABCP Congr. Anual 18th (Sao Paulo), 1:31 (1985).
9. Dinwoodie, J.M., Tappi J. 49(2):57 (1966).
10. Dinwoodie, J.M., Tappi J. 48(8):440 (1965).
11. Barefoot, A.C., Hitchings, R.G. and Ellwood, E.L., Tappi J. 47(6):343 (1964).
12. McKenzie, A.W., Appita 38(4):284 (1985).
13. Nordman, L. and Gustafsson, C., Paper and Timber (Finland) 33(2):36 (1951).
14. Nordman, L., Gustafsson, C. and Olafsson, G., Tappi J. 38(12):724 (1955).
15. Van den Akker, J.A., Lathrop, A.L., Voelker, M.H. and Dearth, L.R., Tappi J. 41(8):416 (1958).
16. Kallmes, O.J. and Perez, M., in "Consolidation of the Paper Web," Tech. Sect. BPBMA, London, p. 779 (1965).
17. Page, D.H., Tappi J. 52(4):674 (1969).
18. Pulp and Paper Manufacture, Third Edition, Vol. 1, Paper Manufacture, Kocurek, M.J., ed. The Joint Textbook Committee of the Paper Industry, TAPPI/CPPA (1983).
19. Seth, R.S. and Page, D.H., PITA/TAPPI/CPPA Intl. Paper Physics Conf., Mont-Rolland, Que., p. 9 (Sept. 14, 1987).
20. Leech, H.J., Tappi J. 37(8):343 (1954).
21. Marton, R. and Alexander, S.D., Tappi J. 46(2):65 (1963).
22. Jayme, G., Tappi J. 41(11):178A (1958).
23. Clark, J. d'A., Tappi J. 45(8):628 (1962).
24. Jones, A.R., Tappi J. 55(10):1522 (1972).
25. Krkoska, P., Misovec, P., Hruskova, K., Blasko, J. and Oroszlanyova, K., Papir Celuloza 41(7/8):v17 (1986).
26. Matolcsy, G.A., Tappi J. 58 (4):136 (1975).
27. Kellogg, R.M. and Wangaard, F.F., Wood Fiber 1(3):180 (1969).
28. Krkoska, P., Misovec, P. and Blazej, A., Papir Celuloza 41(3):v2 (1986).
29. Higgins, H.G., De Yong, J., Balodis, V., Phillips, F.H. and Colley, J., Tappi J. 56(8):127 (1973).
30. Bradway, K., Tappi J. 56(8):118 (1973).
31. Thode, E.F. and Ingmanson, W.L., Tappi J. 42(1):74 (1959).
32. Wangaard, F.F., Kellogg, R.M. and Brinkley, A.W., Tappi J. 49(6):263 (1966).
33. Mason, S.G., Tappi J. 33(8):403 (1950).
34. Luner, P., PIRA Intl. Conf. New Technologies in Refining (Birmingham, UK) Proc.1:26 (Dec. 9, 1984).
35. Pulp and Paper Manufacture, Third Edition, Vol. 2, Mechanical Pulping, Kocurek, M.J., ed., The Joint Textbook Committee of the Paper Industry, TAPPI/CPPA (1987).
36. Pulp and Paper Manufacture, Third Edition, Vol. 4, Sulfite Science & Technology, Kocurek, M.J., ed., The Joint Textbook Committee of the Paper Industry, TAPPI/CPPA (1985).
37. Pulp and Paper Manufacture, Third Edition, Vol. 5, Alkaline Pulping, Kocurek, M.J., ed., The Joint Textbook Committee of the Paper Industry, TAPPI/CPPA (1989).
38. Labosky, Jr., P. and Ifju, G., Tappi J. 55(4):530 (1972).
39. Kibblewhite, R.P., Corson, S.R. and Graham,

K.L., Appita 40(2):121 (1987).
40. Pulp and Paper Manufacture, Third Edition, Vol. 3, Secondary Fibers and Non-Wood Pulping, Kocurek, M.J., ed., The Joint Textbook Committee of the Paper Industry, TAPPI/CPPA (1987).
41. Levlin, J-E., Paper p. 28 (1986).
42. Britt, K.W., Tappi J. 49(5):202 (1966).
43. Hoc, M., Tappi J. 72(4):165 (1989).
44. Wright, B.W., Tappi J. 34(10):133A (1951).
45. Giertz, H.W., Svensk Papperstid. 54(8):267 (1951).
46. Giertz, H.W., World's Paper Trade Rev. 138(19):1451 (1952).
47. McMillan, C.W., Forest Prod. J. 19(7):56 (1969).
48. McMillan, C.W., Wood Sci. Technol. 2:166 (1968).
49. Graff, J.H. and Miller, R.W., Paper Trade J. 109(6):3 (1939).
50. Dillner, B. and Jonsson, P., New Pulps for Paper Ind. Symp. Proc. Brussels (Miller Freeman, San Francisco), p. 25 (May 1979).
51. Koran, A., Veillette, D. and Law, K.N. Tech. Sec. CPPA Conf. Technol. Estivale, Pointe-au-Pic, Que., p. 1 (May 1980).
52. Anon., Paper 192 (4):182 (1979).
53. Dadswell, H.E. and Wardrop, A.B., Appita 12(1):13 (1959).
54. Dadswell, H.E. and Wardrop, A.B., Appita 12(4):129 (1959).
55. Watson, A.J. and Hodder, I.G., Appita Proc. 8:290 (1954).
56. Tocan, M., Fischgold, S. and Constantinescu, O., Celuloza Hirtie 21(1):38 (1972).
57. Clarke, B., Ebeling, K.I. and Kropholler, H.W., Paperi Puu 949 (1985).
58. Clark, J. d'A., "Pulp Technology and Treatment for Paper," Miller Freeman Publications, San Francisco (1978).
59. Kajaani FS-100 Analyzer, Instrument Manual (1982).
60. Einspahr, D.W., Tappi J. 47(4):180 (1964).
61. Giertz, H.W., "The Formation and Structure of Paper", Tech. Sect. BPBMA, London, p. 597-620 (1962).
62. Sastry, C.B.R. and Wellwood, R.W., Tappi J. 54(10):1686 (1971).
63. Sastry, C.B.R. and Wellwood, R.W., Tappi J. 55(6):901 (1972).
64. Leopold, B., Tappi J. 49(7):315 (1966).
65. Kellogg, R.M. and Wangaard, F.F., Tappi J. 47(6):361 (1964).
66. Page, D.H. and MacLeod, J.M., Paprican Rept. PPR 785 (Mar. 1990).
67. Wink, W.A. and Van Eperen, R.H., Tappi J. 45(1):10 (1962).
68. Page, D.H., Seth, R.S. and El-Hosseiny, F., Paprican Rept. PPR 527 (Mar. 1985).
69. Stone, J.E. and Clayton, D.W., Pulp Paper Can. 61(10):T 475 (1960).
70. Cowan, W.F. and Cowdrey, E.J.K., Tappi J. 57(2):90 (1974).
71. Matolcsy, G.W., Pulp Paper Can. 76(5):80 (1975).
72. Gallay, W., "Formation and Structure of Paper," Tech. Sect. BPBMA London, p. 491-535, (1962).
73. Clark, J. d'A., Tappi J. 56(7):122 (1973).
74. Barefoot, A.C., Hitchings, R.G., Wilson, E.H. and Kellison, R.C., Proc. Symp. Growth Acceleration Effect on Wood Props., U.S. Forest Prods. Lab., p. K1 (Nov. 1971).
75. Surewicz, W. and Surma-Slusarska, B., EUCEPA Conf. Torremolinos, Proc. 21(2):3 (May 14, 1984).
76. Retulainen, E., Vaulot, F. and Ebeling, K., Paper presented at 19th Intl. Symp. "Paper, An Engineered Product", Miami Univ. (May 16, 1988).

III

SHEET STRUCTURE CONSIDERATIONS — PAPER AS AN ENGINEERED MATERIAL

G.A. BAUM

Gary A. Baum is Director of Corporate R&D for James River Corporation in Neenah, WI. Dr. Baum received his B.S. degree from Wisconsin State University and his M.S. and Ph.D. degrees from Oklahoma State University. He has gained world recognition for his R&D work in the measurement of fiber and sheet properties, especially for his successful and far-reaching efforts to develop accurate and reliable on-line measurement of mechanical (strength) properties of paper and board. He holds a number of patents and is the author of numerous publications. An active TAPPI member since 1972, he has lectured frequently at TAPPI conferences and seminars. He is presently on the TAPPI Board of Directors. Among his professional honors, he received the 1986 TAPPI Process and Product Quality Division Award and the Division's first Richard S. Hunter Prize. In 1987 he was designated a TAPPI Fellow.

A. INTRODUCTION

In practice, paper can be made from a given furnish to have a broad range of physical properties by changing machine conditions. It is also true, however, that changes in the furnish, without changing machine variables could produce quite different sheets. (Actually, changes in the furnish may force changes in machine conditions in order to make paper at all.) The fact of the matter is that all of the furnish and machine variables are interdependent. It is not always sufficient to separate out just one variable at a time to study. Of course, that makes the situation rather complex and helps explain why the manufacture of paper is still interesting and challenging today, after nearly 300 years of papermaking in the U.S.

This chapter is concerned with paper property development during the manufacturing process. It is fair to say that everything will impact the final sheet properties: the fiber species, pulping and bleaching methods, water chemistry, machine hydrodynamics, pulp consistency and temperature, machine design and related operating variables, and so on. Figure 31 attempts to show the kinds of interactions that may exist to produce a sheet having certain end use performance specifications or converting requirements. While we will touch on most all of these areas, it is not possible to discuss all of them in detail, or all of the possible interactions, in this chapter. Rather, the intent is to discuss the manufacturing process in somewhat general terms, and to ascertain the impact of the various unit processes on the structure of the web. We will dwell mostly on those variables which can be, and often are, manipulated on the machine to yield specific results. In a sense, Fig. 31 provides the outline that we will follow.

This chapter, then, attempts to give an overview of how manufacturing variables may interact to produce a given set of properties and

---

REVIEWED BY

Otto J. Kallmes
President
M/K Systems Inc.
Danvers, MA

Thomas A. Kindler
Energy and Planning Engineer
Mead Corporation
Escanaba, MI

Dan W. Manson
Senior Consultant, Engineering
James River Corporation
Portland, ME

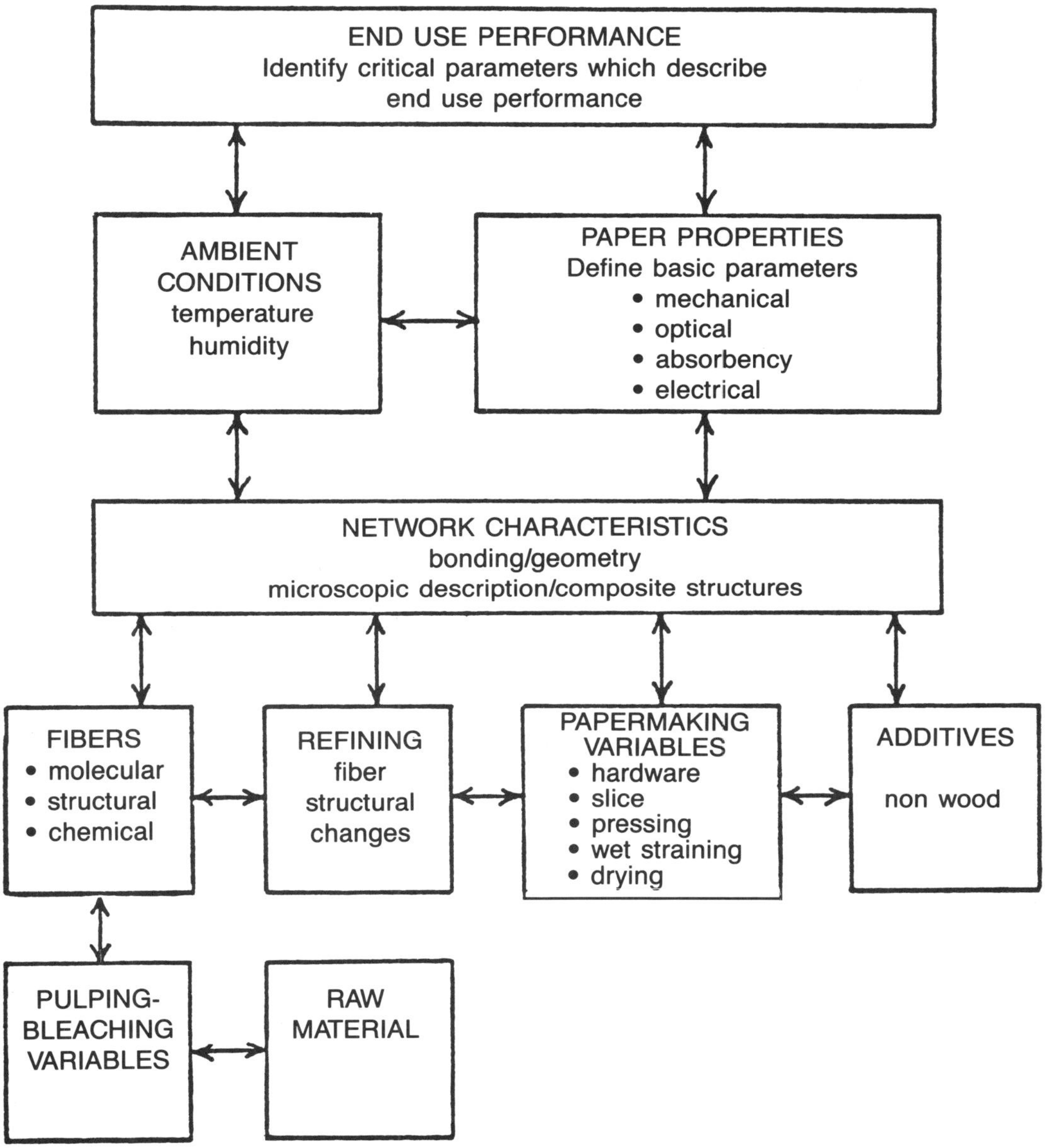

Fig. 31. Schematic diagram showing the relationships between raw materials, manufacturing variables, network and paper properties, and end use performance. The latter includes converting requirements.

furthermore tries to describe how these properties may relate to converting requirements or end use situations. The emphasis is on mechanical properties as developed on the paper machine. The discussions and literature references are not meant to be comprehensive. There are a number of books which cover the subject matter of this chapter in some detail and the reader is referred to them for further information [1-5]. In addition, the published proceedings of the Fundamental Research Committee (London) (the "Oxford-Cambridge Symposiums") also provide specific information on many of the topics discussed here [6a-6h]. The 1989 Cambridge symposium, "Fundamentals of Papermaking" [6h], includes excellent state-of-the-art reviews on stock preparation, forming, chemistry, pressing, drying, surface application, calendering, and controllability. The author recommends that these volumes be studied for in-depth

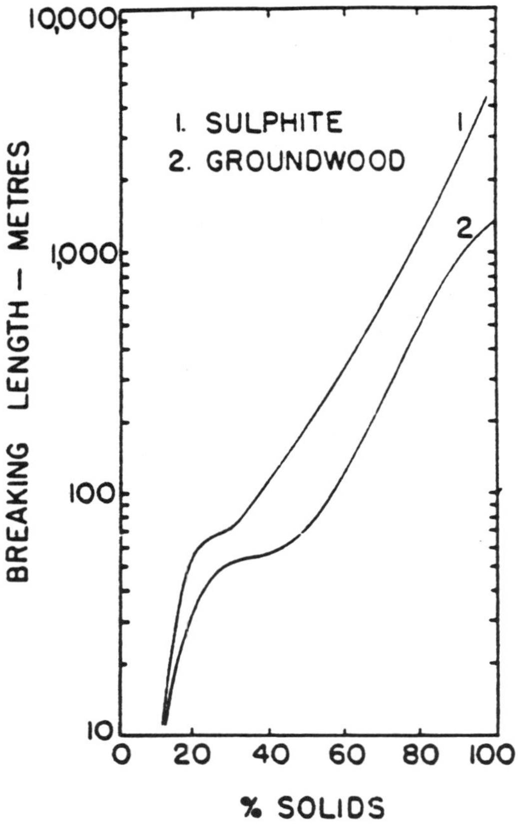

Fig. 32. Development of strength in sulfite and groundwood webs with increasing solids content [7].

discussions of the material presented in this chapter.

B. CONSOLIDATION OF THE WEB

Paper is unique in that the separated wood fibers, of which it is composed, are able to rebond to each other by naturally occurring forces. These are widely accepted to be hydrogen bonds, in which the electron in a hydrogen atom (in a hydroxyl group) is "shared" between two electronegative oxygens. Hydrogen bonds require that the two surfaces (fibers, fibrils, or molecular entities) can come close enough together in order to bond. Surface tension forces are responsible for bringing the wet fibers together so that bonding may occur. According to Lyne and Gallay [7], surface tension forces become increasingly important in the range from 10 to 25% solids, at which point hydrogen bonds can begin to form. This is consistent with experiments done on wet web strength vs. solids in which an inflection point is often seen around 25% solids, as depicted in Fig. 32 for sulfite and groundwood pulps.

As air begins to enter the low solids web, it

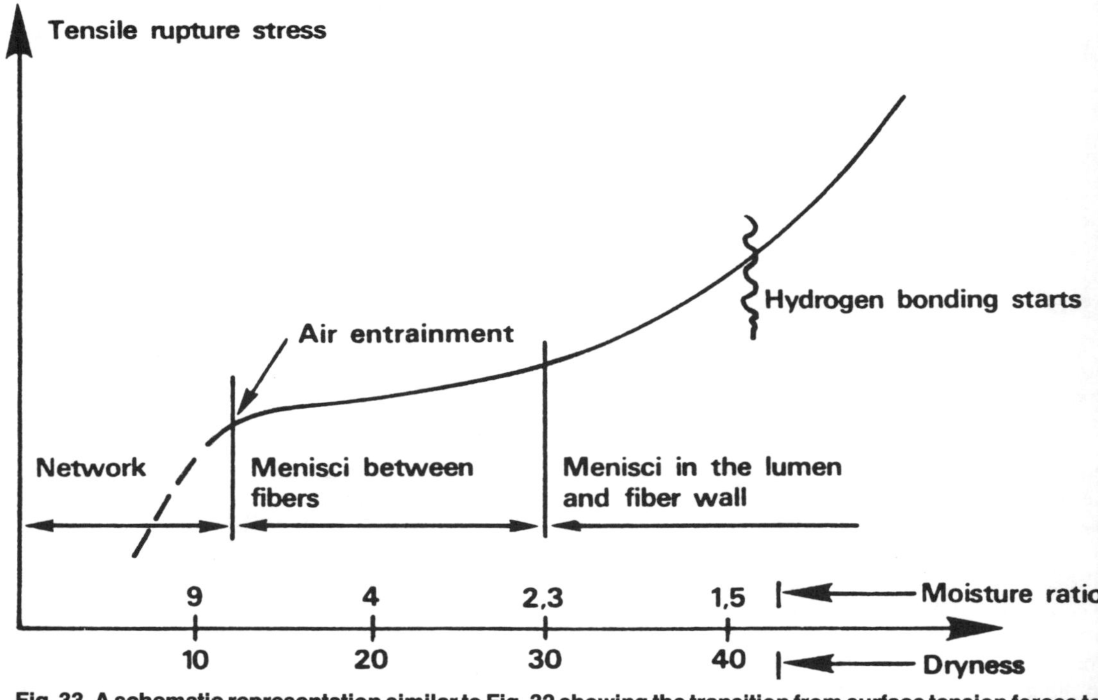

Fig. 33. A schematic representation similar to Fig. 32 showing the transition from surface tension forces to chemical bonding.

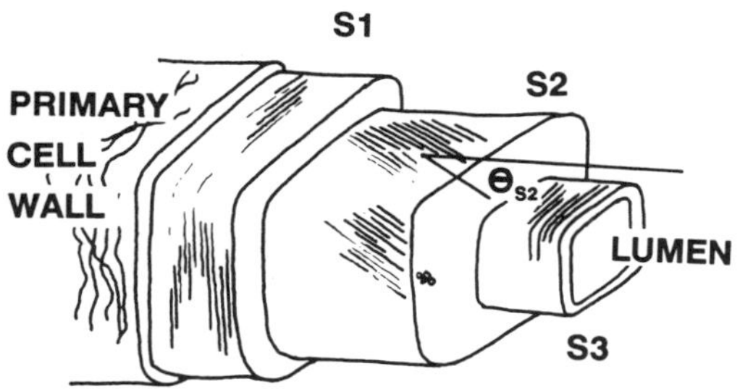

Fig. 34. A schematic representation of the fiber cell wall showing the arrangement of the fibrils in the S1, S2, and S3 layers.

causes a discontinuous liquid (water) film. This gives rise to (attractive) surface tension forces that are dependent on the length of the air-fiber-water contact line. At roughly 25% solids, the surface tension forces gradually change to those which are inversely dependent on the thickness of the water film. The pressure difference between two surfaces, separated by a water film of thickness x, is given by $P = 2s/x$, where s is the surface tension of the water. Decreasing the water film thickness between fibers bring the surfaces close enough for chemical bonding to occur. For hydrogen bonds this would be in the neighborhood of a few angstroms.

Figure 33 illustrates schematically how the wet web strength varies as solids content increases. The initial wet web strength is a consequence of possible mechanical entanglement of the fibers and the onset of the surface tension forces as air enters the structure. Wet web strength increases as the surface tension forces take on the inverse water film thickness form and hydrogen bonds can begin to form. As more and more water is removed, the wet web strength develops rapidly as more and more hydrogen bonds form.

The degree to which this bonding can occur clearly depends on the ability of the two surfaces to conform to each other. Two crossed rigid circular fibers, for example, could only bond in the small area of contact, whereas "flexible" or "conformable" fibers would enable surface tension forces to draw much larger areas together so that chemical bonding occurs over a broader area.

The word conformable is taken to mean a fiber that is flexible both along its axis and perpendicular to its axis. Clearly, the flexibility or conformability of a fiber will depend on the species (hardwood vs. softwood, thick cell wall vs. thin cell wall [ease of fiber collapse], etc.), growth location and conditions (Northern vs. Southern, earlywood vs. latewood, compression wood vs. tension wood, mature wood vs. juvenile wood, core wood vs. slab wood, etc.), pulping method (chemical vs. mechanical, etc.), and refining method and level. As noted earlier, all of these variables tend to be interdependent. For example, the level of refining required to produce a particular "wet fiber flexibility" will differ according to pulping method, growth conditions, etc.

1. The fiber cell wall

This interdependence can be traced, in part, to the nature of the construction of the fiber cell wall (see Fig. 34). Wood fibers in their natural state are hollow filament wound composite structures, composed of cellulosic fibrils surrounded by a matrix of lignin and hemicellulose. The hemicellulose is believed to be the "bridge" between the cellulosic fibrils and the lignin. The lignin-hemicellose matrix distributes loads over the length of the fibril and provides moisture resistance for the fiber. The arrangement of the fibrils in the secondary cell wall provides the necessary structural attributes required by the fibers. Briefly, the innermost S3 layer provides resistance to internal pressures, the S2 layer, with the still cellulosic fibrils oriented at a low angle to the fiber axis, provides the high axial stiffness of the fiber, and the outermost S1 layer provides additional resistance to internal stresses and keeps the entire assembly together.

The above description is the classical description of a fiber-reinforced composite such as fiberglass-epoxy or fiberglass-polyester composites. In the fiber, the fibrils are wound about the fiber axis with distinctive orientations to provide radial and axial strength in the fiber. In a pressurized "fiberglass" vessel, one would arrange the glass fibers in a similar way for similar reasons.

The pulping method modifies the structure of the cell wall. Mechanical pulp fibers are separated without removing much of the lignin from within the cell wall; that is, the lignin which surrounds the fibrils comprising the cell wall. As a result, these fibers tend to be rather rigid and do not readily collapse during drying, since the fibrils are rigidly held in position by the matrix material. During a chemical pulping process, however, the chemicals that remove lignin from the interlamella region between fibers also remove the lignin from within the cell wall. In fact, it is probable that the cell wall lignin is the first to go since the pulping chemicals can move readily through the lumens of the fibers. Removal of the lignin from the cell wall means that the fibrils (in the wet fiber) may now easily move relative to one another, especially when ample water is present in the regions between them to act as a lubricant. Thus, chemical pulp fibers in the wet state tend to be very flexible or conformable and, if the cell wall is not too thick, can easily collapse, all as a result of the ability of the fibrils to readily move relative to one another.

Pulping opens up the cell wall creating more surface area. The mechanical action of a refiner on the pulp fiber has a similar effect. Beating separates fibrils from each other which allows water to enter the volume created. This leads to increased surface area and more flexible fibers. During pressing or drying, water removal from these regions results in high surface tension forces that act between fibers, fibers and fibrils, and fibrils in the cell wall with each other. This leads to many hydrogen bonds and a well-bonded (strong) network. Thus the structure of the cell wall after pulping (and bleaching) and refining to a large extent determines the flexibility of the fiber and the nature of the bonding after the fiber has been dried.

In summary, consolidation of a fiber mat during papermaking depends on surface tension forces bringing the fibers and/or fibrils close enough together for hydrogen bonds to occur (typically a few angstroms). The pulping and refining methods, and interactions between them, determine the extent of the surface area created and available for bonding. In the case of the conformable chemical pulp fibers, more fiber-to-fiber bonds occur and the average cross-section of a bond is larger than in the case of the more rigid mechanical pulp fibers. Surface tension forces are more likely to cause collapse of the lumen in the chemical pulp fibers than in the mechanical pulp fibers. Furthermore, during drying, the cellulosic fibrils and hemicelluloses in the chemical pulp fiber cell wall can also bond. Because refined chemical pulp fibers tend to be more swollen, they will also have the greatest shrinkage during drying. The lateral shrinkage of a fiber bonded to a second fiber causes "microcompressions" in the second fiber, and vice versa. This is discussed in more detail later.

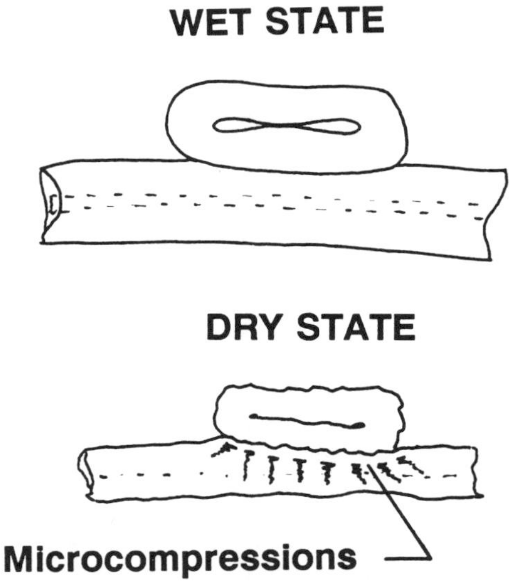

Fig. 35. A schematic representation of two bonded fibers swollen with water (top) and after drying (bottom). The lateral shrinkage of the upper fiber causes microcompressions in the lower fiber.

2. Drying of the network

Thus far we have discussed how the fibers come together to form the network and how chemical (hydrogen) bonds come about. Consider again the case of the swollen chemical pulp fibers and bonds forming between them as the solids content increases. There is something else happening at the same time. As water is removed from the cell wall the fiber begins to shrink in cross-section. This may be typically 15 to 20%. Along the fiber axis, however, there is very little shrinkage because the S2 fibrils, which comprise the bulk of the cell wall, are aligned more or less along the axis. Thus the removal of water from between fibrils results primarily in a lateral collapse of the fiber perpendicular to its axis. The extent of the transverse shrinkage, of course, depends on the extent to which the cell wall was swollen, which in turn is a consequence of the method of pulping and the extent of refining.

Page and coworkers [8] discovered that interfiber bonds form *before* the transverse shrinkage of the web. This creates a rather complex situation at the fiber-fiber interface. In Fig. 35, two swollen fibers are shown bonded together. As the upper fiber shrinks in cross-section, stresses are transmitted into the horizontal lower fiber, causing it to also shrink along its axis. If we now envision that there are many crossing fibers bonded to the horizontal fiber, we conclude that its axial shrinkage would be a sum of all of the separate contributions of the fibers bonded to it. If the average lateral shrinkage was 15% and fiber-fiber bonds occurred all along its length, we could expect a 15% contraction of the horizontal fiber. The compressed regions along the horizontal fiber are referred to as "microcompressions".

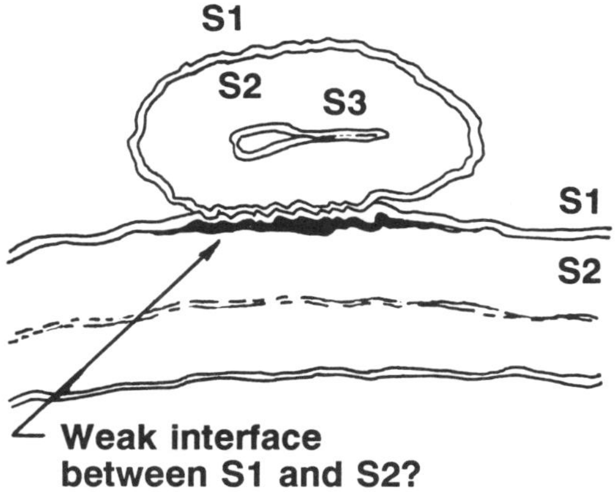

Fig. 36. A schematic diagram illustrating how the axial compression of a bonded fiber could lead to a weak S1-S2 interface. Along the fiber axis the S1 layer is much more easily compressed than the S2 layer.

The web shrinkage stresses can be very large. The extent of the shrinkage, however, is also dependent on paper machine variables which will be discussed later. The shrinkage of the web has important implications in later convertibility or end uses of the paper. For example, the elongation of the web at rupture (stretch) is directly related to the shrinkage during drying. The former is typically 1 or 2% greater (due to the stretch in the network itself) than the latter.

Figure 35 is incomplete in that it depicts events occurring in one direction only. In fact, both fibers are shrinking transversely at the same time with the overall effect of web shrinkage in both directions. There is also a z-direction shrinkage, of course, a direct consequence of the fiber cell wall shrinking as water is removed.

The shrinkage stresses suggest that the contact or bonded area between fibers is not planar, but has a three-dimensional character. If so, the geometric bond area would be an inappropriate measure of the fiber-fiber bond area. We can also argue that the strength of a fiber-fiber bond will depend on circumstances during shrinkage (drying) of the web. For example, an S1 layer bonded to an S1 layer would likely result in a different situation than an S1 to S2 or an S2 to S2 bond. In the first case, the transverse shrinkage of one of the fibers in compressing the S1 layer of the second fiber may result in an interlaminar cell wall failure at the S1–S2 interface in the second fiber since the S2 layer would be much more resistant to axial compressive loads than the S1 layer. Figure 36 attempts to illustrate this situation. Such separation at the S1–S2 interface has been reported in the literature [eg. 9].

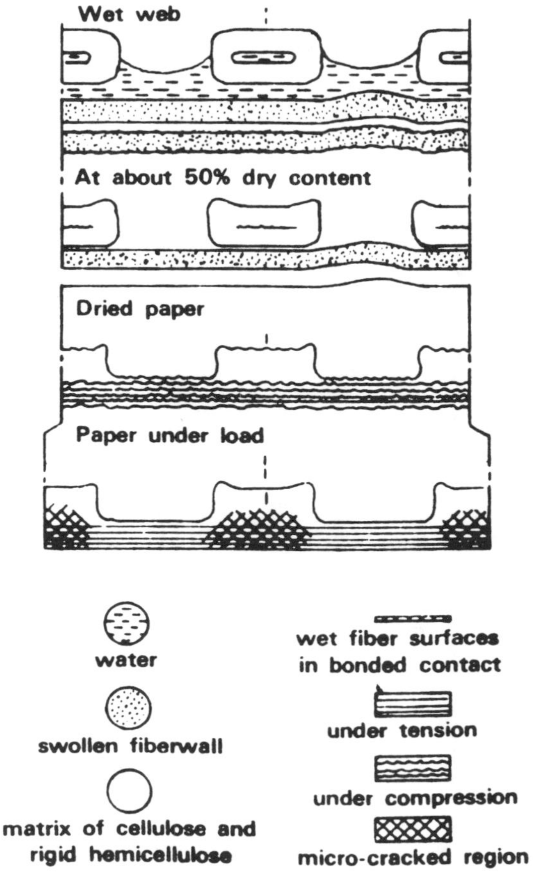

Fig. 37. A schematic drawing of the consolidation of the wet paper web and the phenomena caused by fiber shrinkage taking place during drying [12].

The bonds that hold the fibers together in paper occur naturally as water is removed from the network. Not surprisingly, if water is added back into the structure, it swells and bonds are broken. The continuous dimensional changes that occur in going from the swollen state to the dry state are reversed, although not completely. This lack of *dimensional stability* is a problem in many paper grades. If the moisture entered the network from one side only, that side would begin to expand while the opposite side would not, resulting in *curl*, another undesirable characteristic. The extent to which a web displays these behaviors depends not only on pulping and refining conditions, but manufacturing conditions on the paper machine as well, especially jet-to-wire speed differentials (which govern the extent of fiber orientation), draw ratios, and drying variables [10]. The effects of machine variables on the web structure are

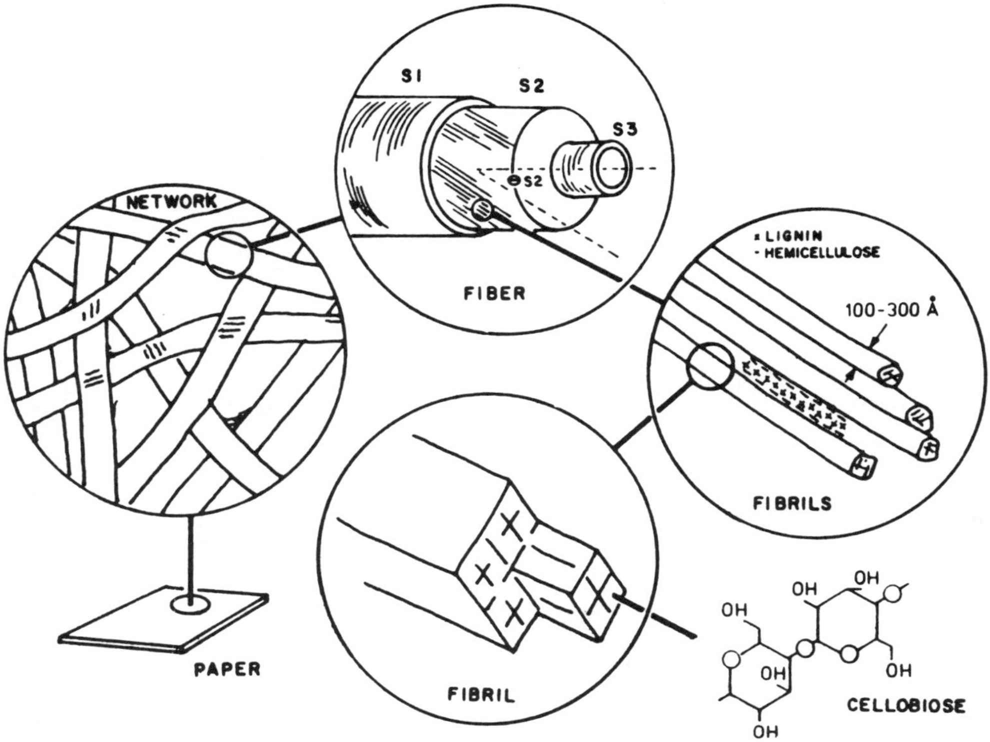

Fig. 38. A schematic drawing showing the "levels of organization" in heterogeneous paper.

dismissed later. Details concerning dimensional stability and curl may be found in Chapter 1 of this volume. An excellent review of the causes of dimensional stability and curl has been prepared by Rutland [11], and appears in Vol. 9 of this series of textbooks.

Giertz [12] has summarized the consolidation and drying of a fibrous network as depicted in Fig. 37. The uppermost part of the figure represents the wet swollen state as surface tension forces begin to act. Next the bonds form and fibers collapse as water is removed from the network. Upon drying, the transverse shrinkage causes microcompressions that result in overall shrinkage of the "network". Finally, if a uniaxial load is applied to the ends of the network, the microcompressions are first "pulled out" and then the fibrous material itself is stressed to failure.

3. Levels of organization in paper

Figure 38 attempts to outline the levels of organization that we must consider when describing fibrous networks made from wood fibers. The far right hand side depicts the cellulose chains that make up the fibrils. These may be arranged in a close packed array (crystalline character) or amorphous array (non-crystalline). As we follow the circles toward the left, we see a schematic representation of the fibrils in the cell wall, the secondary layers of the cell wall (again), and finally the network itself comprising the paper. Thus far we have briefly considered portions of the right side of the figure that relate to pulping and refining and discussed somewhat more thoroughly the impact of the cell wall structure (top figure) on the paper web. We are now ready to examine the network and how we characterize it.

C. CHARACTERIZATION OF THE SHEET STRUCTURE

Paper is made from discrete fibers which have different dimensions and properties and which are arranged in a more or less random fashion. The notable exceptions to the latter comment are that the fibers typically lie in the plane of the paper with little or no thickness or z-direction (ZD) orientation and that conditions on the paper machine give rise to more fibers aligned in the direction of manufacture (machine direction, MD) than in the perpendicular or cross-machine direction (CD). Many papers contain a mixture of hardwoods, softwoods, inorganic fillers, or other non-wood constituents. Altogether, paper can be described as a heterogeneous, anisotropic, composite material. The varied composition and lack of homogeneity make the paper structure difficult to characterize. We attempt to do this, however, in terms of *network parameters* or the macroscopic *physical properties*.

Network parameters attempt to define the structure in terms of the dimensions, physical properties, and geometrical orientations of the fibers and the nature of the bonding between fibers. Such characterization is extremely difficult because of broad distributions in fiber properties and the fact that papermaking variables can change the properties (and distributions) significantly. This, in turn, impacts the bonding between fibers as well as their response to papermaking variables such as refining, pressing, etc.

Some examples of network parameters, aside from terms used to describe the fibers themselves, are fiber orientation, bond area, number of bonds per length of fiber, interbond distance, and bond strength per unit area. These are not necessarily independent. For example, a simple concept like the geometric bond area between two fibers goes from a minimum when the two fibers are at right angles to a maximum when the fibers are exactly aligned on top of one another. Thus bond area and fiber orientation are interrelated. A network parameter which seems to be useful and which combines several of those concepts alluded to above is the *relative bonded area*, defined as the ratio of the total bonded area to the total (surface) area available for bonding. Expressed as a percent, the relative bonded area (RBA) increases as the number of bonds or bond area increases (say by increased refining or wet pressing) but says nothing about the quality (or strength) of the bonds. (The RBA can never be 100% because the outermost fibers cannot be bonded on their outside.)

While fiber-to-fiber bond strength is a straightforward concept, it is a property that is difficult to measure. Much of the difficulty comes about because wood fibers are tiny, but also because it is difficult in the laboratory to reproduce the stresses seen by the fiber in the paper. A single fiber-to-fiber bond tested in the laboratory may not be representative of the state of stress seen by the fibers in the paper. Furthermore, there are always questions concerning the nature of the drying processes in the individual fibers alone or in the paper. Since chemical pulp wood fibers can shrink 15 to 20% perpendicular to their axes if allowed to do so, a fiber in the paper which makes many bonds with other fibers will likely have properties markedly

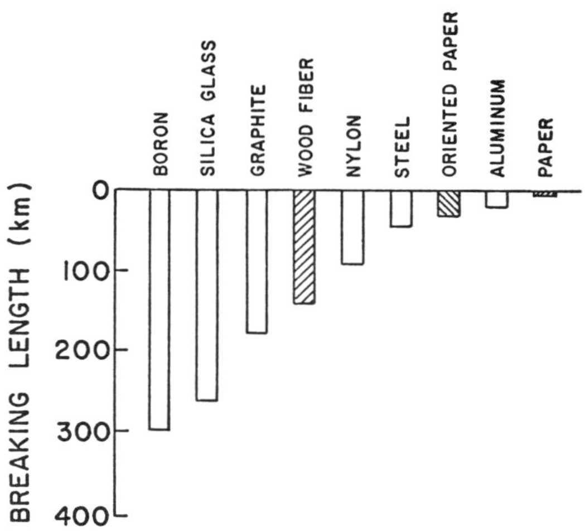

Fig. 39. A comparison of breaking lengths for some common engineering materials. On a mass basis, wood pulp fibers can be stronger than some metals.

different than if dried outside of the network. Hence we cannot be sure that laboratory measurements are representative of conditions in the real structure.

When compared on a mass basis, wood fibers are stronger than many common engineering materials as depicted in Fig. 39. The *breaking length* (tensile strength/density∗acceleration of gravity) of the wood fiber is greater than that for steel or aluminum. While the 140 km breaking length is the strongest the author was able to find reported in the literature [13], even if the average was one third or one half the reported value, we would have a remarkably strong material. Yet the paper we make from such fibers is typically much weaker (right hand side of Fig. 39).

The concepts of *bond area* or *fiber-to-fiber bond strength* imply that the fibers themselves are uniform and stronger than the bonds between them. This is probably not always the case. As we saw in Figs. 34 and 38, the wood fibers themselves are excellent examples of "fiber-reinforced composites" where the fibrils making up the cell walls are the framework material (the "fibers") and the lignin-hemicellulose mixture is the reinforcing (matrix) material. In chemical pulps, some of the reinforcing material is removed. If the failure between two fibers originates within one of the cell walls, as indicated in Fig. 36, the meaning of bond area or bond strength becomes obscure.

In mechanical pulps, where the lignin reinforcing material in the cell wall is still present, the concept of bond area does have meaning. For such pulps, the fibers are less flexible and the contact area (not necessarily bonded area) that results from the surface tension forces will be less than for the case with chemical pulps. The stiffer mechanical pulp fibers also likely result in fewer actual bonds per length of fiber. In addition to fewer bonds and smaller bonded area, the bond strength per area may be different (less) for the lignin-rich surface of mechanical pulps, compared to those of chemical pulps.

Sheet *density* is the mass per unit volume, or basis weight divided by caliper. Density (or its reciprocal, bulk) is often taken as a fundamental network property. In general, as one increases relative bonded area by increasing refining or wet pressing pressure, density increases also since more material (mass) is being compressed into the same volume. This rule of thumb does not work, however, in the case of dry pressing (calendering) where density can increase but relative bonded area may not. Apparently calendering causes some crushing of fibers which decreases bonding while compacting the material to higher densities.

Wet stretching of a web can decrease both density and RBA, as will be discussed later. Shrinkage during drying should have the opposite effect, increasing both density and RBA.

There is a problem with using density as a measure of network structure or an indicator of sheet performance in that density relies on the measurement of a somewhat nebulous property, the caliper. Caliper is defined as the thickness of the sheet under standard conditions of platen size and pressure. This may be perfectly acceptable for monitoring production but is not very good for calculating the sheet density for at least two reasons. First, paper is a compressible material and the caliper, and thus the density, depends on the load applied during the measurement. Second, the surface roughness of paper contributes to the caliper measurement, which introduces an error in the density. This is not acceptable if we wish to use the density as a fundamental network property capable of predicting other properties. The "density" of the surface is apt to be quite different from that in the bulk. To complicate matters further, the compressibility *of the surface* is likely to be very different from that in the bulk.

There have been a number of approaches tried to overcome this problem of surface roughness on caliper. A recent review is given in Ref. [5]. These include a computed "effective caliper", a pycnometric caliper based on mercury intrusion, and a "soft platen" caliper. Each method has drawbacks and advantages but it is probably not appropriate to discuss them here. All of the measurement methods give reasonably similar results over a variety of grades [14]. For any of the methods, the computed densities will be greater than or equal to those computed using the standard hard platen methods. The smoother the surface, the closer they will agree with the hard platen methods.

Perhaps the simplest in concept and ease of use is the soft platen approach. In this measurement, soft rubber platens contact the sheet, conform to the gross roughness features of the surface, and thereby eliminate the surface roughness contribution to the caliper, and hence density. The device is calibrated against precision metal shims of known caliper [14].

There are numerous cases in the literature where sheet physical properties are measured and correlated against density. In many of these the density measurement is based on a hard caliper measurement, making some of the conclusions questionable.

Another property that is often used to characterize the web structure is the light scattering coefficient, S. The light scattering coefficient, determined optically, is directly related to the surface area that can scatter (reflect) light per gram of fiber. Light scattered in the visible part of the spectrum is due to the difference in the index of refraction between fiber and air. Thus, no light will be scattered from the region where two fibers are bonded (or technically in optical contact). The greater the bonded area, the smaller the light scattering coefficient.

Measurements of light scattering coefficient, in fact, are often used to compute the relative bonded area. To do this, one needs to know the light scattering coefficient, S, for the bonded sheet as well as the completely unbonded sheet. Since the latter is difficult to do (although it has been attempted and done), usually S for the unbonded network is obtained by extrapolation to zero tensile strength in a plot of S vs. tensile strength. If Su is the extrapolated value, and S is the value for the bonded sheet, then RBA = (Su − S)/Su.

Rennel [15] studied how refining and wet pressing affected breaking length and scattering coefficient for a groundwood pulp and a bleached sulfate pulp as shown in Fig. 40.

Figure 40 illustrates the differences between mechanical pulps and chemical pulps in terms of strength development and the level of the scattering coefficient. The mechanical pulping process produces many fiber fragments which lead to high scattering coefficients, but the shorter average fiber lengths and lignin rich surfaces give rise to poor bonding and lower strengths. The chemical pulp fibers, however,

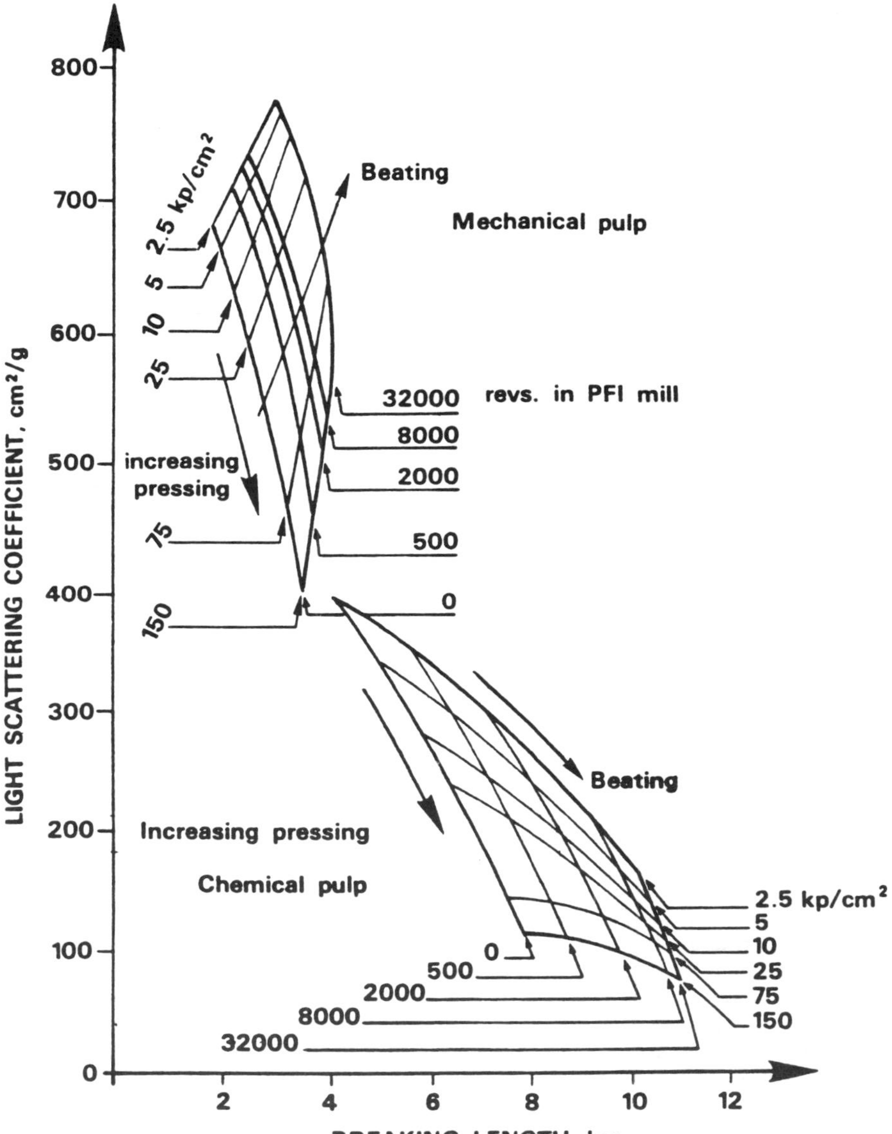

Fig. 40. Light scattering coefficients vs. breaking length for chemical and mechanical pulps [15].

are flexible and more likely to be intact (not cut), leading to greater bonding and thus higher strengths but in doing so produce fewer unbonded sites to scatter light. The response of the two pulp types to wet pressing pressure and beating (in a PFI mill) are different.

For the chemical pulp (lower part of Fig. 40), at constant wet pressing pressure, increased beating causes an increase in breaking length and a decrease in scattering coefficient. The lower the pressing pressure, the greater the effect of beating on increasing breaking length and decreasing S. For the chemical pulp, beating is more effective in producing high strengths and a smaller decrease in S than is pressing.

For the mechanical pulp (top of Fig. 40), however, the situation is slightly different. Increasing the wet pressing pressure still increases breaking length and decreases scattering coefficient just as in the case of the chemical pulp. Beating, on the other hand, increased *both* breaking length and scattering coefficient. One can envision that refining the stiff mechanical pulp fibers produces more fiber fragments by cutting, leading to the increase in S. This is consistent with the normal procedure of refining groundwood as little as possible to prevent loss of tear factor.

D. PHYSICAL PROPERTIES

The physical properties of interest to us in our study of paper are most often those which are typically used to describe any engineering material, including tensile strength, tensile stiffness, elongation at failure, toughness, tearing strength, bending stiffness, etc. These properties, and their measurement, have been discussed in detail in Chapter I. Many tests are unique to paper testing (although similar versions often appear in other fields). Bursting strength and the various sizing tests are some examples. Paper tests are often specific to a particular end use requirement, converting requirement, or both. Paper smoothness, for example, would be important both for printing and the satisfaction of the end user. In some cases the converting requirements and end use requirements for a given physical property may be completely different. For example, the out-of-plane shear stiffness of corrugating medium should be low for the corrugating operation, but as large as possible for good flat crush and box compression in the end use situation.

Paper is a three-dimensional material, and the often-neglected thickness direction properties play an important role in understanding what happens during converting[1] and, in some instances, end use performance. The "traditional" paper physical properties, even collectively, are not always effective in describing the response of the paper to the mechanical stresses that it may experience during converting operations. This is not a shortcoming in the tests, but rather a consequence of the need to describe the stress-strain response of the material in three dimensions. This will be discussed in somewhat greater detail later.

There is a set of fundamental parameters, the specific elastic stiffnesses, that do describe the three-dimensional stress-strain response of paper. Techniques have been developed to measure them non-destructively in sheet-like materials making it easy for us to use them to characterize web structures. In the following paragraphs we briefly discuss the theory behind the measurements.

1. "Converting", as used here, means any process which alters the properties of the web after it has been dried once. Thus, sizing, coating, calendering, etc., would all be converting processes.

The elastic properties of a material describe its deformation when stresses are applied to it. For an isotropic material, one which has no directionality, there would be three elastic properties: a Young's modulus, E, relating axial stress and strain; a shear modulus, G, relating shear stress and shear strain; and a Poisson ratio, v, the ratio of the lateral contraction to the axial extension during uniaxial stressing. Only two of the three elastic properties for an isotropic material are independent. If any two are known, the third can be computed according to $G = E/2(1 + v)$.

For paper, the manufacturing process results in symmetry conditions which require nine elastic properties [16, 17]. These include three Young's moduli (one in each principal direction), three shear moduli (one in each plane), and three Poisson ratios (one in each plane). Six of these parameters are defined in Figs. 41 and 42. The three Poisson ratios also could be determined by taking the ratio of the lateral contraction to axial extension for each of the three uniaxial stress experiments shown in Fig. 41. If the nine elastic properties of paper are known, the three-dimensional response of the paper to applied stresses is known. Such information is valuable in characterizing the end use behavior of paper and for use in modelling containers or other structures.

Procedures have been developed at The Institute of Paper Science and Technology for measuring the nine elastic properties of paper or other sheet materials [18-20]. Seven of these are measured routinely in the laboratory. Measurements have been made on essentially all grades of paper and board, nonwovens, wood, and some plastics. There are limitations, however. The minimum sample size is around 15 by 15 cm, although measurements of elastic properties in the thickness direction can be made on smaller specimens. There is a minimum thickness for the z-direction measurements of about 0.1 to 0.2 mm, depending on the surface roughness. There is no minimum thickness for elastic property measurements made in the MD-CD plane of the paper. As a consequence of these physical limitations on sample size, board samples have been used for most of the three-dimensional work. The in-plane elastic properties measured on thin samples, however, are also useful in understanding the effect of process variables on these properties and have provided improved means to characterize end use performance.

1. Nature of the measurements

The elastic properties are determined by measuring the velocity of ultrasound in the paper. The theory has been described in detail

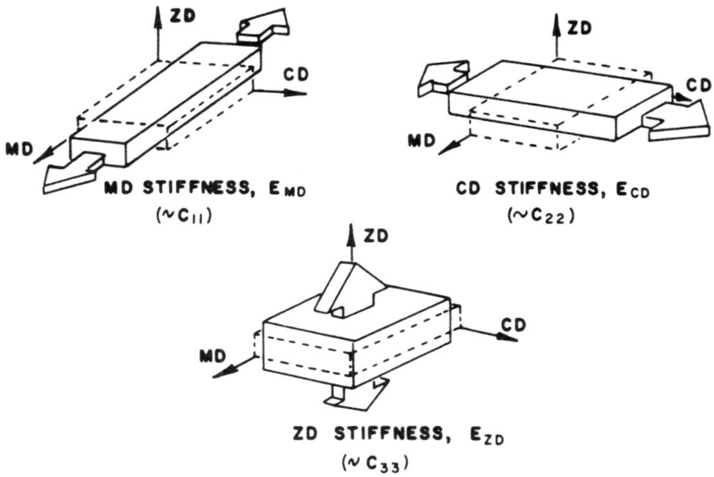

Fig. 41. Three modes of deformation in uniaxial tension. The elastic stiffness (E or C) is the ratio of the applied stress to the resulting strain when the strains are small (approach zero). The Poisson ratio could be computed for each case by taking the ratio of the lateral strain to the axial strain.

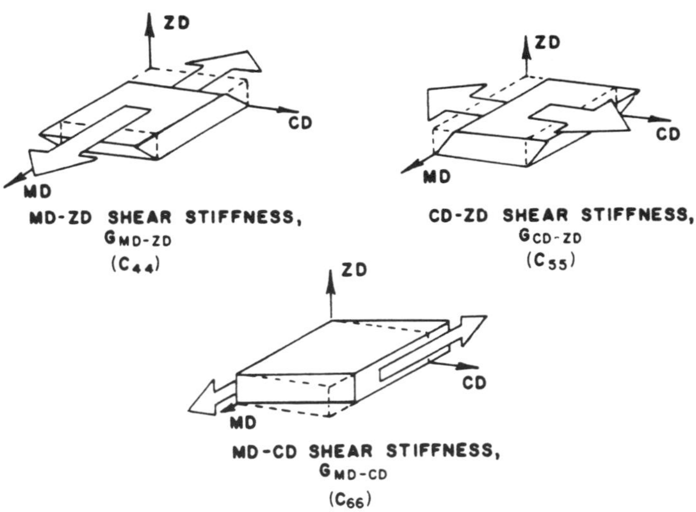

Fig. 42. Three modes of deformation in shear. The shear elastic stiffness (G or C) is the ratio of the shear stress to the shear strain when the shear strains are small (approach zero). For orthotropic materials such as paper, a shear stress is defined for each of the three planes.

elsewhere [17-20]. A brief overview will be given here. The generalized Hooke's Law for a three-dimensional material is

$$\sigma_i = \sum_{j=1}^{6} C_{ij} t_j$$

where i and j have values from one to six and where σ_i is the stress, t_j is the strain, and the C_{ij} are the elastic stiffnesses. The nine elastic stiffnesses are related to the "engineering elastic constants" viz. Young's moduli, shear moduli, and Poisson ratios [e.g., 13].

Three of the stiffnesses are easily determined by measuring z-direction bulk wave velocities:

$$C_{33} = \rho V^2_{Lz}$$

$$C_{44} = \rho V^2_{Sy\text{-}z}$$

$$C_{55} = \rho V^2_{Sx\text{-}z}$$

where

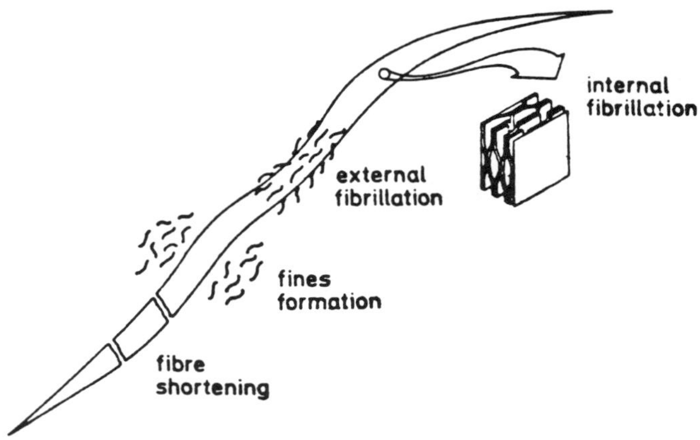

Fig. 43. A schematic illustration of the effects of refining on a fiber [21].

V_{Lz} = velocity of a longitudinal wave in the z-direction
$V_{Sy\text{-}z}$ = velocity of a shear wave polarized in the y-direction
$V_{Sx\text{-}z}$ = velocity of a shear wave polarized in the x-direction
ρ = apparent density.

The constants C_{11} and C_{22} can be determined by propagating longitudinal waves in the machine and cross directions, respectively. The velocities of V_{Lx} and V_{Ly} may then be used to compute C_{11} and C_{22} from:

$$C_{11} = \rho V^2_{Lx}$$

$$C_{22} = \rho V^2_{Ly}$$

The coefficient C_{66} is easily determined by measuring the velocity of a shear wave propagated in either the x or y direction with polarization in the y or x direction, respectively, where:

$$C_{66} = \rho V^2_{Sx\text{-}y}$$

This shear velocity can be measured on either plate or bulk materials[2].

The constant C_{12} is obtained by propagating a shear wave, polarized in the x–y plane, at a direction 45° to both the x and y axes. The expression for C_{12} in this case is:

$$C_{12} = \left\{[2\,V^2_s(45°) - 1/2(C_{11} + C_{22}) - C_{66}]^2 - [(C_{11} - C_{22})/2]^2\right\}^{1/2} - C_{66}$$

where V_s (45°) = velocity of the in-plane shear wave propagated in a direction 45° to the x and y directions. The stiffnesses C_{13} and C_{23} are more difficult to obtain, and at present are not measured routinely.

In many of the following discussions concerning process variables, the impact of a changing variable is often discussed in terms of these elastic stiffnesses, since they permit us to examine the three-dimensional response of the paper.

E. REFINING

Refining is discussed in detail in Chapter II of this volume and in Volume 6. Our purpose here is to understand how refining modifies the fiber and/or fiber cell wall such that sheet properties are impacted.

As discussed above with respect to fiber bonding and light scattering coefficients (e.g., Fig. 40), the impact of refining depends on the nature of the fibers themselves. Chemical pulps, because they have some of the lignin removed, are more easily "refined" than mechanical pulps. That is, it is easier to separate layers of fibrils or fibrils from each other by forcing water between them. This "internal fibrillation" allows the fibrils to move easily relative to one another, making the fibers "flexible" or "conformable" and permits surface tension forces during consolidation of the web to draw fibers together more extensively than would be possible if the fibers were rigid. Figure 43 schematically illustrates some of the results of refining [21].

Prolonged or more extensive refining may

2. A *bulk* material is one in which the boundaries may be considered to be at infinity and thus do not affect the wave propagation. A *plate* material, of which paper would be an example, has one dimension which is finite (paper thickness) and can affect the nature of the wave propagation.

cause fibrils or layers of fibrils on or near the fiber surface to unwrap and separate from the fiber. This is "external fibrillation". External fibrillation also promotes fiber-to-fiber bonding because the surface area of the fiber can be considerably enhanced, allowing more opportunities for surface tension forces to draw fibrils (fibers) together. It is conceivable that fibrils from two or more fibers in proximity with each other could become tangled and provide some mechanical bonding. This is a form of flocculation that may or may not be desirable, depending on whether it occurs during final consolidation of the web or earlier, respectively.

It is highly probable that a fibril hanging from the surface will be broken off by hydrodynamic shear stresses, interactions with other fibers, or subsequent refining. These fibrils become "fines". Fines can enhance bonding in chemical pulps by bridging between fibers. In this sense, their effect on the web structure can be similar to those on the externally fibrillated fiber. On the other hand, these fines may be lost with the white water, representing a material loss. This would occur if hydrodynamic stresses can prevent the fibrillar fines from attaching to fibers or to each other.

Because fibrillar fines or fibrils still attached to the parent fibers represent a large increase in the surface area of the "fibers" in the furnish, it is highly likely that chemical additives or filler pigments will interact with, and possibly attach to, this fibrillar material. This may or may not be advantageous. In the case of fines lost from the system, there may be expensive chemicals or fillers lost with them.

Of course, the refiner, whatever the type, can "cut" or break the fibers. In general, this is undesirable, especially since many sheet properties are directly related to fiber length. In other cases, it may be advantageous to cut long, stiff, fibers into shorter pieces to enhance density or improve smoothness.

1. Effects of refining on sheet structure

"Paper is made in the beater" was a saying used by old-time paper makers. They were implying that desirable paper properties were obtained by how you refined the pulp. This is true, of course, but there are a host of other variables we need to discuss that also strongly influence web structure. We will examine these later. Here we look at how refining, or more specifically the refined fiber, affects the structure.

For a lightly or moderately refined fiber that is internally fibrillated, we have already noted that the increased flexibility gives rise to greater bonded area. Page et al. [22], in a piece of classic research, studied the impact of refining and drying restraints on fiber-to-fiber bonding by directly observing the fiber-to-fiber bonds in 60 g/m^2 papers. They measured the bonded area by measuring the light scattered from the fiber interface. As discussed earlier, bonded surfaces would not scatter light, so the bonded regions would be dark compared to the surrounding area. (Actually, two surfaces in optical contact will not scatter light. The separation distances for optical contact are apt to be two or more orders of magnitude greater than the distance necessary for hydrogen bonding. Nevertheless, because there are no alternatives, we assume that fiber surfaces in optical contact are also bonded.)

Page et al. found that the area of a single bond increased with level of refining treatment. Longer refining times (and one would assume higher wet pressing pressures as well) led to larger bonded area. Drying restraints had little impact on bonded area. Table 4 summarizes some of their results. The degree of bonding is the ratio of the observed bonded area to the geometric crossing area of the two (or more) bonded fibers. A simple bond was one involving only two fibers whereas frequently three or more fibers could participate. Table 4 shows that the degree of bonding or the mean bond area increases as beating time increases while both are insensitive to drying tensions. Page and his coworkers also studied the distance between bonds, using various definitions, and found that as refining increased the interbond distance (by any measure) decreased. Photomicrographs of paper confirm that in most papers there is very little distance between bonds.

As noted when discussing scattering coefficients, the refining of mechanical pulps is less apt to "fibrillate" the fibers than for chemical pulps. Mechanical pulp fibers are more apt to be cut or broken, and these "fines" tend to be different than the fibrillar material produced in refining a chemical pulp. Such differences, as we have discussed, can result in quite different final sheet structures.

For a given fiber and any of the various pulping processes (see Chap. II), one could obtain a more or less continuous range of (paper property) responses to refining in going from high yield to low yield. Of course yield is not the only consideration, because the chemistry and composition of the cell wall after pulping will also have a large influence on the response to refining.

Table 4. Effect of refining and drying tension on degree of bonding and mean bond area [8].

Beating time min	Drying tension g/cm	Degree of bonding %	Mean Bond Area* Simple	Mean Bond Area* All bonds
	0	46.6	772	643
0	35	38.0	656	483
	55	40.3	708	572
	0	71.6	1102	956
20	35	75.1	1199	989
	75	69.0	1003	856
0	All	41.7	712	567
20	All	71.8	1099	932

* Mean bond area has units of square micrometers. Simple bonds involve only two fibers. All bonds include simple bonds plus those involving three or more fibers.

F. STOCK DELIVERY

Paper is typically very directional in its plane. This is commonly expressed as the MD to CD ratio of tensile strengths (MD/CD ratio), or less often as the MD to CD ratio of some other property. For a handsheet we expect a value of one, whereas a sheet from a cylinder machine may be five or greater.

Many incorrectly use the terms "MD/CD ratio" and "fiber orientation" as interchangeable. Actually, the fiber orientation is only a part of the directionality normally found in most machine-made papers. Directionality also results from wet stretching of the web (in an open draw) and/or the drying restraints imposed on the web in the dryer section. These are discussed elsewhere in this chapter (see Sec. H).

Ideally, one would like to have the fibers in the stock suspension delivered to the wire uniformly separated from each other and, in many cases, with completely random orientations in (three-dimensional) space. This is never really the case, however, since hydrodynamic forces and the chemical interactions cause entanglement and flocculation of the fibers. But, even if we had perfect "formation", we would still find that there would be some level of fiber alignment in the final (dried) sheet. As the stock suspension is accelerated toward the slice, the shear stresses tend to cause fiber orientation in the machine direction. In addition, some fibers will tend to align with the papermaking wire when they are deposited into it[3].

If the velocity of the fibers in the stock flow and the velocity of the wire are different, shear stresses can cause fibers to partially align with the wire. This effect is at a minimum when the two velocities are the same, and increases as the velocity difference increases in either direction as shown in Fig. 44.

This phenomenon will occur most readily with those fibers that are in the stock flow nearest the wire during forming. As the mat thickens in moving down the machine, the effect will be smaller and smaller, leading to a gradient in the fiber orientation from the wire side to the top

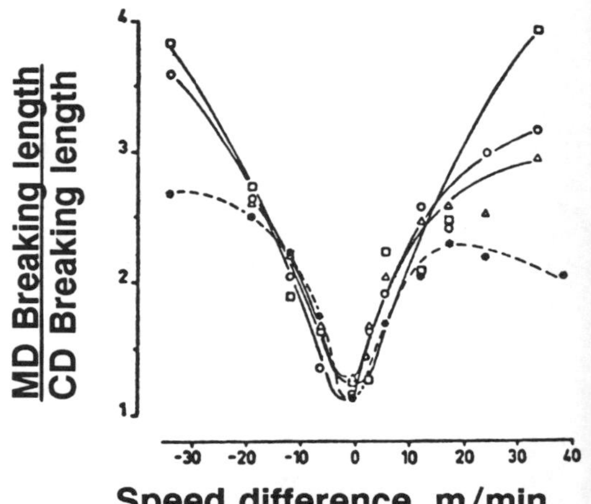

Fig. 44. A graph showing how the MD/CD ratio of breaking lengths varies with the jet to wire speed differential [23]. The speed differential affects the fiber orientation in the sheet. Stretching the web in an open draw or shrinkage during drying can also affect the MD/CD ratio.

3. A fiber randomly oriented in three dimensions in the stock suspension, upon encountering the forming fabric during drainage, to first approximation, may be thought of only losing its z-direction orientation while retaining its angular orientation in the x-y (MD-CD) plane. The latter is altered, however, if there are speed differences between the fibers and the fabric. This tends to orient more fibers parallel to the MD.

side. For a twin-wire machine, with dewatering in both directions, the fiber orientation through the sheet will be more symmetrical and uniform.

The contribution of fiber orientation to the MD/CD ratio can be large, depending on the type of equipment and operating conditions. These can range from a value near one (random fiber orientation in the plane) to values of five or more for a cylinder former. Later we will discuss how the MD/CD ratio can be increased without altering the fiber orientation as a result of wet stretching or web shrinkage.

The MD/CD ratio is important as an indicator of process conditions and end-use performance. However, as in the case of many web properties, the MD/CD ratio may vary by position across the width of the machine. Most often it is observed to be low near the center and higher at the edges. This is a consequence of a non-uniform CD shrinkage of the web during drying. In the MD, the web is under a tensile stress during the entire drying process and shrinkage is prevented. In the CD, however, drying stresses attempt to shrink the web in this direction. This is more apt to happen near the edges of the web as there is nothing to restrain the sheet other than contact with the dryer cans. The shrinkage lessens as one moves toward the center because the material further out helps to restrain the shrinkage. The extent of the shrinkage will depend on the furnish, the coefficient of friction between the paper and metal surface, and felt tensions, among other things.

For example, the non-uniform shrinkage can result in the tensile strength being greater at the center than near the edges. This effect occurs for tensile strengths measured in both the MD and CD, but tends to be greatest for the CD tensile strength. That is, the variation between center and edge is greater for the CD tensile strength than the MD tensile strength, making the MD/CD ratio larger at the edges.

As might be expected, the fiber orientation is not necessarily uniform from point to point across the width of the machine, but this does not seem to have received much attention in terms of sheet properties. The problem is complicated by the fact that MD/CD ratios measured at the end of the machine also contain the contribution due to wet stretching or shrinkage effects.

Another phenomenon related to fiber orientation can occur at the slice. If the stock flow is not exactly parallel to the wire, it is possible to have the fiber distribution maximum lying at some angle to the MD. The transverse component of stock flow does not have to be very large to create a fairly large angle in the fiber orientation distribution. What is important is the *difference* between the MD jet velocity and the wire velocity relative to the transverse component of the velocity. For example, a 1 ft/min (0.30 m/min) transverse flow with a machine direction jet velocity of 2000 ft/min (609 m/min) and a wire velocity of 2010 ft/min (612 m/min) will produce an angle of about 5.7° [arc tan (1/10)]. Such transverse flows are apt to vary from point to point across the paper machine as slice screw settings change, as a consequence of non-uniform manifold flows, or as a result of in-plane shear stresses during consolidation of the web. If the mean fiber orientation angle is too large (angles in excess of 15° have been observed) there may be problems in subsequent converting or end uses of the paper (depending on the product). Diagonal curl and pile lean are examples of this.

G. WET PRESSING

Wet pressing removes water from the web by squeezing it between a roll (or shoe) and a felt backed by a second roll. Once saturation is reached – the point at which the pore volume of the network matches the volume of the water present – continued pressing causes the water to flow out of the sheet into the felt. As the cell walls are compressed, water can flow from fiber pores as well. As the sheet leaves the nip, water may flow back into it from the felt.

Two key variables in wet pressing are the pressing pressure and time. For sheets that have a high flow resistance (very wet, high density, or high basis weight), the dewatering is flow controlled and the press impulse determines overall water removal. The impulse is the integral of the pressure vs. time curve, usually taken as the average nip pressure times nip residence time. For sheets with low flow resistance, dewatering of the fibers becomes more dominant. The press impulse is still an important factor but the pressing pressure alone takes on greater significance.

Conventional wet pressing practices produce a sheet with a solids content of 42 to 50%. This means a substantial amount of water must still be removed in the dryers. This is an expensive proposition, and has led to considerable research dealing with ways to extend the nip residence time (extended nip press) or pressing pressure, as well as a number of new approaches for pressing or drying. The latter, including hot pressing, press drying, and impulse drying, have recently been reviewed [24, 25]. These techniques are also discussed in detail elsewhere in this volume (see p. 73).

Our purpose in this section is to review what happens to the fiber network during the wet pressing operation and to ascertain how sheet properties are impacted.

The compression of the network during wet pressing can cause fiber lumen collapse and high local compressive stresses at fiber crossing points. Lumen collapse leads to more flexible fibers, which when coupled with the stresses at fiber crossings will lead to greater bonded area and a denser network. In a sense, wet pressing helps in bringing the fibers into more intimate contact and thus promotes bonding. This, of course, leads to a denser network and typically to higher values in terms of elastic stiffness or strength values. Since refining leads to more flexible fibers and greater relative bonded area, one might suppose that wet pressing and refining are very similar or even "interchangeable" processes as far as densification is concerned. This is not the case, however, as can be seen in Fig. 45 which shows some strength property plotted against density for a number of pressing and drying conditions [26]. The curve for refining rises faster than that for hard press nips. At a given density, refining results in a greater value for the strength property. The explanation is that refining enhances bonding by its action on the fibrils comprising the cell wall making a more conformable structure and increasing the surface area as well. In contrast, wet pressing physically forces the fibers together without altering the fiber cell wall structure, except perhaps at the crossing points, and by causing fiber collapse.

The above general description, of course, must be modified depending on whether we are discussing chemical pulps or high-yield mechanical pulps since the action of refining or wet pressing on these will be different. For a given pulp, however, while the general discussion may apply, the relative magnitudes of the refining and wet pressing effects for that pulp may differ.

The enhanced density and property development brought about by wet pressing requires that the fibers be saturated with water since wet fibers are compliable and surface tension forces are still required to provide good bonding. Pressing on a dryer web can still provide intimate contact and result in greater network density, but does not necessarily lead to enhanced bonding and development of strength properties. This is evidenced by the line in Fig. 45 for the breaker stack where the significant increase in density is not associated with a concomitant increase in strength. One can envision that the densification comes about because fiber segments are physically forced into void areas without chemical bonding occurring. In the extreme case, where the fibers are "dry", such as in calendering, this action may actually degrade the bonding in the network such that an increase in density is associated with a decrease in strength properties (see Fig. 45).

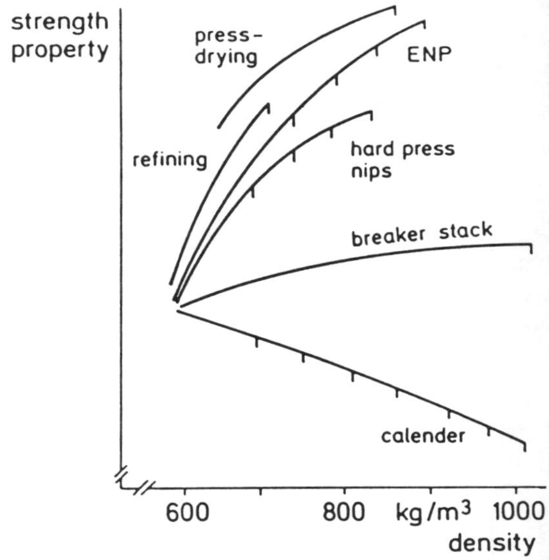

Fig. 45. A schematic diagram showing how strength properties vary with density for different methods of densifying the sheet [26].

In an extended nip press (ENP), the residence time in the nip is increased by providing a "shoe" rather than a roll such that the load is applied for a longer time. This leads to more effective dewatering but also promotes bond development to a greater degree than pressing in a short residence time hard nip press.

The last curve in Fig. 45 will be discussed in the section on drying (Sec. I).

H. DRAWS

Excessive draws tend to pull the sheet apart, causing deterioration of the properties. From this standpoint, open draws are to be avoided, if possible. In the case of newsprint or coating rawstock, minimizing draws is recognized as a key factor in improving the runnability of these grades. On the other hand, draws also can lead to improved paper properties in the machine direction (the direction of the stressing of the web), similar to the effects described earlier for drying under tension. For example, MD tensile or MD bending stiffness can be enhanced by tightening an open draw at low solids, say coming off the couch or out of the press section, as long as the draw is not so tight as to pull the wet web apart.

Perhaps a less well known fact is that an open

draw also affects CD and ZD properties, including caliper, bulk (the inverse of density), and internal bond strength.

As we discussed under "Stock Delivery" (Sec. F), directionality in paper or board (MD/CD ratios) is a result of two different mechanisms. The first is the tendency for fibers to align themselves along the machine direction when they are deposited on the wire and during the early consolidation of the web. The extent to which this occurs is related to the relative speeds of the jet and wire (rush-drag) and thus can be controlled to some degree. The second cause of directionality is related to restraints on the paper in the MD during drying and also the extent to which the wet sheet has been stretched in an open draw. These latter two variables are related in a sense. What seems to be important is the extent of the dimensional change in going from wet paper to dry paper [27]. Paper dried without restraint can shrink up to 15%. If the paper was stretched while wet, the dimensional change would be even greater. The extent of the unrestrained shrinkage depends on the nature of the furnish. As discussed earlier, on the paper machine this shrinkage is largely prohibited in the MD because the web is under tension in this direction. In the CD some shrinkage is likely to occur, especially near the edges. This can lead to a "bell-shaped" distribution across the web of MD and CD mechanical properties such as elastic stiffness or tensile strength, or a "smile" in properties like stretch.

1. Impact on sheet structure

The effect of wet straining in an open draw, then, all other factors being constant, is to increase the dimensional change as the sheet dries. The wet straining elongates the paper, which is thought to give rise to a better distribution of stresses in the MD. This results in the improved MD properties as referred to above when the paper is dried. Intuition suggests that stretching of the wet web would cause an increase in the fiber alignment in the MD, but this is not the case [28]. Wet stretching may result in a decrease in the out-of-plane fiber orientation, however, as discussed below. Increases in MD tensile, ring crush, Taber, etc., resulting from wet straining are offset by decreases in these properties in the cross-machine direction if CD shrinkage during drying is prevented [28]. For example, near the center of the paper web (where CD shrinkage may be expected to be minimal) an increase in MD tensile by wet straining also produces a decrease in CD tensile, that is, the directionality increases. If the product being manufactured calls for maximizing a CD property, such as CD ring crush in the case of linerboard or corrugating medium, it would be desirable to make the sheet as square (nondirectional) as possible. From what has been said here, the way to do this is to minimize wet straining by slackening the draws and to run the jet and wire at the same speeds to minimize the fiber orientation effects. The situation is more difficult if CD shrinkage occurs after wet straining, as it probably does near the edges of the web, because the CD mechanical properties do not always behave in a predictable way in this case [29].

Figures 46 and 47 illustrate how MD and CD tensile strengths, respectively, are impacted by changes in wet pressing, fiber orientation, and wet stretching (draws) [28]. In Fig. 46, MD tensile strength is plotted against density for bleached softwood kraft pulp anisotropic handsheets. The density was varied by changing wet pressing pressure. The different curves result from changes in fiber orientation and levels of wet stretching.

In Fig. 46, as expected, MD tensile strength increases as wet pressing pressure (density) is increased, a consequence of enhanced bonding in the fiber network. The bottom curve represents the case of a random fiber orientation with no wet stretching. As the level of fiber orientation or wet stretching in the MD is increased, the curves are displaced upward to higher MD tensile strength values. The combined effect of going from a random fiber orientation to a "high" fiber orientation (about 3:1) together with a change in wet straining from "low" (essentially zero) to "high" (about 2.4%) is a 200-250% increase in MD tensile strength at a given density.

The consequences of the above changes in CD tensile strength are shown in Fig. 47. In this case, everything is reversed, the uppermost curve is the random fiber orientation – no wet straining case and the lowest curve is that for high fiber orientation and high wet straining. We will comment further on these last two figures in a later section (Sec. I).

The above effects in the MD-CD plane of paper or board have been quite well documented over the years [e.g., 27-29]. The effect of draws or wet straining of the web on the thickness direction of paper, however, has not been explored to the same extent. As it turns out, some interesting things happen during wet straining in this direction. Laboratory studies show that wet straining a long fiber furnish by a few percent will cause an increase in the wet caliper that will persist even when the sheet is dried [30]. A 2% wet strain, for example, can produce an 8 to 10%

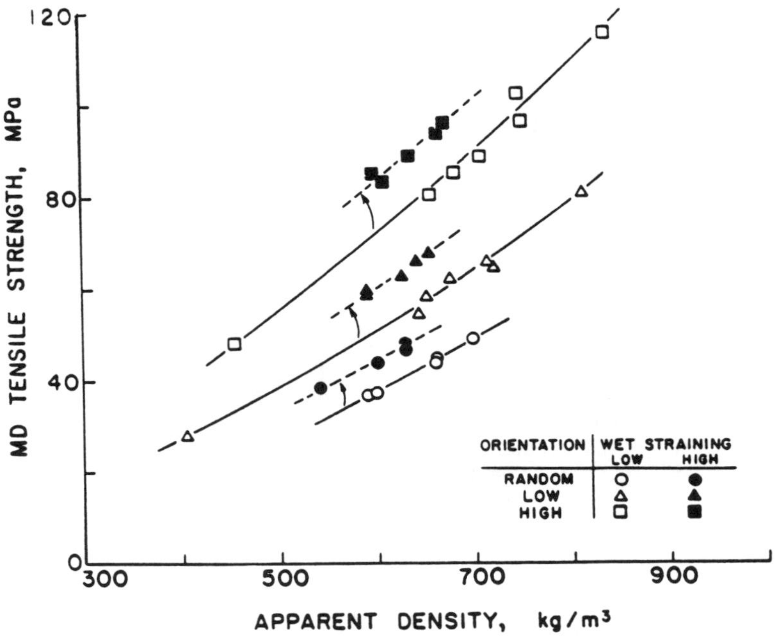

Fig. 46. MD tensile strength vs. density for different combinations of fiber orientation and wet straining. Density was varied by wet pressing [28].

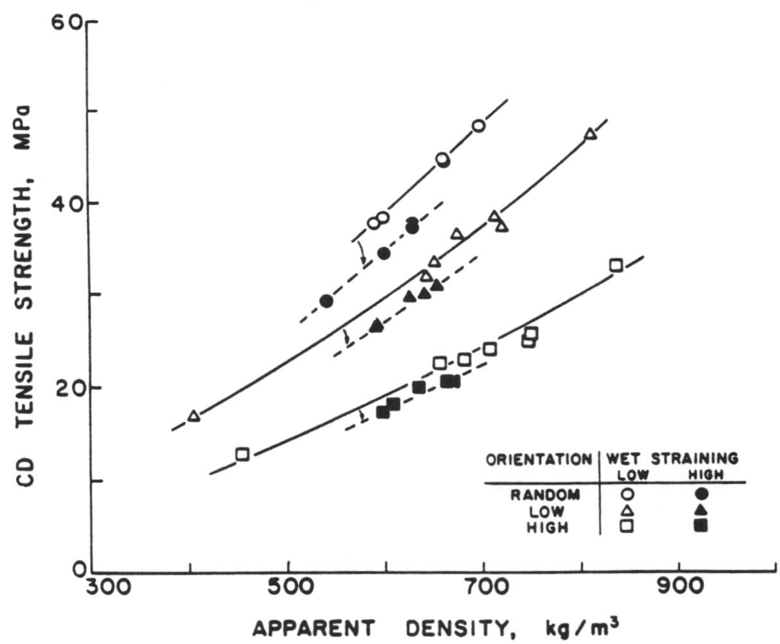

Fig. 47. CD tensile strength vs. density for different combinations of fiber orientation and wet straining. Density was varied by wet pressing. These results are the opposite of those shown in Fig. 46 for the MD tensile strength [28].

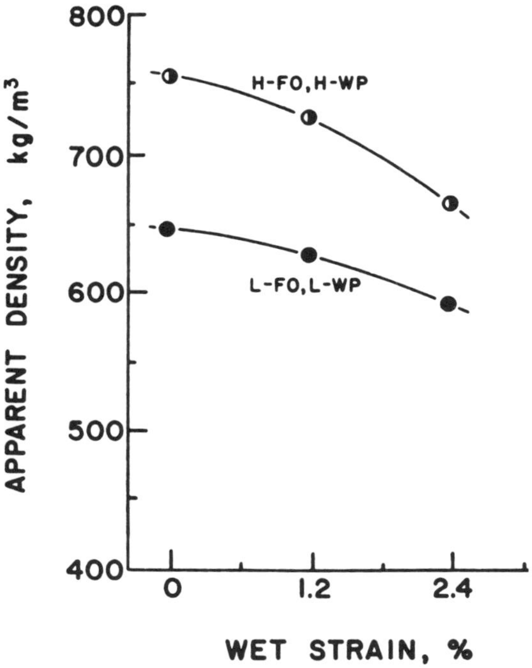

Fig. 48. Change in sheet density with wet straining for samples with a high fiber orientation and high wet pressing pressure (H-FO, H-WP) and a low fiber orientation and low wet pressing pressure (L-FO, L-WP). The density decreases with wet straining because the caliper (of the dried sheet) is increased [30].

increase in caliper. A caliper increase, of course, produces a decrease in density (basis weight over caliper) or an increase in bulk, as shown in Fig. 48. Since many mechanical properties change directly with density, the decreased density would be expected to decrease these properties in all three directions.

Normally when a material is strained elastically in tension, we expect it to become thinner due to Poisson contraction effects. In the wet straining referred to above, of course, we apparently are not straining elastically and some other phenomena must be at work. All of the observed effects seemed to be explained by a fiber straightening model [30], which basically refers to a flattening of the thickness direction (or ZD) undulations of the fibers in tension. It can be argued that the straightening of a fiber under tensile loading causes crossing fibers bonded to it to be pushed apart, thereby increasing the "sheet" caliper.

Such a model also predicts another, perhaps more important effect, in that fiber straightening would likely produce a decrease in the extent of bonding or "connectedness" in the thickness

Table 5. The effect of increase MD wet straining on paper mechanical properties*

	MD	CD	ZD
Tensile strength	+	—	— —
Stretch	—	+	
Tensile Energy Absorption (TEA)	—	+	
Internal bond			— —
Compressive strength	+	—	
Elastic stiffness	+	—	— —
Bending stiffness	+	—	

*Caliper increases and density decreases.
+ = the indicated property increases;
— = decreases; — — = large decrease.

direction. A decrease in bonding in this direction ought to be readily apparent in the internal bond strength of the paper, and it is. In fact, the 2% wet strain mentioned above produces about a two-fold decrease in the ZD tensile strength (internal bond strength) [28] when compared at constant density. The density decrease caused by wet straining would, of course, lower the internal bond strength even more.

Three things seem to be important in the case of internal bond strength: the extent of refining, wet pressing pressure, and wet straining on the paper machine. Increases in the first two factors increase internal bond, but a modest increase in wet straining produces a large decrease in internal bond. In fact, it seems quite probable that increases in strength brought about by wet pressing could be completely overshadowed or lost by running tight draws.

Table 5 summarizes some of the results relating properties to wet straining. All of the results apply to long fiber furnishes. It may be that in short fiber furnishes the effects due to wet straining described above are not as large or not observed. The fiber straightening model would suggest that this could be the case.

I. DRYING

As discussed earlier in this chapter (Sec. B2), the drying of the network has significant impact on the physical properties. In this section we will expand on those earlier discussions and attempt to bring together those factors which strongly interact during the drying process to affect the web properties. Not surprisingly those factors include the nature of the furnish, the extent of fiber orientation, wet pressing conditions, and the extent of the draws as well as a number of drying variables (depending on the dryer type) such as web tension, dryer felt pressures, etc. Chapters XII and XIV of this volume deal with paper drying and Yankee drying, respectively.

If unrestrained, paper will shrink during drying as discussed earlier. The extent of the shrinkage will depend on the wood species and related morphological variables, the type of pulping method and yield, and on the type and extent of refining. The more the cell wall of the fiber swells with water, the greater the transverse shrinkage of that fiber during drying. If this fiber is capable of strongly bonding to many other fibers, the total shrinkage of the web will be maximized as the individual fibers shrink and collectively produce the overall network shrinkage. Such a "freely dried" paper tends to be very rough with both short-range (fiber scale) and longer range variations (e.g., cockle) in roughness. The elastic stiffness or tensile strength of such a sheet will be lower than if it had some restraint during drying, whereas the strain at failure and moisture expansivity will be higher.

If swollen fibers were only weakly bonded, one might expect breakage of these bonds during drying, leading to less shrinkage and a weak network. Fibers with little swelling, perhaps groundwood fibers, would result in little network shrinkage, regardless of the strength of the inter-fiber bonds. (Of course, such fibers do not bond very well in the first place.) The point is that, ignoring other variables for the moment, the extent of the shrinkage for unrestrained drying depends on the nature of the furnish.

The extent of the shrinkage, however, also strongly depends on other factors. For example, if there is an anisotropy due to fiber orientation, the sheet will shrink less in the direction of orientation (MD) than in the cross direction. This is a consequence of the fact that there are simply fewer fiber cross-sections that can undergo shrinkage in the MD. Wet pressing, because it enhances bonding, will partly determine the effectiveness of the stress transfer from the laterally shrinking fibers to the fiber to which they are bonded. Wet straining, as we noted earlier, seems to be the opposite of shrinkage as far as physical properties are concerned.

In a real paper machine, of course, the web structure never really has an opportunity to dry freely. In conventional can drying, the sheet is pressed against the dryer can prohibiting shrinkage in the CD (as long as the shrinkage stresses do not exceed the frictional stresses). In order to move the web through the dryer section, however, it is under tension in an open draw, which can cause a Poisson contraction as well as allow some shrinkage. These effects, acting together at any given point in the web, produce a local drying "restraint" which, together with the other relevant variables mentioned above, determine the physical properties of the dried network at that point including the extent of any graininess or cockle. If either the MD or CD restraints vary with position in the web or with time, the properties will vary as well. We have already discussed the impact of non-uniform shrinkage stresses in the cross machine direction on sheet physical properties.

The interaction between variables can produce some rather subtle differences. An example will help to illustrate this point. Fleischman et al. [28] found in a laboratory study that wet straining a sheet, allowing Poisson contraction to occur, and then drying the sheet under both MD and CD restraint resulted in a direct interplay between MD and CD physical properties as observed in Figs. 46 and 47. As MD tensile strength, for example, increased with increasing wet straining the CD tensile strength decreased concomitantly. Htun [29], on the other hand, did a similar wet straining experiment, but dried the sheet under MD restraint only. He concluded there was no interaction between MD and CD properties as far as wet straining was concerned. One could argue that Fleischman's experiments approximated conditions near the center of a drying web whereas Htun's approximated conditions near the edge of a drying web.

In the laboratory, sheets are usually dried against a plate to prohibit shrinkage and thus provide a way to eliminate the complicated effects on properties due to non-uniform drying. A Yankee dryer provides the same opportunity in a commercial setting, at least for the lighter weight grades, and because the sheet is dried against a smooth surface it also has a restraint on one side.

In practice, the choice of drying systems (besides economic considerations) will depend on the particular grade of paper or the physical properties that are desired. An air dryer, for example, may be desirable if one wishes to have high stretch in the sheet. Press drying is a technique that has received a fair amount of interest in recent years [e.g., 26]. In this process, the sheet is maintained under z-direction restraint during the drying process. This can result in high strength values as depicted in the uppermost curve of Fig. 45.

1. Summary

Table 6 is a general summary of the major effects on sheet properties (for a chemical pulp) of a change in certain process variables. In some cases the direction of change in a property cannot be determined and in those instances a question mark is used. An example of this might be the effect of wet pressing on bending stiffness

Table 6. Effect of increasing the indicated variable on the strength properties and elastic stiffnesses.

	Yield	Refining	MD Fiber Orientation	Wet Pressing	Wet Straining	Super-calendering
MD Tensile Strength	—	+	+	+	+	—
CD Tensile Strength	—	+	—	+	—	—
ZD Tensile Strength	—	+	o	+	— —	— —
MD Compr. Strength	—,o	+	+	+	+	—
CD Compr. Strength	—,o	+	—	+	—	—
MD Bending Stiffness	?	?(E+,I−)*	+	?	+	
CD Bending Stiffness	?	?(E+,I−)*	—	?		
Brightness	—	—	o	—	+	—
Opacity		—	o	—	+	—
MD Elastic Stiffness	—	+	+	+	+	—
CD Elastic Stiffness	—	+	—	+	—	—
ZD Elastic Stiffness	—	+	o,(+)	+	— —	— —
MD-CD Shear Stiffness		+	—	+	o	—
CD-ZD Shear Stiffness	—	+	—	+	—	— —
MD-ZD Shear Stiffness	—	+	+	+	—	— —

*Young's modulus, E, increases, but moment of inertia of cross-section, I, decreases.

or flexural stiffness, EI. Whereas increased wet pressing would increase E, the moment I (which depends on the cube of the caliper) would decrease because caliper is reduced. The table does not show the collective interactions of multiple changes on a property, nor can it give the relative magnitude of the change in the property. The latter would be furnish dependent. As we saw in the case of refining on scattering coefficients (Fig. 40), the effects can be quite different for mechanical pulps and chemical pulps.

J. CONVERTIBILITY

Converting, as used here, means any process that can alter the properties of the web after it has dried once. Thus sizing, coating, calendering or supercalendering, printing, etc. are all converting operations by this definition.

Perhaps the least studied and least understood areas of paper science have to do with sheet properties that are important in a given converting operation or, just as important, how web properties are impacted by converting operations. In many cases, the importance of web properties to successful converting is not known at all. This is probably a consequence of the fact that most converting operations have not been studied in enough detail, especially in fundamental terms, to understand the stresses that may be applied to the web. At the same time, traditional paper tests used to characterize the web do not always provide a complete or proper description of the structure and thus may not be able to define the impact of converting process. Specific elastic stiffnesses, however, have provided new insights into some converting processes and in some instances have led to changes in the process and established new descriptors for the structure.

It is not unusual to find that a particular converting operation has degraded web properties (e.g., calendering as we saw earlier). We can define at least two general classes of operations to which webs can be subjected to degrade their properties. There are *localized* processes, such as bending, folding, scoring, cutting, etc., where the degradation occurs along well-defined lines or regions and *non-localized* processes, such as calendering, decurling, or any nip passage, where the degradation may be over the entire web. Some operations, such as fluting where the corrugating medium is degraded alternately at flute tips or roots, would fall somewhere between the two definitions.

We could also group the degradation of the web into *intentional* (e.g., scoring) or *incidental* (calendering). In many of the operations referred to in this paragraph, the degradation manifests itself most strongly in the z-direction, rather than in the plane of the web.

It would be desirable to establish the properties of the structure that are important to performance in a given converting operation and to establish relationships between those properties and the converting operation. The latter knowledge would lead to superior structure or

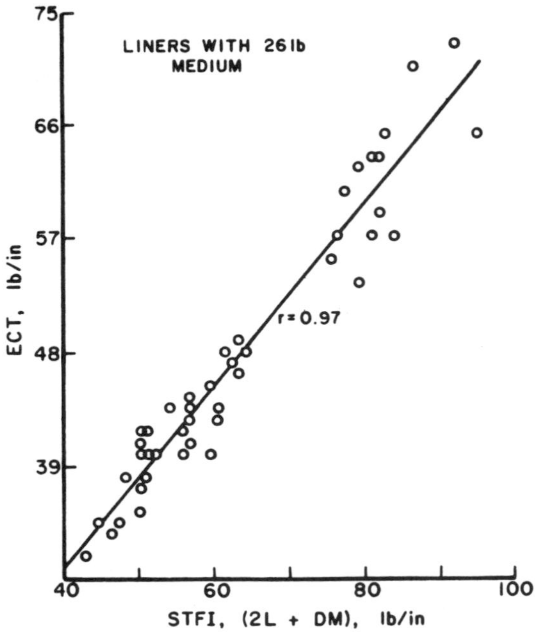

Fig. 49. Combined board ECT (edge crush test) is highly correlated to the STFI (short span) compressive strengths of the components [33]. L and M are the compressive strengths of the liner and medium, respectively, and D is the draw factor (the ratio of the length of the medium in the combined board to the board length).

perhaps adaptation of the converting process to better utilize the existing web structures.

1. Combined board example

Because of the myriad converting operations, it is not possible to discuss them all or even a majority of them. However, it is possible in many cases to group them by their action on the web structure. For example, bending around a small radius, folding, or fluting are all similar operations and might be studied together. In this section we will examine an example taken from the manufacture of combined board.

Box compressive strength is an important end-use performance attribute. In service, corrugated containers must withstand large compressive loads over long periods of time. Compressive loads cause the side panels to bow resulting in the loads concentrated at the corners of the box. The bending stiffness of the combined board is the important attribute that describes the bowing of the sides and the edgewise compressive strength (ECT) of the combined board is important in resisting the concentrated loads at the corners.

Combined board ECT depends mainly on the edge compressive strength of the medium and the liners. This can be described quite well in terms of a simple additivity law [31],

$$ECT = 2L + DM$$

where L and M are the edge compressive strength (STFI) values of the liner and medium, respectively, D is the draw factor (the length of medium for a given length of combined board), and the 2 comes in because there are two liners. A plot of this relationship is shown in Fig. 49.

In a box, the flutes in the corrugated medium run vertically, from top to bottom. Thus box compressive loads are in the cross-machine direction of the medium and liners. Since the above relationship indicates that the combined board strength is directly related to the compressive strength of the components, clearly one would like to optimize the CD compressive strength of both the medium and liners. From our discussions above we know that one way to accomplish this is to make the medium or linerboard structures as "square" (non-directional) as possible during the manufacturing process, because this will tend to increase the CD properties at the expense of the MD properties. Since the former is the important one in box compression, this works to our advantage.

2. Fluting of corrugating medium

While making a square sheet helps to increase box compression, there are other things happening during the converting processes (making the combined board) that must be considered. In order to appreciate these, we need to briefly review the fluting process itself. During the fluting process the corrugating medium is drawn into the labyrinth between the two corrugating rolls as shown in Fig. 50. During this process, several things happen which degrade the properties of the medium, and particularly the CD edgewise compressive strength. The medium is under a "brake" tension which acts toward the left. As the medium moves toward the center of the labyrinth (right side of figure), it is dragged over the flute tips on the top and bottom corrugating rolls. This creates frictional forces which also act toward the left and these increase as the medium moves toward the center, because the angle of wrap increases in moving toward the center. The medium is thus under large MD tensile loads and bending stresses as it is being bent and dragged over the tips of the flutes. At the center of the labyrinth, the corrugating rolls apply ZD stresses that reduce the caliper by as much as 30%. If the stresses get too large,

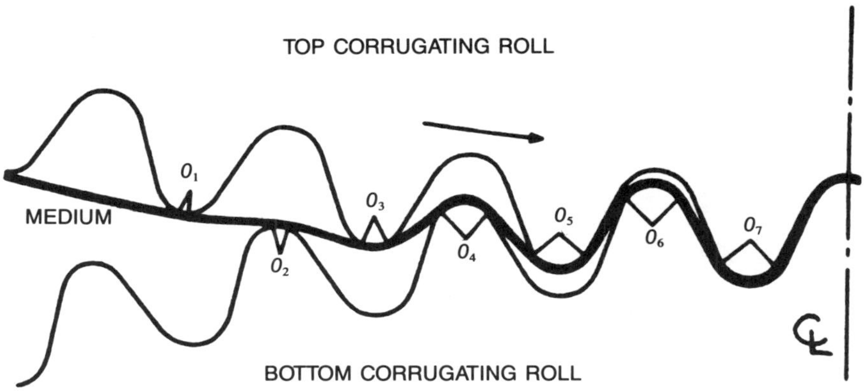

Fig. 50. A schematic representation of the corrugating labyrinth. The medium web is subjected to MD frictional forces as it is dragged over the flute tips of the corrugator rolls. There is a progressive increase in wrap angle over the flute tips as the web is drawn toward the center [33].

fractures at the flute tips or complete web fracture can occur.

Even without complete failure, however, the medium is damaged by the fluting process. Whitsitt et al. [32] found that MD compressive strength was decreased by as much as 30 to 40% and CD compressive strength was decreased by about 20%. This represented a 7% loss of ECT for the combined board. The extent of the degradation in the corrugating medium depended upon position within individual flute tips. Figure 51 shows MD compressive strength vs. position for single-faced board (one side liner only) formed either hot or cold. The dashed line across the top of the figure represents the MD compressive strength for the uncorrugated material. The compressive strength is decreased for all positions along the flute, but is most severe at the flute tips and roots where up to 60% of the uncorrugated strength is lost. The average strength loss on the flute flanks (sides) is around 20 to 25%. Hence there is a sizeable cyclic variation of MD and CD compressive strength as one moves along the MD in the corrugating medium. Whitsitt et al. also found that tensile strength followed a similar behavior with strength losses ranging from 17 to 44%.

Simulated fluting experiments showed that flexing the medium around a small radius is most likely the greatest factor contributing to the loss of strength. This induces bending and shear stresses which damage the sheet and may lead to rejection of the converted product or inferior end use performance. The challenge is to minimize the damage either by manufacturing the board to have different initial properties, by modifying board properties during the converting operation (perhaps by heat or steam), or by modifying the converting operation itself. None of these may be easy to accomplish.

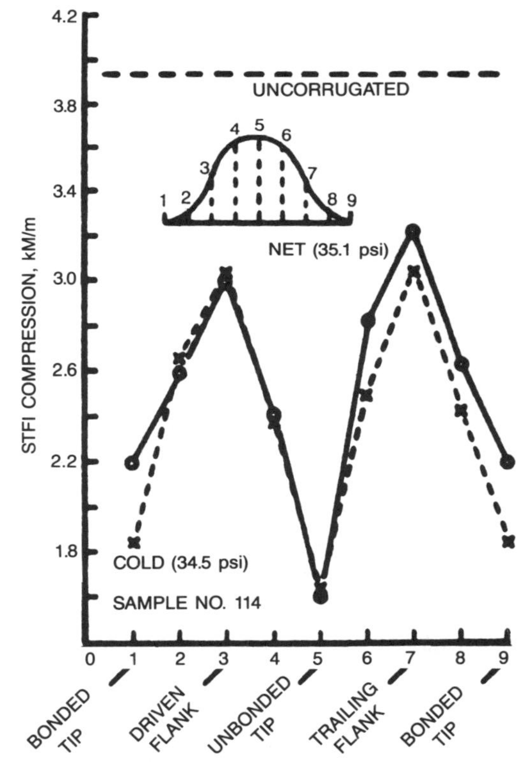

Fig. 51. A graph of STFI (short span) compressive strength vs. position in a flute shows how compressive strength is degraded during the fluting operation [33].

It should be apparent that, while the above example was specific to the manufacture of combined board, the degradation brought about by bending the material around a radius is universal. For example, a similar degradation occurs during decurling or creasing of carton

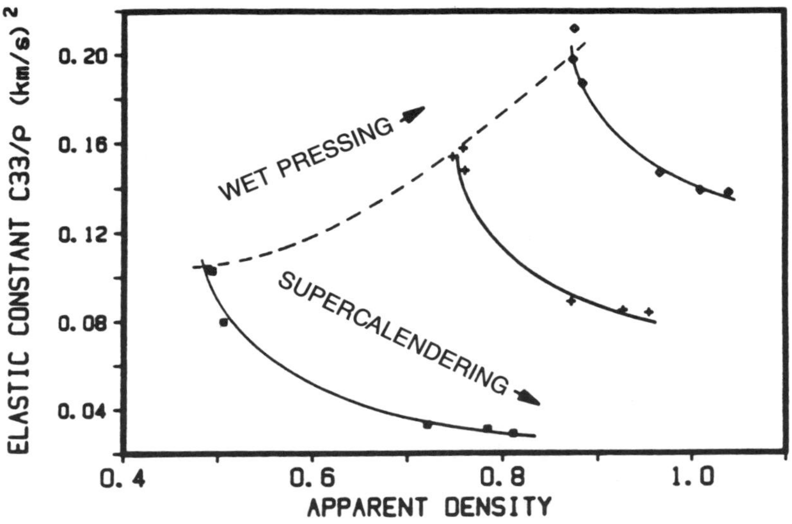

Fig. 52. A plot of specific elastic stiffness vs. density showing how densification by wet pressing or supercalendering impacts the former [34].

board, the folding of magazine pages or envelopes, or forming the top curl in an ice cream carton or paper cup. There are other examples, but the point is that we need to understand better the nature of the degradation that occurs and learn ways to minimize it.

Paper or board may also be degraded by simply passing it through a nip as in calendering or supercalendering. This was evident in Fig. 45 for the case of calendering and was briefly discussed there. Similar effects are found for supercalendering as shown in Fig. 52 [31], which plots the specific elastic stiffness, C_{33}/ρ, versus density. The elastic stiffness often correlates with the z-direction tensile strength, as will be discussed below. This out-of-plane elastic constant increases with increasing wet pressing pressure, as expected, but decreases sharply with increasing supercalendering level. The loss in C_{33}/ρ is greatest for the lower starting densities. At 0.5 g/cm^3 the decrease is from about 0.1 to 0.03 $(km/s)^2$, a 70% loss, while at the higher starting density of 0.9 g/cm^3, there is a 30% loss.

As we noted earlier during our discussions of the impact of draws (wet straining) on sheet properties (Sec. H), the z-direction properties seem to be particularly sensitive to wet pressing pressure (increasing with greater pressures) and to draws and drying (decreasing with greater wet straining levels). Figure 52 reveals that supercalendering (and also calendering) has a large negative impact on out-of-plane properties. This reminds us that, for any paper grades where high ZDT or edge compressive strength is important, we should work to minimize open draws and avoid calendering or supercalendering, if possible.

K. END USE PERFORMANCE

Just as in the case of converting requirements, it is impossible to discuss all of the end use performance attributes or requirements of paper or paperboard. This section attempts only to show relationships between certain of the sheet physical properties, by showing how some of the commonly measured properties relate to the fundamental set of properties, the elastic stiffnesses.

The performance attributes that characterize the product in its ultimate end use are typically those used to characterize the paper during its manufacture or the converting process. For example, a folding carton grade may have bending stiffness and ZDT as important attributes and these are built in during manufacture of the paperboard (although they may be decreased during converting as discussed in the last section). The carton may have bulge resistance as a performance attribute, but to use this effectively it must be related (usually empirically) back to measurable paperboard properties, and these in turn must be related to manufacturing variables (furnish and machine). In some cases there may be little connection between ultimate performance and the physical properties that are measured.

Most grades of paper involve specification of one or more mechanical properties, e.g. tensile, tear, or bursting strength. Since these are

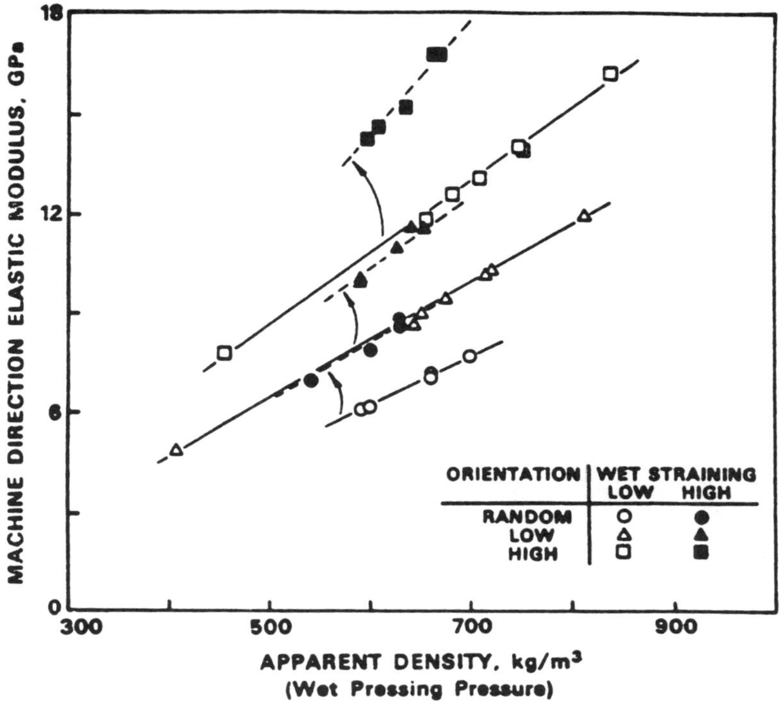

Fig. 53. The MD elastic stiffness vs. density with changing fiber orientation and wet straining. Compare these results with Fig. 46 (same samples).

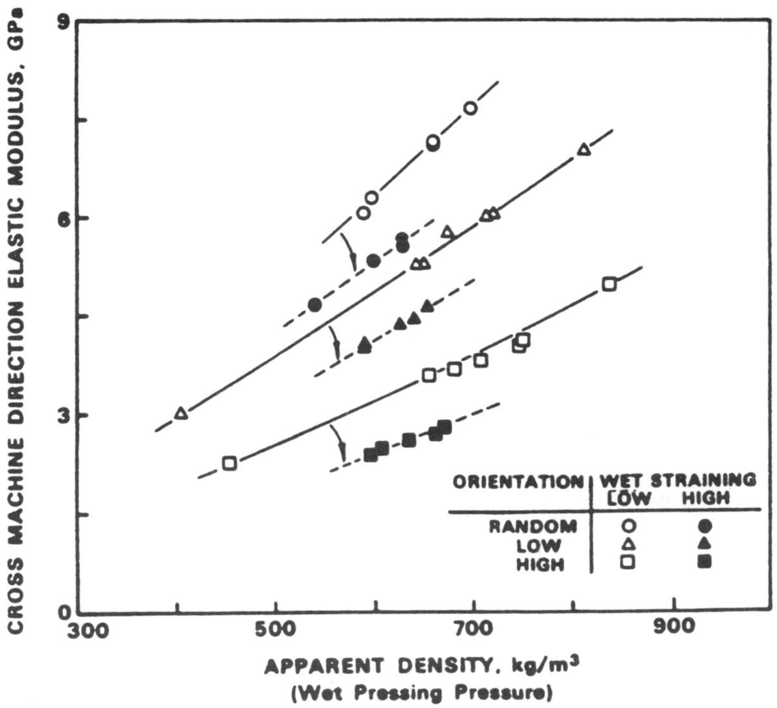

Fig. 54. The CD elastic stiffness vs. density with changing fiber orientation and wet straining. Compare these results with Fig. 47 (same samples).

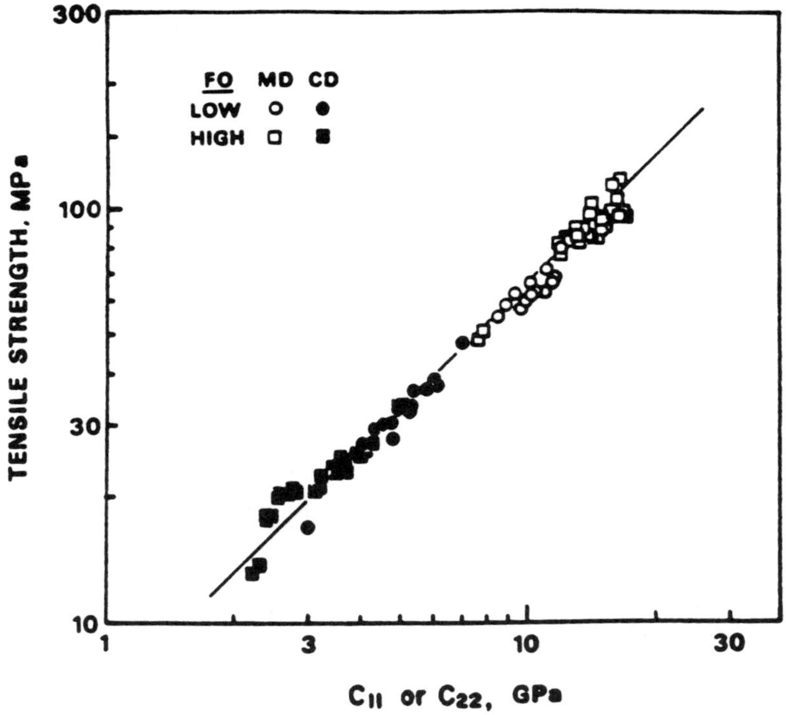

Fig. 55. A graph of MD and CD tensile strength plotted against MD and CD elastic stiffness. Over the range of those variables studied, the impact on stiffness and strength is similar [35].

destructive tests, replication of any given test requires a fair amount of material, and multiple testing requires still more material. Because the point-to-point variability in paper can be fairly high, depending on formation and other factors, test results can often be confusing and inconclusive. This also clouds the nature of the relationships between properties and end use performance.

The specific elastic stiffnesses offer some hope in the desire to relate end use performance to measurable paper properties. They can be measured non-destructively, thus eliminating some of the variability described above. The seven (out of nine) measurable specific elastic stiffnesses can be obtained on a single specimen, providing a three-dimensional characterization of it. In addition, somewhat surprisingly, changes in machine variables such as rush-drag ratio, wet pressing, or draws affect the elastic stiffnesses in a way similar to strength properties. (Much of the material in the following discussions is summarized in Ref. [35].)

Figures 53 and 54 are similar to Figs. 46 and 47 except, in the former case, the MD and CD elastic stiffnesses are plotted against density, respectively. As in the case of tensile strengths, the elastic stiffnesses in the direction of fiber orientation and the direction of wet straining increases, while the tensile in the CD (and ZD) tend to decrease. The close relationships between tensile strengths and elastic stiffnesses can be seen in Fig. 25 which depicts MD and CD tensile strengths against MD and CD elastic stiffnesses [35]. Over the rather broad range of values studied, all of the data fall along a single line. In Figs. 46, 47, 53–55, the sheet properties were varied by changing wet pressing pressure, fiber orientation, and wet stretching. These variables seem to impact tensile strength and elastic stiffness in the same way. Other variables, however, such as formation changes, would not be expected to affect strength and stiffnesses in a similar way.

Similar results are found between z-direction tensile strength and the out-of-plane elastic stiffness as shown in Fig. 56 for a bleached kraft Douglas fir pulp. Similar results have been found for other furnishes. Figure 57 shows MD and CD STFI edge compressive strength vs. a product of two elastic stiffnesses. MD compressive strength correlates with the product $C_{11}*C_{55}$ (the uniaxial and shear stiffnesses in the MD–ZD plane, whereas CD compressive strength correlates with the product of uniaxial and shear stiffnesses in the CD–ZD plane ($C_{22}*C_{44}$). Both of these

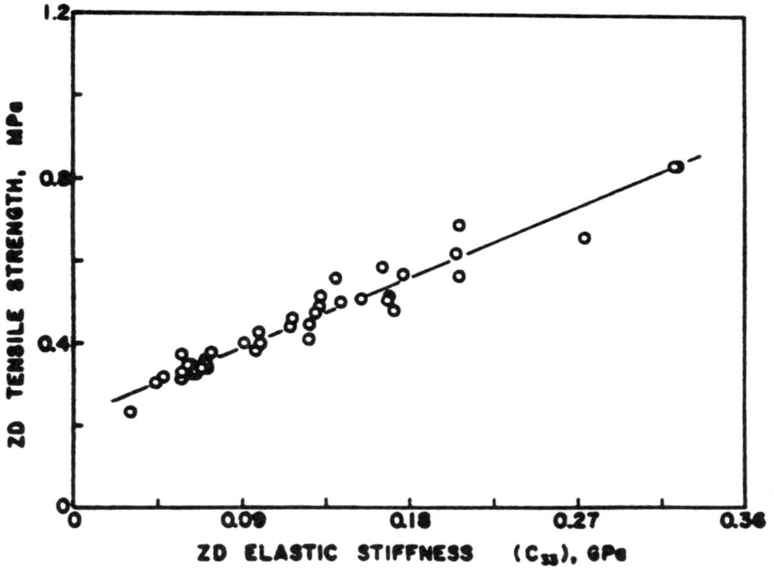

Fig. 56. Z-direction tensile strength plotted against z-direction elastic stiffness. Properties were changed by wet pressing and wet straining. Changing the fiber orientation (in the plane) has no effect on out-of-plane properties.

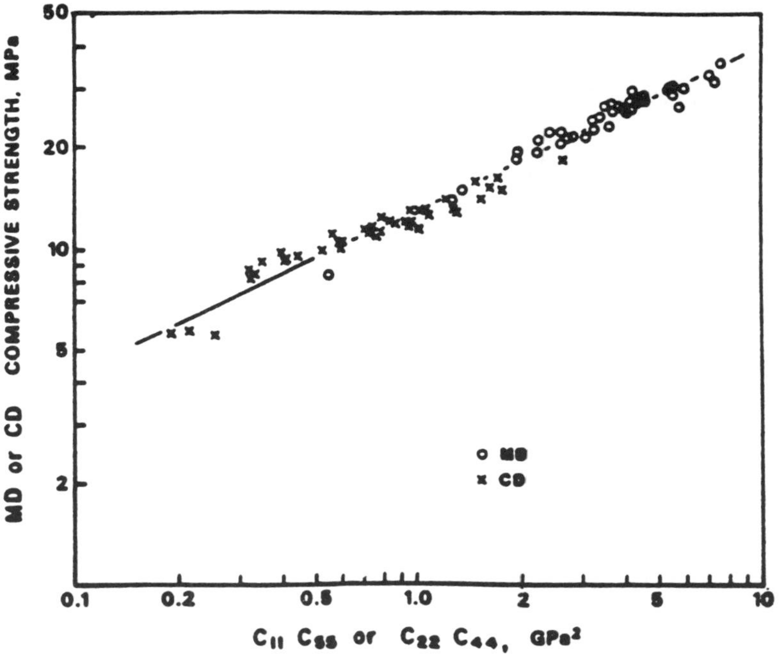

Fig. 57. MD and CD STFI compressive strengths plotted against the products $C_{11}*C_{55}$ and $C_{22}*C_{44}$, respectively. The regression line coincides with the expected behavior [36].

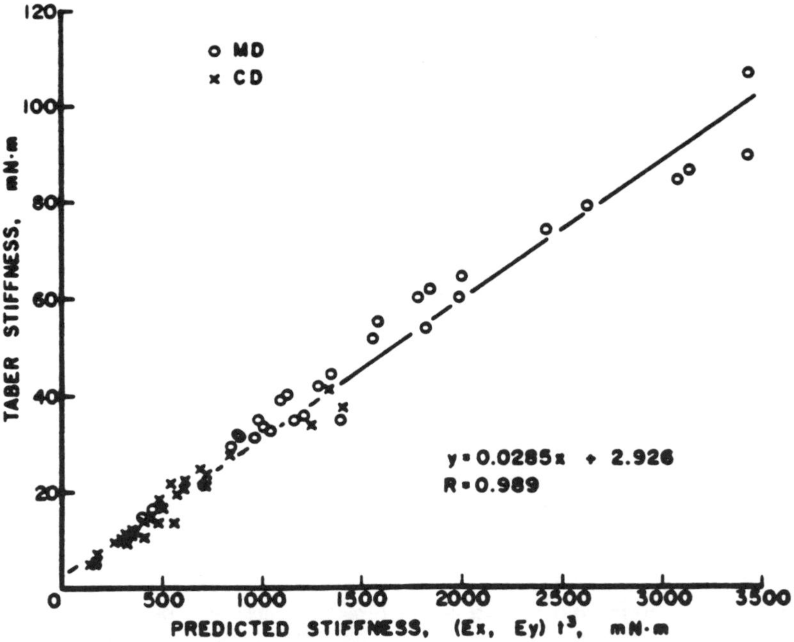

Fig. 58. MD and CD Taber bending stiffnesses plotted against the flexural rigidity predicted from measurements of MD and CD extensional elastic stiffness and caliper, respectively.

results are consistent with a semi-qualitative model of compressive strength [36]. This model assumes an initially curved lamina bending into a reinforced orthotropic medium. Failure occurs when the shear strength between this critical lamina and the medium (the rest of the sheet) is reached, causing delamination and voids.

Figure 58 plots MD and CD Taber bending stiffness against a calculated bending stiffness in terms of the (extensional) elastic stiffnesses and the caliper. Since Taber bending stiffness should be related to the flexural rigidity, EI, and I (the second moment of the cross section of a beam in bending) varies as the cube of the caliper, these results are expected. The results in Fig. 58 are for commercial linerboards, but similar results have been found for a variety of other grades.

Table 7 gives a summary of how some of the important end use tests are related to elastic parameters. It is not realistic to suppose that all of the "customary" properties of interest can be related to elastic stiffnesses, nor is there any reason to attempt to do this. The value of the elastic stiffnesses is, once again, that they collectively characterize the sheet in three dimensions, providing a better description of what happens during a process change, and that they can be measured non-destructively. The latter fact means that the seven measurable elastic stiffnesses can be obtained for a single specimen.

Table 7. End use tests and elastic parameters[a].

Property	Elastic Parameters
MD tensile strength	C_{11} ($\sim E_{MD}$)
CD tensile strength	C_{22} ($\sim E_{CD}$)
ZD tensile strength	C_{33} ($\sim E_{ZD}$)
MD/CD tensile ratio	C_{11}/C_{22}
MD compr. strength	$C_{11}C_{55}$ ($\sim E_{MD}G_{WD\text{-}ZD}$)
CD compr. strength	$C_{22}C_{44}$ ($\sim E_{CD}G_{CD\text{-}ZD}$)
MD bending stiffness	$C_{11} \cdot T^3$
CD bending stiffness	$C_{22} \cdot T^3$
Internal bond strength	C_{33}
Bursting strength	$C_{11} + C_{22}$
Flutability	C_{11} and C_{55}
Combined board performance	C_{22} and C_{44}

[a] T = caliper, C_{ij} = elastic stiffness.

L. CLOSING COMMENTS

This chapter has attempted to describe how the structure of the sheet is developed during the manufacturing process, how the properties of the structure can be modified during converting, and how the elastic stiffnesses, together with other physical properties, might offer a better way to characterize the structure. Of necessity, the treatment has been general in nature. For a given paper grade, we would need to prepare a new version of Table 6 based on the particular furnish used and the papermaking hardware available. In addition to the kind of information given in Table 6, we need to understand the *collective* impact of the important process variables on the properties of interest, before we can really begin to control the process to give the desired web structure. Furthermore, we need to be sure the structure is compatible with subsequent converting operations and the ultimate end use performance.

M. ACKNOWLEDGEMENTS

The author would like to thank Thomas Kindler, Otto J. Kallmes and Dan Manson for their careful and thoughtful review of the manuscript, and the participants of the Pacific TAPPI Seminar during the fall of 1989 for their input.

REFERENCES

1. Clark, J.d'A, Pulp Technology and Treatment for Paper, Miller Freeman Publications, Inc., San Francisco (1978).
2. Rance, H.F., ed., Handbook of Paper Science, 2 vol., Elsevier Scientific Publishing Co., New York (1980).
3. Mark, R.E., ed., Handbook of Physical and Mechanical Testing of Paper and Paperboard, 2 vol., Marcel Dekker, Inc., New York (1983).
4. Fellers, C., de Ruvo, A., Htun, M., Carlsson, L., Engman, C. and Lundberg, R., Carton Board, Swedish Forest Products Research Laboratory, Stockholm, Sweden (1983).
5. Bristow, J.A., and Kolseth, P., eds., Paper Structure and Properties, International Fiber Science and Technology Series/8, Marcel Dekker, Inc., New York (1986).
6. (a) Fundamental Properties of Papermaking Fibres, Transactions Fundamental Res. Symp., Cambridge (Sept. 1957).
   (b) Formation and Structure of Paper, Trans. Fundamental Res. Symp., Oxford (Sept. 1961).
   (c) Consolidation of the Paper Web, Trans. Fundamental Res. Symp., Cambridge, Sept. 1965, F. Bolam, ed., Technical Div. British Paper and Board Industry Federation, London (1966).
   (d) The Fundamental Properties of Paper Related to its Uses, Trans. Fundamental Res. Symp., Cambridge, Sept. 1973, F. Bolam, ed., Technical Div., British Paper and Board Industry Federation, London (1976).
   (e) Fibre-Water Interactions in Paper-Making, Trans. Fundamental Res. Symp., Oxford, Sept. 1977, Technical Div., British Paper and Board Industry Federation, London (1978).
   (f) The Role of Fundamental Research in Paper Making, Trans. Fundamental Res. Symp., Cambridge, Sept. 1981, Mechanical Engineering Publications Ltd., London (1983).
   (g) Papermaking Raw Materials, Trans. 8th Fundamental Res. Symp., Oxford, Sept. 1985, Mechanical Engineering Publications Ltd., London (1985).
   (h) Fundamentals of Papermaking, Trans. 9th Fundamental Res. Symp., Cambridge, Sept. 1989, Mechanical Engineering Publications Ltd., London (1989).
7. Lyne, L.M., and Gallay, W., Tappi 37(12):581 (1954).
8. Page, D.H., and Tydeman, P.A., In Formation and Structure of Paper, Ref. 6(b) above.
9. Sachs, I.B., and Kuster, T.A., Tappi 63(10):69 (1980).
10. Salmen, L. and Fellers, C., J. Pulp Paper Sci. 15(2):J63 (1989).
11. Rutland, D.F., "Dimensional Stability and Curl — A Review of the Causes", In Design Criteria for Paper Performance, STFI Meddelande, A969 (Aug. 1987).
12. Giertz, H. and Rodland, G., "Elongation of Segments-Bonds in the Secondary Regime of the Load/Elongation Curve", In Proc. Intl. Paper Physics Conf., Harrison Hot Springs (Sept. 1979).
13. Kim, C.Y., Page, D.H., El-Hosseiny, F. and Lancaster, A.P.S., J. Appl. Polymer Sci. 19(6):1549 (1975).
14. Wink, W.A. and Baum, G.A., Tappi 66(9):131 (1983).
15. Rennel, J., Pulp Paper Mag. Can., p. T151 (May 16, 1969).
16. Jones, A.R., Tappi 53(5):203 (1968).
17. Habeger, C.C., Mann, R.W. and Baum, G.A., Ultrasonics 17:57 (1979).
18. Baum, G.A. and Bornhoeft, L.R., Tappi 62(5):87 (1979).
19. Mann, R.W., Baum, G.A. and Habeger, C.C., Tappi 62(8):115 (1979).
20. Mann, R.W., Baum, G.A. and Habeger, C.C., Tappi 63(2):63 (1980).
21. Mohlin, U.B. and Lidbrandt, O., In Intl. Symp. Fundamental Concepts of Refining, Sept.

16-18, 1980, Appleton, WI, The Institute of Paper Chemistry, Appleton, WI, p. 61.
22. Page, D.H., Tydeman, P.A. and Hunt, M., In Formation and Structure of Paper, Ref. 6(b) above, p. 171.
23. Svensson, O. and Osterberg, L., Svensk Papperstidn. 68(11):403 (1965).
24. Paulapuro, H., "Pressing and Drying — A Revolution Ahead?", In Proc. XXIII Eucepa Conf., Harrogate, U.K. (May 31-June 3, 1988).
25. Sprague, C.H., "Impulse Drying — A Status Review", In Proc. XXIII Eucepa Conf., Harrogate, U.K. (May 31-June 3, 1988).
26. Back, E., "A Review of Press Drying", STFI Meddelande Series D., No. 224 (Sept. 1984).
27. Setterholm, V.C. and Kuenzi, E.W., Tappi 53(10):1915 (1970).
28. Fleischman, E.H., Baum, G.A. and Habeger, C.C., Tappi 65(10):115 (1982).
29. Htun, M. and Fellers, C., Tappi 65(4):113 (1982).
30. Baum, G.A., Pers, K., Shepard, D.R. and Ave'Lallemant, T.T., Tappi 67(5):100 (1984).
31. Sprague, C.H. and Whitsitt, W.J., Tappi J. 65(12):104 (1982).
32. Whitsitt, W.J., "Relationships between Runnability and Medium Properties", IPC Technical Paper Series No. 238, The Institute of Paper Chemistry, Appleton, WI (May 1987).
33. Whitsitt, W.J. and Sprague, C.H., Tappi J. 70(2):91 (1987).
34. Charles, L.A. and Waterhouse J.F., J. Pulp Paper Sci. 14(3):J59 (1988).
35. Baum, G.A., "The Elastic Properties of Paper: A Review", In Design Criteria for Paper Performance, STFI Meddelande, A969 (Aug. 1987).
36. Habeger, C.C. and Whitsitt, W.J., Fiber Science and Tech. 19:215 (1983).

IV
STOCK AND WHITE WATER SYSTEMS

N. RYTI

Niilo Ryti received M.Sc. (Eng) and Lic.Sc. degrees from Helsinki University of Technology in 1944 and 1960, respectively, and was granted a D.Tech. h.c. degree at the same university in 1981. He worked as paper mill Superintendent and Technical Manager at the Kangas paper mill of G.A. Serlachius Oy during the years 1945–1960. From 1960 to 1963 he worked for Oy Wilh. Schauman Ab, and was in charge of the planning, erection, start-up and running of their kraft paper mill at Pietarsaari. During the years 1963–1977 he held the professor's chair in paper technology at Helsinki University of Technology. He joined the Jaakko Pöyry Consulting Engineers as partner, and served as a consultant until his retirement in 1982.

H. PAULAPURO

Hannu Paulapuro received a M.Sc. (Eng) in 1971 and D.Tech. degree in 1979 from Helsinki University of Technology. From 1970 to 1984 he worked for the Jaakko Pöyry company, where he was responsible for research, technical development and corporate planning. For two years he also served as a process and technology expert for Pöyry-BEK, Inc. in the U.S.A. In 1984 he joined the Finnish Pulp and Paper Research Institute as Vice President, Paper and Board, and since 1989 he has been Deputy Research Director of the Institute's Paper Science Centre. From 1987 to 1989 he was acting professor in paper technology at Helsinki University of Technology, and since January 1990 he has held the chair.

REVIEWED BY

Virgil Holland
Sr. Project Manager
Simons-Eastern Consultants, Inc.
Decatur, GA

Dan W. Manson
Senior Consultant, Engineering
James River Corporation
Portland, ME

A. INTRODUCTION

In the wet end of a paper machine, water is drained from the web during web formation and further by suction boxes and presses. Depending on the type of process, process stage and the paper grade being made, the water contains varying amounts of fiber, fillers and dissolved papermaking chemicals. These substances are recovered as far as possible by recirculating the water—called white water—to preceding stages of the production process.

Part of this water, primarily the water removed on the wire section, is used to dilute the stock entering the headbox. This *short circulation* is also designed to increase the flow of dry solids through the headbox with the object of making the flow of the dry solids of the paper web equal to the flow of the dry solids fed from the refining department to the paper machine. Fines and fillers in particular tend to escape through the wire into the white water. For this reason, one of the tasks of the short circulation is to keep the particle size distribution of the web formed on the wire roughly the same as in the stock fed from refining.

The part of the water removed on the wire which is not used to dilute the stock in the headbox is led to the earlier process stages. The circulation formed in this way is called the *long circulation*. The long circulation, which often consists of several branches, is designed to improve the material and also the heat economy of the system. White water from the long circulation is, for example, used to even out stock consistency variations, or to adjust the consistency for stock preparation. In addition to consumption points with steady water consumption, there are also consumption points with intermittent water consumption. Such consumption points are, for example, broke pulpers which start up in the event of a web break and knock-off showers, which remove the web from the wire to make it fall into the broke pit.

Water is needed in a number of cleaning showers and in pump and refiner seal boxes. These applications often require fresh water. However, white water can be used for example in wire cleaning showers, provided that the fiber content of the water is low enough. Increased use of fresh water also increases the volume of fiber-containing water discharged from the system. Reducing the usage of fresh water is called *closing the system*.

The water discharged from the long circulation is often treated in a saveall. The solids recovered in the saveall are dewatered and returned to suitable points of the process. Part of the water from the saveall is often returned to such points of the process where water with a high solids content cannot be used.

Broke produced during occasional web breaks on the paper machine is reused in the process after repulping and dilution. The paper mill's broke system is an important part of the long circulation.

The design of the paper mill's stock and white water system varies a lot, primarily depending on the paper grade, the product mix and the type of machine used.

B. SHORT CIRCULATION

Figure 59 shows the short circulation of a twin-wire newsprint machine. The water removed in the wire part is discharged into the wire pit. This water is used to dilute the stock fed to the paper machine. The diluted stock is pumped via centricleaners, air removal equipment and screens to the headbox. The system between the wire pit and the headbox is called the *approach flow piping*.

The short circulation is designed so that the fractions with a large fiber content are returned to the short circulation, and the overflow to the long circulation consists of the fractions with the lowest fiber content. On a fourdrinier machine, this is easy to arrange by taking the water from the last suction boxes to the long circulation (Fig. 60). On a twin-wire machine, separating the fraction with the lowest fiber content is more difficult.

Even fairly small changes in the flow volume, dry solids content, fines content or temperature of the short circulation disturb the process, causing changes in the finished product and even breaks on the paper machine. For this reason, the short circulation must maintain its stability under any conditions. A lack of stability shows up at the headbox slice as consistency variations or as flow variations, which both cause variations in basis weight and often also in other paper properties.

Consistency variations can originate at such points of the short circulation in which flows with different consistency are combined. For example, the water drained at the beginning of a fourdrinier section contains more solids than the water removed farther down on the wire. Water removed at the suction boxes, which is used to replenish the wire pit, generally contains less fibers than water removed at the beginning of the wire section. The wire pit must be designed with a view to maintaining a stable flow pattern to prevent consistency variations on the suction side of the fan pump.

The mixing point of the thick stock from the basis weight valve and the white water on the

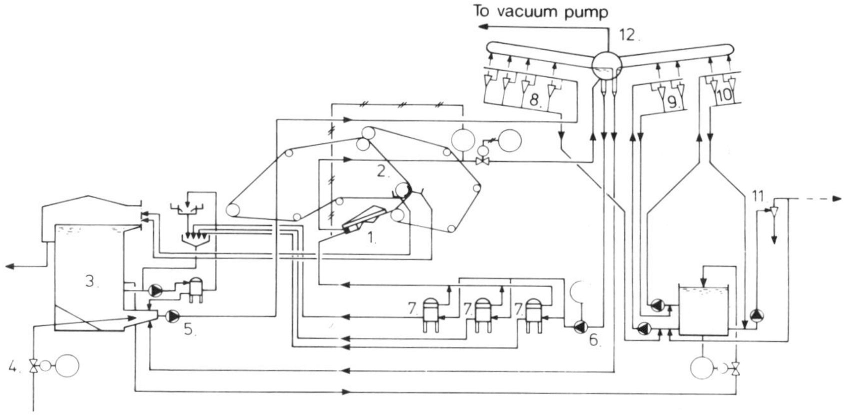

1. Headbox
2. Wire part
3. Wire pit
4. Basis weight valve
5. Mixing pump
6. Fan pump
7. Pressure screen
8. Centricleaners, stage 1
9. Centricleaners, stage 2
10. Centricleaners, stage 3
11. Centricleaners, stage 4
12. Air removal equipment

Fig. 59. Short circulation of a twin-wire newsprint machine.

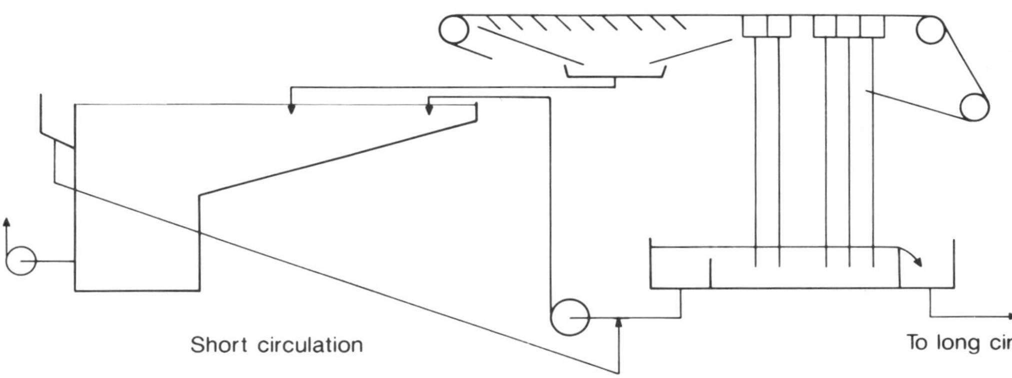

Fig. 60. Separation of water fractions with large and low fiber content into short and long circulation.

suction side of the fan pump should be carefully designed to prevent "clouds" of temporary consistency variations in the suction zone. The thick stock pipeline preferably should be taken right through the white water pit to the suction side of the fan pump, as shown in Fig. 59. It is also advisable to size the pipeline so that the flow velocity of the thick stock in the suction cone of the fan pump is at least five times that of the white water flow velocity [1].

Consistency variations in the short circulation can, in principle, be filtered by taking the stock through an agitated tank with a sufficiently large volume. An air removal unit with horizontal design serves as such a tank. Its volume in relation to the flow through the tank is, however, so small that it effectively evens out only consistency variations with a fairly high frequency.

Consistency variations can also be filtered by dividing the flow into branches, in which the stock is retained for varying lengths of time before the separate flows are recombined. The principle of such an arrangement is shown in Fig. 61. The stock flowing through the left-hand centricleaner of the centricleaner battery has the longest retention time [2].

There may be pressure variations in the approach flow system, which show up in the form of flow variations at the headbox slice. Pressure variations are usually caused by rotating machine parts, such as the fan pump impeller, or screen foils. Flow variations can be dampened by using a vessel with a free surface. The air removal equipment is such a vessel. An air-cushion headbox also dampens flow rate variations but a hydraulic headbox does not. For this reason, a hydraulic headbox is often preceded by a separate flow rate and pressure variation dampener [3]. Horizontally operating hydraulic headboxes are often provided with an air cushion installed in connection with a stilling chamber.

Air entrained in the stock may cause disturbances in the short circulation. In a centrifugal pump, air bubbles in the stock accumulate at the center of the impeller causing flow variations and may in the end cause the whole pump to plug. This is illustrated in Fig. 62. In alternative A, the stock from the level box overflow falls freely into the stock chest. It generates air bubbles, which go to the suction side of the pump, disturbing pump operation. In alternative B, the overflow has been led to the back part of the chest, and directed against the wall of the chest, so that less air is entrained in the stock. In alternative C, the overflow is taken below the stock surface. The overflow level controller keeps the overflow pipe full of stock, so no air can enter the chest.

Air bubbles may also disturb the operation of the basis weight valve. Figure 63 shows three alternative ways of placing the valve. In alternative A, the valve is situated above the surface of the wire pit. Below the valve the hydrostatic pressure is low. It easily collects air bubbles, which cannot rise in counter-flow direction through the valve, causing a varying "empty space" below the valve. Flow conditions in the stock pipe vary and so does the stock flow to the paper machine. In alternative B, the valve is situated below the surface of the wire pit, but on the declining section of the pipe. Air accumulation below the valve is less likely than in alternative A but is still possible. In alternative C, the valve is on the rising section of the pipe. Air bubbles cannot be accumulated at the valve, but pass through the valve.

Air bubbles can also cause disturbances in approach flow pipes containing diluted stock. If air accumulates at a certain point of the pipeline, slime and dirt deposits are likely to form. When

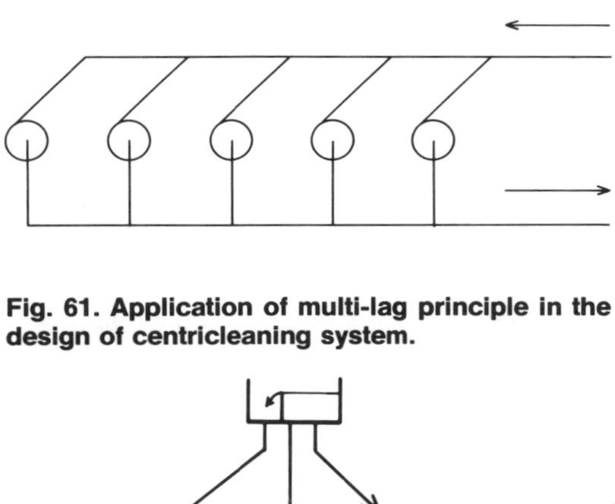

Fig. 61. Application of multi-lag principle in the design of centricleaning system.

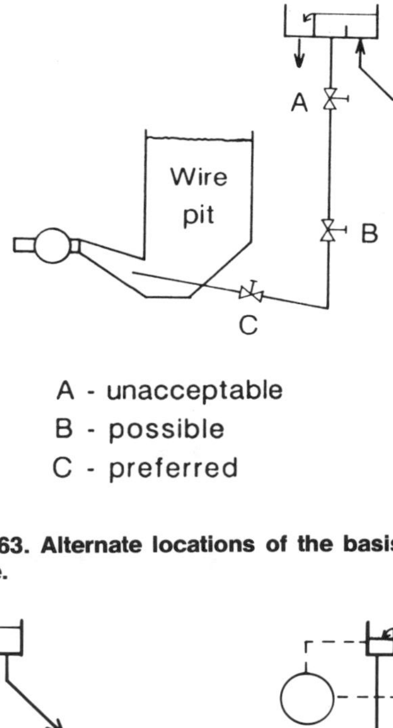

Fig. 63. Alternate locations of the basis weight valve.

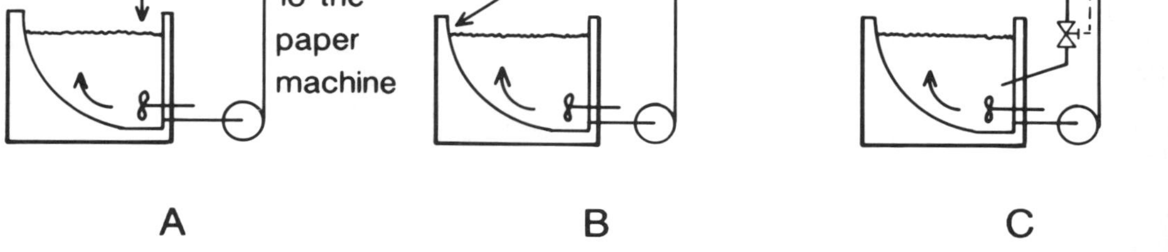

Fig. 62. Different designs of the feedback from the level box to the stock chest (air problems decrease in order A → B → C).

breaking loose and moving with the stock, the slime may cause web breaks on the paper machine and slime spots in the finished paper. To prevent accumulation of air, the pipeline should be designed with a slightly rising slope in the direction of the stock flow [3].

Air entrained in the stock also affects the way pressure variations move in the pipeline, and in this way the air may increase the problems caused by pressure variations. To minimize the amount of air in the short circulation stock, special attention must be paid to pipeline design (cf. above-mentioned examples). Air can also be removed by using chemicals.

The design of the equipment in the short circulation and details of piping are discussed, for example, in reference [4].

C. LONG CIRCULATION DESIGN PRINCIPLES

When assessing the performance of a stock and white water system, the following criteria are regarded as important:

- The system's stability during steady state operation.
- The system's stability during web breaks.
- The system's controllability during grade changes.
- Material losses during steady state operation.
- Material losses during grade changes.
- Time losses during grade changes.
- Heat consumption.
- Environmental impact.

It is impossible to design a system which would perform in the best possible way according to all of these criteria. In practice, these criteria have to be ranked in order of importance, and on this basis strive to find the best design. The system selection depends on the fact whether the paper machine is a single-grade machine, which makes only a limited range of similar paper grades, or whether it is a multigrade machine which makes a wide range of products with a varying raw material composition and possibly even varying colors. It is also significant whether the retention on the paper machine wire is high or low, i.e., whether the dry solids content of the wire and suction box water is low or high. Furthermore, the mill conditions, such as available space, etc., affect the design of the stock and white water system. It is seldom that two identical systems can be found. Thus, the systems discussed in this chapter are only typical examples; other kinds of well-functioning systems can also be found.

1. Broke

The paper machine broke system is an important part of the long circulation. Broke is produced during web breaks. Web breaks usually occur at the couch, in the press section or after the drying section on the machine calender. Wet broke is produced even during normal operation in the form of edge trimming at the wire section, and dry broke is produced in finishing operations. The broke produced during a break must be diluted to allow it to be pumped. A suitable broke pulp consistency for pumping is 2.5-3%. However, broke pulp is often diluted much more, because consistency control is difficult in a discontinuous process like broke handling, and even temporary consistency peaks may cause disturbances in broke repulping and pumping. Diluted broke pulp is pumped to thickeners where excess water is removed. The suitable consistency for proportioning is aimed at on the broke thickener, i.e. about 4% after the thickener and about 3.5% after the consistency control for proportioning. The fiber-containing water from the thickeners is piped to the white water system. However, its dry solids content varies with the operating situation. For this reason, this water preferably should not be led to the sensitive short circulation, but further towards the beginning of the system.

To safeguard the stability of the system during steady state operation, all stock components, including broke pulp, must be proportioned to the paper machine. Since broke pulp is produced discontinuously, but consumed at a steady rate, the system must incorporate a sufficiently large broke storage tower, as shown in Fig. 64 for a single-grade machine. The broke pulps originating from the wet and dry ends of the paper machine are not identical in terms of their papermaking characteristics, but they are usually not proportioned separately into the system.

On multigrade machines, the grade changes create a special problem, because the furnishes of the two grades might not be compatible. A running grade change in such a situation is not possible, but the time-consuming clean-up of the system is necessary. Therefore, on a multigrade machine it is not usually feasible to use a single broke storage tower. At the next grade change, the broke in the storage tower may not be suited for the new grade, so it would have to be discharged. For this reason, wet broke is generally used immediately. At least on small machines, dry broke can be stored and added together with other raw material components at the beginning of the stock system. A broke system designed for a multigrade machine is

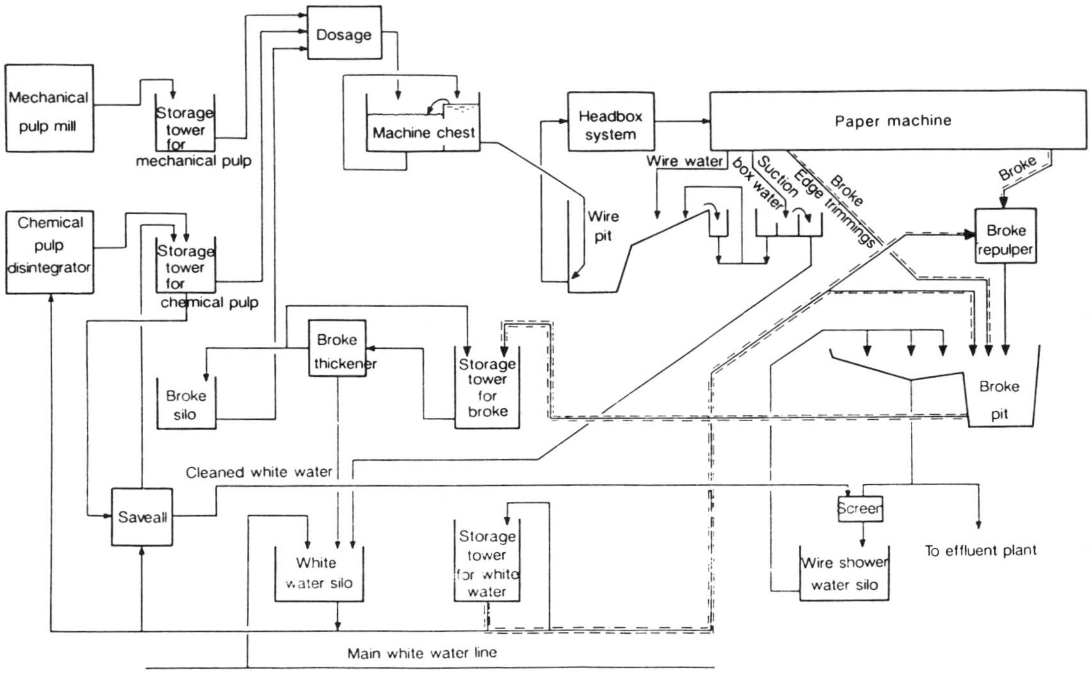

— — — — Flows caused by wet end broke

—·—·—·— Flows caused by dry end broke

Fig. 64. Block diagram of stock and white water system of a newsprint machine. (Block diagram shows the connection principles between the equipment groups, but not the piping details.)

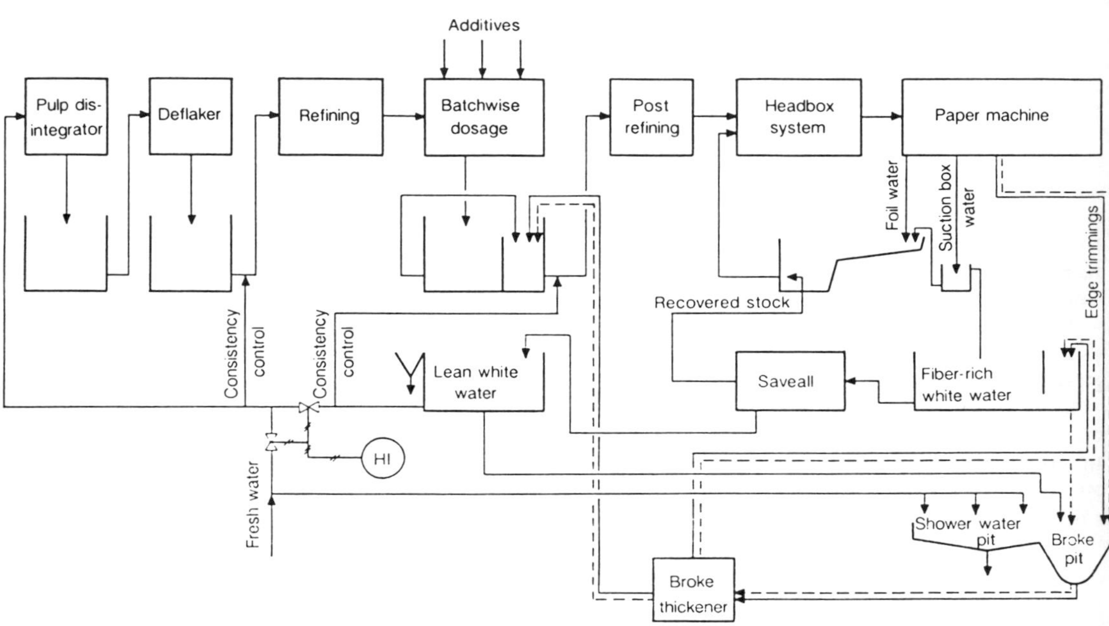

— — — — Flows caused by wet end broke

Fig. 65. Block diagram of the stock and water system of a small fine paper machine with a large variety of grades.

shown in the block diagram in Fig. 65. The flexibility required by a multigrade machine is achieved by sacrificing the system's stability, especially during web breaks. During a web break the proportion of broke in the stock pumped to the paper machine increases, particularly if the break is prolonged. This easily causes new breaks. Therefore, the operating speed of a multigrade machine usually remains lower than the speed of a single-grade machine operating under similar conditions.

2. Savealls

The stability requirement of the system includes the demand for a stable fines content in the stock pumped to the paper machine. Fines are concentrated into the stock recovered in the saveall. The stability of the saveall's operation under varying conditions, the time it takes for the saveall to respond to changes in operating conditions and the point where the recovered stock is returned into the system are all important factors influencing the stability of the whole system. On a multigrade machine the recovered stock should be returned into the system as close to the paper machine as possible. If returned to the beginning of the stock system, recovered stock would cause problems during grade changes. Therefore, on a multigrade machine the saveall must be connected to the system so that its running conditions–and thereby the quality and quantity of recovered material–remain stable in spite of variations in the running conditions of the paper machine. In the system shown in Fig. 65 the volume flow to the saveall during a web break does not change and the fiber flow changes only slightly.

On a single-grade machine, it is not necessary to return recovered stock close to the paper machine. Even if the stock is returned to the beginning of the system, possible variations in the fines content are evened out before the stock reaches the paper machine. Therefore, the requirements for the operation of the saveall need not be so strict. However, the time it takes for the process to reach its equilibrium after start-up and after grade changes is prolonged, but on a single-grade machine this is not very important. Figure 64 shows how thickened stock from the saveall is led to the chemical pulp storage tank at the beginning of the system. The saveall has been assumed to be a filter, with a sweetener stock from the chemical pulp storage tank. This pulp and the fines mixed with it on the filter are returned to the chemical pulp tank.

The operating principles of different kinds of saveall are described in reference [5].

3. Temperature control

The temperature of the stock in the headbox influences the drainage on the wire. For this reason, the system should be designed to keep the headbox temperature sufficiently high and as stable as possible under all operating conditions. The system's temperature tends to change if the volume of cold fresh water added to the warm water system increases. This can happen, for example, in the case when the clear filtrate storage for knock-off shower is too small and make-up fresh water is automatically turned on. In the system shown in Fig. 64, the temperature change due to cold fresh water from the knock-off shower spread quite slowly to the headbox system. In the system shown in Fig. 65, which does not include a broke storage, the temperature change on the paper machine is much steeper.

4. Grade changes

On a multigrade machine, the process should be so controlled during grade changes that the quality of the previous grade remains stable to the end, and so that the new grade quickly reaches the desired quality level. Controlling the system during grade changes is difficult because the stock chests must be allowed to run empty and then filled up again for the new grade. In the system shown in Fig. 65, the broke content in the machine chest increases, as the supply of fresh pulp to the chest is shut off and the level begins to go down. If the stock from the saveall is pumped to the machine chest and not directly to the paper machine, which is desirable with a view to the stability of the process, the fines content changes also in the stock pumped to the paper machine as the level goes down. If a refiner is installed after the chest so that the overflow of refined stock returns to the chest, the degree of refining of the stock in the chest changes due to the overflow until a certain equilibrium is reached. If the level in the chest changes, i.e., if the flow of fresh stock to the chest does not equal the outgoing flow, the equilibrium is disturbed, causing the degree of refining to fluctuate. This problem can be avoided by designing the system so that the refined stock does not recirculate to its feed chest.

To speed up grade changes and to minimize the volume of broke produced during grade changes, stock chests should be small and recirculations short. So, on a multigrade machine, it is not possible to make use of the smoothening effect of large stock chests, which would be useful as a means of improving the stability of the process during steady-state operation.

5. Material losses

Material losses in the stock and white water system are dependent on the volume of water discharged from the system and its fiber, filler and chemical contents. Losses can be reduced either by reducing the fiber and filler content in the water discharged from the system or by reducing the volume of water discharged.

The dry solids content in the water discharged from the system can be reduced by designing the white water system so that water fractions with a high dry solids content are recirculated, and water with a low dry solids content is led to the system's overflow, and by pumping the water through a saveall before it is discharged from the process.

The volume of water discharged can be reduced by reducing the volume of fresh water added to the system. This is achieved by replacing fresh water with white water, in other words by "closing" the system. This is discussed in more detail in a later section (see Sec. D). The solids losses in water discharged via a saveall generally can be minimized quite well.

Rather large amounts of material are often lost through the reject discharge from the last stage of the centricleaner system connected to the headbox and through the water discharged from the paper machine's wet presses. To safeguard the quality of the paper, these water fractions are usually not recirculated but discharged direct from the process. Harmful press felt fibers contained in the water from the wet presses can be removed on a suitable filter, to allow press water to be recirculated to the white water system.

On a multigrade machine, material losses also occur during grade changes. The white water from the paper machine may not be suitable for use in the pulper, in which pulp for the new grade is being repulped. In this case, the water system must be opened up and repulping done with fresh water until the new grade has reached the paper machine. Figure 65 shows an example of such an arrangement for this purpose. Material is also lost when emptying and washing stock chests. For this reason, the chests should be designed to facilitate emptying. The number of chests should also be minimized. On a multigrade machine, large white water tanks and savealls should preferably be avoided, because they cause unnecessary material losses when emptied during grade changes. A large saveall also requires a long time to reach an equilibrium after a grade change.

6. Heat losses

The heat loss of the white water system may also be significant. To promote drainage on the wire the stock temperature is often raised. The simplest way to do this is to blow steam into the wire pit. This causes the mill's steam consumption to increase and results in condensate losses. The energy used by grinders and refiners is converted into heat in the pulp. Accordingly, if the volume of fresh water added to the system is kept low, the temperature of the stock on the paper machine can be raised enough without using steam. However, the time it takes for the system to reach an equilibrium is quite long, especially if pulp towers and water tanks are large. To improve the controllability of the process and to reduce heat losses, it is often advisable to recover steam from mechanical pulp manufacture and to use this steam to heat the wire pit [6]. The temperature of the water used in paper machine wire cleaning showers should be roughly equal to the headbox temperature. This is necessary to prevent the wire from changing its dimensions because of temperature variation, which tends to shorten its lifetime. If white water is used in wire cleaning showers, the temperature of the shower water is always roughly the same as the headbox temperature. If fresh water is used, the shower water temperature must be adjusted. The thermal energy needed for raising the temperature of shower water can be taken either from the hot humid air evacuated from the drying section or from the condensate system. In the event of a wet end break, these sources of heat run out, which requires a large enough warm water tank or the possibility of using steam for heating the water.

7. Process effluent

Even if the internal material economy of a paper mill is arranged in the most effective way, taking into account the restrictions imposed on the white water system, the water discharged from the process does not fulfil the purity requirements for water to be discharged into a waterway. To meet these requirements, the water to be discharged from the process must be purified separately without returning recovered solids to the manufacturing process.

D. CLOSING THE WHITE WATER SYSTEM

Reducing the volume of fresh water added to the system, and thereby reducing the volume of fiber-containing water discharged from the system, is called closing the system.

When closing the white water system, solids losses are reduced. The water to be discharged is as far as possible taken from the fraction containing the smallest amount of fiber, generally the saveall filtrate. The solids content of this fraction does not change much with a change in the degree of system closure, so the solids losses through this fraction are roughly proportional to the volume of water discharged. However, the amount of dissolved and colloidal material in the white water increases as the system is closed.

A decrease in the water discharge volume reduces the need for effluent treatment capacity in the mill.

Closure of the white water system may cause difficulties in mill operation, and may also impair the quality of the paper. Slime formation in pipelines increases. To prevent slime formation, flow rates have to be kept high enough, and pipelines have to be designed to prevent "backwaters" in the system. Appropriate slime control and cleaning procedures also have to be implemented. When changing over to using white water in shower pipes, the pipes should be designed to prevent plugging and allow cleaning during operation. Shower pipes cause mist and the fines contained in this mist tend to adhere onto paper machine structures. Lumps of these impurities falling on the paper web may ruin the end product and even destroy metal wires. The accumulation of such impurities can be prevented by using cleaning showers. When closing the white water system, the concentration of dissolved and colloidal material increases. An increase in this concentration may cause difficulties in controlling the chemical balance of the short circulation. Wood-based material dissolved from wood are generally anionic, forming so-called anionic trash at greater concentration when closing the system. In particular, this anionic material impairs the effect of cationic retention and drainage agents.

Water enters the white water system:

- With raw materials and additives;
- Through fresh water shower pipes; and
- As seal water in pumps and agitators.

In addition, fresh water is used as cooling water, for example, in ventilation systems and in paper machine cooling cylinders. This water can be reused in the fresh water system.

In shower water pipes it is generally possible to use a saveall filtrate, which has a low solids content and particularly a low long fiber content. When closing the system, white water preferably should be used at locations from which the shower water is led direct into the white water system. Such showers on a fourdrinier machine are the knock-off showers for web and edge trimmings, the wire cleaning shower on top of the broke pit, and the couch roll cleaning shower. In other wire cleaning showers fresh water can be used, provided that the water does not get into

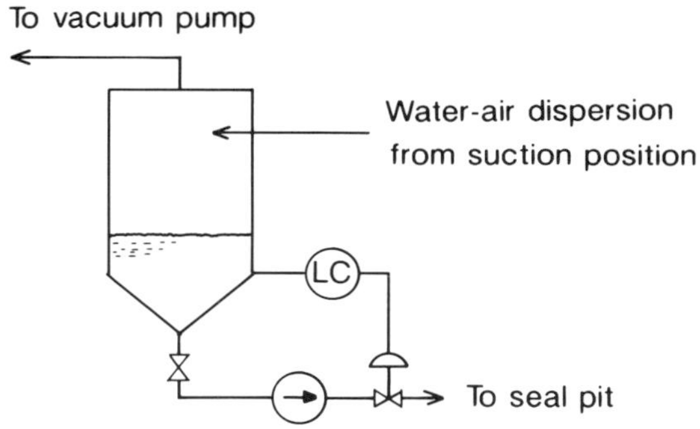

Fig. 66. Water separation in the vacuum system.

the broke pit. An arrangement of this kind is shown in Fig. 65. In the arrangement shown in Fig. 64, white water is used in all wire cleaning showers. Internal shower pipes in wire suction rolls and press suction rolls require fresh water for technical reasons. If Venta rolls are used on the presses, fresh water consumption is lower. White water should preferably not be used in press felt washing showers, because it rapidly leads to felt plugging.

Water-ring vacuum pumps in the paper machine vacuum system consume a lot of water. It is advisable to install a water separator for fiber-containing water on the suction side of the vacuum pump so that the vacuum pump sucks in air only, whereas the fiber-containing water is returned to the white water system, as shown in Fig. 66. The water ring is then created using fresh water only. Vacuum can also be created by using centrifugal blowers, which do not consume fresh water. The seal water in pump and agitator seal boxes should be fresh water. For control of water consumption, a flow rate meter should be installed at each consumption point.

The white water system must be designed so that the system does not open during web breaks or grade changes. A common deficiency is that the white water tank is too small, so the system has to be opened in the above-mentioned cases. This is examined in more detail in Sec. E. If on a multigrade machine white water from the previous run cannot be used on the following run, the white water system has to be emptied for the grade change. This is easier if all tanks are small. In this way, solids losses are minimized and the time it takes for the system to regain its equilibrium is shortened.

Depending on the paper grade and efficiency of the press section, the web carries 1.2 to 2 m^3/t of water to the drying section. If only this volume entered the white water system, there would be no water to discharge. With the conventional technology described in the foregoing it is, in

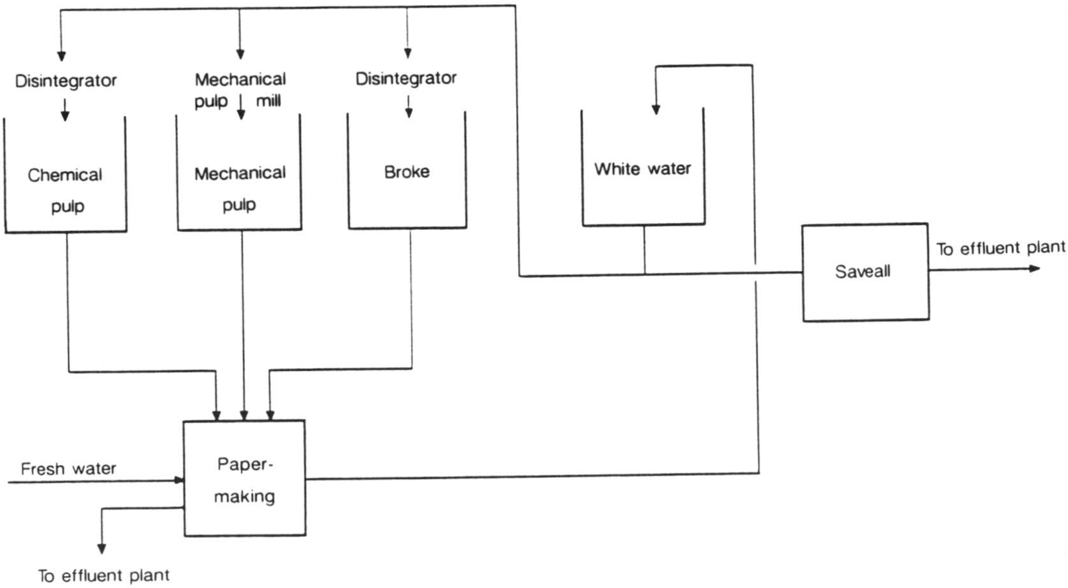

Fig. 67. Simplified diagram of the storage tanks of a newsprint machine.

practice, possible to achieve a fresh water consumption of 10 to 15 m^3/t when making newsprint, fine paper or linerboard [7]. When the system is closed further, the concentration of dissolved and colloidal material increases rapidly, which may cause technical difficulties in the process. If these materials could be concentrated into a small subfraction, which would then be discharged from the process, the concentration of dissolved material in the white water system could be reduced. However, proven technology is not available for this purpose [8].

E. STOCK AND WATER TANKS

The stock and white water system of a paper mill incorporates numerous tanks and chests. The purpose of a tank may be either to act as a buffer or to filter disturbances in the process. In many cases a tank serves both of these purposes.

A buffer tank may be installed between two major mill departments, e.g., a pulp mill and a paper machine. The purpose of the tank is to prevent disturbances or short-term stoppages in one department from affecting the other. The purpose of the buffer tank may also be to change an intermittent process into a continuous one or to act as a start-up buffer between groups of machinery. The buffer tank must have a certain volume, and it must be possible to adjust its liquid volume. A buffer tank does not necessarily have to be agitated.

A tank for filtering disturbances, for example, can be installed between a mechanical pulp mill and a paper mill, in which case its task is to even out temporary quality variations in the pulp produced on the grinders. The machine chest of a paper machine can also be regarded as a filter for dampening disturbances, because its task is to even out variations in stock consistency, degree of refining and fiber composition. The filtering effect is based on the fact that the stock particles leave the tank in a different order compared to the order they went into the tank. For this reason, the tank must have a certain volume and a certain mixing effect. The filtering characteristics of various tanks are discussed, for example, in reference [9].

The size of a buffer tank depends on the process flow and on the duration and frequency of interruptions in it. The various tanks in the system have to be sized according to similar criteria and considering their interconnections.

Figure 67 shows a simplified diagram of the tank arrangement of a newsprint machine. The system has three pulp towers for mechanical pulp, chemical pulp and broke, and a white water tank. The chemical pulp tower is filled up through a pulper, which takes dilution water from the white water tower. The mechanical pulp tower is filled from the mechanical pulp mill. The water needed in the mechanical pulp mill is also taken from the white water tower. In the event of a web break, broke dilution water is taken from the white water tower. The various stock components are proportioned to the paper machine, and the white water produced is returned to the white water tower. Part of the water from the white water tower is pumped through a saveall to effluent treatment. A certain proportion of fresh water is added to the production process.

The system running strategy can be selected in many ways. An apparently sensible running strategy is to keep the discharge flow through the saveall at a certain level as stable as possible. This ensures that the saveall operates under stable conditions, which is favorable for the efficiency of the saveall and the stability of the process. However, the discharge flow must vary slightly, because the inflow of fresh water varies. The discharge flow can be adjusted by using the combined liquid volume of stock and white water tanks as the adjustable control parameter in the control loop and by choosing the control method so that even quite steep changes in the fresh water flow cause only small changes in the discharge flow. For example the following operating situations may occur:

- The paper machine runs without any disturbances and so do the mechanical pulp mill and the pulper. The level in the broke tower goes down, and the level in the white water tower goes up.
- There is a web break on the paper machine. The level in the broke tower goes up, and the level in the white water tower goes down.
- There is a disturbance in the groundwood mill. The level in the groundwood tower and in the broke tower goes down, the level in the white water tower goes up fast.

If the process flows, and the durations of various run situations and their frequencies are known, the towers can be designed to prevent overflows and so that there will be no shortage of any component. The operating strategy of the system can in some cases be changed during operation. If, for example, the mechanical pulp chest is full and the broke chest filling up, and the supply of white water therefore limited, water can be saved by interrupting the pulping process, allowing the level in the chemical pulp tower to go down.

F. PROCESS MANAGEMENT

The connections in a stock and white water system can be drawn in the form of a block diagram, as shown in Figs. 64 and 65. These block diagrams, however, do not show the internal pipelines in subprocesses or the process instrumentation, control points and conditions for achieving an equilibrium. These aspects can be illustrated with the aid of a flowsheet. A flowsheet showing the entire papermaking process is a very large and complicated document, which is difficult to use when striving to explain the operating principles of the system and when looking for alternative designs. For this reason, flowsheets of selected subprocesses are used together with a block diagram of the total system when planning the stock and white water system. An example of a subprocess flowsheet is shown in Fig. 59. It shows the short circulation of a newsprint machine.

When working out a subprocess flowsheet, the process control modes must be considered. The process control points, i.e., the points which allow the process to be manipulated and controlled while in operation so as to influence the state of the process and the quality of the end product, have to be selected and instrumented so that the settings can be repeated. Otherwise, the process should be self-balancing so that the process operator, apart from setting control parameters, does not have to make additional adjustments. A self-balancing process can be achieved by using proper process design or suitable control equipment.

The control points of the paper machine short circulation shown in the example of Fig. 59 are the position of the basis weight valve, the headbox slice opening and the fan pump rotating speed. The position of the basis weight valve influences the fiber flow into the headbox. The rotating speed of the fan pump influences the volume flow into the headbox and thereby the headbox dilution. If the headbox slice opening is not changed, a change in the fan pump rotating speed also influences the headbox outflow speed and thereby the speed differential between the wire and the headbox jet, which has an important effect on web formation. If the headbox slice opening is changed, the headbox jet flow is changed, but the headbox volume flow or headbox dilution does not change much.

These three process control points can be used to adjust the headbox stock flow, the headbox volume flow and the headbox jet speed. Otherwise the process is self-balancing, so the process operator need not make other adjustments, apart from those of the process control points. It is to be noted that other approaches to headbox control also exist.

The volume flow of each stage in a four-stage centricleaner plant must be at a certain level to ensure that the cleaners operate as they should. The volume flow of the first stage is always bigger than the volume flow into the headbox. The excess flow is returned via the air removal equipment overflow to the suction side of the feed pump in the first centricleaner stage.

Though the paper machine short circulation

shown in Fig. 59 is fairly complicated (with a four-stage centricleaner plant, air removal equipment and two-stage screening system between the fan pump and the headbox), its instrumentation is simple, its operation stable and it is easy to control. However, it should be noted in this context that a change in stock temperature causes a change in headbox operation, even though the control points are left untouched. A change in stock temperature causes a change in the air removal equipment vacuum to correspond to the water vapor pressure at the temperature in question. In other words, the pressure on the suction side of the fan pump changes, causing a change in the volume flow produced by the pump. A change in the volume flow into the headbox causes a change in the speed differential between wire and headbox jet, which affects web formation on the wire. This change is thus generated in spite of the fact that the control points of the system have not been touched.

REFERENCES

1. Norman, B. and Tegengren, A., Mixing of Thick Stock and White Water, XXIII EUCEPA Conf. Proc., Vol. I, "The Paper and Board Machine — Today and Tomorrow", May 31-June 3, 1988, Harrogate, U.K., p. 51-56.
2. Cutshall, K.A., Ilott, G.E. and Brooks, B.W., Causes of MD Basis Weight Variation, Pulp Paper Can. 80(6):T193-196 (1979).
3. Pantaleo, S., Stock Piping Principles and Their Effect on Paper Machine Performance, Tappi J. 70(2):81-86 (1987).
4. Reijonen, Y., Optimising and Short Circulation, XXIII EUCEPA Conf. Proc., Vol. I, "The Paper and Board Machine — Today and Tomorrow", May 31-June 3, 1988, Harrogate, U.K., p. 11-33.
5. Doucette, J.A., White Water and Savealls, TAPPI Stock Preparation Short Course, April 13-15, 1988, Chicago, IL, p. 107-116.
6. Paulapuro, H., Valtonen, E., Nyblom, I., Maijala, A. and Ahnger, A., Effect of Headbox Temperature on Energy Economy, Process Operation and Paper Quality in the Manufacture of Wood-Containing Paper Grades, EUCEPA Symp. '87, "Production and Utilization of Energy in the Pulp and Paper Industry", Lisbon, Portugal, Oct. 18-22, 1987.
7. Aaltonen, P., Nyyssölä, M. and von Troil, S., Design Principles of White Water Systems with Special Reference to Effluent Control, XV EUCEPA Conf. Proc., "Harmonizing Pulp and Paper Industry with Environment", May 7-11, 1973, Rome, Italy, p. 336-349.
8. Wearing, J.T., Huang, S., Piggott, A., Ouchi, M.D. and Wong, A., New White-Water Management Strategies for Integrated Newsprint Mills, Pulp Paper Can. 86(5):T139-145 (1985).
9. Ryti, N., Aaltonen, P. and Arjas, A., Utjämnandet av processvariationer med hjälp av massakypar, Norsk Skogind. 20(10):381-389 (1966).

V

ROLL HEADBOXES AND THEIR APPROACH FLOW

D.P. NEWCOMBE

Doug Newcombe's career spans more than 50 years in the paper industry, beginning in 1936 as Plant Engineer for Crocker Burbank and Associates in Fitchburg, MA, for 18 years. This was followed by over seven years with the Orton Corp., also in Fitchburg, and then with Valley Iron Works in Appleton, WI, which is now Voith, Inc., for a period of 17 years. He started Consultants, Inc. in 1980 and has travelled to many parts of the world as a wet end and headbox consultant and has participated in the TAPPI Wet End Operations Seminar since its concept over 15 years ago.

REVIEWED BY

Borje Wahlstrom
President
Borje Wahlstrom, Inc.
Rosemont, PA

A. INTRODUCTION

The headbox is the heart of the paper machine. Its sole purpose is to transform the flow of stock and water from a round pipe to a thin homogeneous flow across the full width of the paper machine but, in doing so, the fiber suspension needs to be without flocculation and streak free. How well this is accomplished depends on the design of the flow approach system and the headbox.

While the headbox does not form the sheet, it has a very significant effect on formation. If the flow from the headbox is not right, or nearly so, there is very little that can be done to correct it on the forming table. In all too many cases great attention has been paid, and continues to be paid, to the forming table with the sole intention of correcting delivery errors from the headbox. Special forming boards, horizontal shear developing units, "stock jumping" elements, formation showers, etc., are being added to existing machines in an attempt to reduce flocculation and streaking problems coming from the headbox. If the delivery from the headbox is right, then the forming table can be used for forming and water removal only. All too much of the forming table area is wasted from its real purpose because of headbox "problems".

B. FLOW APPROACH SYSTEM

To cover the entire headbox operational area, one must start with the flow approach system, starting with the drainage elements in the fourdrinier forming table. Cascading of water entrains air that will cause a foam and dirt buildup problem throughout the system. There is no system on the forming table that can eliminate cascading from the various dewatering elements, table rolls, deflectors or foils, into the collecting trays under the wire or fabric. If the final collecting device is the wire pit with a series of midfeathers directing the flow from the wet end toward the dry end on the outside areas of the pit and then back up the center section to a collecting silo under the breast roll, entrained air will be liberated. However, in some cases, the white water is allowed to cascade into the collecting silo, thereby re-entraining the air and losing most of the benefits of the wire pit collecting area. The collecting silo should be kept full at all times with no cascading from the collecting pit (see Fig. 68).

In other cases, a silo is located on the drive side of the fourdrinier with the collecting trays extended from the fourdrinier to the silo. Here again there should be no cascading from the trays to the silo. A level control device should be

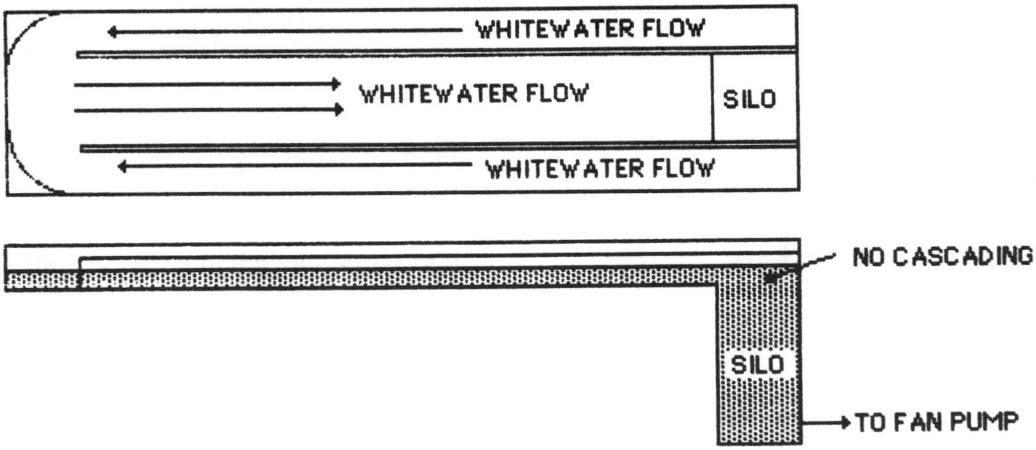

Fig. 68. Wire pit and silo.

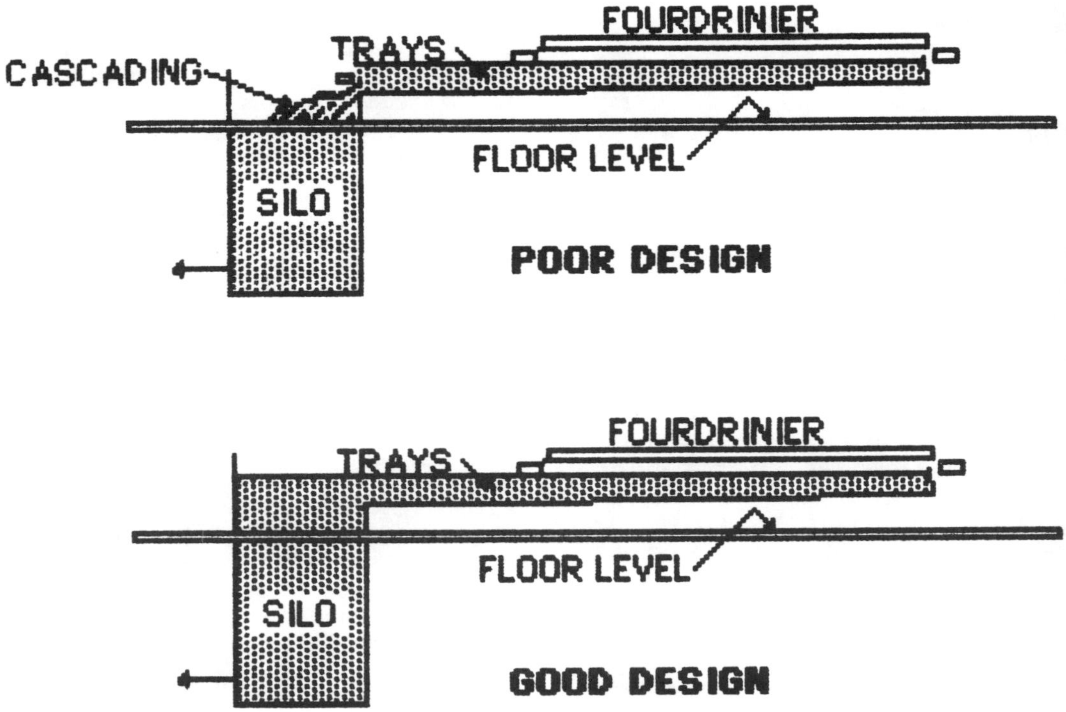

Fig. 69. External silo design.

used to keep the silo full at all times to eliminate the possible cascading (see Fig. 69).

The silo should be sized so that the vertical velocity of the downflow is 0.15 m/s or less to allow any remaining entrained air to be liberated.

Beloit's Stock Piping Recommendations [1] can still be used as a standard of correct piping methods throughout the entire flow approach system. Much of that manual is quoted in the following paragraphs.

All piping should be pitched upward in the direction of flow, if possible, one and one half inches to the foot (15 mm/305 mm) to prevent entrained air from building up on the inside top surface of the pipe resulting in dirt and/or slime

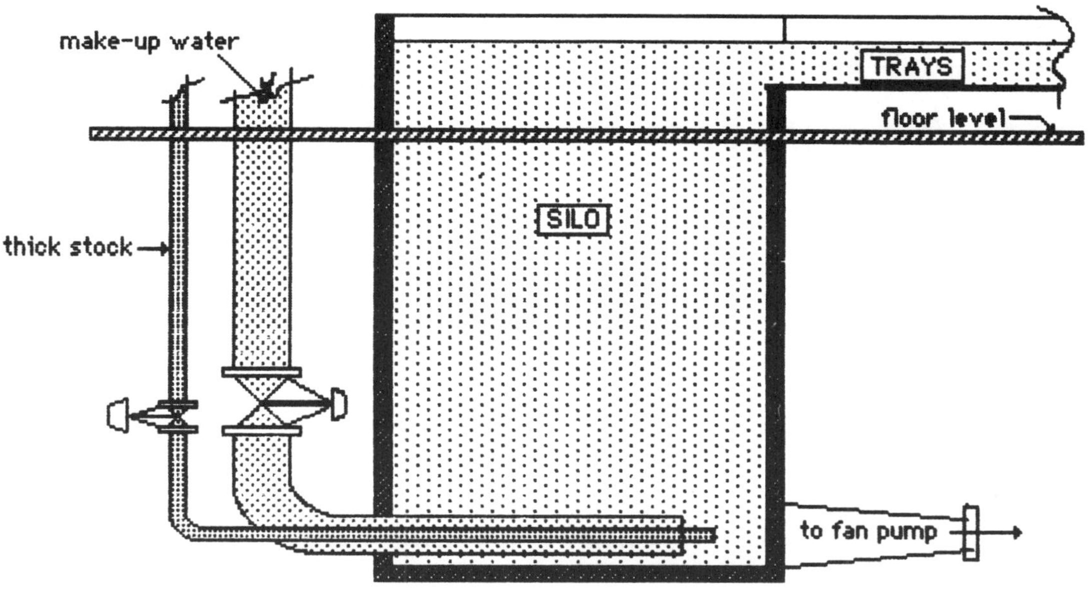

Fig. 70. Silo design.

build-up. If it is not physically possible to pitch the line as stated above, then the velocity should be at least 2.13 m/s.

The ideal stock flow velocity in all piping is between 2.13 m/s and 4.27 m/s. Below 2.13 m/s there is a tendency for entrained air to be liberated and literally cling to the top inside surface of the pipeline if the line is flat with no upward pitch. Velocities above the 4.27 m/s level are not necessarily detrimental to the flow performance if all other considerations are followed but the higher velocities create a high load on the fan pump and other related equipment in the system.

The introduction of thick stock and makeup water into the suction of the fan pump can be critical. Both should be introduced through the wall of the collecting silo and headed toward the eye of the fan pump, as shown in Fig. 70.

In an earlier paper [1], it was stated that both of the feeding lines should enter the connection feeding the fan pump. If this is done there is a danger of air-binding the pump on startup. Both the thick stock line and the makeup water line are empty before startup and the sudden downward surge of fluid will force air along with it to the eye of the pump causing an air bind. The two pipes should terminate at some distance before the pump connection so that the air will evacuate upward and out before entering the fan pump.

All control valves must be placed below the normal level of the liquid in the collecting silo to prevent cascading.

The best system incorporates a variable speed fan pump, automatically controlled, rather than bypass lines around the pump or control valves in the line. A variable speed fan pump will pay for itself in power savings alone as no more power is required than what is needed at the flow rate of the grade being made. A bypass line from the discharge to the suction of the fan pump for coarse control of the flow will create a condition of almost full horsepower requirement at all times regardless of the final flow. In some cases the maximum flow required for a certain grade is also the total capacity of the pump and the bypass line is closed off. This can, and most times does, create a dirt and dewatered stock pocket at both ends of the bypass that may cause problems, especially if there are no cleaners and screens in the system.

There must not be any gaskets in the flanges of the line, especially between the final screen and the headbox manifold. Gaskets will create strings from stock hang-ups. All flanges should be of the metal to metal type with ground and polished welded joints. And, for that matter, all welded joints in the total pipe line must be ground and polished for smoothness to avoid possible fiber hang-ups and consequent strings.

Screens should be located in the basement under the machine so as to keep a positive head on the screen and an even flow to the headbox manifold. If physical space permits, there should

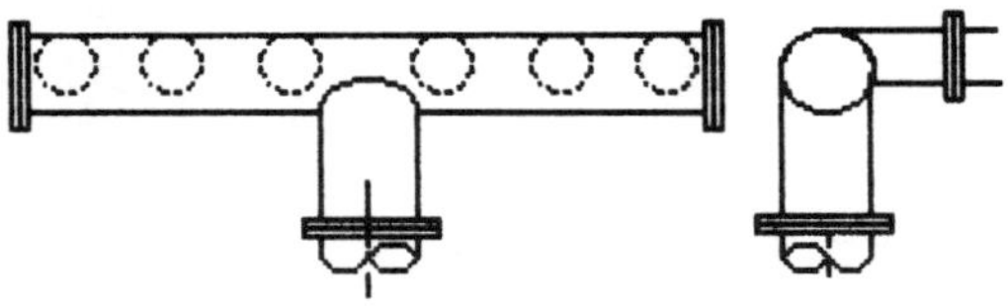

Fig. 71. Straight manifold pipe.

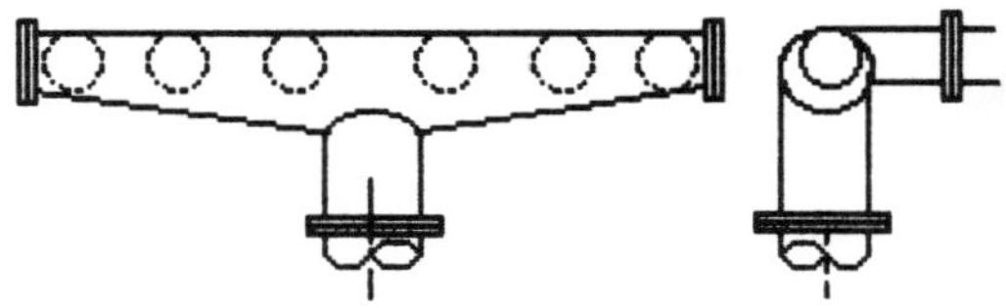

Fig. 72. Tapered manifold pipe.

be a straight length of pipe equal to eight diameters between the last elbow and the manifold transition section. This will reblend laminar seperation before the stock and water slurry enters the manifold. An elbow reacts like a cleaner in that the centrifugal action of the turn throws the heaviest part of the mixture to the outside of the elbow – like clay or titanium dioxide. Unless this is reblended, there could be a difference of filler content from one side of the machine to the other – the side nearest to the elbow having the greatest amount.

If it is not physically possible to have a straight length of pipe between the last elbow and the manifold a hydraulic, or venturi type, elbow must be used to reduce the laminar separation [2].

Much has been said about the value of hydraulic elbows throughout the entire system between the screen(s) and the headbox but, usually, there are straight lengths of pipe equal to more than the required number of diameters between fittings in the feeding line and therefore hydraulic elbows are not really needed at those points.

C. MANIFOLD DESIGN

The first step of the process of transforming the flow from a round pipe to the thin homogeneous flow from the slice the stock and water mixture enters a manifold that, in turn, feeds the headbox. Not much attention was paid to the manifold in past years. The flows were not smoothed out adequately and the older headboxes were designed to take care of this problem. In the early years of headbox development, the entrance manifold usually consisted of a straight length of pipe with a multiplicity of large connections from it to the headbox as in Fig. 71.

This design did not work as the pipes closest to the upcoming connection received most of the flow and the end connections were, usually, "starved". To compensate for this, a series of adjustable gates were installed inside of the headbox at each connection to even out the flow across the machine. It was difficult to make the required adjustments as each had to be changed if there was the slightest change in total flow.

The next design change was tapering the main feed line both ways from the center as in Fig. 72. This was an improvement but there were still problems of unequal flow across the machine.

D. HEADBOX DESIGN

Because of the many problems, the headboxes were made with a series of baffles and distributor rolls to even out the flow at the slice. The most well known design was the Valley three-pass headbox shown in Fig. 73.

This type of headbox is still in use in many mills around the world today. It is most important, if this type is to provide as good a flow as possible, that a differential head be used across the main baffle to create a definite velocity increase under the baffle to help break up flocculated lumps and even out the flow across the machine.

Because of the high volumetric content of this type of unit, flocculation became a serious problem, especially as the machine speed increased, and an additional adjustable auxiliary slice was installed in some units just before the slice distributor roll, again to provide a velocity increase.

Many different types of flow spreaders were tried to even out the flow but none of them really corrected the problem.

As speeds increased, the depth of the pond within the headbox increased to provide the correct jet velocity until headboxes became totally uncontrollable and the air-pad system was designed. The liquid level was kept to just below the top of the slice distributor roll and the difference in the liquid level and the required head was made up by internal air pressure. This was a major step in the reduction of flocculation but the volumetric content was still too large for proper operation.

That a properly designed entrance manifold was required for the elimination of the baffles and other devices to even out the flow was realized only recently.

In the early Sixties, a major study was made in the research facilities of West Virginia Pulp and Paper Co. in Covington, VA, now called Westvaco. The final result of these studies and

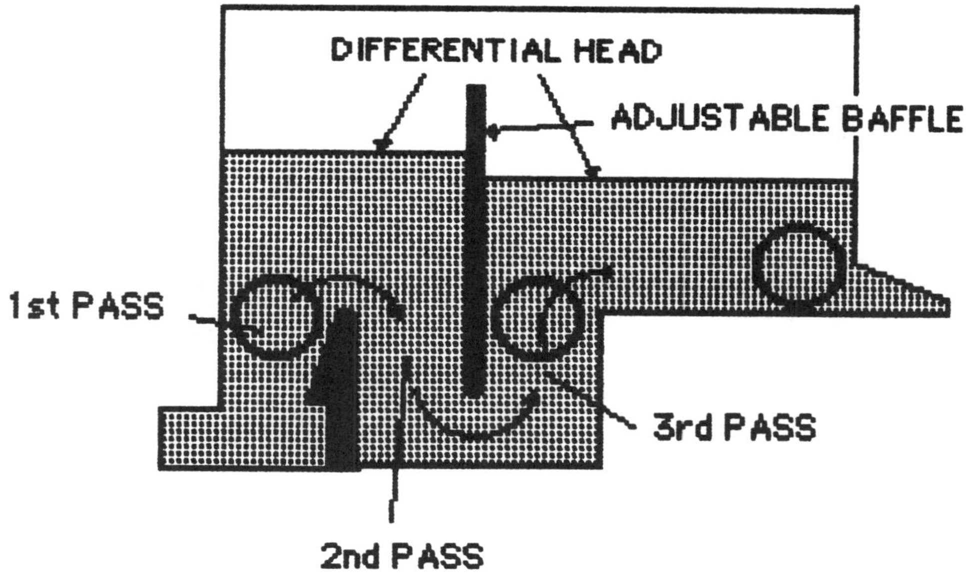

Fig. 73. Valley three-pass headbox.

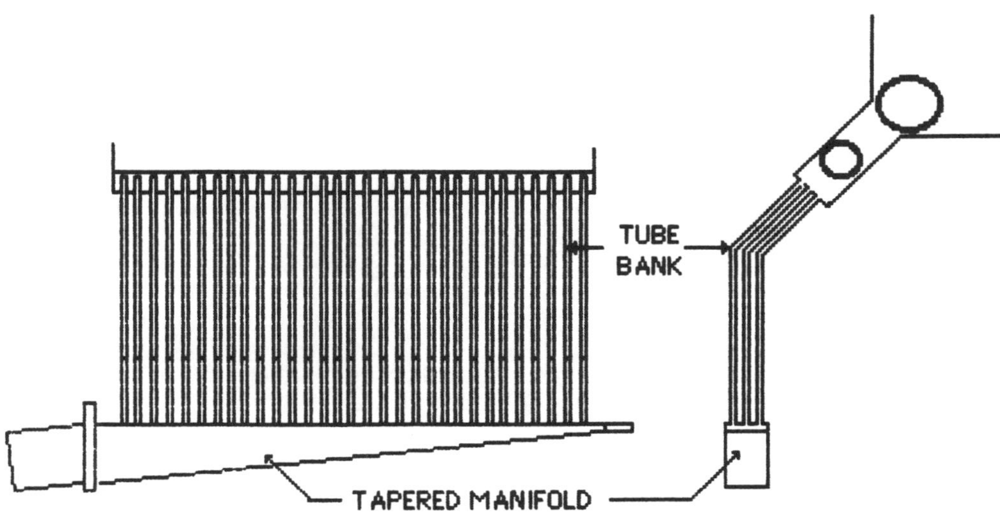

Fig. 74. Multiplex system.

experiments was the Multiplex flow approach system consisting if a tapered manifold, tapering from one side of the machine to the other, usually from the drive side to the tending side, and a multiplicity of long, small-diameter tubes feeding the headbox [3]. Others were working on the tapered manifold system at the time, Jasper Mardon and KMW in particular, but the entire system was patented by Westvaco. The final design was as shown in Fig. 74.

All manifolds today are designed with a taper so that the cross-sectional area decreases in about the same ratio as the flow to the headbox as it travels across the full machine width, plus about 10% for recirculation purposes to balance out the cross-machine pressure within the manifold. Most manifolds are rectangular in shape but there are a few round types, both serving the same purpose. However, in both cases, round or rectangular, the cross-machine direction section has a linearily decreasing cross-sectional area to compensate for the

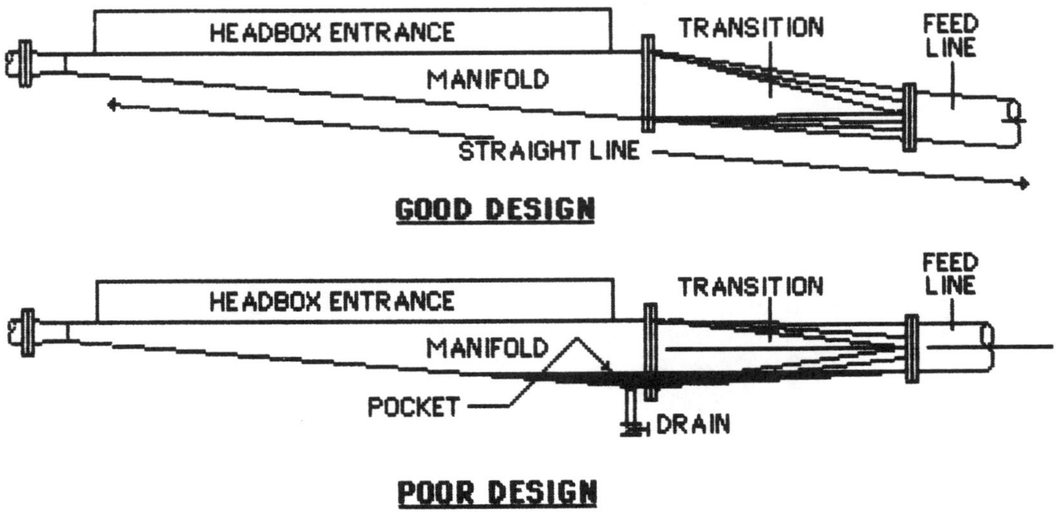

Fig. 75. Manifold system.

frictional loss of the flow.

The recirculation line should be piped to either the silo or the wire pit. A seperate pump in this line should never be used as it could cause an unbalanced condition of the manifold or, in rare cases, the fan pump. There is enough pressure in the manifold to "send" the recirculated stock to just about anywhere in the system.

The long, small-diameter tubes create resistance to provide a pressure drop which, in turn, helps to balance the manifold with equal pressure from one end to the other ensuring an even flow fully across the machine.

It seems logical that the best manifold be rectangular in shape with the sides of the unit parallel to each other, and the bottom, or the surface directly opposite the headbox entrance unit, tapering as the flow is reduced. This puts all of the force of the taper compression directly opposite the tube bank entrance. The rectangular construction is also substantially less expensive to manufacture.

The bottom of the feed line, the round to rectangular transition, and the manifold section should be in a straight line to avoid uncontrollable turbulence (see Fig. 75).

In addition to uncontrollable turbulence caused by the surfaces not being in line, a pocket is formed at the bottom where the manifold and the transition sections join. This pocket is drained at each shutdown but, during operation of the machine, the drain line can fill with dewatered stock and release from time to time resulting in occasional hard stock lumps that will cause a break in the sheet. If such a design is used, the drain line must be left open, at least slightly, to avoid the stock settling out and dewatering. Having the bottom surfaces of the three components in line eliminates the need for a drain at that point.

Balancing of the manifold is important to the operation of the headbox and many devices have been made for that purpose. Sophisticated controls consisting of differential pressure (d/P) cells automatically setting the recirculation valve have been used. Sight glass systems of one kind or another have been used and in a few rare cases a valving system has been incorporated. Figure 76 illustrates the d/P cell system.

In this system, the controller automatically changes the recirculation valve to correct unequal pressures from one end to the other in the manifold.

The sight glass system shown in Fig. 77 is simple. If there is more pressure at the recirculation end of the manifold, the observed flow in the sight glass will be toward the entrance end and the valve needs to be opened slightly—and closed off slightly if the flow is in the other direction. With this system there must be a flow from the pipes feeding the sight glass at all times, except when taking a pressure reading, to prevent the lines from plugging with dewatered stock.

The theory of the straight pipe system shown in Fig. 78 is that the manifold will automatically balance itself without resetting the recirculation valve. If there is more pressure at one end of the manifold than at the other, more flow will occur in that end of the connecting pipe and help to balance the manifold. The constant bleed line at the center, and it must be at the exact center, is

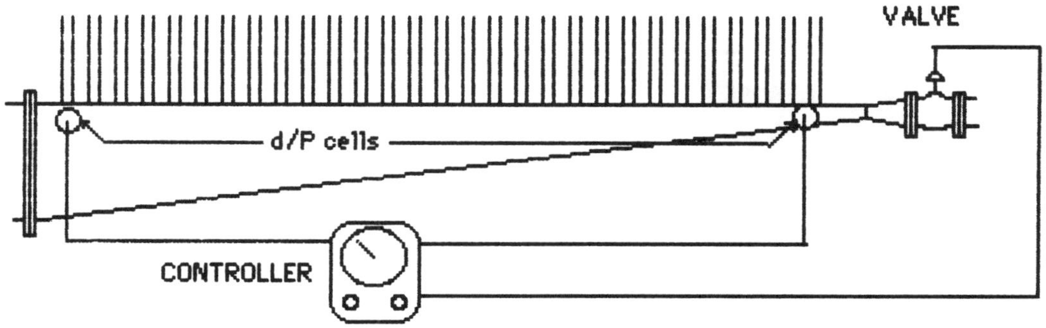

Fig. 76. Differential pressure (d/P) cell system.

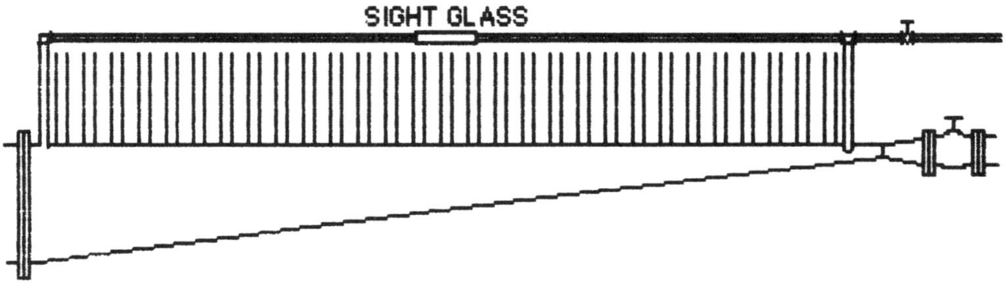

Fig. 77. Sight glass system.

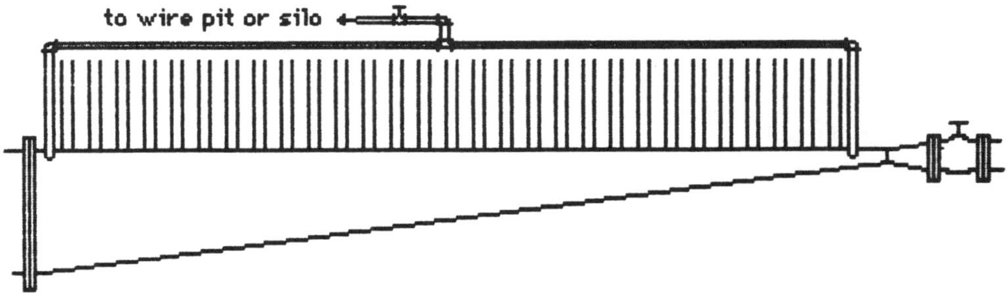

Fig. 78. Straight pipe system.

flow controlled by a valve set in the line. This system, while workable to some small degree, will take care of quite small pressure differences only. It is not recommended.

The simplest way of observing the manifold balance is to observe the profile readout on the computer usually found on most paper machines today. If the profile "curve" shows a downhill slope from front to back, it means that the pressure at the front side of the manifold is higher than at the back and the recirculation valve needs to be adjusted accordingly. The most important part of the operation in the end is the delivery from the slice.

The original concept of the tube bank was to provide tubes long enough to create a pressure drop to assist in the balance of the manifold. The original tubes were 3 m long, or longer. Throughout the years it was found that tubes of much shorter length could be used if the velocity was increased accordingly. This, perhaps, led to the use of a drilled block in place of the tube bank. Today, nearly all headboxes, both the distributor roll and hydraulic types, use the basic tapered manifold with a tube bank or a drilled block.

The drilled block, approximately 150 mm long, depending on the builder, does not seem to

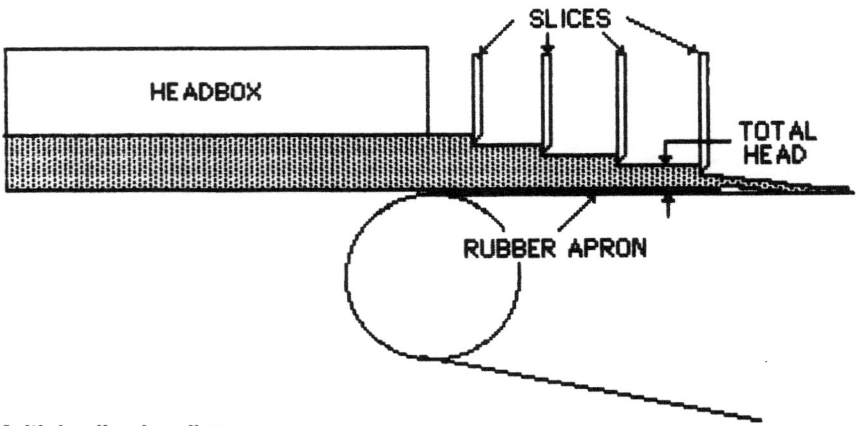

Fig. 79. Multiple slice headbox.

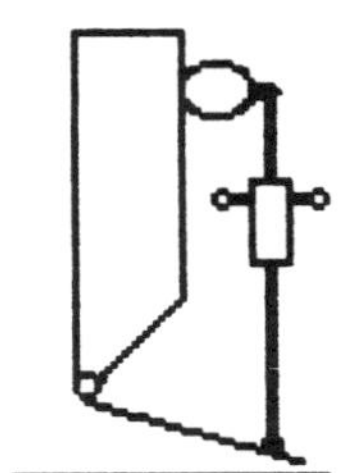

Fig. 80. Converging nozzle with adjusting screws.

be able to cover as wide a range of flows as the tube concept but it has one advantage. If the volumetric through-put of the headbox is increased beyond the velocity design it can be easily replaced with one having larger holes. The tube bank concept is not as easily replaced but it can be done at the expense of considerable downtime.

E. SLICE EVOLUTION

The headboxes of many years ago, when the art of papermaking was simple, were open with a mulitiplicity of slices for basis weight control. Most incorporated rubber aprons out over the breast roll and onto the beginning of the forming table with the slices set, as shown in Fig. 79.

The main purpose of the slices was to create turbulence with a differential head across each slice. The turbulence factor disbursed the fibers to avoid flocculation but there was little or no actual profile control. The slices were fixed at the ends but could be raised or lowered at the center points slightly for a sort of profile control. The head before the last slice body tied in to the speed of the wire. Obviously the speed was limited to physical dimensions of the headbox.

As speeds increased, the under slice turbulence increased and fewer slices were required and, as each slice was removed, the rubber apron was also shortened to suit. All formation at this time was definitely velocity forming on top of the apron.

In the late Twenties, the converging nozzle with profile adjusting screws on approximately 150 mm centers, as shown in Fig. 80, was introduced and became, for the first time, a complete tool for the papermaker [4].

The first converging nozzles were mounted on a cast iron body and the assembly used on wood construction headboxes and proved to be quite successful for the time. Velocity forming was still being practiced.

As time progressed and speeds increased, it became impossible to drain the required water on the table with table rolls and even with deflectors between the rolls. In order to drain the forming sheet it became necessary to pressure form and literally blow the water through the wire as soon as possible—even before the forming board. The forming board soon became a water removal unit and not a forming unit. The difference between velocity and pressure forming is shown in Fig. 81.

Wood, usually cypress, gave way to stainless steel and headboxes became much more durable and, with the new slice, much more manageable but, in order to control the amount of drainage, it became important to have the nozzle adjustable, in and out, for jet impingement. It was still pressure forming of one magnitude or another, however.

Foils were eventually introduced and water removal became easier and more controllable but the habit of pressure forming continued with poor forming results. The trend today is toward a

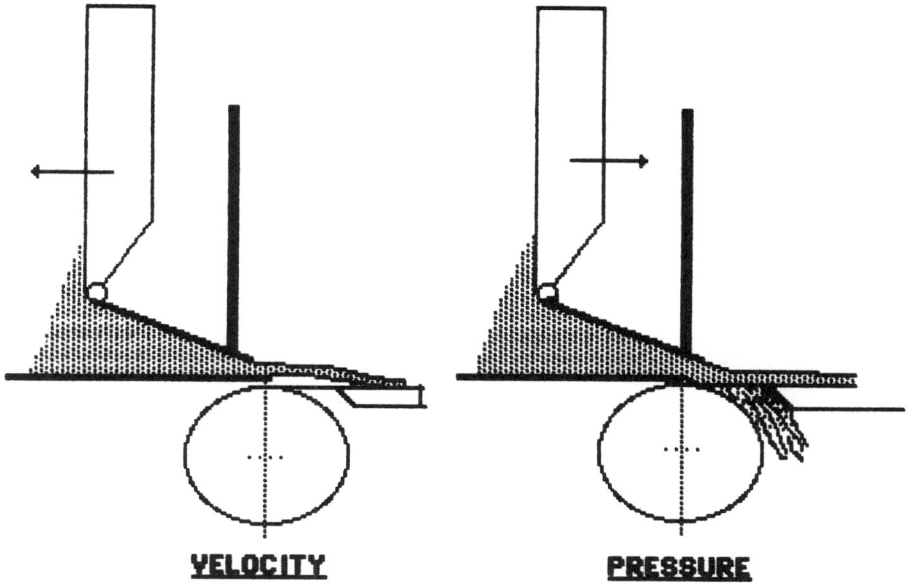

Fig. 81. Velocity vs. pressure forming.

$$d = \frac{K \times H_V^{4/3} \times H_D^{1/2}}{R_D^{1/3}}$$

WHERE:

- d = NO-WAKE NOZZLE LENGTH
- K = FACTOR FROM TABLE
- H_V = HOLE VELOCITY
- H_D = HOLE DIAMETER
- R_D = ROLL DIAMETER

Fig. 82. No-wake formula.

much wider first blade in the forming board and full velocity forming.

It became apparent that, as the top nozzle was moved upstream for jet impingement, streaking became more of a problem. It may have been this fact that led to the study of nozzle lengths and the effect of the drilling pattern of the slice distributor roll on the flow characteristics coming from the slice. Such a study was made by Mardon et al. [5], who developed a formula for determining the "no-wake" length of nozzles on a distributor roll type of headbox. The formula is shown in Fig. 82.

To explain the formula, the K factor is taken from a table in reference [5] and is found by using the percentage of open area of the roll and the diameter of the drilled holes. Generally speaking, the K is 28 for 25 mm diameter holes, 27 for 22 mm holes and 26 for 19 mm diameter holes but only when the open area is 50%. If the open area is smaller than 50%, K changes dramatically. For example, if the roll has an open area of only 40%, K changes to 48, with 25 mm diameter holes making a considerable difference in the required "no-wake" distance.

The hole velocity is a function of the velocity in the headbox and the open area. If, for example, the headbox velocity is 0.50 m/s and the open area is 50%, then the hole velocity will be 1 m/s; with 0.50 m/s velocity and only a 40% open area, then the hole velocity will be 1.15 m/s. If a headbox is designed with a nozzle matching the "no-wake" distance, then the flow will leave the nozzle with little or no internal action and flocculation could be a serious problem. In fact, flocculation will more than likely have started within the confines of the nozzle before discharging out onto the forming table. On the other hand, if the nozzle is too short then the pattern of the drilling of the distributor roll will show up as severe streaks directly in line with the drilled holes. The formula as shown does indeed give the actual nozzle length required to totally dissipate the generated eddy currents from the slice distributor roll, but it must be used with caution. Reference [5] clearly states that there must be a "no-wake" *ratio* between what the formula gives as the "no-wake" distance and the actual distance as measured from the face of the roll to the jet discharge point.

Distributor roll headboxes have been badly maligned over the years as bad "streak makers",

Table 8. Speed — head — pressure table.

Wire speed m/min	Velocity m/s	Theoretical hd. in m	Theoretical hd. in cm	Pressure kg/cm²
15	0.25	0.00	0.32	0.01
30	0.50	0.01	1.27	0.05
45	0.75	0.03	2.87	0.12
60	1.00	0.05	5.10	0.21
75	1.25	0.08	7.97	0.33
90	1.50	0.11	11.47	0.47
105	1.75	0.16	15.62	0.64
120	2.00	0.20	20.40	0.84
135	2.25	0.26	25.81	1.06
150	2.50	0.32	31.87	1.31
165	2.75	0.39	38.56	1.59
180	3.00	0.46	45.89	1.89
195	3.25	0.54	53.86	2.22
210	3.50	0.62	62.47	2.58
225	3.75	0.72	71.71	2.96
240	4.00	0.82	81.59	3.36
255	4.25	0.92	92.10	3.80
270	4.50	1.03	103.26	4.26
285	4.75	1.15	115.05	4.74
300	5.00	1.27	127.48	5.26
315	5.25	1.41	140.55	5.79
330	5.50	1.54	154.25	6.36
345	5.75	1.69	168.59	6.95
360	6.00	1.84	183.57	7.57
375	6.25	1.99	199.19	8.21
390	6.50	2.15	215.44	8.88
405	6.75	2.32	232.33	9.58
420	7.00	2.50	249.86	10.30
435	7.25	2.68	260.03	11.05
450	7.50	2.87	286.83	11.83
465	7.75	3.06	306.27	12.63
480	8.00	3.26	326.35	13.46
495	8.25	3.47	347.06	14.31
510	8.50	3.68	368.42	15.19
525	8.75	3.90	390.41	16.10
540	9.00	4.13	413.04	17.03
555	9.25	4.36	436.30	17.99
570	9.50	4.60	460.20	18.97
585	9.75	4.85	484.74	19.99
600	10.00	5.10	509.92	21.02
615	10.25	5.36	535.73	22.09
630	10.50	5.62	562.19	23.18
645	10.75	5.89	589.28	24.30
660	11.00	6.17	617.00	25.44
675	11.25	6.45	645.37	26.61
690	11.50	6.74	674.37	27.80
705	11.75	7.04	704.01	29.03
720	12.00	7.34	734.28	30.27
735	12.25	7.65	765.20	31.55
750	12.50	7.97	796.75	32.85
765	12.75	8.29	828.94	34.18
780	13.00	8.62	861.76	35.53
795	13.25	8.95	895.23	36.91
800	13.50	9.07	906.52	37.38
825	13.75	9.64	964.07	39.75
840	14.00	9.99	999.44	41.21
855	14.25	10.35	1035.46	42.69
870	14.50	10.72	1072.11	44.20
885	14.75	11.09	1109.39	45.74
900	15.00	11.47	1147.32	47.30
915	15.25	11.86	1185.88	48.89
930	15.50	12.25	1225.08	50.51

Table 8. Continued.

Wire speed m/min	Velocity m/s	Theoretical hd. in m	Theoretical hd. in cm	Pressure kg/cm^2
945	15.75	12.65	1264.92	52.15
960	16.00	13.05	1305.40	53.82
975	16.25	13.47	1346.51	55.52
990	16.50	13.88	1388.26	57.24
1005	16.75	14.31	1430.64	58.98
1020	17.00	14.74	1473.67	60.76
1035	17.25	15.17	1517.33	62.56
1050	17.50	15.62	1561.63	64.39
1065	17.75	16.07	1606.57	66.24
1080	18.00	16.52	1652.14	68.12
1095	18.25	16.98	1698.35	70.02
1110	18.50	17.45	1745.20	71.95
1125	18.75	17.93	1792.69	73.91
1140	19.00	18.41	1840.81	75.90
1155	19.25	18.90	1889.57	77.91
1170	19.50	19.39	1938.97	79.94
1185	19.75	19.89	1989.01	82.01
1200	20.00	20.40	2039.68	84.09
1215	20.25	20.91	2090.99	86.21
1230	20.50	21.43	2142.94	88.35
1245	20.75	21.96	2195.52	90.52
1260	21.00	22.49	2248.75	92.71
1275	21.25	23.03	2302.61	94.94
1290	21.50	23.57	2357.11	97.18
1305	21.75	24.12	2412.24	99.46
1320	22.00	24.68	2468.01	101.75
1335	22.25	25.24	2524.42	104.08
1350	22.50	25.81	2581.47	106.43
1365	22.75	26.39	2639.15	108.81
1380	23.00	26.97	2697.48	111.22
1395	23.25	27.56	2756.44	113.65
1410	23.50	28.16	2816.03	116.10
1425	23.75	28.76	2876.27	118.59
1440	24.00	29.37	2937.14	121.10
1455	24.25	29.99	2998.65	123.63
1470	24.50	30.61	3060.79	126.19
1485	24.75	31.24	3123.58	128.78
1500	25.00	31.87	3187.00	131.40
1515	25.25	32.51	3251.06	134.04
1530	25.50	33.16	3315.75	136.71
1545	25.75	33.81	3381.09	139.40
1560	26.00	34.47	3447.06	142.12
1575	26.25	35.14	3513.67	144.87
1590	26.50	35.81	3580.91	147.64
1605	26.75	36.49	3648.80	150.44
1620	27.00	37.17	3717.32	153.26
1635	27.25	37.86	3786.47	156.11
1650	27.50	38.56	3856.27	158.99
1665	27.75	39.27	3926.70	161.90
1680	28.00	39.98	3997.77	164.83
1695	28.25	40.69	4069.48	167.78
1710	28.50	41.42	4141.83	170.77
1725	28.75	42.15	4214.81	173.77
1740	29.00	42.88	4288.43	176.81
1755	29.25	43.63	4362.68	179.87
1770	29.50	44.38	4437.58	182.96
1785	29.75	45.13	4513.11	186.07
1800	30.00	45.89	4589.28	189.21
1815	30.25	46.66	4666.09	192.38
1830	30.50	47.44	4743.53	195.57
1845	30.75	48.22	4821.61	198.79

Table 8. Continued.

Wire speed m/min	Velocity m/s	Theoretical hd. in m	Theoretical hd. in cm	Pressure kg/cm^2
1860	31.00	49.00	4900.33	202.04
1875	31.25	49.80	4979.69	205.31
1890	31.50	50.60	5059.68	208.61
1905	31.75	51.40	5140.31	211.93
1920	32.00	52.22	5221.58	215.28
1935	32.25	53.03	5303.49	218.66
1950	32.50	53.86	5386.03	222.06
1965	32.75	54.69	5469.21	225.49
1980	33.00	55.53	5553.03	228.95
1995	33.25	56.37	5637.48	232.43

and all too many of them are, but there is a definite reason for this. Original headboxs were designed for certain flows and worked quite well when new. Over the following years, the machines were speeded up and consistencies lowered for better formation until the headbox became badly overloaded. If, for an example, the original headbox was designed for a flow of 15 140 L/min at a speed of 230 m/min and a consistency of 1.0%, it probably operated quite well. Over the years the machine might have been modified and now operates at a speed of 460 m/min; this automatically increases the flow to about 30 280 L/min. If a better forming table with the latest water removal devices has been installed and the consistency has been lowered to 0.5%, the flow has doubled again to 60 560 L/min. Obviously, the headbox is badly overloaded and can result in nothing but streaking. Any headbox will operate extremely well at one point of flow and anything above or below that is a compromise. However, there is a limit of "compromise". The "ideal" headbox is one that has a nozzle that can be changed in length to suit increased, or decreased, flow conditions; this changes the "ideal" flow pattern to one that will bridge a substantial flow change.

Unfortunately, the overloading condition is also a function of internal pressure in an air-padded headbox. The required internal pressure in an air-loaded headbox operating at a speed of 230 m/min is 7.6 kPa but this pressure increases to 28.9 kPa for 460 m/min – four times as much. This may not seem like a high pressure but if it is applied to a headbox having a 2540 mm deckle and a 1524 mm high front plate, the load at 7.6 kPa on the front plate is 2858 kg; at a speed of 460 m/min the internal load becomes over 11 340 kg which can, and often does, blow seals and even the glass ports (see Table 8).

The high pressure applied to old headboxes often results in a problem of head fluctuation. The pressure builds up until the front plate deflects outward, breaking the seal and allowing the pressure to drop. When the pressure is relieved, the front plate seals again and the cycle continues, resulting in as much as three or four inches of head variance on a cyclical basis.

There can be another serious problem with an excessive internal pressure over and beyond the design of the headbox. It is possible, and has happened, that the apron board supporting structure will deflect causing the jet to land unevenly onto the forming table. The center of the landing jet will be upsteam of the edges, resulting in an uneven draining of the sheet as well as a difference in the formation. With the center of the landing jet "hitting" before the leading blade of the forming board and the edges full on the blade, a serious streaking condition could develop, especially in the center of the sheet as shown in Fig. 83.

The excessive internal pressure can also force the front plate to bow outward, further changing the jet landing point and making the condition even worse. This is illustrated in Fig. 84.

In some cases there is a "stop-gap" solution to the streaking problem. If, for example, the slice distributor roll is of an old design and has an open area of 40% with 25 mm diameter holes it can be changed out with a new roll having 19 mm diameter holes and with an open area of 50% making a considerable difference in the "no-wake" distance ratio. In most cases, a good ratio is between 1.75 and 1.25. More than 1.75 will result in a streaking condition and below the 1.25 ratio there will be flocculation. This, however, takes a study for full determination. The angle of the nozzle has a direct bearing on it and the position of the nozzle heel is also important. If the heel of the nozzle is at the center point of, and tight to, the slice roll, there is a different set of circumstances than when the heel is nearly at the top of the roll. Opening up the gap between the roll and the heel point also has a profound effect.

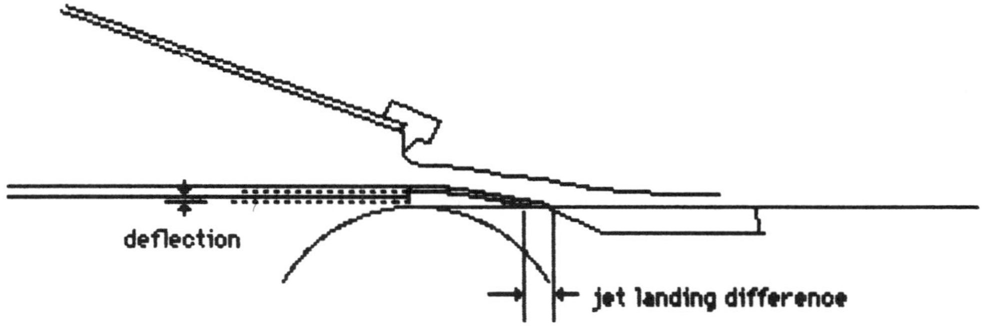

Fig. 83. Bottom apron blade deflection.

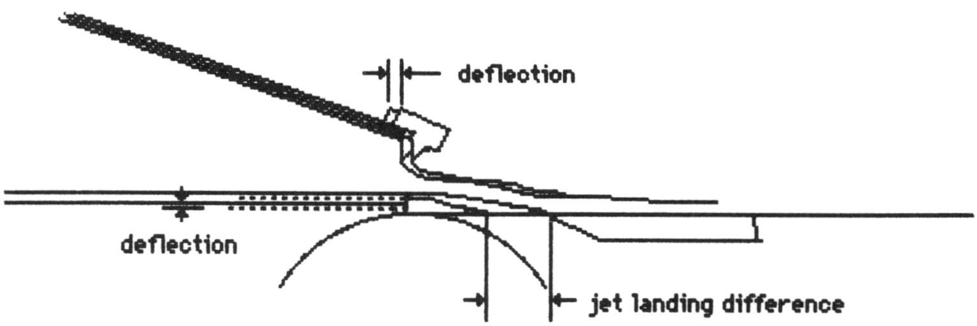

Fig. 84. Bottom and front plate deflection.

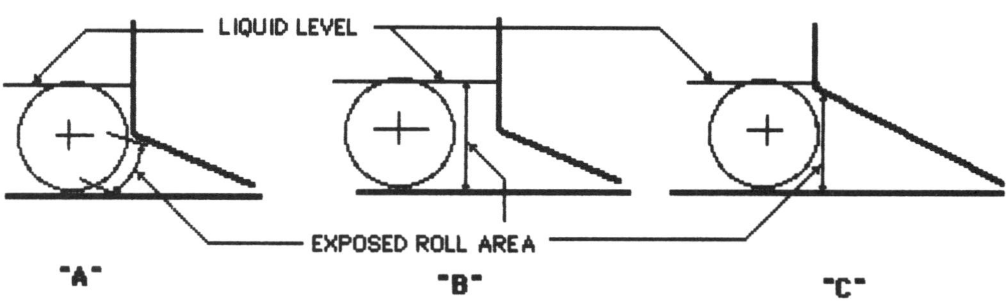

Fig. 85. Slice roll placement.

Figure 85 illustrates this point. In "A" with the nozzle heel point tight to and at the center line of the roll, the exposed area of the roll is approximately one half of the downstream side. This results in a much higher hole velocity than if the entire downstream side of the roll is exposed as in "B". In "C", nearly the same condition as in "B" exists. This illustrates another method of rectifying a short nozzle condition – the roll can be moved upstream, if there is such an adjustment, to further help in correcting a streaking condition. As long as the liquid level is below the top of the roll there will be no bypassing of undisturbed stock between the roll and the front plate.

There are cases where a headbox is undersize either because of over optimistic future expectations or a total misapplication. In cases of this kind, there is usually a condition of severe flocculation between the two rolls in the headbox because of the low velocities. This is not an unsurmountable problem as an auxiliary slice can sometimes be introduced between the two rolls, as shown in Fig. 86, to break up the flocculation.

It is important that there be a head differential before and after the auxiliary slice to create a high velocity under the slice for added turbulence to break up flocculation. Obviously the auxiliary slice needs to be vertically adjustable to

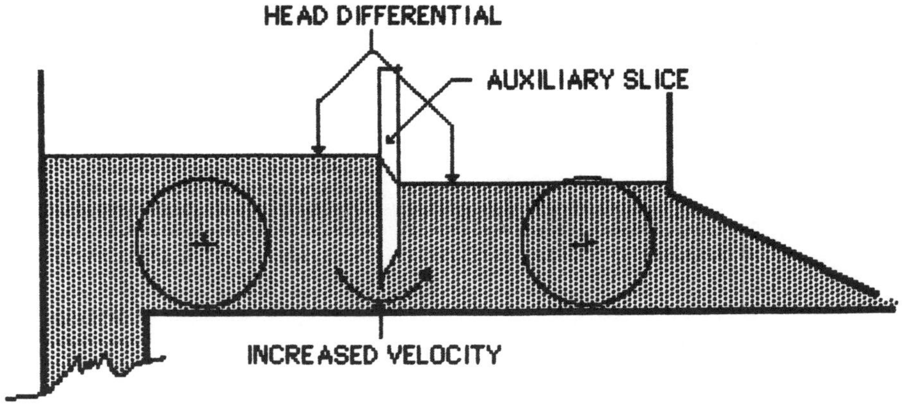

Fig. 86. Auxiliary slice.

accommodate various flows.

Care must be taken when operating with an auxiliary slice. Changing the total flow appreciably will change the head differential across the slice. If the flow is increased substantially without changing the opening under the auxiliary, slice can result in a "spilling" of stock and water over the top. This will create a cascading condition and an erratic flow to the slice. When making a flow change, the differential head must be checked and the auxiliary slice readjusted accordingly.

The velocity of the jet leaving the slice is important to formation as a "rush" or "drag" figure. Unfortunately there is no exact method of determining what the jet velocity actually is under all conditions of flow. In the first place, total head measuring devices usually are placed before the slice distributor roll and the resistance of the roll and the converging slice influence the total head readings. Consistencies and stock makeup also have an effect on the jet velocity. It is possible to measure the actual jet velocity with instrumentation set within the confines of the slice opening but such devices result in a disturbed flow at that point and end up as a sheet defect.

There has been an attempt by many to arrive at a formula for determining the true jet velocity. Some of these formulas are complicated by many variables: consistency, stock characteristics, temperature, through-put volume, slice lip configuration, etc., and, consequently, are not easily workable on the spot. However, a good papermaker can usually tell by observing the sheet and the action on the table as to whether the sheet is being dragged or rushed, or when the jet velocity is very close to the fabric speed. When this has been established, the rush/drag ratio can be determined.

As a starting point, Table 8 shows the theoretical heads and internal pressures required for different speeds in increments of 15 m/min. When using the table, remember that the figures are theoretical and not corrected for the variables listed above.

F. DISTRIBUTOR ROLLS

Originally, most distributor rolls were made of either thin wall tubing or rolled up sheets of metal, usually covered with rubber or some other corrosion-resistant material. Eventually this gave way to all stainless steel as the corrosion-resistant material had a tendency of "peeling" off causing rough areas where fiber collected, causing strings.

Some of the earlier rolls had internal fins spaced on about 150 mm centers to correct a skating action in the headbox from improperly designed manifold and flow entrance systems. This worked for the reduction of the skating action but, when such a roll was used in the slice position, it usually resulted in streaks on the table in a direct line with the internal fins. Finned rolls should not be used in the slice position, especially if the headbox is overloaded.

Many different roll designs are used, ranging from 50% open area down to 35% in some rare cases and with hold diameters from over 25 mm to 12 mm. The open area of the slice distributor roll should be 50% with holes of 19 mm to 25 mm in diameter. Holes over 25 mm in diameter are not effective and holes less than 19 mm can result in land areas so small that stapling usually results, especially on the thick wall (10 mm) rolls more commonly used today.

There is one possibility of headbox design that should be considered when designing a new headbox for a specific application. If the original slice distributor roll is one having a 50% open area and 25 mm diameter holes and the capacity of the headbox is exceeded at some time in the future, the roll can be changed out with one having a 50% open area and 22 mm or even 19 mm diameter holes for a greater working capacity. The "no-wake'" formula (Fig. 82) will result in a different nozzle length of substantial proportions when the different hole diameters are used.

The combination of a different roll drilling pattern and the ability of physically changing the nozzle length will, in most cases, "rejuvenate" an unworkable headbox when the flows are increased beyond the original design and make the unit workable for a longer period of time.

The other rolls in the headbox need to be something less than 50% open area to provide resistance for a better and more uniform flow through the headbox.

The rotational speed and direction of the rolls before the slice position does not seem to be important, as long as they turn to keep clean. The slice roll speed and rotation, however, can have an effect on formation. A "rule of thumb" concerning the rotational speed is to have the surface speed of the roll the same as the stock velocity in the headbox. This is only a starting point, however. There can be times when a higher or lower speed will be of benefit. It is for this reason that most slice distributor rolls have variable speed drives and should be experimented with for the best formation.

The direction of rotation is another optional subject. There are two schools of thought concerning the rotation, both quite valid. One school of thought is that the roll should turn so that the bottom of the roll is turning against the flow to create a turbulence between the roll and the bottom surface of the headbox. When doing this, the top of the roll will then be turning in the same direction as the flow and, if there is a form and/or dirt condition, the holes will scoop the contamination down into the slice nozzle and out onto the forming table with resultant defects in the sheet.

If the roll is rotated in the opposite direction, bottom with the flow, the scooping action will not be there but there will be a danger of undisturbed stock flowing between the roll and the bottom of the headbox, especially if the gap is larger than 6.5 mm. Some headboxes have reversing switches on the roll drive so that the roll can be operated in either direction, depending on the circumstances. There does not seem to be a hard and fast rule.

It is a quite common practice for anyone entering the headbox, for one purpose or another, to use the distributor rolls as a step. This should never be done as it might result in a bent roll. If the roll is bent, two things can happen. The gap between the roll and the interior surfaces of the headbox will open and close during each revolution and the result will be a pumping action, especially at the higher velocities, which will give an uneven flow from the slice. The second possibility is that the roll will rub on the polished surface of the headbox and cause a rough spot that will catch fiber and make strings. There have been cases when the rubbing has been severe enough to wear through the lining and require extensive and time-consuming repairs.

As the sole support of all distributor rolls is the bearing assembly, it is important that the bearings are well maintained. If a bearing fails, the roll will drop and damage the interior of the headbox. With an electrical drive, this condition can go unnoticed for a considerable length of time and has created, in some cases, not only a badly worn condition of the lining but an actual tearing out of the stainless steel sheet and further damage. Bearings are relatively inexpensive and one might be well advised to replace all bearings on a periodic basis to ensure that a failure will not occur.

Nearly all distributor roll types of headboxes have water-sealing glands on the roll journals. It is very important that the sealing water be turned on before the rolls are started and/or the stock introduced into the unit. If it is not, there is a danger of the fiber and/or clay entering the gland resulting in wear of both the gland and the journal. The purpose of the sealing water is to keep the fiber and filler constantly washed out and back into the headbox. Some headboxes have been equipped with an interlock so that the rolls and/or the fan pump cannot be started until the sealing water has been turned on.

G. LIQUID LEVEL CONTROL

When increasing the speed of the machine, one of two things is needed. Either the slice opening has to be reduced or more water added. In either case the liquid level controller signals that more air pressure is required to increase the jet velocity to match the machine speed.

The normal controls used with an air padded headbox are not difficult to understand once the principle is explained. If, for an example, you have an open headbox operating at a speed of 230 m/min, the level of the stock and water must be at a theoretical height of 740 mm to provide the proper jet speed to match the wire, or fabric, speed. To speed up the machine to 305 m/min, the machine operator must either add more

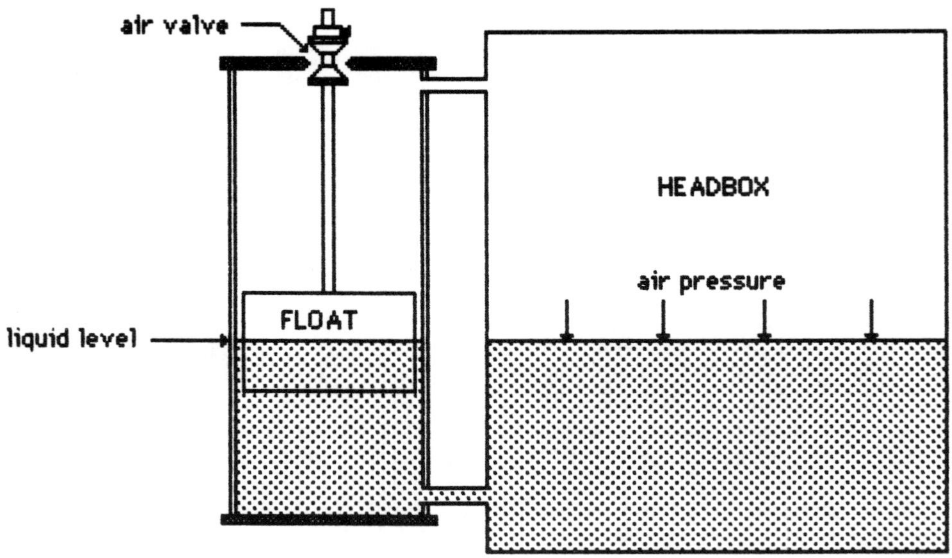

Fig. 87. Float controller (Voith, Inc.).

water or close up the slice to get a greater depth of stock and water to a theoretical head of 1315 mm. A simple procedure. Usually the depth is measured by a scaled sight glass on an open headbox.

When the headbox is provided with a cover and air pressure applied to reduce the amount of liquid volume, the liquid level has to be controlled. This control is accomplished in many ways but the end result is the same. If, for example, the level is at 305 mm above the apron board and the correct amount of pressure is applied to provide a total head of both liquid and air pressure, equal to the speed theoretical total head, the jet velocity will be correct. If the speed is increased, then either more water is introduced or the slice opening reduced. In either case the liquid level starts to rise. The liquid level controller senses this and opens the air valve slightly to increase the internal pressure to keep the liquid level at 305 mm. If, on the other hand, the speed of the machine is decreased, the controller will close off the air valve slightly to lower the pressure.

There is no actual controller other than the flow valve in the feed line and/or the slice opening.

This principle was used in a rather crude form on the early Valley headboxes. The liquid level controller consisted of a float in a plexiglass cylinder mounted on the outside of the headbox. The top of the cylinder was piped through the pond side at the top of the headbox and the bottom at the apron board level. The float stayed at the level in the headbox and was guided by a vertical rod that, in turn, controlled a "V" port air bleed at the top of the cylinder. If the liquid level started to rise the float followed, closing off the "V" port valve, allowing less air to escape and causing an increase in internal air pressure. The liquid level would then lower and the headbox level would stabilize at the required set liquid level depth. Figure 87 illustrates this type of control.

If the liquid level needed to be raised, all that was needed to be done was to raise or lower the "V" port section on the vertical rod.

This system could also be used as a vacuum control for slow machine speed operation where the desired total head in the headbox was less than the required liquid level. The double valve on the vertical rod automatically took care of either a pressure or vacuum system, but it was necessary to change the air inlet valve from the pressure side of the compressor to the vacuum side manually.

It was also necessary to have a manual air relief valve in the line from the compressor to the headbox to allow some of either the pressure or the vacuum to bleed off to keep the valve system of the float within workable limits. If the supplied pressure was too great, then the float would hunt from a full open position to a closed setting on a cyclical basis resulting in a fluctuating head that caused a machine direction profile variation. In fact, if the pressure became great enough the valve would remain fully open with a complete loss of liquid level control.

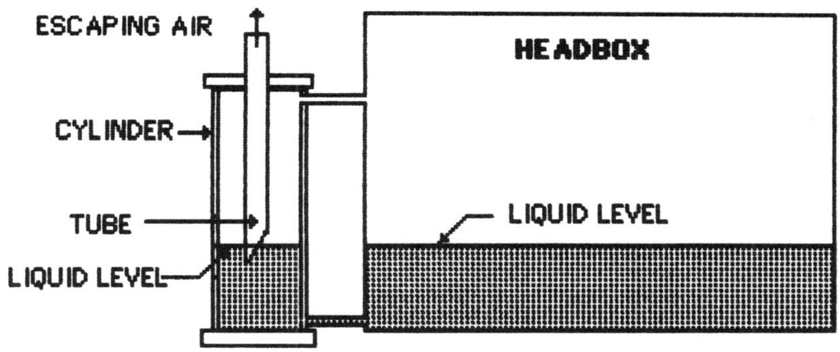

Fig. 88. Syphon controller.

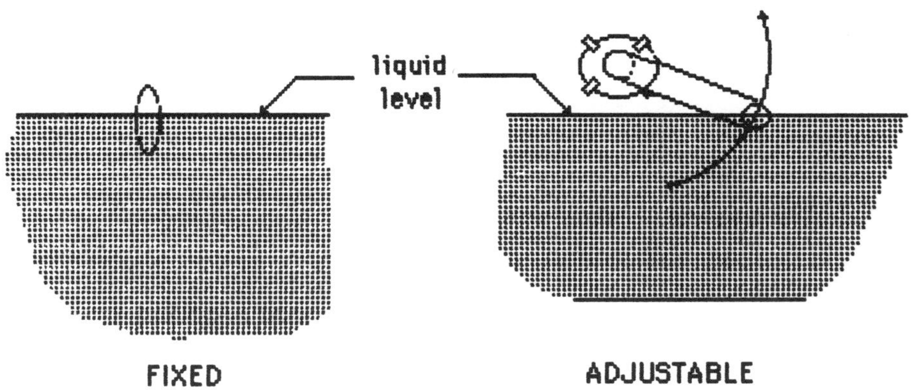

Fig. 89. Hornbostle tube (Beloit Corp.).

The system worked well but had a tendency of filling the cylinder with stock, making it inoperable.

Later on, this system was modified. The float was removed and a tube, called a syphon tube, with a slanted cut off at the lower end was used as shown in Fig. 88. One of the problems with this system was that both water and air were constantly bled off and needed to be contained and piped to some collecting point such as the wire pit or silo.

Another system that was developed was the "Hornbostle Tube" and many are in use today. It probably is one of the most "foolproof" controls developed but it does have drawbacks. The general system is shown in Fig. 89.

In this system, an opening is provided in the pond side of the headbox, either in a fixed position or adjustable. In either case, the outlet is piped to some collecting point such as the wire pit or the fourdrinier trays. The same principle of escaping air applies to this system. As the liquid level raises, the opening is reduced allowing the air pressure to build, forcing the liquid level down until enough of the opening has been exposed to allow the right amount of air to escape. Conversely, the level will lower if the pressure is high, allowing more air to escape. The adjustable system consists of a pipe on the inside of the headbox that can be moved up or down in an arc to raise or lower the liquid level as required.

In all too many cases, the discharge end of the bleed pipe is hidden by either being submerged in the pit or blocked from view by a spray control shield, making it impossible for the operators to see what is going on.

The air supply pressure is critical in all of the systems, however, and it is sometimes difficult to adjust the amount of air entering the headbox. If there is too much air pressure entering the headbox, there will be very little, or no, liquid discharge from the bleed pipe, resulting in a loss of control. If the air volume and pressure are too low, there could little or no air escaping, with the same results of inadequate control. The discharge end of the bleed pipe must be fully exposed so that a visual check can be made at all

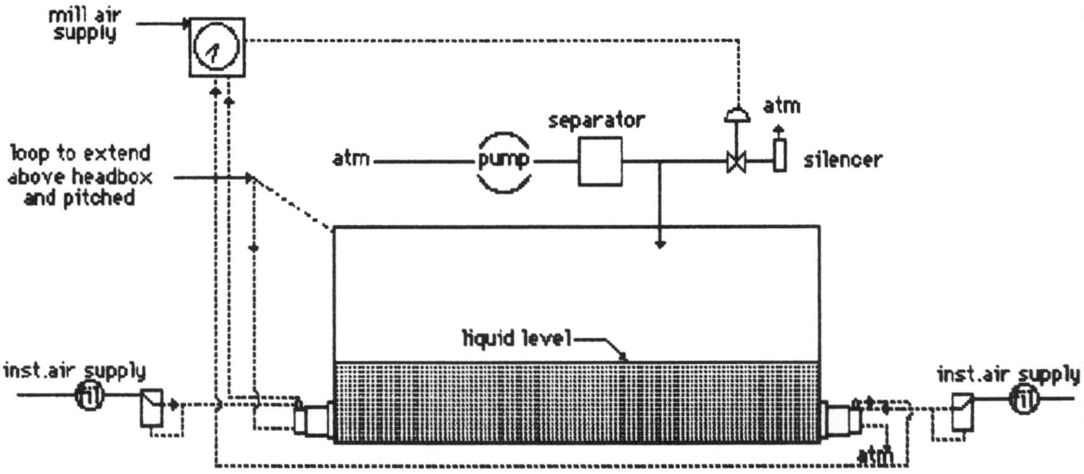

Fig. 90. Liquid level schematic.

times to see if there is both air and liquid discharging from the pipe. Either the air supply to the headbox or an air bleed line must be adjusted to get the correct "blend" of air and liquid discharge.

In all of these systems, it is necessary to have a hand valve for coarse adjustment of the entering air to keep the controllers with working tolerances.

Later, a more sophisticated system of d/P cells and transmitters was developed by the Foxboro Co., among others. In this system, the changing liquid level, detected by the d/P cell, is transmitted to a controller that opens or closes the headbox supply air valve automatically.

Before the development of this sytem, the total head was almost always indicated by either a vertical sight glass or a Meriam manometer. The new system has a second d/P cell for the indication of the total head as a read-out only by transmitting a signal from the second d/P cell to a second pen on the controller. The liquid level is controlled by adjusting the pointer behind the liquid level pen but there is no control of the total head indicating pen. This pen is for indication only. Total head is still controlled by changing the flow valve setting or opening or closing the slice. Figure 90 illustrates this system.

All of the above systems are easy to operate once they are fully understood.

So, in effect, the liquid level controller controls the total head by regulating the amount of air pressure within the confines of the headbox to maintain the desired liquid level, no matter how much the flow or the slice position is changed.

Total head in an air-padded headbox can be controlled with the use of a variable speed fan pump or a flow valve, but the control will still be directly related to liquid level. If the d/P system is used, control will be accomplished through a basis weight and/or machine speed readout. Again, the required total head will be regulated by either changing the flow valve position, the fan pump speed, or the slice opening.

When the hydraulic headbox was developed, it was obvious that, without an air pad, the liquid level control was useless. The outcome is that the total head pressure is measured and a signal sent to do something quite different. Either the speed of a variable speed fan pump, the flow valve opening, or the slice opening is controlled to provide the correct operating head.

H. GENERAL

The interior of the headbox is a highly polished surface for a number of reasons. The first reason is to provide as little resistance to flow as possible. The second, and most important, reason is to prevent corrosion. There is no such thing as completely stainless steel. All stainless steel used today in the manufacture of headboxes is subject to chemical corrosion of one kind or another, but there has to be a place for the corrosion to start. The slightest scratch on the polished surface will provide this starting place. The interior surfaces should be well protected and workers entering the headbox for one purpose or another should remove their shoes to avoid creating the scratches.

If the headbox has a profile bar (sometimes called a "parrot's beak") it should be checked periodically for corrosion on the downstream surface. The interior surface will remain in its original polished condition because of the wash

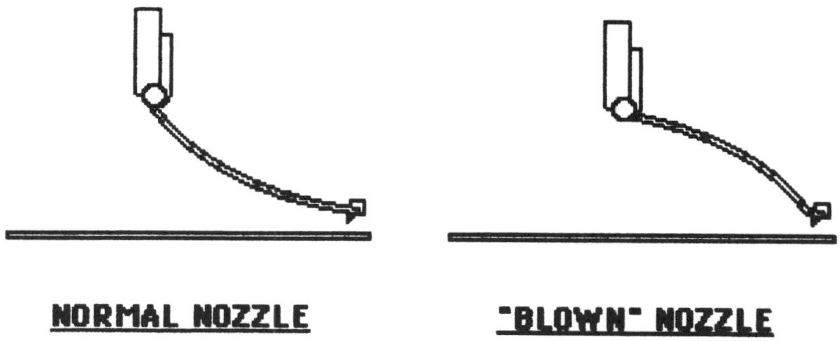

Fig. 91. Damaged nozzle.

of the flow. On the outside surface, the surface not in contact with the stock, the vapor at that point will corrode the surface and create deep pit marks. The pit marks will "grow" until they reach the bottom edge and create a sawtooth edge that will collect fiber and make strings. Usually, if caught in time, the profile bar can be remachined and repolished to its original condition.

One of the "mishaps" that happen to a headbox having a thin nozzle, such as the Valley unit, is "blowing" of the nozzle. This is caused by starting the fan pump before the padding air supply has been turned on, whether that supply is from mill air or a separate compressor. Without the cushion of air, the headbox will fill rapidly with water from the fan pump and, when the headbox is full, the resulting water hammer, similar to shutting of a quick closing valve, creates a sudden shock to the unit. This sudden "water hammer" action will create an extremely high force on the entire headbox. The thin nozzle, being the weakest point, will either bend out and upward or, in rare cases, actually be blown completely clear of the supporting hinge clamp, causing damage, not only to the headbox but to the fabric or wire. There have been many cases when this has happened but has gone unnoticed by the operators. Formation and cross-machine profiles become erratic with no apparent cause. Figure 91 illustrates this condition.

Unfortunately it is sometimes difficult to observe this "blown" condition because of the obstructed view of the area, obstructions caused by the adjusting rods and other equipment.

Figure 90 is not 100% correct, however. Actually, because of reinforcing at the ends of the nozzle, reinforcing by the use of end seal clamps, the ends of the nozzle nearly always hold the original shape of a downward curve and only the center portion of the nozzle bends upward under the water hammer shock. This creates serious cross-flow problems and a changed angle of the profile lip. The only cure for this is to replace the entire top nozzle.

Another problem associated with the thin top nozzle is that, when making adjustments of the profile adjusting screws, the entire nozzle is bent slightly and not just the profile lip. Bending the entire nozzle creates a different flow pattern across the machine and it is for this reason that only one off-profile point should be adjusted at a time. Changing the opening at any one given point with this type of nozzle can change the flow characteristics at some other point without an actual adjustment. Changing more than one profile area at one time can, and often does, result in a much worse profile problem than before the adjustments were made.

When making profile adjustments on any type of headbox with any style of profile bar, one should never go beyond the elastic limits of the profile bar by forcing an individual adjusting rod. A simple rule of thumb is to never turn the adjusting nut, or handwheel, to a hard resistance point. When turning resistance is encountered, rods on either side must be adjusted to relieve the resistance of the adjusted rod. If this practice is not followed, the profile bar will be forced beyond its elastic limit and will have permanent "kink" set in it that cannot be removed except by replacement of the profile bar, or lip.

Forcing the profile lip beyond the elastic limits by the use of a wrench or some other device can also loosen rivets holding the adjusting rod clevis connections on some of the older headboxes with thin top nozzles. The heads of the rivets will then protrude down into the stock flow and create strings.

The newly developed automatic profile control systems where each individual adjusting rod is controlled by either a mechanical or thermal system eliminates this danger. All of the devices have built-in compensating control where the adjusting rods on each side of the one being automatically controlled are also adjusted when the limit of elasticity has been reached to avoid overstressing the profile bar. It is expected that, in the near future, all headboxes, both the distributor roll, air-padded, types and the full hydraulic units, will be equiped with automatic profile controls as a standard procedure.

As the specifications of quality become more severe in allowable tolerances, more attention has to be paid to all design phases of each headbox. It was an accepted fact in the past that the projection of the profile bar down into the stock flow should be about 16 mm but this no longer holds true. It has been found by experience that, the higher the speed, the less projection is required for jet turbulence control. In the latest high machine speed hydraulic headboxes, for example, the projection found to be the best is on the order of 3 mm or even less. It is for this reason that the newer headboxes have "on-the-fly" profile bar projection adjustability for different speeds and fiber compositions.

Distributor roll types of headboxes are excellent units for machines making a variety of grades covering a wide range of flows but, as flows increase, serious consideration should be given to the hydraulic, high micro-turbulence, type. It seems to be a fact that, as flows increase, at least in most cases, the grade structure and consequent flows become smaller in range and fall well within the range capabilities of the hydraulic headbox. In addition to this, the size of the "standard" air-padded distributor roll headbox increases dramatically as the required flow increases. If, for an example, the flow through the distributor roll headbox requires a slice distributor roll of over 457 mm in diameter, a hydraulic headbox might be less than half the size with the same reduction in volumetric capacity, and provide a better delivery out onto the forming area. Any proposed headbox installation should be studied carefully by experts in the field as to the type that should be used. There are distinct advantages of both the distributor roll and the hydraulic, high turbulence types.

REFERENCES

1. "Stock Piping Recommendations — Design Information", Beloit Manual L008-10003 (Aug. 1979).
2. Calehuff, G.L., 'From Fan Pump to Slice Approach Piping', Tech. Sec. CPPA Annual Mtg. (Jan. 1979).
3. Calehuff, G.L., "Distributor System for a Paper Machine", U.S. Patent 3,164,513 (Jan. 5, 1965).
4. Voith, GmbH, Heidenheim, Germany, license to Valley Iron Works (c. 1929).
5. Mardon, J., Monahan, D.E., Manson, D.W. and Smith, K.H., "Perforated Rolls — Design, Structure and Use", Pulp Paper Mag. Can. 67(11):T471–498 (1966).

VI HYDRAULIC HEADBOXES

R.A. REESE

Richard Reese is senior process consultant with CRS Sirrine Engineers, Inc., in Greenville, SC. Mr. Reese graduated from Miami University and majored in pulp and paper technology. He specializes in paper machine evaluation and optimization. His experience includes work on over 100 machines producing linerboard, corrugating medium, fine paper, newsprint, coated paper, foodboard, kraft paper and tissue. He developed paper machine performance and energy consumption standards for major paper grades. He is active in TAPPI and a member of the Water Removal and Papermaking Committees. He is also a member of PIMA and the Pulp and Paper Foundations at Miami University and North Carolina State. He has made presentations at many national and local paper industry conferences and has published over 30 technical articles in industry publications.

REVIEWED BY

John T. Larkey
Staff Consultant Paper
James River Corporation
Green Bay, WI

Douglas P. Newcombe
President
Consultants, Inc.
Appleton, WI

Borje Wahlstrom
President
Borje Wahlstrom, Inc.
Rosemont, PA

A. INTRODUCTION

The headbox takes the stock delivered by the fan pump and transforms the pipeline flow into a uniform, rectangular flow equal in width to the paper machine and at uniform velocity in the machine direction. Design and operation of the headbox is critical to successful papermaking since formation, paper uniformity, and other sheet characteristics are dependent on even distribution of fibers and fillers. Some papermakers feel that the headbox is the most important part of the paper machine.

The functions of the headbox are

1. To provide a uniform and stable jet across the width of the machine without lateral velocities, random components, or streaks.
2. To provide a geometrically stable and defined slice lip geometry independent of temperature, pressure, and slice opening.
3. To produce a well-dispersed fiber suspension with a minimum of flocks.
4. To provide necessary controls of basis weight profile, impact line and angle, and jet velocity.
5. To stay clean and be easy to maintain and operate.

There are three basic types of headbox design:

1. Open headboxes.
2. Air-cushioned pressurized headboxes.
3. Hydraulic headboxes.

Open and air-cushioned headboxes with perforated rolls are discussed in the preceding chapter. Hydraulic headboxes are discussed in this chapter.

Open and air-cushion headboxes are used on machines with moderate and low speeds for applications where large speed and flow ranges are encountered. Hydraulic headboxes are increasingly being used on fourdriniers, basically covering a speed range from 300 m/min upwards and are exclusive on twin-wire formers.

Limited deflocculation capability of roll headboxes requires effective use of fourdrinier dewatering elements to achieve good formation and minimal streaking. On twin-wire formers which set the sheet very fast, there is no disturbance in the forming zone. In these cases, formation and other properties are determined by fiber distribution in the jet, which is controlled by turbulence level, scale, and consistency in the headbox. Streaks in the jet are frozen. Some twin-wire formers have turbulence in the forming zone, which is very beneficial for formation. The special characteristics of twin-wire formers demanded better headboxes and this led to the development of hydraulic headboxes. Elimination of the free surface

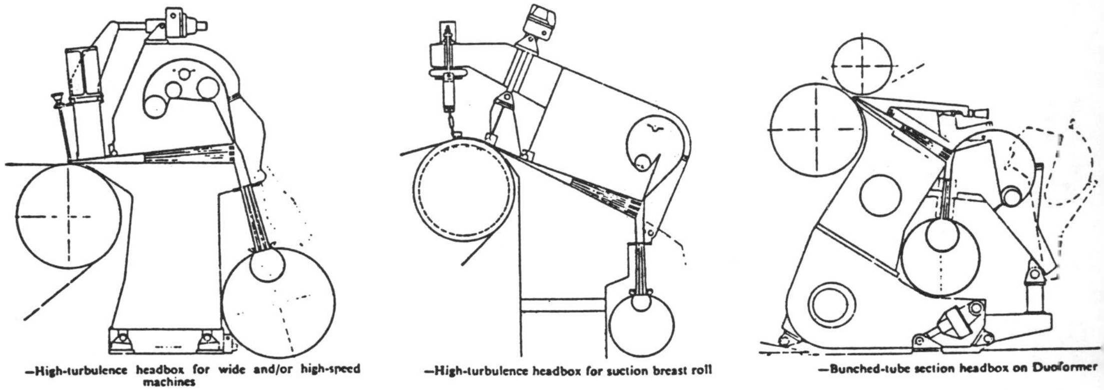

Fig. 92. Hydraulic headbox applied to different formers [1].

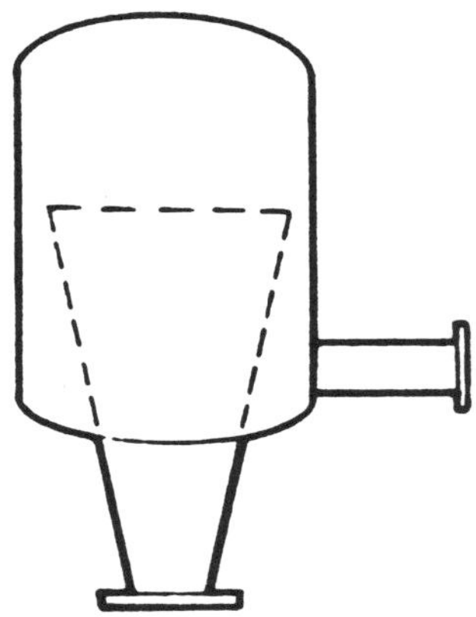

Fig. 93. Surge tank.

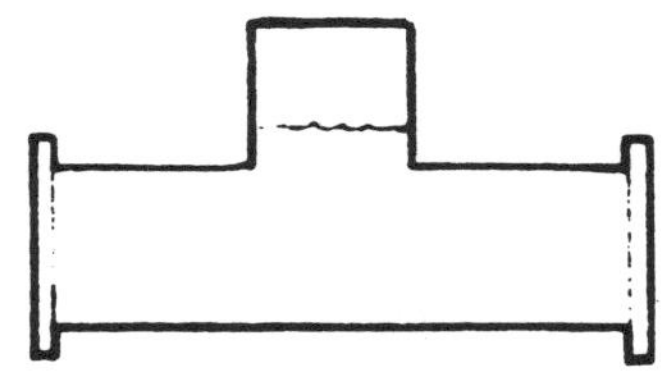

Fig. 94. Surge pipe.

allowed use of stationary elements for flow evening and turbulence generation at much higher flow velocities. The higher velocities, about 1.5 m/s, and use of stationary elements have radically increased turbulence intensity level and reduced turbulence scale which has improved deflocculation. In addition, hydraulic headboxes give much more stable and uniform jets. Because of their improved performance and competitive price, hydraulic headboxes are not only used for twin-wire formers, but are taking a larger share of the headbox market on fourdriniers and top-wire formers [4]. Figure 92 shows the application of one manufacturer's hydraulic headboxes on different applications.

1. Approach flow

Design and installation of approach flow piping between the pressure screen and the headbox is critical to good headbox performance. The guidelines recommended by headbox manufacturers should be followed.

Short-term periodic variations or pulsations are emitted by several wet end sources including:

- Fan pump impeller;
- Pressure screen foils;
- Resonate motion of pipe lines;
- Valve cavitation;
- Structural or equipment vibration.

Some of the short-term variations should be eliminated where possible. However, it is not always possible or practical to eliminate all sources of noise. Air-padded headboxes provide dampening of pulsations around 5 Hz. Hydraulic headboxes tend to pass pulsations through to the sheet.

Air-padded surge tanks (Fig. 93) have been used with hydraulic headboxes but are only as effective as air-padded headboxes. Air-padded surge pipes (Fig. 94) have been applied to systems where a single frequency is troublesome and can be tuned to dampen a single periodic frequency pulse. Diaphragm-type attenuators (Fig. 95) are a preferable arrangement since they attenuate over a broad range of 0.25 to 40 Hz without incurring a reduction in pipeline velocity as with a surge tank or the possibility of entraining air in the stock as with a surge pipe. The effect of a diaphragm attenuator can be seen in Figs. 96 and 97. The figures show machine-direction basis weight variation in the time and frequency domains, respectively.

B. HEADBOX DESIGN

All headboxes can be divided into three parts: the distributor, the evening section, and the slice. The distributor deals with the initial distribution of the flow from the pipe to the full width of the machine. The evening section takes the flow from the distributor and stabilizes it into a uniform flow of controlled, small-scale turbulence. The slice accelerates the flow to a speed close to that of the wire speed and controls the slice opening and impact conditions.

Comments on some of the basic hydrodynamic phenomena that impact headbox design follow.

- Converging flow channels with an accelerating flow stabilize the flow and stretch it in the length direction.
- Diverging flow in an expanding flow channel is unstable over an angle of 8°.
- Instabilities cause eddies, separations, and back-flow.
- Creating a pressure drop across a diverging channel with a perforated roll or a perforated plate allows the diverging angle to be increased up to 20° and still maintain stability.
- A flow that changes direction is subject to formation of Taylor vortices in the direction

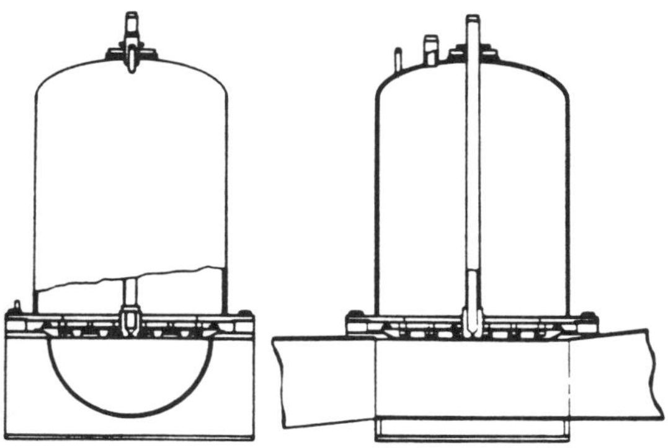

Fig. 95. Pulsation attenuator.

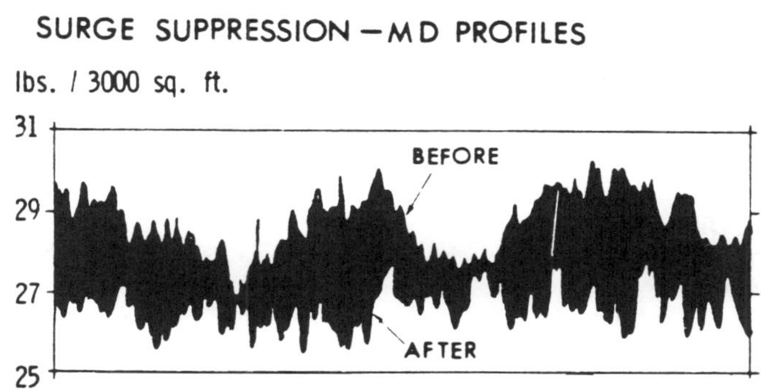

Fig. 96. MD basis weight profile, time domain.

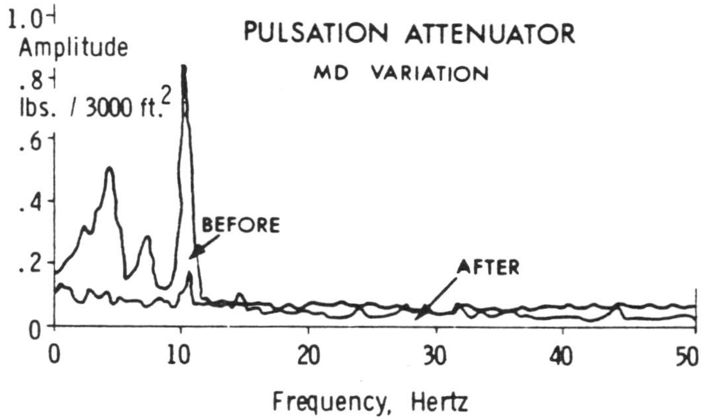

Fig. 97. MD basis weight profile, frequency domain.

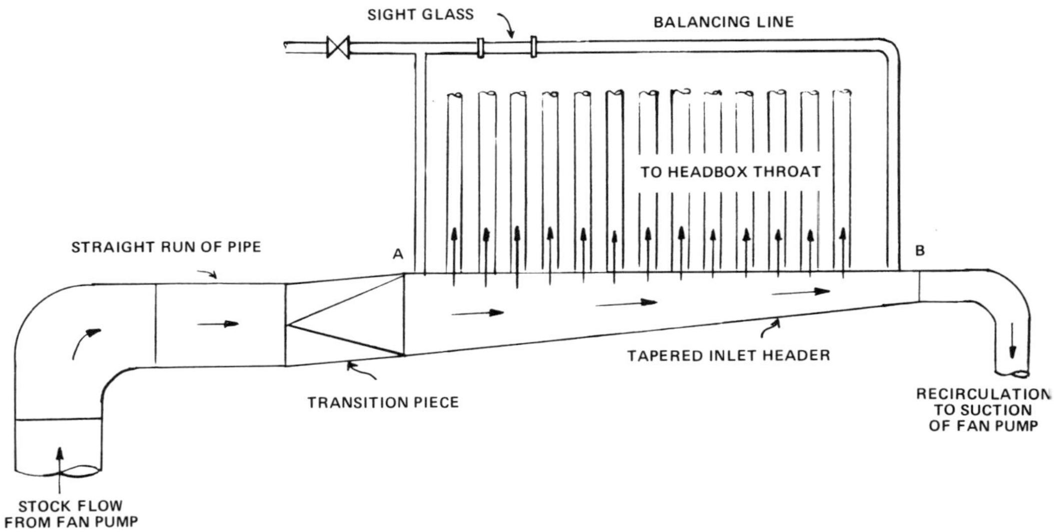

Fig. 98. Tapered inlet manifold with small laterals [3].

of flow. Step changes in the flow channel cause turbulence, separation, and instabilities.

- A non-uniform flow turning at a wall will form vortices, which become more severe the sharper the turn.
- Dividing a flow in a tee is unstable. It can be stabilized by converging the flow through the division.
- A fiber suspension flowing along a wall develops a boundary layer that grows with distance and is free from long fibers. These boundary layers can easily cause streaks.
- Fibers in fiber suspensions have a strong tendency to build dense flocks on protruding surfaces, in gaps between surfaces, and on stationary elements facing the flow, which are less than about twice the fiber length in width if the flow is not sufficiently turbulent.
- High-velocity jets flowing into a slow-moving fluid are inherently unstable and interact to form vortices and cross flows, unless suitably controlled.

1. Distribution system

This distributor is used to spread the flow from a pipe evenly across the full width of the machine for any fiber type, consistency, and flow range. The width of the machines can vary from about 1 to 10 m.

A number of different solutions to this problem have been used. The early ones had a single pipe feeding the backside of the headbox. Before closed screens were developed, feeding through pipes or channels from open screens was the approved method. As machine speeds and widths increased, demand for more uniform and stable basis weight profiles led to development of better systems. The most important of these are manifold systems, flow spreaders, and cross flow distributors. All new headboxes use a one-sided tapered header developed in the late Fifties. All headbox suppliers are using different versions of this basic principle. Virtually all hydraulic headboxes have one-sided tapered headers.

2. One-sided tapered header

Figure 98 shows a typical one-sided tapered header. The flow enters from the side and perpendicular to the machine. This header is designed with a taper and normally with overflow to maintain constant pressure along the header, while turning the flow 90° in the machine direction through pipes or holes evenly distributed across the machine. If properly designed, this ensures constant flow through each pipe and uniform initial distribution from a pipe to the full width of the machine, independent of width. The evening section and the slice will then homogenize and deliver the jet.

An excellent and detailed analysis of design aspects is found in "Design Aspects of Manifold Type Flowspreaders," by Alton D. Trufitt [9].

An ideal taper must be designed to take into account friction losses across the length of the inlet header. Equations to do this and suitable friction coefficients are available, and this approach must be taken when headers without overflow are designed. However, by choosing about a 3 to 1 velocity ratio between the velocity in the manifold and the header and running with about 10% overflow, headers with linearly

Fig. 99. Hole plate design [1].

decreasing cross-section can be used. Rectangular cross-section headers with linearly decreasing cross-section are the most common today. The velocity ratio ensures that small pressure variations in the header will have an insignificant effect on the flow distribution. Maximum velocity in the tapered manifold header is normally 3 m/s. This gives a maximum speed through the manifold of 9 m/s and a maximum pressure drop of about 3 to 4 m water gauge. Round cross-sections are preferred with nonlinear headers. Many header designs use lower pressure drops through the tubes to reduce turbulence level, which otherwise may lead to formation of hard, small flocks. Good distribution is assured by adding another pressure drop downstream.

Hole plates or tube banks are normally used to divide the flow into a large number of small jets. This gives a more uniform distribution and makes it easier to produce a homogenous flow in the evening section. Hole sizes between 13 and 25 mm are normal for hole plates, while tube banks normally use somewhat larger diameters. Stapling at the entrance of the holes can be eliminated by using a washboard design. (Figure 99 shows one hole plate design.)

The length of the holes or the tubes must be sufficient to turn the flow 90° or at least 3 diameters. Hole plates are favored in many designs as they take less room, can easily be replaced, can be made with better uniformity and definition, and eliminate the effect of friction in the manifolds. Both types are used successfully.

Entrance geometry of holes and tubes is critical as it determines the head loss entering the hole. This is a major part of the total head loss across the manifold. A high velocity ratio of about 3 greatly reduces this effect. Tube and hole entrances are normally the same for the whole distributor. Some tube banks are designed with different hole patterns at the ends to compensate for edge effects.

A venturi bend 6 to 8 diameters away from the header is recommended if there is no straight length of pipe between the elbow and the tapered manifold. This ensures that entrance conditions to the header are stable and equal to the flow conditions in the rest of the header. If the flow is entering parallel to the hole plate, a low-pressure zone is produced at the entrance causing uneven distribution. Great care must also be taken with the transition piece connecting the round pipe with the rectangular header.

It is very important for headers with overflow to operate the overflow to give zero pressure differential between front and back. This is especially so with straight-tapered manifolds, which should have automatic control of recirculation.

Headers with linearly decreasing cross-sectional area must be provided with an overflow. The effect of friction on the distribution is minimized by having a sizable pressure drop over the manifold. Distribution systems using this principle can be designed with maximum flow variations of 0.5 to 1.0% over the manifold tubes.

The one-sided tapered distributor is the standard in the industry for all types of headboxes and is used for any width, furnish, or flow range. Different headbox suppliers use different designs. All use overflow, but different approaches for cross-sections, tapers, and manifolds.

The one-sided tapered header with a hole plate or tube bank correctly designed and operated provides a distribution system that is applicable to all conditions and widths and gives uniformity and stability equal to or better than can be profitably utilized. The one-sided inlet tapered distributor is superior to any other known distribution system. It provides a degree of uniformity and stability that meets present and foreseeable requirements.

3. Evening and turbulence-generating section

The evening and turbulence-generating section is located between the distribution system and the slice. Its function is to homogenize the flow to a uniform stable flow without large eddies and cross flows and keep the fibers dispersed through turbulence generation. The flow from a distributor is delivered to the box in discrete jets which must be slowed down and mixed into homogeneous flow without any cross-flow components or large eddies. The better the distribution system, the easier it is to homogenize and stabilize the flow. The elements

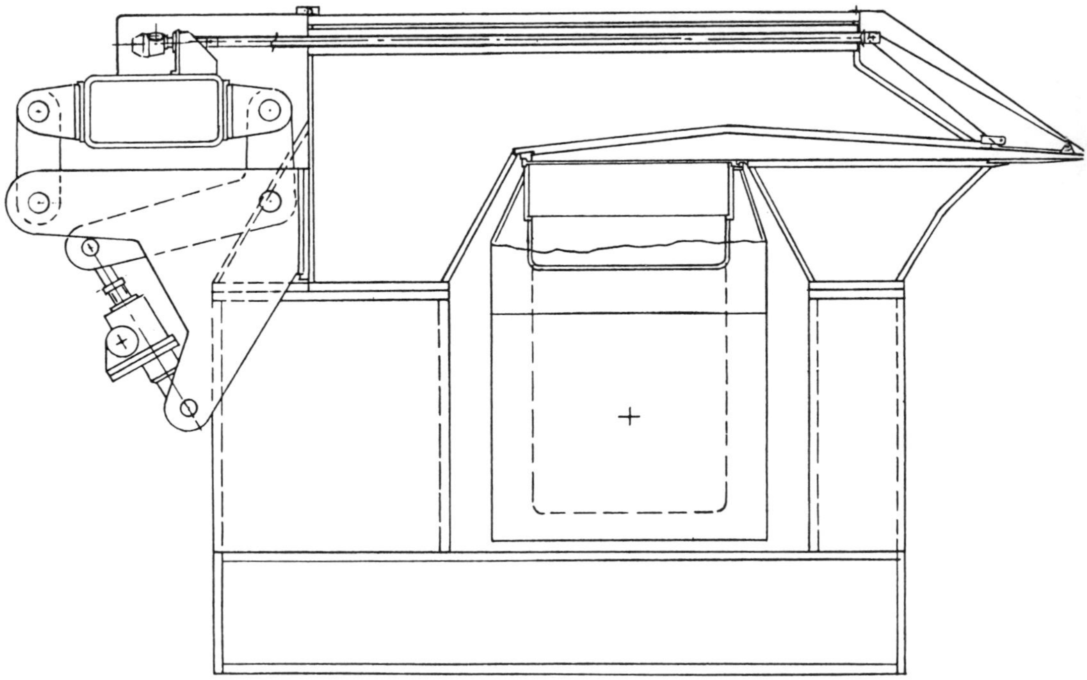

Fig. 100. KMW HTB headbox [1].

used to even and stabilize the flow are also used to deflocculate fibers.

4. Hydraulic headboxes

Hydraulic headboxes use stationary elements to dissipate kinetic energy in high-velocity jets and to produce a stable, homogeneous flow. This, and the generation of suitable turbulence to disperse the fibers, is accomplished using stationary elements. As the same tools are used for stabilization of flow and for turbulence generation, they will be covered together. Hydraulic headboxes are completely filled with fiber suspension and therefore permit use of much higher velocities through the box, enabling turbulence generation using deceleration, step changes in cross-sectional area, change in direction of flow and wall shear. There are three basic principles used for accomplishing this and they are discussed below. They are the flow nozzle, the step diffusor, and the headboxes using different versions of stationary flow evening devices.

(a) Flow nozzles

Figure 100 shows a typical flow nozzle. The jets from the distributor are injected into a narrow nozzle at about a 90° angle. At the entrance, the converging nozzle has a height of less than three to four times that of the manifold jets. The success of this type of nozzle is very dependent on its geometry. If the height of the nozzle is small (3 to 4 jet diameters at the entry of the jets), there will be isotropic mixing of the jets and elimination of major cross flows. If the height exceeds 4 diameters, severe cross flowing results. Due to the necessity of keeping the height of the nozzle small at the entrance of the jets and the lack of elements to stabilize the flow, the flow nozzle can only be successfully applied to small lip openings, below 15 mm. Because of high energy dissipation at the entrance of jets, turbulence level is high which gives good fiber dispersion but creates hard, small flocks. This is an excellent headbox for lightweight sheets such as tissue, toweling, and other lightweight papers made on twin-wire formers or suction breast roll formers. The jet is not stable enough for larger lip openings or for fourdrinier tables. A number of different machinery builders are making headboxes of this kind. The world's fastest tissue machines use a headbox of this type.

A nozzle-type headbox uses a one-sided tapered header either with a hole plate or with a tube bank. The most recent designs use step diffusor hole plates to slow down the jet and thus reduce turbulence intensity which improves formation.

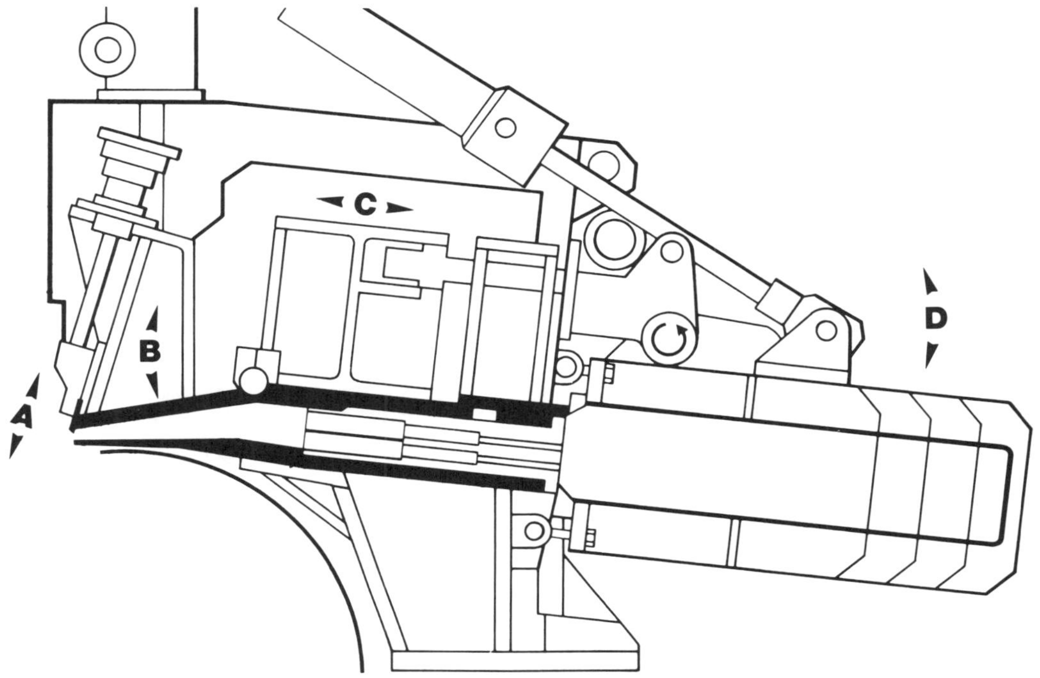

A Setting of slice and fine adjustment

B Adjustment of top lip, depending on flow rate and machine speed

C Horizontal movement of the top lip affects the point of impingement of the jet on the wire which in turn affects sheet formation

D The distributor can be swung clear for cleaning or inspection purposes

Fig. 101. Hydraulic headbox (Escher-Wyss) [3].

(b) Step diffusors

Figure 101 shows a typical step diffusor headbox. It is based on the principles that, after accomplishing the initial distribution through a hole plate or a tube bank, stabilization and turbulence generation are performed through a series of step-wise increases in the diameter of the original branch tubes or holes. From a small open area at the entrance, which gives the necessary pressure drop to give good distribution, the flow is decelerated in these channels with two or more steps until they come out either through a honeycomb with 90 to 95% open area or through holes with an open area of 60 to 70%. Step diffusors produce a stable and uniform flow with a controlled turbulence level that can be delivered directly to the slice over a short rectangular distance. Turbulence is generated by separation at the steps. The length of the tubes between steps is designed to produce a suitable turbulence level high enough to eliminate boundary layer problems. The step diffusor is critically dependent on a good original distribution, as there is no mechanism for evening out flow differences between the different flow channels.

By making these steps about 3.2 and 6.4 mm and dimensioning the length of the tubes, step diffusors combine a stable, uniform flow without streaks and a suitable turbulence level and scale to give good fiber distribution. The last step cross-section can either be cylindrical or square and it is even possible to put square honeycombs in lines without having trouble with boundary layer streaks. This very simple system has given excellent performance.

(c) Stationary flow evening devises

Most hydraulic headboxes use a stilling chamber after the distributor followed by a resistance element which is a type of hole plate followed by a tube bundle, honeycomb, vanes, or foils. The stilling chamber is used to mix the individual jets coming from the distributor and thus creates a homogeneous, uniform flow across the machine. The hole plate acts as a resistor to

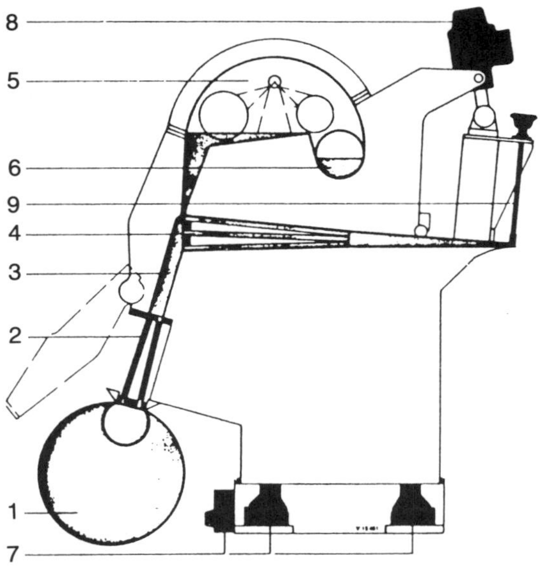

1. Manifold
2. Flow evener
3. Channel
4. Turbulence tube bank
5. Compressed air cushion
6. Overflow
7. Wedge plate gear unit
8. Worm gear unit
9. Indicator handwheels

Fig. 102. Voith W-headbox [1].

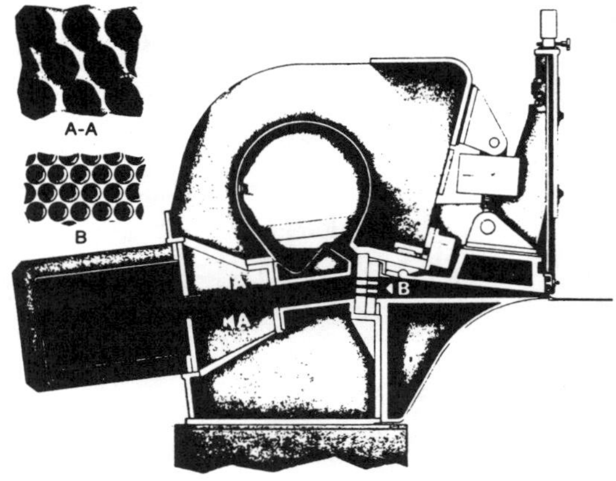

Fig. 103. Valmet Sym Nozzle headbox [1].

further improve uniformity followed by vanes, tubes, or honeycomb sections which eliminate cross flows and develop the desired amount of turbulence. This is achieved in the step diffusor headbox by the length directional partitions that the pipes constitute. There is a certain amount of convergence in the headbox design, but individual characteristics will determine its performance. All major suppliers have headboxes based on these general principles.

Figure 102 shows an example of a hydraulic headbox. The flow from the distributor, which uses a tube bundle, flows upwards through a tapered header design, where the overflow goes on an air cusion and the flow to the slice goes through a tube bundle at about a 90° angle to the vertical header. The slice tube bundle starts with cylindrical tubes that are then expanded and changed into hexagonals that form a honeycomb section at the entrance to the slice. The tubes are about 610 mm long and give extremely good stability to the flow and provide turbulence through wall shear and expansion. Following the tube bundle is a converging slice with a very low angle convergence and straight-through flow which further stabilizes the flow. This design produces an extremely uniform, stable jet without cross flows and an excellent turbulence spectrum for fiber dispersion. The overflow with the air cushion produces some pulsation dampening but, due to the narrow throat, it has a very narrow frequency range and is therefore not satisfactory for most fourdriniers. However, it is very suitable for twin-wire formers. The tube bundle is built around perforated steel plates, which are load carrying which makes it possible to build small, stable headboxes independent of machine width and speed.

The honeycomb tube bundle and the slice are in a straight line which eliminates any possibility of creating Taylor vortices. The end of the slice has a vertical tip for slice profile adjustment.

Figure 103 shows another design for a fourdrinier or a top-wire former. A one-sided rectangular tapered header with a tube bank delivers the flow to a stilling chamber. A double hole plate is used for final flow stabilization and turbulence generation. The flow turns 15° at the hole plate into a straight slice. This design has an overflow and air cushion on top of the stilling chamber for pulsation dampening. The hole plates are load carrying and the slice region was designed to give good stability. Another supplier has a box that is very similar to this design, but without the air cushion.

Figure 104 shows a headbox design for a twin-wire former. This headbox has a one-sided inlet header with the exit through the hole plate in the upper corner of the distributor. The hole plate delivers the jets into a stilling chamber, followed by a hole plate and a honeycomb made by squares, with each square offset half a division. The evening section, which is load carrying, is followed by a long nozzle.

Most modern hydraulic headboxes use length directional elements such as pipes, honeycombs, or vanes to stabilize the flow and create microturbulence. Geometrical diameters or

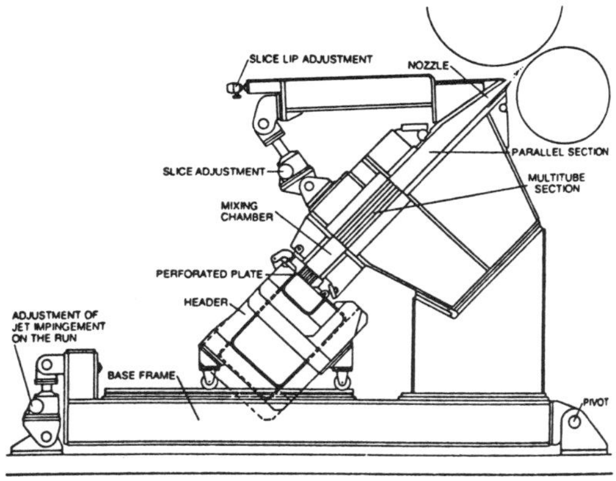

Fig. 104. Dominion Dynaflo headbox [1].

Fig. 105. Beloit Strataflo headbox [1].

Fig. 106. Beloit thin channel Converflo headbox [1].

spacing are in the order of 10 and 30 mm.

Other headboxes use a completely different system of stabilizing the flow and creating turbulence than the previous boxes, as shown in Fig. 105. A one-sided tapered header with expanding tubes built for low velocities is used as a distributor. The flow from the distributor exits into a stilling chamber. At the end of the stilling chamber there is a perforated plate which holds a number of cross-machine foils made out of lexan. These define converging fluid sheets in the slice region. On a headbox for a twin-wire former, these flexible foils extend all the way to the end of the slice. On headboxes for fourdriniers, they end about 100–200 mm from the lip because the slice has to be turned horizontally from its 15° upward angle. The foils, in addition to providing turbulence or shear, define converging flow channels whose exit dimensions are about 2.5–5.0 mm, define the maximum size of flocks, and divide the flow into a number of independent fluid sheets. Therefore, a multi-ply sheet is formed from these individual fluid sheets with a very small floc scale. This explains the fact that a well-formed 161 g/m^2 linerboard sheet can be made from three different furnishes of 0.7% consistency on a twin-wire former. The multilayer headbox approach to design is a major step forward in headbox development.

Some early headboxes had a large cross-sectional area and low flow velocities to keep turbulence levels relatively low to prevent formation of hard flocks. This caused some problems with cleanliness. The headbox shown in Fig. 106 was developed to eliminate this problem and the necessity of terminating the foils 100–150 mm from the slice lip on fourdrinier machines. This headbox turns the flow at the hole plate where the foils are fastened. By doing this, the velocity of the slice is radically increased and thus can be horizontal. This allows the foils to be extended to the end of the slice. Turbulence level can be controlled by adjusting the length of the foils and the shape of the foil tips. This headbox has eliminated problems with cleanliness and improved flow stability. In addition, it has eliminated Taylor vortices which cause streaks and reduced turbulence scale. This headbox has greatly improved the potential of making multilayer sheets on a fourdrinier.

The distance between foils at the end is 2.5–5.0 mm, which defines the floc scale in the z-direction. This has to be compared to the 10–60 mm lip openings that define z-direction floc scale for other types of headboxes. Figure 107 shows a secondary headbox, which is simplified to that extent that the stilling chamber has been eliminated.

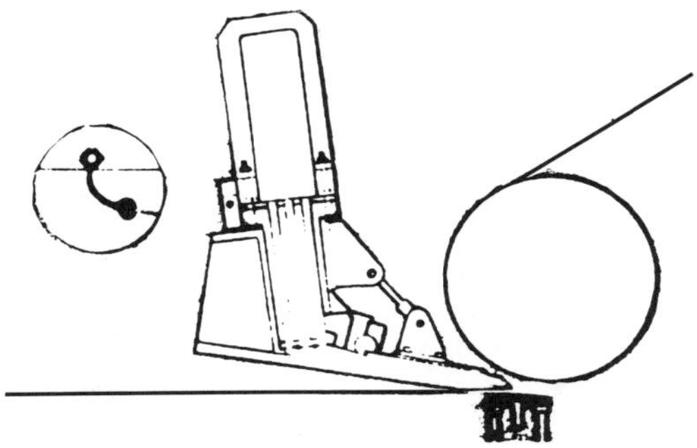

Fig. 107. Beloit secondary headbox [1].

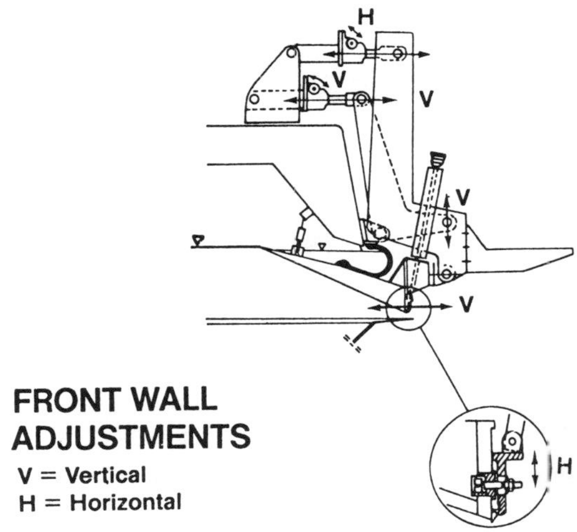

Fig. 108. Front wall design (KMW) [1].

5. Slice

The function of the slice is to accelerate the flow and deliver a stable, deflocculated jet through a well-defined rectangular lip opening. This lip opening must be adjustable but keep its shape independently of temperature, pressure, or total lip opening. In addition to the total lip opening and its shape, the angle and point of impact must be controllable. For a wide, high-speed machine, this is quite a difficult task, as the degree of definition is in the order of 0.01 mm. The slice is the converging part of the headbox taking the flow from the evening section to the delivery of the jet.

(a) Process considerations

From the process standpoint, the slice should have a uniform convergence with a fairly low angle. This reduces the potential of translating ingoing flow nonuniformities into eddies, and thus stabilizes and improves the basis weight profiles. Steep angle slices are very bad due to the formation of eddies.

Due to acceleration, fibers are aligned in slices. However, where turbulence levels are high, this tendency to align is minimal, especially for slices with small convergence. Because of this, it is much more difficult to make directional sheets with hydraulic headboxes than with air cushion or open headboxes.

A vertical metal strip is put on the top lip as a vertical projection into the slice to make the slice less sensitive to its geometry and to make it more adjustable. This metal strip is replaceable and is used for profile adjustment. If it gets permanently deformed, it can be replaced. It helps in reducing streaks caused by slice imperfections or slice deformations.

The slice, preferably, should have a straight convergence. Bends in the slice, which have been used on many hydraulic boxes, can cause Taylor vortices which will cause streaks. The better and more uniform the flow that enters the slice, the shorter it can be made.

The top lip must be movable, both in the vertical direction to control lip opening and thus headbox consistency, and in the horizontal direction to control jet angle and impact point. The importance of this will be discussed under headbox operation (see Sec. D).

Turbulence intensity is reduced in the slice, while the scale grows.

(b) Design considerations

Design of the slice region is critical, as its dimensional integrity and uniformity are necessary for a uniform and stable basis weight profile. The profile must be independent of changes in temperature, pressure, and lip opening. Realizing that the headbox can be regarded as a pressure vessel, the difficulty involved in keeping two parallel lips a distance of 12.7 mm from each other with variations of less than 0.025 mm for a 10 m machine at 138 kPa exemplifies the difficulties. In addition, the slice opening must be controllable, both in profile and total opening. At the same time, the top lip must be movable horizontally in relationship to the bottom lip. Headbox design, especially the slice region, is therefore a difficult design problem.

Air-cushion headboxes must be designed with the four outside walls as the structural elements. Pressure and thermal gradients will distort these boxes and affect lip geometry unless special solutions are used. Pressure and thermally controlled compartments with circulating heated water have been installed between the structural elements and the lips to maintain lip geometry independent of pressure and thermal

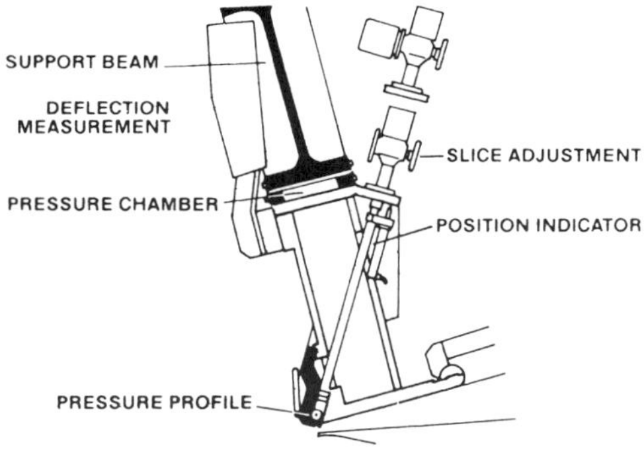

Fig. 109. Lip design (Voith) [1].

gradients. Great care must also be taken in allowing the boxes to expand sideways on their supports without distortion. Figure 108 shows a solution to this problem where the top lip has a very high moment of inertia, takes up small forces and is thermally compensated by an overflow. Hydraulic boxes that use stationary elements for flow evening and turbulence generation are using these elements to take up the forces through internal stiffening. This makes it possible to build these boxes lighter and cheaper with good dimensional stability. However, the slice region which is quite long still has to take up very large forces.

Many new boxes have had trouble with lip distortion due to temperature, pressure, and varying lip opening. This distortion has caused streaks and unstable basis weight profiles. The profiles vary during startup before thermal equilibrium has been achieved and with varying operating conditions. This causes large difficulties with regard to efficient operation especially at high speeds. Therefore, it has been found necessary to build in systems that can compensate and eliminate these effects. Figure 109 shows such a design where the top lip is provided with a strong beam and on top of this beam is another beam with a pressure compartment in between. There is a temperature and pressure controlled fluid in this compartment. A tensioned wire is put inside the lower beam and is used as a reference to measure the position of the top slice lip. This position is used to control pressure and temperature to compensate for deformations caused by the pressure inside the box and temperature of the stock. The metal slice strip provides a vertical projection into the flow. This vertical strip is used to control the slice opening and the profile. This means that the main top lip can be made very strong structurally and machined to good tolerances, as all the deformation is taken up by the replaceable

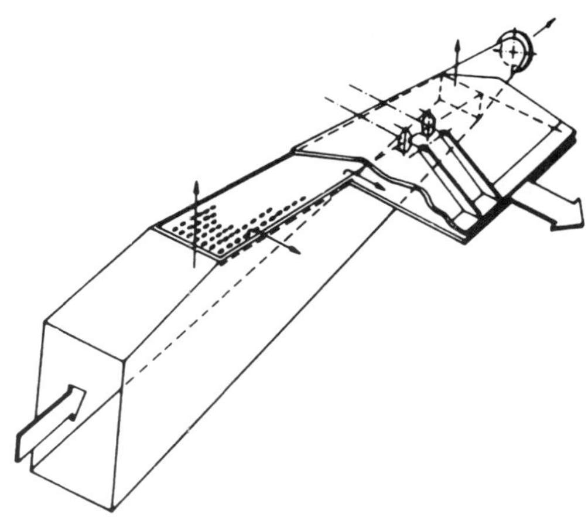

Fig. 110. KWM U-beam slice design [1].

vertical metal strip.

Another method used for automatic compensation for pressure and temperature is shown in Fig. 110. The slice is built as a U-beam and the side walls are only pressed against the ends enough to seal them. Therefore, there are no forces transferred at the ends. The U-beam deforms uniformly across the width of the machine. This becomes a very heavy box, especially at higher speeds, but this is more than compensated for by absence of changes in the slice geometry giving stable, streak-free basis weight profiles. This type of design is used on newsprint, tissue, and towel machines.

Hydraulic headboxes originally did not have horizontally movable top lips in relation to the bottom lip. However, this has become standard as it is a necessary tool to control the impact point on fourdrinier machines. On twin-wire machines, the whole box is normally moved to control jet entrance geometry.

Slice lip controls must be made with a minimum of backlash in the arrangements for changing slice opening or slice profile. This is a very critical area and it is necessary for the slice geometry to be defined and controllable. A recent development using steel bars that can be electrically heated as a means of adjusting the slice profile, has the advantage of eliminating backlash and permitting easy control.

Slice lip adjustment screws have historically had a cross-machine spacing of approximately 150 mm. Higher quality requirements have necessitated narrower spacing to reduce streaking tendencies on lightweight grades. Some current headboxes have spacings as close as 70 mm.

Slice lips preferably should be made in one piece to avoid joints in the slice region. Joints can be the source of buildups of lumps and dirt.

Most headbox manufacturers prefer to use only solid stainless steel materials in headboxes. This eliminates problems with differences in thermal expansion and in the joints. Recently, the inside surfaces of headboxes also have been electropolished to ensure cleanliness. Electropolishing takes away all sharp protrusions and reduces problems with dirt buildup.

The cheeking pieces that define the edges of the jets, the extensions, and the deckles must be fitted perfectly to prevent edge waves. Hydraulic boxes have much better edges than air-cushion boxes because the suspension is treated much more uniformly in a hydraulic box than in an air-cushion box.

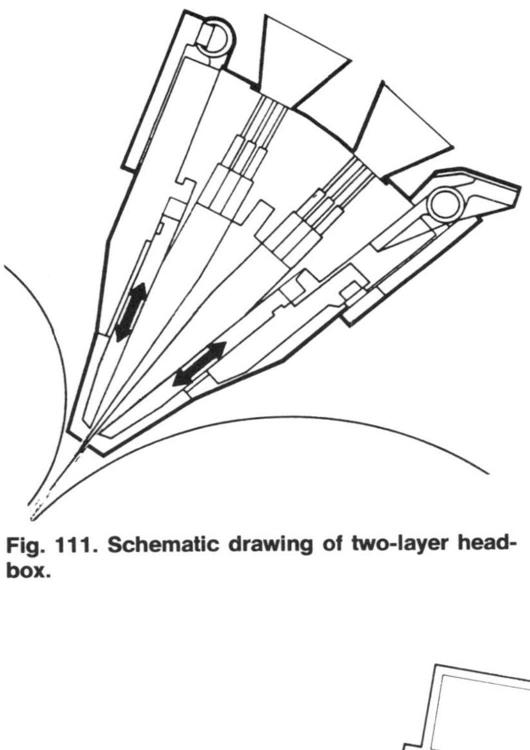

Fig. 111. Schematic drawing of two-layer headbox.

6. Multilayer headboxes

Structuring the paper web in two or three layers is a recognized means of obtaining better tissue quality and higher production economy. This technology is also being used on multi-ply board and linerboard. One of the economic advantages is better use of raw material by positioning the different types of stock with the required properties at the right place in the sheet z-direction. In tissue production, main quality parameters are softness, high bulk and absorbance capacity as well as a certain minimum strength level. The tissue web can be built up such that the surface provides for a "soft feeling" and the inner layer gives the required strength and bulk characteristics.

Some hydraulic headbox designs are well suited for multilayer structures. By dividing the one-sided tapered header, the stilling chamber, and the slice region using flexible foils, it is possible to build headboxes with separate flow channels from the header through the slice. A multilayered sheet can be made by doing this and using different furnishes in these layers. A two-layer step diffusor headbox is shown in Fig. 111.

Another approach has been taken as shown in Fig. 112. Three separate headboxes are combined with two common walls separating the flow channels. These walls are thick and the jets from each layer leave the slice separately, defining air wedges in between. These air wedges are fed with air so that they will be maintained. They are kept long enough so that the jets meet at the entrance of the forming zone, minimizing or eliminating shear between the jets which otherwise cause mixing. Minimizing shear allows the purity of the layers to be better maintained.

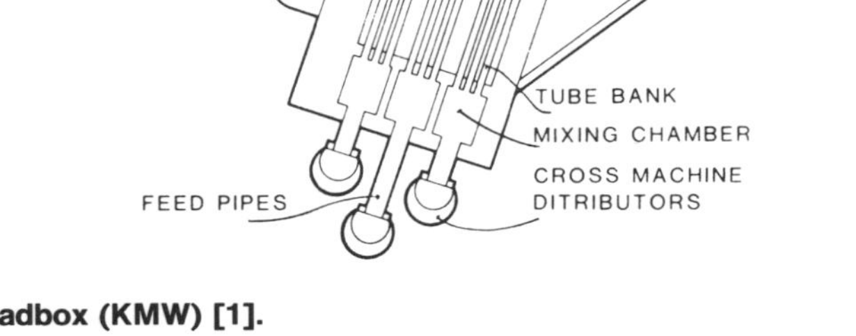

Fig. 112. Multilayer headbox (KMW) [1].

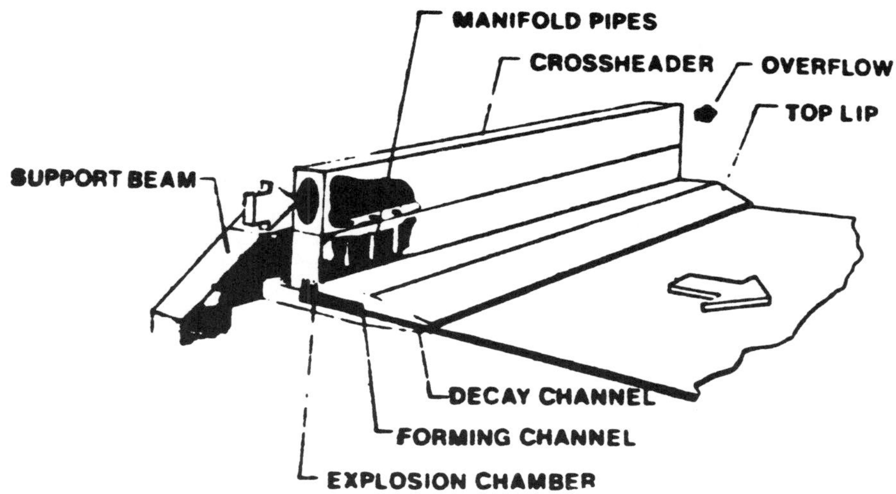

Fig. 113. Formflow headbox — main components.

7. High-consistency headboxes.

The headboxes discussed previously operate with headbox consistency in the range of 0.05 to 1.2%. Development work to permit headbox consistencies of 2.5 to 3.5% was initiated at STFI in Stockholm, Sweden, in the early Seventies.

The principal components of a high-consistency headbox are:

- Inlet crossheader;
- Manifold;
- Explosion chamber;
- Forming channel;
- Decay channel;
- Lip section and precision adjustment screws;
- Rigid support beam;
- Overflow.

Figures 113 and 114 show the major components and a cross-section of a typical high-consistency headbox. The inlet crossheader can be mounted above, below, or in a straight line with the forming and decay channels [5].

High-consistency sheets are characterized by much greater orientation of fibers in the z-direction than sheets formed at lower consistency. This produces a porous, bulky sheet structure. The relatively open sheet structure results in easier dewatering in the forming, pressing, and drying sections of a paper machine. Typical results are a 2 percentage point drier sheet out of the press section and 10% more efficient drying in the dryer section.

Delamination strength (Scott Internal Bond), CMT (Concora Medium Test) and compressive (ring crush) strength are significantly higher. Burst index, tensile index, and surface smoothness are typically lower. These characteristics make high-consistency sheets particularly suitable for pulp drying machines, corrugating medium, and the middle layer of multilayer folding boxboard [6].

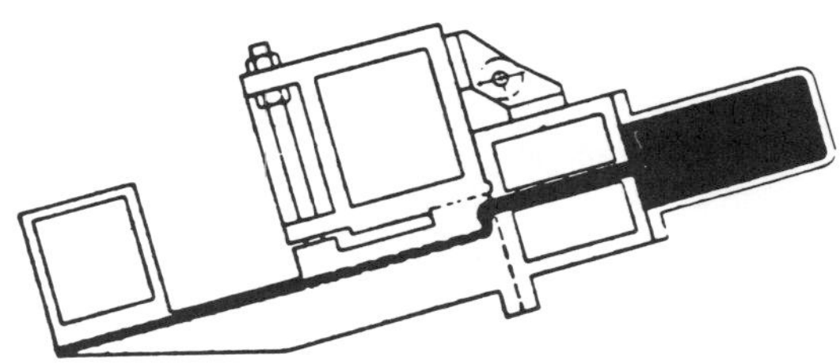

Fig. 114. Formflow — crossheader at rear.

A high-consistency three-ply former is operating successfully on a board machine in Sweden. The wet end is a combination of three fourdriniers with two top-wire dewatering units. The outer layers of the three-ply sheet are formed with conventional low-consistency headboxes. The high-consistency headbox is used to form the center ply on a top fourdrinier and is followed by a top-wire dewatering unit. Use of mechanical pulps in the center ply permit reduction of furnish cost. High bulk of the middle layer improves board stiffness [7].

C. EVALUATION OF HEADBOXES

There is a multitude of headbox designs on the market. It is necessary to evaluate their performance to determine how good the different designs are.

The most critical characteristics are: basis weight stability and uniformity; a streak-free sheet; the effect of pressure, temperature and total lip opening on the basis weight profile; formation, especially any signs of dense flocks or lumps; cleanliness; and control of jet to wire ratio.

Basis weight uniformity and stability for a given type of headbox is best determined by performing variance analysis on a number of installations for the boxes being investigated. Random standard deviation which measures how stable the flow out of the headbox is, is the best number to use in determining the basis weight stability of a given headbox. However, random standard deviation is a function of the cross-machine profile and increases with increasing cross-machine standard deviation. In addition, random variation is dependent on how fast the sheet is set and on the uniformity of drainage in the forming zone. The same box will give a much smaller random number at short forming lengths on twin-wire formers and with new fabrics. This kind of information is important when comparing boxes. It is also important to eliminate the effect of edges, as they contribute a large amount to random and CD standard deviation. Based on a number of such analyses of different boxes, a good comparison can be made with regard to their ability to provide good basis weight uniformity and stability. Machine-direction standard deviation relates to pulsations and the quality of consistency regulation.

A streak-free delivery is very important, especially for twin-wire formers. Whether or not the jet from a given headbox is streak free can best be observed after the slice lips have been cleaned. Streaks can also be observed by viewing the sheet ahead of the calender or viewing large samples of the full width sheet at the reel.

Formation is in most cases determined by headbox consistency, fiber characteristics, and action on the table. However, for formers where disturbance level is low and forming distance short, the headbox has a major impact on formation. This is typical for roll formers. The headbox can also form hard flocks and lumps if turbulence intensity is very high or there are places in the box where hard flocks can form. Careful analysis of papers made by different types of boxes will show differences in this regard. This should be done through visual inspection of papers as well as with formation testers.

Cleanliness is a must for headboxes. This can best be established by visual inspection of operating headboxes when the machine is down and before it has been cleaned. Operators are the best source of this information. Cleanliness behavior can be estimated by calculating flow velocities in different sections of headboxes and looking at the geometry. The headbox should be designed for easy cleaning. There should be no "dead" areas which could cause buildups.

Stability of the basis weight profile as operating conditions change is of utmost importance. The profile must remain the same, independent of pressure, temperature, or total lip opening. Otherwise, the profile will change at startups, at changing machine speeds, or at grade changes. How good a headbox is in this regard is a function of its design and built-in means for automatically controlling the profile. This should be critically evaluated.

As hydraulic boxes are much more sensitive to correct velocities, it is very important to dimension them correctly. Some headbox manufacturers overstate the operating range of their headboxes. Correct evaluation of drainage capacity of the former as a base for determining the maximum flow is critical. Hydraulic boxes are especially sensitive to cleanliness problems due to their roofs toward which stickies, pitch, and other contaminants float.

In addition, the hydrodynamic characteristics, e.g., the velocities in the different sections of each headbox, should be very carefully analyzed to ensure that they meet modern hydrodynamic design principles. The workmanship and materials used should also be carefully checked.

For modern headboxes, the random standard deviation should be below 1% for cross-machine standard deviation of 1%. Under good operating conditions, some of the best headboxes reach random standard deviations of 0.5%.

D. HEADBOX OPERATION

1. Basis weight uniformity and stability — troubleshooting

One key to efficient paper machine operation is stable, uniform basis weight. How do you determine how good a headbox is with regard to basis weight uniformity and stability? Most paper machines have basis weight profile scanners, normally combined with moisture scanners which show the bone dry basis weight profile measured along a diagonal across the width of the sheet. By observing these profiles, it is possible to get a good idea of profile uniformity and stability. If the basis weight profile is very nonuniform and unstable, there is good reason to investigate the sources of these nonuniformities. Modern scanning systems with computers print out statistical numbers on standard deviation or variance in the cross-machine, machine direction, and the random component. The cross-machine profiles that are used to calculate these numbers combine the impact of length directional, stable cross-machine, and random variations. These numbers are not as useful as those that are obtained by taking 10 to 20 cross-machine samples from the reel and running them through a profiler that will measure both basis weight and moisture and calculate bone dry basis weight profiles. Based on this information, the total, length, cross, and random standard deviation can be calculated. Machine-direction variation shows the effect of consistency and flow variations to the box. Cross-machine standard deviation determines the stable, cross-machine profile variation. This is basically a function of the slice lip profile and any stable hydrodynamic nonuniformities. Random standard deviation determines the stability of the flow out of the box and is very much dependent on headbox design. However, random variation for a given box will increase with cross machine variation and with nonuniform drainage. A useful way of viewing these profiles is to make a profile envelope with all the basis weight profiles shown at the same average basis weight. The profile envelope will show the stability of the headbox. The average profile will show the stable basis weight profile.

Standard deviation numbers can be compared to industry-wide numbers to indicate how the headbox compares to other headboxes producing similar grades. To get comparisons, it is normally advisable to exclude the effect of the edges as edge disturbances can vary and are often not characteristic of the headbox itself.

It is important to determine the reasons for poor performance if the variance analysis shows that the headbox is not satisfactory. A number of techniques are available for such determination.

1. Visual observation of the jet. Use of photography or a stroboscope is of great help as these devices allow analysis of streaks and other disturbances.

   The jet should be uniform and stable. However, this is seldom the case. One can normally determine what causes nonuniformities by observing their size, stability, and frequency. Stationary narrow or wide streaks are caused by nonuniformities in the slice area such as deformation of the slice, nicks, or fibers on slice lipos. If ridges are present across the machine and are narrow, discontinuous, and moving, they can be caused by boundary layer phenomena and Taylor vortices. Larger flow instabilities with a lower frequency and wider spacing normally are caused by flow instabilities from the distribution system or the evening system of the headbox. Size, pattern, and distribution normally will give good indications of the source.

2. Observation of the dry line. Observation of the dry line provides an indication of basis weight stability and uniformity. As with the jet, one looks for the size, location, and movement of nonuniformities. There is normally a good correlation between the dry line and the basis weight profile. The dry line looks much more stable at high vacuums than at low vacuums on the dry line suction boxes. Barring is readily observed at the dry line.

3. Observation of the sheet at the calender and the reel. By looking through the sheet ahead of the calender, one can see the types of streaks and non-uniformities. Stationary, stable streaks come from the slice lip or stationary elements in the forming area, while moving streaks are hydrodynamic and characterized by size, dispersion, movement, and pattern. In all cases, location is very important.

   For a more detailed study of nonuniformities and instabilities of the sheet, it is useful to view a large full width sheet. This allows the determination of the size, position, distribution, and stability of the streaks and a characterization of sheet structure. Based on these observations, one can normally determine the sources of nonuniformities and instabilities.

   In addition to observations and measurements of the basis weight profile and its stability, it should be standard procedure to inspect the headbox. This will identify any buildups in the headbox or areas that can cause nonuniformities and/or instabilities. This is especially important if there are problems with cleanliness and if there is a slow and continuous change in the basis weight profile. Stability of the slice region can be studied by measuring how the slice opening varies with lip opening pressure and temperature. This is best done by recording changes in the basis weight profile at startup and grade changes.

   If the headbox is not performing satisfactorily, it is important to determine the cause of unsatisfactory performance. This can be done by a number of experimental techniques that are discussed below.

4. Injection of dye or cold water into selected branch pipes of the distribution system is often used. Exit of the dye or the cold water is then determined in the sheet by visual means, IR photography, or by viewing paper samples. For a good, stable system, the dye should

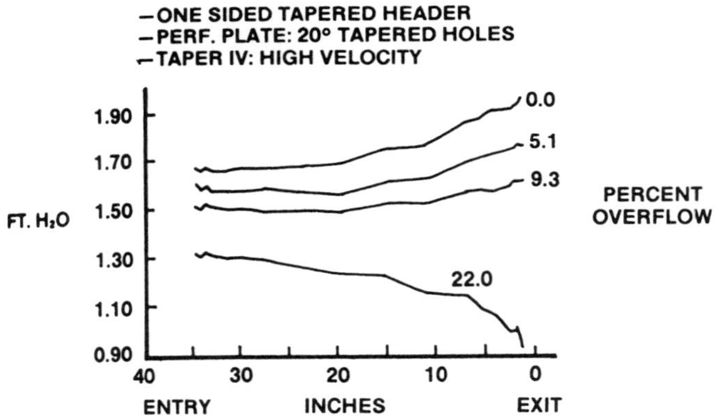

Fig. 115. Figure distribution in one-sided header as a function of overflow (Mardon) [2].

come out in line with the branch pipe, be stable, and show a narrow distribution. However, if the dye comes out widely dispersed in an unstable fashion and offset from the point of injection, there are problems in the distribution and evening systems.

5. Pressure measurements along the pipe, header, and branch pipes will show flow instabilities and are useful in evaluating the reasons for unsatisfactory headbox performance. These measurements can also be used to calibrate and adjust overflow on the one-sided tapered header.
6. Flow measurements in the box are useful if there is reason to believe that flow conditions in the evening section or the pond are a source of instabilities and nonuniformities. This can be done with pitot tubes or propellers which will reveal flow patterns, nonuniformities, stagnant areas, eddies, etc.
7. Pitot tube measurements in the slice lip can be used to show slice flow velocity uniformity. By measuring the pressure differences between a stationary and a movable pitot tube, one can determine the flow instabilities in the slice lip.

Based on this information and a drawing of the headbox, it is usually possible to determine the sources of nonuniformities and instabilities. A short summary is given below.

2. Fixed streaks

Narrow, fixed streaks are caused by nicks, dried fibers, or dirt on the slice lip. Larger, stable streaks can be caused by deformations in the slice lip. These streaks are normally easily observed in the jet and in the sheet. Similar types of streaks, but of a different structure, can be caused by stationary elements in the forming area and by ridges in the wire. These do not show up in the jet.

3. Unstable streaks and an unstable basis weight profile

These are caused by hydrodynamic sources behind the immediate slice area [8].

4. Wake effects

Hydraulic headboxes can cause narrow streaks that move around at relatively high frequencies. These can be caused by boundary layers from vanes, tubes, etc., and by Taylor vortices.

5. Larger instabilities

Larger flow instabilities that are wider, move more slowly and are either varying around a center line or randomly are caused by flow instabilities from the distribution system or the headbox proper. These can be instabilities in the approach to the distribution system; plugged manifold tubes or holes; air; or caused by edges and vortices in the flow through the box. In addition to observations, it is normally necessary to perform measurements as indicated above to identify the cause of these instabilities and nonuniformities.

6. Systematic profile problems

Stable basis weight nonuniformities that show up in basis weight analysis have a variety of causes, some of which are listed below.

– On one-side tapered headers, it is necessary to keep the pressure at the entrance and outgoing sides equal. Figure 115 shows what happens when the recirculation is changed from 0 to 30% for a header that will give uniform pressure at about 10%. This is an important operating variable that should be carefully controlled on one-sided tapered

headers. Not having the flow coming in parallel to the back side of the header can also cause stable nonuniformities.
- Audit the design and installation of the headbox to ensure that there are not design or manufacturing mistakes that cause non-uniformities.
- Dirt buildups in the box can be a cause of stable nonuniformities.
- The slice geometry can cause stable nonuniformities. The slice might not be uniform due to manufacturing errors, deformations of the slice lip, and/or play in the adjusting mechanisms. It is very important to make sure that the slice geometry is stable and independent of pressure, temperature, and total lip opening. Temperature gradients can cause warping.

7. Edge problems

Fit of the cheeking pieces and the edge deckles is very important to give good edges. Any step change will give wakes.

8. Slice profile adjustments

The basis weight profile can be nonuniform due to overadjustment of screws, which causes permanent deformation of the slice. It is therefore very important to check the straightness of the slice periodically and to reset the slice to a uniform opening.

9. Distribution systems

The distribution system is often the cause of large, movable streaks, cross-machine profile instabilities, and large, stable nonuniformities. The appearance of these nonuniformities and instabilities also depends on interaction with the evening section.

10. One-sided tapered header

This is by far the best distribution system. Normally, it has an adjustable overflow. Pressure difference between front and back should be 0 to get a uniform profile.

It is possible to get the hole plates or tube banks plugged at too low flow rates. Plugging will cause a change in the profile with time and make it increasingly unstable. This instability can be overcome by changing the hole plate, plugging holes or tubes, or increasing the flow if possible.

Nonuniformities can be caused by having the approach piping coming in perpendicular to the headbox. A number of design and manufacturing mistakes can cause similar problems. Header geometry, hole entrance geometry, etc., should be checked when stable nonuniformities are found.

11. Unstable profiles

One-sided tapered headers, pulsations and flow instabilities give CD variations. To avoid this, one should eliminate pulsations from rotating elements and design and install the pipeline correctly with suitable attenuators. Bends too close to the header and transition pieces with too large divergence can cause such problems.

Another potential source of instabilities is the stilling chamber, especially if it is used to divide the flow. Using a small number of branch pipes in the distributor will also cause large-scale instability as will poorly designed explosion chambers.

12. Evening section

The evening section is still very important even though the distribution system and the slice are the main causes of hydrodynamic instabilities.

13. Hydraulic headboxes

Hydraulic headboxes use stationary elements in stilling, and explosion chambers for stabilizing and homogenizing the flow. Design of the box is critical with regard to its performance.

Vanes and other elements can cause boundary layer streaks in the jet. These are recognized from spacing and movement of the streaks. The higher the flow rate, the more unstable the flow. There is basically nothing operational that can be done, except by reducing the flow or rebuilding or changing the box.

Hydraulic boxes are much more sensitive to dirt and buildups on the top surfaces. In addition to causing flow disturbances, these also cause cleanliness problems. To keep vanes and roofs clean, velocities must be high 0.9-1.5 m/s. Cleanliness problems show up as a continuous change in profile after wash-up and as an increase in lumps. This can be seen at an inspection. If possible, one can reduce or eliminate these problems by increasing the velocity in the box, by filling in stagnant areas or by increasing the flow rate. Dividing the flow in the stilling chamber can cause flow instabilities. This can only be reduced or eliminated by redesigning the box.

14. Slice region

The slice region can contribute to stable nonuniformities caused by geometry of the slice area and through deformations. The slice can become permanently deformed through exces-

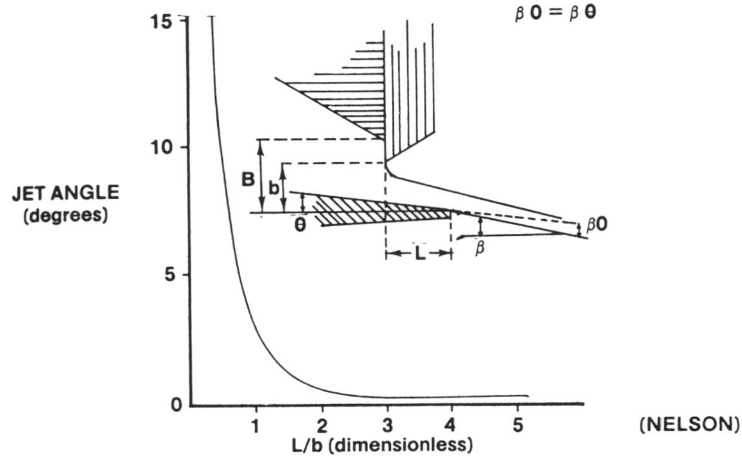

Fig. 116. Jet angle as a function of L/b [2].

sive adjustment and must then be replaced. This is not a big problem for headboxes with small vertical metal strips used for profile control. Slice lips can also be nicked or have dried fibers which will cause streaks. Temperature, pressure, and slice opening can change the profile. This is a very critical problem which has to be dealt with. For good operation, it is very important that the slice profile stay the same, independent of these parameters. It might be necessary to build in pressure and temperature compensation and to eliminate excessive play in the slice adjusting screws to take care of these problems. It is very important that the slice area be checked for accuracy and uniformity.

E. HEADBOX CLEANLINESS — TROUBLESHOOTING

A headbox must stay clean to be acceptable. This has become a much more critical problem with modern hydraulic headboxes than it used to be. Flotation of stickies due to air is often a cause of deposits on the roof and along vanes, mainly on hydraulic headboxes. Velocities above 0.9 m/s will normally keep the headbox clean. Boil-outs, electro-polishing, and keeping surfaces clean and scratch free are the best ways of controlling these problems. Eliminating air in the system is also helpful.

Vanes can become a serious problem if they are too long. Gaps in joints and protrusions into the stock are often places for lump and dirt buildups. These places can be found with careful inspection of the headbox and improved.

Plastic materials in headboxes can sometimes cause deposits because of their affinity to substances like pitch and stickies. Additives can sometimes contribute to cleanliness problems.

Cleanliness problems often occur when headboxes are over-designed, making the velocities too low. This is especially critical for hydraulic headboxes. Boxes should always be cleaned and washed out on shutdowns to prevent fibers from drying onto headbox surfaces. Most modern headboxes are designed for easy cleaning.

F. JET EXIT ANGLE AND POINT OF IMPACT

This is a critical variable both for fourdrinier and twin-wire machines. It determines initial forming conditions and thereby affects formation, wire mark, retention, release, and pinholes. The jet preferably should be set down gently on the wire with a small angle on or ahead of the forming board. It is sometimes beneficial to deposit the jet just ahead of the forming board to control backflow and air entrainment at the point of impact. However, it is difficult to do this with precision along the full width of the machine. Therefore, gently setting the jet on top of the first wide forming board blade is often the preferred method. At high-impact angles, forming pressures are high which tends to give high wire mark, low retention, poor release, and streaky formation. This is accentuated with breast roll discharge. This causes great disturbances as all the nonuniformities in the jet are magnified causing streaks. Very heavy forming pressures further accentuate poor retention, wire mark, and pinholes. Breast roll discharge is sometimes used for coarse grades where these properties are not critical. Breast roll discharge can be used to decrease consistency and thus improve formation in cases of limited drainage capacity. Nonuniform impact angle and position

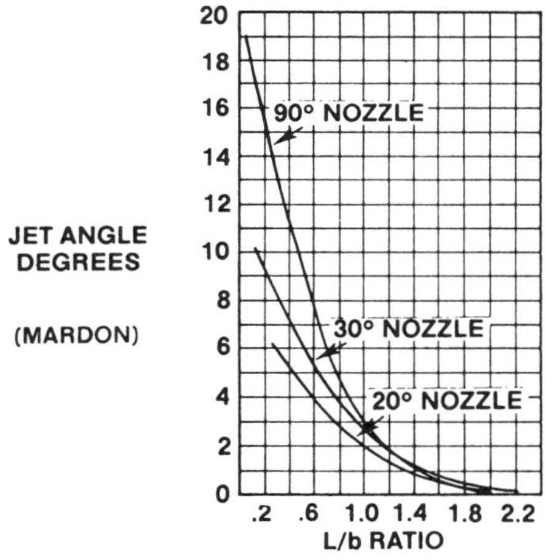

Fig. 117. Jet angle as a function of L/b for varying slice geometries [2].

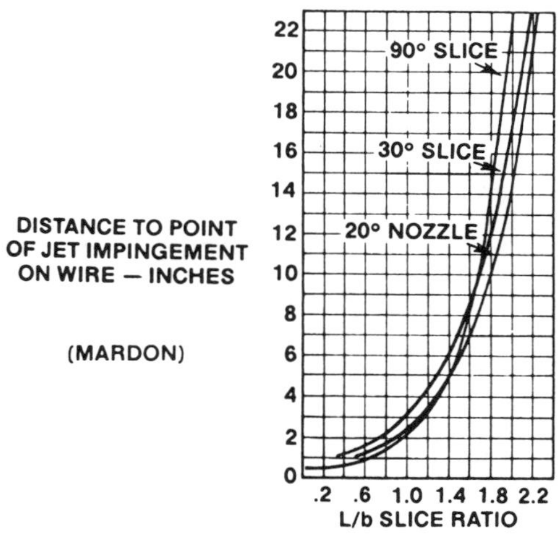

Fig. 118. Jet impact distance as a function of L/B [2].

across the machine can cause large variations in sheet quality across the machine.

Initial conditions are just as critical on twin-wire formers and therefore jet angle and jet entry geometry must be carefully controlled.

Figure 116 shows that the jet angle is a function of L/b, as defined in the figure. Design of the slice has an impact. TAPPI Technical Information Sheets are available to determine the jet angle for different slice configurations. With L/b larger than 2, the jet angle becomes small and less sensitive to varying L/b. Below 1, the angle increases rapidly with lower L/b. The normal operating range for L/b is 1 to 2. Figure 117 shows the effect of slice geometry on the relationship between exit angle and L/b.

The impact line on the wire can be calculated knowing the angle, velocity of the jet, height of the jet above the wire at the exit point, and taking gravity into account. Figure 118 shows the result of such calculations for different L/b's for a given set of conditions. For a given set of conditions, it is important that the position of the forming board in relationship to the headbox is such that the impact point can be controlled. Small variations in L/b have large impacts on the distance that jet travels and thus on the point of the wire that the jet impacts. At small angles, it is very important that variations in conditions across the machine are small.

At low speeds, the jet travels only a short distance before it impacts the wire and L/b provides very poor control. In many cases, especially with large lip openings, it is impossible to avoid breast roll discharge unless the headbox bottom lip extends beyond the center line of the breast roll, or the headbox is provided with an apron or a movable bottom lip.

L/b is an important and critical headbox operating variation which has to be controlled.

G. LIP OPENING

Lip opening determines the effective headbox consistency and flow out of the box at a given speed. It is normally advantageous to form a sheet of paper at the lowest possible consistency that is commensurate with the drainage capacity of the former. Low consistency means better formation and higher strength. However, flow out of the headbox becomes less stable with increased flow and lip opening, which sometimes can restrict the ability to use the benefits of lower forming consistencies due to the limitations of an existing headbox. It is necessary to know the contraction coefficient of the jet to calculate flow out of the headbox based on the head (jet velocity and the lip opening). This is also a function of L/b and slice geometry, as shown in Fig. 119. This is a very useful tool in determining the effective forming consistency, which is the basis weight divided by the jet thickness at the Vena Contracta, divided by the jet-to-wire ratio.

H. SKIMMER OPERATIONS

Some air-cushion headboxes have skimmers. This means that the top layer of the pond flows into a level controlled pocket and is removed from the headbox system. As air, foam, pitch, etc., float to the surface, the skimmer removes these undesirable components. On machines with deaeration, there is no need for skimmers and they are therefore of diminishing interest.

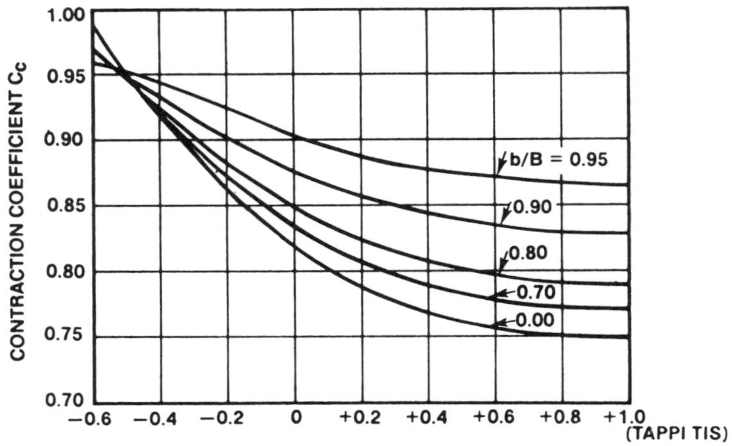

Fig. 119. Contraction coefficient for 45° slice [2].

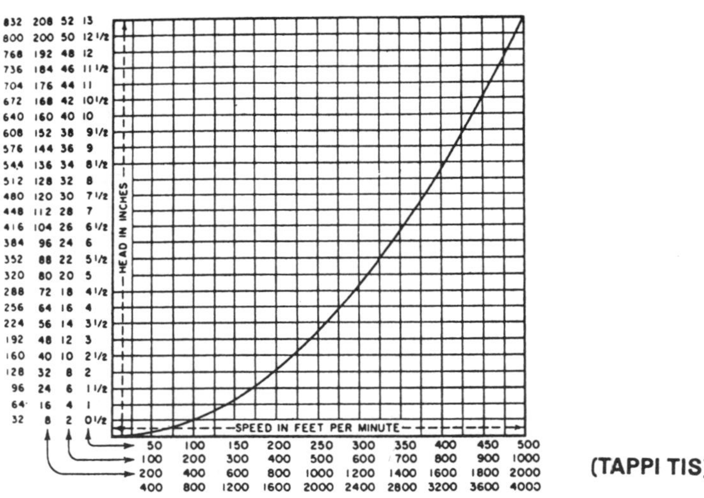

Fig. 120. Jet velocity vs. head [2].

Some hydraulic headboxes have pulsation dampening devices based on free surface with overflow which act as skimmers.

The amount of flow going into the skimmer is normally 1 to 2% of the total flow and can be controlled by level in the headbox in relationship to the height of the skimmer. It is important that this level be controlled well. The outflow from the skimmer, which is also level controlled, must be returned in such a way that it cannot cause basis weight variations.

I. HEAD CONTROL

The most important headbox variable is the total head or pressure that determines jet speed. Figure 120 shows the relationship between total pressure and jet velocity. Jet-to-wire ratio affects sheet properties. Figure 121 shows the effect of jet-to-wire ratio on formation and MD/CD strength properties. The best formation and the most isotropic sheet are made at a jet-to-wire ratio of 1, or very close to 1. As you move away from 1, the sheet becomes more directional in the machine direction and formation deteriorates. This is because there is a speed difference between the jet and the wire which aligns fibers and tends to break up the already formed fiber mat, making a more directional sheet with poorer formation. Because the fibers leaving the headbox have been aligned in the converging slice, sheets are always stronger in the machine direction. However, in hydraulic headboxes

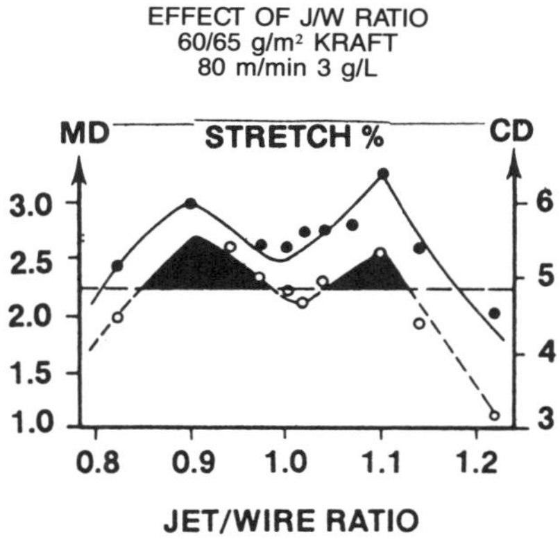

Fig. 121. MD/CD stretch vs. J/W ratio [2].

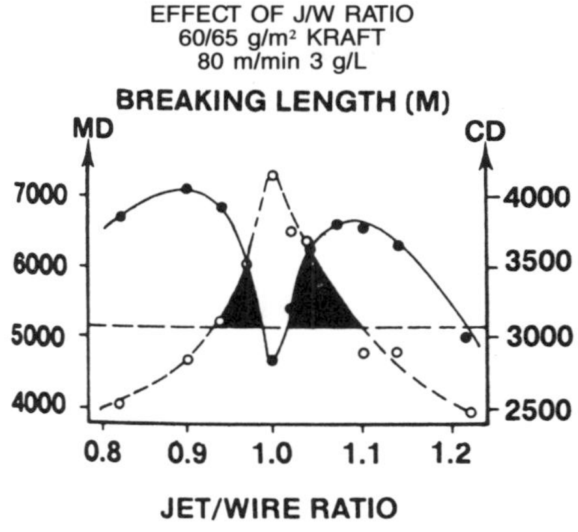

Fig. 122. MD and CD breaking length vs. J/W ratio [2].

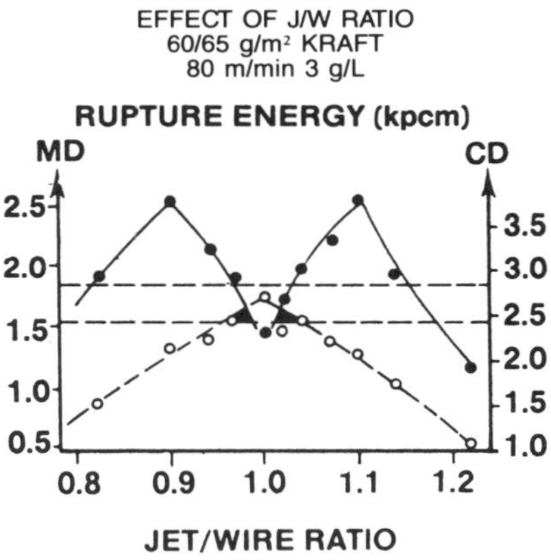

Fig. 123. MD and CD rupture energy vs. J/W ratio [2].

where this acceleration is much smaller and the turbulence is much higher, this alignment is almost nonexistent. Therefore, control of forming conditions are much more important to make machine directional sheets. Figures 122 and 123 show how differently shear works on table roll fourdriniers and wet suction box fourdriniers. It is possible to make directional sheets without loss of formation with wet suction boxes controlling initial forming conditions.

It is necessary to control head very accurately since jet-to-wire ratio is so important. For air-cushion and open headboxes, this is done using the total head in the box. For hydraulic headboxes, it is also necessary to include the velocity head at the point of measurement. On fourdriniers, the jet-to-wire ratio for best formation is 1. For twin-wire formers, it is considerably larger, as the jet must be slowed down sufficiently to build up the pressure needed to balance that exerted by the wire. However, a jet-to-wire ratio relationship identical to that for the fourdrinier exists, taking the slowed-down jet into account.

Modern paper machines with modern control equipment automatically control the jet-to-wire or rush/drag ratio. Because of the importance of the headbox operation, it is advantageous to use on-machine formation testers to keep these variables under control.

J. HEADER CONTROL

Modern headboxes with one-sided tapered headers need O-differential control to maintain a stable, uniform basis weight profile.

1. Basis weight profile

A well-designed headbox should give a uniform profile with a uniform slice opening. An important operating principle is to try to maintain this uniformity and thus to adjust the slice profile only when there are valid reasons. It is advantageous to have automatic systems for recording the slice profile to control the basis weight profile and the slice settings. Such systems are being developed. In addition, there are automatic systems for providing a uniform or preset basis weight profile. It is extremely important that the headbox is satisfactorily designed to maintain a controlled slice opening, independent of pressure, temperature, and total slice opening.

2. Basis weight uniformity and stability

Basis weight uniformity and stability are keys to efficient paper machine operation. The headbox is the key to achieving this. The development of hydraulic headboxes has given a significant improvement in basis weight uniformity and stability and in fiber dispersion.

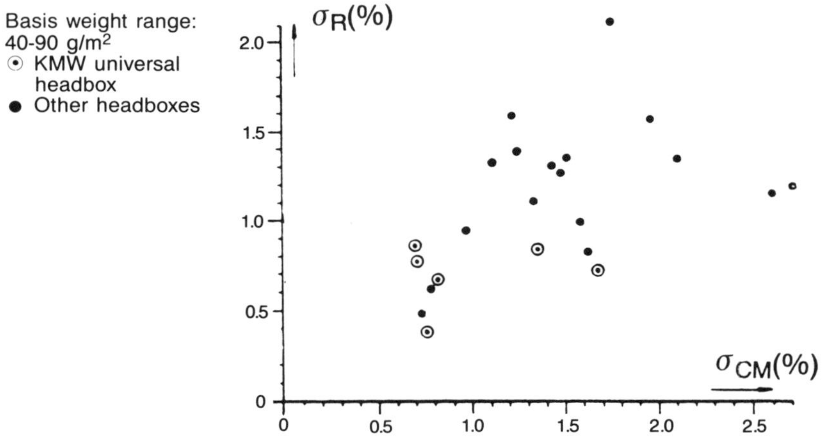

Fig. 124. Random variation as a function of cross-machine basis weight standard deviation [2].

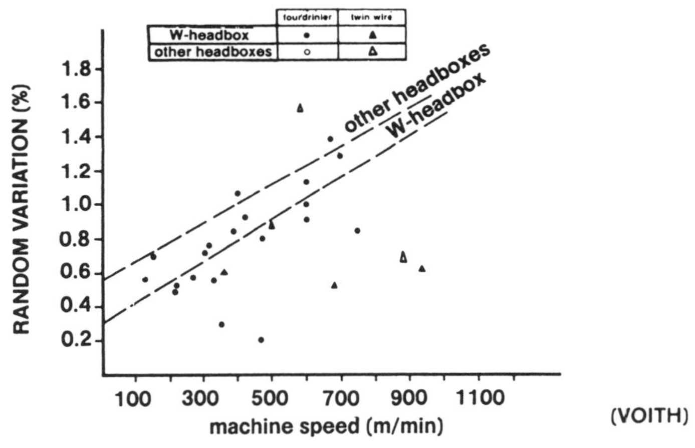

Fig. 125. Random basis weight standard deviation as a function of speed and flow [2].

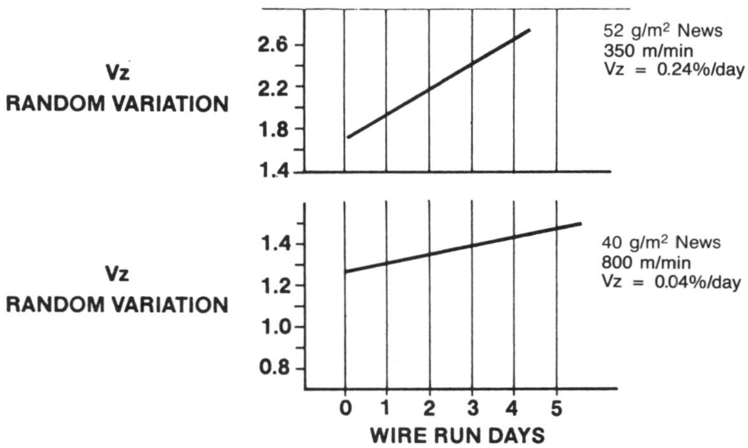

Fig. 126. Random basis weight standard deviation as a function of wire age [2].

Variance analyses of basis weight variations have been performed to characterize the performance of these headboxes. Less than 1% standard deviation in both cross and random variation is being achieved. Very good boxes are down in the 0.5% area for random variation. However, it is important to realize that the operation of the box has an impact on these numbers. Thus, random variation increases as a function of the cross-machine variation, as shown in Fig. 124. This is understandable as a deliberate change in the profile through distortion of the slice opening will increase the effect of instabilities on the basis weight profile. Similarly, increased machine speed and increased flow increase random deviation (Fig. 125). Forming fabric wear and age also add to basis weight instabilities (Fig. 126). It is important to establish comparable conditions and then use a certain level of cross variation at which the random variation is determined.

As the performance of the headbox is very much dependent on how it is operated, it is critical that the relationship between headbox operation, paper properties, and machine performance be well understood.

REFERENCES

1. Wahlstrom, B., "Headbox Design Development, High Speed and Slow Speed", 1989 TAPPI Wet End Operations Seminar, Atlanta, p. 65 (1989).
2. Wahlstrom, B., "Headbox Operation", 1989 TAPPI Wet End Operations Seminar, Atlanta, p. 93 (1989).
3. Smook, G.A., Handbook for Pulp and Paper Technologists, Joint Texbook Committee of the Paper Industry, TAPPI/CPPA (1982).
4. Waller, M.H., "Recent Developments in Headboxes", Tappi J. (Jan. 1987).
5. Stenberg, G., "A New Headbox for Multi-layered Printing Paper", Tappi J. (Sept. 1987).
6. Lehikoinen, E., "High Consistency Forming", Ahlstrom.
7. Bubik, A. and Blum, R., "Systems for Multi-Ply and Multi-Layer Formation", presented at 1987 APPITA Annual Conference, Sydney, Australia.
8. Pantaleo, S., "Stock Piping Principles and Their Effect on Paper Machine Performance," 1989 TAPPI Wet End Operations Seminar, Atlanta (1989).
9. Trufitt, A.D., "Design Aspects of Manifold Type Flowspreaders", Jt. Textbook Committee of the Paper Industry, CPPA and TAPPI (1975).

VII PAPERMAKING CHEMISTRY

W.E. SCOTT

Wiiliam Scott is Professor and Chairman of the Miami University Paper Science and Engineering Department in Oxford, OH, where he directs the academic program, teaches, and carries out research in various areas related to papermaking chemistry. His current research is focussed on the applications of expert systems to wet end chemistry process control. In addition to his academic experience, Dr. Scott has worked for Champion International and served as a consultant to many companies. He is active in various TAPPI committees and is a former TAPPI Director. He has published over 35 technical papers, books, and chapters. Dr. Scott earned his B.Sc. from Miami University and his Ph.D. from The Institute of Paper Chemistry.

REVIEWED BY

Jerome M. Gess
Senior Scientific Specialist, Papermaking R & D
Weyerhaeuser Paper Company
Tacoma, WA

David H. Hollenberg
Research Fellow
James River Corporation
Neenah, WI

A. INTRODUCTION

Papermaking chemistry, or "wet end chemistry", as the subject is more commonly known, is often thought to refer solely to the retention of fine solids during paper manufacture. Indeed, one of the earliest seminars on the topic was entitled just that. However, even a cursory examination reveals that such a definition is much too narrow. This is shown in Table 9, where several common papermaking chemistry phenomena are listed.

Many of the phenomena listed in Table 9 are influenced by actions that occur in the pulp mill, bleach plant, broke reclamation areas, and other parts of a pulp and paper mill. Even the quality of incoming process water can affect papermaking chemistry. While the paper machine wet end is often the point where a papermaking chemistry problem occurs, the solution to the problem commonly lies elsewhere in the mill. For these reasons, the term wet end chemistry will not be used in the chapter. Rather, the broader term "papermaking chemistry" will be adopted.

1. Importance of papermaking chemistry

Table 9 shows that papermaking chemistry involves phenomena that influence both paper quality and paper machine runnability. Chemical additive use efficiency is a third area where papermaking chemistry contributes to the financial picture of a mill. Additives that are lost to the sewer or retained inefficiently will result in increased operating costs. This is particularly true for more expensive materials like titanium dioxide.

The recent, strong interest in alkaline papermaking has placed increased importance on papermaking chemistry due to the high filler levels needed to make the process attractive. Many mills have found it impossible to achieve such filler levels without completely upgrading their papermaking chemistry practices.

Table 9. Common papermaking chemistry phenomena.

1. Fines solids retention
2. Paper machine drainage
3. Fiber flocculation (sheet formation)
4. Sizing agent retention and sizing development
5. Wet strength agent retention
6. Dry strength agent retention
7. Dye retention
8. Deposit formation and deposit control
9. Foaming, foam prevention, and foam control

2. Organization of the chapter

There are two major parts to the chapter. The first part contains a review of colloid chemistry concepts and terms most important to papermaking chemistry. Next, the papermaking chemistry properties of fibers, fiber fines, fillers and flocculating agents are discussed. The third section of part one reviews the papermaking chemistry of sizing agents, wet strength agents, dry strength agents, and dyes. Finally, association colloids are introduced and their relationship to foaming and deposits described. It has been assumed throughout the discussion that the reader has some knowledge of general chemistry, including the basics of organic chemistry nomenclature.

The second part (see Sec. L) reviews papermaking chemistry laboratory and paper machine measurements. Particular attention is paid to the determination of paper machine retention performance. The use of the Britt Dynamic Drainage Jar for laboratory retention investigations is also covered. This section was included because it is recognized that papermaking chemistry is a very empirical area and it is important that papermaking chemists be familiar with the laboratory and paper machine experimental methods and procedures discussed in the chapter.

3. Scope, depth of coverage, and objectives of the chapter

The discussion presented here is broader than previous textbook treatments of papermaking chemistry. Because of this, individual topics are not covered in great depth. Whenever possible, reference to more complete treatments of specific subjects are provided for the reader who wishes to delve more deeply into any given area.

The chapter has two main objectives:

1. To provide the reader with a general understanding of the important chemical concepts underlying papermaking chemistry and how they apply to interactions among the common components of a papermaking furnish.
2. To acquaint the reader with the experimental aspects of papermaking chemistry.

B. COLLOID CHEMISTRY CONCEPTS IMPORTANT TO PAPERMAKING CHEMISTRY

A papermaking chemist must have an understanding of the basic chemistry involved in order to function effectively. Thus, the objective of this section is to introduce the basic chemical concepts that operate in papermaking chemistry and provide a foundation for future study. The discussion first presents a case for categorizing most papermaking chemistry phenomena as hydrophobic colloid systems. Next, various mechanisms proposed for the aggregation behavior of hydrophobic colloids are defined and their relevance to papermaking introduced.

The information presented in this section is applied to various papermaking chemistry phenomena in the following section.

1. Definition of papermaking chemistry

Papermaking chemistry has been defined as the surface and colloid chemistry of papermaking furnish components. To understand the basis for this definition, we will consider some characteristics of a typical furnish.

2. Resident species

Table 10 lists many of the "resident species" of a papermaking furnish. These species are present at various concentrations comprising the composite consistency of approximately 0.3 to 1.0% depending upon grade, speed, etc. They can exist in an environment which is relatively static (i.e., the machine chest) or in one which is quite dynamic (e.g., a high-turbulence headbox).

3. Interactions among furnish components

Furnish components interact with one another in a variety of ways. Some of the more important interactions are listed in Table 11. The interactions listed in this table may be controlled by the papermaker to achieve particular goals, such as sheet formation or fines retention, or they may be beyond his immediate control. In addition, they may be desirable or undesirable. For example, the deposition of an alum/rosin size precipitate onto fibers is desirable, but the formation of an alum scale precipitate is undesirable.

Table 10. "Resident species" of a papermaking furnish.

1. Dissolved electrolytes
2. Suspended fibers
3. Suspended fiber fines
4. Suspended filler particles
5. Water
6. Surface-active molecules (either alone or as aggregates)
   (Examples: detergents, dispersants, wood extractives, defoamers)
7. Dissolved polyelectrolytes
   (Examples: polyacrylamides, cationic starch, wet strength agents, dry strength agents, etc.)
8. Aggregated sizing molecules

Table 11. Important interactions among furnish components.

1. Aggregation of fibers, fillers, and fines
2. Adsorption of dissolved polymers onto fibers, fines and fillers
3. Aggregation of pitch and sizing molecules
4. Adsorption of pitch and sizing molecules onto fibers, fines, and fillers
5. Neutralization of negative charges on suspended and dissolved anionic materials
6. Establishment of equilibria between dissolved inorganic salts and insoluble ionic products
7. Development of micelles composed of surface-active molecules
8. Sorption of water by fibers, fines, and starches

4. Wet end intermolecular forces

Interactions between any two particles involve both attractive and repulsive forces. Chemists usually refer to two types of attractive interactions at the atomic and molecular level: namely *primary* and *secondary* bonding interactions. Primary bonding forces are those which unite atoms into molecules or chemical compounds. These bonds involve energies of 84–840 kJ/mol of bonds formed, operating at interatomic distances of 0.1–0.2 nm.

Secondary bonding forces involve energies of 8–42 kJ/mol operating over distances of 0.2–0.3 nm. *Hydrogen bonding* is a special case of secondary bonding. A hydrogen bond has an energy range of 12–21 kJ/mol and a bond distance of 0.23 nm. It is important in the water-fibre interactions that lead to interfiber bonding in paper. Collectively, secondary bonding forces are referred to as Van der Waals Forces.

In Table 11, the equilibrium between dissolved ions and their insoluble precipitates involve primary ionic bonding effects. Similarly, charged polyelectrolytes may also participate in ionic interactions, but to a lesser degree than salts. The remainder of the interactions in the table involve secondary bonding.

5. Dimensions of furnish components

All the furnish components listed in Table 10 have at least one very small dimension. Several examples of this are shown in Table 12.

Particles having dimensions of the order of those given in Table 12 have large surface area per unit weight. For example, cubes having edge lengths in the 10.0 μm to 0.1 μm range, which are typical dimensions for papermaking furnish components, will have surface areas ranging from 0.6 m^2 – 600 m^2. An informative discussion of the relative dimensions of several papermaking components can be found in reference [1].

Table 12. Smallest dimension of furnish components.

Component	Size	Dimension
Fiber	10-20 μm	Width
Fibril	<1 μm	Width
Nonfibril fiber fine	<1-2 μm	Diameter
Filler particle	0.1-10 μm	Diameter
Dispersed rosin size aggregate	<1 μm	Diameter
Dissolved poly-acrylamide molecule	<1 μm	Length

6. The "colloidal" state

When the surface area-to-mass ratio of a dispersed particle becomes large, then an appreciable fraction of its molecules lie in the interface between the particle and its surrounding environment. These molecules participate in unbalanced secondary bonding interactions with other molecules located in the bulk of the particle and molecules in the surrounding medium. These unbalanced interactions result in excess energy being located at the particle surface, or interface with the surrounding medium. This excess energy, in turn, strongly influences the surface behavior of the particle.

Whenever the surface chemical properties of a dispersed system becomes important to the overall chemical behaviour of the system, the study of the system involves a branch of physical chemistry known as *colloid chemistry*. Thus, papermaking chemistry is really colloid chemistry because of the large surface areas involved and because most papermaking chemistry interactions occur at particle surfaces.

7. Classification systems in colloid chemistry

Having established that papermaking chemistry can be considered to be colloid chemistry, it is convenient to consider papermaking chemistry in terms of common colloid systems. For example, most colloid systems can be classified as either *lyophobic* or *lyophilic*. Table 13 list the principal characteristics of the two systems.

The prefix "lyo-" refers to "solvent". Since most papermaking is carried out in water, the two colloid systems of interest are designated *hydrophobic* and *hydrophilic*. Table 14 lists some typical papermaking examples of hydrophobic and hydrophilic systems.

Two types of hydrophilic colloids are important. The first type includes materials like retention and drainage aids that affect the behavior of hydrophobic substances. The second type are the association compounds like wetting agents, surfactants, defoamers, and dispersants.

Table 13. Characteristics of lyophobic and lyophilic colloid systems.

Lyophobic Systems
1. Suspensions of particles—not solutions.
2. Little attraction or affinity between the solvent and particle.
3. Thermodynamically unstable with respect to the formation of aggregates.
4. Particle-medium interface affects system properties.

Lyophilic Systems
1. True solutions of very large molecules or aggregates of small molecules.
2. Strong attraction between solvent and particle.
3. No true particle-medium interface.

As a group, however, hydrophobic colloids are the most important to papermakers because all retention phenomena can be grouped under this classification. The study of hydrophobic systems is concerned principally with the rate at which they revert to their equilibrium, aggregated state. Indeed, all of the hydrophobic systems listed in Table 14 act through aggregation during papermaking. In practice, the papermaker wishes to control the aggregation process so as to achieve a uniform distribution of materials throughout the sheet, while at the same time retaining the maximum percentage of incoming materials in the sheet.

Further comments are in order about the application of the term "hydrophobic" to the common papermaking systems listed in Tab.e 14. The reader will undoubtedly argue with the application of the term "hydrophobic" to pigments and fiber fines because they are known to be wettable and to interact with water. Indeed, the surface of a fine is covered with a gel formed by solvated hemicellulose and cellulose molecules. Even the rosin size dispersions must interact with water to a certain degree or it would be impossible to form the suspension. Nevertheless, dispersions of these materials do exhibit the behavior described for hydrophobic colloids in Table 13. In addition, they do not have any of the properties listed for lyophilic systems in Table 13. Thus, even though it may run counter to our intuition, it will serve the purposes of this discussion to refer to pigments, fines, etc. as hydrophobic colloids.

The stability of a hydrophobic colloid dispersion is indicated by its ability to remain dispersed and is measured by the time it takes for the particles to settle out of the liquid phase. This phase separation may be the result of gravity sedimentation of the primary dispersed particles, which is affected by the density of the particles and their size and shape, or the aggregation of primary particles into larger aggregates than sediment. The main types of aggregation mechanisms are listed in Table 15.

Table 14. Papermaking examples of hydrophobic and hydrophilic systems.

Hydrophobic
1. All pigments dispersed in water
2. Fines dispersed in water
3. Rosin size in water
4. Dispersed rosin size emulsion

Hydrophilic
1. Starch dissolved in water
2. Gums dissolved in water
3. Hemicelluloses dissolved in water
4. Surfactants, dispersants, wetting agents dissolved in water
5. Retention, drainage, and formation aids dissolved in water

Table 15. Mechanisms for the aggregation of hydrophobic colloid suspensions.

1. Coagulation — The destabilization of a colloidal suspension with salts
2. Flocculation — The destabilization of a colloidal suspension by bonding the particles together with a long chain polymer
3. Agglomeration — The destabilization of a colloidal suspension by low molecular weight, high charge density polyelectrolytes

8. Surface charge on hydrophobic particles

A charge develops at the solid/water interface of solids dispersed in water. In papermaking systems, this charge may arise from dissociated carboxyls and sulfonic acid groups on fiber surfaces and the adsorption of hemicelluloses and dissolved lignin on these surfaces. Mineral fillers, such as clay and titanium dioxide, develop surface charge by ionization at the particle-water interface and by adsorption of other charged substances (hemicelluloses, polyelectrolytes, starches, etc.)

The description of charges at solid/liquid interfaces was first outlined by Helmholtz [2] in 1879 and was elucidated in subsequent years by Gouy [3], Chapman [4], and Stern [5]. Figure 127 depicts the currently accepted representation of the phenomenon [6-8]. The surface potential is determined by the type and surface density of charged groups (e.g., pulp carboxyl groups or sulfonic acid groups). Papermaking components are generally anionic in nature. Therefore, Figure 127 is drawn with a negative charge at the surface.

Ions of opposite charge to that of the particle surface (counterions) are held closely by electrostatic and Van der Waals forces adjacent to the surface. The potential energy drops off

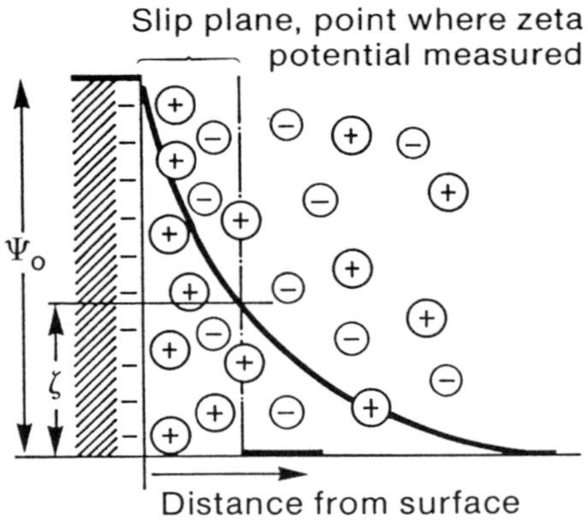

Fig. 127. Representation of the electrical double layer existing at the surface of a hydrophobic particle suspended in water. The zeta potential is taken to be the potential at the slip plane, within which counterions bound to the particle travel with the particle and outside of which counterions are free to move independently of the particle.

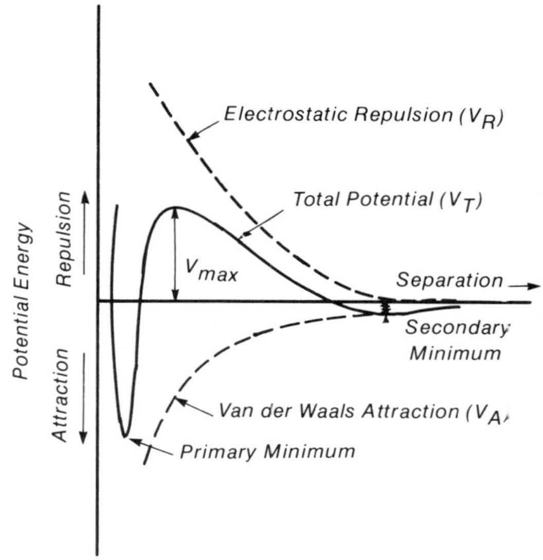

Fig. 128. Representation of interparticle interaction energy as a function of the distance of separation between two particles. The figure illustrates the combining of attractive and repulsive potential energies. The electrostatic repulsion energy, V_R, is represented by the upper dashed curve. The Van der Waals attraction energy, V_A, is the lower dashed curve. The Total Potential Energy, V_T, is the solid curve. The depth of the minimum in the Total Potential Energy Curve is an indication of the stability of the particle-particle aggregate. A shallow minimum suggests that the aggregate can be disrupted more readily than when a deeper minimum exists.

rapidly in this region. A hydrodynamic slip plane exists between the tightly bound counterions and solvent molecules adjacent to the surface (Stern Layer) and the rest of the solution. The less ordered, diffuse portion of the system is called the Gouy-Chapman Region. The potential falls off at a slower rate in this region than in the Stern Layer until zero potential is realized in the bulk of the solution. The potential at the interface plane between the Stern Layer and Gouy-Chapman region is called the *zeta potential*. The common methods of measuring surface charge give values for the zeta potential rather than the actual surface charge. The common name applied to the entire system is the *electrical double layer*. There is no overall net charge in the suspension due to the requirement of electroneutrality. However, local areas of positive and negative potential exist close to the surface of the particles being observed. The strength of these potentials and the distances involved determine the resistance of hydrophobic suspensions to coagulation or flocculation.

9. Importance of the electrical double layer

The stability of a hydrophobic suspension is measured by its ability to remain dispersed over a period of time. This stability depends upon the balance between the relative repulsive and attractive forces existing between suspended particles and how likely the particles are to collide with one another.

Attractive and repulsive forces arise from a number of sources. Van der Waals forces are always attractive for particles of the same chemical composition. Electrostatic forces arising from the interaction of particle electrical double layers are always repulsive if the particles have the same charge. They are attractive for particles of opposite charge. Stabilizing and destabilizing forces can also be developed by adsorbed polyelectrolytes.

The Van der Waals attractive forces and double layer repulsive forces were combined by Derjaguin, Landau, Verwey, and Overbeek into what is now referred to as the DLVO theory [8]. This theory considers the attractive and repulsive forces that exist between two particles and describes the potential energy of the system as a function of distance between the two particles [8]. Figure 128 illustrates this combining of attractive and repulsive potential energies. In the figure, the electrostatic repulsion energy, V_R, is represented by the upper dashed curve. The Van

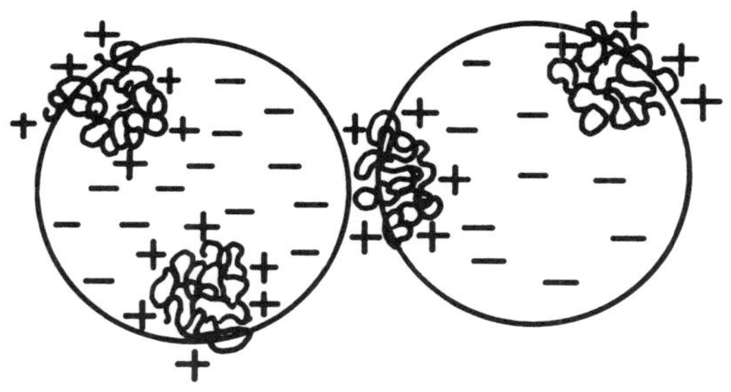

Fig. 129. Schematic representation of the model for patch flocculation behavior. The positively charged patches are attracted to negative regions on the surface of another particle and electrostatic attraction draws them together so that Van der Waals forces can cause them to aggregate. "Soft flocs" are formed by this mechanism.

der Waals attraction energy, V_A, is the lower dashed curve. And the Total Potential Energy, V_T, is the solid curve. The depth of the minimum in the Total Potential Energy Curve is an indication of the stability of the particle-particle aggregate. A shallow minimum suggests that the aggregate can be disrupted more readily than when a deeper minimum exists.

10. Coagulation, flocculation, and stabilization

We will now discuss how coagulation and flocculation can be caused to occur. We will also describe how a system can be stabilized if a dispersed condition is desired.

(a) Charge neutralization

As discussed above, strongly adsorbed counterions and a diffuse region of counter- and co-ions make up the electrical double layer at the surface of particles suspended in water. This layer has a certain thickness for a given set of conditions. There is a reduction in the thickness of the double layer when electrolytes are added to the suspension due to an increase in the availability and number of counterions. This increase in counterion concentration acts to reduce the surface potential and, thereby, the repulsion component of the net interaction potential. Particles are then able to approach each other more easily and coagulation and flocculation are more likely. Coagulation in this manner is referred to as occurring by *charge neutralization* because the addition of excess counterions moves the net potential towards a state of zero charge.

There are two further points regarding this mechanism. First, there exists a critical concentration of salt where the double layer is sufficiently suppressed for coagulation to proceed. This is normally referred to as the Critical Coagulation Concentration, or C.C.C. The second point deals with the effectiveness of various cations in reducing the double layer thickness of anionic surfaces. Shultz and Hardy found that the valence magnitude of the ion charged oppositely to the surface had the principal effect on colloid stability [9]. They found that the higher the valence, the more effective was the ion in reducing surface charge – with the relative ratios of ions of valence +1, +2 and +3 required to achieve the C.C.C. being 4000:100:15 [9].

(b) Patch model

When lower molecular weight (<100 000) cationic polyelectrolytes with high charge density (>4 meq charge/g) are mixed with anionic particles, the polymer molecules are believed to adsorb completely onto the particle surfaces and form a patch of positive charge [10] (Fig. 129). This patch effectively reverses the anionic charge at that site on the particle surface while the rest of the particle remains anionic. (The overall potential may also remain anionic.)

After the polyelectrolyte is adsorbed, collision of the positively charged patch with an uncovered negative area on another particle leads to agglomeration. The aggregation rate is greater than that produced by charge neutralization with a simple electrolyte.

(c) Bridging model

The flocculation of colloidal materials by polyelectrolytes of high molecular weight (>100 000) is presumed to take place by a bridging mechanism [11,12]. In this mechanism, the polymer is thought to adsorb on the particle surface with a series of loops and tails extending into the liquid phase (Fig. 130a). These loops and tails extend well beyond the double layer. Flocculation then occurs by adsorption of the extended loops and tails onto the negative surface of a second particle (Fig. 130b). Bridging flocculation is only dependent on the collision frequency of the particles and the repulsion between the two double layers does not come into play.

The charge density and molecular weight of the polymer are also very important. As you would expect, higher molecular weights promote increased bridging because longer polymer chains extend farther into the surrounding water and away from the particle surface. Polymer charge density influences the attraction of the polymer to the particle surface and the strength

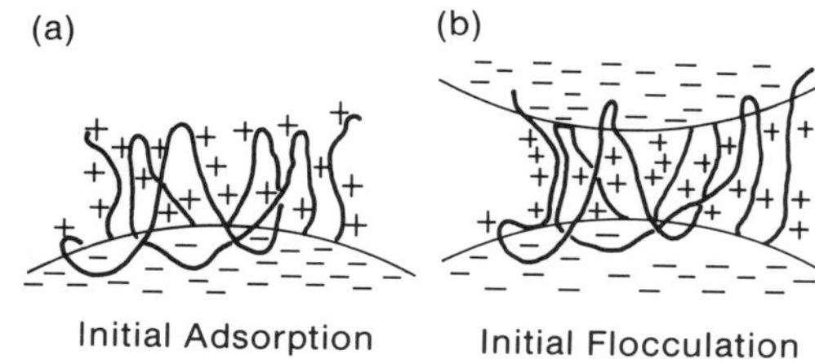

Fig. 130. Schematic representation of the model for bridging flocculation behavior. (a) The positively charged high molecular weight polyelectrolyte partially adsorbs onto the surface of the negative particle. Loops and trains of the polymer extend out into the surrounding medium. (b) The extended loops and trains of the adsorbed polyelectrolyte adsorb onto a second particle, forming a "bridge" between the two. "Hard flocs" are formed by this mechanism.

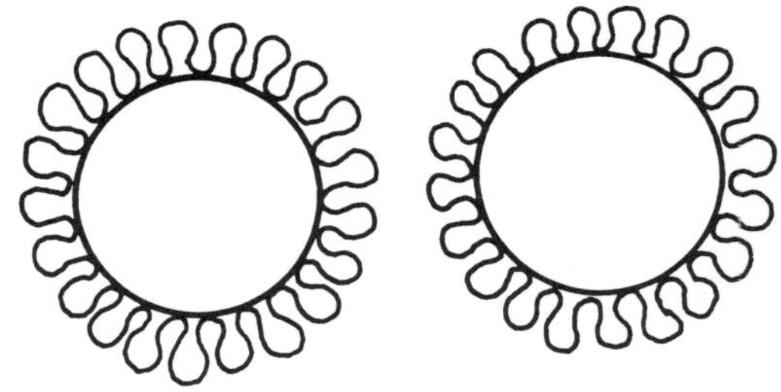

Fig. 131. Schematic representation of the model for steric stabilization. In this instance the surfaces of particles are covered by adsorbed high molecular weight polymer material that extends out past the electrical double layer. When two particles approach one another, the extended polymer coating material provides a steric barrier that holds the particles apart far enough that Van der Waals attractive forces cannot operate.

with which the macromolecule bonds to the surface.

(d) Steric stabilization

In addition to stabilizing a hydrophobic colloid suspension by the development of a high surface charge, a system may also be stabilized when long chain polymers are adsorbed onto particle surfaces in a manner that the loops and tails from one particle interferes with the loops and tails from another particle, preventing them from approaching closely enough to flocculate [13] (Fig. 131). Such polymers are normally nonionic in nature and are of high molecular weight. This phenomenon is often called *protective colloid action.*

11. Summary

This concludes our brief review of relevant colloid chemistry concepts. We began by defining papermaking chemistry as the surface and colloid chemistry of the components of a papermaking furnish. We then explained that this definition applies because the dimensions of papermaking furnish components fall into the range of colloidal particles and the types of chemical interactions occurring in the wet end involve secondary bonding forces characteristic of colloidal systems.

Next, papermaking furnish systems were classified as being primarily hydrophobic in nature with their single most important characteristic being a tendency to aggregate. It was pointed out that the papermaker wishes to control such aggregation during the preparation of a furnish and during the sheet forming process. The basic mechanisms of aggregation were then described.

It now remains to describe how the concepts discussed above are applied to papermaking. In the course of doing this, we will review briefly the papermaking chemistry characteristics of the major papermaking furnish components.

C. APPLICATIONS OF COLLOID CHEMISTRY CONCEPTS TO PAPERMAKING CHEMISTRY

A papermaking furnish consists primarily of fibers, fiber fines, functional additives, and chemical processing aids. The papermaker purchases these materials and blends them into the furnish in the ratios required for a given grade of paper. The process water in which the various ingredients are suspended is another important component. Water plays a direct role in many papermaking steps (dissolving additives, hydrating and swelling fibers during refining, etc.). It also provides the environment for all papermaking chemistry interactions.

Finally, the furnish also can contain a variety of substances, referred to as "cationic demand", "anionic trash", or "interfering substances", that interfere with the action of chemical additives. These substances are introduced to the papermaking furnish as pulp mill and bleach plant carryover, are brought into the system with various types of broke, or introduced with process water. We will discuss these materials later under the general category of *interfering substances.*

We will begin our discussion by reviewing briefly the properties of fibers, fiber fines, fillers, and retention aids that are important to papermaking chemistry. These materials participate in most of the coagulation and flocculation phenomena that occur during papermaking. They also can significantly affect the performance of

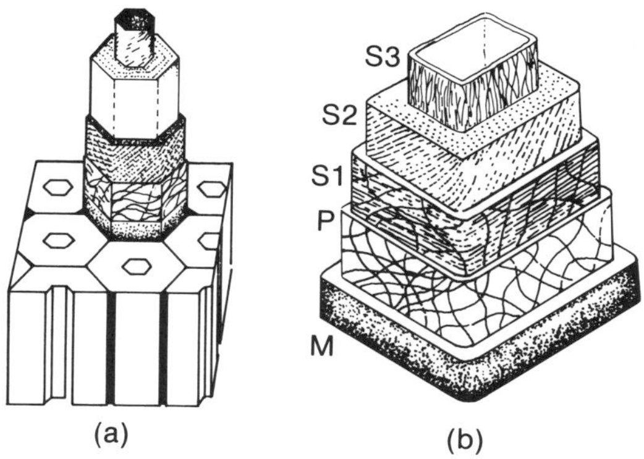

Fig. 132. (a) Cross-section of wood showing relative placement of adjacent fibers. (b) Cross-section of a wood fiber showing the layered structure of the fiber wall. M = middle lamella; P = primary layer; S1 = secondary 1 layer; S2 = secondary 2 layer; S3 = secondary 3 layer. The hole in the center of the fiber is called the lumen.

other functional additives that will be discussed later in the chapter.

1. Papermaking fibers and fiber fines

A detailed discussion of the properties of papermaking fibers is presented elsewhere in this volume (see Chap. II). Therefore, we will concentrate only on those aspects of fiber morphology and fiber chemistry that are important to papermaking chemistry. The characteristics of fiber fines are not described elsewhere in the volume and we will discuss their characteristics in more detail because they are extremely important to papermaking chemistry.

(a) Morphology

The most important morphological characteristics of softwood fibers are fiber length (3–5 mm), which contributes to the tendency of fibers to flocculate when they are suspended in water, and fiber width or diameter (20–50 μm), which lies approximately in the range of colloidal particles and contributes toward the specific surface area of fibers.

Hardwoods are shorter than softwoods (1–2 mm), but still have a strong tendency to flocculate in water suspension. Like softwoods, their diameters lie in the range of colloidal materials (10–40 μm). Unlike softwoods, hardwoods also contain a large percentage of nonfibrous elements. These elements include vessel segments, ray cells, and parenchyma cells. While they have many other influences in papermaking and on paper properties, these elements exert their greatest effects on papermaking chemistry through their behavior as primary fines, in topic to be discussed later in this section.

Figure 132 contains a schematic drawing of the cross-section of a wood fiber. It can be seen that the wood fiber wall consists of different layers; the primary wall and the S1, S2, and S3 secondary walls. The middle lamella is located between the fibers. The lumen is the hole in the center of the fiber.

(b) Fiber chemistry

Many papermaking chemistry properties of fibers and fines are due to their chemical composition. Wood pulp fibers contain cellulose, hemicellulose, and lignin, arranged in a complex mixture in the cell wall. In addition, the wood from which the fibers are derived usually contains organic solvent-soluble resins and fatty materials, designated extractives, that are not completely removed during pulping and can have profound effects on papermaking processes. The relative amounts of each material are 40–50% cellulose, 20–30% hemicellulose, 20–30% lignin, and 1–10% extractives. The average chemical composition of each layer of the fiber wall is given in Table 16. It is evident from the table that the chemical composition varies greatly as you go from the outside of the fiber (middle lamella) to the S3 layer.

The cellulose content goes from 0% to 55%. The hemicellulose content increases from 10% to 40%. And the lignin content decreases from 90% to 5%. Thus, the chemical characteristics of the surface of any particular fiber depends greatly on the mechanical and/or chemical processing treatments that the fiber has undergone. Severe processing may disrupt the outer wall structure

Table 16. Distribution of chemical components in wood fiber [14].

Layer	Thickness, μm	Composition, %		
		Cellulose	Hemicellulose	Lignin
Middle Lamella	1.0 —2.0	0	10	90
Primary	0.03—0.10	10	20	70
S1	0.10—0.20	35	25	40
S2	0.5 —8.0	55	30	15
S3	0.07—0.10	55	40	5

and the chemical properties will be predominantly those of the S2 layer. Conversely, mild chemical or mechanical processing will leave substantial portions of the primary layer intact and it will control the surface properties of the fiber.

Cellulose chains existing at the fiber surface contain hydroxyl groups that participate in fiber-to-fiber bonding in paper. Cellulose hydroxyl groups on the fiber surface and within the fiber wall also interact strongly with water and are important to the swelling phenomena associated with refining and the promotion of fiber-to-fiber bonding.

Hemicelluloses contribute greatly to the swelling wood fibers during beating due to their strong interaction with water. They also promote the development of fiber-fiber bonding in paper.

Residual lignin in papermaking fibers adversely influences the ability of fibers to participate in strong interfiber bonds; provides an inhibiting influence on the swelling of fibers in water; and contributes to the formation of a bulky, rapid-draining mat on the paper machine.

Extractives provide valuable by-products in the kraft pulping process. Their effects on papermaking are all adverse, however. They contribute to foaming, deposit problems, desizing, strength losses, and decreased paper quality.

(c) Influence of pulping processes on fibers

Pulping processes modify the chemical components of wood fibers in a variety of ways, depending on the process selected and yield achieved. Mechanical pulping processes produce few chemical changes in fibers. Consequently, the fiber surface composition and chemistry depend mainly on the degree of structural disruption that occurs during pulping. The inner cell wall region is inaccessible and, consequently, mechanical fibers do not swell much during beating and refining.

Chemical pulping processes produce both chemical and structural changes in fibers. Structurally, the dissolution of lignin and hemicelluloses results in more of the internal fiber wall surface becoming accessible to water and other materials in the papermaking process.

Chemical pulping also results in a number of chemical modifications to the cell wall constituents. For example, cellulose may be degraded by hydrolysis and oxidation reactions, resulting in polymer chains having reduced molecular weight and containing carbonyl and carboxyl groups. Hemicelluloses are dissolved, deacetylated, and hydrolyzed during chemical pulping. Extractives are mostly removed by chemical pulping processes and recovered for their economic value. As was mentioned above, however, sufficient quantities of the extractives are carried on to the papermaking process to cause problems. Much of the lignin is dissolved. The remaining lignin is believed to contain oxidized groups to a greater or lesser extent depending on the pulping process. For example, pulps that have undergone an oxygen stage during pulping or bleaching have greater carboxyl contents than pulp produced by more standard processes.

2. Fiber Fines

Most papermaking furnishes contain fine fiber particles as well as intact fibers. Papermakers have defined fines in many different ways. For example, Jaycock [15] defined all soluble or colloidal material in a furnish as fines. Others [16] have taken all material that passes though a 60-mesh screen as fines. Most recent work [17,18] defines fines to be that part of the furnish passing through a 200-mesh screen – a definition that will be followed in this discussion.

Pulp fines vary widely in morphology, but often are rod-shaped, semicrystalline, fibrillar material having widths of 0.1–0.5 μm and lengths reaching several tens of micrometers. Fines also occur naturally, particularly in hardwoods, where the small rectangular-shaped parenchyma and ray cells represent a substantial percentage of the total weight of a fiber sample.

It is suggested that the following terminology be adopted when talking about fiber fines.

- *Primary fines*: Fines that occur naturally in a wood pulp sample. These include parenchyma and ray cells and some hardwood vessel elements.
- *Secondary fines*: Fines generated by refining. They usually have the geometrical characteristics discussed in the preceding paragraph. They can also include "veil-like" portions of cell wall layers that are stripped off by the refining process.
- *Tertiary fines*: Fines that are generated during the flow of unretained stock around the paper machine white water system. This material may arise as the result of the action of pump impellers, tank agitators, etc. Tertiary fines also arise from the "fractionation" that occurs on the wire during sheet formation.

It is believed that fines have inherently the same chemical composition as their parent fibers in a pure fiber/water system. However, this is not the case in a typical papermaking system, as will be discussed shortly. The fines concentration in any particular furnish depends on factors like the extent of refining applied, the amount of broke utilized, the first-pass retention performance of the paper machine, and the amount of saveall cloudy white water used throughout the system.

3. Fiber and fines characteristics important to papermaking chemistry

The papermaking chemistry phenomena related to fibers and fines can be directly related to their surface and colloid chemistry. The characteristics of importance are sorption capacity, swelling behavior, and ion-exchange properties. These are controlled to a large extent by the surface area, surface charge, and surface chemical composition of fibers and fines.

Because of their small particle size, fines have a large surface area per unit weight. Marton [19] reported that the specific surface area of fines is 5–8 times that of fibers. Table 17 contains sample data to illustrate this point. This high surface area causes fines to play a major role in papermaking chemistry.

Fines also have a major influence on paper machine drainage performance due to their high water sorption capacity (2–3 times that of fibers), which causes them to behave much like a gel [18].

Fines retention is important because fines contribute to paper strength and other properties. Fines retention is also crucial to efficient materials utilization since fines may represent as much as 40–60% of a furnish in some instances. You are referred to the excellent series of papers by Marton [19–22] for a review of the importance of fines to wet end chemistry.

The surface charge on fibers and fines is a complex function of the chemical composition of the fiber (fines) surface, its state of ionization, and the nature and amounts of additional substances adsorbed. For the simplest case, we will consider a suspension of well-washed fibers and fines in very pure deionized water. Under these conditions, the fibers and fines always take on a negative charge due to the ionization of surface carboxyl groups associated with hemicelluloses, oxidized cellulose, and oxidized lignin. If we lower the pH below 3.75, most of the carboxyl groups take on a proton and the charge on the fibers and fines becomes essentially zero. Conversely, if we increase the pH to 7–8, many of the carboxyls are neutralized and the surface charge becomes more negative. This process can be readily followed by measuring the zeta potential or electrophoretic mobility of the slurry.

Table 17. Surface area measurements of pulp fractions [19].

Pulp Fraction	Specific Surface Area (m^2/g)
(BET/N_2—After solvent exchange)	
Fiber Fraction	9.9
Fines Fraction	46.8
(Hydrodynamic)	
Pulp Blend, 410 CSF	2.1
Fiber Fraction	1.2
Fines Fraction (16.5%)	8.0

The "clean" environment described above does not exist under papermaking conditions. Pulps are rarely washed free of extraneous pulp mill or bleach plant chemicals and reaction products. Even when pulp has been well washed, it will adsorb dissolved substances from the water in which it is suspended. These sorption phenomena involve both ionic and nonionic (Van der Waals) interactions. In most instances, such materials are anionic in nature, causing the pulp to take on a greater negative charge. In the case of dissolved salts, the process is called ion-exchange (or base-exchange).

The ion-exchange capacity of cellulose fibers is important to many papermaking phenomena, among which are electrical and electrokinetic properties, adsorption of dyes, aluminum and polyelectrolytes, rosin sizing, and fiber and fines flocculation. The attractiveness of wood fibers for various ions is in the following order:

$$N(CH3)^{+1} < Li^{+1} < Na^{+1} < K^{+1} < Ag^{+1} < Ca^{+2} = Mg^{+2} = Ba^{+2} < Al^{+3}$$

Thus, aluminum binds more strongly to wood fibers than calcium, which binds more strongly than sodium. The charge-neutralizing power of the cations is in a similar order. The adsorption of aluminum ions shifts the fiber charge further in the positive direction than does the adsorption of sodium ions. Indeed, it is possible to neutralize the fiber charge with aluminum under certain conditions.

Studies [23] have shown that the ion-exchange capacity of wood fibers results mainly from carboxyl groups created on cellulose and lignin by oxidation during pulping and bleaching and from the presence of naturally acidic hemicelluloses. Sulfonate groups also contribute to ion exchange reactions.

Rosen [24] studied the ion-exchange behavior of pulps over a wide pH range and concluded

Table 18. Paper properties improved by mineral fillers.

1. Brightness
2. Opacity
3. Whiteness
4. Gloss
5. Smoothness
6. Printability

Table 19. Different types of filler pigments.

Titanium Dioxide (Anatase and Rutile)
Kaolin Clay
Talc
Plastic Pigments
Calcium Carbonate (Ground and Precipitated)
Aluminum Trihydrate
Silicas
Calcium Sulfoaluminates

Table 20. Important physical properties of filler pigments.

1. Refractive Index
2. Particle Size
3. Particle Shape
4. Specific Surface Area
5. Specific Gravity
6. Ink Absorptivity or Hold-Out
7. Brightness
8. Whiteness

Table 21. Filler properties that are important to papermaking chemistry.

1. Particle Size
2. Degree of Aggregation
3. Particle Shape
4. Specific Surface Area
5. Surface Charge
6. Chemical Composition

that carboxyl groups on lignin and carbohydrate fractions were responsible for ion adsorption in the pH range of 3 to 7. In the 9.0 to 11.8 range, phenolic hydroxyls also were important.

It appears, then, that under normal papermaking conditions only the carboxyl groups are involved. However, under the higher pH conditions of bleaching and pulp washing, lignin phenolic hydroxyls also contribute to ion-exchange phenomena.

Other studies [25] have shown that soda-oxygen pulps have a higher affinity for cations than kraft pulps, as a result of the higher carboxyl content of the oxygen pulps.

4. Summary

Pulp fibers and fines participate in papermaking chemistry phenomena through chemical, colloidal, and surface interactions. The most important factors to consider are the chemical composition of the fiber (fines) surface, the size of the particles, and the surface area of the particles. Since they have an exceedingly high surface area per unit weight compared to fibers, fines tend to dominate many papermaking chemistry phenomena through their high adsorptive capacity. A parallel result is that even though fibers and fines initially have similar chemical makeup, the high adsorption of substances by the fines fraction results in a final difference in characteristics between the two fractions. Fines also have a greater specific swelling tendency in water and can exert a significant influence on the drainage performance of a furnish.

Fibers and fines have a naturally negative surface charge that can be modified by changing the degree of ionization of surface carboxyls through a pH change or by the exchange/adsorption of charged species – both cationic and anionic.

5. Fillers

White mineral pigment fillers are used extensively by papermakers to improve a variety of paper properties, most of which are listed in Table 18. Fillers can also be used as fiber substitutes in order to lower furnish cost. Many detailed accounts of filler properties and uses have been published previously [26–29]. It will serve our purposes to review the general characteristics of pigments and then concentrate on those filler properties that influence papermaking chemistry.

There are currently a large number of filler pigments available to the papermaker from several different suppliers. A list of different filler types is given in Table 19. Reference [30] contains an extensive listing of filler types, trade names, and physical characteristics.

The characteristics listed in Table 20 are important for a pigment to enhance the paper properties listed in Table 19.

(a) Fillers and papermaking chemistry

Fillers affect papermaking chemistry through first-pass retention phenomena and through their direct interaction with other wet end additives. Table 21 lists the filler characteristics that are important in papermaking chemistry.

The term particle size is used here to mean the size of the ultimate particles into which a filler can be dispersed. Table 22 lists average particle sizes for a number of commercial pigments. In the table the rightmost column provides an indication of the narrowness of the particle size

Table 22. Average particle sizes for common fillers.

Pigment	Avg. Particle Size μm	Percent Between 0.2 μm and 0.8 μm
Plastic Pigments	0.37	85
Titanium Dioxide	0.37	85
Filler Clay	0.60	33
Precipitated Calcium Carbonate	0.80	43
Ground Calcium Carbonate	0.95	33
Aluminum Hydrate	1.0	26
Calcined Clay	1.1	34
Talc	1.2	27

distribution – the higher the percentage, the narrower the distribution. It is obvious that titanium dioxide and plastic pigments have much sharper distributions than the rest of the pigments. It should be emphasized that the values in Table 22 represent averages over several different products for each filler type listed. For example, a total of 26 different clays were averaged, with the individual average particle sizes ranging from 0.25 μm to 1.7 μm, and most values lying between 0.65 μm and 0.80 μm.

The values in Table 22 represent ultimate particle sizes. In practice, filler particles tend to aggregate into larger structures that the papermaker works with in his daily operation.

The relationship between filler particle size and retention has been studied by a number of researchers [28,31,32]. In all instances it has been found that increased particle size (or aggregate size) leads to increased filler retention. Flocculating pigment particles with a retention aid produces a dramatic increase in retention. Some typical results are presented in Fig. 133.

The difficulty of retaining very fine filler particles leads to their buildup in recirculating white water with the result that there is an increased percentage of fine particles in the headbox compared to the fresh filler pigment introduced during stock preparation [32].

(b) Degree of aggregation

The above considerations would seem to suggest that the papermaker should strive to aggregate filler particles as much as possible in order to maximize first-pass filler retention. This is true if the filler is being utilized for purposes other than to gain opacity. It is well known that the aggregation of filler decreases its light scattering capacity and thereby lowers its opacifying power. Thus, the papermaker is often faced with a compromise between high retention

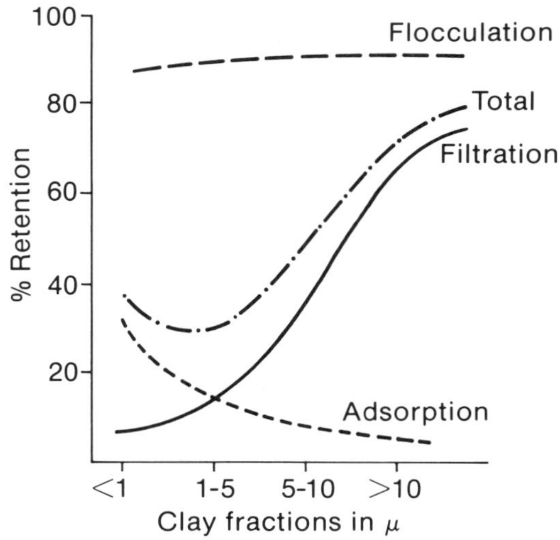

Fig. 133. Influence of particle size on filler retention. The larger the particle size, the greater the net retention [31].

values and high opacity. The challenge is to adjust the papermaking chemistry to achieve the desired opacity at the highest first-pass retention and lowest filler level possible.

(c) Particle shape

Particle shape is known to influence the light scattering performance of pigments, as well as their response to calendering and gloss development. Particle shape also affects the extent to which a filler decreases the air resistance and drying rate of paper. Gill and Scott [29] showed in their work with a wide range of calcium carbonate filler pigments that particle shape can have a strong influence on the particle packing behavior of pigments and, hence, on the properties and papermaking chemistry of aggregated pigments.

(d) Specific surface area

Depending on the particular filler pigment in question, a very wide range of surface areas exist. This is illustrated in Table 23. The surface area of any given pigment depends on its particle size and shape. In general, the smaller the ultimate particle size, the greater the surface area. It is also true that fine materials have a greater tendency to aggregate than coarse materials. This is illustrated by the behavior of the sodium silico aluminates and precipitated silicas, which have extremely high surface areas (20–80 m^2/g), but have average particle sizes in the range of 0.8 μm. Examination of this 0.8 μm material with a scanning electron microscope reveals that the true ultimate particle size for the material is less

Table 23. Specific surface areas of various filler pigments.

Filler Type	Specific Surface Area, m^2/g*
Titanium Dioxide	8—10
Clay	
— Coating	10—15
— Filler	2—5
Aluminum Trihydrate	7—18
Precipitated Silica	30—40
Calcium Carbonate	
— Ground	2—12
— Precipitated	2—10
Plastic Pigment	20—22

*BET specific surface areas taken from technical data sheets.

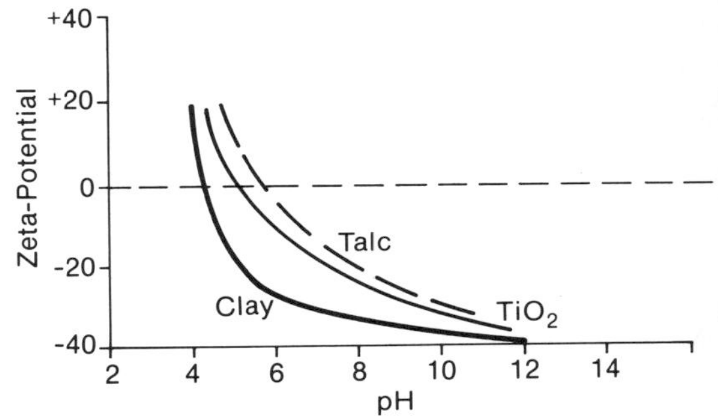

Fig. 134. Influence of pH on the surface charge of filler particles [27].

than 0.1 μm and that the 0.80 μm "particles" are really aggregates. It has not been possible to prepare disaggregated silicas suitable for particle size studies due to their exceedingly strong tendency to form clumps of particles. The particles do not join together through face-to-face contact, however, and the aggregates are very porous, resulting in the high observed surface areas.

Filler surface area affects papermaking chemistry in a number of ways. First, the inverse relationship that exists for most fillers between particle size and surface area suggests that the higher surface area materials will be more difficult to retain. Second, filler surface area influences papermaking chemistry through its relationship to adsorptive capacity.

In our earlier discussion on fiber fines (see Sec.), we stated that fines have the capacity to adsorb more wet end chemicals than fibers due to their (fines) greater specific surface area. Exactly the same thing can be said for fillers. The high relative specific surface area of fillers relative to fibers leads to a greater sorption of wet end chemicals per unit weight of filler than for fibers. The ramifications of this are many. High filler loadings are detrimental to sizing because the sizing material adsorbs onto filler and is rendered ineffective. Dyes distribute themselves more toward filler particles, which increases two-sidedness if the filler is not evenly distributed through the sheet. High surface area pigments require more retention aid to effect a particular retention level under a given set of paper machine operating conditions because the polymer is "diluted" by the large surface area over which it becomes distributed. A recent review of these considerations can be found in reference [33]. It follows, then, that a buildup of filler levels in a paper machine white water system can have adverse effects on a wide range of operating factors and paper properties.

(e) Pigment surface charge

The charge on a filler particle suspended in water will usually be negative unless some other substance is added to reverse the sign of the charge. Several calcium carbonate suppliers have recently developed positively charged calcium carbonate filler pigments for use by the paper industry. The sign and magnitude of the charge play a key role in pigment dispersion and in the retention of dispersed particles by a fiber matrix. By adjusting filler surface charge one can manipulate the dispersing/aggregating behavior of a filler system.

The surface charge on filler particles can be altered by changing the pH of the suspending water, by the addition of soluble inorganic salts (e.g., alum, calcium chloride), by the addition of polymeric dispersing agents (polyacrylates, polyphosphates), and by the incorporation of charged groups in pigment surfaces when the pigment is manufactured. The effect of pH on filler charge is illustrated for several fillers in Fig. 134.

Different commercial pigments exhibit charges ranging from −30 mV to +14 mV [34] when suspended in distilled water at pH 6.5. The more negative products contain polyacrylate or polyphosphate dispersing agents. It should be noted that even though clay exhibits a net negative charge when suspended in water at pH values greater than 2.0, the surface of the clay particle is amphoteric – the edges of clay platelets are positively charged, while the faces are negative. The greater surface area of the faces results in a net negative charge.

Filler charge considerations affect papermaking chemistry in several ways. First, the achievement of a stable, well-dispersed filler slurry that provides good light-scattering performance when the pigment is incorporated in paper requires the use of dispersing agents which directly interfere with the first-pass retention of filler particles on the paper machine. This interference occurs because the filler particles are stabilized in a dispersed state and have little tendency toward aggregation. The papermaker overcomes this situation by neutralizing the filler charge with alum or by utilizing a polymeric retention aid to induce filler aggregation and also cause the formation of filler-fiber flocs.

A second, indirect, way that filler charge can affect papermaking chemistry is through the interaction of pigment dispersing agents with other chemicals in the system. For example, polyacrylate dispersants will react with cationic polyacrylamide retention aids and render them ineffective. Dispersing agents will also interfere with other cationic chemicals in the system. Thus, whenever possible, the papermaker avoids the use of filler pigments that require a high dispersant level.

(f) Filler pigment chemical composition

Most common pigments are inert under normal papermaking conditions. A notable exception is calcium carbonate, which cannot be used in acidic conditions. The restriction of calcium carbonate to neutral/alkaline papermaking conditions has serious implications for the papermaker because conversion from acid to alkaline papermaking conditions requires major alterations in the system chemistry. In spite of this, recent developments in calcium carbonate filler products have encouraged the paper industry to take a fresh look at alkaline papermaking. (A brief discussion of alkaline papermaking is presented later in Sec. K.)

(g) Summary

This concludes our brief discussion of the properties of filler pigments. It should be clear by now that filler pigments play a major role in the papermaking chemistry of any paper machine that produces filled grades of paper.

6. Flocculating agents used in papermaking

Early in the history of machine papermaking, very little was done to promote the flocculation of fibers and fine materials in a papermaking furnish other than to add the alum required by alum-rosin sizing. More recently, the advent of high-speed, high-shear paper machines has dictated the use of polyelectrolyte retention aids that produce stronger flocs than alum.

Alum, $Al_2(SO_4)_3$, is an effective flocculating agent for fines and fillers under certain conditions. It has been shown that the Al^{+3} ion is the active component. The conditions under which alum is most effective are described later in the discussion of alum/rosin sizing.

The discovery of LaMer [35,12,36] and others that low dosages of polymers could be used to flocculate fine particles has been adopted by the paper industry and it is now common practice to utilize polyelectrolytes as retention aids in papermaking. The polymer materials utilized by the paper industry are usually characterized in terms of their backbone structure, molecular weight, charge type, and degree of substitution or charge density.

The backbone structures include polysaccharide materials like starch (see Sec. F), guar gum, and synthetic polymers, such as the polyacrylamides. The average molecular weights of the so-called "high" molecular weight polyacrylamide retention aids are in the range of 1–20 million (higher values for anionic polyacrylamides), while "low" molecular weight materials are less than 100 000.

All effective retention aids have some kind of cationic or anionic groups incorporated into their structure. Figure 135 depicts some of the types of charged groups that are utilized. Typically, 4–10% of the monomer units will carry charges in cationic retention aids while 30% of the monomers carry a potential charge in anionic retention aids. The actual charge density may depend on the pH. For example, anionic acrylamide-acrylic acid polyelectrolytes are 100% ionized at pH values above 6. Below 6 the charge decreases with decreasing pH and, in the typical papermaking pH range of 4.0–5.0, anionic polymers typically contain charges on only 5%–28% of their monomer units. On the other hand, tertiary amines are fully ionized below pH 6, but only partially so above that pH. Quaternary amines are fully ionized over the entire pH range (4–10) of interest to papermakers. A good overview of the chemical properties of synthetic retention polymers is given in reference [37].

7. Polyelectrolyte adsorption

Polyelectrolytes must be adsorbed onto fibers and fines in order to be effective. Adsorption involves Van der Waals forces as well as ionic interactions. The principle variables involved are polymer charge, polymer charge distribution, polymer chain length, polymer conformation, polymer chain flexibility, the availability of hydrogen bonding groups, polymer concentration, contact time between fibers and polymer,

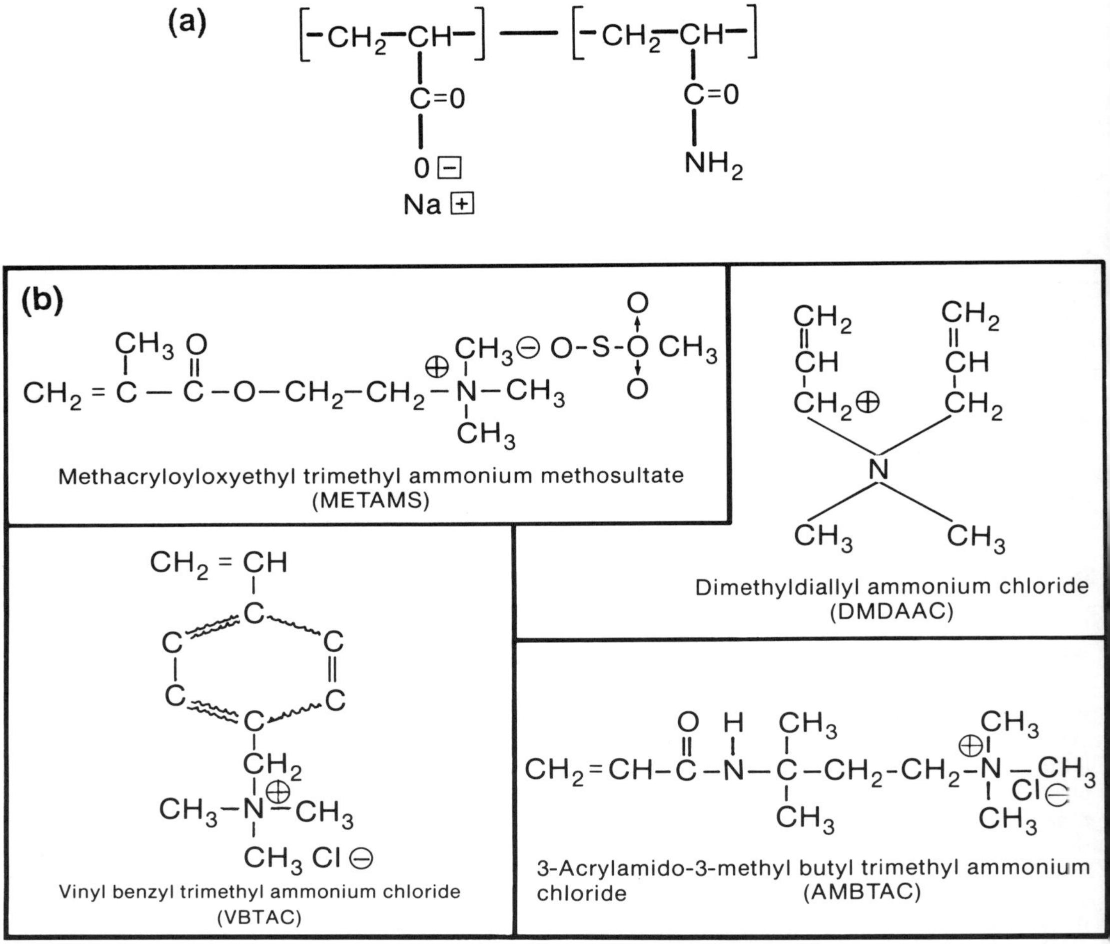

Fig. 135. Charge-bearing monomer units incorporated into retention aids and other polyelectrolytes. (a) Polyacrylic acid/polyacrylamide in anionic polyelectrolytes. (b) Cationic groups.

and the shear forces present at the time of mixing.

D. COLLOIDAL ASPECTS OF SHEET FORMATION

Now that we have briefly reviewed the papermaking chemistry characteristics of fibers, fiber fines, fillers, and retention aids, we will turn to a discussion of how these important wet end components interact on the paper machine immediately prior to, and during, sheet formation. The specific topics of interest here include fiber flocculation, fines flocculation, and drainage phenomena – all three of which are very important in paper manufacture. *Fiber flocculation* primarily influences the "formation", "look-through", and mass uniformity properties of the web. *Fines flocculation* affects the retention of fiber and filler fines in the web as it is being formed and influences both paper properties and the runnability of the paper machine. *Drainage* pertains to the water removal processes occurring in the paper machine forming section. It is well known that drainage is a key factor to the productivity of many paper machines.

It should be understood that the achievement of good sheet formation (the property) and high drainage rates always involves hydrodynamic as well as chemical considerations. The papermaker should optimize the mechanical configuration of a machine before attempting to work with its papermaking chemistry. An excellent description of the influence of machine speed on formation, drainage, and retention is given by Marton [38].

1. Fiber flocculation

Fiber flocculation is the result of the interaction of many parameters. There are two aspects to fiber flocculation. In the first instance, fibers become physically entangled with one another as they flow through pipes and are stored in chests and tanks. The primary approach to minimize this type of aggregation is dilution and the judicious application of shear and turbulence forces. The second type of fiber flocculation behavior derives from the gathering together of fibers through charge attraction (positive to negative) and other attractive forces (hydrogen bonding, etc.)

Much work has been done on the effect of fiber length on the formation ("formation" is used here to denote the sheet property) of the sheet and the physical measurement of this phenomenon [39-42]. "Formation" is often measured by monitoring the attenuation of beta-radiation as the sheet moves past a radioactive source and recording the resulting power spectrum of the transmitted beta particles. Analysis of this power spectrum gives a measure of the size and distribution of flocs. It has been generally found that the longer the fibers, the greater the tendency to form large "cotton-ball" flocs. Longer fiber lengths are also very resistant to mechanical deflocculation. In many cases, substitution of short hardwood fibers for softwood fibers, maximum possible dilution, and the application of turbulence and cross-direction shear immediately prior to sheet formation on the paper machine have been used to improve formation. Synthetic polyelectrolytes, starches, gums, and other chemical additives can have both beneficial and adverse effects. Polymer additives that are utilized specifically for improving formation are called formation aids.

(a) Formation aids

The ability of high molecular weight, viscosity-enhancing materials like the extracts of *Hibiscus manikot* root to facilitate the formation of long-fibered papers has been known and used for centuries in Japan [43]. Naturally occurring products such as guar gum, locust bean gum, and deaceylated Karaya gum have also been used for this purpose commercially [44].

In recent years, synthetic polymers have gained prominence in this area. It has been found that high molecular weight is a primary factor and that anionic polymers are more effective than nonionics. In addition, the active functional groups (carboxyl) must be ionized to be effective [45,46]. Any condition that causes the polymer to expand and increase the viscosity of the medium improves its effectiveness. The mechanism of action in all cases appears to be that the expanded polymer (estimated end-to-end distance = 1 μm) prevents close contact of fiber surfaces so that flocculation cannot occur. Additionally, viscosity increases restrain the motion of the fibers in the suspending water phase and reduce the tendency for fiber flocs to form. There is also a certain amount of fiber-fiber repulsion generated by the anionic charges on the polymers. Along with these desired effects, adverse side-effects can occur, such as sticking on press roll surfaces if excess anionic polymeric material is used. There can also be a large decrease in drainage rate associated with the use of large quantities of formation aids.

(b) Fiber flocculation by retention aids

Retention aids can promote flocculation of the fiber fraction of a furnish. Different materials affect fibers in different ways. Alum and low molecular weight polyelectrolytes seem to form small, weak, easily dispersed flocs (termed "soft flocs") that readily reform after exposure to shear and turbulence forces. These normally do not produce undesirable sheet properties. On the other hand, high molecular weight cationic polyelectrolytes and dual polymer systems can produce tenacious fiber flocs. After exposure to prolonged high levels of shear and turbulence, these flocs (termed "hard flocs") tend to break down and do not readily reform after the shear and turbulence subside [47].

In practice, it is impossible to prevent the occurrence of fiber flocculation when using the high molecular weight polyelectrolyte retention aids required on modern, high-speed paper machines. Excessive fiber flocculation can cause many problems with sheet properties. For example, the fiber flocs result in large differences in mass distribution throughout the sheet. Because of the weak spots that develop between these flocs in the sheet, wet web strength is reduced, crushing can occur in the press section, and the overall strength properties of the final sheet suffer. Appearance properties are also adversely affected.

The papermaker can overcome the potentially adverse effects of fiber flocculation by the use of mechanical (hydrodynamic) action in the headbox and on the forming table, by running as low a headbox consistency as possible within the drainage constraints of the machine, and by taking care that the levels of retention aid utilized are not excessive.

2. Fines retention

As discussed earlier, we will consider fines to be any colloidal material that passes through a 200-mesh screen. This definition encompasses a wide variety of substances found in a papermaking furnish. Fiber fines and fillers are two common examples of fines that were described earlier. Retained fines directly influence sheet structural, strength and optical properties. As was discussed in the Introduction to the chapter, poor first-pass fines retention can adversely affect the runnability, cleanliness, and chemical efficiency of a paper machine. Consequently, the papermaker has a vested interest in achieving good fines retention performance.

(a) Fines retention mechanisms

Fines are retained in the sheet by two mechanisms: colloidal adsorption and mechanical entrapment.

Colloidal interaction is the predominant mechanism and involves the formation of flocs composed of only fines and flocs containing both fibers and fines. In the latter instance, the mechanism involves the adsorption of fines onto fiber surfaces. These fines are then immobilized and incorporated into the mat network along with the fibers on the paper machine forming section.

Mechanical entrapment involves catching the fines in the interstices of the web formed by fibers deposited on the moving fabric. A certain thickness of mat must be formed before substantial entrapment occurs [48]. As the fiber mat builds, more fines are caught in the denser network which provides more available points of entrapment. The papermaker has only a limited control over this type of retention mechanism, which is a strong function of basis weight. Good formation promotes better retention by providing a more homogeneous mat to filter the fines. Likewise, the larger the flocs formed by the colloidal mechanisms discussed below, the greater is the probability that they will be filtered out by the forming web.

As discussed earlier, fiber-fiber flocculation is an undesirable phenomenon that should be prevented as long as possible in the papermaking process. In the case of fines, the situation is more complex. The papermaker strives to promote flocculation between fines and fibers, while at the same time discouraging the formation of flocs containing only fines. It is well known that fines should be well distributed throughout the web in the z-direction. This can only be accomplished if a large percentage of the fines are adsorbed onto fibers prior to sheet formation. If large flocs containing only fines occur, then these fines will be retained primarily by mechanical entrapment, which can only occur after a certain percentage of the sheet has been formed. The first-formed portion of web will contain fewer fines, giving rise to a more pronounced "two-sidedness".

A second reason why uniform fines distribution is desirable is that chemical additives tend to concentrate on fines and it is necessary that the fines be well distributed in order to achieve good distribution of chemical additives throughout the web.

The above discussion obviously presents the "ideal" case. In practice it is impossible to prevent the formation of flocs containing only fines. The realistic goal, then, becomes one of minimizing their formation.

The flocculation of fines occurs in the stock preparation area, the approach flow system, and the headbox. Once flocs have formed, the most important factor in their retention is their ability to resist disruption by the shear and turbulence forces existing in the headbox and forming section of the paper machine. If the flocs are sensitive to these hydrodynamic forces, they will be broken down and pass through the forming media and accumulate in the white water system.

As mentioned earlier, flocs can be characterized as *soft flocs* or *hard flocs*, whose definitions are:

1. Soft floc: A floc that exhibits good fines retention at low levels of turbulence and shear. Upon exposure to greater turbulence, the retention decreases substantially but, when allowed to reflocculate, will return to the original low turbulence retention levels.
2. Hard floc: A floc that shows good fines retention over a wide range of turbulence and shear for brief periods of exposure. Subsequent exposure will cause a decrease in retention that cannot be reestablished after the system is allowed to reflocculate. Upon reflocculation, the system reverts to the behavior of a soft floc system.

The mechanisms postulated for these two classifications are as follows. Soft flocs form when salts or low molecular weight polyelectrolytes are used to induce aggregation. When the system is exposed to shear, the flocs break up but the ions or polyelectrolytes bound to the surface remain in place. When the turbulence subsides, the same degree of aggregation can occur and the original levels of retention are obtained. Table 24 [47] contains data illustrating this phenomenon.

The soft flocs produced by salts occur at system zeta potentials of zero and indicate a charge neutralization phenomenon. The soft flocs produced by low molecular weight PEI (polyethylenimine, 6000 M.Wt.) apparently occur by the patch-type agglomeration illustrated earlier in Fig. 129.

Table 24. Reflocculation phenomena in charge neutralization and agglomeration systems[a].

System	Turbulence (rpm) 0	400	800
NaCl			
—A	67	43	34
—B	80	41	36
$AlCl_3$			
—A	96	58	40
—B	92	57	38
Low M. W. PEI			
—A	93	54	37
—B	91	50	34

[a]Retentions determined on the Britt Dynamic Drainage/Retention Jar as per TAPPI Method T 261

A = Retention of original stock and additives

B = Retention after exposure to 1500 rpm for 1 min followed by a reflocculation period.

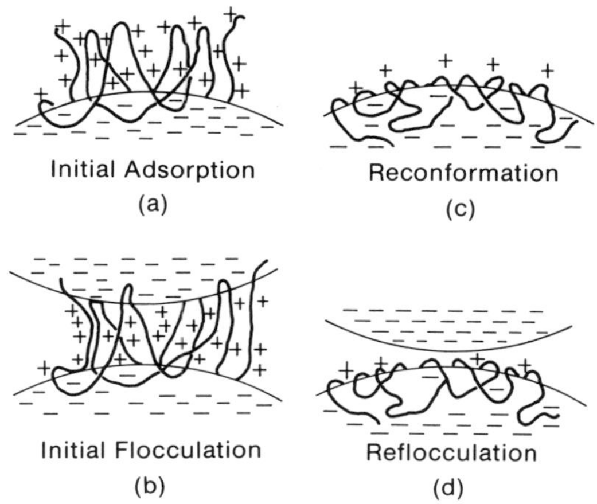

Fig. 136. Hard floc model. (a) Initial adsorption of high molecular weight retention aid onto particle surface. (b) Formation of floc by attachment of second particle to extended loops and trains of retention aid. (c) Reconformation of disrupted and broken retention polymer onto the surface of one of the particles. (d) Attraction of positive patch formed by recondormed polymer to negative surface on a second particle to form a soft floc.

A different mechanism is envisioned for hard floc behavior. When higher molecular weight polyelectrolytes adsorb on a surface, they do so as a series of loops and trains (Fig. 136a), as described earlier (see Sec. A). When flocculation occurs, it takes place via a bridging mechanism (Fig. 136b). This gives high levels of first-pass retention and stability against shear and turbulence. Extended exposure to turbulence breaks the polymer bridge between the two surfaces. Separation occurs with the polyelectrolyte remaining attached to one surface or with chain rupture occurring and parts of the polymer molecule remaining on each surface.

Because the molecule on the surface is undergoing motion due to hydrodynamic forces, and there are adjacent regions of opposite charge on the fines or fiber surface, the molecule can undergo a reconformation (Fig. 136c). This makes the polymer act more like the patch model encountered with low molecular weight polyelectrolytes (Fig. 130) and reaggregation occurs by a soft floc mechanism (Fig. 136d). This effect can be seen in the subsequent flocculation/redispersion/flocculation behavior shown in Fig. 137.

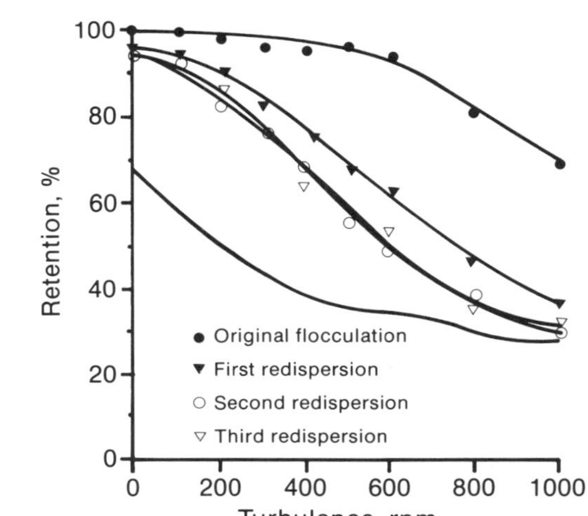

Fig. 137. Flocculation-redispersion behavior of hard-floccing polyelectrolyte systems. The system produces a relatively shear-resistant floc that withstands high stirring speeds in a Britt Jar. Once the strong flocs have been disrupted by stirring and reformed in the absence of stirring, they do not exhibit the same tenacity. Subsequent exposure to increased stirring rates results in behavior characteristic of reversible, soft floccing systems. (The solid line curve in the figure represents the behavior of the system without polyelectrolyte addition.)

(b) Dual additive systems

The use of sequential additions of two oppositely charge polyelectrolytes has been found to produce very tenacious flocculation [49-52]. The sequence of cationic polymer followed by anionic polymer has been found to be most useful. The original work [49, 50] on fine paper systems utilized the addition of an intermediate molecular weight cationic polyelectrolyte followed by a treatment with a high molecular weight anionic polyelectrolyte. The mechanism for producing such tenacious flocculation involves the laying down of a cationic patch of cationic polyelectrolyte (Fig. 138a) which provides an anchor point for the anionic polyelectrolyte. The high molecular weight anionic polymer bonds to the cationic site provided by the patch, but the rest of the chain is repelled by the negative sites surrounding the positive patch (Fig. 138b). This forces the anionic polymer to extend into the surrounding water, facilitating its adsorption onto a cationic patch site on another surface (Fig. 138c) where it acts as a bridge, holding the two surfaces together. The flocs formed are tenacious and fall into the category of hard flocs.

Recently, the combination of cationic starch and colloidal silica [53] has been found to provide good retention and drainage performance on paper machines. The mechanism by which this retention system works is not known at the moment. The good sheet formation achieved at high filler levels suggests that small flocs are formed. Research is currently underway to determine the details of the mechanism. Other polymer-pigment retention systems are also being applied by the paper industry [54].

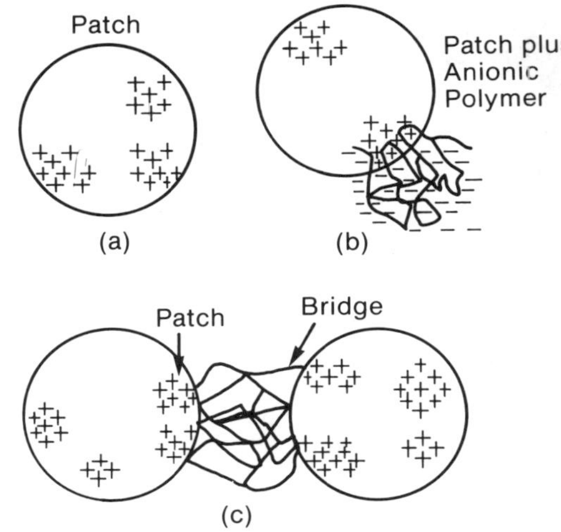

Fig. 138. Model for dual polymer retention aid system. (a) Formation of a cationic patch by a low molecular weight, high charge density cationic polyelectrolyte. (b) Adsorption of high molecular weight, low charge density anionic polyelectrolyte to the positive patch on the particle to form a strong anchoring point. The surrounding negative surface on the particle causes the anionic polymer to extend out into the surrounding water. (c) Attachment of the extended anionic polymer to a positive patch on a second particle forming a strongly anchored bridge between the two particles.

3. Fines distribution

As was discussed earlier (Sec. C), the manner in which fines are distributed in the z-direction of paper depends upon a number of mechanical and colloidal considerations. The mechanical factors have been reviewed by Groen [55]. The principal mechanism at work on the paper machine appears to be self-filtration which can lead to decreased filler and fiber fines content in each hypothetical layer from the felt (top) side to the wire side.

The forming medium is the first structure to filter the stock components. Because of its relatively coarse nature, few fines are trapped and they pass through the medium into the white water. As the sheet consolidates and more of the network is built up, increasing amounts of fines are entrapped in the interstices of the fibrous mat. The fines in the last portion of the stock drained have a greater chance of being filtered. Consequently, there are relatively more fines in the top side of the sheet.

The second mechanism controlling the z-direction distribution of fillers is that of colloidal interaction. Tanaka and coworkers [56] reported that the higher the first-pass retention of filler, the more uniform the distribution of filler in the z-direction (Fig. 139).

Fiber fines and filler are not always retained similarly. For example, it has been reported [57] that, under acid papermaking conditions, the two types of materials behave similarly but, under alkaline conditions, they are retained to different degrees. Insufficient research has been done on this matter to make a general statement on this point.

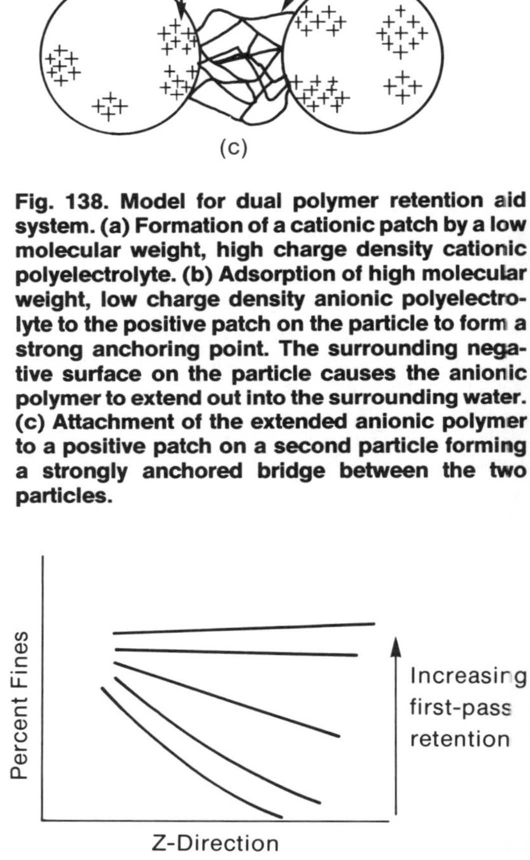

Fig. 139. Influence of first-pass filler retention levels on the z-direction distribution of filler in handsheets [56].

4. Effect on optical properties and sheet formation

Retention aids can greatly increase fines retention and produce a subsequent decrease in white water solids. These positive aspects do not exist without a negative counterpart, a decrease in filler efficiency that often accompanies high retention [58, 59]. When filler retention is increased by the use of chemical additives or polyelectrolytes, a corresponding reduction in the scattering coefficient of the filler fraction can occur due to the flocculation of filler particles. The maximum scattering power of a material occurs when its particle size is in the range of one half the wavelength of the light being scattered. As particles agglomerate and act collectively as one large particle, the number of filler-to-air interface surfaces decrease and their ability to scatter light diminishes rapidly [29]. To this end, care must be taken to guard against overflocculation of filler particles in cases where optical properties are important.

Overflocculation of fines can produce very large flocs that will appear as poor formation when the sheet is viewed by transmittal light. Consequently, excessive fines flocculation is to be avoided.

5. Drainage and vacuum water release

(a) Drainage

"Free" drainage is generally thought of as the water removal taking place between the headbox and a point just preceding the first vacuum box. Drainage occurs during sheet formation when the stock transitions from a random dispersion of fibers to a web network where the fibers have relatively little freedom of movement. (On the paper machine, flow from the stock is enhanced mechanically by forming boards, table rolls, foils and vacuum-assisted foils. This discussion assumes that the operation of these devices have been optimized and focusses on the colloidal chemistry aspects of drainage.)

Both the rate of water removal and the amount of water removed are important factors. The amount of water removal is influenced strongly by the surface area of fines and fibers available to interact with water. The water retention value test [137] estimates how much water can be removed from a stock under "ideal" conditions. The water removal rate is more complex and depends upon such factors as the permeability of the formed web.

There are three main hypotheses used to explain the mechanism of increased free drainage of water by drainage aids [60]:

1. When polyelectrolytes or cations are added to a fibrous suspension, the surface charge is reduced causing collapse of the microfibrils on the fiber surface, thereby decreasing the hydrodynamic specific surface. This results in reduced drag as fluid flows from the forming network.
2. The additives cause collapse of internal as well as external void areas, decreasing the amount of water associated with the fiber.
3. The fines and fibers are flocculated with a polyelectrolyte, therefore the fines are no longer free to plug the interstices in the fiber mat and water can flow freely through the unobstructed fiber network [61, 62].

The third mechanism is generally accepted as the most effective in improving water removal. However, the first two mechanisms have been observed to affect the dewatering process. For example, experiments have been reported where a fines-free pulp suspension showed a drainage increase when treated with cationic polyelectrolyte and alum [63]. This drainage increase also took place beyond the charge neutralization point. Therefore, the cationic polyelectrolyte/alum complex that formed must have affected the fiber surface by either reducing the hydration of the fibers or collapsing the fibrils associated with the fibers [64].

The third mechanism, fines flocculation, depends on a number of physical characteristics of the polymer, the most important of which is molecular weight. If a low molecular weight polyacrylamide or polyethyleneimine is used, the polymer will preferentially adsorb on fines due to their high surface area. This results in the flocculation of the fines onto fiber surfaces, which opens up the fiber network and increases the fluid drainage rate.

Optimum levels of lower molecular weight materials floc fines without producing large fiber flocs. When high molecular weight polyelectrolytes are used, fiber/fiber flocs are formed as well as fiber/fines flocs. This reduces the effectiveness of the drainage aid. Water has a tendency to drain through the thin spots between the fiber-fiber flocs and to be retained inside the flocs.

The charge on the polymer also determines its ability to improve drainage. Anionics alone do not perform well as drainage aids. In most cases, the polymer is not adsorbed and remains free in the water phase, increasing its viscosity. This is detrimental to fluid flow. When alum is added before the anionic polymer, it imparts a positive charge to the fiber surface and acts as an anchor for the negatively charged polyacrylamide [62, 65, 66]. Researchers have observed an improvement in drainage when alum was used in conjunction with an anionic polymer [66]. The combination of cationic starch and anionic polymer was also found to improve drainage [60].

Cationic materials generally improve drainage. Alum seems to decrease the surface charge, allowing fibrillar collapse and the reduction of specific surface area. Cationic polyelectrolytes tend to bind fines to fibers via bridging mechanisms, promoting more open networks. The greater the charge density, the greater the drainage efficiency [67].

Zeta potential plays a role when polyelectrolytes are not used. In systems where salts are used as flocculants, all other things being equal, optimum drainage occurs around the isoelectric point (i.e., zero net charge). When polyelectrolytes are present, there seems to be little correlation between the isoelectric point and optimum drainage [60, 62]. This is most likely true because the bridging mechanism does not require the zero zeta potential required for flocculation via charge neutralization.

Another stock system property affecting drainage is that of entrained air [68–70]. Large air bubbles entrained in a headbox system will manifest themselves as foam, but will not affect water removal. However, drainage is adversely affected by entrained air bubbles. These small bubbles lodge in the fine paper web capillaries and impede the flow of water through the fibrous network. The drainage retardation from this effect can be substantial.

(b) Vacuum water release

As the stock suspension approaches the flat boxes, it reaches consistencies of 4–5%. When the stock passes over the suction box holes or slots, three water removal mechanisms operate [69]. First, the pressure differential causes the fiber network to compact, squeezing water from the sheet. Second, water in capillaries is displaced by air. Third, as the sheet passes the trailing edge of suction box openings, water that has been sucked out of the sheet is scraped off the bottom of the forming media. This decreases rewet after the vacuum zone has been passed [71].

The effect of polyelectrolytes in this zone has been studied to a lesser degree than the drainage zone, but some interesting observations have been made. In mill trials and pilot machine runs, it has been observed that products designated as drainage aids often improved water removal over table rolls and foils, but resulted in a wetter sheet at the couch [72]. A number of laboratory studies have confirmed this [66, 69].

The mechanism that causes this reduction in couch dryness is probably due to a change in sheet porosity [69, 72]. By tying up the fines through flocculation by polyelectrolytes, water and air are able to flow more freely through the sheet. On exposure to vacuum, water is quickly pulled from the capillaries and displaced by air. The air flow then satisfies some of the vacuum, reducing the pressure drop and decreasing the compaction (squeezing) of the sheet.

If the suspension is overdosed so that macroflocs (large fiber flocs) are formed, this condition is aggravated. Water will be removed from the areas between the flocs and air can easily satisfy the vacuum. Water in the flocs is not removed. Therefore, the sheet is wetter at the couch than before the polyelectrolyte was added.

The laboratory investigation of drainage should provide both a measure of the amount of water removed from a furnish as well as the rate of water removal. Unfortunately, drainage testing devices usually measure only the amount of water removed in a fixed length of time. A recently described device provides dewatering rate data and has been used to characterize the dewatering characteristics of furnishes up to the point where the web enters the press section [73].

6. Closing remarks

This concludes our brief review of the papermaking chemistry of fibers, fines, filler, and retention aids. These materials participate in most of the major flocculation phenomena that occur on the paper machine and through these phenomena influence final sheet properties and the runnability of the paper machine.

We now turn to several very important functional additives; namely, sizing agents, wet and dry strength agents, and dyes. With the exception of sizing agents, these materials are utilized in the form of soluble molecules that must be adsorbed onto the fiber, fines, and/or filler surfaces in order to function. As in the above discussion, we will concentrate on those aspects of the subject that pertain most directly to papermaking chemistry.

E. INTERNAL SIZING

The "sizing" of paper refers to the development of a surface that is resistant to penetration by a specified liquid (water, milk, blood, oil, etc.). The penetration of a liquid into a porous structure can be modeled by the Washburn Equation:

$$L^2 = \frac{\sigma R \cos\theta}{2\eta} t$$

where

η = fluid viscosity
R = capillary radius
θ = contact angle

σ = surface tension of fluid
L = depth of fluid penetration in time, t

The equation indicates that penetration is influenced by both the size of the pores in the web and the surface characteristics of the fibers. Of the terms in the equation, only the contact angle, θ, can be changed by altering the surface free energy of the fiber. An unsized (waterleaf) sheet will have a contact angle of $\theta = 0$ ($\cos 0 = 1$). Therefore, a low-energy material must be present on the fiber surface to prevent the penetration and adsorption of a high-energy liquid like water. It is the relative surface energies of the liquid and fibers that determine the magnitude of the contact angle. High-contact angles are achieved by having the surface energy of the fibers much lower than that of liquids with which they come into contact. Most common sizing tests are influenced by both the pore size and surface characteristics. Contact angle measurements are affected primarily by the web surface characteristics.

1. Rosin-alum sizing

(a) Alum

Alum, $Al_2(SO_4)_3$, has been used by the papermaker since the days of hand papermaking. In addition to alum-rosin sizing, alum has been applied as a retention aid, formation aid, drainage aid, pitch control agent, and pH control agent. It is responsible for the acid nature of rosin-sized paper.

The chemistry of alum is very complex and is believed to involve a number of aluminum-containing species under papermaking conditions. Much of the current understanding of the aqueous chemistry of aluminum is attributed to the work of Hayden and Rubin [74] whose concepts have been applied to papermaking chemistry by Arnson and others [75]. Aluminum hydrolyzes in water to form several species by the scheme given in Fig. 140. It has also been postulated that, within the pH range of 4.0 to 5.5, which is of most interest to the papermaker, at least one other species containing several aluminum ions is formed [74]. This is shown by the aluminum-pH relationship presented in Fig. 141.

Arnson [75] showed that the pH range where maximum aluminum is adsorbed onto wood fibers coincides closely with the pH range where $Al(OH)_3$ and $Al_8(OH)_{20}{}^{+4}$ species predominate, as is shown in Fig. 142. As will be explained later, the particular aluminum species important to alum-rosin sizing depends on the type of rosin size utilized.

$$Al^{+3} + H_2O \rightleftharpoons Al(OH)^{+2} + H^+$$
$$Al(OH)^{+2} + H_2O \rightleftharpoons Al(OH)_2^{+1} + H^+$$
$$Al(OH)_2^{+1} + H_2O \rightleftharpoons Al(OH)_3 + H^+$$

Fig. 140. Aluminum hydrolysis equilibrium reactions.

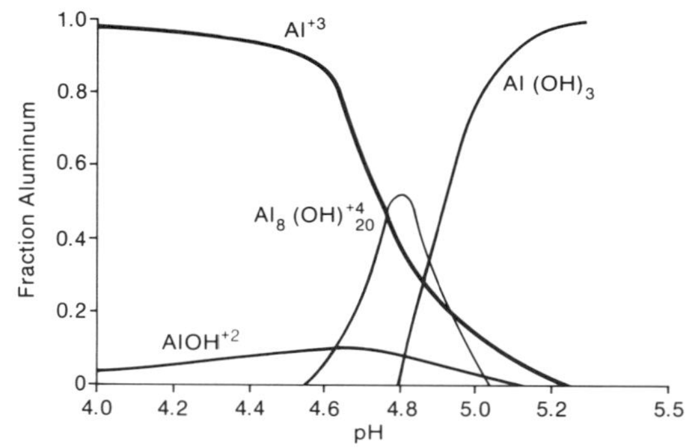

Fig. 141. Influence of pH on the equilibrium species of aluminum present in solution [75].

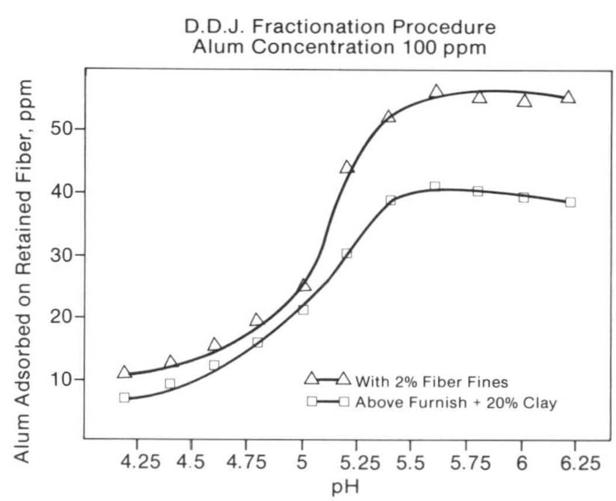

Fig. 142. Alum adsorption as a function of pH and fines content [75].

(b) Rosin size

The most common material used for rendering the fiber surface resistant to wetting by water is rosin. Rosin is an alkylated, tricyclic, unsaturated organic acid derived from a variety of natural sources (tall oil, wood rosin, gum rosin). The two main components of rosin are abietic acid and pimaric acid (see Fig. 143).

These molecules are said to be "amphipathic" because they have a small polar carboxyl group attached to a large, bulky hydrophobic hydrocarbon portion. When these materials are properly adsorbed on the surface of a fiber, the

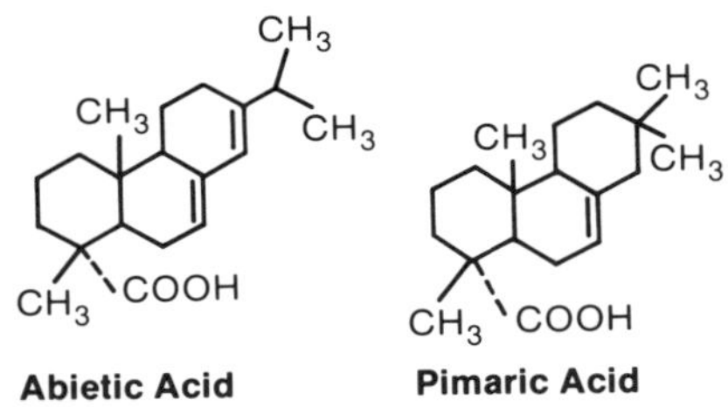

Fig. 143. The two main chemical components of rosin size: abietic acid and pimaric acid.

polar, hydrophilic end is oriented toward the surface of the fiber with the nonpolar, hydrophobic end directed outward toward the penetrating water. The way in which this surface is formed is critical to the successful development of a well-sized paper.

The amount of size theoretically required to cover the surface of fibers is much less than that actually required in practice [76]. The reason for this is related to the way rosin is precipitated and adsorbed onto the surface of the fibers.

Rosin acid is not substantive to the fiber because both fiber and rosin are negatively charged. To make the rosin acid more easily handled, it can be saponified with NaOH, promoting its dispersion in water. When the rosin soap is mixed with water, it forms a colloidal dispersion which must be adsorbed onto the fibers to produce sizing. This is accomplished by precipitation with aluminum ion supplied as alum. The reaction with the aluminum species serves a dual purpose [77].

1. The aluminum reverses the charge of the rosin colloid making it substantive to the anionic fiber surface;
2. The aluminum ion coordinates with the carboxyl group on the rosin acid. Therefore, when adsorbed on the fiber surface, the rosin molecule is oriented with its hydrophobic end outward.

There are some problems with the soap sizing system. The continuity and uniformity of the precipitated soap film will determine the efficiency of sizing. When the rosin soap colloid is exposed to alum, an instantaneous reaction occurs. This, accompanied by the association of the hydrophobic portions of rosin molecules, leads to the formation of large alum/rosin flocs which do not promote the distribution of sizing on the fiber surface and thereby reduce sizing efficiency. In addition, when the floc forms, much of the free rosin remains unreacted with aluminum ion and when this floc adsorbs onto the fiber surface, the free rosin can orient in any direction. If the polar end is oriented outward, it will be free to interact with water. Any rosin molecules that are not fastened securely to the fiber surface will also be free to overturn when the surface is brought into contact with water.

If the system is allowed to reach a pH of 6.0, the rosin-alum precipitate will approach the isoelectric point. At zero charge further agglomeration will occur. Beyond this point, the charge on the particles is reversed and the rosin/alum floc is no longer substantive to the fiber. To avoid this the system pH must be adjusted to around pH 4.5

In order to act as an effective sizing agent, the rosin must be "sintered" in the dryer section of the paper machine, i.e., the rosin complex must be heated to the point where it softens and flows. Soap size precipitates sinter at relatively high temperatures (100–200°C).

When the rosin soap is reacted with alumina in solution, the size will be fixed as it adsorbs on the fiber surface. As the paper dries and interfiber bonding occurs, the immobile rosin-alum precipitate particles can interfere and produce the strength reduction often seen in soap-sized systems.

2. Dispersed size

Because of the difficulties outlined above with saponified size, more size may be needed to reach a given level of sizing than is theoretically required. To increase the efficiency of the sizing operation, "dispersed" rosin size systems were developed where free rosin acid is emulsified. Because they are not substantive to fibers, these small stabilized particles of rosin must be retained by alum or, more normally, by a cationic retention aid or cationic starch.

Alum must be present in the system so that alumina/rosinate complexes can form when the free rosin colloid is heated during drying. Having the rosin react with the alumina during the drying process is thought to have the advantages listed in Table 25 [78].

3. Fortified rosin sizes

When rosin is treated with maleic anhydride, the resulting tricarboxylic acid compond has been found to result in a greater sizing efficiency (see Fig. 144). The main reason for the improved efficiency is thought to be the smaller particle size formed in the size suspension. The smaller particle size arises from stabilization of the particles by an increased concentration of carboxyl groups on the surface of the colloidal particle which hinder coagulation.

A second factor that may contribute to the improved size performance is that maloabietic acid molecules tend to orient more horizontally

Table 25. Advantages of dispersed sizes.

1. The particles of dispersed size are loosely adsorbed on the fiber surface and are free to move. Therefore they do not impede bond formation and result in a stronger sheet than in the case of soap size.
2. Dispersed sizes have a lower melting point than soap sizes due to the greater percentage of free acid present. This material flows more readily during sintering, thereby promoting better coverage and orientation of rosin molecules when the alum-rosin size complex is formed during drying.
3. There is less chance for premature coagulation with dispersed sizes than soap sizes because the alum-rosin complex is not formed until the dryer section. Thus, dispersed sizes exhibit greater sizing efficiency than soap sizes.

on the fiber surface and yield greater surface coverage than the unmodified rosin molecules.

4. Mechanism of action of rosin sizes

The key steps associated with the mechanism of sizing with rosin soap sizes and dispersed rosin sizes were elucidated in an elegant series of experiments reported by Marton [79-81]. In his work, Marton demonstrated that the reactions between alum and rosin are quite different between the two types of sizes. In the case of soap sizes, the neutralized resin acids react with the Al^{+3} form of aluminum to form a positively charged alum-rosin precipitate that is attracted to the negatively charged fiber and fines surfaces and retained there. The aluminum and rosin acid included in the precipitate then react to form a stable size complex on the fiber surface in the paper machine dryer section. Thus, to achieve the most efficient sizing in rosin soap sizing, it is necessary to have the pH of the system maintained in the 4.2–4.5 range where the aluminum exists for the most part in the Al^{+3} form.

In the case of dispersed sizes, the emulsified size particles exist mainly in the free acid form. The size emulsion is stablized by emulsifying agents that do not react extensively with Al^{+3}. Thus, it is usually necessary to use some kind of cationic polyelectrolyte retention aid to promote the retention of dispersed rosin size on the fibers and fines. It is also necessary for alum to be retained on the fibers and fines where it will be available to react with the rosin in the dryer section. From the work of Arnson [75] it is known that the most efficient retention of alum on fibers and fines occurs when the pH of the system is in the 4.5–5.0 range. Consequently, this should be the preferred pH range for sizing with dispersed sizes.

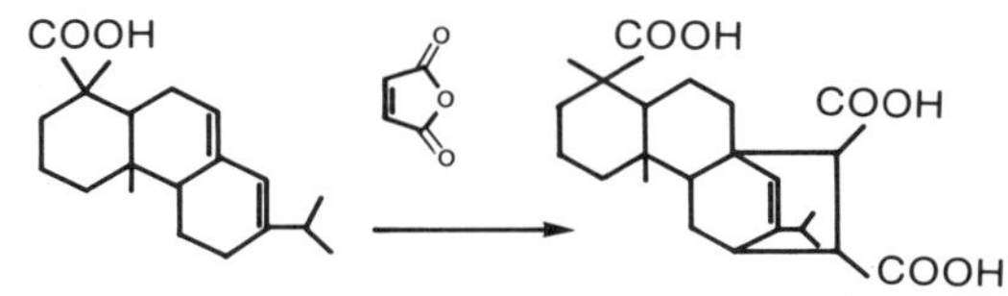

Fortified Size

Fig. 144. Preparation of fortified size by the reaction of abietic acid with maleic anhydride.

In addition to elucidating the different mechanisms by which soap sizes and dispersed sizes are retained and react during papermaking, Marton also showed quite dramatically the importance of retaining the size on the fibers rather than the fines in a furnish. When a furnish containing both fibers and fines was sized, removal of the fines prior to the formation of handsheets resulted in paper that had very low sizing values. Adding back the fines caused the sizing tests to increase to a level very nearly equal to that of paper made from the original furnish. When fines were removed from the furnish prior to sizing and paper was made from the fibers-only furnish, the highest sizing test values were obtained [81].

5. Cellulose-reactive sizes

In recent years, traditional alum-rosin size has in some cases been displaced by cellulose-reactive sizes. These are amphipathic materials that have hydrocarbon chains attached to them that contain at least 16 carbon atoms. The agents are capable of reacting with the hydroxyl groups on cellulose to form covalently bonded ester linkages under the right chemical conditions [82].

These materials also undergo hydrolysis; therefore, there is always a competition between the reaction with the fiber surface and with the water in a papermaking system. These materials are more expensive than alum/rosin, but are much more efficient. Acceptable sizing can be obtained at 0.05%–0.15% dosage rates [76]. Another asset is their ability to react under alkaline conditions.

One potential drawback to the use of synthetic sizes is the "slippery" feel that may be imparted to the sheet when very high levels of the size are utilized. This is likely due to the long hydrocarbon chains that coat the surface and produce the same feel as a paraffin wax if very high dosages are used. Antislip agents are sometimes added to counteract this effect.

There are two cellulose-reactive materials that have gained general acceptance by the paper industry. Alkyl ketene dimers (Fig. 145) have a low melting point and spread more easily onto the fiber surface. They have a low degree of reactivity and are therefore fairly water stable, but are slow to react with cellulose. This means that sizing takes time to develop; often curing occurs in the roll of paper, which can be a problem when on-machine sizing is required. A cationic retention aid is required to retain the alkyl ketene dimer dispersion on the fiber. This material is alum sensitive and is only used above pH 7. The sizing agent is free to migrate on the fiber surface prior to reaction and as a result does not interfere greatly with interfiber bonding.

Alkenyl succinic anhydride (Fig. 146) is a cellulose-reactive sizing material that is very reactive and can be used down to pH 5.0 in the presence of alum. However, optimum performance is achieved under alkaline conditions. The high reactivity also means that the material has a high rate of hydrolysis and aqueous emulsions of the size must not be stored for very long periods of time prior to use. High pH and high temperatures accelerate the hydrolysis rate of alkenyl succinnic anhydrides [83]. Cationic agents are required to ensure high retention rates of the material during papermaking.

As with rosin sizing agents, a large percentage of the added synthetic sizing agents are adsorbed onto the fines material in a furnish. Consequently, it is imperative that good fines retention levels be achieved in order to ensure the efficient use of synthetic sizing agents. As with rosin sizes, the most effective application of synthetic sizing agents requires that the agents be retained on the long fiber fraction of the furnish. Size that is adsorbed on unretained fines will hydrolyze and become ineffective or, worse, cause picking and deposit problems on the paper machine.

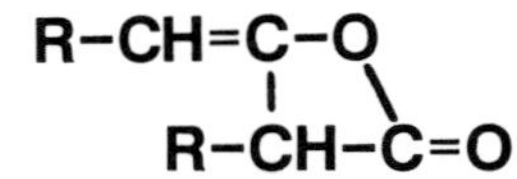

Fig. 145. Alkylketene dimer synthetic sizing agent (AKD).

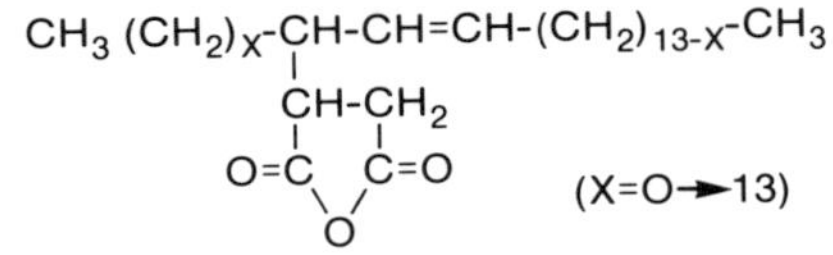

Fig. 146. Alkenyl succinic anhydride synthetic sizing agent (ASA).

F. DRY STRENGTH AGENTS

The "dry" strength of paper generally refers to the relative performance of the paper sheet when specific mechanical properties are tested, i.e., tensile, Mullen, tear, etc., under the normal, dry conditions under which the paper will be used. The main mechanism of attaining strength in paper is through the development of hydrogen bonds between the fibers in the sheet. The strength of the bond between two fibers is directly related to the number of hydrogen bonds formed within the bonded area.

The bonding between fibers can be enhanced by two means. The first means is by refining. Refining increases surface fibrillation and cell wall hydration (which increases the flexibility of the fiber), both of which promote fiber conformability and increase the surface contact between fibers. The second approach is to add a "dry strength" agent, either natural or synthetic, that will enhance and increase the number of bonds between adjacent fiber surfaces.

Dry strength additives are thought to act by several mechanisms. First, they supplement the number of hydrogen bonds that would naturally occur in the bonded area between two fibers [84]. This is a "chemical hydration" of the fiber system. Refining hydrates the fiber, but in the process the fibers are generally weakened and shortened. It is for this reason that the decrease in tear strength that normally accompanies refining can be alleviated by the use of dry strength additives. The increase in bonding from chemical hydration does not affect individual fiber strength. A second mechanism involves improvement of the formation of the sheet. This provides more uniformly distributed fiber-fiber bonds [85]. A third, less important mechanism, involves the improved web consolidation that occurs when dry strength agents enhance fines retention and, hence, sheet dewatering.

During refining, a natural dry strength agent is released – hemicelluloses. These materials produce increased strength when they are retained during the pulping and bleaching process or added after pulp is refined.

1. Natural dry strength agents

As already stated, hemicellulose, with its multitude of hydroxyl groups, is a perfect agent to promote bonding. The commercial materials that have been utilized at one time or another by industry include the natural gums and the natural and derivatized (cationic) starches. Of late, the cationic starches have become the most popular. These materials are thought to provide enhanced interfiber bonding and also increase the bonded area between fibers [86].

(a) Wet end starches

Starch is a white granular substance that occurs widely in the higher plants, frequently in the seed (as in corn and wheat), or in various roots and tubers (tapioca and potato), or sometimes in the stem piths (sago). Its natural function is to provide a reserve food supply for the plant, sustaining it through winter dormancy and nurturing new plants during germination. Starch has been used in the manufacture of paper since the very early development of the industry, even long before the invention of the paper machine. The total spectrum of starch applications in the paper industry now includes wet end addition for filler and fines retention, paper dry strength improvement, surface sizing, pigment coating, and converting adhesive functions. The sum total of these applications represents more than two thirds of the total U.S. industrial starch consumption.

(b) Chemical structure of starch

Chemically, starch is similar to cellulose in that it is a polymer composed of glucopyranose units linked through carbons 1 and 4. Figure 147 is a schematic drawing of starch. The major difference between starch and cellulose lies in the configuration of the glycosidic bond linking glucose units. In cellulose, the bond connecting carbon 1 and the bridging oxygen is oriented in the same plane as the glucose ring (*beta*-link) and the resulting polymer molecule takes on an extended ribbon-like shape. In starch, the bond connecting carbon 1 and the bridging oxygen is oriented perpendicular to the plane of the glucose ring (*alpha*-link) and the resulting polymer molecule tends to take on a more coiled, helix-like shape. This difference in molecular structure between cellulose and starch leads to significant differences in their functions in nature and in their industrial uses.

The structure depicted in Fig. 147 is a linear molecule. In fact, the vast majority of starches occur as a mixture of linear and branched molecules. The linear fraction of starch is called amylose. The branched fraction is called amylopectin. Branching occurs through carbon 6 on the glucose structure most commonly and occasionally through carbon 3. On average, there is one branch point for every 18–27 glucose units. The linear amylose molecules have a greater tendency to associate in crystalline aggregates than the branched amylopectin molecules.

The amylose percentages for several common starch sources are listed in Table 26.

It is believed that the ability of starch to enhance interfiber bonding comes from the readiness with which its free glucose hydroxyls participate in hydrogen bonding with fiber surface cellulose molecules.

The molecular weight of the unmodified starches depends on the source. For example, Harvey [88] reported that potato starches had average molecular weights of $3-3.5 \times 10^6$ for two samples of potato starch while three samples of corn starch ranged from 0.8×10^6 to 1.8×10^6.

(c) Granular structure of starch

Starch is composed of tiny cells or "granules", whose size and shape are characteristic of each botanical variety. The granules are quite small. It has been calculated that one pound of corn starch contains about three quarters of a trillion granules. Each granule possesses a highly complicated internal architecture. The starch molecules are laid down in concentric layers within a granule in an "onion-like" arrangement. It has been found that amylose molecules are associated in ordered crystalline regions in the granule, while amylopectin is found in amorphous areas. A characteristic feature is the appearance of a "Maltese cross" pattern on each granule when it is viewed in polarized light. The existence of this pattern is evidence that the structure of the granule is intact.

(d) Starch dispersion

Unlike many synthetic polymers that are difficult to disperse, starch granules can be slurried easily in cold water without serious intergranular cohesion. To prepare starch for papermaking, the granules must be cooked to bring about either complete or partial disruption.

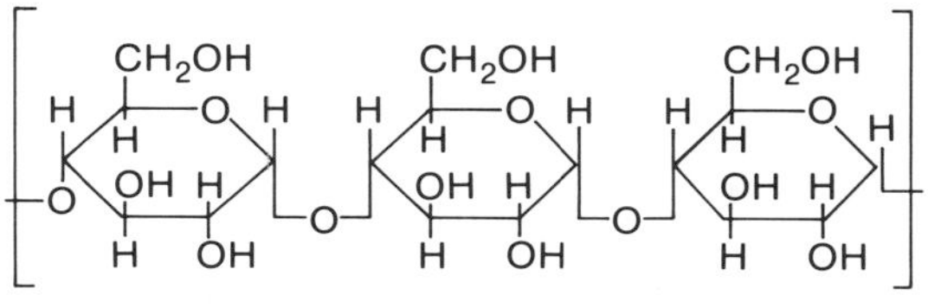

Fig. 147. Molecular structure of the amylose component of starch.

Table 26. Amylose contents of several common starches.

Source	Amylose Content, %
Tapioca	16.7
Corn	24.0
Potato	20.0
Wheat	25.0

When starches are suspended in water and the temperature gradually increased, the granules suddenly swell enormously at the gelation temperature characteristic of each kind of starch. This occurs because the solvent power of the water increases with temperature and the granule is penetrated by the liquid causing it to swell. Concomitantly, molecularly dispersed amylose is leached from the swollen granule. The viscosity of the dispersion increases greatly as the swollen granules come into contact and form a continuum throughout the medium. Although the higher temperature reduces the viscosity of the fluid phase, this is more than offset by the increasing concentration of dissolved amylose and the increasing volume of swollen granules. A point is reached in the dispersion process at which the swollen granules begin to disintegrate and the viscosity then decreases. Subsequent cooling of the paste results in rapid viscosity increases and ultimately leads to gelling (setback). Gelling is related to the amylose content and is a disadvantage in the handling of batches of cooked starch. The degree to which a starch exhibits gelling is related to the molecular weight of its amylose and reaches a maximum when the amylose contains 150–200 glucose units per polymer molecule. Anything above or below this value tends to decrease the propensity toward gelling.

In steam-jet cooking the disruption of the granule is accomplished by a nearly instantaneous temperature increase to about 160°C accompanied by intensive mechanical shearing. Such dispersions differ from batch cooks in that all whole granules are gone, and, in the case of corn starch, the dispersions are remarkably resistant to gelling. Jet cooking can lead to a decrease in average molecular weight of the starch. Since the ability of starch to act as an effective interfiber bonding agent for paper is directly related to both the degree of molecular dispersion and the molecular weight, the balancing of those crucial factors in cooking is of considerable importance for optimizing performance.

(e) Starch modifications

Unmodified starches are available that differ in properties such as amylopectin-to-amylose ratio, granule dimensions, gelatinization temperature, and the molecular weights of the component fractions. In addition to these variables, modified starches greatly extend the number of potential wet-end starch additives available for a given use. Much research has been carried out that has resulted in the development of new applications for starch in papermaking. Physical, chemical, and genetic modifications and combinations of these have increased the useful properties of starches. Table 27 lists some of the methods applied.

Table 27. Starch modification methods.

Physical Modifications
- —Fractionation into amylose and amylopectin components
- —Thermomechanical conversion
- —Acid hydrolysis

Chemical Modifications
- —Oxidation
- —Derivatization — Enzyme conversion

Thermomechanical conversion, acid hydrolysis, and enzyme conversion all lead to a decrease in the molecular weight of the starch and a corresponding decrease in the solution viscosity of starch solutions.

A tremendous number of starch derivatives are available to the papermaker. A 1975 survey of suppliers [87] revealed a total of 39 cationic products, 17 anionic products, and 27 nonionic products that were aimed at wet end application for improving dry strength, fines retention, drainage, formation, stiffness, and dimensional stability.

(f) Cationic starches

Cationic starches are applied in a wide range of paper grades because of their effect on strength, filler retention, and drainage. These starches have the advantage of a high degree of dispersion and high retention on wood fibers. A good introduction to the chemical aspects of cationic starches is provided in an article by Harvey that was presented at the 1987 Advanced Topics in Wet End Chemistry Seminar [88].

Alternatives to cationic starches of fixed cationic charge density are the variable products available from continuous cooking and simultaneous cationization of starch by reactive chemicals, such as various amines and cationic resins.

Starches are cationized with either tertiary or quaternary amine groups. The cationic charge of the former group exhibits a degree of pH dependence, with decreasing charge at pH values above 7.0. Quaternary cationic groups retain their complete charge at all pH levels [88].

The extent of cationization in commercial cationic starches is equivalent to a degree of substitution of approximately 0.02, i.e., there are two cationic groups for every 100 glucose units, on average. This degree of cationization imparts high dispersibility and solubility to the starch. The internal cohesive forces of the granule are diminished and the starch cooks up more rapidly than native starch. Cationic starches differ according to parent starch source, substituent groups, degree of substitution, distribution of cationic groups, and molecular size and shape.

In addition to their applications as dry strength agents and retention aids, cationic starches are also used as emulsifying aids for synthetic sizing materials and as the cationic components of multiple-component retention systems.

(g) Interactions of cationic starch

Marton and Marton [20, 21] studied the adsorption of cationic starch on cellulose fibers and concluded that the material was adsorbed irreversibly on the fiber surface and that adsorption increased with increasing surface area and with increasing carboxyl content of the fiber. The sorption was very rapid and approached 100% retention with time.

Similar to other polymeric additives, cationic starch is preferentially adsorbed by the fines in a papermaking furnish. Marton [19, 20] has shown that the excessive adsorption of cationic starch can lead to a charge reversal on the fines, causing them to be less well retained during sheet formation. The same factor must be considered when cationic starch is being used to retain filler. In general, cationic starches are not as good retention aids as synthetic polymers, especially when higher papermachine speeds and shear rates are encountered.

In practice, cationic starch is added at about 1% concentration at some point that does not subject the material to high shear rates before it reaches the forming section of the paper machine. Cationic starches have been found to be more shear sensitive than synthetic polymers, which can decrease their usefulness on high-speed paper machines. Cationic starch should not be brought into contact with high concentrations of anionic materials if high chemical use efficiency is desired. The anionics will interact with cationic starch and render it ineffective. On the other hand, the presence of excess alum will also decrease the attractive forces between fibers and cationic starch.

2. Synthetic dry strength resins

The use of synthetic dry strength resins began with the introduction of a nonionic polyacrylamide (PAM) to the market. The amide groups on this nonionic polymer are very reactive in hydrogen bonding and serve to increase the number of bonds within a given bonded area. The PAM molecule is ideally suited for use as a dry strength agent because of the broad range of molecular weights and degree of branching that can be obtained and the optional functional groups that can be incorporated in the polymer chain by copolymerization. The normal basis weight range for PAM dry strength resins is 100 000 to 500 000. PAM is more efficient than natural dry strength agents because it can be designed for specific applications. In most cases, a 0.5% or less dosage level on fiber weight is sufficient to bring about a 25–30% increase in tensile or burst strength.

The PAM molecules must be long enough to adsorb on the fiber surfaces and present sites for hydrogen bonding. If the molecular weight is too great, the molecule will be long enough to "bridge" between the fibers, causing excess flocculation. This results in fluctuations in the local basis weight and produces weak areas in the sheet, thereby offsetting the strength gained from the use of the dry strength agent.

The effects achieved from the use of PAM are the same as with the natural gums. An additional bonus in certain instances is increased fines and filler retention.

(a) Anionic PAM

Since PAM is a nonionic polymer that is not very substantive to the cellulose fiber, various anionic and cationic functional groups have been introduced into the polymer backbone. Usually 5–10 mol percent is enough to provide strong enough charges for the polymer to become substantive to fibers.

Anionic PAM was the first charged PAM to be used commercially and is generally a copolymer of polyacrylamide and polyacrylic acid (see Fig. 148). Another way of producing anionic groups is by partial hydrolysis of the amide groups in PAM to carboxylic acid groups. Since both cellulose and anionic PAM are negatively charged, alum is often used to fix the resin onto the fiber surface. A cationic resin may also be used [89].

The use of either of the above two systems can increase the dry strength of a given furnish by 50%. It must be remembered, however, that as refining levels increase, the effect of the resin is reduced. At high levels of refining the fibers

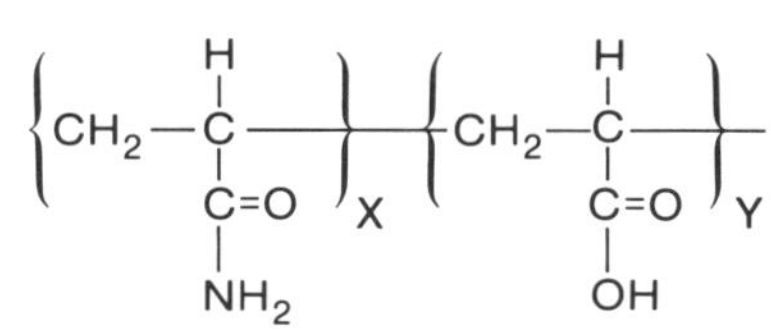

Fig. 148. Anionic polyacrylamide utilized as a dry-strength agent.

become very comformable, resulting in large increases in bonded area. Here, the resin provides some strength improvement, but it is overshadowed by the great increase due to better contact between fibers. The greatest effects of PAM are seen at low-to-moderate levels of refining.

(b) Cationic PAM

Cationic PAM was developed after anionic PAM. Cationic PAMs have the advantage of being substantive without the use of alum. They are useful over a broad range of pHs, well into the alkaline region. Usually some type of tertiary or quaternary amines are inserted into the molecular backbone to impart the cationic charge.

The PAMs, both cationic and anionic, must be well mixed in the stock to ensure that all bonded areas receive treatment and bond strength is optimized. The mixing should not be so violent as to cause degradation of the PAM molecule, which decreases the efficiency of the resin. The resin should also be added after sizing has been set so as not to compete with the size for adsorption sites.

(c) Papermaking chemistry of synthetic dry strength resins

Since the application of dry strength resins requires them to adsorb onto fiber surfaces, it is logical that any factor that strongly affects polymer sorption will be important. This includes the polymer properties, the condition of the fiber surface, and the presence of interfering substances.

The restriction of dry strength resins to a rather narrow molecular weight range means that charge density and sign are the two most important polymer properties when it comes to adsorption, along with the addition levels. The higher the charge density, the greater will be the propensity of the polymer to be attracted to the fibers. This is directly true for cationic dry strength resins and indirectly so for anionic polymers. In the case of the anionic materials, alum must be used to promote polymer retention and anything that blocks alum adsorption sites on the fibers will decrease the effectiveness of the resins.

Cationic demand (anionic trash) will interfere strongly with cationic dry strength resins. It will also interact with anionic dry strength polymers, but to a lesser extent. In the case of the anionic resins, alum probably serves as a charge neutralizer and alleviates the interference.

3. Wet-strength development

When a paper is to be exposed to water during use (e.g., diapers and sanitary products, facial tissues, paper cups, etc.), some kind of treatment is necessary to produce "wet strength" in the paper TAPPI Method T 456 defines the wet tensile breaking strength as the tensile strength of paper when it is completely saturated with water. Complete saturation is important because well-sized papers that are briefly exposed to water before testing will show falsely high wet strength because the sizing prevents wetting of the surface and complete saturation does not occur.

When a waterleaf (unsized) sheet is exposed to water, it loses strength rapidly as the water molecules disrupt and displace the hydrogen bonding that exists between fibers. An untreated paper will be reduced to 3–10% of its original dry strength when wet with water. The strength does not go to zero because some of the interfiber hydrogen bonds that form are inaccessible to the water.

Since the strength of paper is a result of hydrogen bonding, polar liquids other than water will also disrupt the bonds (e.g., polar organic liquids like formamide). Other liquids have an intermediate effect on wet strength. Generally, for paper, wet strength is determined by exposure to pure water.

When a wet-strength additive is applied, an increase of 20%–40% over the untreated wet strength is normal, but the increase can be greater than 50%. The basis for comparison is the percentage of dry strength maintained by the saturated paper. This will not be biased by variables such as basis weight. Obviously, an untreated heavy-weight board will have higher wet strength than untreated tissue paper because the board is a heavier sheet.

4. Wet-strength agents

Wet-strength materials that are added to paper stock can be broken down into two major groups [90]:

1. Aminoplasts: These are amine-based thermosetting polymers that impart wet strength to paper by cross-linking themselves and possibly with cellulose molecules. They tend to bring about little change, other than wet strength, in a paper's physical properties.
2. Elastomers: These are water-insoluble adhesives that bond fibers together. The bonds are resistant to wetting and hold up on exposure to water. These materials change the physical properties of paper and are used to increase flexibility, folding endurance, scuff resist-

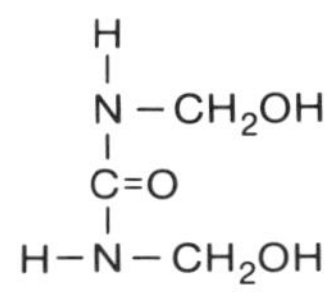

Fig. 149. Dimethylol-urea (DMU). Initial stage of reaction in the use of urea-formaldehyde wet-strength agents.

ance, stretch, etc. Because of this, they are used in specialized situations and we will not treat them further in this discussion.

(a) Aminoplasts

Urea-formaldehyde (U-F)

The first wet strength resin to be used commercially was ureaformaldehyde [91]. The monomeric form is made by reacting urea and formaldehyde under mildly alkaline conditions to form dimethylol-urea (DMU). This is shown in Fig. 149. DMU is referred to as the "A" stage of reaction. When DMU is exposed to acidic conditions and elevated temperatures, polymerization occurs and a water-insoluble polymer forms. This is referred to as the "B" stage. This low molecular weight polymer will continue polymerizing on further heating and cross-link with cellulose, forming a water-stable bond that results in wet strength (Stage "C").

DMU was originally applied by saturating the sheet and then curing. To make the application easier, anionic and cationic functional groups were introduced that made the lower molecular weight B fractions more soluble and substantive. These adsorb onto the cellulose surface and cross-link with additional polymerization when heated in the dryer section. The necessity of exposing these resins to heat to cure them puts them into the thermosetting polymer class.

The "curing" of the resin is accelerated by heat, but the resin will cure at lower temperatures and it usually takes 1–2 weeks in the warehouse for the paper to reach maximum wet strength.

Melamine-formaldehyde (M-F)

Melamine is produced from calcium cyanamide and reacts with formaldehyde to produce mono- or hexamethylol-melamine (Stage A). In stage B, the methylol-melamine undergoes self-limiting polymerization to form a substantive colloid. The M-F cross-links with cellulose to provide wet strength after heat and aging at slightly acidic conditions.

M-F has more functional groups than U-F and is more efficient. Because it is more expensive, cost-effectiveness must be weighed when comparing M-F to U-F.

Polymide-polyamine-epichlorohydrin resins

Both the urea-formaldehyde and melamine-formaldehyde resins adversely affect the absorbency of papers containing them. This condition may arise because the cured resin is hydrophobic or because the alum may react with natural pitch-like materials in pulp to produce a self-sizing effect. Since a major use of wet strength paper is for sanitary absorbent products, this sizing effect is detrimental to the performance of paper. The solution to this problem was provided by the development of polyamideamine epichlorohydrin (PPE) wet-strength resins.

PPE resins are produced by reacting diethylenetriamine with adipic acid (see Fig. 150). The resulting polyamide is reacted with epichlorohydrin, which converts all secondary amine to tertiary amine. The principal feature of this type of resin is its ability to be adsorbed by fibers and to cure to a highly cross-linked insoluble stage under neutral or alkaline conditions. A pH of 6 to 10 is preferred. At this higher pH, paper or paperboard retains more strength upon aging. Absorbency is also improved. The efficiency of the resin is not impaired by the presence of sulfate, chloride, or sulfite anions in the furnish.

5. Other wet-strength additives

A number of other materials have found use as wet end wet strength additives. Some of the more common are listed in Table 28. Each of the agents listed in the table is discussed briefly in reference [91].

6. Papermaking chemistry and wet-strength resins [91]

Experience has produced the following generalizations about the effects of papermaking chemistry variables on wet strength performance [91].

1. Pulp receptivity – Pulps vary in their receptivity to wet strength resins. The literature contains many references to handsheet studies carried out to show the influence of pulps. However, the experimental conditions varied greatly from study to study and it is not possible to draw general conclusions from the reports. It is up to each investigator to

Table 28. Other wet strength additives in use.

1. Urethane Emulsions
2. Cationic Dialdehyde Starches
3. Cationic Dialdehyde Mannogalactans
4. Neoprene Latex Emulsions
5. Polyethyleneimine
6. Formaldehyde Cross-Linking Agents
7. Glyoxalated Polyacrylamide

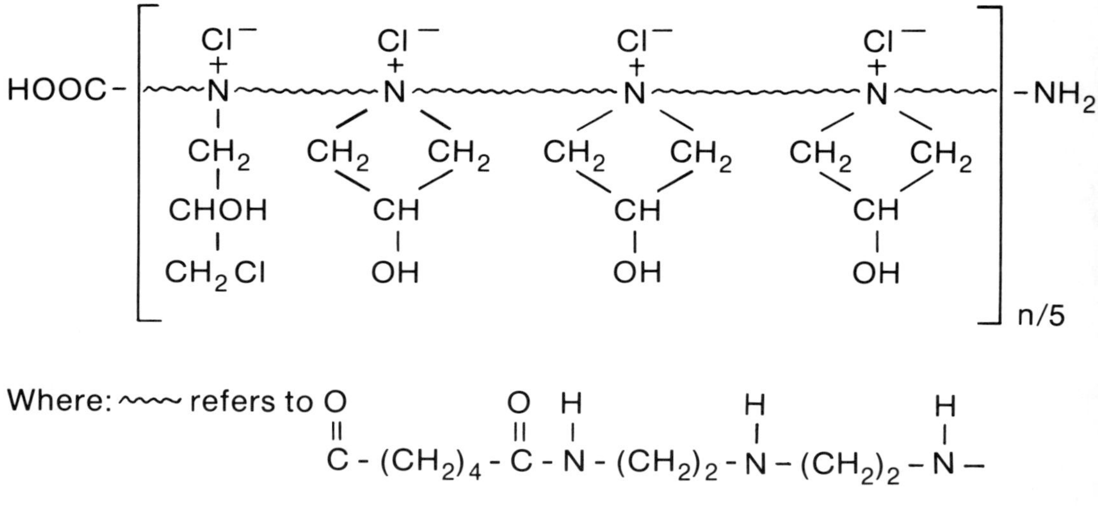

Fig. 150. Polyamine-polyamide-epichlorohydrin wet-strength resin.

establish the relationships that are applicable to a particular situation. The fiber has a limited capacity for wet strength resin. Experience has shown that above, 1.5% addition, the effects of succeeding increments of resin addition diminish rapidly.

2. Curing pH – The pH of the formed sheet is critical for the proper curing of the paper in the dryer section. In the case of ureaformaldehyde and melamine-formaldehyde resins, the pH of the wet end should be around 4.5 and the sheet must be acid for a rapid cure to take place. For neutral cure resins a pH of 6-10 is desired.
3. Additive competition – In order to gain the best efficiency from wet strength resins, care must be taken to avoid interferences from other additives in the system. The adsorption of cationic resins is believed to be promoted by the attraction of the negative fiber for the positive resin. If other substances are present that also adsorb onto the fiber, then they will occupy the adsorption sites utilized by wet strength resins and block their adsorption. Another hinderance to resin retention can occur if the surface charge on the fiber becomes more positive due to the adsorption of substances such as alum, cationic starch, or cationic retention aids. This will diminish the retention-controlling charge gradient that is required for the adsorption of wet strength resins.
4. Fines retention – First-pass fines retention is critical to the efficient use of wet strength resins because the resins tend to preferentially adsorb onto fines. Conversely, wet strength resins themselves can act as retention aids for other chemicals (e.g., synthetic sizes) and fines due to the cationic charge on the resin molecule.
5. Uniform distribution – Resin adsorption is believed to be quite rapid once the polymer and fibers are mixed together. It is critical, therefore, to ensure that very good mixing occurs at the point of addition to the paper machine system in order to avoid nonuniform treatment of fibers with resin.
6. Interfering substances – A high concentration of anionic materials in the system will interfere with the effectiveness of wet strength resins and lead to low chemical use efficiencies. Most chemical suppliers market high charge density cationic polymers ('promoters') that are designed to react with the anionic materials and enable the papermaker to use less wet strength resin for a given level of performance than would otherwise be possible.

This concludes our brief discussion of wet stength resins. It is clear that the efficient use of these materials requires close attention to the papermaking chemistry of the paper machine, both from the standpoint of the retention of the materials in the sheet and the establishment of conditions conducive to the development of good wet strength in the paper machine dryer section and beyond.

G. COLORING MATERIALS

Colorants are used in the manufacture of paper for a variety of reasons, including eye appeal, color coding, and brand identification.

The selection of suitable colorants depends upon a number of factors such as end-use requirements, physical and chemical properties, and handling characteristics. Coloring is done: (1) to produce a paper having a given color or shade; and (2) to produce a white paper having a desired tint or "off-whiteness". To accomplish these ends, the papermaker utilizes a number of different coloring materials that fall into one of the following five categories.

1. Colored pigments;
2. Acid dyes;
3. Basic dyes;
4. Direct dyes;
5. Fluorescent dyes.

These categories differ according to their chemical and physical properties, fastness, affinity for papermaking fibers, effective pH range, solubility, etc. Each colorant has advantages and disadvantages for the papermaker. As a class, the most commonly used materials in the paper industry are the water-soluble direct dyes. Pigments are used sparingly for internal dying, but find applications in the preparation of colored coatings. We will briefly discuss some of the more important characteristics of each of the five categories listed above and then turn to the role of dyes and colored pigments in papermaking chemistry.

The history of the use of dyes and colored pigments dates back to the earliest days of civilization when natural coloring substances that could be readily extracted from plants were applied to fabrics. The synthetic colorant industry is usually related back to 1856 when William Henry Perkin accidently discovered the dyestuff mauve while attempting to simplify the procedure for synthesizing quinine. Good reviews of the technical aspects of dyeing can be found in references [92] and [93].

In this discussion we will concentrate on the chemistry, properties, and papermaking chemistry of the colorants most widely used in the paper industry.

1. Colored pigments

Colored pigments fall into three classifications: natural inorganic pigments, synthetic inorganic pigments, and synthethic organic pigments. As a class, the third category is by far the most important to the paper industry. For reasons that we will discuss later, colored pigments do not find extensive use for the wet end coloring of paper, but are applied more often in coating formulations where they are superior to dyes because they do not have a tendency to follow the water into the sheet when coating is applied.

Table 29. Properties of colored pigments.

1. Excellent lightfastness
2. Low tinctorial value
3. No direct affinity for fibers
4. Contribute to two-sidedness

Table 29 lists some of the more important properties of colored pigments. The low tinctorial value of colored pigments means that high loading levels are required to achieve strong colors, which leads to a weakening of the sheet. The lack of affinity produces poor retention unless retention aids are used. Alum has always been used for this task, although ordinary white pigment retention aids will also work. When alum is used, it is necessary to keep the pH of the system below 5.0 in order to keep the alum in an active form of retention.

2. Synthetic dyes [94]

(a) Acid dyes

This is the name given to a large group of dyes for nylon and wool and one or two for paper. Acid dyes are all water-soluble salts (usually sodium or potassium salt form) of colored aromatic organic acids which dissociate in water to form colored anions. Most acid dyes are azo dyes. A typical structure for an acid dye is presented in Fig. 151. Acid dyes are similar to direct dyes (see Sec. 2c) and the two groups merge together with no clear boundary. Acid dyes generally contain more acid groups which gives rise to their greater solubility in water, compared to direct dyes. Table 30 lists some of the characteristics of acid dyes.

(b) Basic dyes

Basic dyes are the salts (chlorides, hydrochlorides, sulfates, and oxalates) of color acids. Figure 152 contains a typical basic dye. Water solutions of these dyes are acid in character and are particularly sensitive to free alkali. They are soluble in aqueous acid media, and this is often the reason for the use of acetic acid to promote strong concentrated solutions. Basic dyes consti-

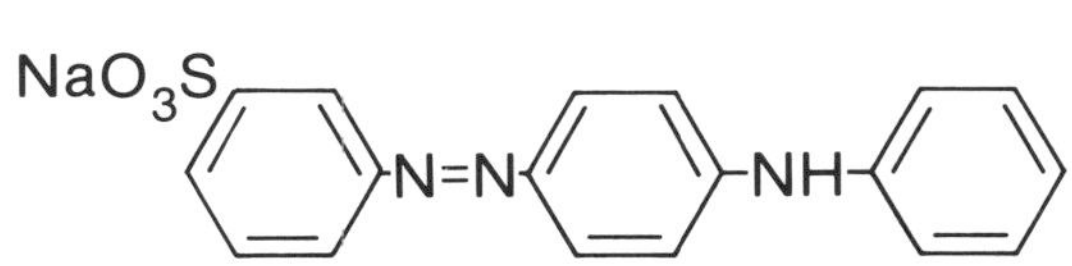

C.I. ACID YELLOW 36

C.I. 13065

Fig. 151. Typical acid dye utilized for paper coloring.

Table 30. Properties of acid dyes.

1. High water solubility
2. Form neutral or slightly alkaline water solution
3. Fair-to-good lightfastness
4. Low affinity for wood fibers
   — dye retention aids must be used
   — the white water will usually be colored
5. Contribute to two-sidedness
6. Sensitive to heat (migrate on dryers)
7. Excellent for calender dyeing

Table 31. Properties of basic dyes.

1. Cationic in nature
2. No affinity for cellulose
3. High affinity for acidic materials such as the lignin present in groundwood and unbleached pulps
   — Retention aids required for retention on bleached wood fibers
   — Can produce "graniting"
4. Highest tinctorial values
5. Brightest of all dyes
6. Poor-to-fair light stability
7. Inferior resistance to acid, alkali, and chlorine

tute the only class of dyes generally applied to paper whose members are soluble to any appreciable extent in methyl, ethyl, or isopropyl alcohols, as well as compounds having similar solvent characteristics, among which are oils and waxes. Table 31 lists some of the important characteristics of basic dyes.

(c) Direct dyes

Direct dyes are sodium salts of dye acids. Chemically, they are similar to acid dyes, and are known as direct dyes because of their affinity for cellulose, which distinguishes them from acid dyes which have no affinity for cellulose. Figure 153 contains an example of a direct dye.

The affinity of direct dyes for cellulose fibers can be attributed in part to the low solubility of the materials. In fact, aqueous solutions of the dyes are often highly colloidal rather than true solutions, and some direct dyes may even gel when a solution is left standing. The classification of acid or direct dyes is often more or less arbitrary and dyes are frequently encountered that have more affinity for cellulose than the average acid dye and less than the average direct dye. These are usually classified as acid dyes.

Recently, cationic direct dyes have been introduced [95]. These dyes retain the planar molecular structure of acid direct dyes, but cationic groups have been incorporated in the structure to increase their affinity for cellulose fibers. An example of one such material is illustrated in Fig. 154. Table 32 lists some of the characteristics of direct dyes.

(d) Fluorescent dyes

These dyes are referred to as "paper whites" or "optical brighteners" and are used by papermakers to brighten white papers. They are very effective when used with highly bleached pulps, but are ineffective when applied to unbleached pulps and groundwood. In some instances they have been used to overcome temporary problems with dirty process water or low-brightness pulps.

Fluorescent dyes absorb light in the ultraviolet

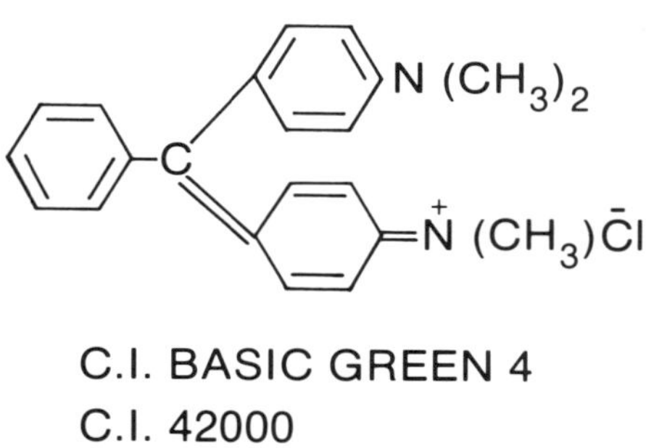

Fig. 152. Typical basic dye utilized for paper coloring.

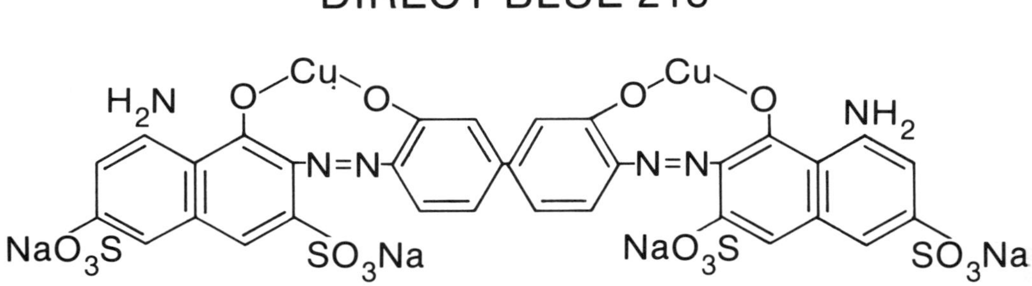

Fig. 153. Typical direct dye utilized for paper coloring.

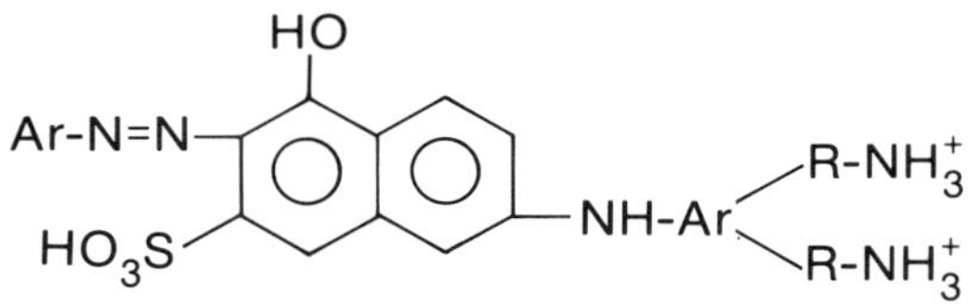

Fig. 154. Typical cationic direct dye utilized for paper coloring.

Table 32. Characteristics of direct dyes.

1. Strong affinity for cellulose
   - Retention can be improved by the use of alum or other retention aids
   - Retention can be improved by heating the stock after adding the dye
   - Retention can be improved by adding a cationic group to the acid dye molecule
2. Superior lightfastness
3. Intermediate tinctorial values
4. Tendency to granite if not handled correctly
5. Amenable to a wide range of pH

region of the spectrum (below 370 nm) and re-emit the light in the visible blue range (peaking at 435 $\pm$ 10 nm), giving a fluorescent effect that in daylight produces bright white, masking the inherent yellowness of the bleached pulp.

Any material that absorbs ultraviolet light will lower the efficiency of fluorescent dyes. Lignin absorbs ultraviolet light and, the higher the lignin content of the pulp, the less the effect from the brightener. Titanium dioxide also absorbs ultraviolet light and will reduce the fluorescence of these dyes. Most clays tend to counteract the fluorescence and reduce the effect of the white dye. Fillers such as calcium carbonate and aluminum trihydrate reflect ultraviolet light, thereby enhancing the effect of the dyes. A high pH (above 5.5) also aids in achieving maximum whiteness. Fluorescent dyes are sometimes added in the size press, also.

3. Dyes and papermaking chemistry

The relationship between dyes and papermaking chemistry can be considered from two viewpoints. First, there are those factors that affect the actual incorporation of the dye in the sheet of paper, i.e., the retention of the dye. Second, other components and conditions of a furnish and paper machine system can influence the development of color by a dye. We will consider the factors related to dye retention first.

(a) Dye retention

Dye retention is closely related to the retention of fines and fillers [96]. This follows from our previous discussion (see Sec. C) of the relative surface areas of fibers, fines, and fillers, and the fact that additives tend to adsorb onto components according to their respective surface areas. As was discussed above, direct dyes have a natural affinity for cellulose fibers due to their (the dye) low solubility. Whenever alum is added to a slurry containing these dyes, a positive charge is imparted to the dye through interaction with the alum.

Other mechanisms assist in the bonding of direct dye molecules to cellulose fibers. You will recall that cellulose fibers have a multiplicity of surface groups that are capable of participating in hydrogen bonds. Direct dyes are all linear and coplanar in structure, i.e, long, flat molecules. They contain groups such as -OH and NH_2 which are capable of hydrogen bonding to cellulose hydroxyl groups. In addition, Van der Waal's forces play a significant role in bonding the dye molecules to the fiber surface.

Cationic direct dyes have a strong cationic charge that can interact directly with negative charge sites on fibers. In addition, dye suppliers are also advocating the combination of cationic and anionic direct dyes to produce "synergistic" mixtures that are reported to be more efficient than the individual components [95].

Acid dyes are similar to direct dyes but lack molecular coplanarity. They also have higher numbers of solubilizing groups that prevent good bonding between the dye molecule and fiber surfaces. The negative charge on the dye is the same as the charge on cellulose and there is no electrostatic attraction between them. That is why cationic retention additives must be used with acid dyes.

Basic dyes carry a positive charge which means that they are electrostatically attracted to fibers. The cationic groups react with any negatively charged acid group present in unbleached pulps and mechanical pulps. Bleached pulps do not have a very high concentration of such groups and they are less attractive to the dyes.

In most instances the papermaker desires to retain as much of an added dye as possible. When a dye is added to a furnish, an equilibrium is set up between the fines and fibers and the suspending water. With light shades, the equilibrium is such that almost all of the dye (direct dye) is adsorbed by the fibers. With deep shades it is often necessary to add something to shift the equilibrium toward the fibers. Alum and/or size will do this. Other actions that promote dye retention include heating the stock, adding salt to the dyed pulp, adding alum to the furnish after dye addition, and adding a dye retention aid (a "promoter").

Numerous dye retention aids are on the market as proprietary products. They are usually low molecular weight, high charge density resins that contain mixtures of constituents like those found in wet strength resins. Wet strength resins and cationic starches also impart a positive charge to a dye and improve its retention during sheet formation.

(b) Interactions that affect color development

The second type of interaction between papermaking chemistry and dyes includes those interactions that affect the color development obtained from the dye. An example of such an interaction might involve interaction with chlorine or chlorine dioxide present in a paper machine system as a microbiological control agent. Unless the residual level of these materials is controlled carefully, they can react with dyes and render them ineffective. Table 33 lists other furnish variables that can affect shade development by dyes.

A discussion of each of these variables is beyond the scope of this chapter. The principal objective here has been to introduce the notion that the performance of dyes on the paper machine is very much affected by the chemistry of the wet end system of the machine.

Table 33. Furnish variables affecting dye performance.

1. Fiber composition
2. Type and amount of fillers
3. Degree and amount of refining
4. Contact time before the sheet is formed
5. Type and amount of sizing agent
6. Alum and aluminum ion concentration
7. pH, total acidity
8. Type and amount of resins and starches

Table 34. Sources of interfering substances.

Fresh water
- A. Calcium and magnesium hardness
- B. Acidity and alkalinity
- C. Organic matter
- D. Suspended solids
- E. Residual chlorine

Pulping residuals
- A. Alkalinity and acidity
- B. Sodium salts
- C. Rosin soaps
- D. Fatty acid soaps
- E. Lignin compounds
- F. Pitch
- G. Calcium carbonate carryover in white liquor

Bleaching residuals
- A. Alkalinity or acidity
- B. Sodium salts
- C. Calcium ion from calcium hypochlorite
- D. Alkali-soluble cellulose and hemicelluloses
- E. Lignin compounds

Broke
- A. Starch
- B. Adhesives
- C. Inks
- D. Miscellaneous contaminants

H. INTERFERING SUBSTANCES ("CATIONIC DEMAND" OR "ANIONIC TRASH")

As mentioned in the introduction to the chapter, a papermaking furnish consists of water, fibers and fiber fines, functional additives, chemical processing aids, and interfering substances. The interfering substances, which are also referred to often as "anionic trash" or "cationic demand", consist of colloidal and dissolved anionic materials that have a variety of sources, several of the most common of which are listed in Table 34.

The materials listed in the table may or may not be present in a given mill. However, these substances can interfere strongly with the action of chemical additives when present in large amounts. Auhorn and Springer discuss the nature of these materials and their effects on papermaking chemistry in two recent articles [97, 98]. A complete session at the 1987 TAPPI Papermakers Conference was devoted to these materials and their effects [99].

When the level of interfering substances is particularly high, the usual practice is to neutralize them with a charge-neutralizing agent like alum or a low molecular weight, high charge density polyelectrolyte prior to the addition of other cationic chemical additives. Thus, alum may be added to enhance a polymer retention aid's effectiveness, even though the alum does not play a direct role in the flocculation chemistry of the polymer.

I. ASSOCIATION COLLOIDS

Up to now, our discussion has dealt with papermaking chemistry topics related to hydrophobic and hydrophilic colloid systems. The third, and final, colloid type found in papermaking systems are the association colloids, whose molecules aggregate spontaneously in a given solvent to form stable particles of colloidal dimensions called micelles. These systems differ from macromolecular solutions because the particles are composed of small molecules held together by Van der Waals interactions.

Most surface active substances used as wetting agents, detergents, and emulsifying agents behave like association colloids. Table 35 lists some typical applications of these materials in the paper industry.

Like rosin size molecules, the molecules of association colloids are *amphipathic*. In other words, they have a "dual" character and exhibit both hydrophilic and hydrophobic traits. The hydrophilic portion of the molecules normally dissolve in water to form molecularly dispersed solutions. The hydrophobic part is chemically similar to materials normally insoluble in water. The compounds in Table 36 are good examples of this. In all cases the hydrocarbon portion of the molecule resembles liquid hydrocarbons or benzene derivatives which are sparingly soluble in water. The salt-like parts resemble sodium sulfate, sodium formate, or ammonium bromide—all of which are water soluble.

The various association colloid materials utilized by the papermaker are often called *chemical processing aids*. These materials are utilized primarily to enhance the cleanliness of a paper machine system and to improve a machine's runnability. These substances are extremely important and their application programs often complex. A particularly good review of chemical processing aids can be found in the Notes from the 1986 TAPPI Chemical Processing Aids Seminar [100]. The objective here will be simply to introduce the various classes of materials.

1. Surface activity

Amphipathic association molecules cause a dramatic, spontaneous lowering of the surface or interfacial tension, even when present in very small concentrations. This occurs because the system tends to decrease its surface energy whenever possible, which can occur if the amphipathic molecule orients itself on the surface with its polar group oriented into the water and its hydrocarbon group extending into the air or oil interface. Materials that exhibit this behavior are called *surfactants*.

2. Micelle formation

As was mentioned above, a characteristic of association compounds is that they form *micelles* of colloidal dimensions. It is thought that micelles have structures similar to that shown in Fig. 155. Micelle formation involves a delicate balance between the hydrophobic and hydrophilic portions of the molecule and depends as well on the solvent properties and concentration

Table 35. Applications of association colloids in the paper industry.

1. Pulp Washing Improvement
2. Pulp Fluffing Improvement
3. Paper Softening
4. Defoaming and Antifoaming
6. Absorbency Improvement in Papers
7. Deposits Control
8. Felt Washing and Conditioning
9. Corrosion Inhibition
10. Yankee Dryer Adhesion Control

Table 36. Association colloids compounds.

Anionic —	Sodium Alkyl Aryl Sulfonates
	Sodium Cetyl Sulfate
	Sodium Oleate
	Sodium Dodecyl Sulfate
Nonionic —	Nonylphenoxypoly(ethyleneoxy) Ethanol
Cationic —	Cetyltrimethylammonium Bromide
Amphoteric —	Dodecyl-beta-Alanine

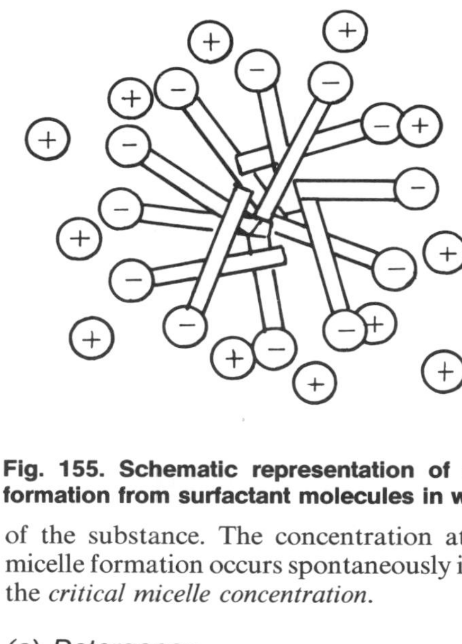

Fig. 155. Schematic representation of micelle formation from surfactant molecules in water.

of the substance. The concentration at which micelle formation occurs spontaneously is called the *critical micelle concentration*.

(a) Detergency

Detergents are used for the removal of dirt for solid surfaces by surface chemical means. Good examples of detergents in the paper industry are the many different felt and fabric cleaning systems available today. A satisfactory detergent must possess the following properties [101]:

1. Good wetting characteristics in order that the detergent may come into intimate contact with the surface to be cleaned.
2. Ability to remove or to help remove dirt into the bulk of the liquid.
3. Ability to solubilize or to disperse removed dirt and to prevent it from being redeposited

onto the cleaned surface or from forming a scum.

The removal of dirt or deposits from solid surfaces involves a combination of mechanical action and surface chemistry. In the absence of a detergent, water is often unable to remove a deposit due to its high surface tension and poor wetting action. When a detergent is added to the water, the hydrophobic parts of the detergent molecules adsorb onto the deposit and the solid surface, thus reducing the adhesion of the deposit to the surface. It is then possible to dislodge the deposit by mechanical action. The dislodged deposit particles are then held in suspension because detergent molecules form adsorbed layers on both the particles and cleaned surface [102].

The most successful detergents are those forming micelles, and this originally led to the opinion that micelles are directly involved in detergent action, their role probably being that of solubilizing oily material. However, detergent action is dependent upon the concentration of unassociated surfactant and practically unaffected by the presence of micelles (other than as a reservoir for replenishing the unassociated surfactant adsorbed from solution). It appears, therefore, that the molecular properties of surfactants associated with good detergent action also lead to micelle formation as a competitive rather than as a contributing process [103].

3. Emulsions

An emulsion is a suspension of droplets of one liquid in another. If a vessel containing two immiscible liquids is shaken vigorously, both liquids are broken up into droplets whose size depends upon the viscosity of the liquids, the interfacial tensions, and the vigor of the shaking. The coalescence of the droplets is retarded in the presence of small amounts of additional components, called *emulsifiers*, and a stable emulsion is formed. Depending upon which liquid is affected most by the emulsifier, either an oil-in-water (O/W) or a water-in-oil (W/O) emulsion is produced. Many retention aids are delivered as water-in-oil emulsions. Synthetic and rosin sizes are also used in emulsion form.

The most important physical property of an emulsion is its stability. Emulsion stability refers to three phenomena: creaming, coagulation, and emulsion breaking by droplet coalescence. In all three instances the emulsion separates into two phases. It is the function of the emulsifying agent to facilitate emulsification and promote emulsion stability. The emulsifying agent forms an adsorbed film around the dispersed droplets which helps to prevent coagulation and coalescence. The stabilizing mechanisms are usually complex and may vary from system to system. Factors which favor emulsion stability are low interfacial tension between the dispersed droplets and dispersing medium, a mechanically strong and elastic interfacial film between the droplets and medium, electrical double layer repulsion between dispersed droplets, narrow droplet size distribution, and a high viscosity in the dispersing medium. The detailed mechanisms involved in emulsification are quite complex. Refer to references [104] and [105] for further treatment of the subject.

4. Foams

Air gets mixed with paper stock at many points in a paper mill and can be a source of significant problems. Among the many troubles related to air are decreased washer efficiency, decreased pumping capacity, excessive foaming, decreased drainage rate, sizing reduction, increased pitch and deposit problems, and pinholes in the sheet.

(a) The nature of foam

A substance capable of adsorbing at the liquid-air interface of bubbles must be present in order for a stable foam to occur. It is well-established that a pure liquid does not foam. Foams are stabilized by two mechanisms:

1. The adsorption of surface-active materials.
2. The concentration of finely divided solid particles around each bubble.

Surface active substances exist in abundance in most paper mill stock as salts of rosin acids and fatty acids. Lignin and other pitchy materials have also been found to be present in surface films on paper stocks. Foam-stablizing, finely divided materials are also plentiful in the form of fillers, fines, alumina floc, etc. These materials do not normally stabilize foams unless excessive aeration occurs or other surfactants are present.

The natural tendency is for the protective layer surrounding a foam bubble to drain and cause the wall to thin and become suspectible to rupture. Foam stabilizers decrease the bubble wall drainage process by increasing the surface and bulk phase viscosities of the liquid in the film. The ability of surfactants and finely divided solids to increase the elasticity of the bubble wall is also important because it dictates the ability of the film to resist mechanical deformation and rupture.

(b) Theory of defoamer action

An excellent discussion of current theoretical ideas about the action of defoamers can be found in reference [100]. For a surfactant-type defoamer to work, it must spread rapidly and thoroughly on the film surface; thereby displacing the liquid in the film and thinning the film to the point of mechanical instability. It is predicted that, as the defoamer becomes saturated with foam stabilizer, the driving force for spreading decreases and the defoamer becomes less effective. This is indeed observed in cases involving tightly closed white water systems. There has not been widespread application of the theory to date. Instead, the formulation of defoamers and antifoams relies heavily on the considerable body of existing empirical knowledge.

5. Summary

We have treated association colloids in a much more cursory manner than the other topics in the chapter because they are treated in more detail elsewhere in the volume. However, it is important that the important contributions association colloids make to papermaking chemistry be recognized. In many respects, the phenomena discussed above represent the most significant challenges facing papermaking chemists today. The chemistry is extremely complex. The systems are difficult to define and characterize. Realistic laboratory simulations are difficult to design. And there are few instruments available that can be used to collect "real-time" data from papermaking systems. On the positive side, many companies and scientists are currently working on projects related to association colloids and the next few years should see the development of a much improved understanding of the mechanisms by which they act and the factors involved.

J. DEPOSIT CONTROL IN PAPERMAKING

Deposit control is a very broad papermaking chemistry topic that includes both inorganic deposits (barium sulfate, calcium oxalate, and calcium carbonate scale; alumina scale; filler deposits), organic deposits (pitch; rosin size; hydrolized synthetic size), and microbiological deposits. A treatment of the subject can be found in reference [106].

It is beyond the scope of this chapter to cover the subject of deposit control in any depth. It should be noted, however, that many inorganic and organic deposits result from poor control of chemical additive addition rates and low values of first-pass fines and filler retention. The production of deposits as a result of overdosing has been a particular problem with alum [106]. Many mills do not maintain close control of alum addition rates. If excess alum is present in the papermaking furnish, exposure to increased pH at wire showers and at other points in the system will result in the precipitation of alumina floc onto forming fabrics and other surfaces. This problem can be so severe as to require the replacement of forming fabrics on a more frequent schedule than desired.

As was emphasized earlier in the chapter, materials that are not retained in the web during sheet formation will remain suspended in the white water and there is a strong probability that they will aggregate and form deposits at some point in the process. Thus, paper machine cleanliness can be adversely affected by poor paper machine retention performance.

1. Pitch problems

Organic "pitch problems" arise with varying frequency and in a variety of forms in every paper mill. Large pitch particles can plug screens and deposit on equipment surfaces. Pitch particles retained in paper may produce dark spots, and cause picking, sheet breaks, and wire and felt filling. These same symptoms may arise from very finely divided pitch. Pitch also produces self-sizing in absorbent and waterleaf papers.

There is still no general agreement as to how to forecast pitch problems reliably, even though many investigations have been undertaken. It is reasonably clear that pitch problems stem from pulp resins (resin acids, fatty acids, fatty alcohols) and, occasionally, sizing agents may also contribute to the problems. However, attempts to relate pulp extractive content with pitch problems have produced inconsistent results and no clear correlation has been discovered.

It has been suggested that colloidal pitch is always present in papermaking systems and causes little trouble. However, alteration of the ambient conditions (pH, temperature, etc.) is often sufficient to induce the aggregation of the very small particles into larger, problem-causing particles. This behavior is typical of hydrophobic colloid systems and colloidal pitch dispersions appear to have the requisite characteristics of such systems.

Salts of the fatty acids and resin acid components of pitch also act as surfactants at low concentrations.

(a) Pitch control

As in the case of defoamers and antifoams, the treatment of pitch problems in the mill is usually approached from an empirical viewpoint. As a result, a very large store of empirical knowledge has been accumulated about procedures to follow when pitch problems arise. The evidence indicates that both hardwoods and softwoods cause pitch problems. The use of unseasoned green wood increases pitch problems. Increased pulp washing and bleaching and highly alkaline pulping conditions reduce pitch problems. Beating and refining also aggravate pitch problems.

There are several schools of thought regarding pitch control. One approach involves using dispersants that stabilize colloidal pitch particles and prevent their deposition. The dispersed pitch particles then leave the papermaking system at water loss points. Another approach involves adding a pigment (such as talc) to induce colloidal pitch particles to precipitate and be carried out with the paper. A third approach is to precipitate the pitch particles onto wood fibers with alum. The particular approach adopted in a given mill will depend upon local factors and past experience.

K. ALKALINE PAPERMAKING

For the past 175 years of machine-made paper, alum-rosin sizing has been the predominant system utilized in the production of sized papers. As a result, most sized papers have an acid pH. It was well known that such papers have a relatively short life and tend to lose their strength rapidly when exposed to high temperatures and other adverse environmental conditions. With the exception of archival papers, library book papers, and a few other grades, the low permanence of acid papers has not proven to be a serious deficiency. About 15 years ago, the U.S. paper industry began to express an increased interest in converting to alkaline papermaking conditions, primarily because a relatively low priced ground calcium carbonate filler pigment appeared on the market. Since it had been known for quite some time that one could incorporate relatively high levels of calcium carbonate filler into paper without seriously degrading its strength, the prospect of substituting a low-cost filler for more expensive fiber became very attractive and a large number of paper companies established alkaline papermaking projects.

The results of these early alkaline papermaking activities were mixed. Several mills successfully converted from acid to alkaline papermaking. Others converted for a time, experienced difficulties, and changed their machines back. A third group of companies ran many trials but, for one reason or another, never reached the stage of attempting a full-scale conversion.

Recently, there has been another resurgence in interest and many more mills have converted. Again the principal driving force was the advent of a new calcium carbonate filler pigment. This time, however, it was a precipitated product that provided the impetus. It is now possible to install a "captive" calcium carbonate plant very near to, or within, any paper mill. The plant will produce a filler-grade precipitated calcium carbonate capable of providing good optical properties in the produced paper.

The above discussion focused primarily on the substitution of calcium carbonate filler for fiber as being the principal driving force for converting from acid to alkaline papermaking. There are a number of other suggested advantages to be gained from making the conversion (stronger paper, higher drainage rates, faster drying, decreased corrosion, increased water recycle, to list a few) and in several instances one or more of these other factors have dictated the change.

1. Alkaline papermaking chemistry

While there are hardware considerations to take into account, this discussion will concentrate on the major papermaking chemistry considerations associated with converting from acid papermaking to alkaline papermaking. The terminology "acid to alkaline" itself indicates that chemistry is a primary issue. Experience has shown that this is the case. It is beyond the scope of this chapter to describe comprehensively the changes in papermaking chemistry required when converting from acid to alkaline papermaking. Entire conferences and many symposia have been devoted to the subject [107–110, 147]. Rather, we will briefly consider some of the more critical papermaking chemistry issues that arise when a conversion is attempted.

It was pointed out above that it is necessary to incorporate high filler loading levels in the sheet in order to gain substantial economic benefits from alkaline papermaking. The exact percentages vary from grade to grade, but it is not unusual for a successful alkaline papermaker to run filler levels of 20–25% in many fine papers. This is to be compared to the 10–12% filler levels common in the same grades made under acid conditions. Experience has shown that the achievement of both high filler levels in a sheet and good paper machine runnability requires high first-pass filler retention. The following argument supports this assertion.

A paper machine manufacturing 270 metric tons/day of fine paper containing 25% filler will have approximately 181 kg of paper leave the forming section of the machine every minute, of which 45 kg will be filler. If the machine has 50% first-pass filler retention, then 45 kg of filler will flow to the white water trays, also. If the first-pass filler retention is 33%, then 90 kg of filler will flow to the trays per minute.

In 1975, Lodzinski published the results of a survey of the retention performance of a large number of commercial paper machines [111]. The results revealed that commercial machines producing paper having a 25% loading level had an average first-pass filler retention of 15%; i.e., in our example, for every 45 kg of filler leaving the forming section with paper, another 300 kg flows to the white water trays! The adverse consequences of having such high levels of filler in the recirculating white water system are obvious. Filler deposit problems are aggravated and the runnability of the machine suffers greatly.

To avoid the scenario described above, it is imperative that high first-pass retention levels be achieved when running highly filled grades under alkaline conditions.

There is also another reason for attaining high first-pass filler and fines retention performance. As was discussed several times previously in this chapter, chemical additives tend to adsorb onto the various furnish components according to their relative surface areas. High furnish filler levels mean that large amounts of chemical additives will be adsorbed on filler surfaces and it is necessary to retain the filler in order to retain the additives.

The above requirement is especially critical where sizing is concerned. The synthetic sizing agents used in alkaline papermaking are sensitive to hydrolysis. Alkenyl succinic anhydride products are particularly so. Consequently, there is a strong incentive to retain these materials in the forming sheet and not allow them to recirculate with white water where they will hydrolyze. Many studies have shown that synthetic sizes behave like other chemical additives and tend to concentrate on fines and filler particles in the system [33]. Therefore, as with other additives, high first-pass fines and filler retention levels are required in order to achieve high size retention levels and prevent size hydrolysis. Failure to accomplish this goal can result in both low sizing efficiency and paper machine runnability problems.

The entire package of functional additives and processing aids must be reevaluated when making the conversion from acid to alkaline papermaking. For example, in addition to utilizing different sizing and filler systems, it is usually necessary to utilize different dye systems under alkaline conditions than under acid conditions. Acid-curing wet strength agents are no longer applicable. The microbiological populations existing in paper machine systems are altered when converting from acid to alkaline conditions, as are the solubilities of various salts. The fact that there have been a number of successful conversions indicates that many papermakers have managed to work through the problems and adapted to the new papermaking chemistry required.

L. BASIC PAPERMAKING CHEMISTRY LABORATORY MEASUREMENTS

The majority of the experimental procedures and methods used in paper machine papermaking chemistry work can be grouped under one of the categories listed in Table 37. Several parameters can be derived from data collected by the primary measurements listed in the table. Foremost are the first-pass retention (single-pass retention) levels of wet end components. Depending on the number of wet end streams investigated, it will be possible to carry out various material balances. The extreme being a complete balance that includes the stock preparation, white water recycle, and broke reclamation areas. If the data points are collected sequentially with time, it will be possible to determine the steady-state or nonsteady-state performance of the machine. The determination of first-pass retention, steady-state and non-steady-state character, and the calculation of material balances will be discussed later. Before doing so, we will consider some of the details of the measurements listed in Table 37.

1. Experimental methods

Laboratory methods for studying component levels, fiber qualities, electrokinetic quantities, and slurry drainage characteristics include the particular determinations that fall under each of these four categories (Table 38). It would be beyond the scope of this discussion to describe each of the determinations listed in Table 38. They cover a wide variety of analytical procedures and methods ranging from rather standard water analysis techniques to very specific measurements unique to papermaking chemistry. Many of the details of the procedures can be found in references [50, 112–137].

Table 37. Categories of papermaking chemistry experimental measurements and procedures.

1. Component levels
2. Fiber qualities
3. Slurry drainage characteristics
4. Electrokinetic quantities
5. Flow rates (liquid, mass)
6. Paper machine operating parameters

Table 38. Specific papermaking chemistry determinations.

- Component Levels
  - A. Suspended solids
    - 1) Percent consistency
    - 2) Percent ash
    - 3) Filler components
    - 4) Percent fines
  - B. Dissolved solids
    - 1) Dissolved inorganic salts
      - a) Specific conductance
      - b) Water hardness
      - c) pH
      - d) Acidity and alkalinity
      - e) Cations and anions
    - 2) Dissolved organic materials
      - a) Starch
      - b) Retention aids and other polyelectrolyte additives
- Fiber qualities
  - A. Fiber composition
    - 1) Hardwood: Softwood percentages
    - 2) Groundwood: chemical fiber percentages
  - B. Fiber length distribution
  - C. Pulp viscosity
  - D. Degree of refining
    - 1) Microscopic inspection
    - 2) Freeness
    - 3) Fines classification
- Electrokinetic characteristics
  - A. Zeta potential
  - B. Electrophoretic mobility
  - C. Cationic demand
  - D. Colloid demand titration
- Slurry drainage characteristics
  - A. Freeness
  - B. Handsheet mold drainage time
  - C. Dynamic Drainage Jar drainage time

Table 39. Papermaking chemistry parameters for which on-line measurements are available.

- Percent consistency
  - A. Thick stock
  - B. Thin Stock
- Thin stock filler levels
- Dissolved solids
  - A. Specific conductance
  - B. pH
  - C. Certain inorganic cations and anions
- Electrokinetic characteristics
  - A. Zeta potential
- Slurry drainage characteristics
  - A. Freeness
- First-Pass retention

Most of the parameters listed in Table 38 are carried out by manual measurement methods. However, a few of them are determined by on-line instruments. Table 39 lists the parameters that can currently be measured on-line.

On-line thick stock consistency measurements have been carried out routinely in the pulp and paper mill for many years. Thin stock consistency sensors have been available for several years but have only recently begun to be utilized. These new sensors are capable of measuring both the total solids consistency and the filler concentrations in the slurries. First-pass furnish and filler retentions can be determined from these measurements.

A survey conducted by TAPPI's Retention and Drainage Subcommittee revealed that pH is by far the most frequent on-line dissolved solids measurement made in the paper industry [139]. Other on-line determinations of dissolved materials are done very infrequently.

The TAPPI survey also revealed that an on-line measurement of cationic demand or some other comparable furnish charge measurement was the "most-needed" measurement as indicated by papermakers and chemical suppliers. Presently, zeta potential is the only electrokinetic characteristic that can be measured on-line. However, research and development efforts are under way to develop on-line cationic demand [140] and on-line streaming current measurement systems [148]. A few of the systems have reached the point where they are being evaluated in paper mills. Once they are fully tested, it is expected that such devices will be rapidly incorporated into the on-line measurement systems of paper machines.

On-line freeness measuring instruments have been available to the industry for many years. Most of these devices have been aimed at the mechanical pulping industry but could be used for measuring the freeness of papermaking stock. Many mills have installed these devices on the wet ends of paper machines at one time or another. However, they have not received widespread acceptance in the industry, perhaps because earlier instruments had reproducibility and reliability problems. Several new instruments have been introduced recently that are reported to have overcome the difficulties associated with earlier versions and papermakers are currently evaluating them.

It is strongly suggested that on-line papermaking chemistry data should be used whenever possible. It is very important, however, that the accuracy and reliability of all on-line data be verified. This means that all sensors must be calibrated regularly and the precision of their output known. A recent paper by Thomas [141] provides an excellent overview of the factors involved in on-line retention measurement.

2. Liquid and mass flow rates.

Process stream liquid and mass flow rates are necessary for carrying out material and energy balance calculations on a wet end system. They are also needed for determining material addition rates and materials loss rates. There are a variety of flow measurement methods available to the engineer – several of which are listed in Table 40.

Several of the methods listed in Table 40 are important enough to deserve very extensive discussion themselves. Indeed, flow rates are the most common measurement made in the paper industry. Here, we will provide only a cursory treatment that should serve to guide you if you wish to pursue any of the specific methods in detail.

Magnetic flow meters and other on-line devices are found in great numbers throughout the paper mill. They provide basic information required by operators. Process engineers rely upon these sensors for carrying out a variety of troubleshooting, process monitoring, and process optimization activities. As with all such on-line instruments, it is absolutely vital that they be maintained and kept in good calibration. This means that a mill should have a routine calibrating schedule for each meter and the routine should be adhered to closely. If this is done and there are sufficient meters in the system, then the collection of flow data will be greatly simplified for the process engineer. Unfortunately, it is a rare mill where the engineer can rely solely on on-line instruments for flow data. Because of this, he or she must be familiar with other flow measuring techniques.

"Bucket-and stopwatch" methods refer to any method where either the length of time required to collect a given quantity of material is determined, or the quantity of material collected in a given length of time is measured. The time-measuring device is usually a stopwatch, while the volume-measuring device can range anywhere from a beaker or graduated cylinder to a barrel or large tank, depending on the magnitude of the flow rate.

This method is usually of limited usefulness around the wet end of a paper machine because it requires collecting the entire flow from a given process stream during a given period and wet end process stream flow rates are far too great to be amenable to such an approach. The method can be very useful for measuring additive flow rates, however.

One method of getting around the high wet end flow rates is to use a chemical dilution method for flow determination. In such an approach, the flow can be determined by introducing a salt solution of known concentration into a stream. By determining the dilution of the salt by the stream, the flow may be calculated with reasonable accuracy.

Table 40. Flow measurement methods.

1. Magnetic flow meters and other on-line devices
2. "Bucket and stopwatch" methods
3. Chemical dilution methods
4. Ultrasonic flow meters
5. Flow balance calculations

Table 41. Stock preparation and paper machine conditions to monitor during a retention aid trial.

1. Couch vacuum
2. Suction box vacuums
3. Machine speed
4. Jet-to-wire speed ratio
5. Dryer section steam use
6. Press loadings
7. Refiner settings

The development of portable ultrasonic flow meters has proven to be a significant advance for the papermaking chemist [138]. These meters provide a convenient, non-intrusive method for measuring flow rates in a wide variety of wet end process streams. Ultrasonic flow meters come in two complementary versions: the Doppler reflective type, and the clamp-on transmission type. The Doppler reflective type of meter is used in streams having suspended solids or air bubbles, while clamp-on transmission meters are applied to streams of clear liquids. Both types are available as portable meters that have proven to be very valuable troubleshooting tools.

Material balance calculations can be used for determining the flow rates of streams that are inaccessible to the types of flow measurements discussed above. The primary requirement is that sufficient data be available to enable the calculation of a material balance that includes the process stream of interest. If this condition is met, then the flow rate in the stream of interest can be computed without making any direct measurements in that stream. This is a powerful technique that can be extremely useful in many instances.

3. Paper machine operating data

Even though we are dealing primarily with papermaking chemistry topics in this chapter, it should be understood that the mode of operation of the stock preparation system and wet end of a paper machine can significantly affect fines retention, sizing, and other papermaking chemistry phenomena. Consequently, it is always advisable to monitor wet end system conditions when carrying out papermaking chemistry work. For example, Table 41 lists some of the paper machine operating data that should be collected during a retention aid trial.

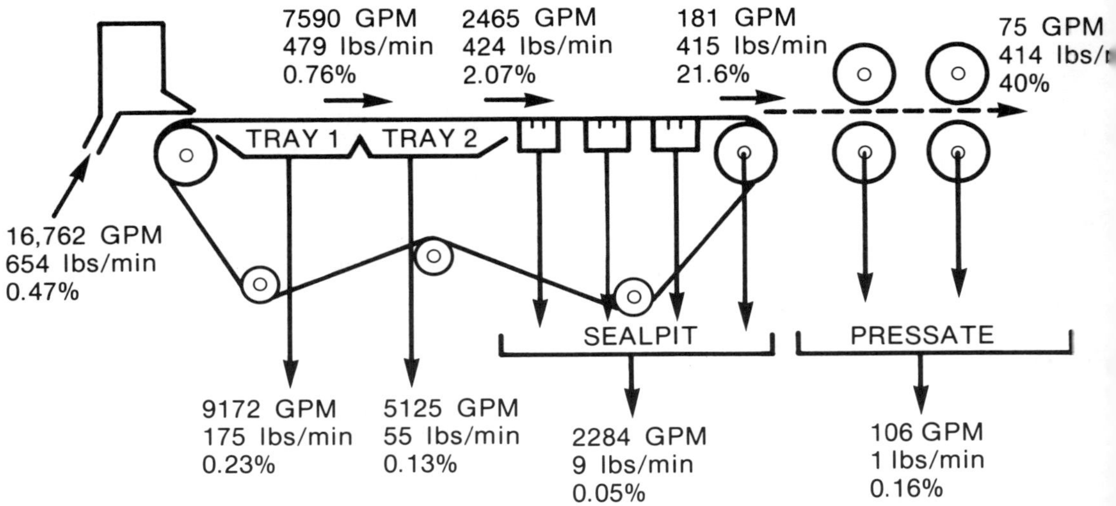

Fig. 156. Sample calculation of water and solids material balance around the wet end of a paper machine.

In addition to the parameters listed in the table, the configuration of the drainage elements on the forming table and the stock jet impingement angle will strongly influence retention and drainage performance.

M. FIRST-PASS RETENTION PERFORMANCE OF A PAPER MACHINE

The retention of a substance is the amount of the substance that remains in the sheet of paper, expressed as a fraction of that in the paper at some earlier stage in the process. The first-pass retention of a material represents the fraction of the material that passes through the headbox slice and is retained during the sheet formation step in papermaking. First-pass retention is of extreme importance to the papermaker because it influences paper machine runnability, material loss levels, and chemical use efficiencies. A detailed discussion of the importance of first-pass retention is presented in reference [142].

First-pass retention values are usually derived from the steady-state material balance conditions existing around a forming section. Since first-pass retention is a steady-state parameter, it is not strictly correct to refer to it when a paper machine is not operating under steady-state conditions. In such instances, it is often more useful simply to monitor the key parameters involved, like headbox consistency and tray consistency.

In this section, we will first present the results of an example calculation of a material balance around the forming section of a paper machine and then show how the balance can be used to derive information about the retention performance of the machine. Next, we will relate the retention parameters calculated from the material balance to those calculated from headbox and tray consistencies and demonstrate the assumptions that underly the latter calculation.

1. Wet end material balance and first-pass retention

Figure 156 contains the results of a typical material balance calculation that can be computed from the flow and consistency data collected during a paper machine drainage performance evaluation. Water and forming table solids are the variables of interest. The most common practice is to measure the headbox flow rate, the consistencies of various white water streams flowing through the forming fabric, and the consistency of the paper on top of the forming fabric. The information contained in Fig. 156 can be used in several ways to describe the performance of the paper machine wet end. We will concentrate on what it tells us about first-pass retention behavior.

If we apply the definition of first-pass retention given above, it can be expressed by Equation (1):

First-pass total solids retention, %

$$= 100 \times \frac{\text{total solids flow in paper}}{\text{total solids flow from headbox}} \quad (1)$$

Equation (1) is a general expression that applies to the entire paper machine or any part of the machine that is related to the headbox by a continuous process flow. The distinguishing feature is where you measure the total solids of the paper – at the reel, after the press section, at the couch, etc. For this discussion we will concentrate on the first-pass total solids retention on the fourdrinier section and, thus, the total solids in the paper of interest refers to the solids in the paper after the couch roll.

From Fig. 156 we see that the total solids coming onto the wire from the headbox is 654 lb/min (297 kg/min). The total solids in the paper leaving the couch roll is 415 lb/min (188 kg/min). Applying Equation (1) results in a first-pass total solids retention on the fourdrinier of 63%.

The papermaker does not usually measure the total solids coming onto or leaving the forming fabric in order to determine the first-pass retention performance of a machine. Rather, the papermaker most often uses Equation (2), which relates first-pass total solids retention to headbox and tray consistencies.

First-pass total solids retention, %

$$= 100 \times \frac{C_H - C_T}{C_H} \quad (2)$$

where C_H and C_T refer to the headbox consistency and tray consistency, respectively.

Utilizing data from Fig. 156 in Equation (2), the calculated first-pass total solids retention is 51% and not the 63% we found earlier. Why the difference? This question can best be answered by considering how Equation (2) is derived. In deriving Equation (2) from the material balance, two assumptions must be made. First, it is assumed that the headbox solids are all divided between the sampled tray and the paper. In other words, the total headbox solids equals the total paper solids plus the total sampled tray solids. The second assumption is that all the water flowing from the headbox flows into the sampled tray, i.e., none goes with the paper.

In practice, the paper leaving the couch roll contains only 1–2% of the total water coming from the headbox. Therefore, the second assumption would appear to be reasonable. However, the situation is not quite so straightforward with respect to the first assumption. The assumption that all headbox solids flow to the sampled tray can be met in at least three ways on a given paper machine. First, if a paper machine has only one white water tray, then most of the solids that flow from the headbox that are not retained in the paper will flow into the tray with the white water and Equation (2) should be applicable. Second, if the machine is configured so that all the white water from various trays is collected and mixed at some point and a representative sample of the mixed white water can be gotten, then Equation (2) will also hold. (Note that, if all white water is not mixed, then it is important that the mixing be done in the ratio of the relative flow rates of the white water to the various trays.) Finally, if the white water flowing to all trays has the same consistency, then Equation (2) will provide a reasonable estimate of first-pass retention.

With these considerations in mind, it is easy to understand why Equation (2) did not give a correct value for the first-pass solids retention of the paper machine in Fig. 156. Referring to the figure, it can be seen that only 73% of the headbox solids and 55% of the headbox water flow into the sampled tray. If it were necessary to obtain an accurate measure of the first-pass solids retention on this machine, it is clear that one would have to measure the white water consistencies of both trays and also have a good estimate of the flow rates to the two trays. With this approach you could account for approximately 95% of the total solids and the calculated first-pass retention would be 65%, a reasonably close estimate to the true value of 63%. In many mills the flow rates are not easy to obtain. Headbox side bleeds can introduce error into the way headbox flows are used and can even enrich white water consistency.

It is not always necessary to collect the flow rates and consistencies required to compute an accurate first-pass retention value. It all depends on the purpose of the work. For example, if the objective is simply to monitor the first-pass retention performance of a paper machine over time, then the primary information required is related to the *changes* in first-pass retention that occur. Thus, the absolute value of the parameter is not so important and headbox and tray consistency data may be all that is needed.

On the other hand, if one is following the results of a retention aid trial, then it is advisable to carry out a more exact retention analysis because the material under investigation can affect retention performance at a number of points on the paper machine – including the vacuum drainage section, the couch roll, and the press section.

First-pass ash retention is usually calculated from the formula presented in Equation (3), which is closely related to Equation (2).

First pass ash retention, %

$$= 100 \times \frac{C_H A_H - C_T A_T}{C_T A_T} \quad (3)$$

where A_T and A_H refer to the tray and headbox ash levels, respectively.

The derivation of Equation (3) involves exactly the same assumptions as those discussed above for Equation (2) and it should therefore be used with the same caution.

Another popular method of calculating first-pass ash retention is to compute the ratio of the ash content of the paper leaving the couch roll to the ash content of the furnish in the headbox. This relationship provides a reasonable value for first-pass ash retention only when the first-pass total solids retention is very high. The lower the first-pass total solids retention, the lower the accuracy with which the couch-to-headbox ash ratio approximates the true first-pass ash retention.

2. Importance of precise consistency measurements

Regardless of which formula is used to calculate first-pass total solids retention or first-pass ash retention, it is absolutely critical that the consistencies and ash levels be determined with good precision. The reason for this is that the retention equations involve ratios of the experimental variables and errors in the measured consistency data can be greatly amplified in the final calculated retention values.

Putting it another way, small errors in the headbox and tray consistency and ash data can produce large variations in the calculated first-pass retention values. For example, it can be shown that a 0.02% deviation in the absolute headbox or tray consistencies can result in a 6% change in the calculated absolute first-pass total solids retention. To avoid these problems, it is absolutely necessary that all weighings associated with first-pass retention consistency and ash content data be made to the nearest 0.001 measurement unit.

N. LABORATORY INVESTIGATION OF FIRST-PASS RETENTION

The paper machine is too expensive to be used for routine or extensive exploratory retention experiments. Because of this, papermakers have always employed laboratory retention investigations as a means of evaluating retention additives and elucidating the effects of various chemical parameters on first-pass retention. The former activity is usually done as a preliminary step to running machine trials with retention aids, while the latter is associated with trouble-shooting and process improvement studies.

Table 42. Devices used in laboratory retention studies.

1. Static settling testers
2. Free drainage testers
3. Handsheet formers
4. Britt Dynamic Drainage Jar

A variety of devices have been employed in laboratory retention studies. Since it is impossible to simulate the conditions on the wet end of a paper machine with any accuracy, most persons have settled for a procedure that will enable them to rank retention aids according to their effectiveness in the laboratory test and also provide input as to whether or not they can be expected to fail completely on the machine. In most cases, it can be stated with fair certainty that any retention aid that is completely ineffective in a laboratory test will also fail under mill conditions.

Table 42 lists several of the devices that have been employed in laboratory retention studies.

Of the devices listed in Table 42, the one receiving the most acceptance in recent years has been the Britt Dynamic Drainage Jar [115]. In addition to the original version, a number of variations of the device have been developed [136, 143].

The Britt Jar allows the direct measurement of the colloidal retention of solids under turbulent conditions independent of fiber mat formation. The measurement of colloidal retention is of practical and fundamental importance because the paper stock system is known to be in a dynamic state at the headbox slice and initial forming areas of the paper machine. In addition, as was discussed earlier, fines retention is a very strong function of colloidal flocculation phenomena.

The basic Britt Jar (Fig. 157) consists of a cylindrical stock sample-holding container that screws into a base designed to hold a perforated plate or sample of paper machine forming fabric. The base has a hole in the bottom into which is inserted a large, one-hole stopper. A small-diameter tube is inserted into the rubber stopper to a point flush with the top of the stopper. A propeller for stirring the stock sample is inserted into the container. The propeller is driven by a variable-speed motor capable of operating in the range of 0 to 3000 rpm. Reference [50] describes the use of the Britt Jar in detail.

Since its introduction in the early 1970s, a

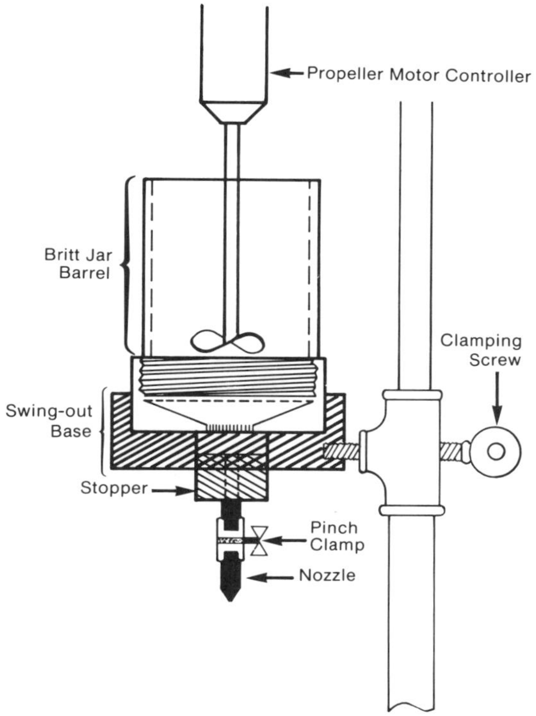

Fig. 157. Diagram of Britt Dynamic Drainage Jar.

Table 43. Uses of the Britt Jar.

Fines Screening
- A. Characterization of stock samples.
- B. Evaluation of first-pass fines retention on the paper machine

Investigation of shear resistance
- A. Additive evaluation
- B. "Calibration" of the Britt Jar against a paper machine

Study of variables that influence retention
- A. Retention aid studies
- B. Investigation of retention interferents
- C. Investigation of mechanical factors and first-pass retention

number of different uses have been developed for the Britt Jar. The most important of these are listed in Table 43. A good introduction to the detailed procedures associated with the uses in Table 43 can be found in TAPPI Method T 261.

O. PAPERMAKING CHEMISTRY PROCESS CONTROL

The final subject that we will cover in this chapter is papermaking chemistry process control; i.e., what is done or should be done in the mill by operators and engineers to keep the papermaking chemistry of a machine within some desired set of conditions. The following three aspects of the topic will be discussed:

1. Papermaking chemistry parameters that are currently controlled.
2. Current measurement and control practices.
3. Suggested future measurement and control practices.

The currently controlled papermaking chemistry parameters can be grouped into three categories. First, there are those phenomena that are closely related to fiber and fines flocculation, such as retention, drainage, and sheet formation. Next, there are those phenomena directly related to component levels, such as pH, total acidity, stock consistency, and the levels of various additives. The third category, paper properties, is not really papermaking chemistry in the usual sense of the term. However, many control actions taken on the wet end of the paper machine to control paper properties do directly influence the papermaking chemistry of the system and, for this reason, we have included them in our list.

While there may be other variables that could have been included in the listing, the parameters listed above cover most of the papermaking chemistry control practiced in the U.S. paper industry today. What, then, are the particular control strategies followed for each of these parameters? We will address this question from both the general and specific viewpoints. First, the general considerations. With few exceptions, papermaking chemistry process control practices employ feedback control loops. In other words, the strategy is to react to observed changes in the parameters under control. This is particularly clear in the case of paper properties where a significant deviation from the desired property value (opacity, brightness, ash content, shade) usually causes the operator to change the addition rate of one or more wet end additives. The change may be made automatically via closed-loop control or the operator may be required to intervene (supervisory control). The control of flocculation-related parameters (retention, drainage, formation) is usually carried out by open-loop, supervisory actions, also. Component level control (pH, consistency, total acidity, etc.) is usually accomplished through closed-loop feedback control.

Many of the control actions described above require operator intervention, which results in the adoption of a conservative, almost passive, approach. Actions are taken infrequently and only when the system is observed to be clearly out of control is a change made. One of the primary reasons for this conservative approach is the paucity of real-time papermaking chemistry information available to the operator. As the TAPPI survey indicated [139], very few on-line papermaking chemistry measurements exist and most papermaking chemistry data available to an operator are based on infrequent manual measurements.

The paucity of real-time papermaking chemistry data also means that operators have not been able to develop reliable control strategies that can be utilized with confidence. Indeed, it is common knowledge that applying the same control response to identical problem symptoms on different occasions often results in two different reactions from the machine. This experience is usually the consequence of having insufficient process information available to the operator.

In spite of the deficiencies described above, a considerable body of empirical and theoretical knowledge exists relative to papermaking chemistry process control. For example, pH and total acidity control is done for the purpose of optimizing the wet end reactions that occur among rosin, alum, and fibers. There has also arisen a clear strategy of separating the control of fiber flocculation from the control of fines flocculation on the paper machine. For the most part, fiber flocculation is controlled by the application of shear and turbulence forces in the headbox and on the forming section of the paper machine. On the other hand, fines flocculation is controlled by the addition of polyelectrolytes and/or alum flocculants. Drainage control is usually done by a mixture of chemical and mechancial means through the use of drainage aids, headbox consistency changes, vacuum box adjustments, foil angle adjustments, and refining changes.

Obviously, the types of control actions listed above vary greatly in the frequency and ease with which they can be taken. Changing the addition rate of a flocculant is relatively simple and can be done as often as you wish. On the other hand, changing the mechanical configuration of the headbox or forming section cannot be accomplished without expending considerable time and energy. Consequently, mechanical changes are not very suitable for use in real-time process control schemes.

It is clear that the papermaking chemistry process control practices described above will not be adequate to meet the needs of the papermaker in the future. Many paper technology developments of the recent past, such as increased machine speeds, high turbulence headboxes, twin-wire formers, increased water recycling, higher loading levels, and alkaline papermaking, have all placed increased emphasis on the achievement of improved papermaking chemistry. It is expected that current trends will continue in the future and there will be an increasing need for the employment of real-time, automatic, papermaking chemistry process control and statistical process control systems.

There is already a strong base of fundamental knowledge upon which to base such control systems. For example, it has been understood for some time that one should attempt to neutralize the interfering substances in the thick stock side of stock preparation, rather than wait until the stock has been diluted. Likewise, it is clear that

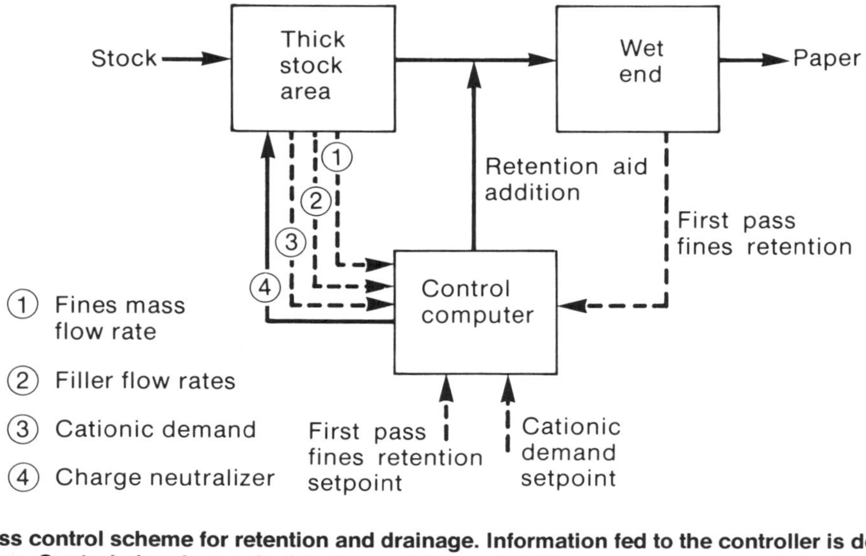

158. Process control scheme for retention and drainage. Information fed to the controller is depicted by dashed lines. Control signals are depicted by solid lines [139].

first-pass fines retention is best controlled within the thin stock loop of the paper machine. Several researchers have suggested control strategies based on these ideas [142, 144, 145]. Figure 158 illustrates one proposed scheme.

The most serious hindrance to the development and employment of papermaking chemistry process control systems is the paucity of on-line sensors that measure papermaking chemistry parameters.

There are reasons to be optimistic, however. Low-consistency sensors have recently been introduced that work well and can be used as the basis of a first-pass solids retention measurement. References [141] and [146] review current applications of these sensors to the on-line measurement of retention. Other researchers are developing on-line cationic demand and charge-measuring sensors. Finally, several mills are applying statistical quality control methods to the control of papermaking chemistry, a requisite step in the development of the algorithms required for automatic control systems.

REFERENCES

1. Linhart, F., Dimensional Analysis — A Novel Approach to Understanding Retention Aids and Their Mode of Action, Proc. 1989 TAPPI Papermakers Conf., Washington, DC, p. 277-285 (1989).
2. Helmholtz, H., Wied Am., 7:337 (1879).
3. Gouy, G., Am. Phys. 7:129 (1917).
4. Chapman, D.D., Phil. Mag. 25(6):475 (1913).
5. Stern, O., Elektrochem. 30:508 (1924).
6. Williams, D.G. and Swanson, J.W., TAPPI CA Rept. 57, p. 47 (1975).
7. Voyutrsky, S., Colloid Chem., Mir. Rub., Moscow (1978).
8. Overbeek, J.Th.G., J. Coll. Interface Sci. 58(2):408 (1977).
9. Hiemenz, P.C., Principles of Colloid and Surface Chemistry, 2nd Ed., Marcel Dekker, Inc., New York (1986).
10. Gregory, J., J. Coll. Interface Sci. 42(2):448 (1973).
11. LaMer, V.K. and Healy, T.W., Rev. Pure Appl. Chem. 13:112 (1963).
12. Fleer, G.J. and Lyklema, J., J. Coll. Interface Sci. 46(1):1 (1974).
13. Napper, D.H., J. Coll. Interface Sci. 58(2):390 (1977).
14. Hale, J.D., "Structural and Physical Properties of Pulpwood", Pulp and Paper Manufacture, Vol. 1, The Pulping of Wood, Chap. 1, McGraw-Hill (1969).
15. Jaycock, M.J., Papers presented at 1st and 2nd Intl. Seminars on Papermaking Chem., Amsterdam (1977) and New York (1978).
16. Hinton, A.J. and Quinn, M., Fiber Surface Characteristics as They Affect Two-Sided Dyeing of Paper, Paper Techn. 5(1):60 (1964).
17. Britt, K.W. and Unbehend, J.E., New Methods for Monitoring Retention, Tappi J. 59(2):67-70 (1976).
18. Htun, A.J. and deRuvo, A., The Implication of the Fines Fraction for the Properties of Bleached Kraft Sheets, Svensk Papperstidn. 16:507-510 (1978).
19. Marton, J. and Marton, T., Wet End Starch: Adsorption of Starch on Cellulosic Fibers, Tappi J. 59(12):121-124 (1976).
20. Marton, J., The Role of Surface Chemistry in Fines-Cationic Starch Interactions, Tappi J. 63(4):87-91 (1980).
21. Marton, J. and Marton, T., Effect of Fillers on Rosin Sizing of Paper, Tappi J. 66(12):68-71 (1983).
22. Marton, J. and Marton, T., Some New Principles to Optimize Rosin Sizing, Tappi J. 65(11):105-109 (1982).
23. Ekwall, P. and Bruun, H.H., Surface Chemical Studies of the Conditions of Rosin Sizing, Tappi J. 37(7):303-306 (1954).
24. Rosen, A., Adsorption of Sodium Ions on Kraft Pulp Fibers During Washing, Tappi J. 58(9):156-161 (1975).
25. Xuan, N.N., Venkatesh, V., Gratzl, J.S. and McKean, W.T., The Washing of Soda-Oxygen Pulps: Sorption Phenomena of Cations, Tappi J. 61(8):53-56 (1978).
26. "Paper Loading Materials", TAPPI Monograph 19 (1958).
27. Bohmer, E., "Filling and Loading", in Pulp and Paper: Chemistry and Chemical Technology, Vol. III, J.P. Casey, Ed., p. 1515-1546, Wiley, Interscience (1981).
28. Beazley, K.M., "Filler Pigments", in Retention of Fine Solids During Paper Manufacture, TAPPI CA Rept. 5060 (1975).

29. Gill, R. and Scott, W.E., The Relative Effects of Different Calcium Carbonate Filler Pigments on Optical Properties, Tappi J. 70 (1):93 (1987).
30. "Register of Coating and Filler Pigments Commonly Used in Paper and Paperboard", in Pulp and Paper Buyers Guide, p. 32-38 (1982).
31. Brecht, W. and Pfretzschner, H., Papier Fabrikant 47:329-333 (1936); Ivanov, S.N. and Ljadowan, W., Zellstoff u. Papier 19(4): 101-105 (1970).
32. Williams, W.R. and Mather, R.D., An Analytical Technique for Recovering Filler to Study Particle Size Fractionation During the Paper Forming Process, Paper Techn. and Ind. 16(3):T150 (1975).
33. Scott, W.E., Fines Management and Control in Wet-End Chemistry, Tappi J. 69(11):30 (1986).
34. Gill, R., Personal Communication (1982).
35. Smellie, R.H., Jr. and LaMer, V.K., J. Coll. Sci. 13:589 (1958).
36. Healy, T.W. and LaMer, V.K., The Adsorption-Flocculation Reactions of a Polymer with an Aqueous Colloidal Dispersion, J. Phys. Chem. 66:1835 (1962).
37. Tarvin, R.F., Hinch, J.D. and Honig, D.S., Retention Aids: Structure-Performance Relationships, 1983 TAPPI Retention and Drainage Seminar Notes, Savannah, GA, p. 15-24 (Nov. 14-15, 1983).
38. Marton, J., Effects of Machine Speed on Formation, Drainage, and Retention, 1989 TAPPI Retention and Drainage Short Course Notes, Washington, DC, p. 19-25 (Apr. 12-14, 1989).
39. Hubley, C.E., Mason, S. and Robertson, A., Flocculation in Suspensions of Large Particles, Can. J. Res. 28(8):770 (1950).
40. Balostis, Y., McKenzie, A.W. and Harrington, K.J., "Consolidation of Paper Webs", F. Bolam, Ed., Vol. 2, p. 659, British Paper and Board Manufacturers Assoc. (1966).
41. Wahren, D., On Three-Dimensional Fiber Networks, Svensk Papperstidn. 13:536-539 (1964).
42. Wrist, P.E., "Formation/Structure of Paper", F. Bolam, Ed., Vol. 2, p. 839, British Paper and Board Manufacturers Assoc. (1962).
43. Hofreiter, B.T., "Natural Products for Wet End Addition", in Pulp and Paper: Chemistry and Chemical Techn., Vol. III, J.P. Casey, Ed., p. 1475-1514, Wiley-Interscience (1981).
44. Yanc, C., Effect of Freeness, Consistency, and Additives on Formation, Tappi 46(10): 169A-172A (1963).
45. Wasser, R.B., Formation Aids for Paper: An Evaluation of Chemical Additives for Dispersing Long-Fibered Pulps, Tappi J. 61(11):115 (1978).
46. Wilharm, M.T., Novak, R.W. and Savino, C.A., Proc. 1987 TAPPI Nonwovens Conf., Orlando, FL, p. 101-104 (1987).
47. Unbehend, J.E., Mechanisms of "Soft and Hard" Floc Formation in Dynamic Retention Measurements, Tappi J. 59(10):74 (1976).
48. Han, S.T. and Chang, N.L., Retention of Fines in Fiber Mats, Tappi J. 52(4):688 (1969).
49. Britt, K.W., Retention of Additives During Sheet Formation, Tappi J. 56(3):83 (1973).
50. Unbehend, J.E., The Dynamic Retention/Drainage Jar — Increasing the Credibility of Retention Measurements, Tappi J. 60(7):110 (1977).
51. Britt, K.W. and Unbehend, J.E., Electrophoresis in Paper Stock Suspensions as Measured by Mass Transport Analysis, Tappi J. 57(12):81 (1974).
52. Moore, E.E., Charge Relationships of Dual Polymer Retention Aids, Tappi J. 59(6):120 (1976).
53. Moberg, K., Microparticles in Wet End Chemistry, 1989 TAPPI Retention and Drainage Short Course Notes, Washington, DC, p. 86 (Apr. 12-14, 1989).
54. Ford, P.A., The Application of a Dual Component Polymer Pigment System in the Manufacture of Multi-Ply Waste-Based Coated Packaging Board, 1989 TAPPI Retention and Drainage Short Course Notes, Washington, DC, p. 95-97 (Apr. 12-14, 1989).
55. Groen, J., The Formation and Structure of Paper, British Paper and Board Manufacturers Assoc., p. 697 (1962).
56. Tanaka, H., Luner, P.L. and Cote, W., How Retention Aids Change the Distribution of Filler in Paper, Tappi J. 65(4):95 (1982).
57. Gess, J.M., Personal Communication (1989).
58. Dobbins, R.J., Overflocculation Reduces Opacity and Affects the Strength of Paper, Paper Trade J. 162(20):42 (1978).
59. Brill, H.C. and Hecklau, F.L., Titanium Dioxide Retention, Tappi J. 43(4):229A (1960).
60. Klungness, J.H. and Exner, M.P., Chemical Additives to Pulps — Effect on Drainage and Strength of Papers Containing Oak Pulps, Tappi J. 63(6):73 (1980).
61. Ellis, W.H. and Foster, W.A., Chemicals Can Improve Drainage Performance on the Paper Machine, Paper Trade J. 152(23):34 (1968).
62. Linke, W.F., Retention and Bonding of Synthetic Dry Strength Resins, Tappi J. 51(11): 59A (1968).
63. Moore, E.E., Drainage and Retention Mechanisms of Papermaking Systems Treated with Cationic Polymers, Tappi J. 58(1):99 (1975).
64. Moore, E.E., The Role of Multivalent Anions in Pulp Systems Treated with Drainage Aids, Tappi J. 56(3):71 (1973).
65. Unbehend, J.E. and Britt, K.W., Unpublished Results.
66. Abson, D., Bailey, R.M., Lenderman, C.D., Nelson, J.A. and Simons, P.B., Predicting the Performance of Shear-Sensitive Additives, Tappi J. 63(6):55 (1980).

67. Urick, J.M. and Fisher, B.D., Factors Affecting the Use of Chemical Drainage Aids, Tappi J. 59(10):78 (1976).
68. May, O.W. and Buckman, S.J., Practical Effects of Air in Papermaking, Tappi J. 58(2):90 (1975).
69. Gavelin, G., Pulp Paper Mag. Can., 55(C) (1954).
70. Scott, W.E., The Miami University Entrained Air Research Program, 1989 TAPPI Retention and Drainage Short Course Notes, Washington, DC, p. 87-90 (Apr. 12-14, 1989).
71. Attwood, B.W., A Study of Vacuum Box Operation, Paper Techn. 3(5):446 (1962).
72. Unbehend, J.E. and Britt, K.W., Retention, Drainage, and Sheet Consolidation, Industrial and Engineering Chemistry Product R and D 21(2):150-153 (1982).
73. Gess, J.M., An Introduction to the G/W Drainage Retention System, Proc. 1989 TAPPI Retention and Drainage Short Course Notes, Washington, DC, p. 49-52 (Apr. 12-14, 1989).
74. Hayden, P.L. and Rubin, A.J., Aqueous Environmental Chemistry of Metals, Ann Arbor Press, Ann Arbor, p. 317 (1974).
75. Arnson, T., The Chemistry of Aluminum Salts in Papermaking, Tappi J. 65(3):125 (1982).
76. Davidson, R.W., The Sizing of Paper, Tappi J. 58(3):48 (1975).
77. Strazdins, E., Mechanistic Aspects of Rosin Sizing, Tappi J. 60(10):102 (1977).
78. Strazdins, E., Chemistry of Rosin Sizing, Tappi J. 64(1):31 (1981).
79. Marton, J. and Marton, T., New Aspects in the Mechanism of Rosin Sizing, Pulp Paper Can. 83(11):T303-308 (1982).
80. Marton, J., Papermaking Raw Materials: Their Interaction with the Production Process and Their Effect on Paper Properties, Trans. 8th Fundamental Res. Symp., Vol. II, p. 785 (Sept. 1985).
81. Marton, J., 1985 TAPPI Advanced Topics in Wet Chemistry Seminar Notes, Atlanta, p. 29 (Apr. 1985).
82. Roberts, J.C. and Garner, D.C., The Mechanism of Alkyl Ketene Dimer Sizing of Paper, Part 1, Tappi J. 68(4):1198-121 (1985).
83. Wasser, R., 1985 TAPPI Advanced Topics in Wet Chemistry Seminar Notes, Atlanta, p. 45 (Apr. 1985).
84. Davidson, R.W., Dry Strength Additives, in Dry Strength Additives, W.F. Reynolds, Ed., TAPPI PRESS (1980).
85. Roberts, J.C., Au, C.O., Clay, G.A. and Lough, C., The Effect of C^{14}-Labelled Cationic and Native Starches on Dry Strength and Formation, Tappi J. 69(10):88-93 (1986).
86. Werdouschegg, F.M.K., Dry Strength Additives, W.F. Reynolds, Ed., TAPPI PRESS (1980).
87. "Commercially Available Chemical Agents for Paper and Board Manufacture", TAPPI CA Rept. 60 (1975).
88. Harvey, R.D., "Properties and Applications of Commercial Cationic Starches", 1987 TAPPI Advanced Topics in Wet End Chemistry Seminar Notes, Memphis, p. 83-90 (Oct. 14-16, 1987).
89. Reynolds, W.F. and Wasser, R.D., Dry Strength Resins, in Pulp and Paper: Chemistry and Chemical Techn., J.P. Casey, Ed., Vol. III, p. 1447-1474, Wiley-Interscience (1981).
90. Britt, K.W., Wet Strength in Paper and Paperboard, J.P. Weidner, Ed., TAPPI Monograph 29 (1965).
91. Britt, K.W., Wet Strength, in Pulp and Paper: Chemistry and Chemical Techn., J.P. Casey, Ed., Vol. III, p. 1609-1626, Wiley-Interscience (1981).
92. Schwalbe, H.C., Pulp and Paper Manufacture, Vol. III, Papermaking and Paperboard Making, R.G. MacDonald and J.N. Franklin, Eds., McGraw-Hill, Chap. 2 (1970).
93. Lips, H.A., Dyeing, in Pulp and Paper: Chemistry and Chemical Techn., 3rd Ed., Vol. III, J.P. Casey, Ed., John Wiley, p. 1627-1666 (1981).
94. Ehrenburg, H., Personal Communication (1987).
95. Brady, B.P., Dye Synergism: A Combined Effort, Proc. 1987 TAPPI Papermakers Conf., Atlanta, GA, p. 17-20 (Apr. 6-8, 1987).
96. Britt, K.W., Some New Approaches to Improving Retention in Paper Sheet Formation, Pulp and Paper 51(4) (1977).
97. Auhorn, W., Linhart, D., Degen, H. and Lorz, R., Anionic Trash: Controlling Detrimental Substances, Tappi J. 70(10):79-85 (1987).
98. Springer, A., Noe, J. and Wegner, T., Fundamental Strategy for Control of Retention and Drainage on a Modern Paper Machine, Proc. 1987 TAPPI Papermakers Conf., Atlanta, p. 137-140 (1987).
99. Proc. 1987 TAPPI Papermakers Conf., Atlanta, p. 141-178 (Apr. 6-8, 1987).
100. 1986 TAPPI Papermaking Chemical Processing Aids Seminar, New Orleans (Apr. 1986).
101. Shaw, D.J., "Introduction to Colloid and Surface Chemistry", 3rd Ed., Butterworths, London, p. 139-142 (1980).
102. Kushner, L.M. and Hoffman, J.I., Synthetic Detergents, Scientific Am. 26-30 (Oct. 1951).
103. Adamson, A.W., "Physical Chemistry of Surfaces", 3rd Ed., Wiley-Interscience (1976).
104. Becher, R., "Emulsions — Theory and Practice", 2nd Ed., A.C.S. Monograph 162, Reinhold (1965).
105. Sherman, P., Ed., "Emulsion Science", Academic Press (1968).
106. Springer, A., Porwal, S. and Proctor, A., Scale Deposits on the Fourdrinier Wire of a Fine Paper Machine, Tappi J. 63(6):67 (1980).
107. Proc. 1988 TAPPI Papermakers Conf., Chicago, IL, p. 33-60, 81-106 (Apr. 11-13, 1989).
108. Daiste, D.P., Considerations for Alkaline Papermaking, Tappi J. 64(4):97-99 (1981).

109. 1985 TAPPI Alkaline Papermaking Seminar Notes, Denver (Apr. 15-17, 1985).
110. Proc. 1989 TAPPI Papermakers Conf., Washington, DC, p. 83-109 (Apr. 10-12, 1989).
111. Lodzinski, F.P., in Retention of Fine Solids During Paper Manufacture, TAPPI PRESS, Chap. 1 (1975).
112. TAPPI Method T 240 (Percent Consistency).
113. TAPPI Method T 413 (Percent Ash).
114. TAPPI Method T 233 (Percent Fines and Fiber Length by Classification).
115. TAPPI Method T 261 (Percent Fines by Britt Jar Screening).
116. TAPPI Useful Method 508 (Analysis of TiO_2 in Process Streams and Paper).
117. TAPPI Method T 627 (Analysis of TiO_2 in Process Streams and Paper).
118. "Methods for Chemical Analysis of Water and Wastes", U.S. Envir. Prot. Agency, EPA-625/6-674-003 (1974) (Sample Preparation Methods for Atomic Absorption Analysis of Process Streams).
119. TAPPI Useful Method 214 (Measurement of pH in Stock and White Water Samples).
120. Avery, L.P., Evaluation of Retention Aids: The Quantative Alum Analysis of a Papermaking Furnish and the Effect of Alum on Retention, Tappi J. 62(2):43-46 (1979) (Analysis of Alum in Process Streams).
121. Aldrich, L.C., A Volumetric Method for the Determination of Sulfates, Tappi J. 57(7): 122-123 (1974) (Sulfate Analysis Methods).
122. TAPPI Method T 419 (Starch in Paper).
123. Allen, L.H., Trans. Tech. Sect., CPPA 3(2): 32 (1977) (Pitch Determination Method).
124. Browning, B.L., "Methods of Wood Chemistry", Vol. II, Interscience Publishers, p. 812-823 (1967) (Methods of Lignin Analysis).
125. Strazdins, E., Factors Affecting the Electrokinetic Properties of Cellulose Fibers, Tappi J. 55:1691 (1972) (Use of Fines in Zeta Potential Measurement).
126. Halibisky, D.D., Wet End Control for the Effective Use of Cationic Starch, Tappi J. 60(12):125-127 (1977) (Colloid Titration Method).
127. Hubbe, M., A Modified Reporting Procedure for Polyelectrolyte Titrations, Tappi J. 62(8):120-121 (1979) (Refinement of Colloid Titration Method).
128 TAPPI Method T 8 (Identification of Wood and Fibers from Conifers).
129. Parham, R.A. and Gray, R.L., "The Practical Identification of Wood Pulp Fibers", TAPPI PRESS (1982) (SEM Microscopy of Wood Pulp Fibers).
130. TAPPI Method T 401 (Fiber Analysis of Paper and Paperboard).
131. Clark, J.d'A., "Pulp Technology and Treatment for Paper", Miller Freeman Publications, Inc. (1978) (Discussion of Handsheet Devices and Other Pulp Treatment Devices).
132. Hergert, R.E. and Waech, T.G., Newsprint Furnish Solids Retention. Field and Experimental Machine Studies, Tappi J. 62(11): 79-81 (1979) (Description of Britt Jar Survey of Paper Machine First-Pass Retention).
133. Pelton, R., Allen, L.H. and Nugent, H., Measuring Fines Retention of Newsprint Pulps, Pulp Paper Can. 80(12):125-129 (1979) (Experimental Variables Associated with the Britt Jar).
134. Werdouschegg, F.M.K., Interrelating a Practical Retention Testing Procedure with Wet End Retention Results — Theory and Practice, Tappi J. 60(7):105-109 (1977) (Use of Minidrinier Retention Tester).
135. Kerekes, J.R. and Harvey, D.M., Drainage Tester Survey, Tappi J. 63(5):89-91 (1980) (Survey of Drainage Testing Practices).
136. Tarvin, R.F. and Westwood, R.L., Proc. 1982 TAPPI Papermakers Conf., p. 243-248 (1982) (Use of Modified Britt Jar for Drainage Testing).
137. LeBel, R.G., Nobleza, G.C. and Paquet, R., Water Retention Value Indicates Machine Runnability of Pulp, Pulp Paper Can. 80(5):64-70 (1979) (Use of Water Retention Value in Papermaking).
138. DiNapoli, L.D. and Weber, H., Paper Stock Consistency and Pipe Size Effects on L&N Doppler Flowmeter Performance, Proc. 1981 TAPPI Engineering Conf., p. 559-577 (1981) (Use of Ultrasonic Flow Meters in Papermaking).
139. Scott, W.E., A Review of Wet End Chemistry Process Control Instrumentation, Tappi J. 67(11):72 (1984).
140. Springer, A.M. and Taggart, T.E., Quantitative Evaluation of Alternative Methods of Determining Total Charge, Tappi J. 69(5):116 (1986).
141. Thomas, K.V., Retention — On-Line Determination and System Factors, 1989 TAPPI Retention and Drainage Short Course Notes, Washington, DC, p. 37-47 (Apr. 12-14, 1989).
142. Scott, W.E., Important Wet End Chemistry Considerations on First-Pass Retention Control on Modern High-Speed Fine Paper Machines, 1983 TAPPI Retention and Drainage Seminar Notes, Savannah, p. 56-69 (Nov. 14-15, 1983).
143. Davison, R.W., Vacuum Pulsation Drainage Procedure for Determination of Fine Particle Retention, Proc. 1989 TAPPI Papermakers Conf., Washington, DC, p. 75-82 (Apr. 10-12, 1989).
144. Springer, A.M., Noe, J.S. and Wegner, T.H., Fundamental Strategy for Control of Retention and Drainage on a Modern Paper Machine, Tappi J. 70(1):43-45 (1987).
145. Lindstrom, T., Rydefalk, S. and Wagberg, L., in New Available Techniques, Proc. World Pulp and Paper Week, Stockholm, p. 492 (Apr. 10-13, 1984).

146. Kortelainen, H., Huttunen, J., Nokelainen, J. and Lehmikangas, K., Application of a New Fiber and Filler Retention Monitoring System at a Fine Paper Mill, Proc. 1989 TAPPI Papermakers Conf., Washington, DC, p. 335-343 (Apr. 10-12, 1989).

147. 1990 TAPPI Neutral/Alkaline Papermaking Short Course Notes, Orlando (Oct. 16-18, 1990).

148. Penniman, J.G., A New On-Line Zeta Potential Instrument, The Zeta Data, Proc. 1988 TAPPI Papermakers Conf., Chicago, IL, p. 21-25 (Apr. 11-13, 1988).

VIII FOURDRINIER PAPERMAKING

D.W. MANSON

Dan W. Manson is a Senior Consultant — Engineering with James River Corporation where he is responsible for technical and engineering support for the paper manufacturing process. His work experience has been involved with all aspects of the papermaking process including technical service, manufacturing, and designing and building paper machines. He has also done research in the papermaking process. He has published numerous articles covering all aspects of papermaking. He received a B.S. in Chemistry from the University of Vermont and an M.S. and Ph.D. from The Institute of Paper Chemistry. He is the Papermaking editor for Tappi Journal. He is a TAPPI Fellow and a past member of the TAPPI Board of Directors.

REVIEWED BY

Terry J. Bambrick
Manager, Mill Service Program
Albany International
Appleton, WI

Borje Wahlstrom
President
Borje Wahlstrom, Inc.
Rosemont, PA

A. HISTORY

Paper was made universally by hand until the invention of the paper machine by Louis Robert in France in 1799. The first paper machine of Robert, shown in Fig. 159, consisted of an endless belt of wire cloth stretched between two rolls. The position of the rolls was adjustable so that the wire could be stretched to keep it flat. The beaten pulp in the vat was thrown up by a revolving paddle against a curved baffle which distributed the pulp and water in an even stream on the moving belt. As the wire cloth traveled slowly forward, the water passed through the wire while the small squeeze roll completed the preliminary dewatering. The receiving roll wound up the wet sheet until a specified length had been obtained. Fifteen meters was the generally accepted practical limit. The roll was then removed and the paper unwound, passed through some press rolls, and hung up to dry. A working model of this machine was made, but changes were necessary in order to make it practical. Robert was granted 8,000 francs to assist him in his studies and experiments. Instead of developing it himself, Robert sold his interest in his patent and his model to his employer Leger Didot.

Didot realized the greater possibility of perfecting the paper machine in a country free from government strife and took his patent and machine to England in the summer of 1800. In England, Didot fortunately secured the assistance of Bryan Donkin, a mechanic well qualified by his training to perfect the machine.

On April 2, 1801, the British Patent 2,487 was granted to John Gamble, Didot's brother-in-law, for an improved paper machine. Further improvements in the design resulted in a new patent, 2708, granted June 7, 1803 for "Improvements and Additions to a Paper Machine for Making Paper in Single Sheets without Seam or Joining from One to Twelve Feet and Upwards Wide and from One to Fifty Feet and Upwards in Length". In the autumn of the year 1803 the first papermaking machine ever to be built and successfully operated was started in Frogmore, England. In the following year another machine was successfully started up and this machine was practically a duplicate of the first. It was put into service in Waters, England.

In 1804, the Fourdrinier brothers Henry and Sealy purchased the remaining interest of Didot and Gamble in the improved Robert machine. Henry Fourdrinier was granted Patent 2,951 on July 24, 1804 for "The Method of Making a Machine for Manufacturing Paper of Infinite Length, Laid and Wove with Separate Molds". On August 14, 1807, an Act of British Parliament

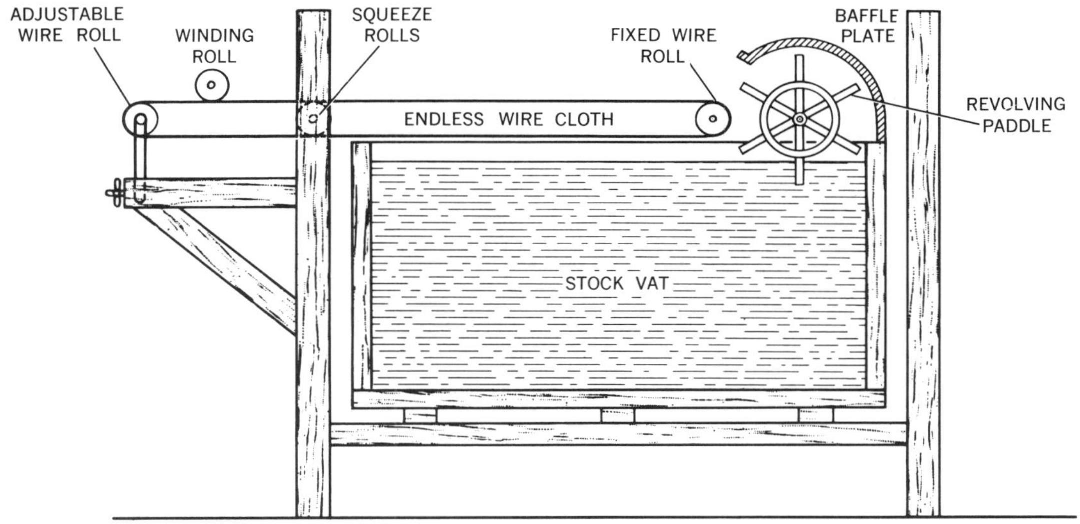

Fig. 159. The first paper machine of Louis Robert.

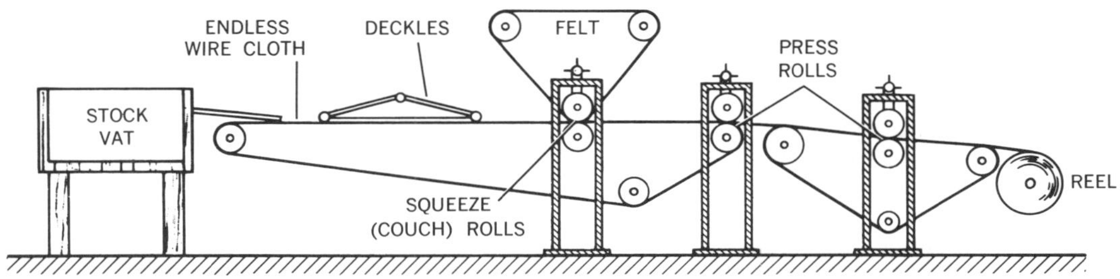

Fig. 160. The Donkin machine.

gave an extension of the patent rights obtained by the Fourdriniers for the invention of making paper by machinery. In this act the machine described by John Gamble in the specifications of his Patents 2,487 and 2,708 together with the added improvements was fully described and illustrated by diagrams.

During the year 1808, John Gamble assigned to the Fourdriniers all his rights in the patents as extended by this Act of Parliament, thus making them the sole proprietors of the patents covering the only successful papermaking machine in existence. So the machine invented by Robert, promoted by Didot and Gamble, and designed by Donkin, and financed by the Fourdriniers came to be known as the Fourdrinier Machine.

The Donkin machine is shown in Fig. 160. The mixture of pulp and water, kept in a state of agitation, flowed from the vat-like modern flow box, through pipes and onto the endless wire cloth and between the endless deckels. The wet sheet of paper was passed between the squeeze or couch rolls as was the case with the Robert machine, but in this case the work was better done because of the traveling upper felt. This felt, the ancestor of the couch roll jacket, also improved the firmness of the paper. The paper traveled through the press rolls and was finally wound up on the reel.

In 1826, Canson in England applied vacuum pumps to the fourdrinier machine to cause a suction beneath the wire on which the paper was formed to assist in water removal. This invention was really an adaptation from the Dickinson cylinder machine which had been invented by John Dickinson in 1809.

It was not until the years 1889 and 1890 that the modern machine was perfected, which, with all its improvements, remained essentially the same machine as the original Fourdrinier and Donkin with the addition of the cone drive and steam-drying cylinders. Originally the paper dryers were headless cylinders with a wood or

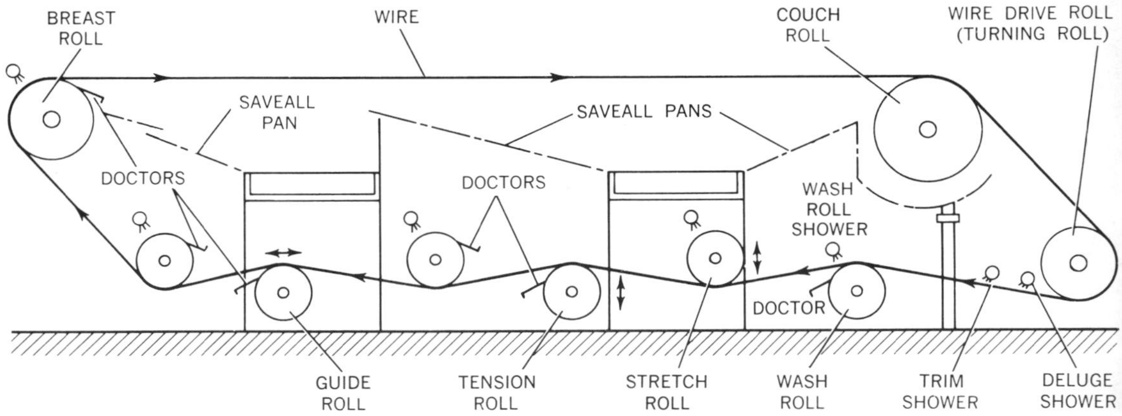

Fig. 161. Characteristics common to most paper machines.

charcoal fire in each one. Steam dryers were first used by Crompton in England in 1823. The dandy roll was invented by John Marshall in 1826.

B. THE MODERN FOURDRINIER MACHINE

The fourdrinier section of a modern paper machine looks strikingly like the original machines. It contains all the parts of the original machines with the exception that a suction couch roll has been substituted for the jacketed top couch roll. The layout of a typical fourdrinier can be seen in Fig. 161.

1. Breast roll

The breast roll is the turning roll for the wire at the headbox end of the section. It is driven by the fabric and must be sufficiently rigid not to deflect under the tension of the fabric. Modern breast rolls are rubber covered. This roll must be doctored so that lumps of stock will not follow around the roll and get between the fabric and the roll, causing damage to the fabric and also disruption to the sheet formation. The doctor usually has an extension on its back to carry the water draining through the wire in that area into the first saveall pan. This prevents water from falling onto the return wire and flooding the wire or filling it with fiber before it goes round the breast roll.

2. Forming board

The space between the breast roll and the first drainage element is quite critical to the formation of the sheet. There is a large span of unsupported fabric in this area and it would sag a great deal under the weight and pressure of the stock flowing out of the headbox. The fabric would then come up sharply over the first drainage element and the change in direction of the stock and fabric would cause hydraulic instabilities that would be detrimental to sheet formation. This is especially true for wide fast machines where the size of the breast roll is quite large to prevent deflection and thus the distance to the first drainage element is quite long. The forming board is inserted in this area to support the wire and prevent it from sagging.

Fig. 162. The forming board (Feldmuehle).

The earliest forming boards were single-slat affairs and played little role in the sheet forming process other than supporting the wire. Modern forming boards use a number of slats as shown in Fig. 162. The purpose of this device is twofold. Not only does it support the fabric, it also inhibits drainage. Modern practice is to have a forming board with about 20 to 25% open area. Earlier models had much greater open area and allowed large amounts of water to be removed at the forming board. In some cases, as much as 30% of the water from the slice was removed at

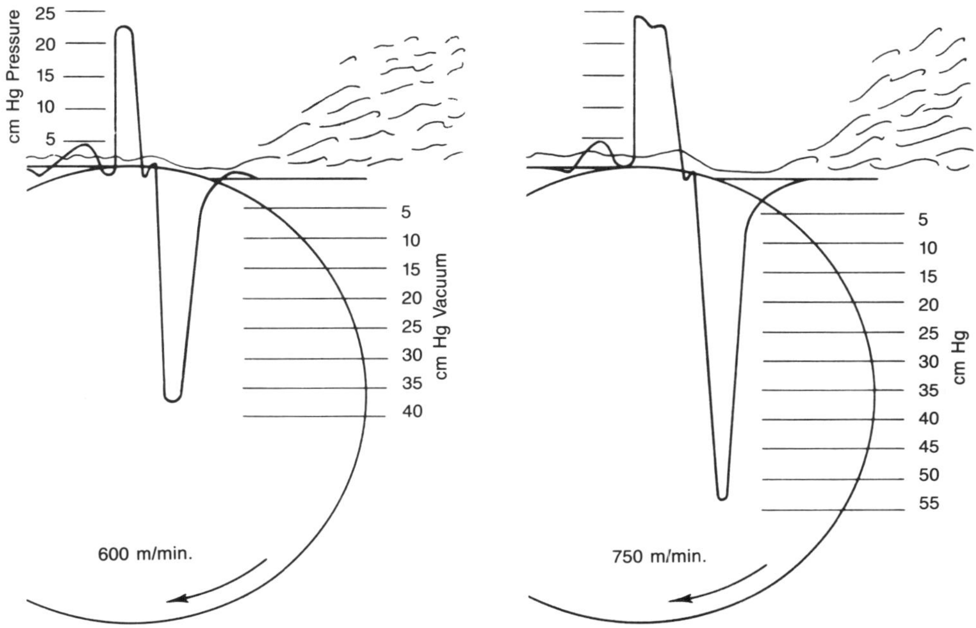

Fig. 163. Table roll action.

the forming board. Because of the diminished amount of water that got to the turbulence generating water removal devices later in the fourdrinier table, their beneficial effect on formation was reduced.

A forming board strip that is worn or damaged can cause damage to the forming fabric. It can also disrupt sheet formation. Therefore the materials of construction of the strips in the forming board are of considerable interest. They must be constructed of wear-resistant material. On slower speed machines, the material is usually high-density polyethylene. On higher speed machines, some form of ceramic material usually is used.

The positioning of the forming board relative to the breast roll centerline is critical to proper operation. The location is determined by the speed of the machine and the relative position of the top and bottom slice lips. It should be positioned so that the slice jet lands just before the leading edge of the forming board.

3. Table rolls and deflectors

Following the forming board are the drainage devices installed to remove water and induce turbulence into the sheet for good formation. These rolls must be straight, rigid and dynamically balanced, since any whip or vibration will disrupt the formation of the sheet and can produce barring matching the frequency of vibration of the rolls. The dynamic stability of a rotating roll is determined by its length and diameter. As the length goes up the stability goes down and as the diameter goes up the stability goes up. From this it is clear that, as machines got wider and faster, the diameter of the table rolls got larger and larger so that the rolls would not reach their critical rotational speed, go unstable, and start to whip.

Research into the action of table rolls has shown that the rotation of the roll in contact with the fabric creates a vacuum at the outgoing nip. This vacuum aids in water removal and it also causes the fabric to deflect over the top of the table roll. The magnitude of the vacuum created is determined by the speed of rotation of the roll and can reach very large values at high speeds. Figure 163 shows the pressure schedule for a table roll at speeds of 600 m/min and 750 m/min. The peak vacuum generated by a table roll has been measured and found to be approximately ½ RV^2 where R is the density of the white water and V is the wire speed [1]. This peak vacuum is not affected by sheet drainage resistance [1]. This means that the table roll pressure profile is independent of such variables as sheet consistency, stock drainage characteristics, and basis weight.

When the downward acceleration created by the induced vacuum at the outgoing nip gets very large, as it does at 750 m/min, it can cause spouting on the wire and seriously disrupt the formation. Spouting can occur at lower speeds, if

Fig. 164. Transition grooved to solid roll 360 m/min.

the headbox flow is unstable and contains streaks or ridges. With unstable headboxes it is possible to have spouting at speeds as low as 390 m/min.

In order to reduce the vacuum produced by the table rolls in the early part of the forming zone, these rolls were often grooved. This did help, since only the zone of the roll in contact with the fabric created a vacuum. The number of these rolls used would depend on the speed of the machine and how far down the fabric the sheet was before it would no longer spout. The grooves were usually about 3–6 mm in width and 1.5–9.5 mm deep. The grooves on adjacent rolls had to be staggered so that the sheet would not show streaks in line with the grooves. A photograph of the effect of a grooved table roll and a plain table roll is shown in Fig. 164 [3].

At higher machine speeds, the water pulled through the fabric did not have time to fall away before the next table roll was reached. This inrush of water from the bottom side would be forced up into the sheet by the next table roll and cause a disruption of formation. In order to obviate this situation, deflectors were inserted between table rolls to remove the water from the bottom of the fabric. Typical examples of deflectors are shown in Fig. 165.

Early deflectors were a single blade set very close to the wire, but not in contact with it. They caught the water and deflected it downward,

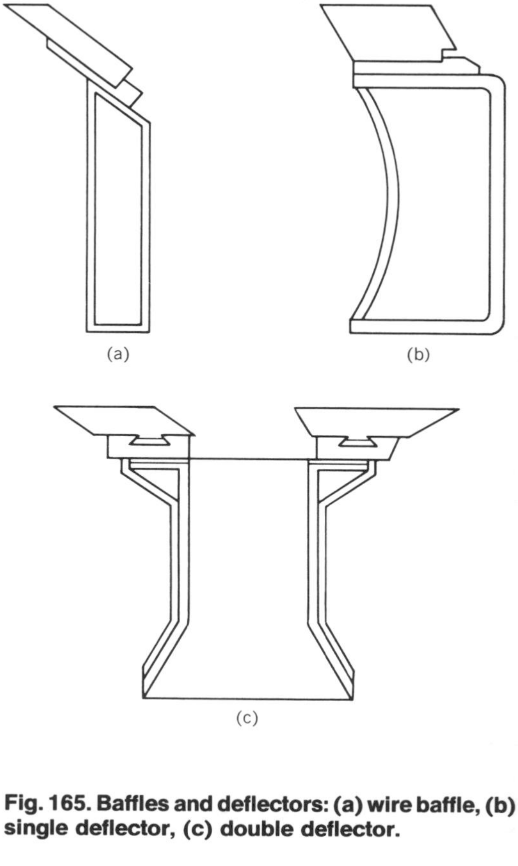

Fig. 165. Baffles and deflectors: (a) wire baffle, (b) single deflector, (c) double deflector.

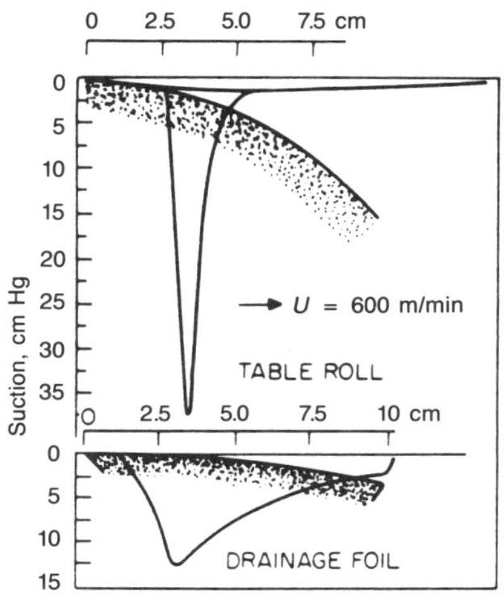

Fig. 166. Comparison of suction profiles over a table roll and foil [2].

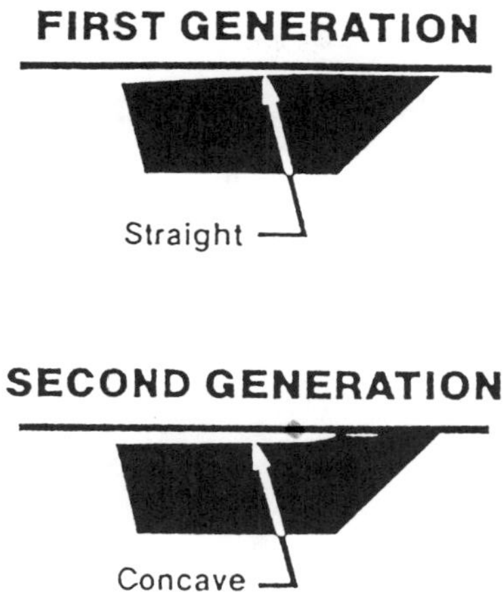

Fig. 167. Examples of straight and curved foil blades.

thus keeping it away from the ingoing nip of the next table roll. This worked quite well until machine speeds increased to the point where these deflectors were not rigid enough and the water passing over the top was too great and caused formation problems. The next step in the development was to go to single-and double-top deflectors. These deflectors were set in contact with the wire to catch and remove all the water on the bottom side of the fabric. The deflectors were usually made in a box section to add stiffness so they would not sag, deflect or vibrate while in service. Any of these defects would cause formation problems in the sheet such as worms and streaks.

4. Foils

The high vacuums generated by table rolls at high speeds, the inability of the operators to control this vacuum, and the resultant disturbances to the forming sheet forced the development of another type of drainage device. The initial development of hydrofoils as drainage devices was done by Burkhard and Wrist and first published in 1956 [1]. These new drainage devices produced a much lower vacuum pulse and therefore created far less distrubance on the wire. A comparison of the pressure schedule of a foil and a table roll is shown in Fig. 166 [2]. It is very clear that a much gentler dewatering takes place with a foil than with a table roll.

The water removed by a single foil is less than that removed by a table roll. However, the space occupied by a foil is much smaller than that of a table roll and so several foil blades can be installed in the space occupied by one table roll. In addition, the angle of the foil blade can be increased or decreased to change the vacuum pulse and thus the agitation on the wire and the amount of water removed.

Early foil blades, sometimes called first-generation blades, had a small level land section followed by an angled surface. Research efforts have shown that water removal was more efficient when the angled surface was curved [3]. A comparison of the two types of foil configurations is shown in Fig. 167. A comparison of the pressure schedules of the various foil types with table rolls at various speeds is shown in Fig. 168 [3]. It is clear from this figure that the table roll vacuum increases dramatically with increasing speed. The three-degree foil blade also shows an increased vacuum pulse as speed increases. The second-generation foil blade shows a lower vacuum pulse over a greater length which gives good dewatering capacity without the large vacuum pulse.

The peak vacuum generated by a foil blade increases with increasing foil angle and foil blade length. As with table rolls, foil-induced vacuum increases in direct proportion to the square of the machine speed. However, unlike the table roll, the vacuum forces generated by a foil blade increase as fiber mat drainage resistance increases. A foil blade located at the dry end of the table will typically exert twice the peak vacuum of an identical blade used right after the forming board [1].

5. Vacuum foil units

A natural extension of the foil technology was to enclose the foils in a box, seal the ends and apply vacuum. The first vacuum foil units were actually evolved from suction boxes used at the

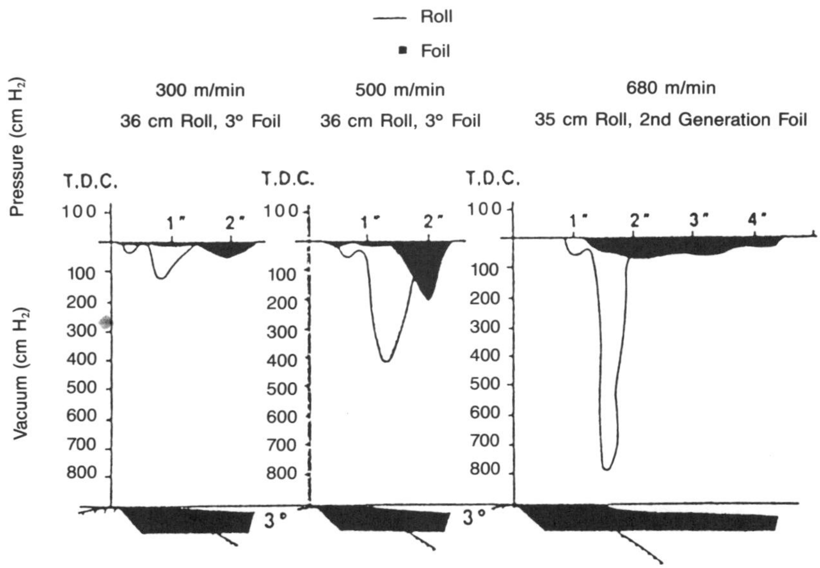

Fig. 168. Comparison of table roll and foil blade suctions. — Roll; ■ Foil.

dry end. The flat blades used in these vacuum-augmented wet boxes do not generate their own vacuum forces. A slight positive pulse is generated over the blade/wire contact surface as shown in Fig. 169. These units had a capacity to remove large amounts of water when vacuum was applied. They were tried near the headbox of the machine and were not too successful as they froze the flow from the headbox without any agitation prior to the water removal. At that time, headbox design was not sufficiently advanced to produce a slice flow of sufficiently good quality to be frozen without further agitation on the wire. Instead, wet suction boxes were used farther down the table for greatly increased drainage capacity. This resulted in higher slice flows and improved formation.

The next step in vacuum-augmented units was to combine the foil unit with a vacuum box. This application has been very successful. The vacuums run are relatively low on the order of 25 – 50 cm water. The water removal is increased significantly without increasing the intensity of the vacuum pulse. Figure 170 [4] shows the actual vacuums measured over a 3° foil unit at different applied vacuums in the box. The actual data, while showing that much of the pressure profile is unaffected by changes in box vacuum, indicate that V_{max} does increase slightly as box vacuum is increased.

A third type of vacuum-augmented foil unit is the step foil. The step-type blade generates a variety of pressure profile shapes depending on such factors as blade length, wire span between

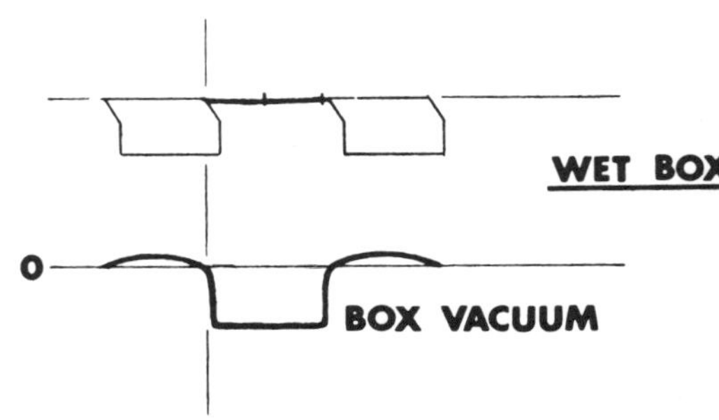

Fig. 169. Wet suction boxes.

blades, step depth and box vacuum. A comparison pressure profile between a conventional foil and a step foil is shown in Fig. 171. The sharp vacuum spike located immediately following the step is characteristic of all step blade profiles. The peak negative pressure is directly proportional to the square of the machine speed. As with conventional foil blades, the peak vacuum generated is greatly affected by fiber mat filtration resistance [5]. It was found that the peak vacuum generated by a step blade is one and one half to two times greater than the peak vacuum generated by a 3° foil measured at the same location on the machine.

Still another vacuum-foil type is the isoflow unit shown in Fig. 172. This unit has foil blades followed by blades set a small distance below the wire level. The fabric is pulled down by the contacting foil blade. The water removed forms a seal with the lower unit that diminishes the

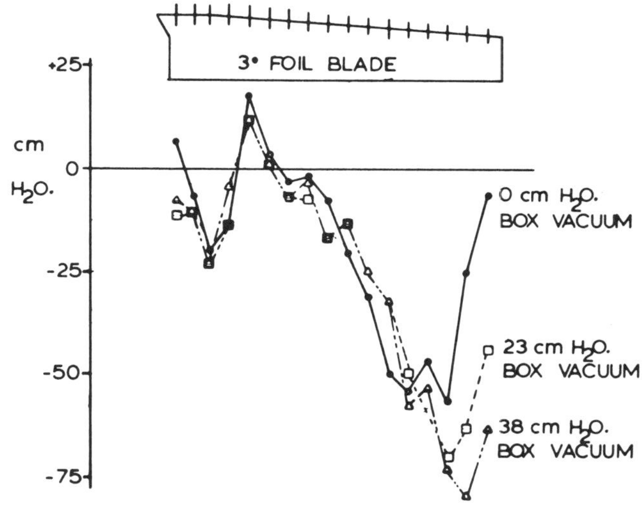

Fig. 170. Pressure profile of vacuum foil blade.

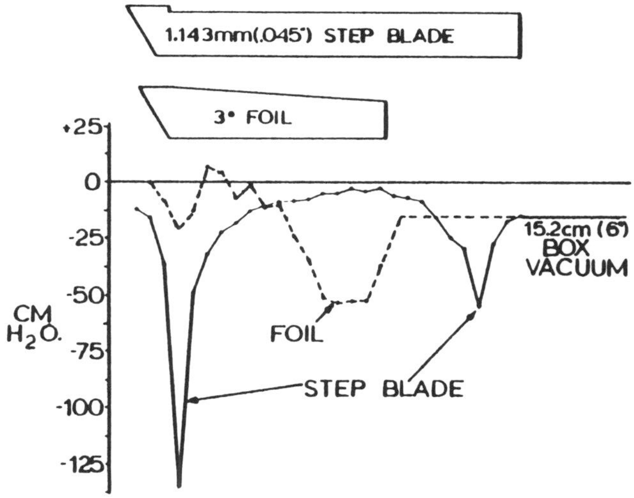

Fig. 171. Pulse shape of 3 in. foil and a 1.14 mm step blade. Blades located mid-table on fourdrinier producing 58 g/m² bond paper at 518 m/min.

pressure pulse and the water removal so that the pulsing is more gentle. The pulse is repeated over each submerged unit. Each of the submerged units removes water at its leading edge so that water removal is gradual at each edge.

6. Vacuum boxes

Depending upon furnish and a number of other operating conditions, it is usually found that somewhere between 2 and 3% consistency the water removal capacity of foils and vacuum foils diminishes considerably. Water removal at this point must be accomplished with vacuum boxes operating in the range of 7.5–25 cm of Hg.

As the wire passes over the boxes, first only water is removed from the sheet, but finally air breaks through and both air and water are pulled into the box. The point that air breaks through the mat is called the "dry line" or sometimes the "wet line". It is the point in the process where the flow changes from single-phase flow to two-phase flow. At this point the air flow aids in sweeping the water out of the sheet and into the box. The exact consistency varies with basis weight and grade, but the breakthrough of air takes place somewhere near 7% consistency.

The flow of air and water must be separated to keep the water in the white water system and the

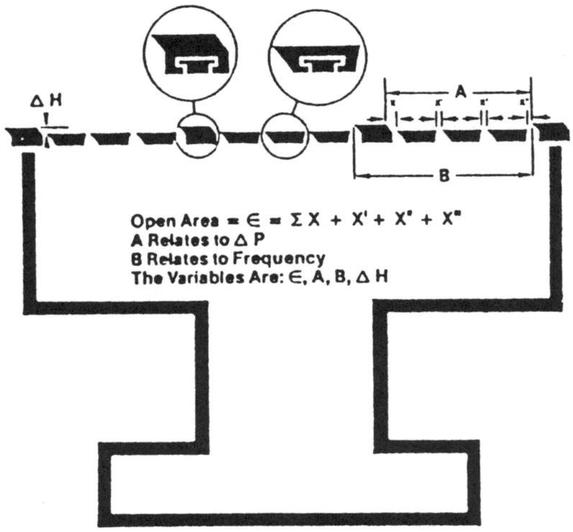

Fig. 172. The universal isoflo type of vacuum foil.

air going to the vacuum system. There are a number of systems that can be used to separate the air and water. The most common of these is shown in Fig. 173. The air-water mixture flows out of the back side of the suction box into a separator unit with a deflector inside of it. The whole flow is deflected downward. The white water continues downward, while the air moves upward toward the vacuum header. There is a valve at the top of each vacuum leg that can be set to control the vacuum in each box. Each of the drop legs from the separators goes into the seal box so that air cannot be drawn up into the system. The length of the drop leg must be greater than the length of water column equivalent to the highest vacuum the system will see.

Normal operation of a flat box system would use five or six boxes between 15 and 40 cm wide with a graduated vacuum. The wet end boxes run at relatively low vacuums of from 5–7.5 cm of

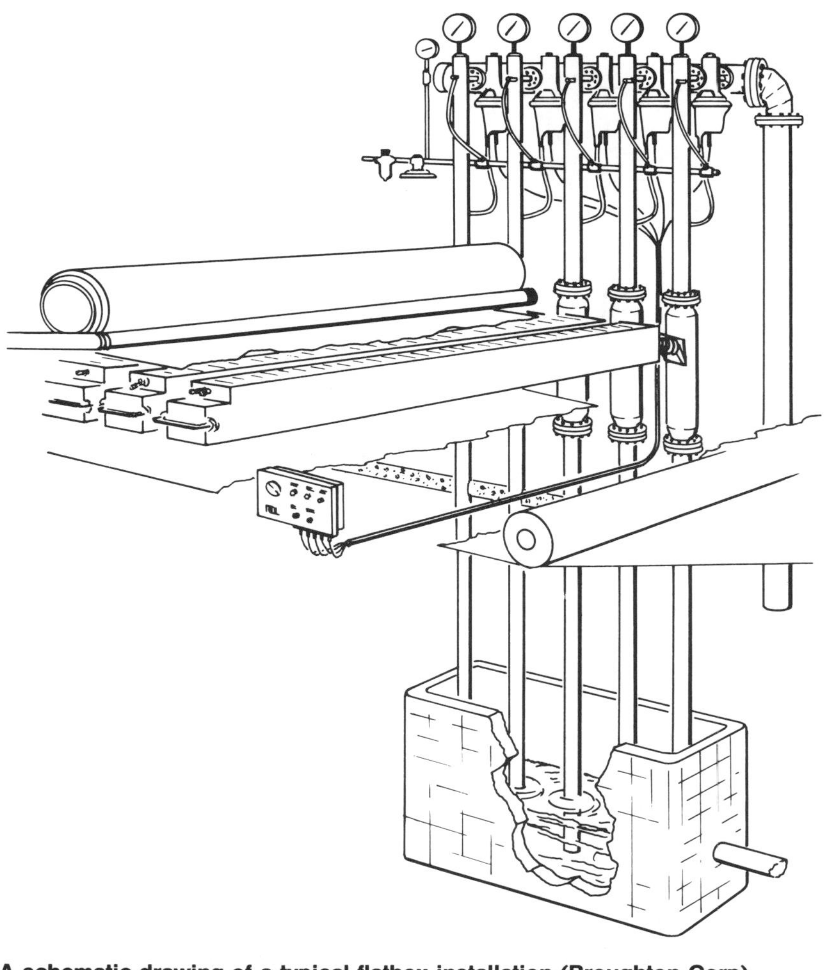

Fig. 173. A schematic drawing of a typical flatbox installation (Broughton Corp).

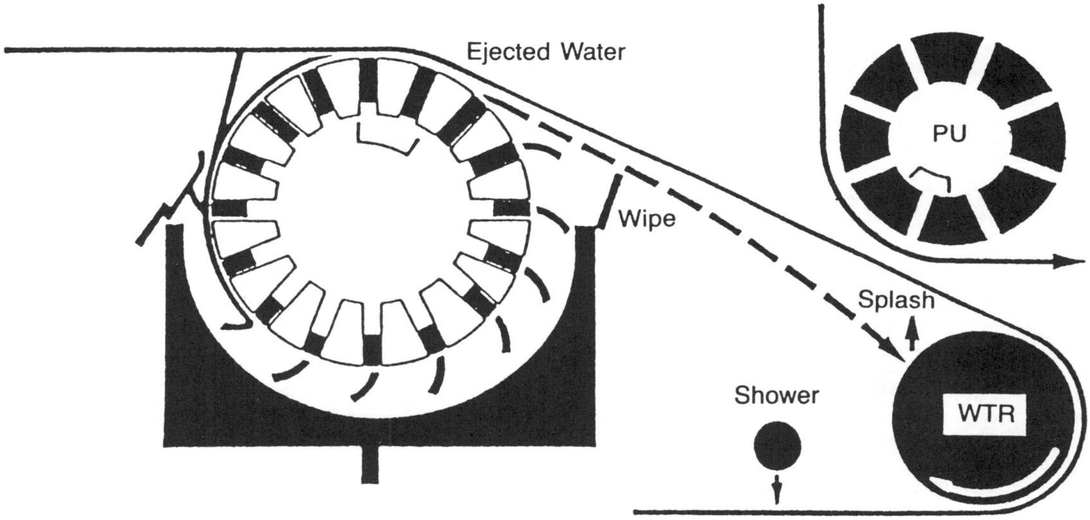

Fig. 174. Water removal at the couch is through the suction box inside the roll.

Hg. The following boxes will be graduated up to 15–20 cm of Hg. A practice that is becoming more common is to use only four suction boxes and run them at rather high vacuums, starting at 7.5–10 cm of Hg. and going as high as 25–30.5 cm of Hg. The water removal is the same or increased and the drag load is reduced.

There are now available vacuum boxes that have multiple compartments in them so that several vacuums can be run in the same box. The argument for this approach is that the vacuum is not lost between stages and therefore the water does not redistribute through the sheet between vacuum stages. This approach has led to increased water removal [3].

The cover material for the suction boxes is normally high-density polyethylene for lower speed machines and a hard material such as silicon carbide or aluminum oxide for high-speed machines. The desire is to have a situation where neither the fabric nor the box cover wear. Shortened fabric life is certainly not desirable and suction box wear can lead to streaks in the sheet that are also undesirable. The covers can have either slotted openings in them or drilled holes. Normally the wet boxes are slotted and the dry boxes may be either slots or holes, depending on the product being made. With fine and coated papers, the preference is for all slotted covers as the drilled covers can put streaks in the sheet that lead to finger ridges in the final product. The slot width is varied from wet to dry with the wet boxes having wider slots on the order of one inch. The intermediate boxes will have a 16 mm slot and the high vacuum boxes may go as narrow as a 13 mm.

7. Couch roll

The couch roll is the final water removal device in the fourdrinier system. This roll is operated at high vacuums on the order of 40–63.5 cm of Hg depending on the basis weight and the product being made. The dryness coming to the couch is normally between 12 to 18% and the couch will raise that dryness to about 18 to 25%.

The couch roll also represents one of the principal drive points for the fourdrinier section. In the case of machines that do not have a wire-turning roll, it is the only drive point. Although on larger machines without a wire-turning roll, the first wire-return roll is frequently driven. When there is a wire-turning roll, then the drive is divided between the couch and the wire-turning roll with the couch being the helper drive and the wire-turning roll the principal drive.

Water removal at the couch is through the suction box inside the roll. At low speed, the water and air are taken into the couch and go into the couch vacuum box. In this case a separator system must be used to separate the air and water so that the white water does not get into the vacuum-generating system. At higher speeds, the air and water are pulled into the couch shell drilled holes and then the water is thrown out by centrifugal force once the holes have left the vacuum zone. In this case, the design of the couch pan and wipe are critical so that the water thrown out does not get back into the sheet on the ingoing side of the couch/fabric nip or carry over the pan and splash on the wire-return roll as shown in Fig. 174 [3].

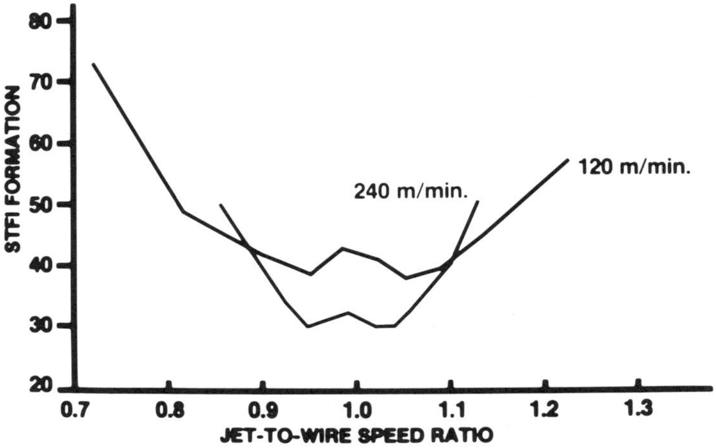

Fig. 175. The effect of J/W ratio on formation.

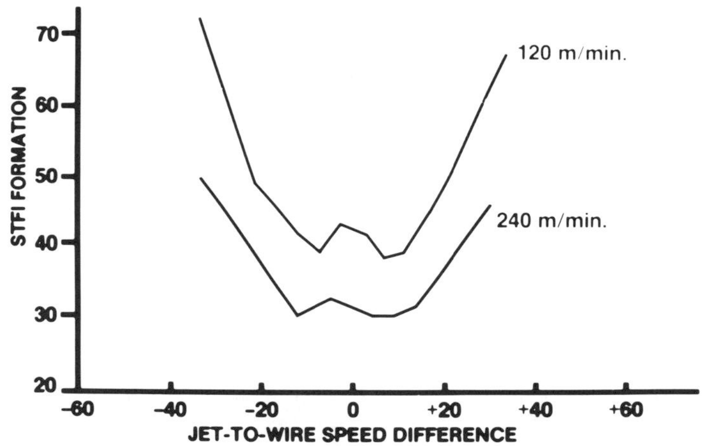

Fig. 176. The effect of J/W speed difference on formation.

Some paper machines that do not have a pickup will run a couch presser roll to improve water removal and increase sheet density at the couch. These rolls are normally covered with a relatively soft rubber that is supposed to be "self skinning", but seldom is for any period of time. To stop the fiber picking, a low-volume fog-type of shower will be run on the roll to keep it wet and help keep it clean.

C. WIRE FLOW AND DRAINAGE

1. Slice delivery

(a) Jet speed to wire speed ratio

The next item on the list of important factors in formation is jet speed to wire speed (J/W) ratio. If you stop and think for a moment, what we are trying to do is deliver a free jet of water and stock onto a moving wire. It would seem apparent that differences between the jet velocity and the wire velocity will produce relative movement between the stock and the wire. Large relative movement should be disruptive to sheet structure and this, in fact, is true. The classic work done by Schroder and Svenson [6] clearly showed that formation is quite sensitive to J/W ratios, and that for practical purposes this ratio should stay between 0.90 and 1.10. They further showed that, for the sheets they were making, the best formation was obtained at J/W of 1.0.

Figure 175 [6] shows two runs from their data expressing formation in STFI units against J/W ratio; the higher the STFI units the poorer the formation. It is very clear that for both operating speeds the optimum formation comes at J/W ratios between 0.95 and 1.05. The reason for the slight hump in the curve is unexplained.

Comparing the two curves, it would appear that formation is more sensitive to J/W as the

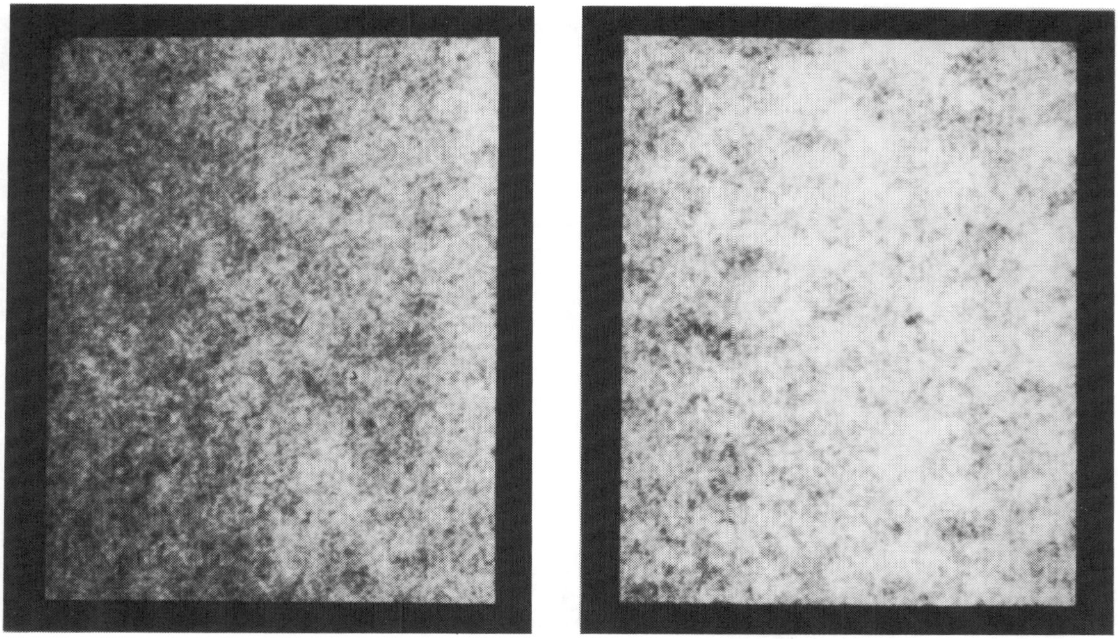

HEAD NORMAL

HEAD TOO HIGH

Fig. 177. The effect of total head on formation.

speed increases and at least for these runs the formation got better as the speed increased. However, if you replot their data using not J/W ratio but difference in speed between jet and wire, the situation looks different, as shown in Fig. 176 [7]. Now the curves take on essentially the same shape. The effect of machine speed is still apparent. What this tells us is that it is not really the J/W ratio that influences formation, but the differences between jet speed and wire speed. This should not be so surprising. Consider two machines, one running at 150 m/min and one at 900 m/min. At a J/W of 1.10 at 150 m/min the jet is traveling 15 m/min faster than the wire, while at 900 m/min the jet is traveling 90 m/min faster than the wire. Clearly the shear and turbulence conditions cannot be the same. This example also gives an indication of the sensitivity of total head control on high-speed machines. If we say that optimum formation is obtained at a jet speed/wire speed difference of 10 m/m, then the allowable total head variation at 900 m/min is 25 cm of water which is 2% of the total head. Therefore the total head control should be controlling to less than ±1% variation to stay within acceptable limits.

What does all this look like as far as the paper is concerned? Figure 177 gives a good measure of the effect of the J/W ratio. This machine normally runs at theoretical head. When the head was raised above this level, the formation clearly deteriorated.

The question of total head should be considered carefully. The total head is really the sum of the static head and the velocity head. With rectifier roll headboxes the manometer tube or D/P cell is usually located in the middle of the headbox where the velocity head is negligible and therefore the static head equals the total head. However, with hydraulic headboxes the velocity head is significant at any point where the manometer can be placed. Under these conditions, the static head is not the total head, but some value lower than that due to the unmeasured velocity head. Normally the headbox manufacturers will calculate the correction and provide a set of tables or graphs to enable you to use the manometer correctly.

The manufacturer's corrections should be checked on the machine. Although the equations are quite good, some of the variables in them are hard to measure accurately and this can lead to calculation errors of 2-4%. We have already seen that control of total head should be better than ±1% for good formation control. A calibration can be accomplished by making small changes in the J/W and taking paper samples at each point. At a J/W of 1.0, the formation should be the best and the tensile ratio should be at a minimum. The formation measurement is preferred for this

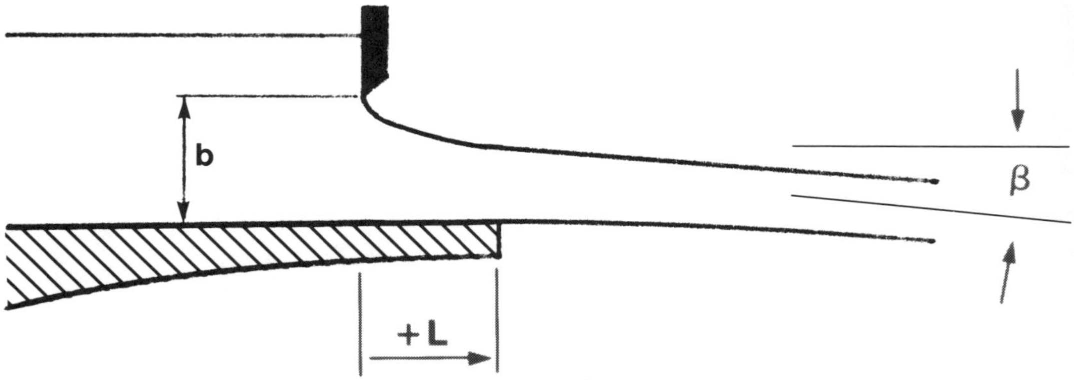

Fig. 178. Slice geometry.

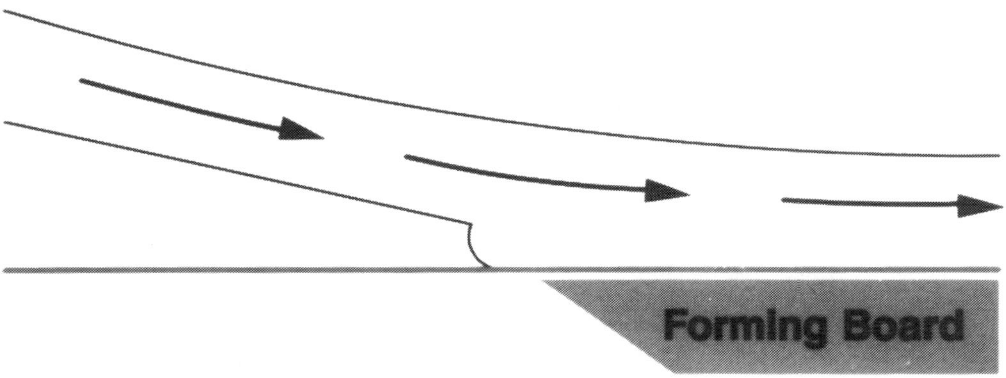

Fig. 179. Slice jet impingement.

test because the tensile ratio can be influenced by other things on the machine, such as draw changes, and the results may be less clear.

Adjustments to the J/W should be made in small increments, when trying to improve formation. The point of best formation is usually within a narrow J/W band and taking large adjustments may pass right by the optimum point. This adjustment is crucial for both fourdrinier and multiwire machines. This is particularly true of twin-wire machines since there is no free wire to use to compensate for improper head adjustment.

(b) L/b ratio

The ratio of the projection of the lower slice lip beyond the upper lip to the slice opening is very important in determining how the slice jet will land on the wire. Figure 178 [7] shows a sketch of this geometry. The ratio of importance is L/b. A low L/b around 0.5 means a steep jet angle into the wire and is frequently referred to as pressure forming. On the other hand, an L/b of 1.0 or greater means a relatively flat jet and is called velocity forming.

Looking first at pressure forming, the sheet is set quickly on the wire as the water is driven through it. This may be helpful with a headbox that has poor delivery, as the defects generally get amplified as they go down the wire and removing the water early tends to minimize this effect. However, with a good delivery, the benefit of the action of the wire in breaking up flocs is lost. In addition, pressure forming frequently causes a breast roll discharge and, if there is not sufficient space around the breast roll to handle the water, the bottom of the sheet will be disrupted. At low L/b ratios, a small change in either L or b will change the jet angle significantly but not change the point of impingement, so forming board position is not highly critical.

During velocity forming, L/b greater than 1.0, the jet angle is low and travels more nearly parallel to the wire. With modern headboxes this is the preferred method of delivery. The jet should be delivered to the wire at a low angle and land just before the leading edge of the forming

Fig. 180. At high speeds with table rolls spouting occurs and sheet formation is destroyed.

board. With this arrangement the bottom of the slice jet gets peeled off by the leading edge of the forming board as shown in Fig. 179 [7]. This is the desired effect since the bottom of the jet tends to flow backwards when it hits the wire and would have poorer formation if allowed to stay with the sheet. Also there is a layer of air traveling with the bottom of the jet that must be eliminated at the forming board or it will also disrupt the bottom side of the sheet. In addition, the low angle of delivery gives low initial forming pressure which is beneficial for good retention, reduced wire mark and improved fines distribution.

The point of impingement is critical to good formation. When velocity forming, a small change in either L or b will change the point of impingement significantly and can disrupt formation by landing the jet directly on the forming board. Adjustments to the L/b ratio should be made in small increments so that the optimum point won't be missed.

2. Flow and drainage at table rolls and foils

Previous discussions in the chapter dealt with the ability of a table roll or foil to generate vacuum and therefore remove water from the forming sheet. The downward acceleration caused by the fabric being pulled down by the vacuum caused by the drainage element causes an equal upward acceleration to the stock on the wire. This acceleration is the source of the turbulence which causes rearrangement of the fibers in suspension. This turbulence is controlled by the speed of the machine, the wire tension and the vacuum produced by the drainage element.

At high speeds with table rolls, the downward acceleration is so great that spouting occurs and the sheet formation is destroyed Fig. 180 [3]. This is clearly an undesirable situation and makes very clear the reason the development of foils became so important.

The use of foils as drainage elements enables water to be removed with gentler vacuum and thus less violent turbulence on the table. This is illustrated in Fig. 181, which shows the wire surface over a 1° foil at 900 m/min. The agitation of the sheet is clear, but the violence is much diminished and will aid in good formation.

The best application of foil technology is to take advantage of the mild pulsing effect of each foil and arrange the foils' spacing so that a regular pulsing of the sheet is accomplished. This regular pulsing of the sheet evidently provides a reinforcing effect on the agitation of the sheet and is most beneficial to formation. This regular pulsing of the sheet can be seen in Fig. 182 [5].

There are a number of different methods to develop and promote the regular pulsing of the sheet. Evenly spaced foils are the most common-

Fig. 181. The use of foils enables water to be removed with gentler vacuum, and thus creates less violent turbulence on the table.

Fig. 182. A regular pulsing of the sheet is beneficial to formation.

ly used system. The spacing is based on experience of foil designers and field engineers. The spacing is determined mainly by the following principles. The higher the speed the greater the spacing; the lower the freeness the greater the spacing; the lower the consistency the greater the spacing [3].

The same effects can be produced by vacuum-augmented foils with the additional advantage of being able to control the pulse intensity by controlling the vacuum. This is very well done with the vacuum-augmented step foils. The same effect is accomplished by the isoflow units where much of the pulsing takes place over the submerged drainage elements shown earlier. In this case, there are many small pulses between major pulses from the blades that contact the wire.

The advantage to the small frequent pulses is that water is removed from the sheet slowly. This prevents sheet sealing on the early part of the wire. If the pulses on the early wire are too strong, the bottom side of the mat will form too quickly and get compacted by the pulses of the foil blades. This compaction increases the drainage resistance of the sheet and makes water removal much more difficult. In addition, this sealing tends to diminish the turbulent agitation of the sheet on the wire and formation improvement becomes more difficult.

D. FORMATION

The quality of a paper product and its suitability for the marketplace is often determined by its formation. Formation not only influences the aesthetic values of a product, such as printability, but it can also influence its physical and optical properties. Formation is therefore a much sought after sheet property and a good deal of effort on both a practical and theoretical basis has been expended in pursuit of improved formation.

The first thing that should be done is to define formation. For this discussion, formation will be defined as the visual appearance of the sheet when held up to the light. Formation really encompasses the structure of the sheet and deals mainly, but not exclusively, with the flocs in the sheet: their size, distinctness and their distribution. Basically, a well-formed sheet will have a uniform fiber distribution with very faint small flocs evenly distributed throughout the sheet. Depending on a number of operating parameters to be discussed, the sheet can look flocculated with large distinct flocs, or very uniform with small flocs, or streaky, or bunchy, or open. The list of descriptive adjectives is endless. The best way to define formation is with the use of an instrument. Most instrumental devices used to measure formation do so by measuring either the spacing between floc centers or the diameter of the flocs. Although they produce numerical results, these are usually related to what the eye sees. So basically, regardless of how formation is evaluated, it relates back to a visual impression. Instrumentation is the preferred method of evaluation as its reproducibility and lack of prejudice yields a more uniform and consistent evaluation.

1. Flocculation and headbox consistency

At this point, it is worthwhile to examine fiber flocs and what we know about their properties and how they behave in the hydraulic turbulence and shear fields that exist in the headbox and on forming sections of the machine.

According to Parker [8], the minimum number of fibers required to form a floc with distinct mechanical properties is four. Therefore a floc can be as small as four fibers and increase in numbers to quite large amounts. At some point in this floc growth path, as the floc gets larger and larger it begins to take on the properties of a network that in fact contains smaller flocs linked together. This is, of course, what happens during the forming process as water is removed from the forming sheet. A network is at first loosely formed and then, as water is removed, the network becomes more and more rigid. As a result the energy required to rearrange the fiber distribution within the network becomes greater and greater. Ultimately the only way to rearrange the network is to rupture it and the sheet is destroyed. However, between the headbox and the point where the sheet can no longer be changed, a lot of opportunity exists to influence the formation of the sheet.

Beginning at headbox consistencies of less than 1%, there exists a degree of flocculation that is dependent upon the fiber properties and consistency and the amount of agitation present. Studies using laser Doppler anemometry [9-11] and visual inspection [12] in special apparatus show that there is a distribution of floc sizes, not just a single size. There is a dynamic equilibrium in which flocs are being created and destroyed all the time with a size distribution dependent on the turbulence existing. If the turbulence level is increased, there is a shift in size distribution toward smaller flocs. If the turbulence level is decreased such as when the jet leaves the slice and moves toward the wire, there is a shift back to larger floc sizes. This change takes place very rapidly, often within hundredths of a second or less. Parker observed that floc size is determined

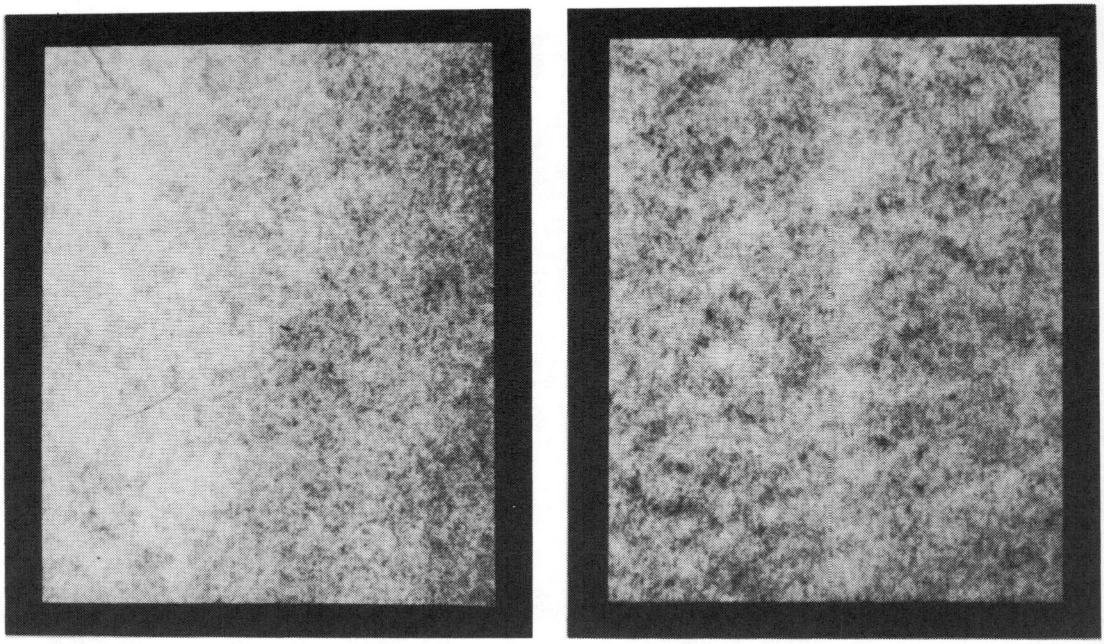

Fig. 183. Effect of headbox consistency on formation. Left — normal consistency; right — lower consistency.

by the scale of the turbulence [8]. In addition, reflocculation is influenced by fiber length and the strength of the fiber network formed in the flowing stream and thus fiber floccing is about the third power of the consistency. We need to be concerned with both the scale and intensity of the applied turbulence when thinking about formation and floc size.

Translating these experiments to the situation existing in the headbox and slice area, it becomes clear that the good done to the fiber distribution by turbulence generated at the slice is soon lost. In fact, it has already begun to deteriorate before the slice jet hits the wire. It is therefore imperative that the degree of flocculation in the headbox be reduced to as low a level as possible. This is done by increasing turbulence levels and decreasing the scale of the turbulence in the headbox by good design of the turbulence-generating elements.

The second factor in controlling flocculation is fiber concentration. The flocculation time goes down very rapidly with increasing concentration. The third factor that influences flocculation time is the fiber length; the longer the quicker. The headbox consistency can be reduced simply by adding more water at the fan pump. The volume flow will go up and the consistency will go down. An added plus to this is that the volumetric flow goes up and the turbulence will also increase.

The consistency run in the headbox for any given grade is normally determined by the gallonage capacity of the headbox and the amount of water that the forming section can handle. Therefore the gallonage run and thus the consistency cannot be changed appreciably unless the system is changed.

The long fiber portion of the furnish is the really important consideration, because fiber length is such an important part of the flocculation equation. The more dilute the long fiber, the better the formation will be. A rule of thumb is to keep the effective long fiber concentration below 0.3%. There are several ways to do this. First, by adding water, the long-fiber stock consistency is reduced. It can also be reduced by substituting short fiber for long fiber in the furnish.

It is hoped the gallonage rate achieved during operation will match the rate for which the headbox was designed. If this is not so, formation defects can be introduced into the sheet. Velocities through the box well below those designed for will allow time for the stock to flocculate. The flocs formed in this way are most difficult to break up with further turbulence and so the resulting sheet always looks heavily flocculated. On the other hand, flow velocities well above the design values can produce severe flow streaking on the wire, which in turn produces streaks in the sheet.

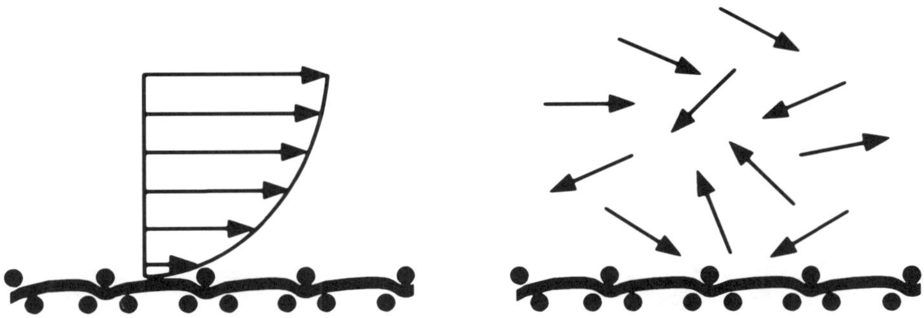

Fig. 184. Turbulence (right) and oriented shear (left) cause deflocculation on the wire.

The effect of exceeding the hydraulic flow limit of a rectifier roll headbox is shown in Fig. 183 [7]. In an attempt to improve formation by lowering the consistency, the hydraulic flow limit of the headbox was exceeded. In addition, the ability of the fourdrinier table to handle the water was also exceeded. The result was a streaky sheet with disrupted formation as is clearly visible in Fig. 183. This is a classic demonstration of what happens in all too many mills. The whole machine gets speeded up, but no one thinks about the effect on the headbox.

2. Agitation on the wire

Going back for a moment to the discussion of flocculation or floc size distribution, it was quite clear that, if the stock slurry is not agitated after it leaves the slice and lands on the wire, the floc size distribution will get rapidly worse. Energy must be put into the draining stock to establish and maintain good fiber dispersion. There are two ways of supplying this energy: turbulence and oriented shear as shown in Fig. 184 [7]. Both of these types are active in causing deflocculation on the wire and maintaining a deflocculated state.

(a) Turbulence

Some years ago, it was generally believed that the sheet on the wire should look like a pane of glass, and some people were even able to accomplish this. Much to their dismay, the resultant sheet looked terrible. It was terribly flocculated. The problem was that, without agitation on the wire, the fibers had adequate opportunity to flocculate. Good agitation on the wire is essential to good formation and is as important as good turbulence in the headbox. Combinations of foil blade angles and table rolls at lower speeds can be used to produce turbulence on the wire. This does not mean that turbulence on the wire will replace good dispersed delivery from the slice, but that a good delivery can be lost by an inactive wire section.

An old/new turbulence generator is now gaining new recognition. This is the so-called Sheraton Roll. This is a sawtooth roll as shown in Fig. 185. It pulses the wire and thereby introduces turbulence into the draining stock. Early work with these rolls met with mixed success. Sometimes a slight improvement in formation was noticed and sometimes nothing happened. Recent work by Kallmes et al. [13] has shown that the roll can be extremely effective if driven at a speed that generates the resonant frequency of the length of free wire above the roll. This is a very new technology and it will be interesting to see how it develops. In the few installations now operating, dramatic improvements in formation have been achieved.

(b) Oriented hydraulic shear

Oriented hydraulic shear is quite beneficial to improving formation. This type of shear is best exemplified by a ridge of stock going down the wire. When it hits a drainage element, the ridge splits into two smaller ridges with a valley between, as shown in Fig. 186. This splitting creates shear which helps to break flocs and keep the fibers dispersed. Obviously the generation of ridges should be carefully controlled. If they are too large (contain too much energy), they will be disruptive to sheet formation when they divide and can cause spouting on the wire. On the other hand, if they are of the correct size, spacing, and energy level, and the splitting is properly done, the resulting improvement in formation can be quite remarkable. The apparent mechanism is for the shear forces to pull the flocs apart and redistribute the fibers.

There are two rather well accepted methods of creating ridges in the flow, the scalloped slice and formation showers [14,15]. In the former, the top slice lip is machined in a scalloped pattern that generates ridges in the slice flow which

Fig. 185. The Sheraton roll introduces turbulence into the draining stock.

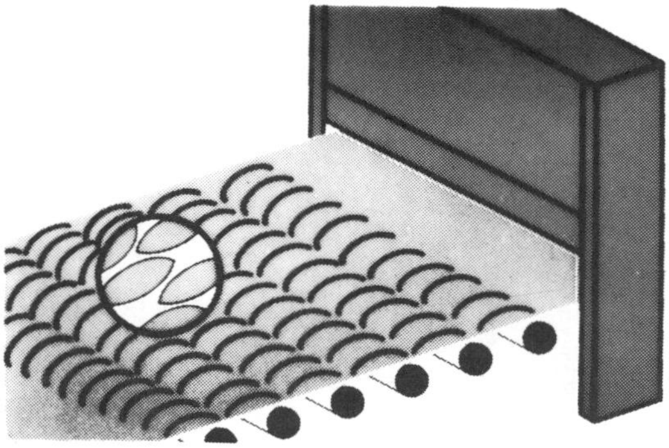

Fig. 186. Oriented hydraulic shear is exemplified by a ridge of stock going down the wire. When it hits a drainage element, the ridge splits into two smaller ridges with a valley in between.

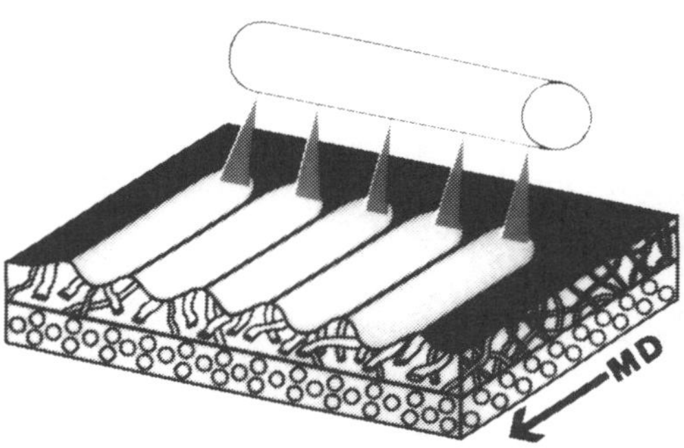

Fig. 187. Schematic representation of a formation shower. After the ridges have died out, they may be regenerated using a formation shower.

subsequently split going over drainage elements on the wire. Each ridge splitting reduces the ridge size to about half its original size. After several splits, the ridges become so small that they really rather disappear and no more splitting can take place. Although the serrated slice can be beneficial to formation, it is not always successful. When the flow out of the headbox is unstable with shifting velocity streaks and turbulent wakes from the rectifier rolls, the effect of the serrated slice ridges tends to be overpowered. In this situation little benefit is seen from a serrated slice.

Another situation where it will not be helpful is when the drainage is so rapid that there is little fluid left to shear a short distance down the wire. The ridges will go through one or two phase shifts and then they will be so small that no further splitting will occur.

After the ridges have died out and if there is sufficient fluid in the sheet, the ridges may be regenerated using a formation shower. The shower shown schematically in Fig. 187 produces a series of fine, closely spaced jets that generate ridges in the stock flow. These ridges are split as they go over the drainage elements as

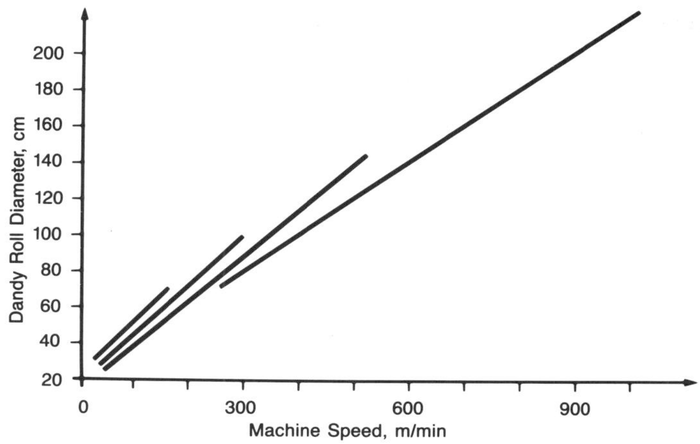

Fig. 188. Relationship between machine speed and dandy roll diameter.

with the serrated slice ridges. This type of shower can be especially helpful on heavy-weight grades.

The position and design of the shower are critical to successful operation. The shower jets must be of small size and coherent, i.e. they must be a stream and not broken up into droplets, in order to produce good ridges. This requires that the shower be positioned fairly close to the wire, a thing papermakers are often reluctant to do. The ridges thus produced do not last many phase shifts before they die out. In some instances several showers are needed to produce a really beneficial effect.

3. Formation aids

(a) The dandy roll

Historically, the dandy roll was used to improve formation of the sheet on slow-speed machines where flocculation of the top side of the sheet was inevitable due to poor agitation on the wire. The dandy roll was placed in the middle of the suction box section where the sheet was just about to pull dry. It was supposed to rework the top side of the sheet and break up flocs. It was very effective and there was a marked improvement in formation.

At operating speeds of 100 m/min or so, the dandy was driven by the sheet and the wire. It ran usually on trunion bearings which were set so that the dandy exerted a certain pressure on the sheet. The original dandies were about 30 cm in diameter or smaller, but as machine speeds increased the shear forces between the small dandy and the wire increased to the point where the sheet was disrupted. Simple drives were installed and the situation improved, but they were still troublesome to run and many were removed from service. The beneficial effects of the dandy on formation remained clear and so efforts were expended to eliminate the operating problems.

As machine speeds increased, it became clear that the diameter of the dandy should increase and the mechanical stiffness had to increase also [16-18]. Dandy rolls that were typically 60–75 cm in diameter increased in size to 125–150 cm. A relationship between machine speed and dandy diameter was developed and is shown in Fig. 188. The key to the correct diameter is to keep the dandy rpm between 120 and 150 rpm.

As diameters increased and machine speeds increased, drives and bearings had to be improved [8, 17-19]. Bearings were changed to large-diameter antifriction bearings and the drive was changed to a timing belt to ensure accurate speed control. The matching of dandy speed to wire speed is important for a number of reasons. First, if the speed difference is too great, the resultant shear forces will disrupt the sheet. Second, the speed match has an effect on the water throw after the dandy. The combination of speed and diameter has a great influence on the water throw. The objective is to have the water drops land on the sheet surface while it is still wet. This relationship is shown in Fig. 189.

The new larger diameter dandy rolls require a somewhat different set-up to operate properly. The amount of free wire under the dandy should be equal to the diameter of the dandy. The dandy should be located with ⅓ of the free wire before the roll and ⅔ of the wire after the roll [8]. The dandy is positioned after the first suction box or in some cases after the last suction foil box. For

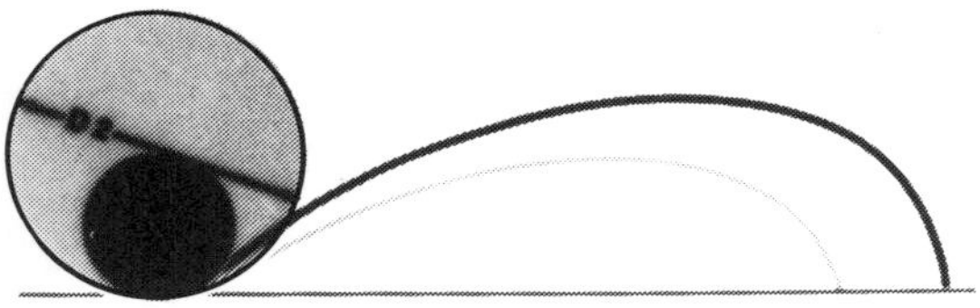

Fig. 189. Water drop path after the dandy roll.

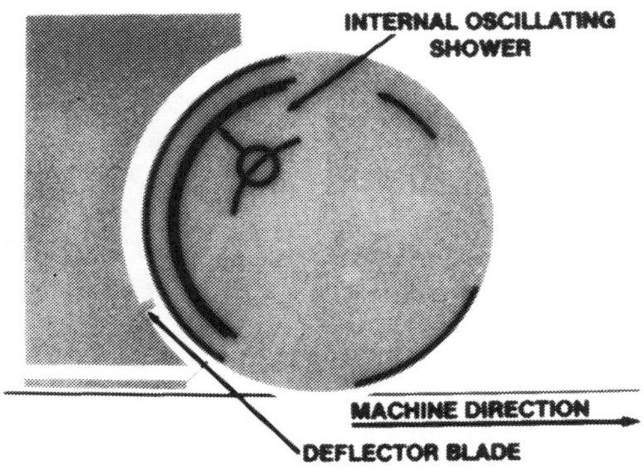

Fig. 190. Typical dandy roll arrangement.

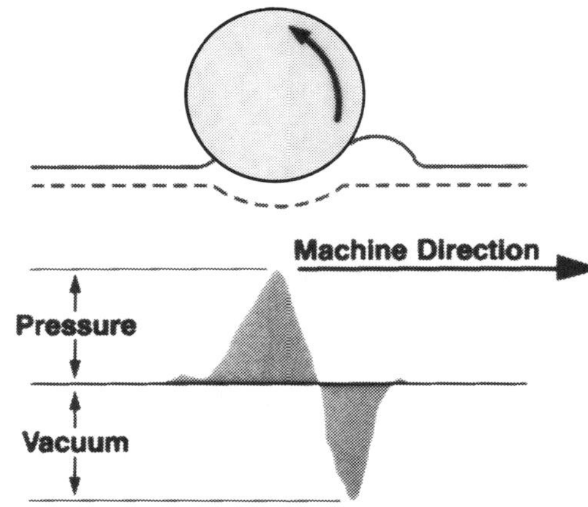

Fig. 191. Pressure profile of the dandy roll.

best action, the moisture content under the dandy should be 2%.

Locating the dandy at this point on the wire makes it a water removal device as well as a sheet structure altering device. In order to provide volume in the dandy to handle the extra water, new double covers have been designed. They have a relatively coarse under wire and then an outer cover of 35 to 40 mesh. The covers have a spiral seam and are almost always made of stainless steel. The extra water held in the mesh of the dandy has to be removed and this is usually aided by the use of a shower. The configuration is shown in Fig. 190. The collecting pan is located on the upstream side of the dandy. The pan has a seal on the bottom to prevent dripping. Drains must be provided to take away the removed water. On very wide machines, there are drains both front and back.

Keeping the roll clean has always been of prime importance to successful operation. There have been a large number of permutations and combinations of water showers and steam showers, all of which were claimed to keep the dandy clean. The most successful arrangement has been to install one shower pointed in the direction shown in Fig. 190. The shower pressure should be 13–14 bar to ensure that the shower streams penetrate both covers and keep them clean. The shower is of special construction such that the nozzles are close to the inner surface of the dandy. It must also oscillate to ensure that the entire dandy is kept clean.

The action of the dandy can be seen by looking at Fig. 191. First of all the wire is deforming in a curve around the dandy. The radius of curvature is determined to a large extent by the radius of the dandy and the tension in the wire. At the ingoing side, there is a compression of the sheet between the dandy and the wire. Water is driven up through the sheet into the dandy and downward through the wire. At this point in the forming process, the water distribution in the sheet is not uniform. All the dewatering has taken place from the bottom and there is gradient of water from top to bottom, the consistency at the top of the sheet being much lower than that at the bottom. Therefore most of the water flow is upward. Past the midnip position, the dandy and the wire are separating and a small vacuum pulse is created. This vacuum is claimed to pull some water out of the dandy and create the spray that is seen on the exiting side of the nip. It is the redistribution of water that causes shear on the fibers and improves the formation [8].

The pressure schedule at the nip is determined by the diameter of the dandy, the machine speed and the wire tension. Larger diameter dandies develop the pressure more gradually so that the pressure pulse is more gentle. The same is true of the vacuum development at the exit side of the dandy wire nip.

The effect of the dandy on formation is mainly in the small-scale floc structure as can be seen from the QNS formation curves shown in Fig. 192. With the QNS formation meter, the lower the number the better the formation. The initial work done was in the speed range of 100–400 m/min and the results are shown in Fig. 192 [20]. The largest effect is in improving the small-scale

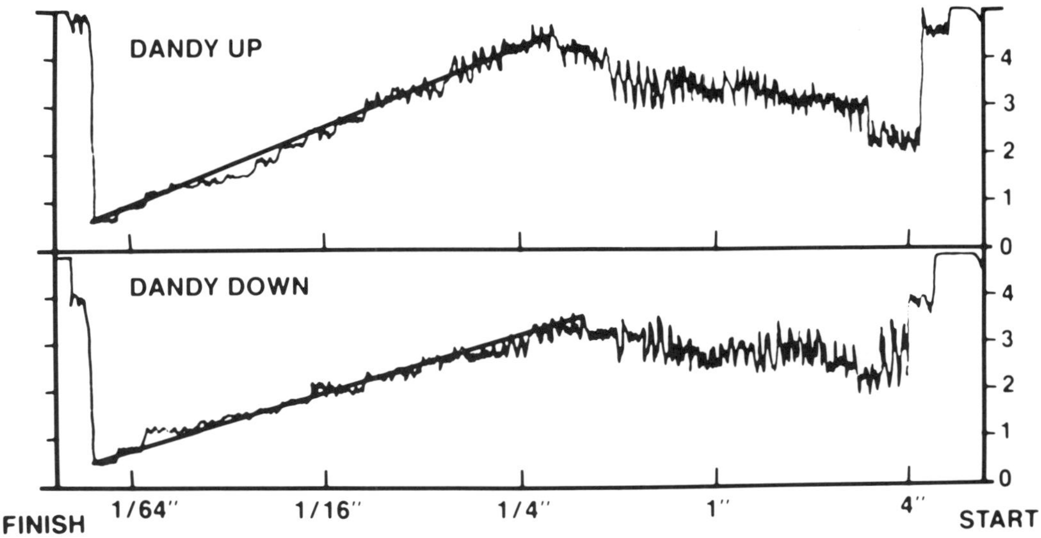

Fig. 192. Effect of the dandy roll on formation.

formation. Later work extended the speed up to 550 m/min with similar results [21]. It is known that there are dandy rolls running at 750 m/min and producing good results.

Consideration has been given to the effect of the dandy on fiber distribution. The effect on filler distribution should also be considered. Since the action of the dandy on the ingoing side of the nip is to squeeze water upward through the sheet, it would seem that any small particle material that was somewhat mobile would travel with the water. Studies done on sectioning sheets show that the filler content of the upper layers of the sheet does increase when the dandy is used [22]. This is shown in Fig. 193.

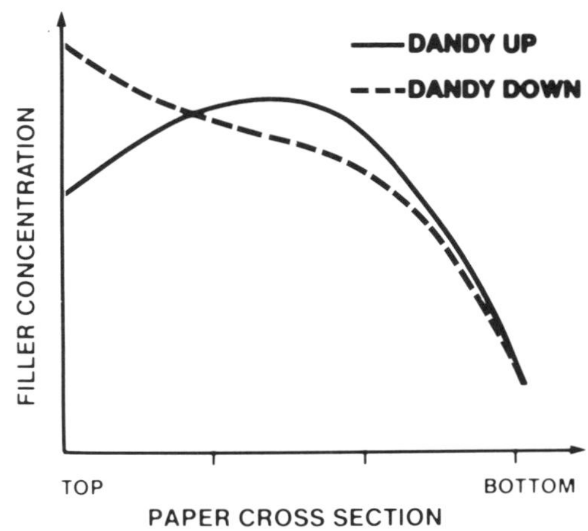

Fig. 193. The influence of the dandy roll on filler distribution.

(b) The shake

The effectiveness of the shake on formation has been the subject of many great debates. To quickly summarize all of this, the following points are generally true. At 600 m/min and above, the shake does little or nothing for formation. There is just too little time for the shake to act on the fibers before they have passed out of the shaken zone. However, high-frequency shakes at speed below 600 m/min and especially with heavy weights at speed of 300 m/min can produce significant improvement in formation.

Investigations have shown that the frequency of the shake is more important than the amplitude. The higher the frequency the more beneficial is the effect on formation. The effectiveness of the shake in improving formation is roughly directly proportional to the amplitude and the square of the frequency and inversely proportional to speed of the machine. The so-called shake number is the product of the amplitude times the frequency squared divided by the machine speed.

Some trials run a number of years ago on several different paper machines making different products at different speeds supported this concept rather well [23]. Figure 194 shows a summary of the results of these trials. The formation number used in this figure is the LIN value from the QNS formation tester; the lower the number the better the formation. All of the results were normalized so that the formation reference was zero at shake number of 250.

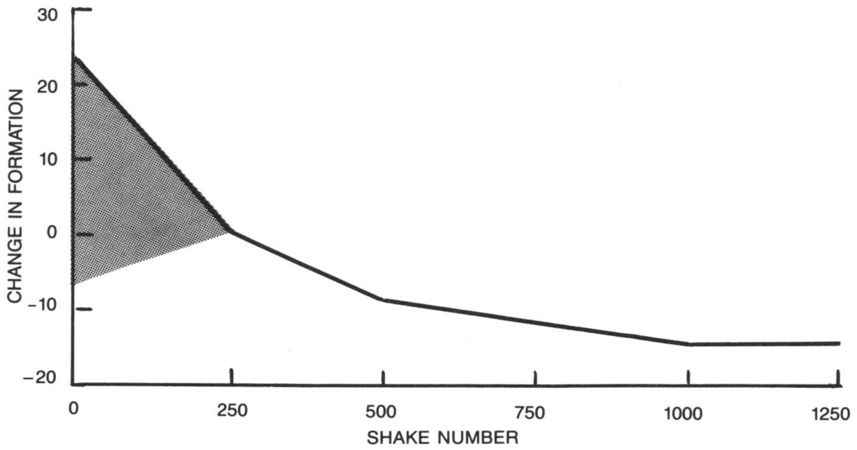

Fig. 194. Effect of shake number on formation.

Formation continued to improve from a shake number of 250 to about 1000 and then flattened out. Below 250, the formation change is much less predictable as can be seen in Fig. 194. Some sheets showed constant improvement with increasing shake number, while others would deteriorate in formation with increasing shake number up to 250. Beyond 250, all sheets improved in formation by a predictable amount.

From the foregoing, it seems apparent that reaching a shake number of 250 will almost always produce an improvement in formation. If we take a mechanically acceptable shake amplitude of 1.2 cm and a frequency of 300 shakes per minute, the calculated speed at which the shake number is 250 is 450 m/min. This seems to fit with experience and indicates that it is possible to predict with some degree of confidence the improvement in formation with adjustments to the shake.

What is done with the shake regarding frequency and amplitude is usually dictated by the mechanical condition of the equipment rather than a desire to improve formation further. The easiest shake system to maintain and also one that puts the least mechanical strain on the fourdrinier is to shake the breast roll and have all other elements on the fourdrinier stationary. This system has the lowest mass to move and shakes the least amount of equipment.

E. AKNOWLEDGMENTS

The author is grateful to T.J. Bambrick of Albany International, Appleton Wire Division and to V.E. Hansen of Atlanta Wire Works for permission to include figures from their papers published in the TAPPI Wet End Operations Seminar Notes.

REFERENCES

1. Burkhard, G. and Wrist, P.E., "Investigation of High Speed Paper Machine Drainage Phenomenon", Pulp Paper Mag. Can. 57 (4) (Mar. 1956).
2. Wrist, P.E., "Formation and Structure of Paper", Trans. of the Symposium held at Oxford, Sept. 1961, Tech. Sec. British Paper and Board Makers Assoc., London, p. 839-888 (1962).
3. Hansen, V.E., "Water Management for the Wet End Part I", Notes 1989 TAPPI Wet End Operations Seminar, p. 117-174 (1989).
4. Bachand, J.P. and Bambrick, T.J., "Effect of Drainage Pulses on Fourdrinier Papermaking", Notes 1989 TAPPI Wet End Operations Seminar, p. 178-185 (1989).
5. Cadieux, S.M., "Influence of Fourdrinier Drainage Elements on Fiber Mat", Intl. Water Removal Symp., Oct. 1982, Vancouver, B.C.
6. Schroder, S. and Svenson, O., "The Effects of Jet/Wire Ratio on Various Paper Characteristics", Svensk Papperstidn. 68(65):25-33 (Jan. 1965).
7. Manson, D.W., "The Practical Aspects of Formation", Notes 1989 TAPPI Wet End Operations Seminar, p. 13-29 (1989).
8. Parker, J.D., "The Sheet Forming Process", A Project of the Fluid Mechanics Committee, TAPPI STAP Series No. 9 (1972).
9. Kerekes, R.J., "Pulp Flocculation in Decaying Turbulence: Literature Review", J. Pulp Paper Sci. 9(3):TR86-91 (1983).
10. d'Incau, S., "Characterization of the Decay of Turbulence in a Pulp Suspension Flow", Proc. TAPPI Engineering Conf., p. 583-592 (1983).
11. Bolon, P., Sabater, J. and Bauduin, S., "Contribution to the Study of the Stock Jet as it Leaves the Headbox", Tappi 66(5): 93-96 (1983).
12. Duffy, G. G. and Nordman, B.G., "Fibre Flocculation in Conical Contractions Simulating the Paper Machine Flow Box Slice", Proc. Intl. Symp. Paper Machine Headboxes, McGill Univ. (June 1979).
13. Kallmes, O.J., Marinari, G. and Perez, M., "A Novel Approach to Optimizing Sheet Formation on the Fourdrinier", TAPPI Papermaker's Conf. (April 1989).
14. Kufferath, W. and Kallmes, O.J., "Optimizing Web Formation on Fourdrinier Machines", Das Papier 35(10A):V159-V177 (1981).
15. Kallmes, O.J. and Thorp, B.A., "New Redispersion System Improves Formation on Fourdrinier", Tappi 66(12):54-57 (1983).
16. Kufferath, W., Paper Trade J., p. 46-48 (June 2, 1969).
17. Planhammar, J., Paper Trade J., p. 36-37 (Dec. 21, 1980).
18. Langdok, R., TAPPI Wet End Operation Seminar (1977).
19. Sinclair, D., Pulp & Paper, p. 26-30 (May 8, 1967).
20. Piggott, R.J.C., Eucepa 79(2):34-47 (1979).
21. Manson, D.W., Tappi 46(8):472-480 (1963).
22. Luhde, F., Pulp Paper Mag Can., 67(10): T381-T392 (1966).
23. Deutschman, H., Zellstoff Papier 3:113-116 (May, June, 1980).
24. Manson, D. W., Gillis R. C., "Effect of the Fourdrinier Shake on Formation", Tappi 49(10):425-429 (1966).

IX

PRINCIPLES OF TWIN-WIRE AND MULTIPLE-WIRE FORMATION AND DRAINAGE

B.A. THORP

Vice President, Engineering
James River Corporation

REVIEWED BY

Dr. Bo Mannstrom
President
Finnish Pulp and Paper Research Institute
Helsinki, Finland

Dr. Dan W. Manson
Senior Consultant, Engineering
James River Corporation
New Gloucester, ME

Dr. Hannu Paulapuro
Deputy Research Director
Finnish Pulp and Paper Research Institute
Helsinki, Finland

A. INTRODUCTION

The idea of forming paper webs between two screens is not new and was conceived at least as early as 1875 [1]. The modern twin-wire former was developed in response to problems that were increasing as machine speeds were increasing. The major problems of fourdrinier forming included:

- Two-sidedness.
- Lack of fine-scale formation.
- Non-uniform profiles (macro variation).
- Length limitations, especially on rebuilds.
- Speed limitations due to "free wire" instability.

There were significant refinements and improvements to fourdriniers from the early 1800s. They included vacuum application, dandy rolls, shake, hydrofoils and synthetic fabrics. Each gave improvement in quality and/or production, but the major limitations returned at higher speeds. Figure 195 shows fines distribution in a fourdrinier-made newsprint sheet at about 760 m/min (2500 ft/min) [2]. The fines poor "wire side" is typical of fourdrinier-made sheets.

Daniel Webster of Consolidated Paper is credited as the inventor of the modern twin-wire former and had a working model in 1953. Figure 196 shows the evolution of various twin- or multiple-wire formers. Types listed after 1985 may be based on opinion as recent trends are not clear. This figure will be used to define the different types of formers. The sections which follow should make each technology more clear.

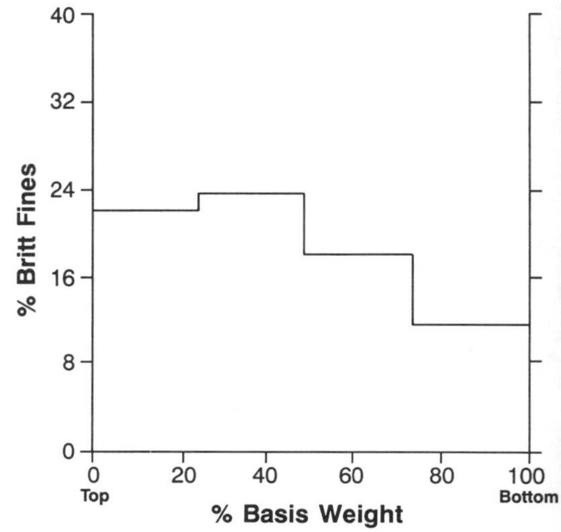

Fig. 195. Fines distribution in newsprint.

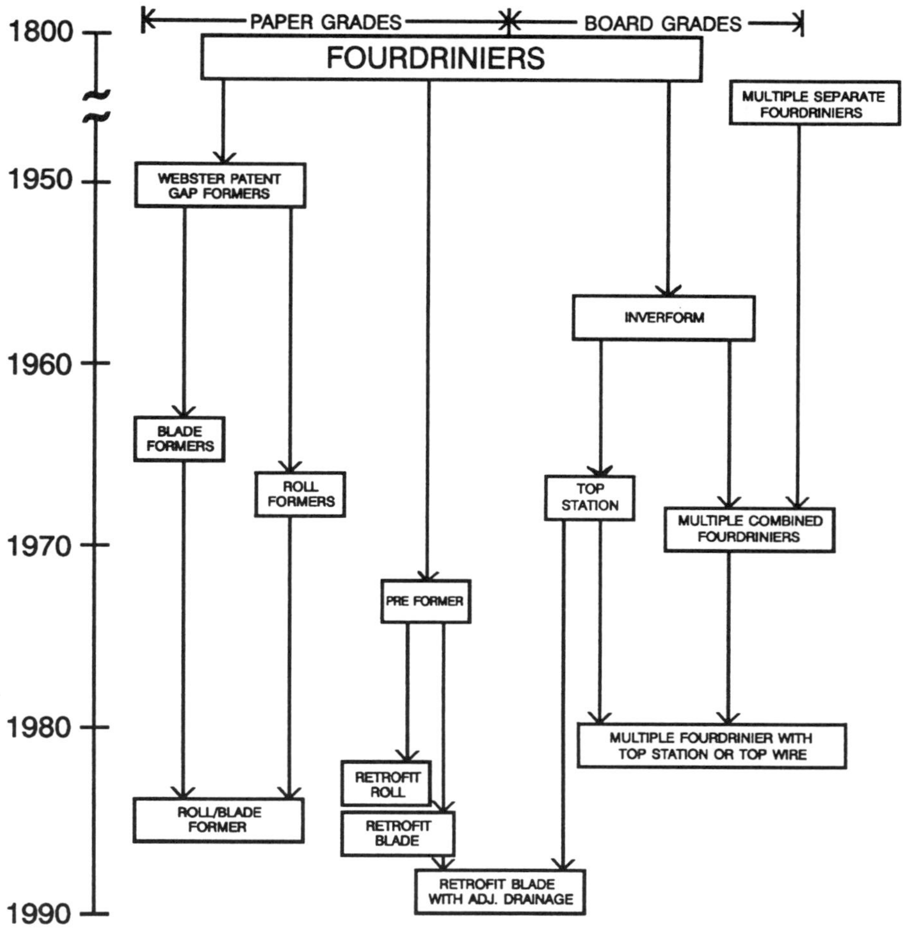

Fig. 196. Genealogy of major types of formers.

Note that Fig. 196 has four major tracks of evolution, two were *mainly* for paper grades and two were *mainly* for board grades. Figure 196 also has an *approximate* time line so one can better understand what was going on at similar times. A more detailed historical perspective is given elsewhere [3].

Perhaps the most critical decision in choosing a paper machine is the selection of the forming mechanism. This will have a major impact on the fundamental properties of the sheet and the runnability of the machine. There does not appear to be one single best choice. Those who have written about choosing a forming mechanism have noted the importance of clearly stating the objectives, then conducting a methodical study [4,5].

B. DEFINITIONS

The "paper side" of Fig. 196 starts with the famous Webster "Gap Former" patent. A gap former is illustrated in Fig. 197 and will mean a former in which the headbox jet is injected between two converging wires. Note that the term gap former does not indicate any dewatering mechanism, just an initial entry condition.

There are three major types of gap formers, namely:

- Gap blade formers,
- Gap roll formers, and
- Gap roll/blade formers.

A gap roll former has roll wrap as the *primary* water removal mechanism. An early roll former is shown in Fig. 198.

A gap blade former has blade action as the *primary* water removal mechanism. A gap blade former is shown in Fig. 199.

A roll/blade gap former is one in which both roll wrap and blade action are water removal mechanisms. The first commercial roll/blade gap former is shown in Fig. 200.

It is important to note that we frequently do

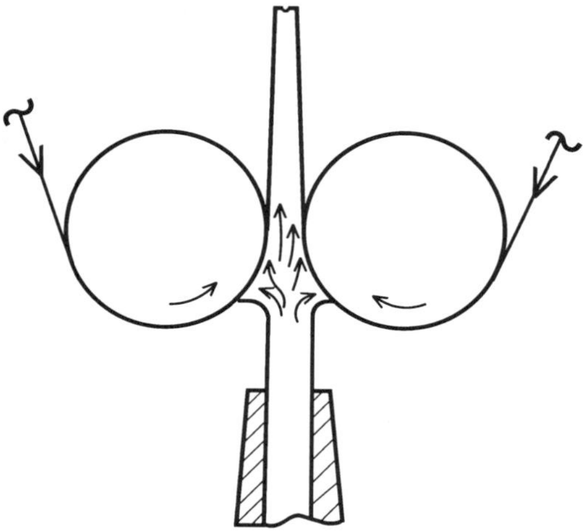

Fig. 197. Gap former.

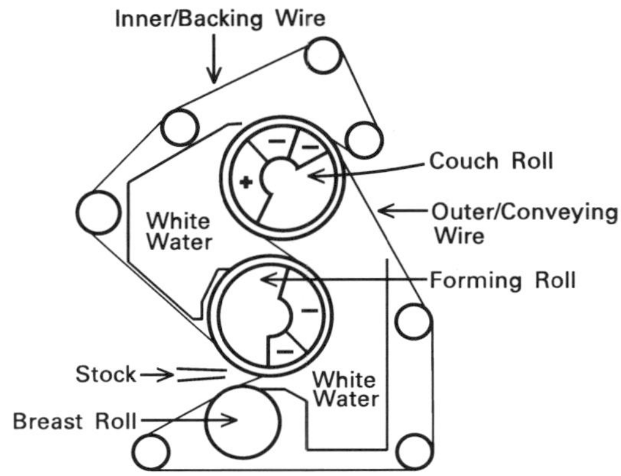

Fig. 198. Gap roll former (Dominion Engineering).

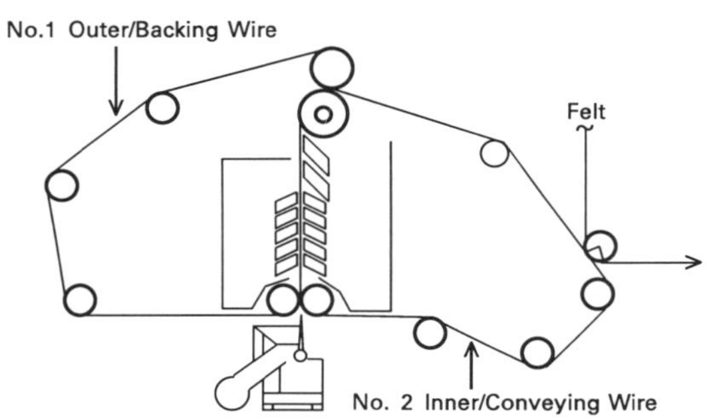

Fig. 199. Gap blade former (Black Clawson).

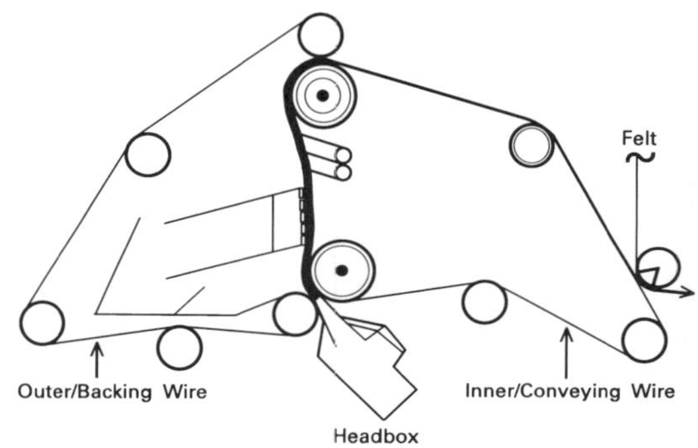

Fig. 200. Gap roll-blade former (Valmet Corp.).

not communicate well when describing formers. For example, we almost never specify the direction of orientation (e.g., vertical or horizontal) and often drop the term "gap". The phrase, "machine No. 5 is a blade former" is common, but not very descriptive.

There is a whole family of top-wire units. They are called retrofits, hybrids, preformers, or "on top" formers. Hybrid is the best description, because not all of them are retrofits. They are all characterized by having a preforming zone of "open wire" papermaking. Figure 201 shows a retrofit type in which roll forming follows the "open wire" papermaking [6].

Figure 202 shows an original machine configuration in which blade forming follows "open wire" papermaking.

It is important to note that the hybrid formers can have roll, blade, roll/blade, or adjustable blade dewatering following "open wire" papermaking. The adjustable feature is of growing importance and will be discussed further in Sec. H.

It is also important to distinguish twin-wire forming, which is typically done for lighter weight sheets, from two-wire or multiple-wire forming, which is typically done for heavier weight sheets. Figure 203 shows one of the less complicated, but more popular designs [7]. This subject will be covered further in Sec. J.

It has taken many words and nine figures to define our terms and gain a basic understanding of the types of units available. There are almost endless varieties of twin-wire formers and it is beyond the scope of this chapter to describe all or even most of them. An attempt will be made to cover the fundamentals of the basic types and use a few figures of some of the latest configurations as examples. The use of these examples does not imply the author's preference.

C. SELECTED SHEET PROPERTIES

In addition to reviewing the basic types of formers, it is worthwhile to review some sheet properties. This will give a common reference for the remainder of the chapter. The properties to

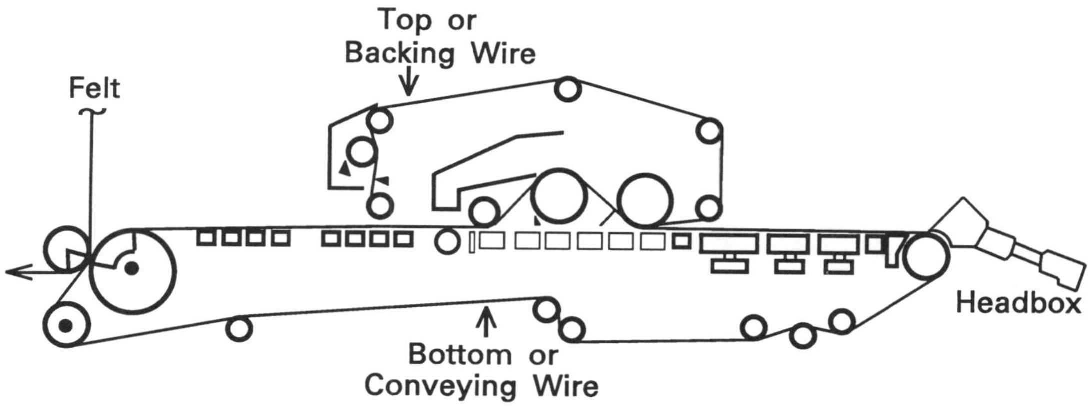

Fig. 201. Hybrid roll preformer (Dominion Engineering).

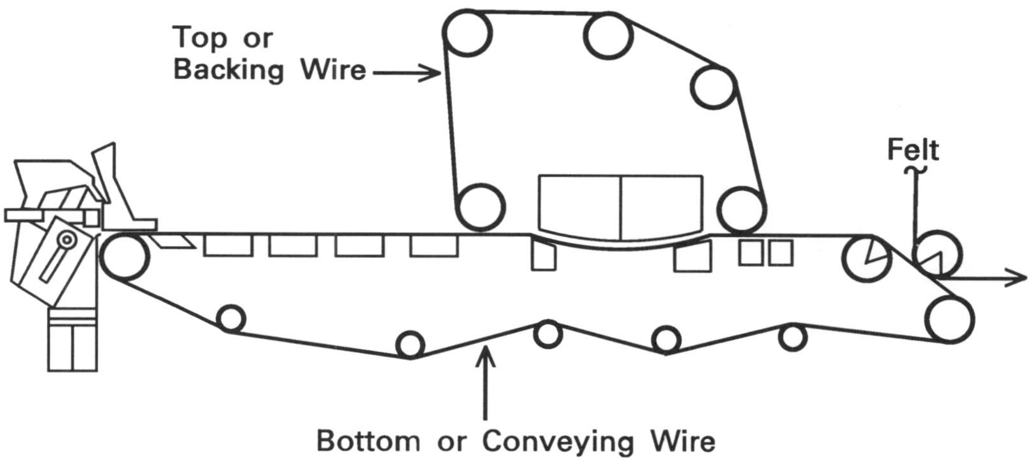

Fig. 202. Hybrid blade preformer (Beloit Corp.)

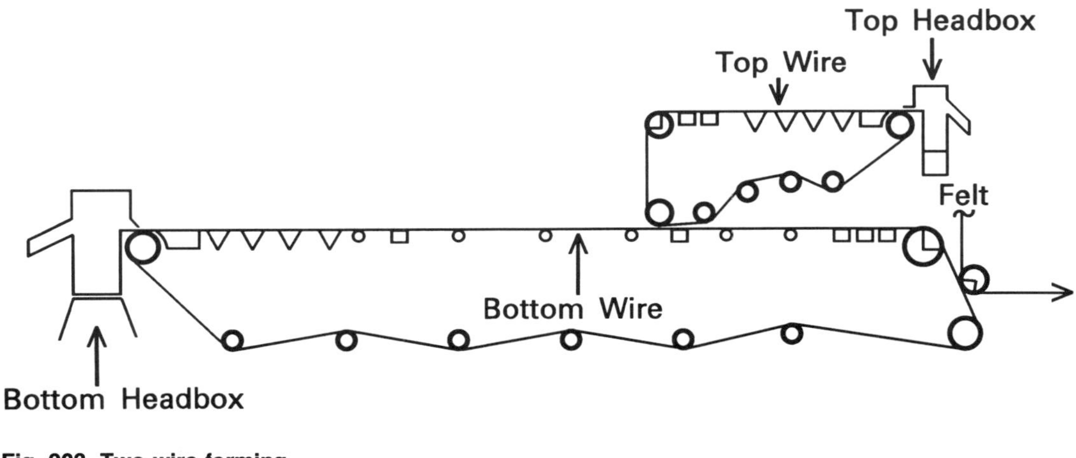

Fig. 203. Two-wire forming.

be discussed are formation, retention, and sheet uniformity.

1. Formation

Formation is perhaps the most fundamental property because it affects most other sheet properties. Formation typically is measured by eye or by meter as an optical variation. However, the most useful concept is mass variation. Micro-mass variation can be measured off-machine with small aperture beta gauges [8]. A new method of indirect recording of electron beam transmission is under development and holds promise for the future [9]. When the mass variation is on a micro scale we use the term formation, and when it's on a macro scale we use the term uniformity. When it's in the z-direction we frequently use the term "two-sidedness".

Mass micro-variation in the sheet is caused by fibers combining into bundles or "flocs".

Fig. 204. Stock discharge from hydraulic headbox.

The tendency for fibers to floc is generally underestimated. Studies have shown that fibers floc in thousandths of a second (milliseconds) [10]. This is critical. One of the more dramatic demonstrations of reflocculation in stock discharge was obtained from a high-turbulence headbox at 427 m/min (1400 ft/min). Figure 204 shows reflocculation about 10 cm (4 in.) from the slice or only 12 ms after discharge [11]. The flocs do not reproduce well in the print, but can be clearly seen in photographs or slides.

The three most important variables in formation are:

– Dilution.
– Shear.
– Turbulence.

These variables are of importance throughout the process because of the rapidity of reflocculation.

Dilution is often expressed as consistency or percent solids consistency. The most important number is not the total, or the average, but the consistency of the long fiber fraction (e.g., softwood kraft). This number is not frequently available and is usually "estimated", often incorrectly. There are a number of examples of shear and they include rush/drag ratios, shake, and convergence into a top wire unit, as will be discussed in Sec. H. Examples of turbulence include headbox induced and table induced. These are not the only formation variables, as indicated by Fig. 205 [12].

2. Retention

Retention is increasingly important as there is a tendency for retention to decrease with decreasing basis weight and to decrease with

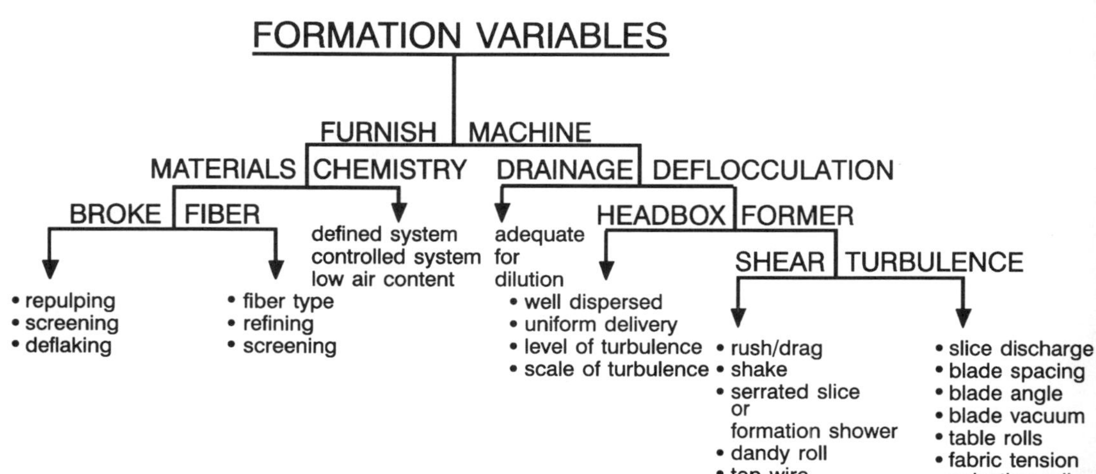

Fig. 205. Major formation variables.

increasing speed, both of which are current trends. The importance of retention is often understated in "closed systems". Retention is of major importance from a water management/operations view. Consider the following: "Cleaner white water means less problem with slime and dirt buildup, which in turn will mean higher machine efficiency. Cleaner white water means less circulating material, and therefore less changes in mix composition with fluctuating retention levels. This will mean a more constant quality of the product. Cleaner white water will mean less material carryover between machines with some connections in the white water systems. Cleaner white water means faster response to grade changes and other process changes, and thus a lower level of broke generation. This is especially important for the long circulation loop" [13].

It is critical that true retention be known and used. There can be significant differences between true retention and the quick method for calculating first-pass retention, particularly on tissue machines [14].

3. Sheet uniformity

Macro-mass variation, or the lack of it, is becoming increasingly important. The importance is being driven by wider, faster machines, and by marketplace demand for higher uniformity. Historically, macro-mass uniformity has been measured by scanning devices which have been manipulated to give CD, MD, and "residual variation". Off-machine profilers have shown that there is much greater variation than that "reported" by the gauges. This difference is due to "time constants", number of "boxes", and the algorithms used. Figure 206 shows that a combination of these historical profiles do not describe paper variation in a meaningful way [15].

There are improvements in modern scanning devices like the zoom feature and other development work is ongoing. In the meantime, these differences and their relationship to the marketplace "fit for use" criteria must be understood in purchasing, optimizing and operating a twin-wire former.

D. SETTING THE JET [16]

The approach flow, headbox selection, and setting the jet are common to all types of gap formers. The gap formers have a limitation in that the operator has fewer papermaking tools or command variables. Approach flow and headbox selection are beyond the scope of this chapter. The critical adjustments the papermaker has with respect to setting the jet in gap forming are typically:

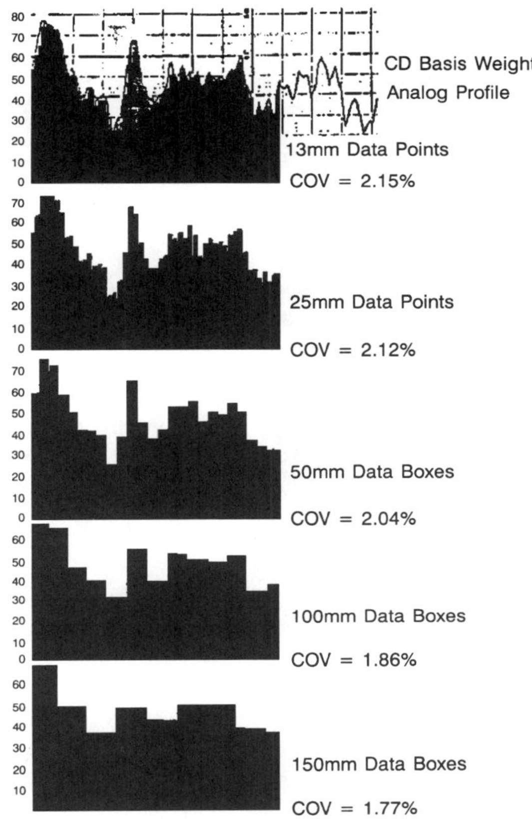

Fig. 206. Variance vs. data box size.

- Free jet length.
- Slice opening.
- Percent impingement on forming roll or blade.
- Angle of impingement into the fabric.
- Forming fabric tension.
- Rush/drag ratio.

The free jet length should be as short as possible and in no case should it be long enough to break apart, or to increase in the z-direction.

The slice opening is determined normally by the speed and lowest consistency possible. At times, this opening can be too large for the free jet length and the jet will begin to break apart. Increasing the consistency and closing the slice could result in a more optimum condition.

The percent of impingement (% jet landed) and angle of impingement are usually particular to the speed, furnish, fabric design, and equipment geometry. A set of organized and carefully conducted trials to determine optimum settings is recommended for each condition or change in condition.

Forming fabric tension is normally selected for drive or dewatering conditions, but can be slightly altered during the trial to determine its impact. Normally, it is a minor variable in setting the jet.

The rush/drag ratio normally follows the classic pattern shown in Fig. 207 [12]. Because of the difficulty in measuring total head in a hydraulic headbox, this curve should be obtained for all gap formers. An on-line formation tester is desirable for these trials and is best obtained as part of the gap former installation.

E. GAP ROLL FORMER

Figure 208 shows a modern twin-wire gap roll former. Note there are no stationary parts, so fabric life is long, allowing some flexibility in selecting fabric design. These formers were popular in the late Sixties and early Seventies on newsprint grades because of their ease of construction and ease of operation. They were less sensitive to changes in furnish than a fourdrinier and could be easily tuned to a set of operating conditions.

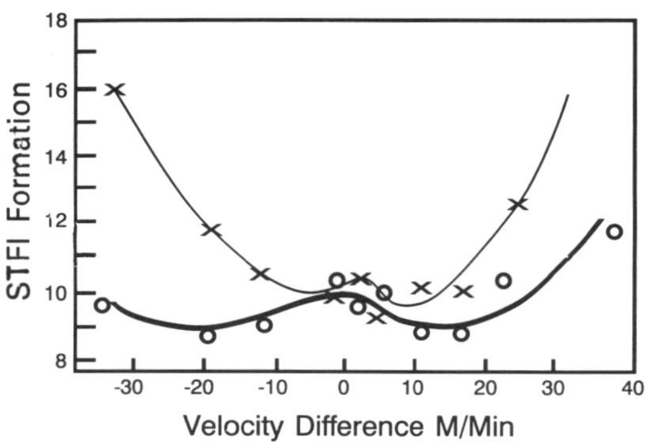

Fig. 207. Formation vs. jet/wire velocity difference.

The generally accepted formula for water removal force is:

$$P = T/R$$

where

P = dewatering pressure
T = tension of the outer wire
R = the radius of curvature of the forming zone [17].

For a wire tension of 5.25 kN/m (30 lb./in.), and roll radius of 46 cm (18 in.), this dewatering pressure would be approximately 0.14 kPa/cm^2 (2 psi). There are other minor forces to be considered, including centrifugal force and the vacuum force [18, 19]. Water removal occurs in approximately five feet. At 914 m/min (3000 ft/min), this equates to 0.1 s, and compares to approximately 0.5 s of forming time on a conventional fourdrinier.

This formula implies that exiting sheet consistency should decrease with increasing speeds due to an increasing quantity of water but constant water removal forces. Experimental results show that exiting sheet consistency increases with speed up to some threshold point. This has led some to believe that momentum is a significant factor which must be considered [20]. Further research is required to develop a comprehensive understanding of the principles of water removal

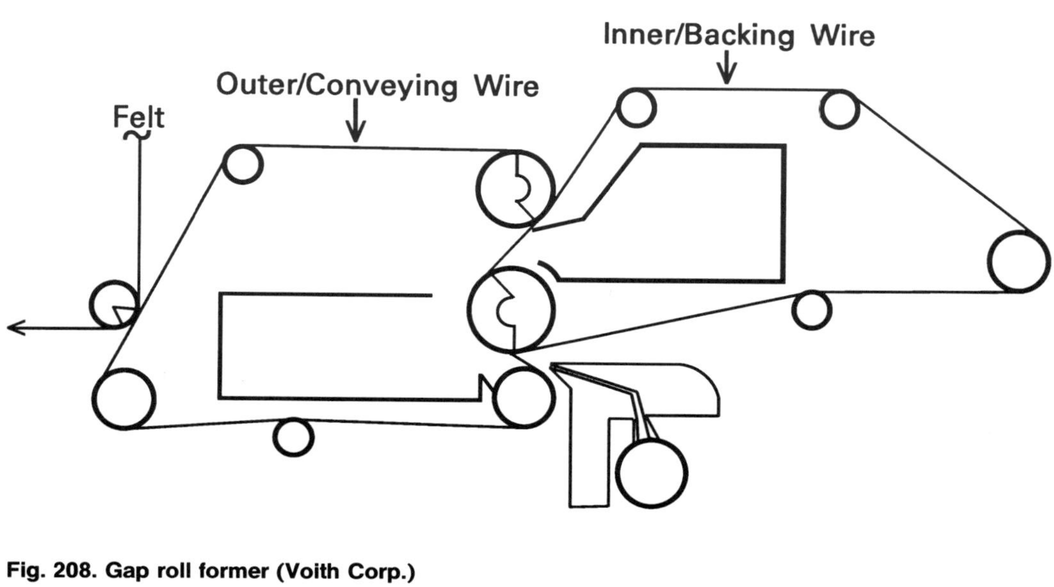

Fig. 208. Gap roll former (Voith Corp.)

on roll, blade, and hybrids.

Figure 209 shows a drainage split or Sankey diagram, which is typical for this type of former [21]. Water removal is much more symmetrical than on a fourdrinier and produces less two-sidedness.

Figure 210 shows sheet ash distribution which is typical for some roll formers [22]. The ash (and fines) content of the sheet side (top) and wire side (bottom) are more equal to each other. When roll formers develop a high ash (and fines) content in the center, the z-directon tensile can be reduced.

Figure 211 shows the type of pressure pulse derived [23]. Note that the main formation improvement mechanisms are headbox selection, dilution, and shear. With rapid reflocculing stock, and no pulsating devices, one would expect soft flocs to develop and the sheet to have a grainy formation.

While the data shown in Fig. 211 fit the formula P = T/R nicely, there are some concerns with the water removal mathematics.

In summary, the advantages of a gap roll former include:
- Low drive.
- Long fabric life.
- Reliability.
- Speed capability.
- Reduced two-sidedness.
- Less linting.
- Improved printability.

The disadvantages of gap roll formers include:
- Higher headbox consistencies.
- More pinholes.
- Lower internal bond.
- Grainy formation.

The formation comparison depends on the degree to which the headbox/foil system has been fine tuned on the conventional fourdrinier. It should be well understood that, with current consistencies in current hydraulic headboxes, the best formation is *not* obtained when the jet is frozen without disturbance. Forming over a curved surface at high speeds can cause Taylor vortices if the conditions are right. This can help explain part of the cause of streaking in roll formers and other formers with curved surfaces.

Presently, gap roll formers are used on high-speed, lightweight sheets where high dilution is possible. The most notable example is tissue.

F. TISSUE FORMERS

Modern tissue formers are gap roll formers where the drainage zone has a "C" or "S" shaped configuration. Consistent with our poor communications, they are most often called "C" formers or "S" formers.

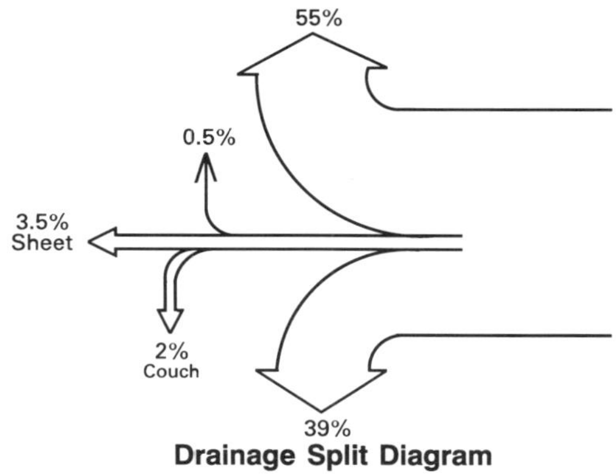

Fig. 209. Drainage split diagram.

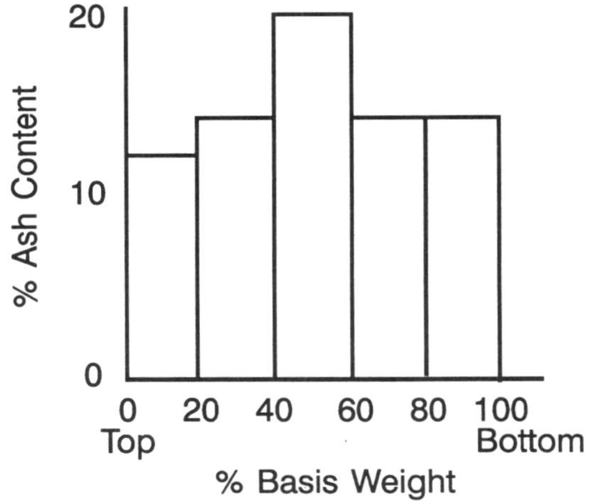

Fig. 210. Ash distribution.

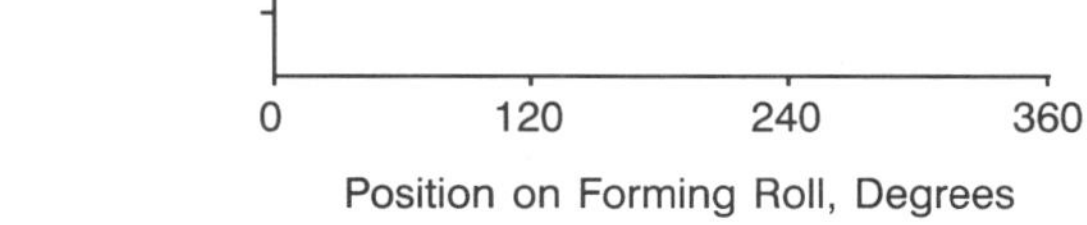

Fig. 211. Pressure over a solid forming roll.

Figure 212 shows a gap roll former with its "C"-shaped drainage zone. The forming roll can be solid for one-sided dewatering or suction to permit some two-sided dewatering.

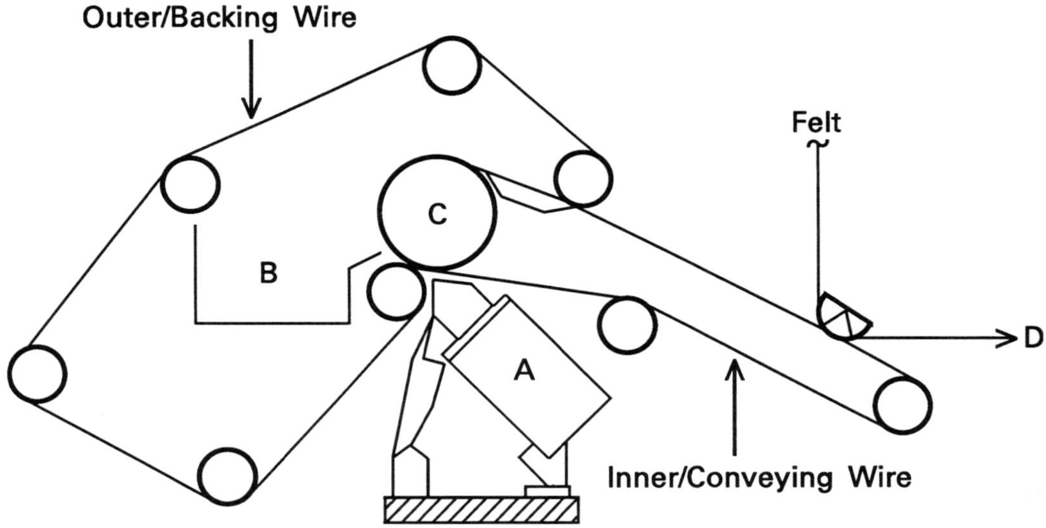

V 94 **Fig. 212. Gap roll former — "C" wrap. (Beloit Corp.).**

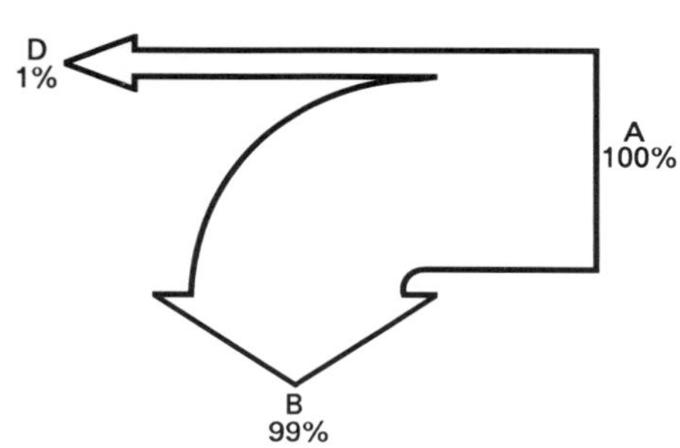

Fig. 213. Drainage split for a solid forming roll (Beloit Corp.).

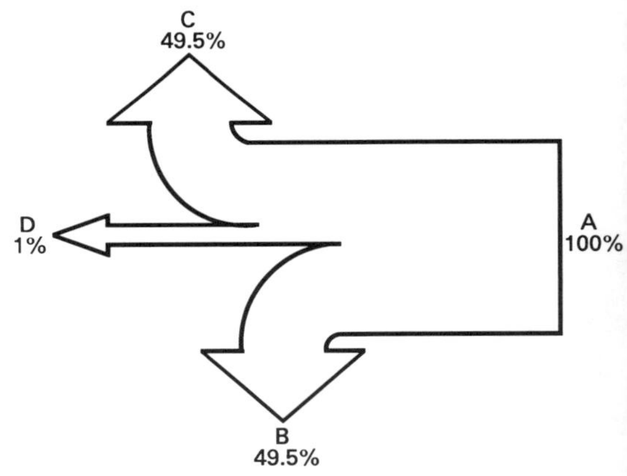

Fig. 214. Drainage split for a suction forming roll (Beloit Corp.).

Figure 213 shows a drainage split diagram typical for a solid forming roll, while Fig. 214 shows a drainage split diagram typical for a suction forming roll.

A consideration when running secondary fiber, or virgin fiber with broke, is contaminants, especially "stickies". The contaminants in many systems tend to stay with the inner/conveying wire (right wire), perhaps due to longer contact time and lower drainage rates. Stickies on the inner/conveying wire do not cause as many pinholes, because it is not the "drainage wire". It should be noted that in some stock systems the contaminants tend to go with the outer/backing wire (left wire). With a suction forming roll there is an opportunity to influence where the contaminants will go. All tissue machines using converting broke are susceptible to contaminants.

Many tissue machines have stratified headboxes which are capable of delivering furnishes of different fiber content. One consideration in the selection of formers is drainage management. In Fig. 212, assume the short fiber furnish is close to the forming roll and the long fiber furnish is away from the forming roll. This means that the short fibers will drain through the long fibers which would give better retention than vice versa. This drainage arrangement also means that the short fiber furnish is in contact with the yankee. If this drainage pattern is not desirable, then an "S"-wrap configuration or a "C" wrap with the headbox on top (see Fig. 217) may be preferred.

Figure 215 shows a gap roll former with its "S"-shaped drainage zone. Two-sided dewater-

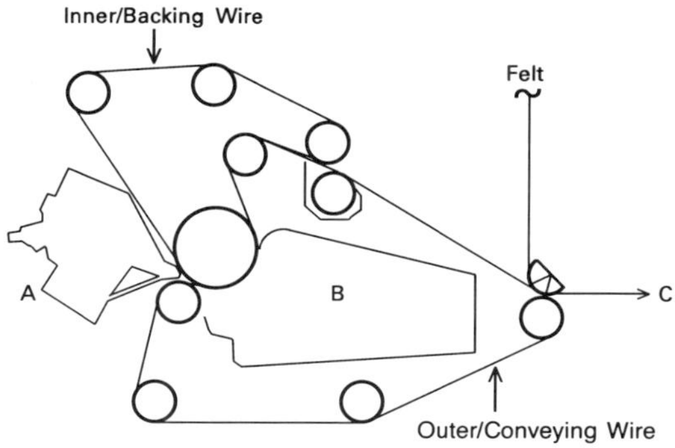

Fig. 215. Gap roll former "S" wrap (Valmet Corp.).

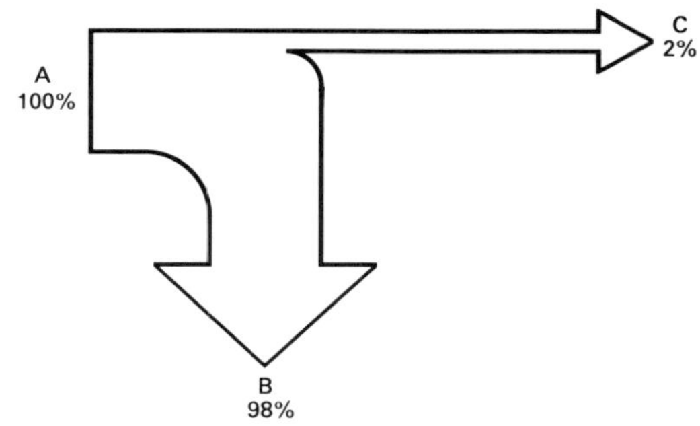

Fig. 216. Drainage split diagram (Valmet Corp.).

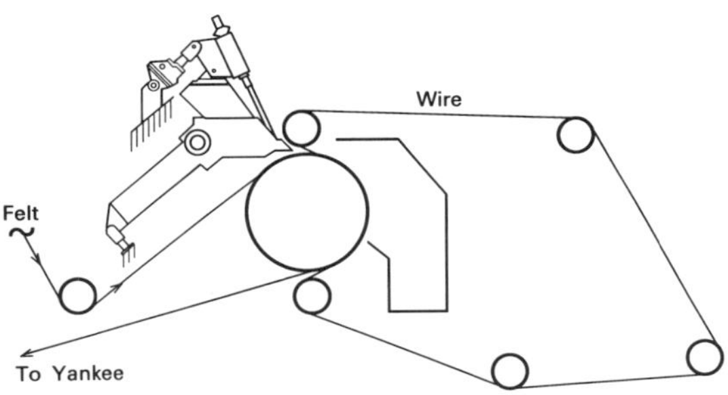

Fig. 217. Crescent former (Beloit Corp.).

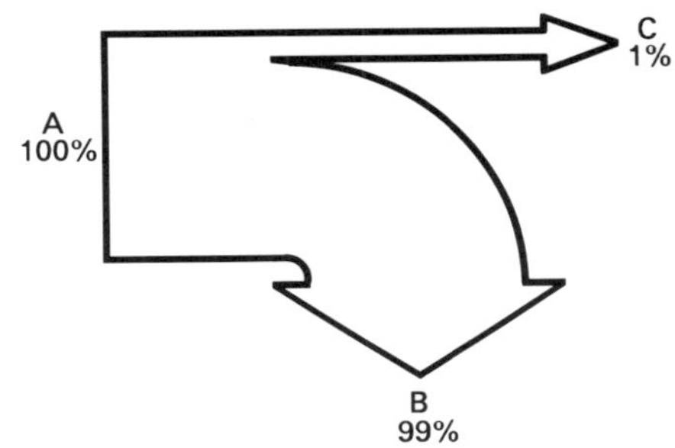

Fig. 218. Drainage split diagram (Beloit Corp.).

ing is not "natural" to this configuration. The solid forming roll could be suction for increased two-sided dewatering.

Figure 216 shows a drainage split diagram typical for a solid roll former of this configuration. Note that the outer/conveying wire (right wire) is also the primary drainage wire. This means that contaminants that want to collect on the outer/conveying wire will create more problems with pinholes because it is also the drainage wire. If the contaminants remain with the inner/backing wire (left wire), then this configuration would have a tendency to "run cleaner".

Another consideration involves stratified forming. If the short fiber furnish runs close to the forming roll and drains through the long fiber furnish, then the long fiber furnish will be in contact with the yankee. If this is not desirable, then a "C" wrap may be preferred.

Figure 217 represents a Crescent former. It is not a *twin-wire* gap former because a felt and wire are used. The original installations date back to the early Sixties which make it the first commercial "gap former". Modern headbox designs and modern clothing designs have added to the original capability and this former is of increasing interest.

The Crescent former is an inverted "C" former. If the contaminants want to stay with a conveying medium then they stay with the felt and have less of a tendency to create pinholes than if they went with the outer wire which is the drainage wire.

Stratified considerations are similar to an "S"-wrap former. That is, if the short fiber stock is close to the forming roll, then it will drain through the long fiber stock, which in turn will contact the yankee.

Figure 218 shows a drainage split diagram typical for a former of this configuration. All of the tissue configurations are very sensitive to clothing design and quality. Small changes in design can usually be detected. Clothing design, selection, and quality deserve the same careful, measured consideration as machine design and machine optimization. Tissue machines can usually be analyzed by the histogram method of charting clothing performance [24]. Clothing cleaning, conditioning and dewatering systems are even more important. The clothing is new only one day out of its life and the cleaning

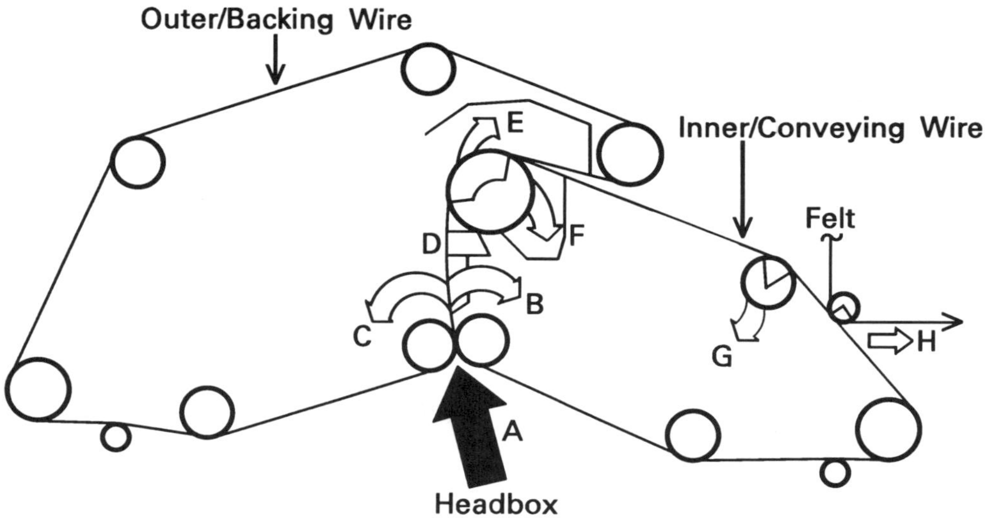

Fig. 219. Gap blade former (Beloit Corp.).

equipment must keep the clothing usable. Inside showers, outside showers, flooded nip showers and cleaning brushes are essential equipment.

G. GAP BLADE FORMING

Figure 219 shows a modern twin-wire gap blade former. The wire and furnish may travel along a large radius as in Fig. 219 or it may have a straighter path as in Fig. 199.

The formula for roll drainage applies to the large [approximately 5.08 m (200 in.)] radius shown in Fig. 219. Also, the water removal takes place over a longer distance [sometimes 3.65 m (12 ft)] than typical roll formers. At 914 m/min (3000 ft/min) the drainage time is 0.24 s compared to the 0.10 s calculated for the roll former. All other things equal, this would give the blade former higher retention. However, things are not equal as the blade formers have a number of stationary elements.

The force which results from stationary elements acting on a wire which is deflected at an angle, a, is:

$$P = T \text{ Sine } a$$

where

T = the wire tension
a = the angle of deflection [25].

If stationary elements are on a radius of 5.08 m (200 in.) with a spacing of 12.7 cm (5 in.), then the deflection angle will be slightly less than 1°. The force for a wire tension of 5.25 kN/m (30 lb/in.) will be less than 0.9 kN/m (0.5 lb/in.). The calculation of pressure distribution is quite complex.

There is considerable water deflection in blade formers. The water which has been removed from the stock will generally adhere to the fabric (as it does on a foilized fourdrinier). Removing this water causes deflection back into the fabric. The maximum local pressure on the deflector *tip* is:

$$P = r/2U^2$$

where

r = density of water removed
U = fabric speed

This results in the magnitude of force shown in Table 44 [where "r" = 1 g/cm^3 (62.4 lb/ft^3)].

Figure 220 shows measured pulses on standard blades and blades with vacuum applied [26]. These pulses break up flocs to improve formation. They also remove fines to reduce retention. The blades in the low-vacuum forming zone are smaller and on closer centers. The total effect is lower pressure pulses to enhance formation and retention.

Figure 221 shows how these forces manifest themselves in the dewatering process [27]. This photo shows the fabric stock sandwich passing upward over the second dewatering box of a blade former with a standard forming zone at 914 m/min (3000 ft/min). The water is being expelled horizontally through the backing fabric. Figure 222 shows a close-up view of the same shot. The above calculations and photos demon-

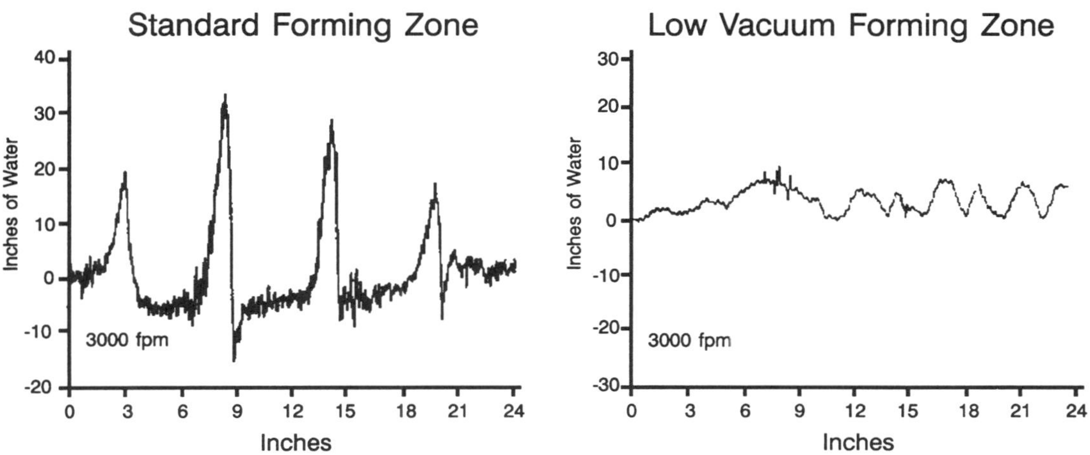

Fig. 220. Pressure over blades with standard and reduced vacuum.

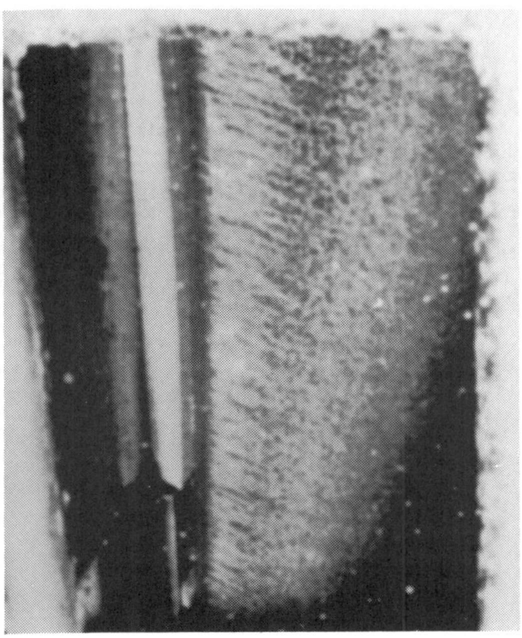

Fig. 221. Blade former at 914 m/min (3000 ft/min).

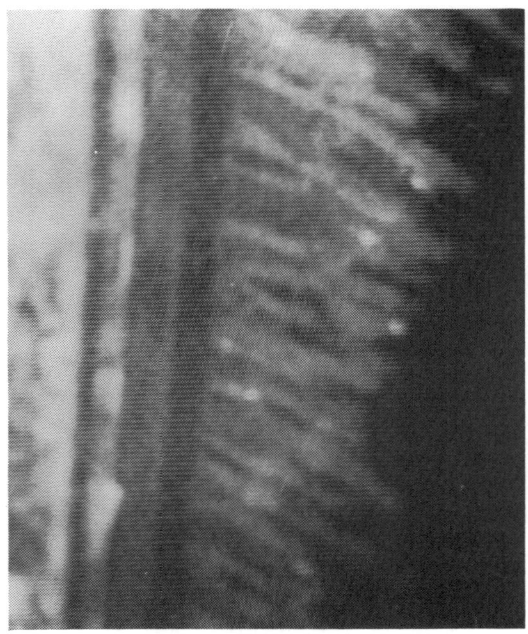

Fig. 222. Close-up of blade dewatering.

Table 44. Tip pressure as a function of speed.

U m/min	100	200	400	600	1000	2000
ft/min	330	660	1320	1980	3300	6600
P kPa/cm^2	0.001	0.056	0.217	0.497	1.4	5.6
psi	0.2	0.8	3.1	7.1	20	80

strate that high-speed drainage on a blade former is *not* a gentle process. These forces have a positive effect on formation due to the shear forces introduced on the flocs, and they have a negative effect on retention.

Figure 223 shows a drainage split diagram which is typical for this type of former. Water removal is quite symmetrical and can be controlled with former design.

Figure 224 shows fines distribution which is typical for this type former [28]. The figure shows that the two-sidedness is reduced and there is not a high concentration of fines/filler in the center.

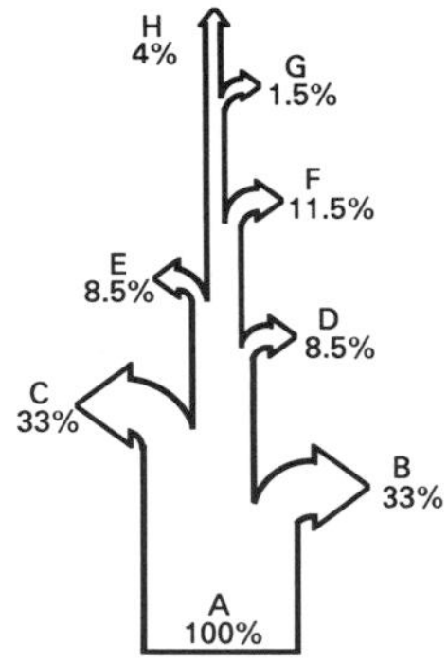

Fig. 223. Drainage split diagram (Beloit Corp.).

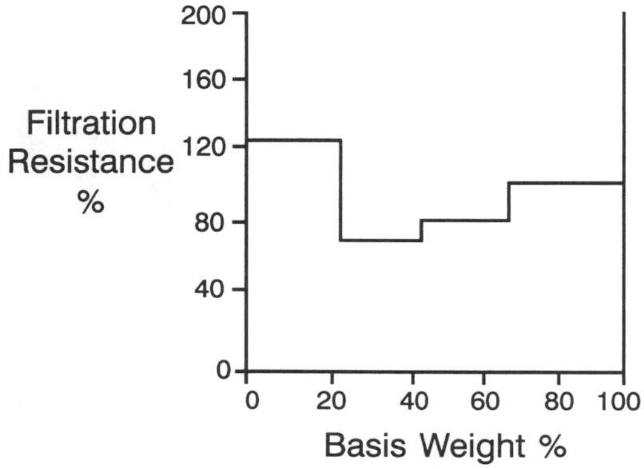

Fig. 224. Fines distribution.

Compared to roll formers, blade formers tend to have:

– Better formation.
– Higher drive requirements.
– Lower retention.

There appear to be some improvements which can be made by additional work on the number and type of blades. It has been pointed out that "twin-wire forming with pulsating dewatering pressure is so far based on empirical results. The need for basic research in this area is extensive and the potential for improvement may be large" [25].

H. HYBRID OR PREFORMERS

The retrofit formers became popular because they were a relatively low cost approach to improve two-sidedness, reduce linting, improve formation and remove the wet end as the production and quality bottleneck. This development was instrumental in expediting the North American switch in newsprint production to all "twin-wire made paper".

The retention, dewatering and formation principles of most retrofit twin-wire formers are combinations of those of a fourdrinier followed by those of a blade or roll former. They will not be reviewed here, but typical results will be discussed. Figure 225 shows a modern retrofit roll former [29]. Figure 226 shows a water split diagram for this former [29], and Fig. 227 shows the before and after ash distribution [29]. Note that the water split diagrams and ash distributions are very similar to those achieved on some gap formers.

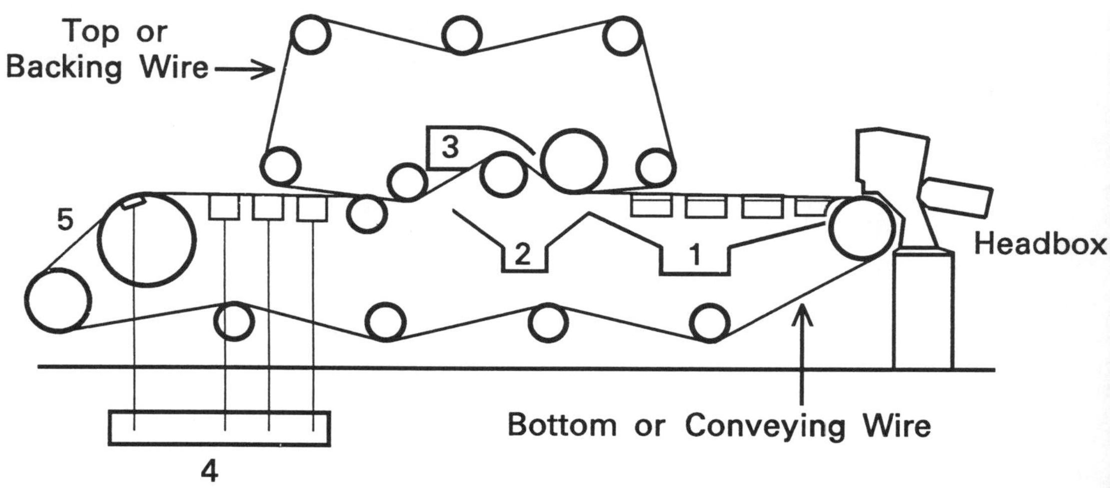

Fig. 225. Modern retrofit roll former (Sulzer Escher Wyss Corp.).

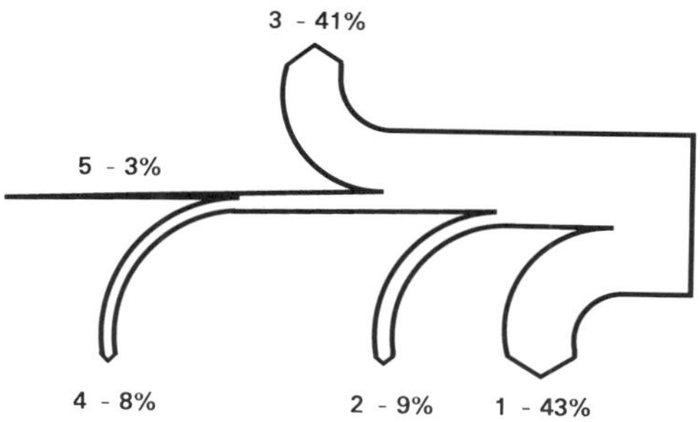

Fig. 226. Corresponding drainage split.

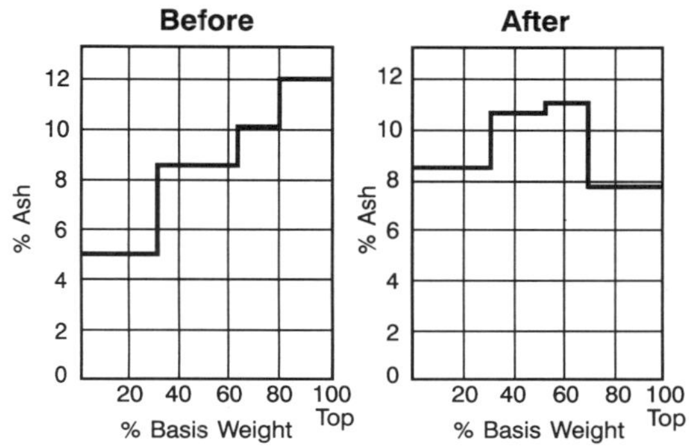

Fig. 227. Ash distribution before and after (60 g/m^2).

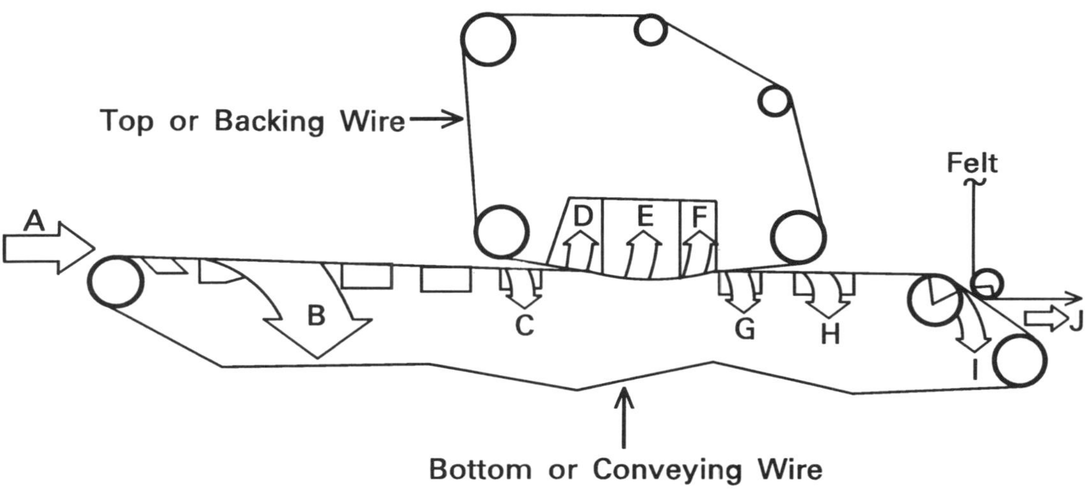

Fig. 228. Modern retrofit blade former (Beloit Corp.).

Figure 228 shows a modern retrofit blade former. This is a schematic of Fig. 202. Figure 229 shows a corresponding drainage split diagram. Figure 230 shows a typical before and after ash distribution for this former.

The ash distribution of the former in Fig. 228 is quite different than the ash distribution of the former in Fig. 225. The difference is characteristic of roll versus blade forming.

The retrofits also gave management and operators a method for trying twin-wire forming on other grades with a fail-safe condition. The retrofits were more successful than predicted by most. They gave reduced streaking, improved formation, improved two-sidedness, reduced linting and retention comparable to a fourdrinier. They provided insight into some basic papermaking principles [30].

The sheet forming process on the fourdrinier is a stepwise filtration process. A partially formed web has a high consistency near the wire and remains close to headbox consistency on the top of the sheet. In the absence of high table turbulence (most machines), the reflocculation of the stock near the top surface is probably more severe and causes the papermaker the greatest problem. A feature common to most of the

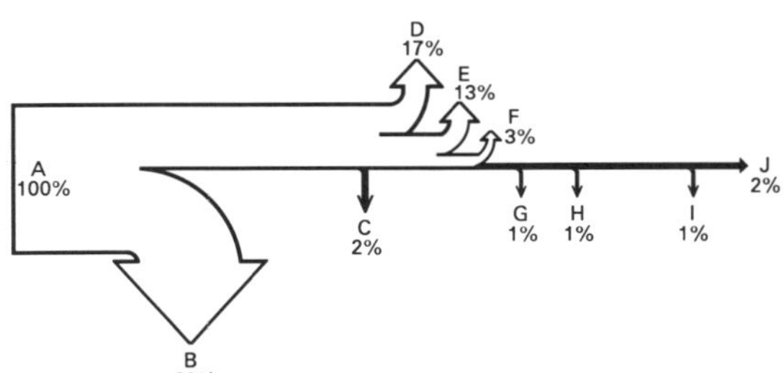

Fig. 229. Drainage split diagram (Beloit Corp.).

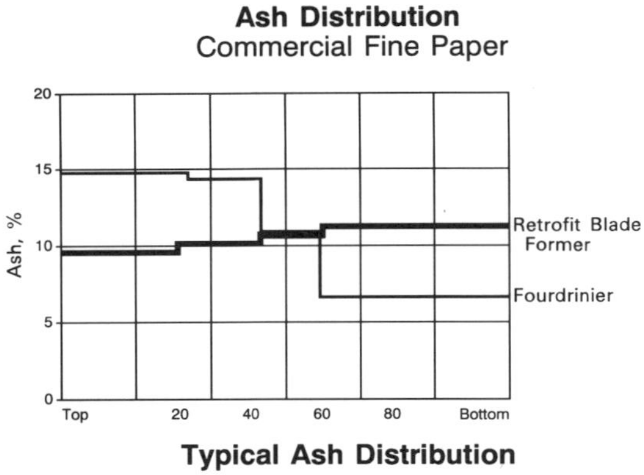

Fig. 230. Typical ash distribution (Beloit Corp.).

preformers is high shear created at the nip where the top wire is introduced. This shear is very long in the MD (like Fig. 211) as the sandwich wraps the "forming roll". The shear is also more effective on the top side where floccing is more severe. This is shown in Fig. 231 which also illustrates why conditions at this nip are critical [31].

Another important feature is that resistance to drainage of the stock at the top of the sheet is comparable to that of the bottom of the sheet over the forming board and foil beams because it is near headbox consistency. This may help to explain why retention on the "top wire" is close to that for the "bottom wire" on a typical preformer.

Headbox streaks are eliminated or reduced because the stock has an opportunity to level out and be "worked" on the forming table. Ongoing investigation is indicating that the headbox is much more of a bottleneck and limitation than previously thought [11]. This is true for even the most modern headboxes. The preformer is simply a way to compensate for <u>some</u> of these deficiencies. In fact, the superiority of the preformers indicates that current headboxes are not good enough for the job at hand. Some preformers have virtually eliminated headbox streaking and very low retention problems. Preformers seem to give most of the benefits of pure twin-wire formers while incorporating few of the negative features. They may have speed limitations due to "open-wire" papermaking. Also, they may preclude high-speed stratification.

Compared to the gap formers discussed so far, preformers have:

- Much less streaking.
- Equal or better formation at improved first pass retention.
- Fewer fabric and unit cleaning problems.

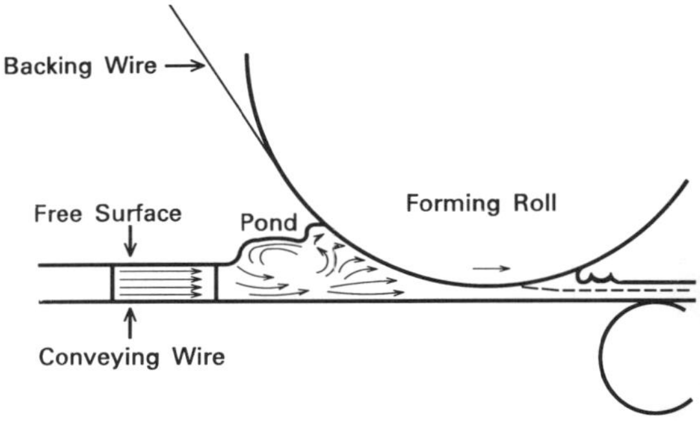

Fig. 231. Shear at top wire forming roll.

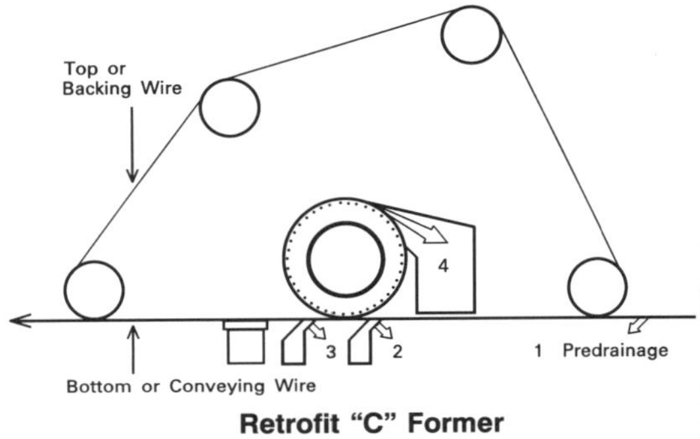

Fig. 232. Retrofit "C" former (Black Clawson).

- An affordable installed cost.
- A vastly improved drainage capability.

While much progress has been made, there are a number of current problems/opportunities. Twin-wire papermaking now dominates lightweight, high-speed grades. But most twin wire formers will not operate at speeds below 1500 fpm, or are too expensive for the older, slower machines.

In the late Eighties, the former shown in Fig. 232 was developed. This unit has the commercial name "C" former, which is mentioned only to further illustrate our communication problem. This is obviously not a gap roll former with a "C" wrap in the dewatering zone like Fig. 212. This former has blade forming on both sides, and a modest degree of roll forming. The preformed sheet enters the nip of two wires, is dewatered by a lower blade, an open roll, and further dewatered and deflocced by an upper and a lower blade. The top blade has an extension that curves around the open roll to allow the former to be run at very low speeds, projected to be 122 m/m (400 ft/min). At low speeds, the open roll and shield act like a "paddle wheel" pushing the removed water around to the catch pan. The unit needs to be driven but it is compact and light so it can fit on top of most fourdriniers with a minimum of structural change. More importantly, it offers the papermaker a great deal of flexibility. The blades and the roll can be adjusted on the run and the whole unit can be raised if some grades require fourdrinier forming. Other configurations of formers are evolving to cover the wide range of papermaking needs.

Figure 233 shows a drainage split diagram for this type of former.

Another problem faced by the papermaker was the decreasing number of adjustments or "command variables". For all of its limitations, "open-wire" papermaking offered a wide variety of variables and quick shutdown changes including forming board adjustments, foil angle

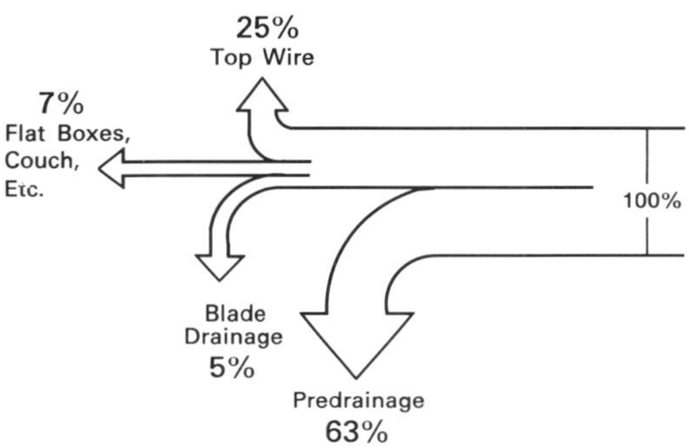

Fig. 233. Drainage split diagram (Black Clawson).

and spacing, level of vacuum, formation showers, Sheraton rolls, Wunderfoils, and even serrated slices.

Some twin-wire machines reduced these choices to jet impingement and wire tension. The preformers limited choices due to the critical sheet consistency requirements entering the top wire unit. In addition to Fig. 232, the machinery builders have responded with formers which have new adjustments.

The former shown in Fig. 234 has formation improvement as one of its functions and has been primarily used on heavyweight board grades. Pilot trials have shown the potential for higher speed grades [32]. The drainage force on this unit is a function of externally applied pressure and is independent of speed, grade, etc. This unit provides the papermaker with another command variable for optimization and responding to grade, furnish, and marketplace changes.

Figure 235 shows a drainage split diagram for this type of former [32]. Figure 236 shows another "on top" unit which achieves the result of adjustable drainage pulses with a different approach [33]. Figure 237 shows a schematic of the former, while Fig. 238 shows a drainage split diagram typical for this type of former.

The unit in Fig. 236 is unique in that there is no downward water removal. The top wire run contains a multi-compartment inverted vacuum blade box with a lead-in auto slice. This box has

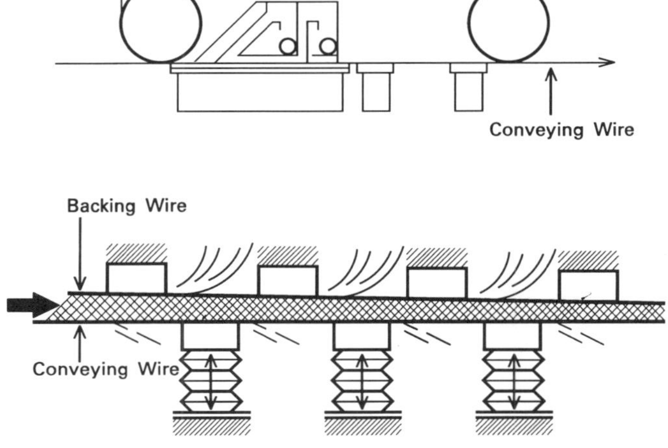

Fig. 234. Top wire unit with adjustable features (Voith Corp.).

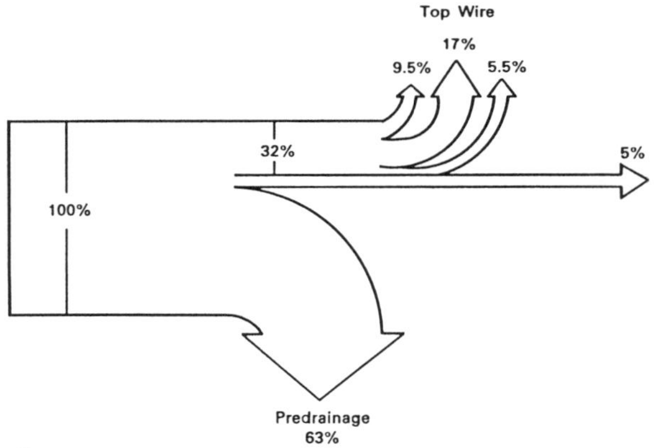

Fig. 235. Drainage split diagram.

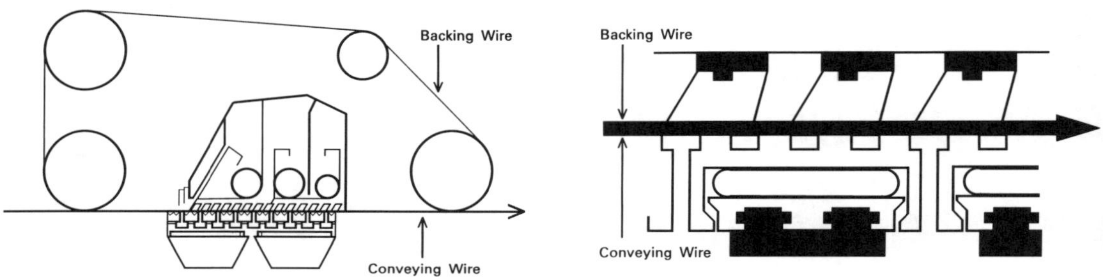

Fig. 236. Top wire unit with upward dewatering and adjustable features (Valmet Corp.).

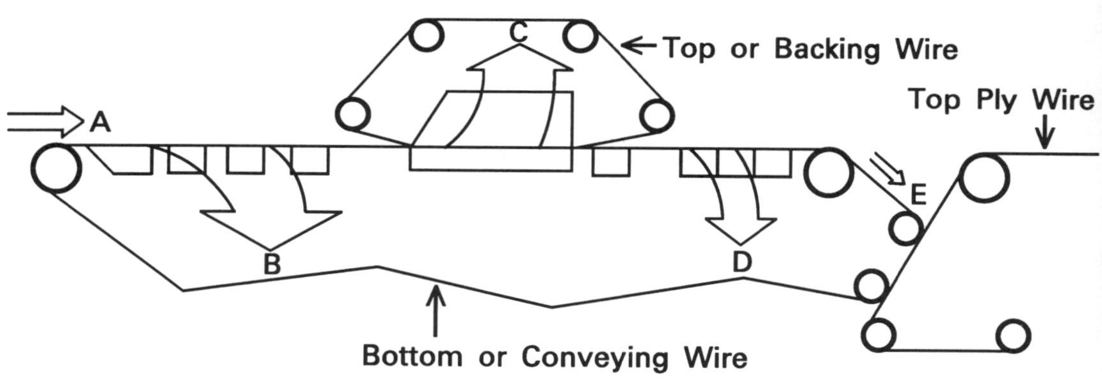

Fig. 237. Top wire unit with upward dewatering and adjustment (Valmet Corp.).

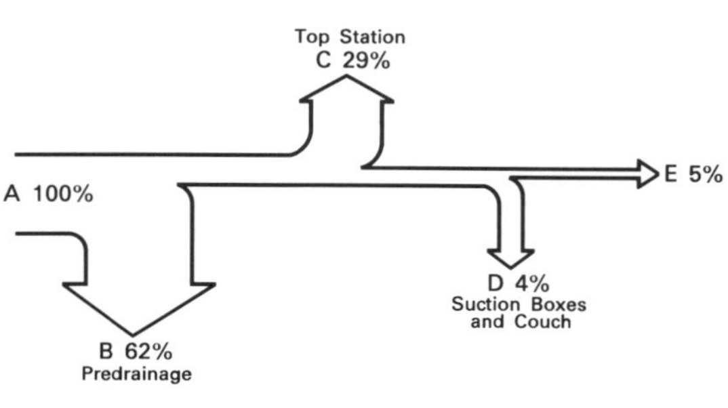

◀ **Fig. 238. Drainage split diagram (Valmet Corp.).**

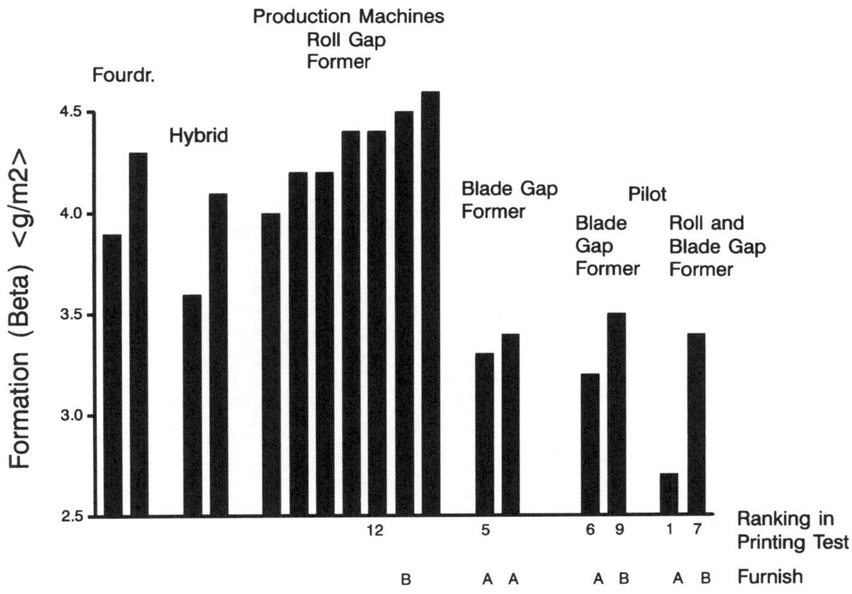

Fig. 239. Formation by former type.

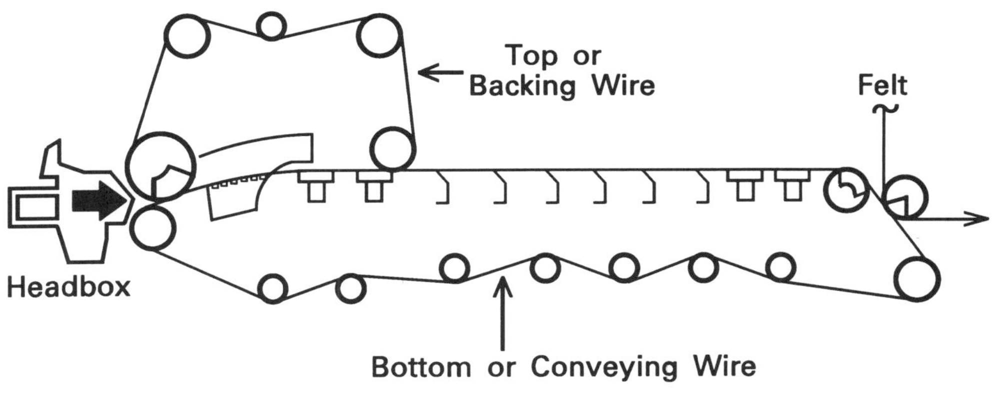

Fig. 240. Horizontal gap blade-roll former (Valmet Corp.).

blades which "mate" to surfaces in the lower box and it is adjustable vertically. The bottom wire run contains multiple cross-machine loading units each individually controllable. These cross-machine loading units have flexible tops about 6 inches in the MD with flat-topped ribs. Mating these surfaces together causes pressure pulses and upward dewatering only. This improves formation and two-sidedness because most of the prior water removal has been downward. These units are used extensively on hard-to-form heavyweight board plys in Scandinavia.

I. GAP ROLL/BLADE FORMING

There have also been problems at the high end of the paper grade speed range. It has been observed that, as speed increases, formation (and sometimes retention) gradually deteriorates [34]. Further, only the blade formers were providing reasonable formation and then at lower than desirable retention and print quality.

Figure 239 shows selected formation data from production and pilot machines. The original figures and the remainder of the article [35] contained significant data on retention, ink

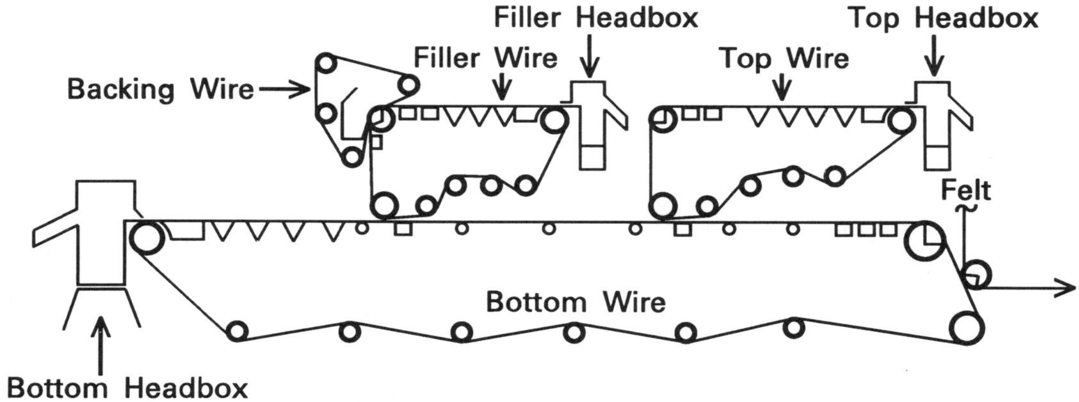

Fig. 241. Multiple wire forming.

absorbency, two-sidedness, and porosity, etc. The readers are urged to refer to this article. The data show that for certain furnishes the formation (and print quality) can be improved by the former shown in Fig. 240 (the vertical version is shown in Fig. 200).

Commercial installations on groundwood grades are giving good high-speed formation, retention and print results [36]. It is likely that formers of this general type will push the speed limit upwards, but probably not too far, especially on free sheets.

It appears that other formers and perhaps other developments will be required for truly high-speed operation which is something like 1524 m/min (5000 ft/min) on free sheets, 2438 m/min (8000 ft/min) on groundwood grades and 3048 m/min (10,000 ft/min) on tissue.

J. TWO-WIRE AND/OR MULTIPLE FORMERS

Historically, multiple-board grades were made on cylinder machines or combinations of fourdriniers and cylinder vats (e.g., Tako in Tampere, Finland – 1932). Cylinder vats were versatile in handling a wide range of furnishes, but gave problems with speed, formation, and plybond. The Scandinavian boardmakers were evolving formers on a commercial basis and, in 1938, at least one mill achieved boardmaking on multiple fourdriniers [37]. These fourdriniers were not connected and often were on separate floors.

These developments led to multiple fourdrinier arrangements as is shown in Fig. 241 [7]. This configuration is more typical for linerboard grades. When fourdriniers are used there is the potential problem of bonding of the "fines rich" top side of one web to the "fines poor" wire side of another web. This event will occur one or more times when three or more fourdriniers are used. It is depicted in Fig. 242 [7].

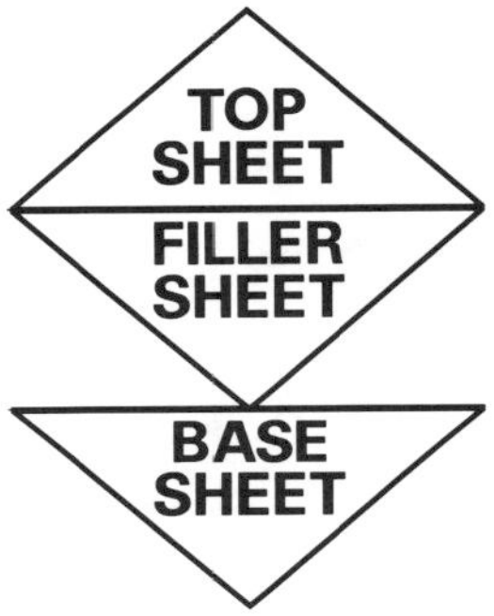

Fig. 242. Potential fines distribution.

Bonding problems are likely to occur at the interface between the filler sheet and the base sheet unless careful attention is given to furnish and former selection. Different former configurations can result in different "wet web" consolidations. With three or more fourdriniers there will always be at least one top side to wire side bond.

A second three-ply configuration is shown in Fig. 243 [7]. This is referred to as a "folded fourdrinier" configuration and is becoming standard for folding boxboard grades in Scandinavia. The middle fourdrinier often has multiple dandy rolls or a formation-improving top wire unit like those shown in Figs. 234 and 236.

In general, the multiple fourdrinier approach is selected to maximize formation and to minimize macro non-uniformity. The furnish and operating conditions must be selected to achieve good plybond.

Another approach was the Inverform unit developed by St. Annes Board and introduced in

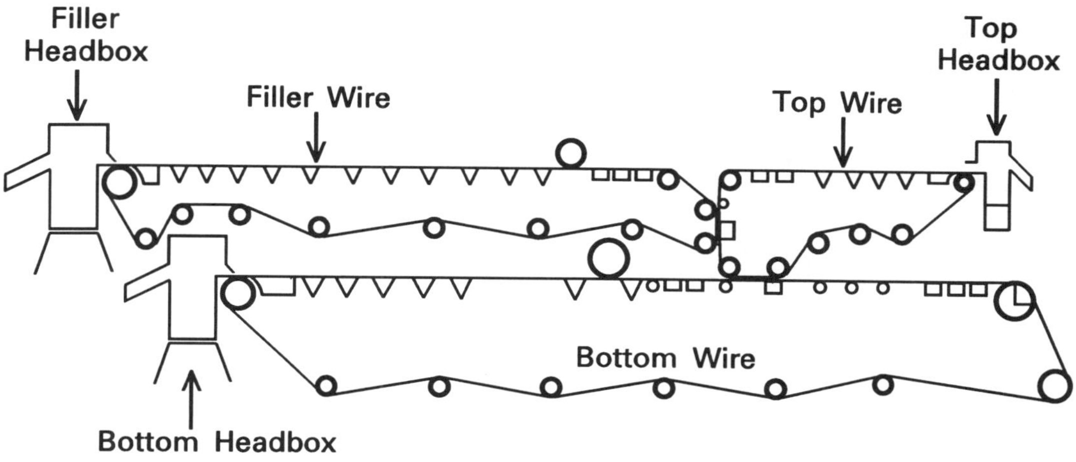

Fig. 243. Folded fourdrinier configuration.

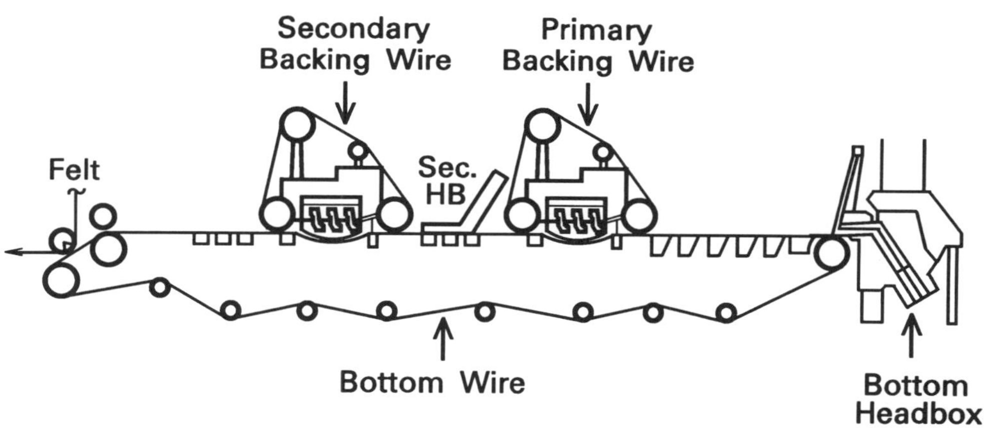

Fig. 244. "Wet on dry" forming (Beloit Corp.).

1957. In some respects this was the first commercial multiple wire unit that was interconnected. This led to the development of a series of top-wire units each with their own headbox mounted on a single long bottom wire as shown in Fig. 244 [38]. This type of former can develop high plybond because of the "wet on dry" application. Some micro non-uniformities are minimized because the ply-to-ply conformance is so complete that surface non-uniformities and defects such as air holes are corrected.

REFERENCES

1. Olding, F.J., British Patent 2477 (July 9, 1875).
2. Publishers Paper Oregon City, OR (Private correspondence, Dec. 1983).
3. Thorp, B.A., "Twin-Wire Forming-Rapid Commercialism Ahead", Proc. TAPPI Wet End Operations Short Course (May 5–10, 1985).
4. Mardon, J. and Vyse, R., "How to Choose a Forming Section", Paperi ja Puu 11:725 (1981).
5. Paulapuro, H., "Aspects of Choosing a Paper Former", Internal Jaakko Pöyry correspondence.
6. Private correspondence with Alex Malashenko, Dominion Engineering.
7. Thorp, B.A., "Fundamental and Commercial Overview of Twin Wire and Multiple Wire Forming", Proc. TAPPI Twin Wire Seminar, Washington, DC (April 12–14, 1989).
8. Norman, B. and Wahren, D., "The Measurement of Mass Distribution in Paper Sheets using a Beta Radiographic Method", Svensk Papperstidn. 77(11):397 (1974).
9. Tomimasu, H., Luner, P. and Hanna, R., "Rapid Imaging of Mass Distribution in Paper by Electron Beams", Papermaking Raw Material, Eighth Fundamental Res. Symp. Oxford, England (Sept. 15–20, 1985).
10. Takeuchi, N., Senda, S., Namba, K., Kuwabara, G., "Formation and Destruction of Fiber Flocs in a Flowing Pulp Suspension", APPITA, 37(3) (Nov. 1983).
11. Kallmes, O.J. and Thorp, B.A., "Back to Basics in Headbox Design and Application", Proc. TAPPI Papermakers Conf. (Apr. 2–4, 1984).
12. Thorp, B.A., "The Two-Sides of Sheet Formation", Pima Magazine (Feb. 1985).
13. Norman, B., "Overview of the Physics of Forming", Fundamentals of Papermaking Ninth Fundamental Res. Symp., Cambridge, England (Sept. 16–22, 1989).
14. Smith, D.T., Miller, D.C. and Murdock, L.D., "Retention Calculation Impact on Flow Rates in Tissue Operations", Proc. TAPPI Tissue Seminar (Apr. 25–27, 1990).
15. Cutshall, K., "Nature of Paper Variation", unpublished paper.
16. Collins, S., "The Response of the Tissue Making Process to Forming Zone Configuration", Proc. TAPPI Tissue Seminar (Apr. 25–27, 1990).
17. Mardon, J., Monahan, R.E., Nelson, J., Meadley, C.K. and Effa, R.D., Norsk Skogindustri 20(10):390–401 (1966).
18. Baines, W.D., Pulp Paper Mag. Can. 68(10):T497–T538 (1967).
19. Kokimie, J., Perkinen, J., Puolakka, H., Schultz, E. and Wahlstrom, B., Paperi ja Puu 54(4a):1–10 (1972).
20. Thorp, B.A. and Barasch, M., "Water Removal Principles on Roll-type, Top Wire Formers", Tappi J. 68(4) (Apr. 1985).
21. Neubert, W. "Duoformer® Review", Proc. TAPPI Twin Wire Seminar, Atlanta, GA (Apr. 8–10, 1987).
22. Wahren, D., Stenberg, G. and Zotterman, C., "Symmetrical Dewatering in Webster-typed Former", Pulp Paper Can. 79(9) (Sept. 1978).
23. Nash, J., "Twin Wire Forming Development", Proc. TAPPI Twin Wire Seminar, Washington, DC (Apr. 12–14, 1989).
24. Dudley, W.L., "An Analysis of Papermachine Clothing Performance on Profits", Proc. TAPPI Papermakers Conf. (Mar. 1975).
25. Norman, B., Neus Entwicklungen in der Zellstoff und Papierherstellung, 17 EUCEPA Konferenz, Wien, Austria, II-23e (Oct. 10–14, 1977).
26. Nash, J.C., "Twin-Wire Forming Development", Proc. TAPPI Twin-Wire Seminar (Apr. 12–14, 1989).
27. Murdock, L.D., "Computer and High-Speed Photography Application", Paper Trade J. (Apr. 1982).
28. Brauns, B., "Beloit Top Wire Formers", Proc. TAPPI Engineering Conf. (1983).
29. Heissenberger, O.L., "Escher Wyss Twin Wire Formers", Proc. TAPPI Twin Wire Conf., Atlanta, GA (Apr. 8–10, 1987).
30. Kallmes, O.J., M/K Systems, Danvers, MA (Private correspondence).
31. Parker, J., "The Sheet-Forming Process", TAPPI STAP, No. 9 (1972).
32. Baumann, W.D., "Duoformer-D–A New Approach to Top Wire Forming", Proc. TAPPI Twin-Wire Seminar (Apr. 12–14, 1989).
33. Case, P., "The Alform-B Three Wire Former and Alform MB Top Former", Proc. TAPPI Twin-Wire Seminar (Apr. 12–14, 1989).
34. Manson, D., James River Corp. (Private correspondence).
35. Ackley, L.F., "Sym-Former: Versatile Forming", Proc. TAPPI Twin-Wire Seminar (Apr. 8–10, 1987).
36. Schneider, M.W., "Valmet Horizontal High Speed Former Paper Mill Operating Experience", Proc. TAPPI Twin-Wire Seminar (Apr. 12–14, 1989).
37. Ahlskog, B., Tako Board Mill, Metsa-Serta Group, Tampere, Finland (Private communication).
38. Reed, F., "Lightweight Linerboard Machine Expansion and Modernization", Proc. TAPPI Papermakers Conf. (Apr. 1986).

X
MULTI-PLY FORMING

B.W. ATTWOOD

Brian Attwood received his technical education at Bristol College of Technology (U.K.) and is a Fellow of the Royal Society of Chemistry. He has spent his working life mainly in the paperboard industry and is currently a consultant working mainly in the field of water removal at presses and dryers, forming and multi-ply technology. His work experience has included various studies appertaining to web forming, pressing and drying and he is the holder of numerous patents. He is a past-chairman of PITA (U.K.) and a TAPPI Fellow. For his contributions to paper technology, he has been awarded the Silver Jubilee Medal and the Gold Award of PITA and the Paper and Board Division Award of TAPPI.

REVIEWED BY

Don Baade
Huyck Felt Company
Wake Forest, NC

Bjorn T. Blomquist
Vice President, Product Technology
Valmet Paper
Karhula, Finland

Multi-layer structured webs until recently have been associated mainly with heavy basis weight materials such as boxboard. Today, other materials, ranging from tissues to printings, are being manufactured with a multi-ply structure. In this chapter the history of multi-ply forming technology will be outlined and the reasons for multi-ply forming discussed. The types of equipment used will be explained in detail, together with information on the range of products made using multi-ply forming technology.

A. HISTORY

An outline of the history of multi-ply forming is shown in Table 45.

The Dickinson cylinder mold machine was invented in 1808 for the manufacture of paper (single ply) and consisted of a suction-forming cylinder and an associated headbox. By 1830, Dickinson had converted his single-ply papermaking machine into a two-ply machine by using two forming cylinders operating together with a common making felt – the basis of the multi-ply cylinder mold board machine of today.

In 1863, J.F. Jones of Rochester, U.S.A., patented the design of a multi-cylinder machine using no fewer than seven vats. The machine described in Jones' patent incorporated vacuum pumps "in order to draw air out of the forming cylinders and to assist the deposition mechanism of the fibers onto the surface of the mould". Instead of using a common making felt, as described in the Dickinson patent, Jones used a separate making felt for each mold stating that "by this method seven separate webs of paper could be produced at the same time".

In 1875, a multi-ply paper machine was described by Ibbotson based upon the use of two separate fourdrinier sections. Incorporated in the machine were two large drying cylinders and these cylinders were mounted on movable bearings so that they could be brought together. Each cylinder was associated with an individual ply and when the cylinders were in contact with each other the two separate plies were combined.

Ibbotson also described a three-ply version in which under one of the drying cylinders was positioned a cylinder mold. The web formed on the cylinder mold was transferred onto the already formed paper web from the fourdrinier wire part. The combined (two-ply) web was then brought into contact with the paper web that had been formed on the other fourdrinier section resulting in a three-ply web. In the patent is also described how the combining of the individual webs could take place at the couch rolls rather than at the drying cylinders.

Table 45. History of multi-ply forming.

50 YEARS	1800–	Development of fourdrinier machine (U.K.). Development of cylinder forming machine. (including multi-ply trials) U.K.
50 YEARS	1850–	Multi-ply (multi-ply cylinder) board machine. Two-wire machine 1875 (U.K.).
50 YEARS	1900–	Millspaugh multi-ply vacuum former machine (U.S.A.). Multiwire and combination machines (Europe). Secondary headbox machines (U.S.A.).
10 YEARS	1950–	Sandy Hill rotary formers (U.S.A.). Inverform (twin-wire) multi-ply machine (U.K.). Stevens former (U.S.A.). First commercial Inverform machine (U.K.).
10 YEARS	1960–	Pilot cylinder mold studies (Muller Rid, Germany). Various cylinder developments (dry vat, half-vat, etc.). Various multi-wire machine developments (e.g., Voith, Europe). Ultra formers (Kobayashi, Japan). Tampella formers (Finland). Manchester formers (U.S.A.).
5 YEARS	1970–	Evolution of "on-top" formers (Europe/U.S.A.). Dry forming of multi-ply structures (pilot plant, Europe). Belvac former (U.S.A.). B.R.D.A./St. Annes formers and headboxes (U.S.A./U.K.). Japanese "up-former" (Japan).
5 YEARS	1975–	Bel-Bond former (U.S.A.; Australia). High-consistency forming; multi-ply (Sweden). Multi-layer headboxes (Strata Flow Beloit, U.S.A.). Wet/Dry concept (St. Annes, U.K.). Contra Flow former (Tampella/Finland). Multi-layer headbox (K.M.W., Sweden).
8 YEARS	1980–	Coanda headbox; multi-ply application (Wiggins Teape, U.K.). Multi-channel headbox tissue machine (Commercial). Multi-channel headbox paper machine (Commercial, Sweden). Multi-ply food board machine with high-consistency middle ply (Finland). Many multi-fourdrinier board machines commissioned (worldwide).
	1986	

In 1884, Escher Wyss and Co. produced a two-ply machine consisting of a main fourdrinier wire part with a second fourdrinier wire part located above the couch press, a design similar to that being used today for the manufacture of a wide range of grades such as test liner, boxboard and wall paper.

The period from then up to the late Thirties was mainly one of "consolidation" of existing techniques. However, there were some significant developments; in Europe multi-fourdrinier machines were developed together with combination machines (multi-ply layers produced on one machine by the use of a combination of cylinder formers and fourdrinier wire parts).

During this period, Millspaugh in the U.S.A. was developing a number of multi-ply forming sections using, as the base for forming, suction roll formers each with its own flow box. With these arrangements many different grades of multi-ply paper and board were produced. In the late Thirties in the U.S.A., the important concept of producing a multi-ply web from a single fourdrinier wire part by the use of a secondary headbox was developed.

Many significant developments took place in the Fifties, for example, the Sandy Hill and Stevens cylinder formers and the concept of twin wire forming. The first commercial application of twin-wire forming was the multi-ply paperboard Inverform machine. These developments stimulated further studies covering cylinder formers, "on-top" formers, multi-layer headboxes etc.

Until recent times, multi-ply structuring has been used mainly for the production of heavy basis weight materials, with the exception in Europe of the manufacture of high-quality and speciality papers using the two wire concept. Multi-ply techniques, however, are now being used to manufacture an ever-increasing range of papers such as tissue and printing grades.

B. FUNDAMENTALS OF MULTI-PLY FORMING

The main reasons for producing multi-ply structures are as follows:

1. To enable high basis weight webs to be produced at speed;
2. Improved fiber economy – by selected distribution of fiber types through the web (cheaper fibers for centre layers);
3. Production of webs with improved characteristics by using the correct selection of furnish for the individual layers;

and in general terms multi-ply webs are produced by:

1. Separately forming the individual layers and then combining them together;
2. Onto an existing layer or layers forming further layers;
3. By using multi-channel (stratified headboxes); or
4. A combination of the above.

C. THE STRUCTURED WEB

1. Improved fiber economy

Peel and Bergstrom [1] explained that in a multi-ply structure total basis weight can be reduced, without impairing web characteristics, by the selective distribution of the fiber furnish. For example, it is possible to produce a multi-ply sheet having a bulky centre with denser and stronger outer layers so that the resulting web has greatly improved bending stiffness. As a consequence, the stiffness of a single-layer sheet can be matched by a structured sheet having a considerably lower total basis weight, hence a substantial fiber saving.

Another method of improving fiber economy is to use a lower cost fiber for the inner plies without adversely affecting sheet properties or sheet appearance. This last technique has been associated mainly with the use of secondary fibers for the middle layer.

2. Improved sheet characteristics

An important reason for producing structured webs is in order to obtain improved sheet properties. The sheet properties can be improved by using a furnish for the outer layers that best meets the requirements for the surface of the sheet, while the inner part of the sheet can be made from a furnish that meets other needs. As an example, the strength properties needed for good runnability of printing papers may be obtained with a long-fibered middle layer whereas shorter fibers may be used for the outer layers, thus providing smoothness and suitable absorptive properties necessary for printing.

Table 46. Contribution of individual plies to the total stiffness.

Layer	% Contribution
Top Liner	38.5
Under Liner	4.3
Filler (Middles)	0.3
Filler (Middles)	0.5
Filler (Middles)	3.1
Filler (Middles)	8.3
Back	45.0

Stiffness is an important property for many paperboard grades. With a multi-ply structure, each ply contributes to stiffness in proportion to the square of its distance from the center of the sheet and its own tensile stiffness. Consequently, the total board stiffness is essentially a function only of the outer ply stiffness values. The center section of the sheet gives bulk and acts as a spacer, thus increasing the contribution of the top and backs liners. This is illustrated in Table 46 which shows the contribution of each ply to the overall stiffness of a typical 7-ply board [2].

Peel and Bergstrom [1] and Seppala and Pöyry [3] have reviewed other aspects of web improvement obtained through multi-ply structuring.

D. MULTI-PLY FORMING EQUIPMENT

Before describing the various forming methods, stock preparation and fiber feed will be considered.

Because of the nature of a structured web and the multi-ply method of forming individual plies, a multi-ply paperboard machine must have a separate stock preparation and stock feed system for each ply. In multi-ply technology, the methods used for stock preparation are similar to those used in paper making.

As a consequence of the above, the paperboard machine beater/refiner house is considerably larger, and more complex, than that associated with a paper machine, each furnish having its own preparation equipment and its own associated storage chests. For example, the liner, underliner, middles and backs may all require individual systems. Often a wide range of products are made on the same board machine and as a consequence its stock preparation plant must be arranged to be extremely flexible.

E. STOCK DELIVERY

The stock delivery system on a multi-paperboard machine receives the stock slurry at approximately 3% consistency after the final defibration stage and delivers this stock in a diluted state to the forming section. If the

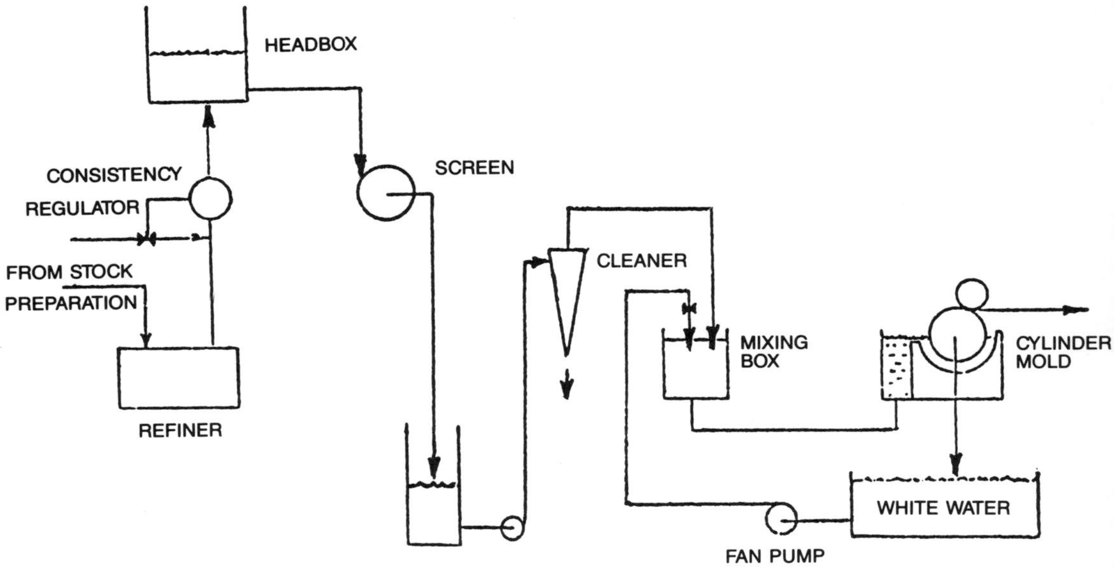

Fig. 245. A cylinder mold stock delivery system.

forming section should have a typical cylinder mold configuration, the consistency is in the order of 0.2% to 0.6%, whereas if the forming section is of the Inverform type it might range from 0.5% for the liner stock up to 2% for the middles plies.

It is in the stock delivery system that controlled dilution, consistency regulation, screening, cleaning and metering take place. Figure 245 shows a typical stock delivery system for a single cylinder mold. A system such as this must be repeated for each layer or ply being manufactured.

The purpose of the consistency regulator is to regulate the white water addition used in diluting the raw stock to a suitable supply box consistency, which is usually in the order of 1.5% to 2%.

The machine supply box, sometimes known as the stuff box, is used to regulate the fiber flow to the individual forming units, its main function being a routing point to the desired vat system. It must provide the flexibility to run different grades of, say, 1, 2, 3 or 4 different stocks to their respective forming sections in order to make up the required multi-ply paperboard structure. Figure 246 illustrates, in diagrammatic form, the layout of a typical cylinder board machine supply box capable of handling a four-stock system. The actual design of the box will differ from mill to mill. For example, some are quite compact units constructed of stainless steel while others can extend over a large area and be of simple wooden construction. Built into the design of the headbox are facilities to minimize foaming problems, the buildup of slime and the accumulation of thickened stock.

It will be appreciated from the diagram that, as well as supplying an individual forming unit with a specific furnish, it is also possible to blend more than one basic furnish being fed to a given forming unit; a chemical pulp liner stock may be blended with a mechanical pulp stock to give a mixture suitable for feeding to the backs former in order to produce a strengthened backs ply.

At the mixing box, final dilution to the required consistency for forming takes place. In modern installations there is a tendency for the mixing box to be eliminated from the system and the feed to the forming unit is often similar to the feed to the headbox of a fourdrinier machine.

F. FORMING EQUIPMENT

Until recent times, the most important multi-ply process was that using molds. Today there are numerous other ways of forming continuously a multi-layer web. For example, the use of a number of separate fourdrinier forming sections arranged in such a way that the webs are combined while still in the moist state, or by the use of twin-wire forming devices such as Inverform/Bel-Bond.

G. MULTI-PLY FORMING METHODS

1. Those based upon the use of cylindrical forming screens:
   a) Cylinder mold machines, Uniflow, Contra Flow and dry vat;

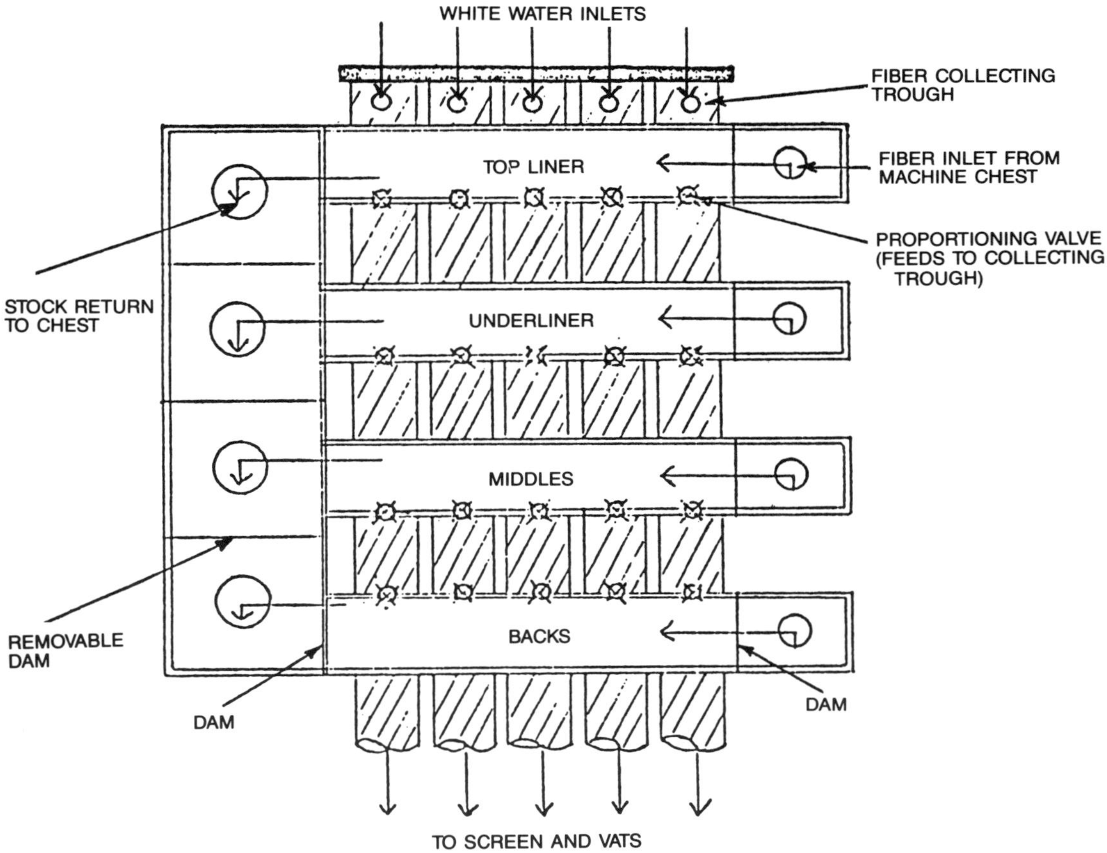

Fig. 246. Cylinder mold stock distribution box.

   b) Suction formers;
   c) Pressure formers.
2. Fourdrinier machines:
   a) Fourdrinier with secondary headboxes;
   b) Fourdrinier with "on-top" formers;
   c) Multi-fourdrinier machines.
3. Twin-wire forming.
4. Miscellaneous, including air laid and high consistency.
5. Multi-channel headboxes.

I. The cylinder mold machine

The principle of web formation on a cylinder mold is as follows. A horizontal cylinder (cylinder mold) having a wire cloth surface is arranged to rotate approximately three quarters submerged in a container (vat) of paper stock so that a small arc of its circumference is above stock level. Water associated with the fibrous suspension drains through the wire cloth with the result that a layer of fibers is deposited on the surface. Drainage takes place because of a difference in level between the stock in the vat and the back water inside the mold. The difference is known as the making head.

It is arranged that a moving felt (mold felt/making felt) is pressed, by means of a roll (couch roll), into contact with the cylinder at approximately the top position. By doing this, the layer of fibers that has formed on the wire screen is transferred to the mold felt which moves away from the forming screen with it. The web having been transferred to the bare wire is washed by means of sprays and re-enters the fiber stock where the fiber layer is reformed.

If a number of these forming units are placed in series, then a multi-ply layer is produced continuously. Each forming unit has its own supply of paper stock and a method for removing the drainage water from its interior so that, in effect, each cylinder mold is a separate web forming machine in itself.

Figure 247 shows a typical Contra Flow vat forming unit where the fiber flow in the vat circle is against the direction of mold rotation. The stock flows into the vat via a distribution arrangement which can be any one of a number of different designs. For example, it can be simply an arrangement of pipes equally spaced across the width of the machine and fed from one main inlet pipe. Other arrangements incorporate taper systems, split manifolds etc.

The stock from the flow distribution arrangement enters the bottom of the vat at A, passes over a weir, then under a baffle, rising again to be fed into the vat circle via wing boards (butterfly) and a making board. The purpose of the wing board is to help correct basis weight levels. When operating a conventional cylinder mold there is nowhere near as much control over basis weight level as there is with a fourdrinier machine and, consequently, the main use of the wing board is, in fact, to offset any general tendency for the web to be lighter or heavier from one side to the other. The making board is where there is the last chance to more evenly distribute the flow of stock across the machine before it enters the vat circle.

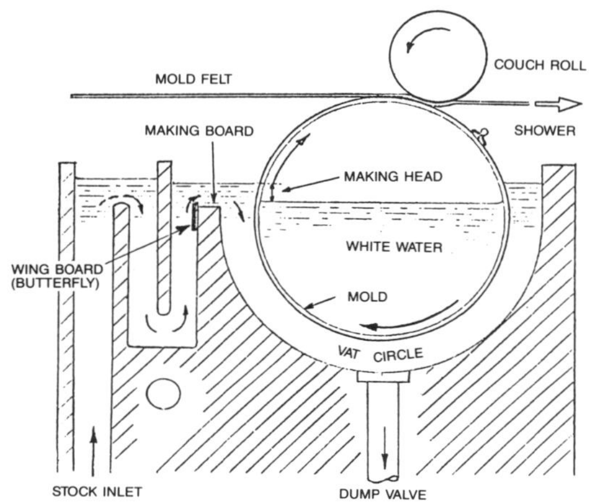

Fig. 247. A Contra Flow vat.

The inside fittings of a vat are often made of wood and the main part of the vat circle is sometimes lined with metal, often stainless steel. Plastic construction is sometimes used. The annular gap between the surface of the mold and the semi-circular section is known as the vat circle. The vat circle has to be carefully constructed and the contour of the circle must be uniform and smooth. Often the cylinder mold is located centrally in the vat and the distance between the surface of the mold and the vat circle varies from machine to machine, ranging from 10 to 15 cm, but in some cases can be as narrow as 5 cm. Some vats are constructed with the mold offset from the center and in this case the inlet gap is wider than the outlet gap.

The cylinder mold itself is of somewhat complex construction, providing the forming surface through which the water drains in order to form the web. The drainage water must be removed uniformly across its width, making sure that no turbulence is created which could disturb the fibers while the web is being formed.

The construction of a typical cylinder mold is

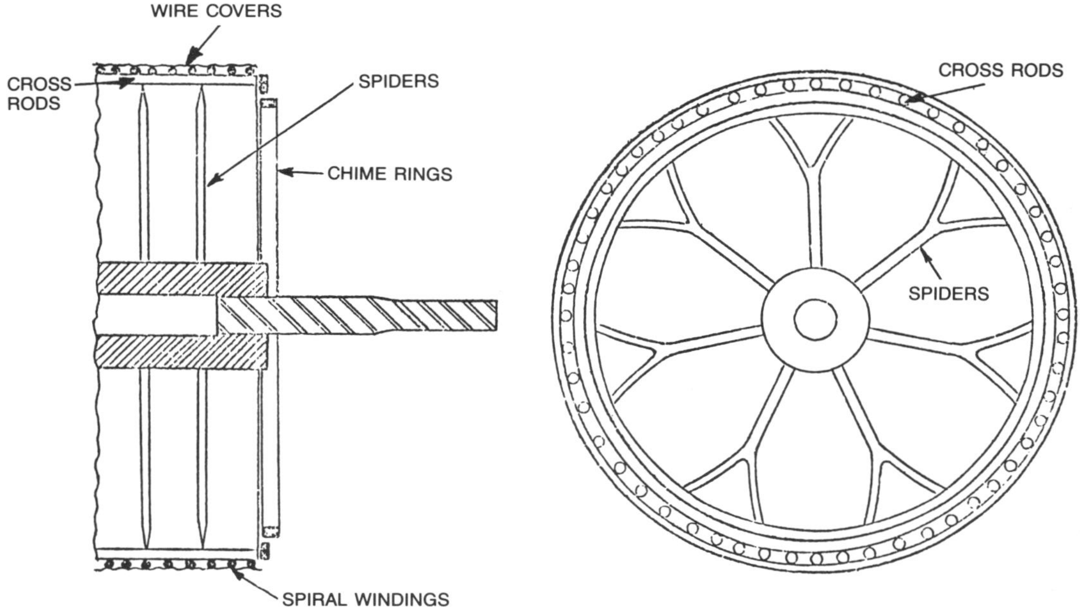

Fig. 248. Typical cylinder mold.

shown in Fig. 248. It is constructed around a cast iron core upon which are secured bronze supporting spokes known as spiders. Spiders can have an approximate spacing of between 10 and 15 cm. The design of the spiders is such that, although they must be strong enough to ensure that the mold is of rigid construction and will not distort, they must be shaped in such a way that they cause minimum disturbance as they rotate in the backwater. The backwater inside the mold must not be violently disturbed as this could upset the fragile web being formed on the mold surface. The spiders support concentric rims, the outside peripherals of which are grooved in order to carry rods, which are approximately 1 cm in diameter and approximately the width of the machine. These rods are on centers of approximately 4 cm and are parallel with the axis of the central shaft. A continuous wire is spirally wound round the cylinder in such a way that there is a gap of about 1 cm between adjacent wires.

On this skeleton is sewn a bronze or stainless steel backing wire having a mesh of approximately six wires to the centimetre (14 mesh). It is over this backing wire that the final wire (forming screen) is stretched and secured. The final assembly must be perfectly round, concentric with the center shaft and rigid enough to withstand pressure from the couch roll without deformation. The diameter of typical cylinder molds ranges from approximately 1 to 1.5 m, 1.2 m being the size most frequently encountered.

Often on machines not required to produce the best quality, the backing wire is eliminated and the forming wire, with a mesh of between 35 and 40, is attached directly to the spiral winding.

The cylinder mold and the vat just described, when put together, become an integral unit and it is necessary to join the mold to the vat by means of a seal. There also exist cylinder molds constructed in such a way as to eliminate the spider struts, resulting in a completely hollow cylinder.

The making head valve can be controlled either by the adjustment of the position of weirs on the front and backside of the molds where the backwater flows out of the mold, or a separate control is used by means of a levelling box situated at the backside of the machine. Figure 249 shows the arrangement for a Uniflow vat and, as can be seen, the basic components are the same as for a Contra Flow vat, but the stock flows with the direction of mold rotation.

The couch roll is usually of simple construction, having a steel pipe body and cast iron roll

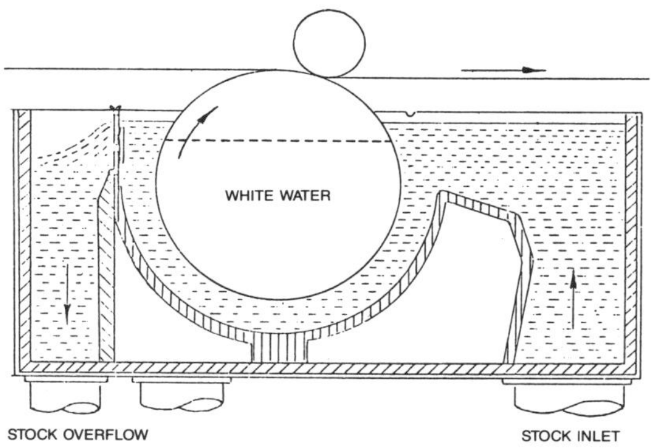

Fig. 249. A Uniflow vat.

ends with pressed-in steel journals. The face of the roll is covered with a thickness of approximately 3–5 cm of soft rubber having a plasticity of approximately 200 P.&.J. Couch roll diameters can range from 40–50 cm, depending on machine width. Various systems are used for weighting the couch roll onto the mold, such as mechanical leverage or a pneumatic system. Occasionally suction couch rolls are used. The pressure applied at the couch roll to the wet web is relatively low, being in the order of 2–8 kg/cm. The couch roll is offset away from the center of the roll towards the direction of felt movement and it is usual for the amount of couching to be approximately a quarter of the diameter of the mold. With this arrangement there is sufficient contact between mold surface and mold felt to ensure adequate drive of the mold through the felt without any slippage. At the same time an initial light pressure is applied to the wet web on the mold surface which brings about some dewatering and allows a reasonable pressure to be applied at the actual couch roll nip without disrupting the surface of the web. Associated with the couch roll are devices to remove buildup of water at the couch roll nips. These devices are known as cylinder slices.

Individual cylinder mold forming units can be arranged in series in order to produce a multi-ply web as shown in Fig. 250(a) and (b). It is usual for the mold felt to pull around the entire forming section, although occasionally individual molds are fitted with separate drives.

A better formation is obtained with a Uniflow vat than with a Contra Flow vat, whereas it is possible to apply more weight with a Contra Flow vat. It is for these reasons that it is often the practice to produce the outer plies by means of Uniflow vats and the middles plies using Contra Flow vats.

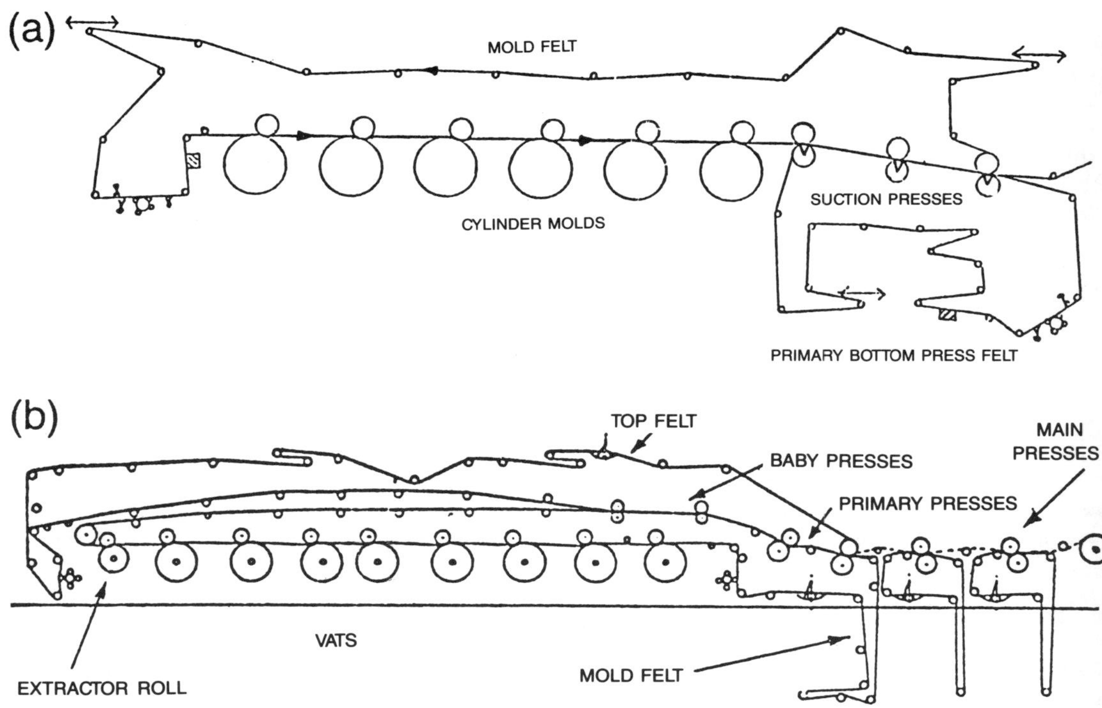

Fig. 250. Typical multi-cylinder board machine forming section.

2. The making felt (mold felt)

The felt which transfers the wet web from the mold surface to the couch roll position is known as the making felt and, as can be seen in Fig. 250(a) and (b), travels the complete length of the forming section of a multi-cylinder machine conveying the web through the first sections of the press part. The felt must have sufficient strength to pull the whole wet end and, at the same time, must be open enough for good water removal at the various presses associated with its run. It must also be fine enough to impart a reasonable finish to the board. Because of the varied duties of the making felt, including the pressing of fairly wet webs, considerable attention is given to felt cleaning and conditioning, including the use of whippers, oscillating showers and vacuum boxes (see Chap. XXII).

3. The mechanism of web formation

The mechanism of forming in Uniflow and Contra Flow vats has been investigated by numerous workers, probably the most detailed by Muller-Rid and Hombsch [4], who used pilot plant equipment. They state that, if the laying of the fibers could take place undisturbed, the wet substance would progressively increase throughout the available forming time, i.e., from entry of the mold into the stock in the vat circle to exit of the mold. In practice, however, the formation is grossly disturbed by considerable speed differences between the stock flow in the vat and the mold surface and this fact causes partial wash-off of the outer layers of the deposited fiber, the amount of wash-off increasing as sheet formation proceeds. Laboratory studies have shown that the maximum rate of deposition of the web takes place over a relatively short distance at the start of the forming process and, although drainage continues for the rest of the immersed forming length, there is little extra fiber deposition taking place because of a continual wash-off effect.

In the case of the Contra Flow vat, the flow pattern is roughly as shown in Fig. 251. The stock enters the vat at considerable turbulence (A), but in a short time becomes less turbulent and moves slowly through the vat towards the opposite side, i.e., the point where the forming surface of the mold enters the stock and where the major part of the web deposition is taking place. It is found that, in this zone, suspension is practically stationary and the stock is in an extremely flocculated state (C). Immediately adjacent to the rotating mold surface a boundary layer (B) is formed which moves rapidly in the direction of

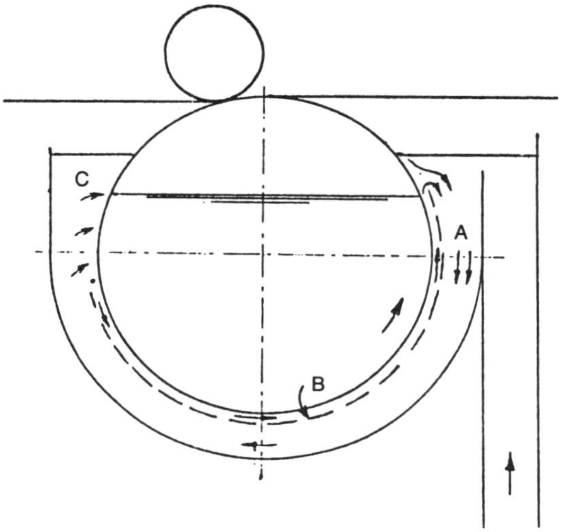

A. HIGHLY TURBULENT FLOW.
B. HIGH-SPEED STOCK FLOW, IN BOUNDARY LAYER.
C. SLOW-MOVING HIGHLY FLOCCULATED STOCK.

Fig. 251. Flow pattern — Contra Flow vat.

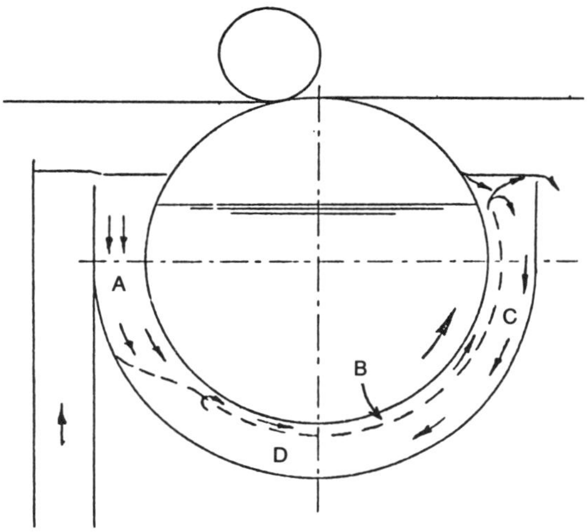

A. HIGHLY TURBULENT FLOW.
B. HIGH-SPEED STOCK FLOW IN BOUNDARY LAYER.
C. SLOW UNSTABLE COUNTERFLOW ZONE.
D. STAGNANT THICKENED STOCK OF VARYING CONCENTRATION.

Fig. 252. Flow pattern — Uniflow vat.

cylinder rotation. This layer can be 1–3 cm thick depending on the consistency of the stock, its freeness and machine speed. It is hardly influenced by the distance between the cylinder and the wall of the enclosing vat except in the case where the distance is very small. Continued drainage without a corresponding fiber deposition leads to the consistency in this layer increasing to become substantially higher than that of the inlet stock. This stream of high-consistency stock follows the cylinder surface to the point where the mold surface emerges. Here it mixes with the incoming stock and is recirculated to the other side of the cylinder, thus increasing the consistency.

Between the two streams of stock mentioned above, an unstable boundary layer is formed and localized velocity changes created in this layer lead to a continuous exchange of stock between the two streams. This in turn leads to a non-uniform flow velocity and a non-uniform consistency across the machine which gives uneven conditions influencing both the web deposition and the stock wash-off at the line of emergence.

In the case of the Uniflow vat, the flow patterns are shown in Fig. 252. At the inlet (A), there is a turbulent flow which extends over the entire vat section, but this turbulence diminishes as the stock flows downwards towards the center of the vat. It is during this first phase at the inlet that the rapid preliminary formation takes place. Some time later, when the flow velocity through the wire has decreased to a certain level, a boundary layer is formed (B) which travels with, and approximately at, the velocity of the mold surface. This layer transports to the side of the vat a sufficiently large volume of stock to cause stagnation of other layers close to the walls of the vat. As in the case of the previously mentioned Contra Flow vat, wash-off takes place and results in the elevated consistency of the boundary layer (B). Where the mold surface leaves the stock, some thickened stock separates from the cylinder, some of this being discharged at the overflow while the rest flows back downwards into the vat at point (C). The consistency of this stock is higher than that at point (B). Counter-flow (C) and boundary layer (B) are separated by an unstable intermediate layer through which thickened stock from the counterflow stream is fed back irregularly onto the boundary layer stream (B). This has negative effects on web formation. Freeness of stock, making head, cylinder speed and amount of overflow control the intensity of the counterflow.

4. Cylinder mold developments

Cylinder mold forming may be summarized as follows. The forming surface of the mold is pulled through the fibrous suspension in the vat circle for longer than is necessary for efficient web formation and in the vat circle stationary and irregular flows are developed. Wash-off and

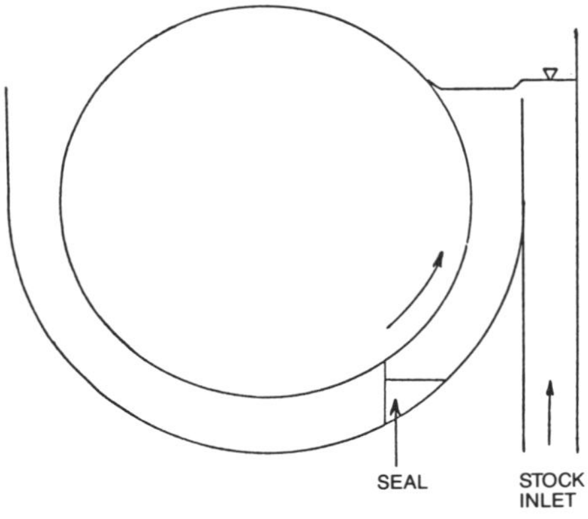

Fig. 253. The dry vat.

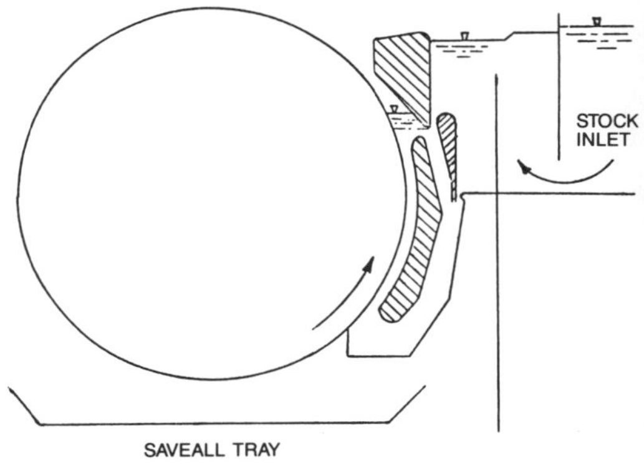

Fig. 254. The restricted flow vat (half vat).

stock thickening occurs and the overall mat deposition is very inefficient.

Another limitation of the cylinder mold machine is that it can be operated only at relatively low speeds. At high speeds, water begins to rim the inside of the cylinder mold and centrifugal force tends to force the water through the mold and lift the mat from its surface. For a 1 m diameter mold, the limiting speed due to this effect has been indicated to be approximately 130 m/min. The use of larger diameter cylinders allows higher speeds to be obtained.

For high-speed operation it is often the practice to use an exhaust system on the cylinder mold. The ends of the mold are sealed and connection is made to a low-vacuum exhaust source. This arrangement results in an improvement in operation and web formation, the drainage rate is increased and drainage water "throw out" problem is reduced.

Figure 253 shows a method used to reduce the forming length. This device is known as the dry vat. A simple seal is introduced into the vat circle so that the fibrous suspension is confined to a shorter section of the vat circle. Because the making length has been decreased in size, this modification reduces the degree of uncontrolled turbulence in the forming zone. At the same time, backwater is allowed to pass unimpeded in and then out of the mold so that buildup inside the mold is eliminated, with the result that the dewatering profile is even across the width of the machine which, in turn, produces a correspondingly improved basis weight profile. The "wash-off" effect associated with Uniflow and Contra Flow vats is greatly reduced and there is no reduction in total forming capacity.

The next stage of development from the dry vat was the evolution of the half vat, sometimes known as the restricted vat which is, in fact, a dry vat with the "unused" half removed (see Fig. 254).

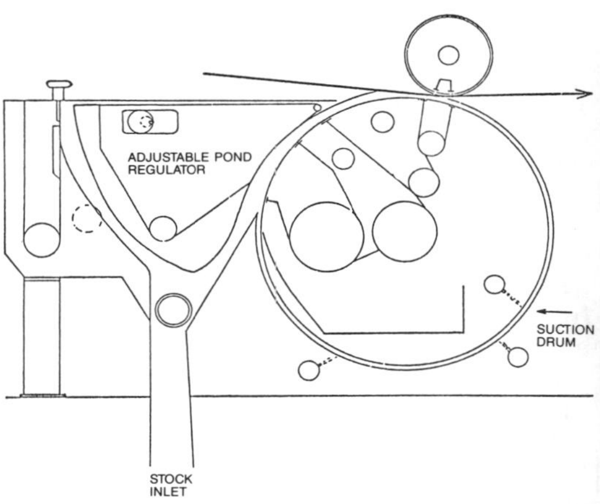

Fig. 255. The Rotoformer (Sandy Hill).

Another cylinder former of importance is the Rotoformer. The Sandy Hill Rotoformer (see Fig. 255) was first used commercially in the Fifties, that is, before the studies of Muller-Rid. It consists of an open-ended perforated suction cylinder which is covered by a coarse backing wire and a fine face wire. Inside the cylinder are adjustable compartmented boxes into which drainage takes place under controlled conditions. There is also an initial draining zone at the beginning of web formation where drainage is by means of gravity. The pond regulator can have its position adjusted in order to influence the stock velocity and pressure applied at the initial forming zone.

Another development is the cylinder suction former, a typical example of which is shown in Fig. 256. It consists essentially of a tapered stock

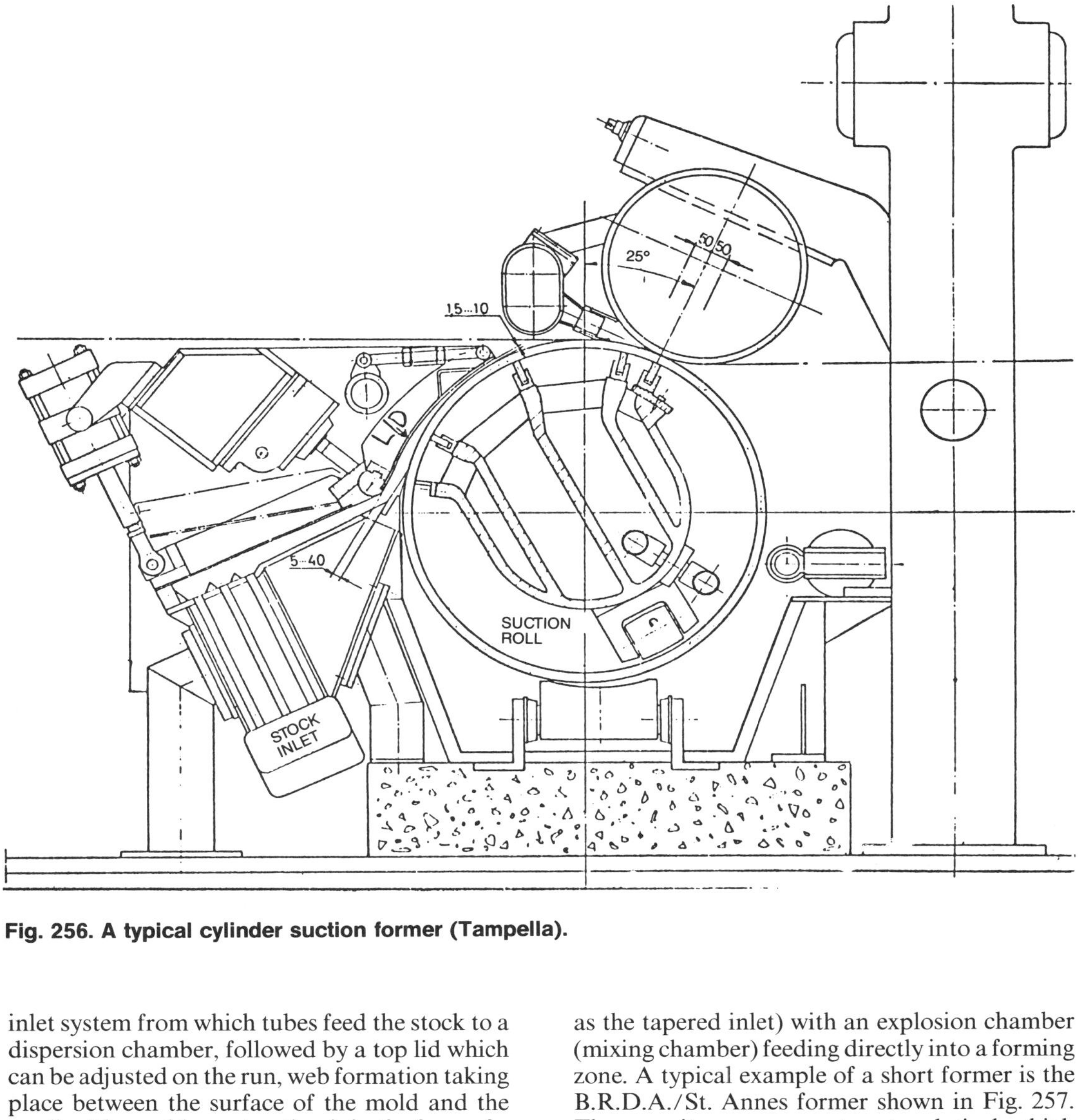

Fig. 256. A typical cylinder suction former (Tampella).

inlet system from which tubes feed the stock to a dispersion chamber, followed by a top lid which can be adjusted on the run, web formation taking place between the surface of the mold and the top lip. The mold construction is in the form of a cast bronze shell inside of which is fitted a suction box having two or more separate vacuum compartments. The position of the suction box can be adjusted on the run.

The use of the tapered stock inlet system described in Chap. VI has significantly improved cylinder mold and rotary former operation by greatly improving basis weight profile.

5. Short or pressure formers

The "short" former or "pressure" former has been developed or, rather, it evolved from the studies associated with dry vat and hydraulic flow boxes. In effect, the short former is the combination of a well-designed stock inlet (such as the tapered inlet) with an explosion chamber (mixing chamber) feeding directly into a forming zone. A typical example of a short former is the B.R.D.A./St. Annes former shown in Fig. 257. These units can operate at relatively high consistencies and often can be installed on existing cylinder molds with the object of improving performance at relatively low cost. In operation, the fibrous suspension passes from a tapered inlet through a series of shear pipes into a small compartment, known as the explosion chamber, where fiber dispersion takes place. Finally, the dispersed fibrous suspension passes to the forming zone where it is confined between a hinged lid, or roof, and the mold surface, formation taking place under pressure. The unit can also be used with suction molds and can also be applied to other cylinder forming devices. The forming length is very short, 10–15 cm. A well-formed web with a good basis weight profile is easily obtained over a wide range of basis weight and operating conditions.

6. Fourdrinier machines with a secondary headbox

Figure 258 is a typical arrangement of a fourdrinier machine fitted with a secondary head box. Machines of this type are used mainly for producing kraft liner board at high rates of production. A base ply, approximately 80% of the total weight of the finished web and having a freeness in the order of 18° Schopper-Riegler is applied at the first headbox. At a suitable position along the forming table vacuum is applied using conventional suction boxes and by means of the secondary headbox a second layer is added to the primary layer. The second layer is of a better quality than the first layer and provides a good printing surface.

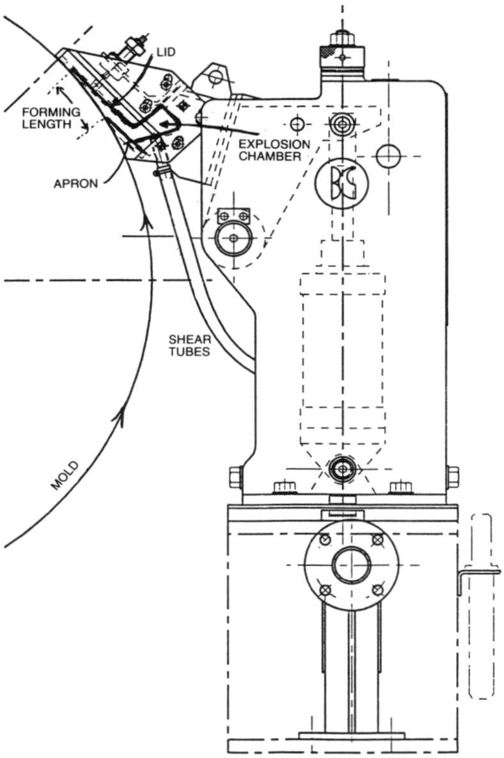

Fig. 257. A short former — B.R.D.A./St. Annes (Black Clawson Bristol Former).

7. Multi-wire fourdrinier machines

Some paper board grades are produced on machines the forming section of which consists of a number of separate fourdrinier wire parts, the webs of each ply being joined together while still moist.

The multi-wire fourdrinier concept is now finding a wide range of applications particularly associated with high production machines producing, for example, linerboard, food board and high-quality folding boxboard. There are many arrangements to choose from and Figs. 259 through 261 illustrate some recent commercial arrangements.

8. Combination machines

Machines where the forming section consists of a combination of one or more fourdrinier wire parts with one or more cylinder molds are relatively common. A typical arrangement is shown in Fig. 262. In practice, the cylinder mold section can be arranged to run in the same or in the opposite direction to the fourdrinier wire and the top liner is often produced on the fourdrinier section.

9. "On top" formers

"On top" forming uses an arrangement in which separate forming units are located above a fourdrinier wire table. A web is formed on the fourdrinier table in the normal way and further webs, formed on the "on top" forming units, are added to it. The previously described method, shown in Fig. 260, is in effect a form of "on top" forming.

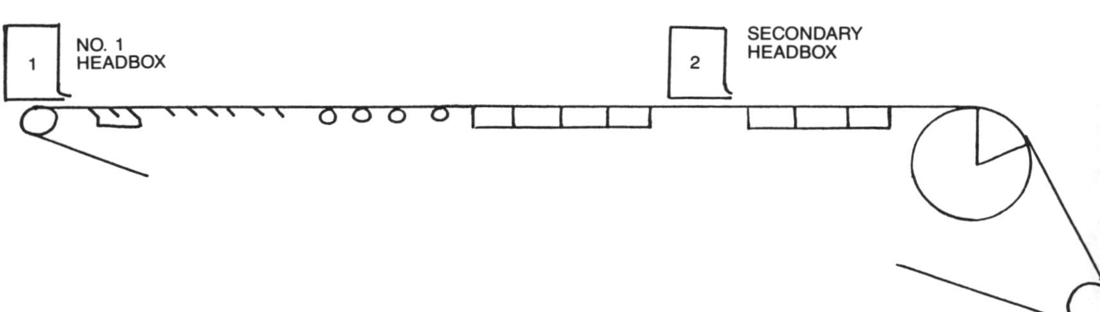

Fig. 258. Secondary headbox application.

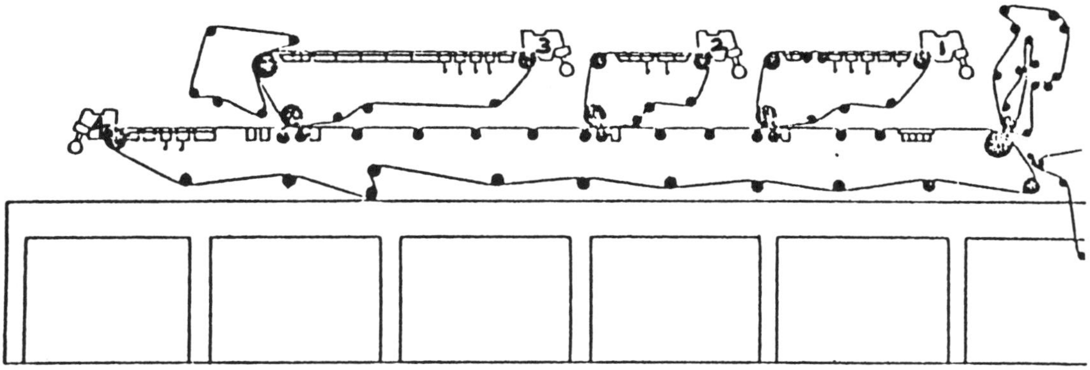

Fig. 259. Multi-wire fourdrinier wet end (Voith).

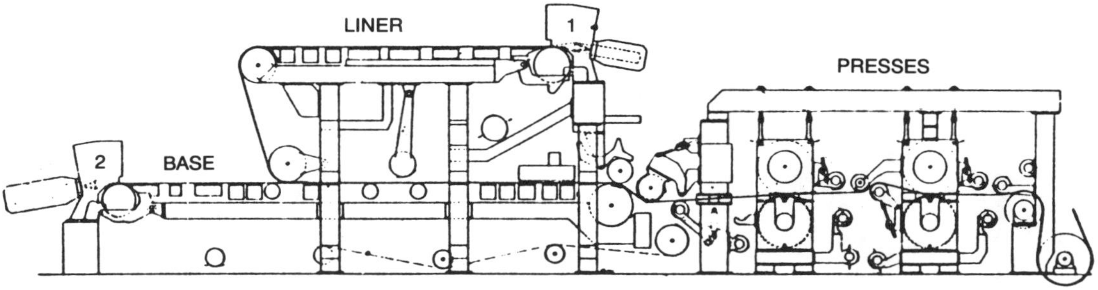

Fig. 260. Multi-wire fourdrinier wet end (Escher Wyss).

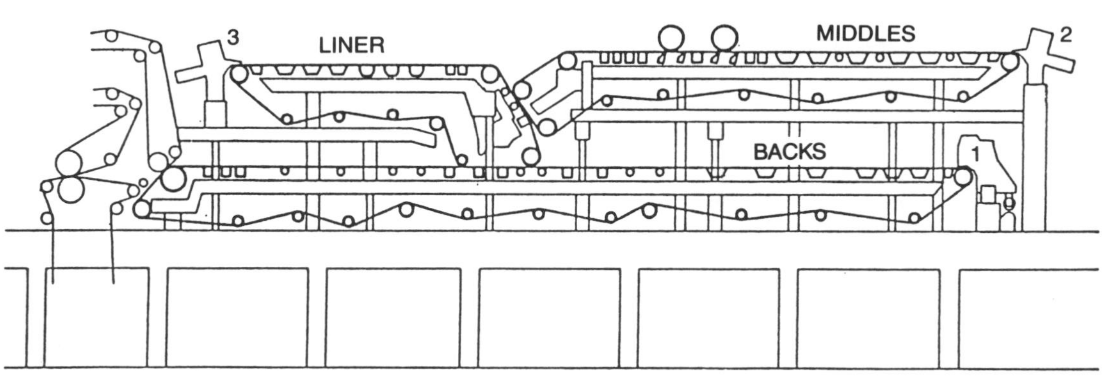

Fig. 261. Multi-wire fourdrinier wet end (Ahlstrom).

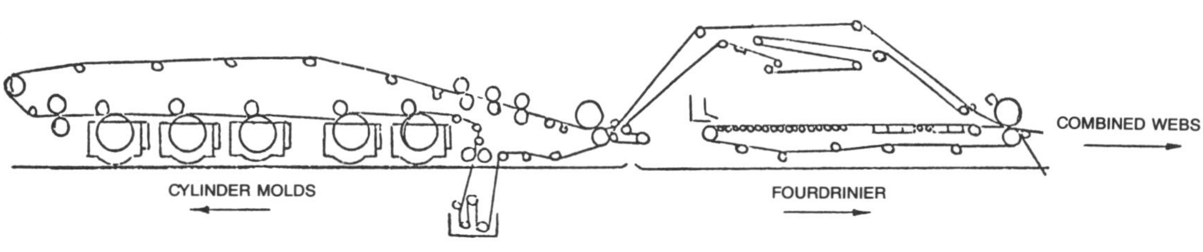

Fig. 262. Fourdrinier/cylinder mold combination machine.

Another example of "on top" forming is the use of the Arcu former as shown in Fig. 263. The suction forming roll does not have a fixed face wire, but instead has associated with it a short forming wire run. In operation, the web is formed on the wire. The first half of the forming zone is kept at atmospheric pressure, but the remainder is at higher degrees of vacuum. As the web and wire move round the suction roll, they come into contact with a short top wire and the web is sandwiched between the two wires where it is subjected to progressively increased pressure. The vacuum on the inside of the suction roll is adjusted to counterbalance the centrifugal forces so that drainage is approximately equal on both sides of the web. This arrangement enhances retention of filler. The web and wires then pass over a suction box following which the top wire returns to the forming roll. The web, now on the main wire of the unit, is added to a previously formed web or webs on, for example, a fourdrinier wire part.

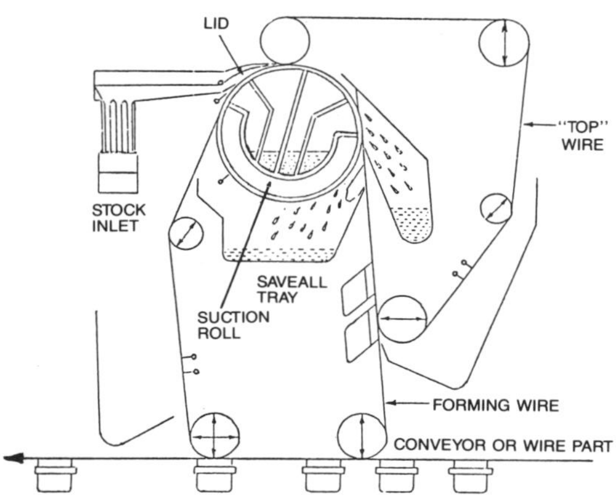

Fig. 263. The Arcu former (Tampella).

10. Twin-wire formers

The Inverform was developed in the Fifties in England at St. Annes Board Mill Co. Ltd. with the object of providing a method of forming high-quality paperboard at speeds in excess of other methods, without the loss of quality, and was the first commercial example of twin-wire forming.

Figure 264 shows the basic Inverform configuration, the operation of which is as follows. A fibrous suspension flows onto the bottom wire immediately before an in-running nip created by a top wire passing round a cellular roll (forming roll) and the bottom wire, which is supported under this roll by table rolls or foils. The two wires, with the fibrous suspension sandwiched in between them, then pass to a rigid doctor blade (auto slice) lightly pressed into the wires. Water is extracted from the fibrous suspension and passes upwards through the top wire, being swept up the sloping blade into a collecting tray from where it is returned to the machine pit. With a single-ply web, or the first ply of a multi-ply web, water also drains downwards. When operating at low speeds, it is necessary to assist the water up the blade of the auto slice. Water removal after the auto slice is carried out by means of inverted vacuum boxes and, in the case of the first ply, by conventional flat boxes.

Final dewatering and web consolidation is often achieved by the use of an assembly, known as a rotary vacuum box, which is a small press assembly with the bottom roll housed in a vacuum box. This assembly also performs the duty of holding the web onto the bottom wire, thus preventing it from following the top wire.

The first ply is carried forward on the bottom wire to the next flow box which deposits on it another layer of pulp. This layer is then formed in another Inverform section. Because of the presence of the first ply, little dewatering takes place downwards. Further plies may be added in

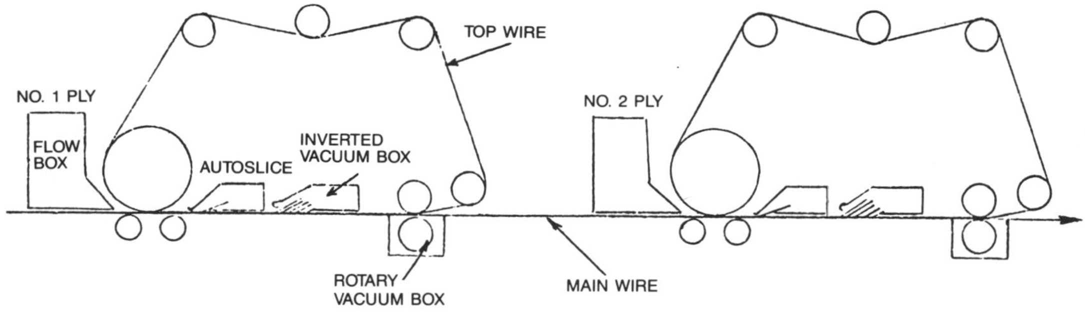

Fig. 264. The basic Inverform configuration (Beloit-Walmsely).

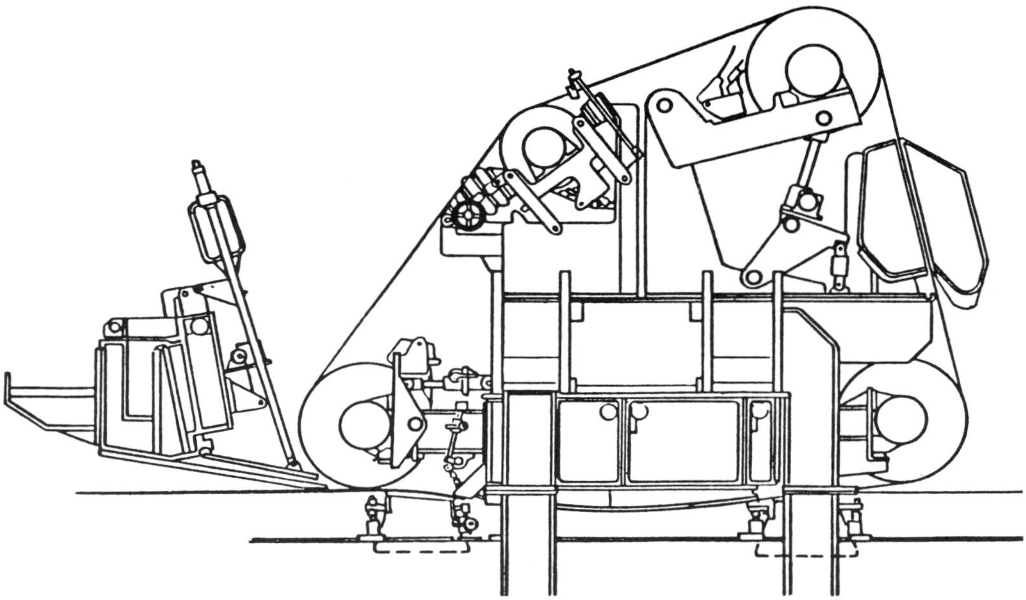

Fig. 265. The Bel-Bond unit (Beloit).

a similar manner. A typical multi-ply Inverform wet end is shown in Fig. 264.

Inverform machines are producing boards at speeds of up to 450 m/min. In the period since the operation of the first commercial Inverform machine, further developments, such as the Bel-Bond development, have taken place.

A typical Bel-Bond unit is shown in Fig. 265 where it is located on top of a fourdrinier table. The Bel-Bond unit provides extra drainage capacity and associated with its operation is a degree of forming and reforming. A secondary headbox is located immediately in front of the Bel-Bond unit.

Since the commercialization of the Bel-Bond system, other equipment manufacturers have developed somewhat similar arrangements. The latest developments incorporate an inverted dewatering section with associated equipment beneath the bottom wire, allowing controlled shear and drainage to be carried out. These arrangements operate over a wide range of basis weight. Figure 266 illustrates a typical arrangement of this type. Suppliers of this type of equipment are Voith, Escher Wyss, and Ahlstrom.

These arrangements are used in a similar way to the Bel-Bond and are now being incorporated in multi-fourdrinier machines in order to enable high basis weights to be produced and 4- and 5-ply structures. Figure 267 shows a typical arrangement.

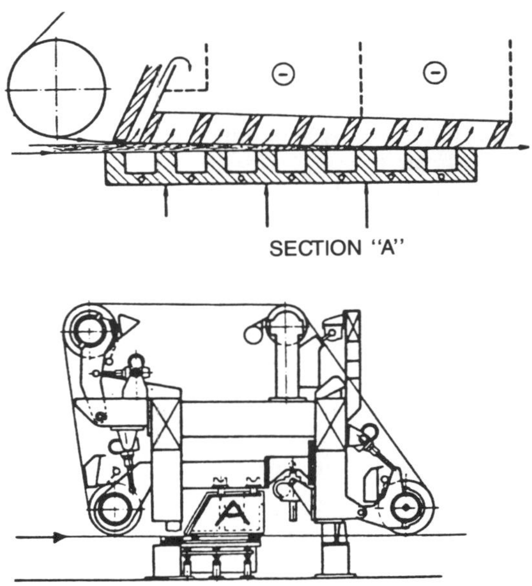

Fig. 266. Voith Duoformer D (Voith).

11. Multi-layered (multi-channel) headboxes

The concept of using multi-channel headboxes is not new and there have been many trial arrangements produced over a number of years. It is now an established commercial operation and has made possible the production of

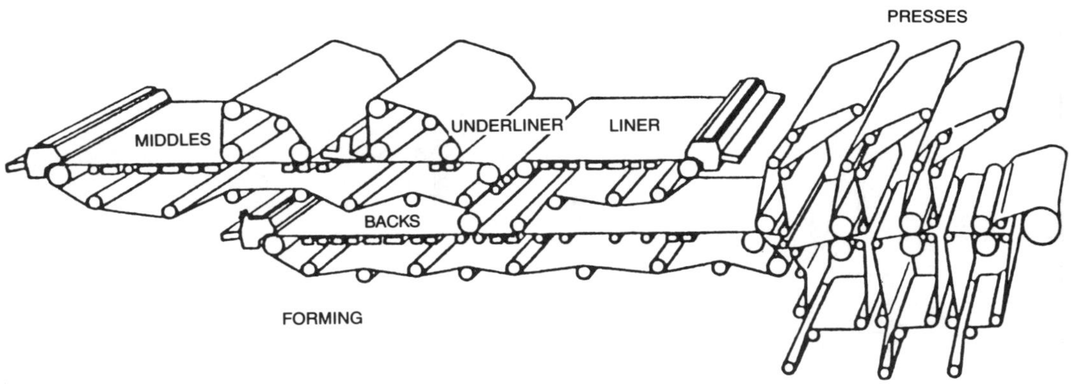

Fig. 267. Modern multi-ply forming section (Valmet/Ahlstrom).

multi-ply lightweight papers.

Figure 268 shows a typical multi-channel layered headbox of the type used for the manufacture of tissue at high speed. The products from this arrangement have a high degree of ply identity, but at the same time do exhibit good ply bonding. Other grades, ranging from printing grades to linerboard, are now produced using multi-channel headboxes. There is also the possibility of using the multi-channel headbox to produce a structure having two outer layers formed at normal consistency with the middle channel layer at high consistency with the aim of producing a high-bulk paper or board.

Multi-channel headboxes are usually operated in conjunction with a twin-wire forming unit, but can also be used with fourdrinier and cylinder formers.

An interesting, if somewhat specific, example of the use of the multi-channel headbox with a twin-wire section is the Contro Flow former shown in Fig. 269. This device with its controlled water removal equipment is capable of handling large volumes of fluid. It can be used to produce a wide range of products including non-wovens containing fibers and heavy basis weight boards.

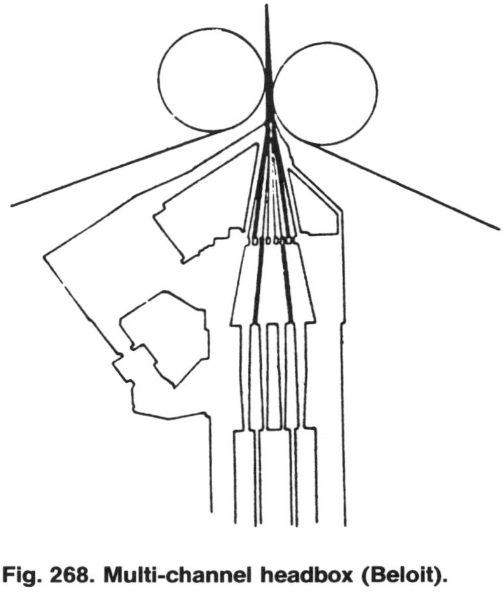

Fig. 268. Multi-channel headbox (Beloit).

12. Ultra former

A development of the Sixties was the Kobayashi Ultra former, which is a type of "on-top" forming. In the case of the Ultra former, the fibrous suspension is dewatered in the pressure wedge between the making felt and the forming wire. Figure 270 shows the basic Ultra former which is in effect a cylinder mold with an associated headbox.

The fibrous suspension is fed by means of a headbox onto the forming wire at the top position of the cylinder mold. This web, sandwiched between the felt and the wire, is subjected to increasing pressure which causes dewatering. As is the case with the Inverform, there is a small turbulent pond developed at the point of maximum dewatering and this is found to assist ply bonding and formation. At point B, the felt with the web leaves the surface of the cylinder and passes onto the next forming unit.

The formation with this basic Ultra former unit is of fair quality and its speed is limited because of the effect of centrifugal forces and rimming. The water inside the forming cylinder runs up to the point of separation causing an excessive pond between the cylinder wire and the top of the sheet. The "washing" of the top liner produces an unacceptable sheet. (In order to overcome this problem, cylinder diameters have been increased with the result that the basic

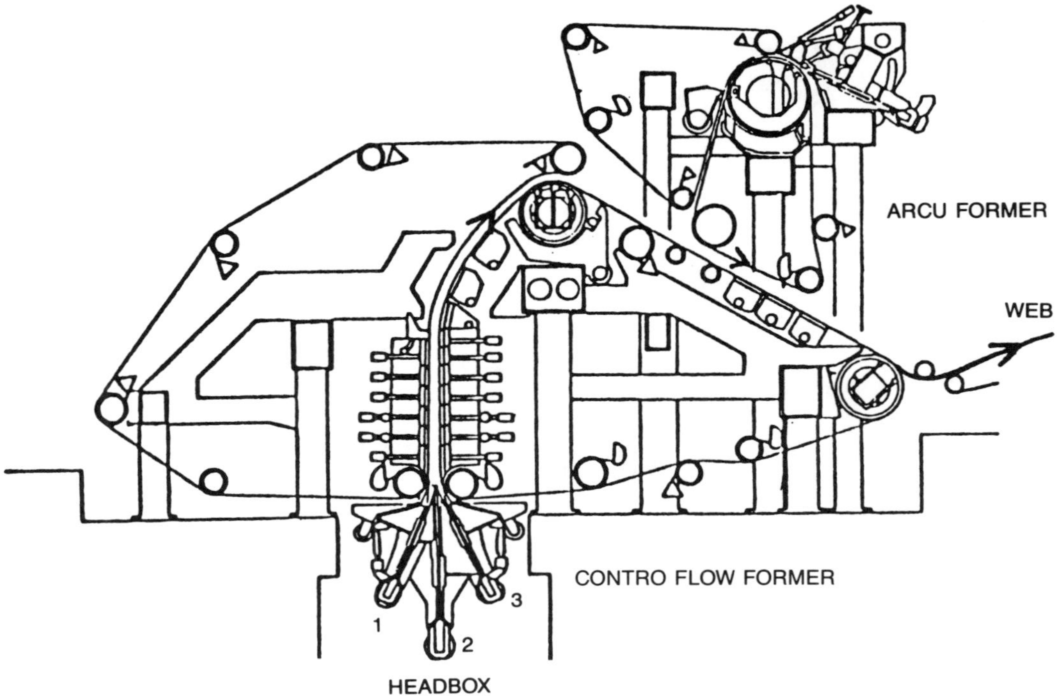

Fig. 269. Contro Flow former (Tampella).

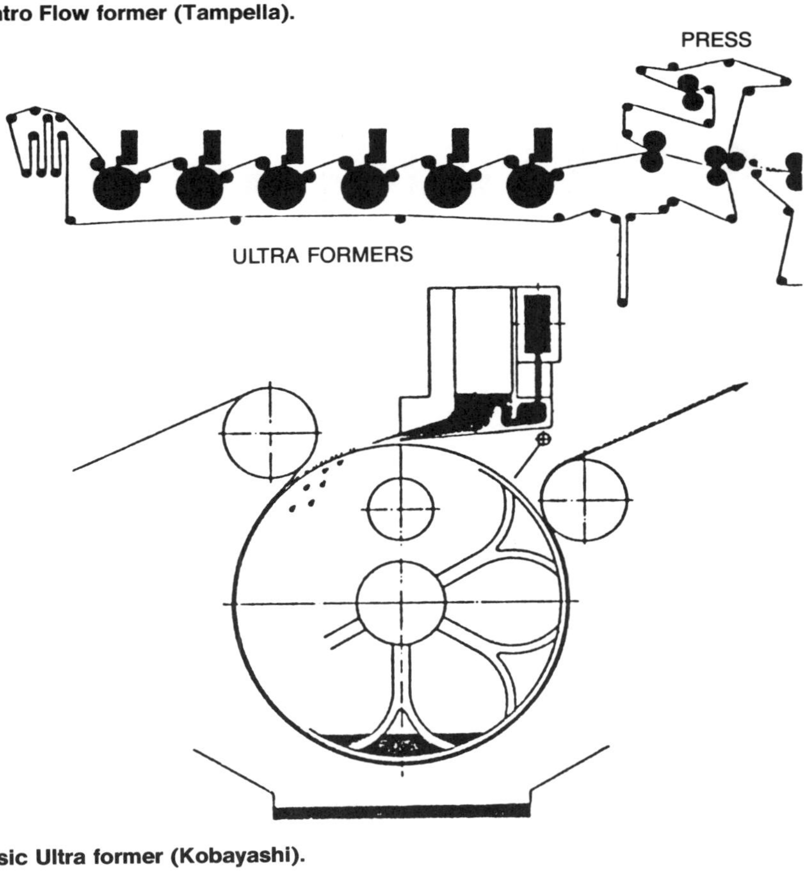

Fig. 270. Basic Ultra former (Kobayashi).

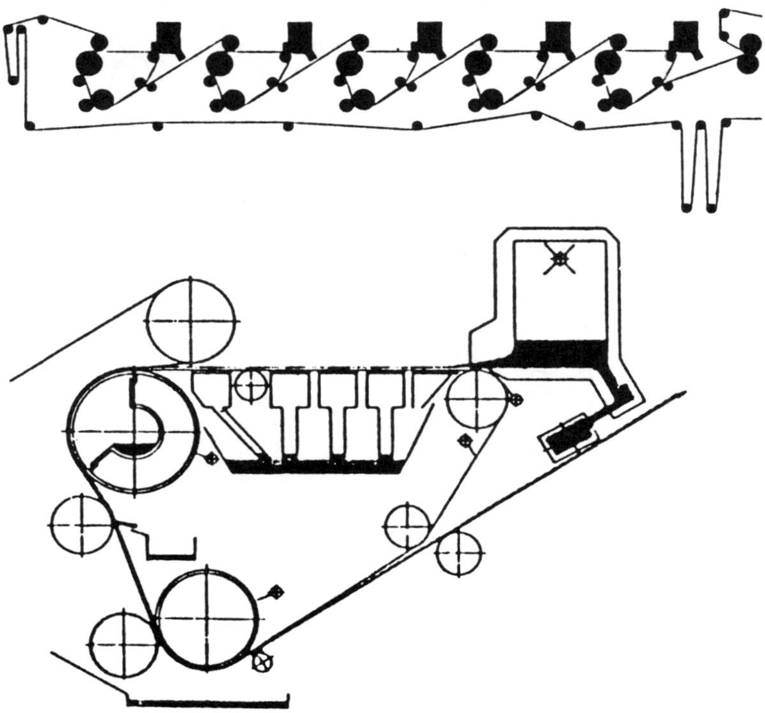

Fig. 271. Super Ultra former (Kobayashi).

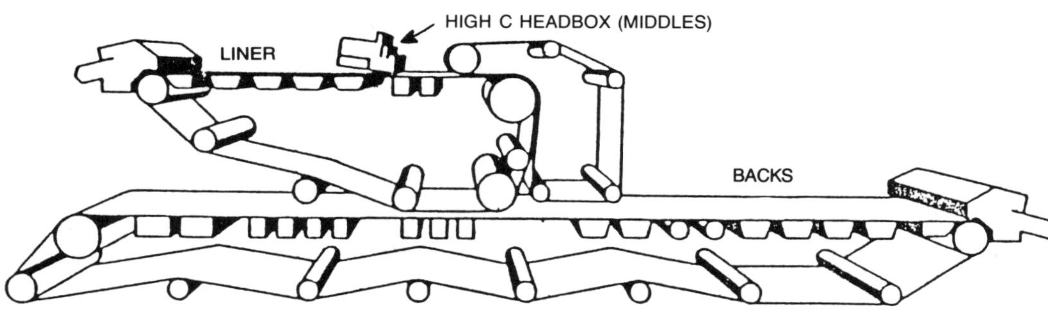

Fig. 272. Application of high-consistency headbox (Valmet/Ahlstrom).

Ultra former can now be found running at speeds of up to approximately 244 m/min).

This limitation has led to the development of a range of Ultra formers, for example the Super Ultra former, which uses, instead of a rotary forming section, a small wire run which can be looked upon as a mini fourdrinier (see Fig. 271). A further development is the Ultra twin former, which is, in effect, an on-top twin-wire rotary former.

H. HIGH-CONSISTENCY FORMING [5]

The concept of producing a fiber web from a high-consistency suspension is still under development, but there is already a commercial application in the field of multi-ply forming. An application, shown in Fig. 272, is the manufacture of liquid carton board. The forming section is essentially a multi-fourdrinier arrangement, but there is also the application of a middle ply using high-consistency forming technology. The resulting product exhibits good bulk characteristics.

The high-consistency headboxes can have applications other than those illustrated. For example, they can be used in conjunction with Kobayashi Ultra formers, as shown in Fig. 273.

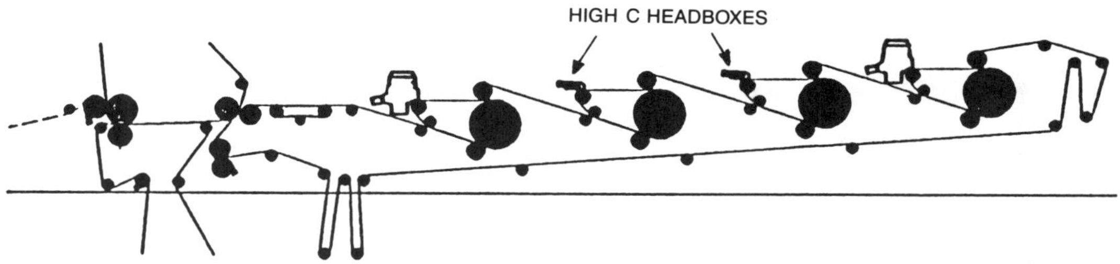

Fig. 273. Application of high-consistency headbox (Kobayashi).

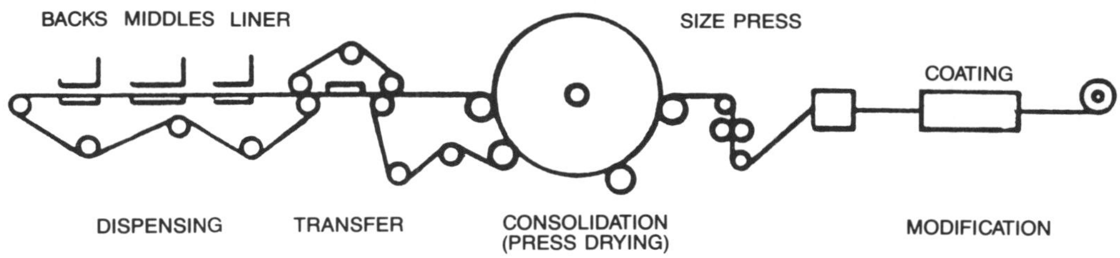

Fig. 274. Multi-ply dry forming machine.

After the forming section of multi-ply paper and board machines, the web is further de-watered by pressing and drying, etc., using the same equipment as that associated with normal paper machine pressing and drying operation.

I. DRY FORMING [6]

A completely revolutionary approach to forming is that using the air laying system. This process is mainly used to produce specialty materials such as wipes, medicals, filters, etc. It can be used to produce these materials having a multi-layer structure, and a typical arrangement is shown in Fig. 274.

A summary of multi-ply forming methods is given in Table 47.

J. MULTI-PLY PAPERBOARD AND PAPER PRODUCTS

1. Paperboard

There is no rigid line of demarcation between paper and board, and board is often defined as thick paper. ISO standards state that paper with a basis weight (grammage) generally above 250 g/m^2 (51 lb./1000 sq. ft.) shall be known as paperboard, or "board". The definition becomes less clear because in some parts of the world board is classed as such when its caliper (thickness) exceeds 0.3 mm (300 μm).

It is also not possible to strictly define paperboard by its structure or by the type of machine used to produce it. For example, paperboard can have either a single- or multi-ply structure and can be formed on a fourdrinier wire part, a single, or a series of molds, or a series of modern formers or sometimes by means of a combination of one or more of the above. For a small part of the market, paperboard is produced by laminating sheets of paper together.

The following terms are in general use in paperboard manufacturing and associated converting industries.

Basis weight. This is the weight of a known area; for example, g/m^2 (grammage) or lb/1000 sq. ft., the weight in pounds of a ream (usually 500 sheets) of paperboard cut to its "basic size".

Caliper (thickness). The thickness of the sheet expressed in μm or thousandths of an inch.

Size of a sheet. The width and length of a sheet of paperboard. The width, always expressed first, is the dimension cut at right angles to the direction of the sheet. Length is the dimension cut in the machine direction. For example, 20 × 30 in. means 20 in. (50.8 cm) cut across the machine by 30 in. (76.2 cm) cut in the direction of the machine. The first dimension is often termed the cross direction and the second direction the machine direction.

Structure. The composition of the web (see Fig. 275).

Ply. A fibrous layer of homogeneous composition.

Topside. The side of the web opposite to the wire side is the normal paper definition of this term. In the case of paperboard topside can also

Table 47. Summary of the main forming methods used today for multi-ply.

Type	Approx. Speed Range m/min	Approx. Basis Weight Range (per unit) g/m^2	Comments
ROTARY BASE			
Contra Flow vat	10 - 150	30 - 150	Commonly used
Uniflow vat	10 - 150	30 - 100	Commonly used
Dry vat	10 - 150	30 - 150	Often used
Restricted flow vat	10 - 150	30 - 150	Often used
Pressure former	20 - 200	25 - 200	Increasing use
Suction former	20 - 500	20 - 300	Increasing use
Rotary top former	30 - 400	25 - 150	Some application
Rotary top former (with wire)	50 - 600	25 - 300	Modern trend
Ultra former (basic)	20 - 200	35 - 100	Many in use
Suction-forming breast roll	50 - 500	20 - 300	Some specific applications
FOURDRINIER BASE			
Basic fourdrinier	10 - 1000	10 - 500	Base of some multi-ply machines
Fourdrinier with secondary headbox	20 - 800	100 - 600 (Total Weight)	Used mainly for linerboard
Two-wire concept	20 - 300	20 - 400	Used for high-quality papers
Mini-fourdrinier (on top) } Multi-fourdrinier }	100 - 700	20 - 220	Used mainly for high-quality paper board
Mini-fourdrinier (under)	100 - 600	80 - 400	Little used
Super Ultra former	100 - 500	40 - 100	Many applications (paper board)
TWIN-WIRE BASE			
Inverform-type	80 - 500	25 - 100	Many installations
"Bel-Bond" types	100 - 600	20 - 150	Increasing use
Vertical twin-wire	300 - 1300	50 - 150	Some applications with multi-channel headbox
Mini-fourdriniers (on top) with twin-wire	150 - 700	50 - 300	Increasing applications
Contro Flow former	100 - 600	20 - 600	Only one commercial application
MISCELLANEOUS			
Multi-Layer headbox	probably 100 - 1200	15 - 300 (total)	Tissues, linerboard printings
Dry former (air laid)	(Max. N/A - 600)	50 - 150	Specialties
High-consistency headbox	300	Approx. 60 - 120	Middle layers

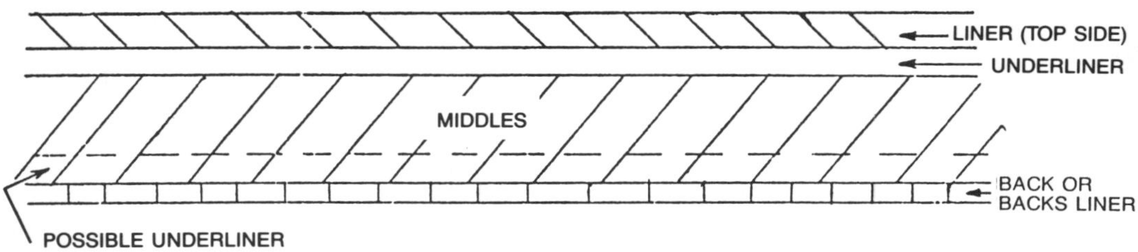

Fig. 275. Structure of multi-ply web.

mean liner side, generally the better quality face of the web. Some grades of paperboard are known as double-lined, in which case the two faces of the web are both of high quality.

Liner. A ply of good-quality fiber (often white) on the topside.

Under Liner. The layer (ply) of fiber between the external layer (topside/liner) and the middle.

Middle (filler plies). The layer (ply) of fiber between the two external layers or between the under liner and an external layer.

Backs (bottom ply). The outside layer (ply) of fiber directly opposite to the liner layer.

Duplex or Biplex. This is a board consisting essentially of two layers of different furnish.

Table 48. Examples of typical paperboard grades.

Board grade	Grammage range g/m^2 (lb/1000 sq. ft.)	Typical furnish	Special requirements (physical)	Typical usage
White-lined folding-box board	200-800 (40-160)	liner: virgin pulp, clean wastepaper; under-liner: mech. pulp; middles: waste; backs: mixture (mech. and chem. pulp)	bending printing plybond stiffness	general packaging cartons
Chipboard	200-800 (40-160)	100% waste	bending plybond	packaging cartons, tubes and stiffeners
Gypsum board	250-800 (52-160)	100% waste	plybond	outer component of plaster board
Test liner	150-300 (30-60)	liner, virgin pulp (kraft): rest, waste	bending bursting strength plybond	outer components of corrugated container board
Linerboard	150-300 (30-60)	from 100% virgin pulp to a mixture of 75% virgin pulp with 25% waste	crush resistant bursting strength	outer components of corrugated board
Food board	200-600 (40-120)	100% virgin pulp (single or multi-ply)	bending printing plybond stiffness	all foods, especially frozen foods (high-quality containers)
Liquid packaging	200-400 (40-80)	100% virgin pulp barrier (coating and/or laminate)	bending printing with plybond stiffness	containers for wide range of liquids, including milk
Fluting medium	90-200 (18-40) 200-800 (40-160)	Typical–100% waste or semichemical hardwood pulp	crush resistance	inner components of corrugated container board
Clay-coated folding carton		100% waste	printing stiffness plybond	soap cartons, cereal boxes, Pampers boxes

Triplex. This is a board consisting essentially of three different furnish layers (external furnish layers may have the same composition).

Multiplex. This is a board with more than three furnish layers. Two or more of the layers can have the same composition. It is also known as a multi-layer board.

Furnish. The constitution of the various materials that are blended in the stock suspension from which the paperboard plies are made. The main constituents are the fibrous material (pulp or secondary fiber), sizing agent, fillers and dyes.

Multi-Ply Paperboard. Table 48 shows a typical range of paperboard grades with comments on their furnish, requirements and end usage. The following definitions describe the most common types of paperboard.

Body or Base Board. This is a board that is ultimately treated by, for example, a coating or a surface application.

White Lined Board (Duplex). A board with a bleached pulp liner and the remainder of the board made up of, for example, a mixture of chemical pulp and mechanical pulp. Often good for packaging.

Kraft Lined Chip Board. An unbleached kraft liner on a wastepaper base used for packaging products, such as electrical and mechanical components.

Test Liner. This is a type of kraft-lined chip board. Sometimes the liner is made from a strong kraft waste furnish. This material is combined with a fluting medium to produce corrugated board.

Double-Lined Board. A board lined on both sides. For example, the outer surfaces (liner) can consist of a bleached pulp and the middles of mechanical pulp. It is used for high-quality packaging of food and cosmetics.

Carton Board (Folding Boxboard). Paperboard of various compositions used for the manufacture of folding cartons and set up boxes.

Food Board. Multi-ply paperboard used for

food packaging. It is hard sized for water resistance.

Liquid Packaging Board. Also called special food board, or milk carton board, this strong board is usually 100% chemical pulp, often plastic coated. It is formed into containers for a wide range of liquids (e.g., milk and other beverages).

Frozen Food Board. Multi-ply paperboard with high moisture and water vapour resistance. It is often single ply, made from bleached wood pulp with a surface coating for high-quality print.

Linerboard. A strong packaging paperboard with a two-ply construction made essentially from virgin kraft pulp and produced on a fourdrinier machine. The top ply is added by means of a secondary flow box. It is combined with fluting medium to form corrugated board and is used in the manufacture of large containers.

Chipboard. A paperboard made from waste paper, used in low-grade packaging, solid fiber and book board.

Machine Glazed Board. Paperboard that in its manufacture has had one face made smooth and glossy by drying on a large, polished, steam-heated drying cylinder (Yankee cylinder).

Tube Board. A paperboard generally unsized and smooth finished. It is slit into narrow widths for winding and pasting into spiral or convoluted mailing tubes, cores etc.

Can Board. Paperboard used for the manufacture of composite cans and fiber drums. The cans are used for packaging a wide range of materials, including liquids and powders.

Coated Board. Paperboards of various grades that have been coated on one or both faces to make the surfaces suitable for high-quality printing.

K. MULTI-PLY PAPER GRADES

1. High-quality/speciality papers

For many years, two-ply grades of paper have been produced using the "two" fourdrinier configuration. These papers are often called twin-wire papers although today the term twin wire is more correctly associated with papers

Two/twin-wire paper is usually produced so that the resulting web has two top sides on which to print – obtained by running together the wire sides of the individual webs. The result is a printing paper with identical surfaces on each side of the sheet.

Other two-ply papers produced by this method use different furnishes for each of the plies, a different color for example. This technique is also used to produce heavyweight papers with good strength characteristics; for example, index papers/board, file paper/board etc.

2. Multi-layer printing papers.

There is now being generated a new range of "printing" with a multi-ply structure. To date the multi-channel headbox technique, used with a twin-wire section, is the favored forming method. Examples are:

1. A "wood-free" paper having a three-layer structure, the outer layers containing some chemical fibers and the center layer mechanical fibers or possibly a degree of waste paper.
2. "Wood-containing" papers with strong chemical fibers in the outer layers and a mechanical/waste paper inner layer. A paper of this type will have improved bending stiffness and a less limiting potential. At the same time total filler content can be increased by adding extra filler to the center layer furnish.

3. Multi-layer tissue.

Three-ply tissue is produced using the combination of a multi-channel headbox and a twin-wire former. Tissues of this type often use premium quality fibers for the outer layers in order to achieve a soft, smooth surface, while the inner layer has associated with it other qualities such as strength, absorbency and bulk. The inner layer could use a more economical furnish resulting in a more competitive price.

REFERENCES

1. Peel, J. and Bergstrom, J., "Fundamentals of Structured Web Forming", Proc. EUCEPA XVIII Intl. Conf., London, Vol. 1, p. 119–130 (1979).
2. Smook, G.A., "Handbook for Pulp & Paper Technologists", Joint Textbook Committee of the Paper Industry, TAPPI and Tech. Sect. CPPA, p. 284, Sec. 19.4 (1982).
3. Seppala, S. and Poyry, J., "Driving Forces to Multi Ply Board", Proc. SPCI Conf. (1987).
4. Muller-Rid, W. and Hombsch, J., "Stock Flow Patterns and Sheet Formation in Cylinder Vats", Paper Technology, 6(6):T159–T164 (1965).
5. Waris, T., "High Consistency Forming", Insko Training Seminar (Rainaus Ja Markapuristus 112-86), Osaaminen Menstykseski, Finland.
6. Attwood, B.W., "The Manufacture of Multi Ply Structures Using the Dry Forming Process", Proc. XIX EUCEPA Conf., Munich, Vol. 11, p. 28.1–28.17 (1986).

GENERAL READING

1. Attwood, B.W., "Multi Ply Forming Past Present and Future", Tappi 68(8):78 (Aug. 1980).
2. Anon., "Multi-Ply Formers", Pulp and Paper (Dec. 1978).
3. "New Technologies in Multi Ply and Multi Layer Structures", Proc. PIRA Conf., U.K. (1987).
4. Various brochures on forming equipment from the major paper machinery companies.

XI PRESSING OPERATIONS

R.A. REESE

Senior Process Consultant
CRS Sirrine Engineers, Inc.
Greenville, SC

REVIEWED BY

Leroy H. Busker
Director Research and Planning
Beloit Corporation
Rockton, IL

Dr. Michael A. MacGregor
Manager Science and Technology
Voith Inc.
Appleton, WI

A. INTRODUCTION

The primary functions of the press section are to remove water, consolidate the sheet, impart favorable sheet properties and promote higher wet web strength for good runnability in the dryer section. The paper web is conveyed from the forming unit on a series of felts, rolls and open draws through one or more presses to the dryer section [1].

The pressing operation may be considered an extension of the water removal process that was started on the forming section. It is more economical to remove water by mechanical means than by evaporation. It has been estimated that the relative cost per ton of paper to dewater the sheet is:

forming section – 10
press section – 12
dryer section – 78

Accordingly, papermakers are always looking for methods to improve pressing efficiency and reduce the evaporative load in the dryer section. Water removal should be uniform across the machine, so that the pressed sheet has a level moisture profile as it enters the dryer section.

Sheet moisture values during pressing are usually expressed as percent dryness. Some typical dryness targets for the sheet leaving the press section for various grades of paper are shown in Table 49 [2]. The targets shown are being achieved on good performing commercial machines. The dryness targets increase as new

Table 49. Sheet dryness performance guidelines — 1989

Grade	Target Press Load, Pli	Target Exit Sheet Dryness, %
NEWSPRINT		
First press	350	33+
Second press	450	38+
Third press	550	43+
BOND		
First press	350	37+
Second press	450	41+
Third press	550	43+
CORRUGATING MEDIUM		
First press	400	37+
Second press	600	41+
Shoe-type presses	6000	47+
LINERBOARD-DOUBLE FELTED		
First press	600	38+
Second press	1200	40+
Shoe-type presses	6000	47+

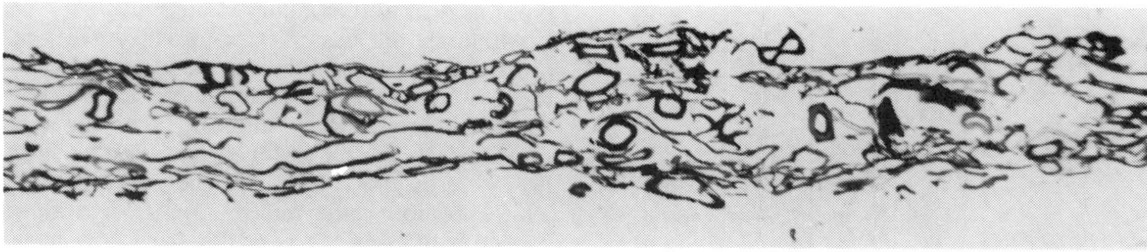

After 2nd press

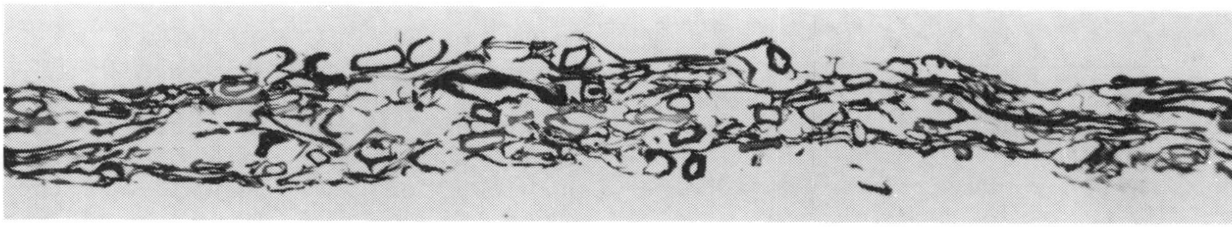

After 3rd press

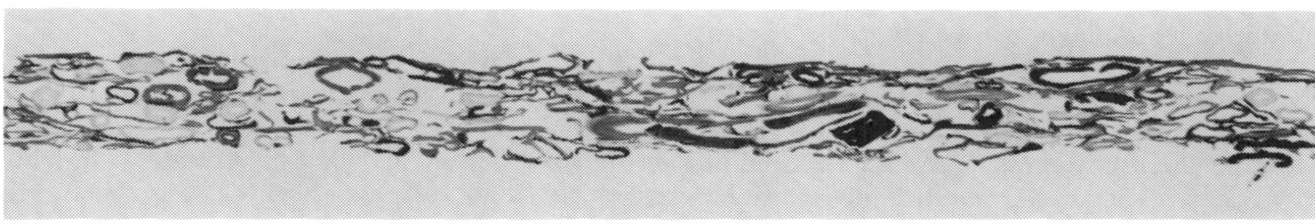

After calendering

Fig. 276. Photomicrographs of newsprint cross-sections illustrating sheet consolidation (courtesy MacMillan Bloedel Research).

pressing technology is developed and implemented.

Changes in percent dryness are not a good indication of absolute differences in moisture content. A better indicator is sheet moisture ratio. For example, a change in sheet dryness from 40 to 42% leaving the presses could be calculated as a 2.5% increase in dryness. In fact, the sheet moisture ratio at 40% dryness is (60 divided by 40) 1.5 and 41%, the moisture ratio is (59 divided by 44) 1.44 so the real difference is 0.06 or 4.0%. The true value of 4.0% lower water content indicates the importance of 1% dryness change during pressing.

Sheet consolidation is a crucial phase of the papermaking process. Fibers are forced into intimate contact during pressing so that more contact points are established. This promotes a much stronger sheet during the subsequent drying stage. Bonds are generally formed in the range of 35 to 65% dryness. The photomicrographs in Fig. 276 illustrate the consolidation between the second and third presses, as compared to a calendered sheet.

1. Pressing fundamentals

The pressing of a sheet of paper appears to be a fairly straightforward mechanical volume reduction process. It involves compression of the sheet to squeeze water from inside the web and out of the fibers themselves. The more the web is compressed, the more water is removed. Wet pressing of paper webs occurs in contact with one or two felts between two rolls in the press "nip".

This picture becomes more complicated when the technical literature on wet pressing is investigated [3, 4, 9–11, 15]. The load applied to press rolls is balanced by the opposite forces generated in the felt and paper web. The area under the total pressure curve must be equivalent to lineal press load. As shown in Fig. 277, the applied press load can be divided into two parts: the hydraulic pressure due to moving water out of the felt and paper; and the mechanical pressure required to compress the sheet and felt. The total pressure at any point in the nip is equal to the sum of these components. The wet press nip has been divided into four phases based on the interaction of the hydraulic and mechanical pressures.

Compression begins in phase 1. Air flows out of both sheet and felt until the sheet is saturated and no air remains. No hydraulic pressure is built up in the web during this phase and little change in dryness of the sheet occurs.

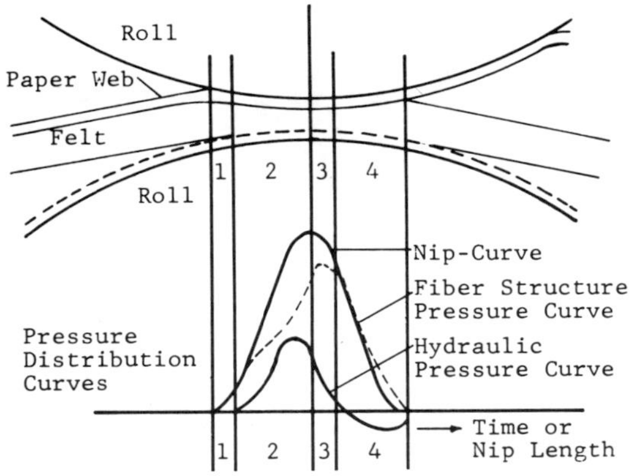

Fig. 277. The four phases of press dewatering.

In phase 2, the sheet is saturated and the buildup of hydraulic pressure in the sheet causes movement of water from the paper into the felt. When the felt also reaches saturation, water moves out of the felt. Phase 2 continues up to mid-nip, where total pressure reaches maximum. It is believed for most cases that the hydraulic pressure reaches a maximum prior to mid-nip.

In phase 3, the nip expands until the fluid pressure in the paper is zero. This is believed to correspond to the point of maximum paper dryness in the nip.

In phase 4, both the paper and felt expand and the paper becomes unsaturated. There is a possibility that water can return to the paper by one or a combination of mechanisms (capillary competition, vacuum in the sheet or separation rewetting).

The possibility of water reentering the sheet in phase 4 is recognized as a limitation to idealized water removal. Various mechanisms for rewetting are postulated, including capillary absorption, mechanical absorption, and film splitting. It is known that additional "rewetting" occurs following the nip unless the sheet and felt are separated immediately. Most modern press designs provide for this separation. Often, the felt wraps the "receptor" roll, while the sheet adheres to the hard smooth roll. Lead-out rolls are generally positioned so the sheet is located midway between top and bottom felts on double-felted presses.

This analysis leads to insights regarding important factors in pressing. For example, the felt must have low enough flow resistance when compressed to allow passage of the water. Addressing these concerns, press felt manufacturers have designed felts which retain void volume longer at high nip loading, and machine builders have provided receptacles of various types in rolls and auxiliary equipment.

Although the basic factors involved in wet pressing appear quite simple, the interactions of these factors can become quite complex. Busker showed some of the complexities of the wet pressing process for grooved or fabric vertical flow presses [14] (Fig. 278). The major factors which promote water removal are the hydraulic pressure difference across the sheet (drainage forces), time in the nip, and uniformity of pressure application. Figure 278 indicates that the major resistance to water removal is the flow resistance of the paper web itself. Sheet moisture into the press and sheet rewetting both have negative effects on sheet water removal. The remaining factors shown usually have only a minor effect on sheet water removal. Plain, suction and blind-drilled presses would have different conditions.

2. Limitations to pressing

Wahlstrom [12, 13] and others determined that there were pressure- and flow-controlled nips. Pressure-controlled nips occur with lightweight/high freeness sheets. These sheets are relatively thin and the sheet structure does not restrict water flow significantly. Water removal is controlled by mechanical compression of the fiber network under the nip pressure and by possible rewetting from the felt to the sheet after the midpoint of the nip. Lightweight sheets are defined as those with a basis weight up to 100 g/m^2. Hard nips where the felt and sheet are separated immediately after the nip generally give the best dewatering performance on lightweight grades.

Flow-controlled nips are formed by heavyweight and/or low freeness sheets. The sheet is much thicker and the sheet structure restricts water flow out of the sheet. Sheet density and temperature and felt void volume all have a significant effect on resistance to water flow. Under these conditions, time in a press nip becomes a controlling factor. High nip loads, soft roll covers, and large roll diameters generally give the best dewatering performance on heavyweight sheets. Heavyweight sheets are those with a basis weight over 160 g/m^2.

Generally, the more readily the sheet loses its water, the greater the pressure that can be applied to the web. Pressing is usually carried out in a series of two to five nips, with increasing pressure in each nip.

Other factors that limit press section water removal include rewetting and pressure uniformity on the sheet. Rewetting is largely a matter of capillary action. The source of the rewetting water is the grooves, holes, or meshes in a

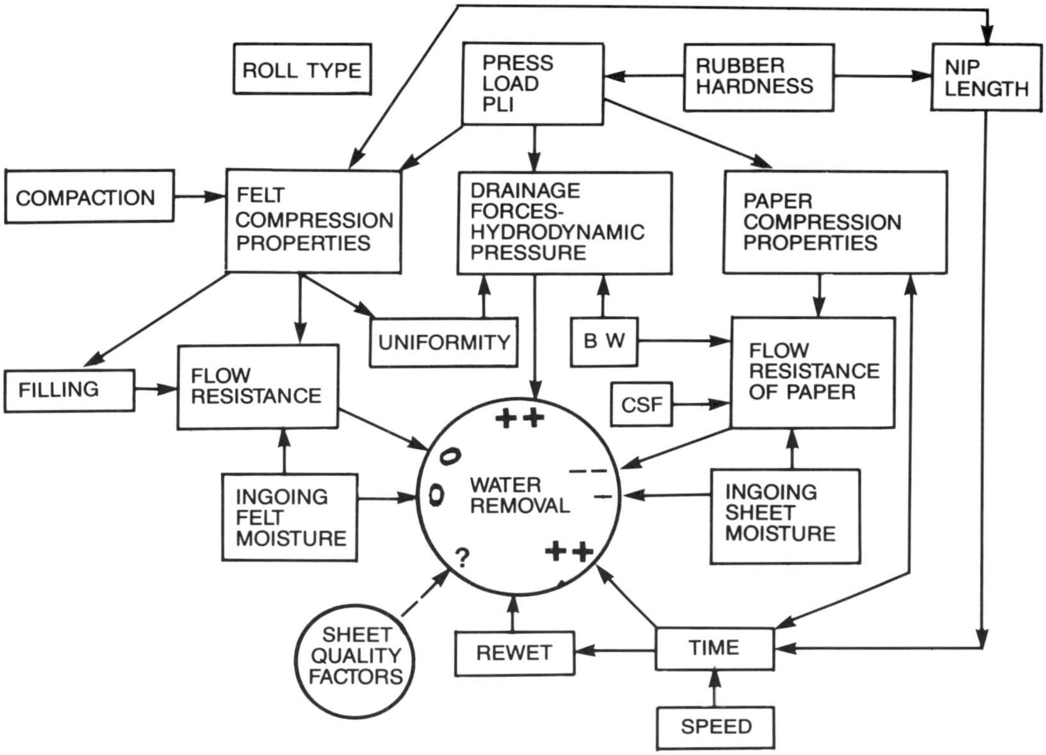

Fig. 278. Basic variables affecting water removal in a press—vertical flow assumed (Busker).

transverse-flow press and the lateral flow in the felt through midnip in a lateral-flow press. The rate of rewetting is determined by the initial recovery and interfacial permeability of the paper and the time during which the paper and felt are in intimate contact. The higher the speed, the less rewetting.

Uniformity of compression is dependent largely on felt structure, and to a lesser extent on the roll surface (land and open areas). Compression is less over the area of the paper which is not sufficiently supported by the felt and the press rolls. Non-uniform compression can be alleviated to some extent by higher basis weight of the paper. High loading tends to worsen the uniformity of compression.

A plain press has the advantages of uniform compression and little rewetting, but cannot be loaded very high due to the possibility of sheet crushing. Sheet crushing is caused by water flow in the plane of the sheet. It occurs when the felt reaches saturation and causes more inplane flow back out the nip entrance than the sheet can stand. Sheet crushing is the main factor limiting press load in a plain press. It is also the reason why several press nips have to be used. Vented nips have large loading capacity but also have potential for non-uniform compression and high rewetting at low speeds.

B. PRESS MECHANICAL ASPECTS

The sheet and felts are pressed between two rotated rolls. Generally, the top roll is mechanically loaded to create the desired pressure within the nip. The lineal nip pressure is the sum of the mechanical loading and the weight of the top roll in vertical press arrangements, divided by the length of the contacting face at the press nip. The weight of mating rolls is not considered in some inverse and cluster press arrangements. The actual pressure developed in the nip follows a curve, as illustrated in Fig. 279. The maximum value depends on several factors, including lineal pressure, roll diameter, roll hardness, and characteristics of the felt. It is difficult to estimate this pressure for many cases.

In single-felted nips, one side of the sheet is usually in contact with a smooth, hard roll, usually granite or "stonite". Sometimes, a solid metal roll with or without hard rubber covering is used. The roll in contact with the felt may be covered with a somewhat resilient rubber cover to "soften" the nip on heavyweight grades or a

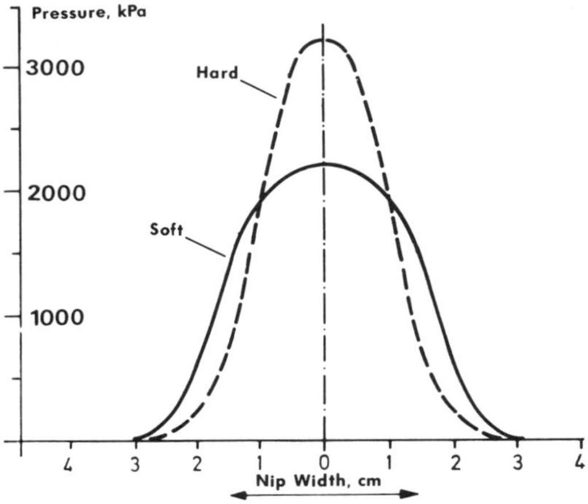

Fig. 279. Press nip pressure distribution, utilizing hard and soft rolls (45 kN/m loading).

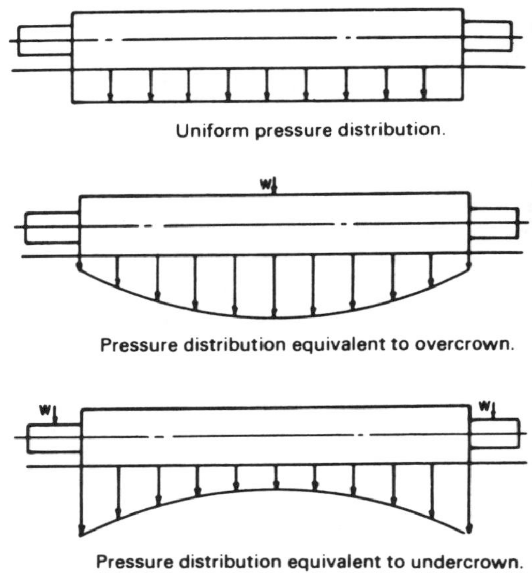

Fig. 280. Although the amount of crown may be only 1.3 mm (0.05 in.) at the midpoint of the roll, small amounts of overcrowning or undercrowning will affect the pressure distribution.

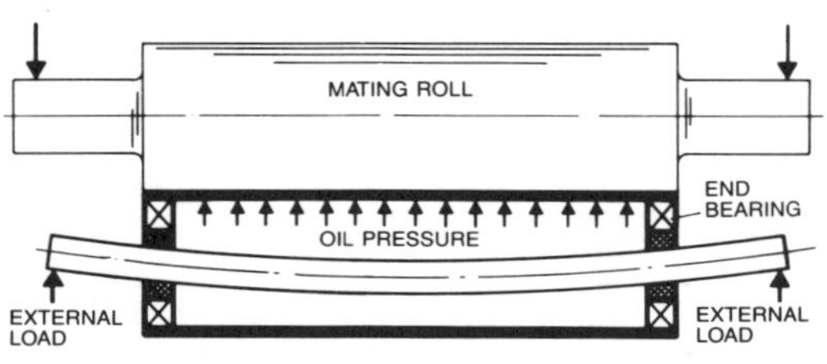

Fig. 281. The swimming roll utilizes oil pressure in the upper cavity to shift deflection from the rotating shell to the fixed shaft.

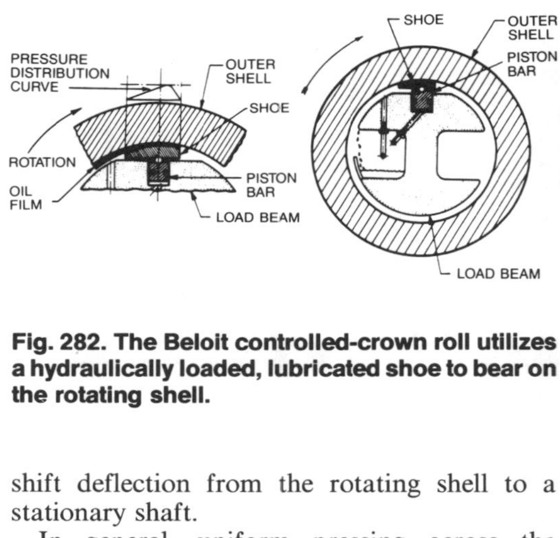

Fig. 282. The Beloit controlled-crown roll utilizes a hydraulically loaded, lubricated shoe to bear on the rotating shell.

bone-hard cover to obtain a narrow nip on lightweight grades.

Because press rolls deflect under loading, it is necessary to camber or "crown" one or both of the rolls to achieve a uniform pressure profile across the contacting face. The amount of crown required is a function of roll diameter, wall thickness, length, material of construction, and applied load. A fixed crown (obtained by grinding the roll) is suitable for only a small range of loadings. The effect of correct and incorrect crowning is illustrated in Fig. 280. Proper roll crowns are more critical with harder roll covers than softer covers. Additional information on press roll crowning is provided in TAPPI TIS 0420-06 "Roll Crown Specifications and Definitions".

Machines manufacturing many different weights and grades of paper usually employ a variable-crown roll design to achieve uniform nip conditions at different press loadings. Several designs of variable-crown rolls are available. Two designs are illustrated in Figs. 281 and 282. Most designs use hydraulic pressure to shift deflection from the rotating shell to a stationary shaft.

In general, uniform pressing across the machine is desired to achieve a level moisture profile into the dryer section. Occasionally, because of non-uniform drainage on the forming section, or other problems, it may be desirable to compensate by uneven pressing action. In this situation, a controlled-crown roll with variable deflection across the machine may be employed. One design is illustrated in Figs. 283 and 284. The width of the "streak" that can be corrected is limited by the bending characteristics of the roll. Sectionalized steam showers can also be used to achieve uniform CD moisture profiles.

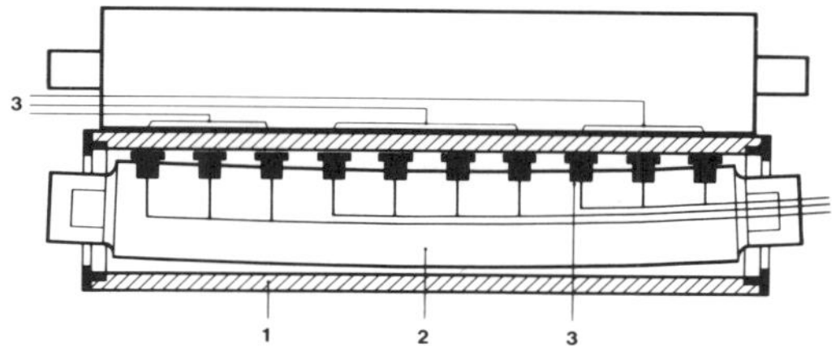

Fig. 283. The Escher-Wyss Nipco roll controls the force distribution by means of individually controlled hydrostatic elements. (1) rotating shell; (2) stationary girder; (3) piston elements.

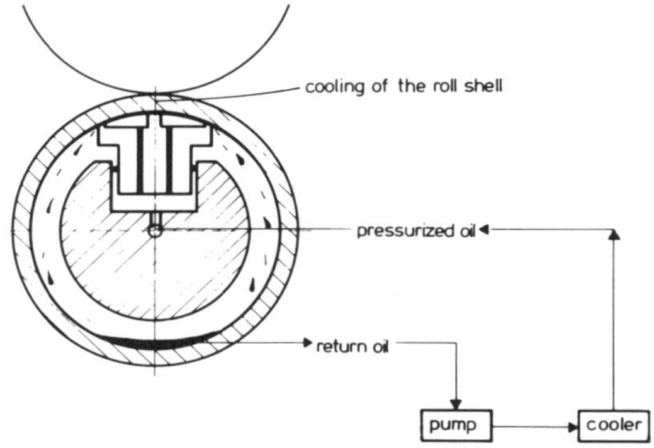

Fig. 284. The operating principle of the Nipco roll is illustrated. Oil under pressure is supplied to the underside of the elements, and through capillary ducts to the top side. As it escapes over the edges of the hydrostatic element, the oil provides a lubricating film between the shell and element.

Fig. 285. Plain press nip.

C. TYPES OF PRESSES

Fundamental research in wet pressing has shown that the most important requirement in press design is to provide the shortest path for water to follow in escaping from the nip. The shortest distance is equal to the felt thickness (usually referred to as the "vertical" direction). Therefore, the main water flow should be perpendicular to the felt, and lateral flow should be minimized. Presses with dominant vertical flow have become known as transverse or vertical flow presses.

The original presses had plain rolls as illustrated in Fig. 285. These presses were severely flow limited because the water could only leave on the entering side of the nip by lateral movement.

The suction press developed in the early 1900s was the first approach toward transverse-flow pressing. The holes in the perforated roll (Fig. 286) provide an easier escape route for the water. Water held in the holes by a stationary vacuum box inside the roll is released on the outgoing side of the nip and thrown clear by centrifugal force at speeds over approximately 300 m/min (1000 ft/min). Most of the water is drawn into the vacuum box at lower speeds and removed through the vacuum system. Some water is able to flow directly to the holes, but most water also must travel laterally as well. The hollow shell construction with perforations limits the amount of pressure that can be applied. Stress fatigue failure of rolls has been a continuing problem over the years. Special alloys have been developed to increase corrosion resistance and minimize suction roll failure. Suction roll application on wide, high-speed machines is the most demanding. TAPPI TIS 0402-10 "Guide for Evaluation of Paper Machine Suction Roll Shells" provides information on monitoring suction rolls.

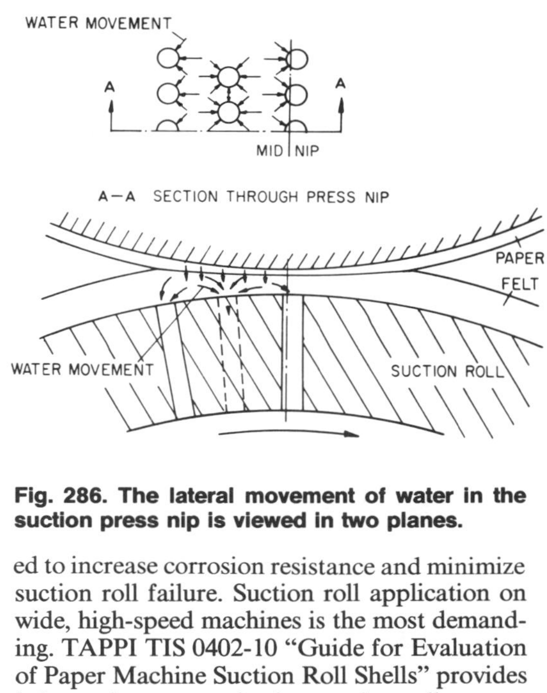

Fig. 286. The lateral movement of water in the suction press nip is viewed in two planes.

The grooved roll transverse-press design was introduced in 1963. As illustrated in Fig. 287, grooves in the roll cover provide easily accessible receptacles for the expelled water. The helically cut grooves are typically 2.54 mm (0.1 in.) in depth, 0.5 mm (0.02 in.) wide on 3.2 mm (0.125 in.) centers (i.e., 8 grooves per inch). The maximum lateral distance for water travel in the grooved press is 1.3 mm (0.05 in.), as compared to typical figures of 5.1 mm (0.2 in.) and 19 mm (0.75 in.), respectively, for the suction and plain

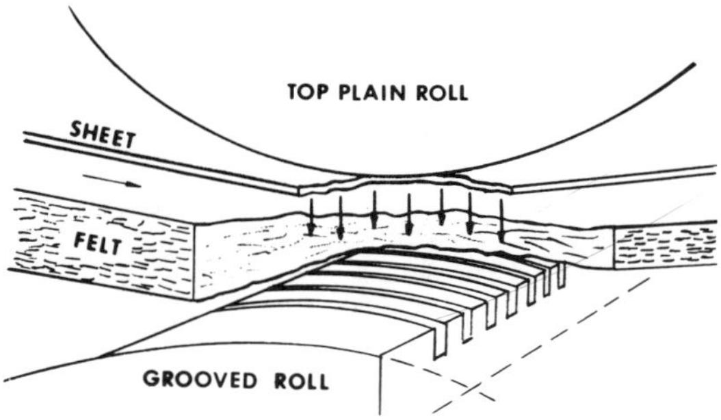

Fig. 287. The grooves provide an outlet for expressed water.

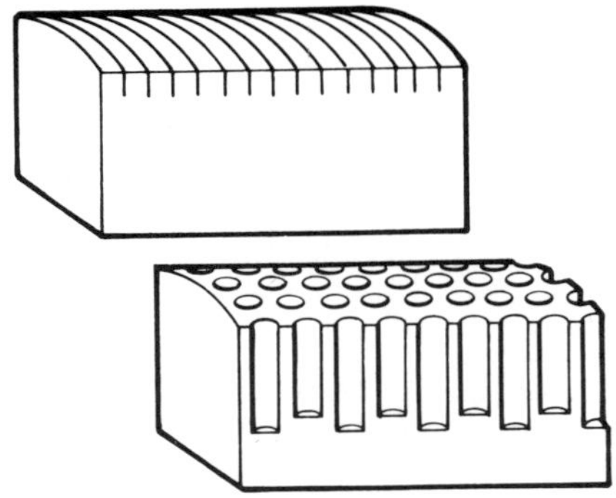

Fig. 288. The greater void volume of the blind-drilled roll as compared to the grooved roll is illustrated by comparing segments.

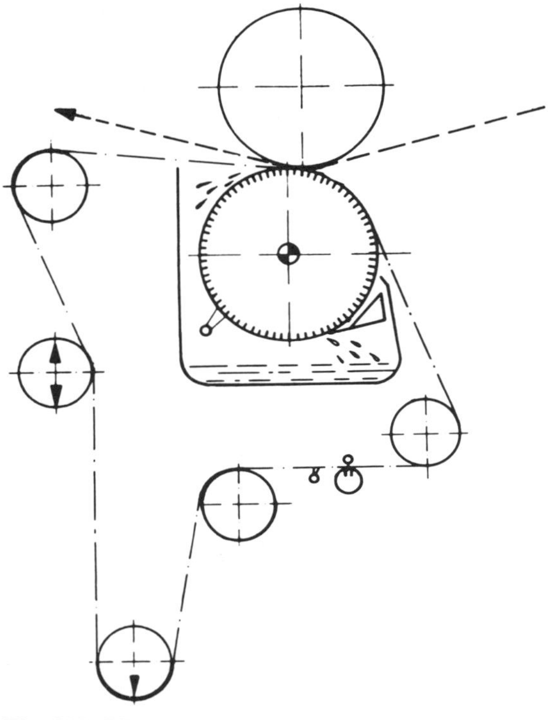

Fig. 289. Blind-drilled press arrangement.

presses. Other groove patterns are listed in TAPPI TIS 0404-37 "Press Roll Cover Application Guidelines". Since the grooved roll has a solid shell, higher pressures can be applied. The water caught in the grooves is thrown off by centrifugal force at high roll surface speeds. Grooves are kept clean with water sprays and doctors at machine speeds below 600 m/min (2000 ft/min). Grooved roll covers must be fairly hard (less than 10 P & J) to maintain groove integrity. The land areas between grooves tend to "mushroom" and hard covers and regular grinding are required to keep the grooves open. Some mills have installed stainless steel grooved rolls to minimize maintenance requirements.

Another innovation in transverse press design is the "blind-drilled press roll". These rolls can be distinguished from suction rolls since only the covers are drilled (Fig. 288). The holes are smaller and more closely spaced, thus reducing the distance for lateral travel. Typical hole sizes and depths are listed in TAPPI TIS 0404-37 "Press Roll Cover Application Guidelines". The wells tend to self-clean by the normal action of centrifugal force. Blind-drilled patterns can be installed in softer roll covers than grooved rolls since open area is higher and there is less tendency to close holes than grooves. A typical installation is shown in Fig. 289.

The fabric press is another development of a transverse press used more in Europe than in North America. The concept is illustrated in Fig. 290. A multiple-weave, non-compressible fabric belt passes through the nip between the felt and the rubber-covered roll to provide void volume to receive the expressed water. Water is removed from the fabric by passing it over a suction box on the return run. Inner fabrics have been installed on plain presses to increase void volume in the nip and on suction presses to reduce the tendency for the cover hole pattern to mark the sheet. Development of modern press felt designs with high void volume has reduced the benefits from fabric presses.

A modification of the fabric press concept is embodied in the shrink sleeve press (Fig. 291). A non-compressible fabric jacket or sleeve is shrunk over the press roll to provide void volume. Shrink shelves are normally installed on plain rolls or suction rolls where shadow marking problems are encountered. Modern press felt designs have reduced the use of fabric presses and shrink sleeves.

One of the most effective presses in the first press position (i.e., for flow-limited presses) is

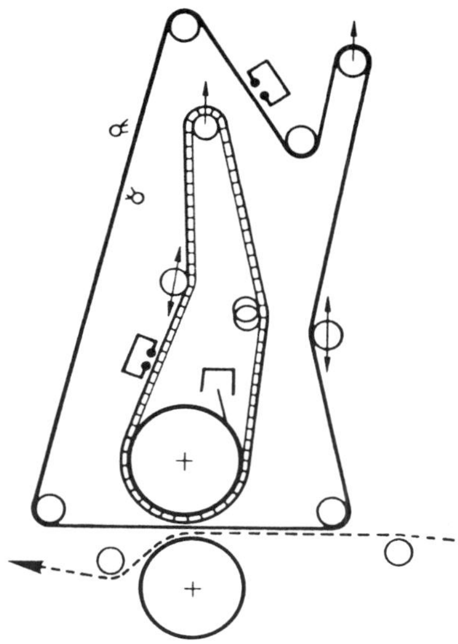

Fig. 290. Fabric press (Albany International Corp.).

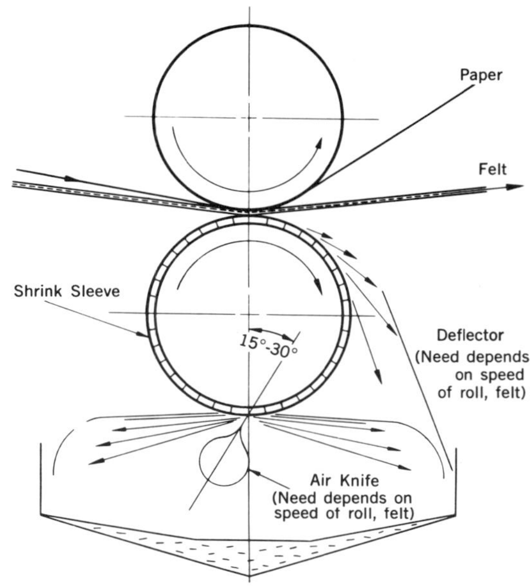

Fig. 291. Shrink fabric press with air knife to remove water from fabric (Albany International Corp.).

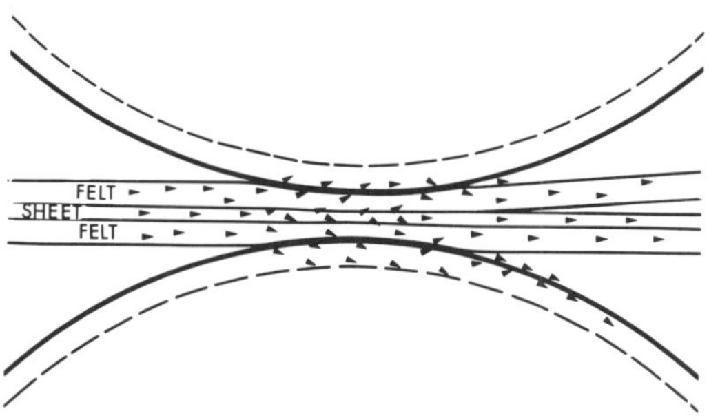

Fig. 292. Double-felted nip with grooved rolls used in both positions.

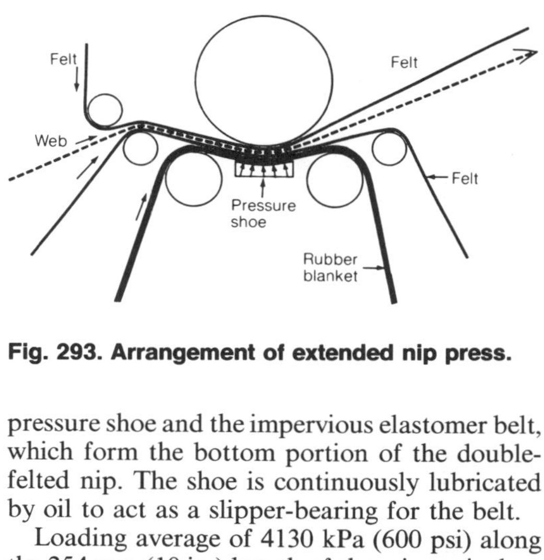

Fig. 293. Arrangement of extended nip press.

the double-felted and double-vented nip, as illustrated in Fig. 292. The distance for lateral travel has been reduced and dewatering can take place in both directions, thus reducing the distance for vertical travel. Double-felting is also useful at second and third presses for sheets above 130 g/m^2 basis weight. Double-felted presses with load capability as high as 350 kN/m (2000 pli) have been installed on linerboard machines.

A totally new type of press was introduced in 1981. The "extended nip press" features a very wide nip to give the sheet a long dwell time at high pressure. This press provides a much drier and stronger sheet after the last nip due to better consolidation of the web structure. The configuration used to obtain the wide nip is shown in Fig. 293. Key components are the stationary pressure shoe and the impervious elastomer belt, which form the bottom portion of the double-felted nip. The shoe is continuously lubricated by oil to act as a slipper-bearing for the belt.

Loading average of 4130 kPa (600 psi) along the 254 mm (10 in.) length of shoe, is equivalent to 1050 kN/m (6000 pli) by the normal method of specifying press loadings. The relatively low pressure level, together with a controlled rate of compression, reduces crushing potential. A dramatic improvement in dewatering is accomplished because the pressure is sustained for a long period of time, up to eight times that obtainable with conventional roll presses (see Fig. 294). Wide nip presses have only been applied on heavyweight sheets (linerboard and corrugating medium) to date. Application of the ENP concept on lightweight grades will require a

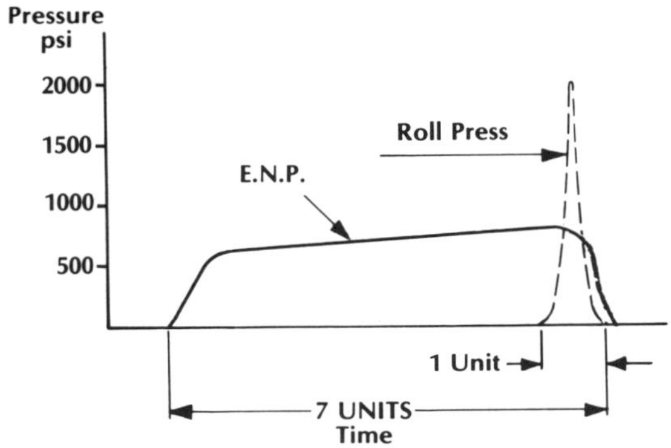

Fig. 294. Nip pressure profiles.

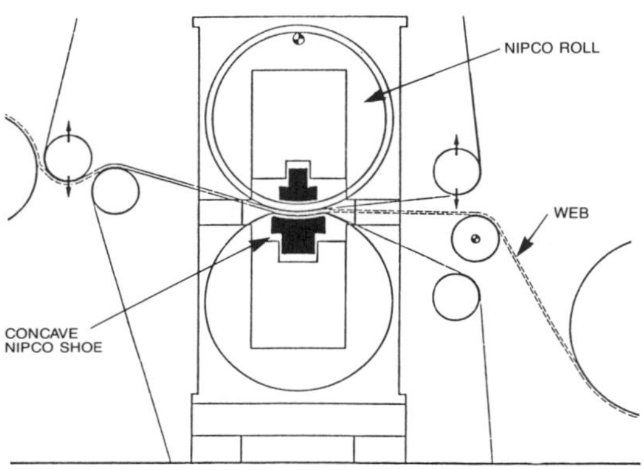

Fig. 296. Intensa press S (Sulzer Escher Wyss).

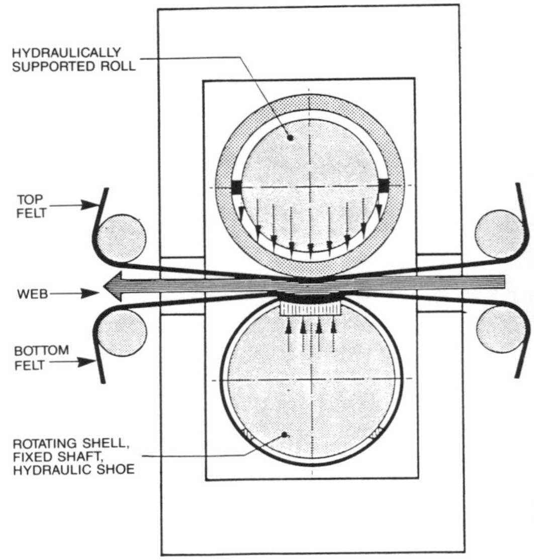

Fig. 295. Flexonip press (Voith, Inc.).

top roll cover with good release characteristics to permit single felting and capability to withstand press loads. Other suppliers have also developed shoe-type presses that achieve high sheet dryness and good web consolidation. One concept is shown in Fig. 295. The bottom roll has a rotating shell, fixed shaft and hydraulic loading shoe. A hydraulically supported roll with anti-deflection control runs opposite the loading shoe. Maximum loadings are equivalent to 980 kN/m (5600 pli).

Another shoe-type press includes a concave shoe inside a flexible roll shell in the bottom position and an anti-deflection roll in the top position. A sketch of this press is shown in Fig. 296.

Some mills have installed two shoe-type presses in series. Tandem shoe-type presses achieve 3–4% points higher dryness than a press configuration where a shoe-type press follows one or more roll presses. Figure 297 shows a press configuration with extended nip presses in the first and second press positions.

A relatively new press design incorporates a large steel roll that is heated with steam internally to improve water removal. The sheet is heated by a 5–9 foot diameter steel roll with a special alloy coating to optimize heat transfer to the web. The large-diameter steel roll can be incorporated in several different press configurations. The center roll in a triple-nip press is a common application.

These presses have been installed on paper machines producing both light- and heavyweight grades of paper. Typical press loadings are 140–175 kN/m (800–1000 pli). Exit sheet consistencies of 52–56% are claimed. High sheet drynesses result from heating the sheet.

Impulse drying is a pressing technique currently under development [7]. The top roll of a press nip is heated to a high enough temperature to generate steam in the nip which pushes water out of the web (Fig. 298). Nip temperatures of 150–480°C, pressures of 275–690 kPa (40–100 psi) and residence times of 100–150 ms are used. Laboratory studies indicate that newsprint at 50% solids can be dewatered to 90% solids in 254–406 mm (10–16 in.) of nip. The combination of press loads, long residence time, and high temperatures provide much greater water removal than is achieved with conventional pressing. Energy requirements are less than half of those required by conventional drying, but high grade energy is used. Problems with sheet sticking to the heated press roll and sheet delamination are currently being investigated.

The term "press drying" denotes the combined application of pressing and drying under

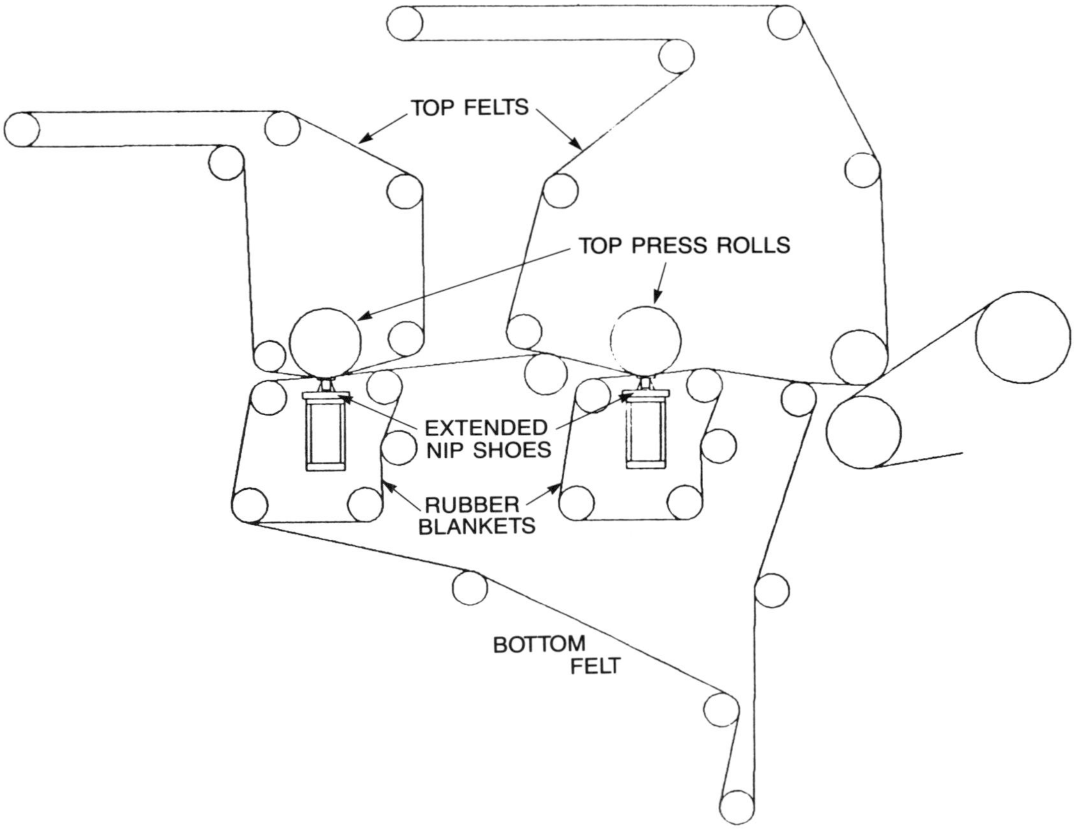

Fig. 297. Tandem extended nip presses (Beloit Corp.).

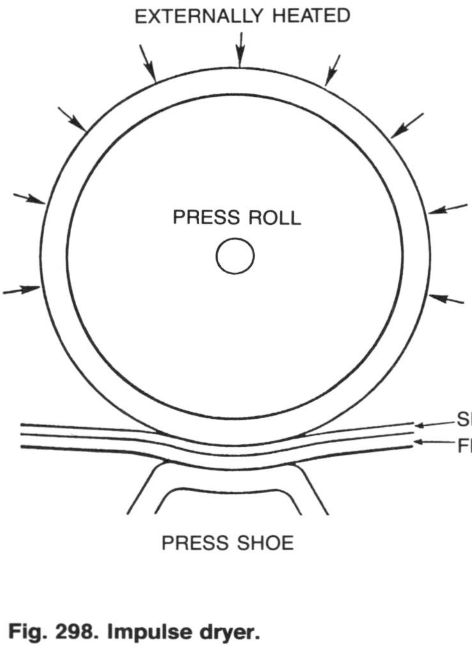

Fig. 298. Impulse dryer.

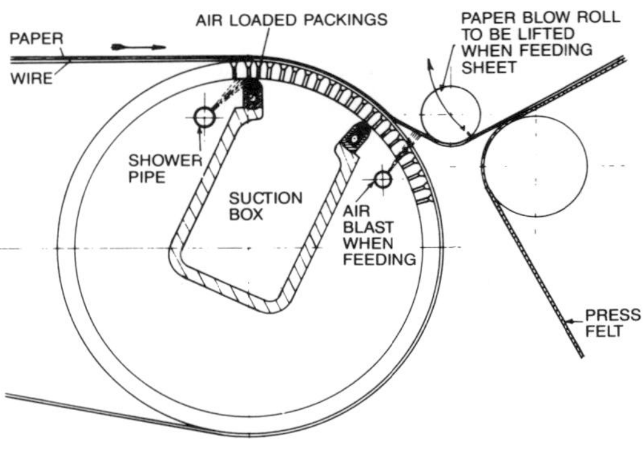

Fig. 299. Open draw transfer. The position of the sheet in relation to the couch roll suction box is typical.

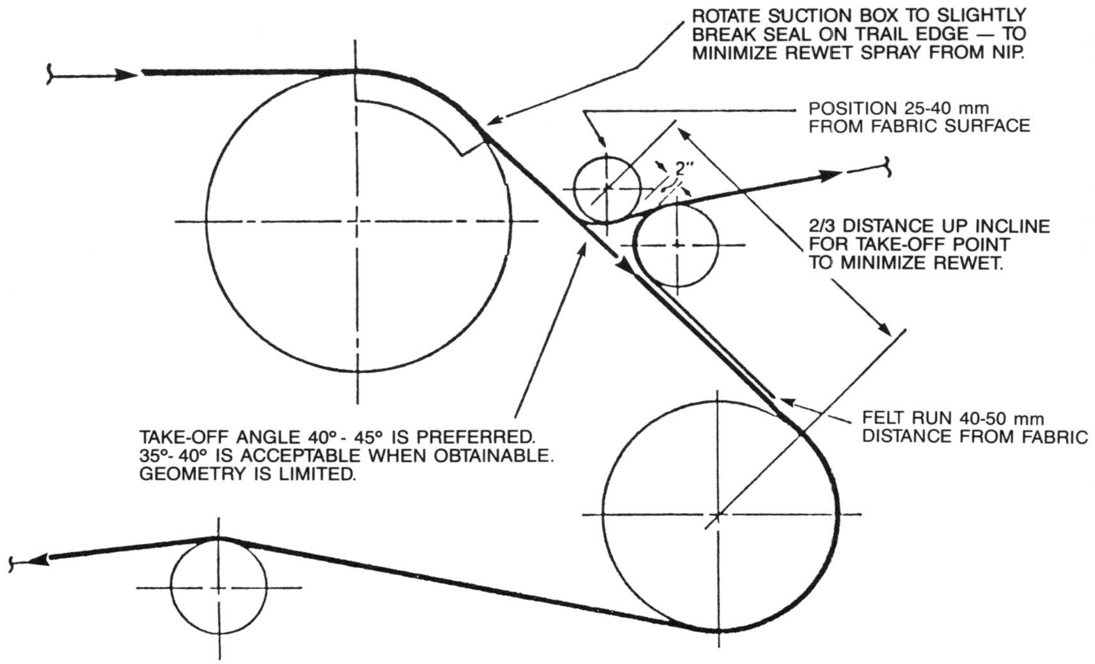

Fig. 300. Open draw transfer arrangement with wire-turning roll (Huyck Formex).

restraint and encompasses a range of technologies. These include compressive pressures and hot surface temperatures sufficient to produce internal sheet temperatures above 100°C. The sheet is dried under restraint to limit in-plane shrinkage. Generally, the conditions are more intense than those used in conventional drying but mild compared to those for impulse drying. Several different press drying mechanisms have been proposed but none is known to be in commercial operation.

D. SHEET TRANSFER FROM WIRE TO PRESS SECTION

For conventional machines, the two basic means of transferring the wet web from the forming section to the press section are an open draw and vacuum pickup (or suction pickup).

Originally, all machines had open draws. The sheet was transferred across the open draw without support onto the first press felt, as illustrated in Figs. 299 and 300. The major difference between the two configurations is that the arrangement shown in Fig. 300 includes a fourdrinier wire turning roll. This configuration is sometimes referred to as a "poor man's pickup". The tension needed to pull the wet web off the wire is provided by a speed differential between the press and wire sections. Open draws are commonly used for heavyweight sheets and older machines operating at speeds below 600 m/min (2000 ft/min). Runnability problems are encountered at high speeds since the required tension for removing the web from the forming fabric increases exponentially with speed. Operation of an open draw is sensitive to changes in stock drainage, quality factors, and sheet formation. Best results are normally achieved with a take-off angle of 40 to 45° and minimal spacing between rolls.

The vacuum pickup concept, illustrated in Fig. 301, was first applied commercially in 1954, and is now an integral part of most press arrangements for lightweight sheets. The sheet is picked up off the wire by a felt which wraps a vacuum roll at the point of contact. The sheet is then carried underneath the felt to the first press nip.

The first open draw on the paper machine occurs in the press or dryer section with a vacuum pickup. The first open draw is always a critical point with respect to machine operation, because the wet sheet is relatively weak and subjected to stresses from a variety of causes. At greater speeds, the need to provide support for lightweight sheets throughout the pressing operation and into the drying section has become paramount. A number of "no-draw" press configurations are offered by equipment manufacturers.

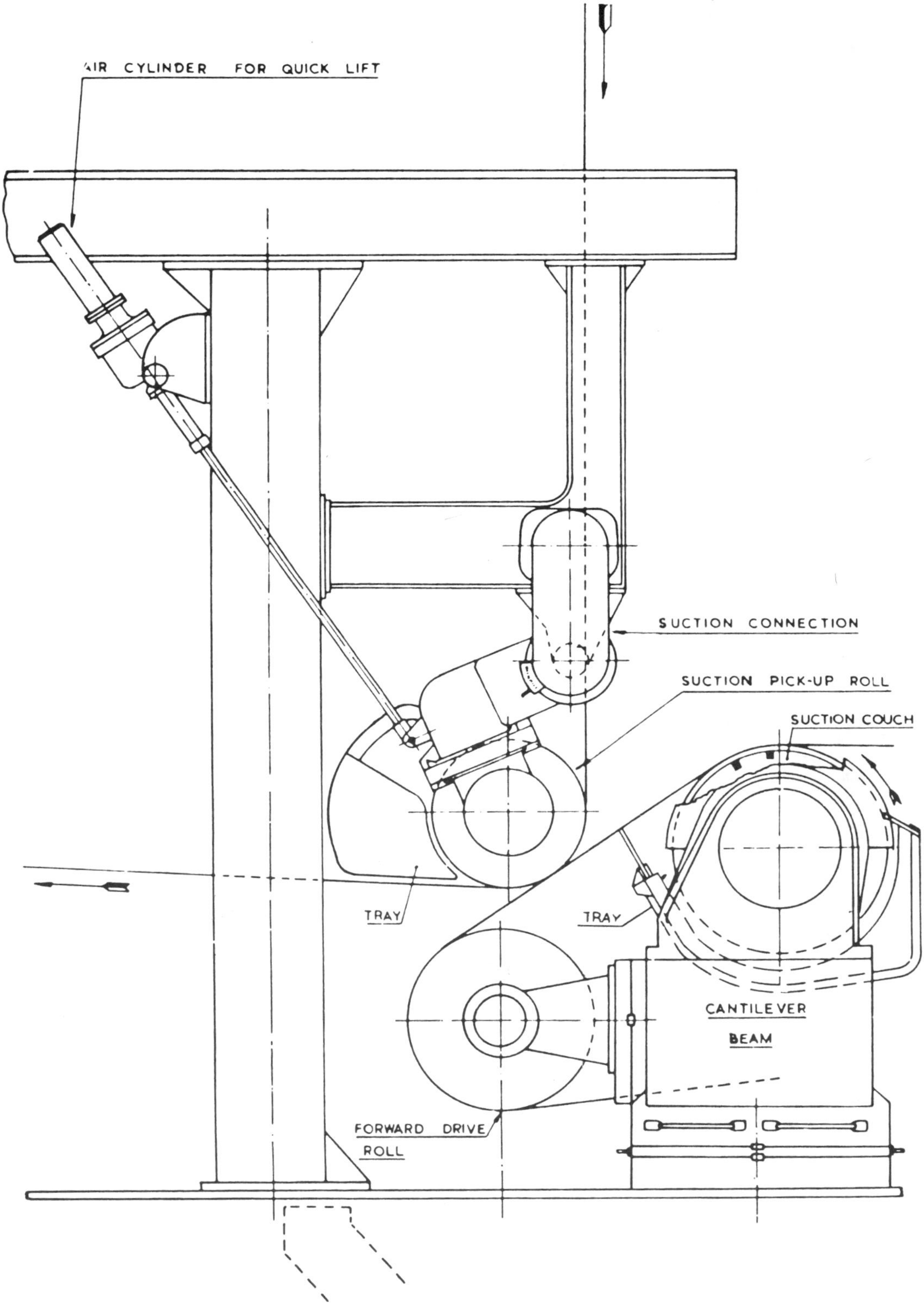

Fig. 301. Typical suction pickup arrangement.

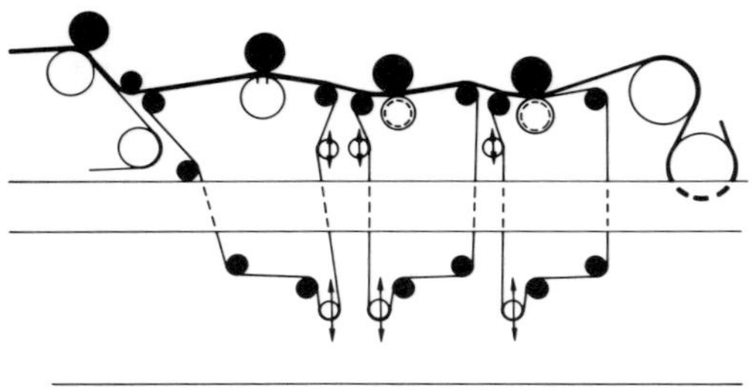

Fig. 302. Straight-through press (Beloit Corp).

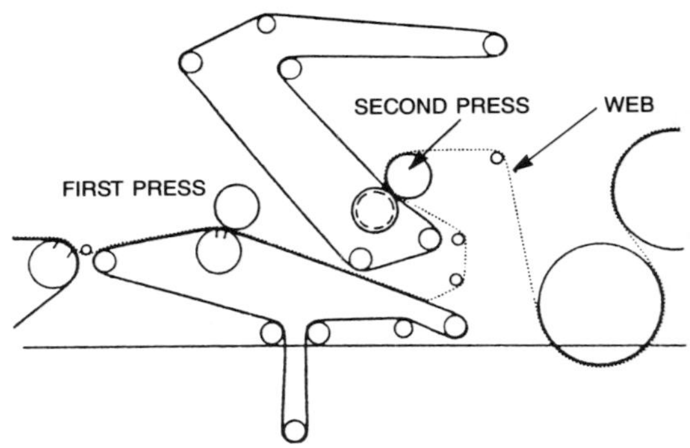

Fig. 303. Reverse press (Beloit Corp.).

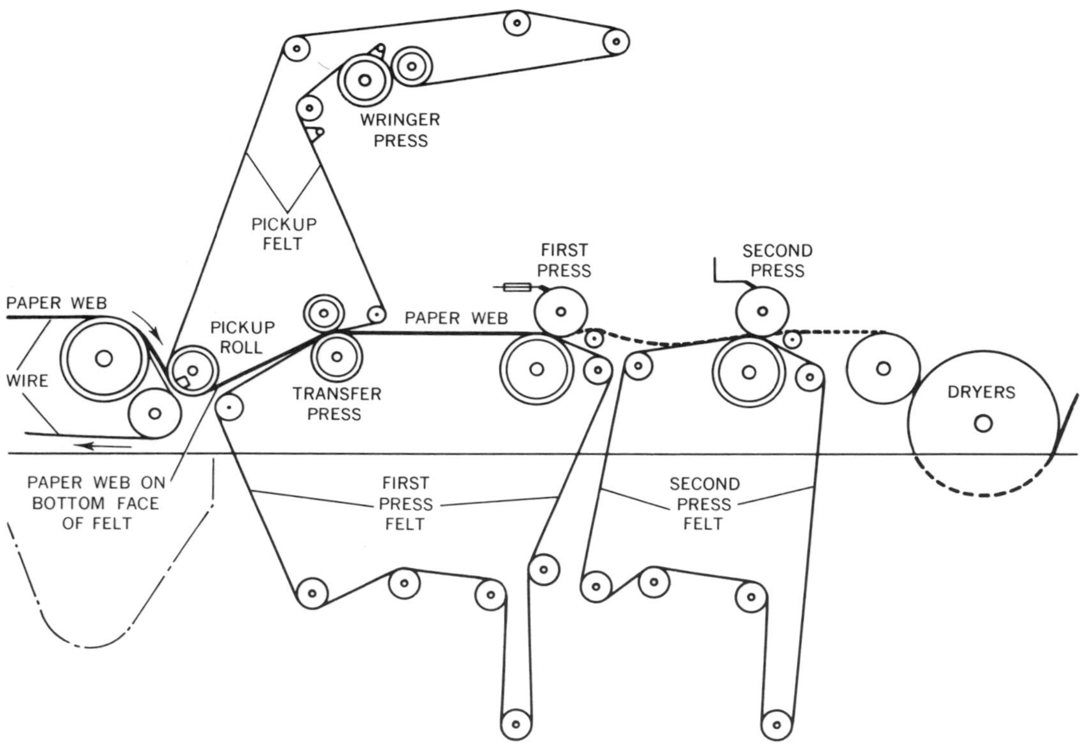

Fig. 304. Transfer press arrangement with suction pickup.

E. TYPES OF PRESS ARRANGEMENTS

The straight-through press (Fig. 302) is the oldest and simplest press arrangement, and is still used on pulp and board machines and many older machines. Originally, each press consisted of a smooth top roll and a bottom felted roll, so that only the top surface of the sheet received smooth roll contact. Later, an inverse second press (smooth roll on the bottom) was utilized so that the wire side of the sheet also contacted a smooth surface. A reverse press is another method to achieve water removal from both sides of the sheet. A reverse press requires that the sheet to be taken under the reverse press, turned around on a paper carrying roll, and brought back to the reverse press as shown in Fig. 303. This arrangement cannot be used at high machine speeds because of sheet runnability problems.

A number of transfer press arrangements have been used on vacuum pickup machines to transfer the sheet from the pickup felt to the first press felt without an open draw (Fig. 304). The older transfer presses were rather inefficient for

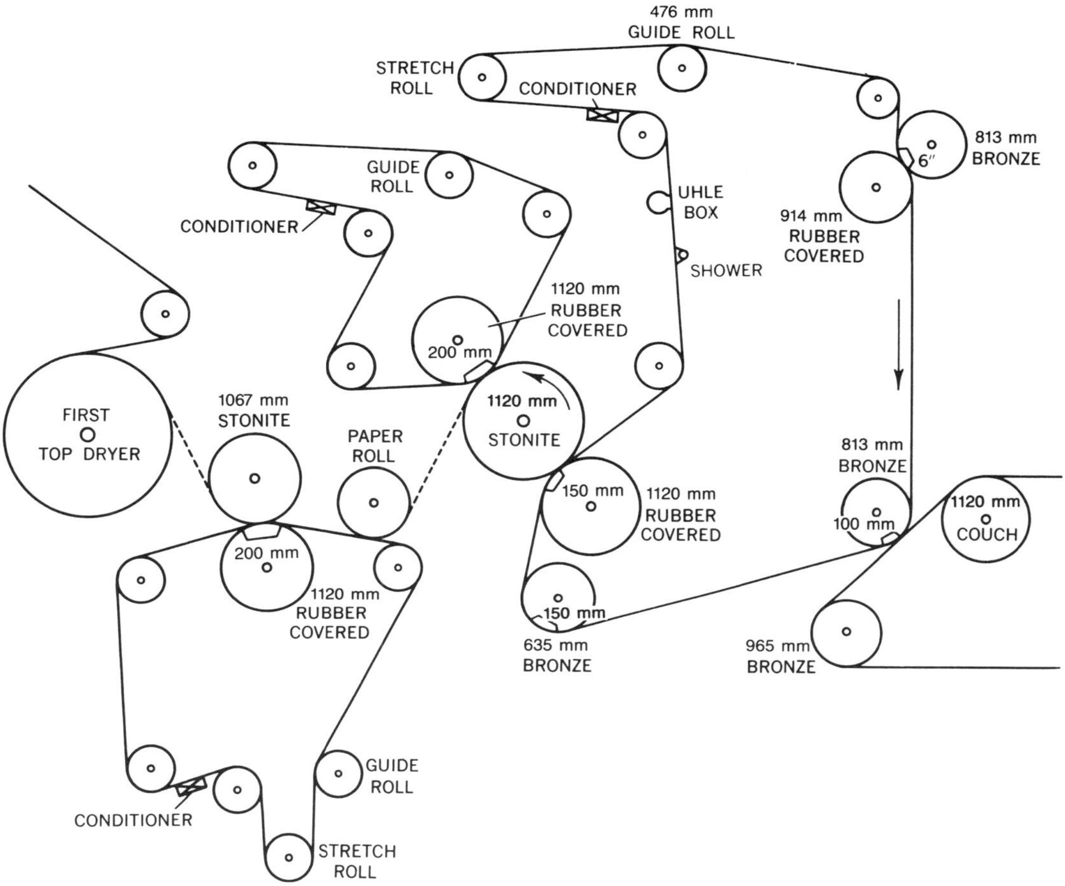

Fig. 305. Twinver press arrangement showing representative roll sizes and suction boxes.

removing water because press loads were low and the suction transfer roll caused high sheet rewetting. However, modern transfer presses with double-felted nips can be very effective [8].

Several arrangements have been developed as an alternative to the transfer press, the best known being the twinver press (Fig. 305). This arrangement has a number of advantages: the draw is eliminated between the first and second presses; the wire side contacts the smooth roll through two nips; only one smooth roll is required for the two nips. A later modification has eliminated the suction turning roll on the pickup felt run by adding a wide suction-holding box to the first press suction roll.

Combination presses are used on some machines to reduce machine direction space requirements and minimize capital cost. Figure 306 shows a configuration where a single suction roll is used for suction pickup and the suction first press nip.

High machine speeds led to development of no-draw presses which provide support to the sheet throughout the pressing operation. Most of these arrangements include at least one double-felted nip, as illustrated in Figs. 307 and 308. Triple-nip, no-draw presses run at machine speeds over 1400 m/min (4600 ft/min) on lightweight grades such as newsprint.

Sheet marking problems and suction roll failures on some wide machines resulted in development of a no-draw configuration using separate rolls for the first and second press nips. The suction roll in triple-nip presses is used for the top roll in the first press nip and is run against the large center roll to form the second press nip. The Tri-Vent configuration uses a separate roll to form the second press nip against the center roll as shown in Fig. 309.

Most of the water is removed from the top side of the sheet in triple-nip presses. This can be a significant limitation on coating base stock and fine paper grades that are printed on both sides.

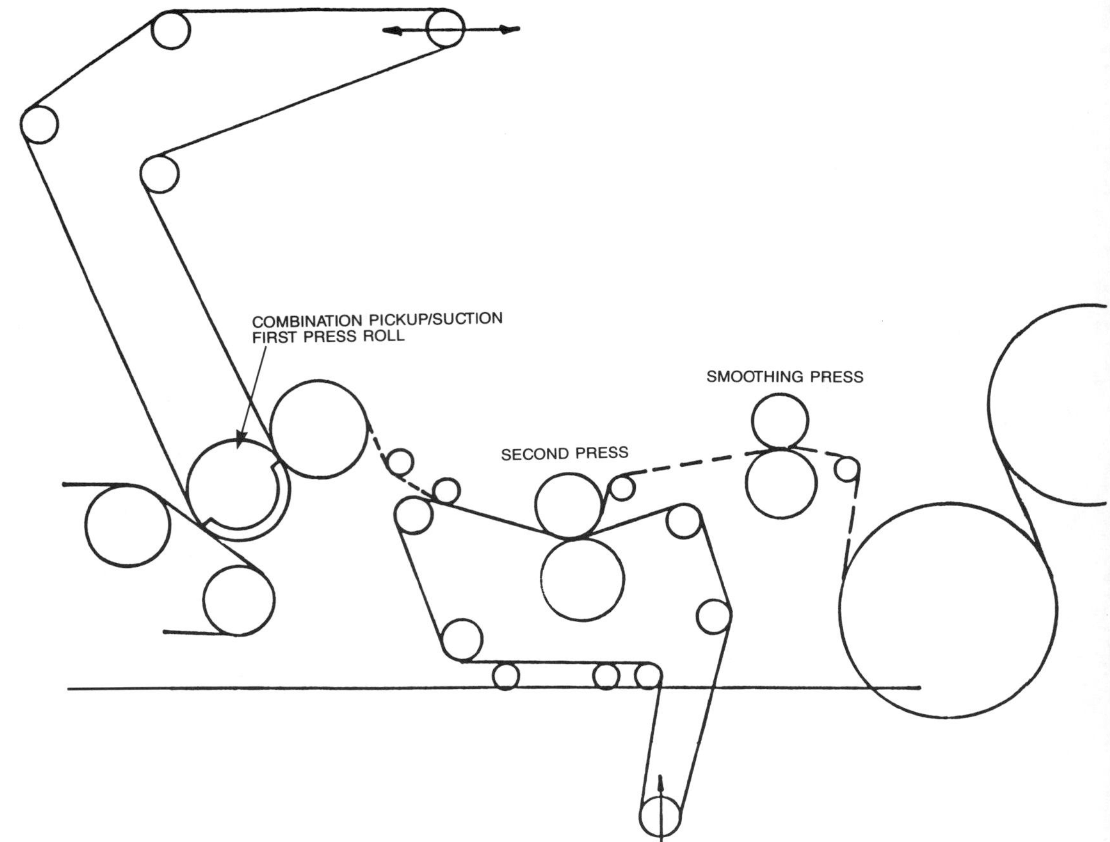

Fig. 306. Combination pickup/suction first press plus second press (KMW).

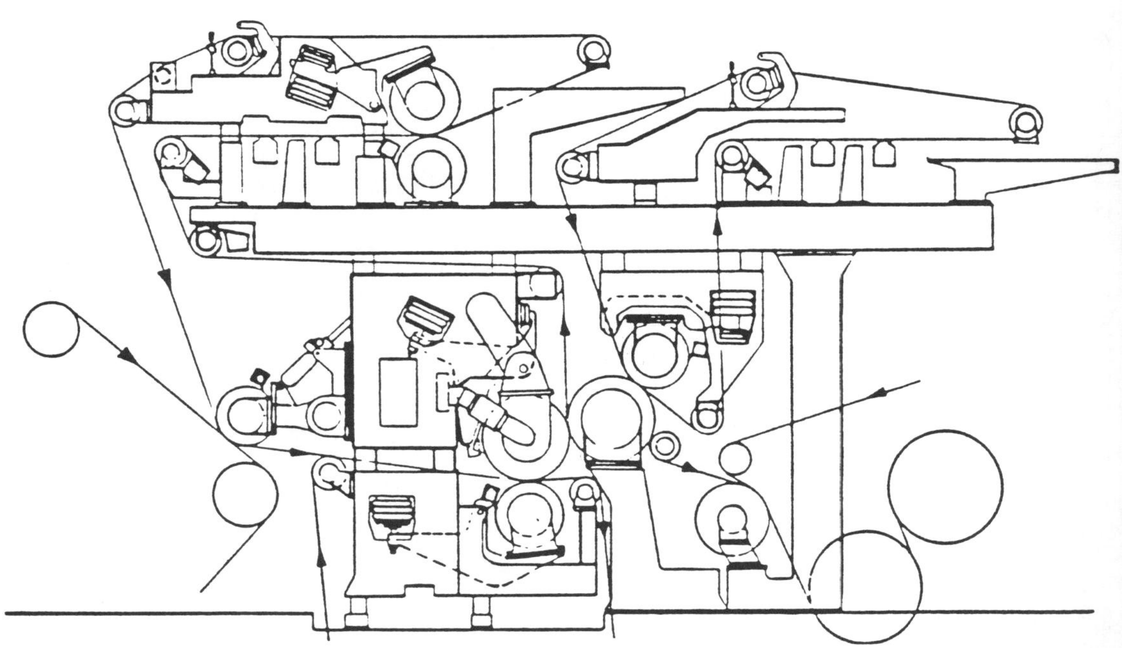

Fig. 307. Beloit tri-nip press.

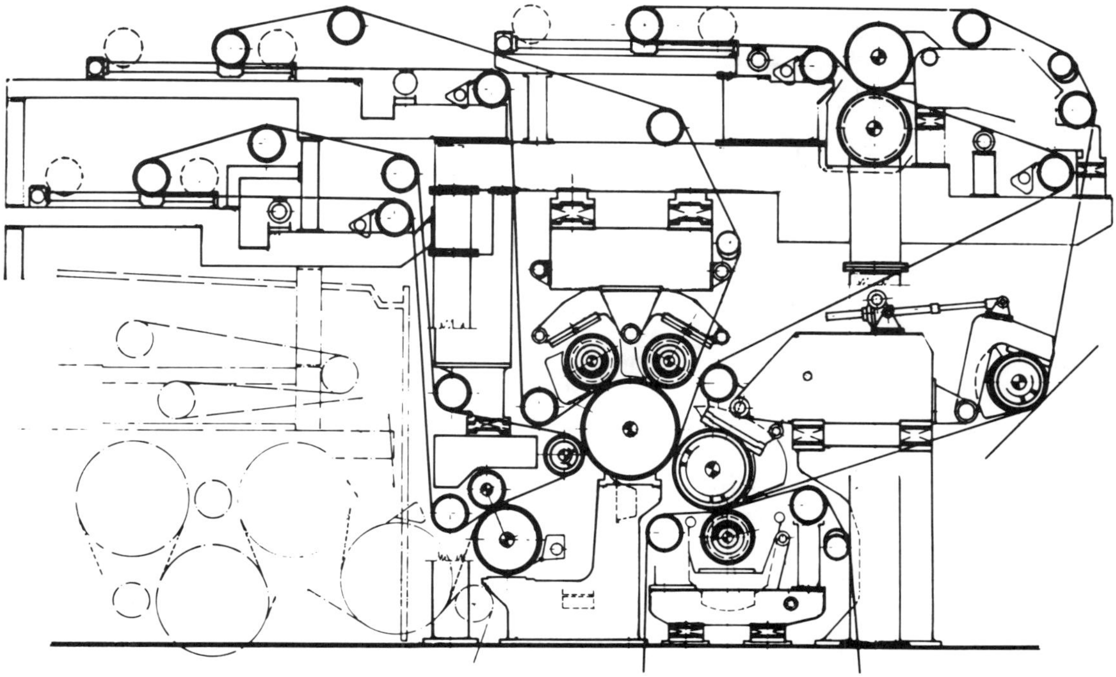

Fig. 308. The Valmet Sym-Press utilizes a closed transfer section to take the sheet off the granite roll and transfer it to the dryer fabric.

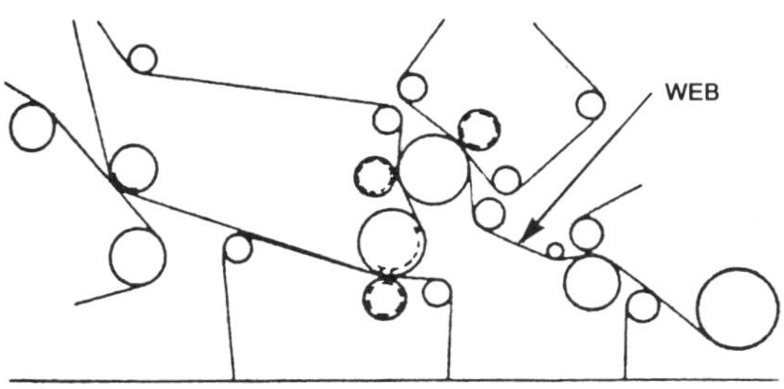

Fig. 309. Tri-Vent press (Beloit Corp.).

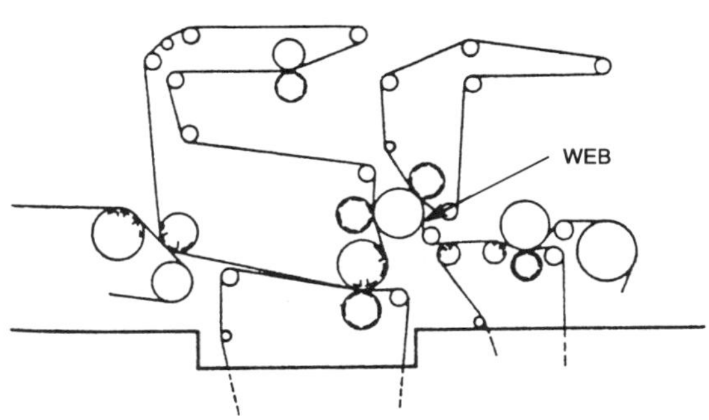

Fig. 310. Tri-Vent plus fourth press (Beloit Corp.).

One-sided dewatering results in differences in coating and ink receptivity between the two sides and can cause linting and dusting problems.

This concern led to installation of separate fourth presses after no-draw three nip presses and press configurations with an open draw between the second and third press nips to permit opposite side dewatering at the last two press nips. Fourth presses increase web dryness to the dryer section in addition to reducing sheet two-sidedness. Sketches showing a fourth press and Bi-Vent plus third press are shown in Figs. 310 and 311.

Some mills install smoothing presses after triple-nip presses to improve smoothness and reduce two-sidedness. Positioning the hard smoothing press roll on the roughest side of the sheet helps reduce smoothness differences between the top and wire sides. Smoothing presses are known to increase drying rates by permitting more intimate contact between the web and dryer surfaces. Smoothing presses also increase sheet density somewhat although no water is removed from the sheet.

Several machines producing linerboard and other heavyweight grades have two double-felted presses to optimize water removal [6]. A typical configuration is shown in Fig. 312. Highest sheet drynesses are being obtained where shoe-type presses are installed in the second press position.

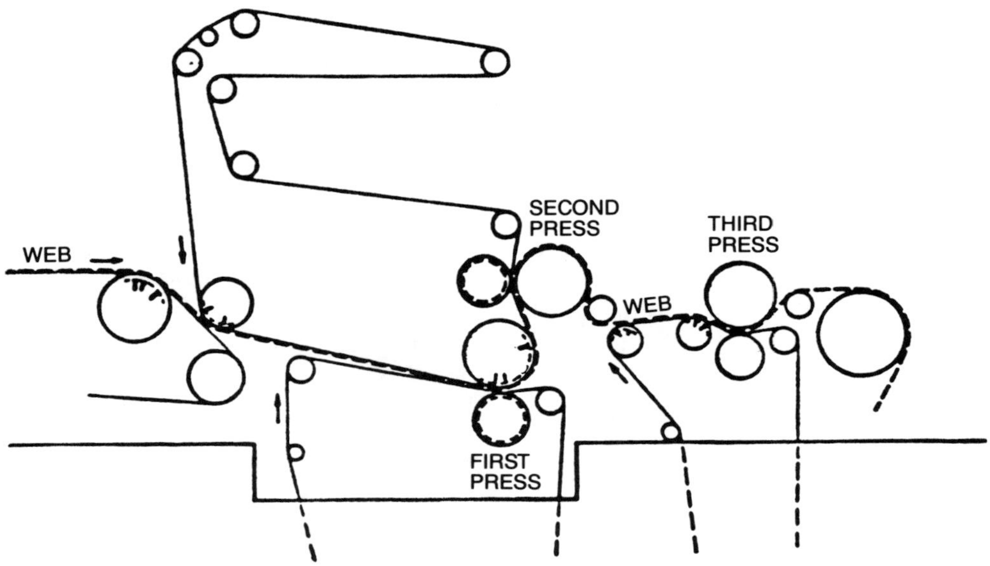

Fig. 311. Bi-Vent plus third press (Beloit Corp.).

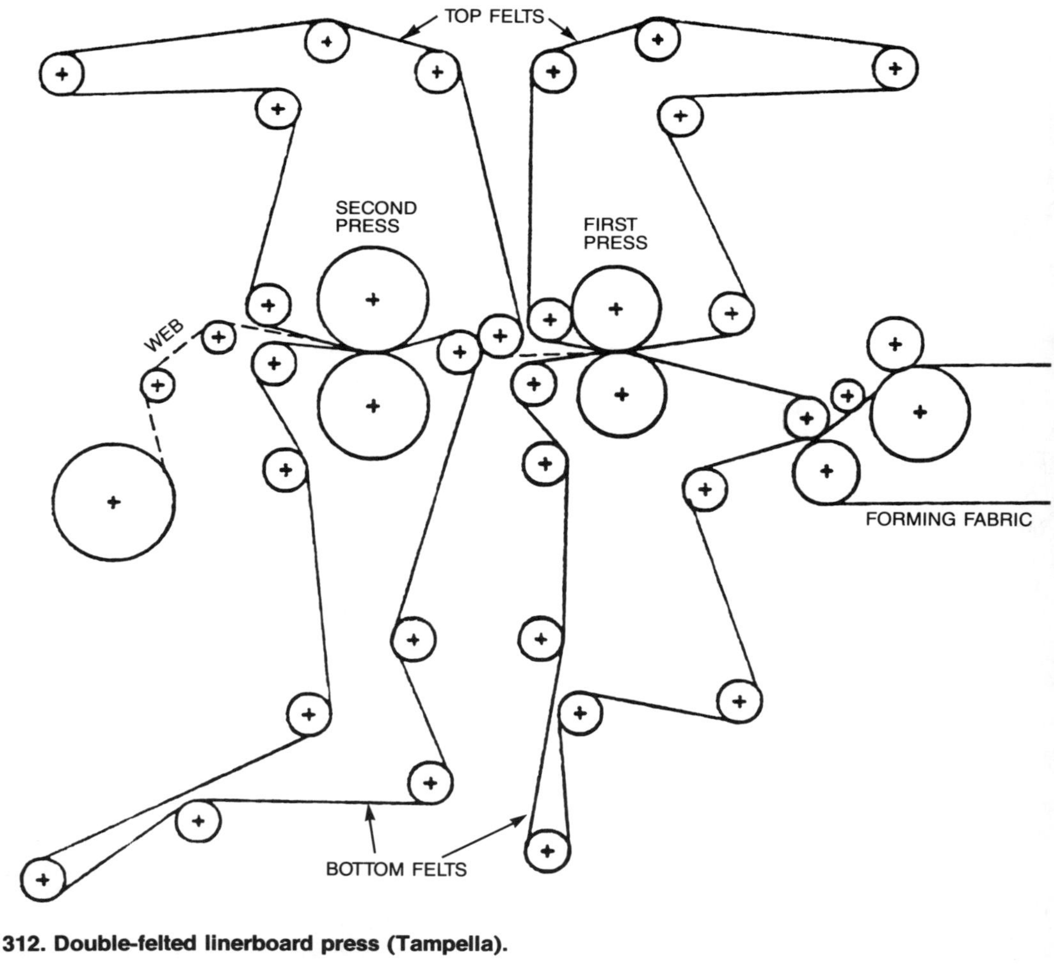

Fig. 312. Double-felted linerboard press (Tampella).

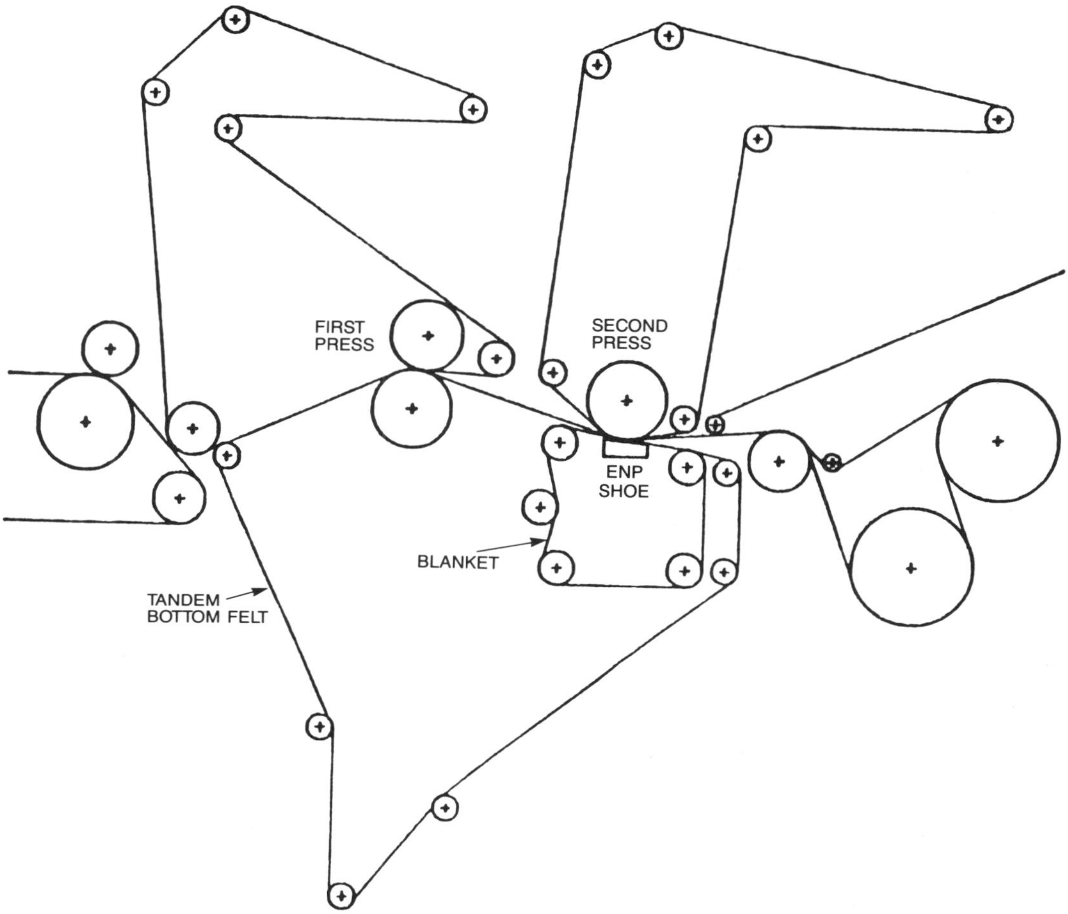

Fig. 313. ENP with tandem bottom felt (Beloit Corp.).

Other machines have installed large-diameter conventional press rolls to permit loading up to 350 kN/m (2000 pli).

Most machines with two double-felted presses have an open draw between the first and second presses. The open draw can cause web breaks, which damage second press felts. Use of a tandem bottom felt as shown in Fig. 313 eliminates the open draw and the potential for second press roll cover and felt damage from wads. Some mills have installed bi-nip presses ahead of the last press to increase sheet dryness at the first open draw.

F. PRESS CONFIGURATION CONSIDERATIONS

There are many factors that should be considered in determining the best press configuration for a particular machine. Some of the parameters that must be considered include:

1. Sheet quality requirements.
2. Machine speed.
3. Sheet basis weight.
4. Stock furnish characteristics.
5. Broke handling.
6. Forming section configuration.
7. Potential for sheet rewet.
8. Felt change requirements.
9. Potential for steam shower installation.
10. Available felt designs.
11. Vacuum requirements.
12. Available space.
13. Roll doctor requirements.

1. Steam showers

Press-section steam showers are an effective way to increase sheet dewatering. Each 10°C increase in sheet temperature increases sheet consistency out of the last press by approximately one percentage point (Fig. 314) [5].

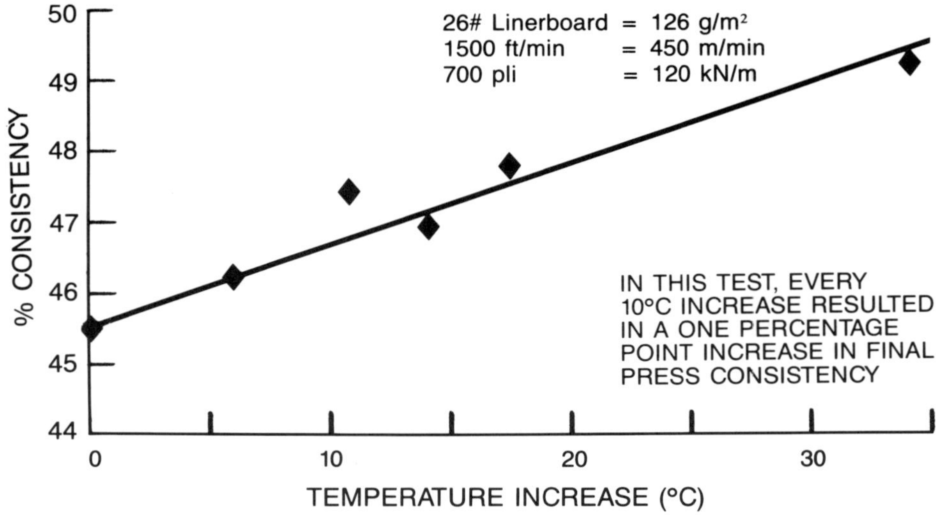

Fig. 314. Sheet consistency vs. temperature increase.

The major benefits of steam showers are reduced dryer-section steam consumption, a drier and stronger sheet leaving the press section, and better cross-machine uniformity. The amount of dryer steam savings depends on the degree of dewatering increase achieved in the press section. Sheet consistency increases up to 3% have been reported on lightweight grades. Since the sheet is hotter entering the dryers, fewer dryers are needed to heat the sheet to evaporation temperature.

Dryer steam consumption is reduced, and more dryers are available for evaporating water from the sheet. Net steam savings of 1.8–2.0 kg steam/kg applied in the press section are typical for well-designed units. Speed increases of 5–25% have been achieved on drying-limited machines.

Many modern steam showers have 150–300 mm (6–12 in.) wide segments across the width of the machine to provide profile correction. Greater application of steam in wet areas provides greater dewatering and helps level CD moisture profiles. The magnitude of correction on a particular machine is dependent on paper weight, moisture content, sheet porosity, and other variables. Good moisture profiles permit raising the average reel moisture content on some grades, which reduces drying energy consumption and can improve calendering. Some profiling steam showers have been integrated with process control computer systems to provide automatic control of moisture profile.

Press section steam showers can also affect sheet quality tests. High web temperature in the press section yields a more consolidated, denser, and better bonded sheet. Installation of a steam shower ahead of the first press nip gives the advantage of more nips working on the hot sheet. Each 10°C increase in sheet temperature softens cellulose fibers by approximately 15%, which increases sheet compressibility. This results in a denser sheet leaving the press nip. Sheet density and bulk changes are minimal if machine speed is increased when sheet temperature is increased prior to the first press nip. The effects of better pressing due to higher sheet temperature are offset by shorter nip dwell time.

G. PRESS CLOTHING

Press felts play an important roll in efficient press sections. Felts convey the web through the press section, should provide enough void volume to accommodate the water pressed from the sheet, and provide uniform pressure application to the sheet. A detailed discussion of press clothing is included in Chap. XXI.

H. PRESS SECTION VARIABLES

Stock and sheet characteristics affect pressing. Important stock and sheet characteristics include:

- Type of furnish.
- Freeness.
- Water retention value.
- Compressibility.
- Basis weight.
- Ingoing sheet moisture content.
- Sheet temperature.

The major variables affecting press section performance are shown in Table 50. Nip pressure is the most significant operating parameter. Nip pressure should be maintained at the

Table 50. Major pressing parameters.

Parameter	Significance	Major Variables	Other Variables
Press load	Water removal Sheet quality	Pressure per unit area Carry through factors Incremental load	Pressure per lineal unit Pressure pulse characteristics Roll diameter Roll hardness Roll crown Felt design Roll venting pattern
Sheet rewet	Water removal Sheet quality	Contact time Nip exit configuration	Sheet basis weight Felt design Entering sheet consistency Press position Sheet compressibility Machine speed
Nip width/ dwell time	Water removal Sheet quality	Roll hardness Roll diameter Machine speed Press load	Felt design Roll cover material Roll crown
Roll cover design	Water removal	Venting pattern Roll hardness	Cleaning showers Roll condition/maintenance Felt design Roll cover material
Felt design	Water removal Sheet quality Runnability	Base construction/weave Batt denier/shape Void volume	Permeability Thickness Weight Surface smoothness/ needling Stiffness/installation considerations Special treatments Compressibility Synthetic content Felt conditioning
Sheet temperature	Water removal Sheet quality Moisture profile	Steam box location Steam box MD length CD control	Air purging Vacuum backing Saturated steam Steam pressure Sheet quality requirements Sheet weight/density Whitewater heating
Carry through factors	Water removal	Press loads Incremental press loads Rewet	Press efficiency Felt design
Double felting	Water removal Runnability	Basis weight Entering consistency Water retention value	Sheet exit geometry Felt design Felt conditioning Roll cover design
Press configuration	Runnability Sheet quality Water removal	Open draws Sheet rewet Number of nips	Suction rolls/boxes Felt/sheet roll positions Felt design Sheet blowing Solid roll replacement Doctoring
Machine speed	Water removal Runnability	Water load/unit time Dwell time Pressure per unit area	Press load Felt design

Table 50. Continued

Parameter	Significance	Major Variables	Other Variables
Sheet quality	Sheet quality	Felt design Side of water removal Press loads Sheet temperature Smoothing press Furnish	Roll venting pattern Sheet consistency Felt conditioning
Vacuum Application	Runnability Water removal	Sheet transfer Felt conditioning Vacuum level	Felt design Vacuum system design Air volume Separators Machine speed Roll cover open area Roll condition Type vacuum pumps
Felt conditioning	Runnability Water removal	Felt design Furnish High-pressure showers Air volume	Slot width Vacuum level Shower pressure Shower oscillation Chemical cleaning Press position Separators Wringer Low-pressure showers Lube showers Uhle box location Shower locations Shower nozzles Water conditions Vacuum system design Machine speed
Roll crown	Moisture profile	Crown compensation rolls Grinding	Load range Dubbing Roll change frequency Felt tension Grinder cams
Felt moisture	Runnability	Felt conditioning Felt compaction/ filling/wear	Felt design
Smoothing press	Sheet quality Drying rate	Load Cover hardnesses Sheet characteristics	Press configuration Draw control

highest level consistent with sheet dewatering and quality requirements. Nip width and dwell time can be related to sheet basis weight and machine speed. It is generally accepted that maximum dewatering occurs with narrow hard nips for lightweight grades and wide, soft nips for heavyweight sheets. Press performance deteriorates as machine speed is increased due to lower residence time in the nip. Sheet rewet is particularly significant on lightweight sheets or with very slow machines using vented presses. Felts and sheets should be separated immediately after press nips unless serious runnability problems are caused.

REFERENCES

1. Smook, G.A., "Handbook for Pulp & Paper Technologists", Joint Textbook Committee of the Paper Industry (1982).
2. Reese, R.A., "Cost-Effective Paper Machine Upgrades", Tappi (Apr. 1986).
3. Wahlstrom, P.B., "Our Present Understanding of the Fundamentals of Pressing", Pulp Paper Mag. Can. 70(10):T349 (1969).
4. Nilsson, P. and Larsson, K.O., "Paper Web Performance in a Press Nip", Pulp Paper Mag. Can. 70(12):T438 (1969).
5. Reese, R.A., "Why Put a Steam Shower in the Press Section", Tappi J. (June 1986).
6. Royo, M.A. and Berggren, C.O., "Double-felting of Second & Third Presses", Tappi 62:3 (Mar. 1979).
7. Sprague, C.H., "Impulse Drying & Press Drying: A Critical Comparison", Tappi J. (Apr. 1987).
8. Wicks, L.D., "Press Design Review, Practical Aspects of Pressing & Drying", Tappi Press (1989).
9. Caulfield, etc., "How Web & Press Parameters Interact to Control Water Removal in the Wet Press", Tappi J. (June 1986).
10. Wicks, L.D., "Press Section Water Removal Principles & Their Application", Southern Pulp & Paper Manufacturer (Jan. 1978).
11. Bliesner, W.C., "Sheet Water Removal in a Press: Time to Review the Fundamentals", Pulp & Paper (Sept. 1978).
12. Wahlstrom, P.B., "Opportunities in Pressing", TAPPI Annual Meeting, Atlanta, GA (1980).
13. Wahlstrom, P.B., "New Felt & Press Section Designs Yield Even Drier, Stronger Sheets", Pulp & Paper (Jan. 1985).
14. Busker, L.H. and Cronin, D.C., "Relative Importance of Wet Press Variables in Water Removal", Water Removal Symp., Vancouver, B.C. (1982).
15. Beck, D.A., "Reexamining Wet Pressing Fundamentals: A Look Inside the Nip Using Dynamic Measurement", Tappi J. (Apr. 1987).

BIBLIOGRAPHY

1. TAPPI Technical Information Sheet 0404-19—Press Section Monitoring.
2. TAPPI Technical Information Sheet 0404-26—Paper Machine Clothing Performance Analysis.
3. TAPPI Technical Information Sheet 0404-37—Press Roll Cover Application Guidelines.
4. TAPPI Technical Information Sheet 0420-06—Roll Crown Specifications & Definitions.
5. TAPPI Technical Information Sheet 0404-32—Felt Uniformity Impression Method.
6. Wicks, L., "The Influence of Pressing on Sheet Two-Sidedness", Tappi J. (Sept. 1982).
7. Reese, R.A., "Strategies in the Press Section: Optimizing Performance"; PIMA (Feb. 1983).
8. MacGregor, M.A., "Practical Effects of Sheet Stratification Caused by Wet Pressing", Tappi J. (July 1983).
9. Reese, R.A., "How Important is Felt Moisture Content?", Tappi J. (Dec. 1983).
10. Reese, R.A., "Working Towards Good Press Felt Performance", Tappi J. (July 1984).
11. Reese, R.A., "How Important is Sheet Rewet After Press Nips?", Tappi J. (Nov. 1984).
12. MacGregor, M.A., "What Happens During Shell & Groove Marking", Tappi J. (Sept. 1985).
13. Pikulik, I.I. and McDonald, J., "Pressing—Induced Two-Sidedness of Paper", Tappi J. (Apr. 1987).

XII
PAPER DRYING

K.C. HILL

Ken Hill is the Vice President of Engineering and Product Management for Valmet-Enerdry. He is a Mechanical Engineering graduate from Lakehead University in Thunder Bay, Ontario. Throughout his career, he has been active in testing and characterizing dryer section performance. He has published a number of papers on the optimization of dryer section performance.

REVIEWED BY

Jeffrey H. Pulkowski
Manager of Drying Research
Beloit Corporation
Rockton, IL

Richard A. Reese
Senior Process Consultant
CRS Sirrine Inc.
Greenville, SC

A. INTRODUCTION

After pressing, the residual water in the sheet must be removed by evaporation in the dryer section of the paper machine. The evaporation process requires a large amount of thermal energy which is normally supplied in the form of steam. The dryer section is typically the largest energy user in the paper machine complex. The transfer of thermal energy from steam to the paper web requires considerable heat transfer surface. In most cases, the dryer section is physically the largest part of the paper machine. In spite of this, the dryer section often receives the least attention by the papermaker. There are often many opportunities to improve the paper making operation in the dryer section.

The objectives of the dryer section can be summarized as follows.

1. The dryers must efficiently evaporate water with the least amount of drying equipment. Drying equipment is large and expensive. The dryer section design must provide high evaporation rates to minimize the equipment requirement. This must be done within the constraints of the paper grade. For example, high evaporation rates on some grades will produce poor sheet quality.
2. The evaporation must take place in a manner that will produce a high-quality sheet. The uniformity of evaporation in the cross-machine direction is the most critical parameter. Any variation in the cross-machine evaporation characteristics will produce an undesirable variation in the final sheet moisture profile. Other properties that can be affected are sheet surface characteristics, cockle, curl and tensile properties.
3. The dryer section and auxiliary equipment must be designed to minimize the energy usage. A good target is to use 2960 kJ/kg water evaporated. There are many machines that use much more energy than this amount. The design of the systems that support the drying operation, such as steam and air handling systems, provide the main influence on energy consumption.
4. The dryer section must be designed to support a high overall paper machine efficiency. Dryer section breaks are often one of the main contributors to lost production time on the paper machine. This problem becomes more of a concern at the high modern paper machine operating speeds.

B. CONVENTIONAL CYLINDER DRYING

The most common drying method uses a series of steam-heated cylinders. Figure 315 shows a typical cylinder dryer section.

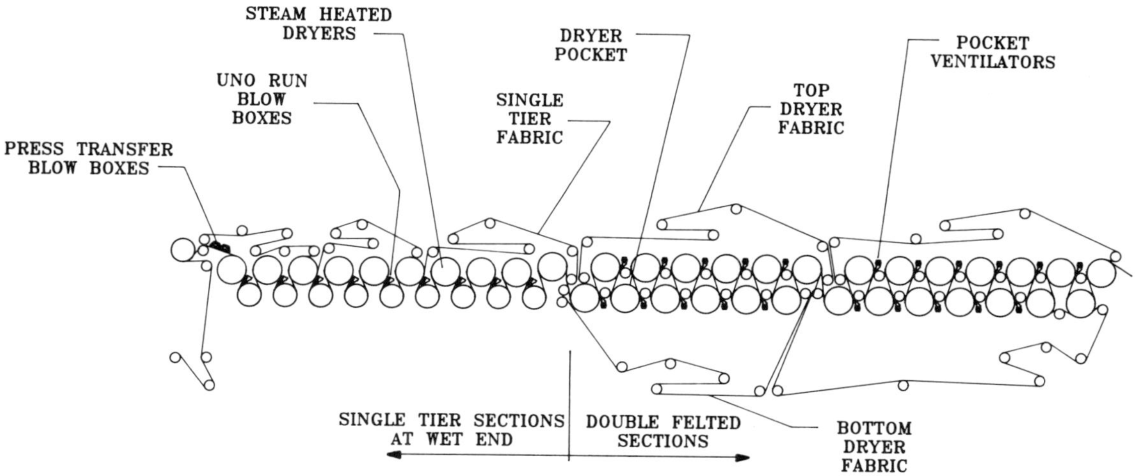

Fig. 315. Typical dryer configuration.

The wet paper from the press section is passed over a series of steam heated cylinders. The cylinders are typically 1.2, 1.5 or 1.8 m in diameter. Modern machines most commonly use 1.8 m cylinders. Steam is fed into the cylinders and the heat from the steam is transferred through the cast iron shell and into the paper web. Steam pressure can vary from sub-atmospheric pressure to 1000 kPa depending on the paper grade. Synthetic fabrics are used to press the sheet tight against the cylinder. This improves the heat transfer by providing better contact between the paper web and the drying cylinder.

Figure 315 shows a single-tier dryer section at the wet end of the paper machine. This type of configuration is commonly used on high-speed, lightweight machines. Another common name for this configuration is a Unorun. A single dryer fabric wraps both the top and bottom dryers which serves to fully support the sheet. This improves the runnability at the wet end where the sheet is still very weak. The disadvantage of this configuration is that the capacity of the bottom dryers is lost since the dryer fabric is sandwiched between the dryer surface and the sheet. Many heavyweight, slower speed machines do not utilize a Unorun section at the wet end.

After the single-tier section there is a more conventional double-felted section. Separate fabrics are used for the top and bottom dryers. The sheet is unsupported between the top and bottom dryers. This optimizes drying capacity but can lead to sheet flutter problems at high speed. It is in the open draw between top and bottom dryers where the majority of the evaporation takes place. The evaporated water is carried away by ventilation air.

C. BASIC DRYING THEORY

The drying process can be broken down into two basic components; a) the transfer of heat from steam into the paper web while on the drying cylinder; and b) the evaporation of water from the paper web into air during the draw between the cylinders. There are a number of variables which govern the process. These variables will be referred to as "Steam Side Variables" (heat transfer to the web) and "Air Side Variables" (evaporation of water from the web).

1. Heat Transfer

Heat is transferred from the steam through the dryer shell into the paper. As the paper web contacts the dryer surface, the dryer fabric presses the sheet against the dryer surface and improves the contact. The fabric also restricts the evaporation while the sheet is on the dryer. Therefore, while on the dryer, the sheet temperature rises as heat is transferred into the sheet and little evaporation takes place.

Figure 316 shows an exploded section of the dryer shell during the heat transfer phase of the drying cycle. Heat transfer into the sheet follows the basic heat transfer formula:

$$Q = U \times A \times (Ts - Tp)$$

where

Q = Rate of heat flow from steam to the paper (kJ/h)

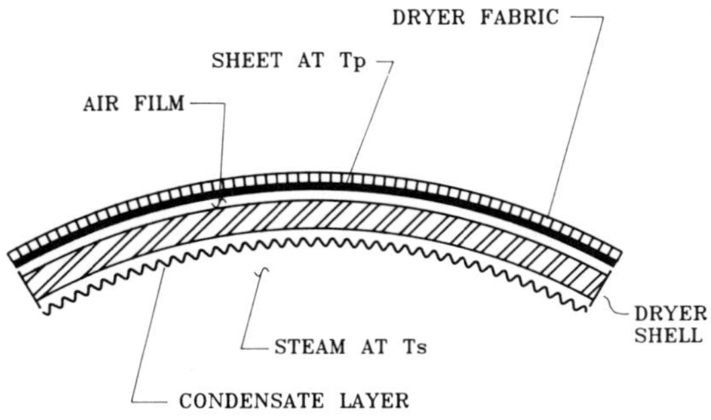

Fig. 316. Heat transfer with sheet on cylinder.

Table 51. Typical steam pressures used.

	Wet end (kPa)	Maximum at dry end (kPa)
Newsprint	−35 to 100	345 to 415
Woodfree Writing Papers	−35 to 0	345 to 520
Woodfree Copier Papers	−35 to 35	520 to 690
Lightweight Coated	−35 to 0	275 to 345
Supercalendered Groundwood	−35 to 0	275 to 345
Linerboard and Corrugating	200 to 520	1030

U = Overall heat transfer coefficient which is a measure of the resistance to heat flow ($kJ/h/m^2/°C$)
A = Dryer surface area in contact with the paper web (m^2)
Ts = Steam temperature (°C)
Tp = Paper temperature (°C)

The temperature of the paper (Tp) is the variable which relates the heat transfer and evaporation process. Increasing the heat transfer will increase the paper temperature as it leaves the dryer which will increase the evaporation in the draw between cylinders. Similarly, an efficient evaporation process will reduce the paper temperature in the draw between cylinders as the water evaporates. This results in a lower sheet temperature as the sheet comes back onto the next cylinder. The lower paper temperature improves heat transfer.

With a fixed number of dryers, the only two variables in the heat transfer equation that can be directly controlled are the saturated steam temperature (Ts) and the overall heat transfer coefficient (U).

Increasing the steam pressure inside the dryers increases the temperature. However, there are practical limitations to the maximum pressure that can be used. Most lightweight printing papers cannot use high pressures especially at the wet end of the dryer section. Pressures must start out low and be graduated to higher pressures as the paper progresses through the dryer section. It is not unusual to see pressures below atmosphere in the wet end dryers of fine paper machines. Higher pressures cause the fibers on the outer surface of the sheet to stick to the dryers. This causes a rough sheet surface and can also induce runnability problems due to the non-uniform release of the paper from the dryer cylinder.

There are similar limitations to the maximum steam pressures that can be used at the dry end of the paper machine. Again, lightweight grades cannot use high dry end pressures due to runnability and quality problems.

Heavyweight board machines are not faced with the same problems as fine paper machines and can use higher steam pressures early. It is not unusual for linerboard machines to use 200–500 kPa pressures in the early dryers and quickly graduate to pressures up to 1000 kPa.

Table 51 lists some typical grades and steam pressures used.

It is possible to use different pressures but sheet quality problems can be a difficulty. A general rule is that "the higher the printing requirements, the lower the wet end steam pressures and the more gradual the increase in steam pressure".

The overall heat transfer coefficient is a measure of the resistance to heat transfer. It is possible to improve this coefficient on many machines and improve the drying efficiency. The resistances that make up the overall heat transfer coefficient are:

1. The condensate layer thickness and turbulence level. This topic will be discussed in more detail in Sec. E. A condensate layer inside the dryer is created as the steam transfers heat and condenses. The layer creates a resistance to heat transfer. At higher speed, the resistance becomes greater as the layer becomes less turbulent. Keeping the condensate layer thin and turbulent will maximize the heat transfer. This is one of the most significant variables in the drying process.
2. The dryer shell metal thickness is a resistance to heat flow. The thickness of the shell is determined by the pressure requirements of the dryer, and little can be done to optimize this variable.

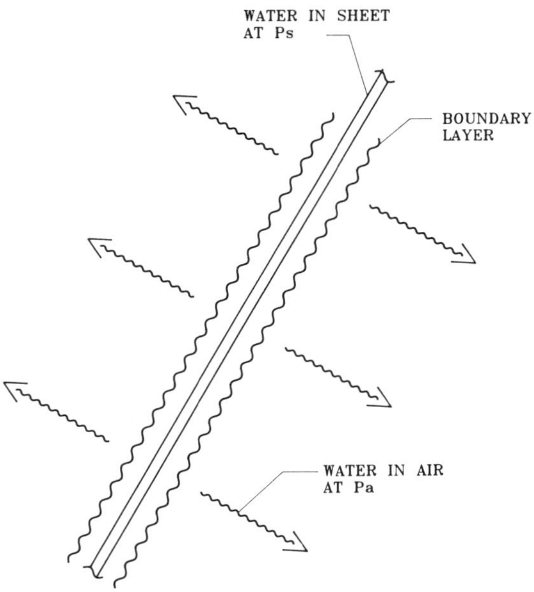

Fig. 317. Evaporation in free draw.

3. Any scale or fiber buildup on the outer surface of the dryers will cause a resistance to heat flow. Dryer surfaces should be kept clean.
4. A thin air film is trapped between the sheet and dryer shell as the sheet comes onto the cylinder. This air film forms a very effective insulation layer. The purpose of the dryer fabrics is to press the sheet against the shell and minimize the air film resistance. The air film is most significant at the wet end where the sheet is impermeable. At the dry end, the sheet is more permeable to air flow and the trapped air can be forced into the sheet which results in good contact to the dryer.
5. The properties of the sheet affect the resistance to heat flow. The water content, thickness, surface roughness, porosity, etc., will influence the ability to transfer heat effectively. Drying rates vary significantly from one grade to another and this is one of the key reasons. It must be recognized that all furnishes do not dry at the same rate.
6. Non-condensibles in the dryer drum reduce heat transfer. There can be air or other non-condensible gases in the steam supply to the dryers which can drastically reduce the heat transfer if they are allowed to accumulate. Proper treatment of the steam supply to the dryers and proper purging of the non-condensible gases is needed to prevent this problem. Normally this is not significant with a well-designed steam and condensate system.

D. EVAPORATION

The heat that is transferred into the sheet while on the dryer cylinder is used to change the phase of water from a liquid to a vapor. For conventional drying conditions, the amount of heat required for this phase change is 2290 kJ/kg water evaporated. The majority of the evaporation takes place in the free draw between cylinders. The evaporation process is governed by a standard mass transfer equation.

Figure 317 shows an exploded view of the paper web in the draw between dryer cylinders. The evaporation from the sheet follows the basic mass transfer equation shown below.

$$\text{EVAP} = \text{K x A x (Ps - Pa)}$$

where

EVAP = water evaporated from the sheet (kg water/h)
K = mass transfer coefficient (a measure of the resistance presented by the air film on surface of the sheet)
A = evaporation area determined by the length of the draw (m^2)
Ps = vapor pressure of the water in the sheet (Pa)
Pa = partial pressure of water vapor in air surrounding the sheet (Pa)

Because of the physical arrangement of a conventional dryer section, it is difficult to reduce the resistance of the boundary layer of air accompanying the sheet (K). No method exists to date which can "scrub" the boundary layer without disturbing the sheet or causing obstructions within the dryer pocket.

The vapor pressure of the water in the sheet is directly related to the temperature of the sheet. The sheet temperature is the link between the heat transfer and evaporation process. Efficient heat transfer will result in a high sheet temperature and a high value for the vapor pressure of the water in the sheet (Ps). As water evaporates from the sheet, the temperature of the sheet drops.

The partial pressure of the water vapor in the air surrounding the sheet is directly related to the absolute humidity of the air. High humidities in the dryer pocket result in a high partial pressure and reduced evaporation. The function of the pocket ventilation system in conjunction with permeable dryer fabrics is to introduce hot dry air into the pocket area and purge the pocket of humid air. The pocket humidity is commonly measured. The guidelines for recommended humidity are given in Table 52.

Generally pocket humidities that are higher

Table 52. Recommended pocket humidities (kg water/kg dry air).

	Average	Peak Reading
Newsprint, SC, LWC	0.18	0.25
Fine Papers	0.2—0.22	0.30
Linerboard & Corrugating	0.2—0.22	0.30

Table 53. Effect of drying variables on drying efficiency.

Variable	% Influence
1. Condensate Removal	30%
2. Furnish, Grade, and Sheet Properties	25%
3. Dryer Fabric Design, Permeability and Tension	20%
4. Pocket Ventilation	15%
5. Hood and Dryer Air Systems	5%
6. Other	5%

than these levels significantly restrict evaporation. Inefficient pocket ventilation or fabrics of low permeability are reasons for high pocket humidity levels. The papermaker must always balance fabric permeabilities against sheet stability. High-permeability fabrics will move too much air into the pocket and cause sheet disturbance at high speeds. Using lower permeability fabrics will block the air flow into the pocket but will raise the pocket humidity and reduce the evaporation rate.

1. Influence of drying variables

There are a number of basic variables that determine the efficiency of a dryer section. These variables assume that the dryer geometry is fixed and the steam pressure is constant.

An estimate of the effects of the different drying variables on the overall dryer efficiency is given in Table 53. The table is meant to give only a general indication of the overall impact of the different variables. It is difficult to give more exact estimates because of the interrelationship between the variables and the many different machine geometries and conditions.

2. Drying zones

The rate of drying is different at different points in the dryer section. Figure 318 shows the typical drying curve for paper.

Initially the sheet enters the dryer section at a temperature lower than needed for evaporation. The dryers must heat the fiber and water in the sheet up to the evaporation temperature. This is called the "sheet heating phase". The sheet heating phase will generally take from one to four dryers depending on the initial temperature, the dryer steam pressure, the moisture content, and the basis weight of the sheet.

The sheet evaporation increases to a peak rate then enters the "constant rate drying zone". The constant rate zone has the highest rate of evaporation. Free water is readily available for evaporation in this zone.

The next drying zone is the "falling rate drying period". In this zone, free water is no longer available at the surface of the sheet. Some of the larger pores in the sheet are empty of water and the effective area for mass transfer is reduced. The evaporation rate is reduced and continues to reduce as more of the pores become empty. The point at which the sheet enters the falling rate drying zone is difficult to determine. Early in the zone the evaporation rates are very high and are at a level close to those in the constant rate zone. There is little distinction between the constant and falling rates zones in this region.

The last small portion of water is very difficult to remove and the rate of evaporation is very low. Many paper machines dry the paper to a lower moisture content than is required. The papermakers will do this if the moisture profile of the sheet is poor. Overdrying the sheet will "flatten" the moisture profile. This practice is wasteful in terms of drying capacity and energy utilization. The water that is removed when the sheet is overdried is the most difficult to remove. Up to 10% of the total drying capacity of the machine can be lost through the practice of overdrying. A better approach is to correct the source of the moisture profile variation.

3. Drying rates for different grades

It is useful to be able to predict the drying rate on a given paper machine and grade. The drying rate will vary with the paper grade and furnish conditions. TAPPI has accumulated drying data from many paper mills in North America. The

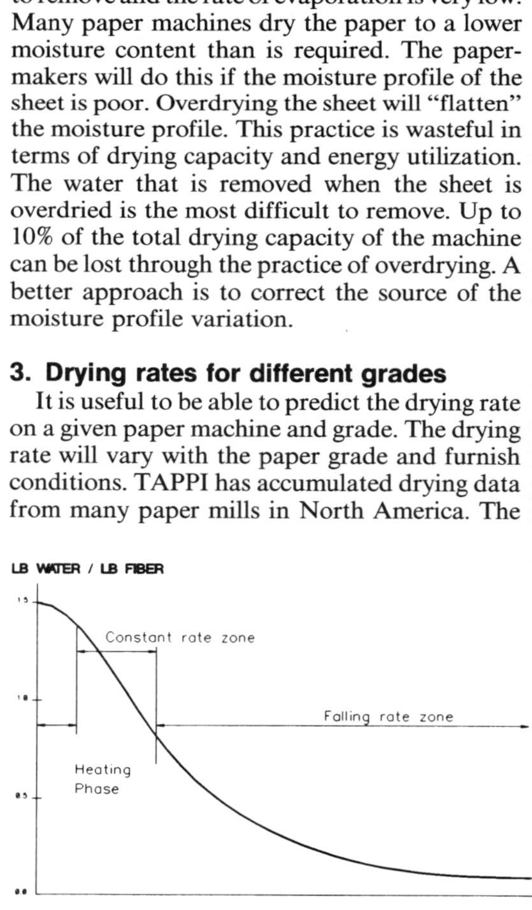

Fig. 318. Drying curve.

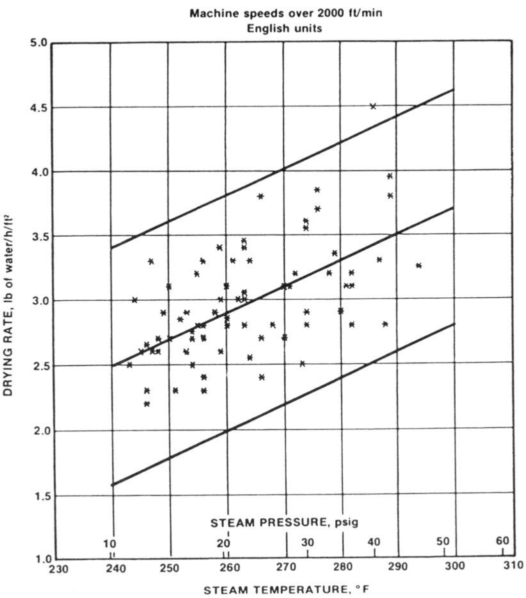

Fig. 319. Paper machine drying rate (newsprint).

data have been sorted by the different paper grades and are plotted on curves. A typical TAPPI drying rate curve is shown in Fig. 319. A full set of curves is available from TAPPI.

The drying rate is evaluated in terms of kg/water evaporated/h/m^2 of total dryer surface. The steam temperature is the weighted average saturation temperature existing in the dryers.

There is a wide variation in the drying rate between the worst and best machines. This is mainly the result of the different machine configurations and the equipment installed on the machines.

4. Dryer energy consumption

A dryer section uses steam for sheet heating, evaporation, air heating, non-condensible bleed, and venting. Table 54 shows the typical heat consumption values for each of these areas.

The heat required for evaporation is essentially fixed and little can be done to improve this value in a conventional dryer section. The heat used to bring the sheet temperature up to the evaporation temperature is determined by the white water temperature used on the machine. Also, if a steam shower is used before the dryer section, the sheet will enter at a higher temperature.

Air heating is often an area where the dryer energy consumption can be optimized. Heated air needs to be supplied to the dryer section to carry away the evaporated water. Air is also

Table 54. Typical dryer energy consumption.

Consumer	Heat consumption (kJ/kg of water evaporated)
Sheet Heating	200
Evaporation	2300
Air Heating	420
Non-Condensible Bleed	45
Venting	0
Total	2965

heated by radiant heat lost from the dryers. The hood and air system design has the greatest influence on air heating energy consumption. A poor design may use up to 700 kJ/kg water evaporated.

Non-condensibles such as air can build up in the dryer section and cause a loss in heat transfer efficiency. A small amount of steam needs to be bled off the system to prevent any buildup. This number can range up to 465 kJ/kg water evaporated with a poor design.

Although a good dryer drainage system will never vent steam to maintain differential pressure in the dryers, there are many machines where venting to a condenser or atmosphere is a normal part of operation. This is extremely bad practice. Venting amounting to as much as 1160 kJ/kg of water evaporated is possible.

E. STEAM AND CONDENSATE SYSTEM

The energy needed to evaporate water comes from steam as it condenses inside the dryer. The supply of steam to the dryer and the removal of condensate is a critical part of dryer section operation. The manner in which this is handled affects drying efficiency, energy efficiency, paper quality, and overall paper machine efficiency.

1. Condensate layer effects on drying

A condensate layer is formed inside the dryers as the steam condenses. This condensate layer has a major influence on drying efficiency and uniformity. One of the major purposes of the steam and condensate system is to efficiently remove the condensate from the rotating dryers.

The condensate layer acts differently inside the dryer depending on the speed of operation. Figure 320 shows the different behavior of the condensate layer.

At very slow speeds, less than 150 m/min, the condensate forms a puddle in the bottom of the dryers. The heat transfer coefficient outside the puddle region is very high. As the speed increases, the puddle begins to climb up the side of the dryer due to friction between the dryer wall and the condensate. Turbulence is also

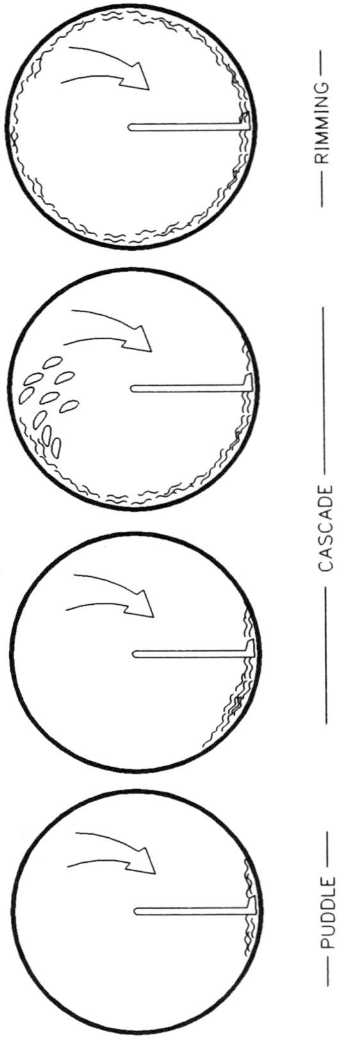

Fig. 320. Condensate behavior with machine speed.

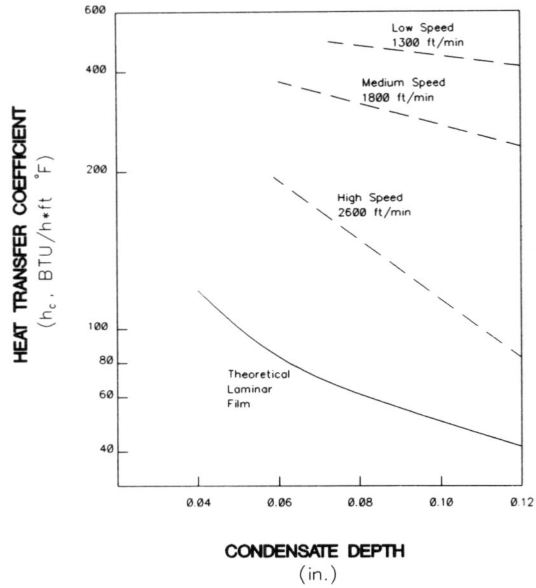

Fig. 321. Variation in heat transfer coefficient with machine speed.

being created inside the condensate puddle by the friction. Heat transfer is enhanced by turbulence inside the condensate layer. At a speed of approximately 300 m/min the condensate is climbing up the wall and cascading back down to the bottom of the dryer. The turbulence level is very high and the heat transfer is good. However, the high level of turbulence places a very high horsepower and torque load on the dryer drive system. Gear failures have been attributed to dryers operating in this cascading condition.

Once the dryer reaches sufficient speed, the condensate will fully rim the dryer. The speed at which rimming occurs depends on the dryer diameter, condensate load, and surface smoothness. Generally, rimming will occur between 300 and 425 m/min machine speed. Even though the condensate layer is rimming, it still has considerable motion relative to the dryer surface. The condensate layer moves with a "sloshing" action. This phenomenon was best described in research work done by Appel and Hong [7].

The sloshing action creates turbulence and enhances heat transfer. The action becomes more stagnant as the machine speed increases. Figure 321 shows the variation in heat transfer coefficient as the machine speed increases. At today's state-of-the-art machine speeds, the condensate layer is approaching a laminar film inside the dryers with very poor heat transfer characteristics. A small variation in condensate layer depth will cause a major variation in heat transfer. Even a 0.5 mm variation in condensate layer thickness can cause a very significant change in the condensate layer heat transfer coefficient. This will cause a variation in the cross-machine drying rate and a moisture profile variation at the reel. It is a difficult task to maintain this uniformity on an 8.9 m wide modern paper machine.

2. Dryer syphons

A piece of equipment called a dryer syphon is used to remove the condensate from the dryer. There are two basic types of syphons used: a rotary syphon which moves with the rotating dryer; and a stationary syphon which is fixed relative to the rotation of the dryer. Figures 322 and 323 show the two designs.

Blowthrough steam is needed to remove the condensate from the dryer. The blowthrough steam entrains the condensate, atomizes it, and carries it out of the dryer. A differential pressure is needed between the steam pressure inside the dryer and the pressure in the condensate piping outside the dryer. The differential pressure creates the driving force for the blowthrough steam flow. The amount of differential pressure, blowthrough steam, syphon pipe size, condensate line size, and syphon design are all critical variables in determining how well the condensate is removed from the dryer.

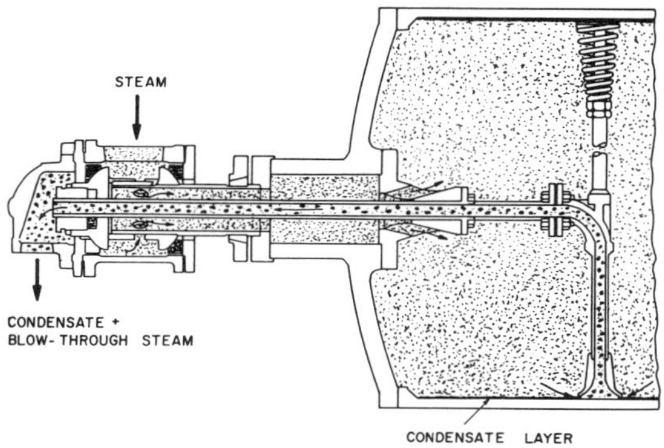

Fig. 322. Rotary syphon.

The main operational difference between a rotary and a stationary syphon is the amount of differential pressure needed to overcome the centrifugal force in a rotary syphon. Extra differential pressure is needed to overcome the centrifugal force of the condensate/steam mixture in the syphon pipe. The amount of extra differential is dependent on the density of the mixture which is dependent on the percentage of blowthrough steam. If the blowthrough steam flow is too low, the centrifugal force becomes very high and the differential pressure requirements are also high. Dryers will stop draining and "flood" with condensate if the blowthrough steam flow is too low. High-speed machines require blowthrough steam flows of approximately 20% of the condensing load to prevent flooding of the dryers. Typical differential pressures for rotary syphons range from 40 kPa for machines operating at 450 m/min to 95 kPa for machines operating at 1050 m/min.

The stationary syphon does not have to overcome the centrifugal force of the rotating mixture, therefore, the differential pressure and blowthrough flow requirements are less. Typically stationary syphons will operate with 35–40 kPa differential pressures. The stationary syphons are more suited to very high operating machine speeds. The disadvantage of a stationary syphon is the mechanical problem of keeping

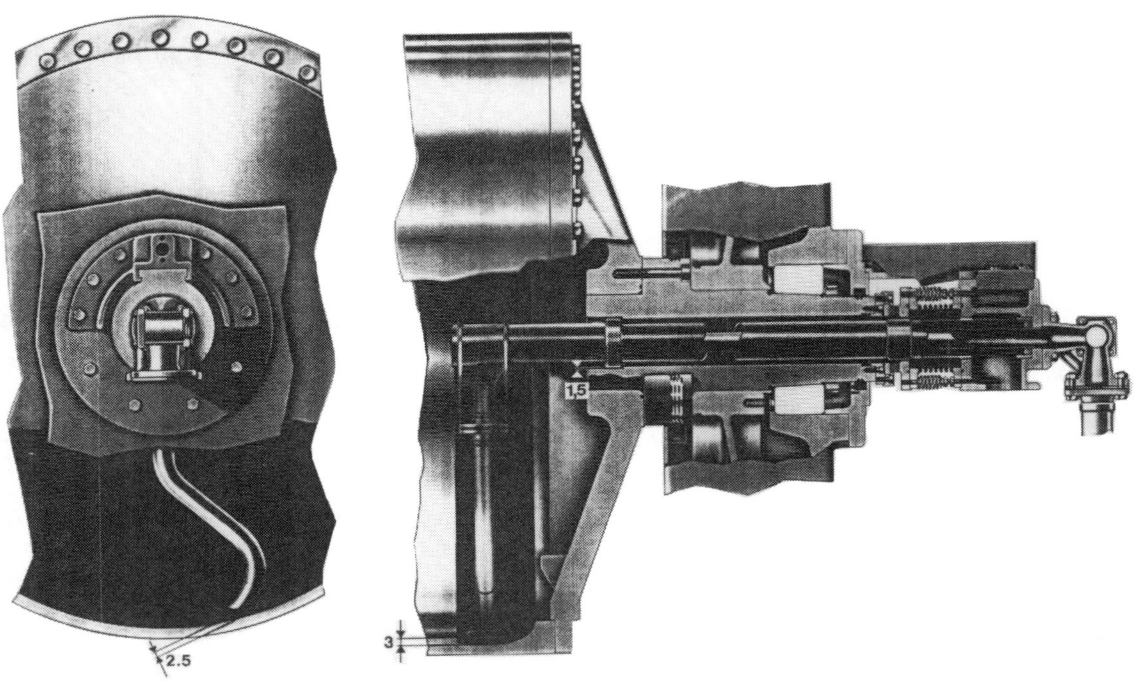

Fig. 323. Stationary syphon.

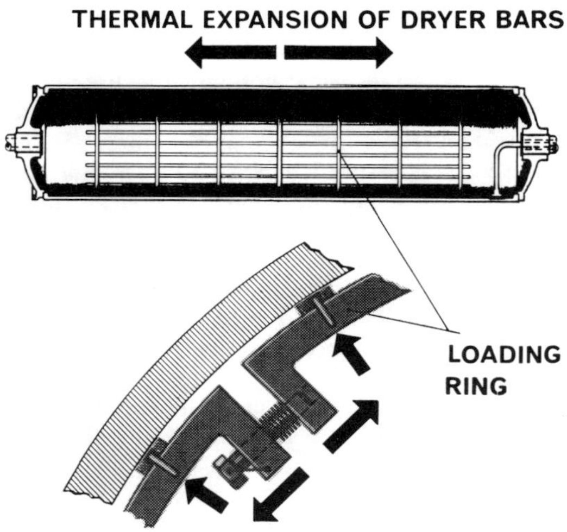

Fig. 324. Attachment of dryer bars to inner surface.

the syphon properly set. The clearance is generally less than 0.75 mm and any misalignment will cause a mechanical failure.

Many modern, wide paper machines use two syphons – one on each end of the dryer. The reason for this is to provide better uniformity of the condensate layer. A gravity head difference is the only driving force which moves the condensate to the syphon. On a wide machine, the head difference is sufficient to cause a variation in the heat transfer across the dryer. The use of two syphons minimizes this problem. It is possible to obtain a uniform moisture profile with a single syphon, however, proper syphon design and adequate blowthrough flows are essential. At speeds over 900 m/min, dryer bars are essential to obtain a uniform moisture profile with single syphons.

3. Dryer bars

There has been an increasing use of dryer bars inside the dryers as machine speeds have increased. Dryer bars are a series of flat bars running lengthwise, mounted on the inside dryer surface. Figure 324 shows dryer bars inside a dryer.

The purpose of the dryer bars is to improve the heat transfer by creating turbulence in the condensate layer. There is some motion in the condensate layer even at high speed. The dryer bars create an oscillation in the condensate layer between the bars. The turbulence in the condensate layer is increased, which results in substantial improvements in heat transfer. Dryer bars are more effective at high machine speeds where the condensate layer has little natural turbulence. At lower speeds, the condensate layer has more natural turbulence and the addition of dryer bars has less effect. Table 55 shows dryer bar improvement potential as a function of machine speed.

Table 55. Dryer bar improvement potential.

Machine speed (m/min)	Drying capacity improvement (%)
150—365	0
365—460	0—5
460—610	3—10
610—760	5—12
760—915	10—15
915+	12—20

The effect on heat transfer is dependent on speed, dryer diameter, condensate removal efficiency, and condensing load.

It should be noted that the variation from machine to machine can be great.

In addition to improving the drying rate, dryer bars also have a great improvement in the uniformity of heat transfer in the cross-machine direction. The heat transfer coefficient is made uniform across the width of the dryer, which minimizes the effect of small variations in the depth of the condensate layer.

4. Cascade system

Differential pressure and blowthrough steam must be created to drain the dryers. The steam and condensate handling system outside the dryers is responsible for creating the differential pressure and handling the blowthrough steam in an efficient manner.

Figure 325 shows a cascade design for a steam and condensate system.

The system consists of three dryer groups. The drying pressure is set in the main control group which has the majority of the dryers. Blowthrough steam and condensate are removed from the dryers. A tank is used to separate the blowthrough steam and the condensate. The blowthrough steam is piped to the next steam section. The condensate is returned to the boiler house.

The pressure for the intermediate steam section is set by the differential pressure requirements for the main drying group. Therefore, the intermediate group operates at a lower pressure than the main group which creates the differential pressure needed to cause blowthrough steam flow from the main group. The intermediate section makeup steam valve is controlled from a differential pressure controller for the main steam group. This valve allows

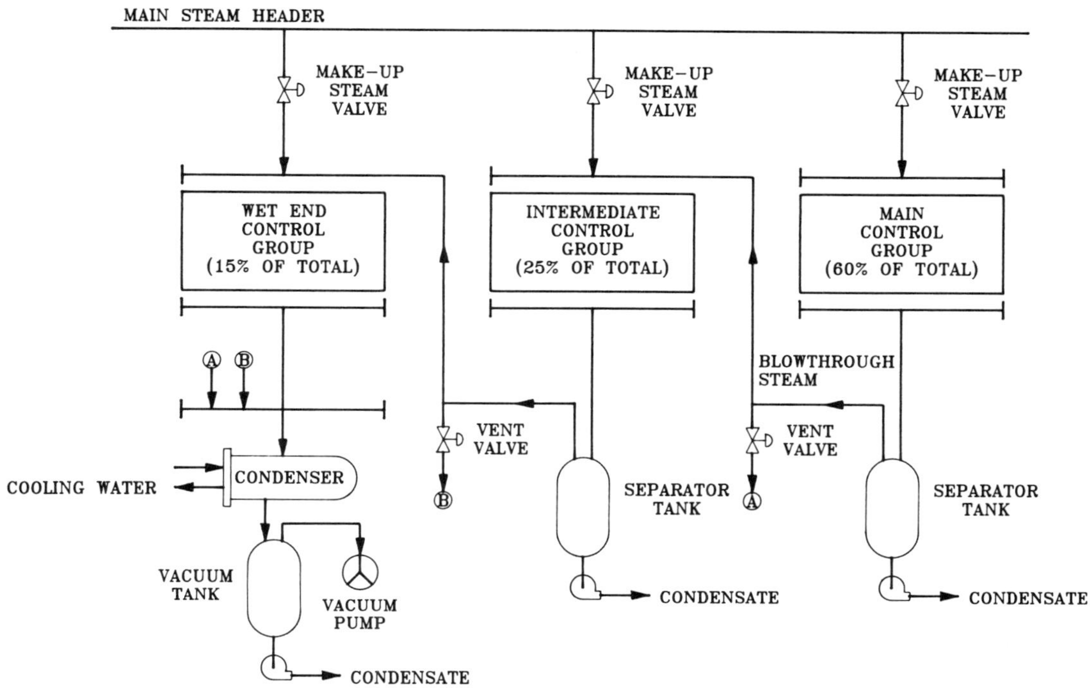

Fig. 325. Basic cascade system.

enough steam into the intermediate group to provide sufficient differential pressure for the main group. If the differential pressure for the main group is not sufficient, then the intermediate section makeup valve will close off and reduce the intermediate group pressure.

The wet end group operates in a similar manner. It has a lower pressure than the intermediate group and its pressure is determined by the differential requirements of the intermediate group. The blowthrough steam and condensate from the wet end group are taken to a vacuum condenser. Only a small percentage of the total steam flow to the machine is taken to the condenser. A vacuum pump is used to remove the non-condensibles from the system. The blowthrough steam purges the non-condensibles from all the dryers which are eventually removed through the vacuum pump.

Vent valves are used on the main and intermediate groups. In normal operation, these valves are closed. However, in the event that the differential pressure cannot be maintained, the vent valves will open to the condenser to guarantee that dryers do not flood.

The cascade type steam and condensate system is widely used but it does have some limitations. These are listed below.

1. There is limited range of operation. The main group pressure cannot be set to low pressures or there will be insufficient differential pressure capability or excessive venting. At high-pressure operation, the wet end pressures may be too high for good sheet quality. Different control strategies need to be used which complicate the system design.
2. The control groups are large. A change is steam pressure will cause a pressure change in all the dryers. When fine adjustments are needed for reel moisture control, this can be a problem. The dryer section becomes too large for sensitive control.
3. The system is difficult to configure to different dryer configurations. The size of the steam groups is determined by the blowthrough flow amount. The system cannot be arranged to match drive groups or easily handle Unorun wet end configurations.
4. The system is difficult to troubleshoot. Since all dryers are related, it is difficult to determine the source of a problem.

5. Thermocompressor system

An alternative to the cascade arrangement is to use a thermocompressor to recompress the blowthrough steam and use it directly back in the same dryers. Figure 326 shows a thermocompressor.

High-pressure motive steam is used to recompress the low-pressure blowthrough steam back to an intermediate pressure which is the drying

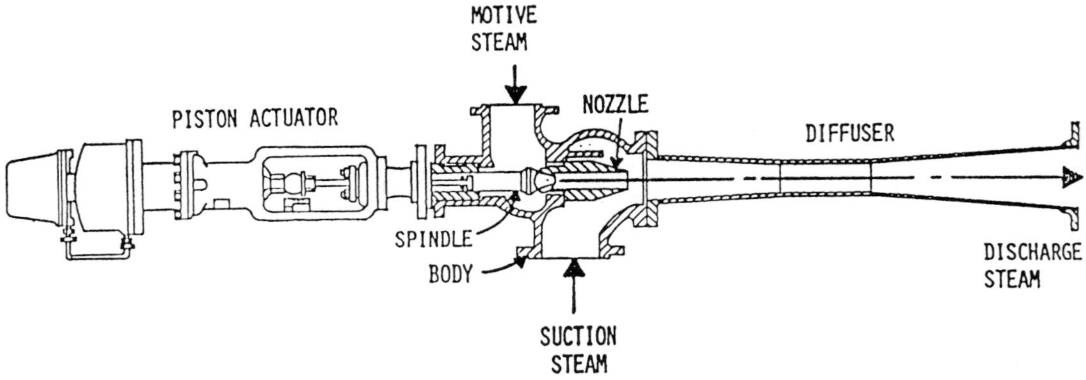

Fig. 326. Thermocompressor.

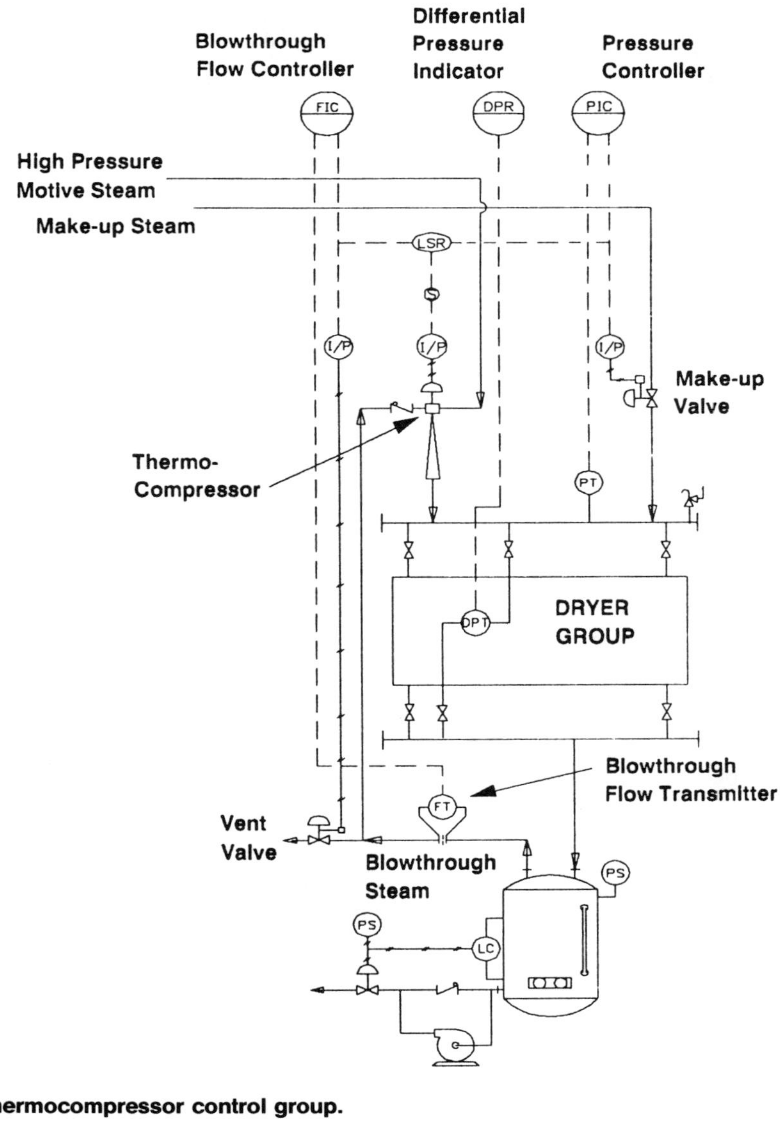

Fig. 327. Thermocompressor control group.

pressure. A nozzle is used to discharge the high-pressure motive steam at high velocity. A momentum exchange occurs between the high-velocity motive steam and the low-pressure blowthrough steam. The diffuser section is used to slow the mixture and recompress it back to the drying pressure.

The amount of blowthrough steam and differential pressure is controlled by an automatic actuator and spindle for the high-pressure motive steam. The introduction of more motive steam will create a higher differential pressure and blowthrough steam flow. The motive steam pressure must be at least 345–520 kPa above the maximum drying pressure for good operation.

Figure 327 shows a thermocompressor dryer group. The total dryer section is divided into a number of independent thermocompressor loops. Many times the steam groups are matched with the dryer drive groups for convenience of operation. Generally the wet end groups are smaller which provides improved wet end pressure graduation. Figure 328 shows a complete dryer drainage control system.

Since each steam group is completely independent from the other groups, the system has great flexibility. The dryer section can be operated over a wide range of pressures. The differential and blowthrough flow settings can be varied depending on the requirement of the dryer group.

The control of the wet end dryers in Fig. 328 is typical of machines producing lightweight grades. The extreme wet end dryers would be on individual pressure control. The dryers would discharge directly to the condenser. This allows the dryers to be operated at low pressures. The pressures can be graduated upward from the wet end.

A small amount of bleed steam should be taken from each thermocompressor group. This ensures that non-condensible gases are purged from the system. The amount of bleed steam should be no more than 1% of the total steam load to the dryer section. The vent valves to the condenser are opened if the thermocompressor cannot maintain sufficient differential pressure. This should not occur in normal operation.

The thermocompressor system has many advantages over the cascade system design. It is the most commonly used system for modern North American machines. The disadvantage of the thermocompressor system is the use of high-pressure motive steam. Typically 25% to 30% of the total dryer steam consumption will be supplied with the high-pressure motive steam. In many mills, high-pressure steam is not readily available in the machine room and is more costly than lower pressure steam. These factors need to be considered when designing the steam and condensate system.

6. Blowthrough flow control

Until recently, the most common control method was differential pressure control. The differential pressure between the steam header and the condensate header was measured and controlled. However, establishing the correct differential pressure was often difficult and the correct setting could change depending on machine operating conditions. "Flow control" is now the most common method to control dryer drainage.

Flow control monitors the blowthrough steam flow from a dryer section and automatically adjusts the differential pressure to satisfy the blowthrough steam flow requirements. This is a more exact way to control drainage since it is the amount of blowthrough steam that determines how well dryers drain.

Flow control offers many advantages over differential pressure control:

1. Simplicity of Operation – Once set, no further operator adjustment is necessary. The system will automatically adjust the differential pressure for changing operating conditions.
2. Minimizes Motive Steam Flow – The system automatically sets the differential pressure at the lowest level which will produce sufficient blowthrough steam to effectively drain the dryers.
3. Reduces Steam Waste – The flow control system will reduce the differential pressure on sheet breaks when condensing rates are low. Substantially less steam is vented to the condenser than with a conventional differential control system.
4. Eliminates Dryer Flooding – The flow control system rapidly adjusts to system upsets which are the main cause of flooding. If dryers do flood, the system will sense a decrease in blowthrough flow and increase differential pressure to drain the dryers.

F. DRYER VENTILATION SYSTEM

1. Objectives of a dryer ventilation system

The objectives of a dryer section ventilation system have evolved since the days of the early paper machines in North America. The earliest machines had absolutely no ventilation equipment. The water evaporated in the dryer section was allowed to simply migrate into the room where it would be removed using room exhaust

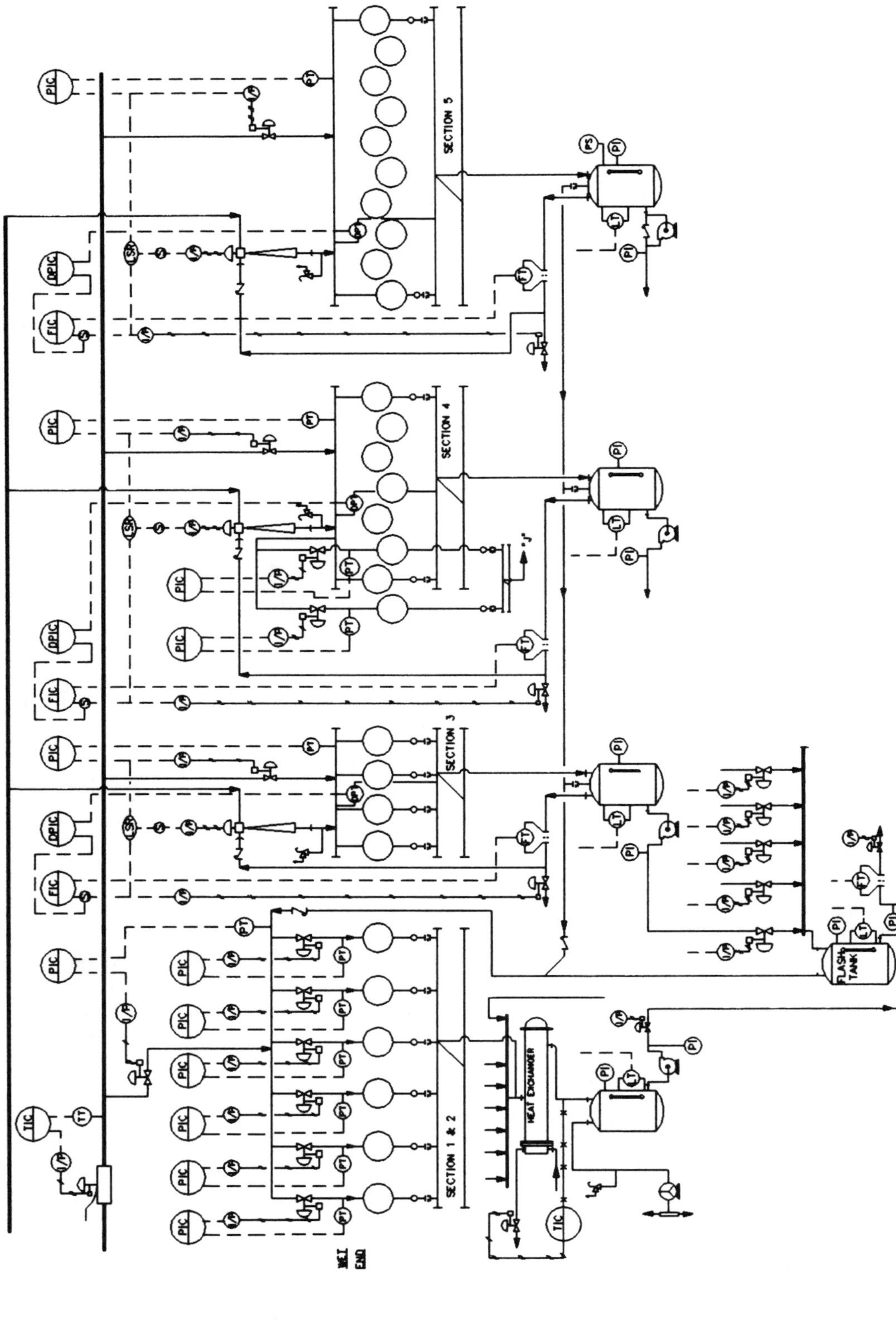

Fig. 328. Complete dryer drainage control system.

fans. Today's dryer section ventilation equipment influences machine efficiency, production capacity, energy consumption, and product quality. Proper treatment of air flows within the dryer section is gaining increasing importance in overall machine operation as machine speeds continue to increase.

The following is a listing of the objectives of a modern dryer section ventilation system.

1. **Capture and remove the vapor from the dryer section**
   This is the most basic function of the dryer section ventilation equipment. The hood and hood exhaust fans are the equipment used to accomplish this objective.
   The dryer section on a modern high-production paper machine will evaporate well over a ton per minute. Room conditions would be intolerable if this vapor were simply allowed to migrate into the room. The hood and exhaust system contain the vapor within the dryer section and remove it from the machine room.
2. **Maximize the drying capacity and machine productivity**
   The importance of pocket ventilation became evident in the 1960s with the introduction of synthetic dryer fabrics. It was found that, if the evaporated water was not purged from the pocket area efficiently, the humidity level would increase to a high level. This would restrict the drying capacity and production of the machine. Proper ventilation of the pocket area is important on every modern paper machine.
3. **Provide a controlled drying environment**
   Just as high pocket humidity levels can influence the drying capacity of a paper machine, a non-uniform pocket humidity level can result in a variation in the drying capacity in the cross machine direction. This will result in a moisture profile variation at the reel. Other paper qualities such as cockle, grainy edges, and curl can be influenced by the pocket humidity level.
   The pocket ventilation system and hood must work together to control the drying environment. The pocket humidities must be controlled at a low and uniform level if the best results are to be achieved.
4. **Stabilize the paper web to maximize machine efficiency**
   The stability of the paper web in the dryer section is becoming an important function of a dryer section ventilation system as machine speeds continue to increase. Many machines are either speed limited or suffer a high number of breaks due to an unstable sheet in the dryer section. Proper handling of the air flows within the dryer section is needed to provide good sheet runnability. There is an entire family of runnability components whose purpose it is to control the air flows within the dryer section resulting in a stable sheet run.
5. **Preserve the machine equipment**
   This is often an overlooked function of the dryer ventilation system. If hood exhaust and supply air flows are not properly designed and balanced, then condensation is likely. Condensation can be a serious problem. Corroded felt rolls, dryer frames, structural members, etc. can be the result of a poor dryer ventilation system. This can reduce the life of the paper machine.
6. **Optimize the energy efficiency**
   Air heating will account for approximately 400-440 kJ/kg of water evaporated. On a machine with a poorly designed hood and dryer ventilation system, this value can increase to 700 kJ/kg of water evaporated. The vast majority of the heat that is released by the steam to dry paper is contained within the exhaust vapor that is carried away from the hood. A well-designed dryer hood and ventilation system will recover as much of this as is practical. The hood system must be designed to maximize the amount of heat that can be recovered and the quality of the recovered energy. A low-humidity open hood will have very poor heat recovery potential compared to a modern high-humidity hood. Both exhaust streams contain the same amount of heat, however, the high-humidity hood is at a temperature level where the heat can be recovered and used effectively.

2. Components of a dryer ventilation system

Figure 329 shows a cross-section through a dryer section. The ventilation system components are illustrated.

1. **Hood**
   The purpose of the hood is to capture the water evaporated from the dryer section and remove it from the machine room. Hood constructions vary from an open uninsulated design to a high technology, high humidity, fully enclosed construction. The type of hood construction determines how the remainder of the dryer air systems are designed.
2. **Hood exhaust fans**
   Exhaust fans remove the moisture from the hood and discharge it outside the building. Axial and centrifugal designs are used. The fan design must isolate the bearings from the

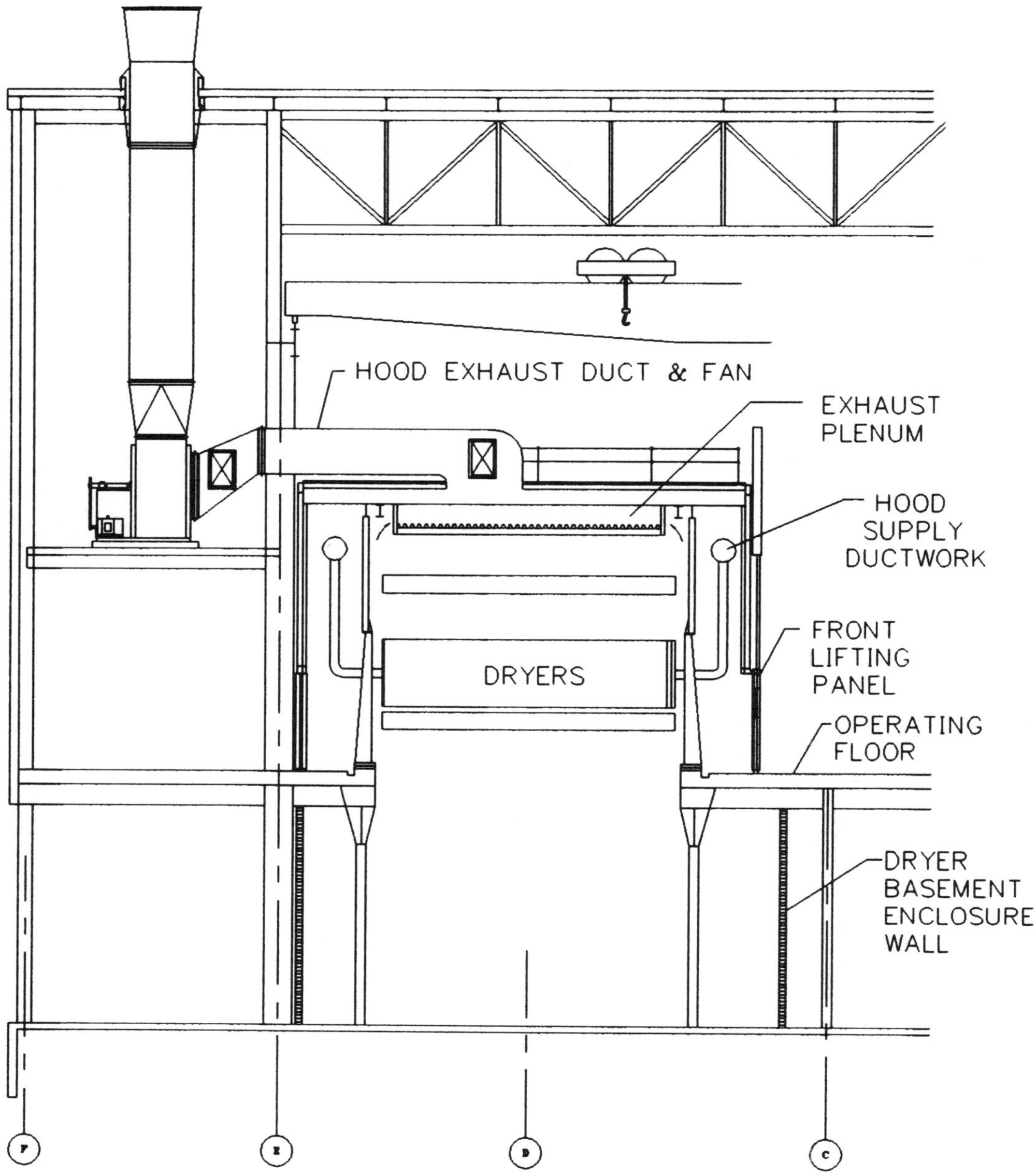

Fig. 329. Hood cross-section.

exhaust stream for good reliability. Stainless steel fans should be used in conditions where the exhaust stream is saturated such as after heat recovery equipment. If the exhaust fans are handling vapor that is not saturated (without heat recovery), then mild steel or aluminum construction is adequate.

3. **Hood supply units**
Heated supply air needs to be introduced into the dryer section to balance the exhaust flow from the hood. Condensation and dryer operational problems will result without the introduction of heated supply air. The hood supply equipment usually consists of heating coils, fans, ductwork, and distribution nozzles. The ideal place to distribute hood supply air is through the pocket ventilation equipment where it is used to enhance drying efficiency and control the drying environment. Hood supply air is sometimes used to

condition dryer fabrics although this is not as effective as pocket ventilation.

The amount of hood supply air needed will vary depending on the type of hood used. A modern closed hood should have 65% to 75% of the air that is exhausted supplied to the dryer section in the form of dry heated air. An open hood should have a 35% to 40% hood balance for good operation. All balance figures are calculated on a mass flow basis.

4. **Pocket ventilation equipment**

   Pocket ventilation nozzles or rolls introduce hot dry air into the dryer pockets to purge the evaporated water and remove the humidity. The drying efficiency is improved with lower pocket humidities. The pocket ventilation system is also responsible for ensuring a uniform pocket humidity in the cross-machine direction. A variation in humidity will cause a variation in the moisture profile at the reel. Increasingly important is the effect of the pocket ventilator design on paper machine runnability. Sufficient air must be introduced into the pockets to lower humidities without excessively disturbing the sheet. This is a difficult task on high-speed lightweight paper machines.

5. **Web stabilizing components**

   There is a group of air-handling devices that can be used to stabilize the sheet. These devices are used to control the boundary layer air flows carried by the dryer fabrics and rolls at high machine speeds. Press transfer blow boxes, unorun blow boxes, drive group transfer blow boxes, and sheet stabilizing pocket ventilators are all examples of sheet stability devices. A separate discussion is included which details their operation.

6. **Heat recovery equipment**

   The majority of the heat that is contained in the steam supply to the dryers is contained in the hood exhaust vapor. The heat exists at a low temperature and quality. Heat recovery equipment is commonly installed in the exhaust stream.

Some of the more common uses for the recovered heat are to preheat hood supply air, to heat shower water for use on the machine, and to heat building ventilation air. The three most used heat recovery equipment types are direct water sprays, air to air heat exchangers, and air to liquid heat exchangers.

The direct water spray system is the simplest. It consists of a chamber in the exhaust stream where water sprays are used to atomize the water in the exhaust vapor. Heat is transferred from the vapor into the water. The water is collected for use on the machine.

Air-to-air heat exchangers transfer heat from the exhaust vapor into separate air stream. Hood supply air can be preheated in this manner. It is common to heat the hood supply air from a 26°C room temperature to 60°C using this equipment. Building ventilation air can also be heated using air-to-air heat recovery equipment.

Air-to-liquid heat exchangers are used to transfer heat from the exhaust vapor into either water or a water/glycol solution. The water/glycol solution is often used for heating building ventilation air. The glycol solution can be pumped to all areas of the mill for heating building ventilation air. This can provide very efficient heat recovery benefits especially in northern mills where building heating is costly.

G. HOODS AND HOOD EXHAUST

The type of hood and its design affect the design of the ventilation system. The main variable is the hood exhaust air humidity. Low-humidity operation requires high air exhaust volumes along with high supply air volumes. Figure 330 shows the relationship between hood exhaust humidity and energy consumption.

1. Open canopy hood

This is the simplest type of hood. A canopy is formed over the dryer section. A high exhaust rate is needed to capture the vapor and prevent condensation. These hoods usually do not provide a tight fit to the dryer section, therefore

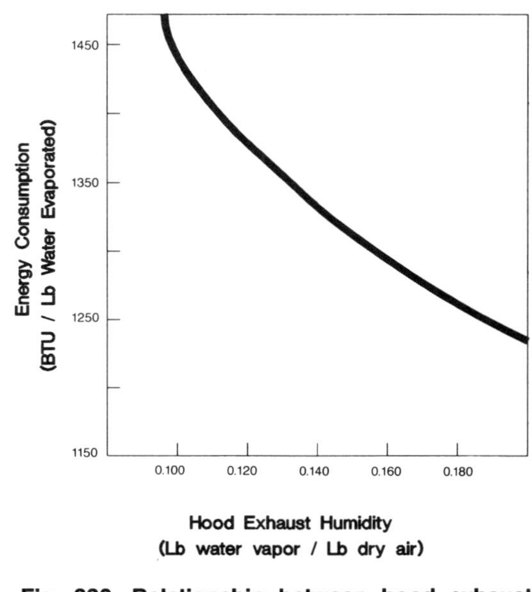

Fig. 330. Relationship between hood exhaust humidity and energy consumption.

spill is a problem. Many canopy hoods are built of single sheet material which does not have a high insulating value. High exhaust rates are needed to prevent condensation. Insulated panel designs are also used which provide better insulation properties and allow slightly higher humidities.

The exhaust humidity design range is 0.04–0.07 kg water vapor/kg dry air. The 0.04 kg value would apply to single sheet canopy hoods. The 0.07 kg value would apply to insulated panel hood designs.

Approximately 13 kg of air are exhausted for each kg of water evaporated. This high air flow results in poor dryer section energy consumption due to the high air flows through the dryer section. The high exhaust rate must be balanced with heated building supply air which increases the building heating load. The low temperature of the exhaust makes heat recovery expensive and in most cases impractical.

Open hoods are very rarely used on new machine installations. Many older paper machines still use open hood designs. These are gradually being replaced with more efficient closed hoods designs.

2. Medium-humidity enclosed hood

This type of hood is by far the most common in North America. The hood consists of an insulated enclosure above the operating floor. Front lifting panels and rear sliding panels are used to provide access to the dryers above the operating floor. The hoods normally include an uninsulated basement enclosure under the dryers. The purpose of the basement enclosure is to provide a control of air infiltration from the basement enclosure. An enclosed hood would be pressurized above the operating floor without a basement enclosure. The natural stack effect within the dryer section causes a pressurization. The basement enclosure restricts the air flow and prevents pressurization.

Early enclosed hoods were designed with exhaust humidities around 0.09 kg/water vapor/kg dry air. Most medium humidity hoods today operate in the 0.11–0.13 kg/kg range. Approximately 10 kg of air is exhausted for every kg of water evaporated. Typical exhaust temperatures for this type of hood would be 80°C dry bulb and 59°C wet bulb.

Heat recovery from a medium-humidity enclosed hood is generally attractive in northern climates. Preheating hood supply air, process water heating, and building makeup air heating are all common uses for the recovered heat.

3. High-humidity enclosed hood

A high-humidity hood is designed to operate in the 0.16–0.17 kg water vapor/kg dry air exhaust humidity range. The air consumption is approximately 7 kg of air per kg of water evaporated. Typical exhaust temperatures are 85°C dry bulb and 62°C wet bulb.

The hood construction for a high-humidity hood is considerably different than the medium-humidity hood. Panels with a high insulation value and tight vapor seal need to be used. Normal tongue and groove insulated panels are unsuitable for long-term service in a high-humidity environment. Careful attention needs to be paid to door seals and penetrations into the hood.

The high-humidity hood offers reduced energy consumption and superior heat recovery opportunities. A main advantage of the high-humidity hood is the reduced air flow through the hood and the influence on sheet runnability. At state-of-the-art speeds, the air flows through the hood have an impact on sheet stability and machine efficiency. High-humidity hoods better control the air flows through the dryer section.

H. POCKET VENTILATION

Modern pocket ventilation systems were made possible with the introduction of synthetic dryer fabrics which have an air permeable construction. This advance occurred in the early 1960s. Prior to this, the dryer clothing was of an impermeable design. Air was introduced into the pockets using low-volume, high-velocity side jets. This method was limited in value especially on wide machines. Humidities in the pockets were high and non-uniform. This resulted in reduced capacity and reel moisture profile variation.

The synthetic dryer fabrics allowed air to be forced into the pocket through the fabric. The combination of permeable dryer fabrics and pocket ventilation resulted in much improved drying capacity and more uniform moisture profiles. On some machines, the increase in drying capacity was up to 20%. The state-of-the-art machine speeds would not be possible without modern dryer fabrics and pocket ventilation. However, the air-moving capabilities of the dryer fabrics has created an entirely new set of problems at high machine speeds. The air movement must be closely controlled or sheet stability problems arise. Fabric permeability, fabric construction, and pocket ventilator design are used to control the air movement.

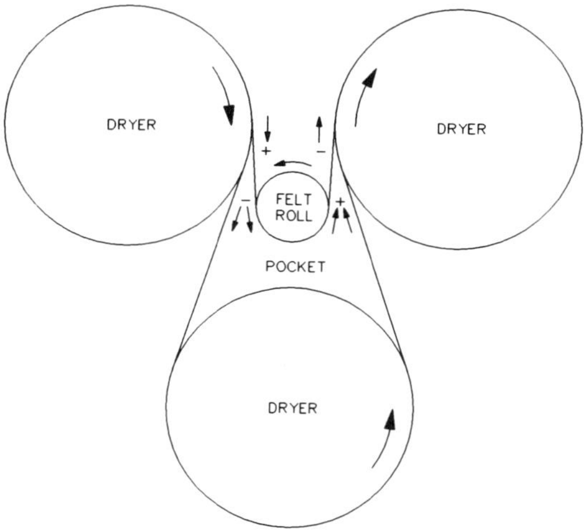

Fig. 331. Pressure induced by moving rolls and fabric.

Figure 331 shows the air movement in the pocket area. The dryer fabric carries a high volume of air with the boundary layer associated with the fabric movement. The converging nip formed by the felt roll and the fabric create a positive pressure area. Air is forced through the fabric and into the pocket. This is a natural form of pocket ventilation resulting from the geometry of the rolls. The amount of air forced into the pocket in this manner is governed by the machine speed, the fabric permeability, and the felt roll location.

On the exit side of the felt roll, a negative pressure area is created as the fabric leaves the felt roll and a positive pressure is created as the fabric contacts the next dryer. Air is forced out of the pocket with this mechanism.

The air that is pumped into the pocket area is humid hood air which creates higher pocket humidities. Another problem with a natural pocket ventilation method is that there is no control over the air flows. High-permeability fabrics will pump an excess amount of air into the pocket. This over-pressures the pocket and forces air out the ends. The excess air creates sheet flutter. The papermaker is forced to reduce the fabric permeability to control sheet flutter. Low permeabilities will restrict the air pumping into the pocket area. This results in high humidities in the center area of the pockets. The edges are normally at a lower humidity than the center with low-permeability fabrics. Reduced drying capacity and a non-uniform reel moisture profile are the result.

The pocket ventilation system must control the natural pumping action of the fabric and introduce dry air into the pocket in a controlled manner. The pocket ventilation system must work in conjunction with the dryer fabrics.

1. Purpose of the pocket ventilation system

The purpose and degree of sophistication of the pocket ventilation system depends on the machine speed and paper grade. A slow-speed heavyweight paper machine can use very open dryer fabrics and use an unsophisticated pocket ventilator design. Hot dry air can be simply uniformly dumped into the converging nip area and the fabrics will naturally pump the air into the pocket. Sheet flutter is not a problem because of the paper weight and machine speed. This type of ventilator will not be satisfactory on a lightweight, high-speed machine. Air flows must be better controlled. The fabric permeability will be lower and natural pumping will be less. The demands on the pocket ventilator will be greater.

The following are some requirements of a high-speed pocket ventilation system:

1. **Balance and control air flows into and out of the pocket**
   Good runnability and moisture profile require that the amounts of air pumped into and out of the pocket are equal. If the incoming and outgoing amounts differ, axial flows are created in the pocket. Axial air flows create sheet flutter.
   The amount of air that is pumped into a pocket exceeds the air pumped out of the pocket if the fabric permeability is high. At low permeabilities, the air pumped out of the pocket exceeds the air pumped in. Figure 332 shows two conditions with lower fabric permeabilities and no pocket ventilation equipment. Humidity levels are high in the center by the incoming air flow at the edges and the sheet moisture profile will be dry at the edges.
   A high-speed pocket ventilator must put the pocket in balance. It must allow the papermaker to use the highest permeabilities possible without excessive sheet flutter. To accomplish this, the incoming boundary layer of air carried by the fabric needs to be blocked. The high volume of humid air carried by the fabric cannot be allowed in the converging felt roll and fabric nip. Figure 333 shows a pocket ventilator design suitable for high-speed operation. The top air nozzle is designed to block the undesirable air flow carried by the dryer fabric. The lower jet blocks the air flow being pumped by the felt roll. A pressure zone is formed between the

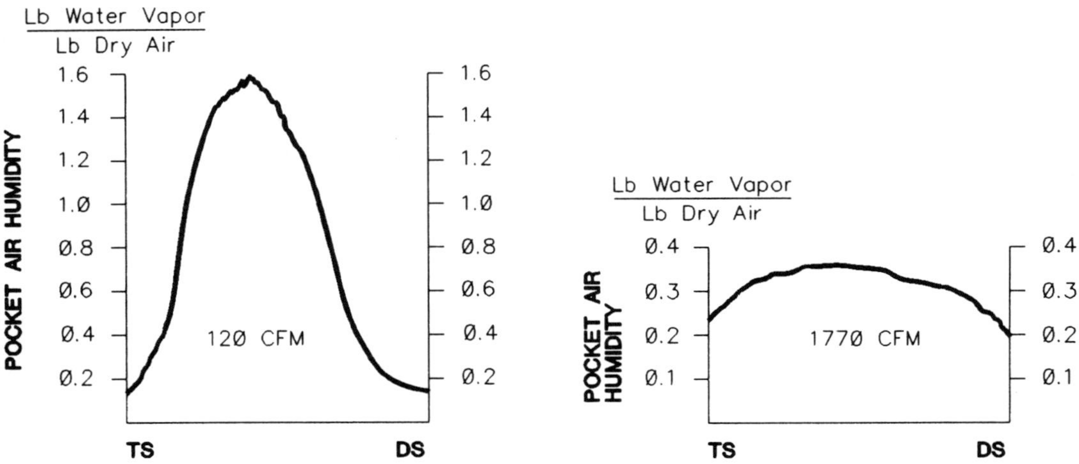

Fig. 332. Pocket air humidity profiles with different fabric permeabilities.

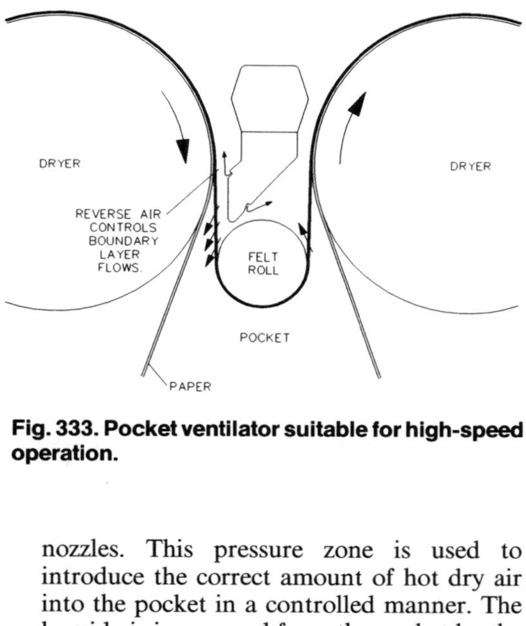

Fig. 333. Pocket ventilator suitable for high-speed operation.

nozzles. This pressure zone is used to introduce the correct amount of hot dry air into the pocket in a controlled manner. The humid air is removed from the pocket by the natural pumping action. The pocket air flows are in balance to minimize sheet flutter. Dry air is supplied to the pockets rather than the humid air from the hood.

2. **Provide low and uniform pocket humidities**
A pocket ventilation system must provide low and uniform pocket humidities. This maximizes the drying rate on the paper machine and helps to provide a uniform reel moisture profile. Typically a pocket humidity level of 0.18 kg water/kg dry air is considered optimum for machines operating with dryer pressures less than 30 psig. A value of 0.2 kg/kg is good for a machine operating at higher pressures. The cross-machine variation should not exceed 0.02 kg/kg best results.
Questions always arise concerning how much air is needed for each pocket to lower the humidity to the levels listed above. The amount of air required will vary depending on the drying rate of the machine and the position in the dryer section. A detailed knowledge of the paper drying process is required to properly size the pocket ventilation air volume. The nozzles should be designed for the highest drying zone on the machine. The amount of air will vary from 12 m^3/min/m of machine width per nozzle for a low-pressure newsprint machine to 18 m^3/min/m for a high-pressure fine paper machine. The majority of today's paper machines require air flows in the 14 m^3/min/m range.

3. **Provide reel moisture profile correction capability**
The ability of a pocket ventilation system to correct the reel moisture profile by varying the cross-machine air distribution is limited. Moisture profiles can be affected by the cross-machine air flows. The amount will depend on grade, final reel moisture, speed, fabric permeability, and average pocket humidity level. The amount of correction capacity that can be expected on any given machine is difficult to predict because of the large number of variables involved. A 1% reel moisture profile correction at the 4% final moisture content is reasonable. A 2% moisture profile correction at the 8% level is reasonable.
Pocket ventilation should never be used as the

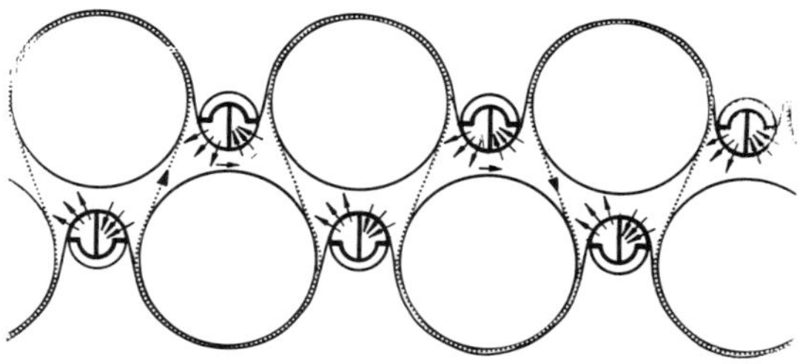

Fig. 334. Pocket ventilation rolls.

primary moisture profile correction device on a machine. Its effect will be limited. It can be used as another tool to help straighten a typically bad profile. Many machines will run with a typical dry front or back edge. Turning off the pocket ventilation air in these areas can help to correct this type of a profile problem.

4. **Provide heated supply air to the hood**
   A hood requires a source of heated supply air. The best place to distribute the air into the hood is through a well-designed pocket ventilation system. The air is supplied to the area where the vapor is generated and where it has the most benefit for the machine.
   Normally if the proper air volume is supplied to the pocket ventilation system it will provide sufficient air for a good hood air balance.

To this point in the discussions, pocket ventilator nozzles which are located above the felt roll have been discussed. Another popular method to ventilate pockets is through the use of a specially constructed felt roll. The felt roll is built to allow air to be passed into the roll, then is distributed from the roll through a series of drilled passages. The air is forced through the fabric and into the pocket. Figure 334 illustrates pocket ventilation rolls.

I. DRYER SECTION RUNNABILITY

Runnability in the dryer section has become one of the major considerations for efficient machine operation at high speeds. Configuring the dryer section for good sheet stability is becoming the major influence in modern dryer section design.

Sheet instability is caused by air movement associated with moving surfaces such as dryer fabrics, dryer cans, and felt rolls. The air carried by the moving surfaces disturbs the sheet. The problems get progressively worse as speed increases. To stabilize the paper web, the papermaker is forced to use speed differences to put tension on the sheet. This causes a high sheet break frequency and a loss in some sheet properties.

Modern dryer sections use unconventional configurations to provide improved sheet stability. There are also methods and equipment available to improve the sheet stability on older conventional dryer sections.

The key to good dryer section runnability is to:

- Shorten or eliminate the unsupported sheet run.
- Control the boundary layer air flows associated with the moving surfaces such as fabrics, dryer cans, and felt rolls.

1. Press to dryer transfer

Transferring the sheet from the press to the dryer section is often the first area where sheet stability problems occur. The press and dryer geometries often dictate that there is a long unsupported sheet run. The wet sheet at this location is very weak.

A speed difference, or draw, is needed between the press and the first dryer. The draw is required to pull the sheet off the last press roll and to stabilize the sheet over the unsupported length. A long, unsupported length requires a high draw, and the draw requirement increases with speed. The sheet stretches with the speed difference. Stretching the weak sheet causes a loss in elasticity resulting in a less forgiving sheet. A high break frequency at the press transfer and further down the machine is the result. The stretching of the sheet at the unrestrained edges is greater than in the center. This causes "floppy" edges and non-uniform cross-direction tensile properties.

The most common solution to this problem is to minimize the unsupported draw length. The turning roll for the first section dryer fabric is brought in very close to the last press roll. The sheet is transferred over a short unsupported length to the underside of the dryer fabric. Special air boxes are used to create a suction on the underside of the fabric, holding the sheet tight and flat to the fabric. The arrangement is shown in Fig. 335.

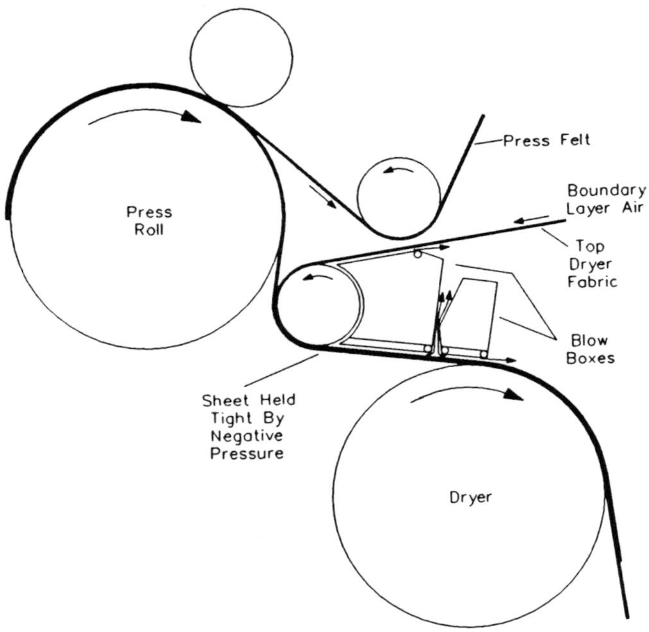

Fig. 335. Low-draw press transfer.

The air boxes are the key ingredient to making this arrangement work properly. Pressurized jets of air are used with a special nozzle configuration. The jets control the boundary layer air carried by the fabric and create the negative pressure zone without drawing the fabric into the boxes. This press to dryer transfer can be adapted to many press configurations. Generally machines operating above 750 m/min can benefit from the use of this improved press to dryer transfer.

2. Single-felted dryer sections

A conventional two-tier dryer configuration has an unsupported sheet run between the top and bottom dryers. Sheet "flutter" in this open draw will cause sheet breaks and wrinkling. The problem is most severe at the wet end of the dryer section where the sheet is very weak.

The initial solution to this problem was to use a single felt which supported the sheet between top and bottom dryers. This is the single-felted or Unorun dryer configuration. The felt supports the sheet and improves the stability.

The problem with this arrangement is that the fabric is sandwiched between the sheet and the bottom dryer. This makes the bottom dryers ineffective in terms of transferring heat to the sheet. However, the sheet stability gains outweigh the loss in drying capacity on most machines. The Unorun concept is used toward the wet end of the paper machine. Once the sheet is sufficiently dry, conventional top- and bottom-felted dryers are used to maximize the drying capacity. The length of the Unorun section depends on the grade, machine speed, and drying capacity.

Even with the fabric supporting the sheet, the sheet can become unstable in the serpentine section. Figure 336 shows the forces that act on the sheet to cause instability.

The fabric carries a high volume of air in the boundary layer that is associated with the fabric. At the ingoing nip on the bottom dryer, a positive pressure is created by the boundary layer air being forced into the lower nip. Some of the boundary layer air is forced through the fabric and is trapped between the sheet and fabric around the bottom dryer. The trapped air will float the sheet off the fabric and stability is lost.

At the outgoing nip from the top dryer, the sheet tends to follow the top dryer can. This will create a bubble in the sheet which can lead to wrinkling. Any air that is trapped between the sheet and fabric is forced out at the ingoing nip to the top dryer. This also can lead to wrinkling.

The solution to the instability problem is to eliminate the boundary layer air before it can be trapped between the sheet and fabric. Air boxes,

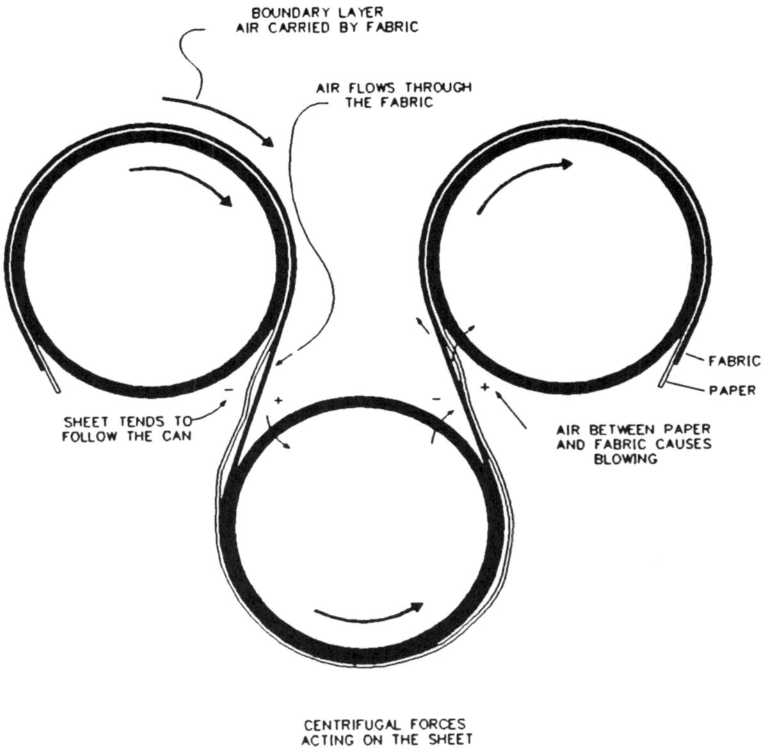

Fig. 336. Instabilities in sheet travel in a single-felted dryer group.

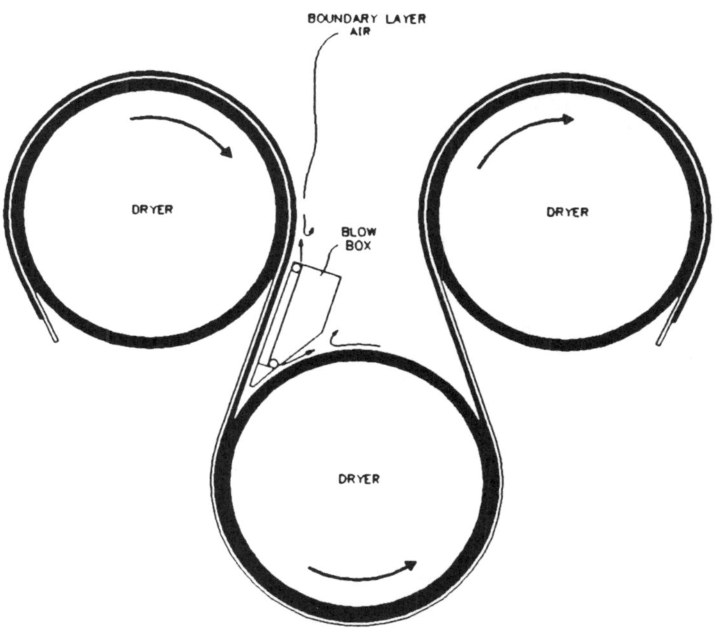

Fig. 337. Serpentine dryer with blow box.

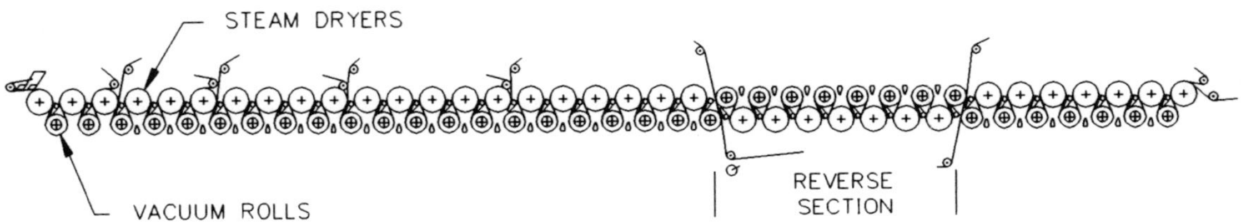

Fig. 338. Complete single-tier machine.

which blow air in a direction opposite to the fabric and dryer travel, remove the boundary layer of air. The blowing action creates a negative pressure to hold the sheet tight to the fabric with no air trapped in between. Modern air boxes span the full height between top and bottom dryers. The sheet is pulled off the top dryers and held by negative pressure until it reaches the bottom dryer.

Figure 337 shows a Unorun arrangement with an air box installed.

Two to three short Unorun sections could be used. The sheet would be transferred between Unorun groups by "kissing" the last top dryer with the fabric for the next Unorun group. The short drive sections allow the papermaker to have a small speed difference between serpentine groups. An air box would be used at the transfer to provide stable operation. The transfer between the last Unorun section and the first conventional double-felted section would use closely spaced felt rolls and an air box.

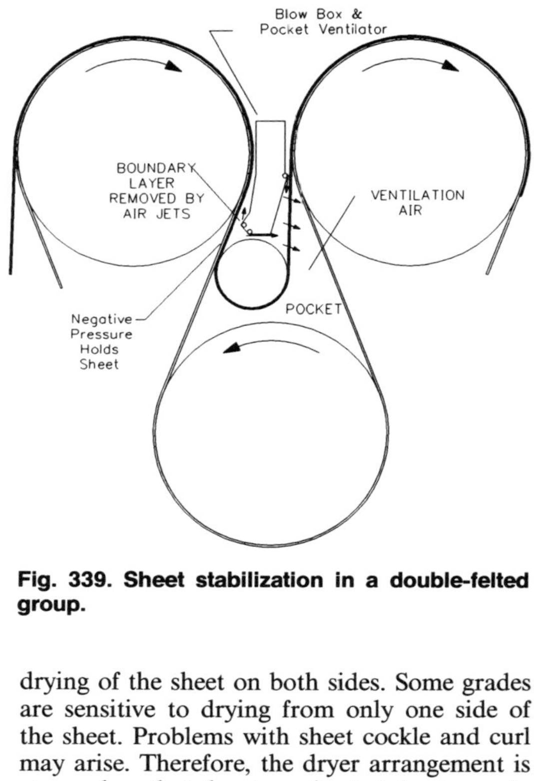

Fig. 339. Sheet stabilization in a double-felted group.

3. Single-tier dryer arrangements

Today's very high speed machines extend the Unorun dryer configuration down the entire machine length but with some major changes to improve operation.

Figure 338 shows a modern high-speed machine which uses a complete single-tier dryer configuration. This type of dryer section would be used for new high-speed machines designs.

The arrangement has only one tier of dryers instead of two. In the conventional serpentine arrangement, the bottom dryers are ineffective as dryers. These dryers have been eliminated and replaced with smaller diameter rolls. The rolls are placed under a negative pressure to improve the runnability by holding the sheet with vacuum. Some designs use the air boxes to remove the boundary layer similar to the unorun configuration. Other designs do not use air boxes and rely on the suction in the vacuum roll to control the boundary layer.

It is necessary to have a number of reverse sections with a complete single-tier arrangement. The purpose of the reverse sections is to provide drying of the sheet on both sides. Some grades are sensitive to drying from only one side of the sheet. Problems with sheet cockle and curl may arise. Therefore, the dryer arrangement is reversed so that the steam heated dryers are in the bottom position. The sheet is transferred to the reverse section using vacuum rolls.

The single-tier dryers sections make more effective use out of each dryer than a conventional two-tier arrangement. The sheet and felt wrap the dryer for a greater distance around the diameter. This improves heat transfer. There is also a greater distance for evaporation as the sheet travels around the bottom roll. Both these factors make each single-tier dryer more effective than a conventional dryer in a double-felted group.

The effect on sheet properties of the single-tier arrangement is a major consideration. The sheet is completely restrained down the entire dryer section and there are no free draws. Drying the sheet under restraint improves the cross-direction uniformity of sheet properties.

The single-tier dryer section is more expensive than a conventional two-tier dryer arrangement. The dryer section is longer which means a longer building. The horsepower consumption is higher due to the vacuum that needs to be generated for the rolls. However, the improved runnability and sheet properties outweigh the higher cost.

3. Double-felted dryer groups

Although single-tier dryer sections represent modern technology for new high-speed machines, older machines and some new machines designed for lower speed operation use conventional double-felted dryer groups. These machines are faced with the problem of sheet instability in the double-felted dryer groups.

The arrangement shown in Fig. 339 is a solution to sheet stability in a double-felted group. The felt roll is offset to the wet end of the machine. An air box is used to create a negative pressure on the wet end side. The negative pressure pulls the sheet off the dryer can and holds it tight to the fabric. The length of the sheet travel is reduced and the sheet is released from the fabric in a uniform manner. Pocket ventilation is performed by pressurizing the dry end of the pocket.

REFERENCES

1. Gavelin, G., Ed., "Drying of Paper & Paperboard", Lockwood Publishing (1972).
2. Smook, G.A., "Handbook for Pulp & Paper Technologists", Joint Textbook Committee of the Paper Industry, TAPPI, Tech. Sect. CPPA (1982).
3. Dreshfield, A.C. and Han, S.T., "The Drying of Paper", Tappi 39(7) (July 1956).
4. Hill, K.C., "Defining Dryer Performance", Tappi 69(4) (Apr. 1986).
5. Han, S.T. and Ulmanen, T., "Heat Transfer in Hot Surface Drying of Paper", Tappi 41(4) (Apr. 1958).
6. Hill, K.C. and Reese, J.R., "Dryer Section Rebuild Requirements", Proc. TAPPI Engineering Conf. (Sept. 1989).
7. Appel, D.W. and Hong, S.H., "Condensate Distribution & Its Effects on Heat Transfer in Steam Heated Dryers", Pulp Paper Mag. Can. (Feb. 1969).
8. Gardner, T.A., "Blowthrough Steam Control Yields Major Reduction in Drying Steam", TAPPI Papermakers Conf. (Mar. 1987).
9. Fishwick, H.P., "Paper Machine Steam and Condensate Systems", 3rd Ed. (May 1986).
10. Gardner, T.A., "Fundamentals of Blowthrough Control", Pulp Paper Can. 87(5) (May 1986).
11. Concannon, M.D., "Condensate Effects on Tongue and Horsepower in Paper Dryers", Tappi J. 63(9) (Sept. 1980).
12. Calkins, D.L., "The Effects of Syphon Clearance on Dryer Performance", The Johnson Corp., Three Rivers, MI (1966).
13. Pulkowski, J.H. and Wedel, G.L., "The Effect of Spoiler Bars on Dryer Heat Transfer", Pulp Paper Can. 89(8) (1988).
14. Wedel, G.L., "No-Draw Drying Restraint", Tappi J. (Apr. 1989).
15. Wedel, G.L. and Palazzolo, "Advances in Dryer Section Runnability", Tappi J. (Sept. 1987).
16. Hill, K.C., "Dryer Sheet Stability for Older Paper Machines", Tappi J. 71(8) (Aug. 1988).
17. Karlson, M. and Vihma, L., "Sheet Runnability and Dryer Section Ventilation Design", Tech. Sect. CPPA Annual Meeting Preprints (1985).
18. Lindstrom, Y., "Dryer Section Ventilation Design and Sheet Runnability", Pulp Paper Can. 86(3) (1985).
19. West, A.O. and McMullen, D.W., "Application of Modern Pocket Ventilation Systems to the Conventional Dryer Section", Tappi J. 59(4) (Apr. 1976).
20. Reese, J.R., "High Humidity Hoods Conserve Energy, Improve Runnability", Proc. TAPPI Engineering Conf., Book 2 (Aug. 1988).
21. Ojala, P., "Analyzing Mill Heat Recovery Systems for Coat Effectiveness in Production", Paper Trade J. (Nov. 1982).
22. Weiner, J., "Drying of Paper & Paperboard, Bibliographic Series", Inst. Paper Chemistry, Appleton, WI (1976).
23. Hill, K.C., "What's New in Glycol Heating & Heat Recovery Systems", PIMA Mag. (Mar. & Apr. 1981).

XIII SURFACE SIZING

C.P. KLASS

Charles P. Klass is Managing Director of Klass Associates Inc., King of Prussia, PA, a consulting firm specializing in coating, surface sizing and development of new paper grades. He has a B.S. in Paper Technology from Western Michigan University and an M.B.A. from Pace University. His three decades of paper industry experience include mill technical and operating positions and technical service/marketing experience with chemical and coating equipment suppliers. He has published numerous papers and has presented his publications worldwide. Mr. Klass is a TAPPI Fellow, recipient of the TAPPI Paper and Paperboard Manufacture Division Award, and served on the Board of Directors. He is also active on the Technical Section, CPPA Coating Committee.

REVIEWED BY

George L. Booth
Product Manager/Coating
Black Clawson Company
Fulton, NY

Dale R. Dill
Director Polymer R&D
Protein Technologies International Inc.
St. Louis, MO

A. INTRODUCTION

Traditionally, the purposes of surface sizing have been to provide resistance to penetration of liquids, to give better surface properties, and to improve certain physical properties of the paper sheet such as surface strength and internal bond. Surface sizing operations have typically utilized starch films to fill in the surface voids in the sheet, reducing the pore radius and therefore the rate of penetration of fluids including printing inks [1]. Over the past decade, however, the variety and complexity of surface sizing operations have grown. This growth has been brought about by several factors:

1. Coated papers have become more important. Surface sizing is a key element in making an improved base stock.
2. Specialty papers demand new and unusual sheet properties. For years, if a property was needed, furnish changes and wet end additives were employed. Now, the papermaker routinely considers surface treatments when he is considering performance requirements and grade development. Many new surface sizing additives and much data have been presented to the industry by suppliers.
3. Everyone has become aware of the need for a more efficient operation, both in cost and in the conservation of raw materials. Surface sizing provides close to 100% retention of added chemicals. This improved retention also reduces the problems of wet end deposits and increases the life of machine clothing.
4. Environmental laws have placed new restraints on paper mills. Compared to wet end addition, surface sizing allows reduction or removal of chemicals from the white water to assist in environmental improvement.

These are only a few of the varied reasons for surface sizing application growth [2].

The contribution of surface sizing to the desired end product properties depends on the type of paper or paperboard being made, and trade-offs may be involved [3]. Some examples may illustrate this point.

In the case of an offset printing paper, surface sizing is used to improve surface strength (to resist picking), to control penetration by water (fountain solution) and ink, and to prevent linting and dusting. The surface size used to achieve these desirable properties may adversely affect optical properties such as opacity and brightness. The choice of surface size application equipment may improve or adversely affect internal bond and smoothness.

The purposes of surface sizing of packaging papers may include resistance to penetration by oils, greases and solvents; improvement of burst and tensile strength; and resistance to abrasion. However, surface sizing may adversely affect tear strength. The bending and score cracking properties of paperboard may be improved by surface sizing.

Specialty surface sizes may be used to obtain specific functional properties such as moldproofing, bacteriostaticity, wickproofing, flame retardance, and oxidation/tarnish resistance.

Since many packaging and functional specialty papers and paperboards are also printed, the optimization of properties sought through surface sizing may be complex. The combination of chemicals used, application equipment, and operating conditions affect not only the end product properties but also the efficiency of the operation.

B. STARCH AS A SURFACE SIZE

The most commonly used surface sizing is starch applied by size press. In common paper mill practice, a relatively low viscosity starch solution of 3 to 9% solids is used at the size press to achieve pickup in the range of 30–50 kg of dry starch per ton of paper produced.

Starch is a carbohydrate synthesized in corn, tapioca, potato, and other plants by polymerization of dextrose units. The polymer exists in two forms: a linear structure of about 500 units and a branched structure of several thousand units. The linear chain polymer, called "amylose", constitutes 27% of natural corn starch, while the branched polymer, called "amylopectin", makes up the remaining 73%. Fractionated starches are available for special uses [4].

Starch is supplied as a white, granular powder which is insoluble in cold water because of the polymerized structure and hydrogen bonding between adjacent chains. However, when an aqueous suspension is heated, the water is able to penetrate the granules and causes them to swell, producing a "gelatinized" solution or paste, depending on concentration. Further cooking causes the solution to thin to a lower viscosity. Cooling this hot solution causes thickening, which is called "setback" or retrogradation. Starch for surface sizing application is "cooked" using either batch or continuous systems [5].

Unfractionated and unmodified starch, called "pearl starch", is "thick boiling" (i.e., viscous) and has a tendency toward gelling or setback, even without cooling. Setback is avoided by using 100% amylopectin, "waxy" starch which forms a clearer paste and is non-gelling; however, a loss of sizing efficiency results because the linear fraction contributes more toward film formation.

Lower viscosity and setback resistance are

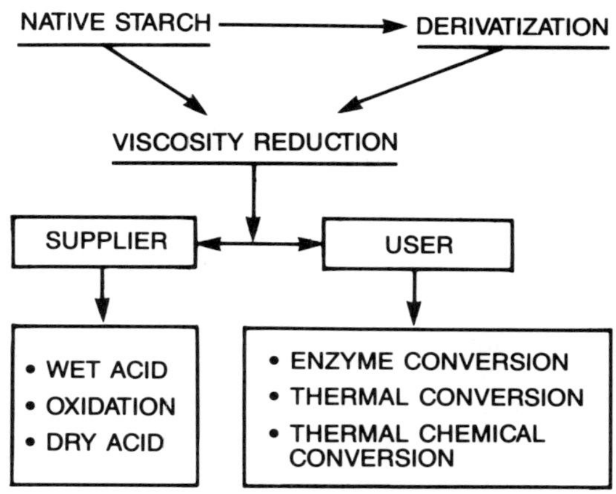

Fig. 340. Starch modification methods.

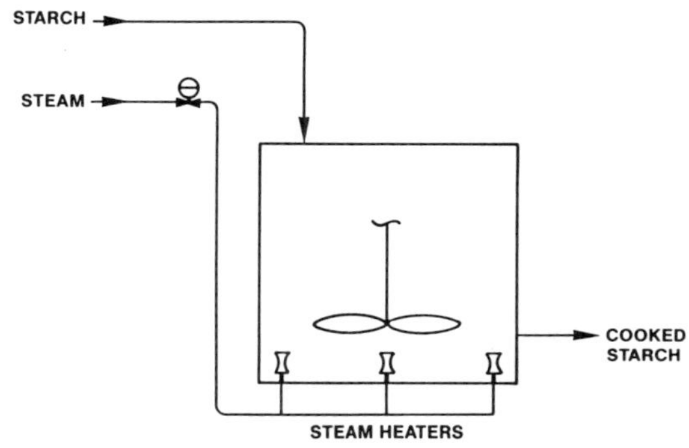

Fig. 341. Batch starch cooker.

achieved by using chemically (e.g., "oxidized" starch) or thermally modified starch. A "thin boiling", low-viscosity starch can be produced by enzyme conversion. Starch modification may be done by the starch supplier or at the paper mill as summarized in Fig. 340 [6].

C. STARCH PREPARATION

As noted above, starch is supplied to paper mills as an insoluble white powder. It must be dispersed in water to form a granular slurry and cooked to make a colloidal dispersion. The cooked form is commonly known as "paste".

Slurry makedown may be either a batch or continuous operation. Starch solids concentration must be controlled by accurate measurement of both water and dry starch. Adequate wetting time and agitation are essential to provide a uniform dispersion for cooking. It is desirable to screen the starch slurry prior to cooking. Since starch is subject to bacterial degradation, it may be necessary to use a microbiocide to preserve the starch slurry in storage.

Starch cooking may be either batch or continuous. As shown in Fig. 341, a batch cooking system involves charging an open vessel with starch slurry and, while agitating, heating with live and or indirect steam to raise the starch temperature to 88°C to 93°C [5]. The batch is held at this temperature for 20 to 30 min. In batch cookers where direct steam injection is employed, it is essential to use turbulent mixing conditions at the point(s) of injection to provide even heating. The viscosity of the slurry increases as the conversion point is reached, so strong agitation is required. The starch slurry should be kept below the boiling point. Boiling of the starch slurry creates foam and a potentially severe safety hazard if paste boils out of the tank. Because cooking temperatures are limited to atmospheric conditions, premodified starches are usually used for batch cooking.

Thermal jet cooking permits the continuous preparation of starch and facilitates cooking at 104°C to 160°C [7]. The starch cooking jet shown in Fig. 342 allows a stream of high-velocity steam to mix intimately with a stream of starch slurry. This causes a rapid temperature rise. Steam injection devices induce turbulence in a controlled manner to ensure dispersion of the steam in the line prior to condensation.

A continuous process for jet cooking is shown in Fig. 343. The process includes a positive displacement slurry pump with a loop line excess return to the slurry tank, a steam regulating valve which is fed by steam, the cooking jet, a back pressure valve, and a flash tank. The amount of steam added to the slurry is controlled to achieve complete condensation. If cooking is to be done above the boiling point, back pressure is used to control temperature.

Since the jet cooker can operate at elevated temperatures, it can be used to cook unmodified starches, i.e., thermal conversion. However, all high-temperature cooked underivatized starches are susceptible to amylose precipitation. Amylose precipitation can be minimized by maintaining the pH in the 8.0 to 8.5 range by adding alkali with effective mixing as soon as possible after the paste is returned to atmospheric conditions.

Some jet cooking systems use "excess steam" in the heating zone to cause mechanical shearing of the starch. This produces a more dispersed starch paste. The amount of excess steam varies from 125% to 300% of the amount of steam required to bring the starch to the desired temperature. This is commonly referred to as "1.25x to 3x steam". The most significant result of this system is the lower apparent viscosity of unmodified starch compared with that prepared by conventional batch cooking or by 1x jet cooking. Excess steam jet cooking may provide cost savings by allowing the use of lower cost unmodified starch instead of a higher priced premodified starch; this is offset to some degree by the cost of the excess steam.

Starch may also be converted by the action of bacterial alpha-amylase enzymes. The enzymes are blended by the manufacturer to provide a standard activity level on a unit basis, i.e., per packet, gram or tablet. Activity of the enzymes on starch is affected by pH, temperature, and the presence of stabilizing ions (usually Ca^{++} ion); all of these variables must be controlled. The temperature effect is critical because higher temperatures simultaneously increase the rate of enzymic depolymerization (deactivation) and the rate of enzyme activation.

While technology exists for continuous processing with enzymes, batch processing is usually used. The enzyme conversion process consists of charging the starch slurry to the cooker, checking and adjusting the pH to the value recommended by the enzyme supplier, and adding the enzyme with good agitation. The slurry is then heated at a controlled temperature rise rate to about 77°C. After holding at this temperature for the specified period, usually about 30 min, the temperature is increased as rapidly as possible to about 93°C. This temperature is usually adequate to kill the enzyme in about 15 to 30 min. The material is then cooled to the desired storage temperature. If the enzyme is not properly killed, reducing sugars can be formed to adversely affect the performance of the starch.

Thermal chemical conversion of starch is a combination of chemical depolymerization and high temperature dispersion. Although numerous soluble oxidants can be used, ammonium persulfate and hydrogen peroxide are the most common. Both function via a combination of oxidative and acid hydrolytic depolymerization to reduce starch viscosity.

The basic thermal chemical conversion system shown in Fig. 344 consists of a positive displacement pump that supplies slurry at a controlled rate to the cooking jet [8]. The slurry is

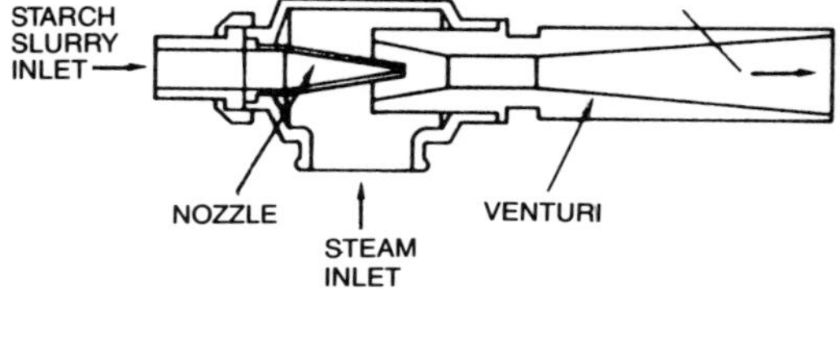

Fig. 342. Starch cooking jet.

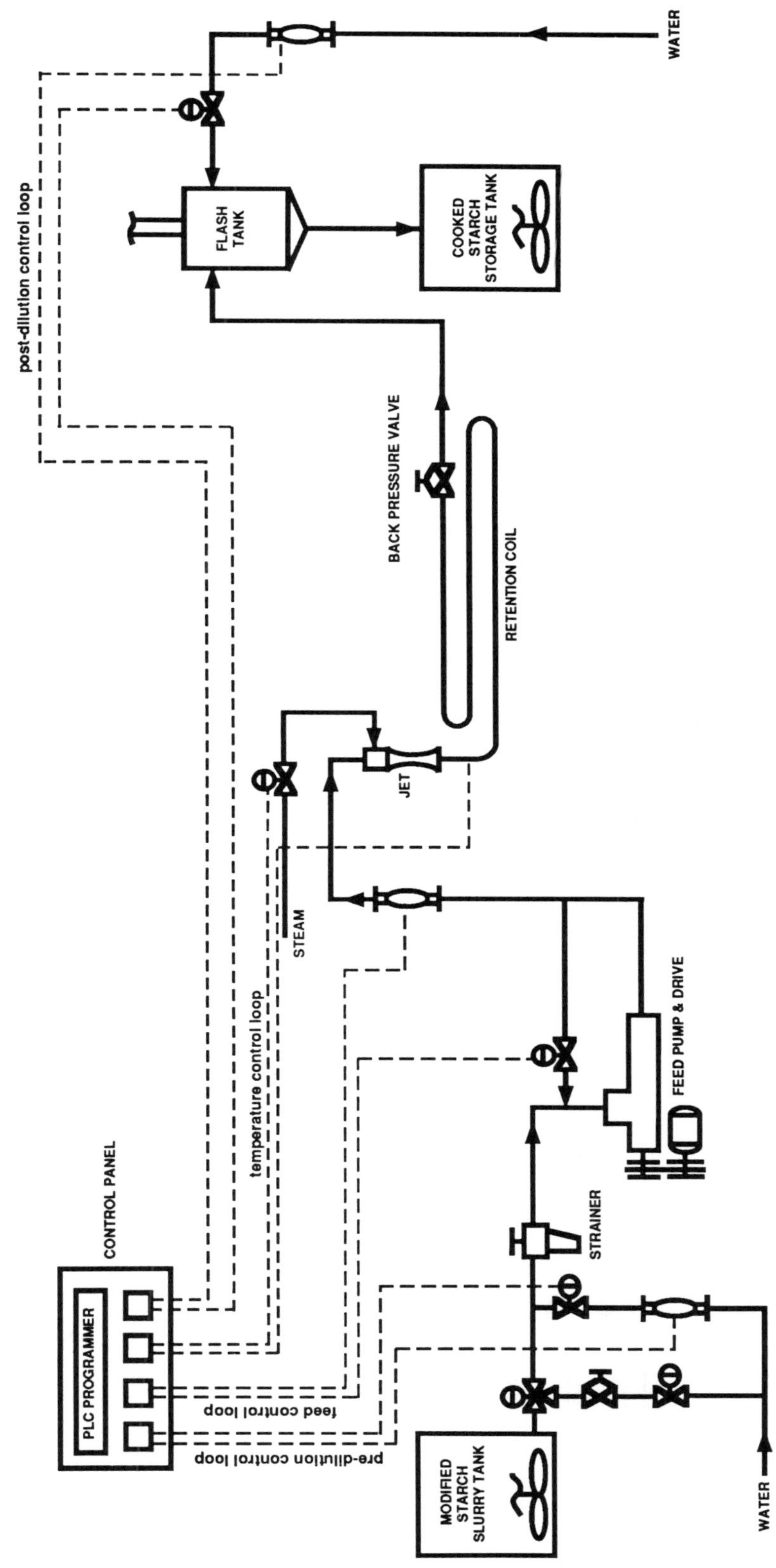

Fig. 343. Continuous process for jet cooking. (Courtesy of Grain Processing Corporation)

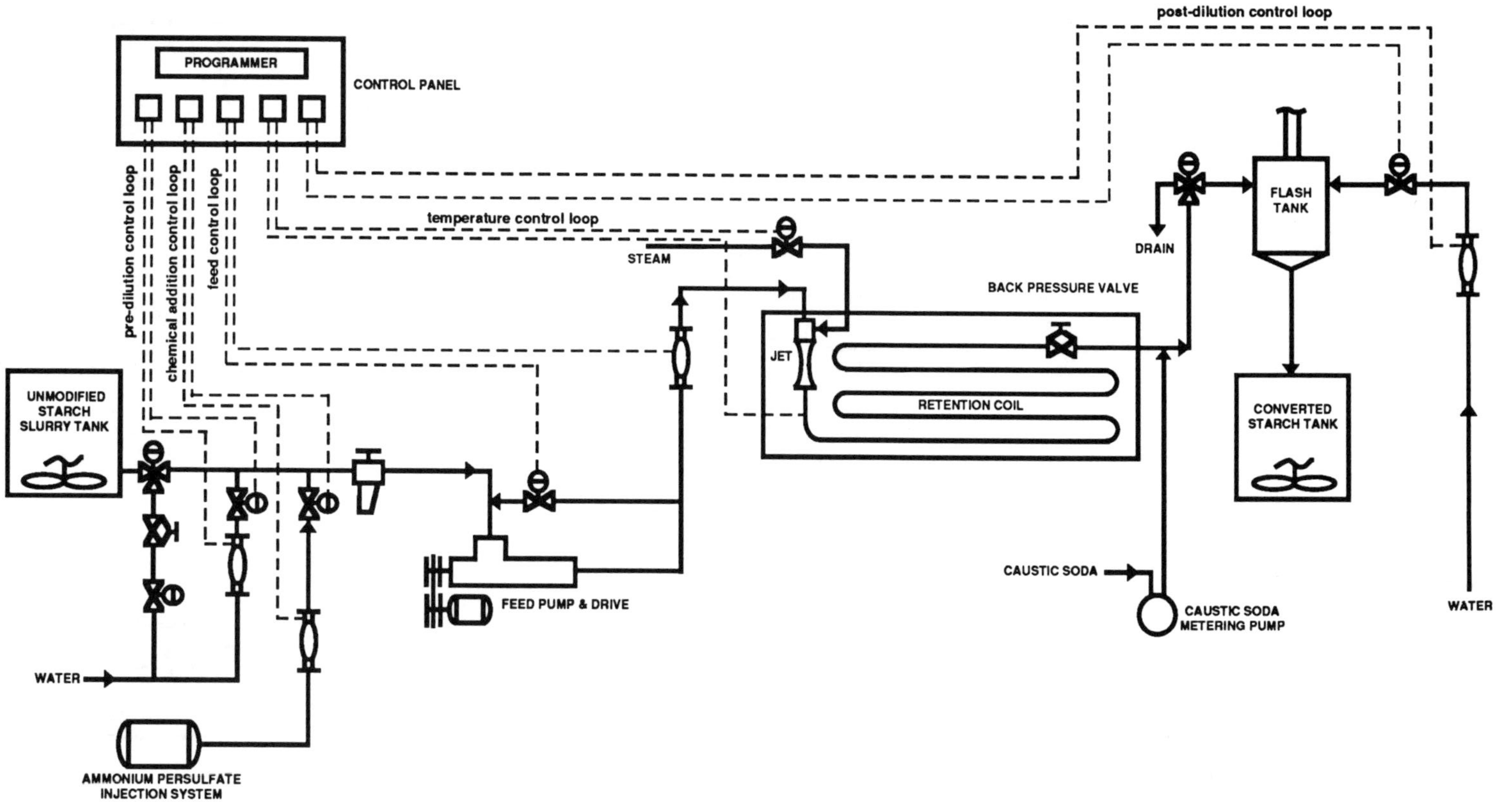

Fig. 344. Thermal chemical conversion system. (Courtesy of Grain Processing Corporation)

heated, usually to about 152°C, in the cooking jet using live steam under closed loop temperature control. The heated starch then enters an insulated retention coil sized to provide about 5 min retention. This results in thorough dispersion and consumption of the oxidizing agent. The pasted starch is flashed through a back pressure valve at the end of the coil into a steam separating chamber and discharged into a starch storage tank. Since most pastes exiting thermal chemical converters are acidic, immediate addition of alkali to adjust and maintain pH at 8.0 to 8.5 is required to minimize amylose precipitation.

One of the potential problems with oxidized starches is their anionic characteristic. This may be undesirable when the starch is reintroduced to the white water system through the use of broke [9]. There may be undesirable effects on wet end chemistry and the mill effluent treatment plant. Some mills have switched to the use of cationic starches for surface sizing. Cationic starch is prepared by reacting starch with a cationic polyelectrolye. Various suppliers market a range of cationic starches with controlled levels of cationic charge in combination with various degrees of starch polymerization. Typically, low levels of cationic charge are used for surface sizing. Some mills prepare cationic starch on site by cooking starch with addition of a cationizing agent, usually a highly branched condensation amine polymer. The cationic sites on the starch are attracted to the anionic sites on the fiber, thus making the starch cellulose substantive. This can result in decreased BOD loading on the paper mill effluent treatment system.

D. FILM FORMERS OTHER THAN STARCH

A number of water-dispersible film formers are used for surface sizing. Among these are carboxymethyl cellulose, methyl cellulose, polyvinyl alcohol, alginate, and synthetic resin polymers. These materials are higher in cost than starch and their use is limited to higher value added papers.

The sodium salt of carboxymethyl cellulose (CMC) is used as a surface size to improve grease resistance, ink holdout, and surface strength. The CMC solution may be used alone or mixed with starch, vegetable gums or proteinaceous polymer. CMC surface sizes may be applied by either size press or calender box. One of the major applications of CMC is in calender box sizing to improve print gloss of paperboard.

Methyl cellulose can be used by itself or blended with starch for surface sizing to produce hard, flexible, and continuous films that resist oils and greases. The film is water sensitive, and it may be necessary to add a crosslinking resin to provide water resistance. Plasticizers such as sorbitol may be added to improve the flexibility of the film. Methyl cellulose solutions will gel or precipitate at higher temperatures. Methyl cellulose can be readily applied by size press. In calender box sizing with methyl cellulose, it may be necessary to circulate and cool the solution to prevent coagulation and resultant picking.

Polyvinyl alcohol is used as a surface size to form films that have very high tensile strength and a high degree of transparency, flexibility, and oil resistance. The water resistance of these films is fairly good, and it can be improved by addition of a crosslinking resin or an alkali stabilized colloidal silica. Solutions of polyvinyl alcohol can be applied via size press or calender box but they tend toward dilatancy, which may cause operating problems at high speeds. These problems will be discussed later in this chapter (See Sec. G3).

Alginate can be used for surface sizing alone or in combination with starch, casein, or vegetable gums. Alginate forms films which are clear and tough but may be brittle unless plasticized. Alginate films are water sensitive, and a crosslinking resin may be needed. Alginate is also sensitive to temperature, and cooling may be required for calender box applications. The primary use of alginate surface size is in grades where a high-density, well-closed surface is required. Included are papers to be used as waxing base and gumming stock and paperboard for high-gloss ink printing.

Animal glue is used as a surface sizing agent on high-quality rag content writing, bank note, and blueprint papers. Surface sizing with glue provides a relatively hard, nonabsorbent surface. If the glue is hardened by passing the sized paper through a bath of alum or a crosslinker, the surface becomes quite resistant to water. Glue sizing also improves the strength properties of paper, particularly the folding endurance.

E. WAX EMULSIONS

Most of the materials used for surface sizing belong to the class of water-dispersible colloids. In some cases, it is desirable to apply water-insoluble materials, such as waxes or resins, in a surface sizing operation, and emulsions offer one of the best means of doing so.

Wax emulsions can be applied either alone or in combination with starch by either size press or calender box. Surface sizing with wax emulsions provides a number of desirable paper properties, including better finish, scuffing resistance, water repellency and blood resistance. In calender sizing, wax can also function as a lubricant to minimize picking.

The increase in water resistance obtained from wax emulsions is not due to the formation of a continuous wax film but rather to coating of the fibers with microscopic particles of wax, thus increasing the contact angle of the fibers with water. The wax particles also decrease the amount of loose surface fiber, thus improving printing properties and scuff resistance. Some of the grades which benefit from wax emulsion sizing include meat wrap (blood resistance), bag papers (water resistance), cup stock (slip), boxboard (foldability), and twisting paper (pliability).

It is desirable to keep the wax on the surface of the paper. This is not a problem when materials such as starch are used with the wax, but excessive penetration may be a problem when wax emulsions are used alone. Drying temperatures and storage conditions of the wax-sized paper can affect the final distribution of the wax in and on the sheet.

F. OTHER SPECIALTY SURFACE SIZES

Several other sizing agents are used to provide functional properties to specialty papers.

The application of starch results in putting a very hydrophilic layer on the surface of the sheet. A combination of starch and a polymeric size, such as styrene maleic anhydride copolymer, can decrease the hydrophilicity of the surface and improve water resistance and edgewick [3]. Various polymeric size compositions are available to provide water resistance with and without film formation. Styrene maleic anhydride copolymers as size press additives can also contribute to improvements in wet and dry strength. Polyurethane emulsion surface sizing additives can provide significantly improved wax pick and alkali resistance. The polymeric surface sizes may be used in combination with other specialty surface sizes to achieve a variety of desired sheet properties.

Fluorochemicals are used to provide resistance to penetration by oil and grease. These chemicals function by making the individual fibers repellent to oil rather than by forming an impervious film. Since the porosity of the paper is not greatly affected, it is possible to make oil-resistant papers which are permeable to gas, e.g., noiseless popcorn bags. Fluorochemical surface sizing is also used to manufacture wickproof papers; use of a conventional film former would not protect against edgewicking of oils. Fluorochemical surface sizes are usually used in combination with starch, polyvinyl alcohol, or styrene maleic anhydride copolymers.

Chromium complexes of long-chain fatty acids, such as myristic acid and stearic acid, are used as water repellants and release agents for surface sizing paper and paperboard [10]. Treated papers are used for water-repellent bags and overwraps. Treated linerboard is used for boxes for wet vegetables, meat and overseas shipment. Release papers surface sized with a mixture of a chromium complex and a film former, such as polyvinyl alcohol, are used for release from plastics and in food packaging.

Ketene dimers are used as surface sizes to provide water repellency [11]. They are normally formulated with starch or CMC and applied to an alum-free unsized sheet via size press or calender box.

Silicone resins in emulsion form are used as surface sizes to provide low surface tension which makes the paper repellent to most materials. This property is desirable in interleaving papers, packaging papers for sticky substances, and release papers for tapes, labels and other adhesive surfaced materials.

Fire-retardant paper is made by surface sizing with ammonium salts, including ammonium phosphate, ammonium sulfamate, polyphosphorylamide, and ammonium sulfate. These materials are dissolved in water and applied at the size press or calender.

Some grades of paper are treated with plasticizers to produce a softer sheet. The commonly used plasticizers include glycerin, sorbitol, mannitol, and glycols–either alone or in combination with starch. Glassine, parchment, and waxing stock are some of the grades commonly treated with plasticizer. Plasticized glassine forms better around sharp corners and has less tendency to spring back; these are useful properties for products such as butter and candy wrapping papers.

G. CONVENTIONAL SIZE PRESS CONFIGURATIONS

In a conventional size press, the objective is to flood the entering nip with sizing solution; the paper absorbs some of the solution, and the balance is removed in the nip. The overflow solution is collected in end funnels or a pan below the press and is recirculated back to the nip.

Conventional size press configurations are generally categorized as verticial, horizontal, or inclined, as shown in Fig. 345. The vertical configuration provides the easiest sheet run, but the pond depth of solution in each nip is unequal. The horizontal size press arrangement solves the

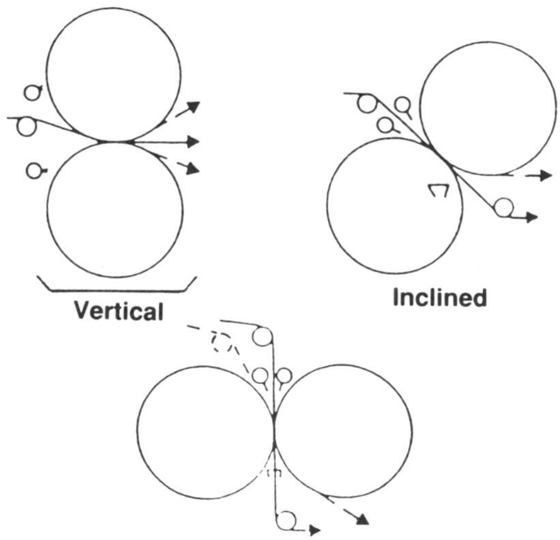

Fig. 345. Conventional size press configurations.

problem of unequal absorption by providing identical pond forms on either side of the sheet. The inclined size press is a compromise, developed to avoid the awkward vertical sheet run of the horizontal size press.

1. The size press as a coater

Pigmented coatings are applied to improve printability. The size press was the first type of coater used to apply pigmented printability coatings. Over the past several decades, the blade coater has become the coater of choice for pigmented printability coatings because of its capability to apply higher coat weights to provide a uniform surface for printing. Today, some pigmented size press coated papers are still produced, primarily for text and cover applications; these grades are usually referred to as "film coated" or "wash coated". Use of the size press as an on machine precoater is common today in mills making some of the highest quality coated papers [12].

2. Size press pickup

Size press operations must be controlled carefully to ensure that the sheet picks up the desired amount of chemical uniformly across the machine. At the same time, the amount of water absorption should be minimized to maintain after dryer requirement at the lowest practical level. The primary factors determining solids pickup at the size press are listed in Table 56.

There are two basic mechanisms for incorporating sizing solutions at the size press [13]. The first is the ability of the sheet to absorb the size

Table 56. Factors influencing size press pickup.

Sheet characteristics
A. Paper substrate (basis weight, density, smoothness, capillary structure, void size etc.)
B. Level of internal sizing
C. Moisture content

Sizing solution
A. Solids content
B. Temperature
C. Viscosity (fluidity)
D. Composition (type of starch, use of additives)

Design and operational
A. Machine speed
B. Pond depth
C. Nip pressure
D. Nip width (affected by roll hardness and diameter)

solution. The second is the amount of solution film passing through the nip and the manner in which the paper and the roll surfaces separate. Factors that favor greater absoption include low solution viscosity (higher solution temperature), low machine speed, and low level of internal sizing. The factors favoring greater film thickness include sheet roughness, faster machine speeds, low nip pressure and higher viscosity.

Of the factors affecting pickup, solution viscosity and solids content are the easiest to manipulate. The solids concentration, however, is usually maintained at the highest manageable level consistent with the desired viscosity in order to minimize the amount of water that must be subsequently evaporated.

3. Conventional size press limitations

In the conventional size press, the size solution is delivered to the nip between a pair of oppositely rotating rolls to form a pond of material above the nip. With high-speed roll rotation and high-speed movement of the web, the pond tends to absorb kinetic energy from the moving web and rolls [14]. As shown in Fig. 346, an excess of liquid is flowing toward the nip, but the nip pressure of the rolls limits the amount of sizing solution passing through the nip, causing the remaining solution to flow back upwards. If hydrodynamic forces are too great, this upward velocity becomes high enough to cause the solution to break the surface of the pond and splash out of the nip. This pond turbulence and "nip rejection" results in uneven pickup of solids across the machine.

Size solution viscosity contributes to the hydrodynamic force in the roll nip. This may make it impractical to use the size press to apply dilatant materials. As machine speeds increase, it may be necessary to dilute the starch solution or go to a more degraded starch to avoid nip

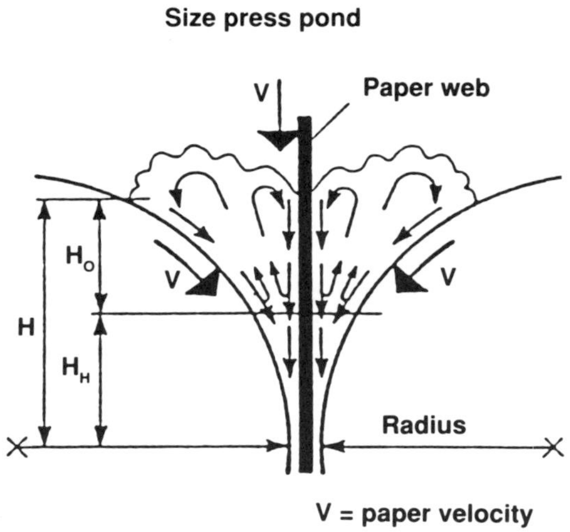

Fig. 346. Hydrodynamic forces in the nip.

Fig. 348. Film split pattern.

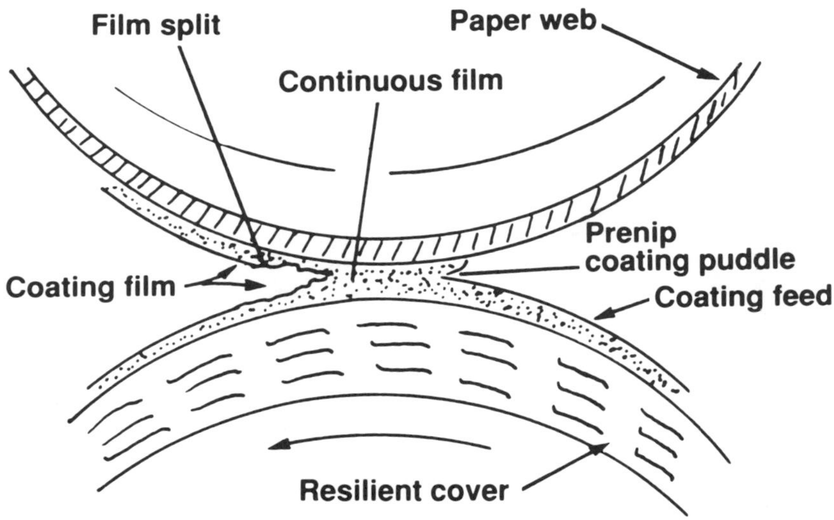

Fig. 347. Development of film split pattern.

rejection. These approaches, in turn, may increase after dryer requirements and also affect sheet properties adversely.

Another problem associated with the size press is film splitting. Although most of the size solution is absorbed into the sheet at and after the nip, excess solution remains on the surface of the sheet between the paper and the roll. At the exit of the nip, this film is split into two layers with part following the web of the paper and the rest remaining on the roll surface, as shown in Fig. 347 [15].

This pattern will appear as heavy rings of fluid across the roll face, as shown in Fig. 348 [16]. The severity of the pattern depends on the rheology of the size, absorption properties of the sheet, speed of the rolls, and volume of material being split. As shown in Fig. 349, the problem becomes worse with higher speeds, higher viscosity, and higher film thickness [13].

Moisture content of the sheet entering the size press has a significant effect on pickup. Higher moisture content promotes absorption; thus a variable moisture content across the machine causes variations in pickup and sheet properties affected by surface sizing. Thus the ingoing sheet is usually dried down to 1 or 2% or less to ensure an even moisture profile and to keep the sizing agent nearer to the surface. This "overdrying" of the sheet is inefficient from an energy standpoint

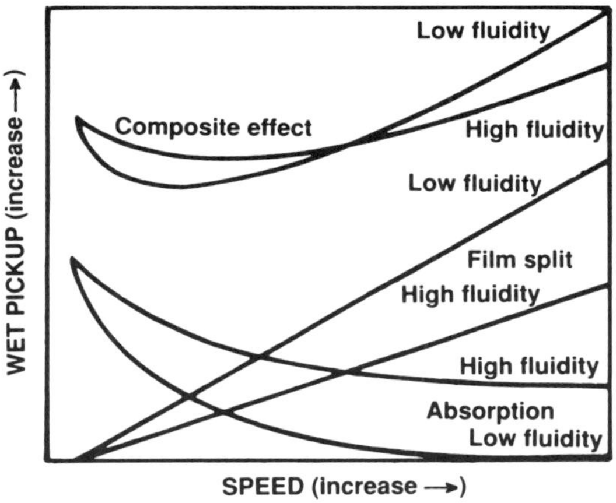

Fig. 349. Factors affecting pickup and film split.

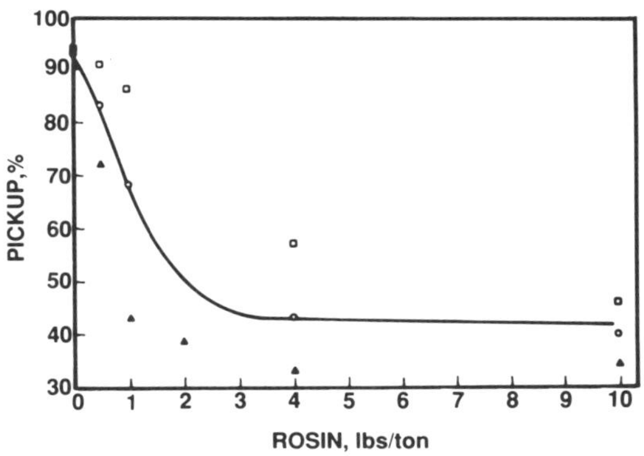

Fig. 350. Size press pickup vs. level of internal size.

and may place a limit on machine speed.

As illustrated in Fig. 350, studies have shown that the level of internal sizing is the greatest factor in size press pickup control [17]. Depending on draw control and basis weight, it may be necessary to have at least a minimal level of internal sizing to run the sheet through a size press and into the after dryer section without breaking.

The required level of internal sizing to provide runnability is inversely related to basis weight and wet tensile strength. Thus it may be necessary to use high internal sizing levels and expensive wet strength agents in order to get lightweight papers to run through the size press. Even with these additives, the size press is usually a major source of lost time on machines making lightweight and/or weak papers.

Since the commonly used reactive neutral/alkaline internal sizes are slow curing, problems are often encountered with size press breaks in alkaline papermaking [18].

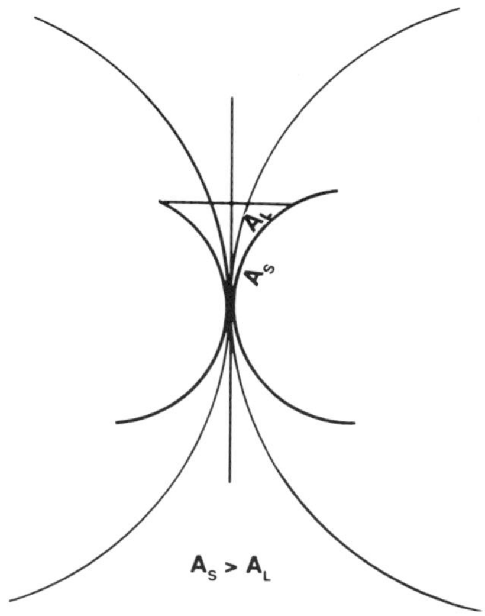

Fig. 351. Larger size press roll diameters reduce relative acceleration lengths.

Another difficulty with the conventional size press is matching the speeds of the two rolls. If the rolls speeds are not matched precisely, "sheet scuffing" or "scrubbing" can lead to paper surface marking and increased web breaks. The difficulty of precision roll speed matching increases as machine speed goes up.

H. SIZE PRESS DESIGN IMPROVEMENTS

As machine speeds have increased, paper machine designers have increased size press roll diameters to help compensate for greater hydrodynamic forces. As shown in Fig. 351, increased roll diameters reduce the proportionate accelerative forces on the sizing solution. Some of the newer uncoated free sheet machines, equipped with 1500 mm diameter size press rolls, operate successfully at speeds approaching 1000 m/min; however, it has been found necessary to use short chain length (degraded) starch and solution concentrations in the 2 to 3% solids range to keep pond turbulence manageable.

Figure 352 shows the apron size press, patented as the "high speed size press" [19]. It is designed to prevent pond splashing by isolating the sizing pond from the high-speed surfaces of the rolls and paper. Since less kinetic energy is absorbed by the size solution, circulation velocities are reduced, and the size solution does not have a tendency to jump out of the pond. The baffles are usually made of sheet plastic material. This size press modification is effective in reducing pond turbulence, but there may be problems in keeping the baffles clean and with

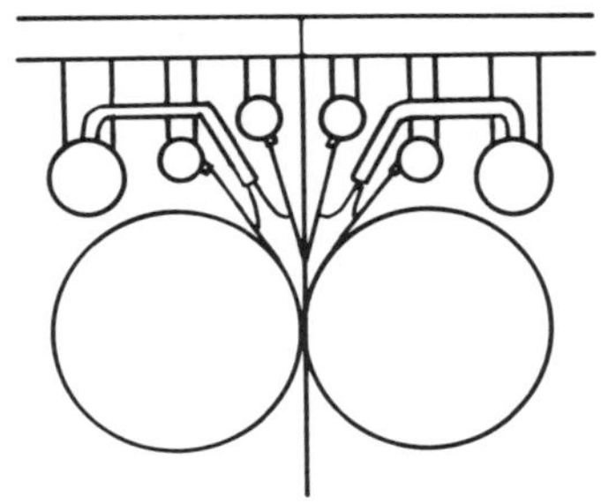

Fig. 352. Apron size press.

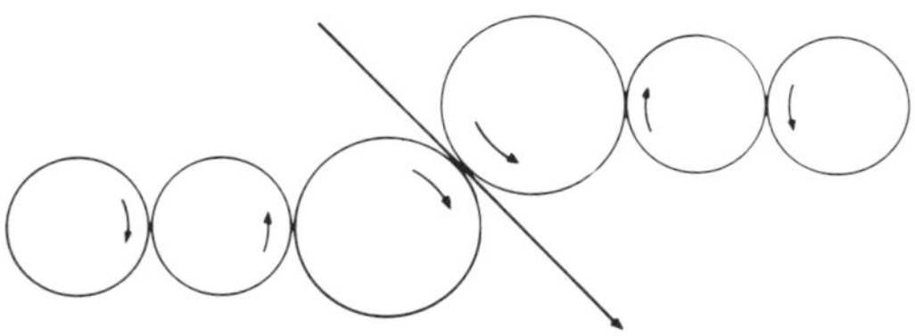

Fig. 353. Gate roll size press.

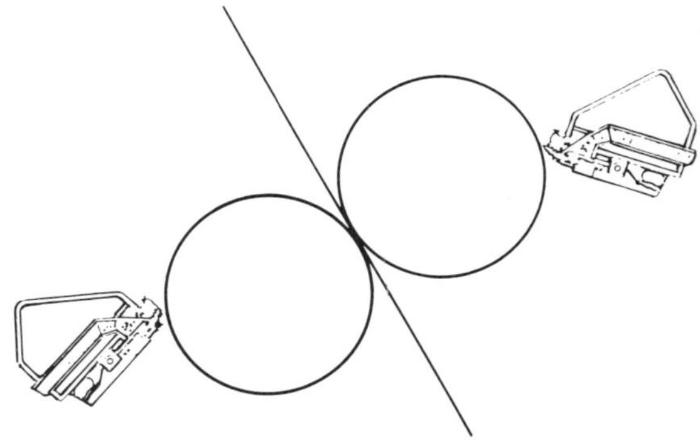

Fig. 354. Blade/rod metering size press.

excessive wear on the dry sides of the baffles. "Icicles" building up on the baffles can create not only streaks but also scratches severe enough to break the sheet. Very few mills have been able to use the apron size press in day-to-day operations.

An improvement on the apron size press concept is to combine its use with restricted flow of sizing solution to the pond. The simplest and most reliable device for this use is a weir and spreader pan. Restricting the flow over the baffle aprons decreases the loads on them and minimizes the dry side wear problem. It also aids in reducing the amount of splashing in the size press nip.

I. MODIFIED SIZE PRESS DESIGNS

The problems with pond turbulence motivated development of modified size press designs which eliminated the pond. The gate roll size press, illustrated in Fig. 353, has an offset pond which is not in contact with the sheet. This offset pond feeds a metering nip which controls the amount of size going to a second nip. The second nip controls the film uniformity. It is essential that the rolls in the gate roll train be run at different speeds to minimize film split patterning [20].

With the gate roll size press, sizing solution solids and pickup rate are independent variables. It is possible to increase starch solution concentration to reduce load on the after dryers, and the "print on" application method keeps the starch on the sheet surface. Nevertheless, the problems associated with film split and fiber picking on the outgoing side are still present and may be increased due to higher starch solution concentration. The high z-direction shearing forces created as the sheet leaves the nip cause sheet roughening and linting. These factors contribute to the "orange peel" sheet surface defects often associated with the gate roll size press.

Substitution of a gate roll size press for a conventional size press involves replacing two rolls with six, contributing to increased initial investment and the potential of significantly higher maintenance costs and downtime for roll changes. It also involves installation and maintenance of six drive units compared to two. It is necessary to crown all six rolls and run them with relatively high nip loads; this causes heat buildup and short roll cover life with a corresponding requirement for a large investment in spare rolls.

The roll maintenance and film split pattern problems of the gate roll size press led to development of the blade/rod metering size press shown in Fig. 354 [21]. Short dwell blade coater heads are used to apply and meter the size to the size press rolls. Blade or rod metering permits operation without a pond or with a very limited one.

With blade metering, the amount of surface size being applied may be controlled and easily changed by varying the blade pressure. This feature makes it practical to control pickup and solids content as independent variables. The ability to control wet film thickness provides the capability to control the degree of penetration of surface size into the sheet, i.e., to locate the size near the surface or to drive it into the sheet with pressure-induced migration.

Use of a bent blade for metering provides the capability to control cross-machine profile of the wet film thickness and thus the amount of starch applied. Although this may provide the paper-

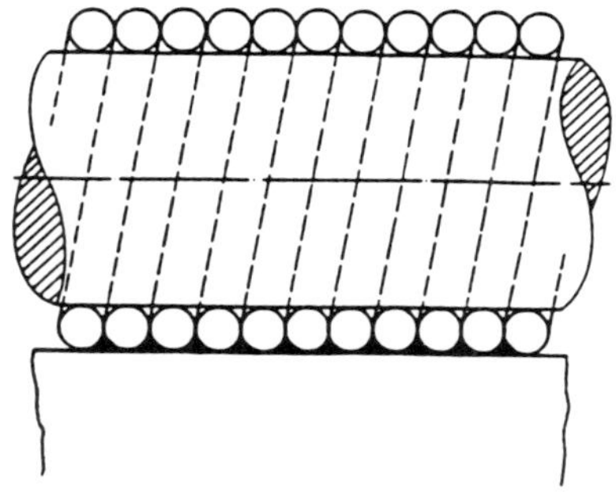

Fig. 355. Wire-wound rod.

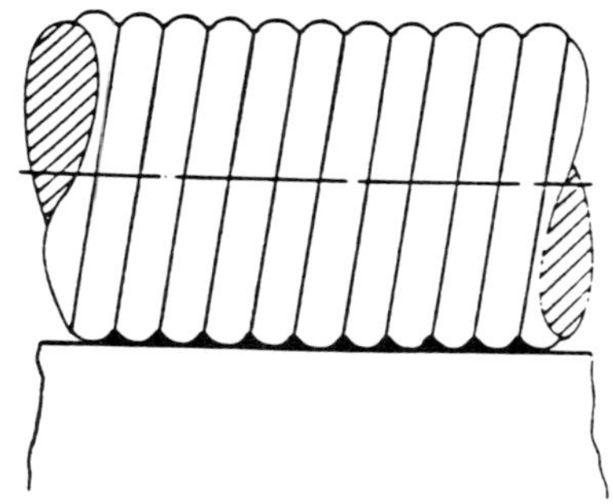

Fig. 356. Profile rod.

maker with a valuable tool, this profiling capability should be used with discretion. It has the potential to mask profile problems that should be corrected at the wet end or in the main dryer section. In addition, application of varying amounts of starch across the machine can lead to differences in printability and other sheet properties.

Rods are also used as metering elements in modified size presses [22]. Initially wire wound rods were used. As shown in Fig. 355, the application of size is not done with hydraulic pressure as in a conventional coater, but is achieved solely by volumetric application. The wet application of size suspension to the size press roll is a function of the wire diameter. The relationship is linear and independent of speed. Pressure differences across the width do not cause variation in the amount of sizing material. Volumetric metering does not subject the material being applied to the same amount of shear as blade metering, and thus the rheology constraints on the sizing material may not be as critical. Since control of wet film thickness is a function of wire diameter, it is necessary to change to a different rod in order to vary pickup.

There is a concern about the use of wire wound rods on high-speed machines, i.e., the potential for damage or injury if the wire unwraps or wears through. These concerns have been addressed by a new profile rod construction which has machined grooves similar to those of the wire wound rod as shown in Fig. 356. It has higher resistance to mechanical loads, compared to rods with thin wires.

In most designs of the blade/rod metering size press, short dwell coater heads are used to supply the sizing material. Standard short dwell coater heads require relatively high pumping rates in the range of 3.0–3.7 L/min per cm of width. This flow rate results in recirculation rates which are considerably higher than those required for a

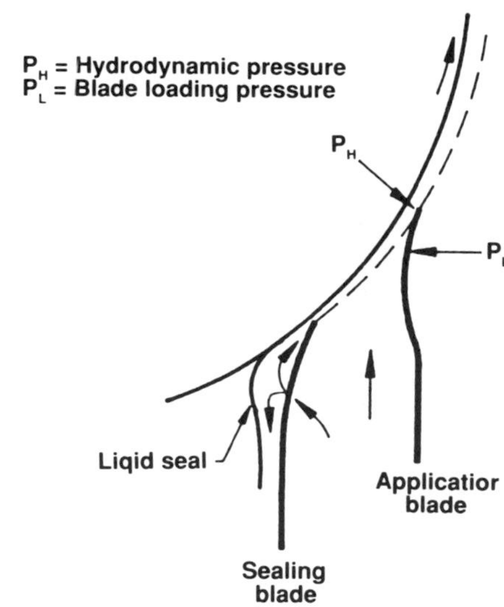

Fig. 357. Sealing blade arrangement.

conventional size press. High recirculation rates necessitate larger filtration equipment, pumps, heat exchangers, and tanks in the surface size supply system. One of the machine builders has a novel applicator design which incorporates a sealing blade in the application chamber ahead of the metering blade [23]. The sealing blade is perforated, which allows the sizing liquid to pass through the blade, with the excess amount recirculated. The puddle before the sealing blade forms an air-tight seal at the tip of the sealing blade and also lubricates the sealing blade. This feature minimizes the potential for skips from air entering the metering zone. Recirculation rates can be kept in the same range as a conventional size press. Figure 357 shows the dynamics of this arrangement.

Another development which combines some of the advantages of volumetric metering and low recirculation rate is the high solids metering roll size press illustrated in Fig. 358 [24].

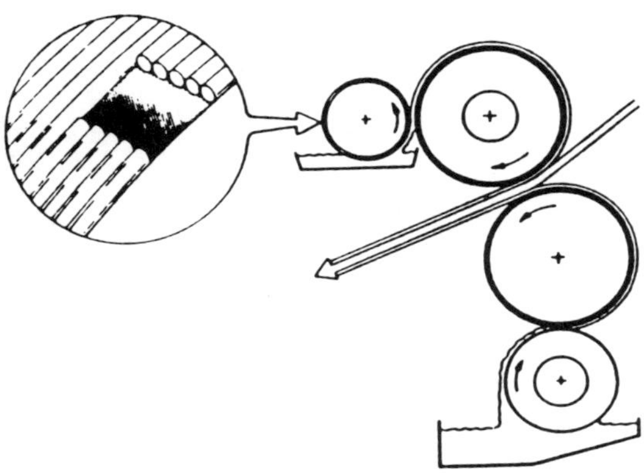

Fig. 358. High-solids metering roll size press.

Large-diameter wire-covered metering rolls are used to apply wet film to the size press rolls. Metering is volumetric, analogous to a wire wound rod metering size press. The shear rate with the high-solids metering roll is not as great as with a blade or rod metering applicator. Thus the rheology requirements for the surface size are not as critical.

J. ALTERNATIVES TO THE SIZE PRESS

The size press is not the only device used for surface sizing. Calender boxes will be discussed later in this chapter (See Sec. N). A variety of coaters–including short-dwell coaters, simultaneous two-side blade coaters and transfer roll coaters–are used as alternatives to the size press for surface sizing [25]. These coater configurations are described in Volume 8 of this series [26].

K. EFFECTS ON SHEET PROPERTIES

The key factors which differentiate the effects of various surface size application methods on sheet properties are:

1. Pressure-induced migration;
2. Dwell time at pressure; and
3. Lifting forces on the sheet surface as it exits the nip.

A combination of the first two factors determines the location and uniformity of distribution of starch in and on the sheet. The second factor is the primary determinant of the amount of fiber swelling. The third factor controls integrity of the starch film and disturbance of the sheet surface. As shown in Table 57, the conventional size press and each of the major alternatives discussed above represent various combinations of the above factors, resulting in differing effects on sized sheet properties.

The size press and modified size presses all have rolls which run at the same speed as the sheet. The sized sheet exits via a diverging nip, which creates film split and lifting forces on the sheet surfaces. These forces are higher in a gate roll coater or blade metering size press than in a conventional size press. In either of the modified size presses, the wet starch film is premetered onto the applicator rolls. Dwell time at pressure is the same as for a conventional size press, but the nip is not full of starch solution. This causes dewatering of the starch solution in the nip and location near the sheet surface. The dewatering causes an increase in viscosity, which adds to the z-direction pull on the sheet as it exits the nip.

Since the severity of film split pattern is dependent on viscosity, it imposes a limitation on the amount of pigmented coating that can be applied via conventional or modified size presses to produce a sheet without objectionable sheet surface marks. In general, the upper limit is about 3 to 5 g/m^2/side.

A starch layer located closer to the surface does not reduce opacity to the same extent as deeply penetrating starch. In cases where titanium dioxide is used to obtain opacity, a significant savings can be realized. It should be practical to control this parameter with the premetering size presses by controlling the degree of starch penetration.

A switch from a conventional to a premetering size press can also help to increase stiffness. This effect is related to location of the starch near the surface, which creates a modulus effect similar to an I-beam.

A higher concentration of starch solution can influence porosity. Premetered starch application provides better film forming and thus reduces porosity. Less refining horsepower may be needed to reach a given porosity. The volumetric premetering size presses may make it practical to apply dilatant materials to produce sheets with high Gurley density.

Table 57. Comparison of key factors for various applicators.

	Conventional Size Press	Gate Roll Size Press	Blade/Rod Metering Size Press
Pressure-induced migration	High	Low	Moderate and controllable
Dwell time at pressure	Long	Long	Medium
Z-direction lifting forces	High	Highest	Moderate and controllable

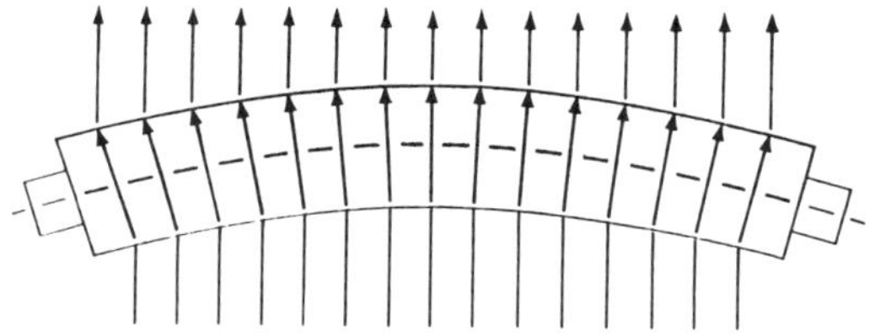

Fig. 359. Curved spreader (bowed) roll.

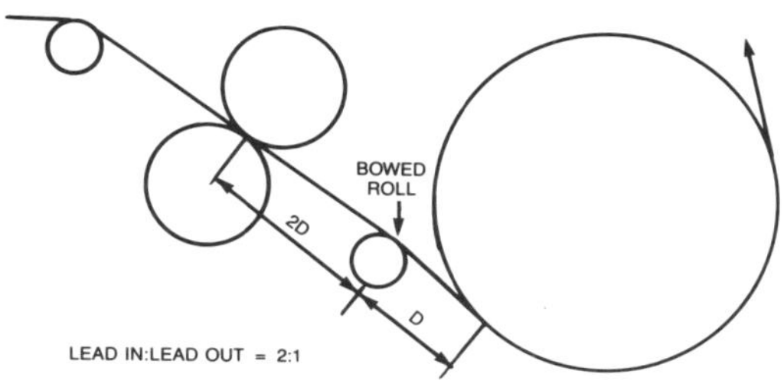

Fig. 360. Sheet run with spreader roll.

Orientation of the starch closer to the surface results in better ink holdout, improved solvent holdout, and higher ink gloss.

The conventional size press forces starch into the interior of the sheet, resulting in an increase in internal bond. With a premetering size press, penetration depth can be controlled, making it practical to obtain internal bond improvement if other quality control parameters favor deep penetration of the starch solution.

Deep penetration of starch into the sheet tends to decrease tear strength. The penetration control capability of the premetering size press makes the tear strength/surface sizing relationship controllable.

Location of the starch nearer the surface increases its effectiveness in providing sheet surface strength. The penetration control of the premetering size press can be used to improve sheet surface strength at equal starch pickup.

L. EFFECTS ON DRYING AND MACHINE SPEED

The decrease in water added to the sheet during surface sizing also reduces the sheet draw to the after dryer section.

With a conventional size press, starch pickup is directly dependent on solution concentration. Thus the primary method of controlling pickup is to vary the solids. With a premetering size press, solids concentration and pickup rate can be varied independently.

With the various premetering size presses it is practical to apply surface size at concentrations 1.5 to 2 times higher than with a conventional size press. Application of surface size at higher concentrations decreases after dryer load proportionately. On a machine which is after dryer limited, a switch to high solids surface sizing with a premetering size press can result in significant machine speed up and added value of incremental tonnage.

M. WEB HANDLING AND DRYING OF THE SURFACE-SIZED SHEET

Surface sizing rewets the sheet. Absorption of water from the surface size causes fiber swelling and sheet expansion. It also causes a decrease in tensile strength. As water is absorbed by the fibers, the solids and viscosity of the sizing material at the sheet surfaces increases, the sheet becomes tacky. This tackiness can cause picking and sheet quality problems. All of these factors must be taken into consideration in designing the sheet run out of the size press and through the after dryers.

One of the biggest potential problems is sheet expansion, which can lead to wrinkles and calender cuts. The most commonly used device to compensate for sheet expansion is the curved spreader roll shown in Fig. 359. A curved spreader (bowed) roll in the sheet run between the size press and the first after dryer, as shown in Fig. 360, will provide a level sheet of uniform tension on the first after dryer cylinder. The spreader roll should be covered with a non-stick surface.

The draw to the after dryer section should be precisely controlled. Too tight draw on the expanding sheet can adversely affect physical strength properties. Too slack a draw may cause wrinkles, cockles, or grainy edges. Modern machines use a load cell mounted on the spreader roll bearing housing to measure sheet tension out of the size press. This tension signal is used as the basis for control of after dryer section speed.

Another option for transporting and spreading the surface-sized sheet is the air turn shown in Fig. 361 [27]. The wet sheet is not contacted by a roll but rather rides on a cushion of air. The ratio of the air cushion to air supply pressure provides a reliable measurement of sheet tension for use in after section drive control.

The after dryer section usually consists of conventional cylinder dryers as discussed in Chap. XII. In most cases, the first two after dryers are run at lower temperatures and covered with a non-stick surface to avoid picking. These first two dryers are usually unfelted. Dryer fabrics used on after sections are usually monofilament fabrics with pin seams to avoid

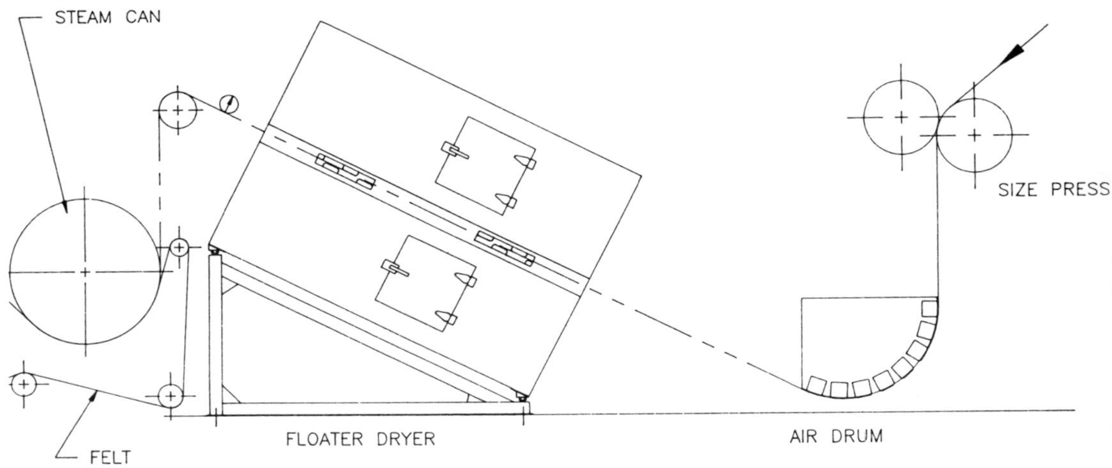

Fig. 361. Sheet run with air turn and floater dryer following size press (Courtesy of Advance Systems Inc.)

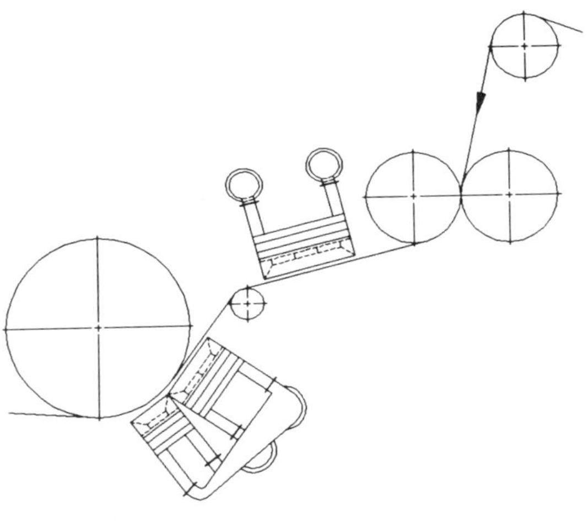

Fig. 362. Sheet run with supplementary infra-air dryer.

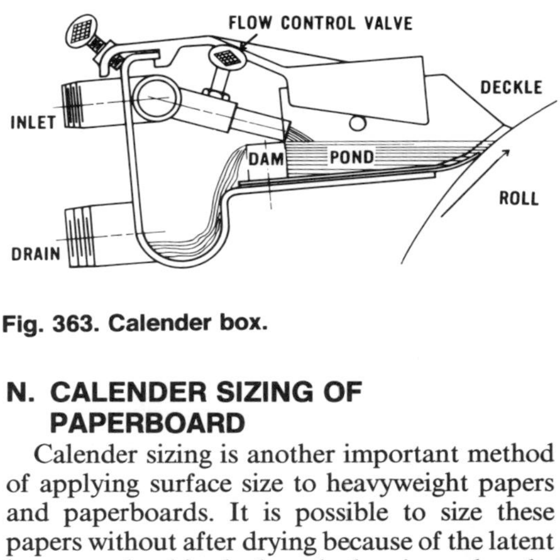

Fig. 363. Calender box.

sheet mark. It is desirable to have the capability to control the temperatures of the top and bottom after dryers separately to control curl.

In some cases, it may be necessary to provide supplementary drying between the size press and the first after dryer as shown in Fig. 362. Infrared or infra-air dryers are usually used in this location becuase they can provide a large input of energy in the short sheet run distance available. When infrared dryers are used, the first two after dryers can be run at full temperature. On some machines, air foil dryers are used in the after size press supplementary dryer position to provide not only supplementary drying but also web guiding, spreading, and tension control.

N. CALENDER SIZING OF PAPERBOARD

Calender sizing is another important method of applying surface size to heavyweight papers and paperboards. It is possible to size these papers without after drying because of the latent heat contained in the board when it reaches the calender stack. This heat causes some evaporation between the calender stack and the reel. The added water that remains can usually be absorbed throughout the entire mass of the board without a significant increase in the overall moisture content.

Figure 363 shows a typical calender box, which extends across the length of the calender roll. The sizing material is applied to the roll and transferred to the sheet surface.

On some machines, water alone is applied by the calender box. In these cases the boxes are usually located high on the stack. The purpose of water application is to soften the sheet surface to improve the smoothing effect of calendering. This is not really a surface sizing operation and is usually referred to as "water finishing".

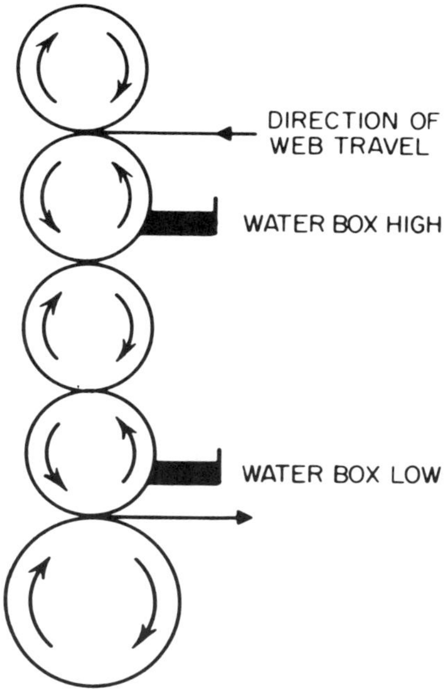

Fig. 364. Calender stack with boxes.

When a calender sizing chemical is used, more than one box may be used as shown in Fig. 364. The pickup and distribution on the sheet are dependent on the surface roughness and moisture content of the paper, temperature and concentration of the sizing solution, location of the boxes in the calender stack, and the number of boxes used. If the box is placed high on the stack, penetration of the size into the sheet is favored, and pickup rate will increase. Placement lower on the stack tends to reduce pickup and hold the size near the surface. Two boxes on the same side of the stack are used when maximum sizing effect is required, such as oil resistance. The first application fills the pores of the sheet, and the second application remains more on the surface to impart most of the oil resistance. High temperature of the sizing solution favors increased pickup of solids, but the absorption of size is more uniform if size temperature is close to sheet temperature. A uniform temperature throughout the size box is important to keep roll temperatures even. If the size temperature is lower than the dew point of the surrounding atmosphere, condensate may form and drip on the sheet.

Depending on the desired result, boxes may be placed on only one or both sides of the stack to treat one or both sides of the sheet. Some mills have two calender stacks, a "wet stack" equipped with calender boxes, and a "dry stack" to do final calendering. There may be a few dryers between the stacks.

Starch is the principal sizing agent used in calender sizing, but other materials are also used as discussed earlier in this chapter. Materials that dry to a tacky surface cannot be used for calender sizing.

Pitting of calender rolls can be a problem with calender sizing. Frequent regrinding and maintenance of high polish on the calender rolls tend to minimize pitting. Buffering the size solution with sodium silicate or sodium nitrate can help to reduce pitting.

REFERENCES

1. Smook, G.A., Handbook for Pulp & Paper Technologists, Joint Textbook Committee of the Paper Industry, TAPPI, Tech. Sect. CPPA p. 262 (1982).
2. Dill, D.R., "Surface Sizing Utility Through Understanding", TAPPI Notes: 1983 TAPPI Sizing Seminar, TAPPI Press, Atlanta, p. 15-20 (1983).
3. Dill, D.R. and Pollart, K.A., "What to Expect from Surface Sizing", 1984 Papermakers Conf. Proc., TAPPI Press, Atlanta, p. 175-177 (1984).
4. Kirby, K.W., "Starch Chemistry and Reactivity", 1985 Papermakers Conf. Proc., TAPPI Press, Atlanta, p. 59-66 (1985).
5. Klem, R.E., "Practical Application of Starch Binders in Blade Coating", TAPPI Notes: 1986 Blade Coating Seminar, TAPPI Press, Atlanta, p. 53-63 (1986).
6. Kearney, R.L., "Starch Binders in Coatings", TAPPI Notes: 1988 Blade Coating Seminar, TAPPI Press, Atlanta, p. 99-102 (1988).
7. Klem, R.E. and Brogly, D.A., "Methods for Selecting the Optimum Starch Binder Preparation System", Pulp & Paper, 55(5):98 (May 1981).
8. Brogly, D.A., "Thermal-Chemical Starch Conversion for Size Press Application", Tappi 61(4):43 (Apr. 1978).
9. Harvey, R.D., "Is There a Better Size-Press Starch?", 1985 Papermakers Conf. Proc., TAPPI Press, Atlanta, p. 83-88 (1985).
10. Fournier, L.B., "Chromium Complex Treatments for Paper", 1984 Papermakers Conf. Proc., TAPPI Press, Atlanta, p. 181-184 (1984).
11. Girardi, F.P., "Alkyl Ketene Dimer — Proper Handling, Metering and Application", 1989 Papermakers Conf. Proc., TAPPI Press, Atlanta, p. 205-209 (1989).
12. Adams, A.A., "Effect of Size Press Treatment on Coating Holdout", TAPPI Notes: 1988 Blade Coating Seminar, TAPPI Press, Atlanta, p. 23-30 (1988).
13. Beals, C.T., "Review of Paper Surface Treatment Applications at the Size Press", Pulp & Paper 54(4):132 (Apr. 1978).
14. Hansson, J.A. and Klass, C.P., "High Speed Surface Sizing", 1983 Papermakers Conf. Proc., TAPPI Press, Atlanta, p. 49-54 (1983).
15. Perry, J.A., "Size Press Coating and Treatment", Pulp & Paper 48(5):144 (May 1972).
16. Eklund, D.E., "Influence of Blade Geometry and Blade Pressures on the Appearance of a Coated Surface", Tappi J. 67(5):70 (May 1984).
17. Dill, D.R., "Control and Understanding of Size Press Pickup", Tappi 57(1):97 (Jan. 1974).
18. Pollart, K.A. and Dill, D.R., "Size Press Sizing Additives on Neutral and Alkaline Base Sheets", TAPPI/PIRA Seminar Notes: New Sizing Methods and Effects on Fibers, Fillers and Dyes, TAPPI Press, Atlanta, p. 1-6 (1982).
19. Justus, E.J., "High Speed Size Press", U.S. Patent 4,340,623, (July 20, 1982).
20. Alheid, R.J., "Operation and Applications of the Gate Roll Size Press", 1978 Papermakers Conf. Proc., TAPPI Press, Atlanta, p. 83-85 (1978).
21. Wight, E.W., "Operation and Applications of the Blade Metering Size Press", 1988 Engineering Conf. Proc., TAPPI Press, Atlanta, p. 549-552 (1988).
22. Sollinger, H.-P., "Speedsizer", 1988 Engineering Conf. Proc., TAPPI Press, Atlanta, p. 553-559 (1988).
23. Finch, K.W., "Sym-sizer: The Film Transfer Size Press", 1989 Papermakers Conf. Proc., TAPPI Press, Atlanta, p. 117-121 (1989).
24. Paczkowski, M. and Engevik, T., "A High Speed Metering System for Billblade and LAS Coaters", TAPPI 90 Proc., TAPPI Press, Atlanta, p. 165-169 (1990).
25. Klass, C.P., "Alternatives to the Size Press", PIMA 70(9):19 (Sept. 1988).
26. Coating, Converting, and Specialty Processes, Pulp and Paper Manufacture Series, Vol. 8, Joint Textbook Committee of the Paper Industry, TAPPI and Tech. Sect. CPPA (1990).
27. DeSanti, E.A., "Circular Flotation Dryer Eliminates Picking", Southern Pulp & Paper 49(5):16 (May 1986).

XIV YANKEE DRYERS

W.G. CORBOY, Jr.

Bill Corboy is a Chief Engineer and Manager of Scott Paper Company's Yankee Dryer Management System in Philadelphia. He chairs TAPPI's Yankee Dryer Safety Subcommittee from where he has overseen publication of numerous committee-sponsored papers, has contributed to and edited TAPPI publications, "Guidelines for the Safe Operation of Yankee Dryers" and "Yankee Dryer Safety and Condition Assessment Guidelines," and has promoted TAPPI's feasibility study into the use of acoustic emission techniques for the non-destructive examination of cast iron dryers. Mr. Corboy received his BSME from Drexel University, an MS in Engineering Mechanics from Widener University and is a registered Professional Engineer.

REVIEWED BY

Erik Danielson
Valmet-Karlstadt AB
Karlstadt, Sweden

Herbert Holik
Head of R&D Paper Technology
Sulzer Escher Wyss GmbH
Ravensburg, Germany

Gerald F. Schreiner
Director, Paper Manufacturing Process
James River Corporation
Norwalk, CN

A. INTRODUCTION

There are a variety of stories in circulation relating the origin of the term "Yankee" dryer. The one with the most currency, perhaps, suggests that a Dutchman, whose surname was Yonke, was involved in the early development of the dryer. When the concept was imported to the United States, the Americanized version of the cylinder associated with his name became "Yankee" [1]. Whatever the case, today, under the definition of Yankee dryer are included all large cast iron drying cylinders employed in the manufacture of creped paper grades, such as toilet and facial tissues, many varieties of paper napkins, towels, wadding for sanitary padding and other special products. These products are all distinguished by the fact that they have been removed from the dryer through the mechanical action of a creping doctor. Similar dryers used in the production of flat grades or machine-glazed papers are called MG cylinders. Instead of being creped from the dryer surface, these products are pulled from it in order that the side against the dryer will take on the very smooth surface characteristic of the polished cast iron shell.

Yankee and MG are generic terms in the industry, depending upon the product manufactured. For the sake of simplicity, and because of their predominant number, "Yankee" will be used herein to represent the category. In general, tissue products are made at higher speeds and steam pressures than machine-glazed papers. The design feature that distinguishes the Yankee dryer from paper or can-type drying cylinders is the centershaft. Because Yankees are typically much larger in diameter than paper dryers, the centershaft is required to stabilize the head and reduce axial stresses in the shell.

The Yankee dryer (Fig. 365) is the heart of the paper machine in the tissue industry because of the high contact drying rate, 10 to 12 lbs. $H_2O/(ft^2)(h)$ vs. 2 to 3 lbs. on conventional felted dryers, based upon full circumference [1], which equates to 49–59 kg $H_2O/(m^2)(h)$ on the Yankee and 10–15 kg on conventional dryers; the elimination of open draws until the sheet is strong enough to support itself; the ability to impart properties associated with the particular product, e.g., bulk, softness, and stretch.

It should be remembered throughout this chapter that comments related to paper machine configuration, dryer design, drying rates, ranges on a variety of physical, mechanical and material properties, etc., are best regarded as merely typical. There are many special products, proprietary configurations and designs, and corporate nomenclatures within the industry. Also, sizes, speeds, and performance capability of equipment are continually evolving, which tends, naturally, to date many such statements.

B. THE TISSUEMAKING PROCESS

In the tissue industry, the wet sheet is adhered to the Yankee on the upstream side at 20–35% dryness with one or two pressure rolls. The pressure rolls, in addition to adhering the sheet to the Yankee surface, mechanically reduce moisture content through water transfer from the wet sheet to the felt. Frequently, vacuum is applied to the first pressure roll to assist in this process and to prevent the sheet from being thrown off the pressure roll at high speeds. A generalized elevation view of a tissue machine is presented in Fig. 366.

Fig. 365. Yankee dryer (Beloit).

Pressure roll loads vary, generally, from 35–90 kg/cm for the first pressure roll and 62–90 kg/cm for the second pressure roll, second roll loading being equal to or greater than the first. Occasionally, a steam shower is installed to heat the sheet and reduce water viscosity, enhancing water removal and improving sheet moisture uniformity at the first pressure roll.

Conductive drying initiates at the pressure roll nip and continues as long as the sheet remains in contact with the dryer surface. Convective and radiant drying begin under the Yankee hood assembly.

The sheet is creped from the Yankee at 55–98% dryness. Machines, where creping is performed in approximately the 55–70% dryness range and further drying accomplished with smaller, can dryers, are referred to as wet crepe machines. Machines which crepe at or above that level are termed dry crepe. A skinning doctor is installed, prior to the creping doctor in the direction of dryer rotation, to remove the sheet while the crepe doctor blade is being changed. And a cleaning doctor is often installed following the creping doctor to remove fiber and maintain a uniform Yankee surface coating.

Finally, chemical spray showers are often

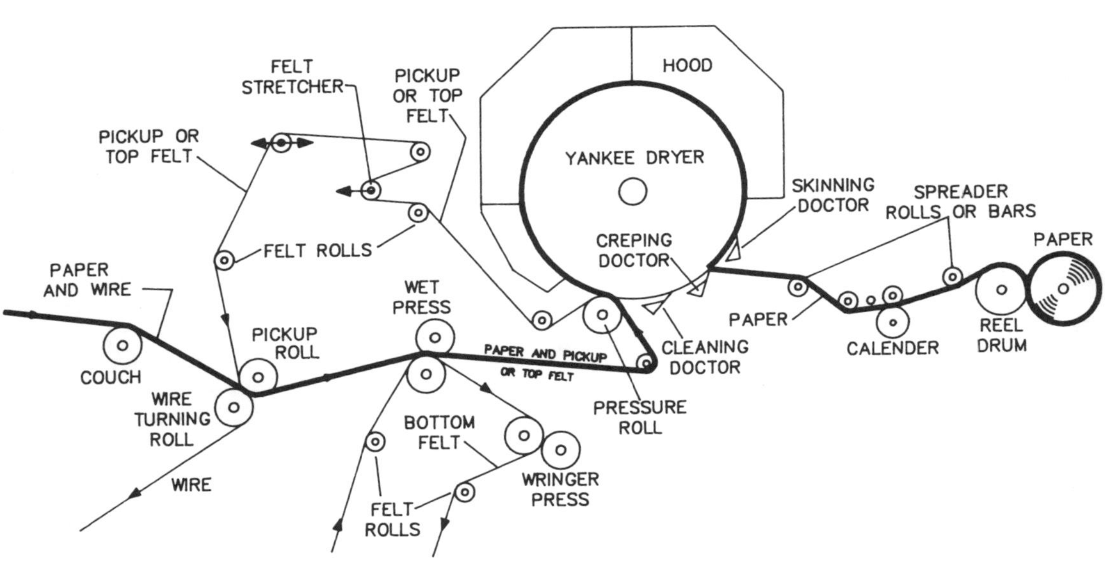

Fig. 366. Tissue machine elevation [1].

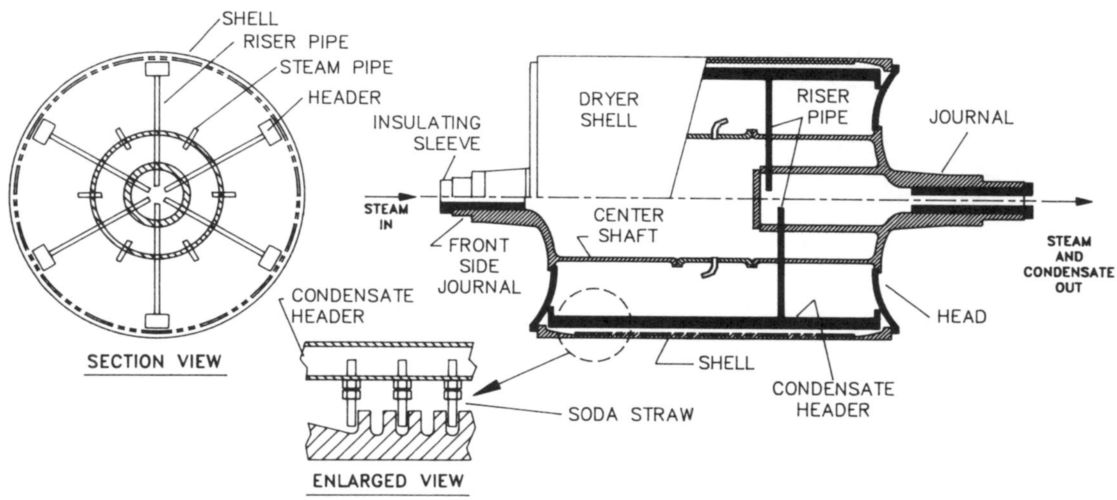

Fig. 367. Standard ribbed-shell Yankee configuration [2].

employed, situated before the first pressure roll, to control the adhesion and release characteristics of the sheet and, in turn, to impart desired physical properties into the product being manufactured.

C. DESIGN AND CONSTRUCTION

Today's Yankee dryers are as large as 5.5 m in diameter and 7.75 m in width, weighing as much as 180 t. Most typically, a Yankee of this size would have a ribbed shell geometry, designed to enhance heat transfer and strength. Speeds in excess of 1850 m/min are becoming commonplace. MG cylinders have been manufactured to 7 m in diameter and 5.25 m face width. Plain bore shells, because the process heat transfer requirement is generally lower, are most often found in MG cylinders.

Modern Yankee dryers are designed according to the prevailing regulations of the jurisdiction in which they are intended to operate. In the United States, this is most usually Section VIII of the American Society of Mechanical Engineers' "Boiler and Pressure Vessel Code". Design consists of the pressure-containing parts–shell, heads, journals, and centershaft–insulating sleeves to protect the bearings from high temperatures, steam joints to provide passage of steam and condensate without leakage where the external stationary piping connects to the rotating cylinder, manholes for entrance, and internal condensate removal equipment. The Yankee is supported on roller bearings that are encased in housings and lubricated through a circulating oil system. The backside bearing is usually "fixed", while the front side is permitted to "float" in order to take up axial expansion that occurs as the dryer is brought to operating temperatures. Figure 367 illustrates a standard Yankee configuration.

The shell is invariably made of ASME SA-278 or equivalent, Class 30 to 60 grey cast iron, with newer Yankees cast in the Class 50 to 60 range only. Heads, journals and centershafts are cast from Class 50, or less, grey iron, ductile iron or steel, or fabricated from steel.

Internal design consists of a method of introducing steam, the heat transfer medium, through the front side journal and into the centershaft, where it is distributed throughout the dryer via steam inlet nozzles. A condensate pickup system is provided, comprising several axially aligned headers having–again in newer designs–a multitude of small-diameter syphon tubes, known as "soda straws". Pressure differential between the steam supply and exhaust and the resultant aspirating action of excess blow-through steam provide the motive force to entrain and remove the condensate at the soda straw orifices, maintaining a thin, uniform film and prohibiting the buildup of condensate in the dryer. Passing up the soda straws, condensate collects in headers. From there, it is again evacuated through riser pipes, connected to the backs of the collection headers, and conveyed into a condensate collection pot for exhaust out the backside insulating sleeve. In plain-bore Yankees operating below rimming speed–the speed at which centrifugal force holds the condensate firmly against the shell wall around the entire internal diameter during a full rotation, such that the condensate forms a thin ring of water within the contour of the shell – scoops are utilized to mechanically gather the condensate from the shell and, as the dryer rotates, transport the condensate to the hollow journal out of which it flows by gravity. Plain-bore Yankees operating above rimming speed, typically of older design, often incorporate only a collection header, set slightly off the shell surface with a series of accurately spaced holes, to syphon the condensate. There are a number of variations on the basic concept of internal condensate removal systems depending upon vintage, manufacturer, and internal shell geometry. However, in modern, high-speed Yankee drying, the rotating syphon design, utilizing excess blowthrough steam to develop a two-phase transport flow, is the one of most interest.

D. STEAM SYSTEMS AND CONDENSATE REMOVAL EQUIPMENT

Steam enters the cavity between the center-shaft and inside diameter of the shell. It condenses and, under sufficiently high speed conditions, forms a water ring of, more or less, uniform thickness. The condensate is drawn from the shell against centrifugal and gravity forces, into the excess, or blowthrough, steam. The mixture of steam and condensate becomes a two-phase flow stream, which passes up the riser pipe and out the backside journal. Depending upon the rotational velocity, dryer diameter, and condensate volume, the condensate ponds, cascades or rims inside the dryer. At low speed, the condensate forms a pool at the bottom of the cylinder and turbulence in the pool maintains the temperature of the inside surface of the shell close to saturated steam temperature. At intermediate speed, the condensate follows the shell some distance around in the direction of rotation before it breaks from the surface and cascades back. This condition results in higher, unstable drive torque [3]. At higher speed the condensate film, under the influence of centrifugal force, follows the shell throughout the rotational cycle. Heat must be transferred through this layer by conduction, if a laminar film is assumed. The speed necessary to create a rimming condition can be estimated from Equation (1) below [4]. Application of that formula suggests a required speed of approximately 427 m/min for a 3.66 m diameter Yankee to maintain a rimming condition when the condensate film thickness is on the order of 5 mm.

$$V_r = \left(5720 - \frac{2610}{D}\right) d_{cond}^{1/3}, \text{ ft/min} \qquad (1)$$

Considerable care in the design, operation, and maintenance of the steam supply and condensate removal systems is required, if they are to effectively perform their designated functions. A typical system (see Fig. 368) might include steam supplied from a boiler at 1.5 or more times the maximum operating pressure of the Yankee, if a thermocompressor (T/C) is included. High-pressure steam flow to the T/C is termed motive flow, which in the prevailing system designs is somewhat less than the maximum condensing load on the Yankee. A thermocompressor design is selected that provides optimized efficiency for that motive flow. A by-pass, pressure reducing valve (PRV), around the thermocompressor, satisfies the remaining condensing requirement and normal system transients that occur due to changes in machine operating conditions. Steam system design should also provide sufficient capacity for a small amount of bleed-off steam to eliminate non-condensable gases. In a steam supply system utilizing only a PRV, the boiler must produce not only motive flow capacity, but sufficient blowthrough steam to evacuate the condensate. Typically this is between 40 and 70% of the motive flow. The blowthrough steam, in the case of a T/C system, simply recirculates, while in a PRV arrangement it must be exhausted to a lower pressure system or to the atmosphere. In systems where blowthrough steam must be exhausted to the atmosphere in order to evacuate a dryer, there is an extraordinary cost penalty to replace, treat, and evaporate boiler makeup water. One or two safety relief valves (SRVs) are situated between the pressure control valves and the Yankee inlet to protect the vessel against overpressurization. Because of the potentially catastrophic consequences of an overpressure failure, redundant SRVs are recommended. Safety relief valves should be sized to prevent pressure buildup in excess of 10% or 20.7 kPa, whichever is greater, above the maximum allowable working pressure of the vessel [5]. Inclusion of a blowdown line and automatic valve permits the rapid reduction of Yankee steam pressure to some preset value during sheet loss or to atmospheric pressure for shutdowns.

The two-phase flow of blowthrough steam and condensate is discharged from the Yankee, usually the backside, and into a separator tank, where the steam and non-condensable gases are segregated from the condensate through a baffling arrangement. Steam and gases exit the top of the vessel – this is a good place to bleed the non-condensables – and the steam is either recycled into the suction side of the T/C or dumped, through a differential pressure control

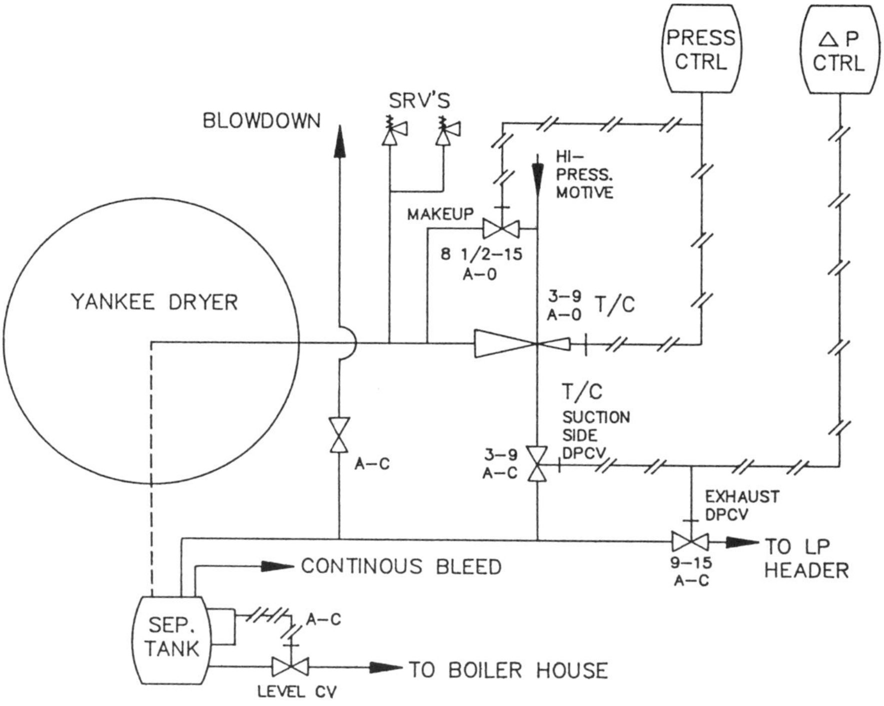

Fig. 368. Yankee steam control schematic.

valve (DPCV), to a lower pressure system. Condensate is discharged from the bottom of the separator through a level control valve, LCV, employed to maintain a relatively constant liquid depth in the tank, and returned to the power house.

Steam pressurization of the Yankee is accomplished through the T/C and/or PRV [6], with the T/C spindle positioning usually being controlled through instrument air pressure, which is split-ranged with pneumatic control pressure to the by-pass valve. In that way, once the T/C has reached a pre-set degree of openness, as determined by the instrument air pressure, the by-pass valve will begin to open. Similarly, differential pressure or blowthrough steam is often controlled by a split-ranged set of valves, one on the suction side of the T/C which acts to choke down differential (flow) created by the T/C and one exhausting to a lower pressure system which increases differential or flow. There has been much debate concerning the relative merits of differential and flow control systems [7], but designers of modern systems appear increasingly to favor the use of flow control.

E. BEARINGS AND LUBRICATION

Roller bearings began to come into general use on Yankee cylinders beginning in the Forties. As greater productivity was sought and Yankees grew in size, weight, speed, and steam pressures, an increase in the load-carrying capacity of bearings was demanded.

Both spherical and cylindrical roller bearings are found in Yankee dryer applications. Spherical bearings are especially suited for the fixed, drive side of Yankees, because they will withstand locational and thrust loads associated with the drive system. Inherent in the spherical bearing's design is the ability to accommodate misalignment between the journal and housing up to 1.5° without loss of load capacity. Cylindrical roller bearings are more typically mounted on the floating, front or operators' side, where they function to take up the axial expansion through the bearing, instead of requiring the bearing to float in the housing.

A Yankee expands axially through the heating-up process, which takes it to operating temperature and pressure. Depending upon width and steam temperature, the Yankee will grow as much as 12 mm in the cross-machine direction, and, in order to eliminate the enormous thermal stress that would result from attempting to resist such growth, a variety of means are employed to permit the cylinder to "float". The use of cylindrical roller bearing has already been mentioned. A second method utilizes ladder bearings on which the housings are mounted to allow linear, axial movement. In a third design, the housings may be mounted on rockers, which are designed to maintain the centerline height of the Yankee while the housing moves axially. Finally, it is possible on smaller cylinders to have a spherical roller bearing on the front side, as well, which slides through the housing on its outer ring as temperatures vary.

The evolution of dryers to higher weights, speeds, and operating temperatures has placed increasing demands on the bearings. Yankee bearings are generally designed for a B_{10} life in excess of 200 000 h. To achieve this longevity under today's operating conditions, great care must be taken in the selection of the bearings and lubricant, in the circulating oil system, and in the handling and installation of the bearings. Due to the higher operating pressures (temperatures) of the modern dryer, bearings are protected by the inclusion of insulating sleeves, which reduce heat transfer from the steam to the journal. Still, startup must be sufficiently protracted to prevent rapid heating and, therefore, expansion of the journal too much before the bearing. Otherwise, there is the risk of preloading the bearing, drastically shortening life, or actually cracking the inner race. To provide some cushion against such occurrences, bearing manufacturers now typically recommend C4 clearances, which permit greater expansion of the inner race before all clearance is eliminated, and carburized inner races, which offer some margin of ductility [8].

Lubricating oils have gotten considerable attention also. To achieve the life desired of a roller bearing in this severe environment, a hydrodynamic film must be maintained between the rollers and the raceways in the load-carrying zone. Hydrodynamic film thickness depends upon the viscosity of the film and the rotational velocity of the bearing rings [9]. To ensure that the necessary film thickness is present, lubricating oils with a minimum viscosity equivalent to ISO VG-220 for dryer pressures up to 7 kg/cm^2 and ISO VG-320 for dryer pressures over 7 kg/cm^2 are a recommended industry standard.

Bearing manufacturers prefer inlet oil temperature near 54°C, with a permissible rise of 16–22°C through the bearing. Customarily, this is accomplished with an oil flow rate of 19–26 L/min per bearing [10]. Prudence would suggest that bearing manufacturers and lubrication engineers should be consulted to determine the proper lubricants, flow and temperature conditions for the particular situation.

F. DESIGN STRESSES

Stress analysis of a Yankee dryer is not a trivial undertaking. It was not until the advent of finite element techniques that acceptable accuracy in the calculation of static stresses became achievable. This is particularily true of stress in the head and the shell corner. As a consequence, dryer manufacturers have generally designed Yankees such that the highest stress tends to occur under normal operating conditions on the outside surface of the shell at the centerline. The stress in this location is a function of the forces due to internal pressure, rotation, temperature gradient through the shell wall, and pressure roll line loading.

Centerline stresses due to pressure and rotational velocity are treated as "membrane-type" stresses in that they are assumed to remain constant across the thickness of the pressure vessel wall. Given that the ratio of wall thickness to shell radius, typical of Yankee dryers, is perhaps 1/30 in the worst cast, the thin shell assumption is generally quite satisfactory; although the ASME Code version of the pressure stress expression given below, UG-27 (c)(1), includes a factor of 0.6 times the thickness to approximate "thick shell" formulations found in textbooks [11].

Pressure stress:

$$\sigma_p = \frac{P(r_i + 0.6\ t)}{t} \tag{2}$$

Rotational stress:

$$\sigma_\omega = \frac{\rho V^2}{g} \tag{3}$$

Thermal stress accrues as a result of the temperature gradient through the shell wall thickness. Accurate determination of thermal stress is, therefore, dependent upon a knowledge of inside and outside surface temperatures. Temperature at the inner wall surface is generally a calculated value, equal to the saturation steam temperature less the temperature drop across the condensate film. Temperature at the outer wall surface is generally derived from a measured value – which depends upon the capability of the instrumentation, and the thickness and thermal conductivity of any coatings on the papermaking surface – or from an estimate of heat flux, based upon condensing rate. Therefore, while the mathematical expression for thermally induced stress is quite simple, it requires knowledge of temperatures and

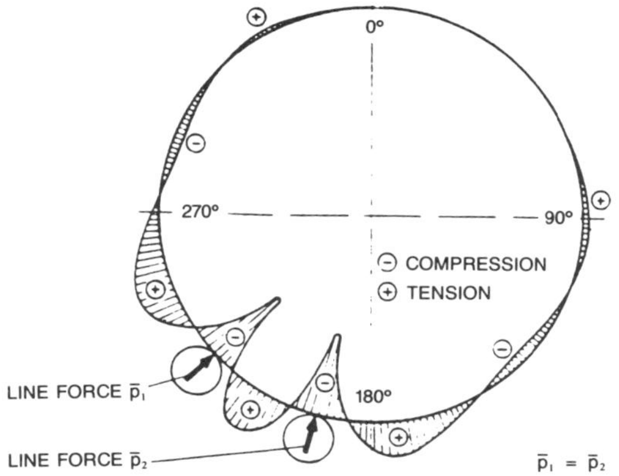

Fig. 369. Circumferential line load stresses (at the centerline due to two pressure rolls) [12].

material properties, i.e., coefficient of thermal expansion, elastic modulus, and Poisson's ratio, which are problematic. The thermal stress is similar to a "bending" stress, in that the stress on the outside surface, being cooler, is the maximum *tensile* stress and that on the inside is the maximum *compressive* stress.

Thermal stress:

$$\sigma_T = \frac{E\alpha\Delta T}{2(1 - \gamma)} \tag{4}$$

Pressure roll nip loading imparts an alternating bending stress to the shell, which, immediately under the nip, is compressive on the outside surface and tensile on the inside. This stress passes through several reversals, eventually dampening out as one progresses, circumferentially, away from the point of loading as illustrated in Fig. 369. The first reversal, typically 30°–50° away from the line load, is significant because it adds to the tensile stress on the outside shell surface and expands the range of cyclic stress produced by the pressure rolls.

Formulae for the resultant stresses and deflections are given by Roark and Young [13] and by Hoff et al. [14]. Mangelsdorf summarized the Hoff et al. solutions to the Donnell equations in a TAPPI paper [15] and, although there are numerous other approximations available, the formulation found in the Mangelsdorf paper is presented here.

Line load stress:

$$\sigma_N = \frac{0.87\, q\, r^{1/4}\, L^{1/2}}{t^{7/4}} \tag{5}$$

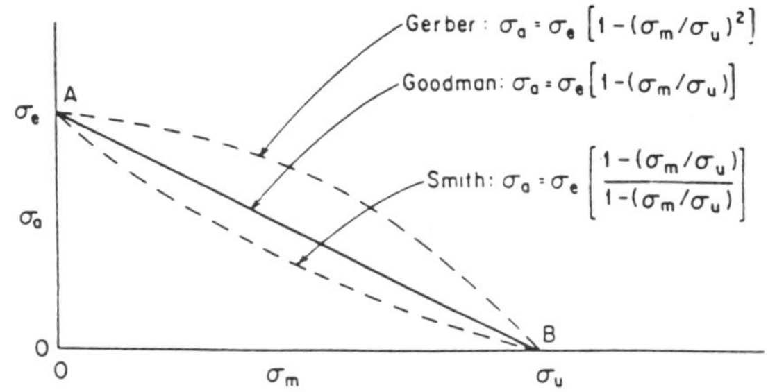

Fig. 370. Alternating vs. mean stress diagram [17].

The magnitude of the reversed circumferential bending stress is calculated to be 30% of the stress at the nip and occurs 51 K degrees away, where $K = (t^{0.25})(L^{0.5})/(r^{0.75})$.

The alternating stress field, attendant to two pressure roll configuration tissue machines, can be determined by superposition of the stresses resulting from individual roll loadings. Further, it can be established that, if properly located, a second roll can reduce the cylinder stress that would have existed under single roll loading.

The failure criterion for brittle materials, which include grey cast irons, is generally that based upon maximum principle stress [16]. In fatigue, cast iron is recognized to follow the Smith relationship [17] on an alternating vs. mean stress diagram, as shown in Fig. 370, along with the Gerber and Goodman formulations for comparison.

The maximum and minimum stresses, acting on the outside surface at the shell centerline, are calculated as follows:

$$\sigma_{max} = \sigma_p + \sigma_\omega + \sigma_T + \sigma_{N_{max}} \tag{6}$$

$$\sigma_{min} = \sigma_p + \sigma_\omega + \sigma_T + \sigma_{N_{min}} \tag{7}$$

The mean stress is then found by combining Equations (6) and (7) in the standard way:

$$\sigma_{mean} = \frac{\sigma_{max} + \sigma_{min}}{2} \tag{8}$$

The alternating stress from Equation (9), the Smith Equation is:

$$\sigma_{alt} = \sigma_e\left[\frac{1 - \sigma_{mean}/\sigma_u}{1 + \sigma_{mean}/\sigma_u}\right] \tag{9}$$

where, σ_e is the endurance limit and σ_u is the ultimate tensile strength of the material. A typical comparison of the contributions of the four loading components is given in Fig. 371.

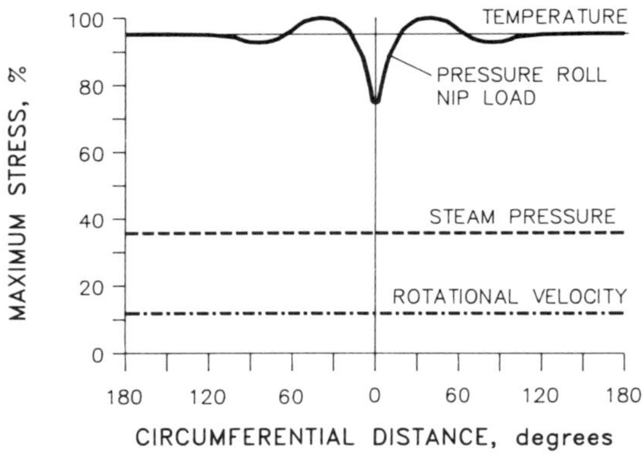

Fig. 371. Typical percentage contribution from four Yankee stress components [18].

Various manufacturers and tissue producers have established safety factors based upon maximum principal and/or fatigue stress criteria to supplement Code requirements, which specify only allowable hoop stress at the shell centerline. The success of these factors is borne out by the exemplary safety record of industry.

Axial shell stress is not typically an issue at the shell centerline, since the axial stress component due to steam pressure is, at most, only 50% of the tangential stress and this is further reduced by the centershaft's ability to carry a portion of the axial load. It is the axial bending stress at the shell corner, added to the axial membrane stresses in that region, that are of interest during "pressure only" conditions.

The above discussion relates primarily to "plain" bore Yankee shells, in contrast to the "ribbed" internal configuration. The latter compounds greatly the difficulties encountered when undertaking such calculations. Variation of internal heat transfer coefficients around the rib, non-orthogonality in shell properties, and non-linearity in temperature drop across the wall thickness demand a finite element approach to attain adequate accuracy in shell stress determinations.

Stresses in Yankee heads, journals, and centershafts are, today, approached by finite element calculations as well [19, 20]. Classical pressure vessel techniques, as described by Harvey [21], are able to account, with only marginal accuracy, for the complexity of the head shape and its varying thicknesses, or for the fixity of the joints. The same is to be said for the accompanying moments, forces, and deformations under the several operating conditions, necessary to determine an accurate crown near the head-to-shell junction.

In summary, dryer and tissue manufacturers have, through time and the utilization of ever more sophisticated techniques, been able to develop procedures for the accurate determination of stresses (and deformations) in Yankee cylinders. Accumulated experience, enhanced knowledge of materials and vastly improved casting quality, combined with more precise calculations, have resulted in an excellent record of pressure vessel performance, even as size and speed have increased dramatically.

G. DRYER SAFETY

Strict adherence to good and prudent engineering practice is essential to the ultimate performance and safety of Yankee dryers. Not only is the Yankee dryer the heart of the tissue-making operation but, by virtue of its size, design, and function, it has the potential for creating enormous destruction should it fail catastrophically. Even non-catastrophic failure, resulting in a loss of functionality, can mean months of idled production. Therefore, it is paramount in the protection of the Yankee dryer—which translates into protection of paper-making assets, profits, jobs, and employees—that safe operating and maintenance procedures be understood and implemented.

In recent years, publications have attempted to provide guidelines for dryer safety [18, 22]. Yankee manufacturers now typically provide manuals with new cylinders, describing safe startup procedures and an envelope of allowable operating conditions for their particular vessels. And, insurance underwriter recommendations have expanded considerably [23]. These represent just a few of the sources that may be consulted. Since it is beyond the scope of this text, the multitude of areas related to safe operation and maintenance of Yankee dryers will not be treated in detail here. However, it is important that the reader have an understanding of the salient elements in any program designed to ensure Yankee dryer safety.

Dryer design has evolved through the years to larger diameters and widths, and faster speeds. This has been made possible through advances in the metallurgy and quality of casting, through improved analytical techniques, and through retained experience. To a major extent this evolutionary process has led to a general uniformity in design, which can be influenced in only a very minor way by the tissue producer. Therefore, the dryer manufacturer is usually responsible for the structural safety of the design and construction. He provides the original documentation for the cylinder. All good safety programs begin with this documentation and continue by meticulous and thorough recordkeeping throughout the dryer's life. Such record-keeping will include, *de minimis*, centerline thickness measured before and after every grind, surface repairs undertaken, such as plugging or metallizing, mechanical history and damage reports, and dryer inspection accounts.

With due care and stewardship, a new Yankee may be expected to deliver decades of productive service. Such expectations demand substantial attention to maintenance, if they are to be realized. Repairs have to be undertaken, consistent with statutory and underwriter requirements and state-of-the-art engineering practices.

Leakage through flanges, pressure-containing walls, and around bolts must be eliminated with all deliberate speed to protect the structural integrity of the assembly. Bolts must be maintained free from corrosion, steam cutting and loss of pre-stress, and they must be of specified Code material and installed in accordance with the initial design. The size and amount of porosity in the dryer shell are variously tolerated by different tissue products and processes but, at some level, defects in the shell begin to degrade operating efficiency and product quality. More importantly, increasing levels of porosity at the highly stressed shell surface also degrade strength. This should be accounted for when establishing safety criteria. The three methods for addressing surface defects–plugging, epoxies, and spray metals–each have their applications and particular installation techniques, which must be closely followed, if a safe and enduring repair is to be attained. Of course, safety relief valves are vital to protection of any pressure vessel. It is critical that they be selected, installed, tested, and maintained in such a manner that their function is preserved; good recordkeeping is absolutely essential to that end.

In Section F above, considerable space was allotted to the individual stresses that combine during the various stages of Yankee operation. The objective of accurate stress knowledge is to establish allowable operating conditions. Through the judicious selection of safety factors, those allowable operating conditions must account sufficiently for transients, minor variations in operating conditions, uncertainty in material properties, and an array of unknown loadings in order to prevent structural damage and failure. Dryer manufacturers and some tissue producers have developed presentations in the form of graphs and charts, which define maximum allowable operating pressure and pressure roll load, given a set of other assumptions such as papermaking surface temperature and speed. An example of one manufacturer's graph is presented in Fig. 372.

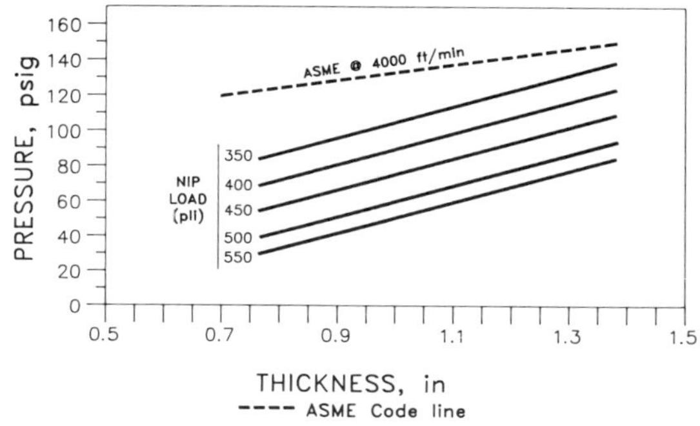

Fig. 372. Typical steam pressure derating curve [18].

Control of temperature transients and gradients is a primary requirement in ensuring dryer safety, since many of the recorded dryer failures may be attributed in large measure to abnormal or unanticipated thermal loading. These can occur through rapid warmup or cooldown, excessive local heating or cooling of the vessel through showers, sprays, hoods, and moisture profiling devices, and over-cooling the shell edges. Temperature differential, through and between assembly components, can easily result in stresses greatly exceeding the load-carrying capacity of the structure, so considerable care should be exercised to avoid conditions which can create excessive thermal differences.

Structural failure is generally attributable to stress, environment, defects, and/or material. An excellent collection of case histories is presented by Ferrari [24], which illustrates contributing factors in nine Yankee dryer failures prior to 1984. Overall, the record of dryer safety in the industry has been commendable but, with increased demands for extended performance and longevity being placed on this equipment, that record will only be maintained–and, hopefully, improved–through adherence to recognized good and prudent engineering practice, based upon accumulated knowledge, experience, and skill in Yankee dryer systems' management.

H. SHELL METALLURGY

Undoubtedly, the greatest advances in Yankee technology have occurred in metallurgy and the foundry practices employed to attain the size and quality cylinders being cast today. Historically, grey cast iron has been the material of choice in drying cylinders and, at this time, there is nothing on the horizon that stands poised to replace it. The characteristics and properties of high-strength grey cast iron, appear to offer the

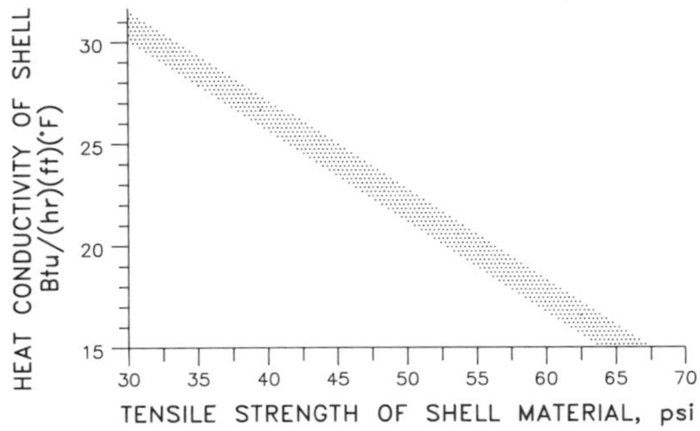

Fig. 373. Thermal conductivity of grey cast iron vs. tensile strength [26].

ideal combination of cost, strength, thermal stability and conductivity, wear and corrosion resistance, machinability, and castability.

The standard grey iron in Yankee dryer shells is alloyed and cast to obtain a pearlitic microstructure, minimal porosity within the anticipated useful shell wall thickness, hardness in the range of 200-260 Brinnell, and strength between ASME Class 50-60. Thermal conductivity is inversely proportional to ultimate tensile strength (see Fig. 373), with values normally between 33 and 47 W/m/°C. Each dryer manufacturer has a unique recipe of alloying ingredients to achieve his particular combination of shell properties, with the usual tradeoffs.

Specifically, lower strength cast irons are usually softer and more thermally conductive, due to a higher graphitic carbon content and lower alloying. Because of the higher carbon content, they should exhibit a reduced tendency for porosity, all other conditions being equal. Also, for the same alloying content, higher strength can be expected with thinner walled castings [28] and the disposition toward shrinkage porosity should be lessened, due to more rapid cooling. Ultimate tensile strength, elastic modulus [28], and hardness [25] tend to be proportional to each other. Tensile strain to failure of high-alloy grey cast irons is usually in the range of 0.004-0.008 m/m.

The fatigue behavior of grey cast iron is similar to that of steel in that it usually shows a specific fatigue limit and a fatigue/tensile strength ratio, or endurance ratio, that decreases with increasing strength [27]. The endurance ratio is reported [28] to be approximately 0.4 for completely reversed bending in Class 55 C.I., but values vary widely due to the sizable diversity in irons and testing conditions [25]. Under push-pull fatigue testing, the endurance ratio is typically lower than under reversed bending [28]. Fracture toughness data are just beginning to be developed for the range of cast irons of papermaking interest [29]. The Design of Iron Castings Handbook [28] suggests a critical stress intensity value, K_{IC}, of between 11 and 19 MPa $\sqrt{m}$.

Perhaps the final word regarding Yankee shell properties and quality is that they are very manufacturer-dependent. Therefore, all handbook and test values should be approached with a high degree of circumspection and, certainly, when attempting to incorporate the values into stress calculations, larger safety factors are required if the specific material's test data are unavailable.

I. DEFORMATION ANALYSIS

The demand for ever-increasing levels of performance in the tissue industry has led to larger diameter Yankee cylinders for enhanced thermal drying capacity, higher pressure roll nip loads for better mechanical water removal, wider and faster machines for greater tonnage, and closer tolerances on machine and cross-directional uniformity for improved quality. Integral to the industry's ability to meet this demand has been the expanded capability in mathematical analysis techniques.

The Yankee dryer deforms in the radial and axial directions under the variety of loadings to which it is subjected – pressure, temperature, line load(s), and rotation. Accurate determination of the deformed shape of a complexly configured pressure vessel under such loadings is no small task, but it is made Herculean by the requirement that a precise contour, machined into the shell surface under the set of grinding conditions, must itself deform to a contour that creates a uniform nip(s) with the pressure roll(s) under the completely different set of papermaking conditions.

To achieve the precision required for the larger, faster, more highly loaded Yankees of today, classical techniques of analysis have given way to finite element methods [19], which are far better able to address the complicated head and shell corner designs, in the determination of deformation – due to changes in pressure, temperature and rotational loadings – between grinding to operating conditions. The head and centershaft are the pressure vessel components that complicate the problem, in that they add rigidity to the cylinder ends, restraining free expansion by virtue of the moments and forces applied there. In theory [30], edge loading of a cast iron shell results in deformation that extends axially a distance

$$a = 2.43 \sqrt{rt} \qquad (10)$$

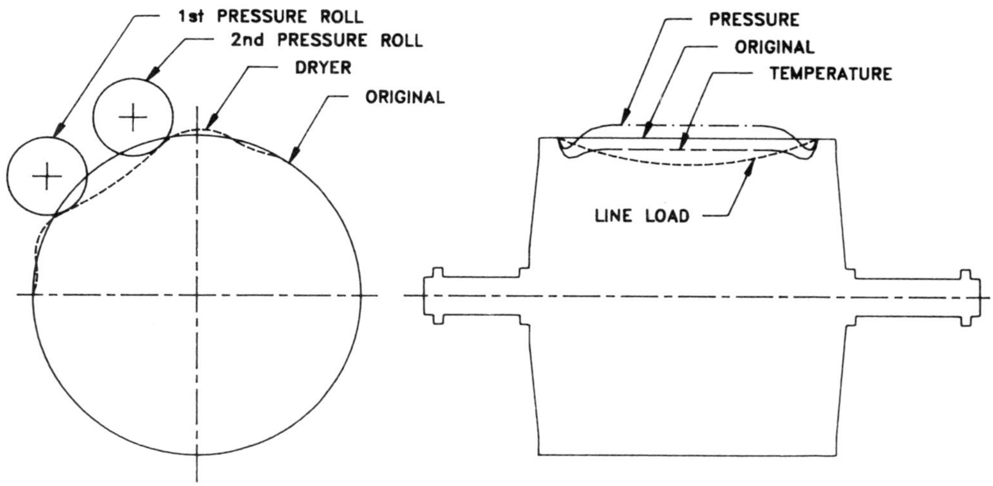

Fig. 374. Deformed Yankee under various loadings [1].

In practice then, edge effects on a 4.6 m diameter Yankee, having a 50 mm thick shell, would dampen to a negligible level some 824 mm from the shell end. This is the approximate extent of the secondary crown, also referred to as end relief or dubbing. The magnitude of the secondary crown is determined by the difference in deformation, in the vicinity of the shell ends, between operating and grinding conditions. The objective of accurately compensating for end effects during grinding is to obtain an operating contour that precisely matches the total deflection curve of the dryer, as a result of pressure roll loading only, i.e., an operating contour that precisely matches the primary crown. A schematic of Yankee deformation under the various loadings is provided in Fig. 374.

Primary crown depends solely upon deflection due to pressure roll loading and is generally calculated under the simplified assumption that the pressure roll and Yankee are free from end restraint. Therefore, the deflected shapes of both rolls can be described by a monotonically increasing algebraic or trigonometric function. The components of Yankee deformation are the result of membrane bending and shear, while the pressure roll includes a beam bending component as well. Shear deformation is a relatively minor term for both rolls and may be calculated, per Reference [31], from Equation (11):

$$y_s = \frac{qx^2}{2\,K'\,A\,G} \tag{11}$$

Membrane bending occurs symmetrically with respect to the line load generatrix. The maximum values of this component for Yankee and pressure roll deflections may be calculated at their respective centerlines using the expression from Reference [13], given in Equation (12):

$$y_m = \frac{0.0305\ B^5\ q\ r^{3/4} L^{3/2}}{t^{9/4}\ E} \tag{12}$$

Beam bending occurs only in the pressure roll and is easily found in any good handbook [32].

$$y_b = \frac{q\ L_f L_B^3}{384\ EI\ (1 - 2z)}[5 - 24z^2 + 16z^4] \tag{13}$$

In addition to the magnitude of the primary crown in both the Yankee and pressure rolls, calculated as above, the shape and the proportion of crown allotted to each roll are also important. Typical roll grinding equipment might use a combination of gears and eccentrics to create the total crown and to impart a crown shape in the form of a partial cosine curve. A cosine curve on the order of 71° (half-angle) has been regarded as most closely corresponding to the "true" deflection curve [33]. Failure to obtain good agreement between system crown, total system deflection and deflected shape, eventuates in non-uniform nip loading. It has been calculated [34] that a nip load "deviation of 10 N/mm (57.1 pli) results in a web moisture difference at the reel of from 0.2% up to 1%". The above paper by Holik and Surat is perhaps the best study to date on the interaction of surface temperature, nip load, and moisture variation.

Correct crowns will result in uniform contact pressure across the entire sheet width, while

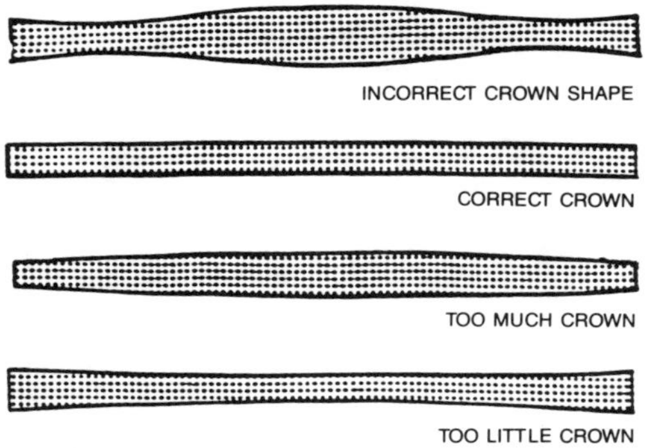

Fig. 375. Nip impression shapes for various crown problems.

incorrect crowns and skewed rolls will cause contact irregularities similar to those represented by the nip impressions shown in Fig. 375. However, because of the difference in dryer contour between operating and grinding conditions, static Yankee/pressure roll nip impressions are of little value, except in highlighting front to backside variances.

When highly crowned rolls mate, papermaking problems can arise if the ratio of their crowns differs significantly from the ratio of their diameters. This is due to the difference in surface speed, the product of radius and rotational velocity, between the center of the rolls and the edges if the crown/diameter ratio mismatch is excessive. Thus, it is frequently recommended that to achieve a perfect speed match across the entire dryer face, a Yankee which is five times the diameter of the pressure roll should be apportioned five times the pressure roll's crown.

Uniform nip loading, the objective of accurate crown calculation and installation, is one vital factor in the complex combination of factors that interact to determine uniformity in creped-sheet properties, moisture profile and coating development, surface wear in the contacting rolls, and, ultimately, productivity of the paper machine.

J. SURFACE WEAR MODEL

Corrosion-wear is the generally accepted model for removal of dryer surface metal under papermaking conditions [35]. In a corrosive environment, an oxide layer quickly forms on a cast iron surface, while the layer's rate of growth rapidly diminishes with time, i.e., the oxide film becomes a self-protecting barrier to further corrosion. In the presence of continued rubbing (sliding) the outer portion of the oxide layer will be constantly removed so that its thickness is reduced and the diffusion of liquid and gaseous

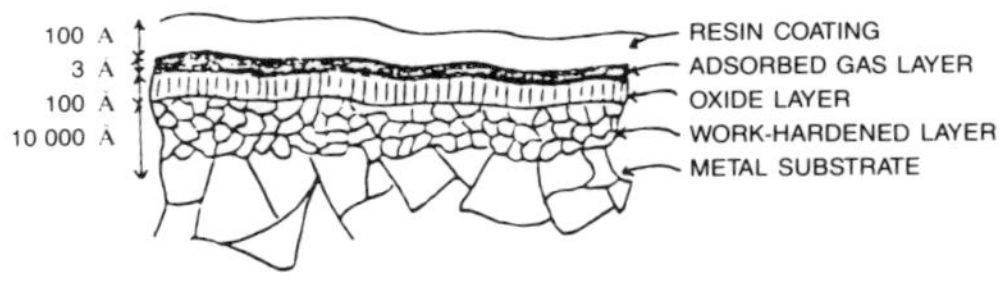

Fig. 376. Schematic of films on a metal surface (not to scale) [36].

corrosive elements to the underlying metallic surface is promoted (see Fig. 376).

External to the oxide film in the papermaking process, a boundary layer (coating) develops, composed of a number of hemicelluloses and derivatives, creping aids, fiber dust and metal oxide particles. The components, thickness and rheological characteristics of this coating depend upon pulp types, system chemistry and pH, creping dryness, the ratio of hood to Yankee drying, and a farrago of associated variables. But the importance of this film in protecting the dryer surface is monumental for, when a coating is stable and uniform, there is a minimum of metal-to-metal contact between the dryer surface (oxide layer) and the doctor blade. The film boundary layer acts as a lubricant and, although it may not totally prevent metal-to-metal contact at the asperities, it limits access of corrosive liquids and gases to the underlying material and produces a low shear strength interface, becoming itself the "wear" surface. This accounts, in part, for the generally lower wear rates associated with dry crepe machines, which exhibit the ability to develop a more substantial coating that is sacrificial and reduces access of corrosive materials to the substrate.

There are any number of ways in which surface wear of a Yankee dryer might be categorized but, broadly speaking, uniform and non-uniform seem to work well enough. Uniform wear may be defined as that which occurs globally to the dryer surface. For example, wet crepe machines tend to wear the papermaking surface more or less uniformally, due to minimal coating maintenance under the moist sheet, leaving the edges high outside the trim. Conversely, dry crepe processes tend to wear the unprotected edges, leaving the well-coated sheet area high. Other factors which affect the wear-corrosion rate in a generally uniform way are: pulp type, degree of bleaching and washing, amount of beating, water hardness, chloride content, system pH, chemical coating, shell material, paper machine configuration, hood type, and doctor loading. The process water system, which tends to be least corrosive, to establish a reasonably thick and stable coating, and to remain relatively consistent over time,

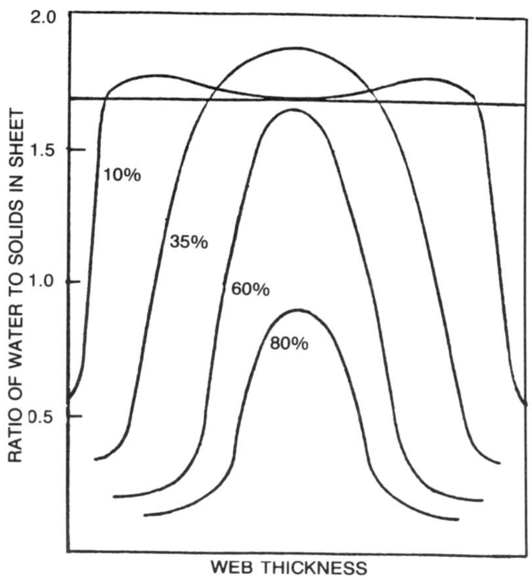

Fig. 377. Moisture profiles through paperboard at various levels of sheet dryness [41].

exhibits those characteristics most likely to deliver wear uniformity.

Non-uniform wear is attributable to those factors which occur differentially around or across the Yankee. They originate from non-uniformity in: pressure roll and doctor contour; sheet and felt moisture profile; hood air impingement velocity and temperature; spray application; condensate removal, and shell metallurgy. To this list should be added the vibratory input from a variety of mechanical sources, which can eventuate in chatter marks randomly located throughout the surface.

K. YANKEE DRYING

Paper drying occurs through the process of evaporation. With paper, which is a thin, porous medium, evaporation proceeds, thermodynamically, through a complex interaction of heat and mass transfer mechanisms. This process has been variously described for a variety of paper grades by a number of authors [37–39]. In summary, when a wet sheet is pressed onto the dryer surface there is a short period of high heat transfer, as surface water is elevated to the interface temperature and gradients are established. This might be denoted as the pre-heating period. Heat is conducted from the hot shell surface to the moisture-saturated paper surface at the shell to paper interface. As that moisture vaporizes it migrates from a location of high temperature and vapor pressure through the sheet until it reaches a lower temperature and vapor pressure, where it condenses. As the moisture vaporizes and moves away from the drying surface, microvoids develop in the paper which result in a reduced vapor pressure region, and so, by capillary action, moisture is drawn back toward the drying surface. Thus a cycle is created of vaporization, vapor flow away from the drying surface, condensation and capillary water migration back toward the evaporation surface. In thicker, heavier grades, where the sheet is heated on alternate sides by a succession of dryer cans, this process results in a fairly symmetrical moisture and temperature profile in the sheet, i.e., high moisture and low temperature in the center of the sheet with the reverse at the inside and outside surfaces, as shown in Fig. 377. This drying phase is one in which surface water is still present at the dryer to sheet interface, and so heat transfer rate remains relatively constant. This phase is suitably termed the constant-rate period. At some point a deficiency of free moisture at the drying surface is reached, when insufficient moisture is available in the sheet to replenish it by capillary flow. The water is now said to be "bound" to the fibers, vapor pressure diminishes from that of pure water, and the drying rate falls; hence the designation, falling-rate period.

Because of the thin, porous structure of tissue and towel webs, Yankee drying differs somewhat from the above [38]. At the end of the pre-heating phase, a short distance after the first pressure roll, sheet temperature reaches a steady-state value around 90°C [40]. It is likely that a portion of the vapor created at the hot surface can pass through the open structure of the sheet directly, without condensing. As a result of the high surface adhesion caused by pressing, heat transfer occurs very rapidly; if the vapor formed is unable to escape, the problems of sheet blistering and blowing may ensue. The ability of sufficient vapor to pass through the sheet without blowing no doubt contributes to the high drying rates attainable with Yankee cylinders. Relative to the "thick-sheet" model, the moisture profile within "thin-sheets", during the constant-rate period, can be significantly altered from that illustrated in Fig. 377, depending upon the split between Yankee and hood drying. This split has major implications, as well, in coating development and control at the Yankee surface when producing tissue products.

The drying model outlined above is expressible mathematically in terms of heat and mass transfer formulae. Heat transfer is governed by Fourier's heat conduction law:

$$Q = A_{surf}\,U\,(T_{sat} - T_o) \qquad (14)$$

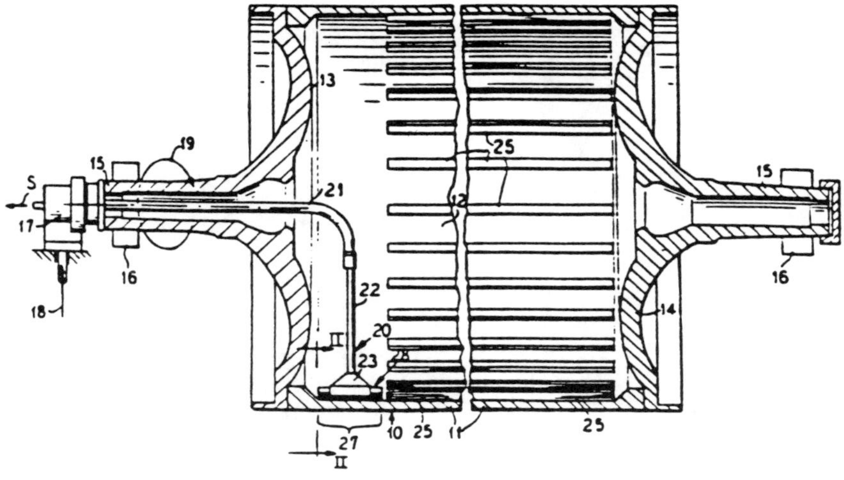

Fig. 378. Spoiler bar installation drawing.

where the reciprocal of conductance for the Yankee's heat resistances is:

$$\frac{1}{U} = \frac{1}{h_i} + \frac{t}{k_s} + \left(\frac{1}{h_e} + \frac{t_p}{k_p}\right) \tag{15}$$

The mass transfer of vapor is likewise governed by the equation:

$$W_e = h_d(C_{wp} - C_{wa})F \tag{16}$$

The factor "F" is the log mean partial pressure difference:

$$F = \frac{p}{p_{wp} - p_{wa}} \ln\left[\frac{p - p_{wa}}{p - p_{wp}}\right] \tag{17}$$

It is seen from the above that drying rate is controlled by the temperature gradient from the saturated steam to the paper and by the vapor pressure gradient from the sheet surface to the air, along with the appropriate coefficients of heat and mass transfer.

The evaporation rate can then be determined through the contributions to heat transfer from the components of conduction, convection and radiation:

$$W_e = \frac{1}{H}\left[U(T_{sat} - T_o) + h_a(T_a - T_o) + Q_r\right] \tag{18}$$

Condensate behavior in a drying cylinder has been described most fully by Appel and Hong [41]. The internal heat transfer coefficient, h_i, is a function of film thickness, dryer speed, the presence of non-condensable gases, shell corrosion, and perhaps high levels of superheat [42]. Considering the condensate film to be laminar and, therefore, that the resistance to heat transfer is simply dependent upon the film thickness and the conductivity of water at the film temperature, a lower bound on condensate film heat transfer coefficient (HTC) can be found utilizing Eq. (1) and Eq. (19) below:

$$h_i \geqq \frac{k_{cond}}{d_{cond}} \tag{19}$$

The conductivity of water at 150°C is 0.684 W/m/°C, which for a 3 mm condensate film thickness results in a lower bound on h_i of 228 $W/m^2/°C$. Because of the movement, "sloshing", of the condensate film relative to the inside surface of the shell, as the shell rotates through a full cycle, more typical HTC values would be in the 2840–4260 $W/m^2/°C$ [37, 43].

Internal heat transfer coefficients can be substantially improved through the installation of turbulence generators, also known as spoiler bars and clips, axially oriented and strategically positioned around the inside diameter to create a "resonant oscillation of condensate between the bars, and a high level of agitation due to formation and collapse of hydraulic jumps and cascading of flow over the bars" [44]. Based upon speed and film thickness, turbulance generating devices may double typical film HTCs. A drawing (Fig. 378) from a Beloit Corp. patent [45] illustrates the spoiler bar installation concept.

Improvement in steam to shell heat transfer may also be gained through a ribbed shell internal configuration. Because of the various heat transfer coefficients on the rib tops, sides and the groove bottoms and the difference in shell thickness through the ribs and the grooves, the calculation of an "effective" internal heat transfer coefficient is a task for more sophisticated mathematical techniques, such as finite element analysis (FEA). Cox [46] estimates that optimized rib geometry may result in an "effective" internal HTC 35% higher than is typical with plain shells.

Shell conductivity was discussed in the section devoted to Yankee metallurgy and was said to be in the range of 33–47 W/m/°C. Cast iron conductivity is a function of temperature, so the conductivity at the average shell temperature should be used in heat flux determinations.

The terms in Eq. (15) for paper-to-shell heat transfer coefficient and sheet conductivity are generally incorporated into a single resistance, due to the difficulty of separating them in empirical studies. Another problem, described above, is that these coefficients vary through the drying cycle, which demands the selection of an average outside HTC to perform the evaporation rate calculation per Eq. (18). Further, the external HTC is dependent upon the type of pulp, openness of the sheet, thickness of the organic and inorganic coatings on the dryer surface, and amount of pressing. Race [41] provides a range for this averaged, external HTC, h_{acc}, of 3833–7666 W/m^2/°C.

It is now possible to calculate a value for overall heat transfer coefficient, U, for use in Eq. (18), and proceed with a determination of the mass flow rate.

Mass flow rate is a function of the difference between the vapor pressure at the outside surface of the sheet and the surrounding air, and of the resistance of the boundry air layer, as described in Eq. (16). The vapor pressure difference may be established from the wet and dry bulb conditions at the two locations. The mass transfer coefficient is related to the air-to-paper heat transfer coefficient at zero mass transfer by the Chilton-Colburn form of the Reynold's analogy:

$$h_d = \frac{h_{ao}}{\rho C_p}\left(\frac{P_r}{S_c}\right)^{2/3} \tag{20}$$

Air-to-paper HTCs are a function of nozzle design (slot vs. hole), nozzle proximity to the sheet, nozzle spacing, air properties and velocity. Spraker [47] offers a series of curves (Fig. 379) based upon the above equations to illustrate evaporation rates for tissue and towel grades in the temperature range of 200–540°C

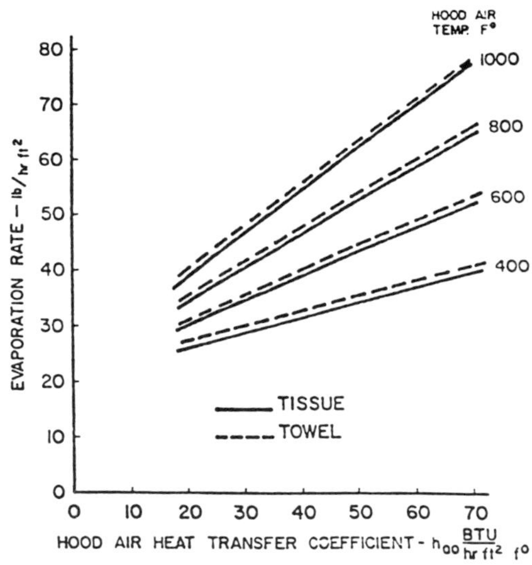

Fig. 379. Drying rate as a function of hood air HTC and temperature.

(400–1000°F) and hood-air HTC, h_{ao}, in the range of 115–400 W/m^2/°C (20–70 BTU/ft^2/°F).

To finally connect the various equations in this section, Silver and Wallis [48] have determined a correction for the effect of mass transfer rate on the hood-air heat transfer coefficient, h_{ao}, which permits the calculation of the convective heat transfer coefficient, h_a, in Eq. (18):

$$\frac{h_a}{h_{ao}} = 1 - \frac{W_e C_p}{2h_{ao}\, Pr^{2/3}} \tag{21}$$

Impingement drying may be divided for simplicity into three general catagories: low velocity/temperature which is typically less than 25 m/s and 150°C; medium velocity/temperature, typically 25–50 m/s and 150–350°C; high velocity/temperature 50–150 m/s and 350–540°C. From Fig. 379 it can be seen that such a range encompasses nearly a four-fold increase in evaporation rate capability of the Yankee-hood system. Unfortunately, high evaporation rate capability is not free. Whittaker [49], for example, calculates that, to raise the minimum nozzle velocity from 35–100 m/s, which achieves a three-fold increase in evaporation capacity, requires a twenty-fold increase in fan horsepower.

Yankee drying is an expansive and complex subject, adequate coverage of which would minimally demand several chapters. It is hoped that this brief overview and the references offered will be of some assistance to the reader in

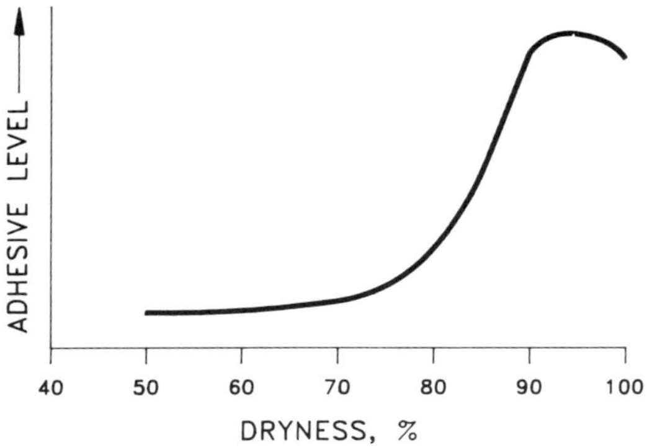

Fig. 380. Typical adhesion force vs. percent dryness curve.

reaching a more complete understanding of the process.

L. THE CREPING PROCESS

In large measure, the tissue-making process culminates at the creping blade; not merely in the sense, at least with regard to dry crepe products, that the sheet is formed, dried, and prepared for finishing and packaging, but that creping is the *sine qua non* of the tissue-making process. It is at the doctor blade that the characteristics of the sheet are, for the most part, established. Therefore, it is both elemental and essential to the understanding of the manufacture of these types of papers on Yankee dryers that a brief overview of the creping process is essayed.

The purpose of creping–in addition to simply removing the web from the Yankee surface–is to make "creped" paper grades softer, bulkier, more absorbent and extensible. This is accomplished through the breakdown of fiber bonds due to the interaction of the adhesive force of the sheet to the dryer, the physical properties of the sheet, the mechanical forces of doctoring, and the geometry of the doctoring system.

The force of web adhesion to the Yankee surface typically varies in a manner similar to Fig. 380. In wet crepe processes, adhesion is controlled by the "viscous deformation and cohesive failure" [50] of the water boundary between the sheet and the cylinder. In the dry crepe process, no such water film exists; therefore, the adhesion, which is usually greatest in the 92–95% dryness range, is a function of the papermaking film (coating) developed on the Yankee surface. At intermediate creping drynesses, coating development is very sensitive to dryer surface temperature and sheet moisture content and, thus, difficult to maintain and control.

The dry crepe tissue process is of most general interest, however, and is the one to be considered here. In order to obtain the sheet properties desired, it is necessary to achieve the optimum level of adhesion between the web and the cylinder at the doctor blade. The mechanism supporting development of an adhesive film on a dryer surface was elaborated in Sec. J above.

Numerous researchers have examined this film in considerable detail, an excellent summary of which is provided by Oliver [51]. Fuxelius [52] has shown that the adhesive layer is composed primarily of hemicellulose, with small amounts of cellulose, lignin, pulp extractives, and dispersed fibrils. There are several types of hemicelluloses found in commercial pulps with differing hemicellulose contents as well. For example, in order of increasing hemicellulose content, pulps were ranked: mechanical = sulfate < high-yield sulfite < semichemical. A similar rank-ordering is found in the degree of adhesion attained. It seems further that glucuronic acid content is an important pulp constituent in its ability to create adhesion because, with the exception of mechanical pulps, an analogous ordering has been observed. In that same seminal paper, Fuxelius states that "The dissolved hemicellulose content of the back water at normal operation with good adhesion was found to be between 30 and 70 mg/L, calculated as xylan" [52]. Other factors found to increase the level of adhesion are: decreased pulping time in sulfite pulps, increased beating, use of slush and unbleached pulps, greater concentrations of salts, e.g., calcium carbonate, and increased alkalinity [51]. Additionally, adhesive agents are frequently utilized to enhance adhesion. These include: animal glues, neutral-setting wet strength resins, dyes, and retention aids. Conversely, adhesion is reduced, i.e., release is elevated, by the employment of emulsified mineral oils, debonders, defoamers, polyethylene glycols, alum, and a range of surfactants. With such a variety of factors influencing the level and uniformity of adhesion, papermakers have turned increasingly to the use of synthetic spray coatings, applied directly to the Yankee surface [53].

The ability to break an adhesive bond is a function of the degree of mechanical and chemical bonding at the interface, the mode of adhesive failure, and the type of stress to which the interfacial region is subjected. The ratio of the adhesive force of the sheet to the dryer in relation to the cohesive forces within the sheet is instrumental in determining creping effectiveness. If adhesion is too small, the sheet will blister off the shell surface before reaching the creping blade; if it is too large, or the cohesive force is too small, the sheet will stick tightly to the Yankee and fracture at the creping blade, causing tears and plugging.

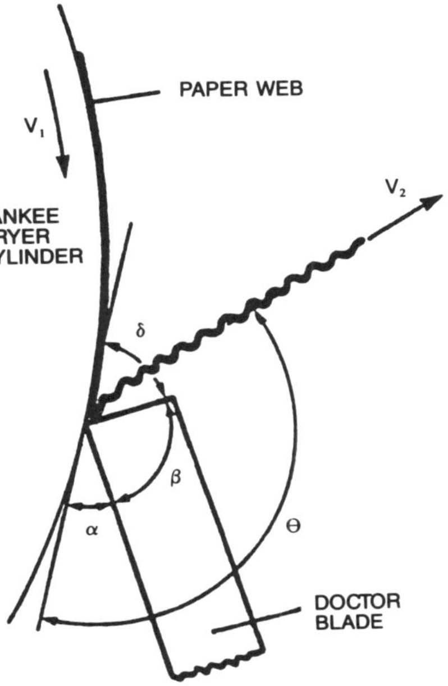

Fig. 381. Creping doctor blade geometry [51]. α **= blade contact angle;** β **= blade bevel angle;** δ **= cutting angle;** θ **= take-off angle;** $(v_1 - v_2)/v_2 \times 100\%$ **= percent crepe.**

Crepe structure has been examined by Hollmark [54] and Falk [55]. The structure of the creped sheet and, therefore, the properties of the paper product are generated through the interaction between adhesion, doctor blade angle, properties of the uncreped sheet, and crepe ratio (the ratio of the speed of the sheet on the Yankee to the sheet speed leaving the crepe blade). Figure 381 illustrates the various angles that are employed to define doctoring geometry.

Hollmark [54], through the techniques of high-speed video, was able to establish the manner in which creping occurs in the vicinity of the creping blade, and to demonstrate the effect of adhesion and doctor blade cutting angle on crepe structure. As is illustrated in Fig. 382, with the proper ratio of adhesion to cohesion, the sheet will detach from the Yankee surface prior to contact with the doctor blade and form a loop, termed a "microfold". As these loops build up on the doctor blade they arrange themselves into a single bundle, or "macrofold". It is possible for several bundles to assemble on the doctor blade before exiting the pocket formed between the blade and the Yankee. Now, the fineness of the crepe, usually associated with enhanced softness, can be improved by increasing the level of adhesion, such as by raising the amount of wet strength resin. This causes the sheet to approach, more closely, the tip of the doctor blade prior to release from the shell surface, which, in turn, causes smaller microfolds. A similar affect is

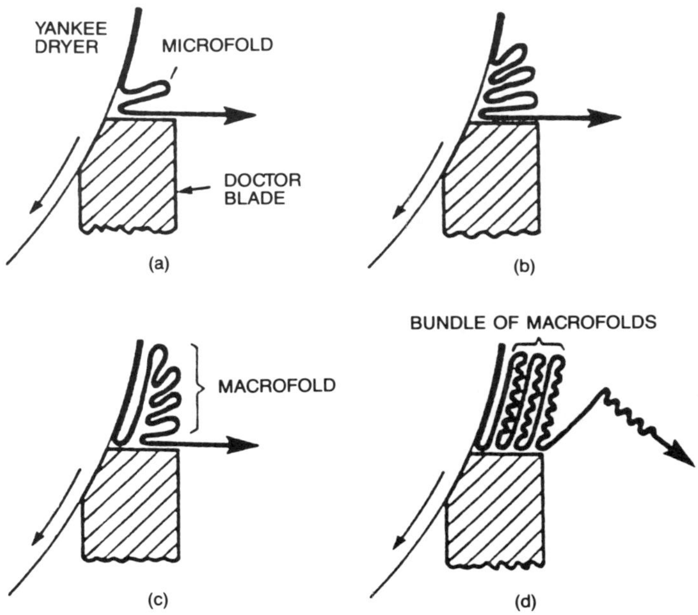

Fig. 382. Diagrammatic representation of the various stages of crepe development in the creping process [51].

obtained by increasing the cutting angle of the blade. At some point, as that angle is increased, microfolds cease to exist altogether, and a fine crepe structure is achieved. The angle cannot be increased indefinitely, however, since a sufficient pocket must be maintained to provide the static friction necessary to cause the sheet to fold into a crepe wave. Additionally, properties of the uncreped sheet, e.g., basis weight, tensile strength and stiffness, fiber aspect ratio and bond strength, are known to influence the crepe structure.

Doctoring uniformity, in loading and angle, is critical to creping effectiveness, as variations result in: variations in blade pressure, which affect the ability to fracture the coating; unevenness in thickness of the coating, which influences heat transfer; differences in the contact and cutting angles, which determine blade loading and fineness of the crepe wave. In the tissue-making process, loadings and contact angles are generally found between 1.75 and 3.5 N/mm and 15 and 25°, respectively. Doctor blade holders range in design from "rigid" to "flexible"; the stated advantage of the flexible being that they more closely mirror the normal topographical variations in the Yankee surface, although this is a subject of considerable debate.

M. SUMMARY

The objective of this chapter has been to proffer an overview of the major component in the manufacture of creped and machine glazed papers, the Yankee dryer. As should be palpable to even the casual reader, because the Yankee dryer is so central to the production of these paper grades, one is immediately led to discourse on an ever-widening range of topics, regarding attendant systems and process fundamentals: hoods, pressure rolls, bearings, steam supply and exhaust, and creping equipment; heat and mass transfer, fluid mechanics, metallurgy, tribology, corrosion and adhesion; deformation and stress analysis, crowning and safety. It is hoped that this summary, however meager, will perhaps pique the interest of those contemplating a field of specialization because, although central to the paper industry, Yankee dryers, and all the systems and processes they encompass, remain rife with opportunity for advancement of fundamental understanding. For those simply interested in attaining an enhanced competence in the manufacture of tissue and MG papers, it is hoped the references listed will prove beneficial.

N. NOMENCLATURE

A = Cross-sectional area, in^2
A_{surf} = Surface area, in^2
a = Journal length, in.
B = $[12(1-\tau^2)]^{0.125}$
C_p = Avg. specific heat of gas, Btu/(lb)(°R)
C_{wa} = Concentration of water vapor in air, lb/ft^3
C_{wp} = Concentration of water vapor @ eff. paper temp. lb/ft^3
D = Diameter, ft
d_{cond} = Condensate depth under rimming conditions, ft
E = Elastic modulus, psi
G = Shear modulus, psi
g = Gravity constant, ft/s^2
H = Heat of vaporization, Btu/lb
h_a = Convective HTC, Btu/(h) (ft^2) (°F)
h_{acc} = Accumulated average external HTC, Btu/(h) (ft^2) (°F)
h_{ao} = Hood-air HTC at zero mass transfer, Btu/(h)((ft^2) (°F)
h_d = Mass transfer coeff., ft/h
h_i = Shell internal film HTC, Btu/(h) (ft^2) (°F)
h_e = Shell external film HTC, Btu/(h) (ft^2) (°F)
I = Moment of inertia, in^4
K' = $2(1-\tau)/(4+3\tau)$, form factor for a thin annulus
k_p = Paper thermal conductivity, Btu/(h) (ft) (°F)
k_s = Shell thermal conductivity, Btu/(h) (ft) (°F)
L = Shell length, in.
L_B = Brg. centerline distance, in.
L_f = Felt width, in.
P = Steam pressure, psig
P_r = Prandtl Number
p = Total vapor pressure above sheet, psia
p_{wa} = Partial press. water vapor in air, psia
P_{wp} = Partial press. water vapor @ effective paper temp., psia
Q_r = Radiant HTC, Btu/(h) (ft^2) (°F)
q = Nip load, pli
R = Gas constant
r = Shell radius, in.
r_i = Shell inside radius, in.
S_c = Schmidt Number
T = Temperature, °F
T_a = Temperature of air adjacent the paper, °F
T_{sat} = Saturated steam temperature, °F
T_o = Temperature of paper surface facing air, °F

t = Shell thickness, in.
t_p = Paper thickness, in.
V = Surface speed, ft/min
V_r = Rimming speed, ft/min
W_e = Evaporation rate per unit surface area, lb/(h) (ft^2)
x = Distance from centerline, in.
z = a/L_B
α = Coeff. of thermal expansion, in/(in) (°F)
τ = Poisson's ratio
ρ = Density, lb/ft^3

REFERENCES

1. Jacobi, E.B., "Yankee Pressure Roll Operation", 1976 TAPPI Pressing and Drying Seminar, p. 2, 5, Fig. 3.
2. "Yankee Dryer Basics", Scott Paper Co., Fig. 6 (1984).
3. Concannon, M.D., "Condensate Effects on Torque and Horsepower in Paper Dryers", Tappi 63(9):69-72 (1980).
4. White, R.E., "Residual Condensate, Condensate Behavior, and Siphoning in Paper Dryers", Tappi 39(4):228-233 (1956).
5. ASME Code, UG-125 (c), 1983 Edition.
6. Fishwick, H.P., "Paper Machine Steam and Condensate Systems", TAPPI PRESS, Atlanta, GA, p. 7-16 (1977).
7. Perrault, R.D., Hardin, L. and Fleming, C., "Flow Control of Dryer Drainage Systems", Tappi (62(11):63-67 (1979).
8. Wilhelm, F. and Kinney, J., "Improving Dryer Bearing Performance for Maximum Productivity", PIMA, 49-53 (June 1985).
9. Bergling, G., "Increased Life of Rolling Bearings for Steam-Heated Papermill Cylinders by Proper Insulation and Lubrication", SKF Publ., p. 6 (May 3, 1987).
10. Good, R.W., "Increased Machine Speeds Could Change Bearing Lubrication Needs", Pulp & Paper, p. 83 (Dec. 1981).
11. Timoshenko, S.P. and Goodier, J.N., Theory of Elasticity, 3rd Ed., McGraw-Hill, New York, p. 71 (1970).
12. Holik, H., Schuwerk, W. and Stumpp, W., "The Effect of Non-Conformities in Material and Geometry of Yankee Dryer Shells on Fatigue Behavior", TAPPI Engineering Conf. Proc., Book 2, p. 428 (Sept. 1985).
13. Roark, R.J. and Young, W.C., Formulas for Stress and Strain, 5th Ed., McGraw-Hill, New York, p. 496 (1975).
14. Hoff, N.J., Kempner, J. and Pohle, F.V., "Line Load Applied Along Generators of Thin-Walled Circular Shells of Finite Length", Quart. J. Applied Mathematics 11(4):420-424 (Jan. 1954).
15. Mangelsdorf, C., "The Applicability of the Donnell Equations to the Stress Analysis of Yankee Dryer Shells", Tappi 49(3):114 (Mar. 1966).
16. Barp, B., Schuch, M. and Povoden, R., "Optimal Layout of Yankee Dryer Cylinders", Tappi 55(2):245 (Feb. 1972).
17. Roark, R.J. and Young, W.C., Formulas for Stress and Strain, 5th Ed., McGraw-Hill, New York, p. 29 (1975).
18. Corboy, W.G., Guidelines for the Safe Operation of Yankee Dryers, TAPPI PRESS, Atlanta, GA, p. 9, 23 (1986).
19. Schuch, M. and Feurstein, G., "Optimal Layout of Yankee Dryer Cylinders from the Viewpoint of Strength", Escher Wyss News, 1/2 (1980).
20. Voll, G. and Fernside, R.L., "Finite Element Correlation with Experimental Stress Analysis Results", 1985 TAPPI Engineering Conf. Proc., Book 2, p. 207-422.
21. Harvey, J.F., Theory and Design of Modern Pressure Vessels, 2nd Ed., Van Nostrand Reinhold, New York, p. 151-159 (1974).
22. Smith, E.C. et al., Safety with Paper Dryers, British Paper and Board Industry Federation, England (1989).
23. Yankee Dryer Data Sheet, 12-29, Factory Mutual Engineering Corp., Norwood, MA (1986).
24. Ferrari, F.J., "Case Histories of Yankee Dryer Failures", 1985 TAPPI Engineering Conf. Proc., Book 2, p. 385-400.
25. Design of Iron Castings, 2nd Ed., The Gray and Ductile Iron Founders' Society, Inc., Cleveland, OH, p. 198, 205, 218 (1971).
26. Herminge, L., "Factors Affecting the Capacity of a Yankee Dryer — The Thermal Conductivity and Cylinder Diameter", Svensk Papperstidn. 65(8):322 (1962).
27. Forrest, P.G., Fatigue of Metals, Pergamon Press, Oxford, England, p. 70, (1962).
28. Design of Iron Castings, 2nd Ed., The Gray and Ductile Iron Founders' Society, Inc., Cleveland, OH, p. 231, 248, 264 (1981).
29. Jordan, C., "A Case Study of an Axial Shell Edge Crack in a Yankee Dryer", 1988 TAPPI Engineering Conf. Proc., Book 2, p. 425-426.

30. Harvey, J.F., Theory and Design of Modern Pressure Vessels, 2nd Ed., Van Nostrand Reinhold, New York, p. 148 (1974).
31. Hopkins, R.B., Design Analysis of Shafts and Beams, McGraw-Hill, New York, p. 66 (1970).
32. Machinery's Handbook, 22nd Ed., Industrial Press, p. 304 (1984).
33. "Calender Roll Crowning — Old Problems, New Solutions", Chap. 2, p. 13, Lockwood Trade Journal Co., New York, NY (1964).
34. Holik, H. and Surat, M., "Origins of CD Moisture Profile Disturbances in Tissue Production and Their Possible Cure", STFI Tissue Symp. (June 1983).
35. Uggla, R., "Studies of the Structure and Build-up of Yankee Cylinder Coatings", STFI Yankee Dryer Symp., Stockholm, p. 18 (Feb. 1976).
36. Rabinowicz, E., Friction and Wear of Materials, John Wiley and Sons, Inc., New York, p. 71 (1965).
37. Gavelin, G., Paper and Paperboard Drying, Lockwood Trade Journal Co., New York, Chap. 1 (1964).
38. Dreshfield, A.C., Jr. and Han, S.T., "The Drying of Paper", Tappi 39(7):449-454 (July 1956).
39. Han, S.T. and Ulmanen, T., "Heat Transfer in Hot Surface Drying", Tappi 41(4):185 (Apr. 1958).
40. Brauns, O. and Jansson, A.L., "Heat Transmission in the Drying of Paper on a Yankee Cylinder", Svensk Papperstidn. 17:627 (Sept. 1957).
41. Appel, D.W. and Hong, S.H., "Condensate Distribution and Its Effects on Heat Transfer in Steam Heated Dryers", 2nd Ann. Symp. on Water Removal, p. T61-T62 (Oct. 1986).
42. Race, E., "Mechanism of Paper Drying on Steam-Heated Cylinders", Paper Trade J., p. 37 (Mar. 1, 1971).
43. Spraker, W.A., "Drying Performance of the Combination Yankee-Dryer Hood", Paper Trade J., p. 48 (Feb. 23, 1970).
44. Appel, D.W. and Hong, S.H., "Optimizing Heat Transfer Using Bars in Dryers", Brit. Paper and Board Ind. Federation, Intl. Water Symp., 109th Conf., London (Mar. 20, 1975).
45. Chance, L. et al., U.S. Patent, 4,538,360 (Sept. 3, 1985).
46. Cox, J.F., Jr., "Optimal Rectangular Rib Dimensions for Heat Transfer in Ribbed Yankee Dryer Shells", Tappi 50(7):371 (July 1967).
47. Spraker, W.A., "Drying Performance of the Combination Yankee-Dryer Hood", Paper Trade J., p. 47 (Feb. 23, 1970).
48. Silver, R.S. and Wallis, G.B., "A Simple Theory for Longitudinal Pressure Drop in the Presence of Lateral Condensation", Proc. Inst. Mech. Eng., 180(I):36-40 (1965/1966).
49. Whittaker, J.D., "Trends in Modern High Velocity Hood Applications", Paper 183(3):132 (Feb. 10, 1975).
50. Hollmark, H., "Replicas of Yankee Dryer Surfaces", STFI, Stockholm, p. 9 (Mar. 1983).
51. Oliver, J.F., "Dry-Creping of Tissue Paper — A Review of Basic Factors", Tappi 63(12):91-95 (Dec. 1980).
52. Fuxelius, F.K., "Adhesion of Paper Web to Glazing Cylinders During Dry Creping", Svensk Papperstidn. 70(5):6, 8 (1967).
53. Sloan, J.H., "Yankee Coatings and Release Agents", Tissue Issues, p. 3, Issue 11, Niagara Lockport Industries, New York.
54. Hollmark, H., "The Adhesion Mechanism", STFI Yankee Dryer Symp., Stockholm, Sweden, p. 4-15 (Feb. 1976).
55. Falk, M., Valmet Tissue Making Seminar Papers, Karlstad, Sweden, p. 39-50 (Oct. 4-5, 1989).

XV
WINDING

K.G. FRYE

Keneth G. Frye is Vice President of Finishing Technology for Beloit Research and Development Group, Beloit Lenox Division in Lenox, MA. He has 40 years experience in the paper finishing area of the paper industry as a winding, sheeting, and printing specialist. An active TAPPI member since 1960, Ken is the First Chairman of theTAPPI Finishing Division, member of the Winding Committee TAPPI Finishing Division, and member of the Sheeting Committee TAPPI Finishing Division. In 1973, he received the TAPPI Finishing Division Award; in 1984, TAPPI's Certificate of Appreciation; and in 1986, the TAPPI Finishing and Converting Division Award. He has published over 20 works.

REVIEWED BY

Donald A. Cumfer
Executive Vice President
Hamilton-Hybar
Richmond, VA

Gerry Forsberg
President
G. Forsberg and Company Inc.
Tigard, OR

A. WINDING DEFINITION

The winding process and all of its ramifications may seem quite simple; but in fact, depending on how deeply one gets into (or is involved in) its technologies, the subject can become quite complicated.

There are many types of winders and winding processes; but the common thread through all of this equipment and winding process is: basic technology – and, the problems of winding. In order to best cover this important subject, it was felt that the only way to do it properly was to divide the subject into categories which, I feel, will describe all aspects of winding and the reason why there are differences in the winding processes, winding equipment and winding problems.

For example: winding theory is a separate category – primarily because, when discussed in terms of pressroom and converting runnability, the basic technology of the pressroom and the mill are pretty much the same, and the problems found in the pressroom and converter really help define winding theory.

Equipment Definition. Any equipment that processes a web of paper into a roll is defined as a winder. The process and the problems encountered further define the type of winder that is used.

Winding Processes. Different types of winding processes use the same basic winding equipment. Variations of the winding equipment are dictated by the winding process; and, again, the problems encountered in that process.

Structural Control Processes. This subject is considered separately – primarily because the only difference in most winding equipment is the method used to build or structure a roll of paper.

Web Control Function. A basic, fundamental function of any process – including winding – that covers control of the web prior to becoming a roll.

Slitting Process. A basic fundamental of winding that covers the separation of the web into different widths – in all winding processes that incorporate slitting.

Unwinding. A basic component that is found in processes that deal with roll-to-roll versus web-to-roll winding.

Productivity and Manpower. This category covers the differences in all of the winding equipment in terms of effect and potential cost of operation.

Definitions of Terms Used. In any dissertation of this kind, terminology is the main deterrent to understanding the subject; therefore, definitions of winding processes and equipment – as well as

Fig. 383. Paper machine winder (reel).

a Glossary of Terms – are given in the following sections of this book.

The first step in understanding this subject is to define it.

1. Winding definition

The winding function is a process which changes a material manufactured in continuous process from web form into roll form for further processing. The first winding process occurs at the paper machine, extrusion coater or non-woven reels (Figs. 383, 384) where the web is wound continuously on mandrels (usually without slitting) for further processing. Subsequent winding processes operate roll-to-roll (unwinding roll to a windup roll – Fig. 385) rather than web-to-roll (web from paper machine to windup reel).

In categorizing the winding process, one finds that the basic fundamentals and technologies of winding, such as roll structure (how tightly the roll is wound) and web control (how to keep the web from wandering in an unwanted direction), are common threads that are found in one form or another in most of the processes of winding. We call these fundamentals the technologies of winding theory.

The basic fundamentals of slitting, and unwinding, are found in most, but not all, of the winding processes; however, this depends more on the process function and in no way eliminates slitting and unwinding as basic functions in the more major categories of winding theory.

The fact is that problems found and documented in the pressroom of a commercial printer and at the converter can now be directly related to similar problems found at the paper machine reel and other in-house winding processes, such as the machine winder, coater (adding additional coating material to the base web or sheet), and the supercalender (adding a glossy finish to a web).

The following sections will address the theory of the winding process, and then discuss some of the fundamentals, equipment, and common problems found in the winding process – all of which will help us to understand and define the complex process of winding.

B. WINDING THEORY

When discussing the process of winding and the proper methods and equipment necessary to wind rolls of paper, one must first discuss the ultimate needs of in-house processes as well as

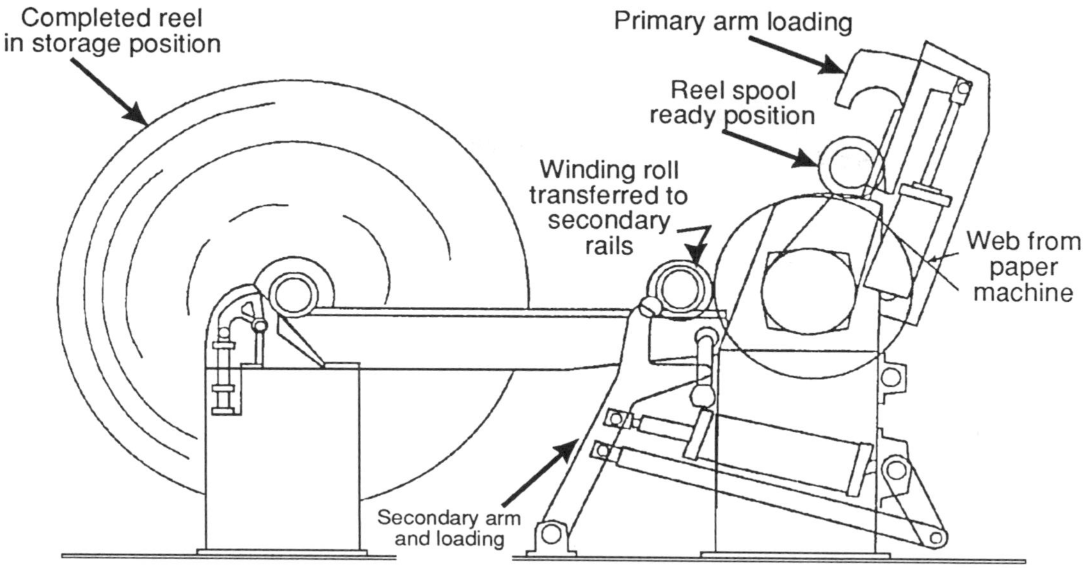

Fig. 384. Web to reel winder (paper machine reel).

Fig. 385. Roll to roll paper machine winder.

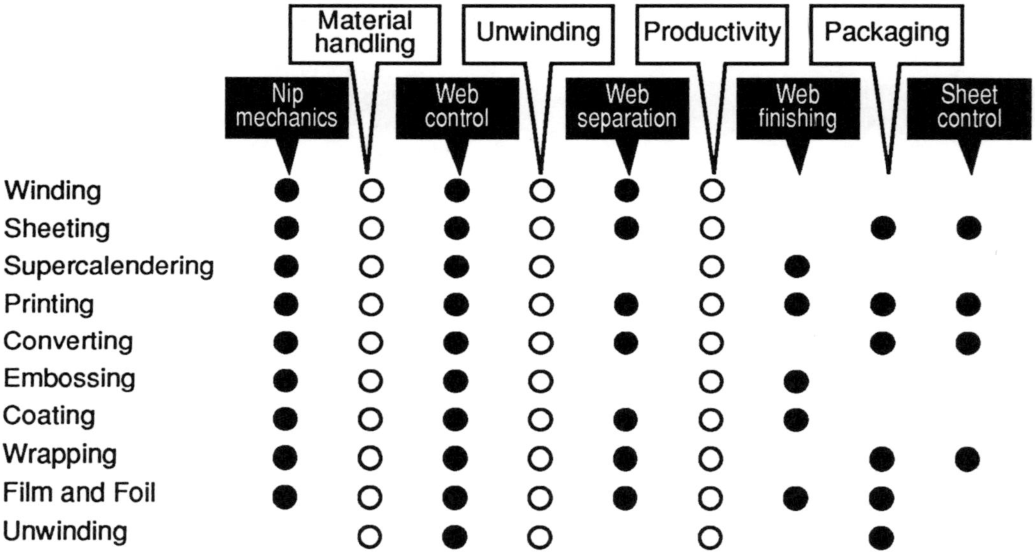

Fig. 386. The basic technologies of finishing.

the final customer in terms of roll quality and press and converting runnability. Consequently, it becomes apparent that those needs for a good roll should be defined in terms of variables over which there is some measure of control.

An important step in this regard was taken by Pfeiffer [12] who put forth a definition, a measure, and the effect of wound-in tension on the structure of a roll. It is now possible to associate improper roll structure with certain defects found in a roll of paper; and is therefore necessary to understand all of the winding variables that affect that roll structure which, in turn, affects the quality of the roll being further processed in the mill or shipped to the customer. Putting it quite simply, roll structure is defined as how tightly and uniformly a roll is wound. Wound-in tension is one measure of that tightness and uniformity.

In an attempt to bring winding variables and shipping roll quality together in proper perspective, the variables affecting shipping roll quality have been divided into four basic categories: the basic fundamental functions of winding which, in turn, are part of the basic technologies of finishing (Fig. 386).

1. Basic technologies

The basic technologies (Fig. 386) are: nip mechanics, which play a major roll in roll structure; web control, which addresses tension control, traction control and spreading; web separation, which covers all processes of slitting and cutting; web finishing, which alters the surface of the sheet and is involved in supercalendering, embossing, coating and printing; sheet control, which is a pivotal part of the sheeting process; unwinding, because every roll that is used in any other process must eventually be unwound; packaging, which is involved in the protective cover for any product; and, finally, production control, which is a fundamental and necessary function in any process. These are the basic technologies of finishing that transcend many of the operations that process web materials such as paper.

(a) Unwind control

The first function of roll-to-roll winding, naturally, is the unwinding process (Fig. 387). This process provides a means of support and braking (tensioning of the web) as well as acceleration and deceleration (increasing and decreasing speed of web) for the "unwinding" roll. Sometimes edge guiding, oscillation and special roll and other material handling equipment are included in this function for the roll being unwound. Edge guiding (Fig. 388) keeps the web edge in a straight line, even though the unwinding roll edge might not have a straight edge. Oscillation (Fig. 389) moves the web in a sideways motion while running to even out cross-machine variation in web thickness during the windup process. This helps avoid corrugations and bursts found in subsequent processes.

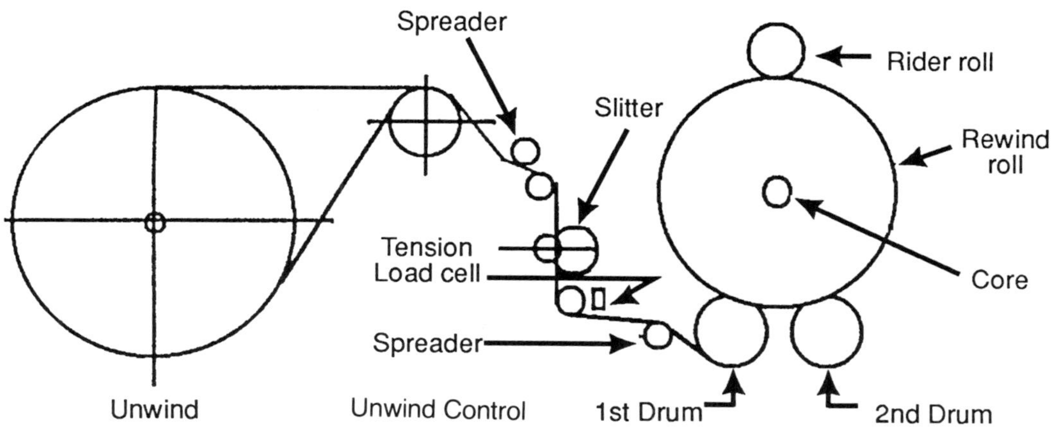

Function	Defects
1. Braking tension a. Regenerative b. Friction c. Emergency stop 2. Acceleration 3. Deceleration 4. Roll support 5. Payout 6. Inertia compensation 7. Edge guiding 8. Oscillation	1. Oscillating offsets 2. Structure inconsistencies 3. Tension offsets 4. Lopping rolls 5. Sheet flutter 6. Edge cracks 7. Wrinkles

Fig. 387. Unwind control.

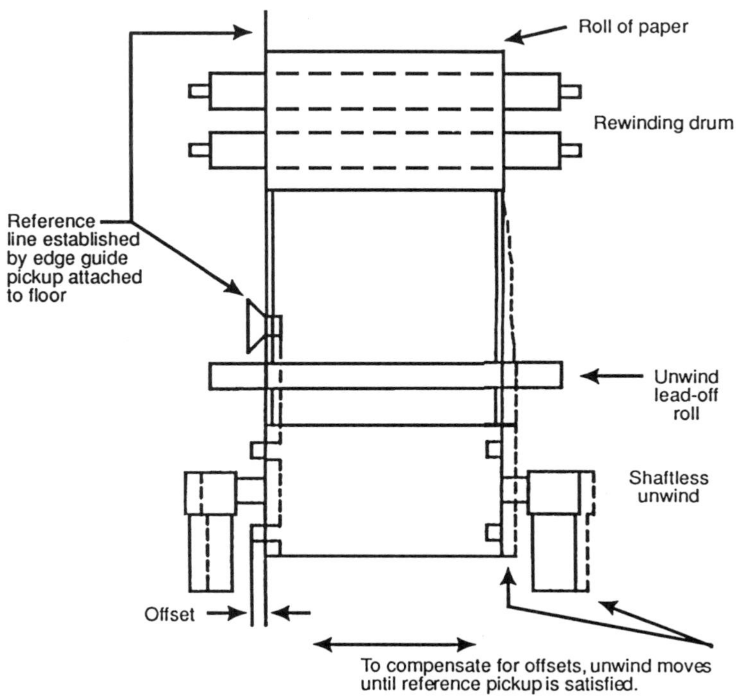

Fig. 388. Edge guiding.

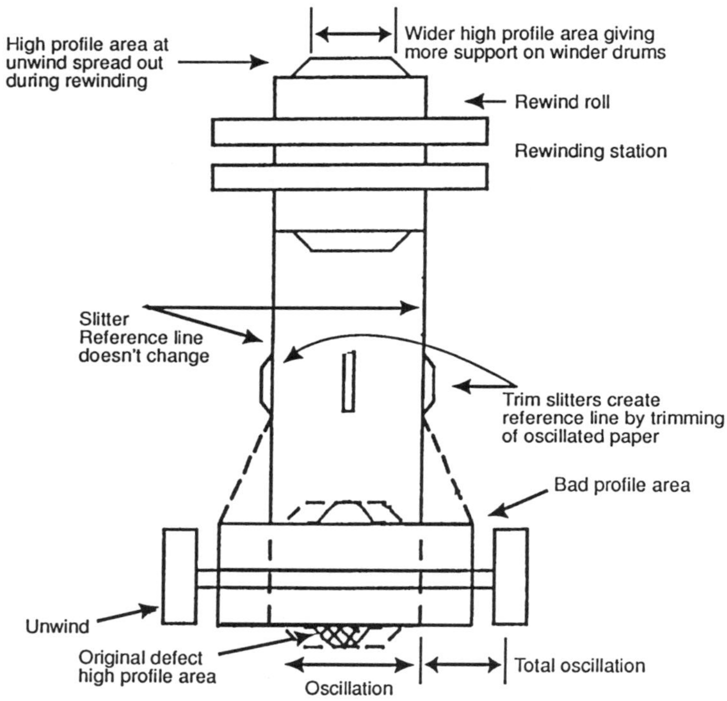

Fig. 389. Oscillation.

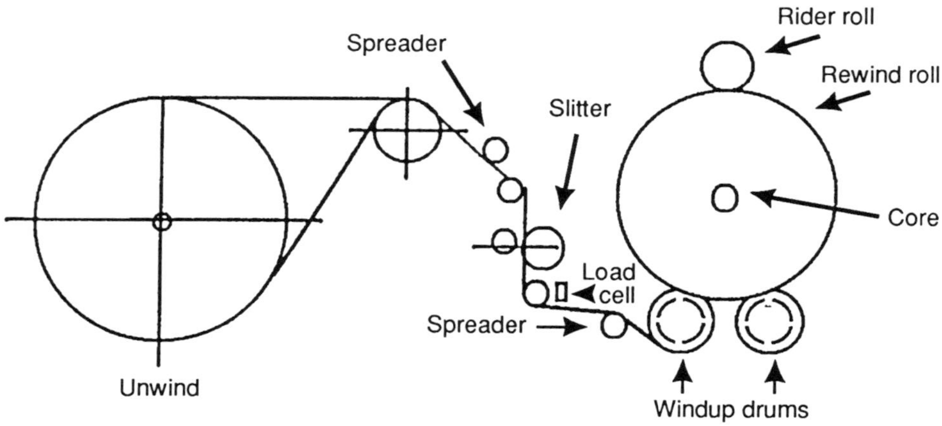

Web Control

Function	Defects	
1. Tension control	1. Acceleration offset	8. Slit offset
2. Traction control	2. Deceleration offset	9. Interweaving
3. Spreading	3. Splice offset	10. Dishing
4. Speed	4. Web ridging	11. Snap - offs
	5. Web fold - over	12. Shinner offsets
	6. Web abrasion	13. Edge cracks
	7. Wrinkles	

Fig. 390. Web control.

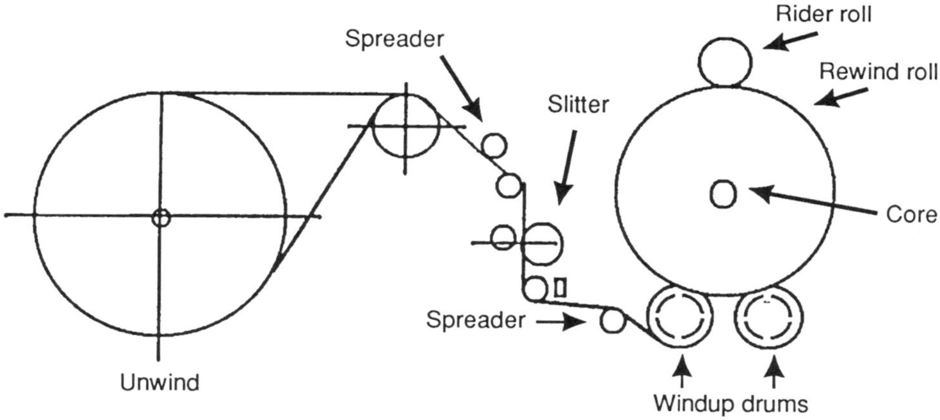

Slitting

Function	Defects	
1. Web separation methods	1. Slitter dust	7. Slitter marking
2. Trim removal	2. Slitter rings	8. Splice offsets
	3. Ragged slit	9. Snapoffs
	4. Trim removal (web break)	10. Shinner
		11. Scalloped edge
	5. Edge cracks	12. Wrinkles
	6. Edge stretch	13. Turnover

Fig. 391. Slitting.

(b) Web control

The second basic function of winding is web control (Fig. 390). This includes tension control, traction control, spreading, and speed. Traction control maintains a constant speed between the web and its supporting members. Spreading keeps the web in a flat, unwrinkled condition, as well as keeping slit webs from overlapping one another in the process of winding. Hence, web control provides the necessary control of transporting the web or webs from the unwinding roll to the roll being wound on the winder.

(c) Web separation

The third basic function is web separation (slitting) (Fig. 391). This technology covers slitting, which is the cutting of the web into specified usable widths, and trim removal, which is the removal of the paper edges left over when the web is slit to specified widths in the winding process.

(d) Roll structure control

The last function of winding, roll structure control (Fig. 392), is, possibly, the most important function of winding. The many variables that affect structure control will be discussed later but, basically, these can be broken down into three categories which we call the TNT of winding. These variables are tension, nip, and torque. The other 18 variables (Fig. 393) all affect these three in some way which, in turn, determine the structure of the roll.

One might ask: Why is roll structure so important to the final customer? To put the needs of the industry in proper perspective one must say that runnability (running paper through any process without problems), printability (the maintenance of print quality), and convertibility (roll and web converted to another form or size) are the categories that most concern the mill and the customer; the problems of runnability many times point to winding equipment in the paper mill and converting processes. Corrugations (Fig. 394), bursts (Fig. 395), bags (Fig. 396), and snapoffs are common defects and problems usually associated with roll structure and winding equipment. (The definitions and runnability in the pressroom will be discussed later.)

2. Roll structure

Before going further, it might be important to first determine what the "structure of a roll" is, in terms of wound-in tension and internal pressures. Figure 397 illustrates the internal pressures and the residual tension found in a roll wound on a nip-type winder. Roll weight and

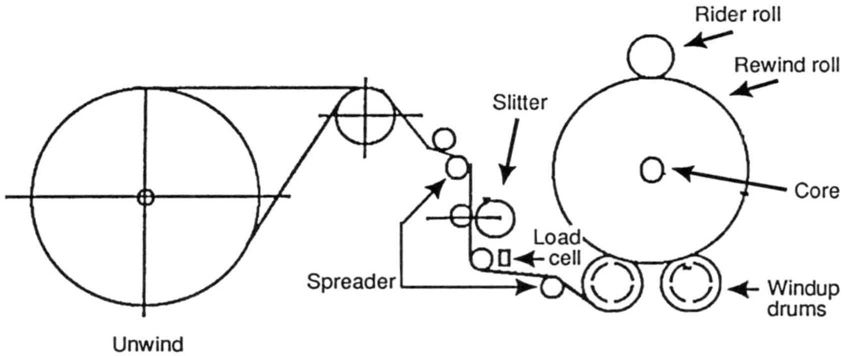

Structural Control

Function	Defects: Nonuniform nip	Defects: Structure
1. Drum diameter	1. Interweaving	1. Full tension burst
2. Rider roll	2. Wrinkles	2. Partial tension burst
3. Core shaft control	3. Dishing	3. Bag
4. Rewound roll diameter	4. Offsets	4. Caliper shear burst
5. Drive	5. Ropes	5. Air shear burst
6. Differential torque	6. Snapoffs	6. Crepe wrinkle
7. Tension	7. Corrugations	7. Loose core
8. Paper	8. Core burst	8. Edge burst
9. Drum coating		9. Core burst
10. Drum grooving		10. Turnovers

Fig. 392. Structure control.

Variable	Tension	Nip	Torque
1. Tension	X		
2. Torque			X
3. Nip		X	
4. Drum diameter		X	
5. Drum arrangement		X	
6. Drum surface			X
7. Paper profile		X	
8. Roll diameter		X	X
9. Core shaft slides		X	
10. Paper grade	X	X	X
11. Fiber roll		X	X
12. Winder drive	X		X
13. Speed	X	X	
14. Core diameter		X	X
15. Spreader	X		X
16. Drum wrap	X	X	X
17. Core type		X	X
18. Vibration	X	X	X
19. Number of slits	X	X	X
20. Rider roll drive		X	
21. Roll width	X	X	X

Fig. 393. Winding variables and their effect on TNT.

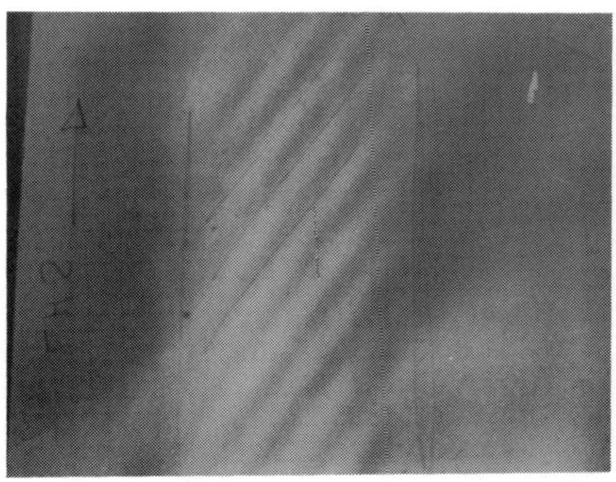

Fig. 394. Corrugation.

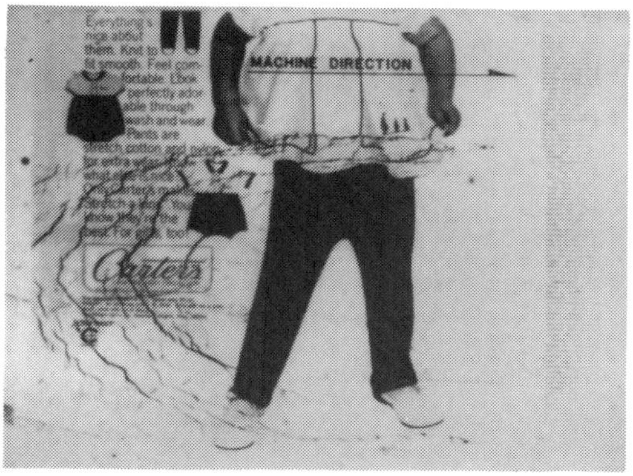

Fig. 395. Burst shear caliper.

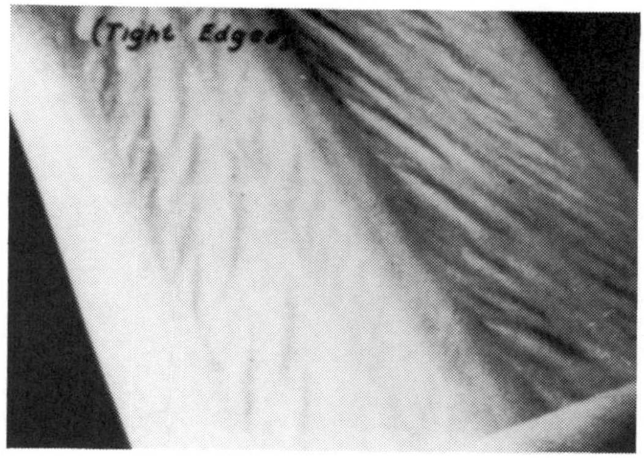

Fig. 396. Bags and corrugations.

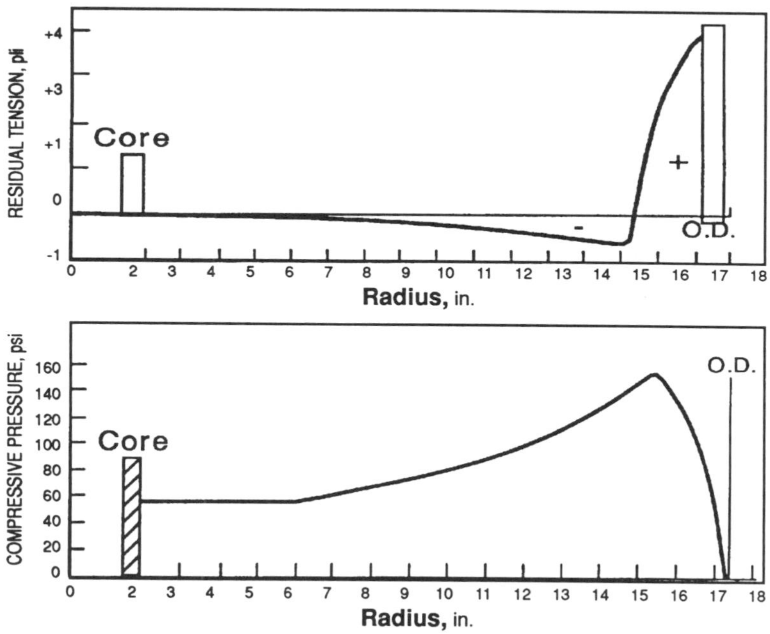

Fig. 397. Internal pressures and residual tension in a roll wound on a nip-type winder.

other mechanical equipment provide nip pressure between the winding roll and some reference surface such as a winder drum.

Wound-in tension and internal pressure can best be visualized by stretching a cut rubber band several times around your wrist. One will note immediately that the rubber band seems to get tighter and tighter on successive wraps, and eventually cuts off the circulation to your hand. This pressure between your wrist and the rubber band is the same as what we call "internal pressure" in a roll of paper.

To best demonstrate wound-in tension: with the rubber band tight on your wrist, now place your wrist tightly against a table, trapping the rubber band between your wrist and the table, and then cut the rubber band. Now, take the two ends of the cut rubber band, without stretching them, and place them again around your wrist. One will immediately find that the two ends will no longer meet, mainly because they were stretched when they were first added to your wrist and then relaxed to a shorter non-stretched length when cut. The residual tension of the rubber band – as it was originally wound around your wrist – is then determined by measuring the force necessary to again stretch the rubber band so that the ends will meet. This is the wound-in tension in the rubber band.

Wound-in or residual tension in a roll of paper

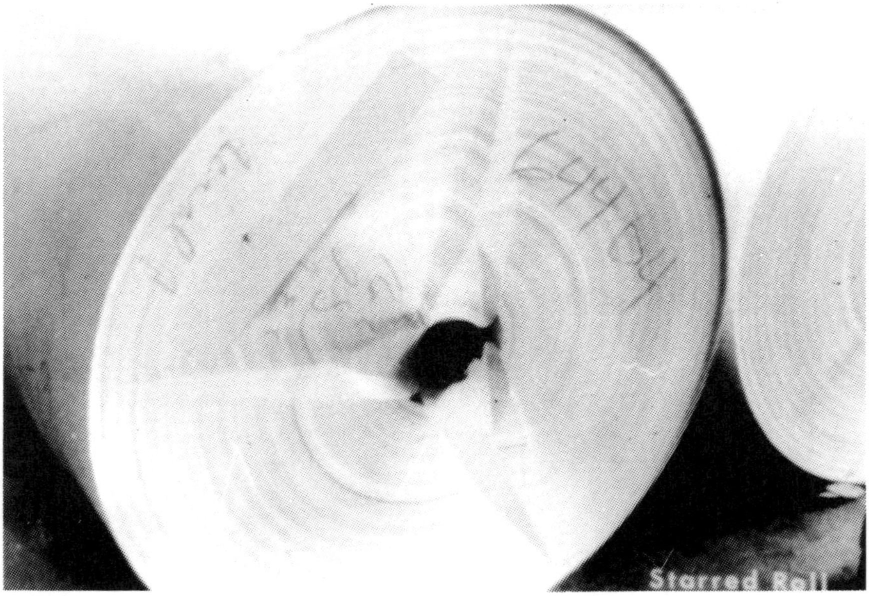

Fig. 398. Extreme starred roll.

can be visualized and demonstrated in the same way. By using a sharp knife to cut the outer layer of a roll sitting on the floor (as was the case with the rubber band), one will find as the knife progresses across the roll that the paper will break before the knife reaches the other edge of the roll. An attempt to tape the web back together would find that there would be a gap between the two broken edges. If this gap is measured and a strip of the same paper is stretched the same amount as the gap, the tension required to stretch the paper to fill the gap is what we refer to as "residual" or "wound-in" tension. The Gap Test was first developed by Cameron Machine Co. and is still recognized as a means for determining residual tension.

Pfeiffer [1] first determined the internal pressure in the roll (Fig. 397) and, from this, the resultant residual wound-in tension. (Note that the only area of positive tension is found mostly in the outer layers of the roll.) The tension band can be anywhere from 1 to 6 inches deep, depending on the type of paper and the equipment on which the paper is run; whereas the area immediately below this band of tensioned paper has decreased tension and much of the paper in the roll has no tension at all, but is in compression.

The familiar starred roll (Figs. 398, 399) illustrates this quite graphically. A star is caused by an external pressure or blow great enough to cause those areas with lower or no residual

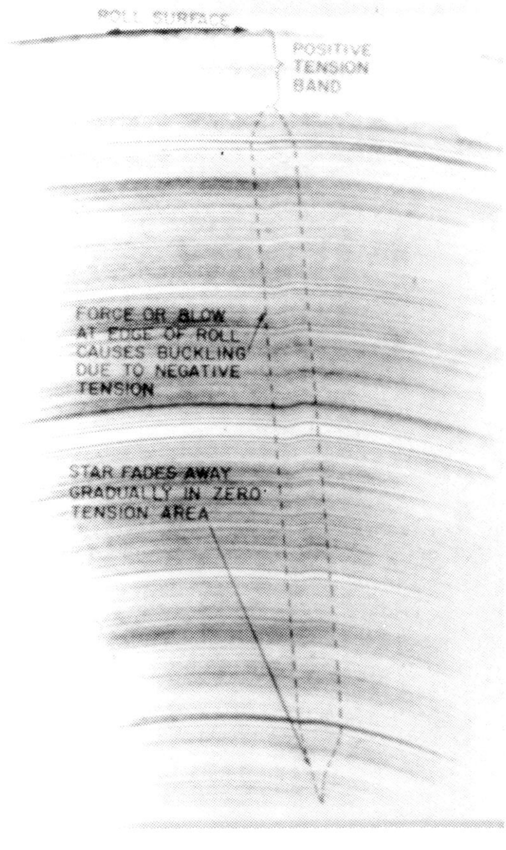

Fig. 399. Normal starred roll caused by external pressure or a blow to outside roll.

tension to buckle. Those areas with higher residual tension near the outside edge of the roll have a certain ability to bounce back from a buckled condition.

When the wound-in tension becomes excessive, however, the positive tension area (Fig. 397), caused mostly by differential and high nip pressures, is the area where many of the roll structure defects occur (bags, bursts and corrugations).

3. Industry-wide problems

To determine the needs and what is best in the development of a quality roll of paper, let us first look to the basic causes of some of the industry-wide problems. Understanding these problems will give us insight into the winding process and better explain why different processes and different pieces of equipment are used.

In a study made in conjunction with Midwest Printer, over 18,000 rolls of paper were monitored through the pressroom. In a related study, over 450 defects in a roll of paper had been identified and their causes determined. Many of the defects were found to be the same as the ones in the pressroom sample. In reviewing the data from this sample, it was determined that 87% of the problems found, originated, or were amplified, in the paper mill winding process. In defining and isolating the defects further, a high percentage of those defects were found to be roll structure and nip-induced type defects and problems.

4. Winding variables

In the winding process, there are many factors that affect roll structure in a roll of paper. Tension, nip, and torque (TNT) (Fig. 400) are the main basic categories; however, Fig. 393 shows other important variables, most of which are directly related to the TNT of winding. Each one of these important variables affects, by increasing or changing in some form or another, one of the three basic variables of TNT. Depending on paper grade, some of these variables are more critical than others – the most significant of the three, in most cases, is the nip.

Figure 393 lists some of the winding variables and how they affect the main variables of tension-nip-torque.

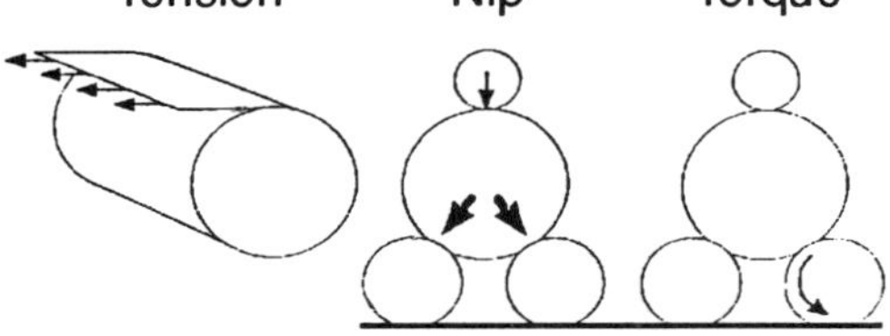

Fig. 400. TNT of winding: tension, nip, and torque.

5. Nip-induced defects

The overwhelming cause of defects in a roll in terms of pressroom or converting problems as related to winding problems is not necessarily nip per se, but non-uniform or differential nip. This is substantiated by the many defects in a roll of paper that have been found to be directly related to nip pressure or, more accurately stated, non-uniform nip pressure. The partial list below shows those defects classified as non-uniform nip-type defects. (For additional definitions and photos of winding functions that cause roll and paper defects, see Ref. [6].)

1. Tension burst
2. Bags
3. Caliper shear burst
4. Crepe wrinkles
5. Loose core
6. Air shear burst
7. Interweaving
8. Wrinkles
9. Dishing – telescoping
10. Offsets
11. Corrugations – ropes
12. Snapoffs
13. Core bursts
14. Edge cracks
15. Turnovers.

By definition, nip requires pressure in pounds per lineal inch of width (pli) or pounds per square inch (psi) between two in-contact components. Nip pressure can be specified in many ways; the most common in the United States is pli. A revealing study of the characteristics and effect of those components that create the nip against the roll of paper is described as follows.

6. Nip and drum-diameter relationship

Pfeiffer has proved that wound-in tension in a rewound roll of a two-drum horizontal winder (Fig. 385) was significantly dependent on two things: nip (pressure) and drum diameter [1]. Through his Instant-Center Theory (Figs. 401–404) he showed that, for a constant drum diameter on a specific grade of paper (Fig. 405), when one doubled the nip pressure, the wound-in tension in the rewound roll increased 1.59 times. Under the same conditions, if the nip pressure was tripled, the wound-in tension in the rewound roll increased to 2.08 times the base

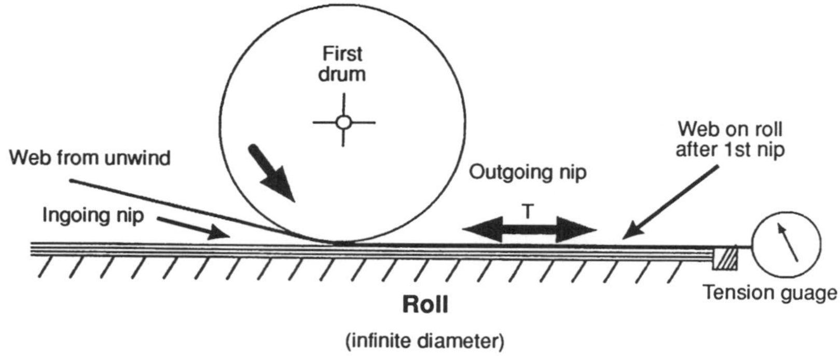

Fig. 401. Rolling nip with a fixed infinite-diameter of paper roll.

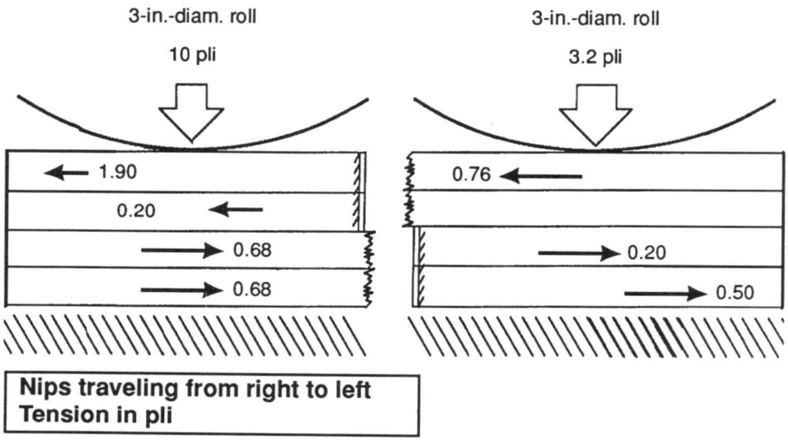

Fig. 402. Nip mechanics.

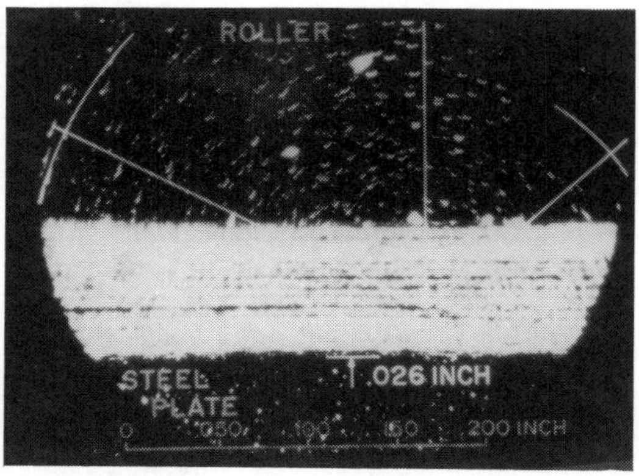

Fig. 403. Nip mechanics — instant center tests by Pfeiffer.

Fig. 404. Nip mechanics — instant center tests by Pfeiffer.

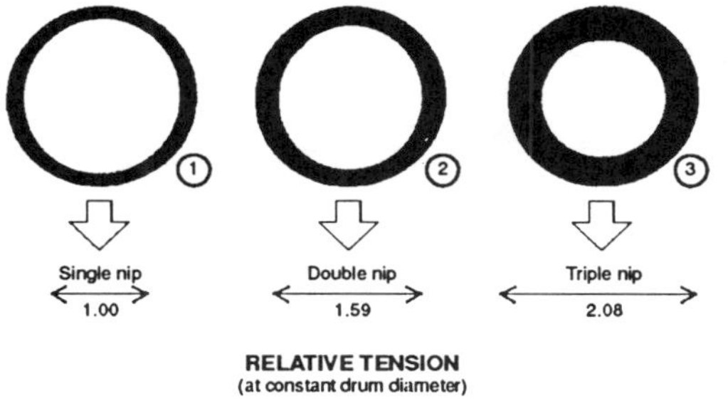

Fig. 405. Effect of nip on tension (WIT) development.

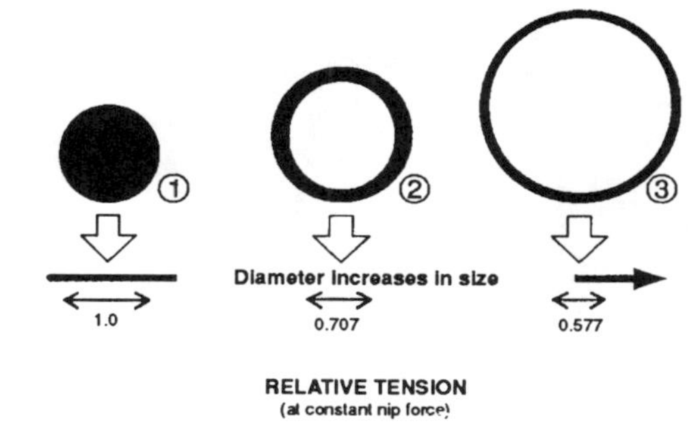

Fig. 406. The effect of diameter on tension (WIT) development.

value of 1 pli (175 N/m). Conversely, when the size of the drum is doubled (Fig. 406) and the nip pressure is kept constant, the resultant wound-in tension in the rewound roll would decrease from the baseline of 1 pli to 0.707 pli (175–124 N/m); and tripling the diameter would decrease the pli further – to 0.577 (101 N/m).

What this means is that, with the non-uniform paper profile, the high or hard areas in a roll of paper (because of nip) will produce more tension in that area in relation to an adjacent softer area. The differential tension between these two areas, if extreme (Fig. 407), will manifest itself in ropes or corrugations – a series of diagonal wrinkles usually in a rope-like pattern found in the machine direction (Fig. 394) and bags (Fig. 396) where the paper has been stretched beyond its elastic limit and does not recover its original form. A larger diameter drum will produce, therefore, for a given nip pressure, much less differential tension and consequently fewer corrugations and defects such as bags.

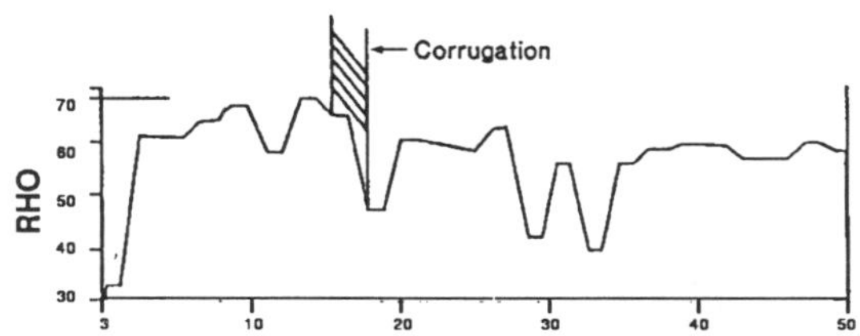

Fig. 407. Rho meter profile shows what a drum nip will see and help predict corrugations.

The "J-line" (Figs. 408, 409), a measurement of roll tightness used by many paper mills, graphically illustrates the motions that occur in the nip. As can be seen, the motion is in two directions, as described by Pfeiffer [1], and the degree of this motion more closely describes many of the problems occurring in a roll of

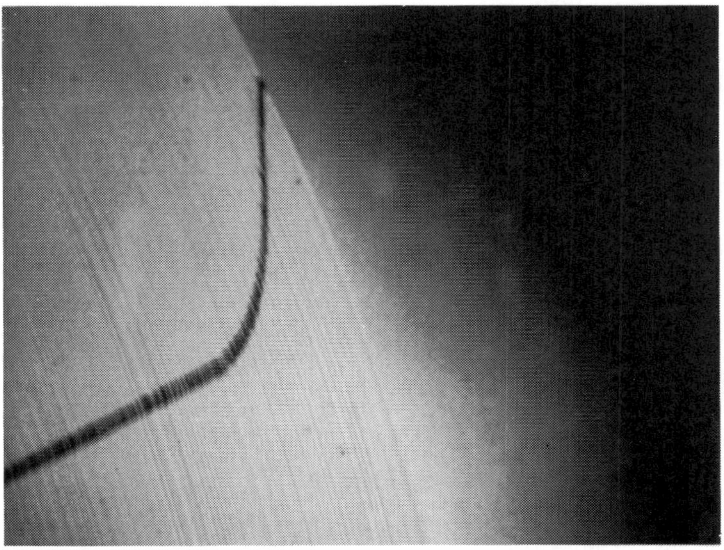

Fig. 408. The "J-line" graphically illustrates the motions that occur in the nip.

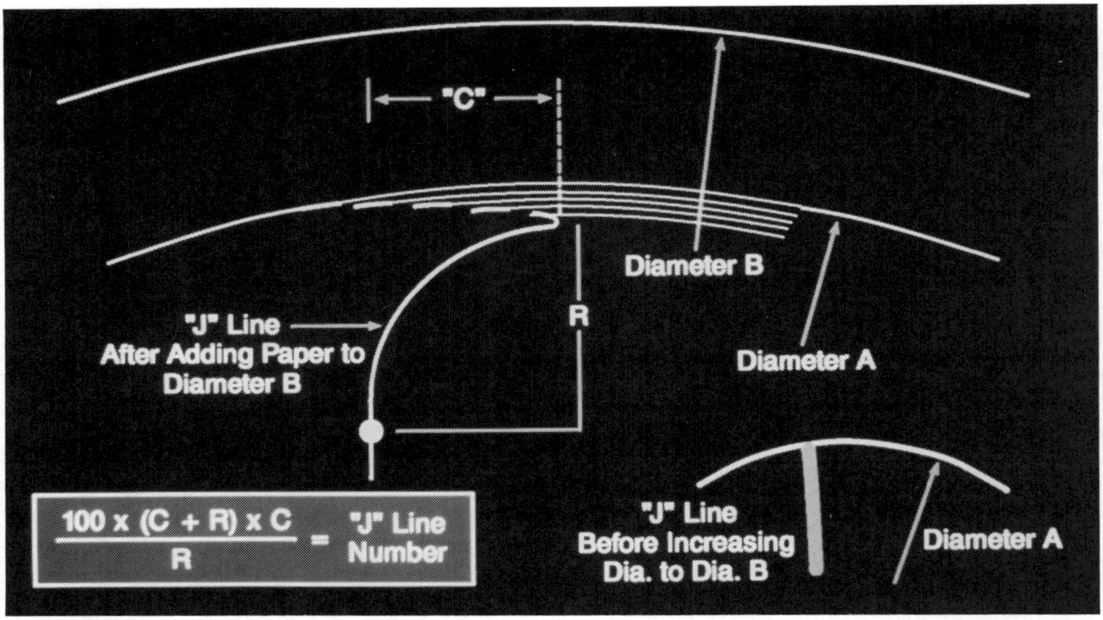

Fig. 409. The "J-line 'C' Dimension" is a measurement of roll tightness used by many paper mills.

paper. A more detailed description of the use of the "J-line test" is found in the Glossary (see Sec. K).

7. Limits of nip-induced tension

If one accepts the preceding, one might ask: Why not go to a very large drum (say, 100 inches) to decrease the wound-in tension even further? To investigate this possibility we decided to find the limits of wound-in tension by paper grade – not only on the different drum diameters, but on all the other variables that affect roll structure. The most recent studies along these lines were done on the effect of multiple nips on wound-in tension by Frye [5].

8. Roll quality measurement devices

Just a word concerning roll quality measurement tests. Many of these test methods, having been researched individually, all indicate some measurement of roll quality. Measurement devices reviewed are the Rho Meter, the Rho Hammer, and the Smith Needle. The Schmidt Hammer, the Tab Pull Test, the Cameron Gap Test, the Bench Nip Test, the Nip Cycle Test, the "J-line", and the WIT WOT Machine are all tests that are presently used in the industry to determine some measure of roll quality.

Unfortunately, many of these tests are misused and misinterpreted. However, those who continually test rolls for the industry and use all of these devices on a fairly consistent basis

Fig. 410. Strain gauge bench test analyzes the effect of multiple nips on WIT.

(mainly to relate to the existing test methods used by any given mill) find that the proper use and the relative value of each test is quite easily determined, depending on what characteristic of roll quality one wants to determine.

To help us evaluate some of these methods, we ran a single roll on a laboratory winder and performed as many of these tests as possible on this single roll so the relative values and resulting curves could be compared and evaluated. The tests used were:

1. A bench test, not unlike the tests made by Pfeiffer [1] (Figs. 401, 402, 410).
2. The bench test results were confirmed on the wound-in tension machine (Fig. 411). This is

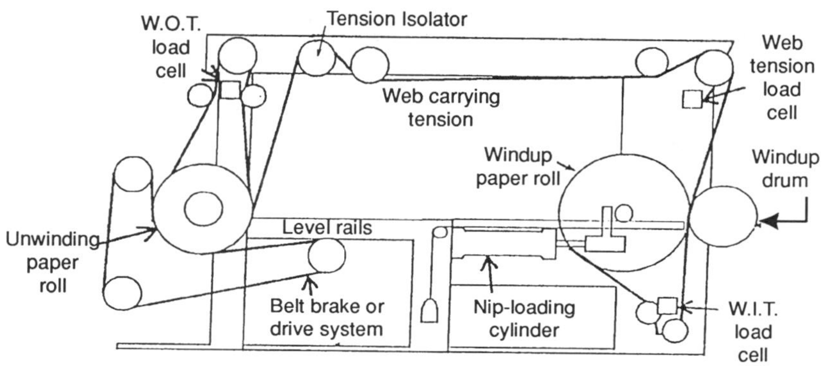

Fig. 411. Wound-in and wound-out tension-roll analysis winder.

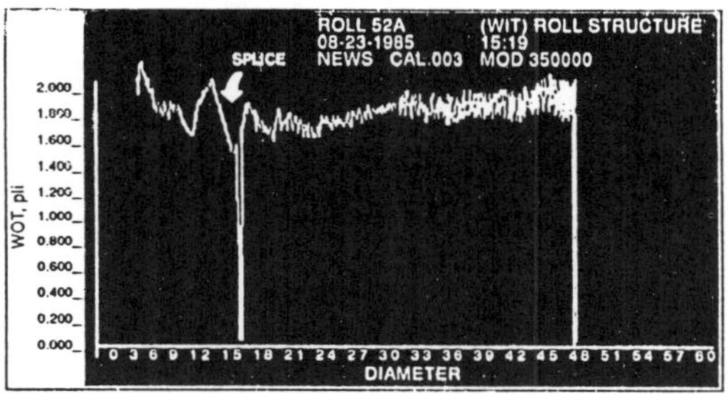

Fig. 412. The wound-in tension machine measures the wound-in tension as the roll unwinds.

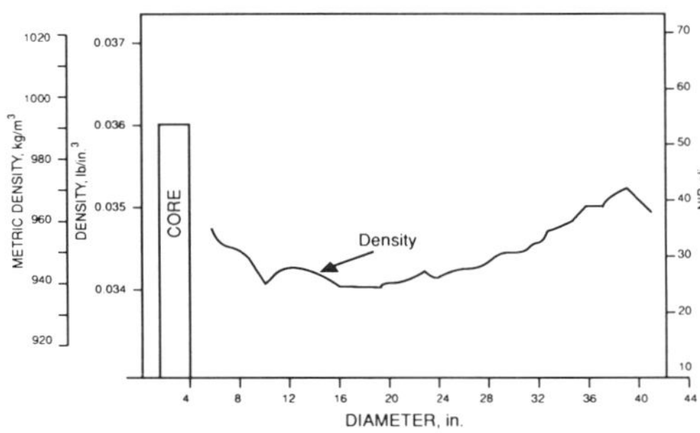

Fig. 413. Beloit roll density profile.

a laboratory test machine that dynamically measures the wound-off tension (WOT) (Fig. 412) as the roll is winding-up, or measures the wound-in tension (WIT) from a previous winding process, as the roll is unwinding. This allows one to analyze mill rolls, and the type of equipment that was used when the roll was being wound, and then to determine those winding parameters that are best for that grade of paper.

3. Running a production machine under similar circumstances (optimum running parameters for that grade) and verifying – with a

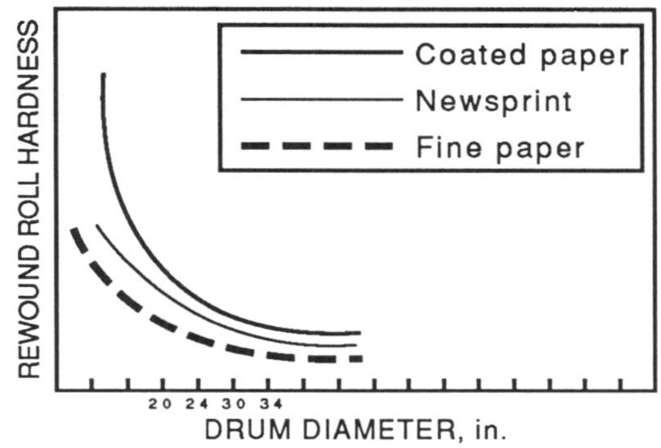

Fig. 414. Roll hardness vs. drum diameter.

properly interpreted Density Monitor (Fig. 413) or a gap test.

The first tests were of the bench test variety (Fig. 410). This test consisted of strain gauges attached to multiple webs which, when placed on a flat table, simulated an infinitely large roll of paper. Multiple drums of actual production size (30 in., 762 mm) were rolled across the web, away from the strain gauges. This test more or less repeated Pfeiffer's work (Fig. 401), with the exception that actual production drum diameters were used.

9. Drum diameter vs. paper grade

Different-size drums were used on one grade of paper to develop a curve of drum diameter vs. wound-in tension. Tests were repeated using different grades of paper, thus developing a family of curves as seen in Fig. 414. In evaluating the test data further, several conclusions become quite obvious:

1. The small-diameter drums produce significantly more wound-in tension than the large-diameter drums (Fig. 414).

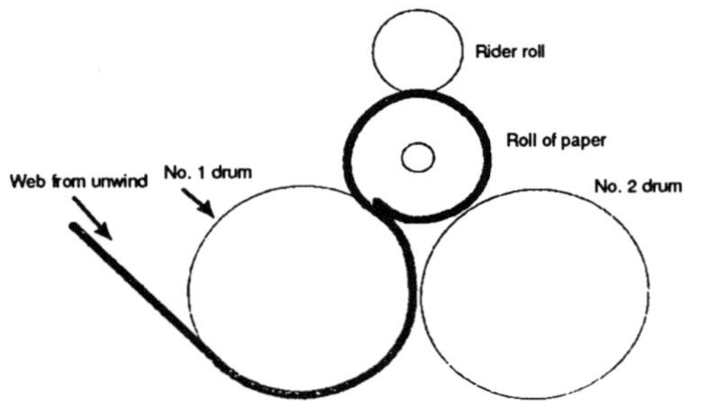

Fig. 415. Three-drum winder (two drums and a rider roll).

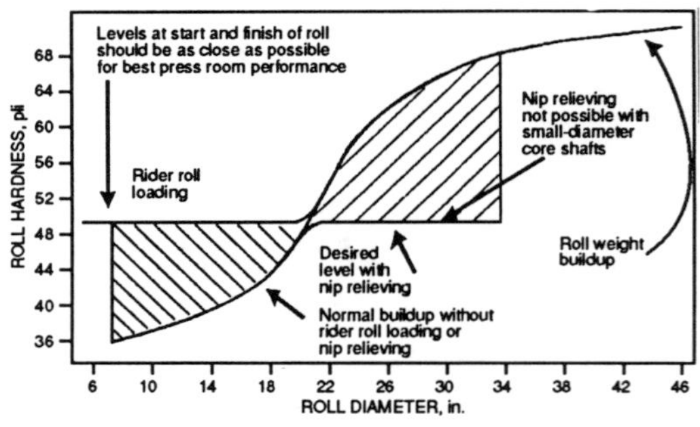

Fig. 416. Rider roll function.

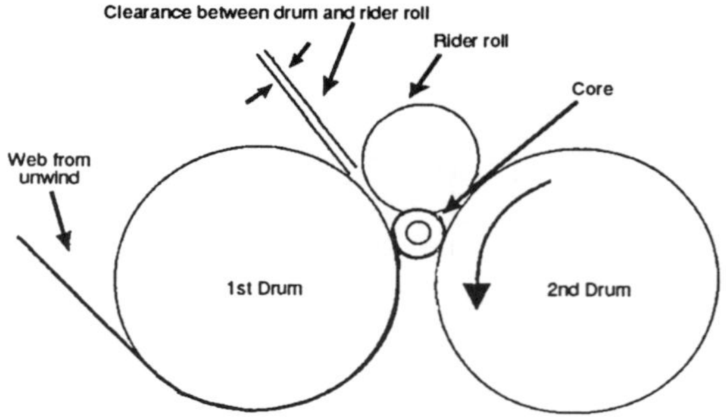

Fig. 417. Rider roll core — drum relationship.

2. There seems to be a diameter whereby any increase in drum diameter [above the 30–34 in. (762–864 mm) diameter range] produces no further improvement in the reduction of wound-in tension.
3. All grades of paper react differently to the nip pressure produced. Based on these limited results one could say that for most fine paper grades, a 30 in. (762 mm) diameter drum, for example, would eliminate the need to consider drum diameter as a significant factor in the selection or the operation of a piece of winding equipment. In fact, this points out that optimizing drum diameter at this level removes it as a point of contention in the two-drum-winder roll quality assessment.

10. Three-drum winder

Unfortunately, even with the elimination of drum diameter as a significant factor, there are still many other variables that figure significantly in roll structure and roll quality. With that statement in mind, it is quite obvious that all our tests were made with one drum, not unlike a single-drum winder. However, a normal bed-roll type of winder (usually called a two-drum winder) (Fig. 415) is really a three-drum winder. The third drum in this case is a roll called a rider roll.

The rider roll is a device on a conventional two-drum winder that provides nip in the early stages of the winding process (Fig. 416). (A more detailed definition of this component is found in Sec. F.)

Out of necessity, the rider roll is usually quite small. As mentioned previously, both drums should be 30 in. (762 mm) in diameter (Fig. 417). And, using the scenario of placing a 4 in. (102 mm) diameter core between 30 in. (762 mm) diameter drums, a small-diameter rider roll is necessary to nip the bare core at the start, without interfering with or touching the drums.

If we continue with the same reasoning we have followed above, we must ask: "What effect does this small-diameter roll have on roll structure?"

I believe the answer is obvious, as Lucas [2] pointed out that crepe wrinkles were caused by too much nip pressure by the small-diameter rider roll at the larger rewound roll diameters on a newsprint winder. The significant fact that the rider roll load is very low at the larger diameters does not minimize the concern on the effect of the small-diameter rolls, even at low nips.

The important point to remember on this subject is that all of the theory above (and most of the literature on winding) neglects the basic fact that the paper being wound on a drum winder is *always non-uniform* (Fig. 407). Because of this profile non-uniformity, paper roll nips wherever they occur (at the core or at the outside diameter), whether the nip pressure is low or high, can be damaging. The lesson to be learned from this and to minimize this possible damage (on sensitive papers) to the rewound rolls, *drums as large in diameter as possible should be used* – at least within the useful limits discussed previously.

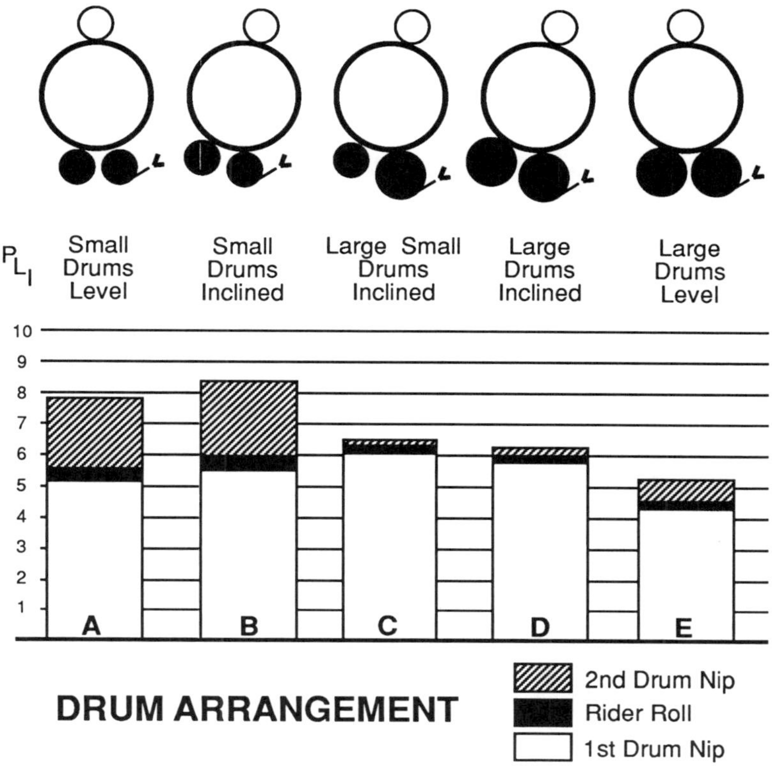

Fig. 418. Wound-in tension development by different drum arrangements.

11. Effect of multiple nips and drum size

To study the wound-in tension effect of multiple nips from different-diameter drums, the same bench test was used. This time, first one, then two, and then three nips were tested (Fig. 410), rolling the actual-size drums that would be used on a production winder across the paper attached to the strain gauges. The final runs used three nips at one time (Fig. 410), simulating the actual pli forces that would occur with a 60 in. (1524 mm) diameter rewound roll.

The results (Fig. 418) show several interesting facts which substantiate the results found by Welp and Schoenmeier [3]. Their tests show that the roll hardness created by two small-diameter drums in the level position is somewhat less than the hardness of two equal-diameter drums on an incline. Welp and Schoenmeier demonstrate that, by substituting a larger diameter drum as the first drum the paper sees in the inclined example, the hardness again returns to a normal level.

Other tests show much the same phenomena in terms of wound-in tension (Fig. 418). However, by simulating the weight of the 60 in. (1524 mm) diameter roll on the inclined drum arrangement, a total of 72 pli (12.6 kN/m) max. load is placed on the first drum the paper contacts. With this same roll weight on a level drum configuration, the load is 42 pli (7.4 kN/m) on both the first and the second drum.

The pli wound-in tension developed by the drum nip on two small-diameter drums shows 7.8 pli (1.4 kN/m) (Fig. 418A). With two inclined drums of the same small diameter, the result is 8.3 pli (1.5 kN/m) (Fig. 418B). This increase is logical when one considers that while unloading one of the drums substantially, a higher load is being created on the other. If the other drum happens to be the first drum the paper sees, it becomes the drum – according to the tests – that produces the most wound-in tension; consequently, the higher nip loading produces less desirable results.

When replacing the small-diameter drum with a large-diameter drum in the first drum position (Fig. 418C), the wound-in tension of 6.2 pli (1.1 kN/m) again becomes substantially less than the small drum conditions. The message,

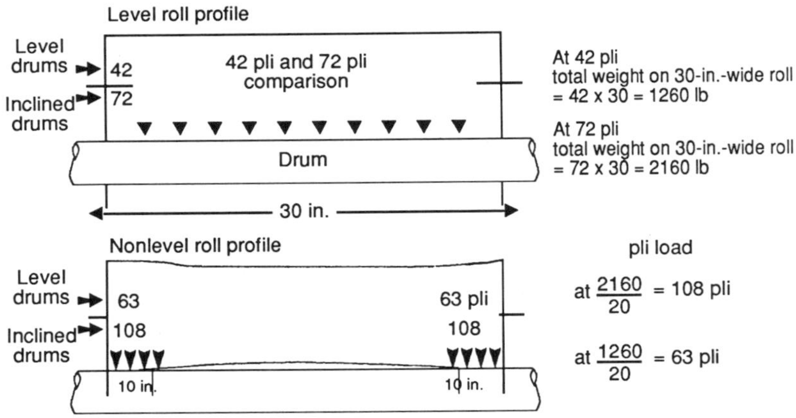

Fig. 419. Effect of non-uniform profile on drum nip.

therefore, remains clear: *the larger the drum, the less the hardness of the rewound roll.*

Carrying this scenario one step further, using all large-diameter drums (Fig. 418D, E), the results still remain consistent. Two inclined, large-diameter drums produced a wound-in tension of 6.1 pli (1.1 kN/m) (Fig. 418D), which is significantly more than the 5.4 pli (0.95 kN/m) (Fig. 418E) produced by two level, large-diameter drums.

12. Large drum and low nip

The results of all of these tests show that large-diameter drums produce less wound-in tension in a roll of paper than smaller drums, and that level drums also produce lower wound-in tension than inclined drums. Therefore, the message is again clear: *To obtain the best results, a large-diameter drum should be run with the lowest possible nip, thus making the better roll. To obtain the loosest possible nip on both drums, the drums should be equal in diameter and level.*

13. Effect of profile

Profile is usually defined as the unevenness in a cross-web direction in terms of moisture, caliper, basis weight, or Rho Hardness nip profile (Fig. 407). As any of these variables are wound into a roll where layer upon layer are wound on top of one another, any variation will therefore multiply and thus amplify the variation.

Because of these variations, the nip across the machine width is therefore never uniform. And one should also agree, therefore, that with an uneven profile across any one roll, that roll will only be supported on the two hardest or highest points, as the softer or lower point would not be in contact with the drum, thus no nip. This is a basic and correct assumption (Fig. 419).

To further illustrate this [and using the same drum loading of 42 pli (7.4 kN/m) and 72 pli (12.6 kN/m) used above] let us calculate the total load on the first drum for a 30 in. (762 mm) wide rewound roll. With a uniform profile, the total load then becomes 30 in. (762 mm) wide × 42 pli (7.4 kN/m) = 1260 lb (572 kg) on the large-diameter, equal-size level drums, and 30 × 72 = 2160 lb (980 kg) on the unequal-size inclined drums. Let us say for the sake of discussion that because of high-caliper areas on each end of the roll we have only 20 in. (508 mm) (10 each end) that are in contact with the drum. This means the pli nip pressures become 2160/20, or 108 pli (18.9 kN/m), on the inclined drums.

It is quite obvious that, if these pli numbers exceed the ultimate strength (breaking strength in pli or elastic limit of the paper), a break, burst, or bag will occur. One can easily go through the mathematics of different rewound roll diameters and their corresponding pli loadings to determine the extent of loadings involved; however, the fact still remains that nip and drum diameters are major factors in wound-in tension and roll quality and, without using the maximum beneficial drum diameter and the lowest possible nip, optimization of the effect of profile and resultant wound-in tensions become difficult, to say the least. Continuing this same thinking, the wound-in tension developed by the small-diameter rider roll might be considered insignificant, and can be discounted. However, remembering that although the rider roll load is almost always decreasing at any given roll diameter from maximum to minimum load, the effect of its small diameter, coupled with a less than ideal profile, can be damaging at any diameter.

From all these tests, one would say that the

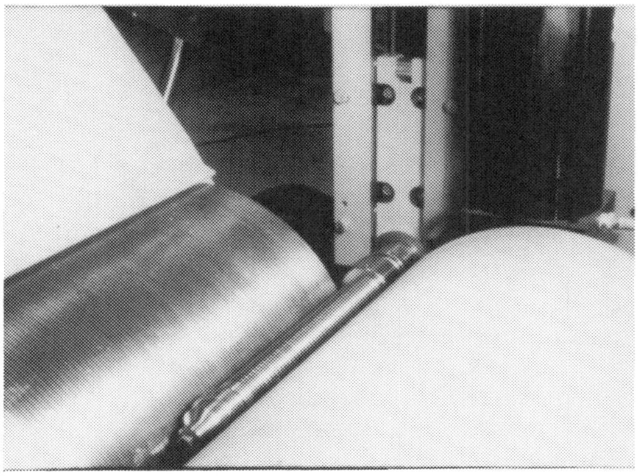

Fig. 420. Carbide-coated drum is a factor in proper torque control.

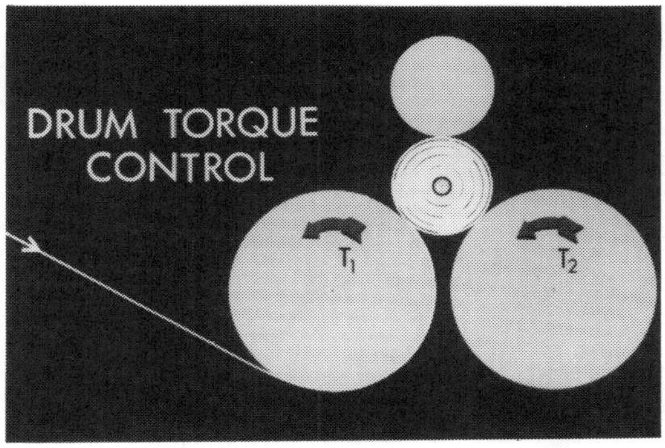

Fig. 421. Drum torque control. The torque-created tension minimizes differential nip tension.

optimum condition for the nip tests conducted would be two 30 in. (762 mm) diameter drums and one 30 in. (762 mm) rider roll. This, of course, cannot be accomplished with the small-diameter rider roll necessary to nip a small core between 30 in. diameter drums. Consequently, the overall nip furnished by the rider roll must be decreased and the necessary wound-on tension developed by some other means.

14. Torque with carbide drum coating

The advent of carbide coating (a rough surface to improve traction) on the front winding drum (Fig. 420), coupled with the ability to provide maximum torque at zero speed on the very first wrap on the core, was a major breakthrough and benefit to the industry as a whole. What this allows is a method to minimize the effect of nip-induced profile oriented tension, substituting torque-created tension (Fig. 421) which, by its very nature, is not nip-induced profile-oriented tension, and, therefore, provides more uniform tension across the whole web. This happens regardless of profile. The thousands of gripping fingers on the rough coating on the drum surface provide a uniform grip which, in turn, through the addition of a higher load on the second drum of a two-drum winder, provides the necessary tension on the very first wrap on the core. Figures 422 and 423 demonstrate the results of this in terms of roll quality during a field study.

A two-set reel was provided for this test. The mill was having corrugation problems (Fig. 422) – specifically, on one end of the roll. The first roll was wound in a normal way and the Smith Needle structure profile and Beloit Density Profiler were used to evaluate the results. The Smith Needle structure is a device which

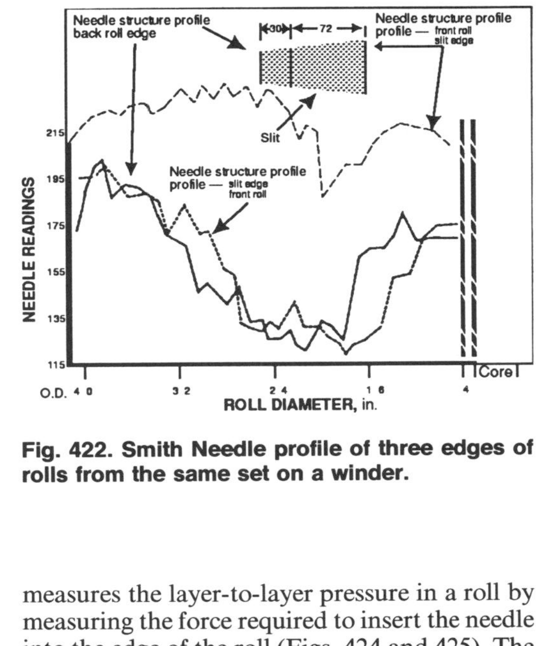

Fig. 422. Smith Needle profile of three edges of rolls from the same set on a winder.

measures the layer-to-layer pressure in a roll by measuring the force required to insert the needle into the edge of the roll (Figs. 424 and 425). The Beloit Density Profiler is a device also manufactured by Jagenberg and Wartsila which measures the density of the roll while winding (Fig. 426). These results (Fig. 422) showed that one end of the set was too hard; corrugations resulted. The other end was too soft, with corresponding soft roll type problems. Increasing the rider roll pressure to make the soft area harder would only aggravate the corrugation on the hard end. To alleviate the problem on the second set, torque was increased on the front drum significantly, with results as seen in Fig. 423. We not only completely eliminated the low tension in the middle diameters or center of the roll (Fig. 422) but also reduced the corrugation problem on the hard end of the roll.

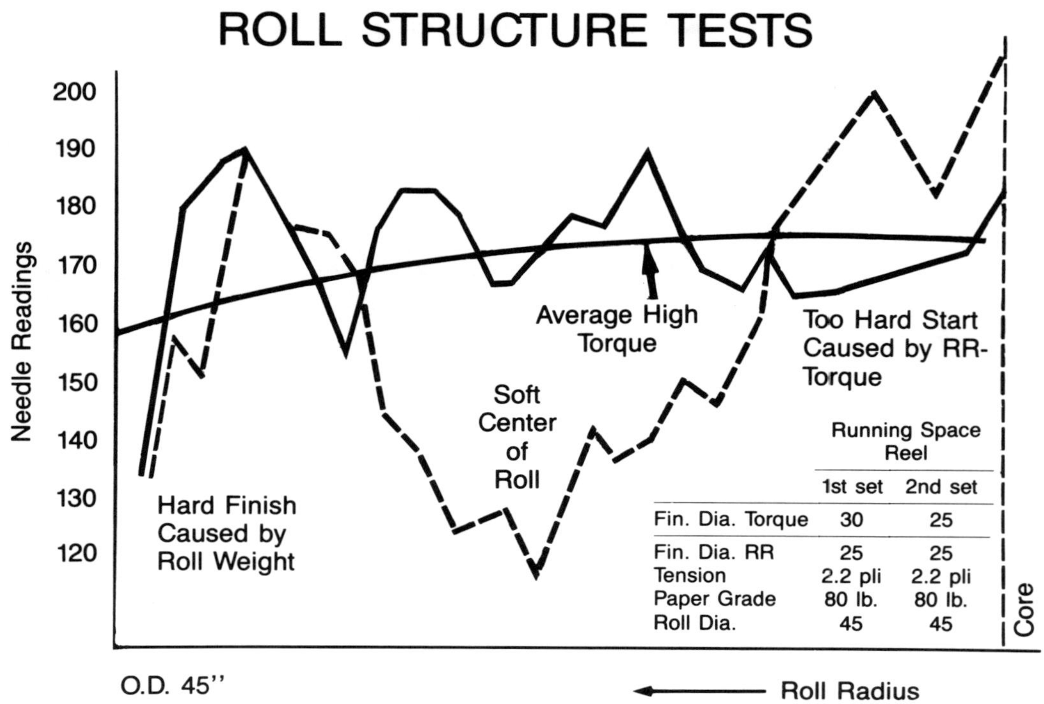

Fig. 423. Effect of torque as shown by a Smith Needle profile.

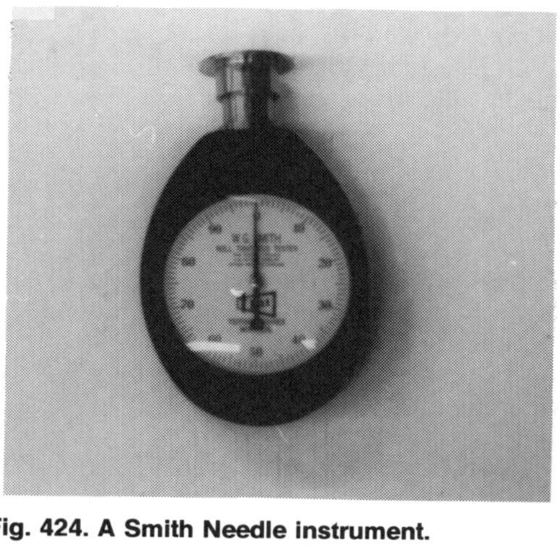

Fig. 424. A Smith Needle instrument.

Fig. 425. Smith Needle instrument in use.

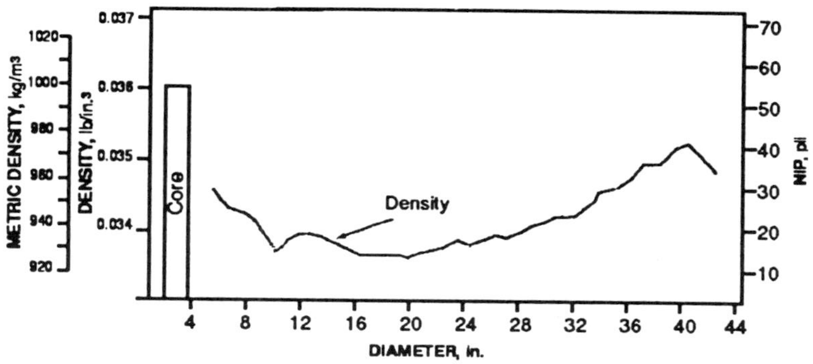

Fig. 426. Beloit roll density profile.

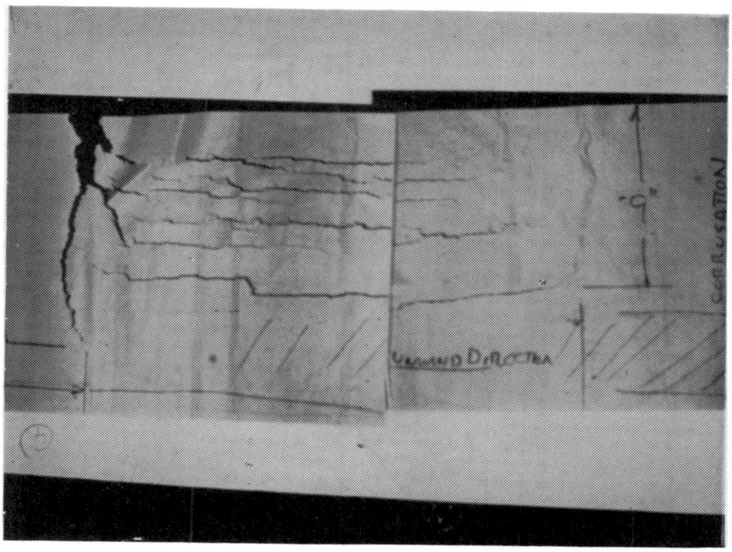

Fig. 427. Core bursts.

15. Core bursts and crepe wrinkles

The above procedure also improves the common industry-wide problems of core bursts (Figs. 427 and 428) and crepe wrinkles (Fig. 429) that occur on wide rolls of lightweight paper near the core. To eliminate slippage or gearing at the core (Figs. 430, 431), which results many times in bursts (Figs. 427, 428), dishing (Figs. 432, 433) at the core, and telescoping, there must be a relationship between core diameter, the weight of the roll, and the necessary pressure on the core (provided by tension on the first wraps) for the paper to maintain its grip on the core(s) (Fig. 430).

This relationship is important because no matter how a roll is wound, whether it be on top

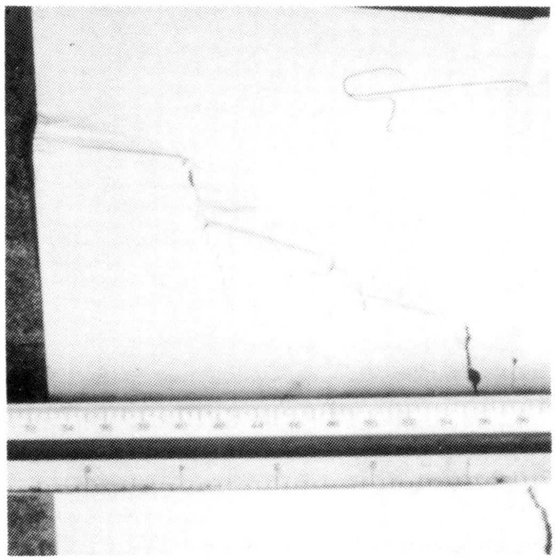

Fig. 428. Core bursts. ▶

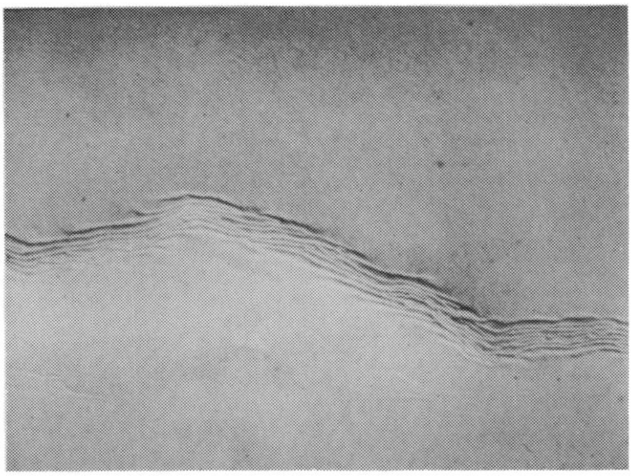

Fig. 429. Crepe wrinkles.

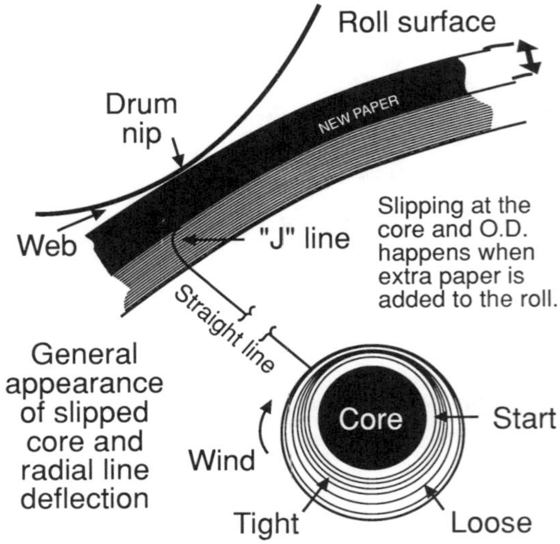

Fig. 430. Gearing at core "J-line" near outer diameter of roll.

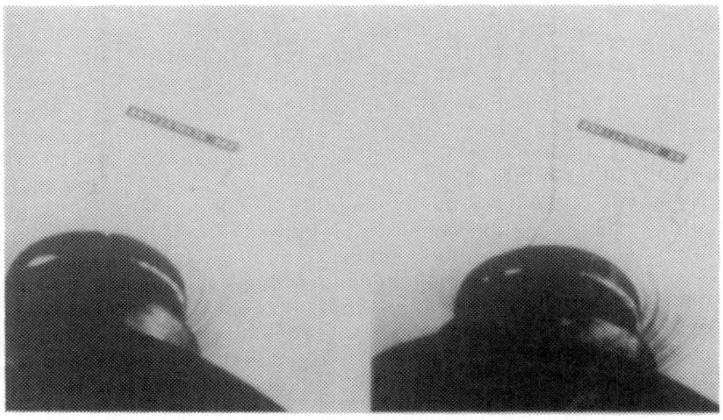

Fig. 431. Gearing, or slippage, at the core.

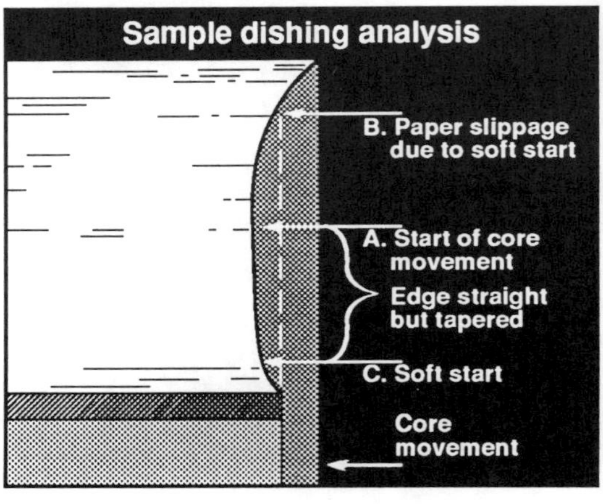

Fig. 432. Dishing edge profile.

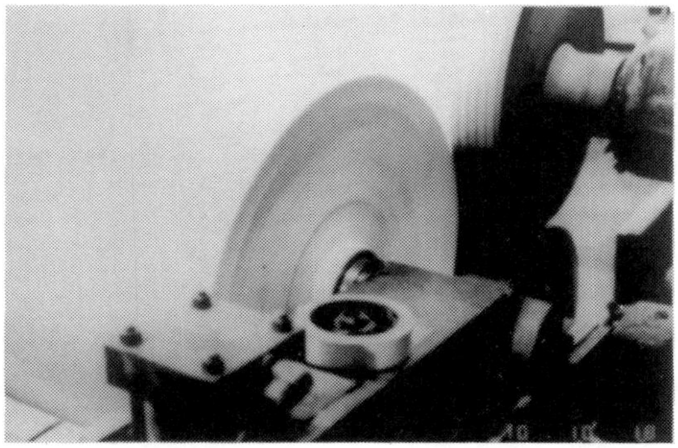

Fig. 433. Dished roll.

of the drum or any other position, on a single drum or between two drums – centerwind or surface wind – the fact cannot be denied that the roll, when unwound, will be supported only on its core (Fig. 435). The resultant core pressures then become the key to roll quality at the core, as all the forces involved with the roll weight will therefore come to bear on the small-diameter core and the specific roll support mechanisms indicative of the pressroom (Figs. 436, 437) and converting equipment. If the initial wraps of paper are not tight enough to withstand the strains of the roll weight, the grip on the core will be lost and internal gearing (layer-to-layer slippage of the paper near the core) and burst edges near the core will occur (Figs. 427, 428, 430).

16. Core pressures—a key

Figure 438 shows different pressures at the core during the winding and unwinding process.

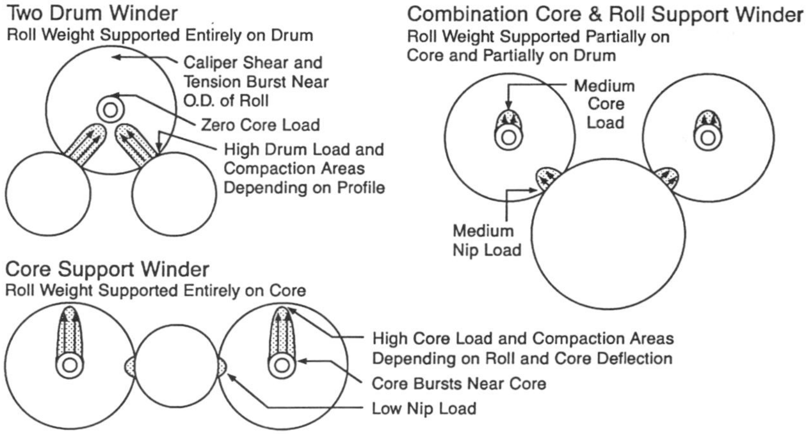

Fig. 434. Compression and decompression of paper during different winding processes.

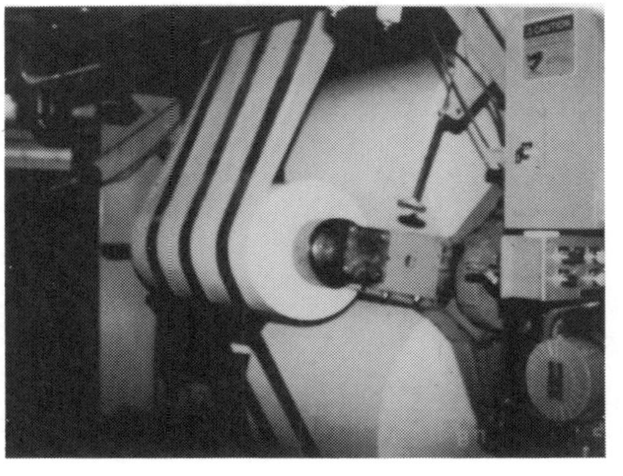

Fig. 436. Pressroom unwinding station.

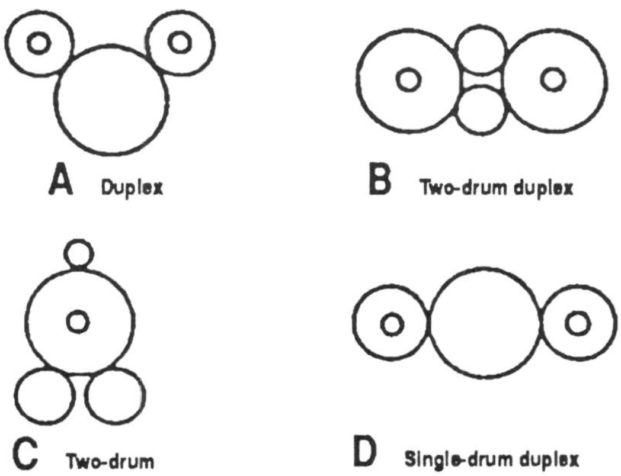

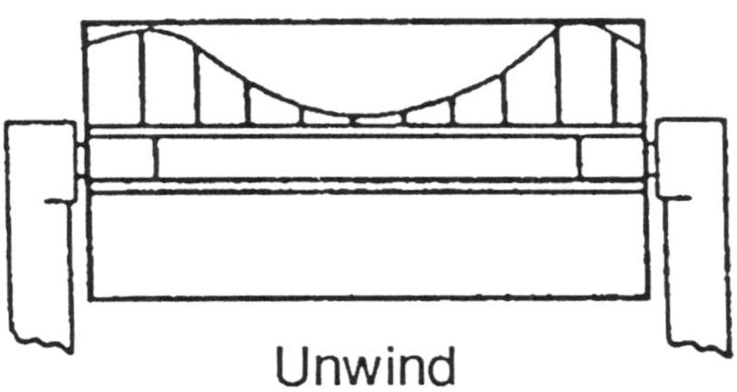

Fig. 435. Core pressure during the unwinding cycle.

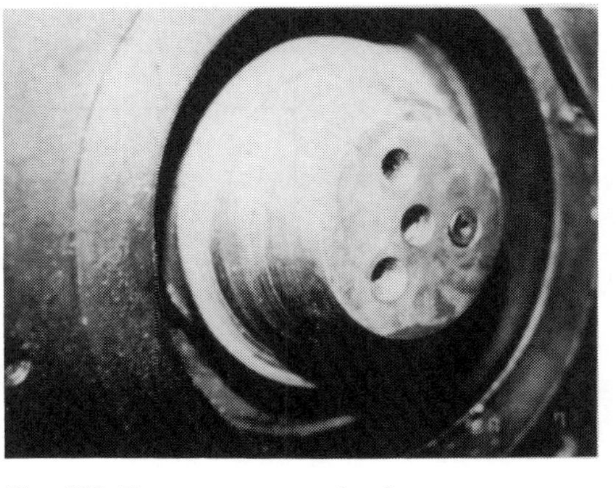

Fig. 437. Pressroom core chuck.

It is obvious that the two-drum winder has little, if any, core pressure during the winding process, as the rewound roll is supported by the drums and not on the core, as is found in duplex winding. This is why core problems on a two-drum winder are first seen in the press room and not during the winding process.

During this winding process, duplex winders (a winder that supports each roll in a set, separately on its core vs. being supported on the roll surface), on the other hand, which support the roll on its core while winding, have the same core pressures that will eventually be seen by the press unwinding equipment.

Significantly, the core pressures at the ends of the roll are the highest, and many of the core bursts (Fig. 428) that occur in the pressroom occur in those core support areas where the pressures are the highest.

Many steps can be taken to reduce these core pressures during the winding process. A few of these steps are: larger diameter cores, smaller diameter and narrower rolls (reducing the roll weight), core inserts, heavier wall cores, different core material, supporting part of the weight of the roll on the winding drum and winding equipment that minimizes core pressures during the winding process (Fig. 434). It should be emphasized on certain types of equipment that paper is sometimes damaged during the winding process if the core and drum nip pressures are not controlled or equalized. The fact still remains, however, that only those factors listed above that will improve the core pressure problem *during the unwinding process* will prove effective in reducing problems at the core.

The fact also remains (Fig. 435) that, no matter how a roll is wound, it will eventually be supported on its core during the unwinding process in the pressroom and, unless proper attention is paid to the first wraps on the core during the winding process, problems at the core are more likely to happen.

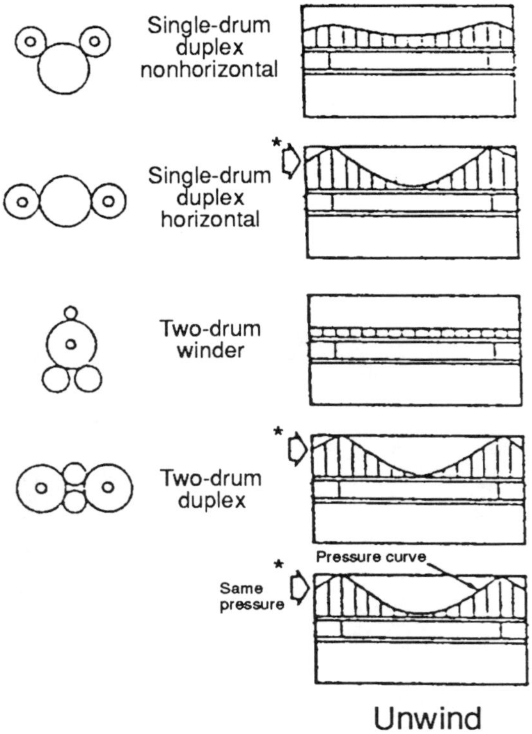

Fig. 438. Core pressure during the winding cycle.

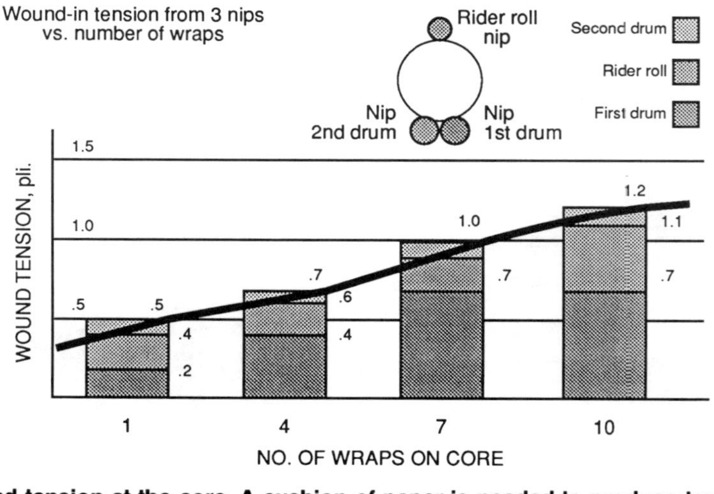

Fig. 439. Nip-induced tension at the core. A cushion of paper is needed to produce tension.

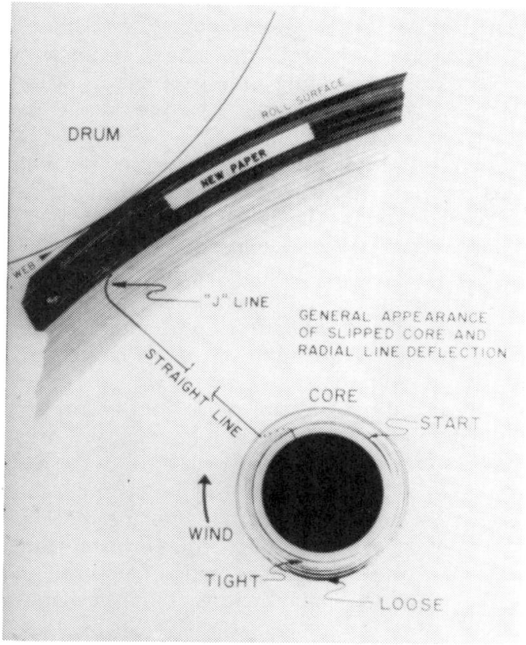

Fig. 440. Demonstration of layer-to-layer slippage (Figs. 440, 441) on a core support winder.

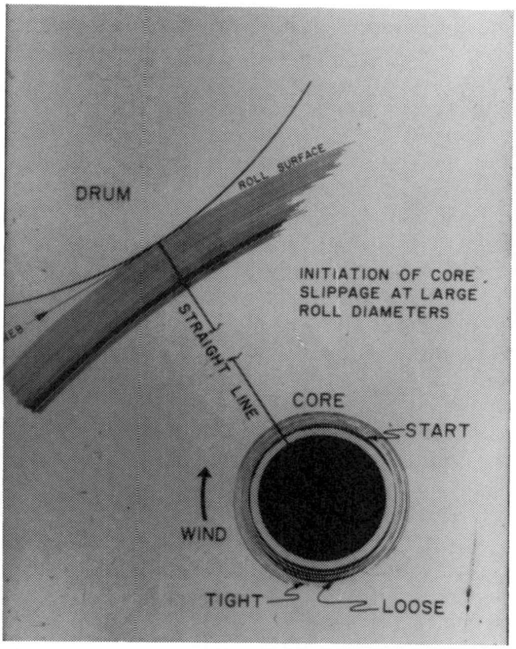

Fig. 441. The break in the straight line drawn in Fig. 440 shows the gearing effect at the core as the roll rotates in Fig. 441.

17. Nip will not make first wrap tight

This attention to the first wraps on the core brings up another important fact that is sometimes overlooked or minimized, as our tests show that nip pressure alone on the core (without the assistance of tension or torque) will produce very little of the wound-in tension required on the so-critical first wraps of paper on the core. To produce any significant nip-induced tension, a cushion of paper is required. This cushion is obviously not available on the first wraps of the core, and thus the nip-induced tension will be very low.

For example, Fig. 439 shows that with a 9 pli (1.6 kN/m) nip load, only 0.5 pli (0.087 kN/m) wound-in tension is developed on a single wrap, which then increases to 0.7 pli (0.123 kN/m) with four wraps, and to 1.0 and 1.2 pli (0.18 and 0.21 kN/m), respectively, as the cushion increases. Unfortunately, by the time there is enough cushion to produce the required tension to support the roll, the damage is already done, as it takes much higher levels to resist the strains caused by the roll weight at or near the core. Most of the testing was done with 30 wraps. The rate at which nip-induced tension increases is dependent on the K_2 or bulk factor of the paper [10] and the very fine line in the relationship of nip to the actual produced wound-in tension is very delicate as too much nip will lock the layers together making torque tension on a single layer ineffective.

18. Single-drum winder without torque

In equating this to a single-drum winder without torque (which includes such equipment as a paper machine reel, some duplex winders, and many of the rereelers found on the market), the first wrap on the core would have only that low tension that can be provided by nip. The nip-induced tension would remain low until such time as enough paper is wound to provide the necessary cushion. Unfortunately, by that time (as stated previously) the damage is done, and when the roll is subsequently unwound the loose, first wraps would have no grip on the core – and gearing occurs. Figures 430, 440 and 441 demonstrate this "gearing" phenomenon where one sees the looseness (Fig. 441) in the bottom half of the roll at the core on a core-supported winder, and the break in the straight line (Fig. 441) drawn previously on the roll showing the "gearing" effect at the core as the roll rotates.

The loose start will cause the core to rotate at a different rate with respect to the outside diameter of the roll. With the core moving ahead of the outside of the roll, the severe pressure on the core end, due to the small length of the core support, locks the layers of paper together, allowing no layer-to-layer slippage in that area of the core support. With a low-tension, comparatively loose start, the rest of the roll continues to "gear", creating first a crepe wrinkle (Fig. 429) and then a burst in the core support areas (Figs. 427, 428).

One might ask: "If center torque or differential torque is not needed on a single-drum winder, why is it needed on a two-drum winder?" It certainly would simplify the whole two-drum winding process if, in fact, only nip pressure could be used to develop tension on the first wraps near the core. Obviously, it is needed; and regardless of the type of winding process, sufficient tension on the first wraps on the core is needed to minimize the core crepe wrinkle and bursting problems that happen during the unwinding process.

19. Summary

In the basic technologies of winding, there are certain fundamental characteristics of the roll of paper itself (to name a few: basis weight, core diameter, roll weight, roll width, roll diameter, tensile strength, tensile modules, compressive properties (K_1, K_2) interlayer friction coefficient and profile of the paper) which dictate the limits of what can and cannot be done with winding equipment without problems.

And, even when using all available tools, there will still be some limits that could be exceeded. Those pieces of equipment that do not provide all of the tools or do not properly use the tools they have must face the limits of roll quality and runnability that will therefore exist.

To determine what equipment works well and what other equipment is lacking, one must go to the pressroom or the converter for the answer.

(a) Runnability is the ultimate test

Runnability in the pressroom or at the converter is really the ultimate and final test of roll quality [6]. Hundreds of thousands or even, possibly, millions of dollars are spent each year by both the pressroom and the converter – as well as the paper mill – when this ultimate test fails.

(b) Problems are cyclic

The statement made so often in the past to a paper mill having runnability problems is: "If your paper won't run, we can always get paper that will – from other suppliers". This, of course, neglects the fact that it will still cost everyone (both the press and the mill) considerable amounts of money, even though a contract is negotiated with another supplier. (This attitude, however, is changing.) The fact that problems in the pressroom are cyclic (and this other supplier could at some time possibly have the same or another type of problem) makes runnability problems never-ending, unless something different and more definitive is done about the problem in the first place.

I believe that both the pressroom and the mills are beginning to realize that a cooperative effort to resolve some of the problems will bring major benefits to all parties concerned.

When one looks at the basic technologies of finishing (Fig. 386), it becomes readily apparent that the printing, winding, and many of the converting and finishing processes have many things in common.

For example, a printing press (Fig. 442) as well as paper mill equipment must be concerned with tension control, web control (control of the web through all of its supporting rolls and other contacts), unwinding, slitting, cutting, packaging, some of the basics of coating (printing) and, of course, the basic process called "nip mechanics" – all of which play an important role in most processes.

The fact is that within the years of experience at both ends of the papermaking and printing processes there are common interests and knowledge that could help and that should be shared by both parties. It is widely believed that the industry is changing very rapidly and that a common and concerted effort is and will continue to be necessary to compete in the world market.

Therefore, our aim must be to continue our investigations into the definition of the problems in order to meet the following objectives:

1. Improve runnability in the pressroom and converter from all sources
2. Improve the communication between paper mill and pressroom or converter and others who can affect runnability
3. Optimize existing mill and press equipment, to minimize known and proven defects and their causes for the best runnability results in the pressrooms as well as in the mill
4. Understand the limitations of the pressroom and converters' equipment
5. Learn the technologies of finishing that pertain to the above objectives that apply, in many cases, to both the pressroom and the paper mill operations.

The learning process of zeroing-in on the problems that resulted in less-than-desirable runnability in the pressroom is ongoing; however, after looking at hundreds of samples from many paper mills and pressrooms, categorization of defects and their causes comes quite easily.

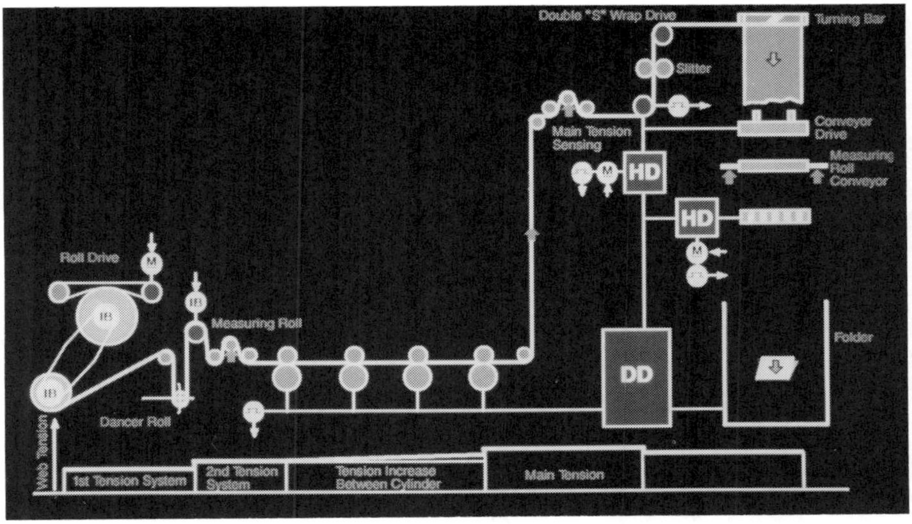

Fig. 442. Pressroom.

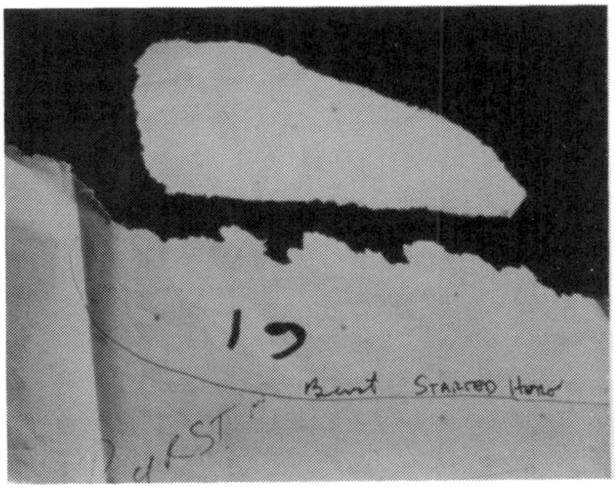

Fig. 443. Burst-tension-edge.

Fig. 444. Burst tear.

(c) Basic defect categories

The fundamental and basic pressroom defect categories, after reviewing hundreds of defect samples, are:

1. Unknown
2. Bursts
3. Wrinkles
4. Cracks and cuts
5. Corrugations
6. Holes
7. Roll condition
8. Miscellaneous.

Unfortunately, the unknown category is the largest, which I feel will be reduced when a better definition of those defects that fall into this large catch-all category that will be further broken down into yet finer cause-oriented categories. When doing this we developed categories in each classification that had different causes. This of course allows us to attack the real cause of the problem. This can be done where the samples show a pattern that allows more detailed classification, as seen below.

Bursts

1. Caliper shear bursts (Fig. 395)
2. Burst-tension-edge (Fig. 443)
3. Partial tension burst
4. Air shear burst
5. Edge burst
6. Burst tear (Fig. 444)
7. Core burst (Fig. 427)
8. Handling damage burst
9. Two-minute burst (newsprint).

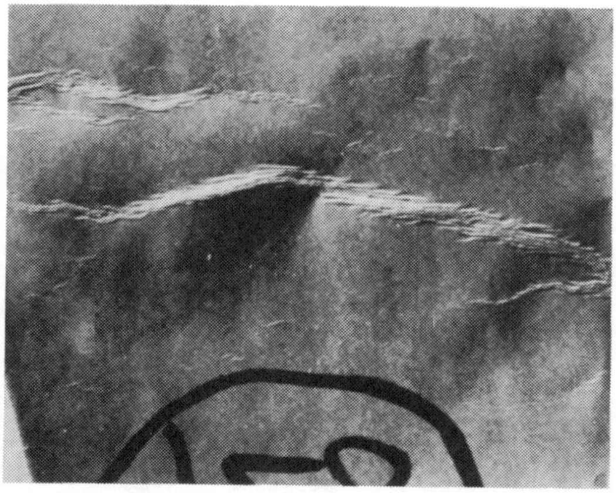

Fig. 445. Wrinkle crepe.

Fig. 446. Cuts.

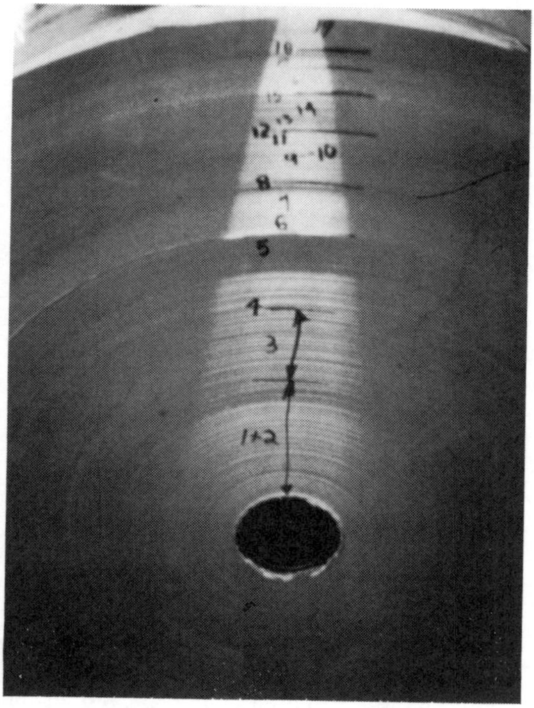

Fig. 447. Roll condition spreader caused offsets.

Wrinkles
1. Hard wrinkle
2. Full crepe wrinkle (Figs. 429, 445)
3. Edge crepe wrinkle
4. Core crepe wrinkle
5. Caliper shear crepe wrinkle
6. Handling damage crepe wrinkle
7. Pucker wrinkle
8. Moisture welts
9. U-shaped blow wrinkle
10. Air crepe wrinkle
11. Two-minute crepe (newsprint).

Cracks and cuts (Fig. 446)
1. Turnover
2. Diagonal edge cut
3. Edge crack (tear)
4. Hair cut
5. Bad start crack
6. Calender cut
7. Core key crack
8. Die cut
9. Diagonal edge tear
10. Edge burst.

Corrugations (Fig. 394)
1. Profile (through all diameters of the roll)
2. Ridges
3. Roll structure (only part of the roll)
4. Bad start corrugation
5. Bags and stretch-outs.

Roll condition (Fig. 447)
1. Offsets (Fig. 447)
2. Off-center core
3. Out-of-round roll
4. Crushed core
5. Dishing
6. Telescoping
7. Wet rolls
8. Core delamination (Fig. 448)
9. Core cap deformation.

Holes
1. Slime
2. Water.

Miscellaneous
1. Roll damage
2. Shives

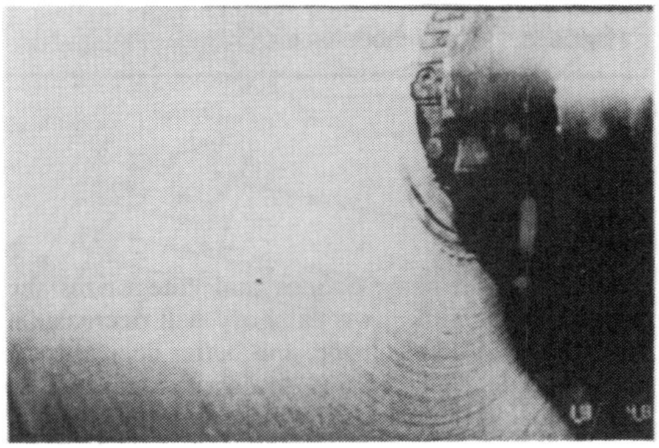

Fig. 448. Delamination of core.

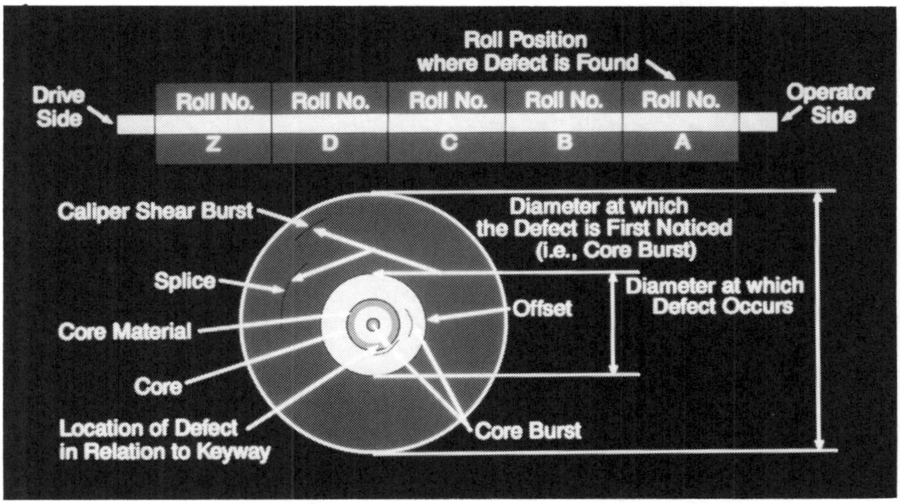

Fig. 449. Location of defect in roll.

3. Surface marking
4. Mis-registration
5. Dust
6. Debris in roll.

These categories were established through a very specific procedure that enabled one not only to describe the defect but also to break it down into smaller increments as a function of the cause, in those cases where this cause could be identified. The procedure used is shown in Table 58.

After one properly defines and categorizes the defects [6], the unknown category will decrease in importance and magnitude, and better decisions on what should or should not be done on equipment that is causing the problem will be made, which will benefit both the paper mill and the pressroom.

Table 58. Procedure for defining defects.

1. Obtain as many examples and samples as possible from the pressroom and converter.
2. Take pictures of the samples.
3. Write a description of the defect (Table 59). Two-minute crepe or break.
4. Define deckle position of defect in relation to paper machine deckle position.
5. Define diameter at which the defect is first observed (Fig. 449).
6. Define the diameter where the defect is found (Fig. 449).
7. Try to identify equipment function or paper machine where the defect can be found and verified in the mill.
8. Speculate as to cause and revise the assessment when new clues are brought forth.
9. Compare runnability records and break frequent records with other equipment and pressrooms.

Table 59. Example: Defect description for a 2 min break or crepe.

— Crepe wrinkle normally present
— Web burst sometimes evident
— Happens after a snap-off
— Happens when cross-over torque drive program is in effect
— Happens more often in the A/Z rolls of a roll set
— Happens more often in first set on reel spool
— Happens if rolls are too soft
— Happens if coefficient of friction is too low
— Happens if tensile modulus is too low

Table 60. Example of cause: possible causes of 2 min break or crepe.

— Gearing problem on paper machine reel
— Drum grooving allowing too much air in roll
— Incorrect torque program
— Non-uniform nip, rider roll level, core shaft slides, or profile
— Cyclic nip
— Inconsistent roll structure
— Too low coefficient of friction
— Too many nip cycles during snap-off
  Stop printing
  Rolling roll on floor in mill and pressroom
— Too soft a roll

We have explored the basic technologies of the structure control of a roll of paper and its effect in terms of runnability in the pressroom. We will now look at some of the different types of winders that are in the marketplace. The first step in this identification process is to define the equipment and its function. "Equipment" that is considered winding equipment is covered in the next section.

C. EQUIPMENT DEFINITION

Since there are many different types of winding equipment which have evolved because of a need, it is therefore important to define each piece in terms of that need. Definitions of the following equipment and components are given in this section:

1. Reel
2. Re-reeler
3. Two-drum winder
4. Duplex winder
5. Center winder
6. Combining (tissue) winder
7. Converting winder.

1. Reel

A nip-type, surface-driven reel is a winder (Figs. 383, 384, 450–452) with a single, large-diameter drum that winds paper on a mandrel called a reel spool, or large-diameter core. This piece of equipment is usually associated with some continuous process that cannot be stopped or started during a reel change (starting a new roll).

To allow the process to be continuous, this equipment (Fig. 450) usually consists of primary arms that hold the new reel spool about to be started on top of the drum and web. Through various means, the paper is automatically transferred from the finished reel to the bare spool. The primary arms then rotate down to a rail where secondary arms provide the needed nip pressure to complete the desired reel diameter. The primary arms then retract and return to the starting position, ready for the next new spool (Fig. 450).

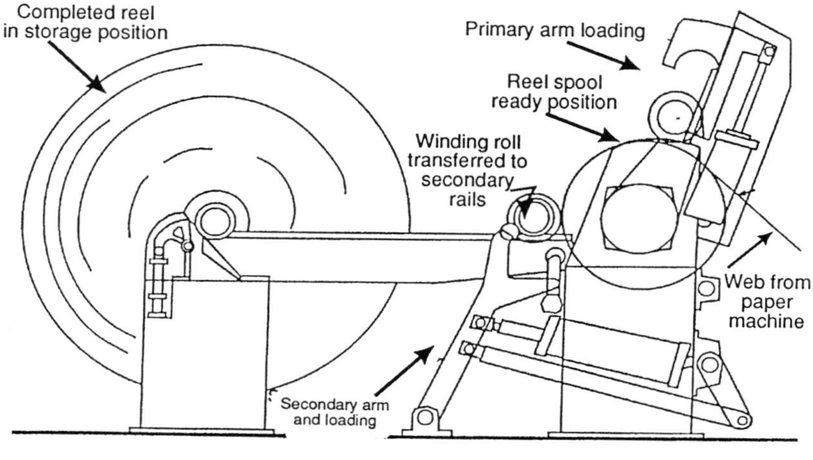

Fig. 450. Web to reel winder.

Fig. 451. Center drive on reel.

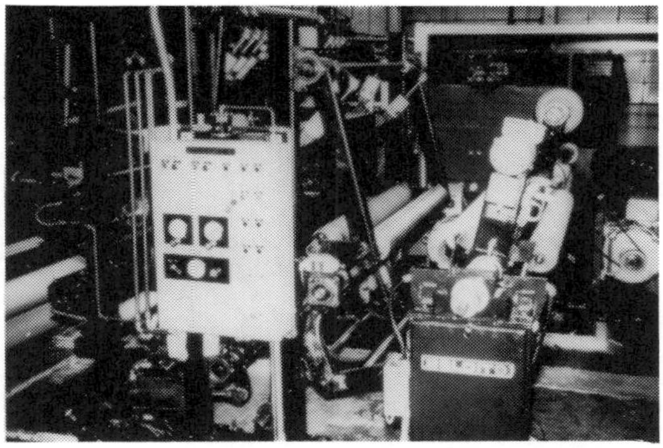

Fig. 452. Envelope-type winder used with the extrusion coating process.

Fig. 453. Single-drum winder.

Fig. 454. Single-drum winder.

A paper machine reel and a continuous coater reel are examples of this type of equipment application.

A centerwind reel (Fig. 451) has the same function as above, with one exception: the material being processed is sensitive to nip. Consequently, the roll structure is provided with center torque-nip combination, or just center torque rather than the nip-type of roll mentioned above.

The winding equipment on pure center-torque winding must therefore be somewhat different to provide the transfer in the continuous process. In the case of pure centerwind, this is usually done with some type of enveloper, as there is no nip to help the web transfer process (Fig. 452).

2. Re-reeler

The re-reeler (Figs. 453, 454) is very similar to the nip-type surface-driven reel. The exception: there are no primary arms, as there is no necessity for a continuous operation. It usually consists of an unwind, trim slitters, or a supercalender, or embosser, or any other process between the unwind and the winder. The paper roll (supported full-width on a spool or large-diameter cores) rides on a level or slightly inclined rail, with secondary arms providing needed nip pressure to structure the roll. In some cases, reels are wound onto smaller cores.

A supercalender windup or a single-drum splice-making winder are examples of this type of equipment. However, any process rewinder which is on the windup end of that process could be called a re-reeler.

3. Two-drum winder

The two-drum winder (Figs. 455–458) consists of an unwind of some sort, a slitter section to cut down a wide web into narrower widths, and a windup station, consisting of two drums that provide the means of support for winding up rolls of paper, side-by-side, on small-diameter cores. The two drums (arranged in a horizontal position, side-by-side) form a natural pocket to support the small-diameter cores. The drums

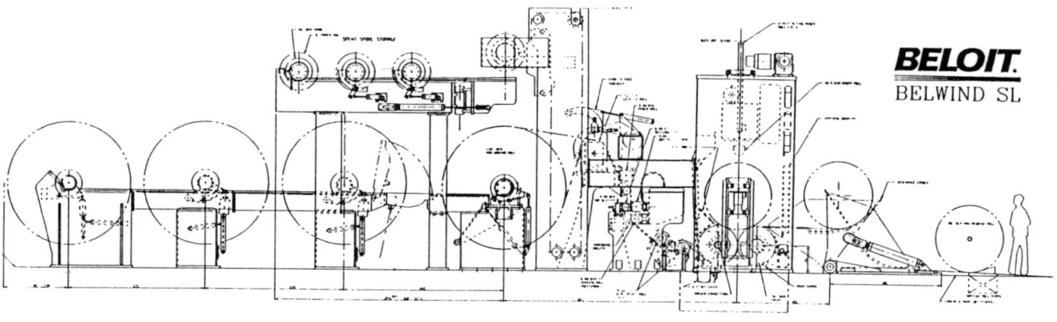

Fig. 455. Vertical slitter section, two-drum winder with rolls and spool handling equipment.

Fig. 456. Two-drum winder.

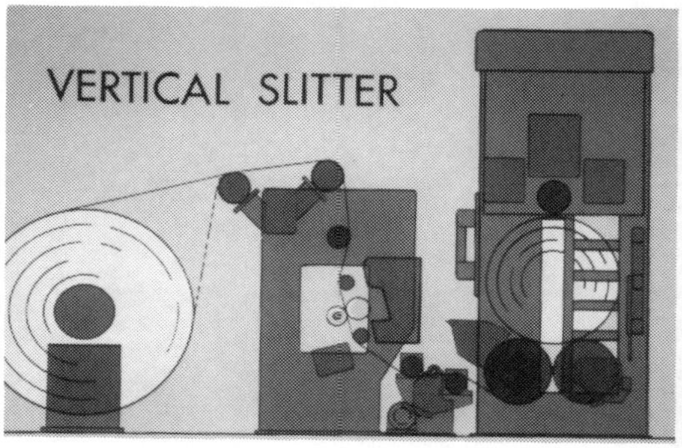

Fig. 457. Schematic of two-drum winder.

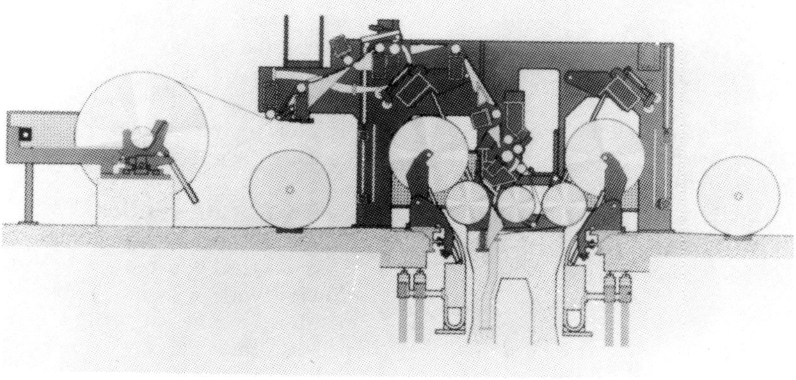

Fig. 458. Valmet duplex winder.

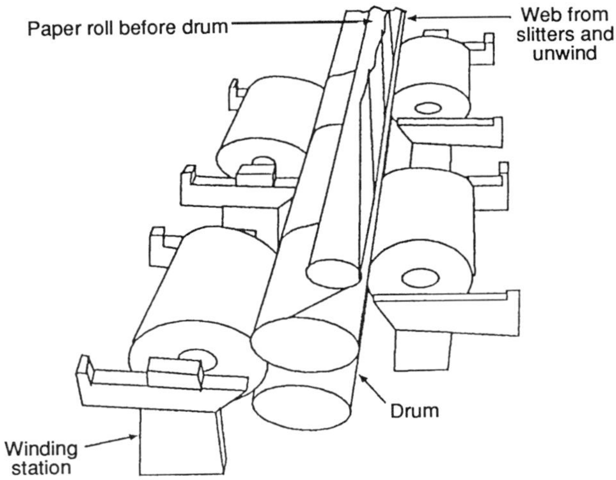

Fig. 459. Duplex core-supported winder.

Fig. 460. Two-drum bi-wind.

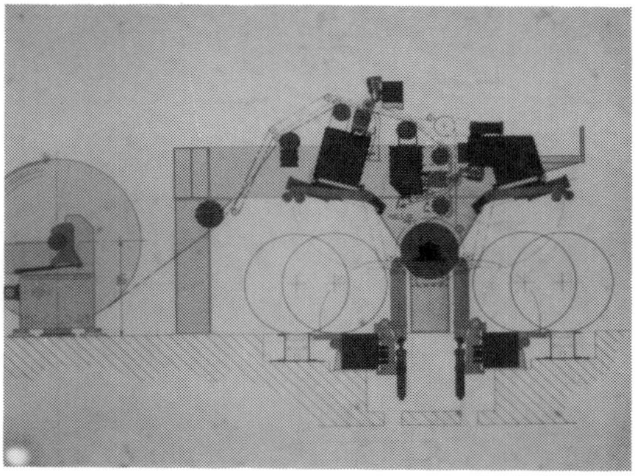

Fig. 461. Beloit model HTC duplex winder.

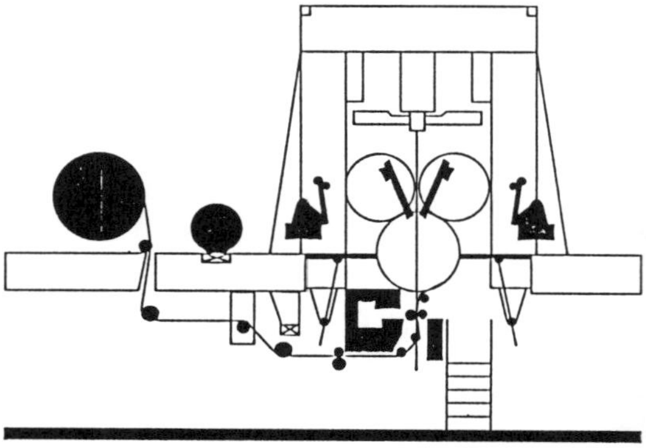

Fig. 462. Jagenberg Vari-top duplex winder.

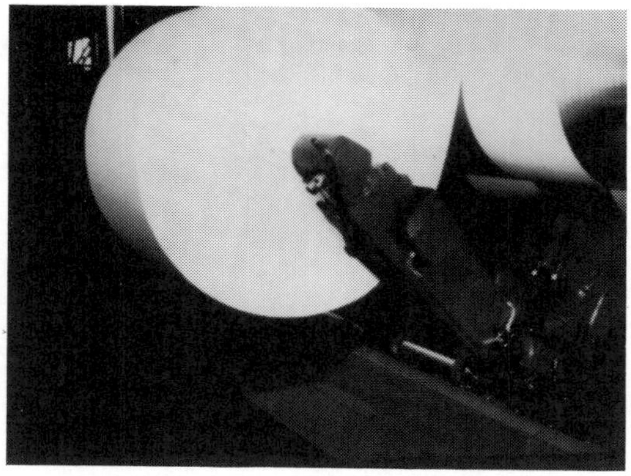

Fig. 463. Center winder.

Fig. 464. Combining tissue winder.

provide the means of support and drive for the finished roll of paper. Variations to this example exist as drums, and sometimes are not of the same diameter or level.

4. Duplex winder

The duplex winder (Figs. 458–462) differs from the two-drum winder above only by the fact that the rolls are partially or fully supported on their cores rather than on the outside surface of the roll. This necessitates that the rolls be wound alternately on either side of a central drum (Fig. 458). This is to allow support for chucks at each end of the individual rolls.

5. Center winder

The center winder (Fig. 463) differs from other winding processes only in the means of driving the rewound roll, and the absence of any nip. As the name implies, the drive is provided to the center of the roll rather than at the outside surface of the roll, as occurs in the conventional surface winder.

6. Combining winder

The combining winder (Fig. 464) winds more than one web at one time, into a single roll or set, as is found in multi-ply tissue. This changes the winding processes mostly in the unwind area, where multiple unwinds and their associated control equipment are necessary.

7. Converting winder

This category is somewhat of a catchall, as any of the above winders could fit into it. There are other specialty-type "converting winders" (Fig. 452), such as automatic paper towel or toilet tissue winders, which still have the same winding technologies but are of a special nature, usually conforming to the needs of a special product.

D. WINDING PROCESSES

Now that we have outlined the specific hardware which could (and does) carry the name "winder" lets look at some of the processes that are considered winding processes.

The following are basic winding process categories describing the type of equipment commonly used:

1. Reeling
   a. Reel
2. Machine winding
   a. Re-reeler
   b. Two-drum winder
   c. Duplex winder
   d. Center winder
3. Reprocessing winding
   a. Two-drum winder
   b. Duplex winder
   c. Center winder
4. Salvage or reclaim winding
   a. Two-drum winder
   b. Duplex winder
   c. Center winder roll doctor
5. Combination salvage and reprocessing winding
   a. Two-drum winder
   b. Duplex winder
6. Combining winding
   a. Tissue
      1. Two-drum
      2. Single-drum
   b. Laminator
      1. Single-drum
      2. Centerwind
7. Converting winding
   a. All types of winding equipment.

1. Equipment process definitions

Since the same as well as different types of winding equipment are used in the many different processes, a definition of each process helps to explain the needs of that process, and the reasons why different types of equipment evolved and are used. Also, definitions give an insight into the specific differences of all types of winders.

2. Reeling

Reeling (Figs. 383, 450, 451) is usually divided into two categories: the primary reeling process, usually found at the end of the primary manufacturing process, like the paper machine (Fig. 383), and the secondary reeling process (Fig. 453) that unwinds and rewinds the full-width web normally on spools or large-diameter cores. The re-reeler coater and supercalender windup sometimes fall into this category.

The reels, in general, wind a web by a continuous (paper machine or coater) or intermittent (supercalender, printing press laminator, etc.) process onto a spool or large-diameter core. Normally this is a core-support versus roll-support (i.e., two-drum winder) type of process.

This process necessitates that the diameter of the supporting spool be large enough to support the full-width roll being wound and that the web be wound at the same speed as the associated or subsequent processes. Webs are not normally slit (other than trim slitting) in this process. Paper machine reels and re-reelers (and supercalender and coater windup reels) are examples of common applications.

3. Machine winding

This process (Fig. 385) uses paper from the primary process machine reel which comes to the winder unwind stand on reel spools or large-diameter cores. Splicing, inspection, slitting, hole patching, and trimming are normally performed as part of this function. Shipping rolls of various and random widths (or subsequent process full-width logs) can also be made during this operation. The width is usually consistent with the associated machine width, whether the machine is a paper machine, coater, or supercalender. This type of winder is usually a two-drum roll-support winder, a re-reeler, or duplex-type core-supported winder.

4. Reclaim or salvage winding

Reclaiming is defined as a winding operation designed to process rolls that have already been wound for shipment or subsequent operations and for some reason (usually defects of some type—i.e., dirt, improper winding, holes or offsets) must be rewound to salvage all or part of the material in the roll.

Reclaiming, under this definition, requires slitting at times and edge guiding (Fig. 388) where the width of the reclaimed roll must not change and trimming cannot be done. The speed of this process is limited to approximately 2500 ft/min (750 m/min), due to the inability to edge guide to compensate for uneven edge conditions in the unwinding roll at high speed. Usually a two-drum, roll-supported winder or a duplex, core-supported winder is used. Roll doctors (Fig. 465) (a centerwind single roll-to-roll winder) are also called reclaim winders and are sometimes used in the reclaiming process.

5. Reprocessing winding

Reprocessing (Fig. 466) is defined as a winding operation that processes rolls (supplied by other functions) making necessary changes that, for one reason or another, are not done on the paper machine winder. Removing defects from a roll is not normally done on this type of winder, as it is usually wider than the so-called salvage (reclaim) winder. This type of winder was normally supplied where there were many

Fig. 465. Stanford roll doctor.

narrow-width rolls needed from the parent log; and due to limitations in spreading, they were not able to process them on a wide machine winder. Or, it could possibly be an inspection winder looking for breaks or other similar defects.

The speed of this winder is normally between 4000 and 6000 ft/min (1200 and 1800 m/min)–much faster than a salvage winder. However, all the necessary equipment, such as edge guiding, slitting, spreading, etc., is available so that either salvage or reprocessing can be done.

Thus the reprocessing function may consist of splitting the roll into specific sizes, changing the core length or diameter, changing the roll diameters, splicing, inspection, hole patching, and trimming. Again, reprocessing (under this definition) is normally done on a two-drum, roll-supported winder with limited application on a duplex-type winder.

6. Reclaim/reprocessing winding

Reclaiming/reprocessing is defined as a winding operation which combines the above two processes. Its major functions include slitting and winding production orders for shipping rolls by reducing the original rolls to narrow widths, as well as changes in core sizes, roll diameters, reversing of wire and felt sides, and any other requirements necessary to fill the customers' needs. This, however, is not a full-time operation, and this winder may double for a reclaim or salvage function as described above.

7. Combining winding

This type of winding (Fig. 464) consists of a process that combines more than one roll of material in the unwinding process and then winds these multiple rolls into one roll on the combining winder. Two- and three-ply tissue and extrusion-coating foil and laminated paper are two examples that illustrate this category.

Fig. 466. Reprocessing winders.

8. Converting winding

This category usually covers those specialty operations that take a roll from the paper mill and convert the rolls into another form. Printing, embossing, winding the printed or embossed paper into a roll, and laminations are examples of operations in the tissue industries. Operations in the film industries are also examples that fit the converting process description.

E. WINDER STRUCTURE CONTROL FUNCTION BY WINDER TYPE

Each of the winding types and processes mentioned previously can be divided into four basic functions: unwind, web control, slitting, and structure control. Although the basics are very similar, the needs of the structure control function in the evolution of the winding process (Fig. 467) caused major changes in the winder configuration dictated by the problems of the ultimate user. Consequently, one must look at the structure control function of each type of winder and investigate what was done in each process to optimize and control these problems (Fig. 392).

In Sec. B, certain needs were outlined in reference to problems of roll structure. There are many types of winder and windup stations that were specifically developed because of (and are directly related to) those specific needs and problems.

These needs, for example, changed the equipment used in winding certain lightweight papers from the conventional two-drum winder (Fig. 455) to the duplex winder (Fig. 459). The problems that caused this change started with the advent of large-diameter, wide rolls needed in the pressroom. Two-drum winders were having major problems on papers that had a low

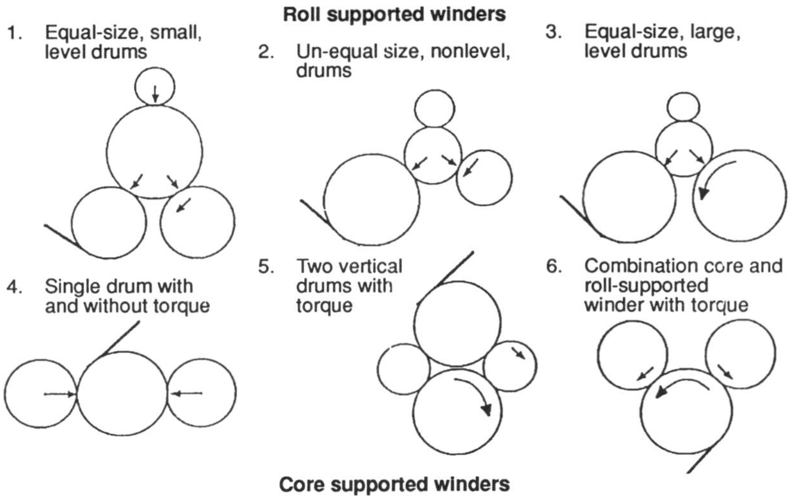

Fig. 467. Winding station evolution.

tensile strength factor, with bursts (Fig. 395) and corrugations (Fig. 394) resulting. (A burst is a break in the paper that happens as the roll is winding on a two-drum winder and, uniquely, the burst happens several layers down inside of the roll while winding; so, it is not seen in the winding process, only in the unwinding process.) The cause of the burst (Fig. 395) is excessive non-uniform nip pressure, creating excessive layer-to-layer slippage and wound-in tension which is caused by the increasing roll diameter on a roll-supported winder.

The industry reacted to this problem by using the type of winder where the roll was supported on its core, and the excessive nip pressure found on the outside of the roll on a conventional two-drum winder was replaced with a controlled nip pressure supplying nip horizontally at the core (Fig. 460).

With the duplex type of equipment, the industry was next plagued with problems at the core: core bursts (Figs. 427, 428), crepe wrinkles (Fig. 429), and delamination of the core (Fig. 448). These problems were caused by too much nip pressure at the core in the area of the windup chucks on the duplex winder (Fig. 438) and the unwind chucks (Figs. 435–437) during the pressroom unwinding process.

To compensate for this problem, the industry next went to a combination core-supported and roll-supported winder (Figs. 460–462), which optimized the problems caused by too much nip on the outside of the roll and the problems at the core caused by roll weight on the core chucks on a core-supported winder.

Thus, three types of winders evolved primarily because of structure control problems (Figs. 392 and 467). A description of these three processes and what was done within each process to optimize its limitations follows:

1. Roll supported on two drums, such as a bed roll or two-drum winder (Fig. 415).
2. Roll supported on its core, such as a reel or many of the duplex winders and re-reelers (Figs. 450, 454, 460, 461).
3. Combination of the above two, which alleviates some of the common problems found in each of the two systems (Figs. 459, 462).

F. STRUCTURE CONTROL FUNCTION

Each of these types of winders has four basic functions: unwind, web control, slitting, and structure control. Even though the basics (unwind, web control, slitting) are the same between each of the different types of winders, the structure control function in each of these three winders is different, primarily because of problems encountered by the ultimate user. Major changes in the winder structure control configuration of the winding process therefore become necessary. What follows is a description of the structure control function in each of these winders.

1. Two-drum winder winding station

For years this winder has been the workhorse of the industry and, in most cases, still is (Figs. 415, 455–458). Its limitations are the rewinding roll weight which, if excessive, causes damaging nip pressures between the roll of paper and the drum that supports the roll. As the diameter gets larger, the damaging aspect of this situation is

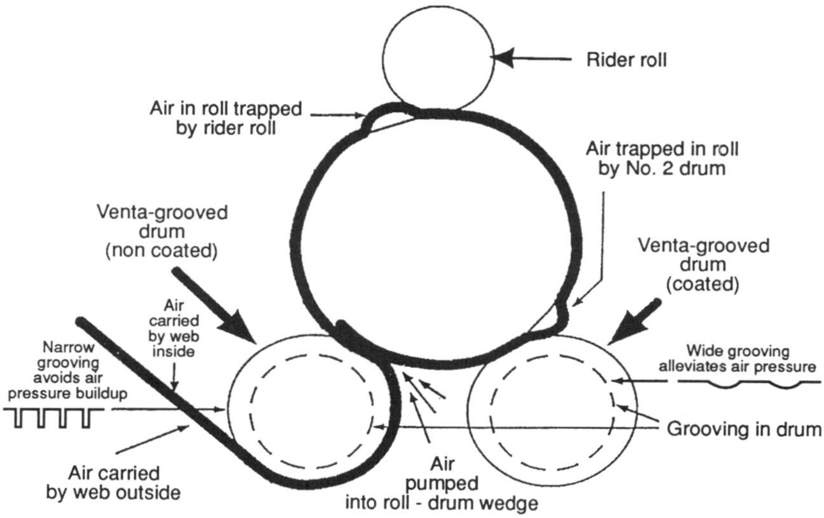

Fig. 468. Drum grooving and coating.

that the nip pressure increases as the roll builds to larger and larger diameters. This nip pressure increases even further because of profile variations (Fig. 407) in the web of paper. The amplitude of these profile differences increases as layer upon layer is added to the roll. This means that more weight of the roll is supported on less area in the nip, a situation causing an increase in wound-in tension in these localized areas of the roll which, in turn, causes defects such as bursts and corrugations.

With larger diameter, wider rolls of lighter weight paper being requested from the pressroom, the two-drum winders had to change to compensate for these market changes. Some of the changes that occurred follow.

(a) Drum diameter

Larger diameter drums (Figs. 414, 418) are replacing the smaller diameters of the past, mainly because the drum diameters are no longer determined by the width of the machine but by the necessity to reduce the nip-induced defects. Thirty to thirty-four inches (762 to 864 mm) in diameter are common sizes mainly because above 34 inches limited benefits accrue, except where deflection problems may dictate larger diameters on wide machines (Fig. 414), as described in Sec. B.

Salvage and reclaim winders also have large-diameter drums so that consistency in roll structure can be maintained between rolls being shipped directly from the machine winder as well as those being shipped from the salvage winder operation.

(b) Drum grooving

The first drum that the paper touches usually has some sort of grooving (Figs. 468, 469). This grooving should be spiral in nature, and its primary function is to vent the air carried by the inside of the web (which is substantial on the new 6000 to 8000 ft/min, winders) so sufficient pressure will not build up and cause the web to skate on the drum. This instability could cause interweaving, dishing and other web control type defects which will be discussed later. "Venta-grooving" (narrow-width groove) is used on this drum to prevent the forming of a tunnel (Fig. 470), which allows air to be pumped into the roll.

Grooving of the second drum is also being done on lightweight papers, primarily to allow the air entrapped in the roll (Fig. 468) to vent itself (one layer down) through the second drum nip, eliminating the bubble and subsequent wrinkle that sometimes occur before and at the second drum nip.

(c) Drum coating

Drum coating (Figs. 468, 469, 471) (usually molybdenum of carbide) has replaced the chevron groove (Fig. 472), which in the past was found on the second drum (sometimes on both drums) in the paper flow direction, with its primary function being to provide traction between the drive components and the paper. The second drum that the paper touches is the drum that is now normally coated, to provide extra traction for the application of the necessary torque mentioned in Sec. B.

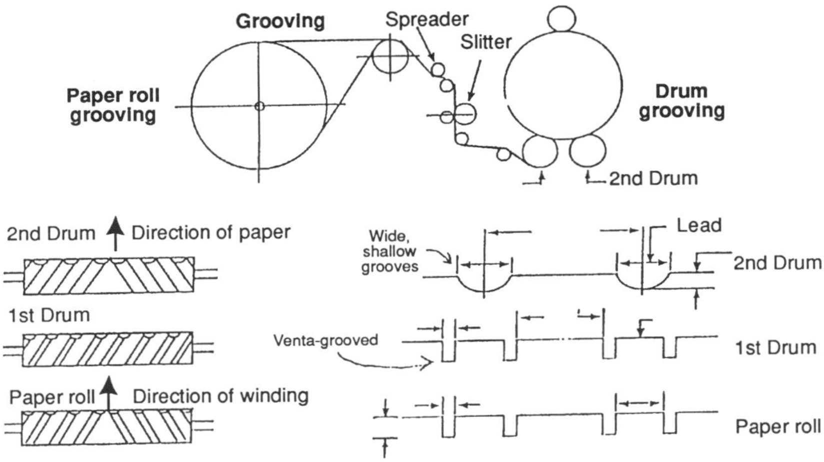

Fig. 469. Drum grooving and coating.

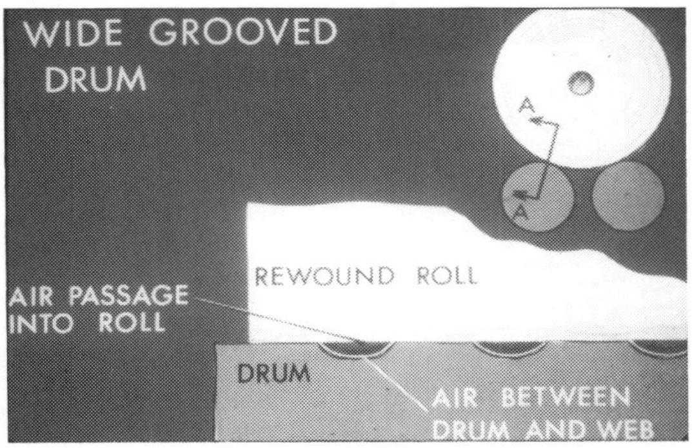

Fig. 470. Tunnel created by deep grooves in first drum.

Fig. 471. Carbide drum coating on second winder drum.

Although certain manufacturers coat both drums, it has been determined that the necessity of coating the first drum is sometimes an indication of inadequate spreading equipment or very high tension found and needed on heavy weights. I must say, however, that this is not true on every grade of paper, and one well might find examples both ways; consequently, I do not believe that this feature (coating both drums) has stood the test of time and should, therefore, be reviewed carefully for your own specific application.

(d) Drum wrap

There are two types of sheet runs addressing the wind-up station of a two-drum winder: a front-drum wrap winder (Fig. 472) and an in-between-drum wrap winder (Fig. 473). The designation of front drum usually denotes, as one stands on the centerline of the paper flow (in front of the winder) and faces back upstream towards the unwind stand, the drum right in front of one is the front drum. The other drum is considered the "back drum".

Fig. 472. Grooved drum on a front drum wrap winder.

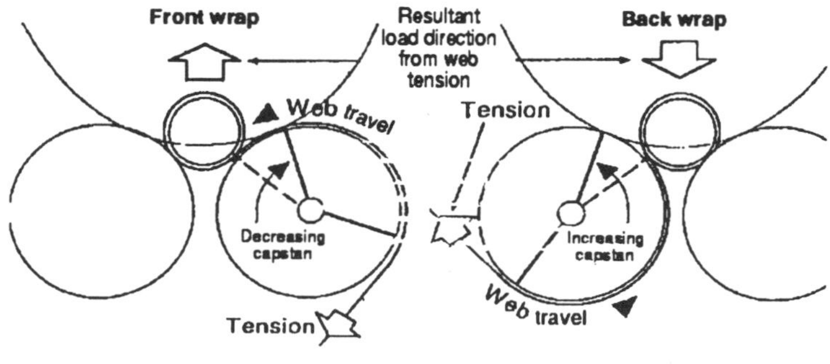

Fig. 473. Drum wrap.

Fig. 474. Anti-friction core-shaft slides.

Fig. 475. Cantilevered-type anti-friction core-shaft slides.

On a front-drum wrap winder, the first drum that the paper touches is the front drum and, although it is easier to make a shipping roll splice on this kind of winder, it does have a tendency to make a harder roll, as changes in tension by the operator (because of a decreasing wrap as the roll builds, Fig. 473) more readily results in wound-in tension changes in the rewound roll. Consequently, on newer applications, to meet the large-roll pressroom needs, the between-drum type of sheet run seems to be favored.

(e) Core shaft slides and shaftless core boxes

The main purpose of the core shaft slide function (Figs. 474–476) is to control and restrict the cross-machine movement of the rewinding roll (to avoid offsets, Fig. 477) without imparting unwanted and concentrated nip to the outside edges of end rolls. Over quite a few years, shaftless winding (Fig. 478) has become necessary, primarily because of roll profile problems. With a core shaft and high areas on your paper profile, one can see that, as the difference in diameter increases as the roll builds up, eventu-

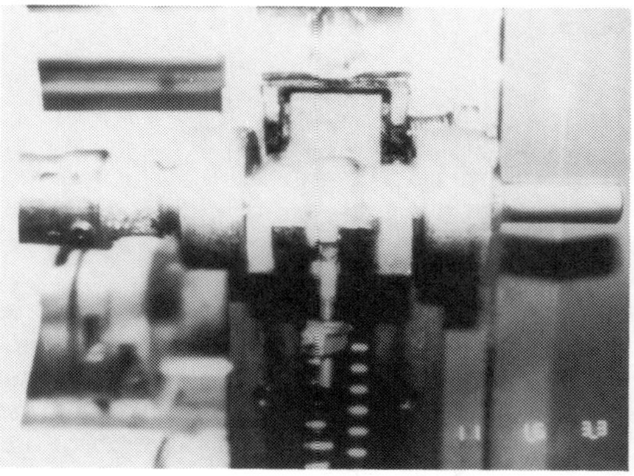

Fig. 476. Friction core-shaft slides.

ally all of the clearance is taken up between the core and the core shaft (Fig. 479), and the center roll is actually lifted from the drum by the shaft in the core. This phenomenon causes a higher nip in the high-profile areas because they are now carrying the weight of the low-profile area on

Fig. 477. Offsets.

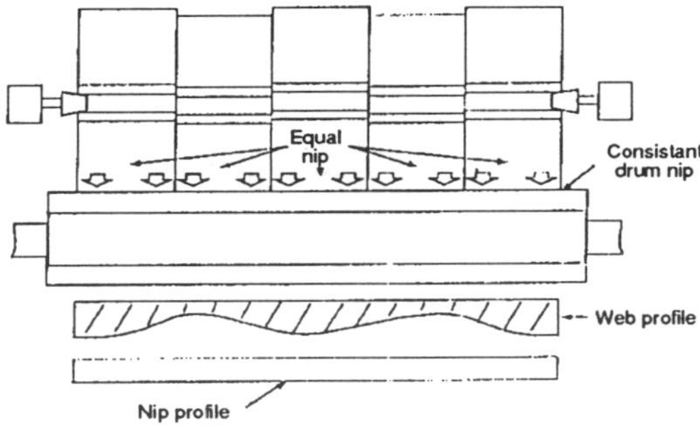

Fig. 478. Shaftless winding.

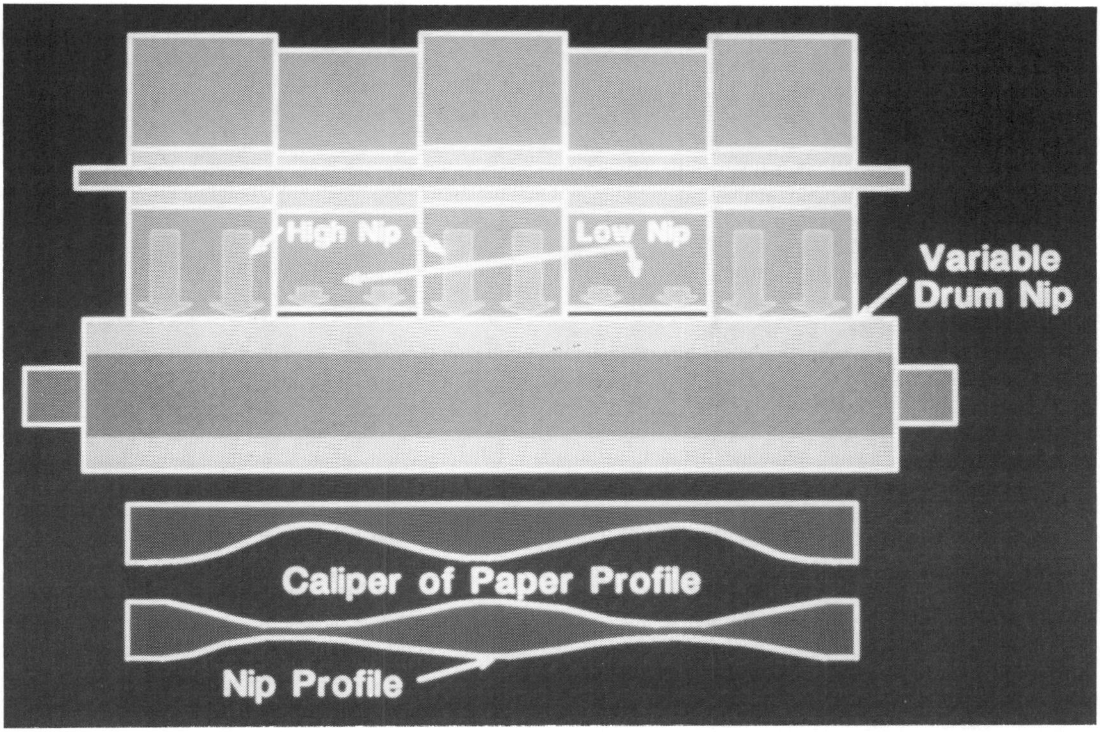

Fig. 479. Core-shaft winding.

each side of the higher profile areas (Fig. 479).

This loss of nip on the low-profile areas (Fig. 479) also causes a loss of control of the web in those areas, creating the strong possibility of offsets (Fig. 477) and interweaving.

With shaftless winding (Fig. 478), on the other hand, each roll is supported on its two high points in the cross-machine profile of only that roll; and the resulting nip pressure is substantially less, thus reducing nip-induced defects. With some nip on every roll, this obviously helps any interweaving problem and loss of control of the web that might exist because of loss of that controlling nip.

A special note on shaftless winding, as this process has a safety problem if not adequately guarded: before installing this process, your winder supplier should be consulted as to the proper guarding.

With the advent of shaftless winding, the core box configuration (Fig. 476) also needed to change, as there are, sometimes, high thrust

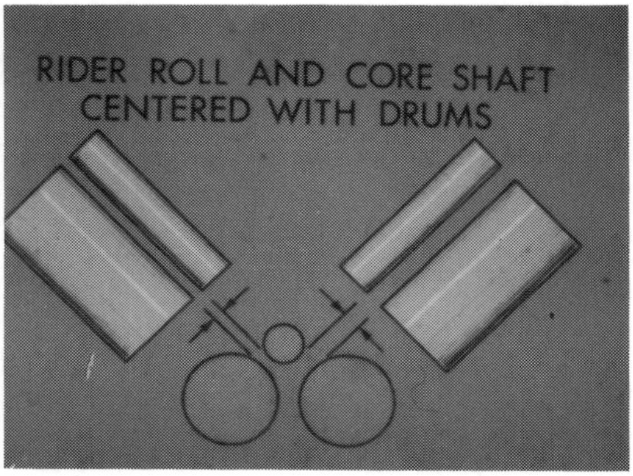

Fig. 480. Core-shaft misalignment with respect to drums.

loads that occur in shaftless winding to keep the roll in line in the cross-machine direction. These high thrust loads necessitated antifriction slides (Figs. 474, 475) (in comparison to the older friction slides, Fig. 475) to ensure that no unnecessary nip load (from the high thrust against the friction in the slides) would be imparted to the rewinding roll of paper, especially at the outside edge of the end rolls. Nonwearing antifriction slides had the additional benefit of maintaining the extremely important dimension integrity (Fig. 480) between the core boxes and the two drums, mainly because there were not any "wearing" surfaces.

(f) Rider roll

The rider roll (Fig. 481) provides the required function of keeping necessary nip pressure at the smaller diameters (Fig. 416) of the rewinding roll, where the roll weight is not sufficient for adequate wound-in tension normally supplied by the nip. It also has a primary function of providing sufficient nip on the second drum, so torque can be transmitted to the roll to make a tight start. The roll can then be unwound without slippage at the core.

Rider roll diameter

The rider roll diameter is usually as large as possible, but is mostly determined by the diameter of the winder drums and the core outside diameter that must be supported by those drums (Fig. 417).

For narrow-width winders, a one-piece rider roll is possible, but with wider machines a rider roll beam is needed with small segmented rolls, to enable them to contact the core without interfering with the drums (Fig. 417).

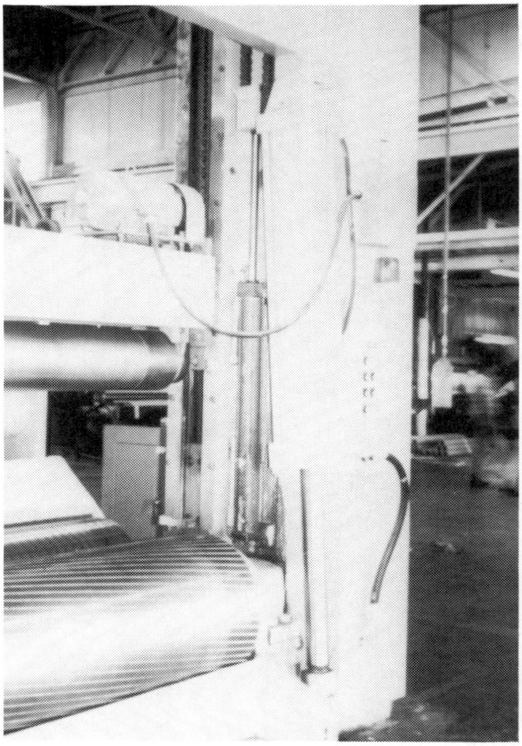

Fig. 481. Rider roll and rider roll drive.

Rider roll drive

The purpose of the rider roll drive is to provide a means of accelerating and decelerating the rider roll without causing a bubble (Fig. 468) or wrinkles during this part of the winding process. Rider rolls are sometimes motor driven, belt driven, or without any drive at all. Drives are especially necessary on lighter weight papers and higher speed winders. Winders that are very slow and running heavy weight papers normally do not need a rider roll drive.

Rider roll control

This control, over the years, has gone from completely manual (Fig. 482C) to a very sophisticated computer control feedback system (Fig. 482B) that basically controls and reduces the nip pressure as a function of the nip on the rewound roll diameter. The purpose is to reduce the nip between the drums and the paper roll as the roll increases in weight (Fig. 416). To control this nip pressure accurately, antifriction slides are needed to ensure that the pressure called for in the control system program will not be altered by unnecessary friction in the slides, causing unwanted, localized nip on the rewinding roll of

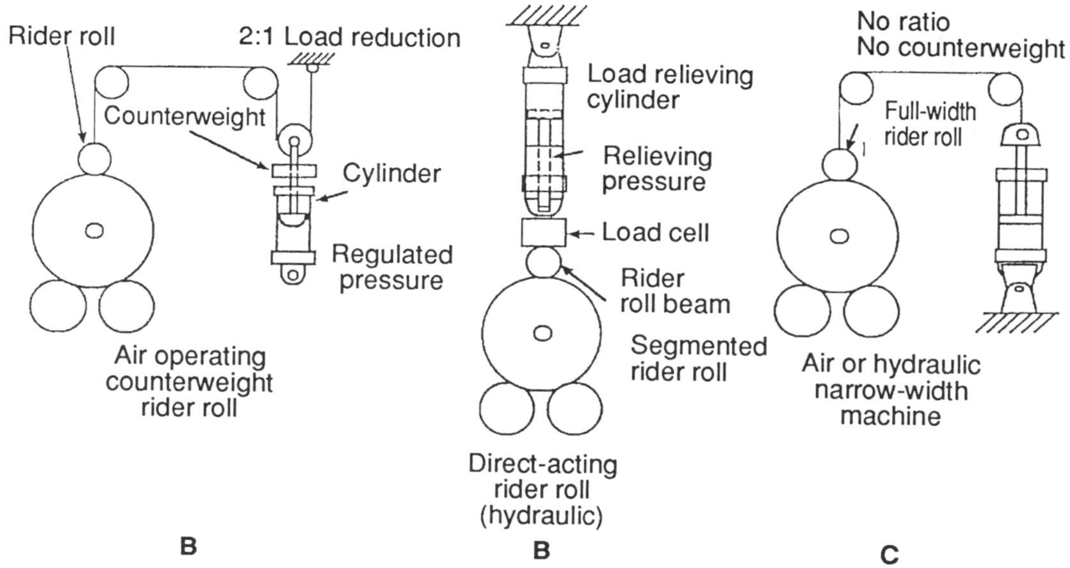

Fig. 482. Rider roll configurations.

Fig. 483. Two-drum winder with two-motor drive.

paper. In many of the older winders, the rider roll hangs on chains (Fig. 482A, C). The chain passes over several sprockets and eventually reaches the actuating mechanism. The friction in the sprocket and the stretch of the chain as a function of time have necessitated that the chain be replaced by direct-acting rider rolls where the rider roll actuating cylinder is directly connected to the rider roll beam (Fig. 482B). This eliminates most of the friction in the system that would affect the nip.

To eliminate the friction factor altogether, some of the direct-acting systems (Fig. 482B) have a load cell mounted directly on the rider roll, thus reading directly the load that the rider roll provides. This obviously allows the closing of the loop in a computerized feedback system. Programming the rider roll to the requirements of the paper grade being run is therefore possible to obtain the proper loading at specific diameters.

Because the profile variances across the face of the roll allow the rider roll to be in contact only on the high-profile area, some manufacturers have in the past experimented with rider rolls that will more closely conform with the profile variations, allowing lower and more consistent loading across the face of the roll.

(g) Two-drum winder drive

The two-drum winder drive (Figs. 483, 484) is usually a two-motor drive. Each motor is directly connected to each winder drum, either through a gearbox or direct-connected (Fig. 484A, B), depending on the rpm needs of the motor and the winder. The drive is programmed for smooth time-rate transition for a change in speed and for different acceleration and deceleration rates – both separately adjustable. A faster deceleration rate is also furnished for safety reasons on "emergency-stop" situations.

Separate power supplies for each motor are normally supplied so that programmed torque as a function of roll diameter can be supplied for optimum roll structure near the core.

Single-motor drives are also furnished, especially on smaller machines (Fig. 484C). In this case, a single motor is direct-connected to the No. 1 drum with some sort of belt tie to the front or No. 2 drum, with speed-changing capability to supply the needed torque program to the second drum.

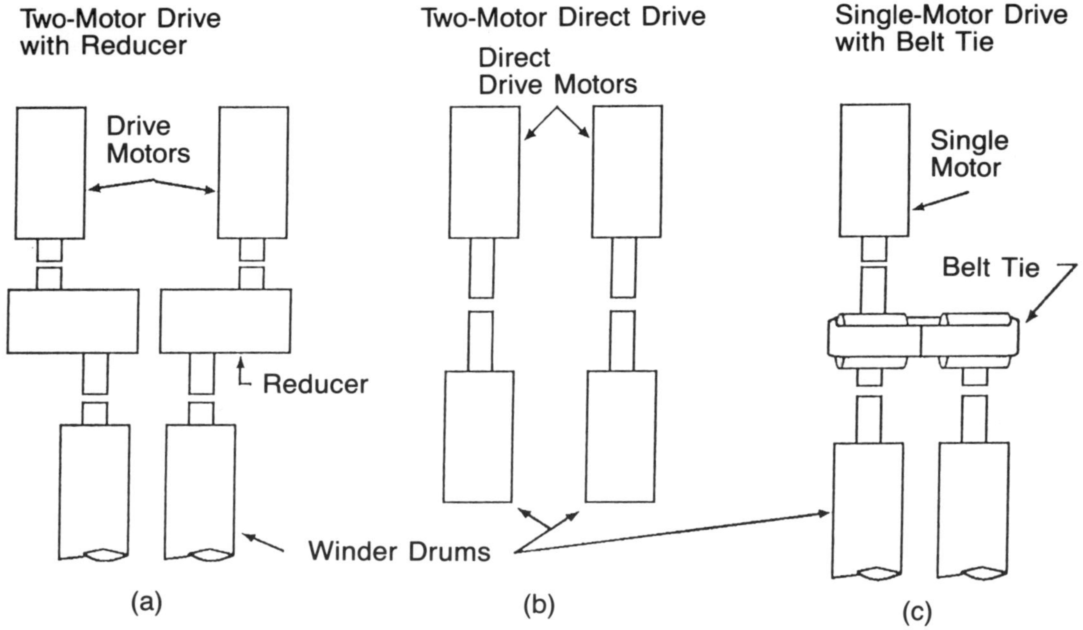

Fig. 484. Two-drum winder drive configurations.

The drive usually is purchased in a package encompassing the regenerative unwind drive, the paper roll drives, the spreader drives, the rider roll drives, and the drum drives. In light of the needs for higher speed and production, the drives include sophisticated feedback systems on each component to ensure tracking during speed changes, "S" curves to ensure proper transition during speed changes, as well as inertia compensation to make tension control more consistent. Auxiliary features such as jog, thread pay-out, and many other production and quality aids are also included.

This basically covers the changes that were necessary to meet the market needs of large roll, lighter weight, and lower strength paper on two-drum winders. Obviously, this did not eliminate the problems; consequently, the industry started using duplex winders that had their own unique problems.

2. Duplex winders

(a) Core-supported winding station

To alleviate the problem of ever-increasing nip as rolls build up on a two-drum winder, the industry has looked for means to reduce this factor by instituting core-supported winding (Figs. 458, 460–462) as found on paper machine reel, re-reeler, and duplex winders. In these cases, instead of the roll weight adding to the nip as the roll diameter increases, the roll weight in this case is supported on its core or core shaft, and the nip pressure between the roll of paper and the drum is not ever-increasing but is accurately controlled. Thus, the main difference between two-drum and duplex winders is in the reduction of the damaging roll weight, nip-induced caused defects usually found on the outside of the roll.

(b) Core diameter

On reels and re-reelers, since the web is not slit, a full-width mandrel is necessary to support the winding roll (Figs. 450, 453).

Shipping rolls, on the other hand, are on small-diameter cores and are narrower in width, necessitating individual support stations for each roll (Figs. 459, 485–489). Obviously, the physical size of the support stations necessitate the winding of each individual roll on alternate sides of the single- or dual-drum winding station. Thus the name duplex winding. Of course, with this arrangement, each station is in fact a separate winder which allows separate control of the nip on each roll (which has mixed blessings). The efficiency and manpower requirements (and material handling arrangements), by the very nature of the equipment, are not as efficient as those found on a two-drum winder. This is why, in many cases, one will find two of these winders following a fast paper machine.

Fig. 485. Bi-wind duplex winder.

Fig. 486. Two-drum bi-wind winder.

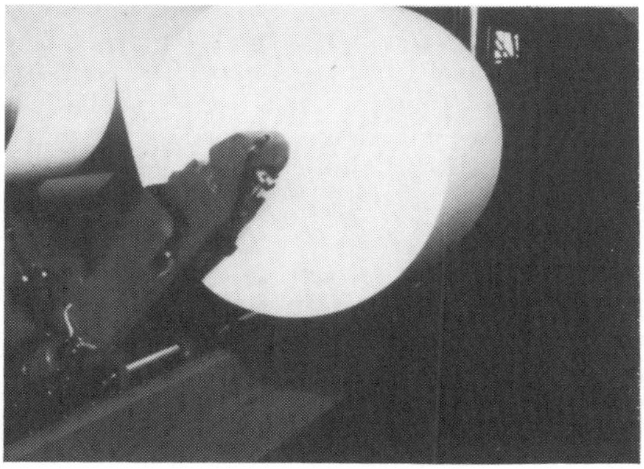

Fig. 487. Ahlstrom center surface winder.

Fig. 488. Jagenberg duplex winder.

Fig. 489. Single-drum duplex winder.

Fig. 490. Wide rolls on a bi-wind winder.

(c) Winding variables TNT

This type of winder requires most of the same winding variables for control as found on the two-drum winder. Therefore, the same TNT of winding applies to duplex winders as well.

(d) Rider rolls

Rider rolls (Figs. 485, 488–491) are still needed on duplex winders to supply the necessary nip at the start, especially on large-diameter and heavier weight rolls. However, the

Fig. 491. Jagenberg Vari-Roll.

rider roll system on a duplex winding station becomes more complicated (Figs. 490, 491) since a separate rider roll (or sectionalized rider rolls for each winding station) is now necessary. And since each winding station has the capability of varying in width, this rider roll must be narrow enough to allow for multiple sections to cover most of the width of the roll. (This is primarily to avoid wrinkles.)

On a single-drum duplex winder, to support the core properly (as is done on the two-drum winder), rider rolls are, in most cases, dual rolls (when the support drum is a single drum) (Figs. 488, 489) to provide core support in this initial stage of the winding cycle. On a two-drum bi-wind (Figs. 486, 490), only a single-roll rider roll is necessary because the two vertical drums provide the core support necessary for a good start.

There have been some winders furnished where no rider roll is used (Figs. 462, 486, 487). This can be done, but it does limit the diameter and the width of the roll that can be successfully wound. Many converting winders do not use any rider roll because the requirements of roll weight, width, and speed are not as severe.

(e) Torque control and drive

Torque control on a duplex winder is provided by two different means:

1. A center drive (Fig. 487) where each station has a separate motor driving the roll at its center, providing the tight start necessary to support the roll in the winding and unwinding process – and this torque is programmed to optimize the roll structure.
2. A two-drum duplex (bi-wind) (Fig. 486), which has two drums arranged vertically (like a two-drum winder turned on its side) where the necessary torque is supplied by the bottom drum very similar to the conventional horizontal configuration.

On some core-supported winders, means of providing a tight start with torque has not been supplied. Again, this is being done in the industry today, but it does restrict the roll length and weight of the rewound roll. Most converting winders do have some means of torque just to pull all of the webs tight in the slitting and winding of narrow-width rolls.

(f) Sheet run

The fact that winding is done on both sides of the drum station necessitates a sheet run (Figs. 458, 460–462, 466) that either goes over or under the inside winding station (Figs. 458, 462). When the web run goes over the inside winding station, special equipment (such as sheet threading arrangements, as well as foot walks) is necessary. When the sheet run passes under the winding station, this requires significant foundation modification and similar threading arrangements, as many such winders are actually located on more than one floor (Fig. 462).

3. Combination core-support and roll-support winding station

This arrangement of winder (Figs. 458, 462) was developed primarily for lightweight, large-diameter rolls of paper. It evolved primarily because it has the ability to limit both the core load and nip pressure between the roll of paper and the drum. In essence, it provides a means of reducing the damage to the core and paper around core and to the outside diameter of the roll, found in the two-drum and core-supported winding equipment. [All of the necessary variables mentioned in Sec. 2(a) above apply to this equipment.]

Most of the earlier converting winders were of this type, but because of the nature of the product (small diameter, weight, and width of rolls) did not require rider rolls and other equipment necessary to minimize damage to the core and the roll. Most of these winders did have the necessary tools of tension, nip, and torque, however.

4. Combining winder — tissue

The combining tissue winder (Fig. 464) winding stations are of three types:

1. Conventional back-drum wrap winder, with rider roll running both shaftless and shafted. This is consistent with the two-drum winder used on other grades discussed earlier.
2. A two-drum winder, front and back drum wrap, without rider roll but having large- and

Fig. 492. Two-drum tissue winder.

Fig. 493. Tissue winder line drive.

small-diameter cores, with core shaft relieving (Fig. 492).

3. A core-support winder, usually the full width of the machine.

(a) Tissue winding theory

Crepe tissue depends on the creping process to maintain bulk. If bulk is lost during the winding process, any bulk variation would create problems in the final packaging of the product. This means, when the tissue is packaged – such as toilet paper and cleansing tissues – and if the bulk changes, either there is too much room in the package, or too little.

Winding on two drums, with the nip on the drums increasing as the roll weight increases, will cause a change in bulk as the roll becomes larger. Obviously this affects the product-package size relationship.

Tissue winder types 2 and 3 above are designed to maintain that bulk by core shaft relieving or core-supported winding (as mentioned previously in Sec. C4). In both of these cases, a large-diameter core shaft is needed so that pli loading will not be exceeded. (Many manufacturers use the figure of 5 pli, 0.9 kN/m.) Any loading above this figure also usually causes wrinkles and corrugations.

The rider roll in type 1 is primarily used with small-diameter cores to keep them from bouncing out of the winder. In these cases, the rider roll is not needed to provide the 5 pli (0.9 kN/m) nip.

(b) Nip or bulk control

When narrow-width rolls are the requirements, a more conventional shaftless winder can be used and, frequently, the pli loading (due to the width and diameter of the rolls) will not exceed the 5 pli (0.9 kN/m) figure; in this case, a conventional two-drum shaftless winder with a rider roll can be used.

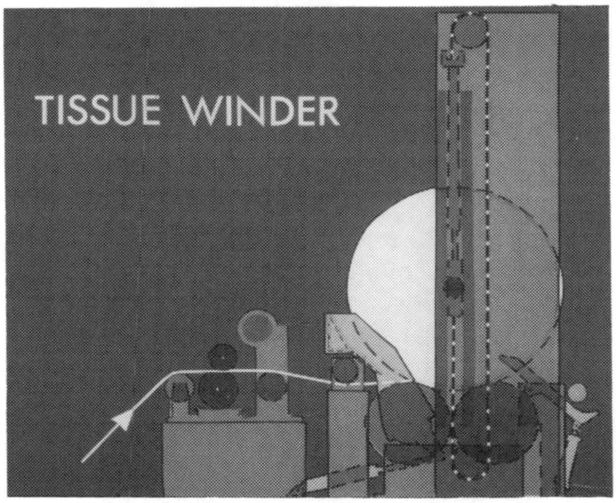

Fig. 494. Sheet run over the back drum of a tissue winder.

(c) Tension control

Because of the nature of crepe, tension control in its normal configuration is not used, as the crepe in the web would be pulled out to develop any tension on a tension-sensing element such as a load cell. Consequently, a draw control system is used to accurately control the crepe pull-out by controlling the differential speed between sections of the total machine (Figs. 464, 493).

(d) Sheet run

With the necessity of special control of the draw between sections, the sheet run on a tissue winder – instead of the conventional wrap of either the front or back drum – has very little wrap and usually presents itself to the rewound roll over the top of the back drum (Fig. 494) so more effective use of the draw can be accomplished to control crepe pull-out in the web.

(e) Drive — tissue winder

The drive on a tissue or combining-type winder (Fig. 493) is one that accurately and, more often than not, mechanically ties the different stations together. The basic requirement is that the draw be controlled without excessive crepe pull-out or any damage to the web.

5. Summary

Structure control of a roll of paper is accomplished through the TNT (tension, nip, and torque) of winding.

Every winding process tries to obtain the most complete and comprehensive control of these variables to optimize the structure of the roll – but, more importantly, to reduce roll defects in the pressroom and converting plants.

The most significant and infamous member of the TNT club is nip. It's infamous primarily because, when it is non-uniform, it causes a high percentage of the problems in the pressroom and the converting plant.

Because of the uncontrollable damaging nip of a conventional two-drum winder and the pressrooms asking for larger and heavier rolls, other winder configurations were developed in an attempt to overcome this problem that seemed to always occur on the outside of a roll of paper.

But in these new winding processes, other problems began to occur – this time, at the core. Consequently, another new series of winder configurations were developed (Fig. 467).

And some problems still exist since the many factors that cause damage at the core and the paper surrounding the core exist not only in the paper mill but in the pressroom, and to alleviate this problem a close relationship of information exchange will need to develop between the paper mill, the pressroom, the core manufacturer, the paper machine manufacturer, and the press manufacturer.

As in any unsolved problem, time and need are good factors that help progress.

Much time has been spent on structure control, and rightly so, as most of the problems relating to structure control usually result in roll defects found in the pressroom or converting plant.

There are, however, many defects that actually happen during the processing of paper in the paper mill and, because they are obvious, they never reach the ultimate user's plant. Many of these defects, such as interweaving, offsets, and snapoffs fall into the category of web control.

Web control, which is the second basic function of the four functions of winding, encompasses most winding processes.

G. WEB CONTROL FUNCTION

Web control (see Fig. 390) is concerned with those factors that control and affect the web of material as it travels along its path to the windup roll or to its first nip or to both. This could be on a paper machine from the previous process (such as calendering) or, in the case of the winder, from the unwinding roll. Some of the factors involved in web control are:

1. Tension control or draw control
2. Traction control
3. Spreading
4. Other
   a. Edge guide
   b. Oscillation
5. Speed.

1. Tension control

Why is tension important?

1. Tension is needed to transport the web from one place to another (Fig. 390).
2. Tension is needed to maintain traction between the web and the paper-carrying components (Fig. 495).
3. Tension is needed to effectively spread (Fig. 390), slit, and unslit webs.
4. Some tension is needed in the process of web separation, commonly called slitting (Fig. 390).
5. Tension assists in building the proper roll structure (Fig. 390).

What are the basic components of most tension control systems?

1. Web
2. Tension controller (Figs. 496, 497)
3. Tension feedback or sensing mechanism (Figs. 496, 498)
4. Tension producers.

(a) Web

The tension necessary to transport a web is dependent on the web itself. Different grades require different tension levels to perform the control necessary. The TAPPI web tension curves (Fig. 499) give at least a starting point to determine at what level the tension needs to be as a function of paper grade. Of course, the operator determines this many times by looking at the draw and increasing or decreasing the tension to reduce bags, corrugation, and other web-type defects. The effectiveness of the tensioning process is determined by the ability of the system to stretch the web. Obviously, on heavier weights, it takes more tension to stretch the web than it does for lighter weights. This is reflected in the TAPPI curve.

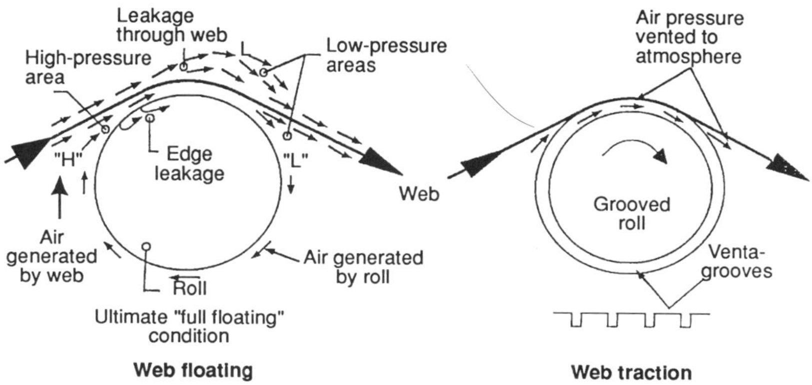

Fig. 495. Air flotation — Venta-grooves — no grooves.

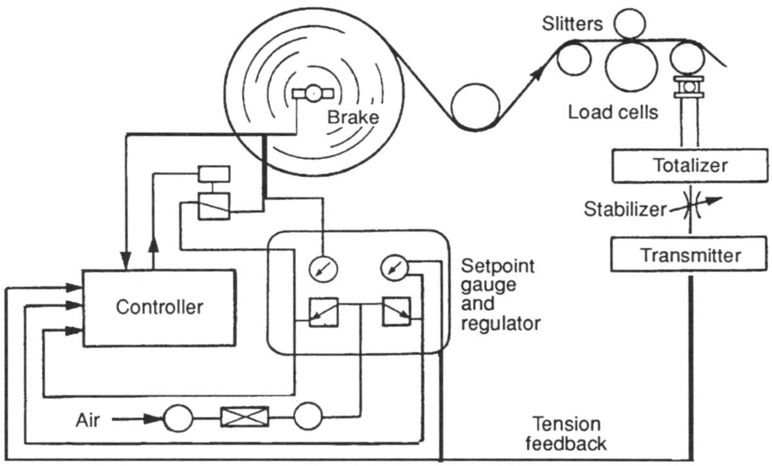

Fig. 496. Tension system.

Fig. 497. Tension controller.

Fig. 498. Segmented tension rolls with multiple load cells.

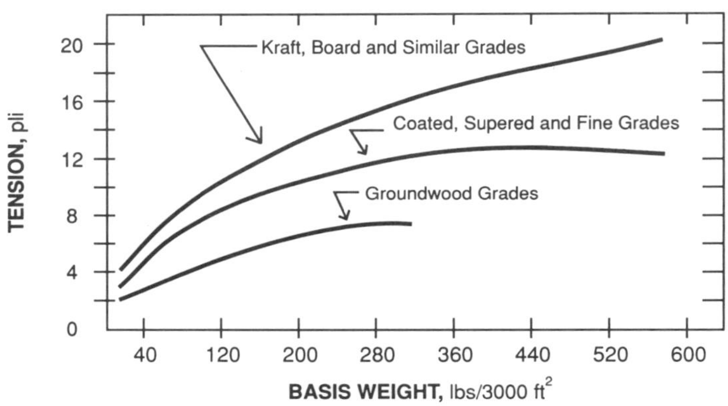

Fig. 499. Tension vs. basis weight.

(b) The sensing element

The tension-sensing elements (Figs. 496, 498, 500) are transducers that provide the function of measuring and feeding back the level of tension in the web to some sort of controller in the system which, in turn, controls or trims the tension-producing braking system. The sensing element transducers are of three basic types:

1. Hydraulic load cell (Fig. 500)
2. Electric load cell
3. Air-bleed system.

The sensing element is usually mounted under a roll (Fig. 500) or a series of segmented rolls (Fig. 498) mounted on a beam which is located in a position where the sheet run will not change because of an unwinding roll or by spreader adjustments (Fig. 390). The sensing element should also be in a position close to the windup drum, so that tension transients from loping out-of-round unwinding rolls will be minimized before reaching the tension transducer. This condition would, of course, cause the tension transducer to become unstable.

(c) Controller

The purpose of the controller (Fig. 497) is to react expeditiously to a tension change with maximum stability during transitory or extreme situations.

The controller can either be pneumatic or electric, and basically receives a signal from the tension-sensing element and conditions it to react to the rate of change of an error in the process control reset. In other words: How often is the signal reset while awaiting the next error – the rate of change to the error signal. How fast the error is occurring. What is the magnitude of the error.

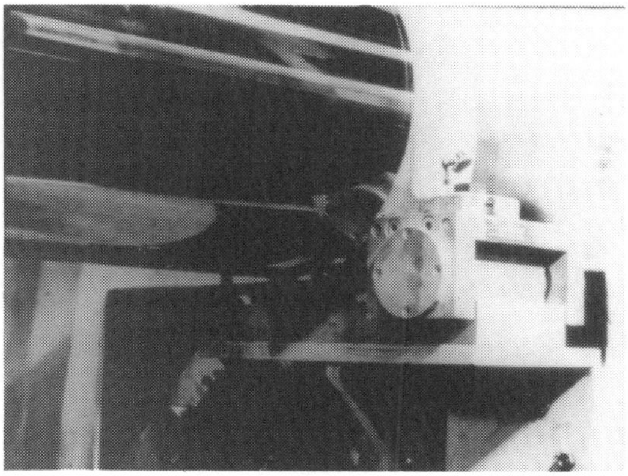

Fig. 500. Hydraulic load cell.

In other words, if there is a large tension change that is very large and abrupt, the controller would sense this and would react differently but with the same controlled results as a tension change that was small and less abrupt.

Some systems control the tension to a braking system directly (as described above) while others, such as regenerative unwind, which programs the necessary braking (based on reel diameter) to automatically maintain a given tension level, use the tension signal to only trim the unwinding paper roll regenerative drive to maintain the desired tension level (Fig. 501).

(d) Tension producers

There are many things that produce tension during the winding process. Some are desirable elements; others, less desirable. It is necessary, however, to recognize and understand these tension producers so proper compensation can be made for the types that are possibly unwanted. The list follows with no comment, as the nondesirable elements are usually compensated for by the equipment manufacturer:

1. Brakes
2. Regenerative drive, acceleration and deceleration, and S curve
3. Eccentricity of unwinding roll
4. Nonrotating elements such as drag boards
5. Spreaders and spreader drive
6. Slitters and slitter drive
7. Paper rolls and paper roll drives
8. Traction control.

The second important function of web control is traction control. It is important primarily because a change in traction can be caused by many elements in the winding process, and this

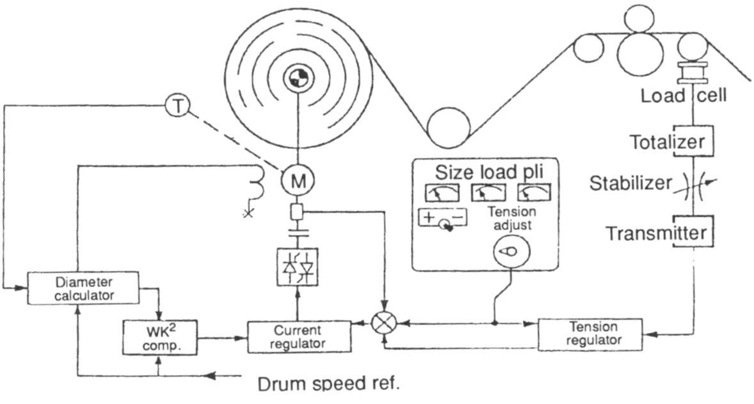

Fig. 501. Regenerative brake tension system.

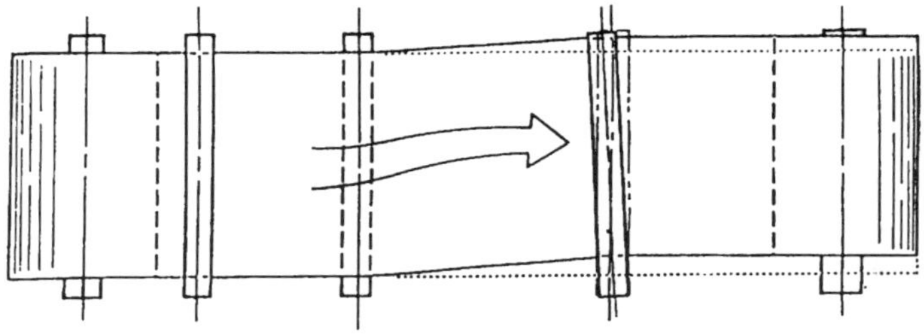

Fig. 502. Misalignment of roll web path.

Fig. 503. Edge profile problems.

change can affect the web and the rewound roll in the form of offsets and snapoffs.

2. Traction control

The reason traction control is so important to any winding system is because a change in traction has a guiding effect on the web as it passes over paper-carrying rolls in any web-carrying system. With improper traction control, wrinkles, snapoffs, offsets, and interweaving all can occur.

It may be fair to say that all rolls in any system, no matter how well they are aligned, are misaligned enough to have a certain amount of a guiding effect on the web (Fig. 502). Due to any given roll misalignment, that roll will guide the web in a given direction until the web passes the next roll in the web run which, with its own influence, could guide the web in still another and different direction. As the web passes each roll in the system, each roll has its own effect on the web.

Combining all of the different directions that the web is being told to go, the web finally finds an equilibrium point that satisfies all of the combination of forces being directed at it. It is not unlike a marble rolling on an uneven sloped cement floor. If started at the same point, the marble will find, basically, the same path of least resistance each time.

If for any reason any one of those rolls should lose their grip on the web, the web would then find another equilibrium point, and a detrimental offset (Figs. 447, 503) on the edge of the rewinding roll will occur. The reason that the offset is detrimental is that many of the rolls are shipped on their ends and, with an offset, the web edge could be damaged or cracked, or both, causing a web break in the next process.

There are many reasons for a change of traction on any web-carrying system.

1. Air carried by the web
2. Change or loss of tension
3. Too little wrap
4. Baggy webs
5. Moisture
6. Speed
7. Out-of-round unwinding roll

Fig. 504. Low-speed nonfloating web.

Fig. 505. High-speed floating web.

8. Too much spread
9. Nonsynchronized paper roll drives
10. Long draws
11. Vibration
12. Paper grade
13. Smoothness of paper
14. Coefficient of friction, paper-carrying roll surface
15. Paper-carrying roll diameter
16. Sheet flutter
17. Uneven profile of web; different length "tracks"
18. Roll bearings.

Each of these items is significant; some are more critical than others, for example air being carried by the web and the air pumped by the roll surface. Creating a floating and loss-of-grip condition are two major reasons for the loss of traction and subsequent offsets on the edge of the roll (Figs. 447, 503).

(a) Air flotation of web on paper-carrying rolls

The effect of porosity and speed was first determined by Daly [7] where he considered the behavior of a web passing over a roll. Daly's statements on the subject have stood the test of time, and after all these years his findings are still valid on the subject of "traction control".

Daly stated that, as a web approaches a roll, both the web and the roll are carrying air on and near their surfaces (Fig. 495). This air is carried into the wedge-shaped nip between the web and roll and is forced to escape the best it can through and out to the sides of the web.

On porous webs, the air can easily pass through the web (Fig. 495) as well as spill out to the edges, so there is little (or no) buildup of pressure. Whereas on the nonporous webs, such as coated paper, or on very fast machines, air cannot escape through the web, so it must all be forced to the edges of the web (Figs. 504, 505).

The back pressure, caused by forcing all the air to the edges, causes air to be forced under the web at the centerline of the machine or the center of any web located in any cross-machine web position. As this happens, that portion of the web is lifted off the roll; the roll is no longer effective in producing traction and, in turn, loses its guiding effect on the web. The web then finds a new equilibrium point and an offset occurs.

Higher and higher speeds pump more air under the web and lift more and more of the web off the roll. Progressively, as a function of speed, the web loses more of its tractive ability and is essentially only supported on a film of air. This is the ultimate fate of any web as the speed is increased. This is quite graphic on a very porous web such as tissue (Figs. 504–506). Snapoffs frequently occur because of this phenomenon, as the tension load is transmitted to the edges of the web and, if weak as in newsprint, can cause snapoffs.

The air pressure required to support the web against the tension is quite small. For instance: at 2 pli (0.35 kN/m) tension, 0.25 psi (1.72 kPa) under the web will lift a web off a 14 in. (356 mm) diameter roll (Fig. 507).

When the web is full-floating on air, it cannot transmit enough torque to the roll to overcome bearing resistance and roll windage, so the roll will slow down and stop. In a full-floating condition, the web is very sensitive to outside forces and can be caused to shift easily, with very small forces, in directions other than web-flow directions. This means that webs at or near a full-floating condition will be more difficult to maintain in stable running condition, even if there are no traction changes taking place.

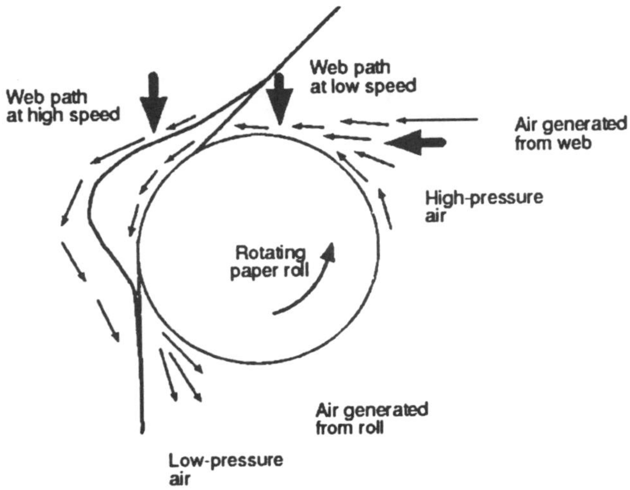

Fig. 506. High-speed floating web.

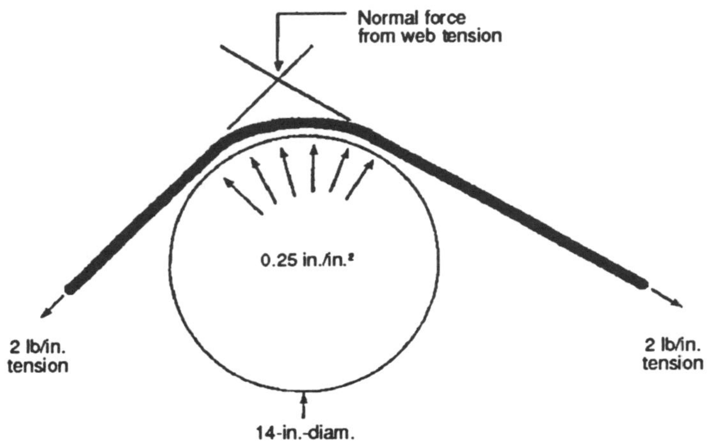

Fig. 507. Traction vs. web tension.

(b) Paper roll grooving

To overcome this problem, a special grooving (Fig. 508) was developed by Daly and Patterson (of Beloit Corporation) for a paper roll which, in effect, provided a means of venting the pressure buildup in the wedge between the paper and the roll. The grooving was first made annular, but problems developed where longer, varying draws and high-tension areas in the web actually pulled the web down into the groove, causing wrinkles.

This was overcome by machining the groove at an angle or lead whereby the web will approach the groove at an angle, thus retaining the venting effect but giving full support to the web as it passes around the roll (Fig. 469).

Fig. 508. Annular groove on a paper roll.

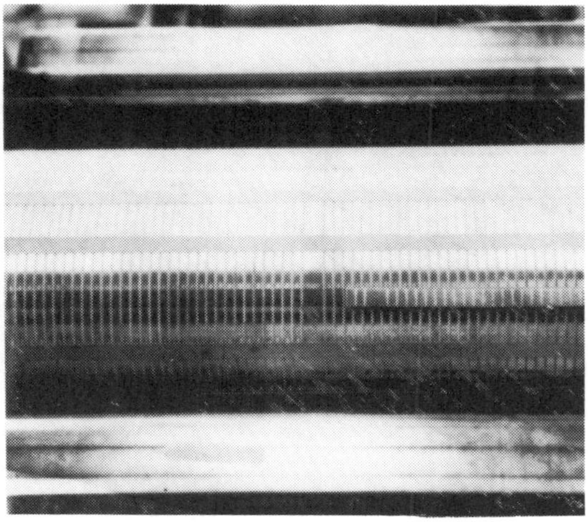

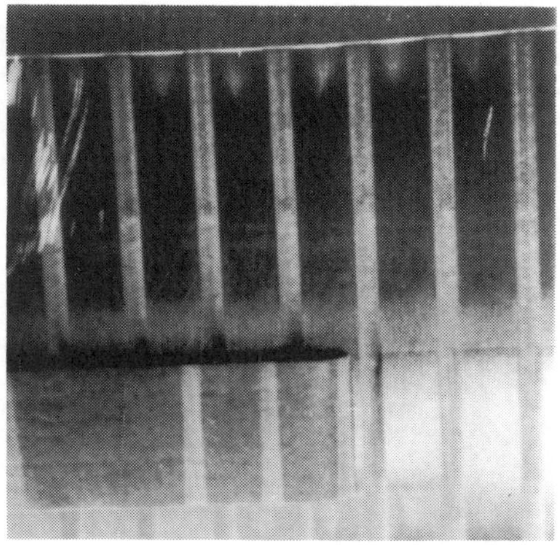

Fig. 509. Venturi effect of Venta-grooved rolls.

An added benefit to this groove was that, as the air was vented and rushed through the groove, a drop in pressure on the downstream side of the groove actually created a Venturi effect (Fig. 509) which sucked the web down on the roll, actually enhancing the tractive abilities between the roll and the paper. This led to the small-diameter, low-inertia rolls (developed by Frye) found on the slitter table of many winders (discussed in Sec. H) which allowed the web to drive the roll without the necessity of providing a drive.

Spreader roll grooving and their drives are possibly the most necessary – and more important than any other paper roll grooving drive. This will be discussed later under Spreading (see Sec. G3).

(c) Paper roll drives

Paper roll drives are necessary on winders to overcome the inertia of the paper-carrying rolls. With full-width paper-carrying rolls, the diameter and the inertia are such that the friction between the web and the roll is not sufficient to maintain traction through all ranges of speeds. As the winder is accelerated to a top speed of, say, 8000 ft/min (41 m/s) – 1000 ft/min (5 m/s) on a coated web, and, possibly, 2500 ft/min (13 m/s) on newsprint – the paper loses its grip on the roll and consequently floats. As the machine is decelerated and the film of air collapses under the paper, the paper now sometimes abruptly grips the roll again and a high-tension transient occurs, sometimes enough to break the web – but more often than not causing an offset. The drive is necessary to overcome these transients.

(d) Salvage winder paper roll drives

On salvage winders, another condition presents itself, necessitating steps to avoid losing traction. It is of more concern on the salvage winder primarily because narrow webs, say 24 in. (610 mm), are wound on a salvage winder whose width is, say, 110 in. (2794 mm). The narrow web in this case does not have enough tractive ability to overcome the bearing friction and the inertia of the wide roll, so it slips during the acceleration and the deceleration process, causing offsets on the edge of the rewound roll.

Drives and/or segmented narrow-width carrying rolls are necessary to accommodate the varying web widths normally processed through a salvage winder. Salvage winder paper roll drives can be relatively simple, using a simple belt drive, either driven from a single motor driving all rolls or a similar belt drive directly connected to the drum.

Many of the newer salvage winders have a combination system: some, segmented rolls that need not be driven; and full-width rolls, where wrap and the many other conditions that effect traction necessitates a drive of some sort.

(e) Tissue paper roll drives

Tissue paper roll drives, because of the nature of the product (low tensile strength), are usually tied to the main-drive draw control system. With tissue being very porous and also having a low tensile strength, the obvious solution to this situation (as it pertains to paper roll drives) is to drive all of the paper-carrying rolls. Due to the enormous amount of air that is carried by the web, the conventional Venta-grooving, in most cases, has been found to be inadequate. Or it seems to look that way.

Fig. 510. Incorrectly located before slitter spreader.

Fig. 511. Correctly located spreader.

As seen in Figs. 504–506, the flotation of the web as a function of speed occurs on the driven grooved roll. It was found that, not only were the grooves inadequate but, even after flotation, the web continued to bellow (blow or float), finally bellowing so much that it actually broke down the web (Figs. 505, 506). As Daly stated previously, the surface air (carried by the roll itself) is an additional source for pumping air into the wedge between the paper and the roll. The solution to this was to drive the rolls at a slower speed, thus alleviating the pumping action of the rotating roll surface and that source of air causing detrimental flotation.

This lesser speed did not have any detrimental effect on the crepe pull-out. Also, the same condition and solution have happened on the very high speed newsprint machines.

3. Spreading

Spreading (Fig. 390), the third important function of web control as the definition reads, causes the web to spread out or lie flat. By the configuration and design of the equipment necessary to spread, traction control is an important consideration in the selection of the equipment.

Spreaders in the winding process have two basic functions: full-width web spreading and a multiplicity of slit web spreading.

(a) Nonslit web spreader

The full-width, nonslit web spreader (Figs. 510, 511) is used primarily to avoid wrinkles as it passes over rolls with a large wrap and long draws. The same condition can occur with small-diameter rolls with a moderate-to-large wrap. The most common usage, however, is spreading just before the slitters (Fig. 511) to ensure that the web lies flat as it passes through the slitters. If the web is not spread properly before it passes through the slitter, it will many times become wider after slitting in the rewound roll, causing an offset. The spreader in these cases is used to alleviate the tendency to wrinkle, and, if incorrectly located, can actually cause wrinkles (Fig. 510). Unfortunately, many times a spreader becomes a crutch to compensate for a problem caused by other things – such as too-long draws.

(b) Slit web spreaders

The slit web spreader, on the other hand, is used, as the name implies, to spread the web after the slitter. The slitter, as stated previously, divides the full-width web into usable widths for the ultimate customer or the next process. To process a multiplicity of webs at high speeds in an orderly fashion, control of each individual slit web, so it does not wander, is extremely important to avoid interweaving and offsets on the rewound rolls.

The first and primary function of this type of spreader is to provide spaces between each web (say, 0.030 in.) (Figs. 512, 519) in the cross-machine direction so that any individual web will not interfere with the adjacent webs, causing snapoffs and interweaving, dishing, and many

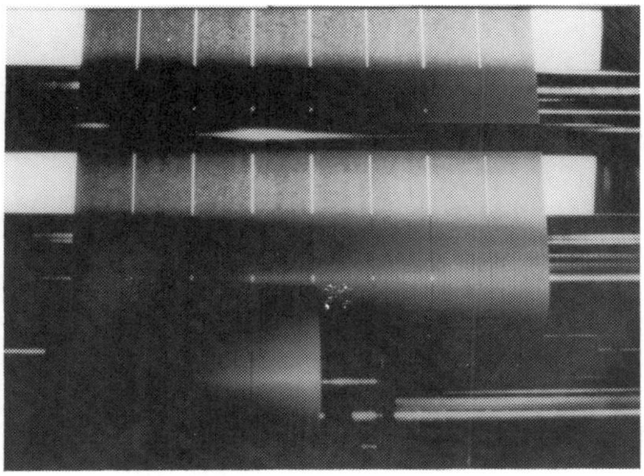

Fig. 512. Split-web spreaders provide spaces between each web.

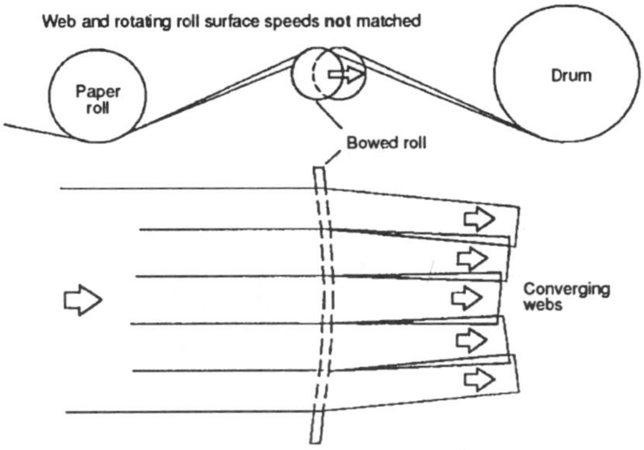

Fig. 513. Webs converge because of slipping on spreader.

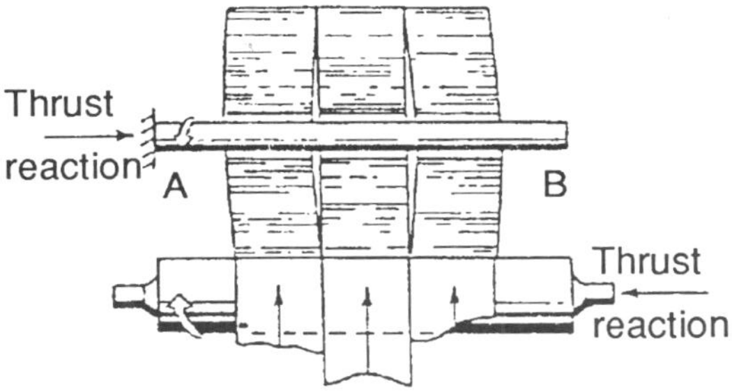

Fig. 514. Symmetrical dishing.

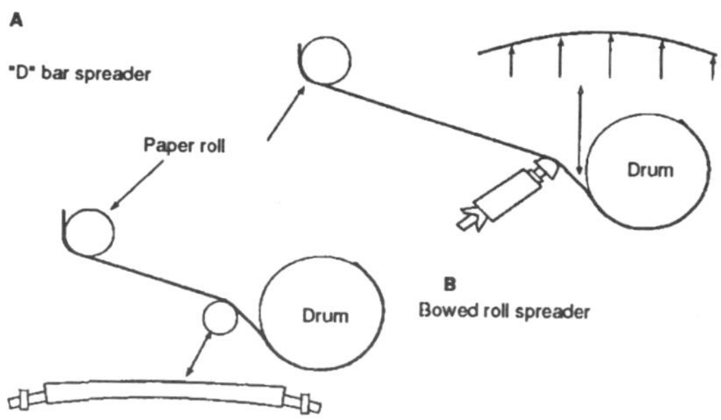

Fig. 515. Bowed-type spreaders.

other winding-type defects. Interweaving (Fig. 513) is the lacing or overlapping of webs from adjacent side-by-side rolls to the extent that the rolls cannot be broken apart easily.

Dishing (Fig. 514) defines where the roll edge becomes the shape of a dish, or any edge that deviates from the norm, or where the edge of the roll is not perpendicular to its face. (Many times this is caused by the after-slitter spreader.)

To provide this proper spacing, the configuration of the spreader roll is usually a bow shape (Fig. 515), causing the web to move in cross-machine directions, symmetrical around the centerline of the machine.

To move the slit webs in a cross-machine direction and to maintain this spacing at any speed requires four basic fundamental criteria:

1. The proper tension
2. The proper traction
3. The proper spreading system
4. A proper nip to hold the other end of the spread web.

(c) Tension and spreading

Tension has already been discussed, but it should be emphasized that, without tension, moving a web in the cross-machine direction, as a spreader tries to do (which is really a direction that puts unnatural and unwanted stresses on the web), would be impossible. By the very nature of a bowed device, intimate and consistent contact is necessary to make this movement possible – or, for that matter, to work at all.

(d) Traction and spreading

Traction has also been discussed but, again, without traction, especially on a rotating spreader, the spreading operation would not function. Most spreader rolls of today are grooved to vent the air carried by the web so the web will not lose contact.

Drives are also used, primarily because a rotating bowed roll has many bearings spaced across the machine. These bearings are sometimes covered with a rubber sleeve or a

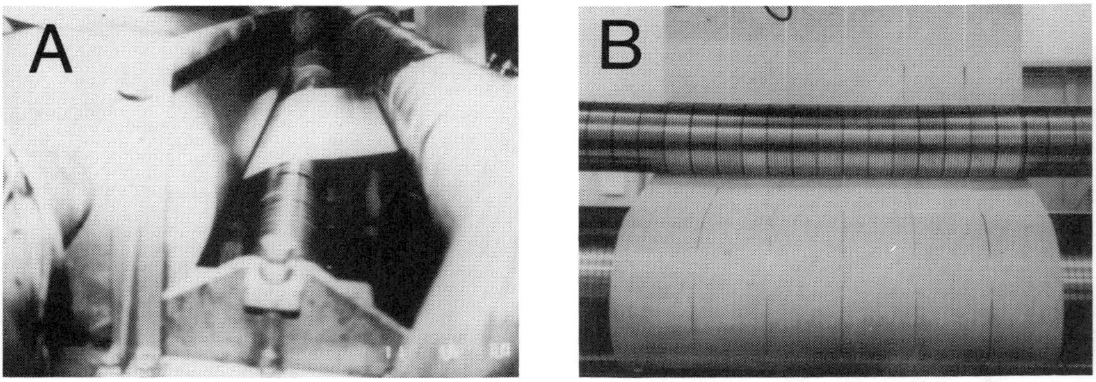

Fig. 516. A: Segmented bowed roll. B: Segmented bowed roll on salvage winder.

segmented steel sleeve – all of which, because of the friction resistance, make it very difficult for the web to consistently drive such a roll without the loss of traction.

If a web is baggy on one edge, with very little tensioned contact with the bowed roll, and the other edge is tight, the tight edge side would obviously have to drive the entire roll. In this case snapoffs and most certainly a loss of traction could occur; therefore, a drive is a necessity.

The only condition where a rotating spreader roll drive would possibly not be necessary is a spreader that is made up of narrow segments (Fig. 516A) allowing each segment to be web driven. The segments in this case are from 4 to 24 in. (102–610 mm) wide, depending on the narrowest web to be spread. Normally, the narrower, the better – except for cost. Each segment is mounted on its own bearings and is usually Venta-grooved and coated with a friction surface. Since that portion of the web in contact with that particular segment of the spreader drives only that segment and not the whole roll, a drive is not necessary, as traction can be maintained. This type of spreader (or a spreader with a drive) is recommended for salvage winder (Fig. 516B) operations that need rotating-type spreaders.

(e) Types of spreaders

There are many different types of spreaders. It is not that one spreader is better than another, but certain spreaders are designed only for specific applications. Therefore, all spreaders, depending on the application, have specific limitations. Understanding the limitations will also define the applications.

There are two basic types of spreaders whereby the name describes the method in which they spread:

1. Fan (Fig. 517)
2. Folding (Fig. 518).

These two basic types can be further broken down:

1. Rotating (Fig. 515B)
2. Sliding (Fig. 515A).

Sometimes these are called friction and positioning spreaders, where in one case (rotating) friction between the web and the spreader roll is necessary to move the web in the cross-machine direction, and in the other case (sliding) the web is positioned or displaced by the spreading device in the cross-machine direction.

Rotating indicates that the spreading elements are rotating at the same speed as the web, and sliding indicates that the spreading elements are at zero speed in relation to the speed of the web.

There are other refinements or choices that must be made in the selection of the proper spreader, such as: Is the web coming to the spreader a fixed width or a variable width? Must the bow be adjustable or fixed? Is localized web doctoring needed? (Doctoring is defined as creating a higher tension in a localized area using a spreading component.)

(f) Fan-type spreader

A fan-type spreader, such as a bowed spreader roll (Fig. 515B), D-Bar (Fig. 515A), or a stationary bowed pipe, spreads the web in a cross-machine direction, not unlike the geometry of a fan (Fig. 517) with all the webs diverging from a central location and symmetrical around that location. The web approaches the fan-type spreader system first, which we will call the paper roll before the spreader, with no cross-machine diverging from the web path at that point. Once the web is in contact with the spreading device, which is the second component in the

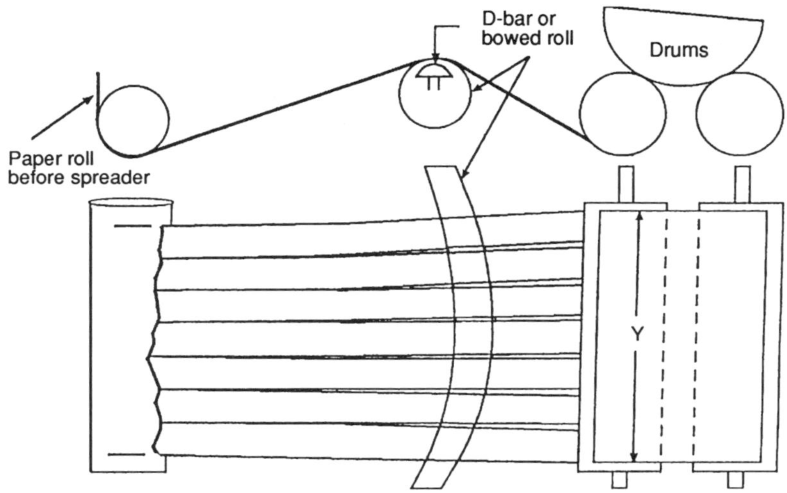

Fig. 517. Fan-type spreader.

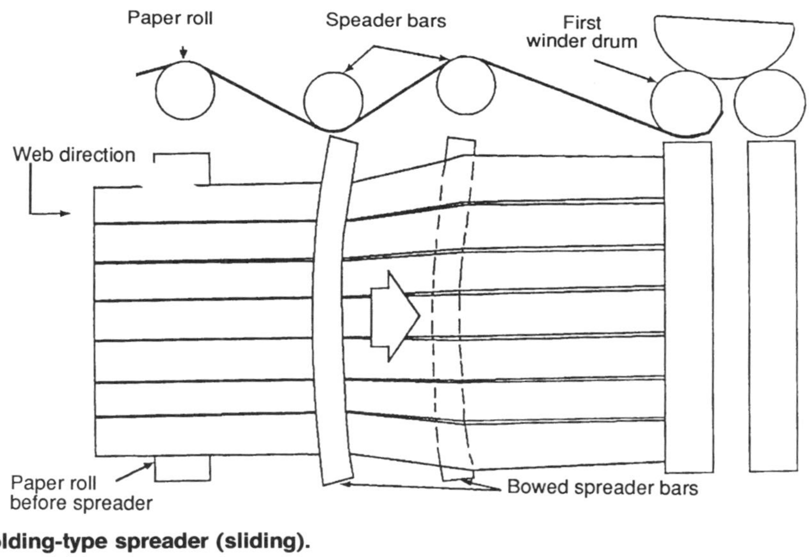

Fig. 518. Folding-type spreader (sliding).

spreading system, it is twisted and takes on a direction that is perpendicular to the axis of that segment of the bowed spreading component.

As one can see (Fig. 519), by increasing the number of slit webs each with, say 0.060 in. (1.5 mm) between them, the outside web must not only move 0.060 in. but also 0.060 in. for each web that is between it and the centerline of the machine. Moving the other way, from the centerline of the machine, the same thing occurs; each slit web must move 0.060 in. to not interfere with the web next to it. Thus, the second web from the centerline of the machine must move 0.120 in. (3 mm), the third 0.180 in. (4.5 mm), etc. In this situation, it is quite obvious that eventually the outside web will have to move so far that a slack edge occurs, the web becomes unstable, and bad offset or interweaving could occur.

From the preceding description, one can see (Fig. 519) the limitation to the fan-type spreader. The more slits, the more the web must be moved in the cross-machine direction. The outside web (having to move the farthest) becomes unstable, and problems occur.

For example, to better describe this so-called slack edge, take a 2 in. (51 mm) wide strip of paper (approximately 10 in., 254 mm, long) and lay it flat on the table, with one end firmly anchored with a piece of tape; now, move the

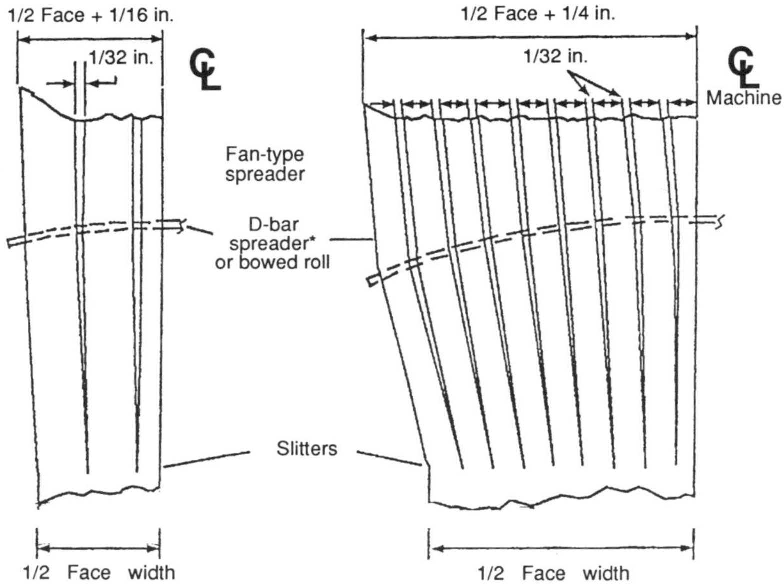

Fig. 519. Displacement of multiple slit widths.

other end of the strip of paper in a direction perpendicular to the long edge of the paper. This is not unlike the situation where the trim is pulling at an angle at the trim slitter, causing the same bubble (Fig. 520). You will see that one edge of the strip buckles, while the other edge seems to be under tension. Tension in the spreading system attempts to tension the web to the degree that the tight edge is stretched a sufficient amount to allow the buckled edge to lay flat. This, of course, must be done within the limits of the strength and the stretchability of the paper, and is why the spreading of very heavy board with this type of spreader becomes difficult to say the least. This is also why this type of spreader is said to be tension sensitive, as a slack edge means that all of the web in that section is no longer in contact with the spreader and that portion of the web that is in contact becomes the only means of stability for the web; and, as the amount of spread increases, this limited contact is eventually not enough to control the web, and instability occurs.

As stated earlier, in this case the need for tension is to stretch the tension edge to the degree that the buckled edge will lie flat as it proceeds over the bowed spreading component. If there are too many slits, the buckle becomes increasingly excessive on the outermost web and the tension level becomes necessarily too high on the tight edge, and snapoffs will occur. On the other hand, if the tension level is lower, it may still not stretch enough to hold the web in this stretched condition; then, oscillation between stretching and slipping occurs, with the web losing a consistent grip on the spreader, thus becoming unstable. Therefore, a limit will be reached regarding the number of slits that can be spread with this type of spreader.

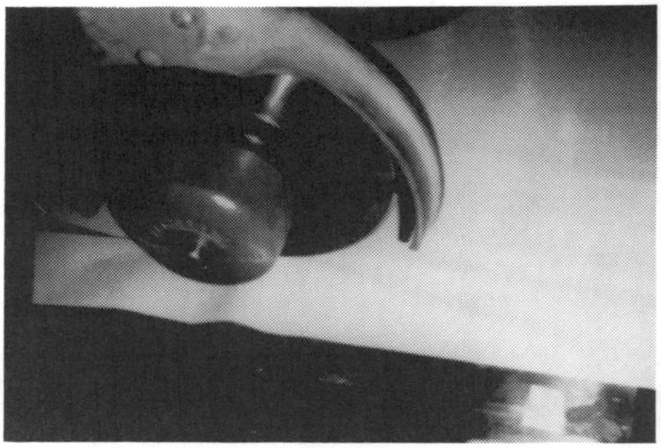

Fig. 520. Web trim being pulled in a direction other than machine direction. Notice bubble on the left and diagonal wrinkle on the right.

During acceleration and deceleration a change of tension sometimes occurs, causing an offset on the edge of the roll that shows graphically how the change of tension during acceleration and deceleration increases and decreases the grip on the spreader (Fig. 447).

Another factor in this instability is the radius of curvature of the spreading element itself (Fig. 521). Radius of curvature can be defined as the radius of the bowed element, which is very

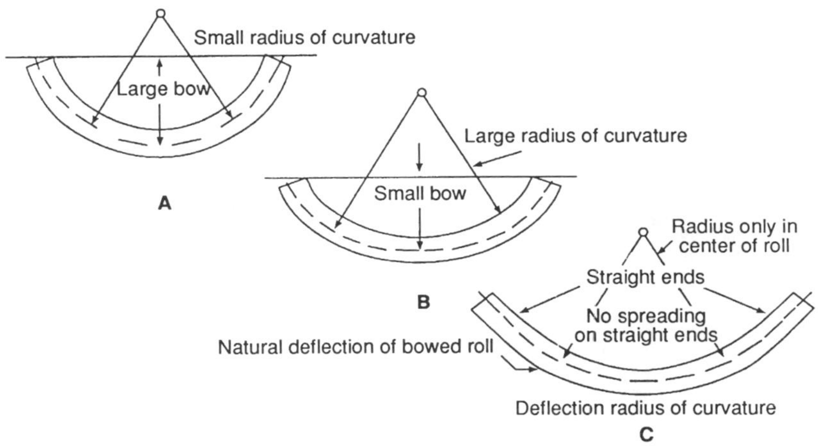

Fig. 521. Deflection-curvature vs. design radius of curvature.

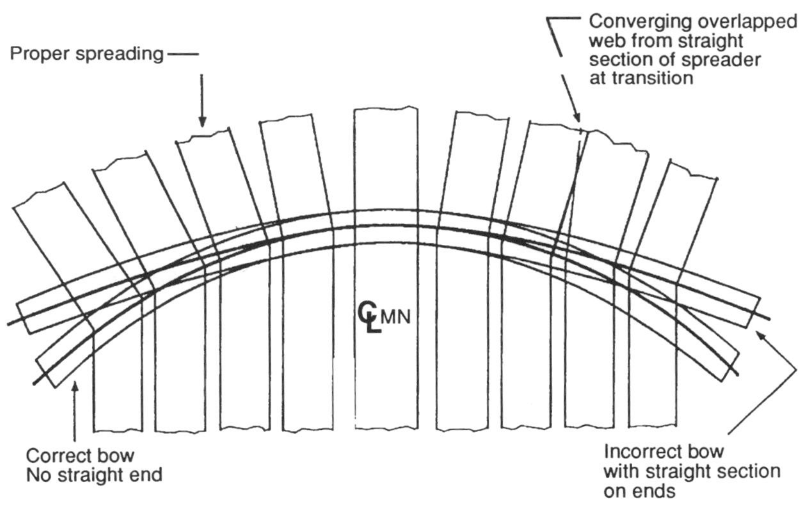

Fig. 522. A parabolic-shaped spreader where radius of curvature is caused by deflection vs. pure radius of curvature.

important in progressively spreading slit webs. One can readily see that the more slits being processed, the less bow on the spreading element is needed to minimize the instability of the outside webs (Fig. 519); thus, a larger radius of curvature means less bow.

This brings another problem to the forefront. With a very small bow, the natural deflection of the spreading element has a major effect on the "effective radius of curvature of the spreading element". If the bow is too much, then the spread on the extreme outside webs becomes too much and instability occurs. If the bow is too small, then the natural deflection causes a parabolic shape (Fig. 521C) with straight ends rather than the circular sheet needed to spread the outside webs. This causes a phenomenon of "interweaving at the quarter points" (Fig. 522), which is really interweaving at the transition between the radius and the straight portion of the spreading element. In other words, even with a minimum number of slits and an incorrect bow or a misaligned spreader (Fig. 523), instability and interweaving can occur.

(g) Folding-type spreader

The folding-type spreader (Figs. 518, 524, 525), on the other hand, is a spreading system that equalizes the strains in the web by folding the web over one bowed bar, causing the web to go in one direction and then unfolding the web back onto the second bar which, in turn, directs

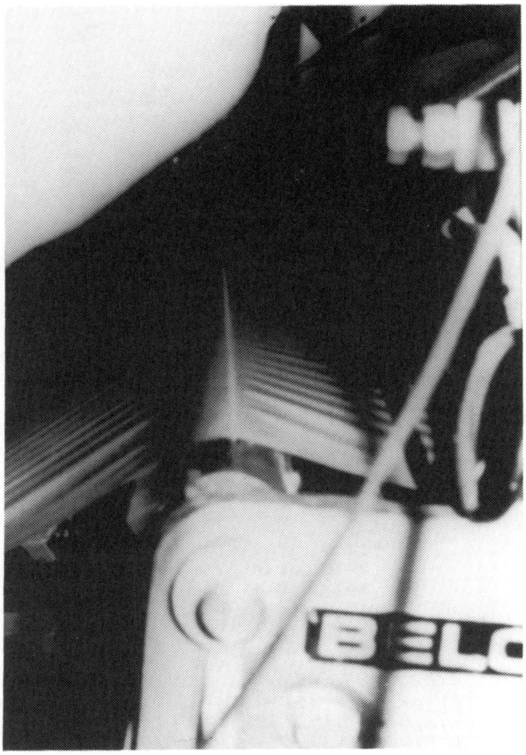

Fig. 523. Spreader with too much misadjusted bow.

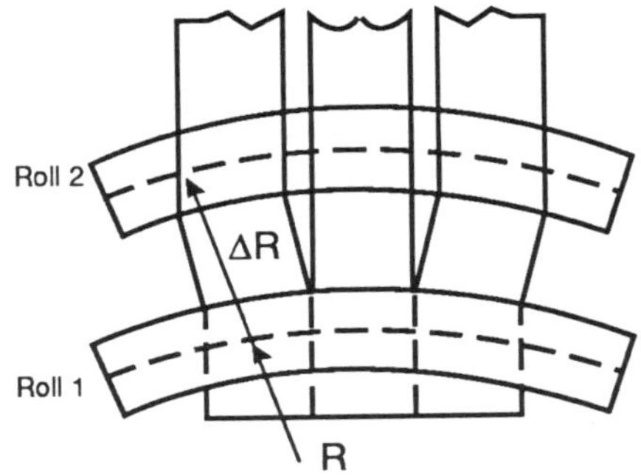

Fig. 524. Two-roll folding spreader.

Fig. 525. Folding-type spreading.

the web back into its original line but in a slightly displaced (cross-machine) position. The degree of unfolding sets the amount of slit separation and also equalizes the strain on the edges of the web.

The over-under web path, the spacing of the bars, the amount of wrap, and the amount of bow all determine the amount of slit separation that can be accomplished. One of the main advantages of (and differences of) this type of spreader and other spreaders is that the web leaves the second spreader more or less straight, rather than at an angle, as it approaches the next roll. In many cases, this next roll is the first drum of the winder. If the web is placed on the next roll at an angle, which is the case for a single spreading element, the unstable condition will be amplified as the web progresses around the drum (Figs. 526, 527), causing other defects such as telescoping and dishing (Figs. 432, 514) to occur.

Thus, the two major advantages of the two-bar spreader are: it minimizes the edge strains (Figs. 510, 523) when a large number of slits must be spread; and as the individual webs approach the drum and the drum nip straight, rather than at an angle, they are in a more natural straight line which, again, lessens the chance for instability.

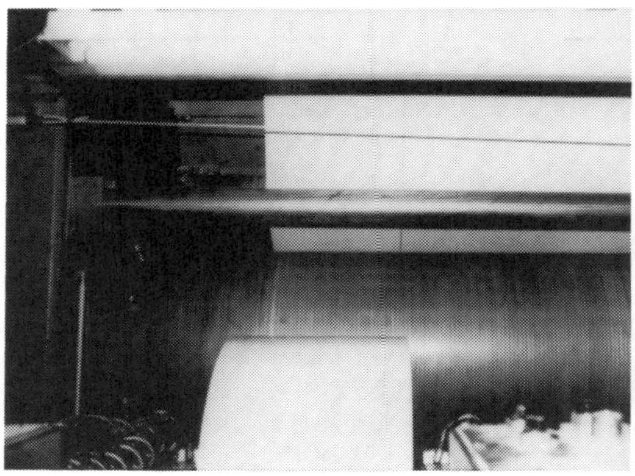

Fig. 526. Interweaving at quarter point.

(h) Rotating vs. sliding-type spreading components

Whether one picks a sliding or a rotating spreader could depend on what the surface of the web being spread can take in the way of abrasion, as well as the tension levels of the web.

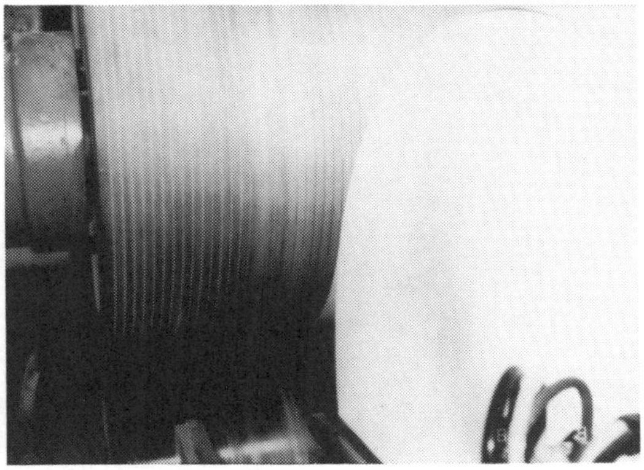

Fig. 527. Web coming onto drum at an angle.

A sliding-type spreader will, of course, have more tendency to scratch the surface of the sheet than the rotating-type spreader. Air-floated turning bars in the pressroom area provide a change in direction of the web, not unlike a spreader on a sliding surface that is air floated with no documented ill effects. Some two-bar folding-and-sliding-type spreaders are also air floated, while other folding-type spreaders are of the rotating variety.

The sliding-type spreader is sometimes called a position-type spreader, where the displacement of the web in the cross-machine direction is a function of the amount of wrap on the bowed bar, whereas the rotating spreader is a coefficient-of-friction-dependent spreader that depends on the coefficient of friction and the pressure between the web and roll to move the web in the cross-machine direction. Obviously, one is very sensitive to tension and the loss of traction (rotating) whereas the other is sensitive to wear and scratching.

Bow direction: one of the major physical differences between the rotating and sliding spreaders is the direction in which the bow of the spreader is pointing. The bow of the rotating spreader is pointing in the paper direction (or downstream) whereas the bow of the sliding spreader is pointed upstream. What makes this important is that when a rotating spreader loses traction, it becomes a sliding-type of spreader, and the bow is pointing in the wrong direction. This wrong direction actually converges the web, causing interweaving (Fig. 513).

Obviously, the sliding-type spreaders impart more tension into the web, which should be compensated for in your drive calculations.

(i) The proper nip

The fourth major criterion in proper spreading is, of course, the process of anchoring of the other end of the web after it is spread – usually done with nip or by a large wrap on a roll. On a slit web spreader, it is usually done with nip.

Unfortunately, there are many causes for the loss of nip in the windup area of any winding process:

1. Paper profile
2. Shafted winding
3. First winder drum grooving
4. Air carried by web and drum surface
5. Bouncing and vibration
6. Narrow-width rolls
7. Speed
8. Rider roll program
9. Core shaft slides
10. Etc.

Two major items that possibly help the most in the present state of the art are shaftless winding and drum grooving.

(j) Shaftless winding vs. spreading

By removing the shaft from a shafted winder and making it shaftless (Figs. 478, 479), one can ensure that each roll is in contact with the drum, regardless of profile. In a shafted situation, as a roll builds up and there is profile variation to the degree that all of the clearance is taken up between the core shaft and the core, many layers of profile variation are added to the roll. As an example, in Fig. 479, you will see a profile of a roll with two high-caliper areas on each end of the roll, with a low-caliper area in the middle section. If the roll is split in three sections, the outside section builds up faster than the center section. The core shaft inside of the core must have some clearance to be able to insert the shaft. As the roll builds, this clearance is taken up in the center section, and the shaft actually lifts the roll from the drum, causing a loss of nip. This loss of nip also creates a loss of stabilization of the web, and interweaving can occur.

With shaftless winding there is no core shaft; consequently, each roll sits on the drum, providing the necessary nip for web stabilization on all three rolls.

(k) Drum grooving vs. spreading

The main purpose for drum grooving (Figs. 468, 469) in any spreading system is to provide web stability. Assuming that, after spreading, the next roll that the web contacts is the first winder drum, the purpose of the grooving on this first drum is to void the air carried by the web between the drum surface and the web, so that instability and web-wandering will not occur (Fig. 468).

This grooving provides a means of stability by removing unwanted air that floats the sheet;

Fig. 528. Photomicrographs of slitter dust from coated publication grade paper.

but – for those applications where the web approaches the drum at an angle – it will still float enough to provide the capability for the web to twist as it would onto the roll of paper. (This is the case with a fan-type spreader.) Surfaces on this first drum – for the purpose of web stability – are many times coated and become too efficient and could have detrimental side effects such as dishing and offsets. Coatings of this type are not to be recommended with fan-type spreaders and are normally only used where the correct spreader (non-fan-type spreader) is being applied.

(l) Summary

Thus the spreading function requires the proper tension, the proper spreader system, the proper traction on rotating spreaders, and a means of anchoring the other end of the spread web. The proper spreader can be either fan (or folding) or rotating (or sliding) types. Special care should be taken to pick the correct spreading system for your application.

The third basic function in any winding process is the web separation function called slitting.

H. SLITTING FUNCTION

The third basic function in any winding process is the web separation function called slitting.

Any industry processing a continuous web of material must, at one time or another, be concerned with the process of web separation which, in terms of winding, is called slitting (Fig. 391).

Obviously slitting is necessary to reduce the wide web of material usually furnished by the paper machine (or other processes) to a usable size for final consumer or some other industry use. Whether the web is paper, steel, plastic or textile, many of the basic fundamentals of severing a web have been found to be the same.

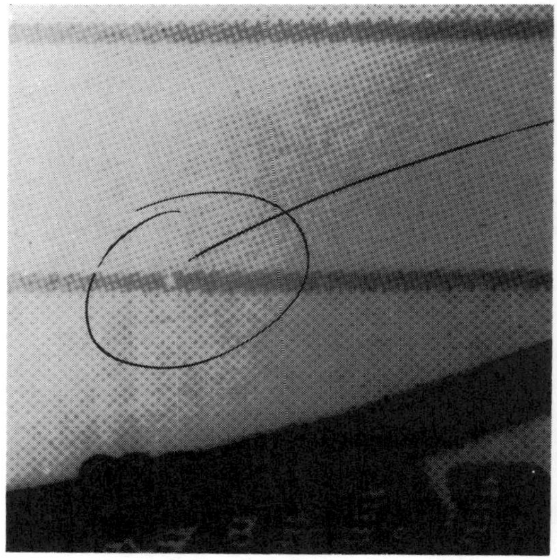

Fig. 529. Slitter dust at the printer. Note the dust in the circled area.

The problems that result from improper slitting might be considered different with different materials; however, with closer scrutiny, they, too, are found to have common threads.

1. Slitter problems and their causes

In the paper industry, the problems were found to be basically the same in processing newsprint, coated paper, tissue, linerboard and glassine. They differed only in degree and severity. Dust, however, was found to be the common denominator for most of the problems (Fig. 528).

Dust is important because, during the printing process, dust interferes with quality printing in the form of white or black specks on the printed sample (Fig. 529). In other words, if the sample has a black or colored background, dust would show up in the form of white specks that are unacceptable to printer and customer.

Many factors and problems affect slit quality – all are important. When evaluating any type of slitter in terms of problems that can occur, there are some general categories into which these problems fall more often than not. These are listed as follows as a general guide and should be looked at first as a baseline for proper evaluation.

Problems that are caused by:

General slitter design and mounting

1. Incorrect toe-in
2. Incorrect knife angles

3. Knife diameter too small or too large
4. Inadequate trim removal system
5. Slitter runout – radial and axial
6. Speed of the process
7. Relative speed of the cutting components
8. Difficulty of changing blade and band
9. Difficulty in adjustment
10. Slitter material incorrect
11. Wear
12. Noise.

Operator adjustments

1. Slitter jump
2. Incorrect penetration (dust)
3. Improper side loading (wear)
4. Slitter not locked in place.

Material path and stability

1. Trim removal, in general
2. Sheet flutter
3. Cutpoint location incorrect
4. Spreader effect and location
5. Unwind oscillation and guiding effect through slitter section.

Paper problems

1. Baggy area
2. Too much bulk or poundage under the knives
3. Grade of material being cut
4. Paper too wet or too dry.

Environment

1. Dust
2. Vibration – slitter
3. Vibration – winder.

Operating and maintenance

1. Slitter nicks
2. Grinding procedures
3. Lubrication
4. Dull slitter blade and band.

2. Theory of slitting

All of these problems are, in some way, directly related to and trace back to the seven basics of slitting (Fig. 530). These are:

1. Toe-in
2. Penetration
3. Side loading
4. Rake
5. Material
6. Blade angles and configuration
7. Band angles and configuration.

To exemplify the effect that these components have on the generation of dust, let us look first at one of the most obvious causes of dust – wear of the slitter components – and determine how wear causes dust.

First, one must set up guidelines of slit quality for different grades of paper. Figure 531 shows three samples of band and blade edges that users agreed were "quality" slits. The band- and blade-side slit edges of a piece of cut paper are shown side-by-side with a light shown from underneath the sample to ensure the edge can be seen clearly.

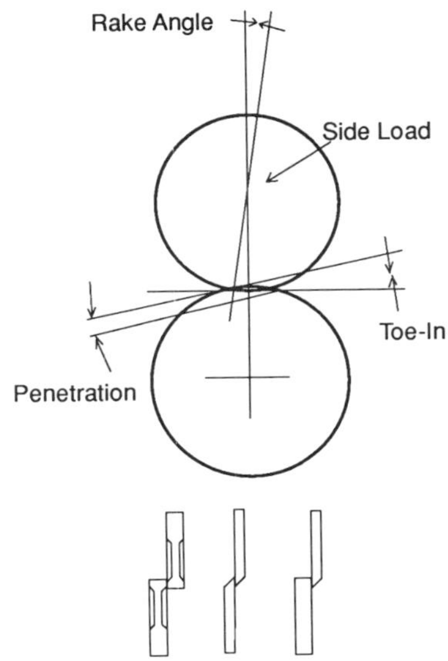

Fig. 530. Slitter basics.

Next, using the same paper samples, we asked the users to define a below-quality slit edge. The comparison (Fig. 532) shows that the edges look quite different, with rough edges and protruding fibers quite visible. These protruding fibers can break off and create dust during the printing process.

It is commonly known that, as a slitter wears, dust becomes more prevalent and eventually the slitter knives must be replaced. In fact, it can be said that a slitter starts to become dull the moment that it is installed and is only replaced when the product quality, in the form of dust, becomes unacceptable.

Therefore, the next logical step in the reduction of dust was to study different slitter knife materials to determine those combinations which would give the longest knife life, thereby maintaining the quality cut as long as possible. Since this is a brute force method of resolving only one small aspect of the slitter dust problem, one must look even deeper into the reasons for wear. This investigation was first done by Daly [7].

(a) Wear investigation

The turning point in Daly's investigation came in cutting polypropylene, as seen in Fig. 532. Unlike newsprint and coated paper, the polypropylene actually extruded along the slit edge.

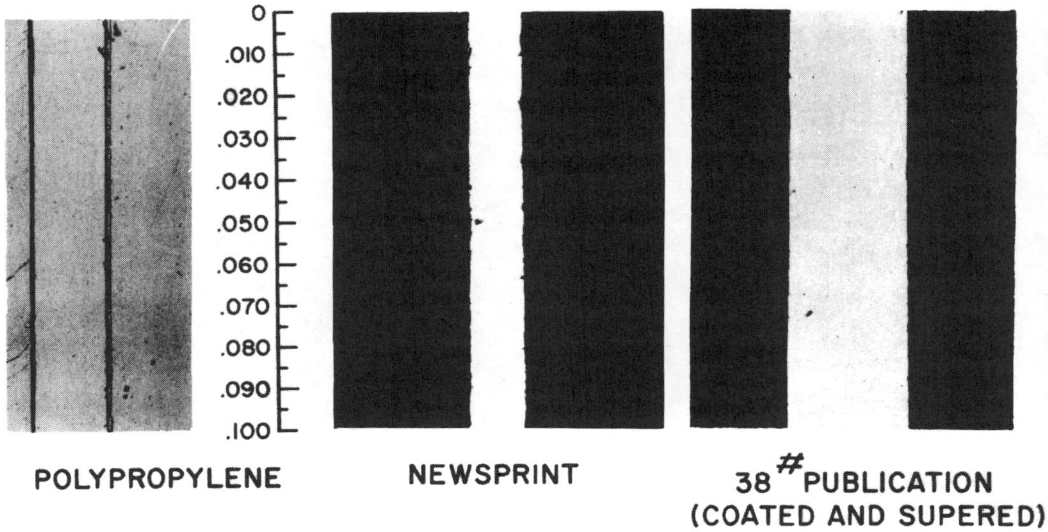

Fig. 531. Comparison of high-quality cuts on different grades.

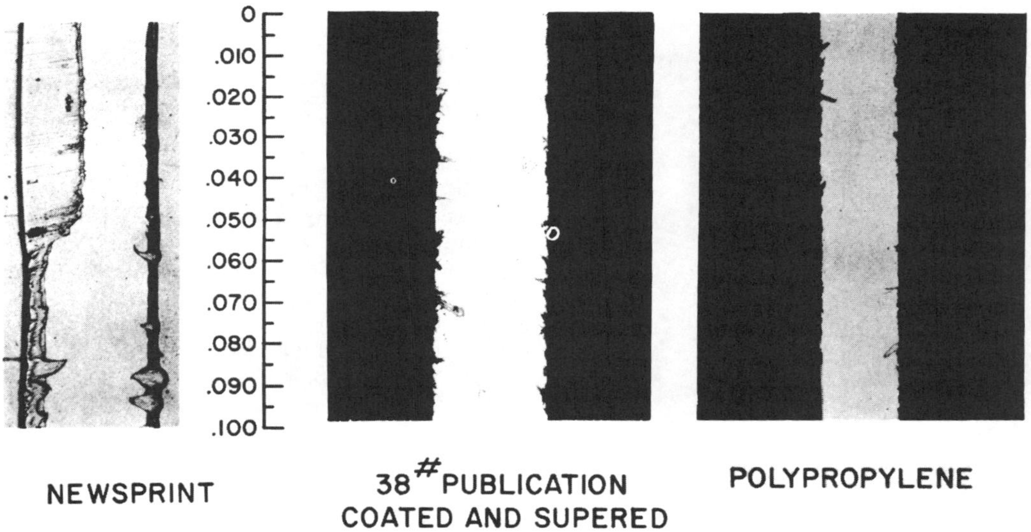

Fig. 532. Slit edges made on different grades by a dull slitter pair.

(This may be the reason this type of material is not usually cut by the shear slitter method.) To determine the reason why this phenomenon occurred, a closer look at the wear pattern of a conventional slitter was in order. We found that, for a given penetration, toe-in and rake angle, the slitter wore at a repeatable and consistent angle (Fig. 533).

In any conventional slitter system, the blade and band (see blown-up views, Fig. 534) are toed-in so only one point of the two components are in contact at one time.

In studying wear patterns, we found that, as a slitter wore, the point of contact ("X") (Fig. 534A) between the two cutting components (or cutpoint) moved downstream (paper flow direction) (Fig. 534B) in the direction of the centerline between the top and bottom slitters (i.e., the paper had to pass the original cutpoint before it reached the point of contact between the top and bottom components of the slitter system). As the cutpoint moves downstream, the original cutpoint location opens up. When it opens larger than one fiber diameter, the paper is pulled through this open cutpoint, thus tearing the web rather than shearing it. Obviously, this is what

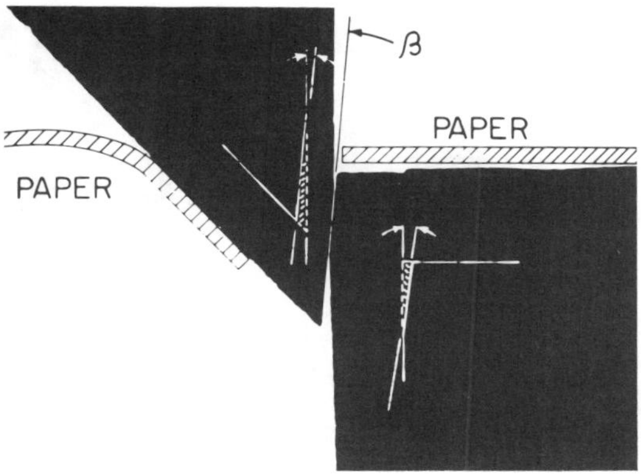

Fig. 533. Slitter knife wear pattern.

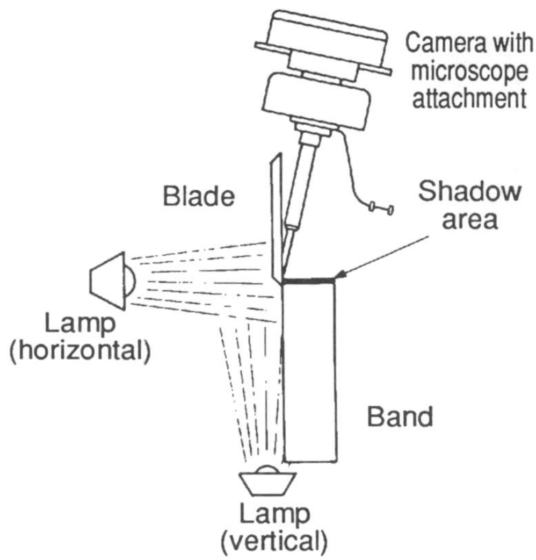

Fig. 535. Equipment arrangement to record pictures as seen in Fig. 537.

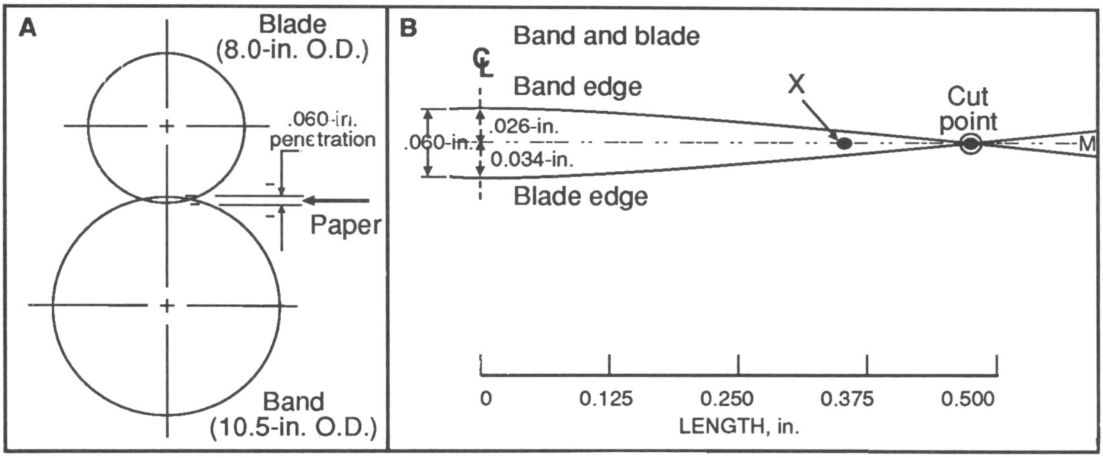

Fig. 534. A: Full side view of shear slitter. B: Enlargement of side view showing area of overlap and theoretical cut point, and propagation of point of contact (X).

happened on the extruded polypropylene sample seen in Fig. 532.

To show this wear more graphically, a top and bottom slitter was set up in a cutting position, as seen in Fig. 535. A horizontal light source was located so that the blade would cast a shadow at the cutpoint, horizontally, across the top of the mating band.

The other light source was located below, so the toe-in or orientation of the blade, with respect to the band, could be observed. A camera was then placed so that the wear pattern of the blade with respect to the original cutpoint could be photographed.

(b) Why a worn slitter causes dust

The resulting pictures (Fig. 536) show first a sharp blade and band toed-in, where the cutpoint shown by the shadow and the point of contact between the blade and band are one and the same. The second picture shows a slitter that was run approximately 8 h. Here one can see that the point of contact has moved downstream from the original cutpoint designated by the shadow. The contact between the blade and band has opened up at the original cutpoint position.

As seen in the third picture, when it is opened up past one fiber diameter, a torn fiber occurs, as

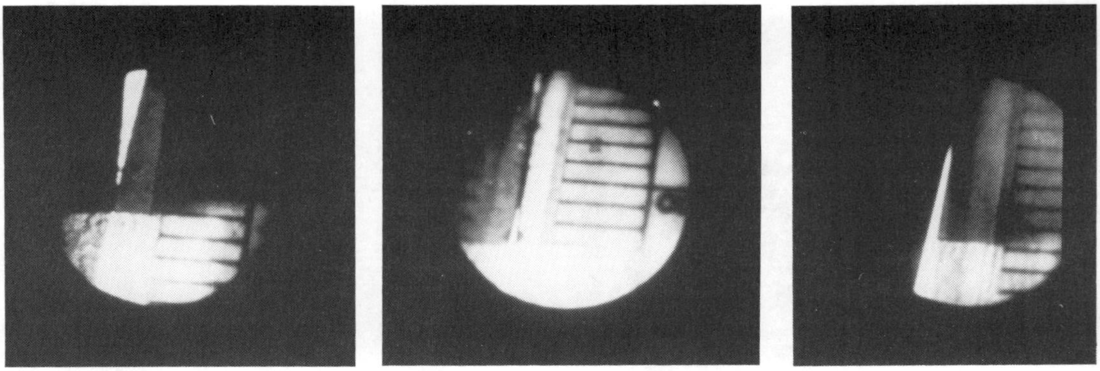

New Worn Grossly worn

Fig. 536. Comparison of cut point openings for three degrees of slitter wear.

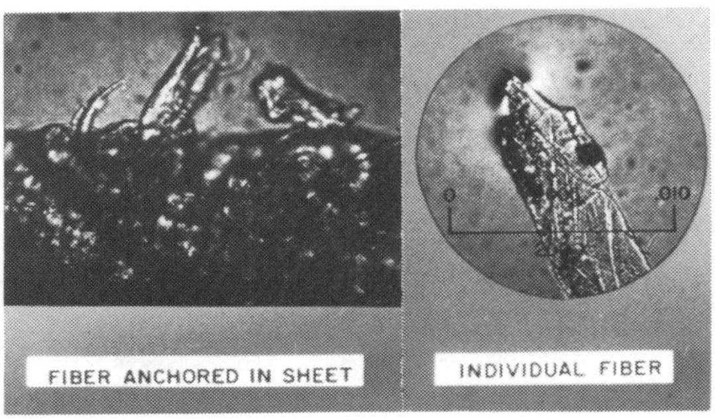

Fig. 537. Torn web fiber.

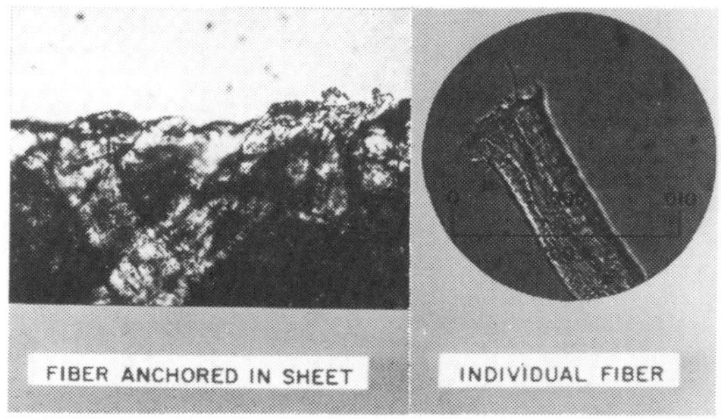

Fig. 538. Mangled web fiber.

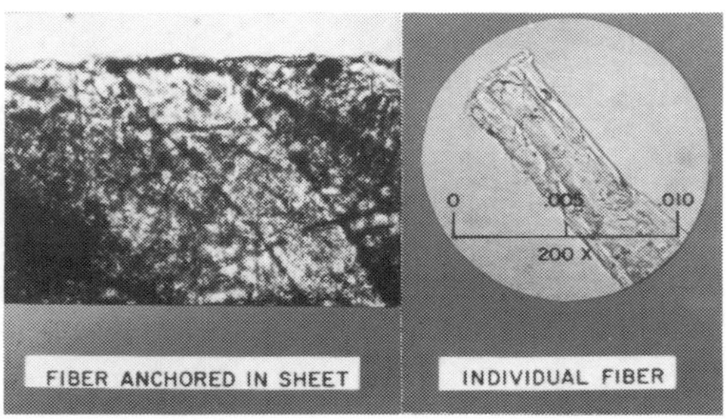

Fig. 539. Properly sheared fiber produced by sharp slitters.

seen in Fig. 537. A slightly worn slitter creating a mangled fiber is seen in Fig. 538, and a properly sheared fiber is shown in Fig. 539.

(c) Other causes of slitter dust

Proceeding with the study of slitting and delving into the area of high-bulk multi-web slitting in sheeting, we found that many of the problems of multiple-web slitting disappeared when we dropped down from six to one web. Subtle slitting problems as seen on a single web were actually amplified by multiple-web slitting. This discovery indicated that the torn fiber shown in Fig. 539 had possible causes other than wear.

One of these causes can be demonstrated by making a short slit with a razor blade in the middle of a piece of paper. Next, insert the conical end of a pencil into the slit (Fig. 540).

If the pencil is moved to one end of the slit, one finds that the strain put on the web in the cross-direction (perpendicular to pencil direction) actually tears the paper prior to physical contact of the pencil at the end of the slit (Fig. 540).

This same thing occurs with a conventional slitter blade and band (Fig. 541), the blade in this case being the pencil. In the same manner, the tear occurs before the paper comes to the cutpoint. When the torn fibers on each side of the slit reach the cutpoint, one side is sheared off – creating dust.

The other side just rubs against the bevel of the blade and is not sheared off. This demonstrates why, on a roll of paper, looking at both sides of

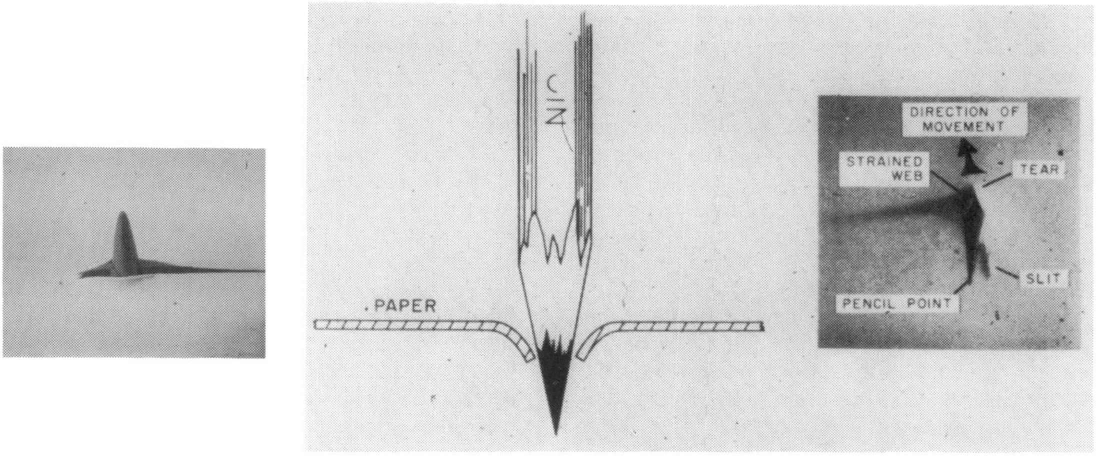

Fig. 540. Pencil penetrating web.

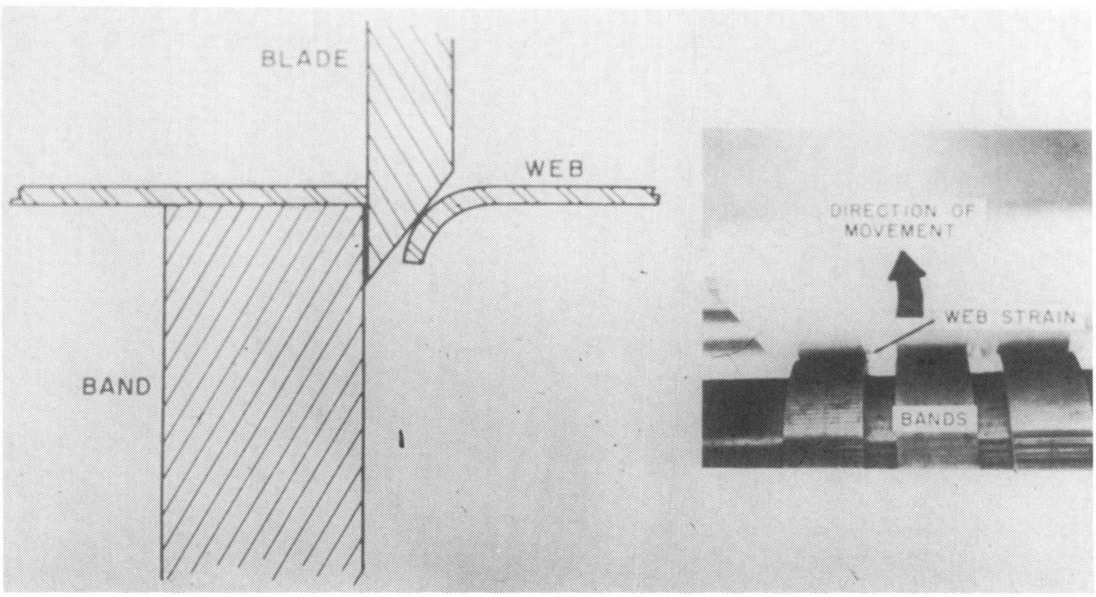

Fig. 541. Plow effect of top slitter.

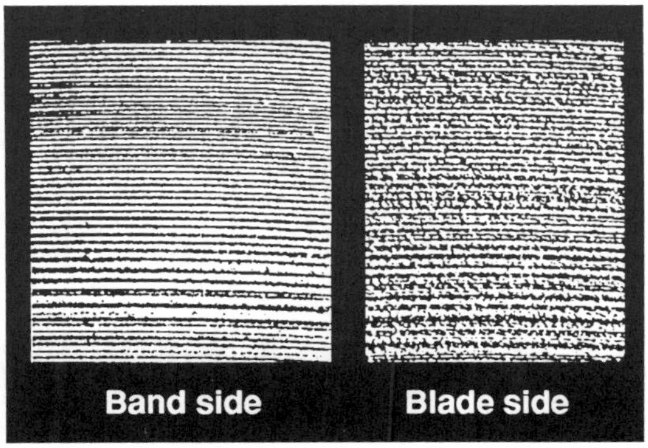

◄ **Fig. 542. Different cut quality of two adjacent rolls.**

the same slit, you will notice a marked difference in slit quality between the two sides (Fig. 542).

3. Causes cannot be ignored

Obviously, there are many things that have been done to improve the dust problem. Our findings, however, point up the fact that there is no one cause for slitter dust. And there are many little things that cause the dust; unfortunately, few conventional slitters, to my knowledge, take into account or compensate for a majority of the factors that can create dust.

Some things have been done, however. As indicated in Sec. H, many of the different types of slitters evolved over a period of time primarily to overcome some of the many problems of slitting and, in an attempt to minimize the worst problem, dust.

The most important thing that has been or could be done to improve slitting is to obtain consistency and repeatability in the slitting process. There are far too many operator adjustments in the conventional slitting process; the industry is slowly reacting to the need in the following ways.

(a) Materials

Material research is done primarily to slow down the migration of the cutpoint, mentioned previously. Actual material hardness and durability have long been a method of getting longer life, and it is obvious that a harder material will wear more slowly and give a longer life. However, there are certain things that should be watched for when choosing materials.

Sometimes the material is too hard and brittle; consequently, instead of giving a longer life, you shorten the effective life by chipping and damage during the slitting operation, such as slitter jump. Some slitter materials actually accelerate the wearing process by galling. Improper chrome content in the slitter material causes galling, and galling causes slitter jump and damage.

Understanding the advantages and disadvantages of all types and combinations of materials will go a long way in optimizing slitter wear. Unfortunately, most mills buy their slitters from a local supplier after the original equipment has been installed, thus making research into the wear problem by the original equipment supplier not very inviting.

(b) Side loading

Improper side loading is probably one of the biggest causes of a high rate of wear. There are many types and methods of loading: spring, air, and mechanical loading.

The key to proper side loading is that all of the slitters be loaded the same to ensure the same wear rate on all slitters; but, more importantly, that the slitters not be over- or under-loaded, as happens quite often with spring- or mechanical-loaded slitters. Unfortunately, when a dusty slit appears, the natural tendency for the operator is to increase the loading of that slitter where actually one would be further ahead if the slitter were replaced with a new or reground slitter.

Heavier weight paper requires higher side-loading, primarily because the web wander on heavyweight has a tendency to unload the slitter. Consequently, a heavier initial loading is necessary.

The baseline normal side loading measured at the point of contact should be approximately 1–3 pounds.

(c) Toe-in

Toe-in (Fig. 543) is one of the major causes of dust, creating a situation where the paper is strained ahead of the cutpoint to a degree that the fibers are torn and dust is created. In this regard, toe-in needs to be much less than 0.5°, with the larger diameter slitters even more critical, as the physical cross-machine displacement of the web as it is forced to pass around the protruding blade becomes greater (in the case of large-diameter slitters), causing more strain at the cutpoint. Unfortunately, manufacturing tolerances pretty much dictate what can and cannot be done to improve upon the 0.5° and be consistent.

Toe-in is extremely important and should be preset by the machine manufacturer, who should make it very difficult to change once the slitter is installed in the machine. The optimum toe-in should be approximately 0.25° and, regardless of manufacturing tolerances, should always be such that the cutpoint is the only point of contact between the top and bottom components, and the toe-in should always be positive (Fig. 543).

(d) Blade angles

Blade angles play an important part in creating too much stress at the cutpoint. If the bevel of the blade is too great, the paper rides up the bevel (Fig. 541), creating a plow effect not unlike the toe-in mentioned above.

There are two extremes in blade configurations: one, the perfectly square slitter (Fig. 544A) where the paper is bent in the machine direction (opposite of cross-machine direction) which, if tried with a partially slit piece of paper, would cause minimal strain; the other extreme is the 45° bevel blade (Fig. 544B) that creates the maximum plow effect of strain, and the paper tries to avoid the thick part of the bevel as it proceeds past the slitter.

Both of these extremes exist: the square slitter is used in the cigarette industry where the paper is extendible and easily strained and damaged, and in the steel industry, where it would be impossible to create cross-machine strain without damaging the slitter components.

Again, the evolution of the needs have dictated variations that are compromises on

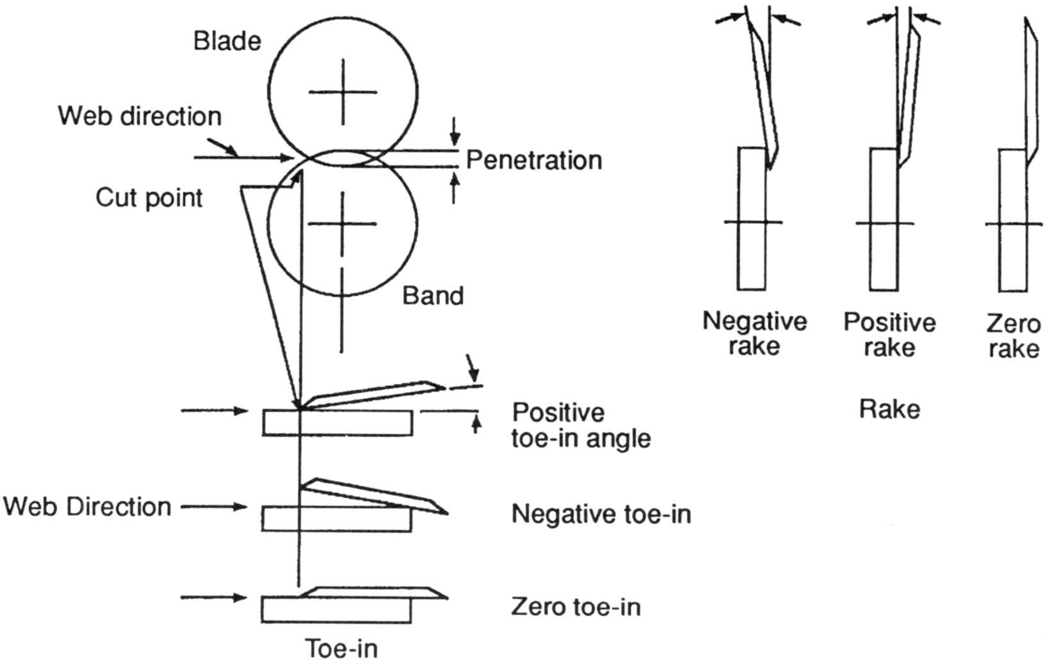

Fig. 543. Toe-in and rake (shear slitters).

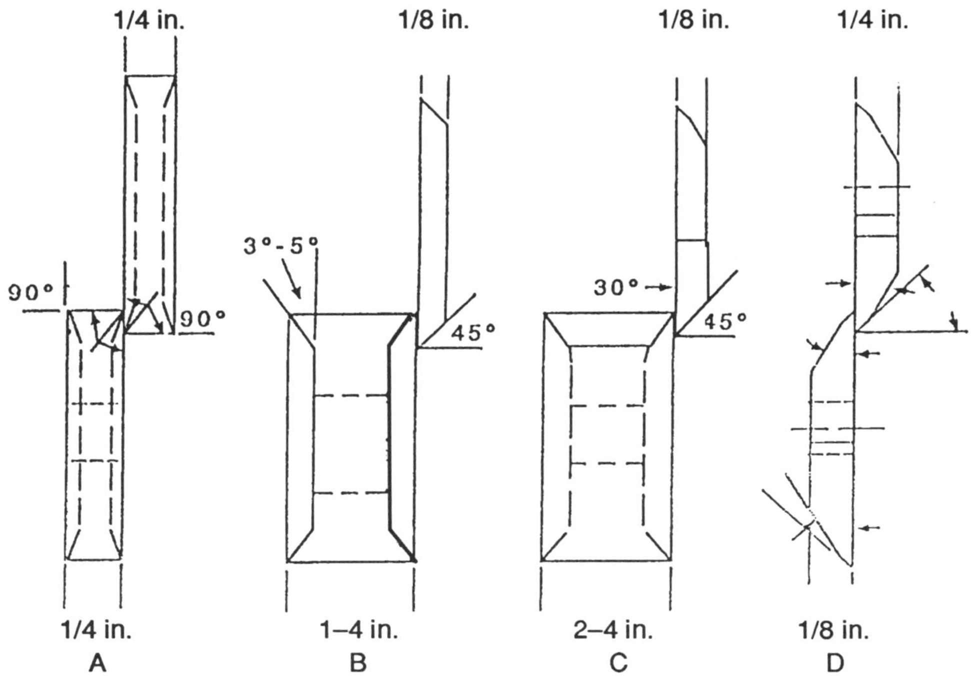

Fig. 544. Slitter cutting configurations.

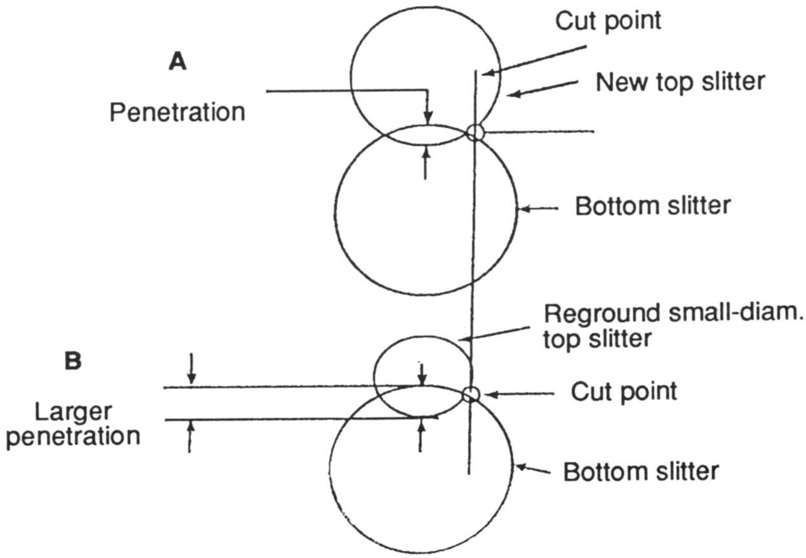

Fig. 545. Penetration change with slitter grinding.

both extremes and which are designed to reduce the unwanted strain while maintaining the high concentration of stress at the cut-point. One such type is the combination-angle blade (Fig. 544C), which is 30° instead of 45° and which lessens the thickness of the blade that causes cross-machine stresses in the paper. This shallow angle creates another problem, as the tip of the 30° angle becomes quite fragile and pliable. And since this is where the slitter side loading comes to bear, it sometimes creates a burr – and, subsequently, dust. To overcome this problem: a very short (0.010 in., 0.25 mm) 45° bevel in machine direction at the very tip of the blade (Fig. 544C) is used to strengthen the slitter, but still maintaining the lesser interference in the cross-machine direction of the web. Other variations of these angles can depend on the product that is being slit, and the machine manufacturer should be consulted for specifics.

(e) Rake angles

Not only are knife angles important, but so is the rake angle (Fig. 543). The rake angle can be defined as the angle at which the top slitter is cocked with respect to the bottom slitter, when looking in the machine direction. Ideally, this angle should be zero, but manufacturing tolerances make this impossible. To compensate for this, a relief angle (Fig. 544A–C) on the bottom slitter band of 3–5° is added to allow the top slitter to pivot around the cutpoint without any interference between the extreme bottom of the top blade and the band. With no relief, the top slitter would pivot around this extreme bottom of the blade which would, in turn, open up the cutpoint and cause a dusty slit.

(f) Penetration

As stated previously, excessive penetration presents a thicker slitter component as the paper rides up the bevel of the blade (Figs. 541, 545), causing more cross-machine strain at the cut-point. If the slitter is set with too little penetration, it has a tendency to jump.

Thus, a compromise situation needs to exist which is demonstrated quite graphically by the state-of-the-art grinding of the dull, top, slitter blade which, in turn, changes the diameter (Fig. 545A, B). As a blade is ground down, in order to maintain the same cutpoint the penetration of the slitter must be increased. If the penetration is increased, more cross-machine stress is created in the paper, causing more dust; consequently, the slitter diameter must be watched closely.

A slitter that does not change in diameter would alleviate the problem of too much penetration (Fig. 546). Again, the cigarette industry and the steel industry were forced to use square slitters and a preset penetration because of the product that they were cutting. The paper industry (in total) has yet to accept the admittedly small but significant benefit of square slitters.

With conventional slitters, a penetration of 0.030 in. (0.7 mm) is a good compromise for quality slitting on many grades, while on high-bulk slitting the need to reduce the

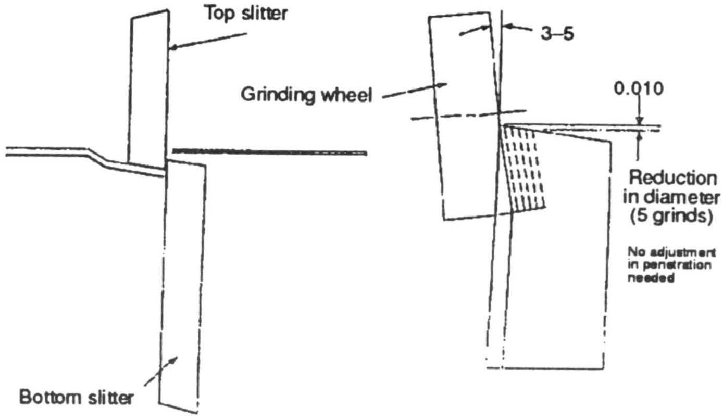

Fig. 546. Grinding of square slitter.

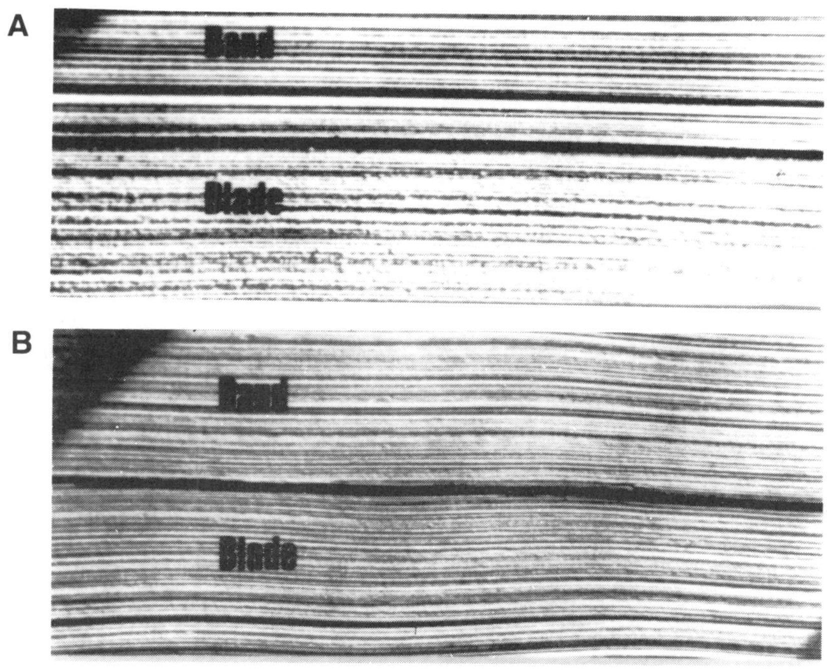

Fig. 547. A: Band side and blade side, conventional slitter. B: High-bulk slitter.

interference even more is graphically demonstrated by Fig. 547A, B, which shows the two sides of a slit web in a stack of cut paper, as would be done on a sheeter. Figure 547B shows the blade and band side of the high bulk slitter to be more or less equal while the quality of the slit, especially on the blade side of the conventional slitter, is less than ideal. Figure 547B demonstrates the contrast and what can be done to alleviate the problem.

(g) Slitter runout

This is sometimes called slitter wobble (Fig. 548) and is the cause of the slitter rings (Fig. 549A-C) seen on the edges of a roll of paper [16].

The concern with slitter runout stems from the condition of the loss of contact of the top and bottom cutting components. If the runout is excessive (above 0.002–0.004 in., 0.5–1.0 mm) and as the slitter goes up in speed, the top slitter

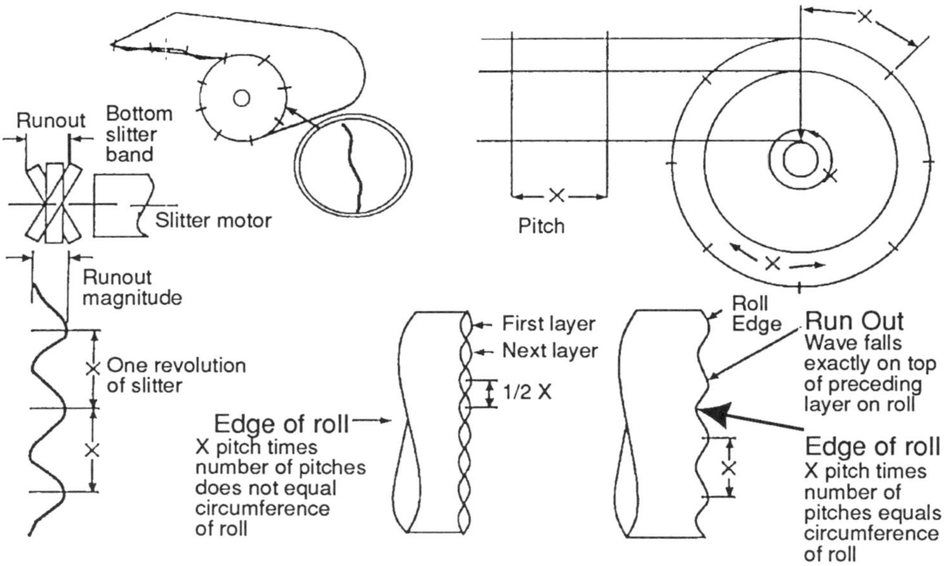

Fig. 548. Slitter runout causing slitter rings.

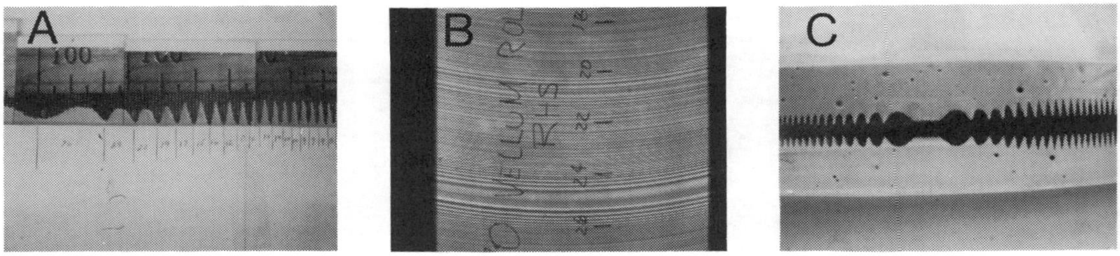

Fig. 549. Slitter rings.

Fig. 550. Dusty edge caused by bottom slitter runout.

Fig. 551. Trim systems.

is thrown away from the bottom slitter; depending on the amount of runout (throw), it does not have time to get back into contact. Thus the paper is torn (Fig. 550A, B), creating dust, rather than being sheared.

4. Trim systems

Trim systems (Fig. 551) are a much neglected but integral part of any slitting system. The trim system, obviously, affects only the outside rolls in the winding process, but from recent investigations into runnability in the pressrooms, outside rolls are the most prevalent in causing breaks on the press. So, attention should be paid to some of the problems caused by the trim system – edge cracks being the most prevalent.

The diagonal wrinkle (Fig. 552, 520) seen in most winding systems just ahead of the cutpoint of the trim slitter is caused by inadequate tensioning of the trim. This occurs when the web of a certain width is tensioned by the braking system at the unwind. At the trim slitters, the web is cut narrower by an amount equal to the width of the trim on each side being removed. Now the total web is narrower than the web coming from the unwind stand, but with the same total tension level. Therefore, the pli tension level is higher and the transition at the trim slitter from one pli level to another causes a diagonal strain, mainly because the trim being removed has no tension.

This diagonal strain can become severe enough to create a bubble (Figs. 552, 520) which (when cut by the slitter) gives the appearance of a scalloped edge on the side of the roll (Fig. 553). Since most trim systems are air systems, to tension the trim properly necessitates air velocities that exceed 2–3 times the winder speed. Then, noise and the problem of getting rid of the air adds more complications to this unsolved and

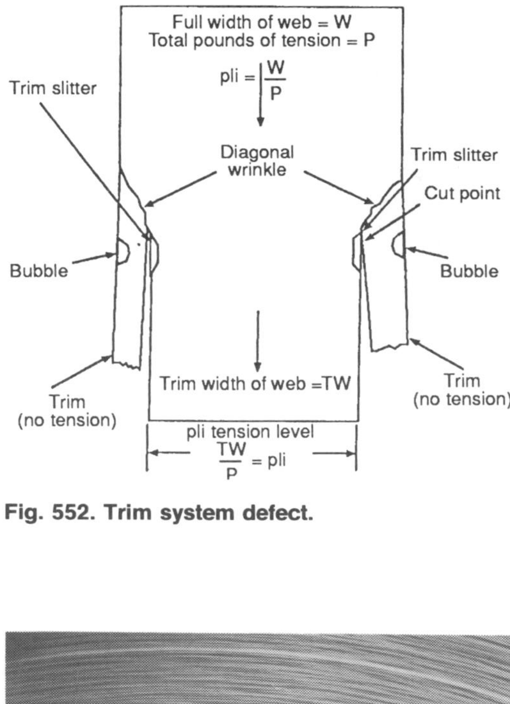

Fig. 552. Trim system defect.

Fig. 553. Trim problem as seen on a rewind roll.

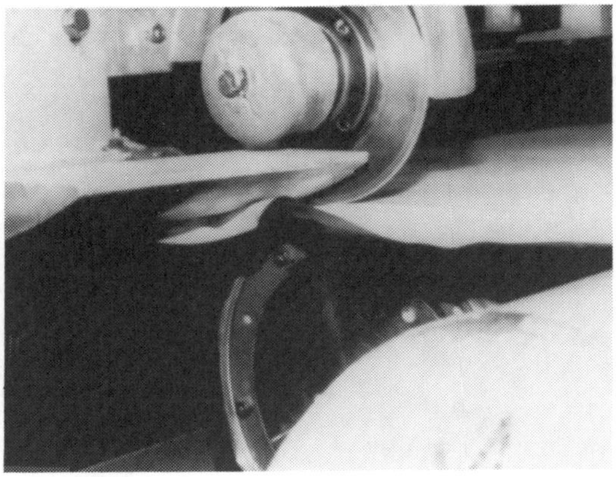

Fig. 554. Trim instability.

Fig. 555. Shear slitter.

Fig. 556. Bottom slitter.

complicated problem of trim removal (Fig. 554).

5. Slitter types

There are seven basic types of slitters used in the web processing industry today. Those that are the most common in the paper industry are the shear and score slitters.

1. Shear tangential slitter (sometimes called a kiss slitter). This is the most commonly used type of slitter in the paper industry (Fig. 555).
2. Shear wraparound slitter. This type of slitter is used mostly on narrow-width, small machines (Fig. 556).
3. Score slitter (Fig. 557). This type of slitter is used mostly on very high density material or on tissue where the edges of multiple webs

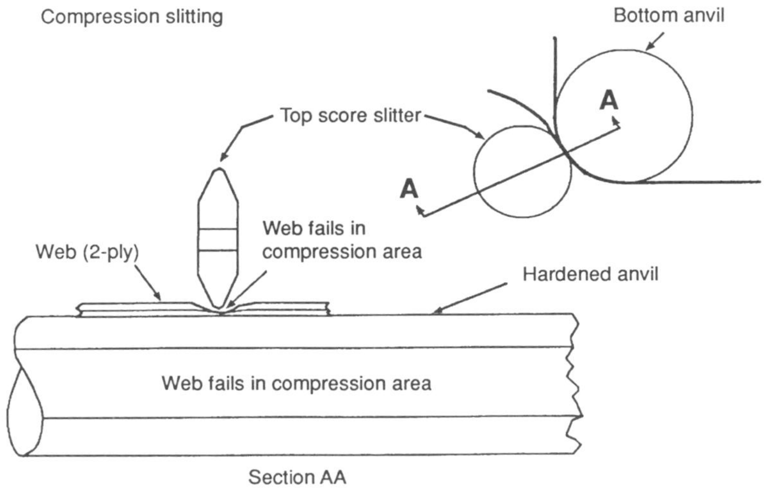

Fig. 557. Score slitting.

Fig. 558. Automated slitter positioning.

Fig. 559. The bottom slitter can be a roll with slitter band mounted to that roll.

must be bonded together in the slitting process.

4. Razor blade and burst slitting. This type is used almost exclusively in the plastic film and packaging industry.
5. Water-jet slitting. This process has limited application, where its use is primarily in the tissue, board, and carton industry.
6. Laser slitting. This type of slitting has intrigued people for years, but because of the large power requirements needed to reach the speed of state-of-the-art winding equipment, this development has not yet flourished.

(a) Shear tangential slitter

The top slitter of the shear tangential slitter (Fig. 555) is mounted on a continuous beam (Fig. 558) spanning the machine and is usually driven by the bottom cutting element commonly called a band. The bottom slitter can be a roll with slitter band mounted to that roll (Figs. 556, 559) – all of which is driven by a motor connected to a journal stickout. Or, the band can be mounted on individual motors (Fig. 560) which, in turn, are mounted on rails that span the width of the machine.

The roll-mounted bottom slitter band system is limited by the width and the speed of the machine, as its diameter must increase as the speed and width of the machine increases. The diameter must be sized to prevent a critical speed vibration. This is possibly one of the main reasons why the individual motor slitters have become a necessity with the machines becoming wider and faster.

The motor-mounted slitter has less vibration; change of a worn band can be done in minutes rather than the hours necessary to remove all of the roll-mounted bands in order to change the dull one. With the motor-mounted slitter, the diameter of the band is not a function of the width or speed of the machine, which is the case in the roll-mounted-type slitter bands. The major drawback of motor-mounted slitter bands is the web stability. The draw across the slitter is quite critical and, unless the slitter function, in particular the slitter table (Fig. 561), is designed properly, it could cause wrinkles and web flutter.

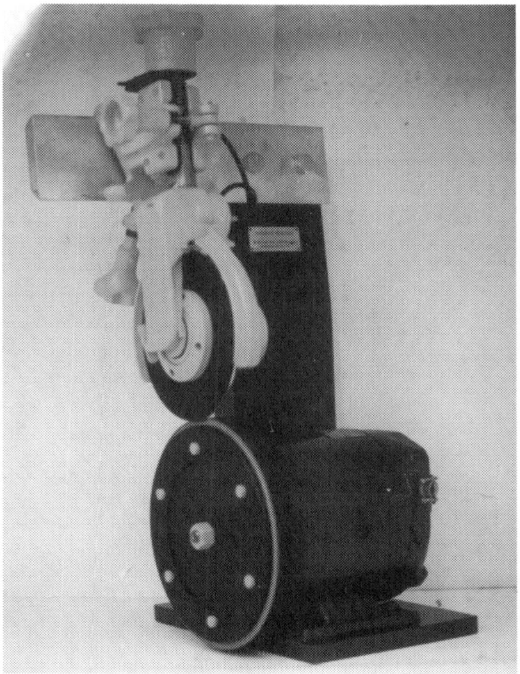

Fig. 560. Individual motor-driven bottom slitter.

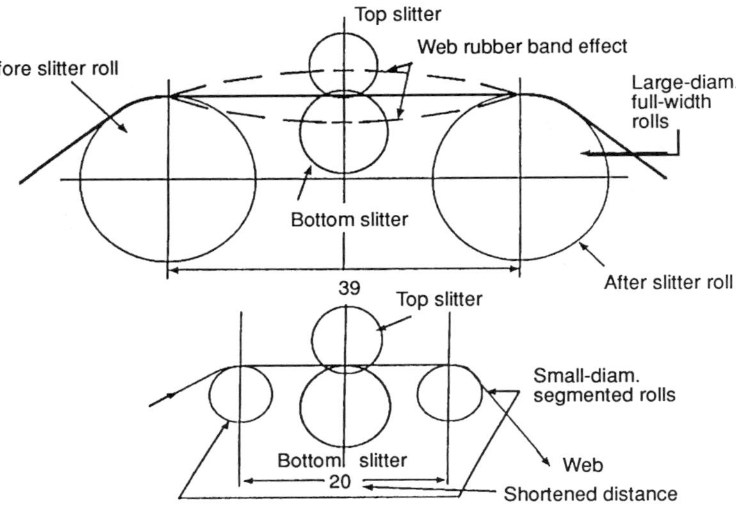

Fig. 561. Slitter tables.

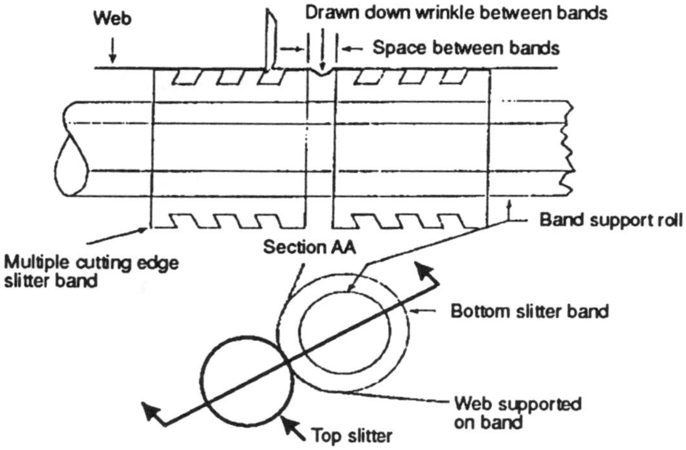

Fig. 562. Wraparound shear slitter.

(b) Wraparound shear slitter

One of the major advantages of the wraparound slitter is web stability (Fig. 556). Multiple bands covering the entire width of the machine actually support the web as it wraps around the bands mounted on the carrying roll.

In this type of slitter, each band has multiple grooves (Fig. 562) which are the cutting edges of the slitter band. These grooves are spaced fairly close together to provide maximum support for the web as it passes around the roll. When one of the edges wears out, the band is moved very slightly to the right or left until another groove comes into position. To do this, it is necessary to provide some space between bands to move the band, thus causing some of the disadvantages of this type of slitter.

Of course vibration and critical speed, width of the machine, and speed of the machine are limits to this type of slitter. The fact that the band must move to reach a new cutting edge necessitates a gap which is detrimental to running lightweight papers, as they have a tendency to pull down between the bands and wrinkle (Fig. 562). The length of time it takes to change a band (especially since the trim slitter bands wear faster because of their use) and the fact that the operator can never remember which edge was dull and many times just runs it because it takes too much time to change, all point to the need for and the evolution to the better system found in the motor-mounted-bands slitting system.

(c) Score slitter

The score slitter (Fig. 557) is a compression-type slitter that causes the web to fail in compression rather than shear. By its very nature it is used in tissue converting because, on multiple tissue, there is a need to bond the edges of the tissue; therefore, a crushed edge is of some benefit. On high-density material such as plastic, where a shear slitter cannot be used because of the nature of the wear (mentioned in Sec. B), a score slitter is also used.

Advantages

One of the main advantages of a score slitter is the simplicity of setting or moving the slitter, as only one slitting component has to be moved in the setting procedure. The top slitter in this case runs against a full-width hardened-bottom mandrel (Fig. 557). This makes adjustment very simple, coupled with the wraparound advantages mentioned above, which covers about the only advantages of this type of slitter.

Disadvantages

The vibration aspects caused by machine width (and speed, mentioned previously) are even more critical on a score slitter, as any vibration of the roll being wound would be transmitted directly to the cutting element edge, causing the loaded blade (which rides directly on the mandrel) to chatter. This chattering rounds off or flattens out the tip of the blade, making the area much larger, so the applied cutting pressure will drop and the slit quality will deteriorate. To improve the quality, the operator increases the blade pressure, which increases the wear, and the cycle starts all over again.

As the speed and the width of the paper machines increase, this necessitates that the speed of a winder with this type of slitter also increase. With this increase, at least in the case of tissue and newsprint (two of the biggest users of score slitters), comes the evolution to shear slitters replacing the score slitter. In the case of tissue, this means the need for other equipment to emboss and seal the edges of the slit web, formerly done by the score slitter.

In any slitting system, the cutting elements mentioned above are only one part of the slitting system. The elements on the downstream and upstream sides of the slitter cutting elements (more often than not being rolls) are also considered an important part of any slitting system. This is sometimes called the slitter table.

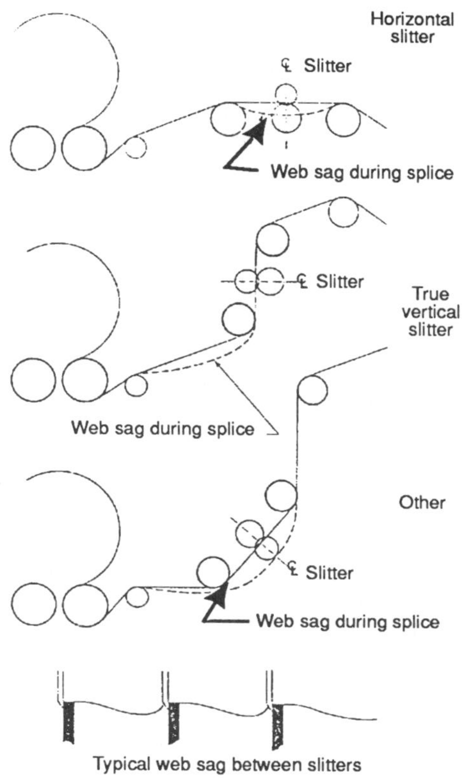

Fig. 563. Slitter offset at splice.

6. Slitter table

There are three types of slitter tables (Fig. 563):

1. Horizontal (Fig. 564)
2. Vertical (Fig. 565)
3. Inclined (Fig. 566).

(a) The horizontal slitter table

The horizontal slitter table arrangement (Figs. 561, 564) gives the best visibility of the web with the least amount of rolls and equipment to transport the web past the cutting components. Small-diameter segmented rolls are used on either side of the slitter to reduce the span across the slitters (Fig. 561).

This is especially necessary on the wider machines where, because of the speed of the machine, the rolls would have to be very large in diameter, making the span across the cutting components very large (Fig. 561). On a large span, the rubberband effect leads to dusty slitting.

Fig. 564. Horizontal slitter table.

◀ Fig. 565. Vertical slitter table.

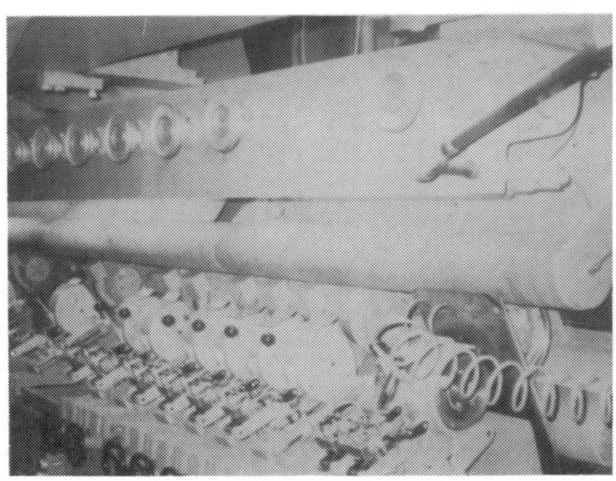

Fig. 566. Inclined slitter table.

In this arrangement, the web must be broken down when the operator needs to adjust or observe the slitters for other reasons.

(b) Vertical slitter section

Vertical slitter sections (Figs. 391, 563, 565), although there is more equipment involved, give the best accessibility to the slitters and spreader adjustment without breaking out the web, which would necessitate a time-consuming rethreading process. Some visibility of the web is lost, especially on the larger machines. On this type of slitter the web goes overhead, so special threading equipment is also required.

The vertical slitter section lends itself to the modular approach (Fig. 455), which isolates the rewinding roll vibration from the slitting function. This type of sheet run also minimizes the sag between the slitter during a splice. The sag on the horizontal and inclined slitter section (Fig. 563) causes offsets at the splice.

Trim removal is said to be the easiest on this type of arrangement, as the trim can essentially drop straight down without having to change direction. I qualified this advantage because the trim flutter in the vertical section still occurs sometimes, creating edge cracks and offsets in the finished roll.

Fig. 567. Reel spool unwind.

(c) Inclined slitter table

Inclined slitter tables (Figs. 563, 566) provide good visibility and accessibility and are inclined, basically, to reduce excessive wraps on the rolls that direct the web into the spreader and the winding station.

Trim removal becomes more difficult with this arrangement, as the trim flutter and gravity more often than not interfere with the cutting process. The sag of the web during splicing causes offsets at the splice [8].

There are so many items that affect slit quality and, as has been said repeatedly, they all are important. Although the subject sounds complicated, understanding and complying with the basics of slitting (Fig. 530) will make equipment selection and problem-solving easier to accomplish.

I. UNWINDING FUNCTIONS

The unwinding process (Fig. 387) provides a means of support and a braking means which, in turn, provides the tension for the process.

Unwinds can be divided into two basic types:

1. Shafted (Fig. 567)
2. Shaftless (Fig. 568).

1. Shafted unwinds

(a) Paper machine winder unwind

Shafted unwinds first appear at the paper machine winder. The shaft in this case is a reel spool that is of sufficient diameter to support multiple sets of paper the full width of the paper machine. A set is defined as a series of rolls wound side-by-side, at the same time, on a two-drum winder to a given diameter and width to satisfy a customer or next process shipping-roll specification.

(b) Flying-splice unwinds

Flying-splice unwinds are found primarily where the process of the full- or half-size reel must be processed continuously. Printing presses, coaters, and sheeters are a few of the processes using flying-splice unwind equipment.

(c) Salvage and reprocessing winder unwinds

Another form of shafted unwind is found on some of the older salvage winders and many of the converting duplex winders. These are being replaced in many cases with shaftless unwinds, primarily because of the necessity of the shaft to be very small to fit into a small (3 in. ID, 76 mm)

Fig. 568. Shaftless unwind on a salvage winder.

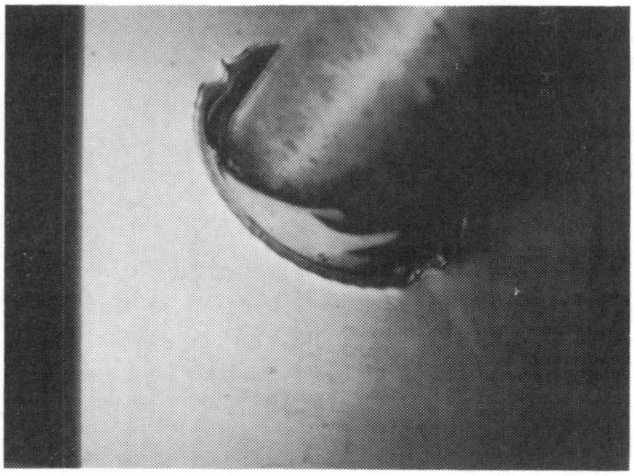

Fig. 569. Core delamination on a shafted salvage unwind.

core. With wide rolls being processed, the deflection of the shaft can (in some cases) be detrimental to the core (Fig. 569), plus causing loping of the rolls, thus limiting speed and subsequently production. Other negatives of this type of unwinding reflect on the need for shaft handling equipment, such as shaft puller and special cranes.

2. Shaftless unwinds

Shaftless unwinds (Fig. 568) are used where the paper coming from a previous process is on small-diameter cores. This usually necessitates a means of picking up the roll from the floor, a chuck that will grip the core so braking and tension can be provided to the unwinding roll through the chuck, and, of course, doing this without damaging the core or the roll of paper.

This type of unwind is usually designed to be a one-man operation customarily equipped with a means for edge guiding (Fig. 570) where a roll must be salvaged without cutting down or changing the width. Oscillation (Fig. 571) is a means of leveling out an extreme profile variation. The unwind has the ability to handle all widths and diameter rolls coming from other processes. Multiple disc brakes (sometimes programmed for the large variety of widths and diameters coming to this type of unwind) attached to each core chuck provide the braking and tension for the salvage and reprocessing process.

3. Braking

Braking is usually provided by two means: friction-type brakes, which can be an air- or water-cooled disc brake (Figs. 572, 573), a hydraulic shear disc brake, or a water-cooled drum brake (Fig. 574) with multiple shoes to provide the necessary tension and a means of stopping the full reel in case of a snapoff.

The main concern with these types of brakes is that the web must provide the forces necessary for the acceleration and deceleration of the unwinding roll. Maintenance, wear, and dust are also negatives for this type of braking.

The other means is an electric regenerative brake (Fig. 574). This type of brake has gained more acceptance mainly because it not only

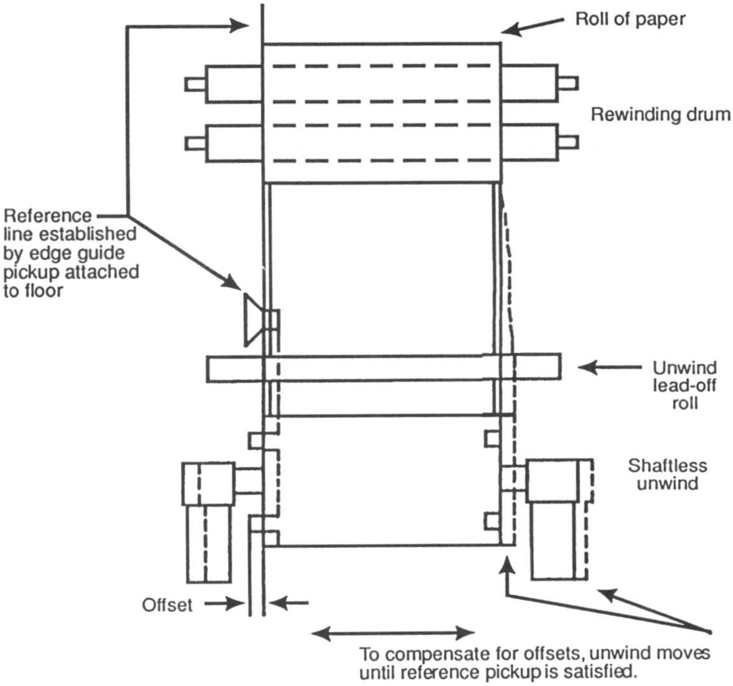

Fig. 570. Edge guiding.

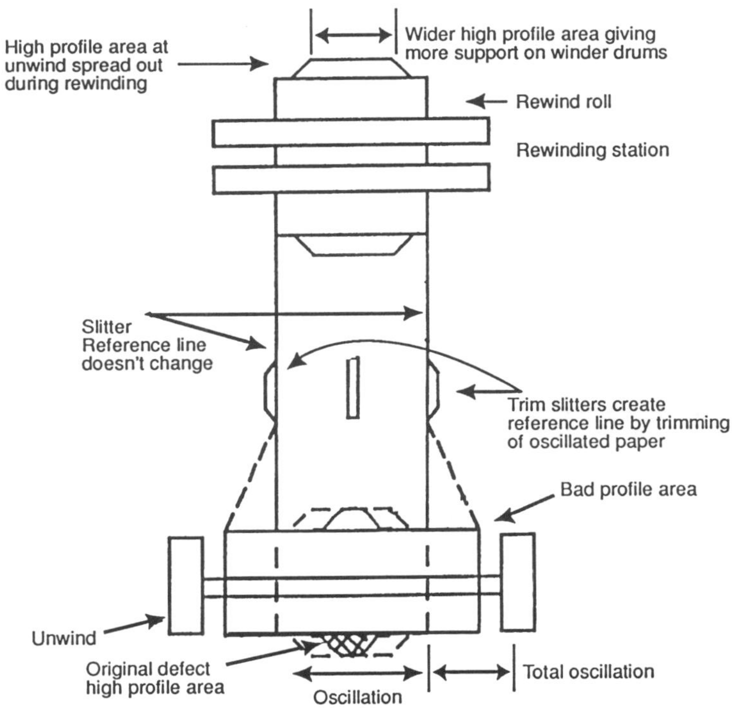

Fig. 571. Oscillation.

Fig. 572. Disc brakes.

Fig. 573. Water-cooled disc brakes on a shaftless unwind.

provides the means of tension control, but it also provides a means of acceleration, deceleration and payout – none of which is dependent on the strength of the web material. Programmed inertia compensation and controlled deceleration are usually part of this drive package.

4. Auxiliary equipment

In the interest of productivity, there is considerable equipment that reduces the downtime in the unwind area. Continuous rails directly from the reel eliminate cranes and the time to handle the reel spool (Fig. 455). Reel-spool ejectors (and other spool and roll handling equipment too numerous to list) also enhance the minimizing of downtime.

Fig. 574. Drum brake on a reel spool unwind which also has regenerative brakes.

J. PRODUCTIVITY AND MANPOWER

Since quality and productivity are the two main motivating forces in any process system such as winding, there should be some discussion about the productivity aspects of winding.

1. Factors that limit productivity

It would be incorrect to discuss productivity of any piece of equipment without first discussing some of the factors that limit productivity, regardless of the type of winder and the automated equipment that one uses.

There are many obvious factors that limit productivity in a winding system. To name a few:

1. Paper quality
2. Small roll diameters
3. Roll quality
4. Snapoffs
5. Product mix
6. Types of equipment
7. Material flow
8. Space
9. Slitter changes
10. Bouncing.

Many of these are obvious; however, the bouncing phenomenon is little understood and should be covered when talking about productivity, especially on grades that historically bounce such as kraft paper and tissue. With the higher coefficient of paper to paper friction that is surfacing on newsprint, this high production grade is now experiencing the bouncing phenomenon.

2. Theory of bouncing

Years of study (first done by Daly [9]) "on vibrating rolls of paper on two-drum and single-drum winders" have revealed at least one type of bounce (bounce that occurs at a certain diameter). The cause of surface winder vibration, and bouncing of the rewinding roll were found to

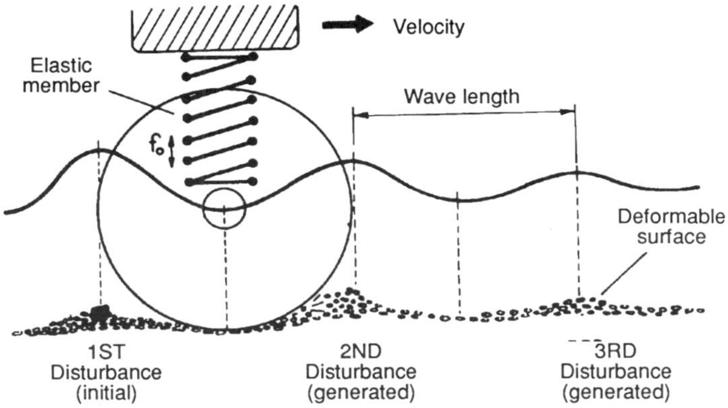

Fig. 575. Generation of a "washboard" road.

be similar to another vibration problem familiar to many motorists – the washboard road. A washboard road is generated by the combination of:

1. A deformable surface (the gravel road) (Fig. 575).
2. A rotatable elastic cylinder supported to a fixed body through a linkage of known natural frequency of oscillation (the tire and wheel suspension of an automobile).
3. Contact between no. 1 and no. 2 with relative motion (the auto traveling over the gravel road).
4. A small disturbance on no. 1 to cause no. 2 to begin an oscillation (a larger pebble in the gravel).

The tire and wheel suspension will go through at least one cycle of oscillation on hitting a disturbance on a road and, on returning to the road surface, scoop out a small depression in the gravel some distance downstream from the initial disturbance. The distance from the initial disturbance will be a function of the natural frequency of the suspension system and the velocity of the automobile. Since automobiles traveling any road have approximately the same velocity and suspension systems, another automobile passing over the disturbance will behave in the same manner, scooping out additional gravel and ultimately producing a second disturbance in the form of a hill.

Thus, an automobile could be subjected to two disturbances precisely tuned to the frequency of the suspension system. Additional disturbances are generated until the road contains a continuous series of precisely spaced "across the road" ridges. Hence, the washboard road (Fig. 575).

The vibration of a surface winder during the winding of paper rolls has the same generating elements as the vibration due to the washboard road. The elements are somewhat inverted, however, in that the rewinding roll of paper becomes both the deformable element and the suspension system of known frequency of vibration (Fig. 576). The winding drum is more or less the fixed member in the system. Vibration can be said to be generated by a combination of:

1. A deformable surface (the roll of paper) (Fig. 576).
2. A rotatable elastic cylinder of known frequency of vibration (the roll of paper).
3. Rolling on a relatively inflexible nondeformable surface (the drum).
4. A small disturbance on no. 1 to cause no. 2 to start an oscillation.

Figure 576 shows the generated disturbances downstream of the original disturbance. These disturbances will continue to be generated around the roll until, unlike the washboard road, they close to form a repeating cycle. If, in closing, the wavelengths pitch out to exactly an even number of cycles per revolution, the condition of amplification of the disturbance will be the worst and may possibly amplify enough to cause the roll to be thrown clear of the drum – hence "bouncing". The winding roll, however, does not require an even number of pitches of vibration per revolution to bounce. Some papers support the deformation of their surface so well that even nonmultiples of cycles per revolution will cause violent bouncing.

In the case of the washboard road, vibration can be eliminated by paving the road to remove the deformable component.

Although this solution cannot be applied to the paper roll, the analogy provides a clue as to why only certain grades of paper vibrate, some to the point of bouncing.

The ability of a roll of paper to become

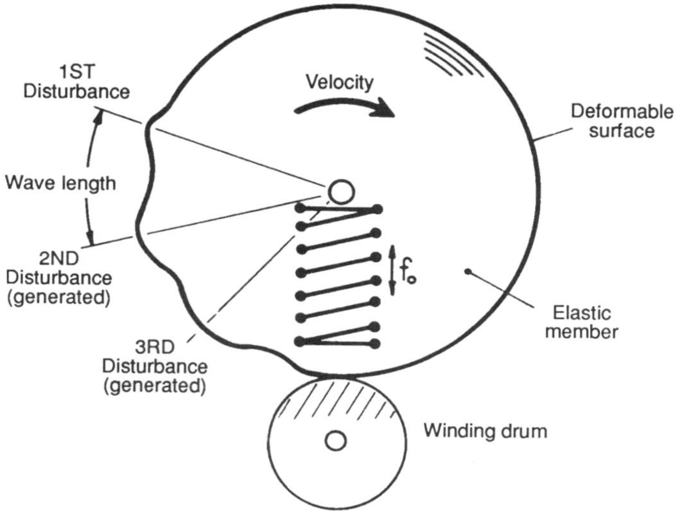

Fig. 576. Generation of vibrating or "bouncing" roll.

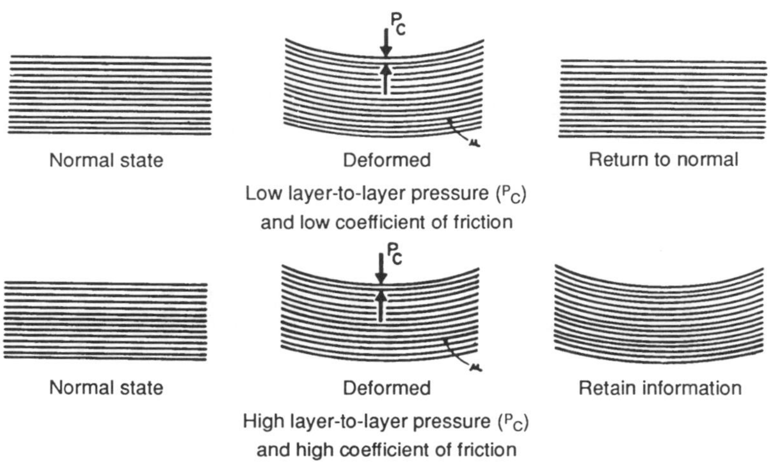

Fig. 577. Factors producing roll deformation.

permanently deformed on its surface is the key to the problem. Those rolls that are easily deformed on their surface and hold this deformation are the "bouncers".

If a deformation can be held for one revolution of the rewinding roll, it then can be amplified by another impact on the drum in the same spot. If the deformation can be pulled out or slips out before the next revolution, the amplification will not take place.

A study of why some papers hold their surface deformations showed the cause to be simply the friction force existing between layers of paper in the rewinding roll. If a deck of cards is bent (as shown in Fig. 577), simulating indenting the surface of a roll of paper, the individual cards are forced to move relative to one another but will easily return to their original state when the bending force is removed. If, however, there is a combination of normal forces, Pc (layer-to-layer pressure) and coefficient of friction, existing between the cards of the deck so as to resist sliding back to normal state, then the deformation will be held. This is to say that papers with a high layer-to-layer friction and a moderate layer-to-layer pressure (or vice versa) are ones which will hold a deformation and are potential vibrators or bouncers.

Measurements of layer-to-layer friction were made on many grades of paper, and a direct correlation was obtained between potentially vibrating papers and this friction. The line drawn

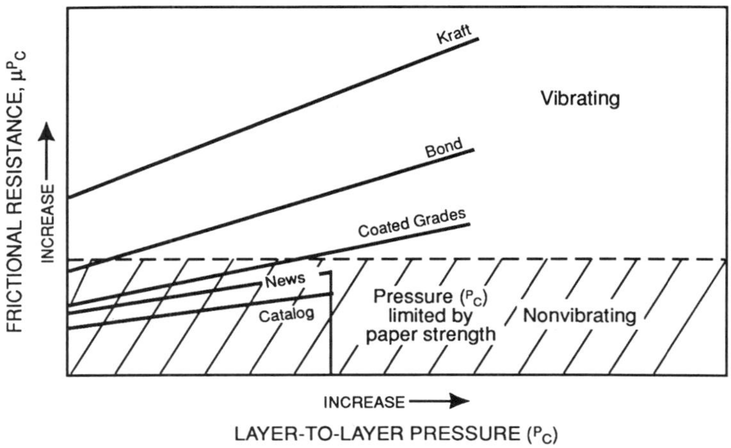

Fig. 578. Vibration potential.

through the family of curves is an imaginary boundary between vibrating and nonvibrating papers (Fig. 578). The curves for newsprint and catalog grades are not extended to the higher pressures because they cannot be wound to these pressures without damage to the paper. However, if the coefficient of friction is increased, the same effect will and does occur.

Note that the coated papers operated just the opposite, even though they have a relatively low coefficient of friction, and also can be vibrators because they wind up with much higher pressures in the rewound roll. The exact boundary definition for vibration of a rewound roll is difficult because all rolls vibrate to some degree. The dividing line indicates where this vibration becomes detrimental to the winding machine or the control of the rewound roll.

It goes without saying that bounce will limit production and, with the 8000 to 10,000 ft/min (2400–3000 m/min) winders that are needed to keep up with 5500 ft/min (1650 m/min) paper machines, already a reality, addressing this problem and understanding it becomes a necessary fact of life. If this is not done, all of the automated equipment in the world will not solve the productivity problem that it causes.

But having solved that problem or not having it in the first place, it is still well to understand the productivity limitations of the different types of equipment and grades now in the marketplace.

3. Paper machine two-drum winder

A rule of thumb on the necessary productivity of a two-drum paper machine winder is that the winder must run two to three times the speed of the paper machine. This has been found necessary primarily because of the downtime needed to change rolls in the unwind, slitter changes, splicing time, paper problems, and for any other reason that the winder is down between sets. The main impetus is, of course, to allow sufficient time so the paper machine will not have to shut down or slow down to compensate for winder productivity problems.

Many things have been done to reduce this downtime. And the simplicity of the two-drum winder makes automation of the many reasons for downtime quite easy.

This is why the efficiency and manpower requirements on the two-drum winder are much better than on a duplex-type winder. One winder can normally keep up with paper machine speeds of 4000 ft/min (1200 m/min). This of course, depends on the sizes (width and diameter) and the grades being run, and the type of production-improving equipment on the winder.

Some of features that have been automated to improve winder productivity include:

1. Automatic roll insertion into the winder unwind (Fig. 455).
2. Spent reel spool ejection and removal (Fig. 455).
3. Auto payout and threading.
4. Automatic programmed acceleration, deceleration and speed, stopping at a preset footage or diameter.
5. Automatic cutoff and roll ejection and roll handling (Fig. 579).
6. Automatic core insertion.
7. Automatic gluing or taping.
8. Automatic sequencing.
9. Automatic roll identification and processing.
10. Automatic selection of TNT for different paper grades.

There are many others, but they are all aimed

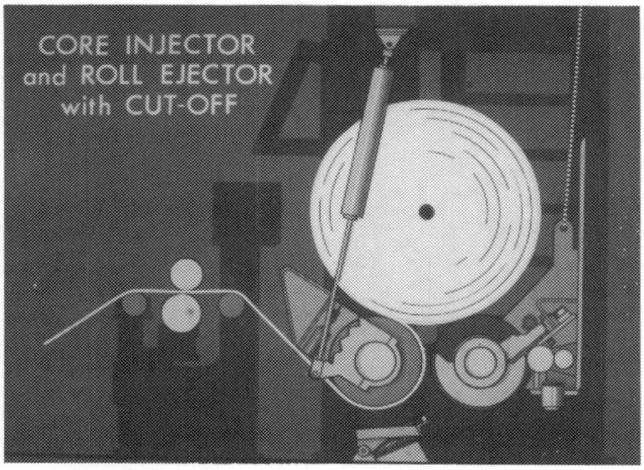

Fig. 579. Winder automation.

at making the winder a continuous, rather than an intermittent process.

4. Duplex winding

The duplex machine winder by its very nature is a much less efficient machine than the two-drum winder.

With separate winding stations on either side of the drum station, automation of splicing, core insertion, web cutoff, taping, gluing, threading and roll handling are much more difficult; the winder, therefore, becomes a man-paced rather than machine-paced piece of equipment. With one duplex winder following a high-speed paper machine, this would necessitate machine speeds in excess of 10,000 ft/min (3000 m/min), which exceeds the present state of the art; consequently, two winders are necessary. This, of course, also increases the manpower requirements on a paper machine winder operation.

Another concern regarding this type of winding equipment is the difficulty in winding narrow-width rolls on a machine duplex winder. This necessitates winding wider rolls and then having them cut down on a reprocessing two-drum winder, which involves another costly operation.

5. Salvage or reclaim winding

Since this operation is basically one that works on defective rolls, productivity is limited by the type of defect one is trying to remove.

For example: if there are any offsets on the edge of the roll (Fig. 477) (which are detrimental to shipping and running in customers' plants), they cannot be removed by slitting. Since this would change the width of the roll being shipped to the customer, edge guiding of some sort is necessary.

Edge guiding (Fig. 388) is that process which moves the unwinding roll of a salvage operation (in relation to some reference line) by controlling the edge of the web as it unwinds with a sensor that establishes that reference line. The need to react expeditiously to remove a single wrap offset necessitates that the winder be run at a slower speed to give the unwinding roll time to react without wrinkling. The reaction time can be improved by guiding the web instead of moving the entire unwind stand; however, the speed must still be limited by the nature of the defect and the process. This obviously affects the productivity of the salvage winding process.

Automation of the salvage operation cannot easily be done, primarily because of the nature of the operation; as the need varies so much for each roll, they must be handled almost as a separate order.

6. Reprocessing winder or a combination

(a) Salvage and reprocessing winder

The productivity of a reprocessing winder is normally the same as a machine winder. The speed must be fast and, with many of the automation features, good productivity can be maintained.

Many times reprocessing is needed because of the inability to perform the operation on the machine winder, such as spreading; however, with the new spreading equipment now found on machine winders, this need is diminishing. Consequently, the combination salvage and reprocessing winder sometimes becomes quite easy to justify – especially when a combination process of reprocessing and reclaiming is needed.

(b) Combining winder—tissue

The productivity limitation of a tissue winder (Fig. 464) was, at one time, the inability to slit with a score-type slitter at high speed. Slitter vibration limited most of these types of winders to 4000 ft/min (1200 m/min) or less. With the advent of shear-type slitters, this limitation no longer exists, and the limit is now the high speed of the paper machine (4000 ft/min, 1200 m/min) small-diameter rewound rolls and how much single-ply tissue is being wound. Obviously when winding 3 ply tissue, three reels from the paper machine necessitate the winder to run only one third of the speed necessary when running single ply, and only one reel from the paper machine.

7. Summary

Again, the two main issues in the winding process are productivity and quality.

Productivity, or the lack of it, is usually driven by factors such as product mix, space, material flow, type of equipment, etc., but roll quality is a force created by the ultimate customer, and must prevail. The evolution of the winding process as described in this chapter substantiates this fact. And, although the winding process seems quite complicated, it has been necessary because of that quality need.

This is why it has become necessary to dissect the problems of the industry and provide winding equipment that meets both the quality in the pressroom and converter, and productivity requirements in the paper mill [9].

If one looks at the basic technologies of finishing and approaches those problems in a systematic way through some of those basics, a better understanding of all of the processes of finishing will enhance not only the winding process but all of the other processes as well.

Problem-solving, as described by Frye [11], demonstrates the process that should be practiced to advance the industry as a whole.

GLOSSARY

AIR SHEAR BURST — This is a machine-direction rupture of the paper that is usually a single rupture occurring 2–12 in. (5–30 mm) from a slit edge. Shear burst is caused by too much air or lubrication being inserted between web wraps in the roll by improper drum grooving on the first winder drum.

ANTIFRICTION SLIDES — Core shaft support mechanism that runs on surfaces that roll on one another. This is in contrast to friction slides that slide against each other.

BAGS — This is where the paper has been permanently stressed beyond its elastic limit. This is, many times, caused by profile variations and is amplified by excessive nip pressure in a localized area.

CALENDER CUT — This is a double-line cut caused by a folded wrinkle passing through calender nip rolls.

CALIPER SHEAR BURST — This is a machine-direction rupture of the paper, usually a "U-shaped" break found near the high-caliper areas of a web of material, and is caused by the differential layer-to-layer slippage as the paper is processed through the nip of a winder.

CAPSTAN — Referred to as the progress of wrap the web takes as it passes around the back drum.

CARBIDE DRUM COATING — Very sharp carbide particles embedded in the front drum surface to provide traction between the drum and the winding roll.

CENTER WINDER — Differs from other winding processes only in the means of driving the rewound roll. As the name implies, the drive is provided to the center of the roll rather than at the outside surface of the roll (as described in the two-drum winder.)

CENTERLINE OF MACHINE OR MACHINE CENTERLINE — An imaginary line drawn the length of the machine that would split the web and the machine exactly in half (in the machine drum).

CENTERWIND REEL — Provides the same function as the nip-type reel, with one exception: the drive is provided at the center of the roll instead of on the surface. Consequently, the roll structure is provided with center torque rather than the nip. The equipment must therefore be different to provide the automatic transfer in the continuous process.

COATER — A machine that puts a coating such as clay or poly onto the (substrate) paper.

COMBINING WINDER — Winds more than one web at one time into a common roll. This changes the winding process mostly in the unwind area where multiple unwinds (and their associated control equipment) are necessary.

COMBINING WINDING — Consists of a process that combines more than one roll of material in the unwinding process and then processes these multiple rolls into one roll on the combining winder. Two- and three-ply tissue and extrusion coating foil and paper laminate are two examples that fall into this category.

CONTROLLER — The controller receives a signal from the tension-sensing element in a tension system and conditions it to react to the rate of change in the process control reset, the rate of change to the error signal, and to the magnitude of the error. The purpose of the controller is to react expeditiously with maximum stability during transitory web control situations.

CONVERTIBILITY — The ability to convert paper into different roll forms without problems.

CONVERTING WINDER — A winder that converts a roll into different sizes.

CORE — The pipe-like form that the paper is first wrapped around during the winding process. This can be made of paper and resin laminate or steel pipe.

CORE BOXES — Usually refers to the mechanisms that hold the core shaft or core on a winder during the winding process. Another name for Core Shaft Slides.

CORE BURSTS — Breaks on the edge of the web inside a roll is found while unwinding, usually near the core. It customarily begins with a crepe wrinkle, proceeds with multiple shears in the machine direction, and ends with an edge crack or burst at the end of the shear.

CORE SHAFT SLIDES — This is a device that holds

the core and/or the core shaft and roll in the winding process from moving sideways and causing offsets.

CORRUGATIONS-ROPES — Permanent web deformation in the form of diagonal wrinkles, either soft or hard, caused by profile variation producing non-uniform nip pressure on the web of material.

CREPE PULL-OUT — Crepe is added into a tissue web by the crepeing doctor on at issue paper machine to produce bulk. Crepe pull-out is referred to as how much reduction in the original crepe occurs during subsequent processing.

CREPE WRINKLES — A condition where the paper bunches up in the machine direction in an accordion fashion, which forms a crepe not unlike that found in crepe tissue.

CRITICAL SPEED — That speed of a rotating component that matches the natural frequency of the rotating component, creating excessive vibration.

CROSS-MACHINE — Refers to a direction that is perpendicular to the paper flow direction.

CUTPOINT — The point of contact between the top and bottom slitting components; the apex of the scissors-like rotating cutting elements.

DECELERATION — The process of going from one speed down to a lower speed.

DENSITY MONITOR — A device that plots the density changes in a roll of paper during the winding and unwinding process.

DIFFERENTIAL NIP — The differences of nip pressure because of paper profile across the web of paper.

DIFFERENTIAL TORQUE — This is the difference in amps or torque between the two drums of a two-drum winder occurring during the winding process.

DISHING-TELESCOPING — Where the edges of the roll are not in one plane and are progressively located outside or inside the theoretical slit line, usually in some form of smooth curve — like a dish.

DOCTORING — In SPREADING, it is using a spreading device to impart localized tension in a specific area of the web to reduce defects such as corrugations.

DOCTORING — In WINDING, it is the process of rewinding a roll with a centerwind winder called a Roll Doctor.

DOWNSTREAM — Refers to a direction that is the same direction as paper flow. In the case of the winder, downstream points to the discharge area of the rolls from the winder.

DRUM ARRANGEMENT — This usually refers to the arrangement of the driving drum or drums in the winding process, i.e., single drum, Bi-wind vertical drum, two horizontal drums, etc.

DRUM DIAMETER — The diameter of the winding drum that usually provides the drive in the nip-type winding process.

DRUM GROOVING — Refers to the different grooving configurations found on the drum or drums of a winder.

DRUM WRAP — The angle of wrap or contact as the web wraps the first winding drum.

DUPLEX WINDER — Differs from the two-drum winder only by the fact that the rolls are partially or fully supported on their cores rather than on the outside of the roll during the winding process. This, of course, necessitates that the rolls be wound alternately on either side of a central drum to allow for support chucks on the ends of each individual roll.

DUST — Refers to the debris found on the surface of edges of a web or sheet of paper which interferes with the printing process by causing specks on the printed papers.

EDGE GUIDING — This is the process of rewinding a roll of paper with uneven edges and guiding the web in a different way to result in a straight-edged rewound roll.

FAN-TYPE SPREADER — Describes a bowed roll, a D-bar, or a stationary-bowed pipe and are spreaders that spread the web in a cross-machine direction not unlike the geometry of a fan, with all of the webs diverging symmetrically from a central location.

FLYING SPLICE UNWINDS — Found primarily where the reel of paper must be processed continuously, and the splice that ties the expiring and the new roll together is made on the fly or without stopping the process.

FOLDING-TYPE SPREADER — A spreading system that equalizes the strains in the web by folding the web over one bowed bar and then unfolding the web back on the second bar, but not quite as far, thus setting the web spacing and equalizing the strain on the edges of the web.

FRICTION COEFFICIENT — This is usually referred to as the resistance to sliding of the web on itself or between the web and other contacting surfaces. It is used to calculate the force needed to move a body by the formula $F = \mu n$ (friction coefficient).

GAP TEST — A test made on a roll of paper to determine the residual tension in the web at that point in the roll.

GEARING — The phenomenon of layer-to-layer slippage near the core during unwinding on a roll that was not wound tightly enough to support its own weight.

HOLES — Refers to a defect or void in the web of paper usually caused by upstream processes.

HORIZONTAL SLITTER TABLE — The paper passing through the slitter in a horizontal fashion supporting the web on either side of the slitter or rolls is called the slitter table.

INCLINED SLITTER TABLE — This type of slitter table is at some angle other than horizontal and vertical with respect to the supporting floor.

INTERWEAVING — When adjacent webs or rolls overlap one another as they are processed or wound up on a winder.

"J-LINE" — This is a phenomenon of layer-to-layer slippage used to indicate how tightly a roll is wound by snapping a straight line on a moving roll, and observing layer-to-layer slippage. The tight individual layer slip causing the line to bend in such a way that it looks like a "J".

LACING — Describes interweaving of two adjacent webs during the winding process.

LAMINATOR — A machine that combines two materials together to form one product.

LASER SLITTING — This type of slitting, using a laser, has intrigued people for years but, because of the large power requirements needed to reach the speed of state-of-the-art winding equipment, this development hasn't as yet taken off. This system uses a laser beam to sever the web.

LOOSE CORE — The paper is loose on the core (no tension). Dishing and telescoping sometimes occur due to this problem.

MACHINE WINDING — Utilizes paper from the primary process machine reel which comes to the winder unwind stand on reel spools or large-diameter cores. Splicing, inspection, slitting, hole patching, and trimming are normally performed during this function. Shipping rolls of various and random widths can also be made during this operation. The width is usually consistent with the associated machine width, whether the machine is a paper machine, coater, or supercalender.

NIP — Refers to the force between the roll of paper and the driving drum or the force applied to paper by nip producing components.

NIP-INDUCED DEFECTS — This term refers to those defects found in the pressroom, paper mill, and converter that are caused by the nip between the paper and other nip-producing elements, such as a winder drum or calender roll.

NIP MECHANICS — The process of creating tension in a web of paper by applying nip between the two components, the drum and paper.

NIP-TYPE REEL — A winder with a single, large-diameter drum that winds paper on a reel spool or large-diameter core. This piece of equipment is usually associated with some continuous process that cannot be stopped and started — to start a new reel of paper on another spool.

NON-UNIFORM NIP — This refers to a difference in nip against a drum because of a difference in profiles of the paper.

NUMBER OF SLITS — This refers to the number of individual webs that are being cut from a wider web.

OFFSETS — Any protrusion or non-uniformity from the slit line or plane of the edge of the roll.

OSCILLATION — The process of moving the unwind stand of a process back and forth to smooth out any high points in the rewinding roll, during the rewinding process.

PACKAGING — The process of covering or protecting of the final product with a protective covering material.

PAPER PROFILE — This refers to the variations in thickness across the web (web width) and can be attributed to variations in basis weight, moisture, hardness, caliper, coating weight, etc.

PENETRATION — Usually refers to the amount of overlap of the top slitter, in relation to the bottom slitter, in a shear slitter.

PLOW EFFECT — The effect of the bevel and toe-in on a top slitter blade as it intersects the plane of the paper during the slitting process. At this point the paper is forced to go around this obstruction so that the blade is said to plow the paper. The paper is plowed sideways.

PRESSROOM — This covers the area where the rolls from the mill are sent and processed through a printing device.

PRINTABILITY — The ability to produce quality printing.

PRODUCTION CONTROL — The ability to monitor and control the production of any process.

RADIUS OF CURVATURE — This is a term used in spreading design referring to the radius of the bow in a bowed spreader.

RAKE — The angle of the top blade in the vertical plane with respect to the bottom slitter component, looking in the machine direction.

RAZORBLADE and BURST SLITTING — This type of slitting is used almost exclusively in the plastic film and packaging industry and reflects slitting with a top knife inserted in a narrow groove with no contact with the bottom support means.

RECLAIMING/REPROCESSING — This is defined as a winding operation which combines the two processes. Its major functions include slitting and winding production orders for shipping rolls by reducing the original rolls to narrow-width slits, as well as changes in core sizes, roll diameters, reversing of wire and felt sides, and any other requirements necessary to fill the customers' needs. This however, is not a full-time operation; this winder may double for a reclaim or salvage function.

REELING — Winds a web by a continuous (paper machine or coater) or intermittent (supercalender, printing press laminator, etc.) process onto a spool or large-diameter core. This process necessitates that the diameter of the spool be large enough to support the full width of the roll being wound. Paper machine reels and re-reelers (and supercalender and coater windup reels) are examples of common applications.

REGENERATIVE UNWIND DRIVE — An unwind and system that uses a motor to brake and drive the unwinding roll.

REPROCESSING — Defined as a winding operation method to process rolls supplied by other functions, making changes needed to meet the ultimate customer or subsequent process needs. Thus, the reprocessing function may consist of splitting the roll down into specific sizes, changing the core length or diameter, changing the roll diameters, splicing, inspection, hole patching, and trimming.

RE-REELER — Is very similar to the nip-type reel, with the exception that there are no primary arms, as there is no necessity for a continuous operation. It usually consists of an unwind, with trim slitters, or a supercalender, or any other process in between. The roll of paper is supported full-width on a spool or large-diameter core which rides on a level or slightly inclined rail, with the secondary arms providing needed nip pressure to structure the roll.

RESET — The term found in an automated tension control system, which indicates that the system will reset itself after a given change in tension.

RHO METER — An instrument for measuring the hardness profile of a roll of paper. It measures the deceleration rate of a mass as it hits a roll of paper.

RIDER ROLL — This is a small roll that rides on the roll of paper in the early stages (at small diameters) of the winding process and provides a means to make a tight start on the winding of a roll of paper.

ROLL STRUCTURE CONTROL — The process that controls the structure or the wound-in tension of a roll of paper during the winding process. The structure is developed by the TNT of winding: tension, nip, and torque.

ROTATING SPREADER — This term refers to the type of spreader that rotates at web speed (such as a bowed spreader roll) and is compared with a spreader that is stationary (such as a D Bar).

RUNNABILITY — The ability to run paper through any process without problems.

"S" CURVE — This term is used in drive terminology and refers to the rounding out of any abrupt changes or directions of speed.

SCHMIDT HARDNESS TESTER — An instrument for measuring the hardness profiles of a roll of paper. It measures the rebound of a mass as it hits a roll of paper.

SCORE SLITTER — This type of slitter is used mostly on very high density material or on tissue where the edges of multiple webs must be bonded together in the slitting process. In this configuration, the top slitter blade rides directly on the bottom slitting anvil in a scoring function.

SET — Describes all of the rolls that are wound side-by-side at one time on a two-drum winder or on alternate sides of the drum on a duplex winder.

SHAFTED UNWINDS — Shafted unwinds are unwinds that support the roll of paper on a bar shaft or spool.

SHAFTLESS UNWINDS — Shaftless unwinds are used where the paper, coming from a previous process on small-diameter cores, needs to be unwound. Since these rolls are of varying widths, the roll is chucked up on its core rather than on a full-width shaft.

SHEAR TANGENTIAL SLITTER — This is sometimes called a Kiss Slitter and is the most commonly used type of slitter in the paper industry. It has the configuration where the paper is sheared by two rotating cutting components with little or no wrap of those components.

SHEAR WRAPAROUND SLITTER — This type of slitter is used mostly on narrow-width, small machines and has the configuration where the web is supported by and wraps the bottom slitting component.

SHEET FLUTTER — The oscillation or rubberband effect of the web in a direction perpendicular to the web plane as it passes through any process such as a slitter.

SHIPPING ROLL QUALITY — Refers to "a measurement of quality" of a roll that could be shipped to a customer.

SIDE LOADING — The load or pressure, between top and bottom slitting components, necessary to keep the top and bottom slitters in contact.

SINGLE-DRUM WINDER — A winder with only one drum used to drive the rewound roll.

SLIDING SPREADER — A spreading component where the web slides over a nonmoving surface which has a bowed shape to induce spreading of the web.

SLITTER BAND — The bottom element of a rotating shear slitter system. It usually becomes the reference for slit width accuracy and is in configuration and is ground in such a way so as to not change the diameter.

SLITTER BLADE — The top element of a rotating shear slitting system which rides against the bottom band. It sometimes has a beveled edge and, when ground, the diameter changes.

SLITTER JUMP — Due to vibration, the normal overlapped situation of a shear slitter is destroyed as the top slitter jumps on top of the bottom slitter, dulling the blade.

SLITTER RINGS — These distinct rings are visible on the side of a roll at different diameters. The ring circumference is a multiple of the circumference of the bottom slitter band, which creates a wavy cut on the web edge due to band wobble.

SLITTER RUNOUT — Radial and/or axial wobble of the referenced bottom slitter cutting component which results in cutting a scalloped edge which, in turn, causes slitter rings.

SLITTER TABLE — Usually describes the area encompassing the web path through the slitter.

SLITTING FUNCTION — Any industry processing a continuous web of material must, at one time or another, be concerned with the process of web separation — which, in terms of winding, is called slitting. Obviously, slitting is necessary to reduce the wide web of material usually furnished by the paper machine or other processes to a usable size for final consumer or industry use — whether the web is paper, steel, plastic, or textile.

SMITH NEEDLE — An instrument that measures the pressure between layers of a roll by inserting a needle between layers at the edge of the wound roll at different diameters. The number resulting will give you a tightness profile of the roll from the core to the O.D. by measuring the resistance to the needle being inserted.

SNAPOFFS — A break in the web causing an interruption of any process such as winding.

SPREADER — A device that creates a separation between webs and rolls being wound side-by-side. This is done by moving the web in the cross-machine direction by a bowed device called a spreader. It is also a device that smooths out a nonslit web.

STRAIN GAUGE — A device that measures a force through the use of a device that outputs a different voltage as a function of a change in force.

STRUCTURAL CONTROL — The factor that controls the internal structure of a roll during the winding process.

SUPERCALENDER — A machine that through high pressure and resilient rolls puts a gloss on a web of paper for better printability.

TECHNOLOGY OF FINISHING — Consists of the technology of nip mechanics — web control, web finishing, web separation, unwinding, packaging, production control, and sheet control.

TENSILE MODULUS — The amount of strain that is developed for a given force as applied to a web of paper. It is the slope of the stress-strain curves at a given tension.

TENSION — The stretching of a material by having equal and opposite forces pulling on the webs in opposite direction, thus creating tension.

TENSION BURST — A cross-machine break in the web that looks like it has exploded. It happens several layers down in the roll during the process of winding a roll. It usually isn't found until the roll unwinds in the pressroom.

TENSION CONTROL — This is the process of controlling tension automatically during the winding process.

TENSION SENSING ELEMENT — The tension-sensing elements are transducers that provide the function of measuring and feeding back the level of tension in a web to the controller in the system which, in turn, controls some form of braking system.

TNT — Refers to tension, nip, and torque in the winding process.

TOE-IN — The included angle between the top and bottom cutting components in a shear slitter system. Its function is to ensure only one point of contact between the top and bottom cutting components.

TORQUE — Referred to as the difference in tension developed by one drum of a two-drum winder being driven faster or with more load than the other drum. This creates an additional tension in the web and is used to make a tight start. Center torque has the same function but develops tension in the web by torquing or cranking up the roll from its center drive.

TRACTION — The ability of the web to match exactly the speed of the roll on which it is wrapped or supported.

TRACTION CONTROL — The ability to control the slippage between a web of paper and its supporting components.

TRIM REMOVAL — The system that takes a very narrow section of the outside edges of a wider web called trim and processes it through a system that chops, bales or recycles it back into the paper processing system. It is the left-over width of the parent web that is not sold to a customer.

TURNOVER — The term used in the pressroom and paper mill that denotes that the edge of the web is turned or folded over as it proceeds through the press or other processes. It is caused by an edge crack or an edge burst.

TWO-DRUM WINDER — Usually consists of an unwind, a slitter section, and a windup station. Its function is to cut down a wide web into narrower widths. The windup station consists of two drums that provide the means of winding up and supporting rolls of paper, side-by-side, on small-diameter cores. The drums provide the means of support and drive for the finished roll of paper.

UNWINDING — Provides a means of support and braking, as well as acceleration and deceleration, for an unwinding roll as the material is transported to another function such as a winder. Sometimes edge guiding and oscillation (and special roll and other material handling equipment) are included in this function for the roll being unwound.

UPSTREAM — Refers to a direction that is opposite that of the paper flow direction. In the case of a winder, upstream points toward the unwind station.

VERTICAL SLITTER TABLE — This type of slitter section is vertical, or perpendicular, with respect to the supporting floor.

WATER-JET SLITTING — This process has limited applications. Its use is primarily in the tissue, board, and carton industry. Water-jet slitting uses a high-velocity water jet to sever the web.

WEB CONTROL — Provides the necessary controls to transport the web from the unwinding roll to the roll being wound on the winder. It usually includes tension control, traction control, speed, and spreading functions.

WEB FINISHING — The process of putting a finish on a web of paper through pressure, such as supercalendering and embossing.

WEB SEPARATION — This technology covers slitting and cutting or the separation of a wide web into many narrow webs — in the winding process.

WINDER DRIVE — The means of providing the power or drive to wind a roll of paper.

WINDING — A process that changes a material that is manufactured in continuous process from web form and winds it into roll form for further processing. A first winding process occurs at the paper machine, extrusion coater, or nonwoven reel where the web is wound continuously on mandrels (without slitting) for further processing. Subsequent winding processes operate roll-to-roll rather than web-to-roll.

WINDING VARIABLES — In the winding process, these are things that affect roll structure in a roll of paper. Tension, nip, and torque (TNT) are the basic categories; however, there are many other variables that affect wound-in tension in a roll of paper, such as drum diameter, core shaft loading, paper grade, paper profile, etc. Depending on paper grade profile, some of these variables are more critical than others, the most significant being "nip".

WOUND-IN OR RESIDUAL TENSION — The amount of tension in the web as found inside of the roll of paper after winding. This tension is developed by three winder tools — tension, nip, and torque.

WRINKLES — Any deviation or variation from the flat plane of a web. Wrinkle can be described as soft or hard, with hard being a fold-over, creased-type of wrinkle.

LITERATURE CITED

1. Pfeiffer, J.D., Tappi 51(8):77A (1968).
2. Lucas, R.G., 1981 TAPPI Finishing Conf. Proc., TAPPI PRESS, Atlanta, p. 91.
3. Welp, E.G. and Schoenmeier, H., Tappi J. 67(4):70 (1984).
4. Pfeiffer, J.D., Tappi 60(3):106 (1977).
5. Frye, K.G., Pulp and Paper 59(2):102 (1985).
6. Frye, K.G., 1988 TAPPI Finishing and Converting Conf. Proc., TAPPI PRESS, Atlanta, p. 139.
7. Daly, D.A., Tappi 48(9):88A (1965).
8. Frye, K.G., Tappi J. 69(7):67 (1986).
9. Daly, D.A., How Paper Rolls on a Winder Generate Vibration and Bouncing, Beloit Eastern Corp., Paper Trade J., pp. 48-51 (Dec. 11, 1967).
10. Pfeiffer, J.D., Tappi 64(4):105 (1981).
11. Frye, K.G., Tappi J. 69(11):98 (1986).
12. Pfeiffer, J.D., Tappi 49(8):342 (1966).

XVI
CALENDERING

J.D. PFEIFFER

J. David Pfeiffer teaches courses in Design and Value Engineering in the Mechanical Engineering Department of McGill University. He has also started a Winding and Nip Mechanics Lab there which has been partially funded by grants from the Natural Sciences and Engineering Research Council of Canada. He has been publishing on problems associated with paper properties and winding as long ago as 1966, and the first paper, "Internal Pressures in a Wound Roll of Paper", is still cited as a classic. Some noteworthy developments in his career include the Rho-Meter hardness tester, web-driven slitters, and the WIT-WOT, or wound-in, wound-off tension machine. These were inventions he made while employed by the Beloit Corporation in the Downingtown, PA, Research Lab. He put forth the replication theory of calendering and supercalendering in 1970, when it was not quite so widely accepted as it is today. In 1978 he received the TAPPI Finishing and Converting Division's award, now known as the Thomas W. Busch Prize.

A. INTRODUCTION

Calendering is an important part of the papermaking process. It is in the calendering operation that the paper passes through the most intense pressure zones that it must encounter in its journey from the slice at the wet end to the reel at the dry end. The incorporation of calenders in the paper machine has been a part of the evolution of papermaking to meet the demands for paper with increasingly better and more uniform qualities. In response to these demands, calenders have been designed into paper machines to a) correct irregularities in sheet formation, b) improve the surface flatness or smoothness of the sheet, and c) compact and densify the paper fibers. The production of calender rolls with a high degree of surface finish and accuracy was not an easy task a hundred or more years ago. It is only the patience and ingenuity of the early craftsmen that allowed them to build steel or chilled iron rolls that could turn out a sheet with reasonable quality as long ago as 1867. Now, it is possible to obtain sheet finishing in a nip where one or the other roll has a non-metallic covering, which gives less peak pressure at the nip impact, and this allows some selectivity in optimizing one particular property without damaging others.

The location of calenders has changed over the years. One could always find a calender stack of rolls just prior to the reel (Fig. 580), and possibly a breaker stack with one or more nips between dryer sections, to finish the sheet at higher moisture level. At the wet end, a lump breaker roll might be found over the suction couch area just prior to the press nips (Figs. 581, 582). With the exception of a size press partway through the dryer section, which is not usually thought of as calendering because it is a wet process, there

REVIEWED BY

Eric R. Ellis
Senior Research Scientist
S.D. Warren Company
A Subsidiary of Scott Paper Company
Westbrook, ME

Kenneth B. Jewett
Research Scientist
S.D. Warren Company
A Subsidiary of Scott Paper Company
Westbrook, ME

Jay H. Vreeland
Vice President of Research and Director
S.D. Warren Company
A Subsidiary of Scott Paper Company
Westbrook, ME

Fig. 580. A six-roll open stack machine calender located before the paper machine reel.

Fig. 581. Section of a fourdrinier paper machine at the first press showing the location of the lump breaker roll D above the suction-couch roll A. Other labeled items include: B, secondary roll; C, wire; E, suction pickup roll; F, suction leg; G, pickup roll lift; H, suction first press; K, pneumatic pressure unit; J, top felt; L, automatic felt guide; M, felt conditioner.

Fig. 582. Breaker stack at entry to dryer section.

might be no other rolling-nip finishing action on an older paper machine.

With the advent of coated publication grades, additional finishing had to be done to the sheet after coating, which required a less harsh type of nip. The supercalender, where roll coverings of compacted cotton fibers or other materials are used in alternate roll positions, filled this need (Fig. 583, 584). Supercalendering, in the early years, was an off-line process, as was coating. Recent trends have put the coating and supercalendering operations in-line with the paper machine as a continuous process (Fig. 585).

Calendering changes the surface and interior properties of the sheet by the action of passing the web through a nip composed of two rollers which may or may not be of equal hardness. The pressures are extreme, and the time that a given section of web spends in passing through the nip is short, especially when considering the high production speeds of today's paper machines. To

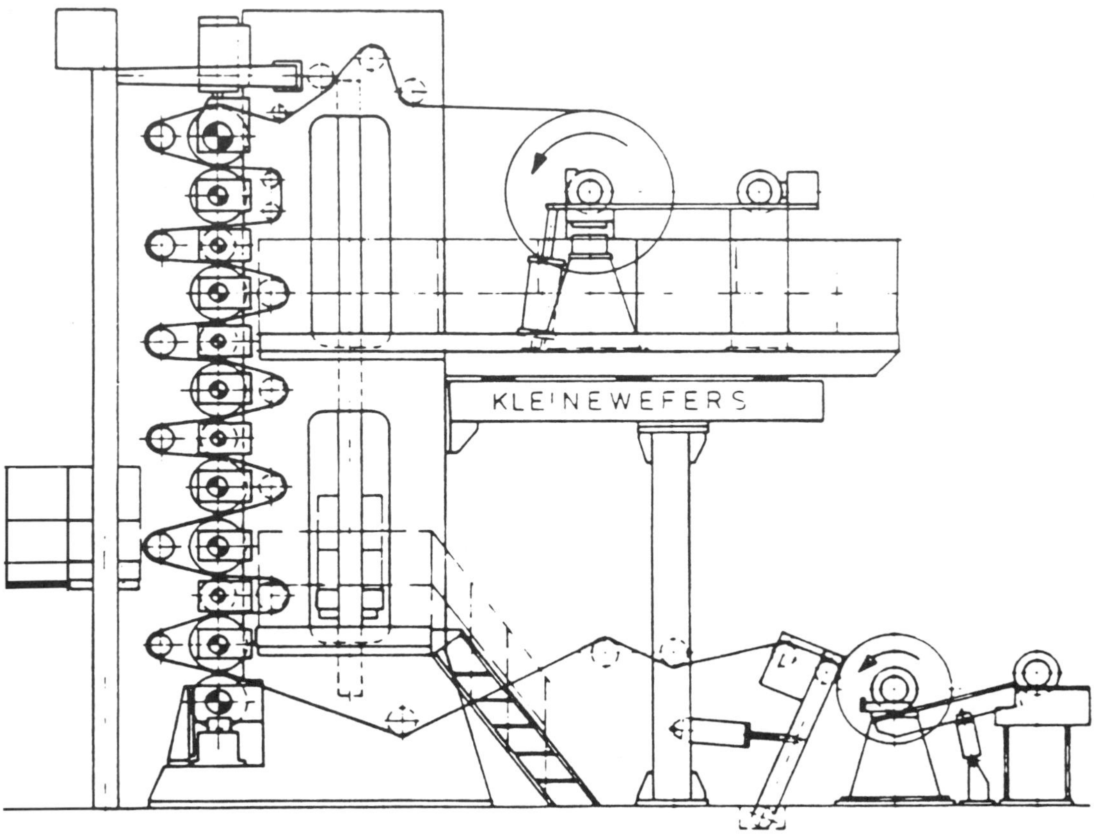

Fig. 583. Sheet run through a typical 10-roll supercalender stack.

Fig. 584. Two 7.6 m wide supercalenders designed to operate at 910 m/min (3000 ft/min). (Photo courtesy Valmet Paper Machinery)

Fig. 585. Valmet SC 2000 soft calender arranged for double-side finishing. First nip (at right) has oil-heated metal roll in top position.

estimate how short this time is, realize that if the web speed is 900 m/min (2950 ft/min or 15 mm/ms), then the widest nip on an off-machine supercalender with filled rolls nearly a meter in diameter might be 30 mm wide. The time spent in the nip would then be 30/15 or 2 ms, while in a 5 mm wide nip between steel calender rolls, the time spent would be only three-tenths of a millisecond, that is, 300 μs!

The nips of the calender are not the only set of nips that pass over the sheet on its way through the paper machine. New developments in high-intensity pressing and extended nip pressing not only do a more effective job in de-watering the sheet, but must have an effect on the sheet properties measured at the surface and throughout the thickness. High-impact presses have been designed with linear loads as high as 350 kN/m, giving a nip width of 90 mm and peak pressure within the nip of 5.5 MPa (800 psi). Another area of current research, impulse drying, promises to further reduce the drying energy required for the sheet. Impulse drying subjects the sheet to short exposures of high temperature at high nip pressures, and will therefore contribute to finishing action in the same manner as calendering.

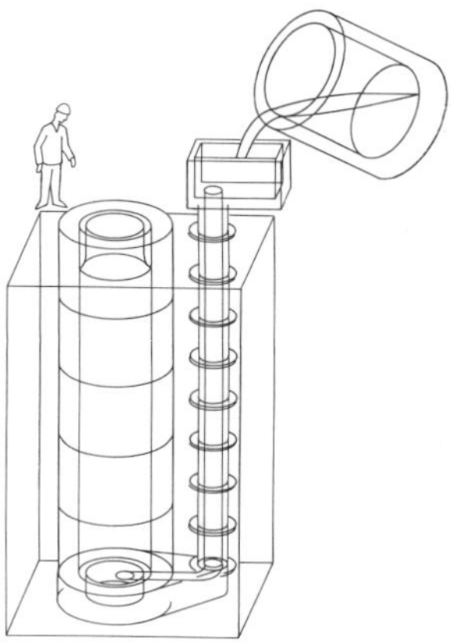

Fig. 586. Casting pit for producing chilled iron rolls loaded with chills. At right, workmen prepare the mold for the pour. (Courtesy SHW, Inc.)

As the technological evolution of the paper machine continues, it may become more and more difficult to point to one section of the machine and say, "There is the calender". New forms taken by calenders, such as the single-nip calender, which is often built as two tandem units with the sheet passing almost straight through (Fig. 585), are not as easy to recognize as the gleaming stack of steel rolls that once stood before the reel.

B. CALENDER ROLL CONSTRUCTION

The surface condition of the metal calender rolls must retain a smooth finish in order to impart the desired sheet properties. For this reason the metal must be hard in order to resist wear by abrasive particles, denting by hard particles, or the occasional sheet wrap-up during a web break. These requirements are met by using chilled cast iron rolls, which are made from iron-carbon alloys which may also contain manganese, phosphorous, sulfur, small amounts of silicon, and sometimes chromium and nickel. The term "chilled" comes about from the rapid removal of heat from the surface of the roll when it is cast. This causes the carbon in that section to be bound to the iron as iron carbides, which are very hard and durable. In the usual method of manufacture, these rolls are cast from a single melt in a vertical pit, where heavy metal rings, called chills, are stacked to a depth sufficient to produce the full roll width (Fig. 586). The thermal mass of these chills takes away the heat of the molten cast iron to produce a chill depth of white iron (the useful hard outer surface) of 10–15 mm in the calender roll. An appreciation of the amount of heat contained in an 80 t pour can be gained from the practice of waiting one week before pulling the roll body out of the pit. The actual surface hardness ranges from about 70 to 78 Shore "C" or 550–660 Vickers. These values are roughly in the range of 517–620 Bhn (Brinnell hardness) or Rockwell C 52–58.

Some regrinding of the roll surface will be needed from time to time in its useful life to remove doctor blade scratches, barring marks due to vibration within the stack or periodic thickness variations in the paper stock, or damage due to contact between a spinning and a stationary roll (thermal shocks) or the passage of a foreign body through the nip. The greater the chill depth, the more regrinding that is allowed. On the other hand, the heat conductivity coefficient of the chilled iron shell is less than half of that of the grey cast iron interior, so where high rates of heat transfer are required, it is desirable to minimize the chill thickness.

C. HEATING SYSTEMS FOR METAL ROLLS

Finishing the sheet between rolls with elevated surface temperature helps to produce a better surface finish with less damage to the web structure than using pressure alone at ambient temperatures. When the required surface temperature is 150°C or less, circulating water or steam may be used for heating. By using heated oil, roll temperatures up to 200°C may be obtained. For temperatures above 200°C induction heating of the roll surfaces is preferred.

At the lowest rates of heat transfer, with surface temperatures below 150°C, a simple through-bore may be adequate, with water or steam entering through a rotary joint on one journal of the roll and leaving by a similar connection at the other end. Where more accurate control of the surface temperature is

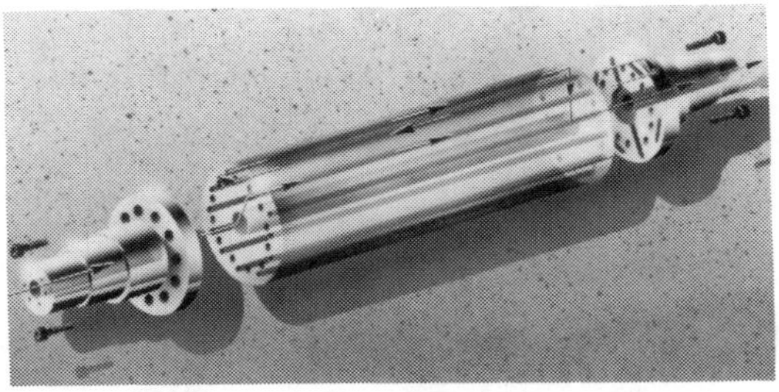

Fig. 587. Construction of SHW Aquitherm®-P Rolls. The Tri-Pass roll is shown, where the heating fluid exits through the right-hand journal after crossing the roll face three times.

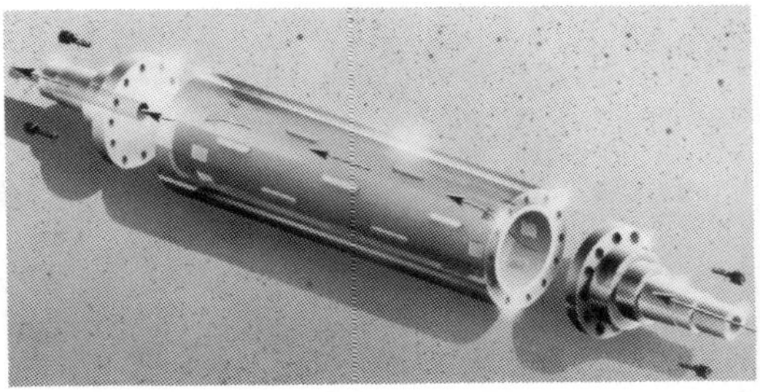

Fig. 588. Construction of Aquitherm®-V Rolls, where the heating fluid flows through the passage between the displacement body and the roll shell.

required, the heated fluid may be conducted through peripheral bores lying close to the roll surface. The flow is redirected to cross the roll face in the opposite direction by channels cut into the roll end headers. This helps to compensate for the temperature drop in the fluid and equalize the temperature at the two ends. In some proprietary drilled roll designs, the flow may cross the roll face three times (Fig. 587). The method of connecting the supply and exhaust tubes to the roll journals also allows for both of these to occupy one journal, leaving the other journal solid for a drive connection.

For long and slender calender rolls the peripheral boring pattern is not suitable and roll balancing is difficult. In this case a displacement body is placed inside the hollow roll shell (Fig. 588), and the fluid flows in the space between the inside of the shell and the exterior of the displacement body. This type of roll is more readily balanced for high-speed operation, although the thickness of the exterior wall through which the heat must flow usually limits the operating temperature to a maximum of about 140°C.

D. HEATED OIL FOR TEMPERATURE AND ANTI-DEFLECTION CONTROL

Years ago, before the advent of anti-deflection rolls, calender stacks had to be designed for operation at one and only one condition of linear force loading across the face of the stack. The reason for this was that the king roll, or bottom roll in the stack, would deflect as a result of the loading received from the rolls above, and the only way to come close to a reasonably constant line nip force load was to grind a crown into the diameter profile. The degree of crown could be correct at only one assumed linear force load: above this point the roll would deflect excessively at the center and the nip force would be highest at the roll edges. At reduced loading, the extra diameter due to the crown at the middle of the roll would cause excessive nip force at the web center. If additional loading was to be applied to the top roll in the stack, then this roll would need to be crowned as well and, here again, the particular crown compensation curve used would be correct for uniform nip at only one condition of loading. Note that the use of the words queen roll to indicate the top roll of the stack is not uniform throughout the industry. In some cases, the term queen roll indicates a major intermediate roll which is fixed in location, and against which other rolls are loaded, and which is often the driven roll of the stack.

Intermediate rolls in the stack are not usually crowned, as they have almost the same loading on their top surface as on the bottom, except for what they contribute in terms of dead weight loading. If the intermediate rolls are properly supported from below and have small overhung loads on their ends, the dead weight contribution will be uniform across the roll face and additional crown compensation is unnecessary. In supercalender construction, fly rolls are employed to lengthen the sheet run between nips, and the method of support for these rolls can cause significant overhung or bending loads on the journals of the intermediate rolls. In this case special attention must be given to the problem of obtaining a uniform pattern of cross-machine nip force.

Various configurations of anti-deflection rolls have been made available since the mid-Sixties, known by trademarks such as Swimming Rolls, Controlled-crown Rolls, and Nipco rolls. These use systems of hydraulic shoes or pistons which are supported in a non-rotating beam inside a roll such as the king roll. The beam ends are spherically journalled to allow the beam ends to deflect under load, while the force to support the nip load is transmitted to the inside of the roll shell by the hydraulic shoes or slippers. The force applied at any one location depends upon the pressure applied to the local hydraulic carrying

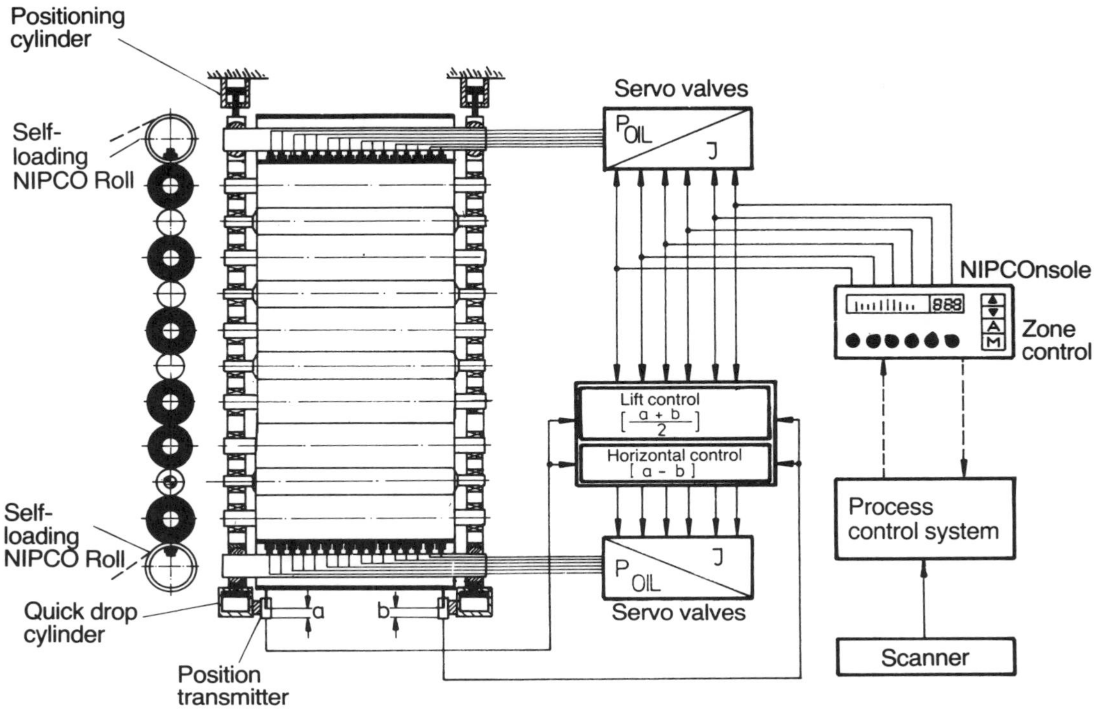

Fig. 589. A floating stack supercalender using two Nipco (T.M. Escher-Wyss) rolls with six pressure control zones.

device. In the Nipco arrangement (Fig. 589), the pistons are grouped into six or eight zones, and each zone may be individually pressurized to allow not only a constant nip, but one which can be profiled over fairly large zones to obtain selective calendering where it is necessary to profile the sheet. By separating the oil fed to each zone and employing isolated temperature control units, the temperature profile of the stack can be adjusted as well.

E. INDUCTION HEATING

The method of heating the calender roll by induction (Fig. 590) for cross-machine caliper control was first developed by Consolidated Bathurst Inc. and is now being marketed under the tradename CalCoil.

This system employs a frequency converter to send a high-frequency excitation to coils, causing a magnetic field which generates eddy current losses in the calender roll [1, 2]. By using an elevated frequency range, the magnetic force of attraction between the work coils and the roll shell is reduced, as is the drag effect that would tend to retard the rotation of the roll shell. The system is intended for caliper control, primarily, rather than the brute force application of heating

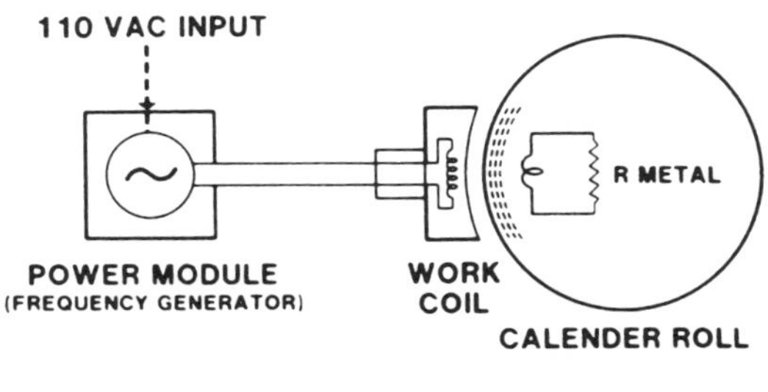

Fig. 590. Principle of high-frequency induction heating of roll surface. Eddy currents generated within the calender roll heat it without unnecessary drag or magnetic attraction forces. (Devron-Hercules Inc.)

power. Nevertheless, it can produce over 40 kW of heating per meter of roll face (Fig. 591).

Some laboratory calenders and other slow-speed stacks are equipped with resistance element heating where the heaters are located within bores in the rolls and the electric power is fed through slip rings located at the roll journals. This arrangement is not suitable for high rates of heat transfer, wide rolls, or where the presence of high voltages and moderate currents at the roll journals presents a reliability or a safety problem.

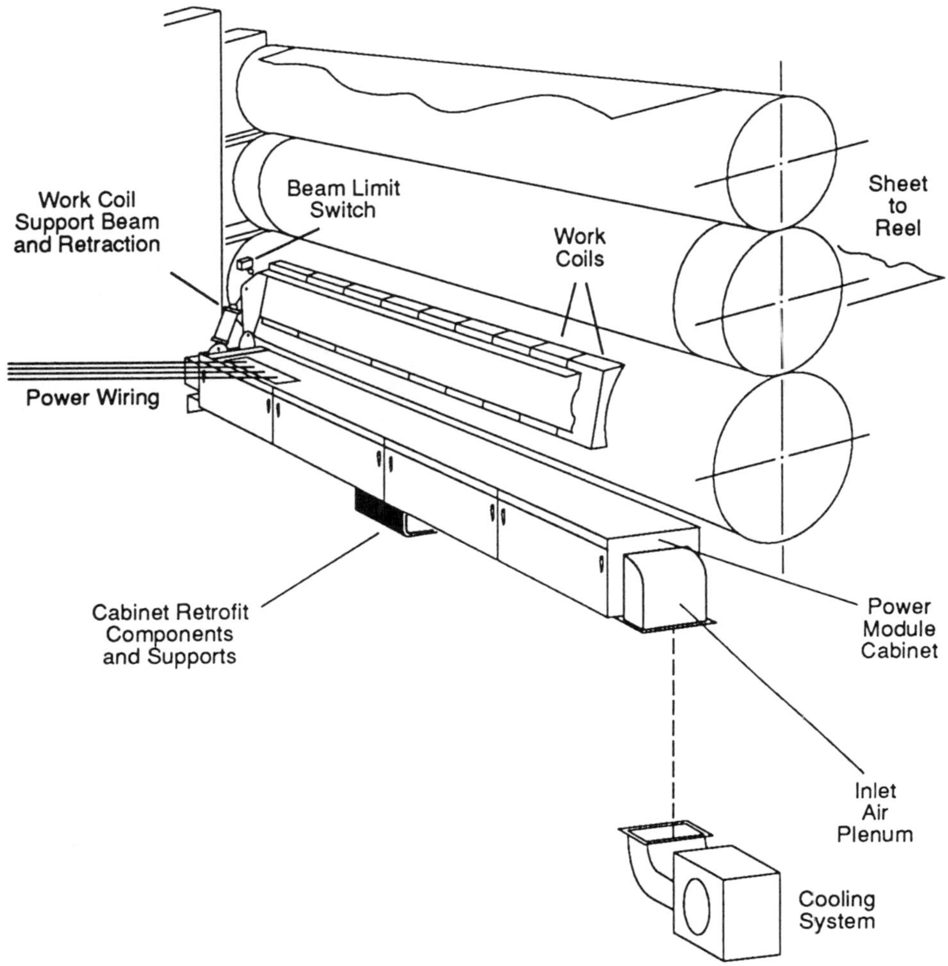

Fig. 591. Arrangement of induction heating work coils and power modules at the king roll.

F. RADIANT HEATING

Radiant heat flux can be used to heat the surface of calender rolls by locating infrared sources close to the surface of the rolls, in a manner similar to that used in many drying applications. The infrared emitters may be powered by electricity, natural gas, or any other power source which delivers heat at sufficient temperature to cause radiant emission. Such a heating scheme is not normally applied to the surface of resilient backup rolls, but the current degree of interest in running the surface temperature of the calender rolls to ever-increasing limits may prompt this approach, if all the problems associated with prevention of overheating can be solved.

G. ROLLS WITH NON-METALLIC COVERINGS

By providing a compressible covering on a roll, the intensity of a steel-steel nip is reduced and the nip width is widened. There is a resultant increase in the time of passage of the material through the nip because of the widening, and the additional resilience reduces the degree of web blackening or web mottle occurring over high spots in the web such as floc centers in the sheet formation. Filled rolls, or bowls as they are called in England, have an external cover or filling made of rings die cut from sheets of paper, cloth fibers, or non-woven mats which have been assembled over the steel core of the roll. The filling is then pressed axially in huge presses capable of exerting thousands of tonnes of force to densify the covering to almost rock-like hardness. As the material compresses, the presses must be opened to allow the addition of more sheets of material. Finally, a locknut is screwed onto threads at the roll end to secure the filling pressure when the press is unloaded. The surface of the roll can then be lathe-turned, ground, or finished using metalworking techniques until it has a high degree of surface finish and a precise diameter. Many different types of filling materials are available for these rolls, including long-fiber cotton, wool, blue denim, and specialty blends. These, in turn, may be pressed to different degrees of hardness, measured on the Shore Durometer 'D' scale, from 78 to 82 for embossing or finishing of nonwovens, 84 to 86 for newsprint and lightweight coated production, to 86 to 89 used in glassine calendering. These are general guideline values, and some variation in values will occur due to mill preferences or the particular combination of roll diameters, speeds, and linear loads employed.

Until recently, the heat required for finishing the sheet between filled roll-chilled iron roll nips was supplied almost exclusively by the heat generated by the cyclic compression of the roll filling material passing through the heavily loaded nip. In general, the higher the hardness of roll filling, the higher the loading of nip force per unit of face width the roll can stand. However, when trying to reach the highest operating temperature, the roll life becomes drastically shortened, and there is a danger of roll burnout, or a blowout, where a large chunk of the filling material may fly off the roll. Obviously, it does not pay to operate close to these limits.

The advent of anti-deflection rolls brought with it the opportunity to use the oil being pumped to the pistons or slippers within the roll as a means of providing heat at a constant temperature to the stack. This in turn meant less production of under-specification paper until the rolls in the stack reached a stable operating temperature. With temperature-regulating controls applied to the heat exchangers in the oil pressure supply, the rolls could be pre-heated to the correct temperature before threading the web through the stack.

The search for a longer lasting roll filling material which was more mark resistant has led to the development of some proprietary composite material roll coverings which are operating successfully in supercalenders in many mills. In general, these are filament-wound matrices with a strong synthetic fiber imbedded in a polymeric material, capable of withstanding hundreds of kiloNewtons per meter of nip force, and operating at temperatures in the same range as ordinary filled rolls, that is, up to about 80°C. Some mills have found that they could run a supercalender stack containing composite rolls at 10–15° lower temperature than normal filled rolls and obtain the same sheet quality. The term synthetic rolls is usually applied to these composite rolls, but the term also includes a solid polycarbonate plastic shell, successfully employed in wallpaper calenders, and Nylon 610 covers, which have seen some use in Europe, but little in North America.

New processes have pushed the search for high-temperature roll coverings to even higher limits. These processes include soft-calendering, gloss calendering, temperature-gradient calendering, impulse drying, and substrata thermal molding. The limits for composite rolls have been pushed to 100–120°C, while a special new cover has been designed for the anti-deflection roll of a soft calender which enables it to operate at 230°C and 440 kN/m nip force. This new technology is said to be based on a synthetic fiber filling that is placed over the anti-deflection roll shell in a manner similar to the construction of standard filled rolls [3].

H. THE OPERATING VARIABLES FOR CALENDER STACKS

Any consideration of the way in which calendering is applied to the sheet must of necessity include a consideration for the ultimate use of the product. Paper mills in regular production of a grade which is utilized in high-volume printing applications are in continuous contact with the quality control people at the printing plant, for instance. This feedback helps to optimize the production of paper to suit the client's needs. Without going into specific detail about any one of the hundreds of different paper grades produced, the fact remains that to specify what is needed usually requires two dozen or more physical tests, such as caliper, bulk, smoothness, brightness, tensile, burst, tear, etc., and even these are not sufficient to completely optimize the paper to its end use. The properties of the sheet can be augmented or diminished by the calendering operation. One must study the structure of the sheet from one surface to the other, paying attention to the properties of the interior fibrous structure as well. In some grades, for instance label stock, release papers, and carbonizing grades, the surface properties may be entirely different on the two sides. Other grades meant to receive printing ink may need to retain equal quality of halftone reproduction on both sides of the sheet and similar gloss.

When dealing with existing paper finishing equipment, some of the variables must be taken as given quantities: the diameter of chilled iron rolls, the maximum attainable loads, the speed of the paper machine, for instance. If a new machine is being designed, or the calender is being rebuilt, any parameter which has an effect on the sheet properties may be considered for modification in the design. At this point in the discussion no differentiation has been made between in-line and off-machine operations. If the finishing operation is off-machine, more variables come into play, such as *how many* off-machine finishing lines are required to keep up with the paper machine output, and at what design speed for the finishing line?

In more general terms, the calendering variables may be categorized in terms of those associated with the sheet: its fiber structure, moisture content and moisture distribution throughout the sheet, and any coating applied to the sheet; the variables of the calender itself: number of nips, linear force loading, roll diameters, roll surfaces, roll coverings, roll temperatures, and speed, which determines the time spent in the nips; and other parameters added around the nips: steam showers, air jets, or even liquids, sprayed or transferred onto the web.

I. SHEET VARIABLES

The selection of fibers in the furnish is not usually adjusted to the operation of the calender. Usually the fiber blend is set by market considerations, the price and availability of pulp, the specifications of the finished product and the price it commands. Noting however that the calender may take away some of the strength or stiffness properties of the sheet, it is not uncommon to fix up the sheet strength after the fact, by making adjustments to the furnish after reports are received that all is not well in the field.

1. Moisture

Sheet moisture is extremely important in calendering. As moisture levels increase the deformations of the sheet and its fibers become more plastic in the nip, the absence of moisture makes the sheet more elastic in the nip, and it will tend to rebound to its original surface finish and have less reduction in bulk. Glassine production in a supercalender is an excellent example of what happens to a sheet with lots of moisture (17–19% is typical) when it hits the finishing nips: the cellulose and hemicellulose flows to the extent that inter-fiber voids are filled in, fiber crossings are welded into a solid interface, and the edges of the fibers which used to reflect and scatter light within the sheet are covered, so that the sheet becomes translucent instead of opaque. It was not too many years ago that all window envelopes had openings covered by glassine paper, through which the address on the letter inside could be read. The effect of high moisture can also be accomplished by high pressure in the nip, and this is what happens to cause sheet blackening in an unrelenting steel-steel nip when either a spot containing too many fibers locally, or fibers with abnormal moisture content come into the nip. An area of the web turns into glassine, and it appears black in comparison to the snow-white reflection from the rest of the web. When the same phenomenon happens to a lesser extent, it is called mottle.

Wet streaks present a problem to the calender. To prevent a permanent loss of caliper where the streak finishes to a lower bulk, the machine calender rolls may be cooled locally with air showers to reduce their temperature and diameter, but this only succeeds in passing the wet streak on the reel, and later to the winder, where more problems may occur.

Moisture variations from the wire side to the felt side of the sheet represent another profiling dimension, the through-the-sheet direction. If the calendering operation is off-line and a time period of one or more hours has elapsed since the last finishing operation, one might assume that the moisture is completely evenly distributed through the sheet. However, the moisture is that which is hygroscopically absorbed into the fibers, free water on the inside and outstide of the fibers, and water which is bound to fillers, coatings or sizing. Although the migration of water may have slowed or stopped, the distribution from one side of the sheet to the other may be quite uneven. Manufacturers of paper used in heat-set web-offset printing have to be particularly careful about trapping sheet moisture behind the coating to avoid blistering of the coating in the heat-set ovens between color application stations of the printing presses [4].

Moisture may be added at the calender, and will be considered separately as a variable which is neither part of the sheet nor of the machinery. However, it should be pointed out that drying the sheet to a greater degree to control a difficult moisture profile is one way to handle problems caused by non-uniform distributions. The moisture deficiency would then be made up at the calender.

2. Coatings, fillers, sizing and binders

The preparation and application of these coatings and fillers are contained in Volume 8 in this series. The point in including them here is to mention that their response in the calender nip is sometimes more important than the response of the paper fibers. They too must flow to allow the surface of the sheet to develop gloss or become smoother. Coating clay platelets randomly oriented in the coating medium must align themselves over most of the sheet in order to develop a high 75° gloss, if this is the target. Different coating formulations may be applied to the wire side and felt side of the sheet to balance their final appearance after calendering. The list of materials that may be added to the paper sheet, which seems to grow longer every year, contains such diverse items as starch, diatomaceous earth, titanium dioxide, ultraviolet converters, coating clays, melamine resins, styrene-butadiene latex, polyvinyl acetate latex, pH control additives, colorants, stearates, satin white, and so forth and so on. The moisture contained in these items affects the overall moisture profile compared to the bone-dry basis weight. It is shared with the paper fibers and affects the rheological flow properties of both the paper and the additives themselves.

While considering how to control the quality of coated products when the coating formulations are so complex, it might be pointed out that a number of problems arise which are only solved by good communications. A situation occurred a number of years ago on an off-machine supercalender line where the coater and supercalender were located far enough apart that the crews for each did not tend to visit each other. For about two weeks, the supercalender crew had been struggling to avoid a situation where a white scale was transferring from the coating to the filled rolls in streaks that ruined the gloss and caused the rolls to go to broke whenever it happened. They tried every change they could think of, cleaning the stack, different loads, speeds, and operating temperatures in the stack, all to no avail. One night, the coater operator just happened to wander by, took one look, and said, "Oh, that. I can fix that with a simple change in the amount of stearate in the coating". And it was done and the problem went away.

Listing the variables brought in with the sheet in approximate order of importance, we have the fiber furnish itself, the moisture and its distribution, and coatings, fillers and sizing. To complete the list, the temperature of the incoming sheet should be added. For in-line steel-stack calendering of newsprint or other lightweight publication paper or base stock, this might seem unnecessary. The heat brought in with the sheet merely serves to reduce the energy requirement to keep the stack up to the desired operating temperature. The temperature experienced by the sheet at the first few nips of the stack can be controlled by adding heat to the rolls which the sheet wraps. However, in the context of temperature gradient calendering, particularly on somewhat thicker products such as foodboard, the aim is to use the thermal mass of the sheet itself to prevent the surface temperature of the calender rolls from penetrating into the central zone of the sheet. Therefore, the lower the temperature of the incoming sheet, the greater the temperature difference that can be developed between the surface of the sheet, which is rendered plastic and deformable by the high temperature of the calender rolls, and the core of the sheet, which tends to stay elastic and resilient because of its lower temperature. This enables temperature-gradient calendering to produce the surface smoothness and gloss associated with finishing at high linear loads and temperatures without as great a reduction in bulk and without reducing the strength and stiffness of the sheet as much as previous methods did. The temperature of the incoming sheet also has an effect on the ability of the sheet to pick up moisture from steam showers, if these are being used.

Finally, one more variable should be put on the list, as it is something which can be altered to affect the net result of calendering. That variable is the boundary layer air that travels on both sides of the sheet coming into the calender. It may seem trivial to include something that we cannot fully get rid of because, if we try to doctor it off, atmospheric pressure will force new air to attach itself to the sheet behind the doctor. But when the addition of steam is considered, the boundary layer air acts as a barrier to the flow of steam, tending to prevent it from entering into the sheet. This is not an unsolvable problem, manufacturers of steam showers have some workable solutions [5, 6].

J. CALENDER MACHINERY VARIABLES

1. Number of nips

Some of the nips may be opened or relieved, and the sheet may be threaded to bypass all but one or two calender nips. Some paper machines are equipped with two four-roll or two six-roll stacks, and one or the other of these can be bypassed. If the sheet is being calendered at all, it will pass through from one nip (soft calendering or gloss calendering, for instance) to as many as 18 nips (release backing paper or glassine supercalender).

Many researchers have measured the change of caliper or bulk of various paper grades caused by multiple passes through nips. Bulk is usually measured in units of cm^3/g, which is the reciprocal of density in g/cm^3. Although the units should be m^3/Kg to follow preferred SI units, the former bulk units may be readily compared to the density of water at 1 g/cm^3. Therefore, any material with a bulk less than one is more dense than water. In general terms, the initial nip causes the greatest decrease in bulk (increase in density), and every successive passage through a nip of roughly the same loading causes the bulk to be reduced by a diminishing fraction. Ultimately a limit will be reached where no further reduction will take place, the material springs back enough between nips to offset the nip deformation. Crotogino has developed a calendering equation [7] which predicts the bulk reduction in each nip of a calender using readily determined empirical constants which measure the effects of initial bulk, nip load, speed, roll radius, temperature, and moisture content. The equation, which combines previous work of Colley, Peel, Kerekes and Pye is of the form:

$$\epsilon = A + \mu B_i$$

where:

ϵ = the relative compression, defined as the change in thickness divided by initial thickness,
μ = the compression parameter defined below,
B_i = the initial bulk, and
A = experimentally determined.

The relative compression may also be defined in terms of the initial and final bulk, since a change of thickness does not affect the basis weight, and bulk can be defined as μm thickness divided by basis weight in g/m^2.

$$\epsilon = (B_i - B_f)/B_i$$

where B_i and B_f are the initial and final bulk.

The compression parameter μ is given by

$$\mu = a_1 + a_L \log L + a_s \log S + a_R \log R + a_\theta \theta + a_M M$$

where:

a_1 = an initial constant, experimentally determined
a_L = a constant factor for the nip load L, kN/m
a_S = a constant factor for the speed S, m/min
a_R = a constant factor for the roll radius R, m
a_θ = a constant factor for the temperature θ, °C
a_M = a constant factor for the moisture content M, %

Once these constants have been determined for the paper being studied, they may be applied to every nip in succession by calculating a final bulk after each nip from the relative compression, then using it as the initial bulk for the next nip.

The total reduction in bulk is important as it tends to measure the amount of finishing that has taken place. For instance, the Parker PrintSurf roughness in μm is usually linearly correlated with the bulk so that, as bulk goes down, so does the PrintSurf number [8]. The question that remains is *how many nips* are required to obtain the bulk reduction. Under some circumstances, it is possible to meet printing standards by passing the sheet through single, heated nips, one for each side of the sheet to be finished. But it is not easy to meet gloss and smoothness specifications under all conditions. It is a long-standing rule that, to obtain the smoothest surface, greatest gloss, and highest brightness, the sheet should be passed through many nips at only moderate speed, and the nips should be wide (resilient) to create as much residence time in the nips as possible. However, rules are meant to be broken, and production costs will come down as the number of nips is decreased and operating speed is increased.

To separate a few of the machinery variables, we might start with machine-calendering and steel-steel nips, where the effect of roll diameter alone on the amount of bulk reduction is only a slowly changing function. Large calender roll diameters give the least bulk reduction, at any one constant linear load. Note, however, that, in practice, higher loads are often applied to the larger diameter rolls. As the calender roll diameter drops, the bulk of the sheet also goes down, and the curve of bulk vs. diameter has an increasing slope toward the smaller diameters, as the nip intensity increases rapidly due to the smaller radius under constant force loading.

Speed is also a variable that does not affect the bulk reduction too strongly, at least in machine calendering. As speed goes up, the bulk tends to go up as well, probably a direct effect of reduced residence time in the nip at higher speeds.

(a) Roll loading

The machine variable with the greatest effect on bulk reduction is the linear load on the rolls. The loading, measured in kN/m or pounds per inch, affects not only the peak pressure which occurs on the nip centerline, but also the nip width, which affects the residence time. The peak pressure and the nip width are related by the formulas for Hertzian contact stresses between two cylinders of similar or dissimilar materials [9]:

Nip width 2b:

$$b = \sqrt{\frac{2\ Fd_1d_2}{\pi}\ \frac{(1-\nu_1^2)/E_1 + (1-\nu_2^2)/E_2}{d_1 + d_2}}$$

where:

b = the half-width of the nip, m
F = the loading force per unit of width, N/m
d = the diameter of each roll, m
ν = the Poisson ratio of each roll material (dimensionless)
E = the elastic modulus of each roll in Pa

The maximum pressure is given by

$$p = \frac{2\ F}{\pi\ b}$$

where F and b are defined above.

The shape of the pressure curve through the nip is elliptical; that is, when the minor axis of an ellipse is placed on the full nip width 2b, the extreme of the major axis represents the maximum pressure, and the pressure rises to this value along the curve of the ellipse. This peak pressure is undoubtedly the predominant factor in calender finishing, although most of the time the linear load is reported, since other factors must be known before the pressure can be calculated. Average pressure in the nip is sometimes mentioned, and is given by the linear load divided by the nip width, F/(2b). In comparison to the maximum pressure, the average pressure is $\pi/4$ times as great.

When the materials of the two rolls are the same, and their diameters are approximately equal, they deform elastically into two nearly flat surfaces along a line of contact 2b wide. If the materials are steel or chilled iron and no web is present, the z-direction stress or pressure (perpendicular to the plane of contact) and the y-direction stress (in the MD) are equal to each other at the center of the nip. However the CD stress will be only 0.6 as large, which will result in a shear stress of magnitude 0.2 times the peak pressure, maximum at an angle of 45° to either of the other two axes. This is mentioned just to set the record right, as there are those who would argue that such a case of pure rolling and deformation contains no shear.

When there is a web present, the calculation of nip width and three-dimensional stresses becomes much more complicated. Carbon paper prints or other recording of nip impression will, in general, show a nip width greater than predicted by the formula above because the material used to take the nip pattern bridges the space between the two rolls just outside the bare contact points and transmits substantial pressure there, enough to mark carbon paper or deform embossed aluminum foil.

Resilient backup rolls have a modulus of elasticity which is typically 500 to 50 times softer than that of a steel roll. When a resilient roll is pressed against a hard steel or chilled iron roll at a given linear loading, the nip width then becomes much wider than with steel-steel contact. The peak pressure drops according to the reciprocal of the nip width, and the surface of contact is no longer a flat plane, but is along the radius of curvature of the steel roll, the steel roll then having a negligible amount (0.2% to 2%) of the total radial compressive distortion. Because so much displacement into the softer roll material is required to reach the maximum pressure, it makes little difference to the final pressure attained whether a sheet of paper between the two happens to be 50% thicker or 50% thinner in a local area; the final pressure will still be reached. This points out the fundamental difference between calendering between steel-steel nips and between a resilient roll and a hard one. In the steel-steel nip, the pressure is not uniform when there are high and low spots in the sheet. When too much material enters the nip at a caliper high spot or floc center, the pressure will peak there and will be reduced in surrounding areas. The net result is that web areas with high fiber density tend to emerge from the nip with the same sheet thickness as areas with low fiber density, which received low-intensity nip pressure impulses. In contrast, when the sheet is pressed into a resilient calender roll cover, the indentation distance of the hard roll into the cover is large enough that changes in sheet thickness do not modify the peak pressure to any great extent. When all elements of the web have been pressed to uniform pressure, they will rebound to a thickness proportional to the original density of fiber distribution; in other words, the thickness will reflect the basis weight

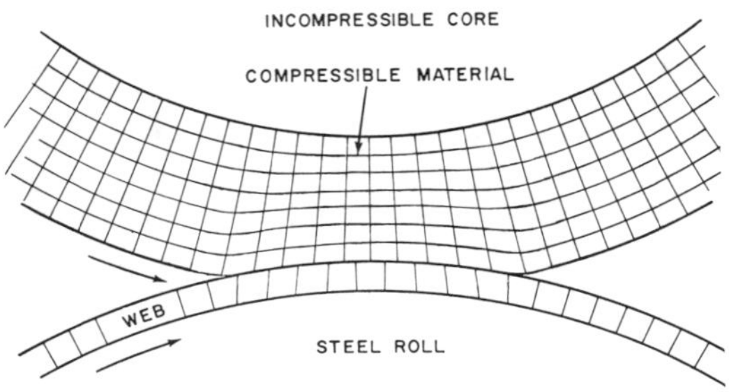

Fig. 592. Deformations at the source of shear in a nip formed between a compressible roll covering material, a paper web, and a steel roll. The amount of nip penetration and the web thickness have been exaggerated for clarity [10].

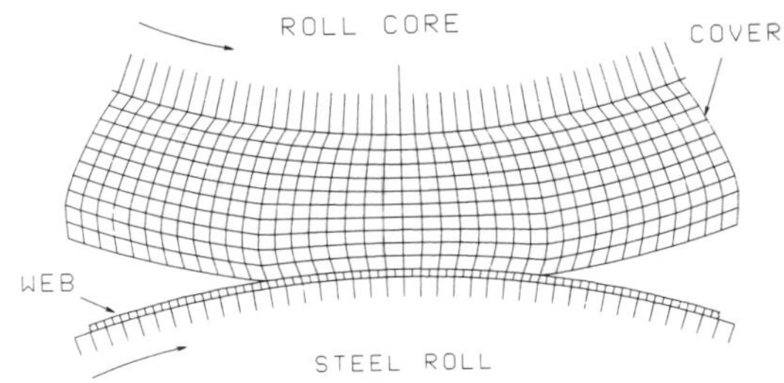

Fig. 593. A computer graphic rendition of the response of an incompressible (Poisson ratio near 0.5) roll cover to being nipped by a steel roller carrying a paper web. Note that the shear directions are opposite to those in the Fig. 592.

distribution. In the case of machine calendering, the product tends to have constant thickness, with variable density (or bulk) and variable basis weight. With supercalendering, the sheet tends to have constant density (or bulk), with variations in basis weight showing up as variations in thickness. Of course, the influence of a variable CD moisture profile may mask these tendencies.

(b) Replication

In calendering against a resilient roll, the nature of the stress within the softer roll will be strongly dependent upon its Poisson ratio. When the Poisson ratio is 0.2 or lower, typical of filled rolls or those compressed from fibrous material, the soft roll material can be compressed in the radial direction without experiencing much MD stress, giving rise to the type of grid distortion shown in Fig. 592, where the amount of shear on either side of the nip centerline is equal but opposite in sense, and is qualitatively shown by the amount of parallelogram tilt in the grid cells. A roll cover which is made of solid elastomeric material which has a small percentage of filler materials compounded into it tends toward a Poisson ratio of 0.5, although any grooving pattern cut into the roll face as in pressing applications will lower the ratio drastically. When the Poisson ratio is near 0.5, the material behaves incompressibly, and the grid cells will tilt the opposite way, as shown in Fig. 593.

Figures 592 and 593 show, in a way that can be interpreted without mathematical formulations, how a cycle of shear can be transmitted to the material passing through the nip at the interface between the backup roll and the hard roll. Shear is the driving force behind flow, and flow is necessary to cause the fibers in the web and the binders in the coating to deform to have the web surface replicate the surfaces that are confining

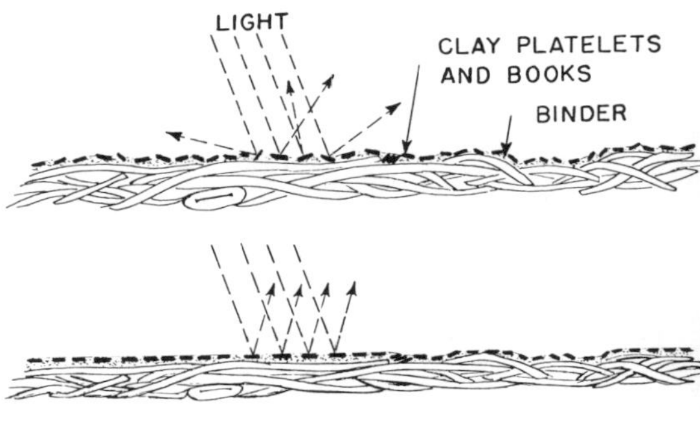

Fig. 594. Schematic cross-section of coated paper before and after supercalendering [10].

the web, the backup roll surface on one side, and the steel roll surface on the other side.

A series of tests were made to prove that replication did exist in both steel nip calendering and supercalendering. This work was reported on at the TAPPI Finishing Conference in 1970, and the findings, which were not part of the 1970 Conference Proceedings, were finally included in paper industry publications beginning with the 1975 Calendering Conference in Manchester, England [10, 11]. A limited set of the original illustrations is included here.

Coated paper before calendering has a surface similar to that of a smooth plaster wall (Fig. 594). That is because the delaminated clay platelets, which have sizes ranging from 2 to 15 μm across, are carried in a binder and deposited on the paper fibers with a low degree of orientation. They follow the contours of the paper fibers underneath as a coating of paint would. In order to make the surface more glossy, the platelets must be leveled out and the binder must flow in order for this to happen, as shown in the lower half of Fig. 594. Note how the drawing indicates that, in some low areas of the web, this

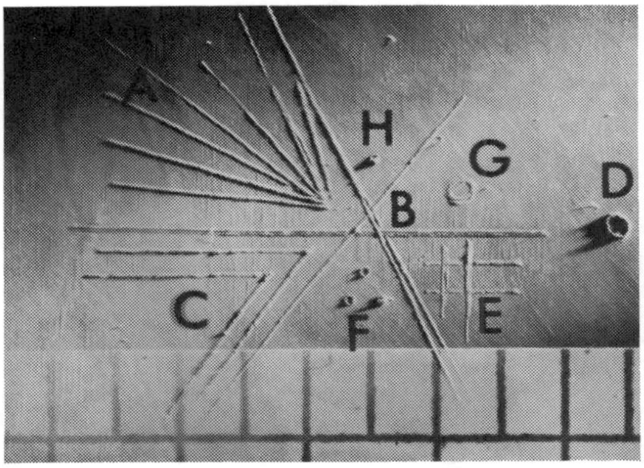

Fig. 595. Silicone replica of engraved marks in the steel roll. Scale superimposed has 1.59 mm (1/16") divisions. Large and small dot patterns at D and F reproduced as hollow craters instead of cone peaks because air bubbles were trapped in the bottoms of the holes when the rubber was poured [10].

orientation will not take place, and that is generally the case.

A small laboratory calender was built to study replication, with three steel rolls 0.203 m in diameter. The center roll could be replaced by a filled roll for studying supercalender action. The maximum nip loading obtainable was 350 kN/m, but lower loads were used in the studies. A pattern was hand engraved on one of the steel rolls by scribing the roll surface with a carbide tool. Then the burrs which were thrown up on either side of the furrow ploughed by the tool were removed by careful stoning with an India stone (Fig. 595). It was felt to be important that any pattern used to prove replication should only descend negatively into depressions in the steel roll; the presence of any positive projections would make it difficult to prove the case.

A grease pencil was used to draw a circle around the exterior of the pattern before the paper was calendered. This facilitated locating the areas of the web to be cut out for study under the microscope. Statistics were kept on the percentage of the total pattern that could be found on the web samples as the linear loading was increased. Up to a linear load of 105 kN/m, steel-steel calendering and supercalendering performed almost identically, no doubt due to the high peak pressure in the more narrow steel-steel nip width. At this point about 75% of the total pattern could be seen on the paper samples from both forms of calendering, as shown in Figs. 596 and 597, depending on the presence or absence of local floc centers which would emphasize various parts of the pattern.

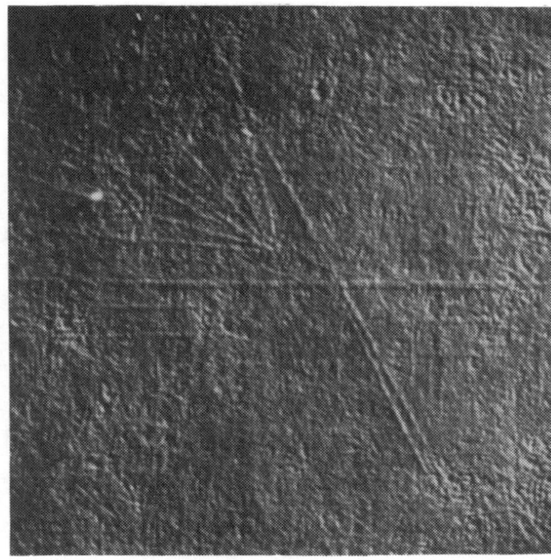

Fig. 596. Sample 600-8 at the same magnification as Fig. 595, showing details of the engraved pattern picked up by steel-roll calendering at 105 kN/m (600 pli).

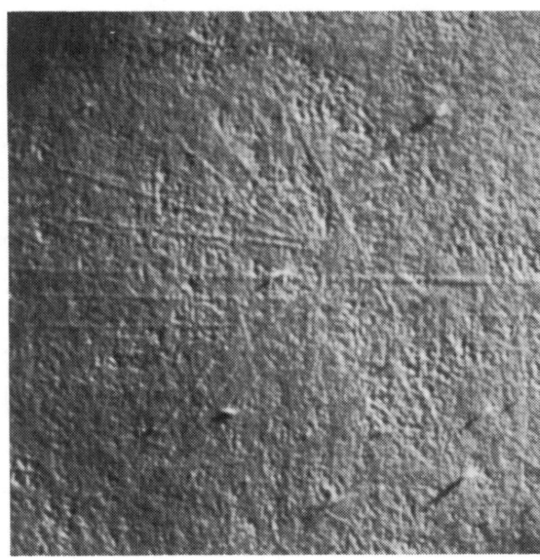

Fig. 597. Pattern replication when supercalendering at 105 kN/m (600 pli). Sample 600-7S [10].

However, a limit was reached at 105 kN/m due to the onset of web mottle, a blackening of the sheet over those areas of intense pressure. Blackening did not occur on the supercalendered samples and, at a loading of 193 kN/m (1100 pli), 90% of the area was replicated on those samples, and in some cases (Fig. 598) even the fine scratches at a 45° angle to the MD, left by the India stone, could be seen replicated on the

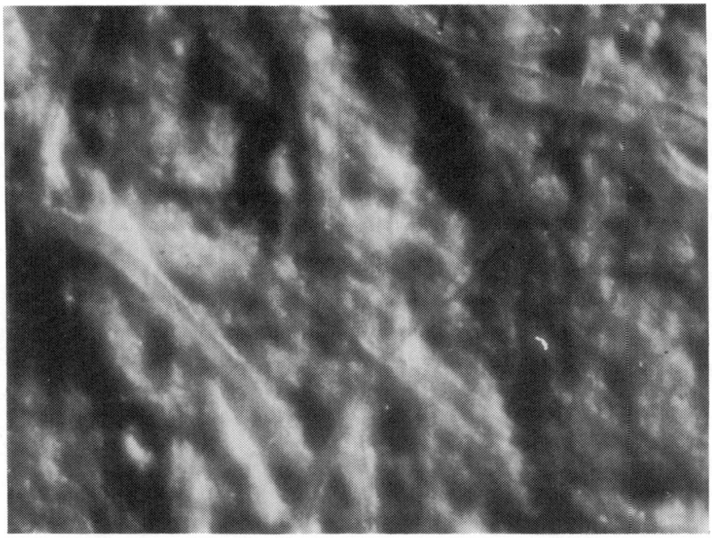

Fig. 598. View at ×16 higher magnification of sample 1100-3S supercalendered at 193 kN/m (1100 pli). Note that in the center of the photograph fine scratches at a 45° angle have been replicated, scatches left by finishing the steel roll with a fine stone.

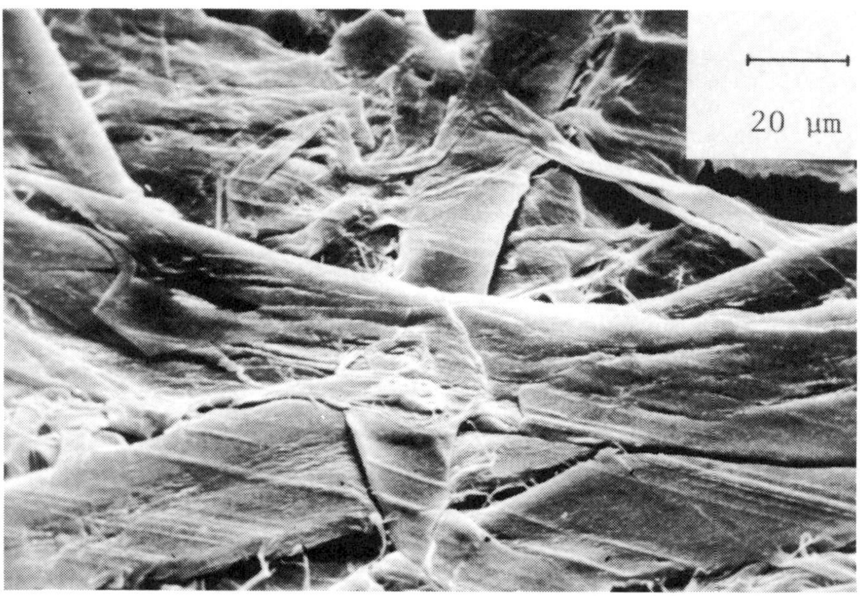

Fig. 599. Replication of scratches from the metal roll on the surface of temperature-gradient calendered newsprint as reported by Crotogino and Gratton [8].

fibers. Any form of slippage within the nip would have wiped out such a delicate pattern, and those who advocated slippage or burnishing within the nip as a means of developing smoothness or gloss began to change their tune.

More than twenty years have elapsed since these first measurements of replication were made using a small laboratory calender under conditions that were not exactly equivalent to those found in paper production. However, the evidence continues to mount that replication does take place in both steel-steel calendering and in deformable nips. Under the power of the scanning electron microscope (Fig. 599), fine scratches may be seen that have been transferred from opposing roll surfaces.

(b) Distortion-energy theory

Is shear required for paper finishing? Rodal argues that fibrous materials such as paper and textiles do not follow the same rules as metals, plastics, and other homogeneous materials [12]. He states that "shear stresses produced by materials with different values of Poisson's ratio are of no major importance in the calendering action. Instead, permanent deformation of the paper sheet occurs by plastic compaction". How can plastic compaction take place without shear? He reasons that porous materials will undergo large, permanent or plastic deformations under high hydrostatic pressures. But in a practical production environment using rolling cylindrical nips for finishing, there is no way in which pure hydrostatic pressures could be imposed upon the material. A tiny sample could be exposed by placing it in an incompressible fluid and elevating the pressure, but with a web passing between rollers there is a necessary constraint on the amount of cross-machine straining allowable, which will prevent the fibers from deforming in the hydrostatic mode. Whenever strains or deformations take place at unequal rates in different coordinate directions, shear will be present. In the rolling nip, the largest initial deformation will be along the z-axis, where the highest negative principal stress develops. Shear stresses will be developed because the MD and XMD stresses are less negative in the rolling nip, and shear flow (the consequence of having shear strains) will cause plastic flow of the web material.

Dealing with three-dimensional stresses in linear, homogeneous materials is difficult enough but, when the material is non-homogeneous and displays a non-linear response to stress, a simple treatment is nearly impossible. However, we may be aided by a theory developed by von Mises and others, called the distortion-energy theory [9]. According to von Mises, the energy to distort or deflect a material comes from the stresses left after subtracting out the hydrostatic components. The differences between the remaining stresses, taken two at a time in the principal stress directions, are squared, added, and then the square root is taken. This resultant magnitude, when multiplied by 1/3, is called the octahedral shear stress. When divided by the square root of two, it is called the von Mises stress. Either way it is a measure of how close the material is to failure in the case of an engineering structure, or to the onset of flow in our case. The remarkable thing about an octahedral shear stress or a von Mises stress is that out of a complicated stress pattern in three-dimensional space, which would otherwise require three normal stress and three shear stress components for definition, only one quantity, a scalar magnitude, results, a number representing the severity of the loading pattern to cause deformation, in terms of the available shear. An important use of the von Mises stress scale in engineering work is to color a display of a machine part to show where it is most critically loaded in finite-element analysis programs, signalling the "hot spots" where the part is likely to fail.

Roll coverings will produce various waves of shear in the entering web according to the Poisson ratio of the cover material, as shown in Figs. 592 and 593. A solid polymer roll covering with a Poisson ratio between about 0.29 to 0.35, and a steel roll should produce shear deformations about midway between those two cases. Nevertheless, the shear cycle experienced by the sheet passing through the nip may be of appreciable magnitude. As the paper first enters the nip, the rapid z-direction compression of the sheet compared to the strain in other directions will cause an initial shear magnitude which may be labeled +, then as the pressure rises the deformation of the backup roll may transmit shear of opposite sign (−) through the sheet to the surface of the steel roll. Since the center part of the nip is symmetrical, the opposite signs of shear will apply to the sheet as it passes center. In this instance, a +,−,+,− shear-cycle would be imparted to the sheet by a single nip. Incidently, the directional angle at which the shear appears to be greatest, relative to a point moving with the sheet through the nip, will not be stationary, but will move about during its passage (like a weathervane before an oncoming storm).

Regardless of what physical laws are being followed, the rolls do function to cause the sheet to deform both plastically and elastically, and the pressures are intense enough to ensure that fibers, coatings, and coated particles which enter the nip stay in one-to-one correspondence with surface features on whichever roll they contact until they leave the roll. Displacements may occur through the sheet, but these are flow displacements, and occur without abrupt slippage. This action causes the replication of roll surfaces onto the sheet surfaces. The side of the sheet which requires the most accurately formed smooth finish should be against the roll which carries the best surface finish, typically the harder one, made of steel or chilled iron. The concept of replication was very quickly accepted in the magnetic tape industry. Magnetic recording medium is produced as a polyester base coated with iron oxide or chromium oxide particles in a binder, and this coated web is supercalendered to set the playing surface as smooth and flat as possible to obtain the best contact with magnetic playback heads used in tape decks, video cassette recorders, and computer floppy disks. When those responsible for magnetic tape quality realized that replication worked on both sides of the web, and that the quality of the tape was affected by the backup roll as well, they switched to resilient composite resin backup rolls capable of being finished to a mirror surface, with a roughness reading of only 0.2 μm. Magnetic tape quality has increased dramatically in the past 15 years, and improvements in the calendering process have allowed a great advance in the state of that art.

2. Hardness of the roll surfaces and roll covers

The result of compressing paper fibers between two chilled iron rolls with modulus of elasticity 145×10^9 Pa each will be predictably different from compressing the fibers between a chilled roll and a backup roll.

3. Temperature of the rolls

The methods used for heating rolls were discussed earlier under roll construction. The control of roll temperature is of great importance for the development of consistent sheet properties. This makes roll temperature one of the key variables for calendering, together with the peak pressure in the nip and the web moisture. It is relatively easy to heat metal rolls, and there will be a temperature drop through the wall of the metal roll in the direction that the heat is flowing. In the case of resilient backup rolls, the operating surface temperature and the rate of heat flow will both be limited by the temperatures that the roll covering can endure at the surface, and at the interface between the cover and roll body, a higher temperature because of the large drop across the roll filling material which will have lower thermal conductivity than metal roll bodies. If the backup roll is unheated, the thermal mass of the web passing by will tend to lower its temperature. The only factors driving up the temperature of the backup roll in this case would be internal losses in the cover due to cyclic loading of the nip, and heat which is transferred across the nip from the metal roll, through the sheet and into the backup roll. If it is necessary to maintain the backup roll surface much higher than that of the incoming web, other methods of heating such as radiant energy or running a heated metal roll against the surface could be employed [13].

4. Use of fly rolls

A fly roll (sometimes called a blow roll) is a turning roll employed to take the sheet away from wrapping contact with a calender roll. The sheet is threaded from one nip to the fly roll, and after a half turn around the fly roll is threaded into the next nip between calender rolls. In supercalenders, fly rolls are commonly utilized on at least the upper nips of the stack, and are frequently employed between every nip. Bowed spreader rolls are often located in the span just before the fly rolls on the upper nips of supercalenders. The combination allows the sheet to expand in the CD between nips. If the expansion were not permitted, hard wrinkles, wrinkle cuts, and fold-overs would occur. Fly rolls are not commonly seen on steel-steel calenders. However, in the newer in-line finishing processes involving heated rolls, it is desirable to avoid prolonged contact with the heated roll when the aim is to provide surface heat to the web only. The web path is constrained by carrying rolls which allow only a limited arc of contact of the sheet with the heated rolls, and these carrying rolls serve a purpose similar to fly rolls.

5. Steam showers

These are devices placed on one or both sides of the web path to allow low-pressure steam to impart moisture to the sheet. If the steam showers are properly constructed, it is possible to utilize economical low-pressure saturated steam without having condensation drip onto the web [5]. If the steam is superheated to prevent droplet condensation, the effect on the web is reduced because the extra heat imparted to the web tends to drive off some of the humidity gained. Steam showers may be arranged in segmented zones across the machine with independent flow controls as a means of correcting moisture imbalance in the incoming sheet, profiling it to the extent allowed by the width of individual showers.

Boundary layer air traveling with the web tends to carry the steam with it, changing its direction so it resists penetration of the web. The interruption of the boundary layer can be achieved by proper design of the lips of the steam shower boxes so that the steam mixes turbulently and partially replaces the boundary layer air. Steaming the side of the web which comes into contact with a turning roll will also result in moisture transfer when the boundary layer air becomes blocked by the obstacle of the roll, and the slight pressure head created at the ingoing contact with the roll will assist in sending the water droplets into the web. The use of steam

Fig. 600. Dahlgren web dampener. A uniform film of water metered onto a specially treated chrome roll (middle) is transferred to the moving web, assisted by pressure from a backup roll (top).

showers on machine calender stacks producing newsprint at 820 to 1300 m/min can be expected to yield a 5 to 20% increase in smoothness, and an increase in caliper reduction of 3 to 7%. The added moisture should result in an increase of reel moisture levels from 0.5 to 1.5%. Tests were made of steam showers added to a 10-roll supercalender stack producing a 70 g/m^2 release grade. The stack had to be limited to a speed of 305 m/min to meet gloss specifications. The steam showers were added just after a fly roll following the first nip, with another unit following the fly roll below the reversing nip, and allowed the same gloss to be produced at 410 m/min, a 35% speed increase. A further benefit was the reduction of CD gloss variation to about 2/3 of the former amount by adjustment of the steam valves.

Dampener rolls are available which literally print a film of water onto the web (Fig. 600). The water is picked up from a hydrophilic, chrome-plated dampener roll as a film 1–20 μm in thickness, applied by a metering roll operating in a fluid reservoir. The film, which generally travels at a higher peripheral speed than the web it contacts, is imparted to the web by pressure with a resilient backup roll, or by wrapping the dampener roll for an angle of about 45°. When using the backup roll, the light nip tends to propel the moisture into the web, and a hydraulic wedge is created where the dampener roll and its fluid film contact the web, travelling at slower speed. The dampener roll is able to operate at speeds up to 1500 m/min without throwing the water from the hydrophilic roll surface, and can remoisturize a web to a level of about 20% moisture. Typical web speeds reported are on the order of 240 m/min.

K. REEL BUILDING

Control of the calender should make it possible to optimize the quality and uniformity of the paper wound onto the reel, so that high production efficiency is possible at the rewinder, with few if any shipping rolls rejected for causes attributable to defects caused by profile or reel structure. If the reel provides base stock for a coater, or off-line supercalender, it should provide the best conditions for those processes.

First, let us consider the CD variations that might be reduced by proper adjustment and operation of the calender. Some degree of caliper profile correction is possible on a machine calender stack. By providing air showers at perhaps 20 to 24 CD locations (Fig. 601), cooling air may be directed at the chilled iron rolls to slightly change their diameters by thermal contraction, allowing an increase in bulk in the zones that are selectively cooled. In one set of investigations on the effect of air showers on newsprint calenders, the maximum correction on the caliper was found to be about 2%. Although this seems to be a small change for the effort, it may afford just enough relief to fill in a soft spot on the reel and prevent roping.

Another method of caliper control which affords a somewhat greater range of correction over roughly the same effective width per unit

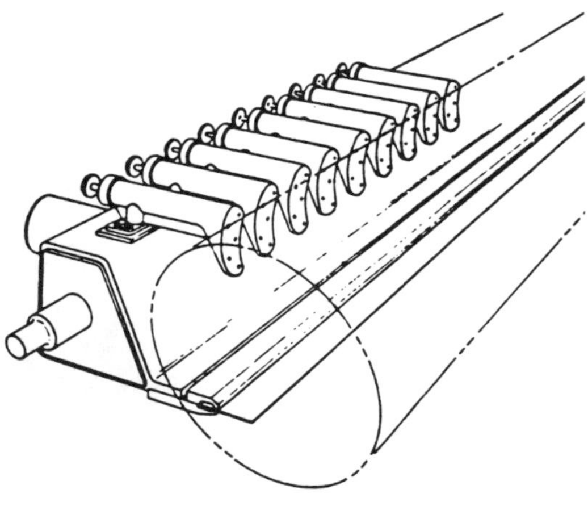

Fig. 601. Calender cooling air nozzles.

(75 to 120 mm) is the previously mentioned CalCoil development, using high-frequency eddy current heating of the chilled iron rolls to locally expand the roll diameters, thereby causing a caliper reduction of up to about 15%, with a rate of correction faster than air showers when changes are desired. The adjustments are made electronically, and are therefore suitable for automated control.

None of the above work very well on calenders with filled rolls or soft nips, as they are not that sensitive to the diameter variation of the metal roll. However, the anti-deflection roll shell arrangements with segmented pressure control zones can be used with these nips to affect the amount of caliper or bulk reduction over comparatively broad zones, on the order of 1 m in width. This will permit a smoothly curved crowned or S-curve correction in sheet thickness to be made, similar to what could be accomplished by regrinding crowned rolls, or using nip relieving on intermediate roll ends [14].

Another aspect of reel building is to develop the proper reel structure in the radial direction, from the turn-up to the final diameter, so that the layers nearest the spool are properly tensioned to support the final reel weight over the initial paper layers. When operating the calender to improve gloss and smoothness, new problems may develop in terms of sheet slippage that did not occur with a rougher sheet. Reels on supercalender windups are often provided with centerwind assist to give an extra dimension in the radial profiling of roll structure. Drives have been added to the reel spools of paper machines to provide additional torque, when this was found necessary.

Finally, MD variations will cause problems from time to time. A careful analysis is required to determine the cause and effect of such things as calender barring excited by patterns of suction roll drilling, or other periodic disturbances on longer centers that might be thought to initiate winder vibration or to cause cores of shipping rolls to end up eccentric to the outside of the roll after winding. The list of possible disturbances includes pressing-induced MD variations, sectional drive stability problems, and even stock surges in chests and tanks. When confronted with such difficulties, one may wonder what possible remedy could be provided by the adjustment of calenders. The answer lies in a review of the effects of changing calender variables. Disturbances caused by periodic MD variations might be moderated by changing the final bulk leaving the calender, within the limits of the specifications imposed for gloss, smoothness, and caliper. By changing temperatures and pressures, the bulk reduction can be made higher in the outer surfaces compared to the inside of the sheet. By subjecting the sheet to a change in the types of nips through which it passes, resilient or non-resilient, or changing the properties of roll covering materials, a balance may be struck between finishing to constant caliper over the sheet surface and finishing to constant bulk. It is not reasonable to expect the calender to eliminate MD problems in sheet formation, but it may be possible to minimize their downstream effects. Here it is appropriate to mention a quote from Crotogino and Gratton [8]: "The most cost effective way to improve the quality of a calendered paper web is quite frequently to improve the quality of the uncalendered web".

L. SUPERCALENDERS: SPECIAL FEATURES

The supercalender is a stack of alternating chilled iron rolls and rolls with resilient covers, usually employed for off-line finishing because of frequent downtime for roll maintenance and limits which may be imposed on speed, making the operation a potential bottleneck in paper production. Because a supercalender contains many additional web-carrying and spreader rolls, it is not as easy to thread following a sheet break as a machine calender stack. All of these reasons are not sufficient to exclude its use as an on-line finishing machine, though, especially if the product contains little or no coating and only a small amount of filler, as do the SC-A and SC-B paper grades.

Most new supercalenders are of open-face construction rather then closed-face, as the changeout of one or more rolls for regrinding or refilling is simplified in the open face design.

Many are equipped with automatic splicing devices on the unwind to speed up production and eliminate the need for rethreading between roll sets. It is possible to pass the splice through the calender without opening the nips if the following conditions are met. The splice pattern should be at an angle to the web (diagonal), the rolls should be mark-resistant or be made with synthetic covers, and (optionally) the stack loading pressure or the line speed may be reduced momentarily while the splice passes.

When a web is threaded through a supercalender stack with chilled iron rolls in every odd position, with filled rolls in between each, one side of the sheet will always contact a filled roll surface at every nip, and the other side will always contact a chilled iron roll. This is the side of the sheet that is most improved in gloss and smoothness, as it is against the smoothest roll surface. To avoid this one-sided finishing action, there are two or possibly three choices: coat the sheet differently on the two sides to favor the production of gloss on the side against the filled rolls; include a reversing nip in the stack to put the other side of the sheet against the chilled iron in the bottom of the stack; or find a new roll filling material that maintains a smooth surface over its life and produces results similar to chilled iron. The last option is not yet available in large roll sizes, but it is a predictable development which will eventually provide a solution.

A reversing nip is formed by placing two resilient roll covers in contact at a position partway down from the top of the stack, leaving three (or five) steel-filled-roll nips above to finish the first side of the sheet. After the sheet passes the fourth (or sixth) nip, the reversing nip, it will enter the next nip with the second side against a chilled iron roll. Note that the reversing nip and all nips below this are doing nothing positive to improve the surface of the first-finished side of the sheet; in fact, the surface is probably degraded. Also there is a nip speed difference caused by the reversing nip which always makes the sheet tension on the ingoing side of the reversing nip much higher than that between previous nips, or at the exit of the reversing nip, and this frequently causes web breaks when a hole, weak place in the web, or edge crack comes along [15].

The speed difference arises from the speed of the top filled roll of the reversing nip being set by the depth of penetration of the steel roll above it. Where the two filled rolls contact each other, the depth of penetration of one into the other is about equal, resulting in a flat zone of contact about half as far below the surface as the steel roll's penetration. This actually imparts a faster surface speed to the web than the steel surfaces do. The percentage speed difference must be taken up by sheet stretch, and this has already been compromised by the action of the first finishing nips, leading to frequent breaks. To minimize the problem it is recommended to run the harder or the smaller of two reversing nip rolls in the upper position. The use of synthetic rolls in this position may afford a little relief.

Other features or devices employed on supercalenders include steam showers, anti-deflection rolls in king roll and queen roll position, ability to adjust the stack loading pressure either differentially between front and back side, or in five or six segmented zones across the face. Roll temperature controls for the metal rolls may employ heated oil from the anti-deflection roll system, or other heating medium. Profiling gloss meters are frequently included in the final sheet run. The reel or windup may have centerwind drive assist and, if there is an automatic splicer on the unwind, there should be a turret windup or other automatic reel spool turnup device on the winder.

M. GLOSS CALENDERS, SOFT CALENDERS, TEMPERATURE GRADIENT CALENDERS, AND SUBSTRATA THERMAL MOLDING

Treating the last subject first will allow an orderly differentiation of the other classes of calendering, thanks to the efforts of Jay Vreeland, Director of Research for S.D. Warren, a subsidiary of Scott Paper in Westbrook, ME, who, together with his colleagues, categorized all types of calendering as functions of temperature and pressure as they sought to develop a means of producing gloss and smoothness on paper surfaces without requiring the many nips of a supercalender, and without extreme densification of the interior core of the sheet [16, 17]. In doing so they have staked out a claim for their own zone of operation, titled it substrata thermal molding, and protected it by two U.S. patents, 4,624,744, Nov. 25, 1986, and 4,749,445, June 7, 1988.

The important keys to calendering finishing action in terms of their development are the average pressure in the nip, the initial temperature of the paper entering the nip, the temperature of the roll contacting the side of the sheet to be finished, the residence time in the nip, and the function of sheet moisture in affecting the glass transition temperature of the coating and the cellulose and hemicellulose in the fibers. So far, this is not greatly different from what has been said already, except for the inclusion of the term glass transition temperature. This is the scientific

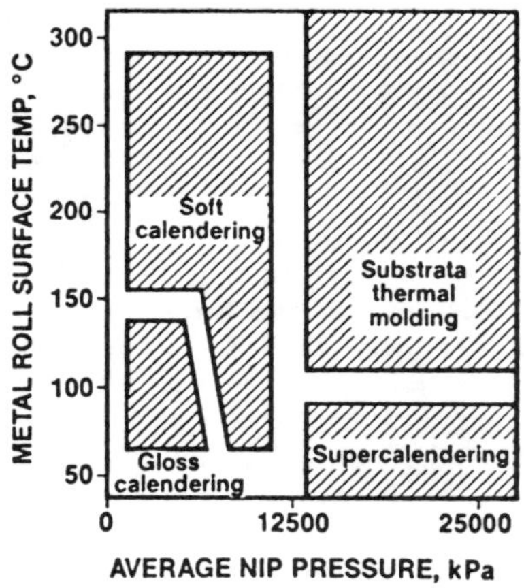

Fig. 602. Regions of resilient roll calendering [16].

measurement of the property that defines when the temperature has been raised to a degree that marks the onset of easy flow of a polymer, and cellulose and hemicellulose are polymers. The instrument most commonly used to measure glass transition temperatures of plastic polymers is a capillary rheometer. In this instrument, the flow of the plastic as a function of temperature is limited by its viscosity in shear when flowing through a capillary tube. Not surprising then that, at or near the glass transition temperature, shear forces applied to the paper sheet easily cause flow of the polymers.

Some new information coming out of the research leading to substrata thermal molding has shown how important and effective water is in reducing the glass transition temperature of cellulose and binders. Water itself has a very low glass transition temperature, −137°C. In small quantities, it depresses the glass transition temperature of 60% crystalline cellulose from 230°C at 0% moisture to less than 70°C at 10% moisture.

Substrata thermal molding is accomplished by having the roll surface very hot, to produce a temperature gradient effect in the incoming sheet, which is initially at a much lower temperature. Little or no angle of wrap is made with the heated roll before the nip, to limit penetration of the heat into the interior of the sheet. The nip intensity is high, and the backup roll is resilient, to increase the dwell time and produce shear forces. The zone of operation defined for substrata thermal molding is shown in Fig. 602, together with the more traditional areas occupied by gloss calendering, soft calendering, and supercalendering. Machine calendering, if it were included, would fill the lower right corner of the supercalendering box, with its high average nip pressure but low metal roll temperature.

Vreeland and associates admit the boundaries are not sharp between the regions occupied by the various processes. For instance, the SC-2000 soft calender shown in Fig. 585 has been reported to operate with a metal roll temperature of 230°C and 440 kN/m loading, which translates into 22 000 kPa average nip pressure, since the nip width is on the order of 20 mm. This is far to the right of the area laid out for soft calendering.

1. Temperature-gradient calendering

The two patents referenced above, assigned to the S.D. Warren Co., mention specifically the use of resilient backup rolls in their claims. On the other hand, the process of temperature-gradient calendering, as applied to board finishing, uses heated steel or chilled iron rolls, nipping the sheet between with no appreciable wrap angle on the rolls (Fig. 603). This causes a temperature gradient to be set up between the surface of the sheet and the center of the sheet [18]. The temperature difference between the center and surface may exceed 100°C if the web travels through the nip at sufficient speed.

The heat from the rolls causes the surface fibers to become more pliable, and they develop the necessary smoothness before the core of the web undergoes a drastic reduction in bulk which would reduce its strength or stiffness. Compared to a conventional machine calender, the temperature-gradient calender can make foodboard at the same gloss but with a 3% higher tensile strength, 10% higher tensile modulus in the MD and CD directions, and about 20% higher Taber stiffness. If the operating zone of the temperature-gradient calender were plotted on Fig. 602, it too would cut across the operating zone of substrata thermal molding, with a roll surface temperature of about 210°C, 50 to 100 kN/m nip loading, and effective nip widths on the order of 4 to 5 mm, causing average pressures of 12 500 to 20 000 kPa. The gloss produced equals or exceeds the acceptable standards for foodboard printing, but is not as high as the substrata thermal molding process would develop, because the residence time in the nip is short compared to resilient roll calendering.

2. Gloss calenders

Gloss calenders were the first types of soft calender developed (Fig. 604). Originally intended to produce gloss on coated boards, their use to finish uncoated grades has been demonstrated. A gloss calender uses a high

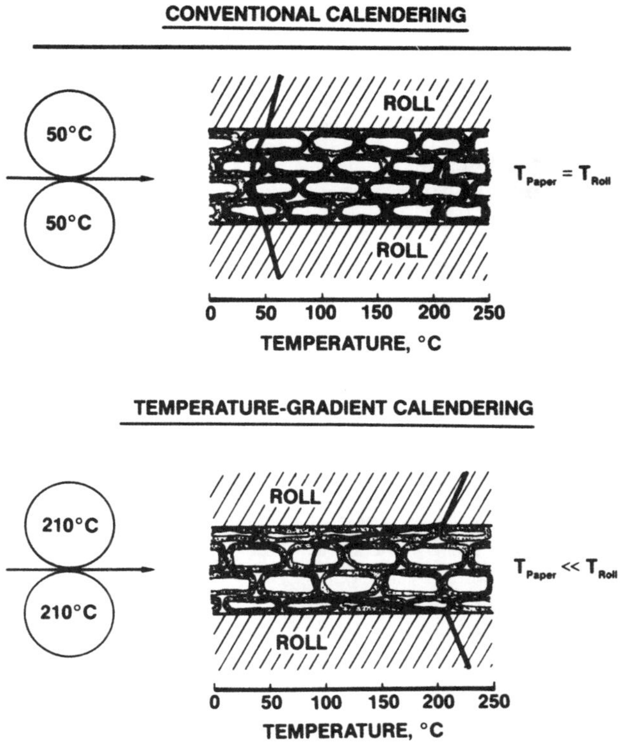

Fig. 603. Roll configurations and temperature profiles through the thickness of a web as it passes through conventional and temperature-gradient calender nips. In temperature-gradient calendering, there is a steep gradient in the temperature profile from the surface toward the center of the sheet [18].

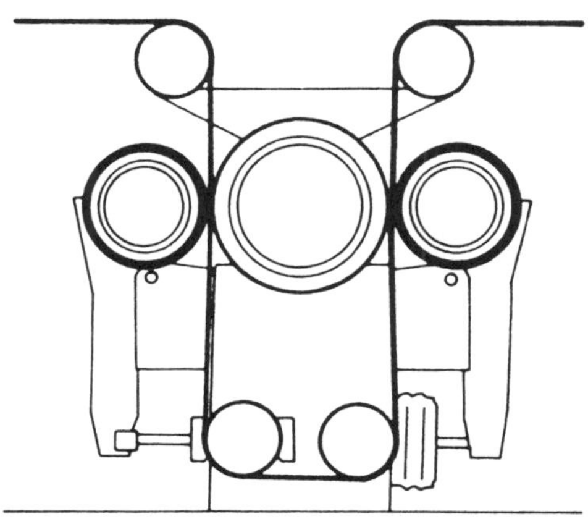

Fig. 604. Gloss calender used primarily to finish coated board. The board is pressed with rubber- or polymer-covered rolls against a central heated iron or steel roll [8].

polished heated metal roll against which one side of the web is nipped by a resilient rubber or synthetic backup roll [19]. The differences between a gloss calender and soft calenders are slight, and seem to be based on some initial roll arrangements and design parameters for the intended application to finish board rather than a functional distinction. They seldom have anti-deflection construction in the backup roll, and rubber backup roll covers where employed have limited operating temperature, loads, and life. If both sides of the sheet are to be calendered, two stations are required, one for each side.

3. Soft calenders

The soft calender normally does have anti-deflection construction in the backup rolls, and the covers are able to withstand higher operating temperatures. On some soft calender arrangements the backup roll is water cooled to extend its life. Again, two finishing stations are required to calender both sides of the sheet (see Fig. 585). The sheet threading arrangement can be as simple as a straight pass through the two nips, making its on-line use attractive. To avoid the danger of damage to the backup roll covers during threading, drive connections may be provided on both the metal roll and the backup roll. As new heat- and pressure-resistant roll covers are developed, the temperature and nip pressure limits on soft calenders will be extended.

4. Substrata thermal molding: integration of calendering variables into one equation for the roll surface temperature

The success of the substrata thermal molding process in achieving multi-roll supercalender gloss readings in one or two passes through heated nips may be attributed to an accurate method of predicting the correct surface temperature for the metal roll under any combination of moisture, web speed and nip pressure.

The process works best when the incoming sheet moisture is between 3–7% and the average sheet temperature is low enough to prevent permanent deformations in the core of the sheet (average temperature well below the glass transition temperature). Even so, it is possible to shift the operating temperature of the metal roll higher to account for a drier sheet. For instance, a Parker PrintSurf reading as low as 2.3 μm can be produced on uncoated bleached hardwood kraft with only 0.37% incoming moisture by raising the metal roll surface to 250°C (nip pressure 27 600 kPa, residence time 17.1 ms). The temperature gradient set up between the heated metal roll surface and the sheet interior is precalculated to do significant sheet molding (or calendering flow involving permanent deformation) in the outer 7.6 μm of the web, which is on the order of 10% of the thickness of a newsprint sheet. When the residence time in the nip is long, with soft backup roll or slow speed operation, the roll surface temperature is reduced to prevent deformation of the zone below 7.6 μm, and when the nip width is short or the speed is high, the surface temperature is raised to produce a rate of heat transfer which will reach to the depth of this zone.

The factors of moisture, residence time, and initial temperature of the web are inserted into a formula which predicts the maximum roll surface temperature required under these conditions. In the range 40° to 20° below this temperature, some surface molding will take place, and this range is mentioned in the patents; however, the critical region of operation lies between the calculated temperature and a point 20°C less than this.

Part of the formula incorporates the glass transition temperature. For moisture levels below 12%, the glass transition temperature is closely estimated by the formula

$$T_g = 234.2 \times e^{-0.131\,m} \quad (T_g \text{ in } °\text{C})$$

where m is the moisture percentage.

The complete formula for determining the surface temperature T_s of the heated metal roll is a function of just three calendering variables: initial web temperature T_i, nip residence time t, and the moisture content m in percent.

$$T_s = \frac{[T_i \times 0.357\,t^{-0.479} - 234.2\,e^{-0.131\,m}]}{[0.357\,t^{-0.479} - 1]}$$

Note that the nip pressure or linear loading force do not enter into the formula for the critical temperature except for the fact that they affect the nip width and therefore the residence time in the nip. The broad range of operating nip pressures has been indicated in Fig. 602 as pressures higher than 13 780 kPa.

The research effort that went into developing the substrata thermal molding process has made a significant contribution to understanding calender finishing action. Whether or not the industry will feel any restrictions on the more widespread utilization of the findings, due to the existence of the previously mentioned patents, is a matter that only time will tell.

REFERENCES

1. Knecht, W., "Fast Response CD Caliper Control", TAPPI 1988 Finishing and Converting Conf. Proc., p. 199-202 (Oct. 1988).
2. Hudson, K.O., "Advanced Finishing Technolgy", TAPPI 1990 Finishing and Converting Conf. Proc., p. 1-4 (Oct. 1990).
3. Tuomisto, M.V., "On-Line Soft Calendering", Pulp Paper Can. 91(1):T42-44 (Jan. 1990).
4. Matsubayashi, H., Miyamoto, K., Takagishi, Y. and Kataoka, Y., "A Study on Blistering by Coating Structure Analysis", Tappi J. 73(5):161-170 (May 1990).
5. Vyse, R.N. and Sawley, D.J., "Practical Aspects of Calender Steam Showers", Tappi J. 71(10):87-90 (Oct. 1988).
6. Bourque, J.A. "Recent Advances in the Application of Steam in the Finishing of Paper", TAPPI 1989 Finishing and Converting Conf. Proc., p. 61-66 (Oct. 1989).
7. Crotogino, R.H., "Towards a Comprehensive Calendering Equation", Trans. Technical Section, CPPA 6(4):TR89-94 (Dec. 1980).
8. Crotogino, R.H. and Gratton, M.F., "Hard-Nip and Soft-Nip Calendering of Uncoated Groundwood Papers", Pulp Paper Can. 88(12):T461-469 (1987).
9. Shigley, J.E. and Mischke, C.R., Mechanical Engineering Design, Fifth Ed., McGraw-Hill (1989).
10. Pfeiffer, J.D., "Finishing Action of the Supercalender Nip", TAPPI 1970 Finishing Conf., and as reproduced in the TAPPI 1982 Finishing Seminar Notes, p. 11-22 (May 1982).
11. Pfeiffer, J.D., "Shear and Finishing in the Supercalender Nip", Univ. Manchester Inst. Sci. Tech. Symp. on Supercalendering (Sept. 1975).
12. Rodal, J., "Soft-Nip Calendering of Paper and Paperboard", Tappi J. 72(5):177-186 (May 1989).
13. Okamoto, K., "New Aspect of Temperature Gradient Calendering", TAPPI 1989 Finishing and Converting Conf. Proc., p. 159-164 (Oct. 1989).
14. Kruska, D. and Stotz, G., "Operating Experiences with the Nipco-Float Supercalender", TAPPI 1990 Finishing and Converting Conf. Proc., p. 189-193 (Oct. 1990).
15. Laumer, E.P., "Troubleshooting at the Reversing Nip of a Supercalender", Tappi J. 70(5):175-178 (May 1987).
16. Vreeland, J.H., Ellis, E.R. and Jewett, K.B., "Substrata Thermal Molding, Part 1", Tappi J. 72(11):139-145 (Nov. 1989).
17. Vreeland, J.H., Ellis, E.R. and Jewett, K.B., "Substrata Thermal Molding, Part 2", Tappi J. 72(12):201-206 (Dec. 1989).
18. Gratton, M.F., Seth, R.S. and Crotogino, R.H., "Temperature-Gradient Calendering of Foodboard", Tappi J. 71(1):81-86 (Jan. 1988).
19. Back, E.L. and Olsson, A.M., "The Effect of Temperature on Gloss Calendering of Board as Evaluated in a Press Simulator", TAPPI 1981 Finishing and Converting Conf. Proc., p. 63-75.

XVII BROKE SYSTEMS

V.K. HOLLAND

Virgil Holland is a Senior Project Manager at Simons-Eastern Consultants, Inc., in Atlanta, GA. His work experience includes several rebuild projects in the paper machine area with this consulting engineering firm. He received B.S. degrees in Chemical Engineering and in Industrial Engineering from the University of Washington. He is a member of TAPPI.

REVIEWED BY

Dr. Hannu Paulapuro
Deputy Research Director
The Finnish Pulp & Paper Research Institute
Helsinki, Finland

Dr. Dan W. Manson
Senior Consultant, Engineering
James River Corporation
Portland, ME

A. INTRODUCTION

Broke is the term used for partly or completely manufactured paper or paperboard that does not leave the machine room or finishing operations as a finished saleable product. The term is also applied to furnish made by repulping broke. Broke is classified as wet or dry depending on its origin. Wet broke is that which is made before the paper or paperboard reaches the dryers. Wet broke includes normal edge trim from the web that drops into the couch pit during machine operation, the full width web that is allowed to run into the couch pit during start-up and upset conditions, and the broke that collects on press roll doctors or drops into the press pit during start-ups or interruptions of normal operations.

Dry broke is produced during start-ups and by web breaks in the dryer section, at a breaker stack or a size press, the calender, reel or winder; at the reel during turnups; at the winder during turnups and as web trim; and trim and paper damaged during finishing and warehousing operations. Dry broke also includes off-quality finished product that is recycled through the broke system to become a part of the paper machine furnish.

An effective and efficient broke system is, therefore, a critical element in the design and operation of a facility for the manufacture and finishing of paper and paperboard products.

Details of the broke system required will depend on the product, source and form handled. The broke system provided must be properly designed to collect, repulp, treat, store and prepare the broke that is generated for its subsequent reuse in the paper machine furnish.

B. BROKE SYSTEM DESIGN

In general, the broke system must be capable of receiving the full machine production during threading and machine breaks. It must also handle normal trim from the former and the dry end. In addition, broke from the reel and the winder as well as off-quality paper and finishing room waste must be handled.

The broke system should be designed to provide a clean uniform furnish to the machine, controlled by the stock preparation blending system.

The characteristics of the dry and the wet broke are different, due to the moisture content and effect of drying on the paper fiber. Dry broke may also be affected by sizing, coating, and other surface treatments applied on the machine or in the finishing operations. The collection and storage of broke from the finishing operations is considered part of the overall broke system. It is very important to protect the complete broke system from contamination.

The satisfactory operation of any broke system is dependent on effective repulping of the broke. Repulping is accomplished in pulpers. In addition, subsequent treatment, such as deflaking, may also be required to ensure that the dispersal of fiber and fillers is complete. Adequate broke storage should provide [4] leveling of the variations and it should accommodate the expected swings in the flow rates.

The design capability of the dry end pulper should be based on the rate required to repulp a full machine break. The dry end pulper will normally receive a break from the last dryer, the calender(s) and the reel. On some machines, breaks at the size press are also handled with the dry end pulper. Accommodating these various sources may be accomplished with conveyors, chutes or multiple pulpers. Smaller, slower machines also rely on manually handling some of the dry end broke. Repulped broke is pumped from the dry end pulper to the broke storage chest.

The wet end broke system consists of the couch pit and, in some instances, a press pit. The wet end broke system must be sized to accommodate a full machine break at the maximum design capability. The wet end system must also handle the web edge trim at the couch pit and shower water during normal operation. Normal production conditions result in a low-consistency stock in the couch pit. When a press pit is used, all wet end breaks are normally sent to the press pit. The couch pit is then mainly part of the white water system. When a paper machine break is sent to the couch pit the stock will quickly thicken.

Three different approaches are used for couch pit system design depending on the product being produced and the rate of production:

- Couch pit broke is pumped to the saveall during normal machine operation and directly to the broke system when thickened during a break.
- Couch pit broke is pumped to a broke thickener at all times.
- Couch pit broke is pumped to the saveall at all times.

C. BROKE SYSTEM CONTROL

The desired result from the control of the broke system is to economically utilize the resource of the available broke and obtain the best quality in the paper produced. A complicating factor is the variations in the broke due to changes in grade, quality and the point in the process where the broke was generated. The system also is used to avoid stock spills during machine breaks.

For an example, consider a broke system where repulped dry end broke is pumped from a single dry end pulper to a broke storage chest and wet broke from the couch pit is pumped to the broke storage chest during the machine breaks after the stock consistency increases above a set minimum. The dry end pulper is also used to repulp winder trim and culled paper (see Fig. 605). This system provides storage of broke in the broke storage chest in the slush form and of rejected paper in the dry form.

This broke system is supplying the stock blending system with the broke portion of the machine furnish. The normal winder broke including trim is usually increased by the addition of any rejected paper available. It is desireable to keep the broke addition to an unvarying percentage and to keep the level in the broke storage chest above a minimum position. The amount of broke added to the stock blending system will need to be changed to allow for the trends in the production of broke. Therefore, if the broke storage chest becomes full because of the need to store the broke from a break, the addition of dry broke at the dry end should be curtailed until the level in the chest can be reduced.

The automatic controls available for the operator to control the system are the consistencies in the system, the levels in the chests, and the addition rate of the broke into the stock proportioning system. The operator also manually controls the rate of addition of dry broke from culled paper or other sources. Variation in the consistency usually results in inaccurate addition of the broke as a portion of the stock furnish. This occurs when the stock proportioning system uses only flow control measurement for control. The consistency also affects the storage capacity in the chests and the pumping characteristics of the broke.

The stock consistency is normally controlled utilizing a consistency sensing element mounted in the pulper pump discharge pipe and controlling the flow rate of the dilution water. The consistency element should be carefully located to achieve accurate measurement. A major factor may be the velocity past the probe, which can be affected by large flow rate changes and insufficient straight piping. The pumping conditions of a single pump will vary considerably from recirculation and transfer of trim and winder waste to a full machine break. Figure 606 shows a typical two-pump dry end continuous extraction

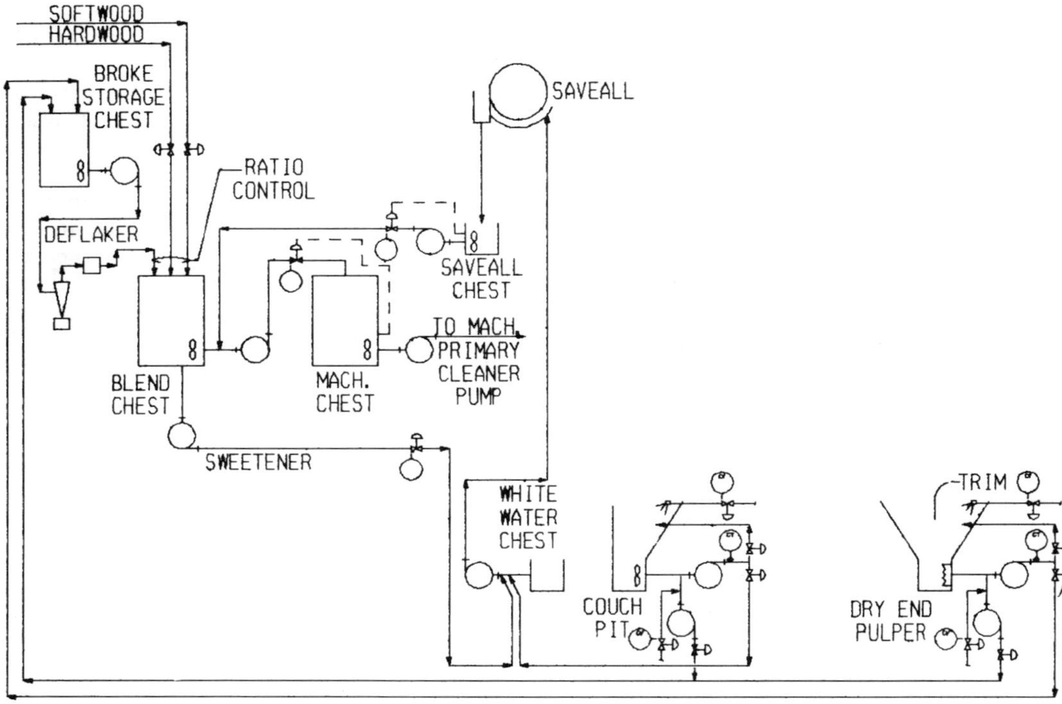

Fig. 605. Typical broke system.

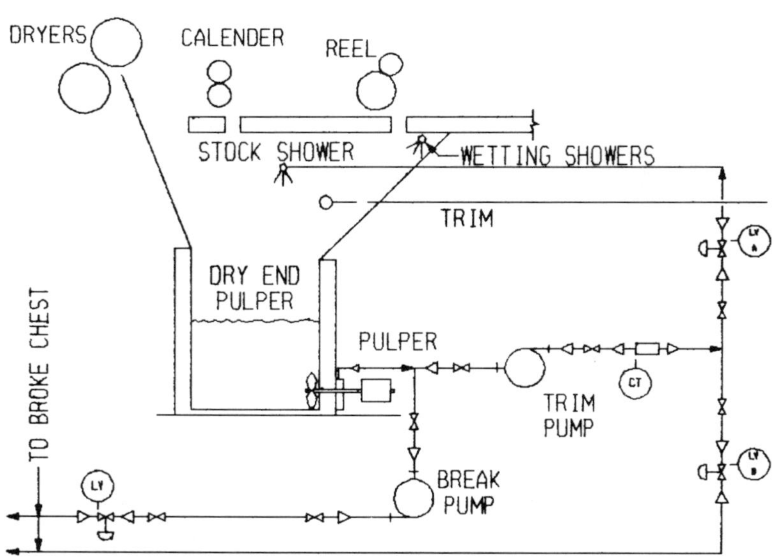

Fig. 606. Typical dry-end pulper.

pulper. In this arrangement, a small pump maintains a stable flow past the consistency element, either recirculating or pumping trim to the broke storage chest. A large pump is used to transfer at the machine break flow rate. The flow rates of the two conditions are a ratio of the total machine trim to the width of trim removed at the winder. The dilution water required will have a similar range of flows. Dilution is required to dilute the dry broke to the consistency desired in

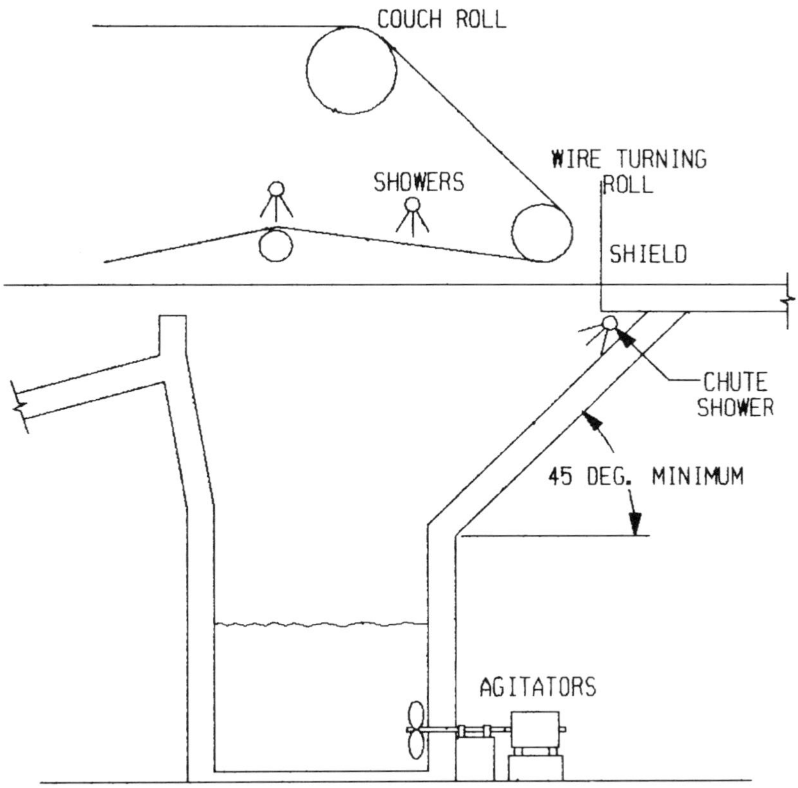

Fig. 607. Typical couch pit.

the broke storage chest, usually 4 to 4.5% bone dry consistency.

The couch pit (Fig. 607) is used to repulp wet end breaks and trim. Water is used to remove the sheet from the fourdrinier fabric and also to clean the fabric. Most of this water enters the couch pit. The sheet is normally trimmed on the wire with the trim entering the couch pit. The result is a stock of low consistency in the couch pit typically around 0.5% or less. This stock is usually sent to the saveall along with the excess paper machine white water for recovery of the fiber and filler. When a wet end break occurs or during start-up of the machine, the full sheet from the wire goes into the couch pit. With this condition, enough dilution water is added to the couch pit so that the consistency of the stock in the couch pit rises to the consistency desired in the broke chest. The control problems in this example are a result of the change in the consistency and the switching of the couch pit from pumping to the saveall to pumping to the broke storage chest [1, 2].

Controls for these systems can best be done with a distributed control system. With the information available to a computer which includes the weight of the sheet entering the couch pit, the couch pit consistency can be calculated. The point of transfer to the broke chest can be controlled and the stock to the saveall can be maintained at a constant rate.

Without a distributed control system, control of the couch pit is dependent upon the ability to measure the consistency of the stock. Because most consistency elements are not effective below 2% consistency, these elements are only useful during the break condition. Under normal running, the couch pit is repulping the trim and the level in the couch pit is being maintained by pumping the pulp to the saveall. On a break sensed by the break detector or the operator, the consistency rises until the consistency element can detect it and then open the dilution valves. The trim pump will go into recirculation only. The break pump will pump out to the broke storage chest when the level in the couch pit rises.

The design of the pulper vat and the couch pit affect the normal level and the ability to vary the level. A dry end pulper is normally operating at a relatively constant consistency. The pulper is designed to operate at an optimum level for proper circulation and submergence of the pulper rotor. A variation in the level may result from the change in the rate of addition during a break, but the controls will return it to the desired point. The couch pit should have some level variation in its operation. The level should be kept at the minimum necessary for the proper operation of the agitator during the trim condition and go to a higher point during a break condition. The volume between the low level and the higher level is used to allow the consistency to increase and come under control before the thick stock is transferred to the broke storage chest.

Variations will be reduced with stock passing through a completely agitated chest. An agitated stock zone of about 15 min capacity at the maximum pump out rate is recommended. The control of the broke can be improved if this leveling function is provided in the broke chests. The broke storage chests also provide the surge capacity required for machine breaks. If it is also to provide for the leveling function, it will require a completely agitated zone of the desired 15 min. Broke storage for a 2.5 h machine break is also recommended to reduce stock spills. For paper machines with a high production rate, this will be a large chest. The agitation zone for 15 min at a nominal addition rate of 10% of the furnish is 1.25% of the total chest volume. This agitation is difficult to design economically. For optimum blending, a large broke storage chest followed by a broke leveling chest is recommended.

When coated products are produced, a separate coated broke system is required for improved ash control. The uncoated broke and the coated broke will have considerably different total ash levels. With a separate system for the coated broke, each can be metered into the stock blending system separately. Within the coated broke there will be a variation in the ash as the ratio of the coating weight to the uncoated sheet weight changes. This variation can be somewhat leveled with storage capacity. The coated broke system may also be fed from pulpers in the supercalender and finishing areas. These systems are normally controlled through level controls.

D. DRY BROKE

1. Dry broke repulping equipment

Repulping equipment used at the dry end of a paper machine is available in many different designs. The purpose of this equipment, which is called a pulper, is to disperse the fiber from a dry sheet of paper into a water slurry for reuse in the paper machine furnish. Repulping is accomplished through agitation which separates the fibers in the broke by the action of high hydraulic shear. Most machine pulpers are of the contin-

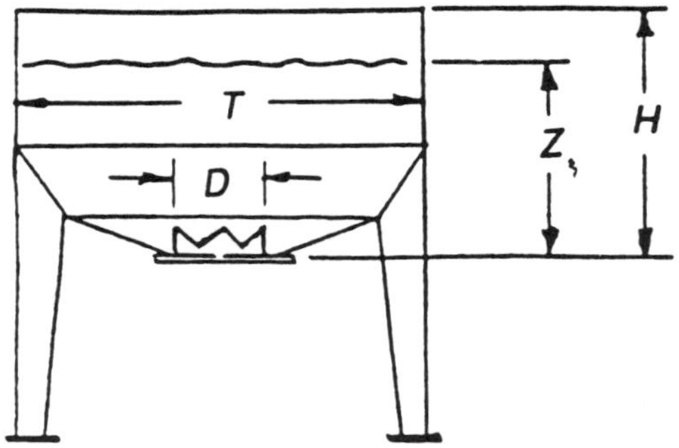

Fig. 608. Vertical bottom rotor pulper: T = tub diameter; H = tub height; Z = stock level; D = rotor diameter [5].

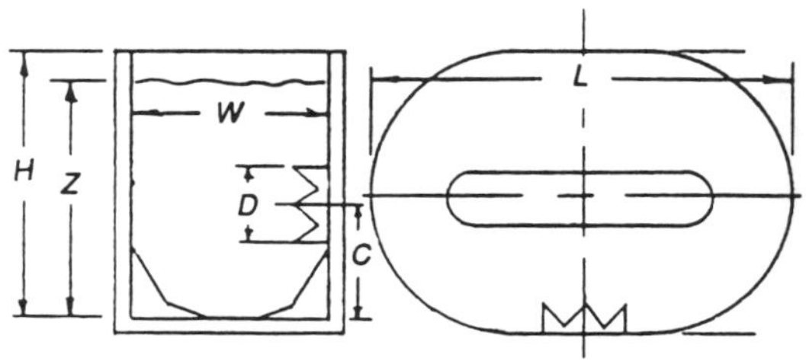

Fig. 609. Horizontal single-rotor pulper: W = tub width; L = tub length; H = tub height; D = rotor diameter; C = rotor off bottom distance [5].

uous extraction type. This is accomplished by the removal of the pulp through an extraction plate under the pulper rotor. Under-the-machine pulpers can have side-entering agitators, bottom-entering agitators or cross-shaft agitators (see Figs. 608–610). For any design to work properly, the sheet of paper must be submerged. Submergence is aided by showers. Showers can be used to add the dilution water required. An additional shower of recirculated repulped broke is effective. The action of the agitator with the shape of the pulper tub creates surface activity which is important in achieving submergence.

Side-entering pulpers are used in multiple units as the width of the sheet increases. Spacing across the machine width and the geometry of the pulper tub should be designed by the pulper vendor to achieve proper agitation and to optimize the power usage.

Bottom-entering pulpers are also used. These pulpers are particularly common in waste and market pulp operations. As the machine width increases, the shape of these units tends to limit their use as a machine pulper. Machine broke slabs and rejected rolls dropped into bottom-entering and cross-shaft pulpers can cause damage to this type of equipment.

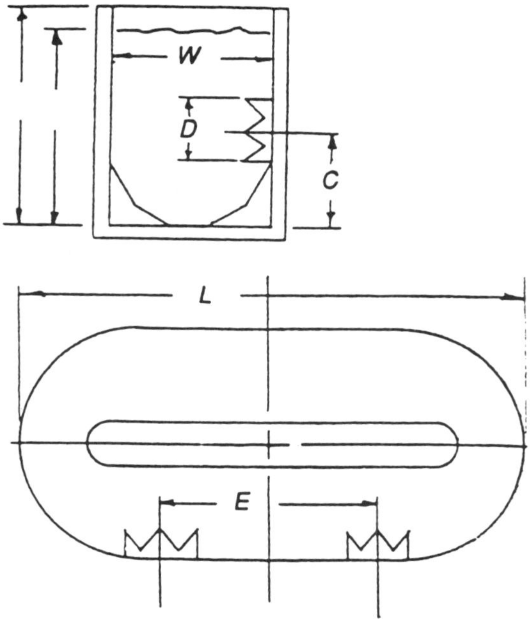

Fig. 610. Horizontal double-rotor pulper: W = tub width; L = tub length; H = tub height; Z = stock level; D = rotor diameter; C = rotor off bottom distance; E = rotor spacing [5].

Older broke systems are still being operated where a beater is used as a broke pulper. The flows in a beater are of a channel type, which does not promote good mixing. The beater tub arrangement does not receive a full sheet break easily. Many beaters were arranged so that the machine broke fell to the basement floor and was manually loaded into the tub. The beater was a batch type of device which leads to the concern that flakes of non-dispersed fiber may be pumped out when transferring broke.

2. Deflakers

The use of continuous extraction allows the pulper to use a control scheme of recirculation with transfer on high level. If the degree of fiber dispersion is incomplete, then further dispersion can be achieved with deflakers. Deflakers are normally installed following the broke storage chest. This allows the capacity to be based on the rate of broke usage rather than the full machine production rate required for processing a machine break. Deflakers are devices designed to place the stock being pumped through them in high shear. This equipment should not further refine the stock. Deflakers usually have a high rotating speed and close clearances. They need to be protected from foreign material, particularly tramp metal. This is most commonly done with a high-consistency centrifugal cleaner. Where maintenance has been a problem, refiners are sometimes used as broke deflakers. Since the broke has been dried, the fibers will refine differently than when they were virgin furnish. This additional refining is probably undesirable.

3. Sheet handling

Controlling the sheet from the paper machine into the pulper during a machine break is the key to good pulper operation. Pulpers can be arranged to take the sheet from multiple sources including the last dryer, one or more calender stacks and the reel (Fig. 605). Pulpers are fed through chutes, direct drops and with conveyors. As the sheet enters the pulper, the time for the sheet to be submerged should be minimized.

Showers are used to help submerge the sheet through wetting. Showers may be the dilution water required for the slurry. The recirculation of broke can also be used to shower the sheet. Using recirculation is recommended for the advantages as follows. First, it is very effective in knocking down the sheet to aid in the submergence. Second, this allows for a continuous flow passing a consistency element which is necessary for consistency control. And third, the pump is running through the continuous extraction plate and is available for transfer without delay when needed.

Pulper conveyors are usually belt conveyors designed to handle the full width sheet. The conveyor speed will be designed to feed the sheet into the pulper considering the speed of the sheet leaving the machine. These conveyors are capable of handling the sheet over a horizontal distance not practical with chutes [3]. Chutes should be inclined so that the sheet will slide easily. The slope required is usually at least 45°.

Winder trim may be blown into the pulper with a pneumatic conveying system. The speed of the trim leaving the winder will be close to the speed of the winder. The trim will be conveyed in air that has a higher velocity than the trim, at least 25% faster. The weight of air required will be two to three times the weight of the trim at the design capacity. For the trim to properly enter the pulper it must be separated from the air. The removal of this air must be considered to avoid interference with the submergence of the trim and to avoid disturbing the sheet on the machine.

4. Safety considerations

A pulper is an extremely dangerous piece of equipment in operation. Accidental entry by personnel into a pulper during its operation must be prevented. Particular attention must be given to pulper conveyors and operating floor broke openings. Personnel must not under any circumstances be able to get on these conveyors with the pulper operating.

5. Control of dry end pulpers

The typical dry end pulper will run at a rate that permits the repulping of winder trim and slab waste during normal operation. Dilution water is added in the proper amount to make down at the desired consistency of the repulped broke. The pulper pump(s) will have a small flow rate of broke to be sent to the broke chest. A pulper trim pump is recommended for the trim flow rates. When a break occurs, the amount of dilution water required increases from the flow required to dilute the trim to that required for the full sheet. Many machines use break detectors which turn on showers to supply most of the additional dilution water. This relieves the consistency controller from having to react quickly enough to avoid plugging the pulper with high-consistency broke.

The source of dilution water should be such that it does not affect any other paper machine system. Pulper dilution is normally supplied by the saveall filtrate system. This system also supplies the paper machine showers and consistency dilution. Pulper dilution should not require the use of cold fresh water makeup into the saveall filtrate. It is recommended that dry end pulper dilution come from an excess filtrate storage.

A typical machine arrangement is for the broke storage chest to be located in the stock preparation area near the wet end of the paper machine. This leads to a long pipeline from the dry end pulper. This line must be sized for the maximum flow for a full machine break. Transferring during normal operation where only a few percent of the machine production is the trim and normal waste leads to a very small average flow rate. The broke needs to be transferred at a minimum flow rate of 0.6 m/s velocity to avoid plugging. If there are multiple pulpers pulping to a common broke storage chest, it is common for several pumps to be connected to a single pipeline. In this arrangement occasionally there may be some problems with overloading the pipe, difficulty in sizing pumps, and avoiding pulper overflows.

6. Other pulpers

Additional pulpers are installed under the dry end to facilitate handling the sheet during threading when justified by the paper machine components. A common example is a machine with an on-machine coater that may have a pulper before the first coating station. This allows threading a full sheet through the dryers before threading the coaters and also allows the drying to stabilize while threading the coaters. Under-the-machine pulpers used only for sheet breaks and startups do not have any broke entering during normal operation. As such, they can be stopped during normal periods and can be started by a sheet break detector or a manual signal.

A separate pulper for the winder is desirable, particularly for higher production rates or where the winder is remote from the machine. A winder pulper handles the winder trim and the winder waste. These pulpers are also sometimes fed with a guillotine roll cutter for repulping rejected rolls.

E. BROKE STORAGE

1. Broke storage chest

The broke system design assigns two functions to the broke storage chest. These are for leveling of variations in the broke and for surge capacity during machine breaks. The break capacity is the difference in the machine production capacity and the pull of the stock preparation system on the broke. It is desirable that the broke chest break capacity be such that the broke chest never overflows. The volume in the chest for good leveling of variations will subtract from the volume available for breaks. This minimum operating level should be set to provide desired uniformity and depends on the degree of variations in the sources of broke pumped to the chest. Sizing of the chest to prevent overflow and to level variations can result in an excessively large sized chest. The agitated zone within the chest will be that which can be effectively blended by the chest agitator(s).

The shape and the size of a broke chest will affect the agitator selection and the power required for operation. The capital cost of a broke chest increases with its size. The operating cost includes the power cost for agitation, the pumping costs and the cost of stock losses from overflow, washup and breakdowns. This cost will reach a minimum as the stock losses approach zero. Optimization of chest size requires an economic evaluation of the various factors projected for the life of the chest and the required rate of return on investment for capital. A size sufficient for a 2.5 h break is recommended.

An alternative to low-density storage of broke is to thicken and store at high density. This allows the broke chest size to be smaller for the same tonnage capacity and perhaps reduces the agitation power requirements. The disadvantage of this arrangement is that the variations resulting from the different sources of broke will not be blended. The capital cost and maintenance costs of additional equipment should also be included in an evaluation. Additional equipment required includes a broke thickener system.

F. WET BROKE

1. Couch pit

The wet broke system includes the couch pit and sometimes a press pit. The couch pit should handle the trim from the wire as well as the full sheet during wet end breaks. The couch pit also may receive press breaks if not restricted by the press configuration. The couch pit will receive water from the sheet knock-off showers and any other wire showers which are above it. This normally results in a low stock consistency in the couch pit during normal operation. During machine breaks, the couch pit will thicken. It can then be controlled to a consistency of 4 to 4.5%, suitable for a broke storage.

The couch pit is designed for the full width sheet to enter from the knock-off position on the wire return (Fig. 606). This point is coordinated with the design of the fourdrinier or former. The design should also consider the farthest the sheet will carry on the wire return in any condition. The couch pit may also extend toward the dry end to the press. The sheet may enter on a sloping pit wall. Wet broke should be assisted into the couch pit with showers. The couch pit pulper designs are usually standard stock agitators or special cross-shaft agitators. Most wet broke is easily repulped. This reduces the concern of broke transfer prior to complete repulping.

The low consistency of the couch pit during normal running varies with the amount of shower water entering the couch pit and the width of the wire trim. The result is usually a consistency of 0.5% or less. This is a rich white water which can be sent to the saveall along with the excess white water from the fourdrinier.

2. Saveall

The saveall system is part of the stock preparation area that recovers the fiber, filler, and heat from the excess white water from the paper machine. The saveall system should be designed to keep the recovered fines from the excess white water in a closed loop within the wet end. Additional stock is added to the white water so there is enough fiber to create a mat on the saveall. This added stock is called sweetener. The sweetener is selected from the alternatives of the various parts of the machine furnish or from the mixed machine furnish. It is desirable that the sweetener not contain expensive fillers and additives that may be lost with the excess white water. The softwood portion of the furnish is usually preferred if it has the proper characteristics to aid in forming a mat and not reduce the capacity of the saveall.

3. Press pulper

Press pulpers are used when the press configuration does not allow the sheet to reach the couch pit during threading. These pulpers may be full width across the machine to accept the whole sheet or be fed by conveyors. A press pulper does not normally run when the machine is making good paper. This leads to this equipment becoming a maintenance problem.

A press pulper is normally installed at the first open draw of a machine with a pickup felt. One way of operating with this equipment is to use the press pulper as much as practical in handling the machine break conditions. The consistency in the press pulper is controlled to the consistency desired in the broke chest. The press pit broke is transferred to that chest. The couch pit can then be sent to the saveall at all times.

4. Furnish control

The control of the machine furnish is affected by the way the couch pit is handled. There are three ways of handling the couch pit broke. These are (1) sending the couch pit trim to the saveall and on a break sending the couch pit broke to the broke storage chest; (2) sending the couch pit broke to the saveall at all times; and (3) sending the couch pit broke to a broke thickener and then to the broke storage chest at all times.

The first method is called a swing couch. The problems with this method are in maintaining a steady saveall operation and maintaining good consistency control in the broke storage chest throughout the swing. During a wet end break, the saveall will run on the excess paper machine white water and sweetener. The couch pit broke will be sent to the broke storage chest. Dilution water for the couch pit should come from the filtered white water system. A large excess white water storage may be required to avoid adding fresh water, depending upon the white water balance and its uses. The amount of water required is equal to the dilution of the broke added to storage less the excess water from the paper machine. The excess water from the paper machine can be calculated from the material balance. Components adding to the amount of excess water are the water contained in the stock prior to dilution by paper machine white water and any fresh water added to the system such as certain showers. Components subtracting from the amount of water are the water in the sheet as it leaves the couch roll and all losses from the system which includes sewer losses.

The second method is sending the couch pit broke over the saveall at all times. During machine breaks, the saveall stock satisfies the total machine furnish demand and this stock is simply recirculated. This creates difficulty with refiner control. Over refining leads to the furnish becoming too slow and losing strength. This results in problems in runnability. Another scheme is to send the saveall stock to broke storage. This is not recommended since the fines in the excess white water will be sent to the broke storage chest. The saveall will need to be designed to handle the change in the flows to the saveall vat. The saveall will tend to be larger than for the flows in a swing operation. The consistency in the vat will need to be kept between a low of 0.5% to a high of 1.25%. During the break condition the sweetener flow is reduced to zero. This method of operation does not require large amounts of excess water for dilution.

The third method is sending the couch pit broke to a broke thickener at all times. This becomes more attractive when the couch pit receives a minimum amount of shower water. The couch broke flow changes considerably. This makes sizing a thickener difficult. One scheme to satisfy this is to recirculate a large broke flow under normal conditions. This is not an efficient use of energy. An advantage can be that the saveall can be run on the excess paper machine white water at relatively constant conditions.

G. FINISHING BROKE

Finishing, converting and coating operations are all sources of broke within the paper mill. This broke results from paper which is being further processed. The operations which produce additional broke include trimming, cutting, splicing, inspecting and removing paper that has been damaged in handling and storage. Some coating and laminating operations produce waste which requires special treatment to recover any useful fiber. Wet strength paper, polyethylene and foil laminates, wax papers, pressure-sensitive and hot melt adhesives, asphalt and other treatments that will contaminate the pulp result in this broke being considered wastepaper.

The collection and protection of finishing broke from contamination is important to the runnability of a paper machine using the broke. Small amounts of contamination can cause large losses due to rejects for defects and considerable lost production time. Separation and storage of some broke may be necessary if it contains any property which will not allow it to be used on all of the machines in a mill.

An example of this would be the handling of colored broke from a finishing room. Printed broke may also present the same problems. Colors may be segregated and stored for use when the machine is making a grade where the color may be used.

The concern of housekeeping to keep such items as garbage, plastic sandwich bags and Styrofoam coffee cups out of the broke collection system may result in strict control of the materials that may be brought into the mill area.

Broke collection systems start with the use of broke carts and include trim conveyors, reject conveyors and the associated systems to automatically bale or repulp the broke as it is generated. A broke system may be designed with conveyors feeding a shredder or paper hog with the shredded paper being conveyed to a separator which feeds the baler or repulper. Such systems should result in restricted access from contaminants. The complete broke system should provide for ease of handling the broke as it is produced or identified. If possible, the broke should be repulped immediately and reused.

Undesirable materials cannot always be removed from a finishing room. Take the example of a white paper mill with a sheet finishing room which uses corrugated boxes on wooden pallets for shipping. Pallet loads may be protected by a shrinkwrap film. All of these materials are objectionable in the broke. Separation of clean broke is mandatory for paper machine operation. Protection from dirt, floor sweepings, grease and oil, and even machine parts is required. The system can remove or contain some of these, if it is designed to do so. The equipment used in reclaiming wastepaper is designed to reduce the contrary items which can contaminate broke.

H. COATED BROKE

Coated paper must be considered separately. The coatings considered include pigments of clay, calcium carbonate, and titanium dioxide in a binder material such as starch or latex. The use of coated broke must be compatible with the wet end chemistry of the paper machine. The pigments used in the coating may be desirable as fillers in the paper. Many coatings are made with pigments of superior brightness and have value as fillers. These can help the quality of the paper. However, some pigments such as calcium carbonate may not be chemically compatible with acid white water systems and wet end additives.

The repulping of coated broke needs to be thorough enough for the coating particle size to be small enough for it to be accepted in the machine cleaning system and to not cause any problem in the quality of the paper. If the coating lumps and debris are plugging the cleaner rejects, further size reduction of the broke must be accomplished.

The coated broke has significant amounts of coating materials containing pigments which will act as fillers when used in the furnish. This makes a separate coated broke system necessary for control. This system is configured with at least a repulper, a coated broke storage chest, and pumps. A deflaker may be desired. The addition of the coated broke to the furnish can then be controlled through the stock blending system.

I. COLORED BROKE

Colored paper is another variable in the usage of broke. Heavy printing may also be considered a color if the print is soluble. Printing with oil-based inks may cause the paper printed with them to be treated as wastepaper rather than broke. Colored broke may be segregated and stored for use when a compatible color is being run on the paper machine.

Many colors are also bleachable. Processing of colored broke by bleaching reduces the storage space required, reduces the cost of carrying the broke in inventory, and reduces the exposure to contamination and damage. Most dyes can be bleached with a single stage. This is normally accomplished in a storage chest constructed of materials resistant to the bleaching agent.

J. WET STRENGTH

There are several different wet strength agents and different ways to apply them. However, most wet strength paper can be repulped by normal methods on the machine as it is being made. Wet broke containing wet strength agents does not present any problem as long as the machine stays on the same grade.

Dry end pulpers may require the addition of chemicals and heat to reduce the repulping time. Heat is usually added by the injection of steam into the repulper. Steam injection requires care to prevent a hazard and to protect the equipment from excessive vibration. Since many wet strength agents increase the difficulty of repulping as they age, waste from finishing operations may not always be useable.

REFERENCES

1. York, J.W., "Paper Machine Broke Considerations", 1986 TAPPI Papermakers Conf. Proc., p. 291.
2. Stevens, W.S., "Stock Preparation — Mechanical Treatment", Pulp & Paper Summer Inst., Univ. of Maine (June 15, 1987).
3. Yackel, D.C., "Under Machine Broke Pulping — Update 1978", TAPPI Engineering Conf. Proc., Bk. II, p. 359 (1978).
4. Thorp, B.A., "The Two Sides of Sheet Formation", PIMA, p. 17 (Feb. 1985).
5. "Recommended Nomenclature for Agitators, Mixers, and Pulpers", TAPPI TIS 0418.03 (1986).

XVIII CROSS-DIRECTION CONTROL

K. CUTSHALL

Keith Cutshall is President of Uniform Paper Inc., a consulting company. Mr. Cutshall's 30 year career has been centered on finding and fixing causes of paper variation. He has worked for Oxford Paper in Rumford, ME; Australian Paper Manufacturers, Melbourne, Australia; Westvaco, Covington, VA; MacMillan Bloedel, Vancouver, B.C.; and Devron-Hercules Inc., North Vancouver, B.C. Mr. Cutshall is a 1962 graduate of the State University College of Forestry at Syracuse University. He is a member of APPITA, the Technical Section, CPPA, IASPM and TAPPI, and has received the Technical Section, CPPA Weldon Medal. He is a TAPPI Fellow.

REVIEWED BY

Donald R. Brewster
President
Profigard Inc.
Montreal, Quebec

Keith D. Otto
Director Control Systems Technology
James River Corporation
Neenah, WI

A. INTRODUCTION

The subject of cross-direction control was born in 1954 when someone other than the backtender officially noticed the problem of basis weight variation. Burkhard and Wrist [1] declared that there would be three components to basis weight variation:

- Machine direction (MD)
- Cross-direction (CD)
- Random (R)

Cuffey [2] gave us a picture of basis weight variation in 1957 by staggering successive CD weight profiles to produce a "topographical" map (Fig. 611). Prior to this, the only methods of sensing CD variation in weight, moisture or caliper were lab tests or the backtender. Control was exercised by chokers on the branched manifold entering the headbox, flow obstructions placed in the headbox, slice spindles, air nozzles in dryer pockets, and air nozzles or friction pads at the calender stack.

The industry discovered how bad CD variation really was when on-machine traversing weight and moisture gauges became commercially available in the early Sixties. Most of the early development effort went into troubleshooting machine faults rather than control actuators.

The first CD weight control system was a joint effort between Taylor Instrument Company and St. Regis Paper Company at Urbana, OH [3]. The system was first tried under analog control [4], but became much more successful under digital control. Profile variation was reduced 6 to 55%, giving a 20% decrease in off-quality, and a decrease in grade change lost time [3].

CD weight measurement systems grew rapidly, but CD weight control remained fairly quiet

Fig. 611. Topographical map of basis weight variation [2].

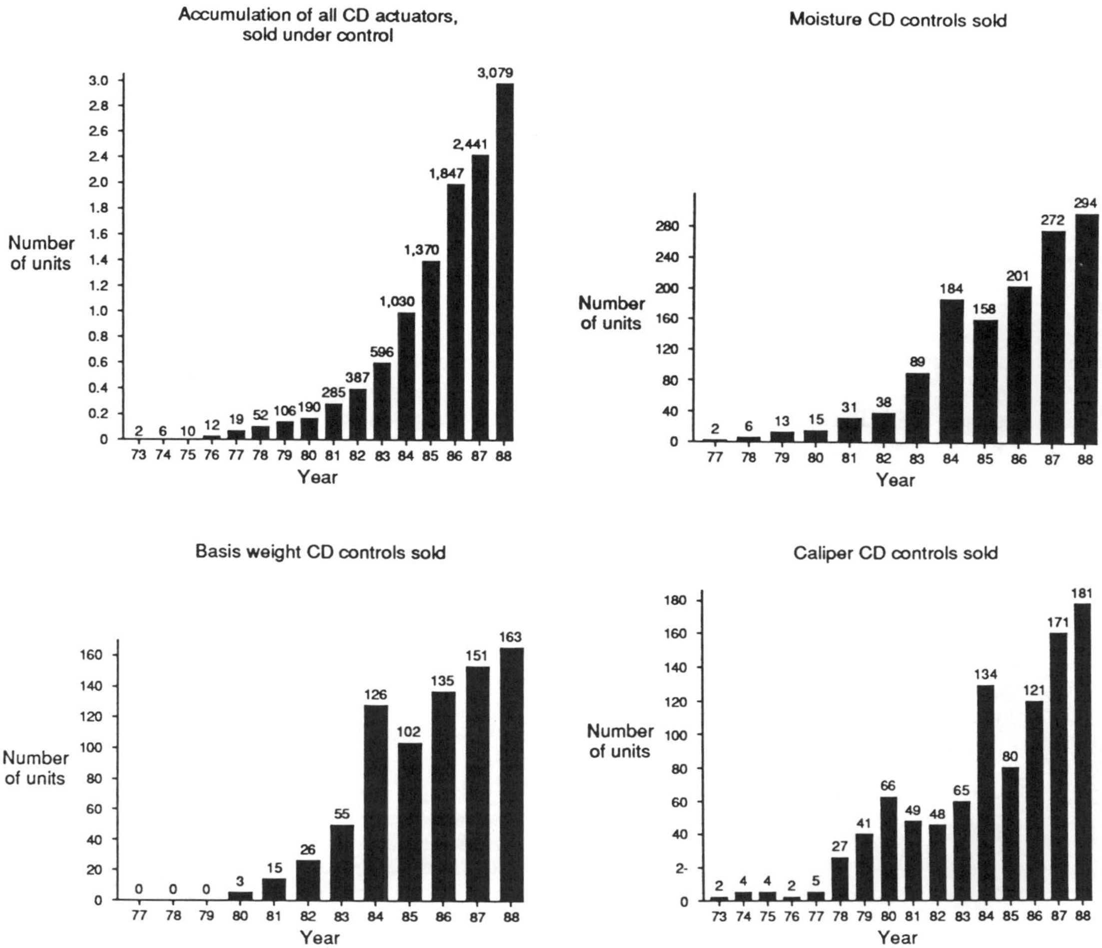

Fig. 612. CD control actuators sold with computer control [6].

until the early Eighties when commercial actuators and control strategies became available. A typical experience is well described in reference [5]:

"Experience with the profile control system has shown that it is not a cure-all for bad machine design or sloppy papermaking. However, in all cases, performance has demonstrated that the system provides minimum profile for any existing conditions on the machine. In order to obtain the optimum profile, good design and operation are still required."

Caliper CD control was the earliest automated system, starting in 1973. Control strategy was simple because of short time lags and simple coupling. Successive developments of cold air showers, hot and cold air showers, induction heating, and very hot confined air showers reduced the cost of the systems and improved results with increasing control capability.

Moisture CD control was the next automated system, starting in 1977. Early control was by water sprays and steam boxes. These systems suffered from neither a lack of power nor from insufficient control quality, but rather an industry reluctance to choose the more expensive automatic control over manual control.

As illustrated in Fig. 612, CD control came together in the early Eighties. By the end of 1988, over 3000 CD systems were on computer control: 42% on moisture, 32% on caliper and 25% on weight. Systems for CD control of smoothness, gloss, and coat weight are also gaining acceptance.

The commercial success of CD control systems is a function of the results achieved and the resulting market demand for improved uniformity. Compare "acceptable" peak to peak CD variation in newsprint between the middle

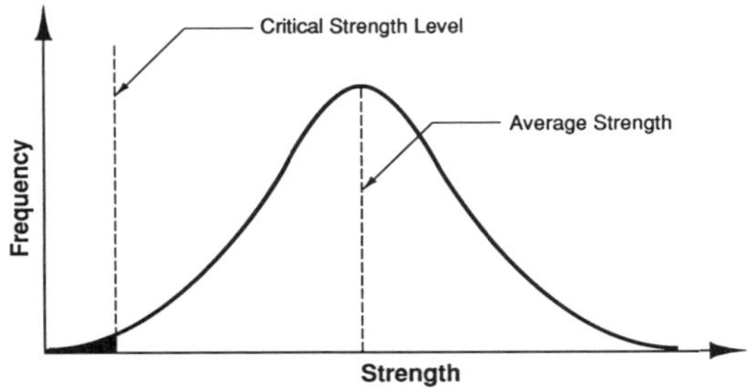

Fig. 613. Strength targets are established to give an acceptably low rate of failure.

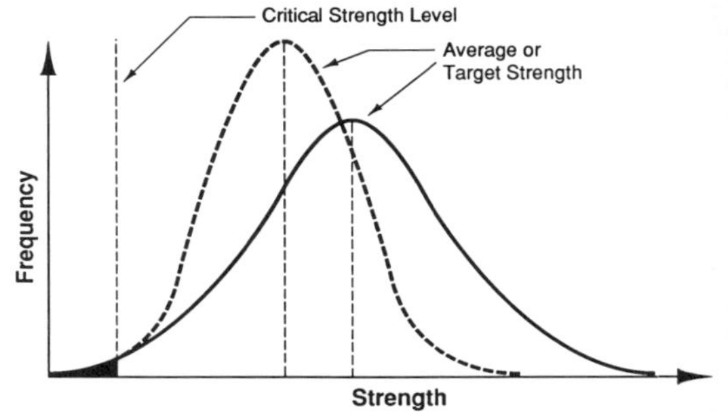

Fig. 614. Reducing strength variation allows the average strength to be lowered while maintaining the same failure rate.

Seventies and 1987. Basis weight peak to peak range has improved from 5% to 2%. Moisture variation has improved from 6 ± 2% to 9 ± 0.5%. Caliper variation has improved from a peak-to-peak range of 5% down to 1%. Very few machines of the Seventies could produce saleable paper in 1987 [7].

The major reason for the improvement is the development of CD control actuators which have the power to flatten any profile. Improved scanners, sensors and computing systems have allowed the more powerful actuators to be fully utilized.

B. BENEFITS OF IMPROVED CD UNIFORMITY

1. Variability costs

Every type of paper has some kind of strength requirement. The requirement is to ensure that a break does not occur in the paper machine, the converting machine or in the end use.

Strength is designed into the paper by using a high strength pulp, by treating the pulp to make it stronger, by addition of chemicals, or by making it heavier. In every case, a stronger sheet is more expensive than a weaker sheet.

Strength targets are generally chosen from experience to give an acceptably low failure rate. Figure 613 illustrates this.

If we were to take 100 samples and test them for strength, we would get a distribution curve as shown in Fig. 613. The average strength will be near the middle of the curve. Somewhere along the weaker tail of the curve, there is a critical strength level. Strengths lower than the critical level are likely to fail; those above are likely to survive. The area under the curve below the critical strength level compared to the total area of the curve gives the failure rate.

Rarely do we know the actual critical strength. The usual case is the strength target, or average, is adjusted until the failure rate is acceptable.

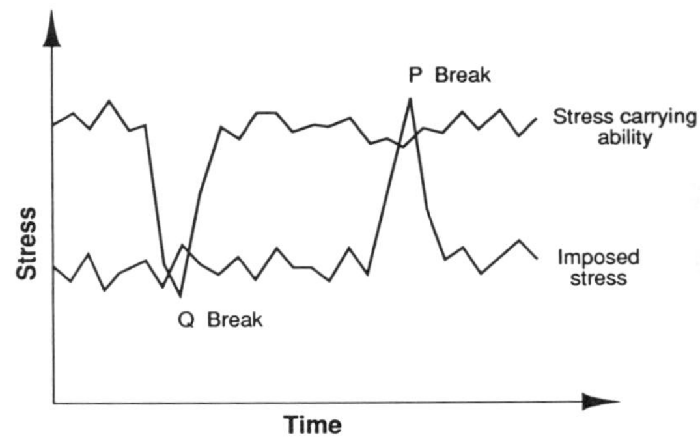

Fig. 615. Both the stress-carrying ability of paper and the stresses imposed on paper vary with time. Breaks occur when the stress-carrying ability falls below the stress level (Q) or when the stress level exceeds the stress-carrying ability (P).

The shape of the distribution curve is caused by the nature of the strength variation. If the variation becomes greater, then (as shown in Fig. 614), the strength target has to be higher to maintain the same failure rate. The higher strength target means higher manufacturing cost.

The critical strength concept used in Figs. 613 and 614 is an over-simplification of reality. Let us look into the occurrence of breaks in more detail. Referring to Fig. 615, imagine a continuous web of paper moving through a machine. The stress-carrying ability of the web varies with time. The stress imposed on the web also varies with time. Sometimes the strength of the web falls below the stress level, and a break occurs. Sometimes the imposed stress exceeds the strength of the web, and a break occurs. Thus, we have two distribution curves as shown in Fig. 616. Where the high stress curve overlaps the low strength curve, breaks occur. The breaks can be reduced by:

– Reducing the average stress level, or

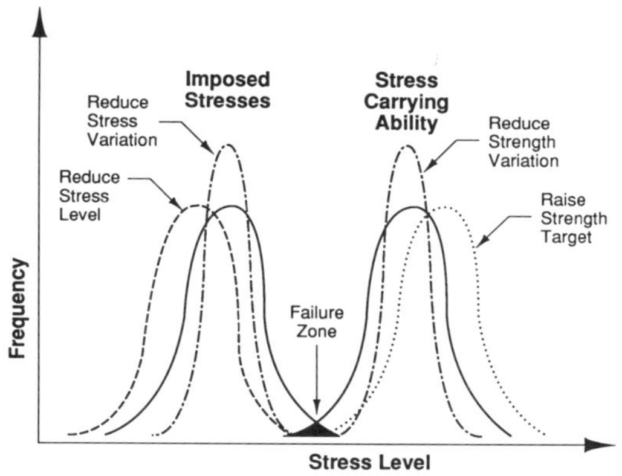

Fig. 616. The rate of failure can be reduced by reducing the stress level, by lowering variation in stress or strength or by raising the strength average.

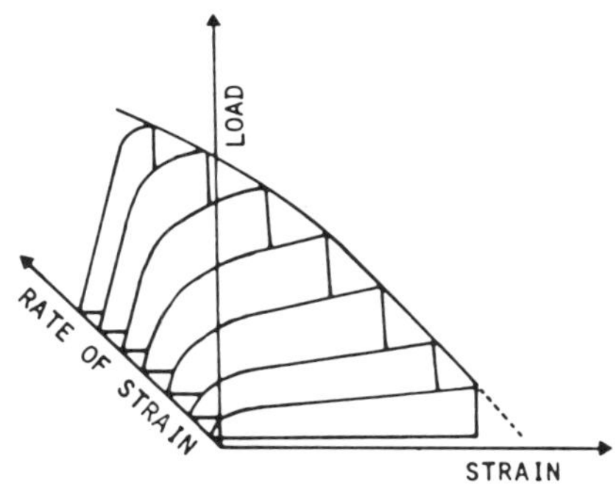

Fig. 617. Stress-strain (load-stretch) curves become steeper and the amount of strain to failure decreases as the rate of strain is increased [8].

- Raising the average strength level, or
- Reducing the strength variation, or
- Reducing the stress variation, or
- A little bit of each.

There are other factors which influence the incidence of breaks. Two that will be discussed briefly are the rate of strain and the repetition of strain. Rate of strain factors explain why quality becomes more critical at high machine speeds and high converting rates. Repetition of strain applies when the number of converting steps increases.

Figure 617 illustrates the effect of rate of strain on the stress-strain (load-stretch) curve. The curve in the foreground shows that, when a load is slowly placed on the web, quite a bit of stretch occurs before failure. However, when the load is quickly placed on the web, failure occurs at a much lower stretch. Machines which operate at high speeds tend to place rapid loads on the web. Thus, high-speed operation requires a stronger sheet or a more finely tuned machine, or a combination of each.

Figure 618 is a diagrammatic illustration of how a web's stress-strain curve changes when a web is stretched, allowed to relax and then stretched again. When the load is taken off, only a portion of the stretch is regained. When passing through the paper machine, it is stressed many times. As illustrated in Fig. 619, each draw which stresses the web uses up some of its stretch.

Whatever the cause or mechanism, breaks are expensive. Much time and effort is spent making the sheet strong to avoid breaks. When breaks occur, money is lost. On a typical paper machine, lost time due to a break can be worth between \$30 and \$150 a minute. Each break can cause 5 or more lost minutes, with averages between 10 and

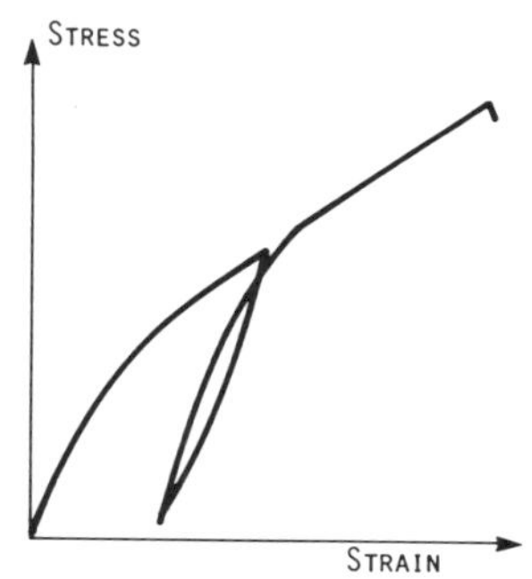

Fig. 618. Only part of the stretch is recovered when a paper web is strained, relaxed and strained again.

20 min common. Thus one break per day is worth between \$106,000 and \$1,065,000 per year in lost production:

Minimum break value =

$$\frac{(\$30)}{\text{min}} \times \frac{(10\text{ min})}{\text{break}} \times \frac{(1\text{ break})}{\text{day}} \times \frac{(355\text{ days})}{\text{year}}$$

= \$106,500

Maximum break value =

$$\frac{(\$150)}{\text{min}} \times \frac{(20\text{ min})}{\text{break}} \times \frac{(1\text{ break})}{\text{day}} \times \frac{(355\text{ days})}{\text{year}}$$

= \$1,065,000

Similar calculations can be made for the various converting operations.

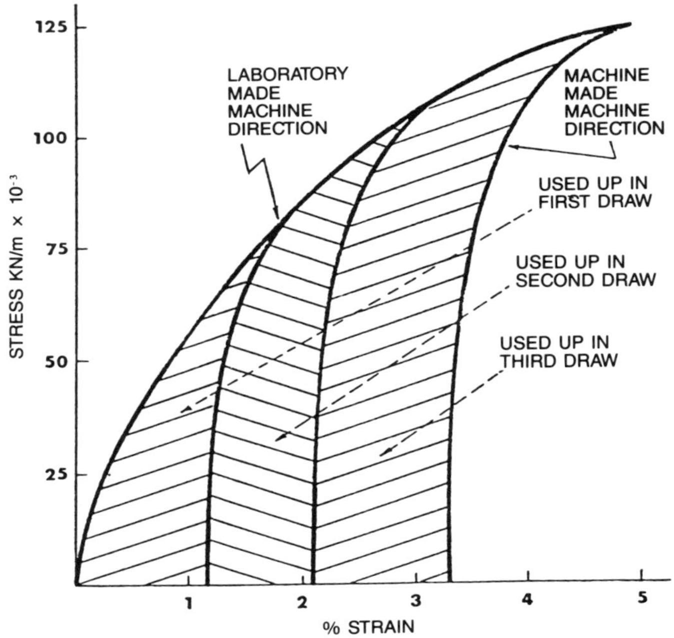

Fig. 619. Schematic diagram of how stretch is used up in machine draws.

To summarize, any form of variation in the paper web contributes directly to poorer production and converting efficiencies by reducing stress carrying ability and indirectly by increasing stress through wrinkles, slack draws, wonky rolls, etc. Measures to increase the stress carrying ability increases material and production costs. In brief, variability costs.

In the following sections, the economic justification for improved uniformity in basis weight, moisture and caliper will be discussed. To a large extent, these benefits are interrelated because improvements in one property will usually yield improvements in one or more of the others.

The most difficult task facing a project engineer is to separately justify improvements in basis weight, moisture and caliper. It is like trying to justify replacing the tires on an automobile one at a time.

2. Basis weight

Basis weight is a fundamental physical property of paper. Variation in weight causes variation in most other properties. Paper strength is related to the sheet's lightest area; all the extra fiber elsewhere is wasted. CD weight variation is more important than MD or random weight variation for paper handled in roll form, because CD variations pile up one on another in roll form, while MD and random variations are averaged out.

Despite the truth of the foregoing, it is surprisingly difficult to economically justify improvements in basis weight variation. One reason is that there are few benefits reported in the literature. Another reason is the difficulty of putting a value on "quality" until sales have been lost. Of course, sometimes the proposed fix is too costly; this can be particularly true on slower, narrower machines.

The first reported CD weight control system installed (at St. Regis, Urbana, OH) reported a 6–55% profile improvement which yielded a 20% reduction in "off quality" paper [3].

Gibson [9] reports that, on the corrugating medium machine at Weyerhaeuser, North Bend, a 68% improvement yielded a flat basis weight profile. This gave a 40% improvement in moisture streaks, which allowed a real moisture increase of 0.6 percentage points, which translated into a 1.5% production increase.

Harke [10] reports the experience at Willamette Industries, Albany, where CD weight is controlled by operators using remote actuators on both the primary and secondary headboxes. Grades are 115–450 g/m^2 (23–90 lb) liner and 325–375 g/m^2 (65–75 lb) kraft bag. The control philosophy is to control CD conditioned weight with the primary headbox, and CD moisture with the secondary headbox. The moisture and weight profiles improved 21–26%, yielding a 0.72% fiber savings due to higher reel moisture. Coincidental with the system implementation, other benefits were noted:

- 10% break reduction.
- 11% break lost time reduction.
- 24% machine downtime reduction.
- 33% cull rate reduction.
- 7% production increase.
- Unspecified energy savings.

Lebrasseur [11] reports results at Great Southern, when both an automated slice and an automated profiling steam box on the Fourdrinier were used:

- CD moisture, 35–50% improvement
- CD weight, 24–36% improvement.
- Production, 3.8% increase
- High moisture flags, 56–93% reduction
- Moisture culls, 54% reduction.

In a companion paper, Jaffe [12] reports the effect of the above improvements on the box plant. The computer-controlled paper gave 41% less warp, which yielded an average 2.4% production rate increase.

3. Moisture

The historical method of obtaining a uniform moisture profile is to overdry the sheet. This method works because, as the dryers work against the sheet's increasing resistance to give up moisture, the relative difference between wet and dry areas decreases. At 0% moisture, there are no relative differences.

The increasing resistance of the sheet to give up moisture below about 20% is due to the removal of water that is chemically bonded within the sheet. In other words, the chemical and physical structure of the sheet is changed. The physical toughness of the sheet decreases. It becomes stiffer and more brittle. Part of this change is irreversible, i.e., suppleness and strength cannot be fully regained by the later addition of moisture.

The physical changes caused by overdrying are not desirable in most grades of paper. The increased brittleness causes lower converting efficiencies. Alternatively, loss of strength is compensated for by using more expensive furnish.

Overdrying is desirable in some grades. For example, the high sheet temperatures that occur with overdrying can be necessary for the cure of resins. Some might also argue that overdrying is necessary to reduce pick-up and penetration of coating in coaters, size at size presses and water at water boxes. Sometimes alternate procedures are available. For example, water boxes at the calenders can be frequently replaced with water sprays before the calenders. By controlling the addition of water, less water can be added, which means that the average sheet moisture entering the stack can be higher. The sheet will calender more easily, not to mention that the dryers have less work to do.

In summary, recent advances in calendering and CD control of moisture have made overdrying an unnecessary expense in all grades of paper except where high temperatures are required for resin cure.

The benefits of CD moisture control fall into three categories:

- Reduced production costs;
- Increased production capacity; and
- Improved product quality.

Most of these benefits are the result of uniform moisture profiles, and apply to whatever method is used to obtain them, except overdrying.

(a) Reduced production costs

The biggest factor in reduced production costs is higher average moisture. Increasing the average moisture by one percentage point means a 1% reduction in raw material costs for grades sold by weight. The second biggest factor is usually improved runnability through the calenders, reel, winders and subsequent converting. A tonne of recovered paper is worth a tonne of extra production at the incremental profit value.

A third production cost factor is drying cost. Not having to dry a tonne of water means a savings of 1.5 tonnes or more of steam.

How much the average moisture can be increased upon obtaining a level moisture profile depends on the major reasons for setting the previous average. If the major reason was to limit peak moistures, then the potential gain is the difference between the old average and the allowable peak moisture.

If the major reason was to reduce the amount of moisture variation, then the potential gain is the difference between the old average and the optimum moisture content for that grade. Although opinions vary regarding optimum moisture levels, experience would suggest the following:

- 10% for toughness-critical grades such as newsprint, sack kraft, towelling, etc.
- 7–9% for stiffness-critical grades such as board and corrugating medium.
- 5–7% for grades needing dimensional stability, such as fine and copy papers.

(b) Increased production capacity

In a dryer limited situation, production can be increased when the dryer evaporative load is decreased. The dryer evaporative load can be decreased in three ways:

1. Drier sheet entering dryers – steam showers reduce sheet moisture content and raise sheet temperature entering the dryers. In general, a 1% moisture decrease is equal to 4% dryer load and a 10°C temperature increase is equal to 1% dryer load.
2. Additional energy source – infrared (IR) units add energy to the drying system. Thus, if the energy capacity of the IR is equal to 4% of the dryer section steam use, then a 4% speed increase can be expected. However, IR can also increase or decrease the capability of the dryers, depending on grade and location.
3. Increasing the average moisture level reduces the load on the dryers, giving the potential for increased production in drying limited situations.

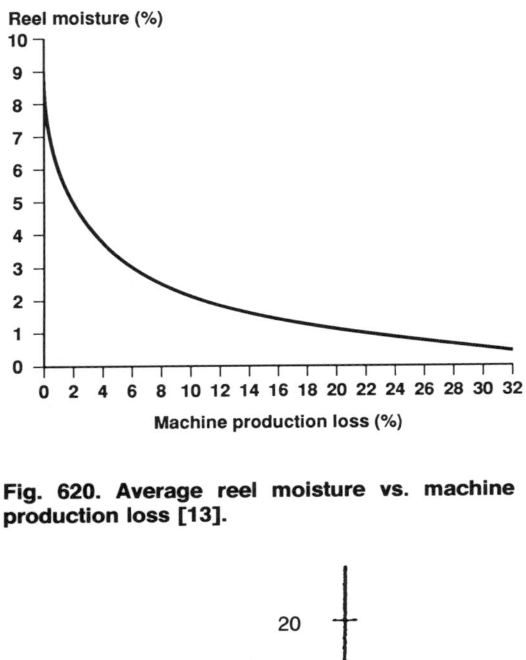

Fig. 620. Average reel moisture vs. machine production loss [13].

Figure 620 illustrates this later relationship. For example, if the average moisture is raised from 3% to 5%, the potential production increase is 4% (6%–2%). However, this is not exactly true in the case of using remoisturizing sprays, because remoisturizing sprays work by filling in the dry areas. The following examples help to clarify this:

	A	B	C
Old Profile:			
– Average	3.0	3.0	3.0
– Peak	5.0	4.0	4.0
New Profile			
– Average	5.0	5.0	5.0
– Peak	5.0	5.0	6.0
Potential Production Gain (%)	0	1.0	2.0

Example A is a moisture peak limiting case and Example C is a moisture variation limiting case. Example B is a blend of the two.

(c) Improved product quality

The improved quality benefits of CD moisture control derive directly from the effect of moisture on paper properties and indirectly from the interaction of converting/finishing operations and paper properties.

A demonstration of the moisture effect on paper properties is in Fig. 621. Newsprint samples were laboratory conditioned to different moisture levels and tested for burst and Elmendorf tear. The burst test, showing bonding properties, peaks at 6–7%. Tear, showing elasticity, is still climbing at 12%. The two tests are

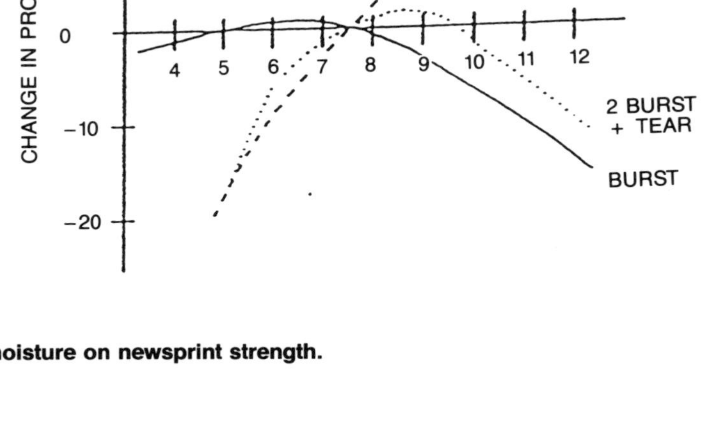

Fig. 621. Effect of moisture on newsprint strength.

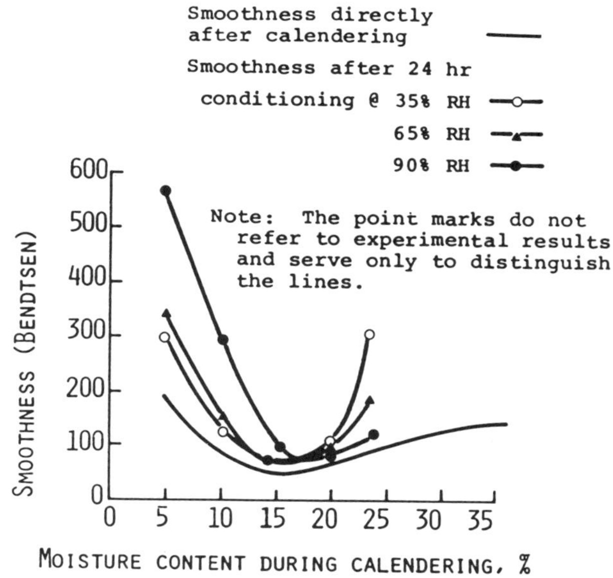

Fig. 622. Variation of smoothness directly after calendering at different moisture contents, and the corresponding smoothness after conditioning for 24 h at three different relative humidities [14].

frequently combined for an indicator of pressroom runnability. This indicator peaks at 8–9% moisture. The validity of conditioned samples to demonstrate the optimum moisture for unconditioned paper is a debatable point. However, the extreme sensitivity of paper strength to moisture content is not debatable. Paper physical properties are more sensitive to moisture variation than any other physical parameter.

The interaction of paper properties and moisture affect the response in all treatments to which paper is subjected. Variable moisture to size presses, coaters and water boxes means variable pick-up and variable absorption, yielding variable sizing, coat weight and finish. Variable moisture to a calender nip means variable finish and caliper yielding variable smoothness, coat weight and reel build. Variable moisture/strength properties means variation in tension during winding and a mixed response in the roll, where interior paper is under compression and exterior paper is under tension. As paper travels through subsequent converting processes and end uses, the original moisture variation coupled with the mixed stress/strain history continues to compound degraded performance. To make matters even more complex, as paper of variable moisture leaves the dryers, the moist areas cool more rapidly, creating a variable temperature profile through calendering and winding, further exacerbating paper quality.

Brecht and Heyn [14] demonstrated the effect of paper moisture on calendering response (Fig. 622). The best smoothness occurred between 15–20% moisture, where the compression resistance is very low. Mitchell [15] calculated the effect of variable CD moisture on average sheet thickness after calendering. A decrease of one point in moisture range will increase average newsprint caliper by about 0.4%.

In sheet-fed converting processes, flatness is of paramount importance. Any deviation in moisture can cause a deviation from flatness. In some grades moisture uniformity and flatness are of sufficient importance to justify an extra processing step of moisture conditioning.

In summary, uniform moisture content is the most important quality property due to its strong influence on paper physical properties. Moisture affects both the final properties and the sheet's response to any intermediate physical treatment.

4. Caliper

Caliper uniformity is important because sheets are often piled one on another. Minor differences in caliper soon become major differences in stack height or roll diameter. Some of the implications of caliper non-uniformity can be shown by working through an example where the sheet being wound onto a 1 m (40 in.) diameter roll has 1% more caliper on one side. When the roll is complete, one side has 10 mm (3/8 in.) more diameter than the other. Thus, the outermost ply must be 31 mm (1 7/32 in.) longer on the big side. Since the same length of paper went into the roll on each side, the long side had to stretch 1% more (see Fig. 619). As the roll is unloaded from the winder, it does not roll in a straight line to the conveyer because it is a truncated cone. So it veers from its path, hits another roll, and both get damaged, sending them both to the rewinder or broke beater. If the roll escapes this hazard, it goes on to the customer where, as it unwinds, it yields 1% more paper on the big side. The extra length from the big side causes uneven draws, tracking errors and misregister. Thus, a variety of benefits are available from improved caliper control, all of which come from the ability to build a good reel and good rolls. Gess and Segre [16] put it nicely:

> "A study to determine how the efficiency of the paper mill finishing room could be improved showed that a key parameter in determining finishing room efficiency was the uniformity of the cross-machine direction (CD) tension profile of rolls as they came from the reel of the paper machine."

Uniform tension profiles require a sheet of uniform rheology being unwound from a roll of uniform diameter.

Other related benefits are:

- Elimination of culls due to roll quality.
- Reduced downgrading of paper due to roll quality.
- Reduced lost production during start-ups and grade changes.
- Higher average moisture due to ability to compensate for effect of moisture variation.
- Reduced calender roll grinding (less downtime) because minor defects in calender rolls can be compensated for by the caliper control system. Also, the old friction pad method of control could damage the roll.

The calender stack is the papermaker's last chance at quality control. After the stack, he can only ship or beater. This helps to explain the early and rapid acceptance of CD caliper control systems.

5. Combined weight, moisture and caliper

Wallace [17] reports a range of improvements from implementing CD control, and the corresponding benefits, as shown in Table 61.

Table 62 is more specific, listing the predicted benefits accepted by the management of a major fine paper mill. The validity of the predictions have not been verified.

Jenkins [18] reports detailed profiles of weight, moisture, coated weight, rawstock smoothness, rawstock burst, rawstock caliper, rawstock porosity, supercalendered gloss, caliper, etc. Note that this was before any automatic CD control. He summarizes:

"1. Virtually every important parameter profile of the sheet from anywhere in the process is overwhelmingly dominated by the base sheet ovendry basis weight profile.
2. The base sheet moisture content profile also bears strongly on many of the parameters studied, its effect often being inverse to that of base sheet basis weight.
3. On-line caliper profiles at the paper machine and at the supercalenders can be invaluable aids for improving cross-machine quality uniformity and improving the mechanical condition of shipping rolls.
4. The high degree of correlation existing between laboratory profiles on single or multiple test strips and on-machine profiles shows minimal interference of machine-direction variations (although present) in characterizing cross-machine profiles.
5. The basic profile shapes display considerable stability day to day and month to month.
6. The more successful a mill is in improving the uniformity of these critical profiles, the more it reduces the strength requirements of the sheet for satisfactory pressroom performance.
7. The use of on-line traversing gauges to measure these parameters and the establishment of limits of variability at the

Table 61. Results and benefits from CD control [17].

Typical reduction in CD variability **Parameter**	**Reduction %**
Caliper	50 – 80
Moisture	50 – 80
Basis Weight	40 – 75
Benefits realized from CD control **Benefit**	**Improvement %**
Reduced Rejects	15 – 80
Fiber Savings	0.5 – 5
Steam Savings	0 – 8
Production Increase	3 – 10
Increased Coater Speed	1 – 4
Fewer Winder Breaks	30 – 75
Improved Converting Throughout	2 – 7
Increased Felt Life	5 – 35

Table 62. CD control benefits.

ASSUMPTION: On a fine paper machine with size press, installation of CD weight, moisture, and caliper control will give the following improvements:

— CD weight	50%
— CD caliper	50%
— CD moisture	30%
— CD ash	50%
— CD opacity	20%

This will yield the following results:

	Weight	Caliper	Moisture
Moisture at reel	+ 0.2%	—	+ 0.75%
Moisture at size press	+ 0.2%	—	+ 0.75%
Steam use	− 0.4%	—	− 1.50%
Speed (over dryer limit)	+ 0.6%	—	+ 2.25%
Culls — Weight	− 50%	—	—
— Condition	− 15%	− 35%	− 50%
Job lot — Weight	− 75%	—	—
— Moisture	—	—	− 75%
Claims — Baggy edges	− 25%	− 20%	− 10%
— Caliper	− 25%	− 25%	—
— Wrinkles	− 25%	− 10%	− 20%
— Runnability	− 25%	− 20%	—
Downtime — 25%			

paper machine will go a long way toward improving overall operation of a mill."

C. DEFINING THE PROBLEM (OPPORTUNITY)

Every problem presents an opportunity. This is fairly true in papermaking, and particularly true in CD control.

Defining the problem (opportunity) is, of course, the first step to a solution (benefit). In the field of CD control, there are four aspects of problem definition:

- Economics (payback and investment)
- Causes
- Cause-effect relationships
- Nature of variation

Section B above discussed the payback aspects of economics. The investment side will be covered in the section on actuators. The causes of each kind of CD variation will also be discussed in the actuator sections. Cause-effect relationships between weight, moisture, etc. will be discussed at the end of this section. The nature of variation is discussed next.

To illustrate the nature of variation in paper, the nature of weight variation will be discussed in detail because:

- It is a fundamental cause of paper variation.
- Other kinds of variation are similar in nature.

2. The nature of basis weight variation

(a) Definitions

The basis weight of paper is the mass per unit area. Basis weight variation is the deviation of the mass per unit area from the mean. It is commonly expressed as percent range, standard deviation, variance, and coefficient of variation. Basis weight variation is synonymous with substance variation and grammage variation.

Total basis weight variation is considered to have three components: machine direction, cross-machine direction and random [1]:

- Machine direction (MD) – variation of the whole sheet in the machine direction, or with time.
- Cross-direction (CD) – variation of the sheet in the cross-machine direction, assumed stable with time.
- Random – variation that is neither pure MD nor CD.

The random variation cannot be measured directly. It has to be estimated by subtracting the MD and CD from the total. Since the random variation is estimated by subtraction, the result contains all measurement and analysis errors. It is therefore called the Residual (R) [19].

The random component of R variation can be considered to have two components: interaction between MD and CD effects and pure random effects. If the MD or CD component is reduced or eliminated, then the interaction component will be reduced or eliminated. However, the pure random component will not be reduced [20].

Thinking of basis weight variation in terms of its directional components is useful in several ways. First, it makes the data amenable to statistical manipulation. Second, it helps to sort out the variation in relation to its cause. For example, consistency variation yields mainly MD basis weight variation; headbox slice distortion gives mainly CD basis weight variation: localized turbulence in the headbox gives random basis weight variation.

It should be noted that many of the measurement system suppliers have taken a different approach to categorizing basis weight variation. Variation that can be controlled by MD loops is called MD. Variation that can be controlled by CD actuators is called CD. Everything else is either filtered out or bumped into a category like "short-term MD". This approach is not unreasonable, because the task of these systems is one of control, not explaining how a paper machine works (or does not work).

In addition to the directional components, basis weight variation contains different wavelengths or frequencies. For convenience, the wavelengths or frequencies have been divided into four ranges [21]:

- Formation – smaller than 100 mm
- Short term – 100 mm to 1 s
- Medium term – 1 s to 200 s
- Long term – longer than 200 s

The 100 mm wavelength bordering between basis weight variation and formation was chosen because techniques used to control formation are effective on wavelengths (or floc sizes) up to about 100 mm. Short-term variation starts where formation leaves off and includes all wavelengths up to the length of paper made in 1 s. For example, if the machine were operating at a reel speed of 600 m/min (10 m/s), short-term variation would include wavelengths from 100 mm to 10 000 mm (0.1 to 10 m), or frequencies from 1 to 100 Hz (cycles per second).

Medium-term variation would include wavelengths from 10 to 2000 m, or frequencies of 0.005 to 1 Hz (or 1 to 200 s per cycle). In like manner, long-term variation includes wavelengths longer than 2000 m, or frequencies whose periods are longer than 200 s.

Each of the frequency or wavelength ranges was chosen because of its relationship to the process. Formation is affected primarily by fiber properties and hydraulic turbulence in the headbox and forming zone. Short-term variation is affected primarily by hydraulic pulsations, hydraulic stability and equipment vibration. Medium-term variation is primarily affected by blending, flow stability and fast control loops. Long-term variation is primarily affected by system stability and slow control loops.

Thus, the basis weight of paper is not uniform. Furthermore, it is rarely uniformly non-uniform. The character of the non-uniformity also changes with time or, in technical terms, the statistical properties are rarely stationary or ergodic. They forever change like the surface of the sea.

(b) Topographical maps

The surface of the sea is a useful way to envision basis weight variation. Or more accurately, the surface of a turbulent river with its longitudinal (MD) ridges and valleys. The first to present basis weight variation in the form of a topographical map was Cuffey [2].

Cuffey floodlighted the sheet and then rapidly traversed the other side with a photocell, relying on the fact that opacity is proportional to basis weight. The successive CD profiles were staggered on a CRT to create the topographical map (Fig. 611). Note the fairly stable ridges at the sides and predominantly random variation in the center.

Britten [21] ran 30 CD strips from the reel of a newsprint machine through an off-machine beta-gauge profiler and staggered the recordings to form the topographical map in Fig. 623. There are three fairly stable ridges on the front side but the rest of the variation is random.

A third topographical map from a kraft machine is illustrated in Fig. 624. The variation is primarily random.

It is worthwhile to spend some time studying these three topographical maps, looking at them

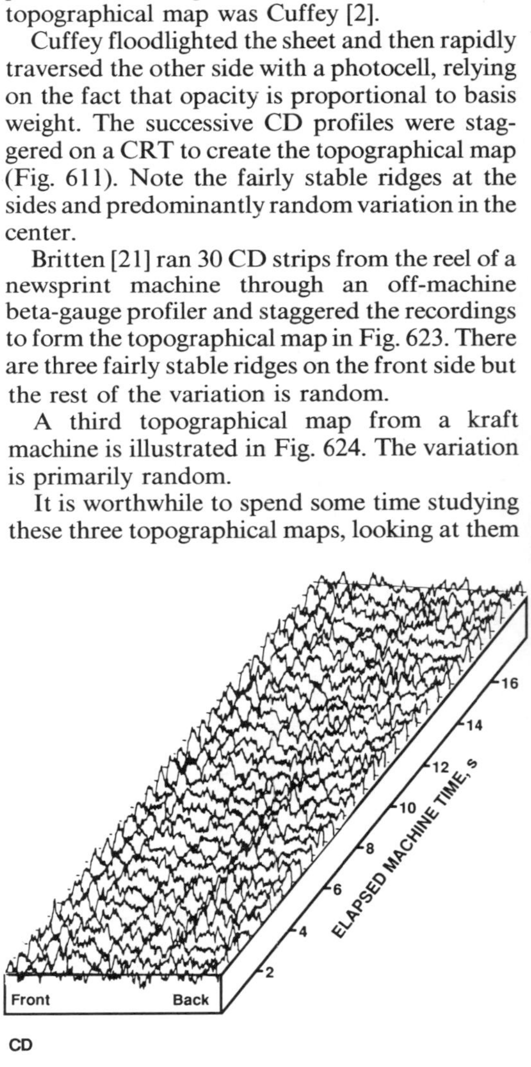

Fig. 623. Topographical map of basis weight variation on a newsprint machine [21].

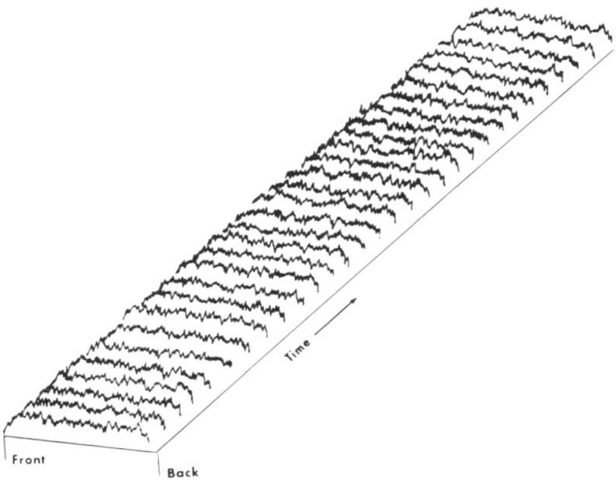

Fig. 624. Topographical map of basis-weight variation on a kraft machine [21].

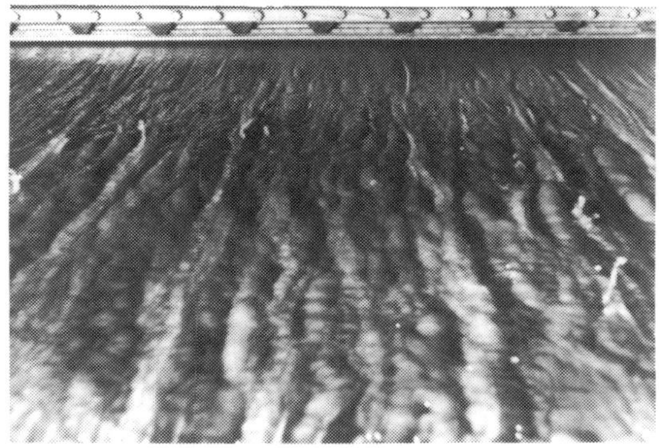

Fig. 625. Photo of turbulence coming from the slice on an obsolete kraft machine [21].

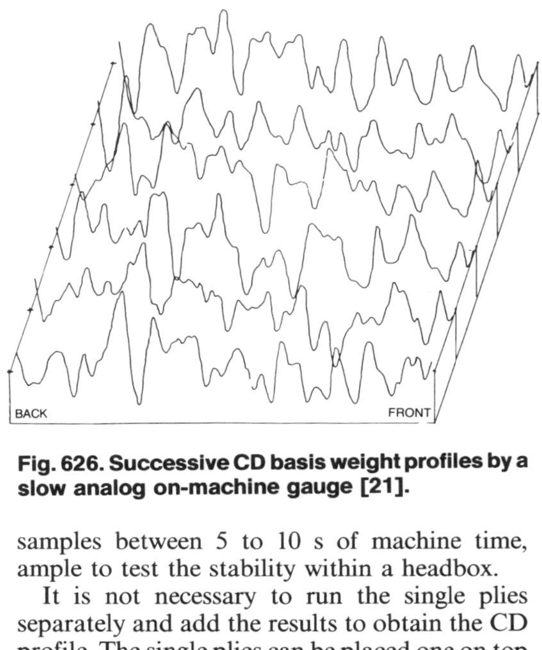

Fig. 626. Successive CD basis weight profiles by a slow analog on-machine gauge [21].

from different angles, until they can be envisioned as the flowing turbulent stock on the wire. Figure 625 shows the turbulence coming from the slice of an obsolete kraft machine.

(c) CD profiles

If a strip of paper is taken off the reel so that its axis is perpendicular to the MD and its length is the full machine width, then it represents a true CD sample. When it is run through a beta-gauge, the result is like one of the records in Fig. 623. The record contains all of the CD and some of the random variation. It does not contain any MD variation because it represents only one point in time.

The random variation has to be eliminated to find the CD variation. This can be done by averaging enough single-ply CD profiles: the random variation, because it is randomly distributed, will eventually cancel out. Ideally, enough single-ply CD profiles have to be used to cover exactly the full period of the longest random variation. Since this is impractical, the number of single-ply profiles used to make the results comparable between similar machines or between different time periods on the same machine is standardized.

How many single-ply profiles to use and how to sample the single plies is not a simple decision. The more plies used, the better the result but the more time consuming and difficult the procedure. Ten plies is about the maximum that can be conveniently handled in most off-machine beta-gauge profilers. The most convenient method of obtaining the ten samples is to peel them from the top of the reel. For most machines, this samples between 5 to 10 s of machine time, ample to test the stability within a headbox.

It is not necessary to run the single plies separately and add the results to obtain the CD profile. The single plies can be placed one on top of the other to form a slab, and run simultaneously.

On-machine traversing beta-gauges do not give a true CD profile, because the sheet travels a considerable distance in the MD as the gauge traverses. Thus the result contains MD variation as well as CD and random.

Figure 626 shows CD profiles obtained with an on-machine traversing analog gauge installed on a high-speed newsprint machine. The six profiles were taken at approximately 4 min intervals; the traversing time for each profile is near 3 min.

The newsprint machines represented in Figs. 623 and 626 have CD variation that is quite similar, but this does not show in the two illustrations for the following reasons:

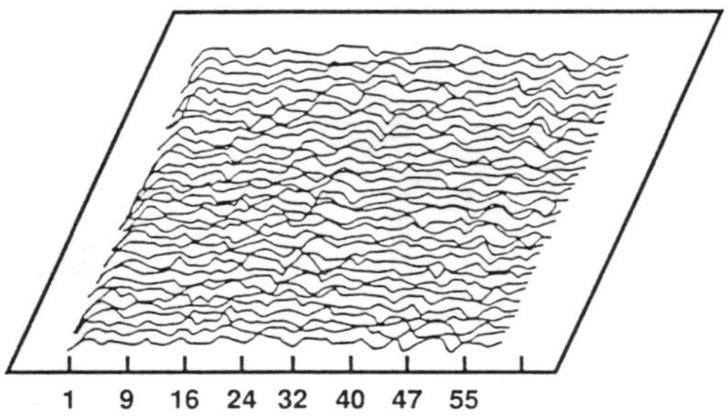

Fig. 627. Successive CD basis weight profiles from a high-speed digital on-machine gauge [22].

- Off-machine vs on-machine.
- Vertical scale exaggerated in Fig. 626.
- Time interval between profiles.
- Response of the gauges; the gauge used in Fig. 623 responds to much shorter wavelength variation.

Figure 627 shows data from a high-speed traversing digital gauge [22]. Each data point represents an area of paper 63.5 mm CD by 9000 mm MD. Each data point is really a data zone or box, each contiguous with the next. Thus, each CD profile would be more accurate if it was shown as a bar chart, but this would be to the detriment of the "3-D" effect. Profiles from all digital gauge systems are contiguous data boxes, even when the display appears to be a continuous analog profile.

It should be noted that modern taper flow headers have confused the concepts of pure CD and MD. Imagine a sudden change in consistency at the fan pump. This could normally be considered to cause a pure MD variation in basis weight. However, when it enters a taper flow header, it hits one side of the machine first (the entry side) and then it works its way across the machine. For example, on a machine operating at 600 m/min (10 m/s), if the flow in the taper header is 3 m/s, the CD/MD diagonal is 3/10. Figure 628 is a diagrammatic illustration of how MD pulses in the taper flow header become diagonal pulses on the wire.

(d) MD profiles

When a beta-gauge is placed on-machine in a stationary position over the sheet an MD single-point profile is obtained. The record contains all MD variation and some random variation, but no CD variation. The random component can be reduced or eliminated by averaging simultaneous MD single-point profiles taken at different CD positions.

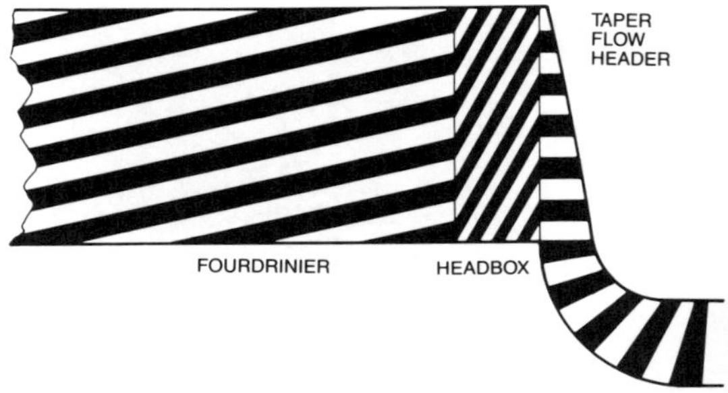

Fig. 628. The mechanism by which a taper flow header can change MD pulses into diagonal basis weight streaks [21].

Table 63. Distribution of variance from single point measurements of opacity (basis weight) on a high-speed news machine [21].

Frequency (Hz)	Period (s)	% of Total Variance*
25–250	0.004–0.04	15
2.5–25	0.04 –0.4	40
0.25–2.5	0.4 –4	10
0.006–0.25	4–167	20
0.0016–0.006	167–600	15

* Average of 8 separate records.
NOTES: Formation is frequencies above 100 Hz; short-term variation is 1–100 Hz; medium-term variation is 1–200 s; long-term variation is longer than 200 s.

A double concept that gives many people trouble is the broad frequency distribution of MD basis weight variation and how a response limited beta-gauge is blind to much of the variation.

Let's take the first point first. Table 63 lists, for five frequency ranges, the variance distribution of a single-point MD basis weight record obtained using an optical gauge, which has a response to over 1000 Hz, and was analyzed using Fourier analysis. The paper machine was a high-speed news machine that exhibits lower basis weight variation than most machines.

Each of the five frequency ranges contains a significant amount of variance; however, the 2.5 to 25 Hz decade contains double the others. The top frequency of 250 Hz falls into the range of formation, being equal to 56 mm (2.2 in.). The 100 mm boundary line between formation and basis weight variation is equal to 134 Hz in this case. However, there was no significant variance above 90 Hz. Figure 629 shows the power spectral density for the top four frequency ranges.

Beta-gauges have a much slower response than optical gauges. It has been found that portable

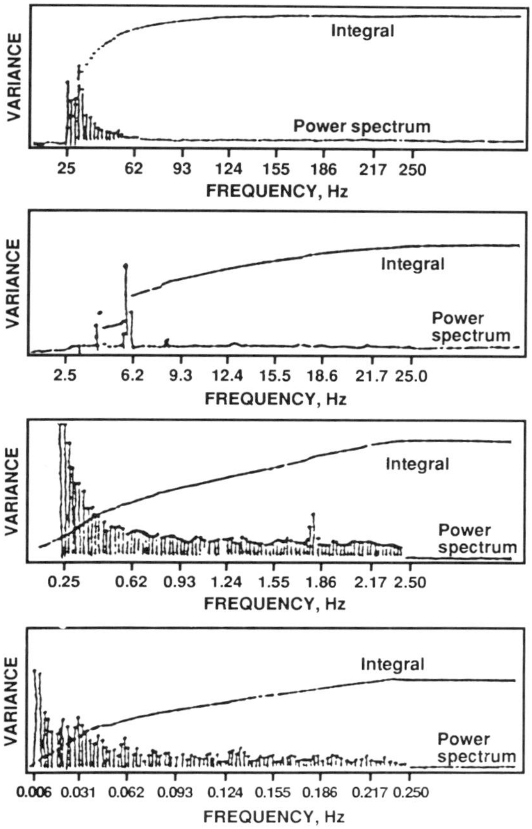

Fig. 629. Power spectra of basis weight variation over four frequency ranges [21].

beta-gauges are useful up to about 1 Hz. Internal noise becomes excessive above 1 Hz. On-machine beta-gauges can have a slower response. In one case, the response was found to start falling off at 0.1 Hz.

When a gauge is response limited, it is blind to variations at higher frequencies than the response limit. For example, the human eye has a response limit of 30–40 Hz. That is why we cannot see a fluorescent light flashing at 60 Hz.

3. Cause and effect relationships

Figure 630 is a diagrammatic illustration of the interactions between CD profiles of weight, moisture and caliper. The table at the top shows which CD faults are present in each sketch below. Sketch A is the most unusual, no faults at all.

Sketch B shows the situation where there is a fault that causes low weight at the quarter points. There are no faults in the dryers, so the moisture profile mimics the weight profile. The caliper

Sketch	A	B	C	D	E	F	G	H
Weight (W) fault	—	×	—	—	×	×	—	×
Moisture (M) fault	—	—	×	—	×	—	×	×
Caliper (C) fault	—	—	—	×	—	×	×	×

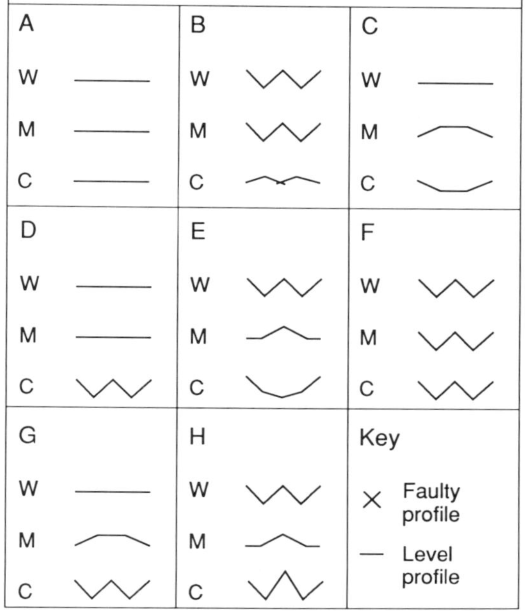

Fig. 630. Diagrammatic illustration of interactions between weight, moisture and caliper CD profiles.

profile will be a function of crown control, caliper control and sheet cooling (time lag between dryers and stack). The caliper profile will show a slight inversion of the other profiles if there is no crown or caliper control, and if there is sufficient sheet cooling to give a mild temperature profile. If the stack were closer to the dryers, such that there is a strong temperature profile (dry areas hot; moist areas cool), then the caliper profile could be opposite to that shown in Sketch B.

Sketch C is for the situation where only a drying problem exists, causing dry edges. This creates a smile in the caliper profile.

Sketch D is for the situation where there is a crown problem in the presses and/or breaker stack, causing low caliper at the quarter points.

Sketch E is where there is both the weight problem and drying problem, which together give a severe smile in the caliper profile. Sketches F, G and H are for other combined problems.

It would be impossible to illustrate all possible combinations of faults and their interactions, because all exist to varying intensities at one time or another, at one location or another. It is

Table 64. Papermaking variation interactions.

Area	Interaction
FORMER	— Heavier areas drain slower — Oriented areas drain slower
PRESSES	— Heavier areas dewater more — Oriented areas dewater much more — Higher load moves more water — Improper crown = uneven load — Plugged/worn felts make wetter sheet
DRYERS	— Exit moisture follows weight — High humidity gives poor evaporation — Thick condensate gives high moisture
SIZE PRESS	— Moist areas absorb more size (size press amplifies moisture variation)
EXITING DRYERS	— Higher average moisture gives more moisture variation — Moist areas cool faster
CALENDERS	— Thick areas carry more load — Hot areas cause more load — Hot areas compress easier — Moist areas compress easier
COATERS	— Rough areas get more coating

necessary to fall back on basic papermaking and physical principles. This is neither the time nor place to review these in detail. The best sources in the references are [2, 8, 13, 14, 19, 23–30]. The last is the most complete.

Not to leave you without any immediate information, Table 64 lists the main general interactions. When reading Table 64, remember that all generalities are invalid.

4. Actuator spacing

Actuator spacing has steadily become more narrow as the need for finer control is being realized. The physical limit for actuator width seems to be about 40 mm (1.5 in.) for CD weight, 15–75 mm (0.6–3 in.) for moisture depending on actuator type, and 75 mm (3 in.) for caliper. However, in most situations, the practical limit seems to be about 75 mm.

The problem behind all of this is that it takes more than two zones to resolve a streak. This is true for both the measurement of a streak and the control of a streak, as shown diagrammatically in Fig. 631.

In the old days of the early Eighties, on-machine measurement computers usually showed the profile in data boxes equal to the slice adjuster spacing (normally 150 mm), or a total of 60 data boxes. Thus, narrower streaks could not be detected on line. As computing power became less expensive, the on-machine systems could be upgraded to narrower data boxes, and better

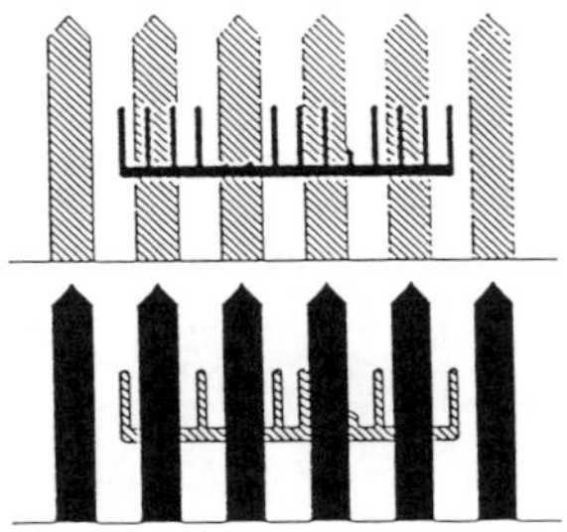

Fig. 631. It takes more than two measurement data zones to see a variation. You cannot count the tines on a rake through a picket fence, but you can count the boards through the rake.

ACTUATOR MATCHING ERROR

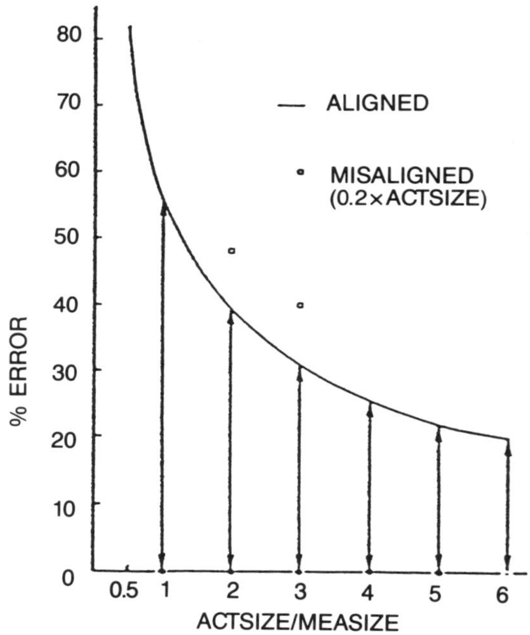

Fig. 632. Mismatching of CD actuator size (actsize) and CD profile data box size (measize) causes considerable control signal error [31].

profile resolution. It was discovered that not only were there narrower streaks, the sheet was just infested with them. So, better measurement resolution has created the need for better control resolution.

In addition to the problem of resolving streaks, there is also the problem of machine actuator size (actsize) versus measurement data box size (measize). Brewster [31] was the first to go public with this thought.

As illustrated in Fig. 632, only when actsize is a multiple of measize *and* there is good alignment

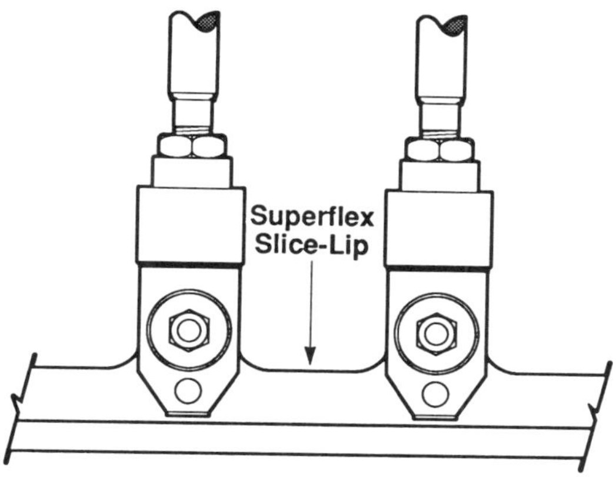

Fig. 633. Scalloping out the slice between attachment points gives less mechanical coupling between spindles and allows a greater control range without endangering the slice by overbending [32].

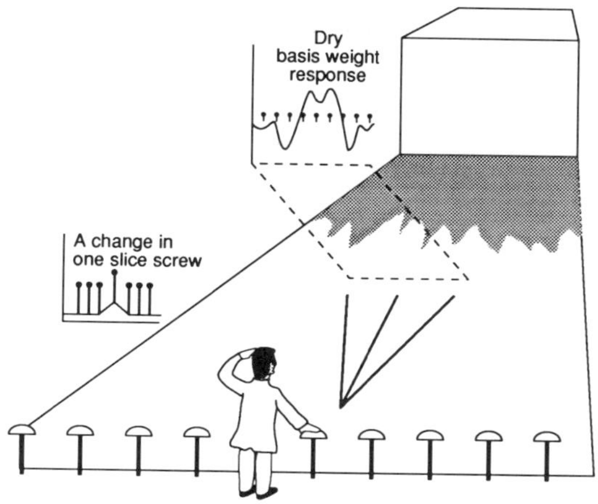

Fig. 634. The effects of water waves on the wire section on many fourdrinier machines are often so dramatic that the machine tender has a very difficult task to make the best possible adjustments of the slice-screws [28, 33].

is the error low. With natural events like sheet shrinkage and sheet wander, constant measize/actsize and perfect alignment is a practical impossibility. The preferred alternative is to provide a small measize so that there are 4 or more data zones per control zone, and to incorporate an algorithm which constantly checks which data box belongs to which actuator zone. It should be noted that alignment problems decrease with decreasing distance between the actuator and the gauge.

The way to think of a CD control system is not to think of a CD control system, but rather to think of a bunch of MD control systems sitting side by side. The successful operation of each MD control loop depends on all the usual things like time lag, signal error, etc. The success of the CD control system depends on the success of all of the MD loops.

The following sections discuss actuators for CD control. Principles of operation, design and application are stressed. Hardware details are generally not covered because these are rapidly changing as competition forces actuator improvements. Supplier contact is recommended for the latest in hardware details.

D. CD BASIS WEIGHT CONTROL

CD weight is controlled via the flow of stock onto the wire. The flow of stock is controlled by adjusting the slice (orifice) opening, normally by flexing the upper lip while the lower lip remains fixed. This requires the rather shaky assumptions that dynamic pressure before the slice opening is uniform, and that the stock stays where it lands on the forming fabric.

Overall, CD weight control is the most difficult of the CD control systems because the time lags are longest, alignment most difficult, and coupling between zones most complicated. Coupling is where action taken in one actuator zone will cause a reaction in other actuator zones. In the case of CD weight control, coupling has two primary causes.

The first is due to the stiffness of the slice lip. Movement at one spindle will cause similar movement at adjacent spindles as well as the opposite movement between adjacent spindles. This is dealt with in three ways:

- Taking appropriate counteraction at adjacent spindles.
- Measuring lip position at each spindle and controlling this position.
- Making the lips relatively more flexible than the spindles (Fig. 633).

The second reason for cross-coupling is wave action on the wire, as shown in Fig. 634 The wave action becomes more complicated as weight increases (Fig. 635), and changes pattern with speed.

1. Slice lip adjustment

Table 65 lists the mechanical parts to the slice adjustment system, and the requirements for these parts.

(a) Mounting beam

The mounting beam to which the slice lip adjusters are attached must above all else be stiff, strong and stable relative to the slice lip.

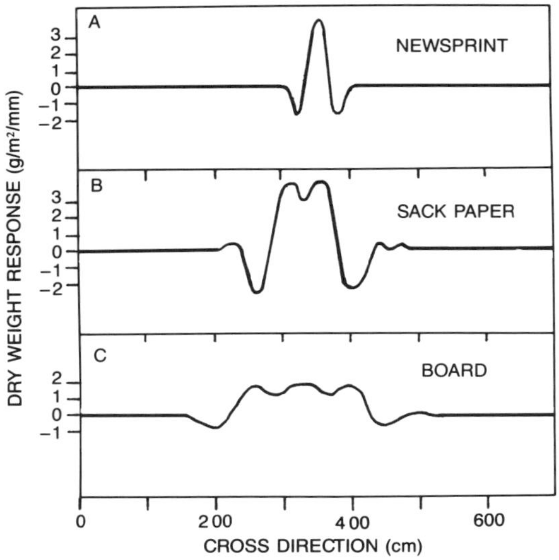

Fig. 635(a). Unit change in the slice-screw affects the dry basis weight of the paper web 75 cm across the web on a fourdrinier newsprint machine. (b) A similar change in a sack paper machine leads to a response with a width of 1.5 m. (c) In extreme cases on a board machine, a change in one slice-screw can have an effect over half the total width of the machine [28, 33].

Table 65. Requirements of mechanical parts for slice lip adjustment.

Part	Required characteristics
MOUNTING BEAM	Rigid
APRON	Straight, level
SLICE LIP	Flexible ZD
	Stiff in MD
LVDT*	Accurate
LINKAGE	Zero backlash
ACTUATOR	Full range
	Fast
	Dependable
SAFETY	Fool proof
	Idiot proof
	Papermaker proof

* Z-direction position measurement.

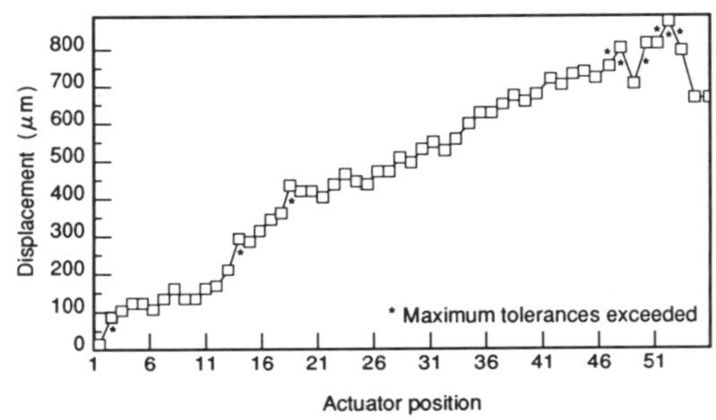

Fig. 636. Results of bottom apron audit [32].

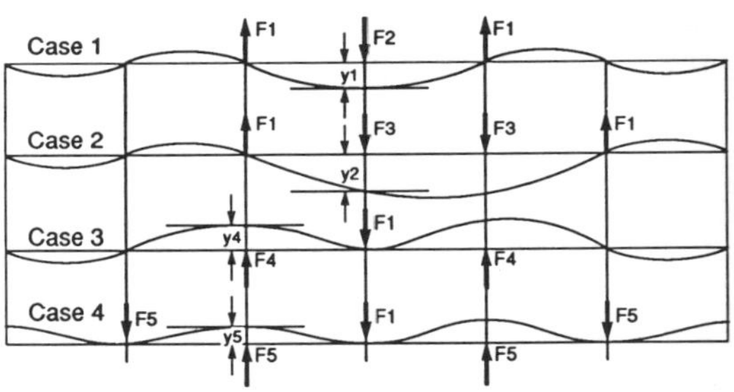

Typical limits for 150 mm (6 inch) spacing

y1 = 335 μm (13.2 mils) y2 = 716 μm (28.2 mils)
y4 = 259 μm (10.2 mils) y5 = 244 μm (9.6 mils)

Fig. 637. Bend configurations for slice lips are illustrated by the four cases. Bend limits are the maximum vertical movement permitted before the slice lip is endangered by permanent kinking [32].

Surprisingly, this is not always the case. Several mounting beams were found to be less stiff than the slice lip, resulting in more movement at the beam than at the slice lip.

(b) Apron

The apron should be straight and level. Maximum recommended tolerances are:

Peak-peak variation = Deckle width/10000

Adjuster-adjuster variation = 0.05%
(75 μm/150mm) (0.0005/in.).

Figure 636 shows an example of a 8.3 m (330 in.) apron that exceeds both tolerances. Usually such aprons can be levelled, and sometimes they can be straightened. If not, replacement must be considered.

(c) Slice lip

The slice lip should be stiff in the machine direction to resist pressure pulsations but sufficiently flexible in the vertical direction so that it can take the shape required for good weight control.

Figure 637 illustrates four possible modes of slice lip bending. The mode of bending determines how much an actuator can move the slice lip before irreversible bending occurs, i.e., the slice is kinked. The harshest bending mode is Case 4 where typical bend limits are less than 250 μm (10 mils) over 150 mm (6 in.). Note that such a slice lip would not be able to conform to the apron shown in Fig. 636 at actuator positions

48, 49 and 54, *assuming* the apron level changes *smoothly* between the actuator positions.

A new slice lip design is shown in Fig. 633. Scalloping the slice between attachment points can increase vertical flexibility and bend limits by two to three times, as compared to conventional designs. This allows a much greater control range without endangering the slice lip. Some mills use slices made of titanium to improve flexibility.

(d) LVDT

The linear variable differential transformer (LVDT), frequently called a linear voltage displacement transducer, is a device that tells us where the slice lip is, not in relationship to the apron (which would be nice, but is impractical), but in relation to where it used to be as well as in relation to each other at each point of attachment. This probably needs clarification.

One starts with a lip that is dead level, dead straight, and mid-range, and then zeroes the LVDTs. This allows the system to tell the lip where to go, to know when it gets there, and to keep it there in spite of what its neighbors are doing. It also allows the system to keep track of the relative differences between positions so as to safeguard against overbending and to model slice lip forces and stresses. Another significant advantage of the LVDT is that it permits the use of preprogrammed lip profiles for fast grade changes.

The accuracy of the positioning depends on the accuracy of the LVDT signal, which in the absence of any electrical noise in the surrounding environment is 2.5 μm (~0.0001 in.). Careful design of the wireways to provide maximum shielding can give an accuracy of 13 μm (~0.0005 in.), giving an acceptable gap error for most applications.

(e) Linkage

The slice to actuator linkage has the simple purpose of making a connection between the actuator and the slice with zero backlash. (If the backlash is small enough so that it cannot be detected by the system, even after years of use, then that is near enough to zero for all practical purposes).

So the connecting rods must be rugged and the couplings tight. The usual couplings on a slice lip seem to be designed to slip after a certain force is applied. This is rumored to be a method of protecting the slice from overbending when operators get overly eager in their cranking. If true, then it is not a very successful design because operators are still kinking the lip. Whatever the cause, actuator connection to the slice is usually poor, and these couplings or clamps normally have to be modified or replaced. The slice lip coupling must also be firmly attached to the slice lip, and not somewhere nearby in the neighborhood. A prime example of the wrong way to attach to the lip is the traditional Valley headbox arrangement shown in Fig. 638A. The better arrangement is shown in Fig. 638B.

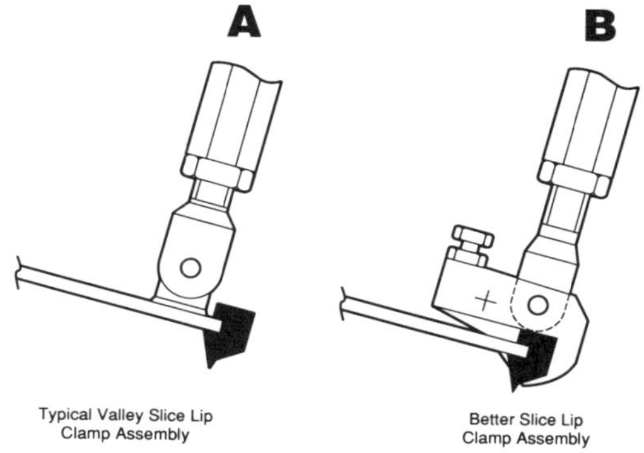

Fig. 638. Traditional method of attaching slice adjuster to slice lip on Valley headboxes (A) permits serious backlash. Better method (B) is to attach directly to slice lip [32].

A nice side benefit of the zero backlash linkage is that it allows placement of the LVDT near the actuator, where it is more easily protected from environmental accidents.

(f) Actuators

The first successful mechanized driver was an electric stepping motor/gear system [4]. The early problem with the electric motors was they did not like the environment. The problem with the gears was wear causing backlash. A series of design improvements in both the electric motors and gear systems have kept such systems alive.

The second successful mechanized driver was the heated rod, using thermal expansion principles on a long metal rod. The primary advantages of the system are freedom from internal backlash, ability to withstand the environment (except on messy twin-wire machines), and a high potential accuracy due to the slow movement. The disadvantages are that long rods do not fit everywhere, the short stroke limits application to light weights, and power failures endanger the slice.

The thermal hydraulic system, where temperature-controlled liquid expansion drives a metal bellows, overcame many thermal rod disadvantages while maintaining the potential for extreme accuracy. The two thermal systems are "active" actuators. That is, control power must always be

Table 66. Response characteristics for various type of automated slice lip drivers [32].

Characteristic	Thermal Rod	Thermal Hydraulic	Air/Electric Harmonic Gear	Electric Motor/Gear
Stroke				
mm	1/m(1)	9	13+	13
in	0.01/ft(1)	0.375	0.5+	0.5
Stroke Rate				
mm/min	0.10/m(1)	0.20	0.8–1.9	2.3
in/min	0.004/ft(1)	0.008	0.03–0.075	0.09
Force Available				
kg	(2)	900	2300	2000
lbs		2000	5000	4400
Effective Accuracy				
μm	13	13	13	64
inches	0.0005	0.0005	0.0005	0.0025

1. Calculated per meter (or ft.) of rod length over operating range of 93°C (167°F).
2. Not possible to calculate due to sensitivity of Euler's buckling constant to alignment. Potential forces well above 1 tonne.

applied to keep them at position or to move them to a new position.

Motorized actuators are "passive" actuators. They receive energy input only when they are moving to their set point. Once at the set point, they passively wait until they receive another command.

In 1986, improvements to motorized actuators were introduced, using either air or electric motors working through harmonic gear reducers. Harmonic gears are the mainstay of the robotics world. They have the advantages of small size, high gear reduction, minimal wear and zero backlash. These new actuators give good accuracy and high response rate, a significant advantage during start-ups and grade changes. The electric motor is hermetically sealed and can operate in wet environments such as twin-wire machines.

Table 66 compares the response characteristics for the various actuator types. The thermal rod values are calculated. Other values are for "standard" supply. Different characteristics can be custom designed into actuators where compatible with space constraints, mill preferences and the laws of physics.

The major point to make in this section is that a variety of actuators are available, and each has its advantages and disadvantages. There is no perfect actuator for all applications. Slice lip automation must be regarded as a system for a particular application, and actuator selection is just one of many choices to be made.

(g) Safety

The system must be fail-safe.

Primarily, the system must be workable whenever the paper machine is running. So if the computer malfunctions, the operator can run the system from a remote operator terminal. If a fork truck takes out the operator terminal, then the operator goes up on the headbox just like in the good old days.

The system must also accommodate computer errors, which means that the system will not self-destruct or damage the lip no matter what the computer tells it to do.

Lastly, the system must be impervious to overly eager gear crankers, fire hoses, acid soaks, and caustic attack.

(h) General arrangements

There are basically two general arrangements for slice actuators. Figure 639 illustrates these. The in-line arrangement is where all the parts are located between the two spindle attachment points. The beam-mounted arrangement is where the "actuator" parts are located above the spindle on a beam, and has the advantage that most of the critical parts can be covered by a canopy. In Fig. 639, the LVDT location on the beam-mounted arrangement is not good. It should be up in the canopy and connected to the spindle below the coupling with a rod.

E. CD MOISTURE CONTROL

From the control view point, CD moisture control is simpler than CD weight control because the time lags are normally shorter and there is usually less cross-coupling between actuators. From the papermaking viewpoint,

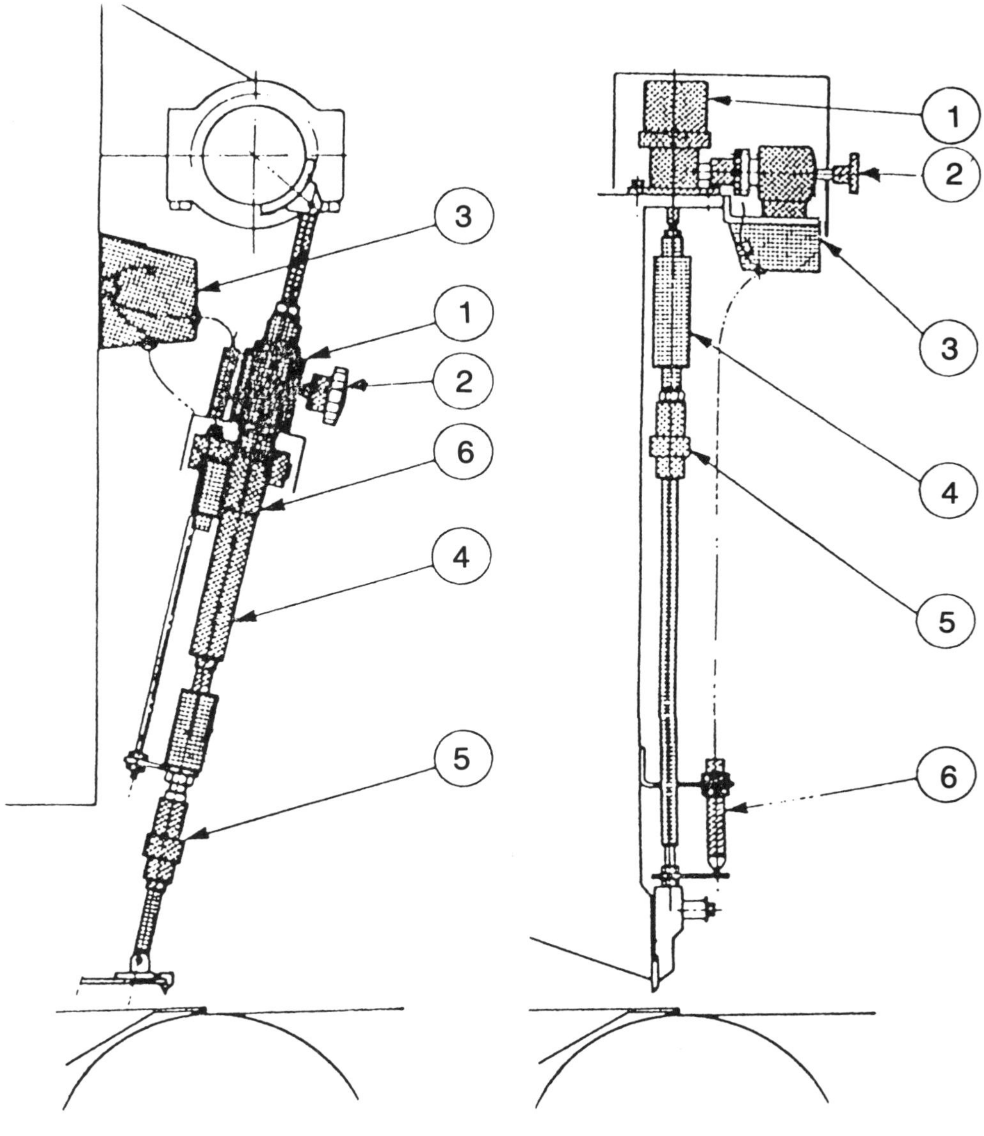

Fig. 639. In-line and beam-mounted arrangements for slice lip actuators. 1. Motor/gear; 2. hand adjust; 3. wire way; 4. force limiter; 5. coupling; 6. LVDT.

however, CD moisture control involves selection from many alternatives.

Firstly, there are different actuator options, each with characteristic pros and cons. Secondly, there are different possible locations for each actuator type, each with pros and cons. Thirdly, depending on actuator type, location and design, there can be significant interference with machine operation and operator comfort.

1. Profile correction

There are several options available for affecting the general shape of the moisture profile: Profile bars, pocket ventilation and crown control rolls.

Profiler bars are devices placed inside the dryer cans to disrupt the condensate film [34–36]. They are effective only where condensate rims the dryer can, that is, at speeds above

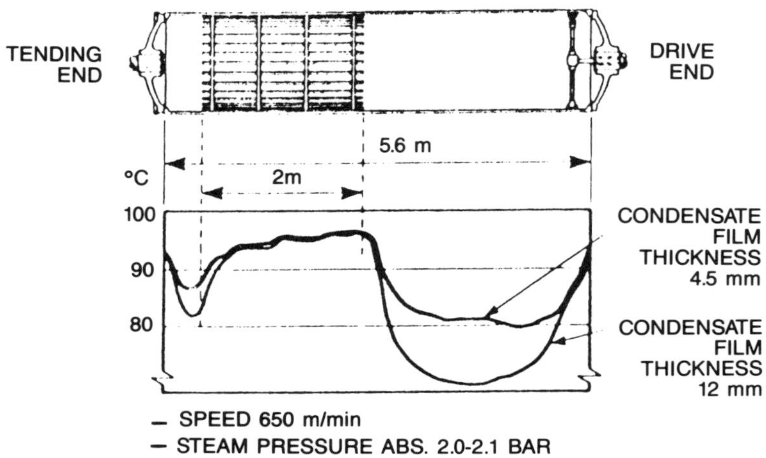

Fig. 640. Effect of profile bars and condensate film thickness on dryer can surface temperature [35].

500 m/min. By disrupting the condensate film, there is better heat transfer to the dryer can inside surface, and this leads to higher outside surface temperatures, as shown in Fig. 640. The number of cans to be fitted depends on the severity of the profile shape to be corrected. The main advantages of profile bars are the low capital cost and the increase in drying capacity. The disadvantages are that they are applicable only to fairly high speed operation and are "permanent" moisture profile correction.

Pocket ventilation compensates for design faults in dryer ventilation by blowing hot dry air into the dryer pockets [37–39]. The air flow to the various CD regions is operator controlled by dampers. The main advantage of pocket ventilation is the increased drying capacity. The main disadvantages are high capital cost and low corrective capability.

Crown-controlled rolls can be used in the press section to alter the moisture profile entering, and thus leaving the dryers [40, 41]. The advantage of the method is that the nip profile can be adjusted with changes in speed and weight to give a flat moisture and sheet density profile leaving the presses. The disadvantage of the concept is its high capital cost.

2. Profile control (grades other than tissue and towel)

Options for *controlling*, i.e. streak shooting, the moisture profile are infrared heaters, water sprays and steam showers. These three options are compared in Table 67.

(a) Streak control

Streak control is the amount that a CD moisture control actuator can lower a wet streak or raise a dry streak. If an infrared heater or steam shower can reduce a wet streak from 8% to 6% moisture, when turned from off to 100% on, then it has a streak control capability of 2%.

Except for hydraulic atomized water sprays, the streak control concept is important for all moisture control actuators because there is a finite amount of water they can remove from, or add to the sheet. In contrast, pressure atomized water sprays have nearly an unlimited capability to add water to the sheet by adding larger flow rate nozzles. The practical upper limit is breaking the sheet, but rarely is so much needed.

The streak control capability of fixed output actuators depends on the average moisture level, actuator type, actuator location, streak % moisture, production rate and grade.

The actuator capability should be matched to future demands. For example, it is common practice to deform the dry weight profile to help compensate for a bad moisture profile. When the moisture actuators are sized they should have sufficient capacity to provide a flat profile at the new higher moisture average, and with a flat weight profile, and at the higher machine speeds.

Moisture profile variation increases as average reel moisture increases. The amount of variation increase can be estimated from Fig. 620.

An average moisture change (exiting the main dryers) from 9 to 7% is equal to a production loss of 0.8%. The same production loss is equal to a moisture reduction of 1.6 starting at 7, or 0.7 starting at 5, or 0.2 starting at 3. Following this logic, the "iso production loss" curves in Fig. 641 were constructed. It has been found that these curves give a reasonable approximation of both

Table 67. Comparing options for CD moisture control (except tissue).

Features	Infrared electric	Infrared gas	Water sprays	Steam showers
Capital cost	High	High	Moderate	Low
Control cost	Moderate	Moderate	Moderate	High
Energy type	Electricity	Gas	Dryer steam	Waste steam
Energy use	High	High	Low	Net saver
Efficiency (%)(1)	17–35	30–50	60–95	50–95
Streak control (see discussion)	Limited	Limited	Unlimited	Limited
MD resolution	Infinite	Infinite	15 steps	Infinite
CD resolution (mm)	20–150	150	100–300	75–300
Maintenance	Moderate	Low	Moderate	Low
Safety hazard	Mod.–high	Mod.–high	None	Low
Operating hassle (2)	Low-high	Low-high	Low-high	Low-high
Location	(3)	(3)	Best applied at draw; sheet moisture above 15%.	Presses or fourdrinier
Principle	Dries wet areas	Dries wet areas	Wets dry areas	Dries wet areas
Comments	Only option after coaters			Gives extra water removal and sheet consolidation in presses

1. How much of the "medium" issued is absorbed in the sheet.
2. The amount of hassle is very dependent on the actuator location and the quality of installation design.
3. Traditionally at dry end for uncoated sheets; after coaters or size press.

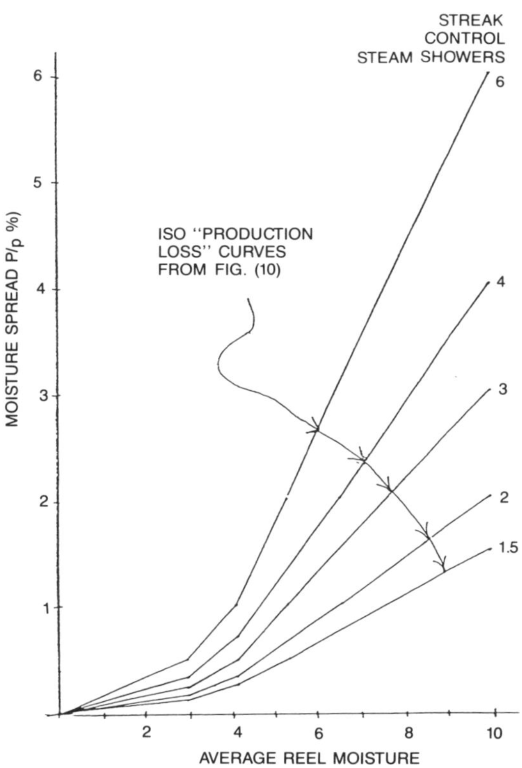

Fig. 641. Moisture profile spread increases as average reel moisture increases. The streak control capability varies in like manner.

the effect of average reel moisture on the profile spread and the streak control capability of steam showers.

If a profile of average 10% with a spread of 4% is reduced to an average of 5%, the spread will be about 1%. The reverse is also true. If a profile with average 3% and spread of 0.5% is raised to 10%, the new spread will be about 6%.

The streak control capability of steam showers has been found to behave in the same manner. A streak control capability of 3% starting at 10% becomes only 0.9% at 5%. This is because steam showers work by establishing a moisture differential leaving the presses, and the dryers attenuate the differential as the sheet passes through.

Infrared heaters with limited energy flux and located at the dry end show opposite behavior. The energy flux input is directly related to an evaporation rate so the streak control tends to be independent of the average reel moisture, decreasing only at moistures below 3%. Infrared heaters thus tend to give the best results in applications where the average moisture is low.

Air-atomized water sprays with a limited flow rate also have more streak control power at dry end locations and low average moistures. Hydraulic-atomized water sprays with "unlimited" flow rates, located in the middle of the dryers, are able to handle the higher average

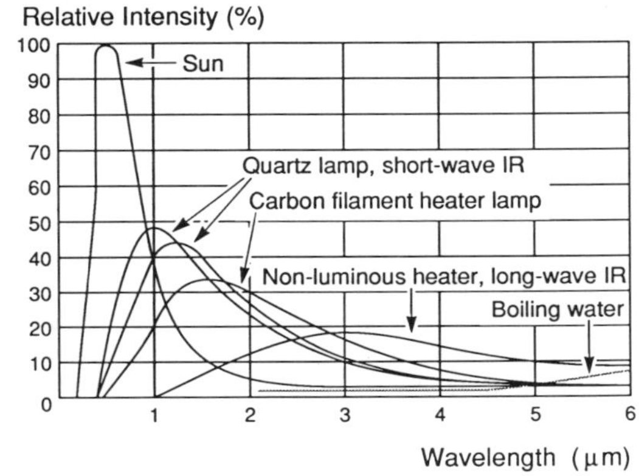

Fig. 642. Energy distribution of infrared sources [42].

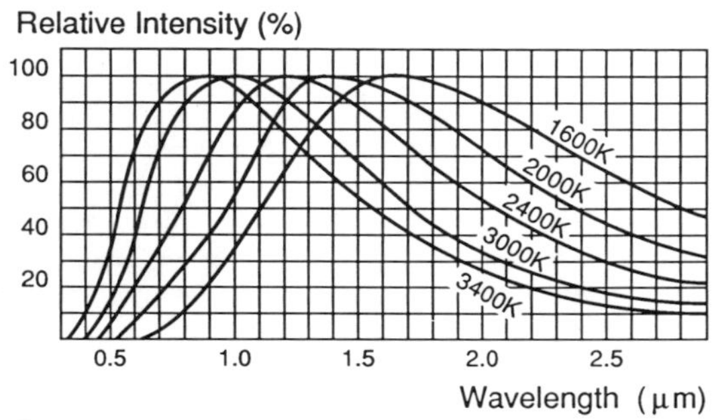

Fig. 644. The distribution of infrared radiation at various temperatures [42].

moisture situations.

Production rate is important to consider with actuators that have a limited flux or flow rate, because as production rate increases the streak control decreases. Steam showers do not follow this pattern, even though the flux rate is limited, because the flux rate is a ratio of the production rate, for example, 0.10 t of steam/ton of paper.

Basis weight affects steam shower capability depending on their location. Located on the fourdrinier, steam showers tend to work better as basis weight increases, whereas the opposite is true for press section steam showers. Infrared heaters work better on the heavier weights because more of the energy is absorbed (less passes through). Brown grades also help infrared absorption.

(b) Infrared heaters

The distinguishing features of electrical infrared heaters are:

- High capital cost.
- High energy cost.
- Dry end location.
- Only system to use after coaters.
- High hazard (fires).

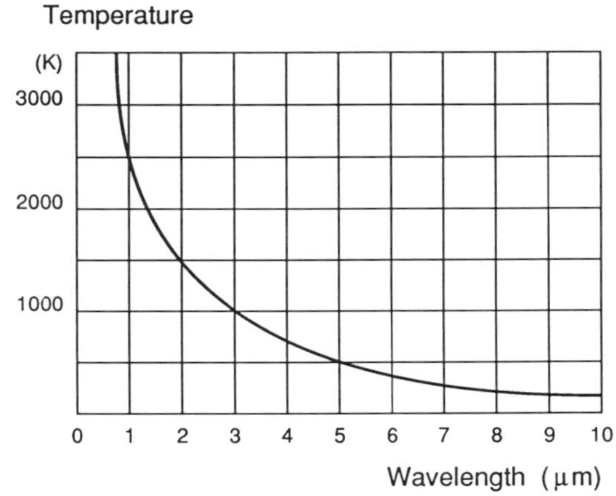

Fig. 643. Relationship of emitter temperature and radiation wavelength at maximum emission, according to Wien's Law of Displacement [42].

Infrared radiation is electromagnetic radiation between the wavelengths of 0.76 μm to 1 mm (0.00003–0.04 in.). The practical range of interest is 0.75–2.7 μm, called near infrared. Visible light has shorter wavelengths, 0.38–0.76 μm.

All sources of heat emit infrared radiation. As illustrated in Fig. 642, the hotter the source, the greater is the intensity of emitted radiation and the shorter is the wavelength. This behavior is described by Wien's law of displacement:

$$\text{Max} \cdot T = 2.8978 \cdot 10^{-3} \text{ (mK)}$$

where

Max = wavelength for maximum emission (m)

T = temperature in °K (°C + 273.16).

The results of Wien's law are illustrated in Fig. 643.

The point of all this is that, if an emitter with a high flux rate is needed so that it does not take up much room on the paper machine, then it has to operate at high temperatures and thus much of the radiation is of short wavelength.

Most of the six or so brands of electrical infrared heaters available for CD moisture control use halogen lamps with tungsten filaments, which operate at a peak 2400°K (2100°C). As shown in Fig. 644, at 2400°K the peak wavelength is 1.2 μm and the lion's share of the radiation falls below wavelengths of 2.0 μm.

Figure 645 teaches us that for radiation of wave length shorter than 2 μm less than 10% is absorbed by water and less than 50% is absorbed by paper. These values vary with grade and basis weight. Nevertheless, most of the radiation is

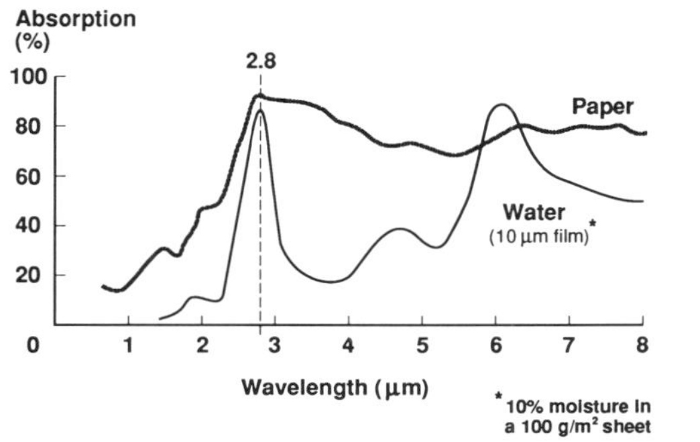

Fig. 645. Effect of infrared radiation wavelength on absorption into water and paper (from [43], who got it in part from [44], who got the water part from [45]. Source of paper curve is not known.)

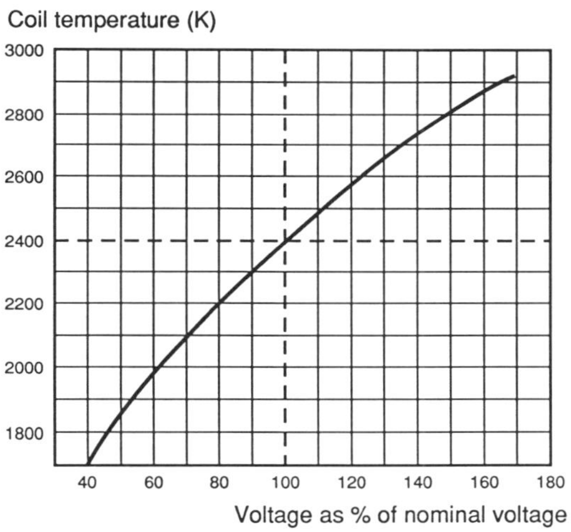

Fig. 647. Relationship between applied voltage and coil temperature [42].

reflected or transmitted. Therefore, much effort is made to trap the radiation on both sides of the sheet and send it back. Some designs use low mass ceramic reflector/emitters which absorb and re-emit the radiation. Other designs use metallic coatings, the best of which is gold, as shown in Fig. 646.

Infrared heaters do not operate at full power all the time, but at some intermediate level depending on the amount of water to be removed. Individual CD modules usually 150 mm (6 in.) wide are controlled individually by voltage. Figure 647 shows the effect of voltage on emitter temperature. For example, at 50%, the temperature is 1850°K, which according to Fig. 644 will shift the peak wavelength to 1.5 μm. According to Fig. 645, a wavelength shift from 1.2 to 1.5 μm will improve infrared absorption

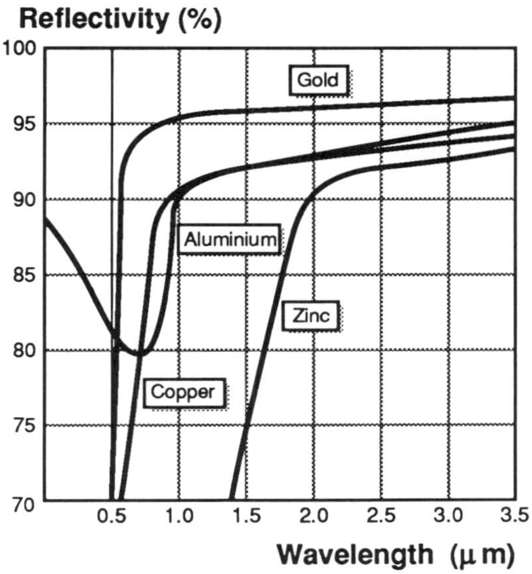

Fig. 646. Radiation characteristics for various reflecting materials [42].

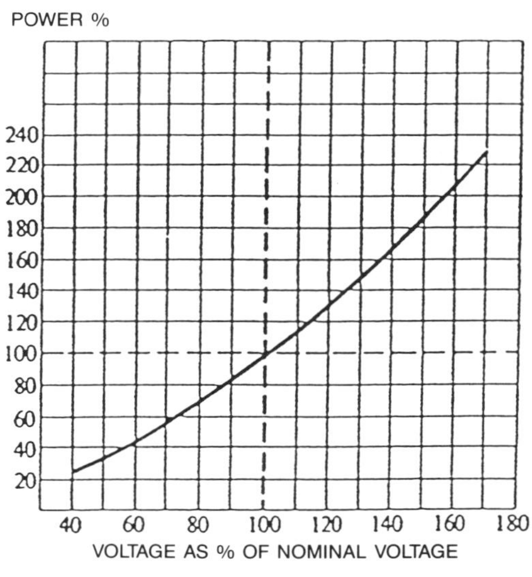

Fig. 648. Infrared lamp: power as a function of voltage [42].

into paper from 20% to 30%. While the radiation emitted at lower voltages is absorbed more efficiently, the power emitted decreases faster than the voltage, as shown in Fig. 648. At 50% voltage, the power is only 35% of nominal. The reduced power compensates for the increased absorption, the net result being a 52.5% first pass retention.

For the record, the water curve in Fig. 645 comes from reference [44], which originally came from reference [45], dated 1958. The percent

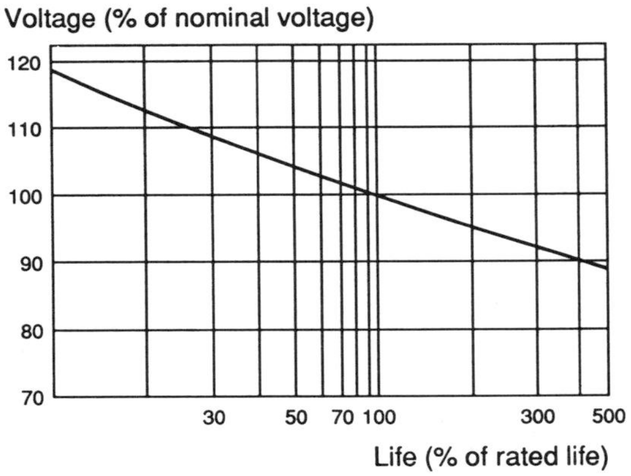

Fig. 649. Theoretical life of the coil as a function of voltage [42].

absorption (A) has been calculated from transmission (T) data assuming reflectance (R) is zero, and using the formula $1 = A + T + R$. The source of the paper curve has eluded efforts to trace it.

One of the inconveniences of electrical infrared heaters is that the lamps burn out. Figure 649 shows a dramatic improvement in coil life for a slight reduction in voltage. This means that average lamp life is much better than the rating of 5000 h at nominal (100%) voltage. It also means that lamps burn out fastest where you need them most.

Most systems have a maximum power output of 120 kW/m (40 kW/ft.). Depending on location, application, design, and who's talking, efficiencies may vary from 17–35%. Efficiency is how much power ends up in the sheet at the point of application, as compared to how much power is used. Of course, the "unused" power does not disappear, but goes to other areas besides the sheet at the point of application. Most of this unused power heats the air that keeps the system cool. This source of heated air can contribute to paper drying, thus improving the overall effectiveness.

Purveyors of electrical infrared heaters emphasize effectiveness rather than efficiency. The results reported in the literature [43, 46–50] and the rapid acceptance by the industry attests to the effectiveness, or profitability, of these systems.

Infrared heaters are most effective on the last heated dryer can because there is the least amount of mass to heat (lowest moisture level), the sheet is already at a high temperature so more of the energy works to evaporate water rather than heat the sheet, and it is at this point that the dryer can is least effective, using more energy to evaporate water than does the infrared heater.

Recent experience has shown that infrared heaters can also be effective at the wet end. At this location they preheat the sheet, thus taking preheat load off of the dryers allowing them to do drying. The wet end location is particularly effective when the hotter sheet allows the early dryer cans to be operated at higher temperatures. On lightweight grades, the heater should be placed opposite a reflector rather than over a dryer can. This is because too much energy goes through the sheet into the can. The external heat source to the can does not raise its temperature significantly, but does reduce steam condensing rates inside the can, thus reducing the condensing capability of the dryer system.

Infrared heaters are the only reasonable CD moisture control system for location after coaters. The features of non-contacting, high-energy flux rate and the deep penetration of radiation lead to rapid drying. The rapid drying reduces binder migration, a welcome relief to mottle problems.

A very distinguishing feature of infrared heaters is their capability to start fires. Early systems caused several disastrous fires. Soon the systems were treated with deserved respect, and alarm bells stopped ringing.

Redundant systems for sheet break detection and flame detection ensure that power is shut off when required. Low mass and high cooling air flow means the heaters cool very rapidly once power is cut off, reducing the fire hazard. Good housekeeping is also essential.

Figure 650 illustrates the design of an electrical infrared heater. Cooling air is supplied across the machine by a plenum. The tungsten emitters (halogen lamps) are the primary source of radiation. The thermal tiles both reflect and absorb to re-emit radiation. Other systems use highly reflective coatings to improve reflection. The dryer can is not a good reflector to send transmitted radiation back through the sheet; however, in many cases there is not room to install a proper reflector.

Recent improvements in gas infrared burners have made them viable alternatives for CD moisture control. The new gas burner systems are approximately the same size and power as the electrical units; the principle of operation is the same; and the results at the reel are nearly the same. The differences are engineering details and energy economics.

The most important difference in engineering detail is that the lower operating temperature produces a peak wave length of 1.9 μm. This means that absorbtion into the sheet can be up to 50% better, reducing the need for reflectors and

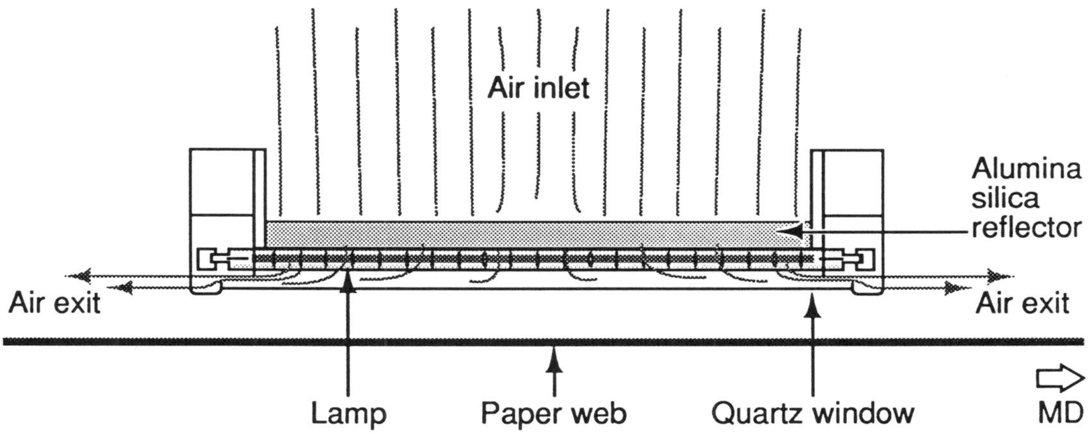

Fig. 650. Design of an electrical infrared heater.

energy input. In many regions gas is a more economical energy source, thus effectively reducing the operating cost of an infrared profiling system by 70% or more. In one mill the operating costs were calculated to be \$0.72/t for gas and \$2.25/t for electric, assuming that the systems operated at an average of 50% of maximum power level.

(c) Water sprays

The distinguishing feature of remoisturizing water sprays is that they increase moisture in the dry areas. Some mills are using moisture sprays in conjunction with infrared to save electrical energy. The infrared works on streaks above the average while the water sprays work on streaks below the average.

Water spray systems were first applied in the early Seventies. The acceptance was poor, partly because it was felt to be immoral to add water after spending money to dry it, but also because there was insufficient design attention paid to making the system reliable. Better engineering helped solve the reliability issue and better profits helped solve the morality issue.

The first successful systems used air pressure to atomize the water into fine droplets. Due to limitations in water flow rates these systems were primarily applied at the dry end. High capital cost and high operating cost slowed acceptance. It was also discovered that remoisturizing a sheet does not recover all of the damage caused by over drying. In subsequent converting operations, the remoistened areas behave differently from non-rewet areas, even though both are at the same moisture content.

Hydraulic atomized systems soon became popular due mainly to the lower cost. These systems are located in the dryers at some location where the sheet moisture is above 15%. The theory is to add water to the dry areas to prevent overdrying. Hydraulic atomized systems have the advantage that water flow rate is essentially unlimited. The larger water droplet does not seem to be causing problems in most applications. Spray pattern and droplet size are very sensitive to nozzle wear, so nozzles must be routinely replaced or be of a special material such as hardened stainless steel, ceramic or tungsten carbide.

Water sprays may or may not use energy. In many cases, the improved moisture profile allows an increase in the average moisture. The set point shift generally more than offsets the extra evaporative load caused by adding water.

Figure 651 shows a typical arrangement for remoisturizing water sprays. The operator or process control computer sends outputs to the actuator control unit which translates this into which nozzles should be operating. The nozzles have flow rates in the ratios of 1 : 2 : 4 : 8 to give 15 flow steps. When the water sprays are located under a dryer as in a Unorun felt arrangement, an elevator is needed to raise/lower the unit from the maintenance position to the operating position. Alternatively, a maintenance platform can be used. Ease of maintenance access is essential to system success.

(d) Steam showers

The distinguishing features of using steam showers for CD moisture control are:

- Net energy savings.
- Increased drying capacity.
- Improves fast and random variation.
- Increased sheet consolidation

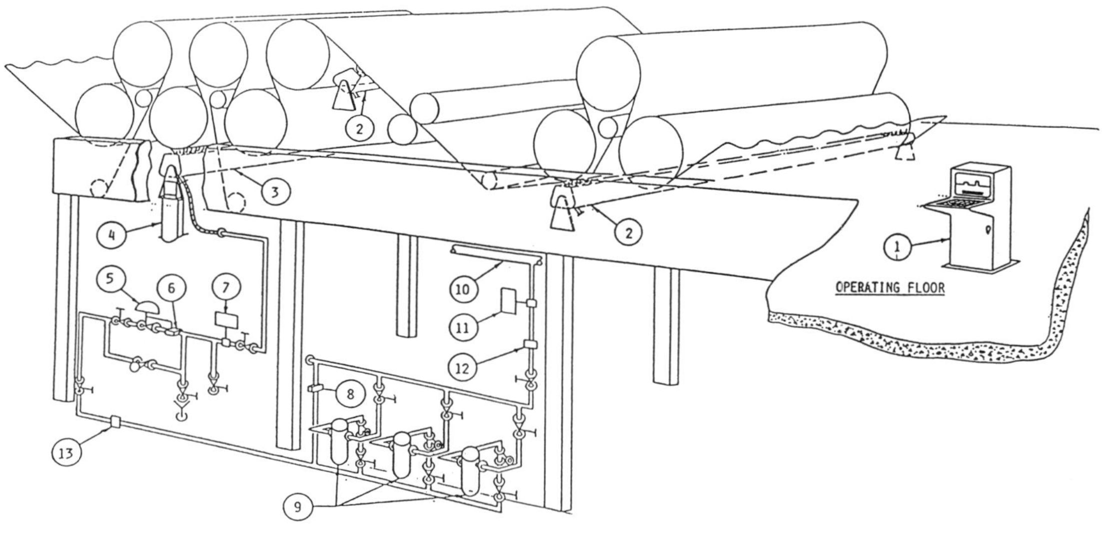

Fig. 651. Typical arrangement for remoisturizing water sprays. Only one of the three locations needed. The interface cabinet translates controller outputs into valve openings on each nozzle bank.

1. **Operator control station**
2. **Interface cabinet**
3. **Remoisturizing spray**
4. **Elevator**
5. **Pressure control valve**
6. **I/O transmitter**
7. **Low-pressure switch**
8. **Pressure differential switch**
9. **Filters (one spare)**
10. **Main water header.**
11. **Temperature sensor/switch**
12. **Solenoid valve for sheet break/high temperature**
13. **Pressure gauge**

Steam showers employ the principle of hot pressing which is improving water removal in the press section by increasing sheet temperature. A 10°C temperature increase will improve sheet dryness leaving the presses by at least one percentage point. Steam showers increase sheet temperature by condensing steam on and in the sheet. Controlling steam application in the CD controls the sheet temperature profile.

The sheet temperature profile controls the sheet moisture profile entering the dryers, which in turn controls the moisture profile leaving the dryers [51].

The improved water removal in the presses reduces dryer loads 5–10%, and thus more steam is saved in the dryers than is used in the steam shower, leading to net energy savings. Another benefit of hot pressing is that it reduces the amplitude of fast and random variation by about one third without any control action being taken [51].

There is a general relationship between sheet dryness leaving the presses and sheet density, an increase in dryness giving an increase in density. The method of obtaining the dryness increase, whether it be time, pressure or temperature, does not significantly alter this relationship. An exception to this is long nip presses. Another minor exception is location of the steam shower; located before the first press it gives much more consolidation than before the last press [51].

Due to the press dryness/sheet density relationship, there is a potential disadvantage to controlling the dry end moisture profile with a press section actuator. Distorting the CD moisture profile out of the presses to compensate for dryer system faults can create a CD density profile variation. However, in most cases the resulting CD density profile variation is less severe than that originally caused by moisture variation through the calender stack. The best papermaking practice is to correct the moisture profile variation wherever it occurs.

Figure 652 shows steam shower locations for light weight grades while Fig. 653 shows locations for heavy weight grades. Steam showers are not normally placed on the fourdrinier for lightweight grades due to excessive sheet cooling over the couch. The fourdrinier is the best and easiest location for heavier grades from the water removal/hot pressing viewpoint, but raises the consolidation aspect mentioned above. Press applications can be very effective, but the arrangement must be carefully designed to

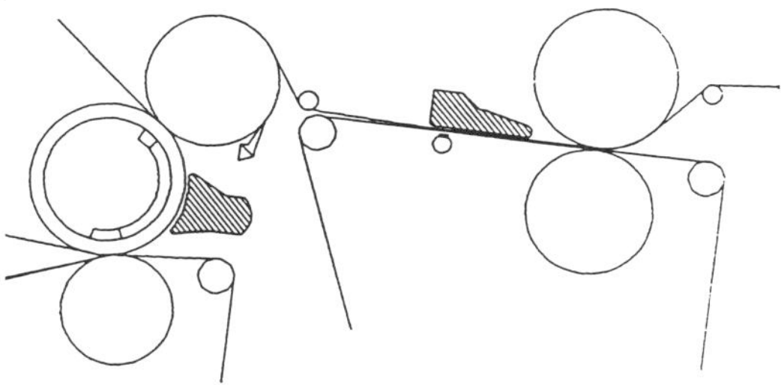

Fig. 652. Steam shower locations for light-weight grades [51].

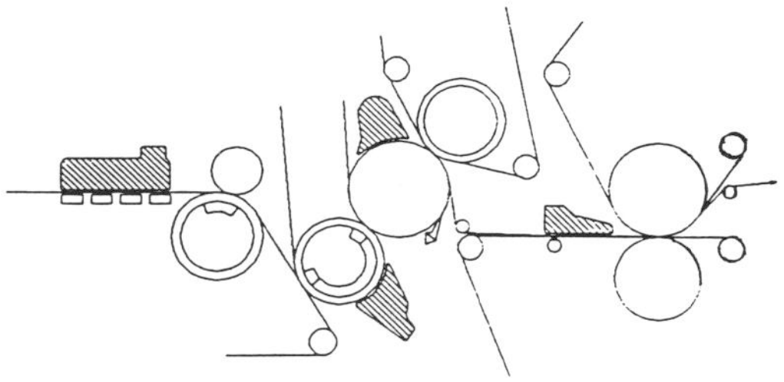

Fig. 653. Steam shower locations for heavy-weight grades [51].

maximize heat transfer while minimizing operating difficulties [51].

Increased heat in the press section can give rise to a number of operating problems. Most of these can be remedied by good engineering and operating practices. The most severe problem is that steam showers have been linked to failure of a certain design of granite roll. Details are reported by Pikulik [52].

3. Profile control (tissue and towel grades)

CD moisture control on Yankee tissue operations can yield dramatic improvement in operational efficiency and converting efficiency. CD moisture control can even improve the uniformity of the Yankee coating leading to fewer creping doctor changes, and therefore more tonnage.

Table 68 compares the major features for three control options:

- Induction heaters.
- Electrical infrared heating.
- Steam showers.

Induction heating of the Yankee has experienced limited acceptance, perhaps due to the high capital cost. The principle is to place variable electromagnets near the shell. Dryer rotation creates "eddy currents", which creates heat. Energy for the heat is supplied by the Yankee drive. Figures 654 and 655 explain the technique.

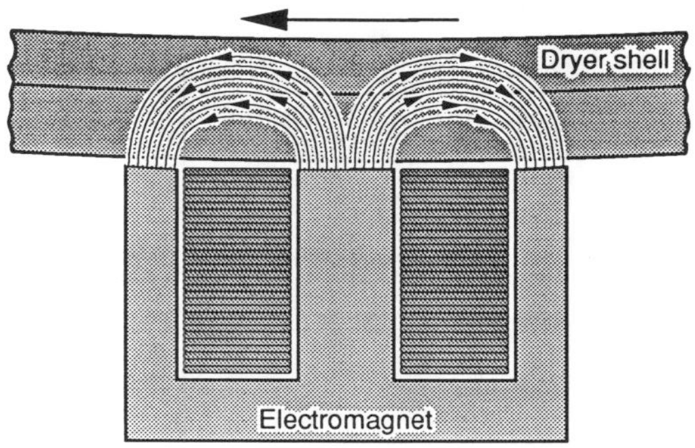

Fig. 654. Induction heating for moisture profiling on Yankee cylinders [53].

The infrared heater is applied between the pressure rolls on those machines with two rolls. The wet environment reduces the fire hazard.

Table 68. Comparing options for CD moisture control (Yankee tissue and towel).

Features	Induction heaters	Electrical[2] infrared heaters	Steam showers
Capital cost	High	High	Low
Control cost	Moderate	Moderate	High
Energy type	Electricity	Electricity	Waste Steam
Energy use	High	High	Low
Streak control (% H_2O)	1–2	1–2	2–3
MD resolution	Infinite	Infinite	Infinite
Maintenance	Low	High	None
Safety hazard	None	High	None
Operating hassle[1]	Low	Low-high	Low-high
Location	On Yankee	Between pressure rolls	At or before pressure roll
Principle	Dries wet areas	Dries wet areas	Dries wet areas
Comments	Energy via Yankee drive		Housekeeping needed

1. Depending on location and quality of installation design.
2. As of late 1989, there were no reports of gas infrared heaters installed on a tissue/towel machine.

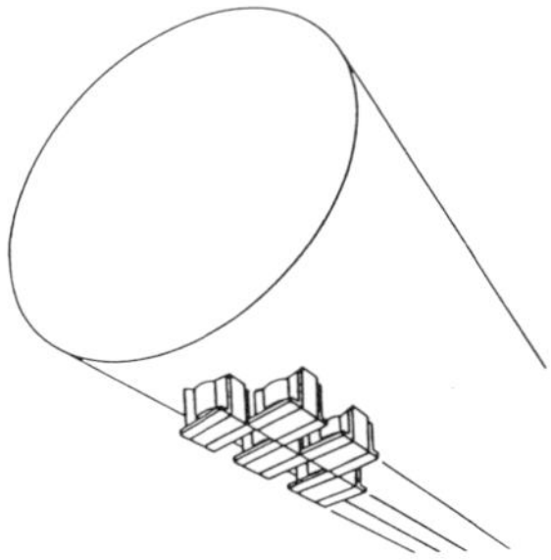

Fig. 655. Mounting configuration of inductors on Yankee cylinder [53].

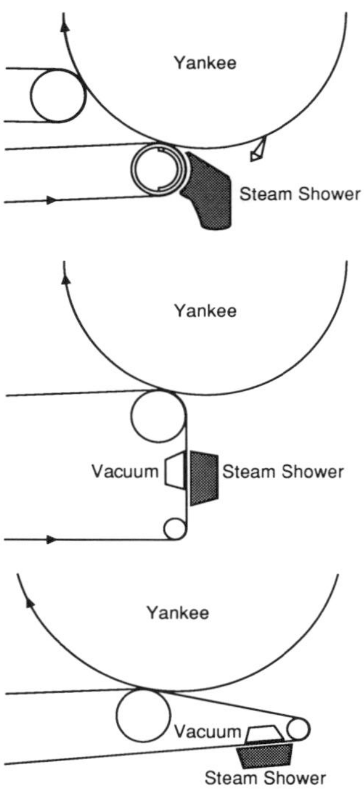

Fig. 656. Alternative arrangements for steam showers on Yankee tissue machines.

There has been no public feedback on the effect of the wet environment on emitter maintenance. Again, acceptance by industry has been limited.

With over 80% of the market share, steam showers seem to be the industry preference. Located at or before the pressure roll, steam showers give 5–15% increases in drying capacity plus the profile control. Figure 656 shows several alternate arrangements for steam showers on tissue machines.

On a per ton basis, steam use is rather high, so net steam savings rarely result. However, savings of the higher cost gas to the hood has given some economic benefits.

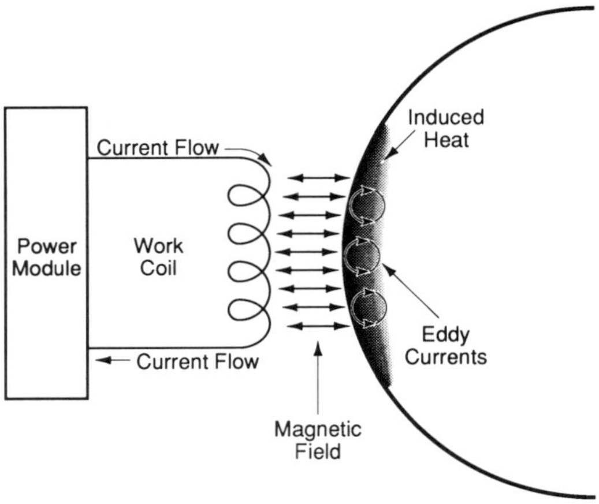

Fig. 657. AC induction — principle of operation. A high-frequency AC magnetic field passes through the roll surface and the resistance to the current induced causes roll heating [55].

The major problem has been housekeeping on "messy" machines. Where pressure rolls fling a lot of stock, the operating face of the steam shower can become covered, leading eventually to plugged holes. On machines where there are glue sprays and lots of dust, the combination of dust and glue can accumulate on top of the steam shower, sending the occasional lump through the nip. Minor changes in design and operating procedures, plus periodic cleaning has overcome the plugging and dust buildup problems.

F. CD CALIPER CONTROL

CD caliper control is done by controlling the CD temperature profile of one or more calender rolls. The roll temperature affects the roll diameter, and thus the nip pressure at any given point. Controlling the CD temperature, therefore controls the CD distribution of load. It does not affect total load.

Early CD caliper control was done with air nozzles, using hot and/or cold air. While these systems did improve the profiles, they did not always have enough power to provide complete control of the profile. Induction heating using high frequency AC supply generally overcame this limitation [54]. Profile improvements of over 90% were reported by mills.

Figure 657 illustrates the principle of AC induction heating. Each time the magnetic polarity changes, due to the AC direction change, the ferrous molecules in the roll surface material are stressed; this causes the heat buildup. A major advantage of induction heating is the energy efficiency, which is above 80%. Another advantage is that induction heating can

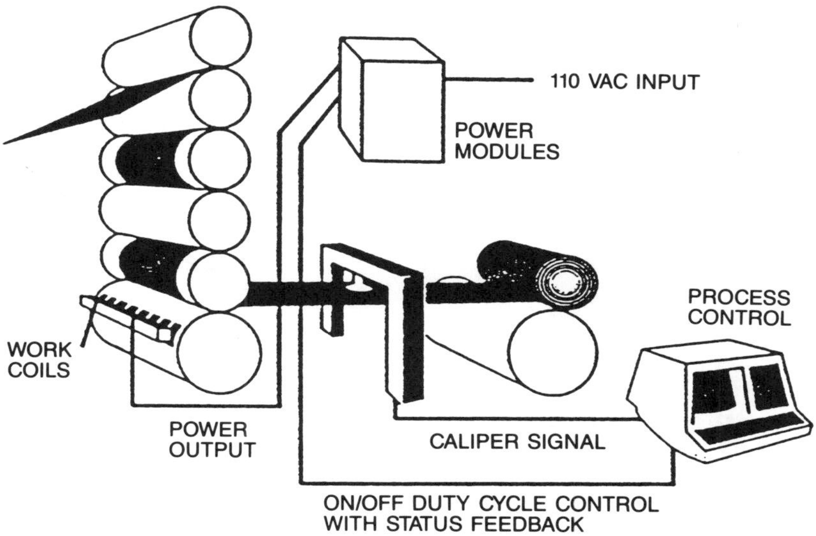

Fig. 658. Arrangement of components to an induction heating caliper control system [55].

be applied through the sheet, giving more options for actuator location. Figure 658 shows the parts and arrangement to a caliper control system.

Improved designs of air showers, called confined air showers, were soon able to match the results of the induction heating designs by keeping the hot air close to the roll surface and raising the air temperature [56]. Figure 659 illustrates the principles.

A fringe benefit from both systems is the trend to calendering at higher temperatures. Small but noticeable improvements to average smoothness have resulted. A recent entry to the caliper control field goes against this high temperature trend by using air cooled by water evaporation [58].

As the demand for better calendering forces calender roll temperatures to climb toward and beyond 200°C, the confined hot air showers will lose effectiveness faster than the induction heaters because induction heating does not rely on the temperature differential to transfer heat. At the same time, the cool air showers will gain effectiveness.

There are several factors to consider when choosing the location of a CD caliper control actuator:

- Lower rolls, where there is more load on the sheet, will give more effect for any given roll diameter change.
- Intermediate rolls, where the actuator affects two nips, give good results (a rare case of one stone hitting two birds).
- Thin-shelled rolls, e.g., crown controlled, react more for a given energy input.
- Locating units under the sheet (wet end side of stack) to avoid creating a spot where dust can settle and then fall on the sheet.

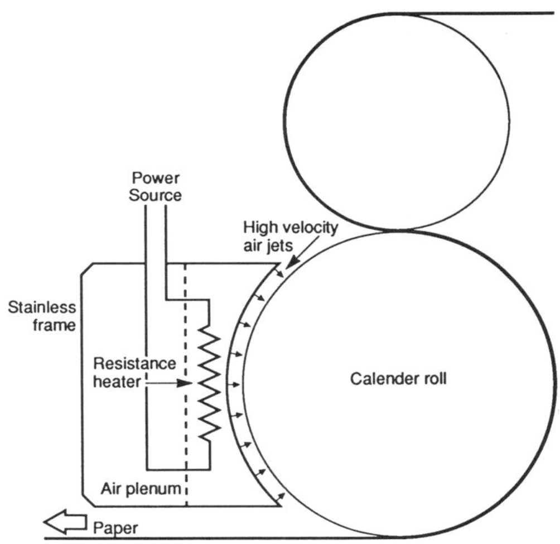

Fig. 659. Principle of high-temperature confined air caliper control system.

Thus the usual preferred location for hot air systems is the third roll or crown-controlled king, while for induction heating the preferred location is a crown-controlled queen in the second roll location.

G. CD COAT WEIGHT CONTROL

CD coat weight control is the same as CD caliper control, except it is applied to the

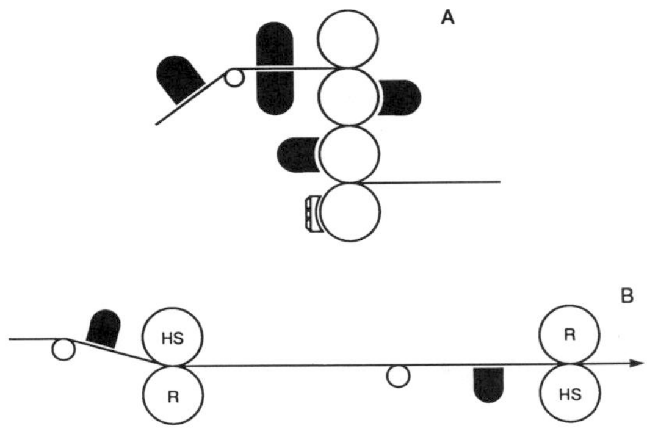

Fig. 660. Potential locations for steam showers on calender stacks. (A) Multiple-nip stack; (B) single nip with resilient roll (R) against heated steel roll (HS) [61].

calender located before the coaters. This generally applies only to machines with on-machine coaters. The principle is that a smooth surface picks up less coating. Higher nip pressure gives a smoother surface. Therefore, increasing zone temperature decreases coat weight.

The problem in all this is that, on machines without a calender stack and caliper control before the reel, caliper control is sacrificed for coat weight control. What these mills tend to do is have programs for both caliper control and coat weight control on the same actuator. If they are having reel build troubles then they operate the caliper control. Otherwise, the system is on coat weight control.

H. CD SMOOTHNESS/GLOSS CONTROL

Improved steam shower technology has permitted the reintroduction of steam showers to calender stacks [59–61]. Steam showers improve calendering by both heating the sheet and increasing the surface moisture. It is not clear which, if either, of these effects dominate.

The CD profiles of smoothness/gloss can be controlled by controlling the CD application of steam. Furthermore, smoothness/gloss can be controlled independently from caliper because the steam application can shift the caliper vs. smoothness curve [59]. The smoothness/caliper effect is very sensitive to the distance (time) between the point of steam application and the last nip.

If the steam shower is located near the last nip, then there is minimum effect on caliper and all of the smoothness effect occurs on the side of the sheet to which the steam is applied. When the steam shower is located well before the first nip, little effect is seen on the slope of the smoothness caliper curve, and the smoothness effect is balanced on both sides of the sheet. Thus, there is the opportunity to affect the sheet two-sidedness as well as the smoothness/gloss profile. Figure 660A illustrates various potential locations for a steam shower on a multiple-nip calender stack, while 660B shows the arrangement on single-nip resilient stacks.

Unfortunately there are several limitations to the process. Steam cannot be applied to many coated sheets because the additional moisture causes scaling. The steam cannot condense on hot sheets, so the sheet has to be cooled to 70°C or less by methods such as sweat dryers.

CD gloss profiles are being controlled on super calenders by using controlled deformation rolls and steam showers [62] or by standard caliper control actuators on a steel roll. A problem with using the controlled deformation rolls or caliper actuators to control the nip pressure profile is that the soft roll soon takes the shape of the steel roll. Therefore, the system works best on “slow unstable” profiles. Stable profiles have to be made unstable by reversing alternate logs end for end. Another technique is to have the controlled nip between two steel rolls. Such hard nips also give more control range for any level of energy input.

Another gloss control technique rapidly becoming popular is to replace calender stack water boxes with remoisturizing sprays. This gives the options of CD moisture control or CD gloss control. Mills are also finding they can come to the stack with a higher average moisture because the sprays add less water than water boxes. The higher moisture means energy savings or more production (see Fig. 620). Best of all, the technique gets rid of water boxes!

I. ACKNOWLEDGEMENT

The support of Devron-Hercules Inc. during the preparation of the manuscript, and for doing the artwork for the majority of the illustrations is gratefully acknowledged.

REFERENCES

1. Burkhard, G. and Wrist, P.E., "The Evaluation of Paper Machine Stock Systems by Basis Weight Analysis", Pulp Paper Mag. Can. 55(13):188 (Dec. 1954).
2. Cuffey, W.H., "Some Factors Involved in Basis Weight Uniformity", Tappi 40(6):190A (June 1957).
3. Anon., "New System Controls CMD Basis Weight Profile Automatically", Paper Trade J. 34 (Nov. 8, 1971).
4. Beecher, A.E. and Bareiss, R.A., "Theory and Practice of Automatic Control of Basis Weight Profiles", Tappi 53(5):847 (May 1970).
5. Carey, E.W., Bietry, C.R. and Stoll, H.W., "Performance Factors Associated with Profile Control of Basis Weight on a Paper Machine", Tappi 58(6):75 (June 1975).
6. Smith, K.E., "Survey Shows More than 3,000 Cross-Direction Controls Now Installed", Pulp Paper 63(2) (Apr. 1989).
7. Norberg, L.O., "Experience from CD Control of Basis Weight on the Fine Paper Machines in the Modo, Husum Mill", Pulp Paper Can. 90(5):T188 (May 1989).
8. Steenberg, B., "Behaviour of Paper Under Stress and Strain", Pulp Paper Mag. Can. 50(3):207 (1949).
9. Gibson, K., "Profile Control System at Weyco Mill Cuts Sheet Weight Variations", Pulp Paper 68 (Feb. 1984).
10. Harke, D.E., "Remote Manual Profile Control on a Three-Ply Liner Machine — A Mill Experience", Proc. 1987 TAPPI Papermakers Conf., p. 197 (Apr. 1987).
11. Lebrasseur, J.G., "Technical Advances to Improve Linerboard Moisture", Tappi 67(10):60 (Oct. 1984).
12. Jaffee, R.L., "Quality Control and Moisture Uniformity for the Corrugated Box Plant", Tappi 67(10):62 (Oct. 1984).
13. Perrault, R.D., "Trouble Shooting Design and Operation of the Dryer Section", TAPPI Notes, 1990 Practical Aspects of Pressing and Drying, p. 249 (Mar. 1990).
14. Brecht, W. and Heyn, D., "The Smoothness of Calendered Paper and its Dependence on Atmospheric Humidity", Woch fur Pap. 86(23/24):1037 (1958).
15. Mitchell, J.G., Private Communication (Nov. 1980).
16. Gess, J.M. and Segre, G., "Cross-Machine Direction Profile Measurements and Their Use by the Papermaker", Tappi 60(8):117 (Aug. 1977).
17. Wallace, M.D., "Cross-Direction Control: Profitability and Pitfalls", Tappi 69(6):72 (June 1986).
18. Jenkins, D.F., "Cross-Machine Variability in Coated Publication Paper Production", Tappi 56(2):89 (Feb. 1973).
19. Mardon, J. et al., "The Struggle for Uniformity", Tech. Sect. CPPA, Monograph (June 1971).
20. Burns, J.W., "The Use of Analysis of Variance to Characterize Paper Web Non-Uniformities", Tappi 57(12):2143–146 (Dec. 1974).
21. Cutshall, K.A., Ilott, G.E. and Rogers, J.H., "Grammage Variation — Measurement and Analysis", Tech. Sect. CPPA, Monograph (1958).
22. Barber, L.R., "Predicting Quality Improvements from Cross-Direction Controls", Proc. 1987 TAPPI Papermakers Conf., p. 179 (Apr. 1987).
23. Mardon, J. et al., "Control of Basis Weight Formation of Coating Rawstock", Tappi 48(6):76A (June 1965).
24. Rocheleau, A., "A Practical Approach to the Reduction of Machine Direction and Cross Direction Basis Weight Variations", Tappi 148(9):48A (Sept. 1965).
25. Mardon, J. et al., "Analysis of Paper Machine Stability and Performance by Means of Basis Weight Investigation", Tech. Sect. CPPA, Monograph (May 1973).
26. Kenworthy, I.C., "Investigation of Cross-Machine Uniformity on Paper and Board Machines", TAPPI Jt. Papermakers/Testing Conf., p. 63–71, Milwaukee (Sept. 25–27, 1973).
27. Crotogino, R.H., "Temperature-Gradient Calendering", Tappi 65(10):97 (Oct. 1982).
28. Karlsson, H., Haglund, L. and Hansson, A., "Optimal Cross-Direction Basis Weight and Moisture Profile Control on Paper Machines", Pulp Paper Can. 86(8):T213 (Aug. 1985).
29. Koskimies, J., "Leaning Behaviour of Form Paper Stack", Tech. Sect. CPPA, 72nd Ann. Mtg., Preprints p. A123 (Jan. 1986).
30. Wahlstrom, B., "Fundamentals of Web Consolidation", PIRA Intl. Conf., New Technologies in Web Consolidation and Drying, Brighton, Sussex, U.K. (May 19–23, 1986).
31. Brewster, D.B., "Interactions in CD Control", Control Systems '86, EUCEPA, Stockholm (May 1986); PIMA 28 (Dec. 1986).
32. Cutshall, K.A., "Precision Lip Control", Proc. 1987 TAPPI Wet End Operations, p. 407 (May 1987).
33. Karlsson, H., Lundquist, I. and Ostman, T., "Spara Pengar Optimera Traersprofilstyrringen", Svensk Papperstidn. 7:30 (1983).
34. Schiel, C., "Improved Dryer Section Performance by Selective Application of Thermo Rims", BPBIF Seminar, The Theory of Water Removal, Birmingham (Dec. 12–13, 1979).
35. Saastamoinen, M., "Improved Operation of the Dryer Section by Using Valmet Dryer Bars", SPCI 84, Stockholm, p. 228 (Apr. 1984).

36. Pulkowski, J.H. and Wedel, G.L., "The Effect of Spoiler Bars on Dryer Heat Transfer", Tech. Sect. CPPA, 73rd Ann. Mtg., Preprints p. A35 (Jan. 1987).
37. Soinen, M., "Pocket Ventilation at Different Geometrics of a Multi-Cylinder Dryer Section with Mesh Screens", Paperi ja Puu 11:735 (Nov. 1970).
38. Marshall, H.G., "Optimizing Newsprint Drying", Pulp Paper Mag. Can. 73(3):40 (Mar. 1972).
39. Grott, R.W., "The Economics of Pocket Ventilation System", Am. Paper Industry 58(4):14 (Apr. 1976).
40. Holik, H. "Improving Paper Quality by Optimum CD Profile", Control Systems '86, EUCEPA, Stockholm (May 1986).
41. Holik, H. and Brugger, H., "Improving CD Moisture Profiles in Tissue Production", Proc. 1986 TAPPI Engineering Conf., p. 301 (Sept. 1986).
42. Anon., "The IR Handbook", U.D.C. No. 621.365.46, Svenska AB Philips.
43. Roth, R.C., "Electromagnetic Radiation for CD Moisture Control", SPCI 84, Stockholm, p. 432 (Apr. 1984).
44. Hultgreen, O., "Raising Quality and Production Through High-SP Drying", Tappi 45(6):178A (June 1962).
45. Borchert, R. and Jubitz, W., "Infrarottechnik", Veb Verlay Tecnik, Berlin (1958).
46. Higham, J.D., "New Techniques for Cross Direction Moisture Control on Paper Machines Using Infrared Radiation", SPCI 84, Stockholm, p. 422 (Apr. 1984).
47. Roth, R.C., "Electromagnetic Radiation for CD Moisture Control", Tappi 69(1):57 (Jan. 1986).
48. Sharma, R., "Effective Use of Electric Infrared for Moisture Control", Proc. 1986 TAPPI Engineering Conf., p. 295 (Sept. 1986).
49. Swanson, D., "Quality Benefits from Cross-Direction Moisture Control", Tappi 69(11):86 (Nov. 1986).
50. Higham, J.D., "Benefits of IR Moisture Profiling on Fine Paper Machines", Tappi 70(4):85 (Apr. 1987).
51. Cutshall, K. and Hudspeth, D., "Hot Pressing", Proc. 1987 TAPPI Wet End Operations, p. 411 (May 1987).
52. Pikulik, I.I., "Granite Roll Failures in Canada", Tech. Sect. CPPA, 74th Ann. Mtg., Preprints p. B427 (1988).
53. Pulkowksi, J.H. and Wedel, G., "Moisture Profiling by Induction Heating", Proc. 1986 TAPPI Pressing and Drying Seminar, p. 214 (Mar. 1986).
54. Hilden, K.K., "Improved Caliper/Hardness Profile Control Using Induction Energy Calender Roll Heating", SPCI 84, Stockholm, p. 416 (Apr. 1984).
55. Hilden, K.K., "Caliper Profile Control Results Using Calcoil Induction Heaters in a Canadian Paper Mill", Tech. Sect. CPPA, 71st Ann. Mtg., Preprints p. A217 (Jan. 1985).
56. Kan, C.E., "A New Caliper Control Actuator — Its Evolution Process and Application Results", Proc. 1986 TAPPI Finishing and Converting Conf., p. 69 (1986).
57. Swanson, D., "Effective Use of Moisture/Caliper Control", S. Pulp Paper 43 (May 1985).
58. Taylor, B.F., "Evaporative Cooling — A Unique, Effective Approach to Caliper Control", Proc. 1987 TAPPI Papermakers Conf., p. 219 (Apr. 1987).
59. Dunfield, L.G., McDonald, J.D., Gratton, M.F. and Crotogino, R.H., "Gravure Printability of Steam-Treated Machine Calender Newsprint", J. Pulp Paper Sci. 12(2):J31 (Mar. 1986).
60. Hilden, K.K. and Sawley, D., "Calender Steam Showers — A New Effective Means of Hot Calendering", Tech. Sect. CPPA, 73rd Ann. Mtg., Preprints p. A343 (Jan. 1987).
61. HIlden, K.K. and Sawley, D., "Calender Steam Showers", Tappi 70(7):87 (July 1987).
62. Thielbor, R.B., "Supercalender is Logical Extension of Cross-Direction Profile Control", Pulp Paper 89 (June 1985).

BIBLIOGRAPHY

1. Church, F., "Transverse Moisture Profile Control", Tech. Sect. CPPA, Ann. Mtg., Montreal (Jan. 1963).
2. Swanson, W.W., "On-Machine Measurement of Basis Weight and Moisture", Paper Trade J. 41 (July 22, 1963).
3. Kahoun, J.B., Zurowitch, W. and Newby, L., "System for On-Machine Measurements and Control of Paper Caliper", Tappi 48(7):60A (July 1965).
4. Nykopp, E.A., "Moisture Profile Correction with an Electrically Heated Dryer", Paper Trade J. (Feb. 22, 1971).
5. Church, F., "Transverse Profile Control of Basis Weight/Moisture", Pulp Paper 48 (Apr. 1973).

6. Lyne, M.B., Haglund, L. and Bjelkhagen, H., "Control of Machine Calender Roll Diameter with Air Showers — An Experimental Investigation", Svensk Papperstidn. 79(8):251 (Aug. 1976).
7. Rutland, D.F. and Kerr, R.B., "Analysis and Correction of Headbox and Approach Flow-Related B.W. Problems", Pulp Paper Can. 79(2):T49 (Feb. 1978).
8. Mitchell, J.G. and Sheahan, D.K., "The Control of Reel Building with Calender Stack Air Showers", Tech. Sect. CPPA, Pacific Coast-Western Branch Jt. Conf., Jasper, Alta. (Apr. 1979).
9. Dove, N.F., "New Technique is Used to Control Cross-Machine Moisture Profiles", Pulp Paper 54(4):174(Apr. 1980).
10. Thomas, C.E., "Flute Formation on the Singlefacer", Tappi 68(8):67 (Aug. 1980).
11. Ostrow, K.P., "New System Selectively Dries Out Wet Streaks Across Paper Machine", Pulp Paper 56(11):130 (Nov. 1980).
12. McDiarmid, R. and Cutshall, K., "Steam Showers Increase Newsprint Machine Press-Section Efficiency", Pulp Paper 56(8):110(1982).
13. Duffy, G.G. et al., "The Drying of Paper Webs Using a New Infrared Generator", Appita 37(2):131 (Sept. 1983).
14. Smith, D.W.P., "Thermal Hydraulic Actuators for Automatic Slice-Lip Control", Tech. Sect. CPPA, 70th Ann. Mtg., Preprints p. A117 (Jan. 1984).
15. Adams, W.L., "Profile Manager, The Application of a Self-Tuning Regulator to CD Controls", SPCI 84, Stockholm p. 405 (Apr. 1984).
16. Cutshall, K.A., "Update on Hot Pressing Results", SPCI 84, Stockholm, p. 199 (Apr. 1984).
17. Elliott, R.M., "Model-Based Cross Direction Control", SPCI 84, Stockholm, p. 396 (Apr. 1984).
18. Nilson, H.E., "Quality and Capacity Improvements with New Infrared Drying System for Moisture Profile Control", SPCI 84, Stockholm, p. 427 (Apr. 1984).
19. Pfeifer, R.J., "Process Measurement Requirements for Cross Machine Control", SPCI 84, Stockholm, p. 410 (Apr. 1984).
20. Smith, D.W.P., "The Thermal Hydraulic Approach to Automatic Slice Lip Control", SPCI 84, Stockholm, p. 401 (Apr. 1984).
21. Radwan, M.N. and Nayar, R., "Application of Hot Pressing on Domtar Paper Machines", Pulp Paper Can. 85(7):T189 (July 1984).
22. Cutshall, K.A., "Cross-Machine Moisture Control Via Hot Pressing", Tappi 67(9):72 (Sept. 1984).
23. Wilhelm, R.G., Jr., "On the Controllability of Cross-Direction Variations of Sheet Properties", Proc. 1984 TAPPI Engineering Conf., p. 621 (Sept. 1984).
24. Silar, S.J., "Cross Machine Basis-Weight Control: Machine Considerations", Tappi 67 (12):52 (Dec. 1984).
25. Daunais, R. et al., "A Comparison of Steam and Infrared Heating for Hot Pressing", J. Pulp Paper Sci., 11(1):J12 (Jan. 1985).
26. Powell, R.I. and Cutshall, K.A., "Press Part Steam Showers — The Extra Nip", Appita 38(1):49 (Jan. 1985).
27. Higham, J.D., "The Changing Face of Paper Machine Controls — Cross Direction and Beyond", Proc. 1985 TAPPI Process Control Symp. (Mar. 1985).
28. Boyle, T.J. and Hamby, H., "Cross-Direction Weight Control: Pro and Con", Tappi 68(6):56 (June 1985).
29. Hadra, J.J., "Selecting Actuators for Cross-Machine Control", Tappi 68(6):60 (June 1985).
30. Leboeuf, J. and Ruel, R., "Automatic Control of Moisture and Caliper Profiles at Papeterie Reed Ltee", Pulp Paper Can. 86(8):49 (Aug. 1985).
31. Fortin, R., "Cross-Direction Moisture Control", New York-Canadian PIMA Mtg., Lake Placid, NY (Sept. 1985).
32. Fournier, J.R., "Cross Machine Direction Control of Moisture", New York-Canadian PIMA Mtg., Lake Placid, NY (Sept. 1985).
33. Pelletier, L., Crotogino, R.H. and Douglas, W.J.M., "CD Calender Control with Air Jets — An Experimental Study of Impingement Heat Transfer", Tech. Sect. CPPA, 72nd Ann. Mtg., Preprints p. A183 (Jan. 1986).
34. Bell, N., "Remoisturizing in the Dryer Section", Proc. 1986 TAPPI Engineering Conf., p. 315 (Sept. 1986).
35. McDonald, J.D. and Stevens, R.K., "Measurement of Calender Nip Pressure", Tech. Sect. CPPA, 73rd Ann. Mtg., Preprints p. B315 (Jan. 1987).
36. Taylor, B.F., "The Thermal Electron Steam-Foil for Curl and Finish Control", Tech. Sect. CPPA, 73rd Ann. Mtg., Preprints p. A165 (Jan. 1987).
37. Audas, J., "Quality Improvements Using V.I.B. Spray Dampening Systems for Moisture Profile Control", Proc. 1987 TAPPI Papermakers Conf., p. 305 (Apr. 1987).
38. Cutshall, K.A., "Moisture Control with Steam Boxes", Proc. 1987 TAPPI Wet End Operations, p. 407 (May 1987).
39. Flading, J., "Cross-Machine Control", Proc. 1987 TAPPI Wet End Operations, p. 349 (May 1987).
40. Spitz, D.A., "Effect of Cross Direction Caliper Variations in Winding", Tappi 52(6):1168 (June 1969).
41. Metcalfe, W.K., "What Pocket Ventiliation Systems Can Do — And What They Cannot", Pulp Paper, p. 95 (Feb. 1970).
42. Anon., "Cross Machine Moisture Uniformity — Its Importance and Control", Paper Trade J., p. 67 (Mar. 30, 1970).
43. Rhorer, C.R., "Diagnosis and Correction of

Moisture Profile Problems on Paper Machines", Tappi 54(1):43 (Jan. 1971).
44. Emmanuel, A., "On the Verification of On-Machine Sensors for Reel Building", Appita 41(1):51 (Jan. 1988).
45. Cutshall, K.A. and Hamel, R.G., "Slice Lip Automation", Pulp Paper Can. 89(1):T1 (Jan. 1988).
46. Sharma, R., "Using Infrared Selectively for Improving Paper Quality and Production", TAPPI Seminar Notes, 1988 Practical Aspects of Pressing and Drying, p. 173 (Mar. 1988).
47. Smith, D.W.P., "The Changing Requirements for On-Line Sensor Data", Pulp Paper Can. 91(3):T118–122 (Mar. 1990).
48. Anon., "Infrared Moisture — Profile Control System Improves Machine Speed", Pulp Paper Mag. (Sweden), p. 40 (May 1988).
49. Anon., "Gas or Electric?", Paper 18(7):19 (July 1988).
50. Pulkowski, J.H. and Wedel, G.L., "The Effect of Spoiler Bars on Dryer Heat Transfer", Pulp Paper Can. 89(8):61 (Aug. 1988).
51. Higham, J.D., Stockford, D.A. and Egierski, J., "Locating Infra-Red Drying Equipment for Moisture Profiling", Appita 41(5):398 (Sept. 1988).
52. Francis, K., Kleinsmith, C. and Stenbak, M., "The Stationary Bone-Dry Weight Sensor", Proc. 1988 TAPPI Engineering Conf., p. 561 (Sept. 1988).
53. Knecht., W., "Fast Response CD Caliper Control", Proc. 1988 TAPPI Finishing and Converting Conf., p. 199 (Oct. 1988).
54. Vyse, R.N. and Sawley, D.J., "Practical Aspects of Calender Steam Showers", Tappi J. 81(10):87 (Oct. 1988).
55. Walton, L., "Thermal Rods and Process Modeling Help Control Tissue Weight Profile", Pulp Paper 62(11):110 (Nov. 1988).
56. Kenyon, P., "Michigan Paper Mill Improves Its Production with Infrared Profiling", Pulp Paper 63(2):78 (Feb. 1989).
57. Cutshall, K.A., "Cross Direction Control Gives Major Improvement in Paper Uniformity", TAPPI Seminar Notes, 1989 Wet End Operations, p. 265 (May 1989).
58. Bourque, J.A., "New-Design Dry End Steam Showers Improve Sheet Quality, Runnability", Pulp Paper 63(5):100 (May 1989).
59. Slowik, A.M., "Cross-Machine Control Experiences", Tappi J. 72(6):57 (June 1989).

XIX PAPER MACHINE SHOWERS AND DOCTORS

M.M. CUTTS

Murray M. Cutts is a Graduate of the Montreal Institute of Technology and also of McGill University's Executive Development Program. In 1969, Mr. Cutts joined Harland Engineering, a manufacturer of variable speed drives for pulp and paper machines, as Sales and Application Engineer. Since 1972, he has been employed in various positions by Albany Engineered Systems, a manufacturer of specialized accessories for pulp and paper machines. Mr. Cutts has been an active member of TAPPI, the Technical Section, CPPA, and PIMA and has presented papers at a TAPPI Wet End Operations Short Course, a TAPPI Pressing and Drying Short Course, and a TAPPI Tissue Seminar.

REVIEWED BY

Richard A. Reese
Senior Process Consultant
CRS Sirrine Inc.
Greenville, SC

Ron Green
Manager Product Development
Johnson Foils Inc.
Springfield, MA

A. INTRODUCTION

This chapter will deal specifically with the design, application, operation, and maintenance of paper machine showers and doctors, excluding formation showers and sheet moisture showers.

These most important accessories are primarily used for the cleaning and conditioning of paper machine fabrics (covered in detail in Chap. XXII), cleaning and conditioning of paper machine rolls, sheet knock-off and lubrication of wear surfaces that interface with rolls and paper machine fabrics.

The importance of showers and doctors cannot be overemphasized. The correct choice and application of equipment significantly improves paper machine fabric life, production efficiency, and lowers production costs. However, the greatest benefit results from a more uniform and consistent product quality.

Since both the effectiveness and efficiency of paper machine showers depends on the type and quality of water used, this chapter will also deal briefly with the treatment of mill fresh water and mill recycled white water.

B. SHOWERS

Basically, three types of showers are generally used in the paper industry for a multitude of applications: 1) stationary showers with fan nozzles; 2) oscillating showers with either fan and/or needle jet nozzles; and 3) a single nozzle shower which is designed to traverse the paper machine. These three groups can be further sub-categorized into: A) showers designed with fixed orifice nozzles; and B) showers designed with cleanable or purgeable nozzles.

1. Stationary showers

Stationary showers are shower pipes that are fixed in a certain position on the paper machine and do not oscillate back and forth. These showers are designed and manufactured with fan-type nozzles. The orifice size of the nozzle is chosen based on the water volume requirement of the application while the angle of the fan is chosen to provide uniform distribution of water based on nozzle centerlines and distance of the shower from the sprayed surface. Figure 661 illustrates this relationship.

2. Oscillating showers

Oscillating showers are those units which are designed with a mechanism to move the shower back and forth on the paper machine at a constant speed over a specific distance (commonly referred to as stroke length). Oscillating showers are supplied with either needle jet

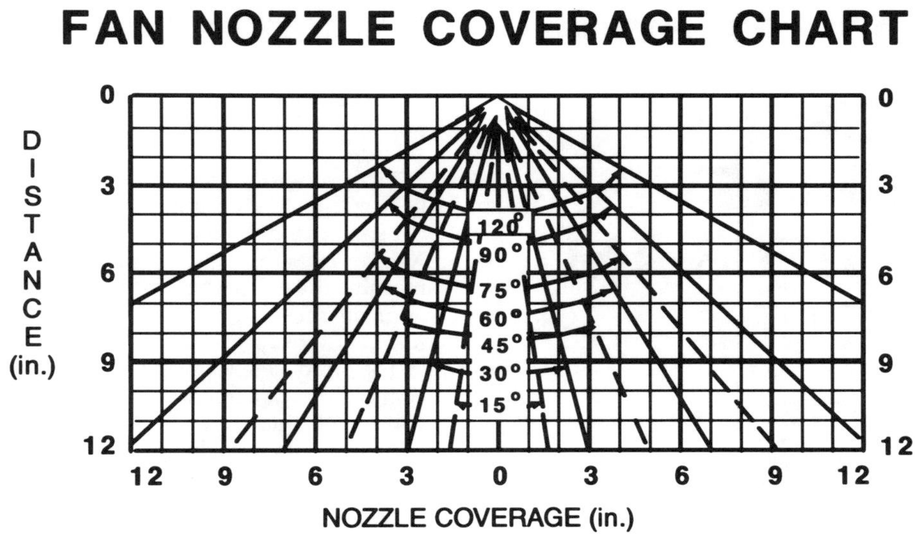

Fig. 661. The vertical axis of this chart shows distance from sprayed surface, and the horizontal axis gives nozzle coverage.

nozzles or a combination of needle jet/fan jet nozzles. When used in this configuration, a needle jet is usually followed by a fan jet, then a needle jet, then a fan jet and so on. This alternating needle/fan configuration is becoming more popular and will be discussed in detail later.

The main function of a shower oscillator is to ensure uniform coverage of water applied to the application. In view of the impact on moisture profile, this function is extremely important and warrants detailed review.

Oscillators are available in the form of an integral electric motor/gear reducer/crank arm assembly, hydraulically or pneumatically operated reversing cylinders, or a linear velocity/low-speed electromechanical arrangement. The electric motor gear reducer combination with a crank arm is a non-linear velocity device and therefore causes sinusoidal wear patterns in paper machine clothing. Because of this, this device is considered practical only for non-critical applications where surface marking is not a consideration.

The two most important operating criteria of any oscillator are stroke length and speed. Generally, it is recommended that the oscillator be designed to stroke at a distance equal to twice the nozzle spacing centerlines. In other words, nozzles mounted on 150 mm (6 in.) centerlines would have an oscillator stroke of 300 mm (12 in.). This arrangement results in double coverage so that, if one nozzle becomes blocked, coverage will still be maintained, although reduced. Where it is not possible or practical to stroke the oscillator for double coverage, then single coverage will suffice.

The adjustment of oscillator stroke relative to nozzle spacing is critical. As shown in Figs. 662, 663 and 664, short stroking or overstroking causes non-uniform coverage. Uniform coverage, resulting in optimum moisture profiles can only be achieved through the accurate relationship between stroke and nozzle spacing.

The speed of an oscillator generally can be adjusted within certain limits. It has been proven that most effective cleaning of forming fabrics and press fabrics occurs when the oscillator speed is set to advance a shower one nozzle diameter per fabric revolution. The formula and an example calculation of this is shown in Fig. 665. When calculated, the resultant velocity or speed is generally much less than is capable of being maintained using the traditional hydraulic or pneumatically operated reversing cylinders. More appropriate for these applications is the electromechanical low-frequency linear velocity type of oscillator employing a ball screw or similar mechanism, as shown in Fig. 666.

The higher speed reversing cylinder oscillators are quite adequate for such applications as suction roll and grooved roll cleaning.

3. Single-jet traversing showers

For specialized applications in press sections where high volumes of water cannot be tolerated because of extreme nip loadings and also for treating fabric streaks, a single-jet shower with programmable traversing features is often used.

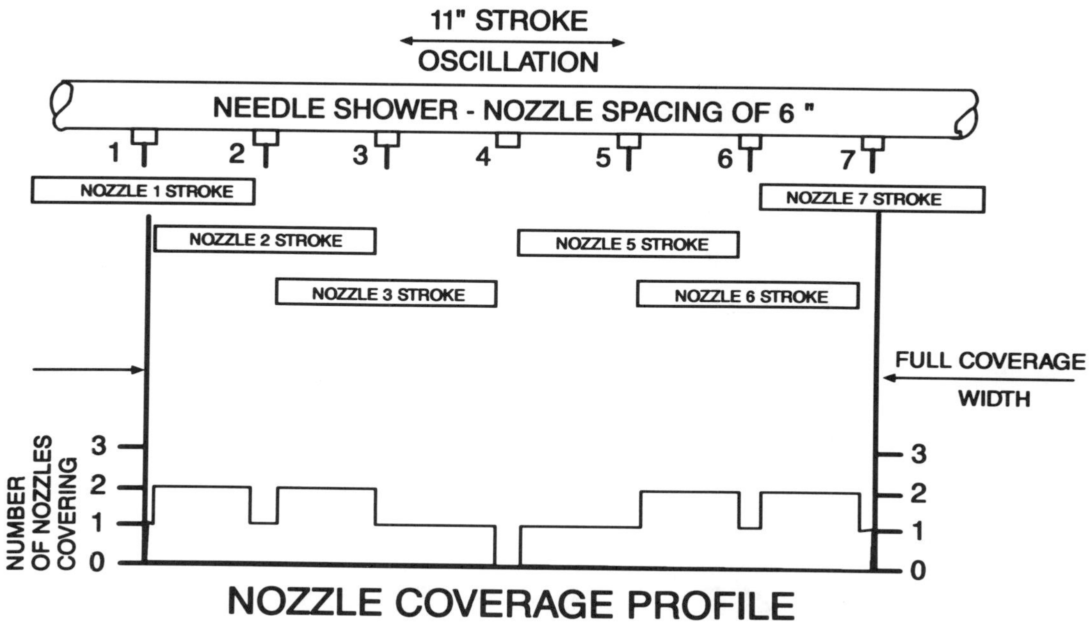

Fig. 662. This profile illustrates a short stroking oscillator.

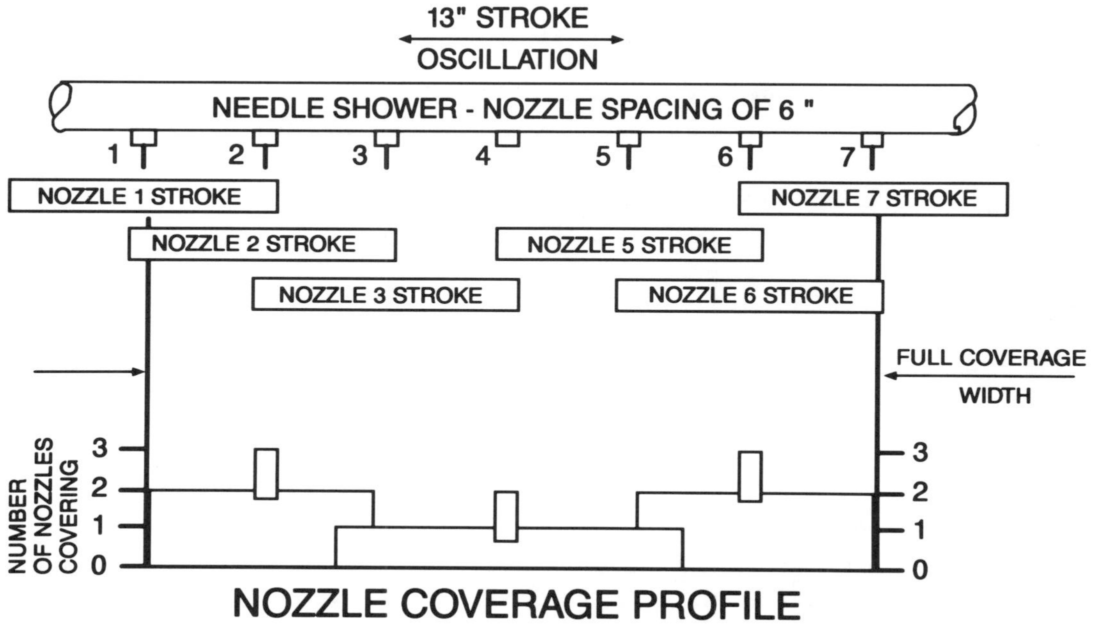

Fig. 663. This profile shows an oscillator that is over-stroking.

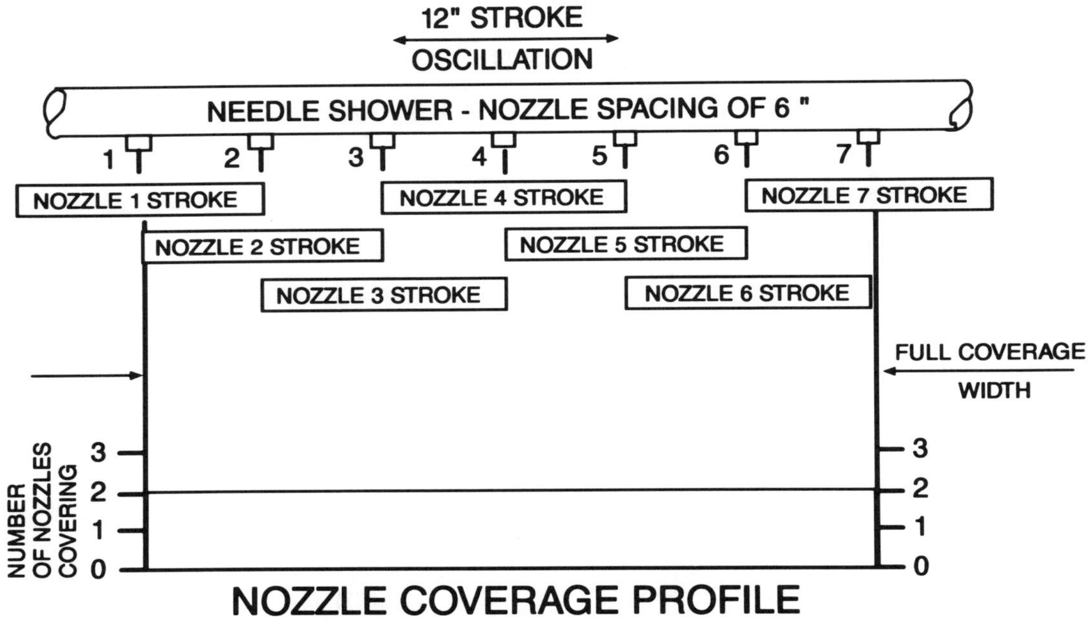

Fig. 664. This profile illustrates an oscillator with correctly set stroke.

OSCILLATION SPEED CALCULATIONS

LOW FREQUENCY OSCILLATORS
OPTIMUM TRAVERSING SPEED (.040" JET)

$$\frac{\text{FPM} \times .040}{\text{FT.}} = \text{TRAVERSING SPEED (INCHES/MINUTE)}$$

FPM = MACHINE SPEED (FEET/MINUTE)
FT. = FABRIC LENGTH (FEET)
.040 = WATER JET WIDTH (INCHES)

———— //———— //———— //———— //

EXAMPLE: MACHINE SPEED: 2000 FPM
FABRIC LENGTH: 65 FEET

$$\text{OSC. SPEED} = \frac{2000}{65} \times .040 = 1.23\text{"/MINUTE}$$

Fig. 665. Calculation of oscillator speed.

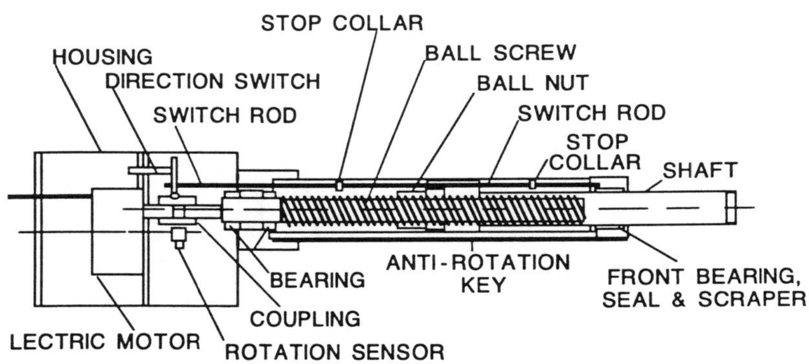

Fig. 666. A typical low-frequency electromechanical oscillator.

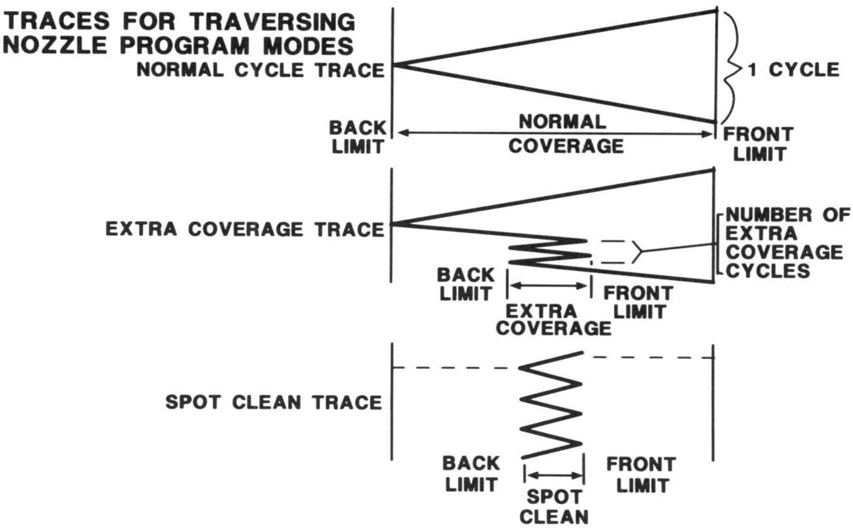

Fig. 667. Typical coverage patterns of single-jet showers.

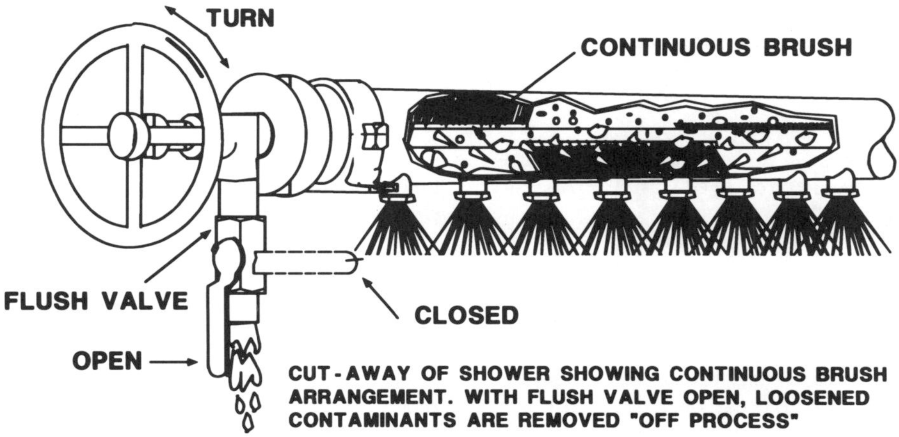

Fig. 668. A typical shower with an internal brush used for cleaning nozzle orifices.

This style of shower can also be used for continuous cleaning of dryer fabrics.

As the name implies, this shower has a single jet which traverses the paper machine according to the needs of the application. As shown in Fig. 667, they can be programmed for normal coverage, extra coverage cycles or for spot cleaning. This shower provides a high degree of flexibility for difficult and special need applications.

4. Fixed-orifice showers

The sub-category of fixed-orifice showers available either as stationary or oscillating units is designed and manufactured with a nozzle that has a drilled orifice. Once drilled, the size of the orifice is fixed and cannot be altered for the purposes of cleaning. If the nozzle becomes blocked, the only means of cleaning it is to flush the shower header, run a brush or go-devil through the shower header, or remove the contaminant with a pointed or pin-type of tool. This limitation of fixed-orifice showers make them applicable for use only with relatively clean water. Guidelines relative to this will be discussed later.

To overcome the problem of nozzle blockage associated with fixed-orifice showers, units are available with a built in brush as shown in Fig. 668. This design employs a continuous brush inside of the shower header, which, when rotated manually, sweeps contaminants from the shower

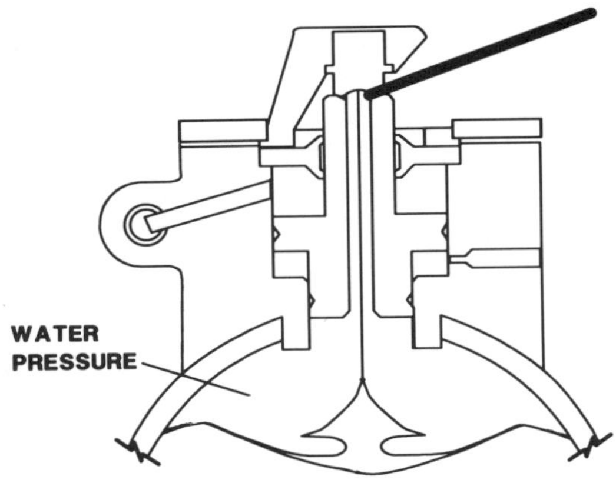

Fig. 669. A purgeable-type shower in its normal operating mode.

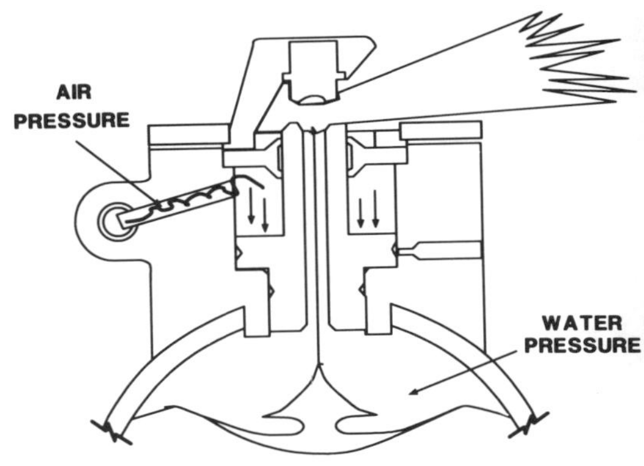

Fig. 670. A typical purgeable-type nozzle in the purge mode.

Table 69. Shower water quality.

Solids loading	Application of water strained with 100 mesh screen
0-50 ppm	Equivalent to filtered fresh water
50-75 ppm	Usable in 1 mm (0-0.40 in.) and larger fixed orifice nozzles with minimal problems
75-100 ppm	Usable in 1.5 mm (0.055 in.) and larger fixed orifice nozzles with minimal problems
100-200 ppm	Usable in 3 mm (0.125 in.) and larger fixed orifice nozzles with minimal problems
200-up ppm	Purgeable showers recommended

pipe and the face of the nozzle orifice. These contaminants are then removed from the process through a flush valve.

For applications where contaminant loadings are relatively high and where a brush-type shower is not practical, a self-cleaning purgeable nozzle is desirable. This type of nozzle is designed to duplicate a fixed orifice when in its operating position. When the nozzle becomes contaminated, the orifice can be increased in size for the purpose of flushing the orifice and the associated spray button. Figure 669 illustrates a typical nozzle in its operating position while Fig. 670 illustrates the same nozzle in a purging condition. With this type of nozzle, it should be noted that the contaminants are discharged directly into the process.

When to specify or apply a fixed-orifice nozzle versus a brush-type shower or purgeable shower depends on the process shower water quality relative to its solids loading. Table 69 provides guidelines for nozzle choice relative to solids loading or amount of contaminants in the process water.

C. PROCESS WATER SYSTEMS

For a given application, the process water system is as important as the selection of shower. Either fresh filtered water or recycled mill white water are used for these applications. In either case, it is desirable to have the shower water heated to the same temperature and pH as the process. This will assist in preventing thermal and chemical shock.

Fresh water, generally recommended for press fabric and dryer fabric cleaning, must be sufficiently free of contaminants to prevent nozzle blockage and excessive erosion of orifices. To protect against these conditions, it is highly recommended that a pressure filter be installed directly ahead of each shower. The filter media should be designed to remove any contaminants larger than the orifice size of the shower divided by six. This industry-accepted rule of thumb seems to provide adequate protection against nozzle blockage. If a common source of fresh water is being used to supply several showers, then a multi-bank of pressure filters can be used rather than individual filters ahead of each shower. In this situation, the bank of filters should be installed as close as possible to the showers being serviced.

Reuse of mill white water as a source of water for showers requires considerable attention to application detail and water treatment. A typical white water shower system as shown in Fig. 671 and in Ref. [1] would involve the use of a gravity-type strainer to separate fiber from the process water. When using this water with fixed jet type nozzles, it is important to include also an

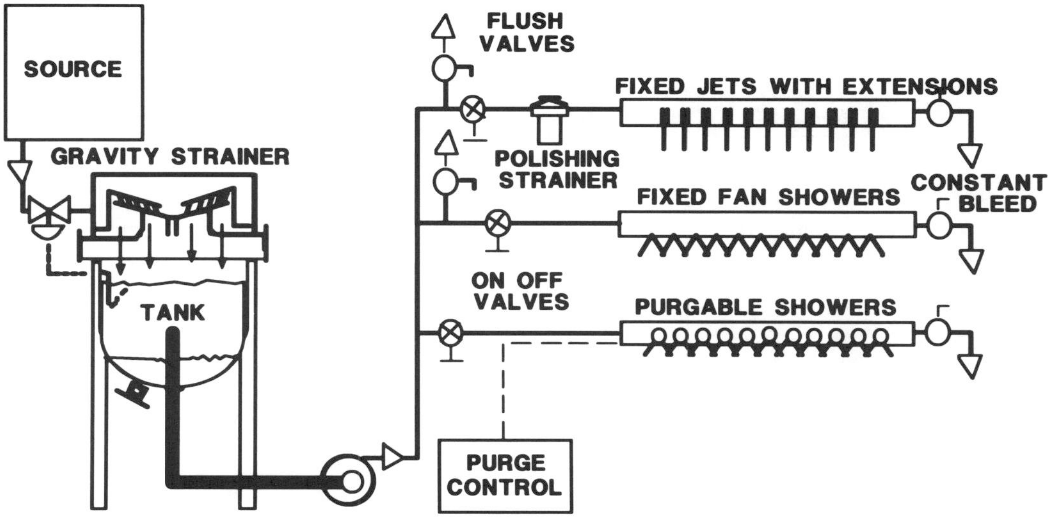

Fig. 671. Equipment associated with design of a white water shower system.

inline pressure filter which acts as a polishing strainer. When this water is used as a source for fixed-fan showers, this precaution is not necessary as usually the orifice size is larger. Similarly, a polishing strainer is not required with purgeable showers.

In all cases, a bleed valve installed on the end of the shower opposite the feed line is recommended. These valves should be left open during operation to increase the velocity of water flowing through the pipe. The minimum recommended velocity is 2.5–3 m/s (8–10 ft/s). This increased velocity tends to break up fiber flocs, thus reducing the incidence of nozzle blockage.

Also, flush valves located ahead of each shower are recommended. These provide a means to flush the pipes prior to start-up and following mill shutdowns.

Mill white water, processed as just described, is usually acceptable for use on most forming section applications.

D. SHOWER APPLICATIONS

This portion of the chapter will not deal with specific choices of shower hardware, but rather the function of the shower and its contribution to paper machine operation. Choice of actual equipment should be based on process water quality and other design criteria previously described.

1. Forming section

The forming section of paper machines can be either conventional fourdriniers, twin-wire formers, or a combination of both commonly referred to as a hybrid former. In all cases, the shower applications are common. With fourdriniers, there is only one forming fabric to clean, while with a twin-wire or hybrid former there are two fabrics to clean. The criteria are the same except that the number of showers required are in proportion to the number of fabrics.

Table 70 describes the various applications of showers in the forming section relative to their function, shower type, nozzle spacing, operating pressure and water applied in gallons per minute per inch of machine width. Following is a brief explanation of each application.

1. Headbox showers are used in both open- and closed-type headboxes for defoaming and prevention of stock buildup. In open-type headboxes, the shower is designed to swing 180° while, for closed headboxes, the shower is designed to rotate a full 360°. For either type of shower, 3 mm (0.125 in.) diameter fixed-orifice nozzles are suitable for use with filtered fresh water or clarified white water providing it meets the qualities shown in Table 69. It is important to note that modern hydraulic headboxes do not require such showers.
2. A solid breast roll shower is used for the purpose of filling the void volume of the forming fabric just prior to the discharge of the headbox. This type of shower usually is supplied with either fixed-orifice nozzles, brush-type nozzles or purgeable nozzles, depending on the quality of the process water. As stated in Sec. E8, a doctor is also normally required.

Table 70. Forming section showers.

Application	Function	Shower type	Nozzle spacing mm	(in.)	Operating pressure kg/cm^2	(psi)	Flow L/min/cm	(gal/min/in.)
Headbox	Foam killing	Rotary	25-50	(10-20)	1.5-3	(20-40)	0.1-0.3	(0.075-0.204)
	preventing stock buildup	Swing	25-50	(10-20)	1.5-3	(20-40)	0.1-0.3	(0.075-0.204)
Breast roll (solid)	Fill void volume	Sta.	150	(6)	2-3	(30-40)	0.25-0.3	(0.167-0.197)
Breast roll (slice)	Cleaning	Sta.	150	(6)	2-3	(30-40)	0.018-0.2	(0.12-0.133)
Dandy roll	Cleaning	Osc.	38	(1.5)	7-10	(100-150)	0.36	(0.239)
Lump breaker roll	Cleaning	Sta.	25-30	(10-12)	1.5-2	(20-30)	0.04-0.065	(0.0285-0.0427)
Suction couch roll	Cleaning	Osc.	75	(3)	25-50	(350-700)	0.30-0.46	(0.202-0.313)
Knockoff	Easy	Sta.	75	(3)	7-10	(100-150)	0.33-0.48	(0.22-0.32)
	Normal	Sta.	75	(3)	7-14	(100-200)	1.2-1.5	(0.75-1.05)
	Difficult	Sta.	75	(3)	10-18	(150-250)	1.5-2.7	(1.09-1.77)
	Flooded nip	Sta.	75	(3)	Calculate R.V.V. (refer to Fig. 672)			
Forming fabric H.P. cleaning	Fourdrinier single-layer fabrics	Osc.	150	(6)	14-32	(200-450)	0.125-0.165	(0.083-0.110)
	Multi-layer fabrics	Osc.	75	(3)	14-32	(200-450)	0.25-0.33	(0.167-0.220)
Wire return roll	Lubricate	Sta.	75	(3)	2-3	(30-40)	0.23-0.25	(0.15-0.163)

Osc. — Oscillating; Sta. — Stationary

3. A breast roll shower used for cleaning the headbox apron is installed so that water will be directed over the top of the breast roll onto the apron of the headbox. Its function is to prevent stock buildup which could possibly fall off in large accumulations, thus affecting the quality of sheet formation and cross-machine direction profile. This type of shower can be designed with either fixed-orifice nozzles, brush-type nozzles or purgeable nozzles, again depending on the quality of the process water.
4. A dandy roll shower is installed inside the dandy roll for the purpose of maintaining its open area by removing fiber from the screen. This shower is typically designed with needle jet nozzles on 38 mm (1.5 in.) centerlines and is oscillated to maintain uniform coverage. Most often, filtered fresh water is used for this application. A "non-drip" catch pan is normally required to prevent sheet rewetting.
5. The lumpbreaker roll shower is a stationary shower used for cleaning the surface of the roll and to prevent "picking". This low-pressure, low-volume shower applies a thin film of water which puts into suspension any paper fines or fiber that adhere to the roll.
6. The suction couch roll shower is installed on the outside of the roll and is used to keep the drilled holes free from contaminants and blockage. These showers are supplied with needle jet nozzles mounted on 75 mm (3 in.) centerlines and an oscillating mechanism designed to stroke 150 mm (6 in.). The nozzles can be either fixed orifice, brush type or purgeable depending on the quality of the process water.
7. The knock-off shower is probably the most important shower in the forming section. Its purpose is to separate the sheet from the forming fabric in the event of a break. The design of this shower varies relative to flow and pressure depending on whether a paper grade is easy, normal or difficult to knock off. This shower is generally installed between the forward-turning roll and the first wire return roll.

When certain types of double layer fabrics are used [2], sheet knock-off is not effective using conventional means. To deal with this, the knock-off shower is relocated between the couch and the forward-turning roll. In this location, the shower is referred to as a flooded-nip knock-off shower. It directs water into the area where the forward-turning roll and the forming fabric converge. The resultant hydraulic forces pump water into the void volume of the forming fabric causing the sheet to release. The correct amount of water for this application is determined by calculating the running void volume of the forming fabric. This calculation is illustrated in Fig. 672.

8. The forming fabric high-pressure cleaning showers are used to keep the void volume of the forming fabric free from contaminants. This function is absolutely necessary to maintain the drainage characteristics of the forming fabric throughout its normal life. As can be seen in Table 70, the nozzle spacing configuration depends on the style of forming fabric used.

For general cleaning, a shower should be located on the sheet side of the fabric and adjusted to spray at a slight angle into the direction of "run". For more difficult cleaning, a second shower located on the inside of the fabric must be used. Additional information regarding the cleaning of forming fabrics is found in Chap. XXII.

9. Showers are used on wire return rolls in conjunction with doctors for the purpose of lubricating the interface between the roll and the doctor blade. Furthermore, the wire return roll shower provides a film of water which suspends the foreign particles so that they can be more easily removed by the doctor. In this application, the shower and doctor complement each other. Used individually they have little effect. Either fixed-orifice nozzles, brush-type nozzles or purgeable nozzles can be used for this application depending on the quality of the process water.

Figure 673 illustrates where forming section showers are typically installed on a fourdrinier-type machine while Fig. 674 illustrates where they would be installed on the top wire of a hybrid former. These are provided to help visualize the normal forming section shower applications.

$$\text{R.V.V.} = \text{g.p.m.}$$

$$\text{R.V.V.} = \frac{C \times W \times S \times V}{19.25}$$

C - CALIPER OF FABRIC IN INCHES
W - FABRIC WIDTH IN INCHES
S - SPEED IN FPM
V - VOID VOLUME (USE 0.6)

$$19.25 = \frac{231\ \text{in}^3/\text{GAL}}{12\ \text{IN/FT}}$$

Fig. 672. Calculation of running void volume.

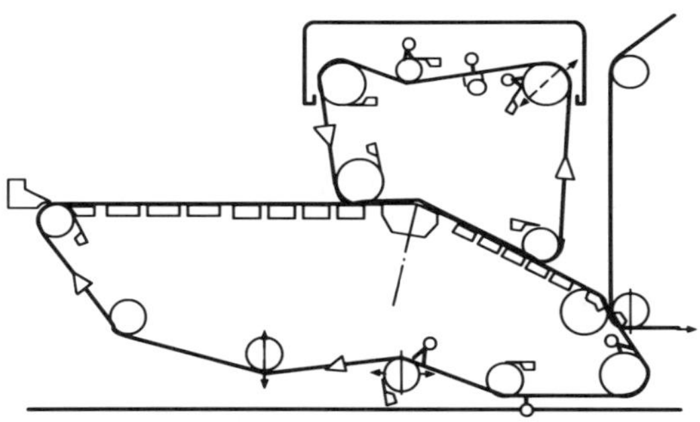

Fig. 674. Location of showers on the top wire of a hybrid or twin-wire former.

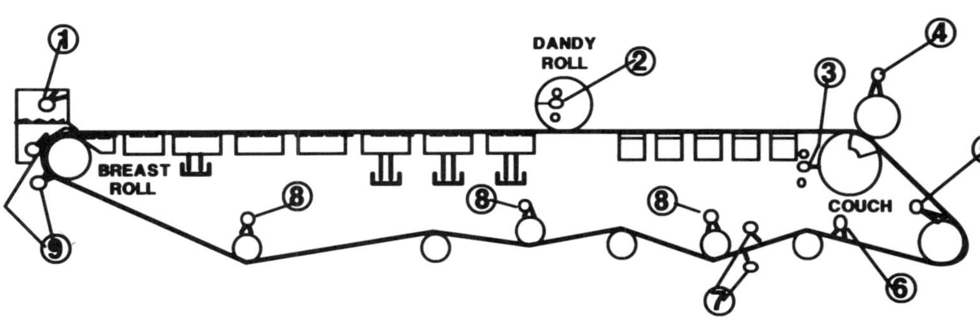

1. HEADBOX
2. DANDY ROLL
3. COUCH ROLL
4. LUMP BREAKER
5. FLOODED NIP
6. KNOCKOFF
7. WIRE CLEANING
8. WIRE RETURN ROLLS
9. BREAST ROLL

Fig. 673. Location of showers on a fourdrinier.

Table 71. Press section showers.

Application	Function	Shower type	Nozzle spacing mm (in.)	Operating pressure kg/cm^2 (psi)	Flow g of water/ g of felt	Flow (lb. of water/ lb. of felt)
Felt	High-pressure cleaning	Osc.	150 (6)	14-25 (200-350)	1st press 0.1	(0.1)
	Flooding	Sta.	75 (3)	4.2-5.6 (60-80)	2nd press 0.08	(0.08)
	Cleaning assm. lip lube	Sta.	175 (7)	2-3 (30-40)	3rd press 0.06	(0.06)
Grooved roll	Low-speed cleaning	Sta.	75 (3)	4.2-5.6 (60-80)	0.74-0.85	(0.49-0.56)
Venta-nip, etc.	High-speed cleaning*	Osc.	150 (6)	14-35 (200-500)	0.13-0.19	(0.083-0.125)
Suction rolls	High-pressure cleaning	Osc.	150 (6)	25-50 (350-700)	0.10-0.14	(0.063-0.096)

Osc. — Oscillating; Sta. — Stationary
* — over 425 m/min (1400 ft/min)

2. Press section showers

Table 71 shows the various applications of showers in a typical press section. As with Table 70, the application, function, shower type, nozzle spacing, operating pressure and recommended flow are shown.

1. For each felt in a press section, a high-pressure needle-jet cleaning shower, a flooding shower, and a suction box lubricating shower normally are used. The function of the high-pressure cleaning shower is to remove contaminants from the surface of the felt. The flooding shower is used to put these contaminants into suspension so they can be removed by the felt suction box. Also, a flooding shower can be used for wetting the felt prior to start-up and for applying chemical for those applications requiring such treatment. The felt suction box lubricating shower provides a lubricant between the interface of the felt and the wear surface of the suction box.

   The amount of water applied to the felt by these three showers should not exceed 0.1 g (0.1 lb) of water per 1 g (1 lb.) of felt in the first press position, 0.08 g (0.08 lbs.) of water per 1 g (1 lb.) of felt in the second press position, and 0.06 g (0.06 lbs.) of water per 1 g (1 lb.) of felt in a third press position. These flows are predicated on the assumption that an appropriate vacuum system is installed for the purposes of water removal.

   Cleaning and conditioning of press felts is a detailed subject. The recommended approach varies from grade to grade and within specific grades based on furnish, fabric design and other operating conditions. This subject is discussed in detail in Refs. [3-5]. Reference should be made to these publications for detailed examples of water flow calculations, vacuum requirements and other criteria.

   When dealing with contaminants that can only be removed using chemical treatment, the correct approach and recommendation should be sought from a consultant specializing in this field.
2. Grooved roll showers are used primarily to keep the grooves void of water and solid contaminants. On machines that operate slower than 425 m/min (1400 ft/min), a stationary shower with nozzles on 75 mm (3 in.) centerlines is recommended. The distance from the sprayed surface and alignment of this shower is critical for optimum operation. For machines running faster than 425 m/min (1400 ft/min), an oscillating needle-jet shower is used to clean solid contaminants from the roll grooves. For either shower, a fixed-orifice nozzle, brush-type or purgeable nozzle can be used depending on process water quality.
3. Suction roll showers are high-pressure oscillating needle-jet showers. These are installed outside of the roll and are used to dislodge contaminants from the drilled holes so that the void volume of the roll can be maintained. Usually these showers are supplied with fixed-orifice nozzles intended for use with filtered fresh water.

3. Dryer fabric cleaning

The cleaning and conditioning of dryer fabrics is necessary to maintain their permeability, and surface smoothness. How permeability affects efficiency is given detailed attention in Chap. XXII, Sec. C, and in Ref. [6].

Cleaning and conditioning of these fabrics is usually done by the intermittent operation of either a stationary fan-type shower or an

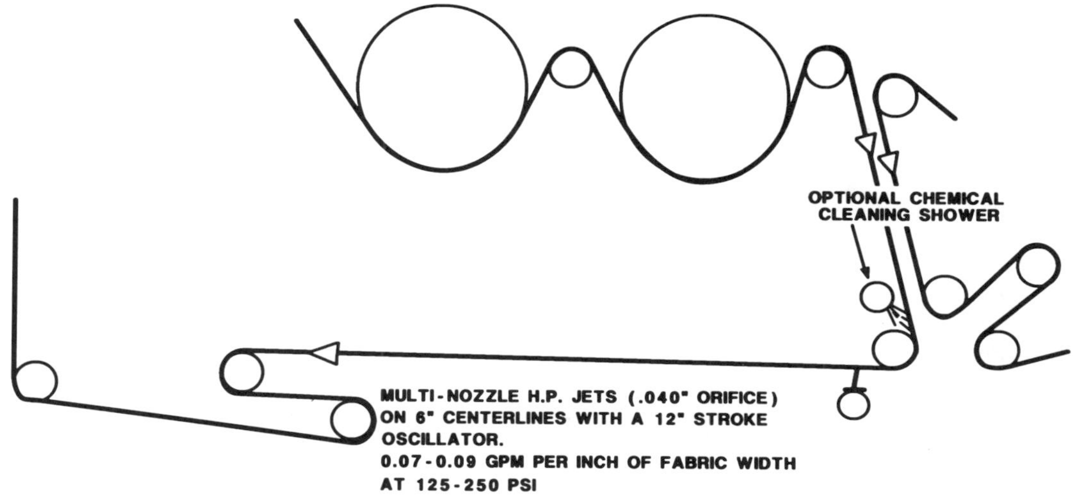

Fig. 675. Recommended location of dryer fabric cleaning showers.

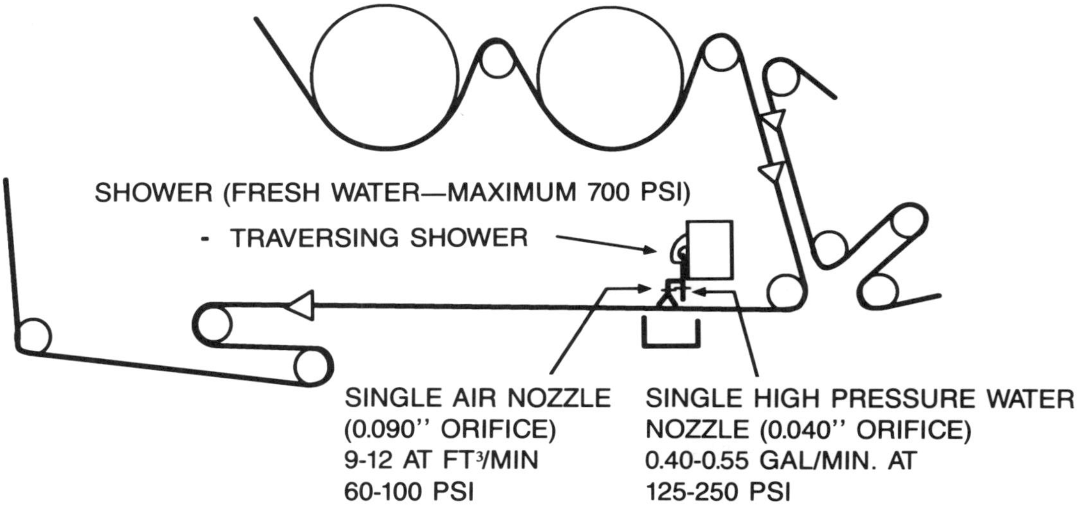

Fig. 676. Recommended location for single-jet showers applied to dryer fabric cleaning.

oscillating high-pressure shower designed with alternating needle-jet nozzles and fan nozzles. In this application, intermittent use refers to cleaning during a break or other downtime. Becoming more popular for certain grades of paper and contaminants is the use of a single-jet traversing shower used on a continuous basis from the time the fabric is installed new.

Figure 675 illustrates a typical intermittent dryer fabric cleaning installation while Fig. 676 illustrates a continuous dryer fabric cleaning application. These illustrations are intended for general information only as the cleaning of dryer fabrics varies from grade to grade and type of contaminants involved. A complete publication [7] is available on this subject.

4. Shower maintenance

Maintenance of paper machine showers is a relatively simple matter. During every shift, any filters associated with showers should be checked and thoroughly cleaned. Approximately every year, a shower should be removed from the machine for complete nozzle maintenance. This usually involves removing the old nozzles and replacing them with new ones. This will ensure the integrity of the fan pattern and flow capacity.

Approximately every six months, the oscillating mechanism on oscillating showers should be rebuilt to factory specifications. During this procedure, all bearings, wear sleeves and guide rods should be checked and repaired if necessary. The final procedure should be verifying proper alignment of the shower so that it moves freely within the dimensions of its stroke.

E. DOCTORS

A doctor is a mechanical device designed to maintain a blade in a fixed position under uniform pressure against a roll for the purpose of cleaning contaminants and water from the roll and, in certain applications, such as top and center press rolls, removing the sheet from the roll in the event of a sheet break. When used in conjunction with a shower on wire return rolls, the doctor can be an effective tool for maintaining the cleanliness of forming fabrics. Most contaminants on paper machine clothing will transfer from the fabric onto a smooth roll. A doctor used to remove these contaminants will prevent them from building up and recontaminating or damaging the forming fabric.

1. A doctor assembly

Figure 677 illustrates the various components of a doctor assembly.

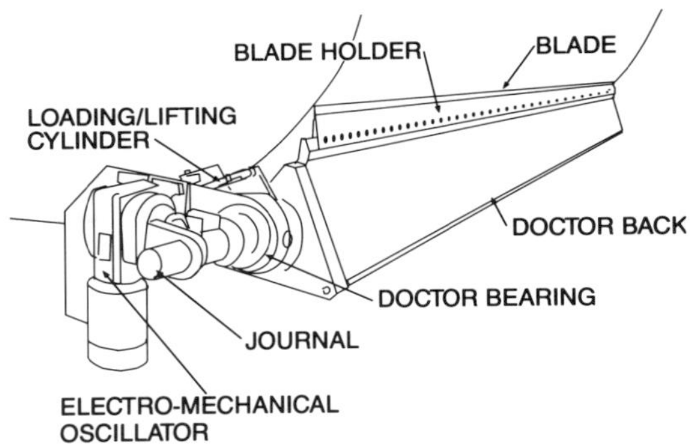

Fig. 677. The various components that constitute a doctor assembly.

2. Doctor back

The doctor back must be designed and manufactured to demanding specifications of rigidity. It is this structure which supports the blade holder and blade in position. For wet-end applications, the doctor back must be clad in high-quality stainless steel to prevent corrosion. For dry-end applications, a painted mild steel structure may be sufficient.

Doctor backs are designed with various cross-sections dependent upon the machine width. Anything from a simple angle to a complicated three- or four-piece weldment is necessary to ensure overall deflection of no greater than 0.13 mm (0.005 in.). For applications such as wire return rolls, some press roll applications and for some dryer applications, special accessories such as deflectors, scoops, white water collection trays and water down spouts must be designed as an integral part of the doctor back.

3. Doctor blade holder

The doctor blade holder is that portion of a doctor which retains the blade in its operating position. The holder must be designed for quick and easy insertion and removal of the blade.

There are two types of doctor blade holders. One is a rigid-type holder which when adjusted to a particular roll does not have the ability to automatically change should a change in the roll surface occur.

A second type of doctor holder is a flexible or self-conforming type of holder. This holder when adjusted to a particular roll can automatically compensate for a change in the roll deflection due to nip loading changes or if used in conjunction with a variable crown roll. These holders employ a tube which can be either hydraulically or pneumatically loaded. The tube applies sufficient pressure to maintain good doctoring while maintaining sufficient flexibility to conform to roll surface changes.

4. Doctor blade

The doctor blade is that portion of a doctor assembly which comes in contact with the roll surface and actually performs the ultimate function of a doctor. Doctor blades are available in many different metallic and non-metallic materials. Each type of material varies in hardness and is applied according to the application and the material of the roll surface. The best doctor blade for the application often is determined through a method of trial and error. This procedure can be expedited by consulting an expert in this field.

5. Journals

The journal is the component of a doctor assembly that supports the doctor back in a bearing arrangement located on the tending side and drive side of a paper machine. The journal must be designed to withstand not only the weight of the doctor back, but also the forces applied during operation. The location of this journal is extremely important. A doctor, without any external loading forces, should fall by gravity onto the roll with a force no greater than 9–13 kg/m (0.5–0.75 pli). For this reason, it is difficult to relocate a doctor from one position to another without relocating or redesigning the journal arrangement.

6. Bearings

The doctor bearings support the doctor in position by retaining the journals. Bearing design varies drastically and therefore it is important to choose the correct bearing for the application. The correct selection of bearing will depend on whether the doctor is stationary or oscillated, applied on the wet end or dry end, or applied in a high ambient temperature dryer section.

7. Loading and lifting

The mechanisms used for loading and lifting a doctor can vary from a simple gravity load cam lift-off lever to a spring-loaded manually operated barrel assembly or a more sophisticated pneumatically operated cylinder arrangement. It is important to note that loading/lifting mechanisms are used in conjunction with rigid blade holders only. The self-conforming pneumatic tube type of holder requires that the doctor back be rigidly fixed by turnbuckles during operation. If this precaution is not taken, the self-conforming holder will oppose the forces of the loading mechanism, thus defeating its purpose.

8. Oscillators

The most common types of doctor oscillators are pneumatic which employ air ride type devices, hydraulically operated cylinders or electromechanical systems consisting of a motor gear reducer combination driving an eccentric cam. The purpose of the doctor oscillator is to move the doctor back and forth at a controlled rate of speed. This movement results in a cutting or skiving action which assists the blade in removing contaminants from the roll.

Also, doctor blades wear with a feathered edge. Oscillation constantly removes this feather and keeps the blade edge well honed. This feathering phenomenon applies to all doctor blade materials except high-density, high molecular weight polyethylene. On applications such as wire return rolls where this type of blade is used, it is not necessary to oscillate the doctor as the edge wears smooth.

F. DOCTOR APPLICATIONS

1. Forming section doctors

Whether the forming section consists of a fourdrinier, twin-wire former or hybrid former, doctors are used on the breast roll, inside wire rolls and outside wire rolls. As shown in Fig. 678, a breast roll doctor requires a deflector to ensure

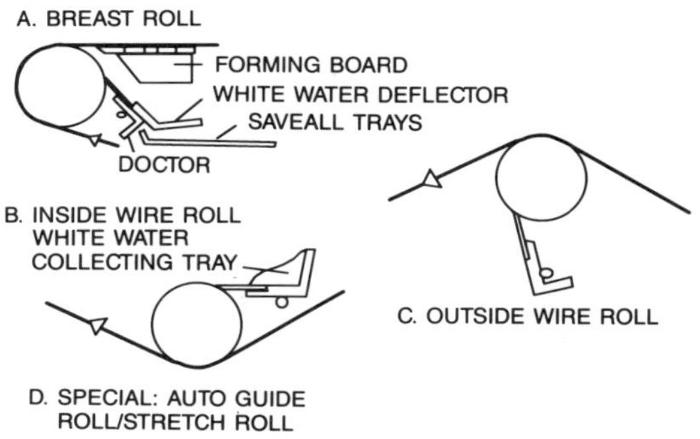

Fig. 678. Typical forming section doctor applications.

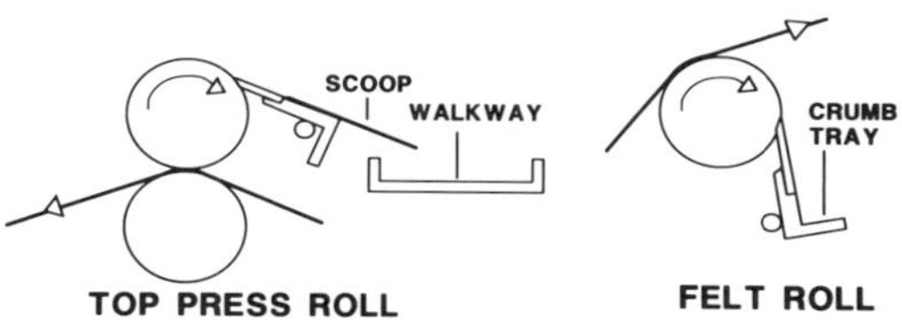

Fig. 679. Press section doctor applications.

that the white water removed by the doctor is directed into the saveall tray. For inside wire rolls, a white water collector tray with down spouts is necessary to direct the flow of water from the doctor to either side of the forming fabric. For doctors applied to the outside wire roll of a fourdrinier, no special water handling mechanism is necessary as the water can flow directly into the wire pit. On twin-wire and hybrid formers, this application requires specially designed saveall trays to collect water and other debris removed from the roll.

When either the inside or outside wire roll is an auto guide roll or stretch roll, special brackets are required to ensure that the doctor maintains its position relative to the roll at all times during operation.

2. Press section doctors

In the press section, doctors are applied to top press rolls, face-side felt rolls and center press rolls.

Figure 679 illustrates a doctor located on the top press roll of a straight through press. The main function of this doctor is to remove the sheet from the top roll in the event of a break. The operation of this doctor is critical because, if the sheet is not removed from the roll and returns to the nip, it will cause wad burns on the press fabric rendering it unusable. This sheet removal function applies much greater forces to the doctor compared to a doctor used for cleaning purposes only. Therefore, the doctor back must be designed to be exceptionally strong and should also include a scoop to direct the sheet into either a walkway or a sheet conveyer.

One of the greatest opportunities to complement cleaning and conditioning of press fabrics is by installation of a doctor on a sheet-side felt roll. Contaminants such as paper fines, fiber and other solids have a tendency to transfer from the press felt onto the surface of a smooth roll. Doctoring this roll removes these contaminants, thus preventing them from building up and recontaminating the press fabric. A doctor used in this location normally includes a crumb tray which is required to prevent debris from falling into the press section.

Figure 680 illustrates a doctor applied to the center roll of a press section. This application is similar to a top press roll in that the doctor must be designed to remove the sheet in the event of a break. The nips on these types of presses are heavily loaded which makes sheet removal extremely difficult. This doctor is probably the most critical of any doctor used on a paper machine. Because of the difficulties experienced with this application, many doctor suppliers are presently designing new units with two holders rather than one. This approach further ensures removal of the sheet which otherwise will cause damage to the press rolls and press fabrics should it not release.

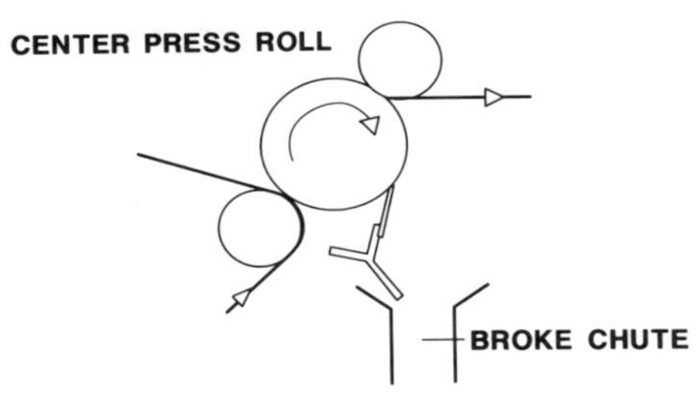

Fig. 680. Center press roll doctor application.

Figure 681 illustrates a grooved press roll doctor application. In this application, the doctor is used in conjunction with a shower for the purpose of removing contaminants and water film from the roll surface. The mechanical demands on this doctor are not as great as the top press roll nor center press roll, and therefore lighter structures can be used. Oscillating this doctor could cause damage to the grooves in the roll and therefore the doctor should be designed for stationary operation.

3. Dryer doctors

Typically, dryer doctors are installed on the lead in and lead out can of each dryer section. However, when the grade of paper and the type of furnish cause excessive buildup on intermediate dryer cans, doctors can also be applied for the purpose of keeping them clean, thus improving heat transfer between the dryer surface and the sheet of paper.

Figure 682 illustrates the typical application of doctors in these three positions. When a doctor is installed as shown in "A" of Fig. 682, the doctor back acts as a crumb tray to collect debris thus preventing it from falling back into the dryer section.

When a doctor is applied on the last dryer as shown in "B" of Fig. 682, it is designed and installed to direct the sheet into the broke chute.

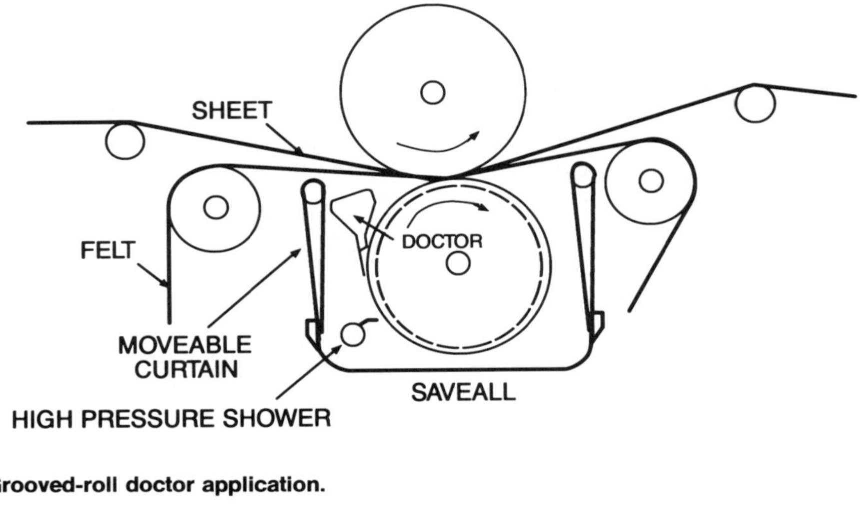

Fig. 681. Grooved-roll doctor application.

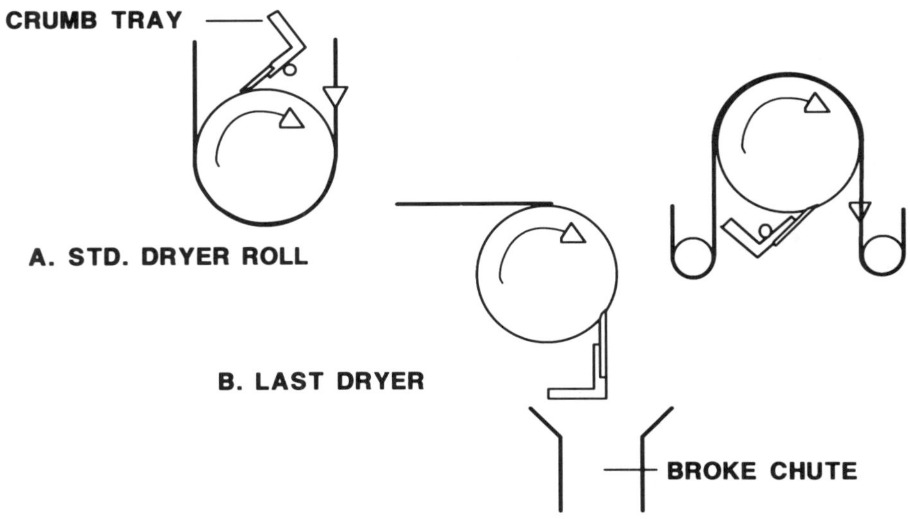

Fig. 682. Dryer section doctor applications.

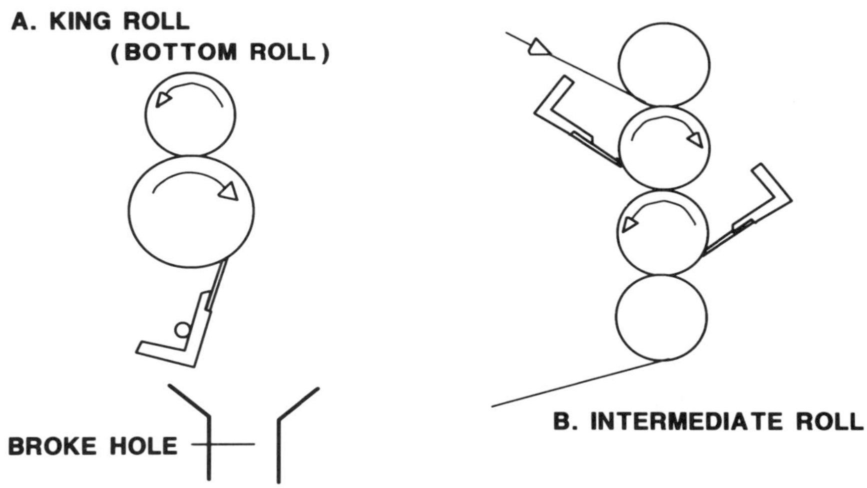

Fig. 683. Calender roll doctor applications.

4. Calender doctors

Figure 683 illustrates the application of doctors in a calender section. When installed on the bottom or king roll of a calender stack, the doctor's function is to remove the sheet in the event of a break and direct it into the broke chute. Again, because of the sheet removal function, the doctor back must be of heavy duty design to perform effectively.

The intermediate roll doctors are primarily used to keep rolls clear of contaminants and therefore can be of a lighter design.

5. Reel doctors

Figure 684 illustrates a reel doctor. In this application, the function of the doctor is to remove the sheet from the reel in the event of a break. Because of the sheet removal function, a heavy duty doctor assembly is required.

G. DOCTOR INSTALLATION

Most start-up and operating problems relative to doctors are associated with incorrect or incomplete installation. A doctor is a precision piece of mechanical equipment which requires meticulous attention relative to installation and alignment. The following list highlights the various procedures relative to installation of doctoring equipment.

1. When handling doctors, they must be slung from their journals. Slinging a doctor from its center could cause damage to the doctor back structure and/or doctor blade holder.
2. The doctor brackets must be located exactly according to the manufacturer's drawings. Careful and accurate measurement before drilling the paper machine frames is critical. Ensure that, when brackets are installed, there is no interference preventing their movement which is necessary for alignment.
3. After installation of the brackets, install the bearings on the journals and then lower the doctor and bearings onto the bracket. At this point, do not tighten the bolts. Rather just make them snug so that the doctor can be moved during the alignment procedure.
4. Alignment gauges should be installed in the doctor blade holder. Following this, a level gauge as shown in Fig. 685 is used to level the doctor. Readings are taken at each end and then a shim of appropriate thickness is placed under the low end for the purpose of raising the doctor until it is level.
5. The next step is to adjust the angle of the blade relative to the roll based on the manufacturer's specification. Figure 686 illustrates the use of an inclinometer (angle

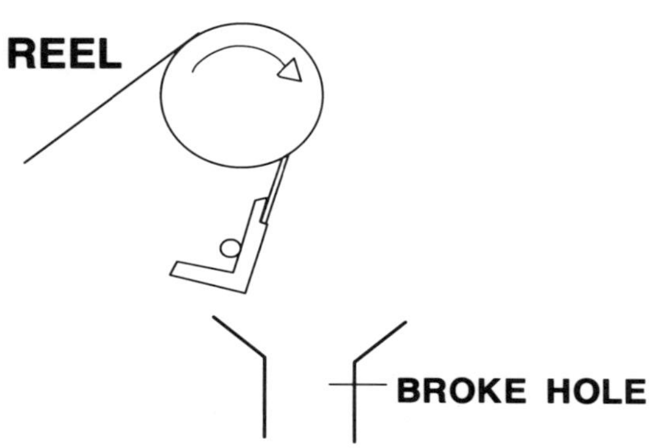

Fig. 684. Reel doctor application.

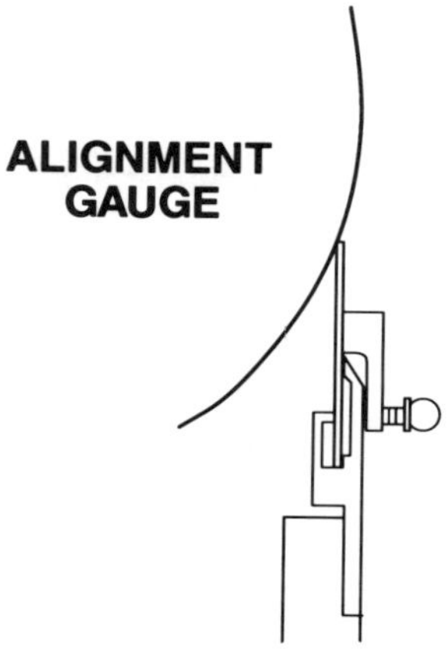

Fig. 685. The use of a level gauge for leveling doctors.

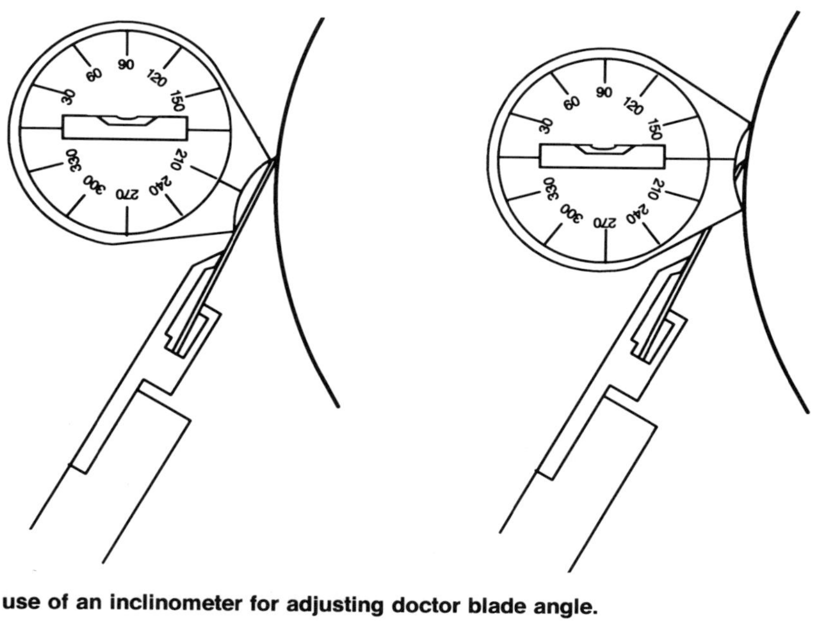

Fig. 686. The use of an inclinometer for adjusting doctor blade angle.

BLADE WEAR PATTERNS, CAUSE AND CURE

SYMPTOMS	BLADE APPERANCE	PROBABLE CAUSE(S)	REMEDY
UNEVEN WEAR FREQUENT BLADE CHANGE	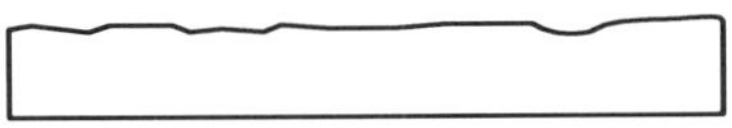	A. POOR ROLL SURFACE B. DAMAGE OR DIRTY BLADE HOLDER C. DEPOSITS	A. USE MORE RLEXIBLE BLADE B. CLEAN, REPAIR OR REPLACE HOLDER C. USE ABRASIVE BLADE
EXCESSIVE WEAR IN CENTER OF BLADE OR AT ENDS	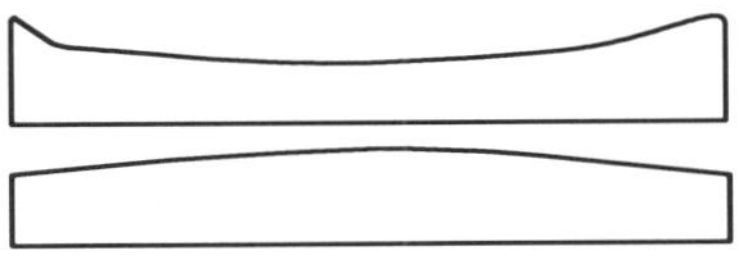	A. DOCTOR NOT MACHINED TO MATCH ROLL CROWNS B. OLD DOCTOR SAGGING INTO (OR AWAY FROM) ROLL	A. REFIT AND SHIM HOLDER TO DOCTOR BACK AS NECESSARY (RETURN NEW DOCTOR FOR REPLANING IF MISMATCH IS EXCESSIVE B. REPLACE OLD, SAGGING DOCTOR
EXCESSIVE WEAR ON ONE END	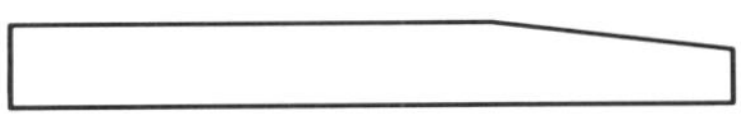	A. DOCTOR MISALIGNED- NOT PARALLEL TO ROLL AXIS	A. SHIM, OR MOVE BEARINGS & BRACKETS AS NECESSARY TO ALIGN DOCTOR PARALLEL TO AXIS
EVEN WEAR BUT ROUGH EDGE	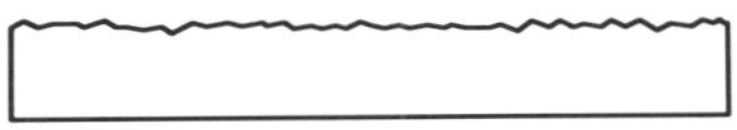	A. DEPOSITS ON ROLL B. EXCESSIVE PRESSURE ON BLADE C. BLADE MATERIAL TOO HARD	A. USE ABRASIVE BLADE AS REQUIRED B. REDUCE PRESSURE C. USE SOFTER BLADE
EXCESSIVE WEAR IN SHEET RUN	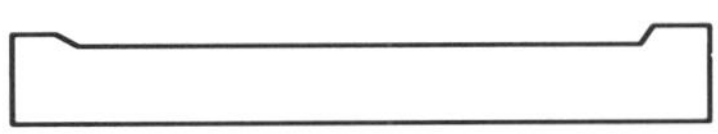	A. DEPOSITS LEFT ON ROLL BY SHEET	A. USE ABRASIVE BLADE AS REQUIRED B. TRIM BLADE EDGES FREQUENTLY C. USE END SLOTTED BLADES TO PREVENT BUILD-UP OF EXCESSIVE PRESSURE ON ROLL ENDS
EXCESSIVE WEAR OUTSIDE SHEET RUN	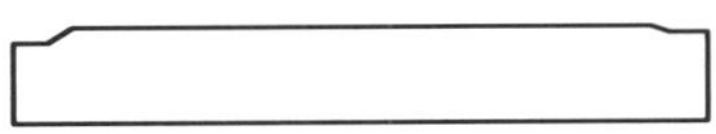	A. SHEET LUBRICATES ROLL	A. SLOT BLADE ENDS TO RELIEVE PRESSURE
LOCALIZED PITTING	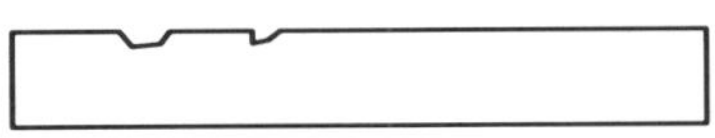	A. BLADE ATTACKED BY 1. ELECTROSTATIC DISCHARGE 2. ELECTROLYSIS 3. HEAT	A. INSULATED DOCTOR FROM MACHINE FRAME: OR USE NON-METALLIC BLADE
BLADE DEVELOPE FEATHERED EDGE	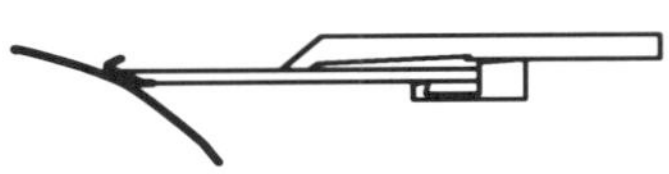	A. BLADE MATERIAL TOO SOFT B. EXCESSIVE PRESSURE C. BLADE ANGLE TOO FLAT	A. USE HARDER BLADE B. REDUCE PRESSURE C. MOVE BEARINGS OR BRACKETS TO INCREASE ANGLE. CHECK FOR CORRECT ANGLE WITH GAUGE
HOOKS ON BLADE OUTSIDE ROLL SURFACE	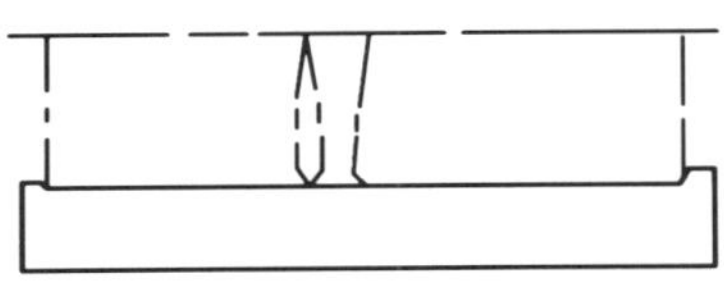	BLADE OVERHANGS ROLL A. NOT CENTERED B. TOO LONG C. NOT OSCILLATING PROPERLY	A. CENTER BLADE B. TRIM LENGTH C. ADJUST OSCILLATION STROKE
DENT IN BLADE EDGE	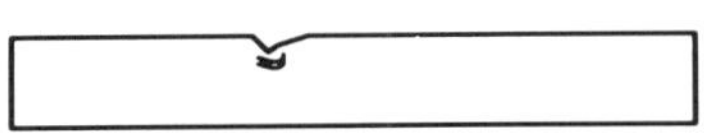	A. BLADE CAUGHT SCAB ON ROLL AT START-UP. (MOST COMMON ON OLD DRYERS.)	A. CHANGE BLADE AND REMOVE SCAB

Fig. 687. Observing the wear pattern of doctor blades is a useful approach to troubleshooting.

gauge) for the purpose of making this adjustment. The doctor should be designed so that it can be easily moved back and forth until the specified angle is achieved.
6. Now it is necessary to install the doctor blade and lightly load the doctor against the roll. The blade holder should now be adjusted in finite increments to ensure even pressure between the doctor blade and the roll. A sense for uniform pressure can be achieved by sliding a thin feeler gauge between the doctor blade and the roll.
7. At this point, any auxiliary equipment such as loading mechanisms and oscillators should be installed according to the manufacturer's drawings and specifications. When complete, these mechanisms should be tried to ensure that there is no malfunction such as binding or mechanical interference.
8. When complete, the final and most important procedure is to drill and pin or dowel all brackets to the frames, bearings to the brackets and oscillators to the brackets. This procedure is the only way to ensure that the doctor will maintain proper alignment during operation. Also, if the doctor must be removed to perform maintenance on the roll, it can be relocated exactly as previously installed.

H. DOCTOR MAINTENANCE

Once put into service, doctors are often neglected. A few simple routine maintenance procedures can ensure continued efficient operation of this equipment.

The first and most obvious routine is to change the doctor blade at frequent intervals as necessary. As a doctor blade wears, the angle of contact with the roll changes, thus making the doctor less efficient. When a doctor blade has worn down from its original width by 6–12 mm (0.25–0.5 in.), it is time to change it.

The next most important component to monitor is the doctor blade holder. Any damage to the holder will affect the uniformity of pressure between doctor blade and roll. A damaged holder should be replaced immediately.

During every shift, a doctor should be visually inspected to ensure that the loading mechanisms and oscillators are operational. When found to be defective, they should be repaired as soon as practical.

One of the most useful methods of troubleshooting doctor problems is a visual inspection of the doctor blade. Figure 687 illustrates various blade wear patterns relative to probable causes and possible remedies. This method of troubleshooting is most useful.

I. SUMMARY

Showers and doctors, when applied to the paper machine either individually or collectively, can result in improved paper machine fabric life and have an extremely positive effect on product quality and paper machine efficiency. The main considerations in selecting equipment for these applications involve not only the correct choice of hardware but also the correct location and operation of each component.

The need for routine maintenance cannot be overemphasized.

REFERENCES

1. Green, M.R., "White Water Showers: Optimum System Design", S. Pulp Paper (July 1985).
2. "Cleaning of Forming Fabrics", Appleton Wire Division, Albany International Corp., Publication CL2 200-43 (Mar. 1985).
3. Cutts, M.M. and Smith, J.C., "Recommended Felt Conditioning for Coated and Uncoated Fine Paper Grades", Albany Felt and Fabric Facts 26(12).
4. Cutts, M.M. and Smith, J.C., "Recommended Felt Conditioning System for Kraft Grades", Albany Felt and Fabric Facts 28(8).
5. Mikkelson O.P. and Salls, D.J., "Recommended Felt Conditioning System for Tissue Machines", Albany Felt and Fabric Facts 34(1/2/3).
6. Dickens J.H. and Khan A.Q., "Dryer Fabric Cleaning: A Way to Reduce Steam Consumption and Improve Sheet Quality", Tappi J. 97(9):71 (1988).
7. "Cleaning and Conditioning of Dryer Fabrics", Albany International Corp. (Aug. 1989).

XX PAPER MACHINE AND WINDER DRIVES

R.P. DERRICK

Robert Derrick received a B.E. degree in electrical engineering from Vanderbilt University in 1954. His entire career has been in the pulp and paper industry serving in the Industry Engineering Department at Westinghouse Electric Corporation, Process Control and Electrical Engineering Design Group Manager at Simons-Eastern Consultants, and Corporate Senior Staff Electrical Engineer at Union Camp Corporation. He holds National Certification as a Professional Engineer and is a registered engineer in eight states. Mr. Derrick has been an active member on both TAPPI and IEEE pulp and paper electrical engineering committees and is a past chairman of each committee. He is a Fellow Member in both TAPPI and IEEE and has been the recipient of the division awards in each association. He has been editor and contributing author to several books, written over 40 technical papers and reports, and is the author of several patents. His field of specialization has been the application of advanced design systems for paper machine and winder drives. His biography has been published in "World's Who's Who in Commerce and Industry" as well as "Who's Who in America".

Drives for paper machine and winder applications have demanding requirements. They must be:

- Highly reliable
- Highly accurate, and have
- Low maintenance requirements.

In addition, the drives must:

- Provide adequate power under all normal operating conditions,
- Provide the necessary power for acceleration and deceleration,
- Provide the necessary speed ranges,
- Provide the necessary operational functions and features.

A. TYPES OF DRIVES

1. Direct current

Since about 1960, the adjustable voltage, direct current (dc) electric sectional drive has been the most common type of drive on machines and winders. This type of drive operates on the principle that a dc motor's speed varies almost proportionally with the applied voltage. The basic drive is configured as shown in Fig. 688. Silicon Controlled Rectifiers (SCR) are used to convert conventional alternating current (ac) to dc. The dc voltage is controlled by the "gate" of the SCR from the regulator.

Up until about 1982, the dc drives utilized analog speed regulators: a reference dc voltage, proportional to the desired speed, was compared to the voltage produced by the tachometer on the motor shaft. The error difference in voltage between the desired speed and the actual speed was then amplified by the regulator and used to control the gate firing of the SCRs to adjust and control the voltage applied to the motor. The accuracy of these dc analog speed regulators was approximately $\pm 0.1\%$ of top speed under steady-state operating conditions. In the later types of analog regulators, a master frequency was compared to the pulses produced by a pulse

---

REVIEWED BY

John M. Bentley
Principal Engineer
Allen-Bradley Company Inc.
Milwaukee, WI

Patrick T. McNulty
Vice President Engineering
Simons-Eastern Consultants Inc.
Decatur, GA

Thomas O. Trueb
Chief Engineer
Dean Oliver & Associates
Duluth, GA

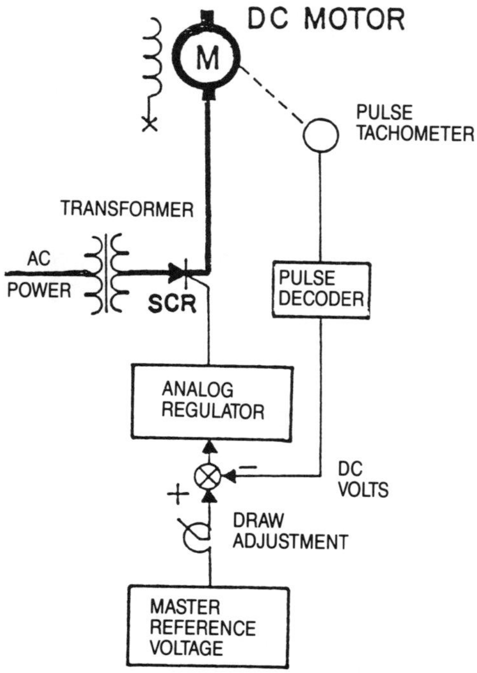

Fig. 688. Pulse analog drives, 1965–1980 era.

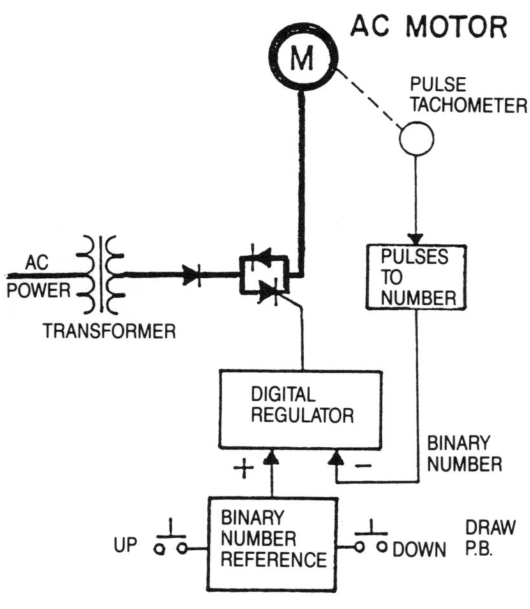

Fig. 689. Adjustable frequency numerical AC drive.

digital tachometer. This design reduced the draw control and tachometer maintenance required and improved the accuracy of the regulator to approximately ±0.01% of top speed under steady-state operating conditions.

By the mid-Eighties, the analog technology was replaced by the numerical digital regulator using microprocessors. The pulse signal from the speed sensor was converted to a binary number and compared to a binary reference signal in the microprocessor. The entire regulator was now digital using algorithms to produce the necessary regulator functions. These types of regulators are usually more stable, reliable, and easier to maintain than analog regulators. Digital regulators also interface more easily with mill-wide computer systems. By 1990, all new paper industry regulators were of this type.

The accuracy of the digital regulator under steady-state conditions is limited to approximately ±0.01% of top speed primarily due to the mechanical components of the machines. These include the stiffness of shafts, couplings and roll shells as well as by the back-lash in gearing and universal joints. Under transient conditions the accuracy is limited by the torque available, the section inertia, and the tuning of the regulator parameters.

2. Alternating current

In the early Eighties, adjustable frequency alternating current (ac) sectional drives were developed in Europe and Japan. This technology was successfully introduced into the United States in the mid-Eighties and has been applied to coaters and supercalenders as well as winders and paper machines. This type of drive operates on the principle that an ac motor's speed varies almost proportionally with the applied frequency. The basic drive is configured as shown in Fig. 689. Power semiconductor devices of several types are used to produce variable frequencies from the mill's fixed frequency electrical service. This is accomplished by first rectifying the mill power into dc power. The dc power is then switched on and off, and in opposite directions, by semiconductor devices to produce adjustable frequency ac power for the motor. This switching is controlled by the numerical digital regulators in much the same way as this is accomplished on dc drives. The prime advantage of the ac drive is the far more reliable ac motor which requires less maintenance and repair than dc motors (Fig. 690). Other principal advantages of an ac drive are:

- Lower installation costs since motor ventilation ducts are not required.
- Lower installation costs since smaller and more standard motor cables may be used.
- Lower operating costs due to better power utilization, i.e., improved efficiencies, power factor, harmonics, etc.

Fig. 690. AC motors in the drive aisle of a large kraft paper machine. The totally enclosed, fan-cooled, cast iron frame AC motors are the prime advantage of AC drives.

The ac drive has achieved rapid early success in European, Japanese, and North American mills and is expected to become the drive of choice in the Nineties. Most major drive manufacturers now offer ac drives to the industry.

B. TYPES OF REGULATORS

Several types of regulators are required to serve the needs of winders and paper machines:

- The basic drive regulator is for the regulation of speed as previously described. It is functionally illustrated in Fig. 691. Note that, within the regulator, speed is controlled by adjusting the output torque of the motor.
- Tension regulators are used to adjust the speed of a section to control the tension in the sheet ahead of that section. As machines become faster, the application of tension regulators on dry end sections will increase.

Absolute simple rules cannot be stated for regulator type selection, but the following general rules usually apply:

- Each section will always have a speed regulator to control the speed of that section.
- Sections with more than one motor will usually have a torque regulator to control the

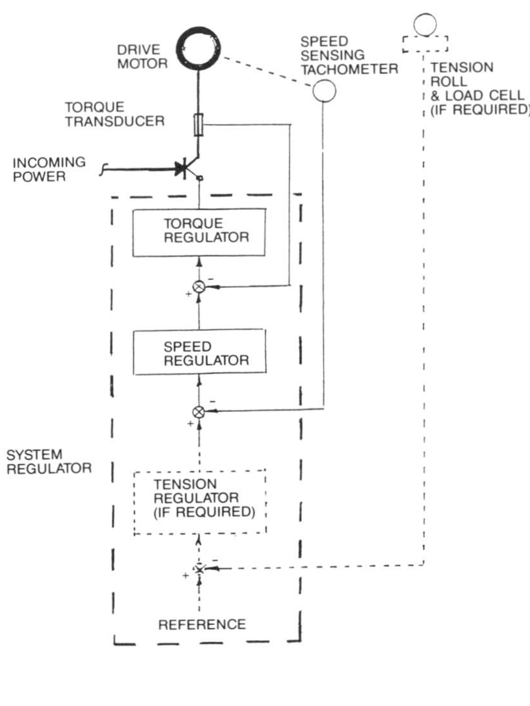

Fig. 691. Basic drive regulator.

Fig. 692. Multiple drive line-ups on a machine drive.

division of load torque between motors.
- Electric backstand winders, reel sections, as well as some other dry end sections, will usually require tension control. Tension control is also applied to size press and coater areas of the paper machine.

The section regulators and power electronics are usually housed in individual cabinets (one or more for each section). The cabinets then are mounted in multiple drive line-ups with their associated interconnecting electrical bus and wiring (Fig. 692). A safety electrical disconnect device is located in each drive for "lock-out". Often each lineup is served by a single power distribution transformer which reduces the distribution voltage (e.g., 13 800 V) to the utilization voltage[1] of 480 to 600 V. These transformers are rated 1000 to 2500 kVA depending on the rating of the drives in a particular line-up. The drive line-ups are located in electrical equipment rooms which are air conditioned and may be humidity controlled. Charcoal filters are usually applied for corrosion protection of the sensitive microelectronics and electronic components, wiring, and connectors.

C. OPERATOR'S CONTROL FUNCTIONS

The paper machine drive will require a means of setting the machine master speed. Machine master speed changes are typically about 1 m/min/s (3 ft/min/s). This may be a function of the process distributed control system or a separate control panel. In addition, a "hard-wired" Master Machine Emergency Stop push-button or switch is frequently applied.

Each section of the machine is usually controlled from a local panel, but may also be controlled from the process distributed control system. Typical section acceleration rates are 6–8 m/min/s (20–25 ft/min/s). The section control may include any or all of the following functions:
- Run and Stop buttons.
- Jog (forward and reverse) buttons for clothing changes [this is usually 15–20 m/min (50–65 ft/min) and functions only so long as the button is held depressed].
- Crawl button [this is usually 15–30 m/min (50–100 ft/min) and is a maintained function. It is released by the Stop button].
- Slow Run button operates the section at approximately 90–150 m/min (300–500 ft/min) to "run-in" clothing, warm-up dryers, etc. (It is released by the Stop button.)
- Slack Take-up button increases a section's speed momentarily as long as the button is depressed. This is *not* usually a progressive function. The increase in speed is dependent upon the length of open draw and is typically 1–4 m/min (3–12 ft/min).
- Draw Mode and Draw Increase and Decrease buttons adjust the section's speed for draw. Most modern machines fix the speed of the forming section and provide progressive draw for downstream sections. Typical draw adjustments are 0.5 m/min/s (1.5 ft/min/s).
- Match Speed button matches the section's speed to that of the preceding section for thread-up. Draw control is then restored by depressing the Draw button.
- Tension Mode and Tension Increase and Decrease buttons adjust the tension from the preceding section. In the past, swing rolls were used to control sheet tension. Direct tension load cells are now increasingly being used on dry end sections. The section can be returned to the Draw Mode by depressing the Draw Mode button. Some load cell controls are capable of automatically returning the section to the Speed Mode upon a sheet break. Some drive systems "remember" a tension and will return to the previous tension value when placed into the Tension Mode.
- Indicating lights, or other graphics, should be provided for diagnostics such as Ready,

[1] In the future, it is expected that ac machine drive utilization voltages will increase to several thousand volts as components become more available.

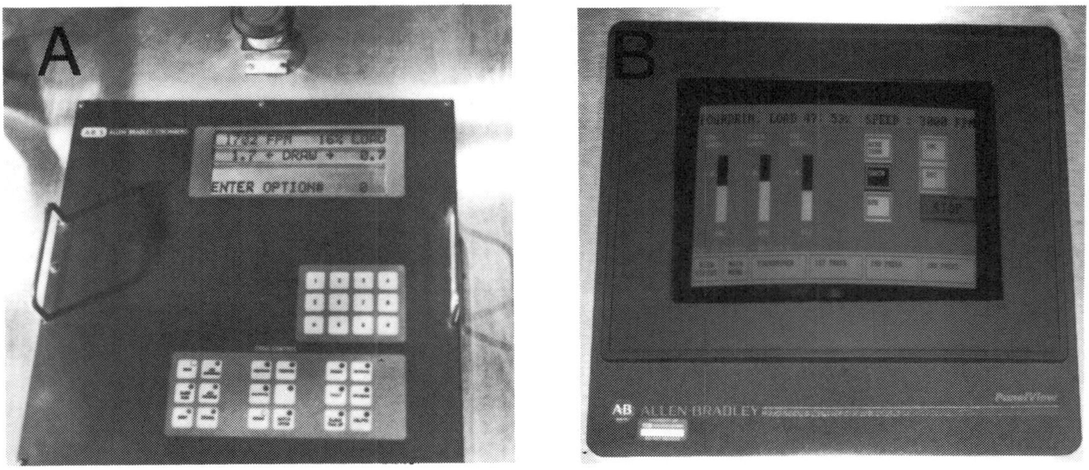

Fig. 693. Typical modern operators' controls. A) Operators' drop-in control station, mounted in the benchboard in the tending aisle. B) "Smart" CRT drop-in control station.

Process Fault, Electrical Fault, etc. Some graphics are capable of precisely describing the fault that prevents the drive from running or causes the drive to stop.

- Numerical displays to indicate the section's speed, percent load, draw from the preceding section and draw to the following section, and section Mode (i.e., Tension, Torque, or Speed).
- On multi-motor sections, Load Share buttons may be provided to control the division of torque between motors. This may also be an automatic feature on some drives as a function of total load, nip pressure, or other process variables such as roll hardness or diameter on a winder.
- In addition to the above operator controls, winder drives will also provide Pay-out and Take-up control of the backstand. Modern winder drives usually also have controls to automatically wind a roll to a preset length or preset diameter.

In the past, operator control devices were often discrete devices mounted on operators control consoles. In the mid-Eighties, "drop-in" operators' panels were developed with all the devices on a single panel (Fig. 693) which, in the event of trouble, can be easily and quickly changed out as a complete unit. These "drop-in" panels were linked to the drive control cabinets by a digital link in a single cable. In 1990, color graphic "touch screen" CRTs were developed as "drop-in" panels and are expected to be applied as "smart" operators' panels in the future. They are referred to as "smart" in that they change their screens depending upon the operator's needs such as clothing change, startup, normal operation, sheet-break, normal shutdown, diagnostics, etc. In this way, the operators are presented with only those displays, controls, and data which are actually needed at a given time thus reducing the opportunity for operator error. The "smart drop-in" panels also provide diagnostics such as "why the section won't start" and "why the section stopped".

Interlocking of the drive to other process equipment is usually accomplished in either a programmable logic controller (PLC) or in the distributed control system. The required interlocking varies widely between machines and can include any of the following:

- Fan Pump to shut down in the event the fourdrinier stops.
- Fourdrinier to shut down in the event that the couch or turning rolls slips on the fabric.
- Presses to unload and the suction pickup roll to raise in the event that the fourdrinier stops.
- Machine to stop, after a very short time delay, in the event of a lubrication failure.
- In the event of a sheet break, the proper sections stop to drop the sheet into the available broke chest.
- On multi-motor sections, all motors to stop if an individual motor "trips out".

Some machines provide for automatic startup and shutdown of all equipment within a process subsystem.

D. RELIABILITY AND MAINTENANCE

Electrical drive systems applied since the Fifties have typically enjoyed an excellent record for reliability. These simple drives typically had a single motor per section, about 10–12 sections,

Table 72. Drive lost time causes. Lost time: min/month/drive motor.

Cause of downtime	Type of Drive System: DC analog pulse-frequency	DC digital (numerical) regulators	AC digital (numerical) regulators
Motor	1.0	1.0	0.3
Tachometer	0.5	0.5	0.4
Power supply	0.3	0.2	0.3
Regulators	1.4	0.6	0.6
Operator's & logic control	0.7	0.5	0.4
Master & reference control	0.3	0.2	0.2
Wiring & grounds	0.5	0.3	0.3
Trip-outs & unknown causes	0.9	0.6	0.6
Total:	**5.6**	**3.9**	**3.1**

and used motor-generator sets and analog vacuum tube type of regulators. Nevertheless, a 1954-55 TAPPI Survey noted an annual average downtime of about 4.6 min/day per machine or approximately 11.6 min/month per motor. This translates to an availability[2] of 99.68% of the machine drive system and an availability of an individual section of 99.98% of the time! With such a high reported reliability, it is not surprising that the variance among mills was greatly dependent upon the reporting procedures, preventive maintenance, availability of spare parts, and the qualifications of the maintenance personnel.

With the introduction of solid state regulators and SCR power devices in the Sixties, availability of the individual drives improved significantly. However, the number of drives on the newer, larger, faster machines also increased. It became common to have multiple drives on all wet end sections and as machine speeds increased so did the number of dryer sections. Thus the overall machine availability did not improve and, in fact, actually decreased on some machines! In the Seventies, additional diagnostic devices were provided on the drives which did help improve the overall reliability somewhat but perhaps most of the improvement was realized by better training of the electricians responsible for drive maintenance. In the Seventies, a modern drive (with frequency-to-pulse regulators) had an average lost time of 5.6 min/month per motor. Thus a typical machine with 20 drive motors had an average lost time of 112 min/month or an overall availability of 99.74%.

The introduction of the digital regulator in the Eighties further reduced the average lost time to about 3.9 min/month per motor. A machine with 20 drive motors had an average lost time of 78 min/month or an overall availability of 99.82%.

The improved up-time of ac motors on adjustable frequency drives further reduced the average lost time to only about 3 min/month per motor. A machine with 20 drive motors could be expected to experience only about 60 min/month of downtime, i.e., 12 h/year. This is an overall machine drive reliability of 99.85%. Some machine drives can expect to experience better availability and some somewhat worse. In a 2.5 year period at least one machine drive had an ac drive reliability of above 99.95%.

Reliability in this class can, and often is, completely destroyed by a single instance of a major failure, especially if spare parts and capable trained personnel are not readily available. A typical breakdown of the causes of drive downtime is illustrated in Table 72. It can also be expected that, as this modern sophisticated equipment ages, downtime may increase, especially if it has been subjected to hostile environments.

In the mid-Eighties, personal computers were introduced for maintenance, tune-up, and diagnostic capabilities for digital regulators. With suitable computer programs, and in the hands of well-trained personnel, this equipment should further aid in rapidly determining the cause of drive difficulties. Some of the equipment and programs are capable of communicating with trained factory service personnel over telephone lines using computer modems.

E. POWER REQUIREMENTS

1. General

In North America, standard electric motors are currently rated in horsepower (746 W = 1 hp). In the rest of the world, standard electric motors are usually rated in kilowatts (kW). The standard sizes are shown in Table 73 and are *not* equivalent. In the following discussion, the

[2] Availability is defined as:

$$\% \text{ Availability} = \left(1 - \frac{\text{lost time}}{\text{total time}}\right)(100)$$

Table 73. Standard motor ratings.

NEMA US horsepower			ISO/IEC metric kW		
10	100	700	11	100	710
15	125	800	16	125	850
20	150	900	20	160	1000
25	200	1000	25	200	1120
30	250	1250	31	250	1400
40	300	1500	40	315	1600
50	400	1750	50	400	1800
60	500	2000	63	500	2000
75	600	2250	80	600	2240

Motor kW = 0.746 (Motor Rated hp)
Motor hp = 1.34 (Motor Rated kW)

STANDARD TRANSFORMER RATINGS
(Base kVA Ratings for 13.8 kV : 480 & 600 V)

15	75	225	750	2000
30	112.5	300	1000	2500
45	150	500	1500	3000

calculation methods presented will allow the selection of standard drive motors rated in horsepower or kilowatts.

Paper machine and winder drive power requirements are generally considered to be of a "constant torque" characteristic and the *power constants* are expressed as hp per inch of width at 100 ft/min or kW/m of width at 1 m/min*. Thus, each machine has a *machine drive constant*, K_m, based on width and maximum rated speed, i.e., "gear-in" speed. The product of K_m and the proper *power constant* determine the expected drive power required:
English units:

$$K_m = (0.01)\text{ (inches width) (max. ft/min)}$$

Multiply by 0.7740 to obtain K_m in metric units (meters width) (m/min).
Example: A 300 in. wide machine with a rated top speed of 2500 ft/min would have a K_m of:

$$K_m = (0.01)(300)(2500) = 7500$$

Appropriate *power constants* have been determined from analysis of actual load studies for the most common sections of paper machines and winders. These studies are published as TAPPI Reports and condensed in the appropriate TAPPI Technical Information Sheets (TIS). For other sections of machines, the general data shown in Table 74 may be considered. Two power constants have been established: The Normal Running Load (NRL), and the Recommended Drive Capacity (RDC)*.

* Multiply hp/in./100 ft/min by 0.9636 to obtain NRL and RDC constants in kW/m at 1 m/min.

2. Normal Running Load (NRL)

The NRL is the expected running load of equipment when operated under rated conditions with typical vacuum, felting, doctors, dryer drainage, sheet tensions, acceleration/deceleration, etc. The suggested values are for modern, well-maintained machines, properly aligned, adequately balanced, and equipped with anti-friction bearings. The NRL *power constants* have been developed from several sources, including:

- Critical review of published current and past load data from available sources.
- Mechanical/process/mathematical analysis of the equipment to establish theoretical power requirements.
- Acquisition and analysis of raw data from numerous machines under typical normal operating conditions.
- Consideration of the current recommendations of machinery builders and others knowledgeable in the field.

Engineering and mathematical (usually statistical) analyses of all data so acquired are conducted and preliminary recommendations proposed for comment from all concerned individuals and TAPPI Committees. The Committee's published final report provides the basic development and data for the suggested NRL and RDC constants and may be viewed as the prime reference source.

After all pertinent comments and additional information are reviewed, the suggested power constants are published as TAPPI Technical Information Sheets.

Table 74. Typical power requirements.

Equation 1

Typical roll NRL power constants (hp/in. @ 100 ft/min) for usual "friction" loads:

a. Small rolls (>20″ dia): 0.00025 to 0.00040, e.g., table rolls, paper rolls, & Mt. Hope
b. Medium rolls (20″ to 30″): 0.00040 to 0.00050, e.g., wire rolls, coater rolls, calender, reels, etc.
c. Large rolls (30″–72″): 0.00050 to 0.00100, e.g., breast, wire turning, press, etc.
d. Suction rolls (30″–72″): 0.00100 to 0.00200, e.g., couch, suction press, dryer, etc.

Add "work" loads of Eq. (2), as applicable and check acceleration load by Eq. (3), as required.

Equation 2

Typical NRL power constants (hp/in. @ 100 ft/min) for usual "work" loads

a. Stationary suction box loads (e.g., fourdrinier flat boxes): $0.0015\ \mu \cdot wv$ where μ, the effective coefficient of friction (varies 0.1 to 0.2 depending upon materials) and w is box width, inches and v is vacuum in inches of mercury.
b. Nip loads, e.g., presses & calenders

$$\begin{pmatrix}0.00005 \text{ to}\\ 0.00006\end{pmatrix}\begin{pmatrix}\text{nip}\\ \text{pli}\end{pmatrix} + \begin{pmatrix}\text{roll \& possible}\\ \text{hydraulic loads}\end{pmatrix} + \begin{pmatrix}\text{felt, wire \&}\\ \text{carrier roll loads}\end{pmatrix}$$

c. Sheet tension loads: $0.003\ (\Delta\ T)$ where T is tension in pli.
d. Doctor blade and similar blade loads: $0.003\ \mu \left(\frac{\text{load}}{\text{pli}}\right)$ where μ is the coefficient of sliding friction (varies 0.15 to 0.30, dependent upon materials)

Equation 3

Typical roll RDC power constants are at least 115% to 125% of the NRL and the RDC for a "high inertia" rolls and sections is usually determined by the acceleration/deceleration requirements.

a.
$$\begin{matrix}\text{hp/in.}\\ \text{@ 100 ft/min}\end{matrix} = \frac{1}{T_{pu}}\left\{\begin{pmatrix}\text{NRL hp/in.}\\ \text{@ 100 ft/min}\end{pmatrix} * \left[9 \times 10^{-4}\left(\frac{WK^2/in}{D_o^2}\right) R\right]\right\}$$

b.
$$\begin{matrix}\text{hp/in.}\\ \text{@ 100 ft/min}\end{matrix} = \frac{1}{T_{pu}}\left\{\begin{pmatrix}\text{NRL hp/in.}\\ \text{@ 100 ft/min}\end{pmatrix} * \left[2 \times 10^{-7}\left(\frac{D_o^4 - D_i^4}{D_o^2}\right) R\right]\right\}$$

* Sign is plus (+) for acceleration and minus (−) for regenerative deceleration.

Where T_{pu} is per unit torque available for acceleration/deceleration, WK^2/in. is inertia per unit width, lb-ft^2/inch; R is acceleration rate, ft/min/s; D_o and D_i are outside and inside roll diameters, inches.

Equation 4

Duty cycle loads, e.g., winder, off-machine coater, supercalender, etc.:

$$\text{RMS HP} = \sqrt{\frac{\Sigma\,(hp_i^2 \cdot t_i)}{t_t}}$$

Where $\Sigma\,(hp_i^2 \cdot t_i)$ is the sum of the individual values of $hp^2 \cdot t$ for each portion of the duty cycle and t_t is the total cycle time, including idle time.

	Unwind	Windup
Acceleration	$hp_a - hp_t$	$hp_a + hp_t$
Deceleration	$-hp_d - hp_t$	$-hp_d + hp_t$

where hp_a and hp_d are the actual acceleration and deceleration horsepower [e.g., Eq. (3)], except omit the $1/T_{pu}$ factor and hp_t is the tension horsepower [Eq. (2c)].

3. Recommended Drive Capacity — RDC

The RDC is the suggested drive *minimum* rating for equipment operated under maximum anticipated load conditions. The difference between RDC and NRL reflects several considerations:

- A minimum application margin of 15% as a "safety factor" for the machine which requires *slightly* more power than the typical machine studied due to vacuum conditions, felt or sheet tensions, clothing and material characteristics, and *minor* maintenance difficulties (e.g., lubrication, tight vacuum or steam joints, minor misalignment, etc.). The typical RDC/NRL factors for application margin range from 1.15 to 1.33 for typical machine sections with *well-defined* and *known* operation conditions.
- Some machine sections (e.g., dryer) require additional power for the acceleration of inertia, i.e., the NRL is low compared to the acceleration requirement. The RDC constant for these sections provide typical acceleration power plus the normal running load *without any* application margin applied.

Unless otherwise specified, the suggested *power constants* assume drive equipment with 1.0 service factors and 150% torque – 60 s ratings (200% torque – 60 s ratings are specified for dryer, calender, and most winder acceleration loads). Drives with 1.15 service factors may provide an additional margin of power for temporary unusual service conditions, for above the standard ambient temperature of 40°C, or abnormal load conditions. Applications with acceleration/deceleration times in excess of 60 s will require derating of the standard 60 s rating or specially designed equipment for the acceleration/deceleration period.

Applications such as winders, supercalenders, and off-machine coaters involve operating duty cycles and dictate that the average motor heating load (RMS) and peak loads (e.g., acceleration and deceleration) be determined. The calculation method is shown in Table 74, Eq. (4). Unless definite application knowledge indicates otherwise, it is suggested that the applied motor rating be *at least* 115% of the calculated RMS rating and capable of the short time peak rating for acceleration and deceleration periods.

Motor locations subject to accumulations of pulp and/or coating material should consider derating the allowable temperature rise of motors since these materials are excellent thermal insulators. For example, a standard motor is rated 105°C rise above a 40°C ambient at 1.0 or 1.15 service factor. It is common to specify only 90°C rise at 1.15 service factor in critical applications subject to accumulation of pulp and/or coating materials.

Standard ratings for motors and transformers are given in Table 73. Currently sectional drive motors are being applied up to about 1800 kW (2000 hp). This is expected to increase as larger and faster machines are developed.

4. Transformer and cable ratings

The power distribution transformers applied to paper machine and winder drive line-ups are rated for only a 30°C ambient and usually have two ratings: self-cooled and force-ventilated. Special transformer designs are required for ambient temperatures above 30°C or installations at elevations above 1000 m.

Paper machine drive transformers should be selected such that the self-cooled rated kVA is not less than 133% of the sum of the NRL kW (100% of NRL horsepower). The force-ventilated rated kVA should not be less than 133% of the sum of the RDC kW. The drive's incoming line equipment, including bus work, should be rated to match the transformer rating and braced for the maximum available short circuit current.

Winder drive transformers should be similarly selected except that consideration must also be given to the duty cycle of the winder. The heat transfer characteristics of transformers are quite dependent on the type of transformer and its individual design. The winder duty cycle should always be specified to the transformer manufacturer's design engineers.

In recent years, cast coil design transformers have found favor as distribution transformers for drive systems as well as in other areas of the mill. This transformer's encapsulated design has a proven high reliability and very low maintenance requirements when all exposed bus work is tin plated and spray coated with insulation. The construction is environmentally sound and suited to withstand harsh airborne contaminants and high humidity characteristics of many paper mills.

The required power and control cable for drive systems should be selected and installed completely in accordance with local codes as well as the National Electric Code requirements for wet areas. Two and four conductor cables should be used for dc circuits with the positive and negative conductors in the *same* cable. Some designs for large ac motors use more than one inverter and thus require an even number of motor cables. Control, signal, and communication cables should exactly follow the recommendations of the drive system design engineer and are usually installed in steel conduit for magnetic shielding. Careful attention to grounding design is required to prevent interference in signal circuits. (It is expected that fiber-optic signal and communication cables will see increased application in the future.)

5. Speed reducing gears

Appropriate service factors for pulp and paper industry applications are suggested in the TAPPI Technical Information Sheet "Service Factors for Gears Used on Major Equipment in the Pulp & Paper Industry". These service factors, thermal and mechanical, are applied to the electrical drive *nameplate rating* and must encompass the entire operating speed range of the drive at constant torque, continuous duty, loads. Machines with wide speed ranges will require additional design consideration of the gear thermal rating and *may* require external lubrication systems, e.g., a central lubrication system, if the ambient temperature expected is above 30°C (86°F). Typical ambient temperatures in the drive aisle of many paper machines are at least 40°C (104°F) and many may approach 50°C.

Detailed application data for applying speed reducers in the paper industry have been presented at TAPPI Engineering Conferences and are recommended reading for those involved in drive systems.

Three basic considerations are involved in selecting a speed reducer for a paper machine drive:

1. The recommended Thermal Service Rating (hp or kW) which may be described as that continuous power rating at which the gear lubricant reaches a constant operating temperature in the specified ambient air. The thermal rating of the speed reducer must be at least equal to the motor rating (at *maximum* gear-in speed) and encompass the entire operating speed range. If the motor provides a 1.15 service factor rating, the speed reducer should also provide a 1.15% thermal rating.
2. The recommended Mechanical Service Rating (hp or kW) which may be described as that maximum continuous power rating which will not damage the gear teeth or cause excessive wear. The mechanical rating should be specified at the motor *base* speed (i.e., maximum torque) and should be at least 1.25 times the motor nameplate rating for 1.0 service factor motors or 1.44 times the motor nameplate rating for 1.15 service factor motors, i.e., 1.15 × 1.25 = 1.44.
3. The required gear tooth ratio should be specified as between 105% and 113% of the motor base speed.

Table 75. Available gear ratios. (Both are available from most U.S. suppliers.)

U.S.A. American Gear Manufacturer's Association	Intl. Standards
1.84	2.03
2.25	2.49
2.76	3.05
3.38	3.74
4.13	4.57
5.06	5.60
6.20	6.86
7.59	8.40
9.30	10.29
11.39	12.61
13.95	15.44
17.09	18.91
20.93	23.16

Coupling service factors should not be less than twice the motor nameplate rating.

It should be noted that gear units, unlike motors, *do not* have standard thermal and mechanical power ratings.

Example: Consider a speed reducer for a 38 in. diameter press roll on a 2500 ft/min machine driven by a 400 hp, 1150 rpm motor with a 1.15 service factor. The motor is operated from an individual power supply and provided with fixed field excitation.

a. The minimum thermal rating should be (1.15) (400 hp) = 460 hp at the gear-in speed determined in (c) below. The gear thermal rating must be suitable for operating continuously at constant torque over the entire speed range of the paper machine.
b. The minimum mechanical rating should be (1.44) (400 hp) = 576 hp at the motor base speed of 1150 rpm.
c. The roll speed is:

$$\text{rpm} = \frac{12\ (\text{ft/min})}{\pi D} = \frac{(12)\ (2500)}{38\pi} \text{ or } 251.3 \text{ rpm}$$

For a motor gear-in speed between 1200 and 1300 rpm, the gear ratio *must* be between 4.775 and 5.173:1 and a standard ratio of 5.06:1 (Table 75) will provide a satisfactory motor gear-in speed of 1271 rpm.

6. Mechanical considerations

The optimum performance of a modern drive system is usually limited by the response obtainable within the mechanical components. The mechanical equipment must be kept in good repair, properly aligned, balanced, and lubricated. The natural frequency (f_n) of the components *will limit* the response time (t_r) obtainable from the drive system. The period of the mechanical natural frequency should be one order of magnitude (i.e., 10:1) greater than the desired drive response time if it is not to affect the regulating system, i.e., f_n must be greater than 10 divided by t_r. Thus for a response time of 1 s from a speed regulator, the mechanical system *must* be designed for an f_n *greater* than 10 Hz. Backlash in gearing and universal joints accentuates any torsional problem which may be present and should be kept to a minimum for optimum performance. Modern drive regulators usually have available optional "torsional compensation" circuits (or programs) which should be specified for high-performance applications.

Torsional calculations are always required on the high-inertia sections and are quite advisable for all sections on a high-speed, high-performance paper machine. A simplified calculation method for calculating the torsional frequency is illustrated in Fig. 694.

Note that this method assumes that the motor and gear unit shafts, as well as the coupling, are *significantly* stiff compared to the jackshaft. Note also, that *only* rigid couplings should be used on high-performance drives since flexible couplings will significantly degrade the drive performance. Double-engagement gear tooth couplings are the most common recommendations of the drive manufacturers. The required torsional frequency should always be specified to the machinery builder and/or the designer of the jackshafts and universal joints.

7. Forming sections

Fourdriniers, and most other type of forming sections, consist of two types of load elements: rotating and stationary elements.

1. The rotating rolls of the forming section have NRL constants of about 0.00028–0.00095 kW/m at 1 m/min (0.00030–0.00100 hp/in./100 ft/min) as illustrated in Table 74. The suction couch roll will have an NRL constant of about 0.00120 plus (0.00005 times v for each suction box) where v is the vacuum in inches of mercury (Hg).
2. The stationary elements consist of forming boards, foils, and flat boxes:
   - Forming boards, deckle, and deflector losses have been lumped with and distributed within table roll losses. For machines without table rolls, add 10–15% to the calculated NRL to accommodate these losses.
   - Plain foils have friction losses of about 0.00038 to 0.00042 kW/m at 1 m/min (0.00040–0.00044 hp/in./100 ft/min) *per blade*. Vacuum foils have losses of about 0.00043–0.00053 kW/m at 1 m/min *per blade*. (Board grades have the larger values.)
   - The major friction loads on many paper machines, especially those producing the heavier grades, are flat boxes. These NRL constants for these losses are well represented by:

     $[0.0015]\ (u_e)\ (vw)$ hp/in. at 100 ft/min

     where v is the operating vacuum in inches Hg, w is the effective box width in inches, and u_e is the effective coefficient of friction. For ceramic, metal oxides and carbides box covers and 1990 monofilament fabrics, typical values of u_e are 0.12 to 0.14 with a mean of about 0.125.

Example: Consider a machine with a k_m of 7500 manufacturing up to 60 lb/3000 sq. ft. kraft papers up to 2500 ft/min. The machine has 14 table rolls, 5 wire return rolls, a breast and a wire-turning roll, and a suction couch roll operating at 20 in. Hg. The stationary elements consist of 4 plain-5 blade foils, 5 vacuum-7 blade foils, four 12 in. flat boxes at 5 in. of Hg and two 12 in. flat boxes at 7 in. of Hg. Thus the NRL constants for this machine would be:

Rotating elements:	
– 14 table rolls @ 0.00035 ea:	0.0049
– 5 wire return rolls @ 0.00050 ea:	0.0025
– Breast & wire-turning roll @ 0.00100 ea:	0.0020
– Couch @ 20 in. Hg	0.0022
Subtotal rotating elements	0.0116
Stationary elements:	
– 20 plain foil blades @ 0.00042 ea:	0.0084
– 35 vacuum foil blades @ 0.00048 ea:	0.0168
– 48 in. flat boxes @ 5 in. Hg:	0.0450
– 24 in. flat boxes @ 7 in. Hg:	0.0315
Subtotal for stationary elements	0.1017
Total NRL for this fourdrinier (hp/in. @ 100 ft/min)	0.1133

Note that the flat boxes were 90% of the total. Thus this machine's fourdrinier would have a total NRL of (7500) (0.1133) = 850 hp. The RDC should be at least 115% of the NRL or 977 hp. Typical control strategies would be to apply 60% of the total RDC to the wire-turning roll and 40% to the couch:

– Wire-turning roll RDC: 586 hp
– Couch roll RDC: 391 hp

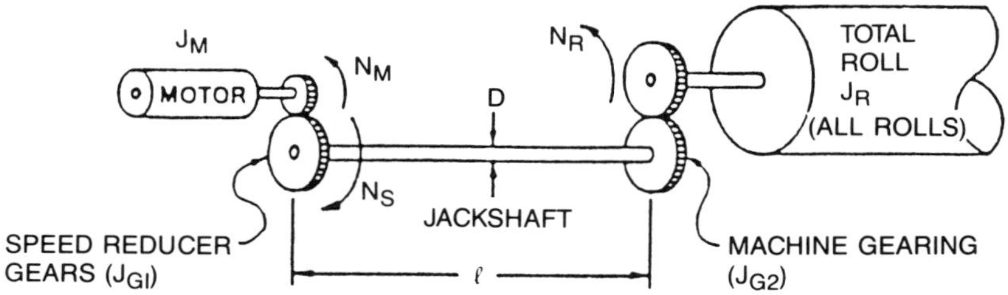

$$J_1 = J_{G1} + J_M \left(\frac{N_M}{N_S}\right)^2 \qquad J_2 = J_{G2} + J_R \left(\frac{N_R}{N_S}\right)^2$$

FOR HOLLOW SHAFTS, $D^4 = D_o^4 - D_i^4$

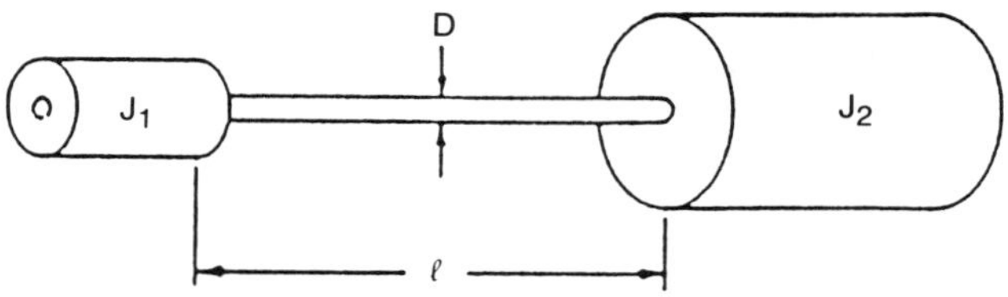

$$fn = \frac{1}{2\pi}\left(\frac{k}{J_e}\right)^{1/2} \text{Hertz} \qquad \text{WHERE} \quad J_e = \left(\frac{J_1 J_2}{J_1 + J_2}\right) \text{ OR } \approx J_1 \text{ WHEN } J_2 \gg J_1$$

k IS THE SHAFT SPRING CONSTANT IN UNITS DEFINED BELOW.

ENGLISH UNITS

$J = \frac{WK^2}{g}$ lb-ft-s² where g is 32.174 ft/s²

$K = \left(\frac{\pi \cdot E_S}{384}\right)\left(\frac{D^4}{\ell}\right)$ lb-ft/rad where D & ℓ are inches

E_S is $\approx 11.5 \times 10^6$ psi for cold rolled steel

$k \approx 0.094 \times 10^6 \cdot \left(\frac{D^4}{\ell}\right)$ lb-ft/radian

For a Specified Frequency, f_n, the required shaft diameter, D, is approximately:

$D \approx \left[(4.2 \times 10^{-4})(\ell \cdot J_e \cdot f_n^2)\right]^{1/4}$ inches

METRIC UNITS

$J = \frac{GD^2}{4g}$ kg-m-s² where g is 9.8066 m/s²

$k = \left(\frac{\pi \cdot E_S}{32 \times 10^9}\right)\left(\frac{D^4}{\ell}\right)$ kg-m/rad where D & ℓ are in mm

E_S is $\approx 8.08 \times 10^9$ kgf-m² for cold rolled steel

$k \approx 0.8 \left(\frac{D^4}{\ell}\right)$ kgf-m/rad

For a Specified Frequency, f_n, the required shaft diameter, D, is approximately:

$D \approx (50 \cdot \ell \cdot J_e \cdot f_n^2)^{1/4}$ mm

EXAMPLE

A dryer section with 16-60 in. (1525 mm) rolls has a total (including gearing) WK^2 of approx. 2 413 050 lb-ft² (GD_2 of 406 840 kg-m²) referred to the indrive shaft speed, N_S. Thus J_2 is 75 000 lb-ft-s² (or 10 372 kg-m-s²). The motor and tachometer have a WK^2 of 115.63 lb-ft² (GD^2 = 19.5 kg-m²). The speed reducer has a 10:1 ratio and a WK^2 of 593 lb-ft² (GD^2 = 100 kg-m²) referred to the low-speed shaft, N_S. Thus J_1 is 378 lb-ft-s² (or 52.26 kg-m-s²) and J_e is 376 lb-ft-s² (52 kg-m-s²). The jackshaft is 6 in. (152.4 mm) in diameter and 72 in. (1829 mm) long with a spring constant, k, of 1.692×10^6 lb-ft/radian (2.36×10^5 kg-m/radian). Thus the system torsional frequency is $\approx$10.7 Hz. To obtain a frequency of $\approx$15 Hz would require a shaft diameter of about 7.1 in. (180 mm) whereas a 10 in. (255 mm) shaft would be required for 30 Hz. Note that, since $J_2 \gg J_1$, $J_e \approx J_1$; thus if a much larger spare motor (larger J_1) were installed, the torsional frequency would actually *decrease*.

Fig. 694. Natural torsional frequency.

At least a 400 hp drive should thus be applied to the couch roll and a 600 hp drive should be applied to the wire-turning roll.

8. Wire drive ability

In addition to calculating the power required by the fourdrinier, it is also necessary to determine the maximum power any given fourdrinier roll can transmit without slipping. With these data, one can intelligently divide the load between the couch and wire-turning rolls as well as select the required rating for any driven wire-return rolls. The maximum amount of power a roll can transmit is affected by:

1. The amount of wrap made by the fabric, the slack side tension, and the coefficient of friction between the fabric and roll:

$$hp = K_m\left[\frac{T_s\,(e^{\mu\alpha} - 1)}{330}\right]$$

   where T_s is the slack-side fabric tension (pli), α is the angle of wrap in radians (57.3° = 1 radian), and e is the base of Naperian logarithm, 2.7182818.

   Typical values of the coefficient of friction (μ) between various roll materials and monofilament fabric are given in Fig. 695.

2. The amount of vacuum (v), size of vacuum box (w), and coefficient of friction (μ) between the roll and fabric:

$$hp = \frac{0.491\ \mu\ v\ w\ K_m}{330}$$

3. If a nip roll is used, e.g., lumpbreaker, the nip also adds to the drive ability:

$$hp = \frac{P_n\ K_m}{330}$$

   where P_n is the nip pressure (pli).

For example, consider the following machine:

- K_m = 7500, as in previous examples.
- Couch has a coefficient of friction of 0.18, a 12 in. wide suction box and a *minimum* operating vacuum of 18 in. Hg. The angle of wrap is 0.78 radians (about 45°).
- Wire-turning roll has a coefficient of friction of 0.30, and an angle of wrap of 2.37 radians (about 136°). The slack side *minimum* fabric tension is 30 pli.
- The lumpbreaker roll has a *minimum* operating nip pressure of 60 pli and a coefficient of friction of 0.18.

ROLL MATERIAL	MONOFILAMENT FABRIC
BRONZE OR STAINLESS STEEL	μ = 0.18 to 0.20
FIBERGLASS ROLL COVERS	μ = 0.22 to 0.25 (DEPENDS UPON FINISH)
RUBBER ROLL COVER	$\mu = 0.004\ (P\&J)^{1.22} + 0.24$ (SEE CURVES BELOW)

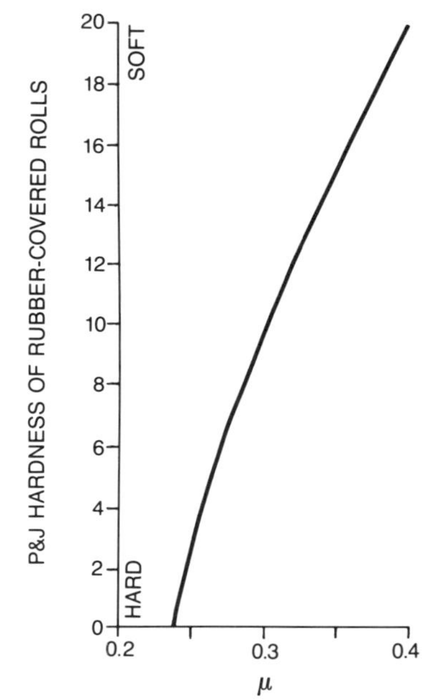

Fig. 695. Coefficient of friction for monofilament fabrics.

Note: $K_m/330 = 22.73$

The wire-turning roll maximum hp is:

$$\begin{aligned} hp &= (22.73)\,[30\,(e^{0.3 \times 2.37} - 1)] \\ &= 706 \text{ hp} \end{aligned}$$

The couch maximum hp is:

$$\begin{aligned} hp &= (22.73)\{30[e^{0.3 \times 2.37}\,(e^{0.18 \times 0.78} - 1)] \\ &\quad + (0.491)\,(0.18)\,(12)\,(18) + (0.18)\,(60)\} \\ &= 889 \text{ hp} \end{aligned}$$

While these values plainly allow the application of the 400 hp couch and 500 hp wire-turning roll drive, it should be noted that a rubber-covered (10 P & J) turning roll was necessary, i.e., coefficient of 0.30, for a bronze turning roll would slip at only 364 hp.

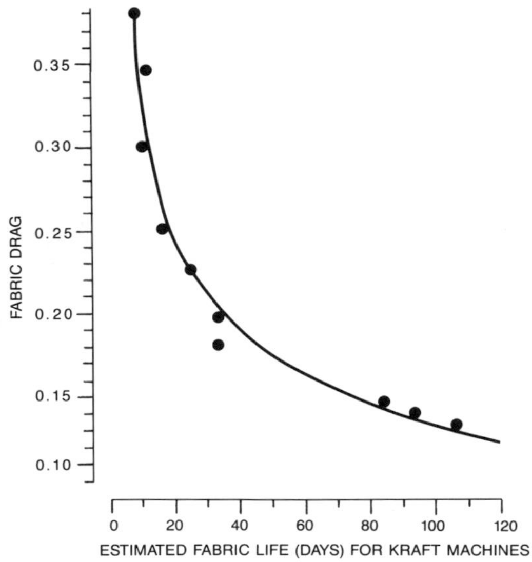

A recent survey* of monofilament fabric life on various machines indicated an average life of 92 days. The higher speed machines typically had the shortest fabric life. The fabric life also varies considerably with the grade:

	Life (Days)
Market pulp	185
Specialty	115
Linerboard	97
Kraft paper	95
Coated papers	73
Tissue/towel	63
Corrugating medium	57
Newsprint	50

* "Paper Machine Clothing", K.L. Patrick and K.H. Ferguson, Pulp Paper Mag. (Aug. 1990).

Fig. 696. Effect of excessive fourdrinier power on expected fabric life.

9. Fourdrinier power effect on fabric life

After considering the power required for the fourdrinier, and the ability of the driven rolls to transmit this power, one must then consider the effect of this frictional drag on fabric life. This is illustrated in Fig. 696. Note that in the previous example an NRL constant of about 0.113 would predict a theoretical fabric life of 120 days. If all 1000 hp available is utilized (RDC of 0.133), the fabric life would be reduced to about 80 days. On board grades especially, fabric life must be considered when selecting drainage elements and applying the required drive power.

10. Multi-wire and multiple fourdriniers

The principle illustrated for calculating the component values may be used for calculating the drive requirements for advanced type formers. At no time, should power be transmitted through the sheet–each forming group *must* provide its own drive power!

11. Regulator control and strategy

The usual control strategy is to speed regulate the wire-turning roll which has a drive capable of 60% of the total fourdrinier. Without a sheet on the machine, the couch roll is regulated to follow the speed of the wire-turning drive. The presence of the sheet is sensed by couch vacuum and the couch is then torque regulated to carry 40% of the total fourdrinier load. The load division between the two drives may be adjustable by the operator within limits, i.e., the turning roll must always carry at least 50% load to avoid slack fabric between the couch and wire turning roll.

Note that, should the fabric start to slip on the wire-turning roll, its load would decrease sharply, shifting the entire fourdrinier load to the couch. Since the couch drive is regulated to share a percentage of the wire-turning drive's load, the couch drive would then have insufficient torque to run. The couch and fabric would thus stall and stop while the wire-turning roll would continue to slip and run within the fabric. This condition will most likely destroy the fabric. Modern drive regulators have speed difference controls to detect this condition and shut down the fourdrinier and fan pump.

The speed regulator for the fourdrinier must be stable under all operating conditions with a fast response time and be very accurate. The current generation of digital regulators have a response time of just under 1 s and a steady state accuracy of $\pm 0.01\%$ of top speed. The regulators are typically tuned for a critically damped response (Fig. 697).

F. PRESS SECTIONS

The dominant load factor for press sections is the basic nip load of each nip. The NRL constant (hp/in. @ 100 ft/min) is well represented by: [(0.00006) (pli) + 0.00015] for roll-type presses.

Other elements are also present, however, including:

- Friction load losses for each roll which are not proportional to nip load but are dependent upon roll type.
- The load associated with each felt and its felt rolls.
- The load associated with each felt conditioning box.

The typical component values for these loads are illustrated in Table 76.

The minimum ratios of RDC/NRL suggested are:

Less than 500 pli:	1.30
501 to 1200 pli:	1.20
1201 to 1800 pli:	1.25

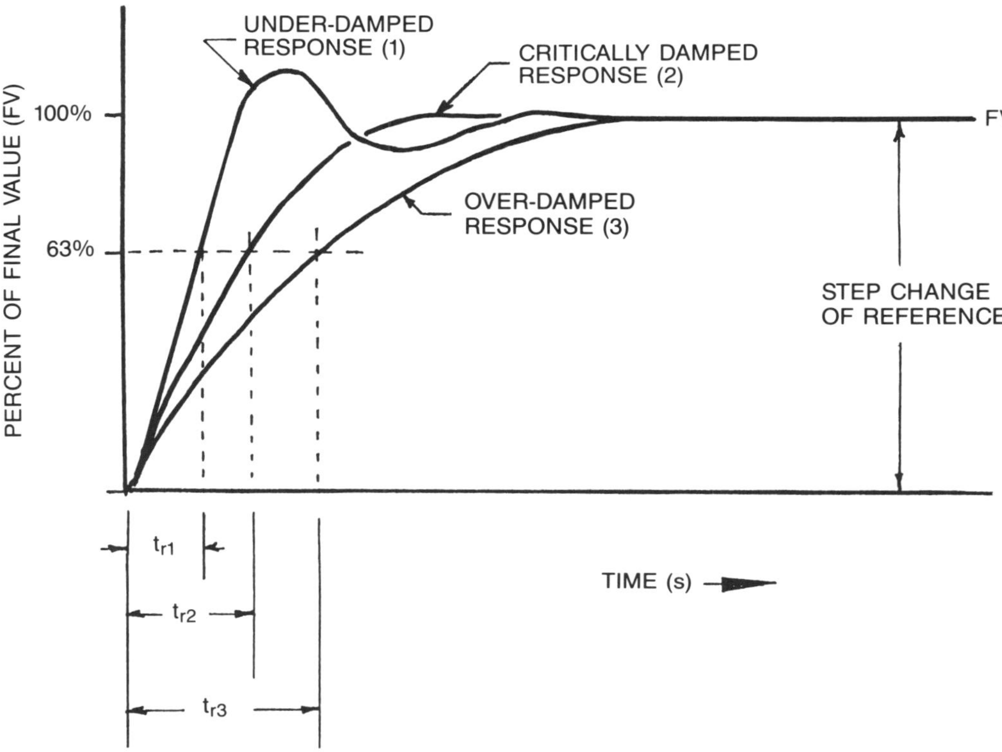

Fig. 697. Speed regulator response to a step change on reference.

Table 76. Suggested NRL power constants, paper and board machine press sections.

Load Element	hp/in. @ 100 ft/min
1. Basic nip load of each nip	[(0.00006) (pli) + 0.00015]
2. Main roll loads for each roll	
a) Grooved roll	0.00100
b) Plain roll — Hard	0.00120
or — R.C.	0.00140
c) Suction roll	0.00200
d) Press roll doctor blades	0.00050 to 0.00100
e) Adjustable crown rolls:	[(0.000 004) (pli × ft/min)$^{0.5}$ + (0.0060)]
3. Felt load	[(0.000 003) (pli) + (0.0007) (No. of felt rolls)]
4. Felt suction conditioning boxes	0.005 per box plus 0.002 for each slot in excess of two

The use of hydraulically controlled adjustable crown rolls on press sections is quite common. The additional load that these rolls represent is quite dependent upon the individual designs of the various manufacturers. The addition load represented by the hydraulic shoe on roll type presses typically is of the form:

$$\text{NRL Constant} = [(0.000004) \times (\text{pli} \times \text{ft/min})^{0.5} + 0.006]$$

Example 1

Consider a 500 pli press (K_m of 2500) with a bottom synthetic felt, 8 felt rolls, and 2 felt conditioning boxes. The bottom roll is a suction roll, the top roll is rubber covered with a doctor blade:

Nip load:
[(0.00006) (500) + 0.00015] = 0.03015
Top roll load: 0.00140 × (1) = 0.00140

Bottom roll load: 0.00200 × (1)	= 0.00200
Doctor blade load: 0.00100 × (1)	= 0.00100
Felt load:	
[(0.000003) (500)] + (0.0007) (8 rolls)	= 0.00710
Felt boxes: (0.005) (2 boxes)	= 0.01000
Total NRL constant	0.05165
RDC content = 1.3 (NRL)	= 0.06715

Thus the NRL is 387 hp and the RDC is 504 hp.

It is suggested that presses above 500 pli and all felted presses have drives on both rolls. Since it is desirable *not* to transmit power through the nip, the loads should be divided as follows:

Top roll drive

½ nip load: ½ (0.03015)	= 0.01508
Top roll load	= 0.00140
Top roll doctor	= 0.00100
Total top roll NRL constant	= 0.01748

NRL hp = 131 hp,
RDC hp = 170 hp

Bottom roll drive

½ nip load: ½ (0.03015)	= 0.01508
Bottom roll load	= 0.00200
Bottom felt	= 0.00710
Felt boxes	= 0.01000
Total bottom roll NRL constant	= 0.03418

NRL hp = 256 hp,
RDC hp = 333 hp

An examination of the NRL constants would indicate a good application would be 200 hp on the top roll, 300 hp on the bottom roll. An alternate application might be 150 hp on the top roll and 400 hp on the bottom roll.

Example 2

Consider a three-roll bi-nip press with all three rolls driven. The first nip has a felted, grooved roll rated 400 pli. There are 6 felt rolls and 1 felt conditioning box on this roll. The center roll is a hard, plain roll with a bottom felt, 8 felt rolls, and 2 felt boxes. The second nip roll is an adjustable crown roll rated 800 pli.

1st roll drive

Nip load:	
½ of [(0.00006) (400) + 0.00015)]	= 0.01208
Top roll	= 0.00100
1st felt:	
[(0.000003) (400) + (0.0007) (6)]	= 0.00540
1st felt boxes: 1 @ 0.00500	= 0.00500
Total NRL constant	= 0.02348

NRL hp = 176 hp,
RDC hp = 1.3 (NRL) = 229 hp

Center roll drive

½ of 1st nip load	= 0.01208
½ of 2nd nip load:	
½ [(0.00006) (800) + 0.00015]	= 0.02408
Center roll – hard	= 0.00120
2nd felt:	
[(0.000003) (800) + (0.0007) (8)]	= 0.00800
2nd felt boxes: 2(0.005)	= 0.01000
Total NRL constant	= 0.05536

NRL hp = 415 hp,
RDC hp = 1.2 (NRL) = 498 hp

2nd roll drive

½ of 2nd nip load	= 0.02408
Adjustable crown roll:	
[(0.000006) (800 × 2500)$^{0.5}$ + 0.0060]	= 0.01166
Total NRL constant	= 0.03574

NRL hp = 268 hp,
RDC hp = 1.2 (NRL) = 322 hp
In this example one would apply:
250 hp 1st roll
500 hp center roll
400 hp 2nd roll

In a multi-nip press, a modern control scheme would be to speed regulate the lead roll, i.e., center roll in the example, and control the torque on the other two rolls proportional to their individual nip pressures, i.e., sense nip pressure as a reference. The speed regulator should typically be tuned for a critically damped response and the response time matched to that of the fourdrinier (Fig. 697).

The various single-roll belt type presses [e.g., Beloit ENP (TM)] are dependent upon the individual designs of the various manufacturers. On this type of press, the dominant nip load is represented by [(0.0125) (nip pli/1000)]. The adjustable crown hydraulic load is typically of the form:

$$\text{NRL Constant} = [(0.0005)\,(\text{ft/min})^{0.5}\,(\text{ID/OD})^{1.5}]$$

where ID and OD are the inside and outside roll diameters. The additional load represented by the shoe and blanket currently is identified as

$$\text{NRL Constant} = [(0.0007)\,(\text{number of rolls}) + (0.003)\,(\text{nip pli}/1000)]$$

The felt load on this type of press is estimated as:

[(0.0007) (number of rolls) +
(0.001) (pli/1000)]

Example 3

Consider a 6000 pli single-roll press for a 2500 ft/min machine (K_m = 7500). The adjustable crown roll has a ID/OD of 0.85. The felt has 6 felt rolls and 4 felt suction boxes. The blanket has 5 rolls.

	NRL constants
Nip load	= 0.0750
Adjustable crown roll load	= 0.0196
Blanket and shoe load	= 0.0215
Felt load	= 0.0102
4 suction boxes @ 0.005 each	= 0.0200
Total NRL constant	= 0.1463
Total RDC = 125% NRL	= 0.1829

Thus the RDC hp is 1372 hp and two 700 hp drives should be applied to this press.

G. DRYER SECTION

While the power requirements for dryer sections have been well known for many years, they are perhaps the *most* misunderstood of all sections! The normal running load for non-felted dryers is about 0.0006 to 0.0008 hp/in. per roll per 100 ft/min. The recommended value of the NRL constant is 0.0008 for 72 in. diameter rolls and 0.0007 hp/in. per roll per 100 ft/min for all smaller rolls. On felted dryers, each felt roll will require about 0.00055 hp/in. per 100 ft/min. Doctor blades require more power than a dryer roll (0.001 to almost 0.002 hp/in. per 100 ft/min) and are well represented by:

$$\text{hp/in. @ 100 ft/min} = \frac{\mu \times \text{pli}}{330}$$

where μ for metal doctors is 0.25 to 0.30; μ for plastic doctors is 0.17 to 0.25; and μ for fiber glass doctors is 0.22 to 0.27.

The dryer drive rating is established by the *acceleration* requirements of the dryer section – *not* the normal running load! The acceleration requirement results in an RDC of at least 2 to 5 times the typical NRL of the section. It can be shown that the required RDC of a dryer section is:

$$\text{hp required} = \text{NRL hp} + \frac{WK^2\,(\Delta \text{rpm})^2}{1.613 \times 10^6 (t_a)}$$

where t_a is the acceleration time in seconds and the inertia is expressed as WK^2 in lb-ft^2 referred to the shaft rpm. Since rpm is determined by machine speed and roll diameter, the relationship can be expressed as:

$$\text{hp/in. @ 100 ft/min} = \frac{1}{T_{pu}}\left[\begin{matrix}\text{NRL hp/in.}\\ \text{@ 100 ft/min}\end{matrix} + 9.045 \times 10^{-4}\left(\frac{WK^2/\text{in.}}{D_o^2}\right)R\right]$$

where T_{pu} is the per unit accelerating torque, R is the acceleration rate in ft/min/s and D_o is the outside roll diameter in inches.

It can also be shown, for machines of usual widths, that approximately:

$$\text{hp/inch @ 100 ft/min} = \frac{1}{T_{pu}}\left[\text{NRL hp/in @ 100 ft/min} + 2 \times 10^{-7}\left(\frac{D_0^4 - D_i^4}{D_0^2}\right)R\right]$$

where D_i is the inside roll diameter.

This equation can be used to calculate the RDC constants. The typical acceleration rate (R) for dryer sections is 20-25 ft/min/s. Thus any machine above 1200-1500 ft/min will require more than 60 s to accelerate and the drive must either be of a special design or derated for the acceleration period.

Example: Consider the previous paper machine with a K_m of 7500 and a top speed of 2500 ft/min. The first dryer section is unfelted with twelve 72 in. dryer rolls (with 1.5 in. shell) and two metal doctors.

NRL	=	(12) (0.0008) (7500)
plus		(2) (0.0010) (7500)
or	=	(0.0116) (7500)
NRL hp	=	87 hp

The RDC constant for a 25 ft/min/s, 200% torque, is estimated as:

$$\text{RDC constant} = \frac{1}{2}\left[(0.0116) + (12)(2 \times 10^{-7})\left(\frac{72^4 - 69^4}{72^2}\right)(25)\right]$$

$$= 0.030$$

$$\text{RDC hp} = 225 \text{ hp}$$

A 250 hp drive, specially rated for 200% torque

for the acceleration time of 100 s, should be applied. (This is in excess of a standard drive's rating.)

Consider the second dryer section on the machine, duplicate of the first, except with two felts. There are 25 felt rolls, each 24 in. in diameter (1 in. shell).

$$\begin{aligned}\text{NRL Constant} &= 12(0.0008) + 2(0.0010) + 25(0.00055)\\ &= 0.02535\\ \text{NRL hp} &= 190 \text{ hp}\end{aligned}$$

Note that the two felts more than doubled the running hp. The RDC constant is the same as before (0.030) plus the power required to accelerate the 25 felt rolls:

Felt roll RDC constant

$$= (25)\left(\frac{1}{2}\right)\left[0.00055 + (2 \times 10^{-7})\left(\frac{24^4 - 22^4}{24^2}\right)(25)\right]$$

$$= 0.0175 \text{ or } 131 \text{ hp}$$

$$\text{Total RDC hp} = 131 + 225 = 356 \text{ hp}$$

A 400 hp drive, rated 200% torque for the acceleration time of 100 s, should be applied. The available drive torque should *always* be specified to the machinery builder.

Any change in the required acceleration time will have a major influence on the required drive rating. Note also that, since the required RDC greatly exceeds the NRL, for economical reasons there are *no* service factors recommended on dryer sections and *no* margin recommended for *any* water in the dryers! It is of the utmost importance that the drive engineer be well aware of the expectations of mill operations.

As can be noted from Fig. 698, even 1% water in dryer rolls will overload the drive system. At slow speeds, the water "puddles" in the bottom of the dryer increasing its weight and bearing losses. As the dryer's speed increases, the water "rides" up the side of the roll and cascades. This greatly increases the drive load and the impact loading usually results in serious drive regulator instability. If the speed continues to increase, the water will "rim" due to centrifugal force. The load will drop dramatically but will remain above normal due to the increased weight on the bearings. It is considered economically impractical to select drive systems which allow more than about 0.5% water in the dryer rolls.

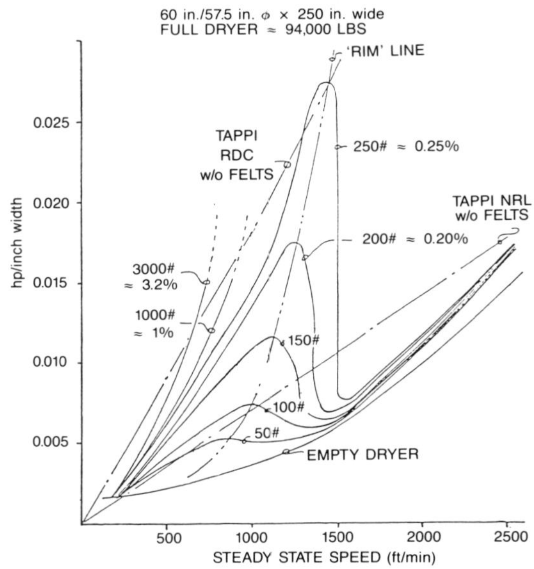

Fig. 698. Dryer acceleration tests with water loads.

Felt-driven dryer sections are currently being introduced into the industry. Early indications are that the normal running loads are somewhat lower due to the lack of machine gearing losses. The power required for acceleration (RDC) however remains essentially unchanged. On felt-driven dryers there is an additional application consideration: the amount of power that the driven roll can impart to the felt without slipping. This power is calculated by:

$$HP = K_m \left[T_s \frac{(e^{\mu\alpha} - 1)}{330}\right]$$

where T_s is the slack side felt tension (typically about 8-10 pli), μ is the coefficient of friction between the driven roll and the felt (typically about 0.18), and α is the angle of wrap in radians. Typically about every third roll will need to be driven to avoid slippage. The typical maximum allowable (tight side) tension is about 30 pli.

On conventional as well as felt-driven dryers, multi-motor drives are torque regulated from a common speed regulator. The speed regulator should be applied at the drive with the tightest felt tension. Dryer section speed regulators are often tuned slightly underdamped to improve their response time (Fig. 697). The response time of all dryer sections should be matched so far as practical. Typically the response times of dryer sections are limited by the available power to accelerate the inertia to about 1.3–2 s. Normally the dryer sections speed regulators have "stiff" speed regulators, i.e., flat speed vs. load characteristics. However on some heavy board machines, a "drooping" speed vs. load characteristic may be beneficial on the last several dryer sections if excessive tension is run going into the calender.

H. CALENDER SECTION

The load elements for a modern calender, or pull stack, consist of:

- The nip load: On hard nip calenders with steel rolls operating less than 600 pli, the work in the nip is insignificant and is neglected in drive calculations. Above 600 pli, this nip load is estimated as (0.000 005)(pli-600) hp/in./100 ft/min. On soft nip calenders, the work in the nip *is* significant and similar to a press:

  $[(0.00006)\ (\text{pli}) + 0.00015]$ hp/in. @ 100 ft/min

- Roll loads: On plain rolls the friction load is in the order of 0.0009 hp/in. @ 100 ft/min for each roll. Adjustable crown rolls have an additional hydraulic load, typically in the form of:

  $[(0.000\ 004)\ (\text{pli} \times \text{ft/min})^{0.5} + 0.0060)]$ hp/in. @ 100 ft/min

- Doctor blade loads as previously listed for dryer sections with a coefficient of friction of 0.25 to 0.30 and observed loadings of up to 3 pli.
- Sheet tension loads: The differential tension across a calender stack is estimated as follows:

  Kraft: $T = (0.05 \times BW)^{0.6}$

  Lightweight coated, bond, tablet, and CRS grades:

  $$T = (0.03 \times BW)^{0.5}$$

  News, groundwood, directory grades:

  $$T = (0.013 \times BW)^{0.3}$$

  In each case, the basis weight, BW, is expressed in lbs/3000 sq. ft. The power required for this differential tension is then:

  HP = (0.003) (T) hp/in. @ 100 ft/min

  The required differential tension across the calender should *always* be carefully discussed with mill operating management! Some mills often desire to run much higher tensions than the typical values.

The recommended RDC/NRL ratio is 1.33. On two-roll stacks, 150% torque-60 s rated drives are considered adequate. Four, or more, roll stacks should have 200% 60 s rated drives. Regenerative braking, for rapid stopping in the event that the sheet wraps a calender roll, is often desirable, especially on high-speed machines.

Example: On the 2500 ft/min kraft machine, with $K_m = 7500$, previously discussed, consider a four-roll calender stack for 126 lb/3000 sq. ft. sheets. The calender has two adjustable crown rolls and is rated for nip pressures up to 800 pli. The other two rolls are plain chilled iron rolls. Each roll has a metal doctor rated 2 pli blade pressure.

NRL Constant:

Two plain rolls at 0.0009 *each*	= 0.0018
Two adjustable crown rolls, 800 pli *each*: $(0.000\ 004)\ (800 \times 2500)^{0.5} + 0.006$ or a total (2 rolls) of	= 0.0233
Doctor loads, four at: $\frac{(0.3)\ (2)}{330} \times$ (four)	= 0.0073
Sheet tension load: $T = (0.05 \times 126)^{0.6} = 3$ pli	
hp = (0.003) (3)	= 0.0090
Nip loads (*three* nips): $[(0.000\ 005)\ (800 - 600)]\ (3)$	= 0.0030
Total NRL constant	0.0444

Thus the NRL is 333 hp and the RDC is 443 hp. A 500 hp, 200% 60 s drive may be applied. Note that some calender designs may use two motors on a four-roll stack.

The regulator's speed response should be fast and critically damped (Fig. 697). Modern calender stacks are usually operated under tension control from direct tension measuring load cells and a tension control paper roll. During thread-up, the calender is operated under speed control. The transfer from tension-to-speed-to-tension may be automatic or manual.

I. REEL SECTION

The drive power requirements for *surface* type reels are straight forward and consist of the sum of three basic factors:

1. Bearing, windage, and friction load of the reel

drum, spool, and the small nip load between the two is in the order of 0.003 hp/in. @ 100 ft/min.

2. Doctor blade pressure load which may range in the order of 3 pli, using the relationship previously given for dryers, a coefficient of friction of 0.3, and 3 pli blade pressure, the doctor blade load is calculated as approximately 0.0027 hp/in. @ 100 ft/min.
3. The sheet tension load at a reel is up to about 70% of that observed of winders and is calculated by the relationship hp = (0.003) (T). The tension, T, is dependent upon sheet basis weight and grade and can be calculated as follows:

Kraft: $T = (0.185 \times BW)^{0.60}$

Lightweight coated, bond, tablet, CRS and similar grades:

$$T = (0.12 \times BW)^{0.53}$$

News, groundwood, directory grades:

$$T = (0.13 \times BW)^{0.33}$$

In each case, the basis weight is expressed in lbs/3000 sq. ft.

The recommended RDC/NRL ratio is 1.25. Typically 150% torque, 60 s drives are applied to reel sections but, if it is desired to accelerate a part roll, a 200% torque drive may be required. An acceleration calculation similar to that applied to dryer sections is required.

Example: Consider a surface type reel for the typical kraft machine previously discussed. The maximum basis weight to be considered is 42 lb *board*, i.e. 126 lbs/3000 sq. ft. and doctor blade loading up to only 2 pli.

NRL Constant

Friction load: $= 0.00300$

Doctor load: $\dfrac{0.3 \times 2}{330} = 0.00182$

Tension load:

$T = (0.185 \times 126)^{0.6}$

$= 6.6$ pli

$(0.003)\ (6.6) = 0.01980$

Total NRL $\quad 0.02462$

RDC = 1.25 NRL $= 0.03080$

NRL hp = 185 hp,

RDC = (1.25 × NRL) = 231 hp

A 250 hp drive should be applied. It should be noted that, in this example the tension load is about 80% of the total NRL! Since some mills often desire to run higher than typical values of tension, the required tension value should *always* be carefully discussed with mill operating management.

The regulator's speed response for a reel drive should be fast and critically damped. Under normal operating conditions, a reel is operated under tension control using the torque algorithm of the speed regulator. For roll "turn-up", and tail thread-up, the reel is operated under speed control. The transfer from tension-to-speed-to-tension control may be manual or automatic. Regenerative braking, for rapid stopping, may be desirable on high-speed machine reel sections.

1. Reel spool drives

Reel spool drives are recommended on all machines with K_m greater than about 5000 and on any machine running lightweight sheets—i.e., sheets less than about 35 lb/3000 sq. ft.

The friction load of the spool, tire, and hub is in the order of 0.0012 hp/in. @ 100 ft/min. The acceleration load may be calculated using the equations of Table 73 (Eq. 3).

Example: Consider a spool of 20 in. diameter, with 1 in. shell (D_i = 18 in.). The desired acceleration time to 2500 ft/min is 20 s at 200% torque, R = 125 ft/min/s

$$RDC = \frac{1}{2}\left[0.0012 + 2\cdot10^{-7}\left(\frac{20^4 - 18^4}{20^2}\right)(125)\right] \text{hp/in. @ 100 ft/min}$$

$$RDC = 0.00232,\ hp = 17.4\ hp$$

At least a 20 hp drive rated 200% torque should be applied.

The regulator's speed response should be fast and critically damped. The spool starter should be speed matched to the reel drum.

J. WINDER DRIVES

Modern electric backstand winder drives are among the most sophisticated and complex drive systems in a pulp and paper mill. [Note that supercalender and off-machine coater drives are similar to winder drives in their complexity.] The backstand drive unwinds the sheet under tension control while the windup and its auxiliary drives operate under speed control. The load division of the windup drives may be under torque control to affect the quality of the rewound roll. The

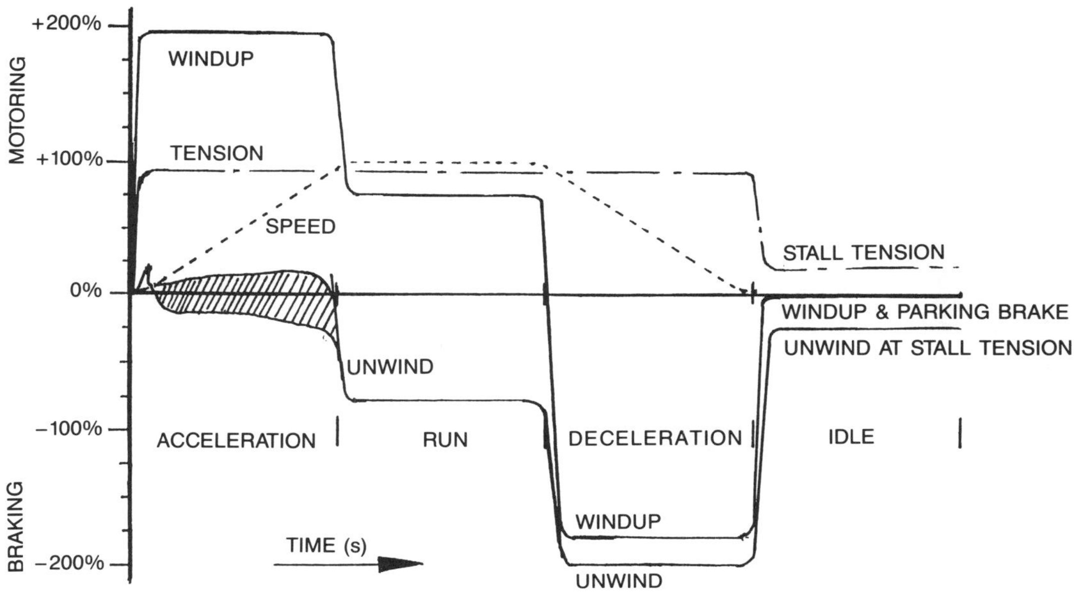

Fig. 699. Winder duty cycle.

winder drive then also operates on a duty cycle[1], as illustrated in Fig. 699:

- Acceleration
- Run
- Deceleration
- Idle

The backstand and windup drives behave quite differently during the duty cycle of a winder, as illustrated in Fig. 699.

1. Unwind backstand drive

The unwind backstand drive may motor or brake slightly during acceleration, depending upon sheet tension. The small irregularities at the beginning are due to the required breakaway, or starting, torques.

In the run portion of the winding cycle, the unwind drive will be in the braking mode to maintain the required sheet tension.

During deceleration, the unwind drive will be braking significantly into its overload rating. On Emergency Stop, the unwind drive will also brake significantly into its overload rating and usually will be assisted by a "power failure" mechanical brake.

Stall tension is typically 25–30% of the running tension value and is used to maintain a tight sheet in the winder while the winder is idle and stopped. The winder drives are stopped (zero torque) with their parking brakes set and the stall tension is provided by the unwind drive.

The *core*-driven unwind must be capable of constant torque operation over the speed range of the winder (e.g., zero to 8000 ft/min) *and* then constant hp over the unwind's range of diameters. For example: Consider a 1000 hp unwind on an 8000 ft/min winder with a 25 in. reel spool and a 100 in. diameter machine parent roll. At 8000 ft/min, the 100 in. parent roll would be running 306 rpm, the empty reel spool would run at 1222 rpm. Thus the unwind drive must be rated 1000 hp, 306/1222 rpm constant hp, and provide constant torque from zero to 306 rpm.

---

[1] Typically, winders accelerate and decelerate at 120 to 140 ft/min/s. The "run time" (seconds) can be determined from the following equation:

$$\text{Run Time} = \frac{981.75(d)(D_m)^2}{(\text{MaxBasisWt})\ (\text{ft/min})} - \frac{t_a + t_d}{2}$$

Where

d	= sheet density (typically 40–60 lbs/ft^3)
D_m	= minimum shipping roll diameter
Basis weight	= lbs/3000 sq. ft.
ft/min	= the maximum winder speed
t_a and t_d	= acceleration and deceleration times in seconds.

On high-speed winders the run time is often very nearly zero, especially for the heavier weight sheets. This equation, with Run Time Zero, can be used to determine the maximum usable winder speed.

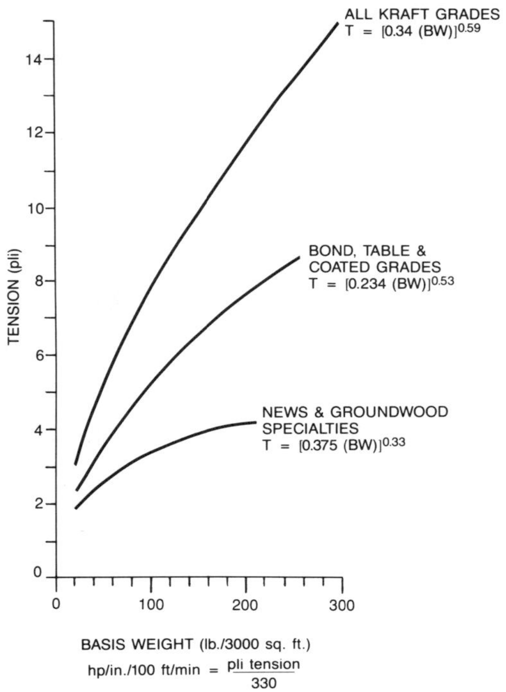

Fig. 700. Typical winder tension.

2. Windup drives

The windup drives include the drum drives, rider roll drives, and guide roll drives all of which are speed regulated and may be torque controlled also. These drives will motor into their overload ratings during acceleration, motor within rating during the run cycle, and will be braking significantly into their overload rating during deceleration, or emergency stopping, cycles.

The windup drives *surface* drive the sheet, i.e., the sheet speed is the surface speed of the roll. They are thus considered constant torque drives. With the usual direct connected motor, the motor and roll speed is:

$$\text{rpm} = \frac{(12)\ (\text{ft/min})}{\pi\ (\text{dia})}$$

Thus on an 8000 ft/min winder with 31 in. windup drums, the roll speed would be approximately 986 rpm. Two 600 hp drum motors then must be capable of constant torque operation over the zero to 8000 ft/min speed range (0–986 rpm). They would not have any required constant horsepower operating range.

It will be noted that winder drive ratings are significantly affected by the required sheet tension (Fig. 700), the various inertias of the rolls and paper, and the desired acceleration and deceleration rates. The drive rating will be determined by:

- Sheet tension (Fig. 700) and duty cycle RMS hp from Eq. (4) of Table 73; or
- The peak acceleration or deceleration hp, including tension and inertia considerations, or
- The required emergency stopping peak hp.

TAPPI has developed a detailed calculation method for winder drives which should be carefully followed when selecting the various drive ratings. The selection of winder drive ratings should *never* be less than those calculated using the TAPPI TIS 403-11 method. The successful application of a winder drive requires the complete coordination of mill operations, winder manufacturers, and the drive supplier.

Both AC and DC drives have been applied to electric backstand winders. Because of the frequent starting and stopping of winders under peak load conditoins, the AC motors have a distinct advantage over the DC motors in this application.

K. SUMMARY

The data presented in TAPPI Technical Information Sheets, as well as other published data, may not be complete or totally exact. Nevertheless, they are considered worldwide to be the best information available. The data presented should be carefully applied and analyzed critically. On any specific application, the recommendations of the machinery builder, consulting engineer, and mill operating management should each be carefully considered.

USEFUL REFERENCES

1. TAPPI Paper Machine Drives Seminars (latest issue).
2. TAPPI Technical Information Sheets for Paper Machine and Winder Drive Requirements (latest issue).
3. Derrick, R.P. and McNulty, P.T., "Why Replace Your Paper Machine Drive?", Tappi J., p. 120 (May 1983).
4. Derrick, R.P., "A Method for Calculating the Power Transmitting Ability of Fourdrinier Rolls", Tappi J., p. 78 (Oct. 1977); p. 109 (Dec. 1977).
5. Concannon, M.D., "Condensate Effects on Torque and Horsepower in Paper Dryers", 1980 TAPPI Engineering Conf.
6. Kessler, N. and Van Lieshout, R., "An AC Gearless Dryer Installation", 1990 TAPPI Engineering Conf.
7. Baker, W.F., Cortes, T. and Derrick, R.P., "The Selection, Installation, and Startup of a Large Adjustable Frequency Paper Machine Sectional Drive", Tappi J., p. 57 (Sept. 1988).
8. Eikenberry, M.L. and Brink, K.D., "An AC Winder Drive System", Tappi J., p. 183 (Apr. 91).
9. Derrick, R.P. and McNulty, P.T., "Energy Cost Considerations for Electrical Backstand Winders", Tappi, p. 57 (Oct. 1981).
10. Derrick, R.P., Martiny, W.J. and McDonald, W.J., "Motors Covered with Pulp: The Effect on Motor Life", Tappi, p. 57 (May 1987).
11. Many of the problems encountered by engineers can often be easily solved if a ready source of reference material is available. Technical papers presented at the various annual conferences of TAPPI, IEEE, and the Technical Section, CPPA, form an ideal base for such a library. A bibliography of electrical engineering papers presented during the past ten years, with a subject index, provides this base. The two most recent published bibliographies were published at the 1977 and 1987 TAPPI Engineering Conferences and are available from TAPPI PRESS.

XXI
PROPER DESIGN OF PAPER MACHINE VACUUM SYSTEMS

M.C. JOHNSON

Mickey Johnson is Manager of Pulp and Paper Process Engineering for Rust International Corporation in Birmingham, AL. He directs the work of the engineering group responsible for the process design in the pulp mill and paper machine areas on all Rust engineering projects. His experience includes 16 years in the consulting field and nine years in mill operations. He has a B.S. in Mechanical Engineering from the University of Alabama. He is an active member of TAPPI, a regular speaker at the Wet End Operations Short Course, and has published several papers.

REVIEWED BY

David C. Farnsworth
Vice President
Nash Clark & Vicario Inc.
Clearwater, FL

Darryl L. Getman
Vice President
Albany International
Glens Falls, NY

A. INTRODUCTION

The proper selection of vacuum equipment and the proper installation of the vacuum system are vital elements in the successful operation of a paper machine. Many existing machines have inadequate vacuum, combined with high energy consumption, resulting in lower speeds and poor operating efficiencies.

The purpose of this chapter is to cover some of the basic fundamentals of vacuum terminology, examine the different kinds of vacuum generating equipment that are available on the market today, review some aspects of proper system design, and examine vacuum requirements for various elements of the paper machine.

B. BASIC FUNDAMENTALS OF VACUUM SYSTEMS

Most paper mill engineers and operators are reasonably comfortable when dealing with positive pressure, non-compressible fluid flow systems. If someone mentions a 1000 gal/min pump operating at 90 feet of head (feet of water), there is a "feel" for what that pump can do. The 90 feet can be converted to 39 psig and one has a system operating at slightly less than normal mill water pressure.

However, when dealing with negative pressure, compressible fluid systems, most engineers and operators have more difficulty. When someone mentions a press felt suction tube requiring 8000 cfm at 12″ Hg, what size vacuum pump is really being discussed? If a press section has three felts, and the first felt suction tube is operating at 6″ Hg, the second is operating at 10″ Hg and the third is operating at 14″ Hg, how much vacuum pump capacity is required for each felt?

Vacuum is a negative pressure indication, generally measured in inches of water column (wc) or inches of mercury (Hg). Figure 701 shows

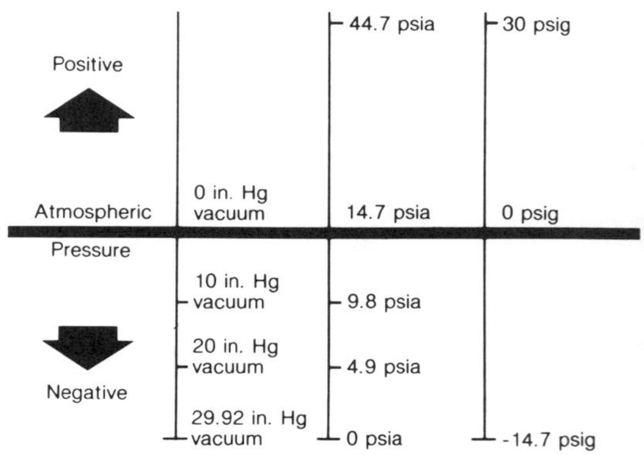

Fig. 701. Relationship between gauge (psig) pressure and absolute (psia) pressure.

the relationships between gauge pressure and absolute pressure in psi and in inches Hg.

One serious mistake that is sometimes made in vacuum calculations is the addition or subtraction of vacuum volumes which are at different vacuum levels. One thousand cfm at 10″ Hg cannot be added to 3000 cfm at 8″ Hg without converting to the same vacuum level. Familiarity with gas laws is essential in making the necessary conversions.

Boyle's gas law states that the volume of a given weight of gas varies inversely with the absolute pressure, under constant temperature conditions. Boyle's law shows the relationship between pressure and volume at one condition and pressure and volume at another as follows:

$$P_1 V_1 = P_2 V_2$$

Charles' law states that the absolute pressure varies directly with the absolute temperature when volume is held constant. The relationship between pressure and temperature at different conditions is expressed in the following equation:

$$\frac{P_1}{T_1} = \frac{P_2}{T_2}$$

These two laws are combined and arranged as follows for a fixed weight of gas having both a temperature and pressure change in passing from condition 1 to condition 2:

$$\frac{P_1 V_1}{T_1} = \frac{P_2 V_2}{T_2}$$

Another form of this equation, called the ideal gas law, can be used to define any one unknown of a particular condition if the other factors are known:

$$PV = WRT$$

It is essential to remember that pressure and temperature must be given in absolute units in all the above equations. Normal units are as follows:

P = Absolute pressure, lb/sq ft = (psi gauge + 14.7) × 144
or
(14.7 − in. Hg gauge vac/2.03) × 144

V = total gas volume, ft^3 (m^3)
W = weight of gas, lbs (kg)
T = absolute temperature, °R (°K)
(°F + 460 = °R)
(°C + 273 = °K)
R = gas constant, lbf × ft/(lb-mol × °R)
N × m/(kg-mol × °K)
Ra (air) = 53.3 (287 for SI units)
Rw (water vapor) = 85.8 (462 for SI units)

Two other frequently encountered terms are "scfm" and "acfm". SCFM is standard cubic feet per minute while acfm are actual cubic feet per minute. SCFM is the flow rate of air at standard conditions of 60°F and 14.696 psia.

Figure 702 illustrates the volume and pressure relationship in a system [1].

Using Boyle's law, the volume at different vacuum levels is calculated as follows:

Felt suction tube service specifies that vacuum must be designed for 4000 cfm at 14″ Hg. A liquid ring pump (constant volume device) is furnished for the specified conditions, but pipe line friction results in an operating vacuum at the pump of 18″ Hg. What is the actual volume at the felt suction tube?

$$V_2 = \frac{P_1 (V_1)}{P_2} = \frac{\left(14.7 - \frac{18}{2.036}\right) \times 144}{\left(14.7 - \frac{14}{2.036}\right) \times 144} \times (4000)$$

$$V_2 = 2995 \text{ cfm}$$

Or a simplified approach, since the pressure units cancel out:

$$V_2 = \frac{P_1 (V_1)}{P_2} = \frac{29.92 - 18}{29.92 - 14} \times (4000)$$

$$= 2995 \text{ cfm}$$

Using 30″ Hg as absolute vacuum, rather than 29.92, is normally adequate for sizing paper machine vacuum systems.

As another example, consider a couch roll requiring 20,000 cfm at 20″ Hg. The stock temperature in the headbox is 140°F. Assume that the cooling effect of the sheet moving down the wire, plus the evaporation which results from the expansion as water is drawn into the couch suction box, lowers the temperature to 110°F in the suction box. Condensing water is sprayed into the vacuum line just preceding the liquid ring pump, lowering the temperature to 90°F. To

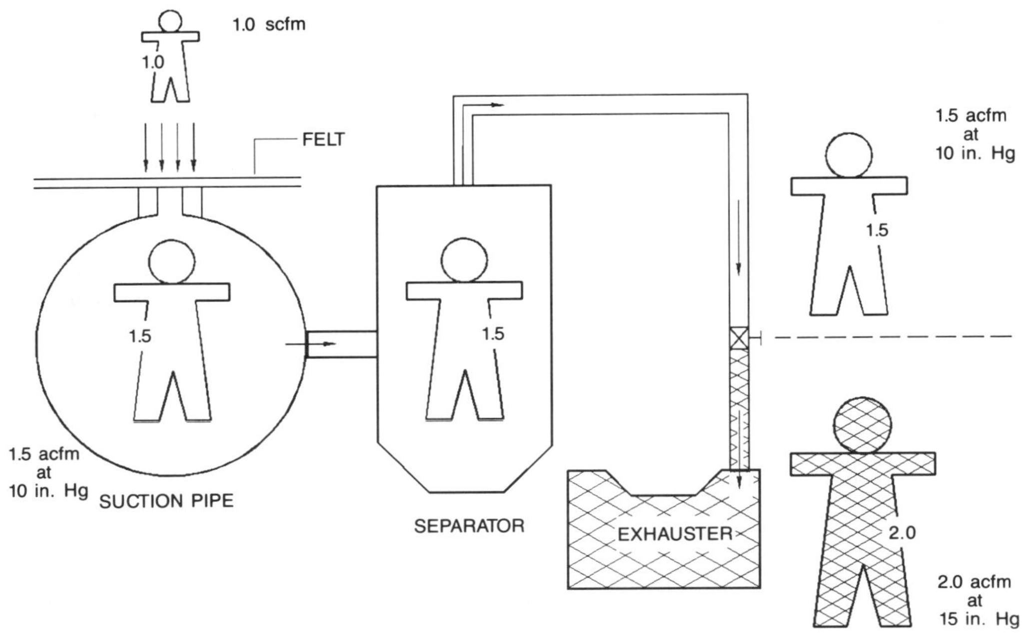

Fig. 702. Vacuum system airflow expansion diagram.

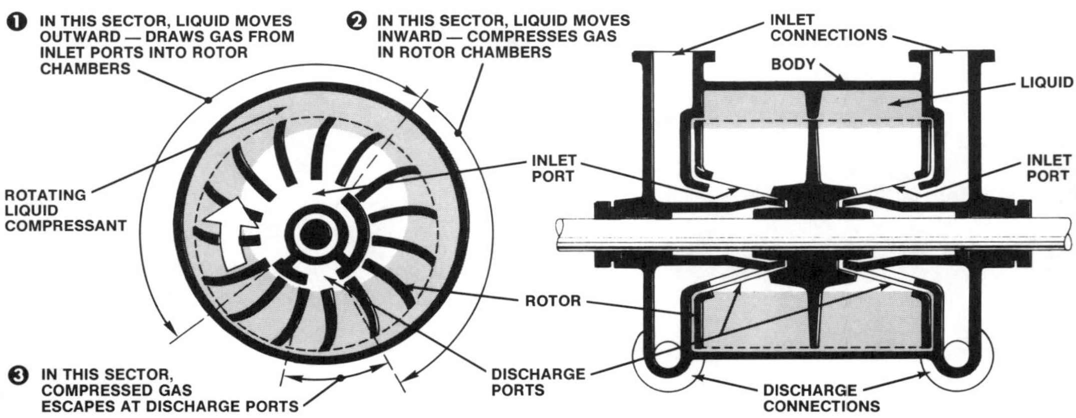

Fig. 703. Liquid ring vacuum pump schematic (Nash Corp.).

satisfy the couch vacuum requirements of 20,000 cfm at 20″ Hg, what is the required pump capacity operating at 90°F and 21″ Hg?

$$\frac{P_1 V_1}{T_1} = \frac{P_2 V_2}{T_2}$$

$$\frac{(30 - 20)(20{,}000)}{110 + 460} = \frac{(30 - 21)(V_2)}{90 + 460}$$

$V_2 = 21{,}443$ cfm

C. VACUUM GENERATING EQUIPMENT

Two basic types of vacuum generating equipment are used in the paper industry: the liquid ring pump, a constant volume device; and the centrifugal exhauster, a variable volume device. Centrifugal exhausters can be divided into two categories, vertically split casing and horizontally split casing.

1. Liquid ring pumps

Figure 703 shows a cross-section of a liquid ring pump and describes its operation. The rotor

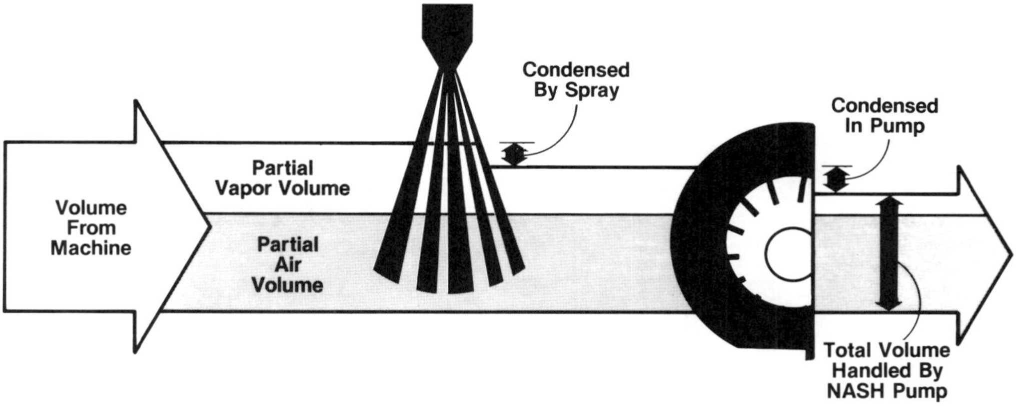

Fig. 704. You can visualize the bonus effect of condensing moisture out of incoming air by comparing widths of the block arrows on this conceptual diagram (Nash Corp.).

keeps a ring of water rotating in the pump body or casing. The water advances into and recedes from the rotor chambers because the rotor axis is offset from the body axis.

This motion sets up a piston-like action which develops the vacuum. As the schematic diagram shows, water almost completely fills and then partially empties each rotor chamber during each revolution. The vacuum is drawn in at each end of the pump through the inlet ports or cones [2].

Some water flows out of the pump with the discharge air. Makeup water is introduced continuously as a replacement and also serves as a coolant and helps condense a portion of the vapor carried by the air stream.

Figure 704 shows the condensing effect that occurs in the pipe prior to the pump, where a spray nozzle injects cooling water into the line [2]. The spray water and the seal water both condense some of the incoming warm vapors, reducing the volume the pump must handle. Caution must be used to not overestimate the benefit from the condensing effect. The final temperature of the water vapor will be higher than the temperature of the seal water. Also, the temperature of the vapor in the suction box on the paper machine is lower than the headbox temperature.

A typical piping arrangement for liquid ring pumps is shown in Fig. 705 [2]. Vacuum is distributed to the respective pumps using a vacuum header. Use of the header gives improved flexibility for future modifications as well as sharing pumps if it is necessary to remove a pump from service. The flanges shown on the vacuum header can either be blank or spectacles to either isolate pumps or connect them.

It should be noted that the different services are isolated in the vacuum header. To achieve maximum efficiency, it is important that systems not be connected together in which the change in operation of one will adversely affect the operation of the other. This is particularly true in the press section, with each felt suction tube isolated from the others. Also, the flatboxes, or wire boxes, are segregated from the couch vacuum because of the difference in vacuum levels required by the two systems. Tying the two systems together would result in a high energy loss if wire boxes were throttled from the 8–12″ vacuum level up to approximately 20″ vacuum level.

Liquid ring pumps are normally installed with a bottom discharge into a below-ground discharge trench and sump for separating the water and the air. (Above ground, individual separation for each pump is also available.) The sump must be large enough so that the flow velocities are reduced and its shape must be designed to induce the right flow pattern so the entrained water will not be discharged through the air exhaust stack. The velocities should be limited to the values shown in Fig. 706.

The air flow used in determining the velocities is calculated under startup conditions, assuming no vacuum is being developed by the pumps. That is the worst condition, and also the easiest to calculate, and results in the highest air flow through the sump and trench. For example, a 9000 cfm pump has inlet conditions of $P_1 = 14.7$ psia, $V_1 = 9000$ cfm; since the discharge to the trench is $P_2 = 14.7$ psia, then $V_2 = 9000$ cfm.

Operating performance of these systems has been very good. The reliability of the low rpm pumps has been exceptional. A liquid ring pump

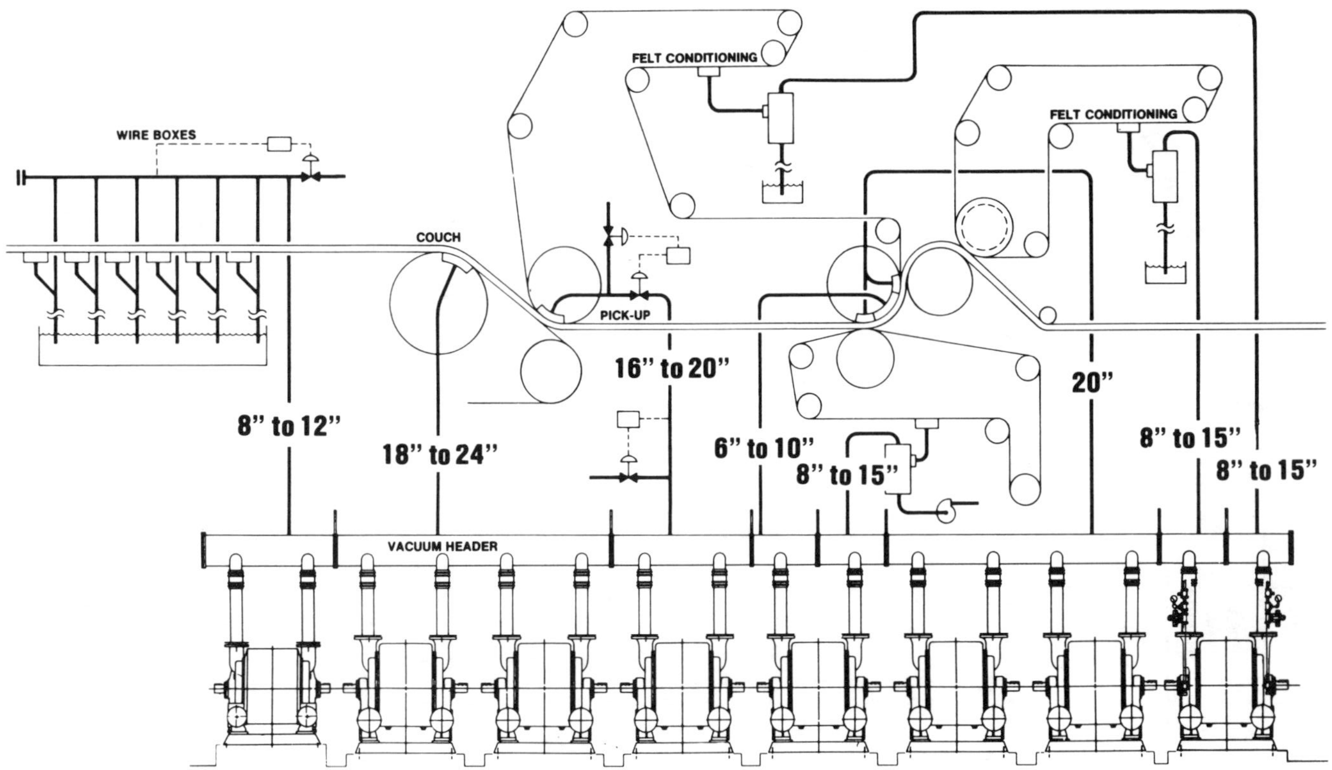

Fig. 705. A typical piping arrangement for liquid ring pumps. Typical vacuum ranges are shown for different services. (Courtesy Nash Corp.)

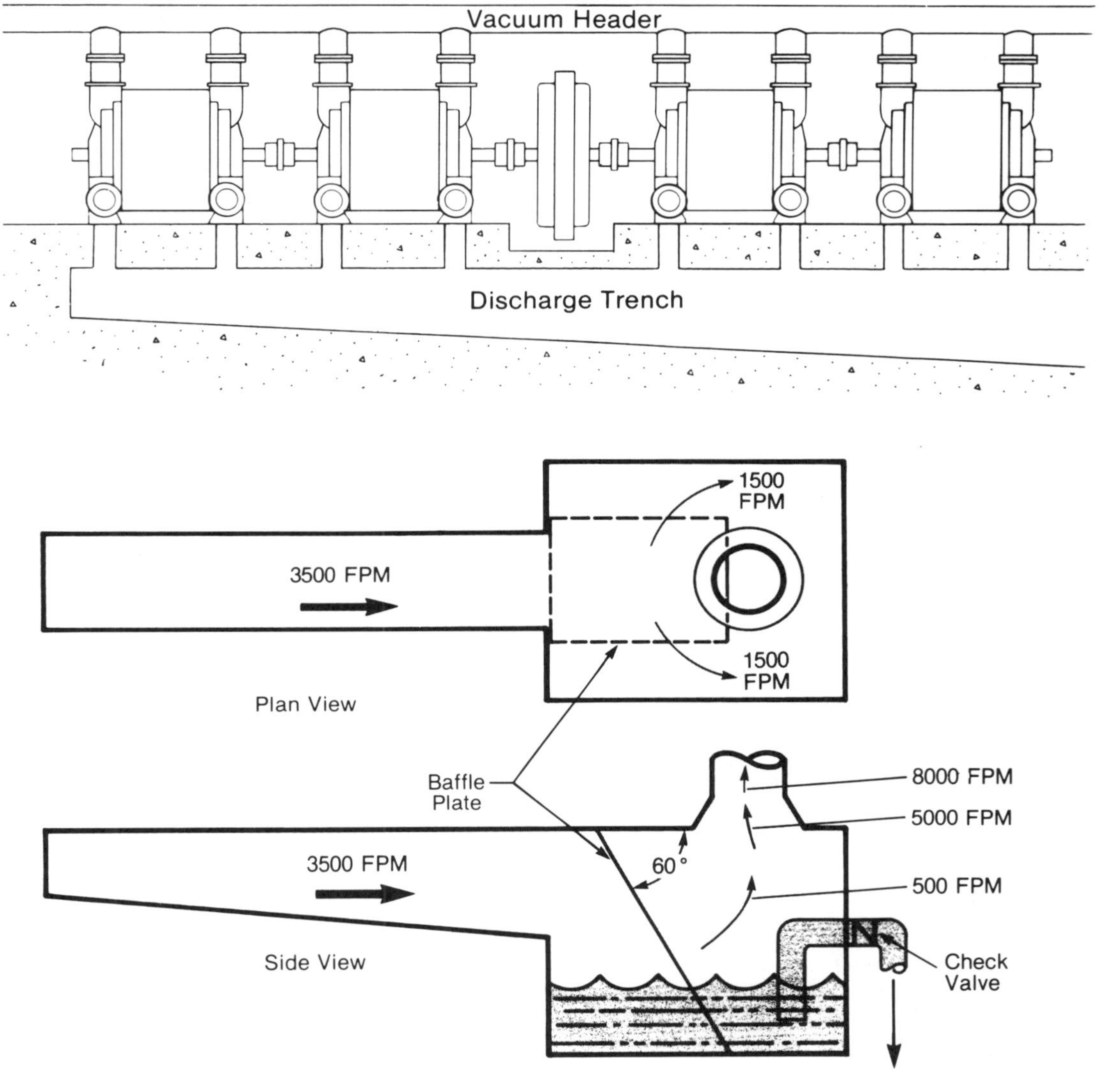

Fig. 706. Routing pump discharge streams downwards into a trench eliminates virtually all discharge piping, and the concrete trench roof provides an excellent machinery foundation. Flow velocities drop to low levels at the end of the trench. Water carried by the air stream then comes out of entrainment and settles in the sump instead of being discharged up the stack. (Courtesy Nash Corp.)

vacuum system requires minimal controls.

The operating costs of a liquid ring pump system can be reduced by recovering the warm seal water. There are a number of different systems that can be used to reclaim this water. The selection of the system must be influenced by the total mill water requirements.

The most simple system is to pump the water from the vacuum trench sump to the paper machine white water storage tank. However, in a fully integrated mill this arrangement will only increase the overflow of valuable white water from the white water storage tank. In mills which purchase most, or all, of their fiber, this system may prove satisfactory, but a complete flow balance of the paper machine system is needed in order to make an accurate assessment.

Another system that has been successful is to use the seal water for felt showers as shown in Fig. 707. The seal water must be filtered over a gravity-type filter, then this warm water can be used for felt showers, replacing fresh water. The shower water system should be designed using good design principles for white water shower

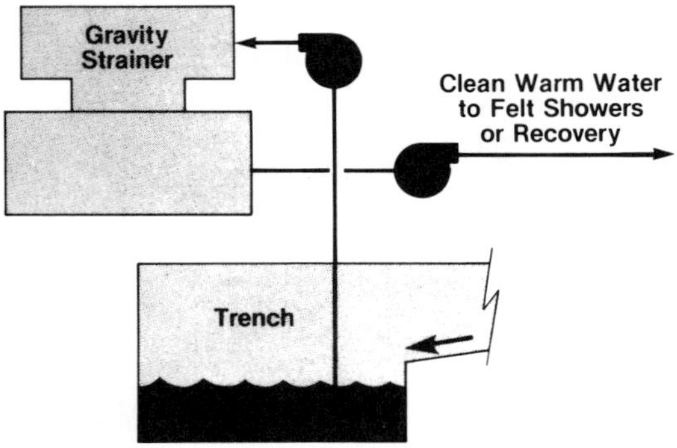

Fig. 707. Discharge water from liquid ring pumps can be filtered to remove short fibers and reused in felt showers. Both water and heat are conserved this way. (Courtesy Nash Corp.)

systems. A word of caution on machines producing a highly filled sheet. Unless very good air water separators are provided at each vacuum connection on the machine, filler can get into the vacuum seal water system, pass through the water filter, build up in the shower system and/or be showered onto the felts or other service.

A cascade seal water system is another method of reducing water consumption. As shown in Fig. 708, a portion of the seal water is recirculated back to the pumps and reused. Cool water is added to the recirculated water to control temperature. A variation is to use the recirculated water exclusively on the lower vacuum level pumps.

Cooling towers can be used as a method of water conservation, as shown in Fig. 709.

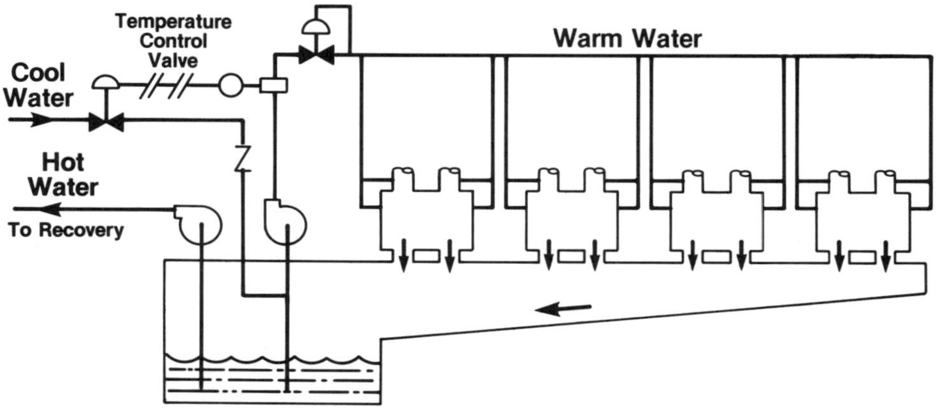

Fig. 708. A cascade seal water system. This is a recirculated system with fresh-water makeup flow based on temperature regulation. Heat added to the system loop is substantial, so that there will always be a demand for cool incoming makeup water. Its flow rate, depending on its temperature, may range from about 35% to virtually 100% of a once-through system's water rate. (Courtesy Nash Corp.)

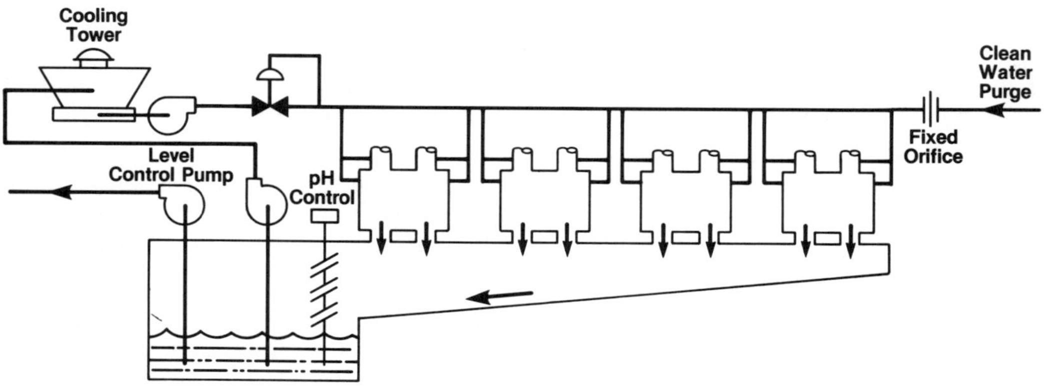

Fig. 709. Cooling towers can be used to conserve water. Saving as much as 90% of once-through water is feasible when fresh-water makeup is added to a recirculated system loop at a constant rate. Heat is dissipated in a cooling tower. (Courtesy Nash Corp.)

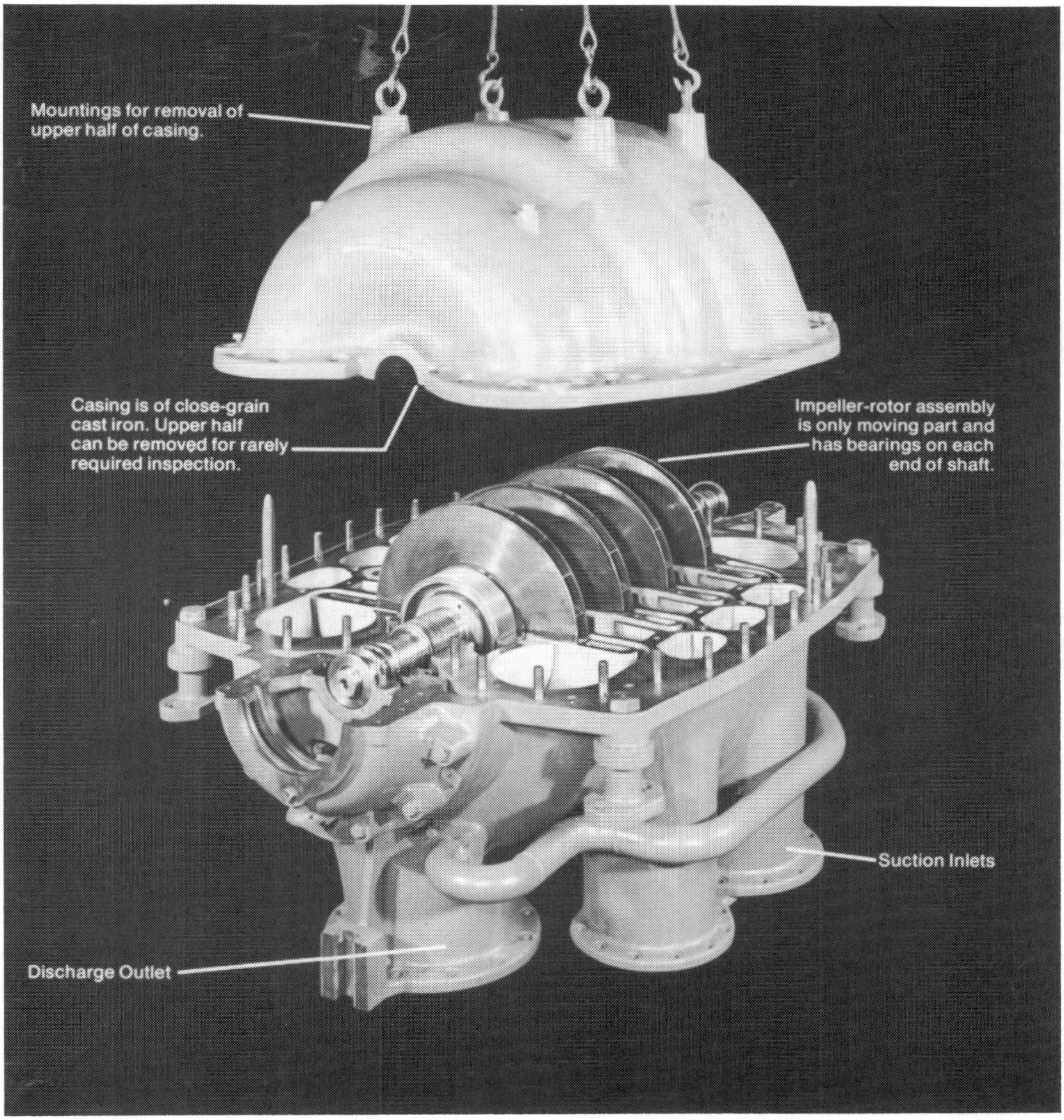

Fig. 710. A horizontal split centrifugal exhauster or typical four-stage Turbair exhauster. (Courtesy Dresser Industries, Inc.)

However, there are a number of inherent operating problems with these systems. Contaminants will build up in the system unless there is a controlled blowdown. The water must be treated for biological growth. Also, the cooling tower must be designed for this type of service so that pulp fines will not build up and damage the tower structure.

2. Centrifugal exhausters — horizontally split

Figure 710 shows a photograph of a horizontal split centrifugal exhauster. Figure 711 is a schematic cross-section of this multi-stage exhauster. The exhauster uses a number of multi-stage impellers within a casing and diffuser system. As the impeller rotates, air enters the eye of the impeller and is accelerated and pressurized as it is moved radially outward. The second stage then accepts the air from the first stage and develops vacuum as it pumps to the third and successive stages. On the multi-stage units shown in Figs. 710 and 711, air can be drawn in at various stages, and hence different vacuum levels, to meet the system requirements of the paper machine [3].

A typical schematic for a paper machine is shown in Fig. 712. Contrasted with the liquid ring pump system, it will be noted that several parts of the paper machine, requiring approximately the same vacuum level, are connected to a common section of the concrete water separator.

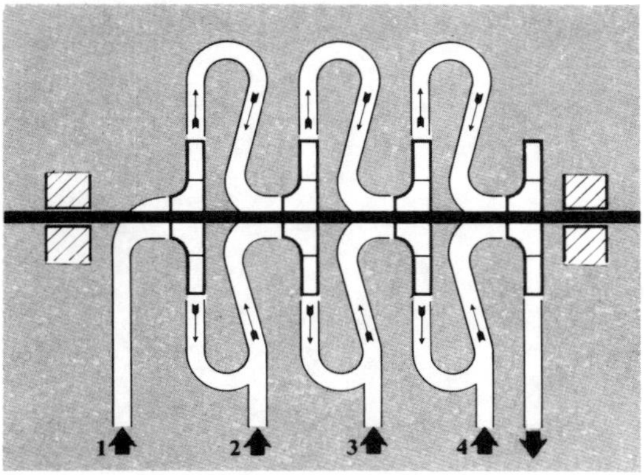

Fig. 711. Cross-section of the multi-stage exhauster shown in Fig. 710 or Airflow diagram of the Roots Turbair exhauster. (Courtesy Dresser Industries, Inc.)

The schematic shows the four-stage vacuum exhauster with a low vacuum level for vacuum drainage elements, a second level for wire boxes, the high vacuum for the suction rolls and the intermediate vacuum level for felt conditioning. The large concrete separator is essential for protection of the vacuum unit from carryover of entrained water. The separator has internal stainless steel baffles, and is sized for maximum air velocity of 10 ft/s.

The installation of this unit requires a heavy foundation because of its very high operating speed and heavy weight. However, it is very compact when compared to liquid ring pumps and does not require as much floor space.

It can be driven by a steam turbine or an electric motor with a speed increasing gear unit. The drive horsepower ranges from 400 to 3000. The operating speed of these units is approximately 5000 rpm. It can be furnished in a single stage up to 15″ Hg vacuum, a three-stage up to 22″ Hg and a four-stage up to 24″ Hg. The flow range is approximately 10,000 to 60,000 cfm for multi-stage units. Single stage units up to 200,000 cfm are available.

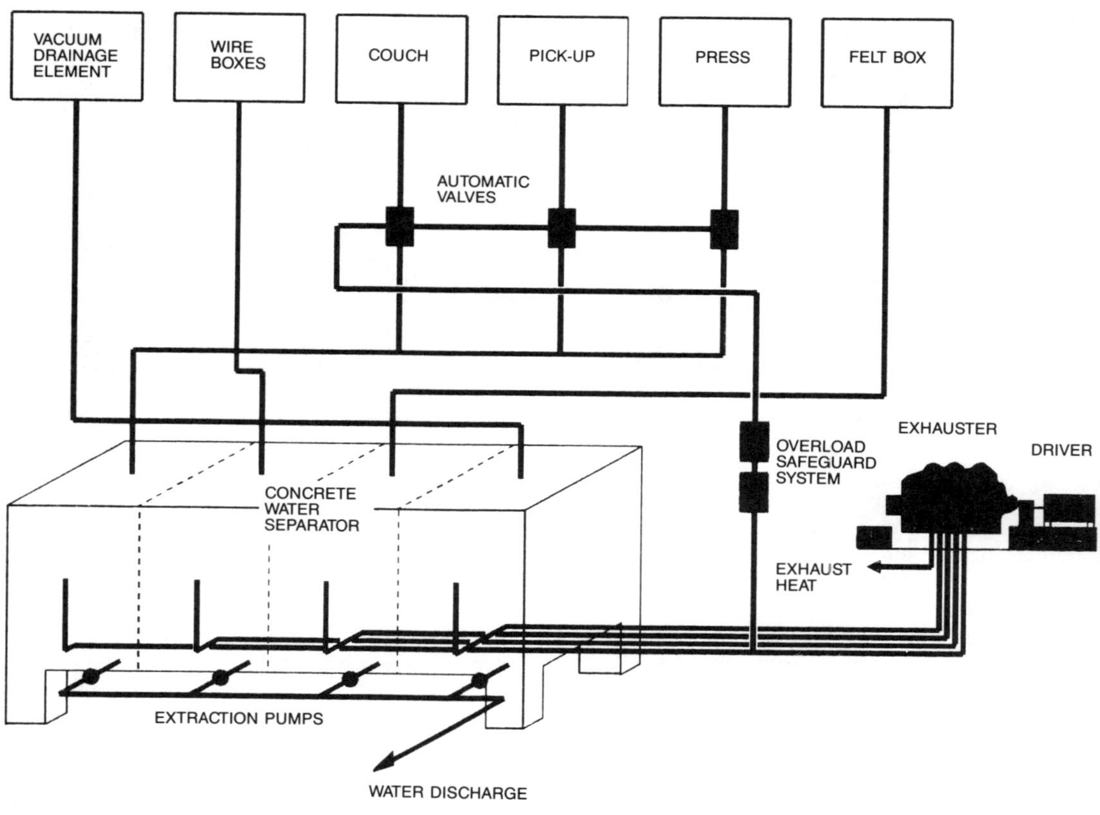

Fig. 712. Paper machine vacuum system schematic for multistage exhauster.

A real advantage of the centrifugal exhauster, both horizontally and vertically split casings, is that no sealing water is required, as with the liquid ring pump. Depending on the particular environmental conditions, this feature can be significant.

Another important feature of the multi-stage exhauster is the potential for reclaiming the exhaust heat from the units. The exhaust air is in the range of 250°F to 350°F and approximately 2% relative humidity. It is normally suggested that the exhaust air be discharged to the intake of the air system for the machine hood pocket ventilation (PV). However, in many installations this humidity gets too high and the full flow of exhaust air cannot be used. If the air at 300°F and 2% relative humidity increases to 5%, the absolute humidity goes from 0.06 to 0.19 lb water per lb of dry air. The air would then be too wet to supply all the PV air. Auxiliary coils will be needed in the PV system to heat normal machine room air if it were necessary to discharge a portion, or all, of the exhaust to the atmosphere. A bypass pipeline should always be provided, so that the exhauster discharge can be sent to the atmosphere rather than the PV system.

The performance of these systems has been very reliable. Very little maintenance has been required on the systems that have been in operation in the industry for a number of years. Care must be exercised in designing the concrete separators; however, the pump vendors can provide good assistance.

3. Centrifugal exhauster — vertically split

Figure 713 shows a photograph of a vertically split centrifugal exhauster. The operating principle for the vertically split unit is the same as the horizontally split unit. One of the main differences, however, is that the vertically split units are generally smaller in capacity and are not multi-ported. There is one inlet and one discharge point. As with the horizontally split exhauster, the vertically split unit operates at a high rpm, typically in the 3000 to 4000 range.

Fig. 713. A vertically split centrifugal exhauster (Courtesy Albany Intl.).

One advantage of the vertically split exhauster is the smaller size compared to the horizontally split unit. Exhausters can be supplied as individual units to meet an individual requirement on the machine. Heavy foundations are not required and the units can be installed on the machine room mezzanine. The flexibility of installation can often be used to real advantage in minimizing vacuum piping. However, the smaller size and multiple units result in more floor space requirements compared to the horizontally split unit.

Very good air water separators preceding the exhauster are essential. These separators should be purchased with the exhausters to ensure good system compatibility.

D. SYSTEM REQUIREMENTS

To determine the capacity required for the vacuum generating equipment, it is necessary to analyze each section of the machine that uses vacuum, to determine the vacuum flow and vacuum level required at each point. It is also necessary to design the total system for most efficient operation. It is not uncommon to find modifications to the existing vacuum system on a paper machine which have been done for the sake of expediency, that result in both poor vacuum control and wasted vacuum horsepower.

TAPPI publishes two Technical Information Sheets (TIS) that are useful in determining paper machine vacuum requirements. TIS 0502-01 provides vacuum design factors, for most areas of the machine, based on operating conditions of actual machines. TIS 0404-27 provides equations for calculating vacuum requirements for press felt conditioning.

1. Deaerators

Although a deaerator is not listed in the TAPPI TIS for vacuum factors, it is a piece of equipment that does require vacuum. The vacuum requirements will be determined by the deaerator vendor, based upon data supplied by the mill or engineering firm in the deaerator specification. A normal vacuum range is 27 to 29″ Hg.

The purpose of the vacuum system is to sufficiently lower the pressure inside the deaerator vessel to create a boiling condition. Consequently, the operating temperature of the system, stated in the specification, becomes very important. If the temperature is stated higher than the machine actually operates, the vacuum pump(s) will be too small. The temperature is easy to determine for an existing machine, to which a deaerator is to be added, but is more difficult to predict for a new machine.

The vacuum system consists of a condenser, or condensers, followed by vacuum pumps. Some systems also use steam ejectors to generate part of the vacuum. However, if the energy in the steam is converted to horsepower, it becomes obvious that ejectors are extremely inefficient.

The condenser reduces the vapor volume as discussed in Sec. C1. The condensers can be either direct or indirect contact type. Direct condensers are not adversely affected by carry-over or flooding of the deaerator, but the cooling water becomes contaminated. The mill water balance should be considered, as well as the condenser operation, when making a selection between the two types.

The vacuum pumps are almost exclusively liquid ring. Liquid ring pumps have the advantage of generating the necessary very high vacuum levels at low flow rates.

2. Vacuum augmented drainage elements

Vacuum capacity required by vacuum drainage elements will be specified by the vendor. The vacuum level will normally range from about 4″ water (wc) to 40″ of water. The vacuum nearest the headbox will be at the lowest level with progressively increasing vacuum levels down the table [4].

Because of the low vacuum levels required, single-stage blowers (or exhausters), rather than liquid ring pumps, are normally used for the vacuum equipment. A blower should be selected that has proper materials of construction for the damp environment. The blowers are small and inexpensive compared to the main machine vacuum system. Consequently, the selection is often not given proper attention, resulting in a poor selection and installation, followed by high maintenance requirements shortly after startup. Poor and unreliable operation of the vacuum blowers results in poor performance of the drainage elements. The low vacuum requirements of the table will generally require two blowers, one low vacuum level and one of higher level.

Vacuum blowers must be adequately protected from any water carryover by installing a good separator ahead of each blower. Do not rely upon getting good separation in the drainage elements. Figure 714 shows a good separator arrangement for a vacuum drainage element.

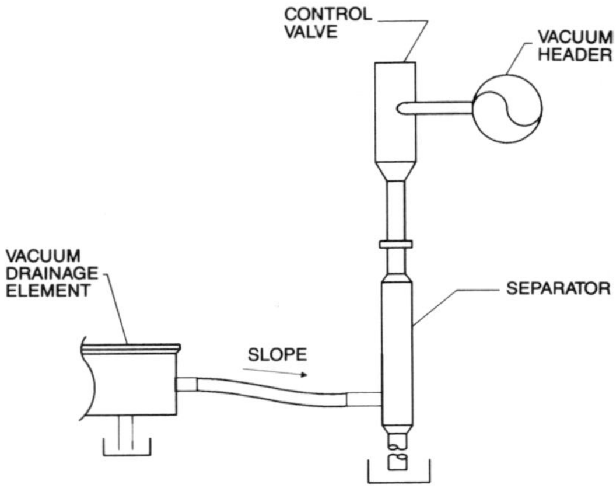

Fig. 714. A good separator arrangement for a vacuum foil or drainage element.

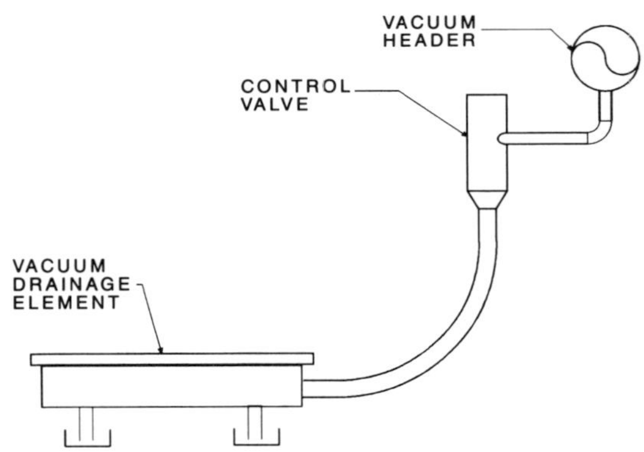

Fig. 715. Improper installation of vacuum drainage element.

The separator is normally supplied by the vendor of the drainage equipment, and should normally be sized for 500 to 800 ft/min vertical velocity. Note that the vacuum line going into the vacuum header goes into the side of the header, not the bottom where water may be present. Always avoid the arrangement shown in Fig. 715.

Figures 716 and 717 show typical vacuum header designs. Figure 716, which shows a separator between the header and the blower, provides the best protection for the blower. Figure 717, with a drip line discharging to a seal pot (or wire pit), has also been used successfully, provided that each individual drainage element has a well-designed separator as shown in Fig. 714. The arrangement shown in Fig. 718 should always be avoided because water will be trapped in the vacuum header, restricting the air flow and often causing vacuum surges.

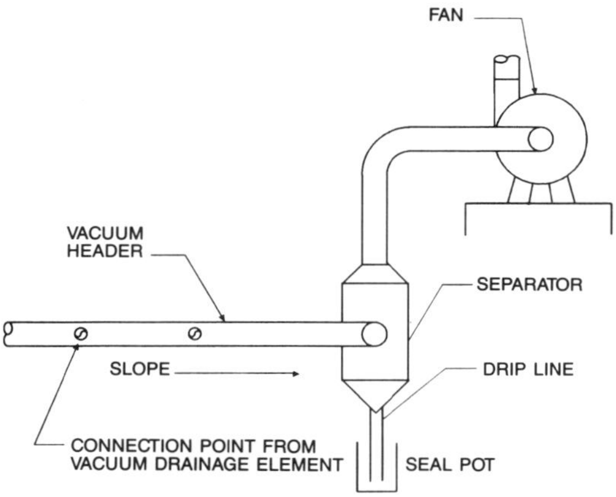

Fig. 716. This vacuum header design provides the best protection for the blower.

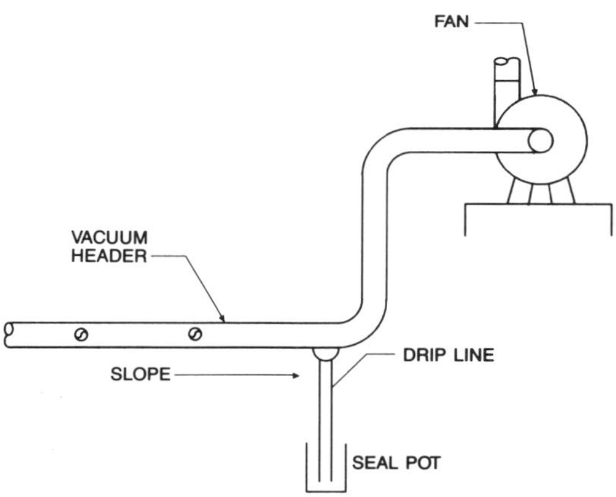

Fig. 717. This vacuum header design has also been used successfully.

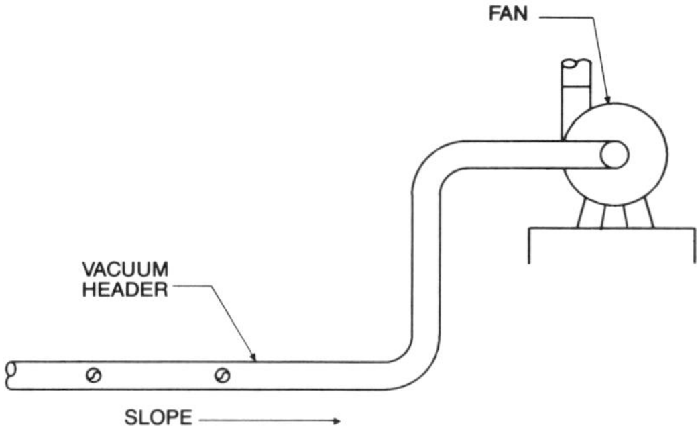

Fig. 718. This vacuum header arrangement should be avoided.

3. Wire boxes

Vacuum requirements for wire boxes (or flatboxes) have a wide range as shown in TIS 0502-01, with the heavier, stronger sheets requiring higher cfm. The vacuum level for individual boxes ranges from about 2″ Hg to 12″ Hg, with the level progressively increasing down the table, as with vacuum drainage elements.

The vacuum volume requirements are affected by the grade of paper, the basis weight, freeness, machine speed, and forming fabric. With double-layer and especially triple-layer fabrics, it will be necessary to provide more volume to handle the air that will leak in at the leading and trailing edges of the box.

When determining the total wire box vacuum flows, it is important to remember that the flows for individual boxes cannot be added together because each box is at a different vacuum level. The acfm for each box must first be converted to the acfm at the vacuum level of the header, then the flows can be added, once they are all at the same vacuum level.

4. Couch rolls

Moving from the slice down a fourdrinier table, or through a twin wire unit, the vacuum level requirements progressively increase with the couch requiring the highest level, normally designed for 20″ Hg for most grades but can range from 12–22″ Hg. On machines with double couch boxes, the first box is generally designed for 10″ Hg.

Laboratory trials were run to determine the change in sheet consistency with changes in couch vacuum level. Results on a sheet weight of 92 lb/3000 sq ft, running at 2000 ft/min, showed that a vacuum change from 14″ to 20″ Hg resulted in a consistency increase from 20.5% to 21.7%, indicating that the higher vacuum level is not significantly beneficial [5].

However, actual field experience does not appear to support this conclusion. Some machines, particularly linerboard, have shown significant improvements in runnability with higher vacuum levels and the corresponding improved dryness. On linerboard, much of the dryness is attributed to the lumpbreaker, which today is commonly designed for 250 pli, compared to 75 pli a few years ago. Other grades have shown improved operating efficiency, even without significant increases in sheet dryness. The operating efficiency has been attributed more to "sheet hardening" than to improved dryness [6].

External air water separators are essential on the stationary drainage elements of the forming section, such as wire boxes and vacuum drainage elements, to prevent water carryover to the vacuum blowers or liquid ring pumps. But the couch roll itself serves as a separator, on machines operating above 2000 ft/min. Above this speed, but also depending on the diameter, the centrifugal force is sufficient to prevent the couch vacuum from pulling the water into the suction box. The vacuum holds the water in the holes of the roll, balancing against the centrifugal force on the water. This separation of water from the air allows the design engineer to size the couch vacuum piping for dry conditions rather than wet, as discussed in Sec. E2. However, it must be recognized that total separation does not always occur. Fines and some water will be drawn into the suction box, the amount varying with paper grade and vacuum level.

The volume required for the couch is made up of two components: the air flow through the sheet and the flow needed to evacuate the couch shell. The volume required by the couch can be determined from the vacuum factors in TAPPI TIS 0502-01. The factors include some allowance for shell hole evacuation. The volume required for hole evacuation can be calculated as follows:

$$\text{CFM} = (V)(b)(s)(E)(M)$$

where:

- V = Machine speed, ft/min
- b = Roll shell face width, feet
- s = Hole depth, feet
- E = % open area of shell
- M = Expansion factor, $\left(\frac{P_2}{P_1}\right)^{0.9} - 1$
- P_2 = Ambient pressure, in Hg absolute
- P_1 = Suction box vacuum, in Hg absolute

As machines get faster, are built wider and have thicker roll shells, it becomes more important to consider the effect of air pumped into the suction box by the shell holes [3].

5. Press suction rolls

Vacuum is applied to press suction rolls to increase void volume in the press nip, providing a place for water to move away from the sheet as it is pressed in the nip. Hydraulic pressure forces the water to flow from the sheet into the felt and the vacuum allows the water to flow into the shell holes.

On slow machines, normally below 1000 ft/min, but also dependent upon roll diameter, the centrifugal force on the water is low and the vacuum pulls the water into the suction box. But at higher speed the vacuum holds the water in the holes, allowing the water to sling out into the saveall pan (or back onto the felt), when the holes cross the suction box seal [5].

The effect of air pumped into the suction box by holes in the shell is the same as discussed in Sec. D4 for the couch roll. Vacuum requirements for suction press rolls are listed in TIS 0502-01.

6. Felt conditioning

Felt conditioning on modern machines today is accomplished by use of felt suction tubes, often called uhle boxes. Felt tubes normally have slotted covers, although herringbone pattern covers continue to be used, but in a limited number. Some felt tube covers have adjustable slot widths to give the proper air velocity through the slot and the proper dwell time. Dwell time is normally in the range of 3 to 6 milliseconds for most machines in the speed range of 1500 to 4000 ft/min. For tubes with double slots, use the total width of both slots for calculating dwell time. Individual slot widths less than 0.5″ or greater than 1″ should be avoided. Slots less than 0.5″ have a greater tendency to plug while slots wider than 1″ will increase felt wear.

Felt conditioning air requirements are influenced by the felt construction, felt tube slots, moisture content of the felt (showers plus water from the sheet) and the pressure drop across the felt [7]. Air flow rates, shown in TIS 0502-01, are listed for both new felt conditions and old felts.

Air flow required for felt dewatering can also be calculated using the DeCrosta equation [8, 9]. This equation, which considers the variables affecting felt dewatering, and is the basis for TIS 0404-27, is listed below:

$$V^1 = \frac{0.069(P_{sp})^{0.476}\,(td)^{0.110}\,(V^*)^{0.916}}{(M_{p,1})^{0.628}}$$

$$M_{P,2} = \frac{1.23\,(M_{p,1})^{0.819}}{(V^1)^{0.024}\,(P_{sp})^{0.124}\,(td)^{0.096}}$$

where:

- V^1 = Specific air flow through the felt at the suction tube, scfm/in^2
- P_{sp} = Pressure drop across the felt at the suction tube, in. Hg.
- td = Dwell time of a particle of felt at the suction tube, milliseconds (ms)
- V* = Felt permeability, cfm/sq. ft. at 0.5″ H_2O
- $M_{p,1}$ = Felt moisture content approaching suction tube, lb H_2O/lb of felt
- $M_{p,2}$ = Felt moisture content leaving suction tube, lb H_2O/lb of felt

The felt permeability of a new felt can be readily obtained from the manufacturer. The permeability will decrease through the life of the felt to about 20% of the original value. With that information, one can calculate the new and old felt vacuum requirements. Typical inlet moisture values are 0.8 lb H_2O/lb felt for the first press, 0.7 for the second and 0.6 for the third.

Air water separators should be provided for each felt tube, or for each pair of tubes if on the same felt. The separators protect liquid ring pumps from cleaning chemicals and reduce contamination of the vacuum pump seal water. Separators are essential for protection of centrifugal, vertically split exhausters. Central vacuum systems using centrifugal, horizontally split casings, use large concrete separators, as stated in Sec. C2. However, individual separators have also been used to further reduce the chance of water carryover from the concrete separator.

Separators can be drained to a seal pit using a barometric leg if the separator elevation (set by the felt tube elevation) is high enough. On felt tubes for bottom felts, removal pumps are required. The drain line from the separator should be sized for 3 ft/s velocity and the pump discharge sized for 7 ft/s.

E. PIPING DESIGN

After determining the vacuum requirements of the machine and selecting the proper vacuum equipment, it is still necessary to avoid the pitfalls of improper piping design.

1. Throttling valves and vacuum control

Avoid excessive pressure drops across throttling valves. A lot of energy can be wasted by taking a low vacuum level requirement system, such as the wire boxes, and piping through a throttling valve to a high-level system such as the couch. Match the vacuum equipment with the system requirements. Section B discusses the effects of system losses on the performance of the vacuum system.

2. Vacuum pipe sizing

Vacuum lines should be sized to avoid excessive friction loss. Excessive line losses have the same effect as a throttling valve, the air flow at the paper machine is reduced for constant volume flow at the pump.

Traditionally, piping has been designed using standard line velocities of 5500 ft/min for "dry" air from separators and from suction rolls. For wet air with entrained water, the industry has traditionally used 3000 ft/min, from sources such as felt suction tubes, wire boxes, etc. Some vacuum pump suppliers have pipe sizing charts available that show air velocity versus flow rate for various pipe diameters. Selecting the proper pipe size will help minimize line friction losses.

A much better approach is to size vacuum pipe based on friction. It is recommended that vacuum pipe be sized on the basis of 3 to 4 in. water column drop per 100 ft of pipe, with a maximum drop of 7 to 14 in. water for the pipe system. As vacuum systems get larger, and as stainless steel pipe is being used more commonly, economical selection of vacuum pipe is critical. Sizing on the basis of friction will often result in smaller vacuum pipe than would be used when sizing on the basis of velocity.

3. Pipe routing

As stated above, it is important to select the proper line size, and design a good route for the line. Locate the vacuum pumps as near the machine as reasonably possible to minimize the length of vacuum lines.

Avoid pockets in the pipeline at all costs. Once the vacuum line reaches its lowest elevation, it cannot be turned back up. Water will collect in the pocket, resulting in high velocity or vacuum surging. There are cases in which the elevation of the vacuum pumps is set by the allowable pipe run. Refer to Sec. D2 for recommended piping to vacuum drainage elements. The same principles used in vacuum piping to drainage elements apply to general vacuum piping.

4. Piping materials

The materials for construction for vacuum service piping must be selected to meet the operating conditions of the vacuum system. Corrosion resistance, pressure requirements and cost are all factors to consider when selecting piping materials for use in a vacuum system.

Thin-wall stainless steel pipe, though somewhat expensive, offers excellent corrosion resistance to the conditions usually encountered in the wet end environment. However, when using thin-wall piping, close attention must be given to the ability of the pipe to withstand full vacuum. Stiffening rings installed along straight runs of piping are the most common solution to reinforcing vacuum piping against collapse. These rings are made of rolled angle or flat stock and are then welded to the outside of the pipe at regular intervals.

REFERENCES

1. Salls, D., "Total Approach to Vacuum System Design Has Efficiency Advantages", Pulp Paper (Apr. 1983).
2. The Nash Engineering Company, "Vacuum Systems for the Paper Industry".
3. Hen⌐ L.A., "Increasing Operating Profit by Optimizi⌐g the Paper Machine Vacuum System", Proc. 1980 TAPPI Engineering Conf.
4. Hansen, V.E., "Water Management for the Single Wire Machine", Proc. 1986 TAPPI Wet End Operations Seminar.
5. Salls, D., "Vacuum Needs are More Stringent on Today's Fast Paper Machines", Pulp Paper (Apr. 1986).
6. Bell, N.E. and Phibbs, D.P., "Couch Vacuum: How to Optimize a Neglected Dewatering Element", Paper Trade J. (Nov. 30, 1979).
7. DeCrosta, E.F., "Air Flow Requirements for Conditioning Press Felts at Suction Pipes", CA Report No. 4410, Tappi 63(5) (May 1980).
8. DeCrosta, E.F., "Experimentally Determined Equations for Water Removal in the Press Section", Tappi 56:11 (Nov. 1973).
9. DeCrosta, E.F., "How to Estimate Air Flows in Felt Dewatering", Pulp Paper Can. 81(1):97 (1980).

XXII PAPER MACHINE CLOTHING

B.A. STOWE

Bruce Stowe is the Technical Director of the Appleton Wire Division of Albany International in Menasha, WI, where he is responsible for product development, process development, product quality, and the laboratory for forming fabric manufacturing. Mr. Stowe has a B.S. in Chemical Engineering and a Pulp & Paper Certificate from the University of Maine. He has written several papers on forming design/application and has been a speaker on this subject at TAPPI seminars. His work experience includes research, design, application, and product quality within Albany International Corp. and has specific experience with clothing for all three sections of the paper machine.

C.B. EDGAR

Clem Edgar is Technical Services Manager for the Dryer Fabrics Division of Albany International Corporation. He has held positions of Research Director, Design Manager, and Marketing Manager in his career with Albany which spans ten years. Prior to that, he was Design Manager for Asten Group — Dryer Fabrics Division. His early career was with Scott Paper as a Research and Project Engineer in Paper Manufacturing. He holds masters degrees in Engineering from both Dartmouth College and Widner College.

R.H. ROOT

Richard Root is a graduate of the Philadelphia College of Textiles, having majored in textile science. His career spanned the time of great evolutionary change from wool felts to engineered synthetic fabrics. Mr. Root began his career with Albany International as a Special Projects Engineer in the Technical Department, and later held various positions, recently retiring from the position of Manager of Technical Support Services after 32 years of service. Mr. Root was an active member of TAPPI, publishing several articles as well as being an instructor at the Pressing and Drying Seminar.

REVIEWED BY

Benjamin A. Thorp
Vice President, Engineering
James River Corporation
Richmond, VA

Charles A. Lee
Charles A. Lee Associates Inc.
Knoxville, TN

A. FORMING FABRICS

1. Introduction

Paper machine clothing (PMC) refers to the "engineered fabrics" used to support the web being made on tissue, paper and board machines. Frequently referred to as wires, wet felts and dryer felts in the forming, press and dryer sections, respectively, current nomenclature favors the use of forming fabrics, press fabrics, and dryer fabrics. The reasons for these changes reside in the needs of the papermaker for the greater productivity provided by faster machines running wider webs, while trying to satisfy the needs of the consumer for higher quality paper products. Hence, the evolution of wires and felts into the engineered fabrics used to clothe the forming, press and dryer sections of today's modern, high-speed machines.

The demands for greater fabric stability, strength and resistance to thermal and chemical degradation led to the use of polymeric materials in the form of fibers, monofilament yarns, and multifilament yarns. These polymeric materials are referred to, generically, as polyamides, polyesters, acrylics, aramids, and so on. Modern paper machine clothing uses spun yarns (made from short fibers), monofilaments (a single, long yarn having a relatively large cross-section which is usually circular, but may be rectangular or square), and multifilaments (a bundle of continuous, relatively small diameter fibers). Only monofilaments are used in the production of forming fabrics. When making press fabrics, spun yarns, monofilaments and multifilaments are arranged in intersecting or overlapping patterns to form woven and nonwoven "bases". A "batt" of fibers ("combed" on a carding machine) is needled onto one or both sides of the base fabric. Chemical treatments may be applied to the fibers, the spun yarns, or the multifilament yarns before weaving or to the completed fabric after the batts have been applied to the base fabric. Dryer fabrics are manufactured in various forms, similar to forming and press fabrics.

Fabrics can be made having one, two or three layers of machine-direction yarns. Depending on the application, the nature of the MD (machine direction) and CD (cross machine direction) yarns will change in type, polymer make-up, and size.

Figure 719 shows the difference in fabrics with MD layered yarns that are used in pressing and dryer fabrics. Forming fabrics of single and double layers are similar to those of pressing and drying except that the double layers are typically in the CD rather than MD. Triple-layer forming fabrics differ from pressing and drying as they

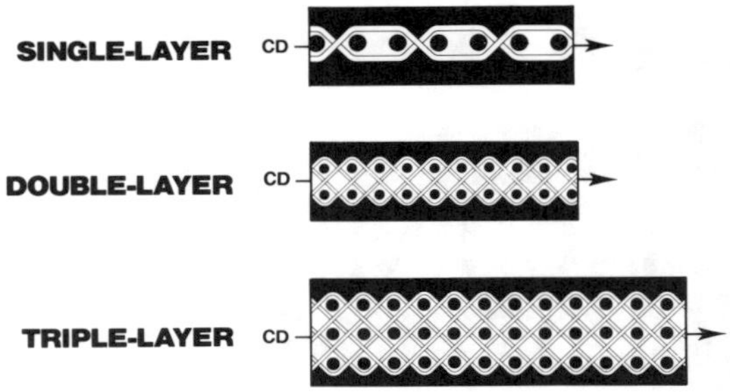

Fig. 719. Layered paper machine clothing is made by weaving stacked machine-direction or cross-direction yarns to produce a single-, double-, triple-, or more multi-layer fabric. These schematics show machine-direction yarns stacked in single-, double- and triple-layer forms. Variations on these constructions are used in the manufacture of all types of paper machine clothing (PMC).

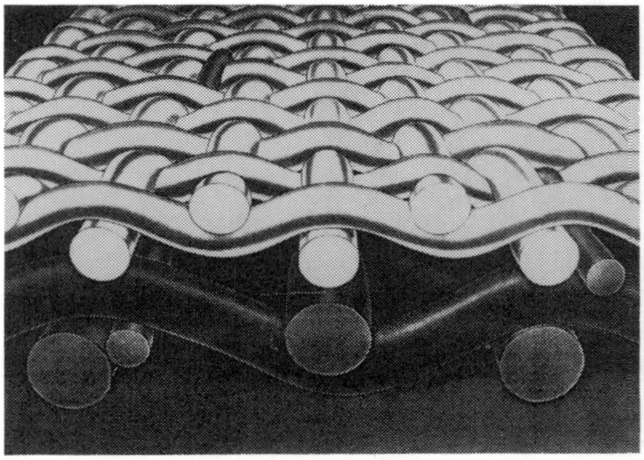

Fig. 720. Typical triple-layer forming fabric.

are better described as two single-layer fabrics held together with a binder yarn (third layer) (see Fig. 720).

This section deals with the traditional forming fabric woven from monofilament yarns using either the flat-woven or endless-woven techniques. The information given here and in the references will provide the reader with a working knowledge of modern forming fabrics enabling that person to have meaningful discussions with suppliers and other papermakers on the subject. The progression from metal to plastic materials of construction is depicted in Fig. 721; how papermakers evaluated the progression to plastic forming fabrics is described by Lewyta [1].

2. Functions of a forming fabric

The functions of a forming fabric can be described best by referring to three concepts:

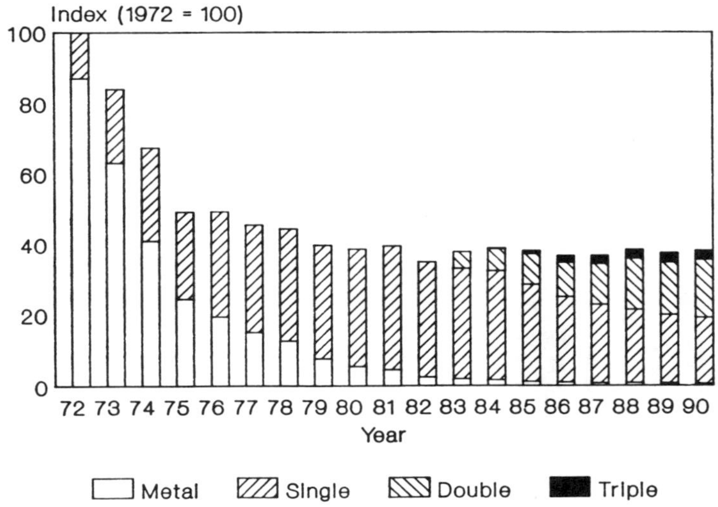

Fig. 721. Product mix distribution on the U.S. market (1972–1990).

drainage, sheet support and mechanical stability.

(a) Drainage

This is the most frequently discussed attribute of a forming fabric. It refers to the high volume of water coming from the headbox which is passed through the forming fabric and into the drainage equipment on the paper machine. Drainage equipment consists of the forming board, table rolls, foils, vacuum-augmented foils, flat boxes, curved shoes, and rolls especially on twin-wire machines. Not only is the total drainage of the forming fabric important to the papermaking system, but the "drainage rate" must also be considered. As it implies, the drainage rate describes how fast the water is removed from the pulp slurry. Depending upon machine configuration and paper grade requirements, it may be desirable to have a fast or slow drainage rate. The typical terms used to describe forming fabrics with respect to drainage are open area, permeability, void volume and void volume distribution. Reference [2] gives a more detailed description of these terms. Figures 722 through 724 illustrate the concept of void volume distribution.

The figures show cross-sectional slices at five different locations through the thickness of the forming fabric. As a general statement, the higher the open area, air permeability, void volume, or bottleneck, then the higher the initial drainage rate. The subsequent drainage rate may or may not continue at the rapid rate with a more open fabric, because plugging of the forming fabric with fibers may occur. This is referred to as "sealing the sheet".

(b) Sheet support

The ability of the forming fabric to retain the fibers and other desirable components of the furnish on top of the fabric and not let them pass through is referred to as sheet support. In this respect, the forming fabric is acting like a screen to filter out the papermaking fibers. Sheet support has to be understood with the simultaneous understanding of drainage of the forming fabrics. Low levels of sheet support would either allow the fibers to pass completely through the fabric or to plug up in the mesh of the fabric. In either case the drainage and papermaking aspects could be ruined. Beran [3] has proposed the following formula for fiber support index (FSI):

$$\text{FSI} = (2/3)\ \{(\text{A}^*\ \text{mesh}) + (2^*\text{B}^*\ \text{count})\}$$

The constants A and B are determined from the exact weave pattern and structure of the forming fabric. Mesh and count refer to the number of machine-direction and cross-direction strands of the papermaking surface. Helle [4] characterizes the support function of forming fabrics by looking at the "unsupported" areas in the fabric and refers to them as "frames". These can be mathematically described in shape, orientation and frequency once the weave pattern, mesh, count and material diameters are known. An easier method to characterize fiber support is simply to count the number of support points per unit area of fabric. Only the mesh, count and weave pattern need to be known for this calculation. Figure 725 compares fiber support points for fabrics of different designs.

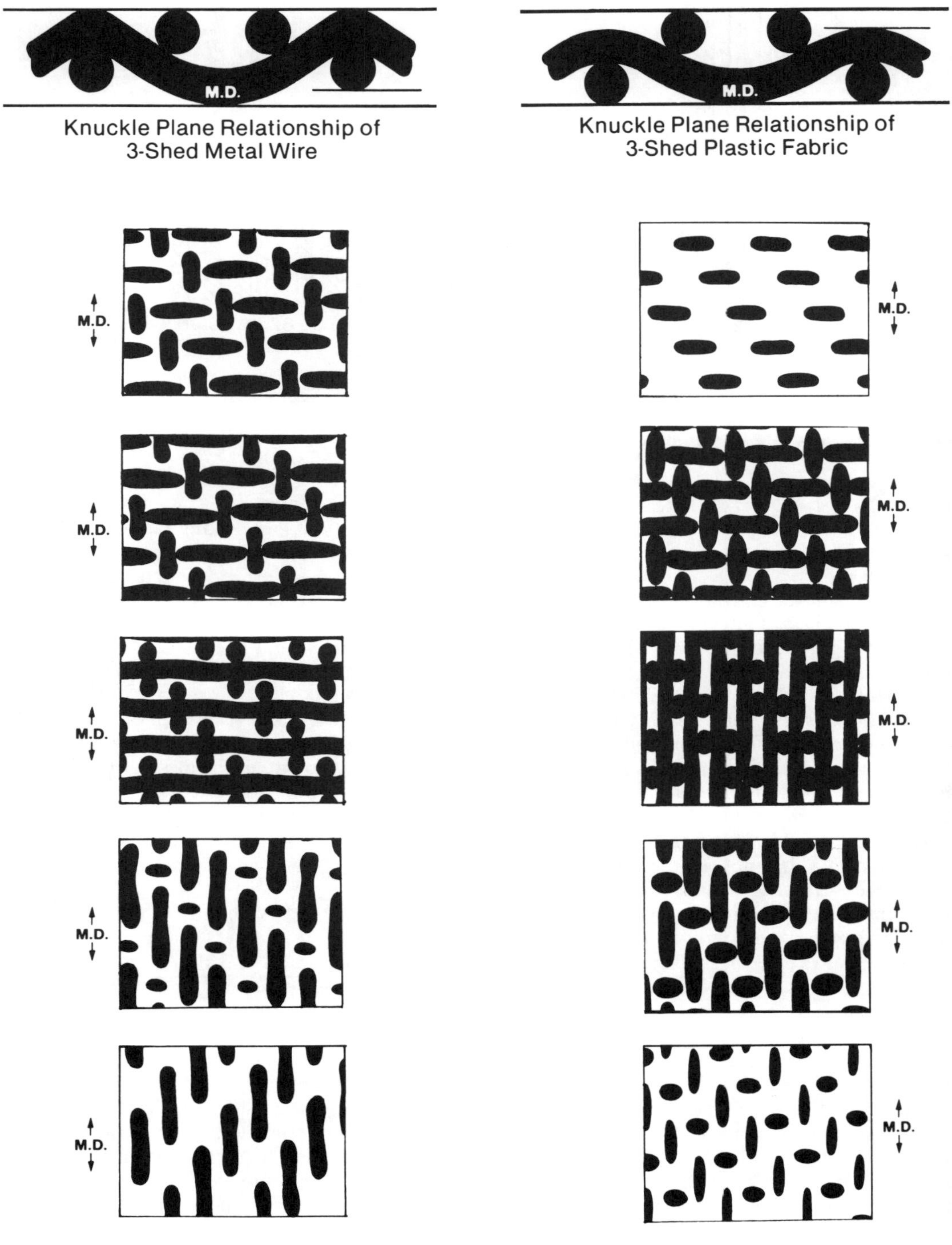

Fig. 722. Cross-sections of a three-shed metal wire and a three-shed plastic forming fabric are shown to illustrate how the void volume available at successive depths through the thickness of the fabrics varies. The area occupied by the yarns in the successive planes is indicated by the black areas. The open channels, which relate to the available void volume, are represented by the white areas.

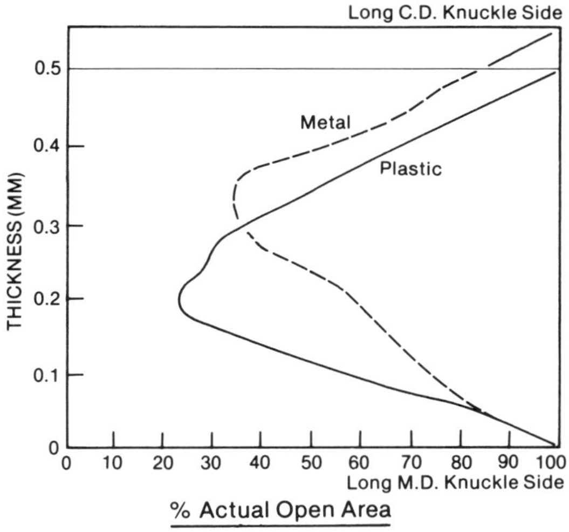

Fig. 723. Void volume distribution of a three-shed plastic fabric vs. that of a three-shed metal wire.

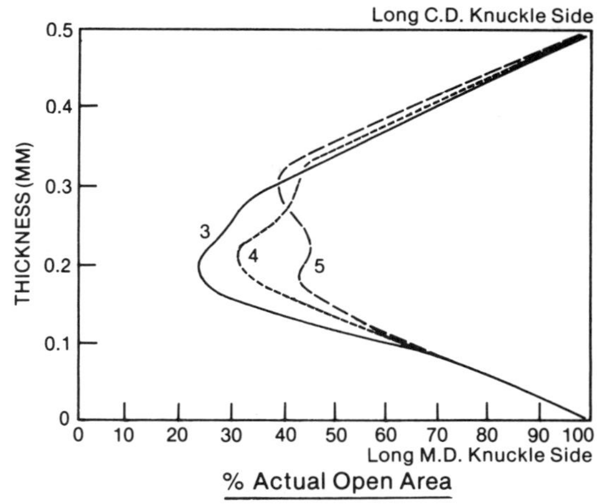

Fig. 724. Void volume distribution curves for the three-shed vs. four-shed vs. five-shed plastic fabrics.

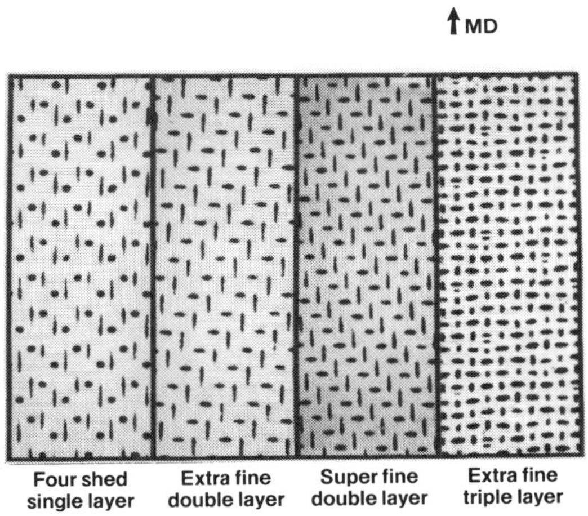

Fig. 725. Fiber support patterns for single- vs. double- vs. triple-layer fabrics.

(c) Mechanical stability

Mechanical stability refers to the physical characteristics of a forming fabric. The forming fabric has to have strength enough to run on high-speed, wide machines without excessive stretching or wrinkling. In addition, the fabric must give suitable life from a cost justification standpoint. By using more strands, and/or larger strands, the mechanical stability can be improved. Weave pattern can have a large influence due to the number of yarn crossover points or intersections of yarns. The three major functions of the forming fabric are often at odds with each other. For example, the use of a relatively few monofilaments of small diameter would increase drainage. For more sheet support, the use of a greater number of monofilaments with small diameter is indicated. Maximum stability requires a large number of large diameter monofilaments. In order to optimize the combination of these three functions of a forming fabric, papermakers must rely on the accumulated experience of their forming fabric suppliers.

3. Terminology

Some of the words and terms used to describe forming fabrics need definition to minimize confusion in communications. Table 77 lists the major terms.

4. Material

Most of the material currently used for forming fabrics is polyethylene terephthalate (PET) and often referred to as just polyester. Through experience, the PET has proven best from the standpoint of weaveability and runnability on paper machines. The specific properties of the polyester in regard to molecular weight, elongation, shrinkage, etc., are all utilized by the forming fabric supplier to manufacture the different weave patterns to obtain the necessary properties for papermaking. For example, high molecular weight cross-direction strands will be utilized to improve abrasion resistance. In addition to polyester, polyamide (nylon) is used to improve abrasion resistance of forming fabrics. The biggest drawback using polyamide is its lack of stiffness and stability (dimensional

Table 77. Forming fabric terminology.

Term	Description	Typical Range Metric	Typical Range Imperial
Caliper	Cloth thickness (mm or in.)	0.30 to 1.15	0.012 to 0.045
Count	Number of cross-direction strands per cm or per in.	16 to 63	40 to 160
Mesh	Number of machine-direction strands per cm or per in.	17 to 80	45 to 200
Open Area	Projected openings in a cloth, normally used with single-layers fabrics (%)	12 to 30	12 to 30
Monoplane	Reference to plane difference when machine-direction and cross-direction strands are exactly in the same plane		
Permeability	Air flow through a fabric when the pressure difference across the fabric is 10 mm water or 0.5 in. water (m/s or cfm/sq ft)	1.0 to 3.5	250 to 800
Plane Difference	Topographical knuckle difference on paper or machine sides. Normally, paper side difference is very low, whereas the wear side has a high knuckle difference (mm or in.)	0.00 to 0.35	0.000 to 0.014
Shute	Cross-direction strand, weft or filling yarn, for flat-woven fabrics		
Stacking	Vertical alignment (registration) of shutes in double-layer fabrics		
Super	Reference to plane difference where cross-direction strands are higher than machine-direction strands		
Twill*	Reference to the plane difference when machine-direction strands are higher than the cross-direction strands		
Twinning	Pairing of adjacent strands of either machine- or cross-direction yarns. Expect a ratio of larger spacing divided by smaller spacing		
Warp	Machine-direction strand for flat-woven fabrics		
Weave	Description of crimp in a monofilament, and also pattern of placement of these crimped monofilaments [1].		

*The word "twill" has three different meanings in regard to forming fabrics. The one used above is in reference to the plane difference of the monofilaments. Twill is also a description of the weave pattern, for example, semi-twill, full twill, in which filling yarns pass over one and under two or more warp yarns. Twill also describes the prominent angle of the knuckle pattern on the paper surface. The twill angle can be left to right or right to left. In addition, the twill can be prominent or the twill can be broken (satin).

change from wet to dry state). Normally, when polyamides are used, they are limited to the wear side of multilayer fabrics alternating with polyester materials.

Monofilaments used in forming fabrics range in diameter from 0.10–0.50 mm. As can be expected, the finer diameter monofilaments are used in the finer mesh fabrics and for the higher paper surface property requirements. Finer diameter strands are more easily damaged by high-pressure showers, foreign objects, wear elements, etc. The normal shape of monofilaments used is round. Both square and flat monofilaments have been used to obtain specific fabric properties. For example, flat monofilaments would allow a reduction in plane difference on the papermaking side to enhance sheet support.

5. Fabric application

Fabric application is a very complicated topic that requires a complete understanding of forming fabrics and machines. The most important items to consider are grade, furnish, speed and machine configuration.

Grades of paper can first be ranked from coarse to fine with pulp and linerboard being on the coarse side and lightweight tissue and one-time carbonizing on the fine scale. As a general statement, the coarse-to-fine forming fabrics can be laid out in the same order as coarse-to-fine paper grades. Lewyta [1] expands on this discussion.

Furnish considerations normally focus around fiber length and additives. The shorter the fibers, the more sheet support is needed for good operations. Headbox loading of inorganic material is used to enhance paper properties and needs to be retained in the forming process. These fine particles will easily pass through a forming fabric, so it is very important to control the drainage rate and the initial fiber mat

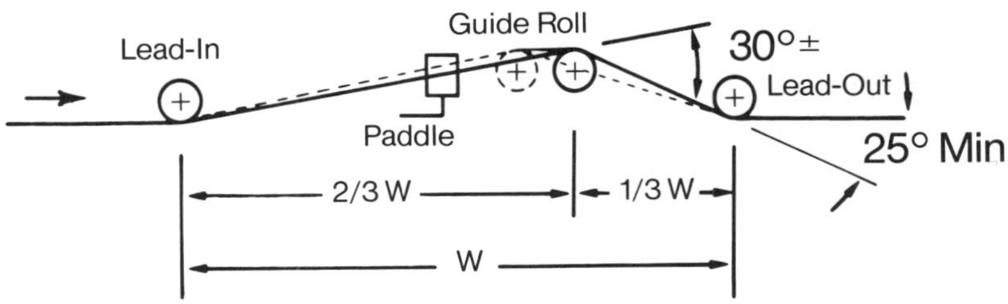

Fig. 726. Recommended relationship of lead-in and lead-out rolls to the guide roll and guide paddle.

formation to help improve this retention of inorganic materials.

Machine speed affects the drainage pulses and drainage forces imparted through the forming fabric to the sheet. It also affects the time available for water to be discharged from the forming fabric into saveall receptacles. On twin-wire paper machines, as the speeds have continued to increase from 1200 to 1500 to 1800 m/min, the fabric drainage capability was increased without the sacrifice of sheet support or stability. On the other extreme, paper machines below 300 m/min must be careful in using multilayer fabrics due to the pulse dampening when using complicated weave structures. Table rolls at this low speed would not present a problem as they generate good pulses whereas low-angle foils might not generate enough drainage pulses.

Twin-wire machines take special consideration as there are numerous configurations with different requirements. Stowe [5] gives a more detailed description. Twin-wire machines can be broken down into two major categories and several subcategories:

1. True twin-wire
   - Blade former
   - Roll former with suction
   - Roll former without suction
   - Combination
2. Hybrid former
   - Blade former
   - Roll former
   - Roll former with suction
   - Combination

True twin-wire refers to the slice jet being trapped simultaneously by two fabrics where the hybrid former has a typical fourdrinier preformer before a "top-wire unit."

6. Fabric operation

Tension on forming fabrics normally ranges from 3.5–7.0 kN/m (20 to 40 pli) when it is measured on the return run. Actual fabric tension at the highest point in the loop of the fabric could be twice that of the return run tension. Fabrics that are run too slack can be hard to guide and may show a tendency to depart from flatness. Fabrics run too tight can stretch or skew the fabric and possibly give ridging problems. For each paper machine there is a proper running tension, and forming fabric suppliers will manufacture fabrics to run at that tension.

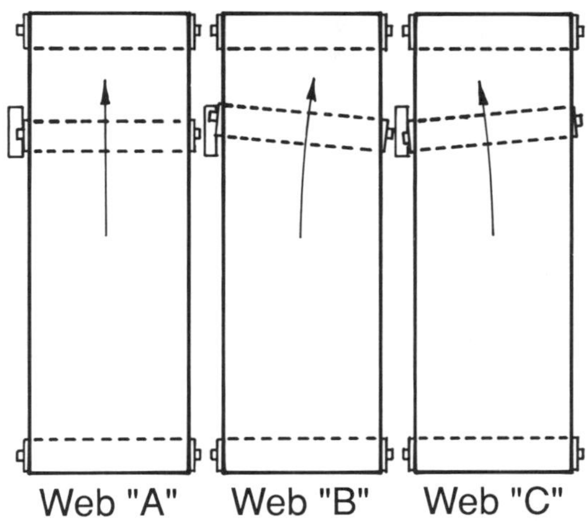

Fig. 727. Guide roll movement governs fabric displacement as illustrated in this schematic. The fabric will tend to move in a direction perpendicular to the long axis of the moveable guide roll.

Guiding of forming fabrics has become more and more critical as machine widths have increased and machine speeds increased. Fabrics with a strong diagonal pattern tend to have a preferred tracking direction. Refer to reference [6] for detailed description of proper guide roll settings (see Figs. 726, 727). Some of the key points for proper guiding are:

- Use rubber-covered guide rolls with 20 P&J

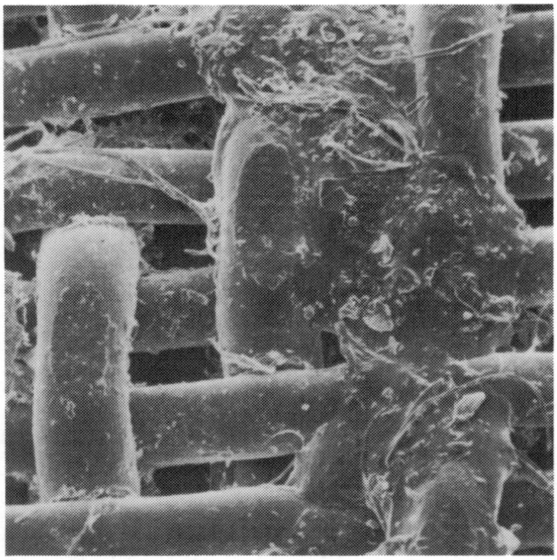

Fig. 728. An example of fabric contamination.

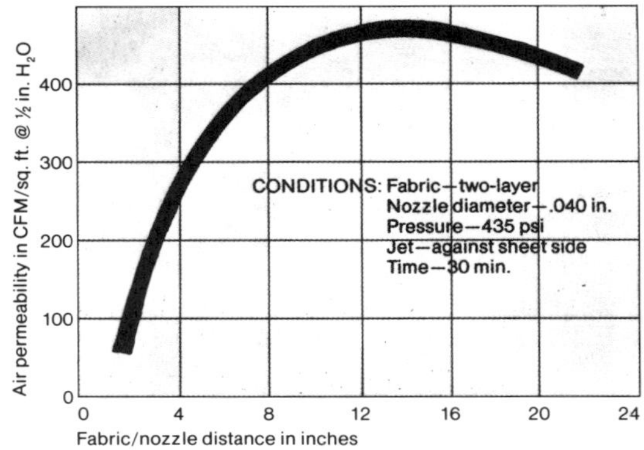

Fig. 729. Air permeability vs. fabric-to-nozzle distance after cleaning a fabric with a high-pressure shower under the conditions noted in the figure indicate that the positioning of the shower is very important. Under any set of shower and machine operating conditions, an optimal separation of shower nozzle to fabric exists, with respect to cleaning efficiency.

- Use an evenly divided wrap of approximately 25°
- Adjacent return rolls are to be ⅓ to ⅔ of the roll length on either side of the guide roll
- Do not locate the stretch roll adjacent to the guide roll as fabric length will change angle wrap
- Roll displacement should be at least ±1% of roll face

7. Cleaning the fabrics

Cleaning of forming fabrics is critical, as illustrated by Fig. 728. High-pressure oscillating needle cleaning showers are generally used to clean modern forming fabrics. These showers can be located either inside or outside the forming fabric depending upon machine configuration and type of contaminant being cleaned. If the contaminant is mainly on the papermaking surface, then the showers should be placed on the papermaking side. If the contaminants are further into the fabric, the showers may be placed on the inside of the fabric run. For face side showers the distance should be 20–40 cm from the fabric with pressures of 25 to 40 bar. Inside showers should be 10 cm from the fabric with pressures of 20 to 30 bar. See Fig. 729 for a comparison of cleanability of the fabric versus distance of the shower from the fabric.

The test results shown here indicate that forming fabric cleanability can be optimized as a function of shower-to-fabric separation. The needle shower jet breaks up about 15 cm from the nozzle. Thus the optimum separation is found after the breakup of the needle jet. The fabric is then subjected to vibrating pulses from the shower. Oscillating the shower one or two nozzle jet diameters once per revolution of the forming fabric gives maximum coverage of the forming fabric in the shortest amount of time. Normal spacing of showers are 7.5 or 15 cm centers with 15 or 30 cm strokes. See reference [7] for a more detailed description.

An additional way to continuously clean a forming fabric is that of using a "brush" on the forming side of the fabric in the return run. This brush can be stationary, similar to a stiff bristle broom, or it can be rotating like the "beater bar" on a vacuum cleaner. These "brushes" physically dislodge fiber and contaminants from the fabric. The brush is subsequently cleaned in either a continuous or a batch mode with appropriate showers and/or solvents. The bristles are made from polyethylene, polyester, polyamide, and other polymeric materials. Often, the stationary brush can be rotated out of place while a new brush is put against the fabric to enable batch cleaning of the brush.

8. Fabric maintenance

Synthetic forming fabrics can be patched with either an adhesive or ultrasonic welding to bond the patch to the fabric. Key points to remember when patching forming fabrics are to thoroughly clean the fabric around the hole to be patched and to follow all of the instructions for the technique that is being used. Patches have been known to last the full life of fabrics if they have been properly applied.

If a fabric has been trimmed, then it is necessary to repair the edge to prevent further raveling and loss of fabric width. First the fabric should be trimmed square and at least one machine-direction strand raveled to ensure a straight edge. If there is a choice of sides to trim, select the edge away from the guide palm. The edge can be sealed with a hot soldering iron or ultrasonic tool and/or epoxy applied to prevent raveling. Epoxy can normally be purchased at a local hardware store and applied through a plastic cup with a slit in its side.

Predicting the usable life of the forming fabric is done by monitoring its caliper or thickness, and tracing the historical performance of previous fabrics. For example, if a fabric starts out at 0.61 mm and through experience is known to be worn out at approximately 0.48 mm, then monitoring caliper would help to predict usable life. The loss of caliper is a non-linear function of running time during the first few weeks. The decrease in thickness is relatively rapid at first. The wear rate gradually reduces and finally becomes essentially constant after about two weeks. If no history is known for a particular design, then the best method to predict usable life is to allow 60% of the wear strand diameter as the maximum amount of caliper loss. For example, if the wear strand is 0.22 mm then 60% of this strand is approximately 0.13 mm. This would be the estimated caliper loss for that fabric. However, it is only an estimate because different fabrics can be worn to different levels and machine requirements may be different. Pitt and Condon [8] give a detailed description for examining fabric wear.

9. Summary

Forming fabrics are an integral part of the paper machine. As the name implies, the sheet is made or formed on a filtering medium referred to as a forming fabric. Forming fabrics have evolved from single-layer constructions made with metallic wire to the modern, engineered constructions having one or more layers of woven monofilaments. These changes were and are being made because of the trend in paper machine design toward wider, faster machines. This trend places increasing demands on the forming fabric for controlled drainage rates and improved sheet support while providing the mechanical stability necessary to ensure desired productivity at acceptable quality levels. For any specific machine, this implies the use of historical records of previous forming fabric performance to guide the subsequent changes in forming fabric design. Equally important, the people involved in the selection and evaluation of the forming fabrics must have a thorough understanding of the principles of sheet formation and the functions of the forming fabric. The process is evolutionary and involves the cooperation of both parties, papermaker and forming fabric supplier.

B. PRESS FABRICS

1. Overview

In the press section, two important functions are simultaneously occurring: paper web support/conveyance and paper web dewatering. In a press nip, externally applied mechanical forces subject the web and press fabric to increasing and decreasing pressure. Both mechanical and hydraulic pressure gradients exist resulting in water flow from the paper web. The press fabric is subjected to cyclic changes in pressure, up to millions of cycles in its lifetime, depending upon machine speed and fabric length. Dimensional stability and durability are prime requirements.

Water expressed from the paper web and the water in the press fabric can move in all three dimensions: lengthwise (MD), crosswise (CD) and vertically (z-direction) through the fabric. In plain presses, lateral-flow water movement occurs in the direction opposite fabric travel. In vented presses used on most modern machines, water will flow efficiently in all three directions. Water retained in the press fabric is removed by suction pipe conditioners to allow the fabric to accept the sheet water during its next pass through the press nip.

An excellent paper on the theory of press nip dewatering was authored by Wahlstrom [9]

(a) Dewatering phase

In Fig. 730, it is assumed that both felt (press

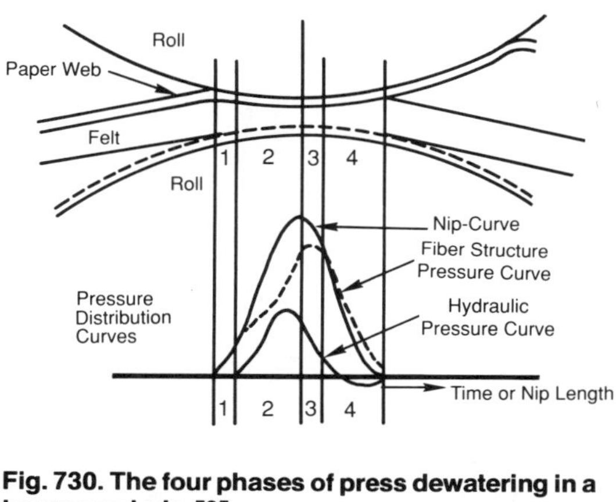

Fig. 730. The four phases of press dewatering in a transversal nip [9].

fabric) and paper are unsaturated on entering the nip. Both contain a sufficient amount of water to enable them to become saturated before reaching the middle of the nip. The geometric form, pressure distribution, water transport mechanism and graphs stating the thickness of paper and felt (press fabric) are given for the whole nip width. The press nip has been divided into four phases.

- Phase 1 starts at the entrance of the press nip where the graph for the specific pressure starts and reaches up to the point where the paper is saturated. The felt (press fabric) is shown unsaturated during the whole of Phase 1. No hydrodynamic pressure differential is generated at this stage.
- Phase 2 extends from the point of saturation to the mid-nip or, to be more precise, to the peak of the total pressure graph for the nip. At this stage the felt (press fabric) also reaches saturation.
- Phase 3 extends from the peak of the press nip graph to the point of maximum dryness in the paper. This point corresponds to the maximum point of the graph which describes compression in the sheet and by the zero point for hydrodynamic pressure in the sheet. In this section of the nip, where the expansion takes place, the felt (press fabric) passes zero point for hydraulic pressure and once again becomes unsaturated.
- Phase 4 covers the point where the paper begins to expand and becomes unsaturated. The felt (press fabric) is already unsaturated and it expands further.

Because of the dramatic increases in speed engineered into paper machines in recent years, the increased rate at which the mechanical and hydraulic pressures build up in the press fabric places relatively stringent demands on it. These demands are being met by using manmade fibers and yarns in engineered fabric construction that tend to optimize the often conflicting requirements described in the following text. Figure 731 shows the appearance of the different yarns used in press fabrics.

2. Functions of the press fabric

Regardless of paper grade or machine configuration, press fabrics are engineered to satisfy two functions: (1) to remove water from the sheet efficiently and economically; and (2) to convey the most uniform sheet possible to the dryers. Associated requirements to accomplish the above are:

- Accept sheet water.
- Provide uniform pressure distribution.
- Provide desirable sheet finish.
- Support the sheet through the nip.
- Transfer the sheet through the presses.
- Act as a transmission belt.

3. Press fabric properties

To perform the preceding functions, press fabrics have to be designed to include certain characteristics. Many terms are used to describe the required properties such as stability, durability, drainage, surface, cleanability, and pliability. These result from the engineering that goes into fabric development. Some definitions follow.

Mass—Press fabric weight per unit area.

Caliper—Thickness measured under some applied force.

Batt/Base Ratio—Mass of the batt fiber divided by mass of the base fabric.

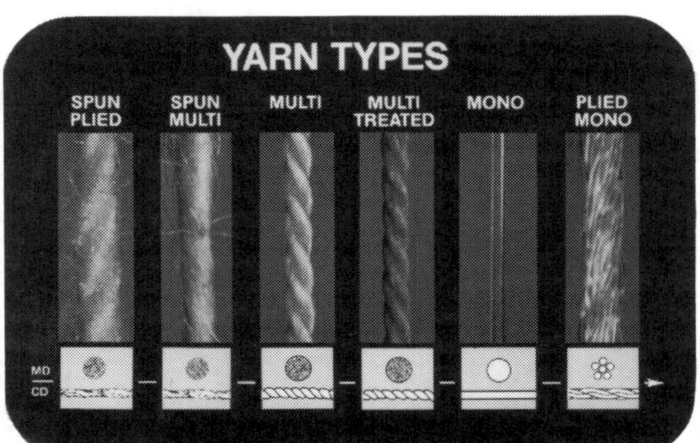

Fig. 731. The appearance of the yarns used in press fabrics is characteristic of their type. Notice, for example, how the treated multi-yarn is more compact than the untreated multi-yarn shown above. Treatments often impart color changes to the yarns treated.

Permeability—Volume of air that can pass through the fabric per unit time, at a specified pressure differential across the fabric.

These terms, in addition to the fabric class, e.g., "two-layer batt-on mesh" generally described the press fabric. The following terms are parameters which must be optimized for a given set of machine operating conditions such as speed, press load, press type, fabric conditioning equipment available, in addition to considerations relating to paper grade.

Void Volume—A measure of the amount of water that a press fabric can accept. Properly defined under load, it is the press fabric's maximum water-handling capacity.

Flow Resistance—This measures the resistance of the flow of water through the fabric in all three dimensions; machine direction (L), cross-machine direction (X) and vertical (Z). The lower the flow resistance, the easier it is for the fabric to accept water in the press nip, and release it at the fabric conditioners.

Compressibility/Resiliency—Compressibility is a measure of the ease with which a fabric is compacted under pressure. Resiliency is a measure of the fabric's ability to recover from the compressed state. A "compaction resistant" press fabric can be one which is either extremely incompressible, or which is quite compressible but resilient.

Uniformity of Pressure Distribution—Uneven transmission of pressure from the press rolls or the fabric to the sheet can lead to uneven sheet moisture profiles, and other sheet quality problems relating to printability. Pressure uniformity is achieved by the use of fine base fabric yarns, a smooth surface weave pattern, and optimally fine batt fibers.

These factors are all inter-related and must be optimized to achieve efficient press fabric performance.

4. Factors influencing press performance

While there are a great many factors which have an influence on water removal from the paper web, it is accepted, generally, that the most important variables are the applied pressure, dwell time in the nip, temperature, moisture content of the web and the ability of the press fabric to handle the water expressed from the sheet. These factors can be varied to a greater or lesser extent through machine design, operating conditions, fabric construction and fabric conditioning.

With respect to sheet water, the furnish makeup, the head box, and the type of former dictate the sheet consistency and quality delivered to the press section. The papermaker has control over many variables such as sheet freeness, machine adjustments to the forming section and the selection of forming fabric design.

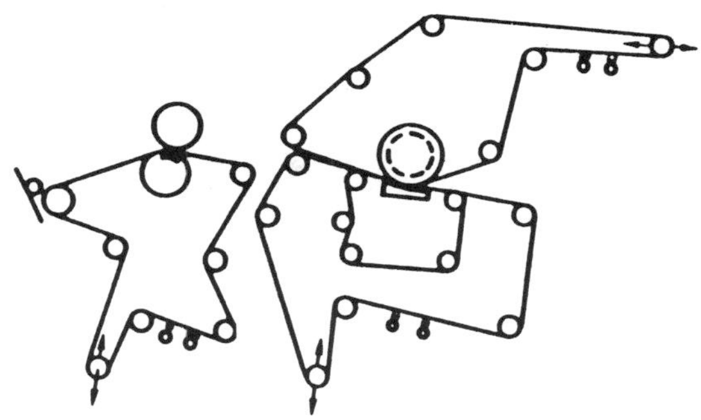

Fig. 732. Standard double-felted shoe press.

With respect to press fabric water content, similar criteria establish the amount to be handled. Design of the press configuration, roll covering, and manipulation of press loading coupled with press fabric design are the controllable variables that alter its water content. Fabric conditioning equipment, used for cleaning and dewatering the fabric, probably has a large impact on the amount of water carried by the press fabric.

The pressure gradient in the press nip is most strongly influenced by machine design, secondly by roll cover selection, and finally by press fabric constructions best suited to the operating conditions so that maximum water removal from both web and fabric is optimized.

For a given machine speed, dwell time in the press nip is basically controlled by roll diameter, cover material and applied loading. Secondary benefits can accrue by using a two-fabric press as a shoe press or with two large-diameter press rolls and different press fabric designs (Fig. 732).

5. Press fabric constructions

Because press fabrics are an integral part of the paper machine, their evolution from "conventional felts" to engineered fabrics had to parallel changes in paper machine design. Conventional felts were made from spun yarns consisting essentially of wool, but which could contain some polyamide fiber. The next major type of felt to evolve was the "batt-on-base" felt which incorporated layers of carded fibers (wool, or wool plus polyamide) "needled" (mechanically bonded) to a base fabric. This change allowed the use of higher percentages of synthetic fibers, increasing the fabric durability and resulting in a smoother surface to enhance sheet finish. A major variation on the batt-on-base construction used no CD yarns or very fine CD yarns in order to minimize the accumulation of paper fines and dirt in the felt structure. In the recent past, felts were removed from machines because they were filled with contaminants, worn out, or because they marked the sheet resulting in quality problems.

(a) Batt-on-mesh press fabrics

The development of the "batt-on-mesh" fabrics followed as monofilament and multifilament yarns became available. Modern engineered fabrics are made by selecting appropriate yarns for use in the MD and the CD which are interconnected in a woven base fabric. It is referred to as a mesh fabric because it somewhat resembles a screen. The yarns selected for constructing the fabric can be any of the available types depending on the fabric design: e.g., spun, spun/monofilament combined, multifilament, monofilament, or plied monofilament. One layer or multiple-layer weave patterns may be used. The type of batt selected is determined by the particular press position being considered. Layered paper machine clothing is made by weaving stacked machine-direction or cross-direction yarns to produce a single-, double-, triple-, or more multi-layer fabric (see Fig. 719). These schematics show machine-direction yarns stacked in single-, double-, and triple-layer forms. Variations on these constructions are used in the manufacture of all types of paper machine clothing.

(b) Planar press fabrics

The constructions employed in multi-layered fabrics occasionally lead to a more pronounced "knuckle" effect at yarn crossover points. On some grades of paper, this effect can cause marks to show on the finished sheet because of the high pressures exerted at these points in the press nip. To eliminate "knuckles" in the fabrics used on sensitive press positions, fabrics were developed in which a grid of CD and MD yarns were laid perpendicular to each other without the benefit of any interweaving. The "planar" press fabrics offer greater sheet-side support because the top-side MD yarns can be closely grouped to improve uniformity. Figure 733 shows the construction of a planar fabric.

(c) Seamed press fabrics

While the development of new yarn forms and multi-layered fabrics solved the problems of compaction and yarn marking, they created

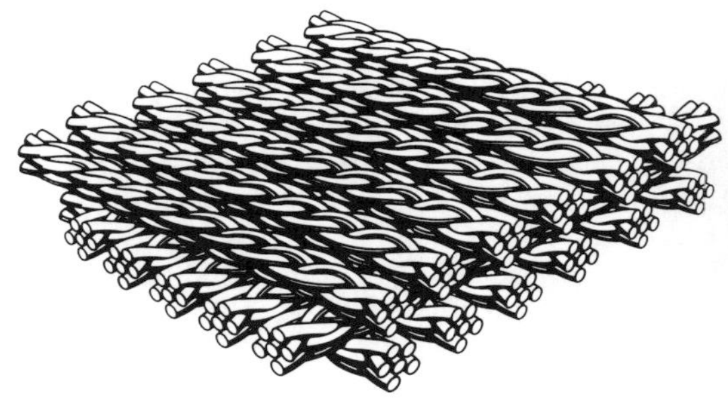

Fig. 733. This triple-layer planar base fabric is composed of plied monofilament yarns in both the CD and MD. The MD yarns can have any combination of multifilament or plied monofilament yarns.

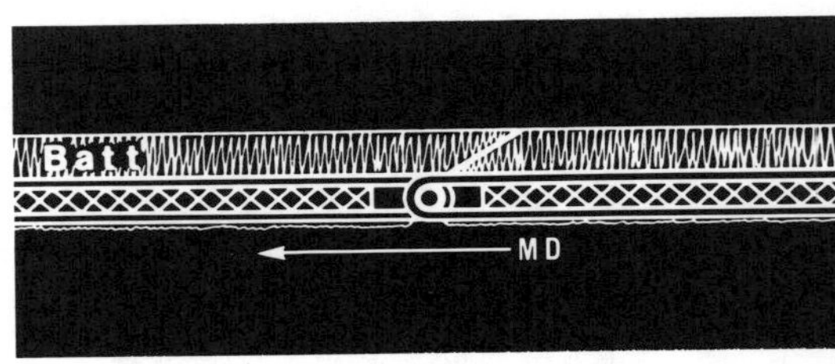

Fig. 734. Seamed press fabrics are used to clothe positions where fabric installation would be difficult because of the stiffness of the fabric.

handling problems during installation. The newer fabrics, being heavier and thicker, are also stiffer and more difficult to bundle and force between narrow machine frame clearances. The solution was to engineer an open-ended fabric which could be seamed on the machine. On-machine seamed press fabrics have found widespread application on many types of press and pickup positions. Heavily loaded presses employing large-diameter rolls or shoe-type configurations designed to provide a wide nip in the kraft market are especially suited for this type of fabric. A schematic of a seamed press fabric is shown in Fig. 734. It shows MD monofilament yarns formed into loops which are meshed together and joined with a compatible monofilament connecting pin (pintle).

(d) Laminated base fabrics

Laminated base fabrics are made by combining the technologies associated with multi-layer and planar base fabrics into a single engineered structure. This tends to optimize the fabric parameters needed to ensure proper performance while extending fabric life. Figure 735 shows one of the possible base fabric constructions. Many possibilities exist in the choice of

Fig. 735. Multi-layer laminated base fabrics are made by combining two base fabrics together by means of the needling operation. In this figure, a single-layer base is combined with a two-layer base to create a three-layer base fabric.

yarns and weave patterns in these multiple-layer fabrics.

(e) Batt fibers

While the genealogy of generic felt (press fabric) types describes the base construction, the web portion of the press fabric (batt) has undergone an evolution of its own.

The original "batt-on-base" base fabrics were made from fibers having the same length and diameter in all the layers. This gave way to blending various diameters and lengths. The batts used in modern press fabrics are composites of "stratified batts" that can be made from different lengths and diameters. Normally, the batt with the coarsest fibers is placed next to the base fabric, while the batt having the finest fibers is placed on the sheet side of the fabric. Special applications requiring heat resistance employ a top stratum of heat-resistant fibers. Other special-purpose fibers such as fusible types may be employed to provide improved bonding.

(f) Chemical treatments

Chemical treatments are applied to enhance certain applications such as startups on tissue and other high-speed lightweight grades. They may be chemically linked to the molecules of the fibers to improve compaction resistance, or provide a coating to protect the fabric from abrasion and fiber shedding. Some treatments may simultaneously provide more than one beneficial parameter, such as improved drainage and resistance to filling.

6. Specifying the press fabric

The specifications of the press fabric to be used on a particular position of a paper machine are based mostly on the experience of the fabric supplier and the papermaker. Usually, it results from a process of trial followed by design modification. For a new machine, analogies are made to like positions on similar machines. After the fabric has been run, it is inspected and analyzed and, if necessary, modifications are made to the original design.

The papermaker is usually interested in fabric life and low fabric cost per ton of paper produced. A trend can be seen toward emphasizing overall cost reduction. TAPPI Technical Information Sheet TIS 0402-22 provides a useful approach to pressing economics [10].

Ultimately, press fabric selection depends on the grade of paper made, its weight and finish requirements, the machine type, the severity of the conditions under which the fabric will be used, and the fabric conditioning equipment available. To these ends, the fabric designer may vary weight, caliper and permeability by modifying yarn count, yarn form and weight, weave patterns, batt distribution and fineness, chemical treatments, overall synthetic content and other manufacturing processes where applicable.

(a) Cylinder or multi-ply board machines

Many paperboard grades such as containerboard, folding boxboard, non-folding boxboard and solid bleached board have a top liner, underliner, filler as well as backliner and are produced on machines having more than three cylinder vats. Most machines commonly have

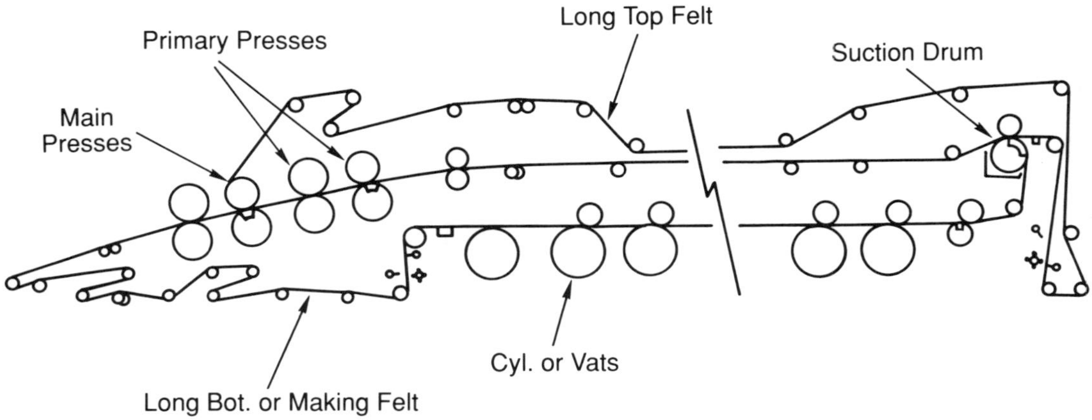

Fig. 736. Conventional multi-ply board cylinder machine.

seven or eight cylinders and may go up to ten cylinders. A former can be used to replace a conventional vat, usually for the top liner or underliner. These formers are advantageous because they give better formation and, in many cases, less liner stock can be used.

Multi-cylinder machines are labeled with a commonly used nomenclature system. Each cylinder is numbered from the beginning of the run. The primary presses are so designated. They are often called baby or helper presses depending on roll size and mill terminology. The primary presses are followed by what is usually called the press or first press. The fabric that contacts the cylinders is called the bottom (or making) fabric. Most modern multi-cylinder machines have two top positions, the suction drum and the primary.

The first top fabric is called the suction drum fabric; the second top fabric is called the primary fabric. It is not uncommon for the main press to be followed by a separate press section, or a combination of straight-through presses. These positions are normally designated second and third presses, etc. Instead of the three or four plain primaries commonly used, some mills are replacing these with combinations of plain, grooved, and suction rolls, which can be more heavily loaded. On the main press, a trend is seen toward a three-roll combination of suction, plain, and grooved rolls.

Multi-cylinder fabric applications

The bottom fabric on the cylinder machine, which picks up the web plies, must be open to allow for drainage, smooth enough to impart a desirable finish, and be strong and stable because it is the press section drive belt. The top fabric is very much like the bottom fabric in construction, but is usually much lighter. Finish is more important, thus yarns are smaller. The top fabric must be open to prevent crushing of the paper web. Tension requirements for it are not as severe as those for the bottom fabric. For split top machines, the first suction drum fabric must be open (i.e., more permeable) to allow for the passage of water. The second top, or primary fabric, must also be open, but it has a higher finish requirement.

Machines with long top and bottom positions (Fig. 736) usually run single or one and one-half layer mesh fabrics that carry a minimum of water and are very open to facilitate good drainage. The split top machine (Fig. 737) with heavier loading at the suction drum and main press positions is ideally suited for multiple-layer mesh products, combination fabrics and on-machine seamed fabrics in the top positions, with one and one-half layer mesh or seamed press fabrics in the bottom position. For the press positions, higher loaded presses, mainly grooved and/or blind-drilled, are increasing in use. Therefore, heavier weight, multi-layer, batt-on-mesh fabrics are needed to accept more water from the sheet without crushing, and to impart good finish.

(b) Tissue machines

Tissue machines are a special technology characterized by a singular large-diameter "Yankee dryer" (typically 10 to 20 feet) with the larger diameters utilized in the higher speed applications. The grade structure on Yankee machines includes toilet, facial, napkin and toweling.

The "two-felt" (pickup and bottom press fabric) machine configuration has given way to the "single-felt" (pickup fabric only) machines

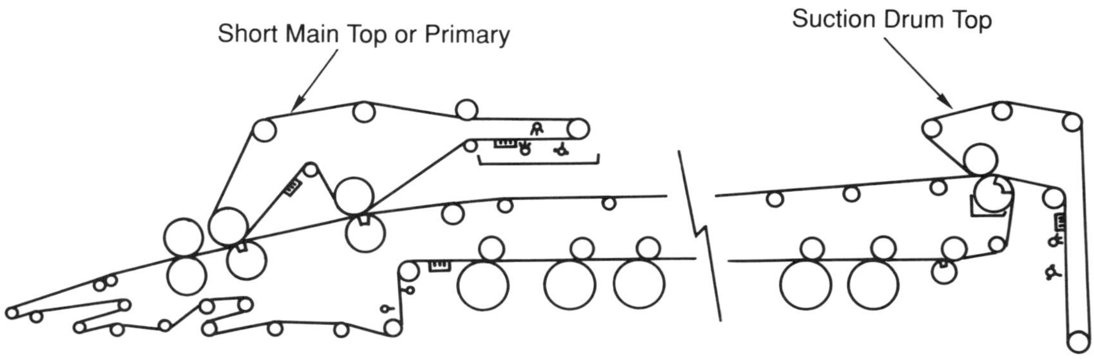

Fig. 737. Split top configuration of a multi-ply board cylinder machine.

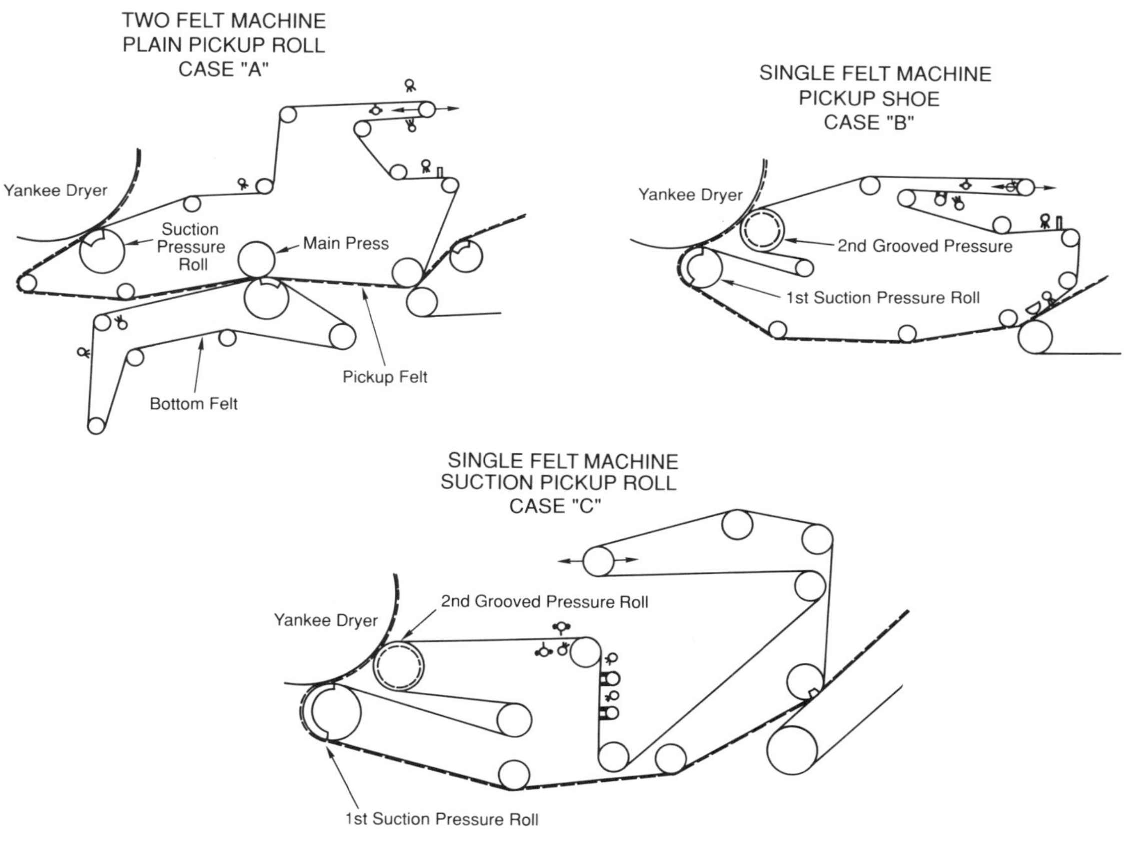

Fig. 738. Modern configuration of high-speed tissue machine press section.

which are the fastest of any grade structure: 1067–2000 m/min (3500–6500+ ft/min).

Yankee machines can be equipped with either a solid pickup roll, suction pickup roll, or a pickup shoe.

The Yankee machine with two felts (press fabrics) has traditionally been predominant, but now there is a trend toward the single fabric configuration because it eliminates the bottom felt (press fabric) with its accompanying equipment and maintenance requirements. Moreover, sheet stealing is eliminated, the noise level is considerably lowered, and economic gains are realized in decreased water requirements and power needs. Figure 738 identifies common configurations.

Tissue felt (press fabric) applications

Tissue pickup fabrics require smooth surfaces and uniform pressure distribution. The trend is toward the use of two- and three-layer fabrics. Many machines use two fabrics with a main press position, referred to as a pickup fabric and a bottom fabric. The bottom fabric accepts the water expressed at the main press. It is usually a single-layer batt-on-mesh design.

(c) Press section terminology

Yankee and multi-cylinder machines are unique in both their machine configurations and the products they produce. All other major grades of paper are made on a broad classification of machines encompassing many configurations, including open draw (Fig. 739) and suction transfer (Fig. 740).

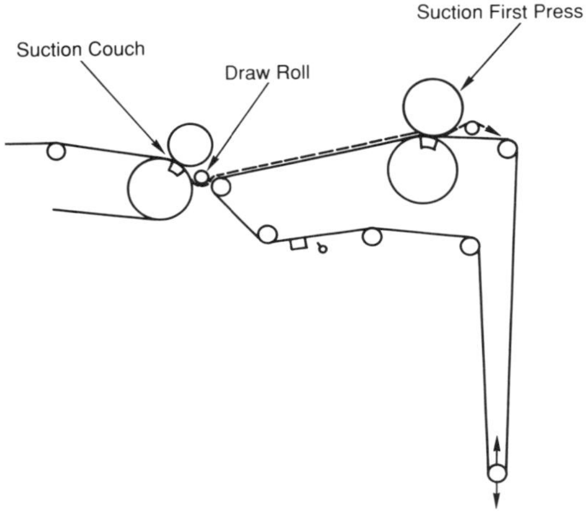

Fig. 739. Simple open draw suction press with "poor man's pickup".

Presses are numbered according to position, with the press nearest to the forming section designated as the first press in open draw machines, or as a suction pickup. Press fabrics are named according to the press positions that they clothe, i.e., the second press fabric clothes the second press. If the first press position is a suction pickup, the press fabric is usually designated as a suction pickup. In those cases where two fabrics pass through a common press, the one wrapping the top roll is usually referred to as a top fabric. The press clothing wrapping the bottom roll is designated as the bottom fabric.

(d) Fine paper (printing and writing) machines

Fine papers is a term that generally includes coated and uncoated publication, tablet, bond, litho, copying, business, cover stock, photographic, machine-glazed (MG) tissue, one-time carbon and numerous other fine, specialty grades. Many two-press and three-press open-draw press positions are commonly used for the production of these fine grades. Plain, grooved, suction, blind-drilled, and reversed presses are used.

In the early Fifties, the demand for fine paper grades exceeded the capability of open-draw machines (Fig. 739), which had speed limitations due to sheet breaks resulting from the open draw. The suction transfer came into use and provided a better transfer with fewer sheet breaks; higher

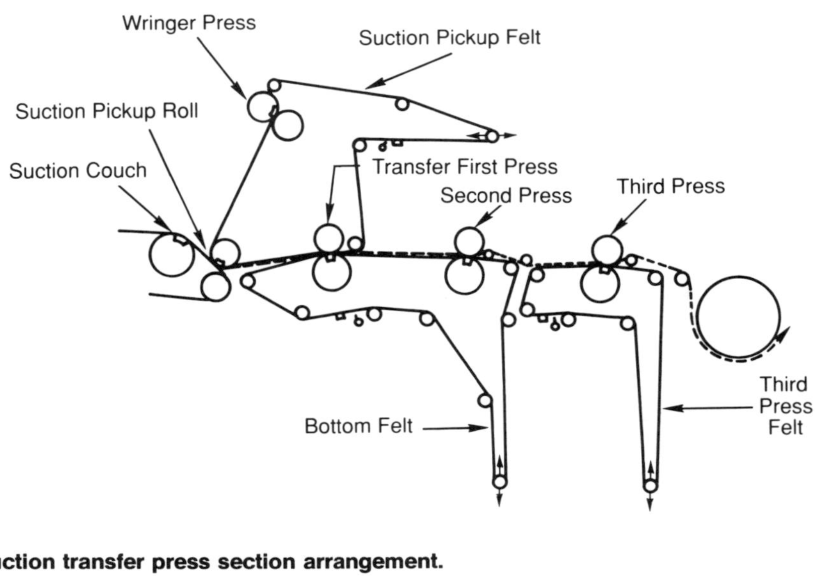

Fig. 740. Suction transfer press section arrangement.

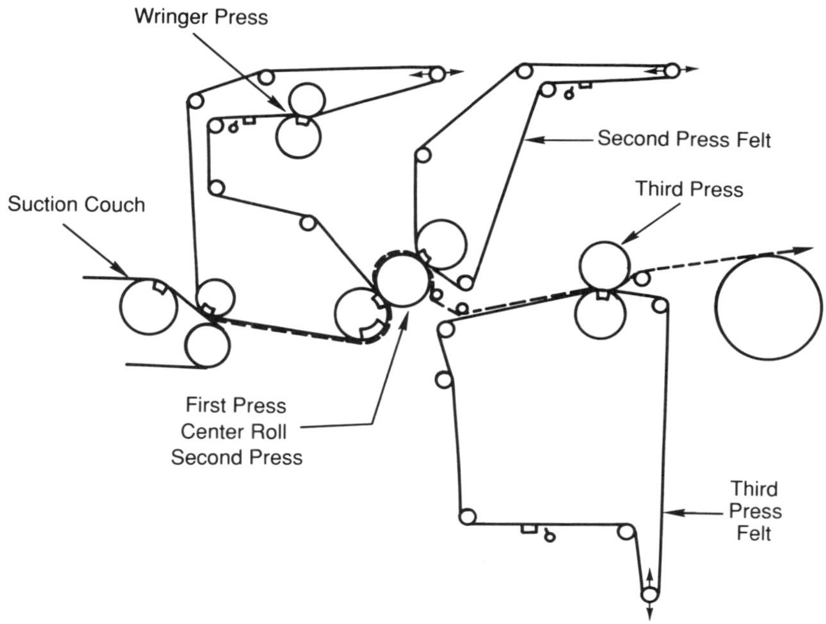

Fig. 741. "Twinver" press section with open-draw third press.

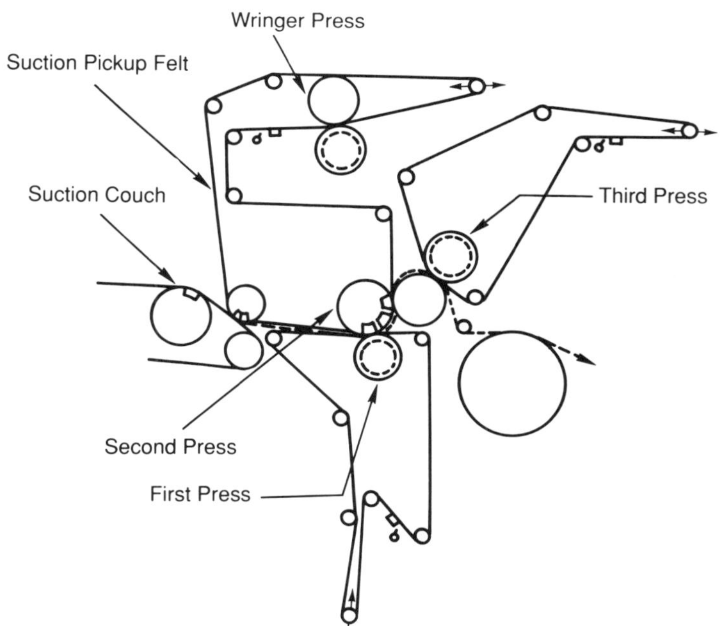

Fig. 742. Modern three-nip "no-draw" press section.

speeds were therefore possible. The first press of a suction transfer straight-through machine is "double felted" (two press fabrics) (Fig. 740), a principle that is now receiving renewed interest both here and abroad.

The next developments in the area of fine paper machines were the inverted first press and three-roll inclined press (Fig. 741).

Today, the latter is the most widely used configuration, and only recently has it been surpassed by other machine designs for the lighter weight publication grades. These configurations use a suction transfer but without the double-felted first nip (Fig. 742).

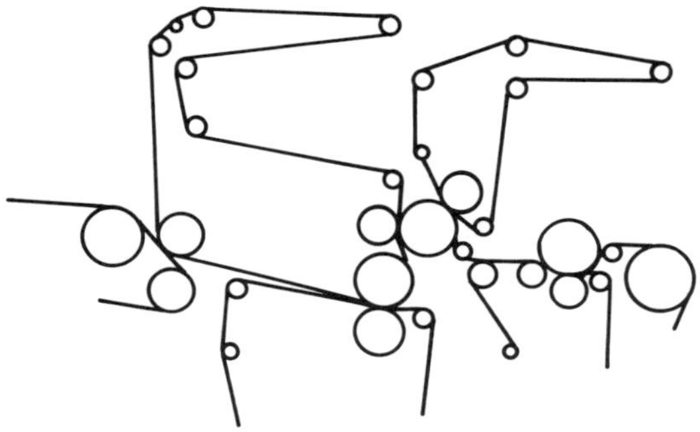

Fig. 743. Three-nip no-draw "divided" press section with open-draw fourth press.

The newest machine applied to fine papers, a three-nip no-draw press, plus an open draw fourth press is currently being introduced to the United States papermaker. The use of two press fabrics with double venting of press rolls is again becoming important and is being viewed as the most efficient way of pressing water from the sheet. Because of the multiplicity of machine types, all fabric constructions (batt-on-mesh, etc.) have an application for all fine paper grades. The trend on modern, high-speed machines is to use multi-layer press fabrics (Fig. 743).

(e) Newsprint machines

Newsprint is generally a lighter basis weight sheet compared to the average printing paper grades and runs at higher speeds. The newest machines run at 1067–1220 m/min (3500–4000 ft/min) and usually are 1.0–1.5m (40–60 in.) wider than comparable fine paper machines.

Stock for newsprint is primarily composed of groundwood, PGW, TMP or CTMP pulps. It is characterized by short fibers and a percentage of long-fibered chemical pulp for added strength and brightness. Secondary fiber (deinked) is fast becoming an important furnish component. Machines used for these grades are similar to those producing fine paper, with the three-roll inclined press (Fig. 741) and three-press, no-draw (Fig. 742) configurations predominating. The newest and fastest machines are now operating with the open-draw fourth press to improve quality and reduce moisture content.

Newsprint fabric applications

Pickup felts (press fabrics) for newsprint should be very permeable and not susceptible to filling; hence, multi-layered batt-on-mesh designs are the preferred applications, with weight and caliper commensurate with the applied loading. Compared to fine paper, these fabrics can be slightly lighter on the later presses due to less chance for sheet marking.

(f) Kraft machines and fabric applications

This category includes kraft papers, bag, as well as corrugating and kraft linerboard, which are considerably heavier. Kraft grades generally use softwood sulfate pulp, which is characterized by long fibers.

Machines for kraft papers and bag are similar to those producing news and fine paper, and machine speeds are comparable with fine paper machines. Base fabric constructions are similar to those used for fine paper: pickup fabrics must be open enough to resist filling, and finish must also be considered; fabrics at the second and third presses must have more batt for uniform pressure distribution and to prevent base strike-through.

The heavier kraft linerboard and corrugating grades were produced primarily on a three-press open-draw machine with higher press loadings and lower speeds compared to kraft papers and bag. Many of these machines have been rebuilt to take advantage of double felting and wide nip pressing which are best suited to the high-speed commercial grades of linerboard. Most new machines and rebuilds incorporate the newer concepts of press configurations.

Because of the extremely high press loadings, both in single- and double-felted presses, the application dictates very heavy, thick fabrics in three-layer batt-on-mesh or laminated constructions. A high percentage of mills are using on-machine seamed fabrics in all positions.

(g) Pulp machines and fabric applications

The pulp sheet is rarely used as a finished product and is generally converted to other grades. Pulp sheets used outside of the paper industry result in varied finished products, including rayon yarn, cigarette filters, cellophane, photographic and X-ray films, plastic articles, lacquers, explosives, and nonwovens.

Most pulp machines are three-press machines, with a first suction press, a second suction or grooved press, and a third grooved press. To aid in drying, a short section of four to eight dryers sometimes precedes the third or "hot" press. High press loadings are also typical, particularly on the third press, which might have a loading as high as 268 kg/cm (1500 pli). In some cases, up to 50% dryness out of the press section has been achieved.

It is particularly important that fabrics used on pulp machines provide good drainage, hence, such fabrics are coarse and open. Pulp fabrics must also be bulky to withstand the heavy press loadings and hard nips. The majority of machines are using heavy multilayer batt-on-mesh fabrics and on-machine seamed designs, especially on the modern machines where the principles of "double felting" and extended dwell time in the nip are being applied.

7. Press fabric records and storage

Minimum recorded data on press fabric use should consist of the following: supplier, fabric style, number, size, weight [total weight and g/m^2 (oz./sq. ft.)], percent synthetic and cost. It should also include the date installed, date removed, days life, reason for removal and tons of paper produced (to determine fabric cost/ton). The overall view of machine operation can be more realistically observed if the following information is also tabulated: average machine speed, steam pressure (or other measure of steam consumption) to dryers and machine efficiency in terms of a quotient (actual machine production divided by scheduled machine production time). Reasons for machine downtime should be recorded. These records help measure the economics of press clothing.

Most papermakers maintain an inventory at the mill site of several press fabrics for each position. They should be stored in a cool, dry, well-ventilated area, out of direct sunlight and away from heat sources. They should remain in their original container until installation. It is advisable to use them in the order in which they are received from the manufacturer.

8. Press fabric installation and operation

The press fabric should be placed on the machine and spread out evenly and flat with as few wrinkles as possible over its entire width. The stretch roll should be brought out until the fabric is snug but not tight. With a straight trade line at slow speed and the guide roll correctly functioning, the press fabric can be wet up with a shower until it is thoroughly saturated.

(a) Operating tension

After the fabric is uniformly wet up, the stretch roll can be adjusted until it is running at its length marked on the tag by the manufacturer. When running at its recommended length, it will be at its designed tension and should be at its optimum point for length, width, and drainage.

(b) Roping

Since most press fabrics used today are of synthetic construction, they should not be roped-up; but if it is necessary, then only during installation. Even then, roping should be kept to a minimum and excessive creasing should be avoided.

(c) Shutdown recommendations

When the press section shuts down for extended periods, i.e., in excess of 4 to 6 h, the jack should be slacked off 3 to 4% of the fabric running length to prevent excessive tension buildup. If the machine is to be idle over a weekend, it is good practice to wash and rinse the fabrics thoroughly just before shutting down.

9. Press fabric conditioning

Press clothing must be kept clean of contaminants or filling materials that accumulate in the fabric structure. Most fabrics are removed from the machine because they are filled or compacted to the point where they no longer handle water uniformly and efficiently. Generally, all raw materials and additives used in papermaking are potential contaminants. The materials most commonly responsible for filling a press fabric can be divided into five main categories.

(a) Alkaline (caustic) solubles

Materials that can be removed from a fabric by dissolving in a strong alkaline solution are called alkaline solubles. These are normally organic materials, such as lignin from wood pulp or furnish additives such as starch and rosin.

(b) Paper fines

Small fiber particles that become trapped in the press fabric are called paper fines. They cannot be dissolved by any chemical that does not damage the fabric. Paper fines are best removed by the mechanical action of a high-pressure shower and suction pipe.

(c) Inorganic materials (ash)

The residue left behind when a used fabric sample is burned at high temperature is called ash. This material is composed of inert mineral materials such as titanium dioxide, clay, sand, and other acid-soluble substances such as alum, calcium carbonate, and other hard water metallic complexes. Removal of these fillers normally is aided by cleaning with acidic solutions.

(d) Extractables

Resinous or polymeric materials soluble in organic solvents are called extractables. Pitch, tar, asphalt, latex, and inks are some examples referred to as extractables.

(e) Resins

Synthetic polymers in two categories, acid and neutral cure, used to improve the wet strength of paper are known as wet-strength resins. Removal of these resins before they are cured is usually aided by an alkaline wash system.

The fillers and additives used in papermaking act as binders and adhesives and tend to stick the batt fibers together. This action, together with the compressive action in the press nip, results in compaction of the fabric. Even though adhesives or fillers are absent, most fabrics will ultimately compact, reducing available void volume so that drainage is retarded and pressing becomes inefficient.

10. Fabric conditioning systems

Conditioning can be done by mechanical and/or chemical means. With the emphasis on greater use of recycled fibers, consumption of lower grade pulps, increased use of fillers, and the need to "close up" white water systems in the mill to meet ecological demands, both conditioning methods might be necessary.

(a) Mechanical conditioning

All press fabrics should be mechanically conditioned continuously, from start-up to removal. The ideal mechanical cleaning system uses hydraulic forces from high- and low-pressure showers to loosen the contaminants from the fabric structure. After the filling material has been loosened from the fabric fibers, it is removed by the action of suction pipes. The suction pipe also dewaters the fabric so that it has the capacity to accept all water pressed from the sheet in the nip. When selecting the design and open area of a pipe cover, it is necessary to consider fabric construction, machine speed and dwell time at the suction pipe slot. TAPPI TIS 0404-27 [11] provides more information.

(b) Chemical conditioning

Modern press clothing can be conditioned adequately in many cases by a properly designed mechanical system, but this does not eliminate the need for chemical cleaning. Chemical cleaning, on a continuous, intermittent, or shutdown basis, is often needed to keep the press fabric operating efficiently.

An analysis of the furnish in conjunction with analyses of returned fabrics will indicate what contaminants are present. Once the fillers have been identified, appropriate chemical cleaners compatible with existing furnish and pH conditions can be selected to help remove the contaminants from the structure at the suction pipe.

(c) Chemical cleaners

These cleaners contain a heavy-duty detergent to which other components and solvents may be added. Such cleaners are formulated to remove specific filling materials. Inhibitors are also added to minimize corrosion of machine parts contacting the solution. The three major cleaners widely used today are acid, alkaline, and solvent cleaning solutions.

Acid cleaning solutions usually have a non-ionic detergent base, with either muriatic, sulfuric, phosphoric, or sulfamic acid added. They are very effective in mills such as those making fine paper, which have a high inorganic content in their furnish.

Alkaline cleaning solutions usually have a non-ionic detergent base but may be anionic to promote formulation stability. The builder may be strongly alkaline, such as caustic soda, sodium hydroxide, potassium hydroxide, or mildly alkaline like soda ash and potassium carbonate. They assist in the removal of rosin, uncured resins, and alkaline soluble materials.

Solvent cleaning solutions contain detergents and emulsifying agents mixed with an organic solvent. They are used primarily to remove heavy concentrations of resinous fillers and other solvent-extractable materials.

Care should be taken to ensure that the cleaning solutions are compatible with the furnish and pH conditions found on the machine. The cleaning solution should not cause a degradation of the fabric structure.

11. Repairing the press fabric

The composition and strength of today's press fabrics makes them quite resistant to punctures, tears, and edge cracks. The damage that does occur in many instances is repairable and can be justified versus the downtime required to change a fabric. Repair kits are available from fabric suppliers and contain the necessary yarn, sewing needles and adhesive to mend small damaged areas.

12. Problems: causes and solutions

Press clothing and mechanical components are closely related and both should be considered when operating problems on the paper machine such as shadow marking, crushing, and sheet following occur. A press fabric is often blamed for an operating condition that is really caused by a combination of both components. Many of the more common problems encountered are described below with probable causes and solutions.

(a) Fiber shedding

In its simplest aspect, fiber shedding is the adhesion of batt fibers to the sheet surface. Many factors can cause batt deterioration and fiber shedding. Two principal causes are abrasion from machine parts and chemical damage. More frequently, the batt surface fiber can be torn loose by showers operated at excessively high pressures. Sometimes the press fabric batt is not necessarily responsible, since fibers may come from the pulp or even from contamination of the system by outside sources. When objectionable fiber shedding occurs at the startup of a new fabric, modifications in design may help correct the situation. If a fabric sheds for its full life, excessive mechanical or frictional wear is probably the cause.

(b) Yarn marking

Proper yarn selection, yarn count and weave patterns in modern engineered press fabrics have reduced the problem of marking of the sheet that was so characteristic of the older designs. However, an improperly designed or worn fabric can still induce yarn marking. Machine-direction marks appear as continuous lengthwise ridges and valleys, uniformly spaced, parallel to the sheet edge and very straight. Cross-machine direction marks are generally similar, but are harder to detect. Needled fabrics have introduced other marks in the sheet not experienced with conventional woven fabrics. A needle tracking pattern is caused by the needle depressions in the batt. These are sometimes severe enough to cause a pattern of machine-direction marks.

(c) Shadow marking/groove marking

Shadow marking is caused by the differential pressures over the hole or groove and land areas at the interface between the fabric and suction roll. Hydraulic flow from sheet to fabric to void area in the roll is directly related to differential pressures, and causes shadow or groove marking in the sheet if the pressures become too extreme. Often the problem can be corrected with changes in operating conditions such as reducing the press loading, decreasing the size of holes or grooves, grinding out the dished or cupped holes, or reducing the fabric moisture content before the nip. A heavier, multi-layer press fabric will minimize this marking by providing more cushion in the nip. Press fabrics must also be designed to provide uniform pressure distribution in the nip and must have low hydraulic resistance to flow in order to minimize pressure differences.

(d) Crushing

A crush in the sheet occurs when the press/press fabric system cannot prevent an excessive build-up of hydraulic pressure which disrupts the sheet. It can be caused by a machine condition such as plugged holes in a suction roll or streaks from a faulty shower nozzle. Other contributory factors can be a higher than normal sheet moisture entering the press or the addition of an excessive amount of shower water to the system. Fabric-associated problems can also be the cause of crushing. Generally, these can be attributed to low void volume and/or high flow resistance caused by: (1) an improper design that is too fine, too thin, or too dense; (2) a dirty fabric which does not handle water properly; and (3) a fabric that has become thin because of abrasive wear and thus cannot properly accept water.

(e) Blowing

Blowing is a localized accumulation of air from the stock or the fabric that causes an air bubble to form between the fabric and sheet. This can occur at the cylinders of a multi-cylinder machine or at the entrance to a press nip. It is usually associated with a new press fabric. The use of anti-blow rolls, reducing the amount of shower water on the press fabric and eliminating face-side showering until after the fabric is broken in, should keep start-up blowing to a minimum. Pre-compaction of the press fabric to reduce void volume, densifying the fabric and engineering a smoother surface usually eliminates the problem. If the press fabric operates satisfactorily for a while and then starts to blow, it may be too dirty or too compact. In such a case, the blowing is caused by minute amounts of air that cannot escape through the fabric. To cure this type of blowing, the fabric design should be changed, or fabric cleaning methods must be improved.

(f) Sheet pickup (dropoffs)

Pickup problems may be defined as problems associated with the transfer of a web of paper from the forming fabric to a pickup fabric that carries the web to the presses. Poor formation on the forming fabric produces a sheet that is difficult to pick up. The sheet that has a non-uniform basis weight with heavy and light streaks is difficult to pick up from the forming fabric, particularly if the edges vary in basis weight. It is essential to have the trim squirts properly set to make a clean cut, thus ensuring that the trim will break away clean and not befoul the transfer to the pickup fabric. Drainage rate and degree of fiber stapling onto the forming fabric influence the release of the sheet to the pickup fabric. Too coarse a press fabric surface or a fabric that is too open (not at the required optimum density) may cause this problem.

(g) Sheet following

The press geometry, water distribution, and smoothness of the surfaces contacting the sheet dictate which surface will attract and hold the wet web after a press nip. The sheet will always tend to follow the smoothest surface with the most uniform film of surface water whether it be a roll cover or a fabric. Sheet stealing is a condition wherein the sheet does not follow the designed path of travel. For example, in the manufacture of tissue on a Yankee machine, the sheet may follow the bottom fabric on the out-going side of the suction press nip. In a dual fabric press nip, the problem is basically the water distribution between the fabrics and is influenced by the press arrangement and the surface characteristics of the press clothing. If sheet stealing does occur, cleaning of the fabric to offset compaction and filling is the best means to combat this condition. In addition, adjustments such as reducing fabric water content, relocating the suction box in the suction press roll, reducing the press load, and changing the sheet/fabric geometry on the exit side of the nip may help to alleviate this condition.

(h) Press vibration

On high-speed machines, presses may severely vibrate. This may be related to machine conditions or variations in mass density of the fabric. Machine conditions that cause press vibration include: improperly balanced press rolls; a malfunction in the pneumatic of hydraulic press-loading apparatus; a soft spot on a rubber-covered roll; a flat spot on the press roll caused by leaving the press loaded when the machine is not operating; or by the occurrence of a natural frequency of vibration at a given machine speed. Sympathetic vibrations related to another roll or press section can be transmitted through press fabrics. Fabric-related press vibration may be alleviated by changing its moisture content, the press loading, machine speed and, in some instances, cocking the trade lines.

(i) Wad burning

Wad burns are caused by a wad of paper becoming lodged at the in-going press nip. This creates large frictional forces that melt and fuse the synthetic yarns on the surface of the fabric. The two most frequent causes of wad burning are: (1) sheet breaks when threading up; and (2) poorly maintained roll doctors which allow the build-up of stock that forms wads. The wad burn damage can be minimized by altering poorly design press configurations and installing devices on existing presses that either deflect the sheet or reduce the press load almost instantaneously when the sheet breaks in the nip. The extent of wad damage may also be reduced by using a press fabric with high temperature resistant fibers and/or wool on the surface. While high temperature resistant fibers or wools alleviate burned streaks, they do not have the abrasion resistance of polyamide and will wear at a faster rate.

(j) Blistering

Blistering is usually associated with drying problems on a Yankee dryer at the transfer point between the pickup fabric and the dryer. It is often caused by a dryer surface that is too hot or a web sheet or fabric that has wet streaks. Blisters are created when steam forms between the sheet and dryer, forcing small sheet areas away from intimate contact with the dryer. On MG finish, the sheet may have a pock-marked appearance, since the blisters do not have the glazed surface where the sheet closely contacted the dryer. Blistering may be caused by a non-uniformly coated Yankee dryer or one operating with profile problems. It may also be caused by a fabric that is filled or experiencing wet streaks.

13. Summary

The optimization of the press section operation is an important part of papermaking because it provides the last opportunity before the dryer section to mechanically remove large amounts of water from the sheet. Large amounts of energy are consumed by dryer sections and costs are relatively high. At the same time, the press section presents a possibility for imparting unwanted quality faults in the sheet. Understanding the technologies of papermaking and press fabric applications gives both the papermaker and the clothing supplier an opportunity to have a large impact on the quality and cost in the manufacture of paper products.

C. DRYER FABRICS

1. Overview

Increasing the speed and width of paper machines results in two major problems in the dryer part that reduce the expected increase in productivity, especially when the lighter grades of paper are being made. These problems provided the impetus for the transition from "conventional" cotton, cotton/asbestos and cotton/synthetic dryer felts to the modern, open-mesh, screen-like, engineered fabrics used today. The first of these problems is the development of a non-uniform moisture profile in the cross direction (CD). It is associated with the poor ventilation of the dryer pockets resulting from the increased width of the sheet. Two approaches to the problem have eliminated it, for the most part. The first was the introduction of the open-mesh dryer fabric, and the second was use of the pocket ventilating (PV) systems. Both solutions work synergistically to level the moisture profile. The second problem is that of web instability, usually referred to as sheet flutter. It is associated with the unbalanced air flows on either side of the sheet as it passes through the draws from roll to roll in the dryer section. This second problem can be moderated by using open-mesh dryer fabrics designed with specified permeabilities, i.e., controlled permeability dryer fabrics. Another approach is that of changing the machine configuration or geometry to eliminate the open draws where the sheet is not supported by a dryer fabric. This latter solution has resulted in the single-fabric dryer section, and the single-tiered dryer part. In the dryer part, dryer fabrics are subjected to high temperatures and high humidities which tend to limit their useful life. Optimizing the performance of the dryer part requires an understanding of the technologies associated with papermaking and paper machine clothing design and manufacture.

2. Function and requirements

To operate on a paper machine, a dryer fabric must meet certain requirements for strength and dimensional stability; it must be resistant to degradation under the ambient conditions of elevated temperatures and high humidities; and it must operate at a specified permeability. To maintain these requirements, the fabric must be resistant to abrasion and flexing, and be cleanable.

Moreover, dryer clothing must perform one or more of these functions:

1. Support and guide the sheet through the dryer section.

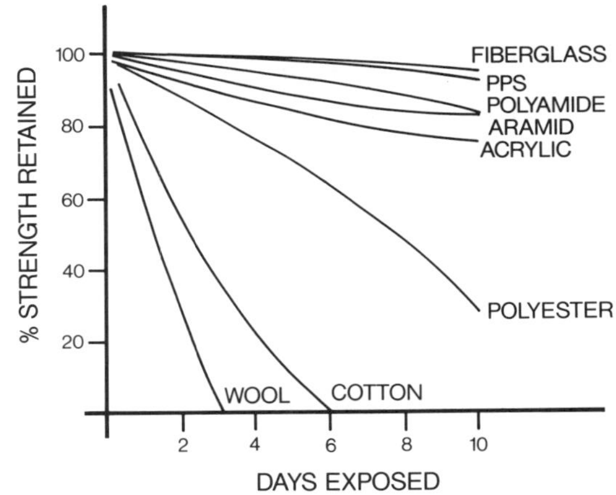

Fig. 744. Moist heat degradation of different yarns made from the materials indicated is shown here. The breaking strengths are measured after exposure to relatively severe conditions of high temperature and high humidity in an autoclave for the time indicated. PPS = polyphenylene sulfide.

2. Aid in drying. The tension of the clothing improves the contact between the cylinder and the sheet, causing increased heat transfer.
3. Prevent cockle. Cockles may be caused by uneven or too rapid drying. The clothing should serve to diffuse the heat and supply the sheet with a uniform condition for drying.
4. Control shrinkage. The tension of the clothing helps to control shrinkage in the sheet.

3. Materials and construction

An important consideration when selecting a material for dryer clothing is degradation from moisture and/or heat. Figures 744 and 745 show comparative data from laboratory tests. Synthetic fibers show a distinct improvement over cotton and wool fibers.

Air permeability, an important property of dryer clothing, is shown in Table 78 for different types of dryer fabrics. Note that, with monofilament and multifilament dryer clothing, it is possible to achieve much higher permeabilities than with cotton felts, which were the "workhorse" of the industry long ago. Monofilament synthetic fabrics also provide the widest range of permeabilities, from 30 to 1000 cfm/sq. ft. at 0.5 in. water pressure.

Overall, polyester has the best balance of characteristics for modern dryer clothing. However, its susceptibility to moist heat degradation sometimes decreases running time under extreme conditions. Materials more resistant to hydrolysis, such as acrylic, polyamide, and

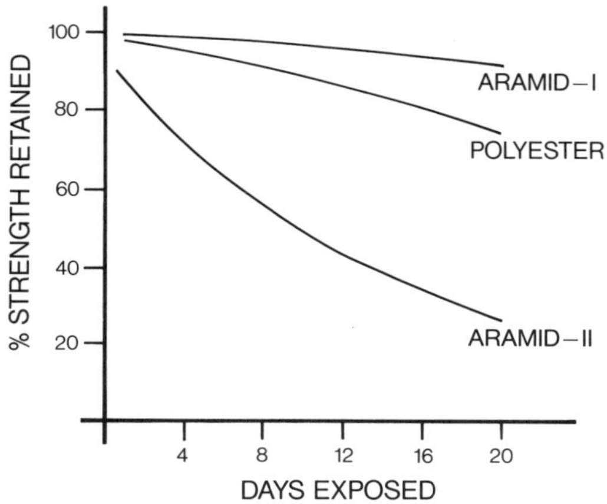

Fig. 745. The dry heat degradation of yarns is shown here. The breaking strength of yarns is expressed here as a percentage of their initial strength after exposure to elevated temperatures in a forced draft, hot-air oven. That relative breaking strength is shown as a function of the time of exposure.

Table 78. Air permeability of dryer clothing.

Type	cfm/sq. ft.	m/h
Cotton felt	0-2	0-33
Needled fabrics	40-100	650-1630
Open-mesh dryer fabrics	30-400	490-6500
Monofilament dryer fabrics	30-1000	490-16 300

aramid are being used in such cases. Monofilament fabrics must be heat set and may be chemically treated to add stability.

4. Dryer fabric classifications

Dryer clothing can be classified into the following categories: conventional dryer felt, needled dryer fabric, open-mesh multifilament dryer fabric, and monofilament dryer fabric.

(a) Conventional dryer felt

Included in this group are cotton and cotton-synthetic felts. These felts, with their lower initial cost, are still used. Spun yarns, often reinforced with a multifilament yarn for strength, are used in these constructions, in two-layer or three-layer weaves for stability.

(b) Needled dryer fabrics

In these constructions, 100% synthetic multifilament or monofilament yarns are used in the machine direction and monofilament yarns are used in the cross-machine direction. The base weave is usually a plain weave or two-layer weave. A batt of synthetic fibers is needled into both sides of the base fabric, adding stability, strength, and a smooth surface to the base as well as protecting it from physical and thermal degradation. Since needled dryer felts are most commonly made of 100% synthetic fibers and yarns, they are heat set for stability. A variety of seams, metallic and nonmetallic, are available.

(c) Open mesh dryer fabric

Designed to be more open than the conventional or needled constructions, open mesh dryer fabrics are manufactured for controlled permeability. Multifilament or spun yarns are woven in the machine direction, spun yarns or fiberglass content yarns are woven in the cross-machine direction.

Fiber content varies from low synthetic to 100% spun aramid for machine-direction yarns, and combination fiberglass is often used in the cross-machine direction for stability. A two-layer or three-layer weave is frequently used, and open-mesh fabrics are supplied with clipper or multifilament seams. The aramid in this construction is used for the very high temperature applications. Open-mesh fabrics are also selected for their less expensive synthetic content where such is applicable.

(d) Monofilament dryer fabrics

Like the open-mesh dryer fabric, this fabric can be designed for specified permeabilities, a factor which has led to the remarkable success of this clothing. Monofilaments are used for machine-direction yarns and monofilament/multifilament combinations are used in the cross-direction yarns to control permeability. A one-, two-, or three-layer weave of synthetic monofilaments and/or multifilaments is used. The yarns are made from polyester and polyamide resins. Polyester is recommended for dry heat and acid stock conditions; polyamide combined with polyester yarns in the machine direction have achieved a longer operating life for moist heat conditions.

Fabrics made only from monofilament yarns are usually not treated chemically, but occasionally treatments may be used to lower the fabric permeability. All fabrics are heat set for dimensional stability. These dryer fabrics can be supplied endless, with clipper seams, or with nonmetallic seams. The latter, particularly the pin seam, have proven to be most advantageous.

5. Application

The following factors should be considered when selecting the appropriate dryer fabric.

(a) Definition of desired advantages

The first step is to clearly define the application. One must define the advantages to be gained and be certain that the dryer fabric being selected has helped give desired results to similar applications. Machine conditions should be checked for compatibility with anticipated performance. If, for example, an increased production rate is desired, one must verify that mechanical limitations or speed limitations other than drying do not exist.

(b) Paper grade

When considering paper grade, one must evaluate whether or not the fabric used will mark the sheet or fill up. Marking must be considered for grades requiring a fine finish. High caliper or high bulk grades are more susceptible to marking than hard finished grades.

Open-mesh dryer fabrics and monofilament dryer fabrics have few limitations with regard to paper grade. Monofilament dryer fabrics are applicable to most paper grades, with finer yarns used for finer paper grades. For some applications the monofilament knuckles of these constructions can be sanded to present a smoother surface to the sheet. The pin seam has eliminated seam marking on sensitive grades. Dryer fabrics with a very fine surface and a pin seam can be used for high bulk grades and for extremely fine specialty grades.

In most applications, a monofilament dryer fabric has the most resistance to filling and is more easily cleaned than the combination-type fabrics. On after-coater applications, or where stock contains contaminants, the best dryer fabric to use is the all monofilament fabric.

(c) Cylinder temperature and air moisture

The emphasis on high production rates in the paper industry has resulted in a trend toward increased temperatures in the dryers. All dryer clothing, whether conventional cotton, or synthetic, tends to degrade and is weakened by high cylinder temperatures and high humidity in the dryer pockets. New heat-resistant materials (i.e., polyphenylene sulfides and aramids) and fabric constructions have been designed to greatly increase fabric life compared to that of conventional cotton felts.

(d) Operating life and economic considerations

Economic considerations other than fabric life should be made when considering installation of a dryer fabric. Experience has shown that a filled dryer fabric may reduce the effectiveness of a pocket ventilating (PV) system and lower the paper drying rate. Dirty clothing can also cause wet streaks in the paper.

Because dryer fabrics absorb very little water or condensed steam from the paper, felt dryers can often be by-passed, thus the running size of the fabric can be shortened and common sizes can be interchanged, reducing the fabric cost by 10 to 30%. Appreciable steam savings are made possible by eliminating the felt dryers. Pocket ventilation fabric carrying rolls or forced hot air ducts are often used to evaporate condensed water from dryer fabrics on wet positions.

The higher drying rates made possible with modern dryer fabrics, by leading to increased production, often pay for the higher fabric costs many times over. Open-mesh and monofilament dryer fabrics also assist in leveling the cross-machine moisture profiles of the web. This allows for the production of more uniform parent rolls which can have higher moisture content. Production is increased and steam savings are realized. Increased production resulting from less downtime associated with fewer dryer fabric changes is also possible. All of these factors have been recognized by the paper industry as shown by the acceptance of the open-mesh and monofilament dryer fabrics.

(e) Machine speed

A major concern associated with increased machine speed is whether or not air pumping will be a problem, with resultant sheet fluttering. This problem primarily occurs when producing lightweight sheets such as newsprint and publication grades. The first and second sections and bottom dryer positions are most critical. Lower permeability fabrics are often used on these sections, with increasing permeability used as the strength of the sheet increases toward the dry end of the machine.

(f) Pocket ventilation

There are many pocket ventilation installations throughout the paper industry, and dryer fabrics of all types have run successfully with them. The fabric manufacturer will recommend a permeability compatible with machine requirements, including the type of pocket ventilation system. The decision about permeability is determined by machine speed and the weight of the sheet, since blowing and sheet flutter must be considered. The ability to control permeability in a wide range for all-monofilament fabrics has been particularly advantageous. Moreover, monofilaments lose relatively little permeability during their machine life.

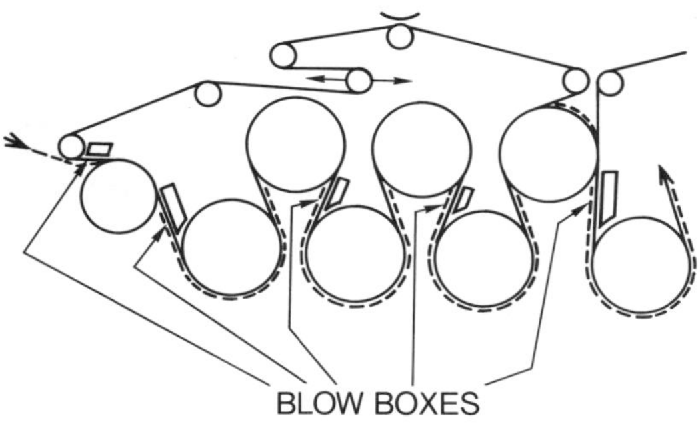

Fig. 746. A single-felted dryer section. This schematic shows how a single dryer fabric clothes the top and bottom positions in a dryer section. "Blow boxes" are positioned as shown to help control the tendency of the web to separate from the fabric in the positions where they are located.

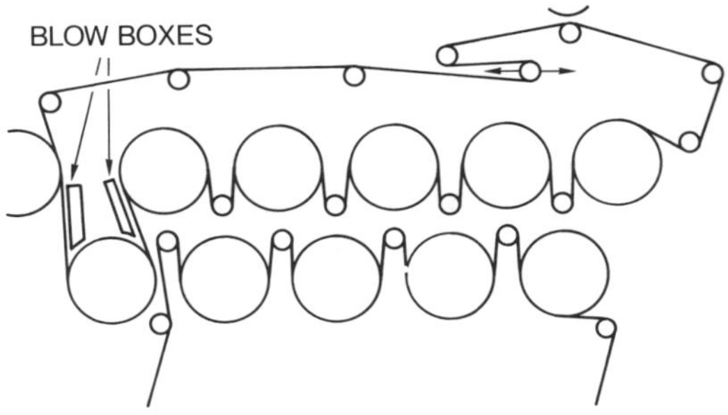

Fig. 747. This schematic of a "pistol grip" dryer section shows the position of "blow boxes" used with higher permeability fabrics because of the presence of dryer pockets after the first dryer cylinder. The first dryer cylinder constitutes the "pistol grip".

(g) Felt drying equipment

The following are general-application rules and should be reexamined when a dryer felt position is actually surveyed. With modern dryer fabrics it is recommended that at least one fabric dryer on first positions be clothed and all fabric drying equipment on other positions be by-passed. With open-mesh, spun yarn content felts, the usual recommendation is to clothe a fabric dryer on all positions but to by-pass all Feeney dryers.

(h) Tension

Tension should be uniform across the machine, to ensure uniform sheet drying. In recent years, dryer fabric tensions have increased to the 1.5–2.5 kN/m range in order to achieve faster drying and control of sheet cockle. Limits to higher tension are fabric roll journal breakage and increased roll deflection causing bowing and narrowing of the dryer fabrics. Fabric seams are also more susceptible to failure as a result of dryer wraps and wads running through the section.

(i) Single-fabric dryer sections

One result of higher speeds and lighter basis weights has been the development of single felting with blow boxes installed behind the fabric to maintain sheet/fabric contact. The term blow box is really a misnomer because the action of the blowing air really creates a *suction* to attract the sheet to the face or sheet side of the fabric. This configuration, usually located in only the wet end portion of the paper machine dryers, dictates a different role for the dryer fabric. Here the sheet must cling to the fabric surface (Fig. 746) and pocket ventilation is not a factor since there is no pocket. Seam design is critical since the seam is in contact with the sheet throughout the section; this is not the case with a conventional top/bottom dryer section configuration.

Spun yarns are sometimes used on the dryer fabric surface in an effort to attract the sheet to the resultant smoother surface. Permeabilities range from 40 to 65 cfm/sq ft for single fabric positions without blow boxes and 50 to 120 cfm/sq ft for single fabric positions with blow boxes.

Pistol grips, partial single felts (Fig. 747), present a special problem since the single felt portion of the run may dictate a lower permeability to attract the sheet whereas the top/bottom portion may need a higher permeability to permit adequate pocket ventilation. The use of blow boxes here may allow for the use of a higher permeability pistol grip fabric.

(j) Single-tier sections

The latest innovation in sheet support technology is the single-tier section which allows for both sides of the sheet to contact the dryer surfaces (see Fig. 748). Vacuum rolls and blow boxes located on the upruns and downruns assist in maintaining sheet/fabric contact. This equipment is required for speeds of 1200 m/min and up in lightweight grades.

A further advantage of the single-tier dryer section is the reduction in width shrinkage of the sheet with this process. The relatively restrained drying allows for more uniform sheet properties across the width, especially cross-direction stretch and tensile ratio. Less deckling of the

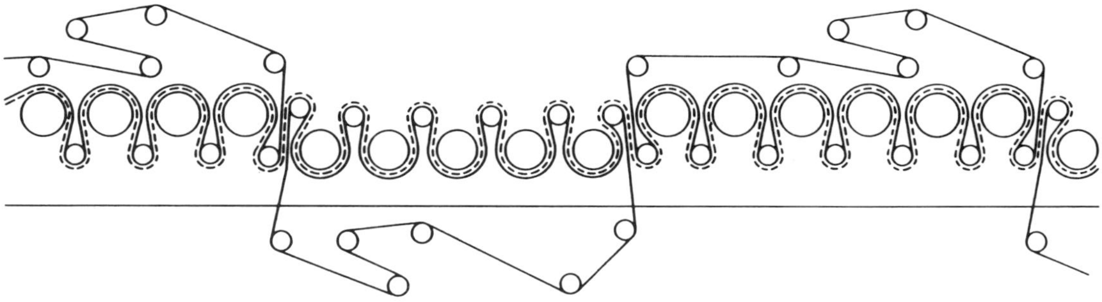

Fig. 748. A single-tiered dryer section has only one tier of dryer cylinders. This contrasts with the two tiers of dryer cylinders associated with the conventional dryer section. The single-tier arrangement eliminates the dryer pockets and the open draws between upper and lower tiers that contribute so greatly to sheet flutter. Notice that the fabric runs on alternate sides of the sheet in the single-tier arrangement to ensure proper drying and web uniformity.

sheet is required for uniform basis weight which results in reduced edge graininess and cockle. The dryer fabric properties noted in Sec. (i) above are applicable here.

(k) Other factors

There are other factors and questions which should be considered when selecting dryer clothing.

Is the present seam running straight? One type of fabric can accommodate seam distortion better than another style. How many drying sections are there? Results with fabrics can be different between a two-section and multiple six or seven-section machine. What are the guiding indications? Is the exhaust system operating at rated capacity and properly located? Is the position being considered one that always experiences seam marking? Is less or more than normal tension used on the position? And has running length and width been checked, as dryer fabrics are often manufactured narrower than conventional felts?

6. Seams

Dryer fabrics are usually not manufactured endless and must be joined on the machine. Hence, it is important that the two ends be joined carefully so that the seam resembles the rest of the fabric as closely as possible. This means that seam permeability should closely match that of the fabric, that the seam should be thin and not protrude above the fabric surface. The seam should be flat, strong, and nonmarking.

Figure 749 shows five of the most common types of dryer fabric seams.

(a) Clipper seams

When used in modern dryer clothing, clipper seams often limit the performance capabilities of

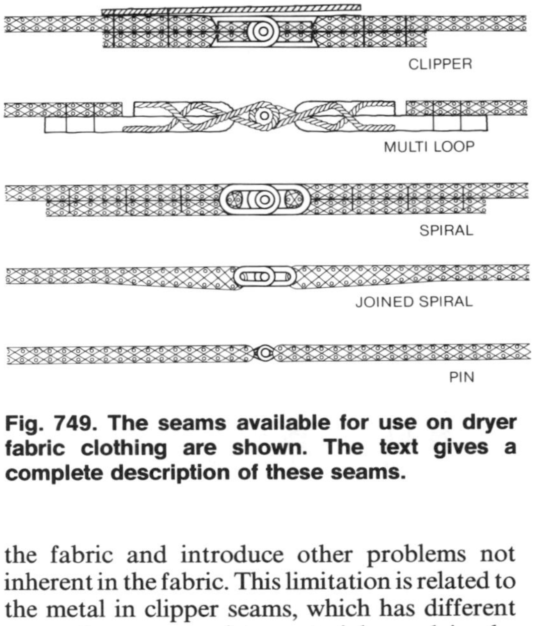

Fig. 749. The seams available for use on dryer fabric clothing are shown. The text gives a complete description of these seams.

the fabric and introduce other problems not inherent in the fabric. This limitation is related to the metal in clipper seams, which has different properties compared to materials used in the dryer fabric. Compared to the body of the dryer fabric, a clipper seam is hard and inflexible. This may intensify marking and air pumping problems. Flexing of the seam as the fabric runs causes abrasion of the metal hooks and cable. Unlike the fabric, metal hooks are also subject to corrosion. In addition, the permeability of a clipper seam is quite different from that of the fabric. This can also cause drying problems.

(b) Multifilament seam

The material which makes the seam consists of a webbing with woven loops. The webbing containing the loops is sewn to the body of the fabric. Both loops and webbing are made of high temperature resistant yarns. There is no metal in the seam area so that many of the disadvantages of a clipper seam are not present. The advantages are as follows: it is more flexible; there is no metal to fail, either in the loops or in the cable; and, it is stronger than a clipper seam.

(c) Spiral seam

The spiral seam is basically constructed by raveling out cross-direction strands in the fabric, creating a machine-direction fringe. A precoiled synthetic spiral, usually of polyester, is then slipped into the fringe area and the fabric folded over and sewn. The spiral forms “loops” which can engage a similar spiral on the other end of the fabric. When joining the two ends of the fabric, a perfect match loop for loop is not necessary. However, alignment marks placed on both ends of the fabric should be made to coincide if possible. An important feature in the construction of this seam is that the sewing used to fix the fold-over should be buried in the fabric. If not, it will wear prematurely and the seam will fail. Also, machine-direction strands near the base of the spiral must mesh correctly or high spots are created which mark the sheet. The high spots can wear prematurely.

(d) Joined spiral seam

This seam also uses a spiral to form the “loops”. However, instead of raveling out cross-direction strands to create a fringe and folding over the fabric and sewing, the area adjacent to the spiral is woven in. Machine-direction strands of the fabric are used to weave behind the spiral and anchor it into the fabric. The “woven back” area behind the spiral is normally 40–60 mm in the machine direction and just slightly greater than the thickness of the fabric itself. This area helps protect the spiral loops from abrasion and early failure.

(e) Pin seam

This seam is only constructed in monofilament dryer fabrics. The monofilament machine-direction yarns are formed into loops and woven back into the fabric. In seaming, the loops on each end are interconnected by inserting a plastic pin. This seam matches very closely the thickness, permeability, and flexibility of the fabric. There is essentially no difference in thickness across the “woven back” area adjacent to the pin loops. Usually the pin loops are slightly thinner than the fabric, thus protecting them from abrasion. The pin seam performs well on high-speed machines where flex fatigue and abrasion can cause early seam failure.

(f) Connecting spiral and pin seams

It is relatively easy to join the two ends of a dryer fabric with a pin seam. Actual mill experience shows that the ends of a 760 cm (300 in.) wide dryer fabric can be connected in about 30 min. The correct procedure is as follows.

Beginning at the guide paddle edge of the fabric, the first few loops are meshed so that the edge is aligned. The first set of alignment marks are matched and the loops are meshed between the edge and this first set of marks. The alignment marks across the full width of the seam ensure correct meshing of the loops.

Next, the steel leader wire is passed through the loops. After connecting the seam for about 0.9 m (3 ft.), the leader is pulled through the meshed loops. The plastic monofilament, attached to the steel leader wire, is then pulled through the 0.9 m (3 ft.) of meshed loops until about 0.3 m (1 ft.) remains beyond the connected edge of the fabric.

The steel leader wire is then reinserted into the next 0.9 m (3 ft.) of meshed loops. In effect, we are “sewing” the pin seam together. This procedure is repeated until the fabric has been connected across its entire width. After the monofilament cable has been inserted through all the loops, its ends are cut so that approximately 10 cm (4 in.) of the cable extend from each edge of the fabric. The cable ends are then bent back into the open loop area and tucked tightly against the fabric on the guide paddle edge. On the other edge of the fabric, the plastic loop cable should extend about 2.5 cm (1 in.) beyond the edge to allow for movement in the connecting cable.

Alternatively, a pintle may be used which has a full-length steel leader attached to the monofilament pintle. Here, the steel leader is pushed the full width of the fabric without any need for “sewing”. A problem with this technique is that, on wide machines, fabric tension cannot be relieved completely across the width and it is not possible to push the metal leader through the loops without binding. This often results in kinks in the metal leader which make it even more difficult to push across. In such cases, a compromise method is sometimes best which involves “sewing” the pintle in two or three places across the width rather than every 0.9 m (3 ft.).

7. Installation

The first step in installation is to position the rolled dryer fabric properly so that, when installed, the sheet side of the fabric actually contacts the sheet. Normally, the roll will be placed at the dry end of the section and run onto

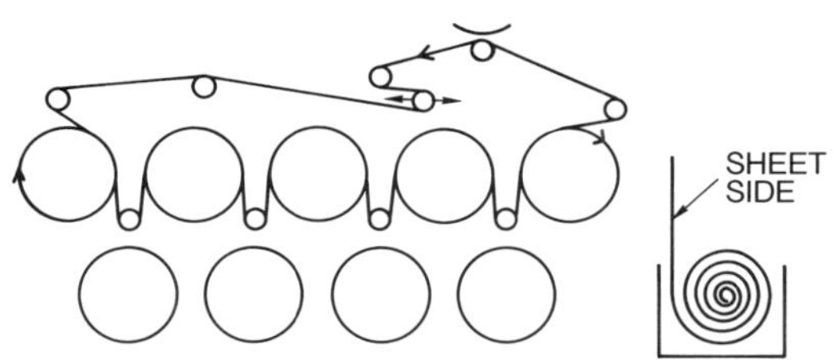

Fig. 750. Positioning the dryer fabric for installation on the paper machine. See the text for details.

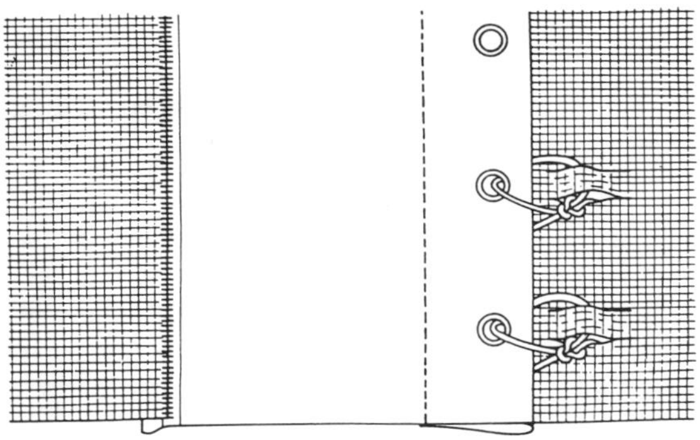

Fig. 751. The installation leader is used to pull a dryer fabric onto its section by attaching it to the end of the fabric being removed.

the machine return run first (see Fig. 750). This assumes that the machine will be running forward during installation. For reverse installation, the fabric will run onto the machine opposite to the directional arrow printed on the fabric. Also, it will be necessary to place the roll at the wet end of the section. If the seam has a flap, the fabric should be specially packaged for reverse installation and these instructions must be supplied to the manufacturer.

The face of the fabric, or sheet side, is always marked and usually on the inside of the roll. A leader (an extra piece of fabric) is provided so that the old fabric may be tied to the leader without damage to the dryer fabric, as shown in Fig. 751.

It is important that the leader be tied to the old fabric at several evenly spaced points, across its width to provide a uniform pull. The dryer fabric must be run on straight.

When joining the seam, the alignment marks are used as a guide. Except for the pin seam, the cable is run through the entire width of the dryer fabric beginning at the guide paddle edge. A check should be made to be certain that gross mismatching of the loops has not occurred. With spiral, multifilament, and pin seams, however, some mismatch can be tolerated without adverse performance; the alignment marks are more important than perfect loop-for-loop match.

8. Operation

Before startup, some preliminary checks should be made. Since dryer fabrics guide best when warm, heat should be supplied to the dryer cylinders as soon as practical after installation and before running a fabric. The guiding equipment should be checked for proper operation. Tension should be lower than normal, ranging from 0.7–1.0 kN/m. The fabric should be tracked slowly and speed gradually increased. It should be run for at least 5 min at full speed under careful observation. The machine should then be shut down and the body and the seam examined. Any necessary corrections should be made at this time.

Dryer fabrics may move rapidly in response to hand guide corrections; initially, these corrections should be gradual until the operator learns how the fabric will react on a specific position.

Treated fabrics are bound together at yarn crossover points making them very stable with low accommodation to diagonal distortion. Seam distortion should be minimized for best performance. Nontreated fabrics, on the other hand, derive their stability from their monofilament construction. The crossovers are not bound and the fabric can yield to diagonal stresses. As a result, it will quickly adapt to machine misalignment and roll deflection. Although a higher degree of seam bowing can be tolerated with untreated fabrics, these conditions should be minimized for best operating life.

9. Cleaning

Sometimes contaminants fill up monofilament or multifilament dryer fabrics and reduce drying efficiency and sheet quality. A major advantage of these fabrics is that they can be cleaned. Their open construction and stability make possible the use of showers and certain cleaning methods.

(a) Contaminants

There are four basic groups: paper dust (normally composed of short fibers and fines); pitch, tar, asphalt and latex (found in the stock); water-soluble sizes (typical for fine papers); and solvent-based or heat setting coatings (for coated papers). The cleaning approach for each contaminant is different.

Paper dust is usually removed by an air shower. The fabric can be cleaned easily with a

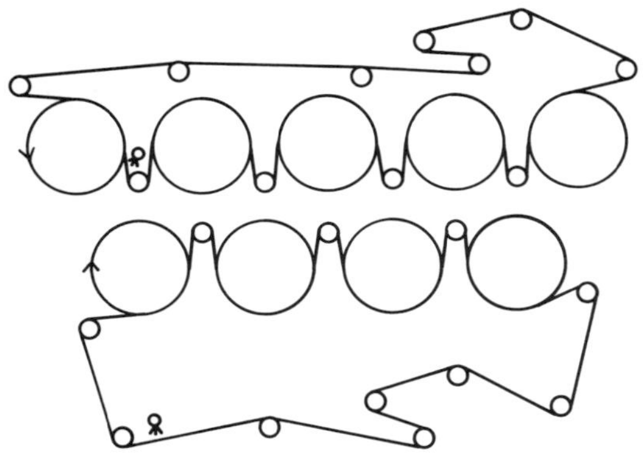

Fig. 752. The recommended location for dryer fabric cleaning showers.

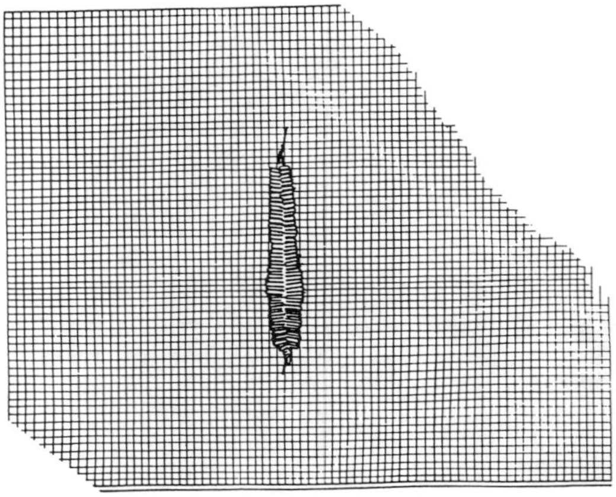

Fig. 753. The baseball stitch used to mend torn fabrics.

shower during breaks or shutdowns. A bristle brush will also aid cleaning.

Pitch, tar and the like adhere strongly to the fabric. Off-machine cleaning of the dryer fabric may be necessary if these contaminants are present.

For water-soluble sizes, a steam hose, steam shower, or hot water may be applied directly on the fabric during shutdown. The rate of contamination build-up will determine the cleaning frequency.

Solvent-based or heat setting coatings should be removed before they harden. A high-pressure hot water or steam shower on the inside return run is effective. The shower should be started as soon as a coating carryover occurs. This will maintain the openness of the fabric. If the fabric becomes plugged, a special chemical cleaning recommendation for each type of coating and off-machine washing is usually required. Recommendations are based on laboratory analyses of plugged fabrics returned from the field.

(b) Shower design

A typical shower installation for shutdown cleaning of dryer fabrics uses fan nozzles positioned for an overlapping spray pattern. The shower should be located so that it impinges on the side of the fabric not exposed to filling agents. Figure 752 shows suggested shower locations.

For the top section, the shower is placed inside the fabric just before or after the carrying roll preceding the last cylinder. This position is best because contaminants will drop down at the end of the bottom section. Also, at this point, the fabric is the warmest. Heat acts to soften some filling agents and thus aids cleaning. If, as in Fig. 752, the top section overhangs the bottom, then the drop-off is directly into the pit or basement. If the reverse is true, the drop-off will fall on the last bottom cylinder and be carried by the bottom screen to the bottom section shower to be discharged to the basement. The bottom shower should be located on the inside of the felt run and installed so as to purge the fabric directly into the basement or pit.

In general, chemical cleaning agents should only be used when normal steam or hot water methods are ineffective. Petroleum solvents, non-ionic detergents, and emulsions of petroleum solvents have been used and are usually applied in dilute solution by bucket and through showers.

It is very important to rinse the fabric thoroughly after using cleaning agents. If chemicals are left on the fabric, they will concentrate through evaporation and may hasten fiber degradation and shorten the useful life of the fabric.

10. Repairs

The most frequent repair made is to the edge of the seam where damage and deterioration usually occur first. An excellent idea here is to reinforce the edge with sewing when the fabric is new, long before degradation and wear occur. The reinforcement will prolong the life of the fabric. The most commonly used stitch is the baseball stitch shown in Fig. 753. Aramid multifilament thread is the best material to use for these repairs.

Frequently, a heat-resistant adhesive is used to fix the sewing and keep it in place. The baseball stitch may also be used for other cuts and tears in the fabric. The sheet must tolerate the extra pressure caused by sewing thread in the area of repair, otherwise it is best simply to trim off loose threads and leave the tear unrepaired. This can usually be done only when the tear or cut is primarily in the machine direction.

11. Summary

Dryer fabric designs and construction have been changing in the recent past to keep up with need for increased production rates demanded by papermakers. Those demands resulted in faster machines and wider paper webs. This led to problems with sheet moisture profiles, and to the need for pocket ventilation (PV) systems. PV rolls, PV ducts and open-mesh dryer fabrics were forthcoming from suppliers. The increased machine speeds led to decreased web contact times with dryer cylinders. The suppliers' response was to produce single- and multi-layer dryer fabrics capable of operating at high speeds and high fabric tensions under the demanding ambient conditions found in a closed dryer hood. As a consequence, the papermaker/supplier team must keep in mind the three parameters of efficient and effective dryer fabric use: permeability, dryer fabric strength retention, and operating conditions.

12. Concluding remarks

Because paper machine clothing is an integral part of the paper machine, it must be properly specified in order to act synergistically with the machine parts. Once the desired level of performance for both the machine and the fabrics to be used has been designated, clothing suppliers can specify changes in fabric construction designed to meet the needs and desires of papermakers. The process of evolutionary development can be slow, however. Typical useful lives for paper machine clothing can range up to several months for forming fabrics; press fabrics can have useful lives measured in terms of weeks; and dryer fabrics can have useful lives approaching one year. Although paper machine clothing costs represent only about 1% of the total cost of manufacture, poorly selected and applied clothing can have dramatic effects on the overall performance of the machine. Drainage rates in the forming section, water removal rates in the press section, and drying rates in the dryer part can be affected adversely. Web stability, in and between sections, can be a function of bound layers of air associated with the clothing. Product quality is somewhat dependent on clothing performance. For these and other reasons, papermakers work closely with the suppliers of paper machine clothing to ensure optimal results.

REFERENCES

1. Lewyta, J., "Recent Developments in Paper Machine Clothing: Forming Fabrics", Tappi J. (Feb. 1987).
2. Chiu, K., "Forming Fabric Characteristics Affecting Drainage and Formation", Vol. 25, No. 10, Albany International Felt and Fabric Facts.
3. Beran, R., "The Evaluation and Selection of Forming Fabrics", Tappi J. (Apr. 1979).
4. Helle, T., "On the Evaluation of Forming Characteristics of Papermaking Wires", Norwegian Institute of Technology (1982).
5. Stowe, B., "Forming Fabrics for Twin-Wire Machines", TAPPI Twin-Wire Seminar (1987).
6. Brown, H., "Location and Operation of Automatic Wire and Felt Guides", Gilbert & Nash Publication (1970).
7. "Cleaning of Forming Fabrics", Appleton Wire Division, Albany International Corporation, Publication CL2 200-43 3/85 (1985).
8. Pitt, R. and Condon, M., "Wear of Forming Fabrics", TAPPI Wet End Seminar (1987).
9. Wahlstrom, P.B., "Our Present Understanding of the Fundamentals of Pressing", 2nd Intl. Symp. on Water Removal at the Presses and Dryer Technical Section, Tech. Sect., CPPA, Mont Gabriel (Oct. 1968).
10. "Press Section Economic Evaluation", TIS 0404-22, TAPPI PRESS (1984).
11. "Estimating Air Flows for Suction Pipe Felt Conditioning", TIS 0404-27, TAPPI PRESS (1983).

BIBLIOGRAPHY

1. Race, E. and Hindle, T., "The Principles and Practices of Permeability Measurement in Dryer Felts", Tappi.
2. Soininen, M.A., "Air Flow and Pocket Ventilation in the Dryer Section", Paper Trade J. (Mar. 1971).
3. Hay, K.D., "How to Improve Performance with Open-Mesh Dryer Fabrics", Pulp Paper (May 1973).
4. Hill, K.E., "Dryer Sheet Stability for Older Paper Machines", Tappi 71(8) (Aug. 1988).
5. Wedel, G.L., "No Draw Drying Restraint", Tappi (Apr. 1989).

XXIII
PAPER MACHINE AND SUCTION ROLL CORROSION

D.A. WENSLEY

Dr. Angela Wensley heads the Materials and Corrosion group at MacMillan Bloedel Research in Burnaby, B.C., where her responsibilities include consulting on materials selection, corrosion protection, fabrication and inspection of equipment, failure analysis and corrosion research. She obtained her Ph.D. in Metallurgical Engineering in 1977 from the University of British Columbia, and has been with MacMillan Bloedel ever since. She has chaired the NACE Corrosion in the Pulp and Paper Industry committee, the NACE Recovery Boiler Fireside Corrosion task group, and the TAPPI Continuous Digester Cracking task group. She has written numerous technical papers on corrosion in the pulp and paper industry.

M.D. MOSKAL

Mr. Max D. Moskal is a Project Manager in Metallurgical Engineering at Stone Container Corporation in Burr Ridge, IL, where he is responsible for corrosion and material engineering in the Containerboard and Paper Division. Mr. Moskal received his B.S. degree in Metallurgical Engineering from the Colorado School of Mines in 1958. He has authored more than twenty papers on corrosion in the paper industry and has contributed to several books. Mr. Moskal is past Chairman of the TAPPI Corrosion and Materials Engineering Committee and was the recipient of the Engineering Division's 1990 Leadership and Services Award.

REVIEWED BY

Dr. Andrew Garner
Director
Paprican
Vancouver, B.C.

The publishers acknowledge with thanks the help of Dr. Andrew Garner, Paprican, B.C. in arranging with ASM INTERNATIONAL for the publication of this material and for its review.
This chapter is reprinted (with updating) with kind permission from *Metals Handbook*, Vol. 13, 9th Ed., "Corrosion", ASM INTERNATIONAL, Metals Park, OH, 1987.

A. PAPER MACHINE CORROSION

Corrosion problems in paper machines are most severe in the wet end and in ancillary equipment handling white water. Areas susceptible to corrosion are those exposed to the white water environment by immersion, splashing, vapor mist, or in the damp crevices beneath pulp pads or other deposits. Acid white waters can cause rapid corrosion of bare carbon steel, copper alloys and cast irons, as well as localized corrosion (pitting and crevice corrosion) of stainless steels.

This section is divided into three parts. The first describes the major paper machine components with respect to their materials of construction and specific corrosion problems; the second part deals with the composition and corrosive nature of white water; the third part is a more detailed discussion of corrosion mechanisms affecting stainless steels.

1. Paper machine components

Most wetted paper machine surfaces are constructed from austenitic stainless steel alloys [1, 2]. Types 304L (UNS S30403) and 316L (UNS S31603) have traditionally been employed in low-chloride (<20 ppm) and high-chloride (>100 ppm) white waters, respectively. Type 316L contains molybdenum (Mo) as an alloying addition, which confers improved resistance to chloride pitting and crevice corrosion, as well as to thiosulfate pitting. Alloys with even higher Mo contents may be necessary for particularly corrosive white waters, such as those resulting from high degrees of system closure [3]. Specification of a corrosion-resistant alloy is not in itself sufficient to ensure freedom from corrosion. Great care must be taken during fabrication and cleaning of equipment to avoid contamination of stainless steel surfaces. The presence of embedded iron, iron oxides, heat tint, weld spatter, slag, and dirt can impair the corrosion resistance of the underlying metal. The misuse of carbon steel wire brushes for weld cleanup may cause serious pitting problems initiating at iron particles embedded in the stainless steel surface. Iron contamination is best removed by passivating with a nitric acid solution, which dissolves the iron and leaves the stainless steel intact. Pickling solutions, which typically contain fluoride in addition to nitric acid, are much more aggressive, and are used to remove heat tint in addition to surface iron. Pickling should be done only as a final cleanup step, and on equipment for which a very smooth surface finish is not required.

(a) Stock piping

Paper machine stock piping is typically thin-gauge (schedule 10) type 304L or 316L austenitic stainless steel. Corrosion is generally not a problem in stock lines if the correct stainless alloy, weld filler metal, welding process, and cleaning procedure have been chosen. Surfaces that have become roughened, or which have weld projections that have not been removed, provide sites for stock hang-up. Crevice corrosion or microbiological attack is liable to occur under deposits [4]. Fatigue cracking problems in stock piping are associated with welding defects such as incomplete penetration. Great care must be taken to ensure that good quality, full penetration welds are obtained. For root passes, inert gas purging is to be preferred over so-called anti-oxidation fluxes.

(b) Headboxes

The headbox environment is considerably more severe from a corrosion standpoint than that inside stock piping. This is because of the existence of crevices inside the headbox, the presence of an air/liquid interface, the potential for internal mechanical damage (rectifier roll failure), and even the possibility of stray current damage [5]. Clean, smooth surfaces inside the headbox are essential. Stock hang-up on rough or corroded surfaces may result in the release of pulp pads, causing product quality problems or even sheet breakage. Pickling, passivation, and buffing with abrasives are commonly used to provide a smooth surface finish. Electropolishing improves the finish for surfaces of 0.4 μm (15 μin.) and above [6]. Calcium deoxidized machinable grades of stainless steel have greater freedom from macroinclusions than conventional grades, and are often used for construction of paper machine headboxes because of their superior electropolishing properties. Machinable grades, however, have less corrosion resistance than conventional grades of equivalent molybdenum alloying content.

Headboxes are often constructed from alloys more corrosion resistant than type 316L as additional insurance against corrosion of internal surfaces. Overmatched weld fillers (filler metals with higher Mo content than the base metal) have been used to provide additional corrosion resistance for weldments. Headboxes may be constructed from higher Mo grades such as the Swedish grade SIS 2353 or type 316LM or from type 317L (UNS S31703) stainless steel. Some headboxes have even been constructed with UNS N10276, a nickel-base alloy that is usually used in much more severe corrosion applications. The reasoning behind this material

Fig. 754. Severe pitting corrosion of a type 316L stainless steel slice lip.

selection is that immunity to corrosion means freedom from maintenance.

Headbox apron and slice lips are particularly prone to pitting corrosion attack (Fig. 754). Corrosion pits initiate outside the liquid flow, either on bare surfaces or beneath deposits. These pits grow in size and eventually reach the edge of the lip, where they then have a detrimental effect on sheet formation. The use of highly pitting resistant stainless steels such as UNS S31254 has been successful in combating pitting of apron and slice lips.

(c) Paper machine wires

Before the advent of synthetic plastic mesh fabrics, paper machine wires were typically phosphor bronze screens. Bronze wire life was limited by wear and corrosion attack [7]. Sacrificial galvanic corrosion of the wires resulted in cathodic protection of bronze couch rolls. Corrosion inhibitor additions were often required to obtain acceptable service lives for bronze wires [4]. Wires coated with tin, nickel, or plastic gave superior service, although the use of synthetic fabrics has now eliminated wire corrosion problems. The resultant loss of cathodic protection appears, however, to have contributed to increased corrosion of bronze couch rolls [8].

(d) Wet end structures

Wet end carbon steel structural members must be protected either by paint coatings, or with stainless steel cladding [9]. Where coatings fail or where white water can gain access beneath stainless cladding (at incomplete seal welds or perforations), the underlying structural steel may corrode away rapidly. Corrosion rates of 1.6 $mm{\cdot}y^{-1}$ (64 mils per year) have been observed inside hollow carbon steel couch beams used as part of the vacuum system. Internal, as well as external, cladding with stainless steel is required in such cases. Industry experience with wet end coating systems involving inorganic zinc-rich primers and epoxy-polyamide top coats has not been good. Immersion service coating systems such as organic zinc/amine adduct-cured epoxy and inhibited epoxy/amine adduct-cured epoxy systems both appear to offer promise of longer term service [10].

(e) Vacuum pumps

Liquid ring vacuum pumps typically have ductile cast iron rotors and gray cast iron casings. Corrosion of the rotors (Fig. 755) and casings results when corrosive seal waters are used. From a corrosion standpoint, it is desirable to use cool, fresh seal water. Paper machine white water is considered unsuitable for use as seal water because of its high temperature and low pH. Even fresh water may be corrosive if the flow rate is so low that the temperature increases inside the pump, or if it becomes contaminated with white water carryover. A discharge seal water temperature of 50°C (122°F) and/or a pH less than 4.5 can cause accelerated cast iron corrosion. Corrosion rates may be lowered by adding a small amount of caustic to the seal water. High discharge seal water conductivity (in excess of 200 μmho) may indicate excessive white water carryover. Accelerated impeller corrosion can occur at rotor tip speeds above 27 $m{\cdot}s^{-1}$ (80 ft/s) [11]. The presence of sand or grit in the water will also shorten liquid ring vacuum pump life because of erosion.

As cast irons corrode, a residue of graphite is left behind (Fig. 756). Pearlitic ductile iron forms a more protective graphite surface layer than ferritic ductile iron because the three-dimensional carbonaceous residue left behind from pearlite dissolution acts as a binder for the graphite (Fig. 757). Rapid liquid ring vacuum pump corrosion can be expected if environmental, velocity, or erosion conditions are such that the protective graphite film is removed. In severely corrosive environments, liquid ring vacuum pumps have been built with stainless-lined bodies and stainless steel rotors.

(f) Roll journals

Corrosion fatigue results from the combination of cyclic stresses and a corrosive environment. Corrosion fatigue produces earlier failure than that which would have occurred either by pure mechanical fatigue or by corrosion alone. Paper machine roll journals in wet end environ-

Fig. 755. Corroded ductile iron rotor of a liquid ring vacuum pump.

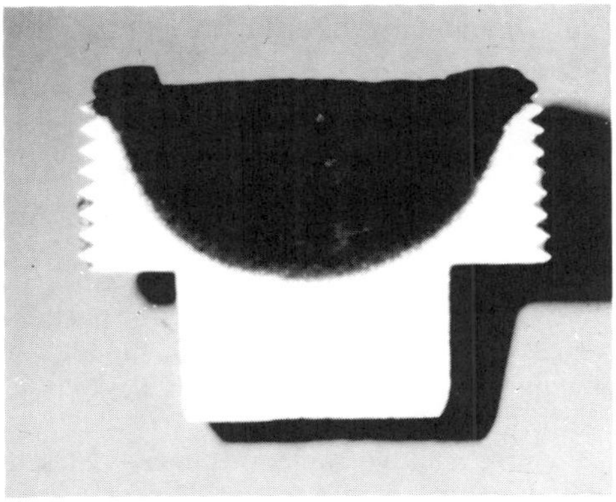

Fig. 756. Cross-section of a corroded gray cast iron pipe plug removed from white water service, showing the graphite residue left behind as the iron matrix dissolved.

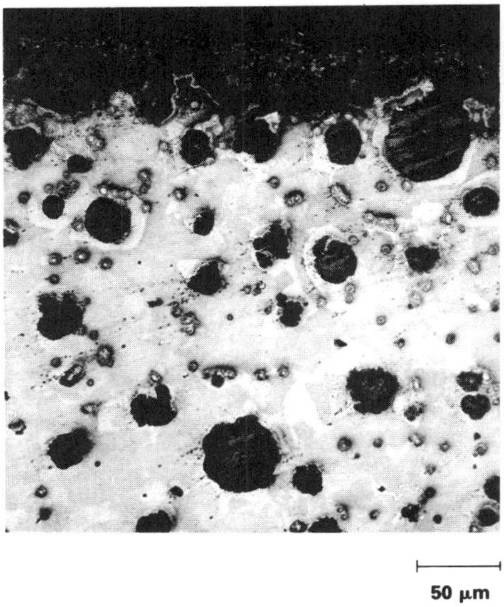

Fig. 757. Micrograph of a cross-section of a pearlitic ductile iron coupon exposed for 1872 h in vacuum pump discharge seal water.

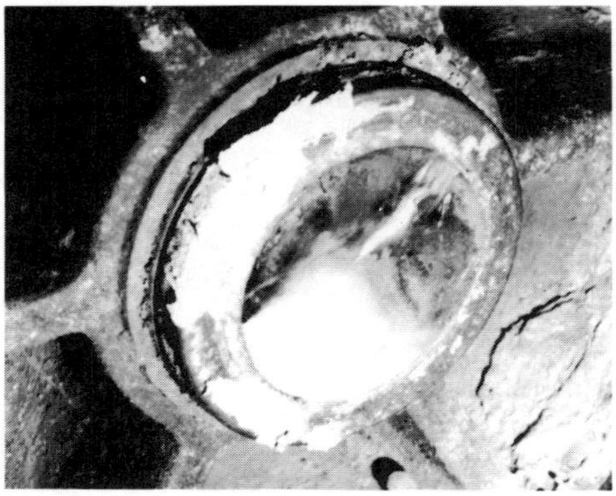

Fig. 758. View of the fracture surface of a felt roll journal that failed by corrosion fatigue. Material of construction is AISI 4140 low-alloy steel quenched and tempered to 300 HB.

ments can suffer corrosion fatigue failure (Fig. 758). Cracking often develops at abrupt changes in journal section size. The environment promoting corrosion fatigue may involve splashing of white water and/or the formation of damp pads of pulp on the journals, or may simply be the humid paper machine atmosphere. Felt roll journals may also be exposed to felt cleaning chemicals, some of which are acidic.

Roll journals are constructed from a variety of materials, including medium- or high-carbon steels, low-alloy steels such as AISI 4140 (UNS G41400) and AISI 4340 (UNS G43400), cast irons, and various grades of stainless steel. Improved corrosion fatigue performance can be realized by protecting journals from the environment, by redesigning to increase their diameter as well as removing step changes in section size [12], by polishing to remove surface flaws, and by selecting a material with superior corrosion fatigue resistance. Fatigue strength is generally proportional to tensile strength; however, no roll journal material is immune to corrosion fatigue because, unlike pure mechanical fatigue, there is no fatigue limit below which indefinitely long operation could be guaranteed. Austenitic stainless steels have relatively poor corrosion fatigue resistance, and are not commonly used for roll journals in wet end environments. Duplex (such as UNS S31803) and precipitation-hardening (such as UNS S17400, condition H1100) stainless steels, however, may offer much better corrosion fatigue resistance than the more conventional journal materials.

Wet end roll journals can be physically protected from exposure to white water. Stainless steel shields and coatings, ranging from paints to thermally applied metal spray, have been employed for journal protection. The disadvantage of physical barriers is that they hinder inspection of surfaces for cracking. Integral roll heads and journals made from ductile cast iron having a generous radius between the head and journal may also offer superior service. Dry end (reel spool) journals are usually made from high-strength low-alloy steels in the quenched and tempered tempered condition such as AISI 4140 or AISI 4340. The latter steel is deeper hardening, and is to be preferred for larger section sizes.

(g) Suction rolls

Corrosion-related failures of suction roll shells are the most serious materials and corrosion problem in modern paper machines. More detailed information can be found in Sec. B of this chapter.

2. White water

White water is the name applied to the aqueous media which carry the fibers, and from which fibers are separated in the papermaking process. White waters inevitably contain appreciable amounts of dissolved organic and inorganic compounds, some of which contribute significantly to white water corrosivity.

(a) Closure

Closure of papermaking systems involves the recycle of white water, either directly or after treatment. The degree of closure can be expressed as the volume of water consumed per tonne of paper produced [13]. Closed paper mills may discharge less than 5 m^3/t, whereas an open paper mill may discharge over 40 m^3/t.

Closure results in an increase in the concentration of dissolved inorganic and organic solids, a decrease in pH, and an increase in temperature, all factors known to increase system corrosivity. Greater degrees of closure can be tolerated in a mill where the white water was not particularly corrosive to begin with, because the corrosivity of a given white water ultimately depends on its composition.

Closure is universally considered detrimental from a corrosion standpoint, even though there is no clear parallel between increasing closure and increasing paper machine corrosion [13, 14]. Accelerated corrosion may be expected for materials that undergo general corrosion attack, such as carbon steels, cast irons and copper alloys. For stainless steels the situation is not as clear. Increasing closure may not result in stainless steel corrosion, but rather only in a decreased margin of safety for corrosion-free service with a particular grade of stainless steel. Localized corrosion may also occur when a certain critical white water composition has been reached. Spontaneous pitting of the less corrosion-resistant stainless steels such as type 304 (UNS S30400) may occur after only moderate closure has been achieved. The Mo-alloyed grades of stainless steel will be more resistant to corrosion under increasing closure. There is a concern that the Mo content of modern type 316L stainless steel is usually only 2.00–2.10% Mo which is at the low end of the standard 2–3% Mo specification for 316L [15]. Recently purchased equipment made from 316L may thus be less corrosion resistant that earlier equipment made from older renditions of this alloy which had higher Mo contents.

(b) Composition

Paper machine white waters can vary widely in composition from mill to mill, within a given mill, or even with time in a given paper machine system. The corrosivity of white water depends primarily on the pH, temperature, and the concentrations of aggressive inorganic anions, such as chloride and thiosulfate. The kinds of dissolved inorganic compounds found in white waters will depend on whether the wood was seaborne, on the pulping process used, on whether or not bleaching or brightening was carried out, and on the chemicals used in the wet end of the machine, such as fillers, sizes, retention aids, defoamers, slimicides, and dyes. Large quantities of corrosive sea salt will be absorbed by logs during ocean transport in booms. Kraft pulping will introduce sulfate. Chemithermomechanical pulping may add sulfite. Chlorine, chlorine dioxide and hypochlorite bleaching will also introduce chloride ions; further, unreacted oxidants such as chlorine dioxide can be carried over onto the paper machine with bleached pulp. Sulfur-containing ions in white waters also include decomposition products of sodium hydrosulfite brightening of mechanical or thermomechanical pulps. A typical white water composition for a West Coast newsprint mill using a furnish of semibleached kraft and hydrosulfite-brightened chemithermomechanical pulp and groundwood is given in Table 79.

The contribution of each of the above parameters on white water corrosivity is discussed below.

(c) Temperature

Increased white water temperature due to closure may contribute to energy savings, and also help to improve drainage in the wet end of a paper machine, thus permitting an increase in machine speed. Higher temperatures are detrimental from a corrosion standpoint because the rates of most chemical reactions (including corrosion reactions) typically double for each 10°C (18°F) increase in temperature. This is likely the case for general corrosion reactions involving carbon steels, cast irons, and copper alloys. If a certain critical temperature is exceeded, pitting corrosion of stainless steels will initiate in white waters where previously no pitting occurred. This critical temperature is lower for white waters which have higher chloride concentrations.

Table 79. Composition of a typical white water from a west coast newsprint mill.

Temperature	118°F (48°C)
pH (initial)	4.6
pH (on standing)	3.0
Sodium chloride (NaCl)	311 ppm
Sodium sulfate	855 ppm
Sodium thiosulfate	34 ppm
Sodium sulfite	75 ppm
Conductivity	1749 μmho

(d) pH

Although there are both acid and alkaline papermaking processes, most papermaking uses the acid process, where the typical white water pH is in the range 4 to 6. The pH is controlled on the acid side by alum additions and/or souring with sulfur dioxide, sulfite, or sulfuric acid.

Below about pH 4, the general corrosion rates of many non-passive alloys are accelerated. This is because hydrogen ions become so plentiful that hydrogen ion reduction to form hydrogen gas replaces the reduction of dissolved oxygen molecules as the predominant cathodic half reaction in the corrosion process (see Sec. A3 below). In acid solutions, corrosion is thus no longer limited by the rate of diffusion of dissolved oxygen to the metal. Decreased pH also lowers the stability of passive films on stainless steels; this makes the steel more susceptible to pitting initiation and increases the probability that a pit, once initiated, will continue to grow.

The pH of a white water may decrease by 0.5 to 2.0 units within approximately 1 day of sampling. There are many reasons for this. The continued oxidation of hydrosulfite and bisulfite generates free hydrogen ions. It would be expected that white waters containing these ions may become more corrosive when kept in storage or during shutdowns. Instantaneous corrosion rate meters are often used to monitor white water corrosivity. It has been observed that the instantaneous corrosion rate of carbon steel often correlates well with the magnitude of the pH decrease subsequently observed after white water sampling. White waters which experience large decreases in pH are those in which higher instantaneous carbon steel corrosion rates were originally measured.

(e) Chlorides

Coastal mills that rely on the transportation of logs in seawater may have chloride concentrations ten times higher than those found in inland mills. Chloride ions are very soluble and will tend to persist through the pulping process, eventually ending up in the white waters. Chloride concentrations increase as closure proceeds. Chlorides are particularly insidious because they can support pitting and crevice corrosion of austenitic stainless steels. Chlorides are also agents for the stress-corrosion cracking (SCC) of stainless steels, but usually at temperatures above 60°C (140°F). SCC is therefore not a serious problem in most white waters. Chlorides accelerate the general corrosion of carbon steels; however, they do not appear to produce significant acceleration of the corrosion of copper-base alloys.

(f) Sulfates

Sulfate ions ($SO_4^{=}$) appear in white water as a residual chemical from the kraft pulping process, from seawater, from alum or sulfuric acid additions, and from the oxidation of other sulfur-containing ions, such as bisulfite (HSO_3^-), thiosulfate ($S_2O_3^{=}$) and hydrosulfite ($S_2O_4^{=}$). Sulfate is the thermodynamically stable form of sulfur in white waters.

Sulfate is not an aggressive anion; indeed, it can act as an inhibitor for chloride-related pitting. The inhibiting effect is particularly pronounced if sulfate is present in (molar) excess over chloride ions [15, 16]. On the other hand, sulfate ions make a significant contribution to white water conductivity and therefore indirectly tend to increase the rates of electrochemical corrosion reactions by facilitating charge transfer in the white water electrolyte. Sulfate ions are used by sulfate-reducing bacteria, which in turn cause microbiological corrosion of paper machine equipment. Further, in white waters containing thiosulfate, sulfate can enhance thiosulfate pitting [17].

(g) Thiosulfates

Mills employing hydrosulfite brightening will have appreciable amounts of corrosive thiosulfate ions in their white waters. A major thiosulfate source is anaerobic decomposition of warm hydrosulfite solutions during storage. Under non-oxidizing conditions hydrosulfite is very rapidly, and stoichiometrically, converted to thiosulfate. By contrast, air oxidation of hydrosulfite does not produce thiosulfate. In practice, additions of hydrosulfite in excess of that required to meet target brightness levels serve only to introduce even more thiosulfate into white water systems [18]. The kinetics of subsequent oxidation of thiosulfate to the more stable sulfate form are so slow that appreciable thiosulfate concentrations can build up.

Thiosulfate concentrations in white water can be reduced by ensuring that the concentration in the original source is as low as possible. Storage of hydrosulfite solutions at 2 to 9°C (36 to 48°F) and/or the addition of an alkaline stabilizer to maintain a pH of 10 or more will minimize thiosulfate formation [19]. Brightening at a pH over 5 will also help keep thiosulfate levels down. Thiosulfate ions have been found to be particularly aggressive pitting agents for stainless steels (see Sec. 3(b) below) [17, 20]. Pitting can occur at much lower concentrations (for example, 10 ppm $Na_2S_2O_3$) than that required for chloride pitting (>200 ppm NaCl). Thiosulfates also increase general corrosion rates of copper-base alloys [21].

(h) Sulfites

Sulfite additions are employed in chemithermomechanical pulping, and also made elsewhere to control pH or to brighten pulp stock. Bisulfite (the stable form of sulfite in white water) is also produced during the decomposition of hydrosulfite brightening solutions. Sulfites are readily oxidized to sulfate in the presence of dissolved oxygen. Sulfite is not considered to be aggressive; in fact, sulfites are often used as oxygen-scavenging corrosion inhibitors in other aqueous systems, such as boiler feedwater. Sulfites should also have inhibitive properties in paper machine white water. Very high bisulfite levels (>200 ppm SO_3^{2-}) in white water may result in sulfur dioxide evolution, leading to severe atmospheric (dew-point) corrosion problems, such as attack of aluminum paper machine hoods.

(i) Conductivity

White water conductivity depends on the concentrations of all dissolved ionic species, both inorganic and organic. Higher conductivity is an indication of increased corrosivity of white waters toward carbon steel, cast iron, and copper-base alloys. This is not the case for stainless steels as long as they are passive and are not undergoing localized corrosion attack.

The corrosivity of paper machine white water can be correlated with individual white water

Table 80. Correlation coefficients for various parameters affecting the corrosion rate of carbon steel in white water.

Property	r^2
Conductivity	84
pH (on standing)	74
Na_2SO_4	72
NaCl	66
pH (on standing)	64
pH (initial)	18
Temperature	2

parameters by linear regression of the instantaneous corrosion rate of carbon steel with analysis data obtained at the same time as each corrosion rate measurement (Table 80).

The best regression coefficient for white water corrosivity is obtained with conductivity. These data imply that white water corrosivity is not a strong function of temperature; however, there was only a small temperature variation over the course of this particular test.

3. Corrosion mechanisms

Corrosion is an electrochemical process involving anodic (oxidation) and cathodic (reduction) half reactions. For example, the corrosion of iron in waters of near-neutral pH is the sum of an anodic half reaction (iron dissolution) and a cathodic (dissolved oxygen reduction) reaction:

$$2Fe \rightarrow 2Fe^{2+} + 4e^- \quad \text{(oxidation reaction)}$$

$$O_2 + 2H_2O + 4e^- \rightarrow 4OH^- \quad \text{(reduction reaction)}$$

$$2Fe + O_2 + 2H_2O \rightarrow 2Fe^{2+} + 4OH^- \quad \text{(net reaction)}$$

This reaction rate is limited by the solubility of oxygen in water and the rate of diffusion of oxygen to the metal surface. In acid waters (pH <4), hydrogen ion reduction replaces oxygen reduction as the cathodic half reaction:

$$Fe \rightarrow Fe^{2+} + 2e^- \quad \text{(oxidation reaction)}$$

$$2H^+ + 2e^- \rightarrow H_2 \quad \text{(reduction reaction)}$$

$$Fe + 2H^+ \rightarrow Fe^{2+} + H_2 \quad \text{(net reaction)}$$

This reaction is not diffusion limited, and can proceed at a rapid rate.

A review of corrosion problems in the paper industry is covered in Volume 9 of this series.

(a) Chloride pitting and crevice corrosion

Stainless steels rely on the formation of a stable passive surface film for immunity to corrosion in paper machine white waters. The metal is protected from corrosion by forming an oxide film which is highly resistant to dissolution. Oxidizing conditions must prevail for the passive film to form and be maintained. Dissolved oxygen from the air is usually sufficient to maintain stable passivity. Stainless steels can even withstand slightly reducing conditions in flowing white waters without suffering appreciable attack. Excessively oxidizing conditions (caused by oxidizing agents such as chlorine dioxide or chlorine) are dangerous because they promote the localized breakdown of passivity, leaving the metal vulnerable to pitting or crevice attack.

For a given white water there is a certain safe range of oxidizing conditions within which a particular stainless steel alloy is stably passive. In the presence of aggressive anions such as chlorides, however, this safe range is narrowed. When the oxidizing potential of the solution exceeds the breakdown potential characteristic of the particular white water/stainless alloy combination, chloride ions can attack the stainless steel surface. The attack manifests itself as pits because passive film breakdown occurs only in isolated locations, such as weak spots in the film due to defects in the underlying metal. Pitting attack is favored by the following conditions:

- High chloride concentration.
- High temperature.
- Oxidizing conditions.
- Low pH.
- Stagnant or low-velocity conditions.
- Low molybdenum content in the stainless steel.

A newly initiated chloride pit tends to repassivate unless the favorable conditions for initiation are maintained. Figure 759 is a schematic diagram of the pitting process. Once a pit has become established, the driving force for its growth is the development of an oxygen concentration cell, where the metal within the pit becomes anodic with respect to the surrounding metal. The majority of the metal surface remains passive because it has ready access to dissolved oxygen for the reduction half of the net corrosion reaction. The exposed surface thus becomes cathodic to the pit. Since the surface area of a pit is very much smaller than that of the surrounding metal surface, anodic dissolution of metal from the pit occurs at a very high rate. Subsequent hydrolysis of dissolved metal complexes inside the pit results in the generation of free hydrogen

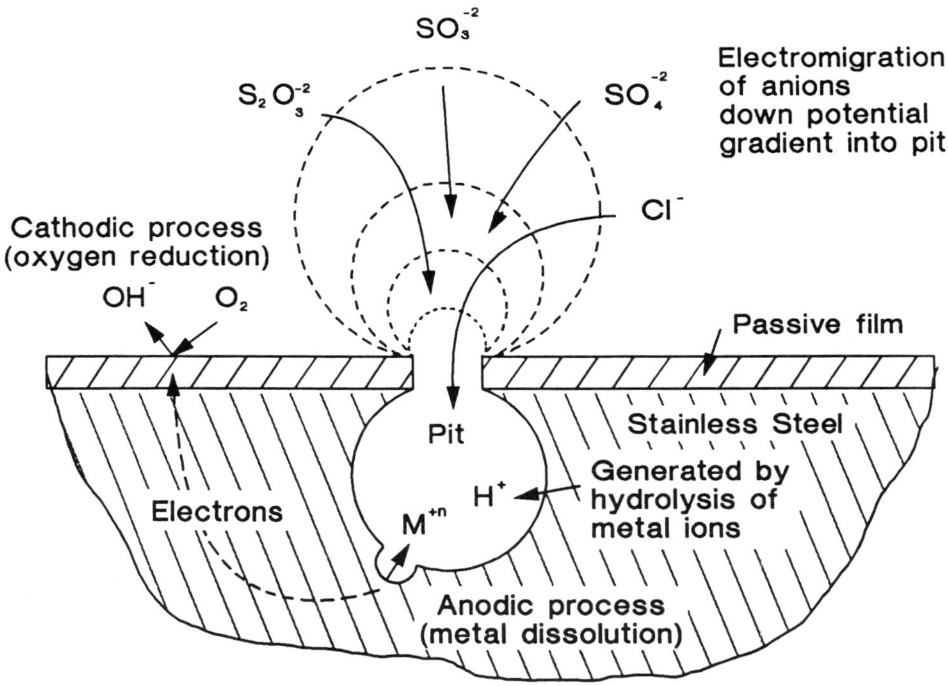

Fig. 759. Schematic diagram of pitting corrosion of stainless steel in white water.

cations. Additional anions diffuse into the pit to maintain charge neutrality. Anions concentrate in the pit by electromigration. Eventually, continued metal ion hydrolysis and electromigration of chlorides may result in the solution inside a growing pit resembling more hydrochloric acid than paper machine white water. Pit growth may then continue as acid corrosion.

Crevice corrosion is a form of pitting corrosion which occurs beneath deposits or in other areas shielded from direct contact with the white water environment (Fig. 760). Crevice corrosion initiates more readily than pitting attack because the conditions for an oxygen concentration cell already exist. The metal surface inside a crevice has difficulty maintaining passivity because of reduced access to dissolved oxygen, whereas the surrounding metal has ready access to dissolved oxygen. Passive film breakdown within the crevice is facilitated by the natural tendency of the crevice to become anodic and the surrounding metal to become cathodic. Once crevice corrosion is initiated, its growth mechanism is the same as that for pitting corrosion.

The primary corrosion concern with white water system closure is that the critical chloride concentration and/or temperature for stainless steel breakdown will be exceeded. Under such conditions, pitting corrosion will occur spontaneously. This is a particular concern with type

Fig. 760. Crevice corrosion of a type 316L cheeking plate located adjacent to a headbox apron. Corrosion developed under pulp pads that formed despite the highly polished surface.

304L stainless steel because this alloy does not contain molybdenum. Stainless steels are chosen for service in chloride pitting environments on the basis of molybdenum content. It is common to select the alloy with the minimum molybdenum content required for resistance to pitting. A hierarchy of austenitic alloys, representing increasing resistance to chloride pitting and crevice corrosion, can be listed in increasing

Table 81. Molybdenum contents of stainless alloys.

Alloy	Molybdenum (typical minimum) %
UNS S30403 (Type 304L)	0
UNS S31603 (Type 316L)	2
(SIS 2353, Type 316LM)	2.5
UNS S31703 (Type 317L)	3
UNS N08904	4
UNS S31254	6
UNS N10276	15

order of their minimum molybdenum contents (see Table 81).

(b) Thiosulfate pitting

Thiosulfate is an aggressive pitting agent, especially for stainless steels that do not contain molybdenum. Thiosulfate pitting can occur only below the thiosulfate reduction potential (unlike chloride pitting, which occurs only above the passive film breakdown potential) [17, 20]. Reduction of thiosulfate in the presence of hydrogen ions produces an adsorbed sulfur monolayer on the metal surface. The adsorbed sulfur activates the anodic dissolution of the metal and hinders repassivation. Excess hydrogen ions must be present for acidification of the pit; further, there must also be a larger amount of inert ions (sulfate and chloride) that can be transported into the pit to meet charge transfer requirements and take part in metal-complexing hydrolysis reactions. The worst case of thiosulfate pitting occurs within the molar concentration ratio:

$$\frac{Na_2SO_4 + NaCl}{Na_2S_2O_3} = 10 \text{ to } 20$$

Above the range represented by the ratio, there is insufficient thiosulfate to reach the pit nucleus. Below this range, there is too much thiosulfate reduction to bisulfate, which prevents acidification of the pit.

Thiosulfate corrosion particularly affects those grades of stainless steels that do not contain molybdenum. Once formed, pits are very stable and are not subject to spontaneous repassivation. Scratches encourage the initiation of pits. Few, large pits tend to form rather than many, small pits, as in chloride pitting. Type 304 stainless steel is particularly susceptible to thiosulfate pitting when it becomes sensitized, such as at weld heat-affected zones. Type 316L stainless steel is the minimum grade that should be used for white water service where high

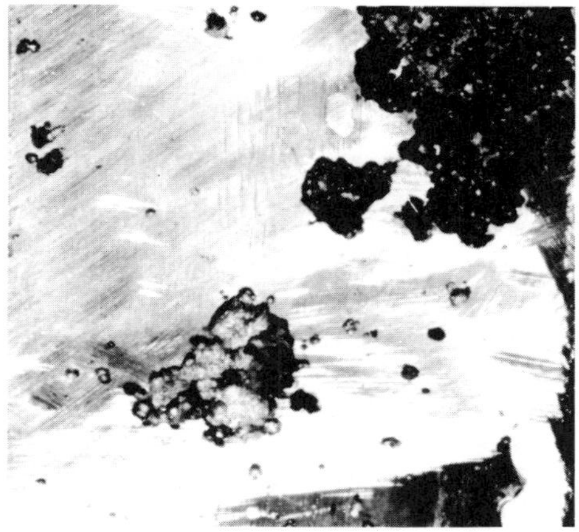

Fig. 761. Section of type 304L stainless steel plate removed from a tapered header used to deliver stock to a headbox showing severe microbiological corrosion.

thiosulfate levels may exist. It appears that in practice corrosion problems can be mitigated if thiosulfate levels are controlled below 5 and 10 ppm for equipment made of type 304 and 316 stainless steel, respectively.

(c) Microbiological corrosion

Paper machine white waters contain nutrients that can sustain bacterial growth. Microbiological growth thrives in near-neutral pH environments. White water temperatures are also usually within the favorable range of 40 to 50°C (105 to 120°F). Although higher temperatures may prevent the growth of some forms of bacteria, increased temperatures can increase the metabolism of those bacteria that can adapt to heat. The result is the formation of slimes [4, 22].

Stock and white water flow systems are designed to minimize slime accumulations. Surfaces are polished and weld projections removed to prevent hang-ups. Wherever slime deposits can build up, however, microbiological corrosion can occur [23]. Once a deposit has grown to a sufficient thicknesss to exclude oxygen, a colony of sulfate-reducing bacteria (*Desulfovibrio desulfuricans*) can become established. Enzymes produced by these anaerobic bacteria catalyze the reduction of sulfates to form free sulfide ions. Chemically reducing conditions quickly develop, resulting in the depassivation of the stainless steel surface beneath the deposit. Active corrosion in the form of pitting then proceeds. Corrosion due to *Desulfovibrio desulfuricans* is characterized by the presence of large, shallow pits covered with a black crust (Fig. 761). Perforations through

Fig. 762. External view of the type 304L stainless steel tapered header in Fig. 760 showing leakage occurring at small perforations.

stainless steel equipment are usually small because the entry of oxygen at a leak will stop the activity of sulfate-reducing bacteria (Fig. 762).

B. SUCTION ROLL CORROSION

Suction rolls are used to remove water from paper at the wet end of the paper machine. One way to accomplish this is by passing the paper web through a roll nip, one roll of which is the suction roll. The suction roll is drilled to an open area of about 20%, and a vacuum is applied to the inside of the roll. A typical cross-section through a suction roll is shown in Fig. 763. A rubber cover is often used on the suction press roll or mating roll to give a desired nip pressure relationship. Many machine designs have been used in the wet end of paper machines throughout the years. A typical paper machine may use three or more suction rolls designated as couch, wringer, pickup, or press rolls (Fig. 764).

The critical component in the suction roll is the drilled shell, which is subject to corrosion and fatigue cracking. During the past 30 years, changes have evolved in design and machine environment that have increased the cracking and corrosion tendency of the suction shell. For example, demands for increased machine production and speed have been resolved by increasing roll nip pressures. Wider machines have also been designed. These new designs have required heavier and stronger shells to meet the increased stress that is imposed. The corrosive environment in the paper machine wet end has also become more severe as water systems have been closed, water temperatures increased, and chemicals used to clean paper machine felts. Numerous changes have also occurred in a search for materials and designs that better resist shell corrosion and corrosion-assisted cracking.

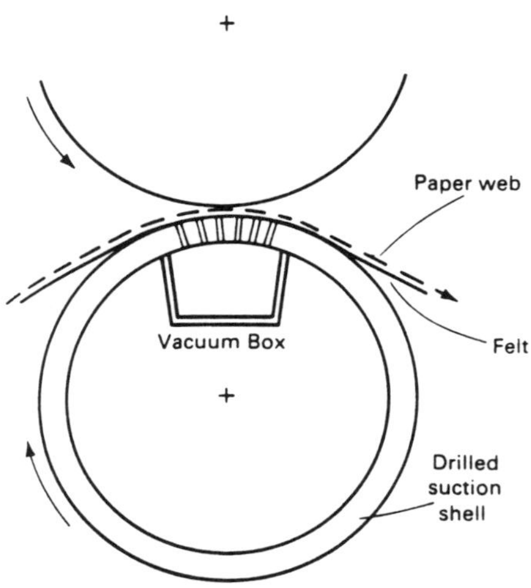

Fig. 763. Cross-section of suction roll configuration.

This section will describe the history of suction roll and materials development, the operating environment, corrosion fatigue of suction rolls, shell manufacturing methods, and corrective measures.

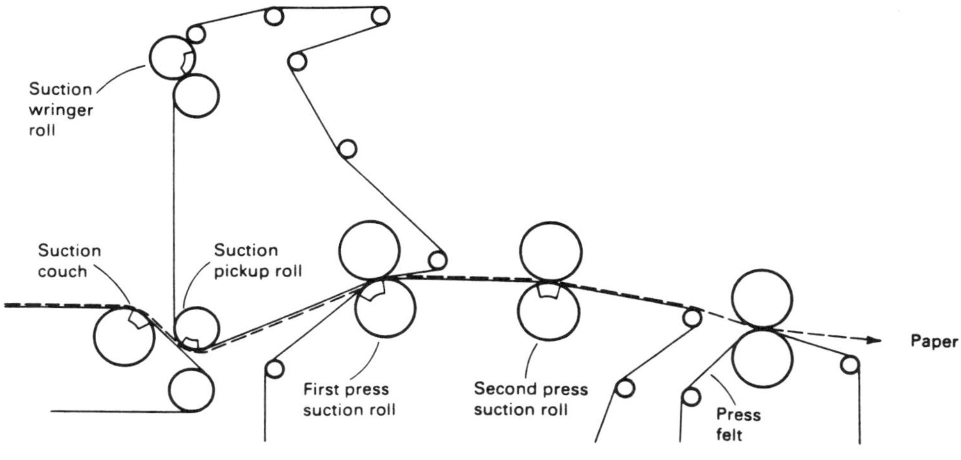

Fig. 764. Schematic diagram of wet end of paper machine showing types of suction rolls.

1. History of suction roll alloys

Prior to about 1950, suction shells were cast from bronze having a composition of approximately 85% Cu, 5% Pb, 5% Zn, and 5% Sn. Bronze has many advantages, such as castability, machinability, corrosion resistance, and relatively low cost. However, bronze lacked the strength needed to resist the higher press loads that were introduced at that time, and it lacked the stiffness necessary to maintain shape and roundness on larger and wider paper machines. In 1964, it was noted that suction roll nip pressures had increased from the original levels of 26 kN/m (148 lb/in.) up to 79 kN/m (451 lb/in.) [24]. Designs with nip pressures to 131 kN/m (750 lb/in.) were proposed in the Sixties and were actually achieved in the Seventies. Also, paper machine widths increased from typically 3.8 m (150 in.) in the Fifties to over 6.3 m (248 in.) in the Seventies, with maximum machine widths achieved in excess of 10 m (394 in.).

The first stainless steel materials to be widely used were forged type 410 martensitic alloy and centrifugally cast CF-3M and CF-8M austenitic alloys. The austenitic alloys were deemed especially desirable for low-stress, high-corrosion applications, such as the couch position.

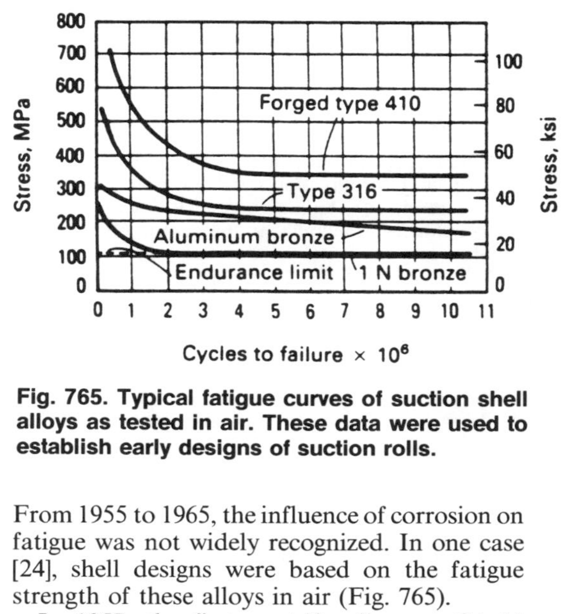

Fig. 765. Typical fatigue curves of suction shell alloys as tested in air. These data were used to establish early designs of suction rolls.

From 1955 to 1965, the influence of corrosion on fatigue was not widely recognized. In one case [24], shell designs were based on the fatigue strength of these alloys in air (Fig. 765).

In 1957, the first centrifugally cast CA-15 martensitic stainless steel shells were cast. Also during this early period, shells were cast from alloy A-63 stainless steel and aluminum bronze. Tables 82 and 83 list the nominal compositions of the alloys used for suction shells.

Table 82. Composition of copper-base suction shell alloys.

Material	Composition, %							
	Cu	Sn	Pb	Za	Al	Ni	Fe	Mn
1N bronze(a)	85.4	4.7	4.6	4.6	—	0.6	—	—
GC-CuSn5ZnPb(b)	84-86	4-6	4-6	4-6	—	—	—	—
GC-CuA19.5Ni(b)	81.5-83.5	—	—	—	9.0-9.5	4.0-4.5	3.0-3.5	0.5-1.0

(a) Typical composition. (b) Composition range

Table 83. Nominal composition of stainless steel suction shell alloys. Materials are categorized according to structure.

				Composition, %[a]			
Material	**C**	**Cr**	**Ni**	**Mo**	**Cu**	**Mn**	**Si**
Austenitic							
CF3M	0.02	17.7	13.8	2.3	—	1.3	0.8
CF8M[b]	0.05	17.7	13.8	2.3	—	1.3	0.8
PM-3-1811MN	0.015	16.5	13.5	2.1	—	1.6	0.5
Martensitic							
C-169 CA-15	0.07	12.4	0.6	0.5	—	0.5	0.6
PM-4-1300	0.12	12.5	0.4	0.4	—	0.7	0.5
DSS-69[c,d]	0.04	12.4	4.0	0.7	—	0.7	0.6
A-70[d]	0.03	11.9	4.0	1.5	—	0.8	0.5
Duplex							
A-63[d]	0.05	21.8	9.4	2.7	—	0.8	1.3
A-170[d]	0.07	23.3	10.7	2.1	—	0.7	1.5
A-171	0.07	22.2	8.3	1.2	—	0.8	1.1
A-271	0.06	24.6	4.3	0.7	—	0.7	1.3
3RE60[e]	0.02	18.5	4.7	2.8	—	1.5	1.7
A-75	0.02	26.0	6.8	—	—	0.8	0.5
A-86	0.02	26.0	6.0	—	2.8	0.8	0.7
VK-A378[f] A682[g]	0.06	20.0	5.0	2.0	4.5	0.6	0.8
ACX-100[h]	0.02	24.0	5.7	2.4	0.5	0.7	0.7
PM-3-1804M	0.06	17.9	4.0	2.0	—	0.6	0.6
PM-2-2505	0.07	26.0	4.0	0.8	—	1.2	1.3

(a) Remainder of compositions is iron. (b) Alloy has been discontinued for suction shells in North America. (c) Details of composition not published, but similar to ASTM A-296, alloy CA-6NM (discontinued specification). (d) Obsolete alloy. (e) Contains nitrogen. (f) Contains nitrogen and tungsten. (g) Alloy A-682 has a composition similar to that of VK-A378, only reduced chromium and nickel. (h) Contains nitrogen and cobalt.

Between 1969 and 1977, a host of new centrifugally cast stainless steel alloys was introduced for suction shells. These were DSS69, C169, alloy 70 (martensitic), and alloys A170, A171, A271, and A-75 (duplex/ferritic-austenitic). Also during this period, rolled and welded stainless steel shells were produced in Sweden from type 316 and 3RE60 stainless steels. Continuous-cast bronzes GC-CuSn5ZnPb and GC-CuA19.5Ni also emerged during this period.

In 1971, the reduced fatigue strength of bronzes and stainless steels in synthesized white water was reported [25]. The importance of residual stresses on fatigue cracking was also more widely recognized at this time.

Between 1978 and 1989, additional new stainless alloys were introduced. These included centrifugally cast precipitation-hardening duplex alloys VK-A378 and KCR-A682 and duplex alloys A-86 and ACX-100. In 1978, the argon-oxygen decarburization process was introduced in the United States for suction shells. This permitted more economical melting and casting of stainless alloys with a very low carbon content. Centrifugally cast CF-3M has since been produced instead of CF-8M. Several forged alloys were also introduced during this period: PM-4-1300 and PM-4-1300M (martensitic), PM-3-1811MN (austenitic), PM-3-1804M, and PM-2-2505 (duplex).

During the Eighties, emphasis was placed on stainless steel development with greater corrosion and fatigue resistance and low residual stresses.

2. Paper machine environment

A detailed discussion of the paper machine corrosion environment is given in Sec. A above. However, considerations specific to suction roll corrosion will be covered in this discussion.

Suction rolls are subjected to a variety of corrosive environments. The severity of the environment depends primarily on the type of paper produced on the machine, the source of water, and the degree of closure of the mill. The corrosivity of the environment depends primarily on pH, temperature, dissolved solids content, chlorides, and the presence of sulfur compounds. The area around the suction rolls is subjected to splash from stock and white water, and deposits of paper fiber are present in crevices. Bacteriological growth can also occur, especially in areas that are difficult to treat with biocide chemicals. Suction rolls are also subject to corrosion from

Fig. 766. Corrosion damage to a CA-15 stainless steel shell caused by muriatic acid felt cleaning.

chemicals and waters used in showers around the rolls and from felt treatment chemicals.

Laboratory studies have shown that $Na_2S_2O_3$ in stock and white water systems results in severe pitting corrosion of the less resistant martensitic stainless steels, and thiosulfate is believed to be important in corrosion and cracking of suction roll shells. Sodium thiosulfate, residual sodium hydrosulfite ($Na_2S_2O_4$), and the sulfide ion are also highly corrosive to bronze.

Biological corrosion has also been observed within the drilled holes of suction shells. The hole area is highly favorable to the growth of microbes because of the presence of dissolved organic materials, dissolved inorganic salts, and a favorable temperature range of 40 to 50°C (105 to 120°F). Anaerobic sulfate-reducing bacteria thrive in crevice locations and under salt and fiber deposits. Therefore, because of the concentration of ions and the localized pH reduction, the chemical environment beneath deposits can become far more aggressive than the bulk stock or white water chemistry.

The showers and chemicals used on the paper machine also influence the environment around suction rolls. Showers are used to clean machine felts, to remove deposits from drilled holes in suction rolls, and to lubricate roll seal strips. A wide variety of shower cleaning chemicals are used for felts and rolls, including acids, aliphatic and aromatic solvents, or animal and vegetable oils. Lubricating and needle showers are often supplied with white water at temperature ranges of 40 to 50°C (105 to 120°F); this white water is more corrosive to the suction roll than freshwater. Attack from chemicals has also contributed to failure of elastomeric covers on suction rolls [26]. Figure 766 shows damage to a CA-15 martensitic stainless steel suction shell from muriatic acid used for felt cleaning.

Off-line chemical cleaning has been used to remove deposits from drilled holes in suction rolls. There have been numerous cases of severe damage to both shells and rubber covers when corrosive cleaning chemicals were used [27].

3. Corrosion fatigue testing

In the Sixties, several investigators recognized the need to obtain corrosion fatigue data for suction shell materials using a synthesized paper machine white water environment. Corrosion fatigue testing of suction shell materials was first described by R. Thompson, who used a conventional R.R. Moore rotating-beam machine equipped with a corrosion chamber around the specimen [28]. Liquid was continuously dropped onto the stressed portion of the specimen. Corrosion fatigue tests were also conducted using a reverse bending plate with a single hole drilled in the stressed regions [29]. Crack growth rate tests were employed to evaluate the corrosion fatigue characteristics of suction roll materials [30, 31]. These investigators used compact-tension specimens subjected to cyclic loading in an electrohydraulic fatigue machine.

Corrosion fatigue test data have been published largely by shell producers and paper machine equipment manufacturers. A variety of test data have been shown for all of the alloys used for suction rolls. Because different test methods and a variety of environments have been used for tests, direct comparison of corrosion fatigue strength data is very difficult.

Rotating-bending fatigue test data for several alloys in white water at pH 3.5 and varying chloride and sulfate contents are shown in Fig. 767. Rotating-bending fatigue test data by a different investigator for alloys in a highly aggressive white water are shown in Fig. 768.

In 1985, the TAPPI Corrosion and Materials Engineering Committee summarized survey results on the correlation among corrosion fatigue testing methods. The survey included data provided by ten investigators on the type of testing used, specimen configuration, test conditions, and environment [32]. The TAPPI committee also published documents describing standard methods and test environments for suction shell alloys [33, 34].

Investigations in near-threshold fatigue crack growth testing have indicated that mean stresses could have a significant effect on the threshold for crack growth in duplex stainless steel alloys [31]. These data correlate with the concept that high residual tensile stresses in suction rolls decrease fatigue life in service.

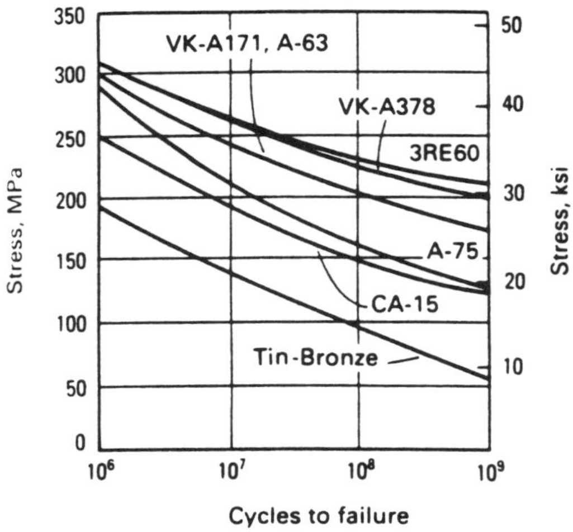

Fig. 767. Fatigue curves in white water based on the data of material suppliers. 1500 to 1750 rpm rotating-bending test, pH = 3.5, Cl^- = 20–400 ppm, SO_4^{2-} = 250–1000 ppm.

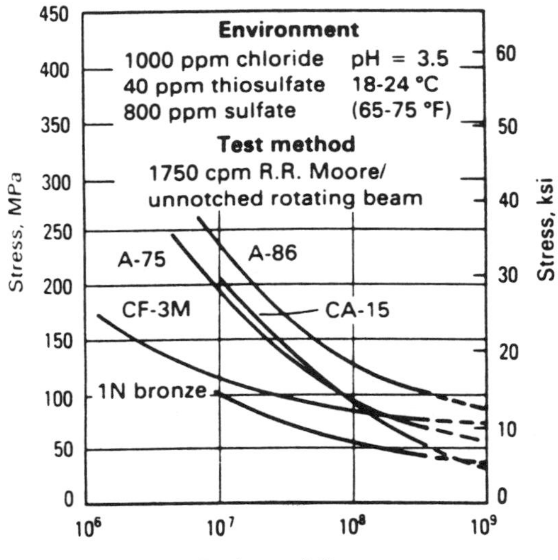

Fig. 768. Fatigue curves for suction shell alloys in an aggressive environment. (Courtesy of Sandusky Foundry & Machine Company)

4. Roll failures

Corrosion and cracking failures have been reported for couch, press, and felt wringer rolls. There has been no widespread problem with the lightly loaded suction pickup rolls. Failure by cracking is frequently unexpected, and is usually the most cause for concern. When cracks progress a significant distance around the shell, the suction roll begins to vibrate and must be

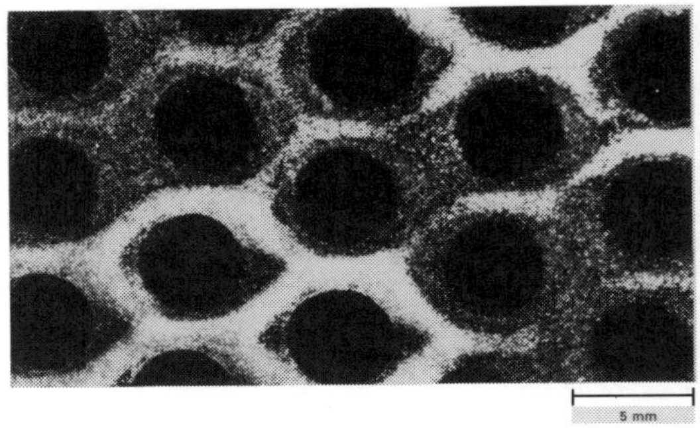

Fig. 769. Thiosulfate corrosion created surface Roughening (dark area) on this uncovered bronze roll.

immediately removed from the paper machine. Deterioration of shells by general corrosion and pitting is another important means of failure. Severe roughening of the outside diameter surface of bronze couch rolls is known to reduce the life of fourdrinier wires. An example of outside diameter surface corrosion in bronze is shown in Fig. 769. General corrosion in drilled holes results in hole enlargement, decreasing the land area between holes to a point at which corrosion fatigue cracking can occur (Fig. 770).

Most suction rolls that are reported to have failed have been removed from service because of cracking. Cracked shells of bronze and martensitic stainless alloys almost always exhibit a visible condition of surface corrosion or pitting in the drilled holes. Higher alloyed duplex and austenitic stainless steels seldom show visible corrosion, although fatigue failures in these alloys are known to be corrosion assisted.

(a) Failure experiences

Two separate surveys covering roll failures have been published since 1979 [27, 35]. The surveys indicated that most failures occurred in bronze and martensitic stainless steel shells. However, it was also noted that there are far more bronze and CA-15 shells in service as compared to other alloys. Cracking was also more prevalent in the higher speed and more corrosive newsprint machines than in fine paper, pulp, or board machines.

(b) Corrosion fatigue failures

Fatigue cracking is first observed in shells as single cracks in the land area between two drilled holes or as cracks that begin at a hole and have not yet progressed to a neighboring hole. Cracking then proceeds across the land regions connecting several holes, eventually joining and

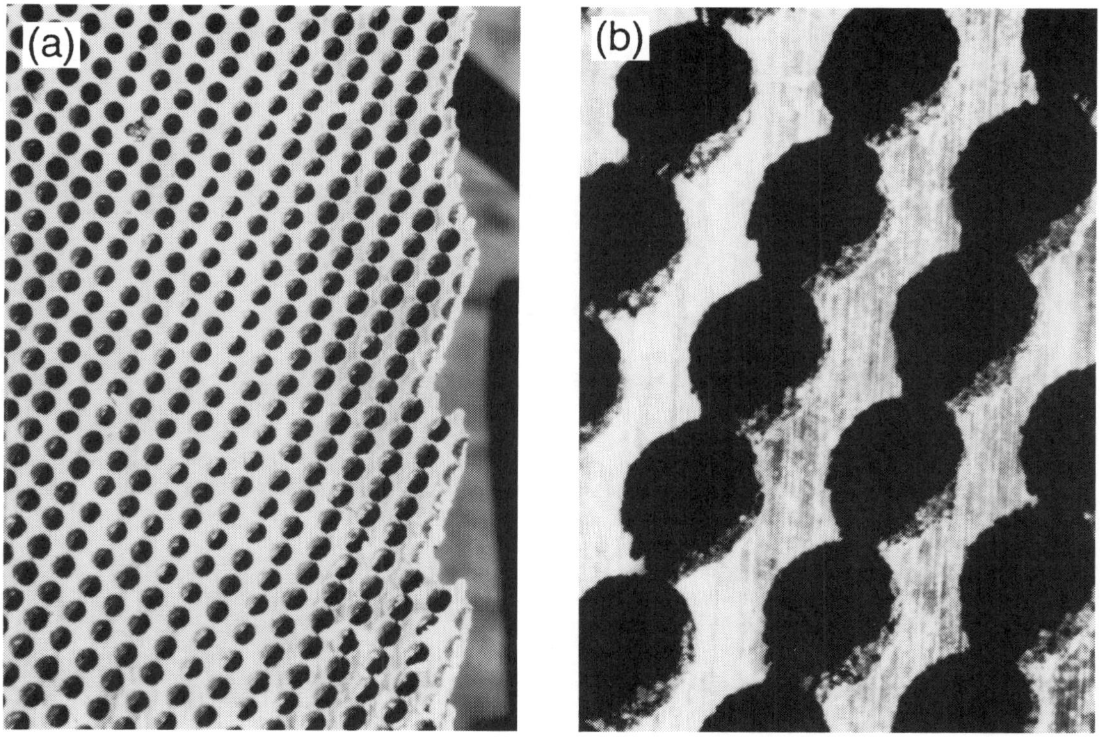

Fig. 770. General corrosion in drilled holes. (a) Inside surface of bronze shell adjacent to fracture. The hole enlargement varied periodically along the length of the shell and corresponded to shower nozzle spacing. (b) Close-up view of hole corrosion. (Courtesy of MacMillan-Bloedel, Ltd.)

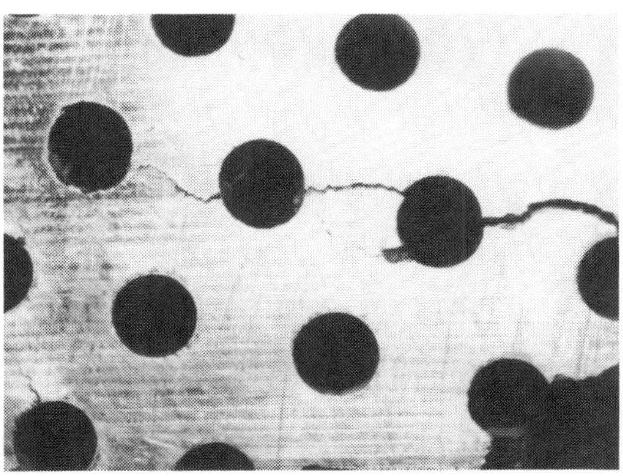

Fig. 771. Typical cracking between drilled holes on the outside surface in a stainless steel alloy shell. (Courtesy of The Institute of Paper Science and Technology)

proceeding circumferentially around the shell (Fig. 771). Cracking usually manifests on the roll inside diameter surface near the mid-span, where operating stresses are the highest. Martensitic stainless steels show many fatigue initiation sites through the cross-section of the fractured surface (Fig. 772). Bronze and duplex stainless steels seldom show well-defined fatigue initiation sites.

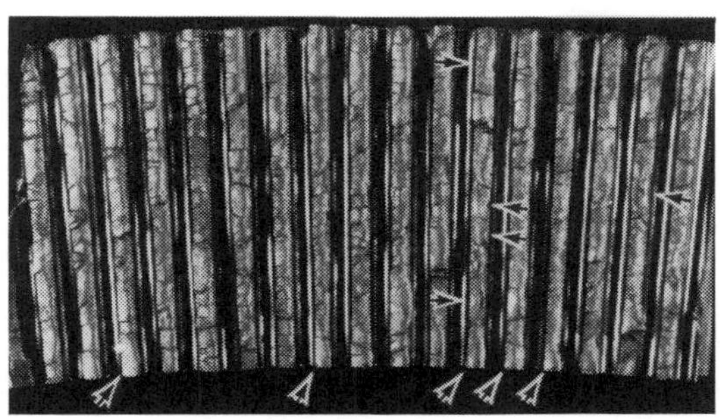

Fig. 772. Fracture surface of CA-15 stainless steel suction shell showing several fatigue nucleation sites (arrows). See also Fig. 773.

Longitudinal cracking is often observed near the mid-span of austenitic and duplex stainless steel shells, and this has been attributed to overloading or residual stresses from heat treatment. Longitudinal cracks eventually turn

Fig. 773. Photomicrograph of specimen in Fig. 772 showing a fatigue crack initiating at the base of a corrosion pit. Etched with Marbles reagent. ×75.

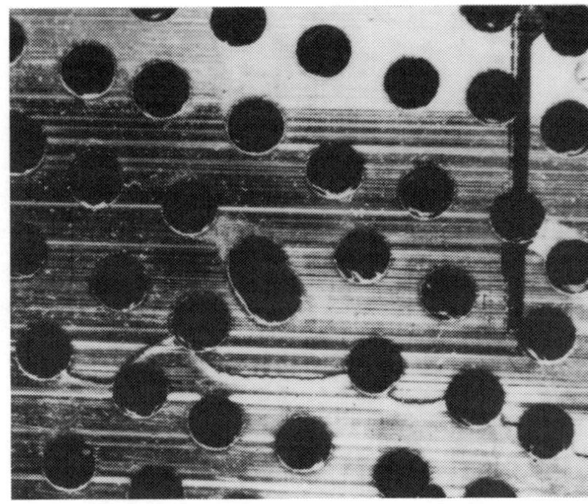

Fig. 774. Fatigue cracking of CA-15 stainless steel suction roll due to poor drilling quality.

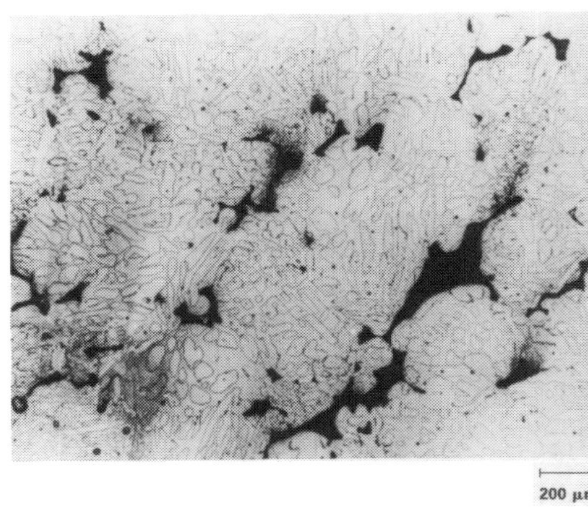

Fig. 775. Casting porosity in duplex stainless steel ball.

to the circumferential direction and proceed around the roll to ultimate failure.

Rolls may develop only a single crack that progresses to failure, but more often, they will develop tens or hundreds of cracks before final failure occurs. Bronze and martensitic steel shells usually exhibit many small cracks before a larger crack develops. Duplex stainless steels develop fewer cracks or a single large crack.

Microexamination of the cracked region frequently shows cracking associated with corrosion pitting (Fig. 773). Cracking has also been attributed to manufacturing defects, poor drilling quality (Fig. 774), excessive porosity (Fig. 775), welding repairs, and electrical discharge machining used to remove broken drill bits. Intergranular cracking has been observed in austenitic stainless steel that was sensitized during heat treatment and in bronze shells that were exposed to mercury and ammonium contamination.

(c) Rubber cover failures

A 1985 Canadian survey showed that the cost of premature rubber cover failures on suction rolls was about as large as the cost of shell failures [35]. In the survey, 38 mills were questioned about rubber cover failures on suction and nonsuction rolls for the year 1983. A total of 70 cover failures were experienced in the 38 mills surveyed.

A significant cause of rubber cover failures is deterioration from chemicals used on the suction rolls, both on-line and off-line. Organic solvents have been found to be highly detrimental to rubber covers, especially when used in a continuous mode as opposed to batch treatment. High-pressure hydraulic cleaning can produce mechanical damage and detachment of rubber covers.

5. Shell manufacturing methods

Most suction roll shells are produced by the centrifugal casting method, followed by continuous-casting (bronze), rolled and welded (stainless steel), and forging processes (stainless steel). Each method of manufacturing claims certain advantages. After production of the rough shell, it may be heat treated (depending on the alloy), machined to near-final shell dimensions, and drilled. The shell is then finished on the inside diameter surface and head seats machined. The outside diameter surface is finished by turning, belt sanding, or grinding. A smooth outside surface finish is not required if the shell is to be subsequently rubber covered.

(a) Castings

The process of centrifugal casting is the oldest and most versatile method of shell production. It is applicable for both bronze and stainless steel alloys. Metal is poured into a spinning mold that is horizontally positioned, and the spinning motion is continued until the metal shell has completely solidified. In this process, the last metal to freeze is at the inside diameter surface of the casting. Non-metallic and intermetallic impurities concentrate on this inner surface, which is later machined away.

Continuous casting is a relatively new method of producing cast shells. It is applicable only to bronze alloys. In this method, the molten metal is poured through holes in a distributing vessel into a vertical collar-type mold. Metal flow, speed of extraction from the mold, and cooling are coordinated so that the casting process is continuous. An advantage of continuous casting is that the uniformity of metal properties through the wall can be readily controlled.

(b) Weldments

Welded stainless steel shells have been produced in Sweden since the early Seventies. Most production has been limited to 3RE60 duplex stainless steel alloy, which is hot formed from plate into semicylinders. The semicylinders are subsequently welded longitudinally to form short cylinders, and several short cylinders are welded at the ends to form the suction shell.

(c) Forgings

Forged suction shells are produced by first upsetting and punching a hollow cylinder from a solid cylinder, then enlarging and forging the shell on a pipe-forming mandrel. The process minimizes the porosity and segregation that may be found in castings, but can be subject to inclusions and other internal flaws.

(d) Heat treatment

Heat treatment of suction shells has proved to be a critical operation affecting the corrosion and the residual stress properties of shells. Current practice in heat treatment is limited to stainless steel suction shell alloys. Although aluminum bronze is heat treatable, shells from this material are currently furnished in the as-cast condition.

Each stainless steel alloy requires a specific heat treatment to provide the best combination of corrosion resistance and low residual stresses. Slow cooling from the final heat treatment produces lower residual stresses, but (depending on the alloy) may produce undesirable intermetallic phases and a low corrosion fatigue resistance [36, 37]. Austenitic alloys may be given a solution anneal treatment, but martensitic alloys are quenched and tempered. Duplex alloys may be subjected to a solution anneal treatment or, in the case of copper-bearing duplex alloys, may be given one or two additional subcritical aging treatments.

(e) Drilling

Suction shells are drilled using multiple-spindle machines. Couch rolls are frequently drilled with larger hole sizes of up to 6.3 mm (0.25 in.), but press roll drilling is typically 4.0 mm (0.16 in.). The drilling process is critical in that hole spacing must be maintained to within close dimensional tolerances.

Gun drilling was first developed in Europe for drilling stainless steels and is particularly applicable when drilling long, straight holes in thick-wall stainless steel shells. Twist drilling of these alloys sometimes results in converging of drilled holes at the inside diameter surface (Fig. 774). Gun drilling also has the advantage of producing a smooth surface finish within the drilled hole. This reduces the tendency toward the initiation of corrosion and deposit formation. Gun-drilled holes have been measured to have a surface finish of $R_a = 0.5$ μm (20 μin.) as compared to $R_a = 5.0$ μm (200 μin.) for twist-drilled holes, where R_a is the arithmetic average roughness. A process of reaming after twist drilling has also been used to produce a good surface finish in the drilled holes.

(f) Shell quality

Several types of manufacturing defects have been attributed to the failure of suction shells, as discussed above. In an analysis of 52 suction roll failures, about 50% of the failures were attributed to defects in manufacture or design, 35% were due to improper material selection for the environment, and 15% could not be attributed to any cause [38].

6. Corrective measures

Three factors influence suction shell resistance to corrosion fatigue: stress, corrosion environment, and the corrosion fatigue strength of the alloy. A favorable change in any of these factors will produce an increase in shell life.

Bronze shells are occasionally removed from service because of general or pitting corrosion. The factors that influence these failures are the corrosion environment and the resistance of the alloy to the environment.

Good judgment in roll design, material selection, and control of the paper machine environment all contribute to the long life of suction rolls. Good quality control during the manufacture of the shell and timely in-service cleaning and roll inspection also contribute to a favorable service life.

(a) Operating stresses

Both the applied stress and internal residual stresses inherent in the shell material contribute to the stress in operation of the suction roll. The responsibility for the level of applied stress rests chiefly with the roll designer who selects the limit of design stress for the material selected. Each shell material will have a limit of stress that will give long life in a particular environment. The magnitude of applied stress is a function of roll configuration and applied loads. Drilled hole patterns must also be favorably designed to provide good distribution of the applied stress. The most important factor influencing the stress in the roll is shell thickness; of course, stress decreases as the shell thickness increases. A stress calculation procedure has been developed by TAPPI that can be used as a guide for determining stresses in suction rolls [39].

The residual stress that is inherent in the alloy and the heat treatment are also important factors contributing to roll life. Shell materials with known high residual stresses should be designed more conservatively or, preferably, not used at all.

Gun-drilled or reamed holes for improved surface finish will reduce corrosion and improve the cleanliness of drilled holes in service. Gun drilling or reaming is usually not performed on bronze shells, but would be helpful in the maintenance of the shell in service.

(b) Corrosion environment

The papermaker often has little control of some factors influencing the corrosion environment, yet there are many conditions that are within his control. For example, thiosulfate-ion contamination and residual hydrosulfite ($NaHSO_2$) in newsprint paper machine white water can occur in $NaHSO_2$ brightening; practices for reducing this problem are addressed in Ref. [35]. Freshwater showers will provide more effective cleaning and reduce corrosive effects. Good maintenance of the roll is also an important factor in minimizing corrosion. Drilled holes must be kept clean of deposits by the use of needle showers and periodic off-machine cleaning. A careful program of biological control should be maintained. The papermaker should also avoid the use of cleaning chemicals (such as muriatic acid) that are known to be damaging to shell alloys and roll covers. The use of a corrosion inhibitor such as sodium mercaptobenzotriazole should be considered to reduce the corrosion of bronze rolls. The use of inhibitors for couch rolls is described in Ref. [8]. An example of severe pickup roll corrosion due to thiosulfate that was corrected with inhibitors is discussed in Ref. [40].

(c) Material selection

When a new or replacement roll is contemplated, the question of material selection is of primary importance. Past operating experience with existing suction rolls should be used as a guide in selecting the new shell material. Reference [41] describes conditions that contribute to roll failure and will aid in new shell material selection.

Duplex stainless steel has the highest resistance to fatigue cracking, followed by martensitic stainless steel and bronze. Austenitic stainless steel has good corrosion resistance to most environments, but has relatively low fatigue resistance and moderate-to-high residual stresses [36, 37]. Duplex stainless steels have the highest fatigue strength and generally have moderate-to-low residual stresses; bronze has low residual stress, as does CA-15 martensitic stainless steel.

Selection of a new alloy should involve an evaluation of the condition of the existing roll and the factors contributing to its deterioration. The changes in the corrosion environment that have occurred on the paper machine should also be evaluated. One is tempted to select an identical alloy if the previous material served for 15 or 20 years. However, it is more likely that changes in the machine environment have occurred at a recent time and that the life of a new roll of the same alloy would be substantially reduced. This factor accounts for the recent popularity of the stronger corrosion-resistant duplex stainless steel alloys.

Suction shell material cost is also a consideration. In North America and Europe, bronze is the lowest cost shell alloy, followed by martensitic stainless steel, austenitic stainless, and the duplex alloys. In Japan, bronze is seldom used because of availability and cost.

(d) Manufacturing quality

There is little published information relating to suction shell manufacturing quality. Shells should be inspected using liquid penetrant prior to drilling to detect such flaws as porosity, slag, and cracks. Skillful control of melting and casting procedures can minimize these defects. Weld repairs should be used with discretion, and only before shell heat treatment. Type 1N bronze should not be weld repaired, because of hot shortness, which develops in the alloy. Careful attention should be given to drilling and machining quality and dimensional control. Industry-wide standards giving acceptance criteria for casting flaws and drilling quality have not yet been developed.

(e) In-service inspection

Periodic inspection of the suction roll is a critical step that will help to extend its life and minimize unexpected shutdowns. Rolls should be removed from the machine, disassembled, and thoroughly cleaned and inspected at least annually. Inspection should include an evaluation of drilled hole cleanliness, fatigue cracking of the shell, and rubber cover deterioration or detachment. Water-washable liquid penetrant has been successfully used to detect cracks on the inside and outside surfaces. A careful visual examination should also be made for corrosion and cracks.

When cracking is observed during an annual inspection, immediate consideration must be given to replacement because of the long lead time required for shell manufacture. Consideration should also be given to inspecting the roll more often than once per year to observe the rate of crack growth. Thus, a careful record must be made whenever cracks are observed.

REFERENCES

1. Muhonen, J.M., Corrosion of Stainless Steels in Whitewater, Pulp and Paper Industry Corrosion Problems, NACE, 1:75 (1974).
2. Kurkela, M., Suutala, N. and Kemppainen, J., On the Selection of Stainless Steels in Bleach Plants and White Water Systems, Pulp and Paper Industry Corrosion Problems, Can. Pulp and Paper Assoc., 5:127 (1986).
3. Rushton, J.D. and Kelly, P.A., White Water Corrosion Data from an Integrated Newsprint Operation, TAPPI Engineering Conf., p. 41 (1984).
4. Piluso, A.J. and Nathan, C.C., Chemical Treatment to Control Corrosion in the Wet-End Operations of Pulp and Paper Mills, Pulp and Paper Industry Corrosion Problems, NACE, 1:14 (1974).
5. Mueller, W.A. and Muhonen, J.M., Pitting Corrosion of Stainless Steels in Six Paper Machine Headboxes: Mechanism and Prevention, Tappi J. 55(4):589 (1972).
6. Ewald, J.L. and Hundley, D.P., Surface Finish Requirements of the Headbox, Tappi J. 63(11):121 (1980).
7. Gerhauser, J.P., Corrosion of Fourdrinier Wires, Tappi J. 43(4):207A (1960).
8. Nathan, C.C. and Piluso, A.J., Wet End Corrosion Problems in Paper Mills, Pulp and Paper Industry Corrosion Problems, NACE, 2:126 (1977).
9. Danielsson, E., Corrosion and Corrosion Control in the Paper Machine Wet End, Pulp and Paper Industry Corrosion Problems, NACE, 3:191 (1980).
10. Bennett, D.C. and Magar, I.J., Protective Coatings for Wet Ends of Paper Machines, Pulp and Paper Industry Corrosion Problems, Finnish Pulp and Paper Res. Inst., 6:348 (1988).
11. Thorpe, P.H., Corrosion in Paper Machines — An Overview, Pulp and Paper Industry Corrosion Problems, NACE, 3:184 (1980).
12. Blakey, B.M. and Thorpe, P.H., Failure and Redesign of Press Roll Ends in Paperboard Machines, Appita 33:45 (1979).
13. Sund, G. and Strom, S., The Consequences of System Closure for Corrosion in Swedish Pulp and Paper Mills, Pulp and Paper Industry Corrosion Problems, Can. Pulp and Paper Assoc., 5:51 (1986).
14. Bowers, D.F., Changes in Water Properties and Corrosivity with Closure, Tappi J. 66(9):103 (1983).
15. Wensley, D.A., Localized Corrosion Resistance of Stainless Steels in White Waters, Materials Performance 28(11):68 (1989).

16. Bennett, D.C. and Federowicz, C.J., Prediction of Localized Corrosion of Stainless Steels in White Water, Materials Performance 21(4):39 (1982).
17. Newman, R.C., Pitting of Stainless Alloys in Sulfate Solutions Containing Thiosulfate Ions, Corrosion 41(8):450 (1985).
18. Barton, R.W., Thiosulfate Generated by Excess Application of Sodium Hydrosulfite, Pulp Paper 59(6):108 (1985).
19. Garner, A., Thiosulfate Corrosion in Paper Machine White Water, Corrosion 41(10):587 (1985).
20. Newman, R.C., Wong, W.P., Ezuber, H. and Garner, A., Pitting of Stainless Steels by Thiosulfate Ions, Corrosion 45(4):282 (1989).
21. Butterfield, W.S. and Glogowski, P.E., Determination and Corrosivity of Thiosulfate Ion in Paper Mill Systems, Proc. TAPPI Engineering Conf., p. 61 (1984).
22. Gorelov, V.V. and Talybly, A.K., Corrosion Protection of the Equipment in Closed and Reduced Water Recycling Systems, Pulp and Paper Industry Corrosion Problems, Swedish Corrosion Inst., 4:113 (1983).
23. Thorpe, P.H., Microbiological Corrosion of Stainless Steel in Paper Machines and its Causes, Pulp and Paper Industry Corrosion Problems, Can. Pulp and Paper Assoc., 5:169 (1986).
24. Stirling, J.B., Modern Suction Roll Design and Construction, Proc. Technical Section, CPPA, p. D74-D79 (1964).
25. Vadas, R.M., The Suction Roll — Its Problems and its Future, Pulp Paper Can. 72(11):66–70 (1971).
26. McNamee, J.P., Influence of Suction Roll Covers on Shell Corrosion, Pulp and Paper Industry Corrosion Problems, NACE 1:94–98 (1974).
27. Wooster, F.S., Moskal, M.D. and McNamee, J.P., Tappi J. 62(9):71–74 (1979).
28. Thompson, R., The Suction Roll — Its Problems and Its Future, Part II, Pulp Paper Can. 72(11):70–76 (1971).
29. Dahl, C.B. and Neubauer, E.P., Modern Practices Can Present Hazards to Stainless Steel Suction Rolls, Paper Trade J. 155(34):46, 48, 50 (1971).
30. Kelley, C., Vestola, J., Sailas, V. and Pelloux, R., Corrosion-Fatigue Crack Propagation of High-Strength Stainless Steels Used in Suction Rolls, Tappi J. 58(11):80–85 (1975).
31. Yeske, R.A., Prediction of Suction Roll Performance from Laboratory Testing, Corrosion '86 Conf., NACE, Paper 147 (1986).
32. Bowers, D.F. and Butterfield, W.S., Corrosion Fatigue Data For Paper Machine Suction Roll Alloys — Its Significance, Proc. TAPPI Engineering Conf. (1986).
33. Plate Bar Bending Fatigue Testing in a Corrosive Environment, TAPPI TIS 0402-08 (1984).
34. Rotating Bar Bending Fatigue Testing in a Corrosive Environment, TAPPI TIS 0402-09 (1984).
35. Garner, A., Suction Roll Failures in Canada, Pulp Paper Can. 86(3):81–89 (1985).
36. Moskal, M.D. and Eggeman, G.W., The Effect of Residual Stresses on Corrosion Fatigue of Suction Rolls, Pulp and Paper Industry Corrosion Problems, NACE, 2:94–99 (1977).
37. Dahl, C.B. and Rainger, C.W., Stainless Steel Suction Roll Performance Design and Materials, Pulp and Paper Industry Corrosion Problems, NACE, 2:105–110 (1977).
38. Sailas, V., Critical Components of the Paper Machine: Damage Caused by Metallic Corrosion, Pulp and Paper Industry Corrosion Problems, NACE, 2:100–104 (1977).
39. Guide for Evaluation of Paper Machine Suction Roll Shells, TAPPI TIS 0402-10 (1985).
40. Butterfield, W.S., Private Communication (1986).
41. Suction Roll Shell Service Report, TAPPI TIS 014-56 (1982).

XXIV PAPER MACHINE AND ROLL VIBRATION

J.A. CHARLES

Dr. John A. Charles earned a Bachelor's Degree in Aerospace Engineering from the University of Texas in 1970, and a Master of Science degree in Mechanical Engineering from the University of Oklahoma in 1972. He earned his Ph.D. in 1976, with thesis work in the area of material fatigue failure prediction. Prior to joining Crown Zellerbach (later part of James River), Dr. Charles taught Mechanical Engineering for five years at both Montana State University and the Ohio State University. He helped develop the Advanced Design Method Laboratory for computer-aided design at Ohio State. In 1981, Dr. Charles joined the Central Research Division of Crown Zellerbach, with responsibilities in the areas of CAD and vibration analysis, and in 1987 he became Manager of the James River Technical Services Paper Machine Analysis Group. Dr. Charles is a member of TAPPI and ASME. He has published over 20 articles covering such diverse topics as radiant heat transfer, material fatigue, computer-aided engineering and teaching techniques.

A. INTRODUCTION

Vibration is a phenomenon with which papermakers must constantly contend during the normal course of papermaking operations. Vibration is always present to some degree in any mechanical or hydraulic system as a result of imperfections in equipment and abnormalities in operating conditions. In the pulp and paper industry, vibration has been both an enemy and a friend of the papermaker for years. Vibration has been the papermaker's enemy for the mechanical failures it induces and because of vibration-related degradations in paper quality which can result from running papermaking machinery in poor condition. On the other hand, when used properly vibration has proven to be an effective tool for improving paper properties, particularly formation. In addition, vibration can be highly useful as a diagnostic tool for machinery condition monitoring.

There are numerous reasons why vibration should be of interest to the conscientious papermaker. First and foremost, the papermaker wishes to maintain the highest possible level of operating efficiency. To do this requires maximizing mechanical reliability and efficiency, which is not possible without an understanding of vibration and its control. To complicate the issue, many paper machines today are running at speeds significantly above their design speeds, regardless of their construction date. This condition is likely to aggravate the vibrational behavior of the equipment and make mechanical failures even more likely. Often, it is the vibrational behavior of the paper machine which limits the production speed of such equipment. As speeds increase, vibration amplitudes may escalate to the point that the machinery becomes unsafe for normal operation. The remedies for such high vibration situations are usually very expensive and often require major rebuilding of machine sections to allow the machine to achieve production goals.

Unwanted vibration in the paper machine can result in several other potential problems for the papermaker. Among these are reduced life

---

REVIEWED BY

Michael A. Smetana
Maintenance Superintendent
Potlatch Corporation
Lewiston, ID

John Yolton
President
Safematic Inc.
Kennesaw, GA

expectancy of equipment and clothing, production limitations, and increased unscheduled downtime to repair vibration-induced problems. The reduction of equipment and clothing life can be costly from both the capital and operating expense standpoints. Rolls for paper machines are costly to manufacture and maintain, and excessive vibration can lead to premature degradation of roll quality and early bearing failures. Clothing, particularly in the press section, is expensive and critical to optimum operation; felts can quickly be ruined by vibration in the nips of the press.

Production limits may arise when rolls are unbalanced and induce high levels of vibration. In such cases, the paper machine speed must often be reduced until a repair can be affected. Although such action can cause a severe reduction in machine productivity and is therefore undesirable, neglecting the condition can lead to very serious (sometimes catastrophic) machinery failures such as broken journals and roll shells. Often, the correction of problems before failure must occur during unscheduled machine downtime. In the event that a problem goes uncorrected and a failure results, another unscheduled down is necessary. These times are undesirable since they are usually not as well planned or efficient as scheduled downs and thus are detrimental to machine productivity.

1. Advantages of vibration

While the effects of vibration discussed above are undesirable, there are some ways the papermaker can use vibrations advantageously. For example, it is not uncommon to induce a controlled amount of vibration into the wet end of the paper machine in order to improve forming conditions and produce a higher quality sheet than would be possible otherwise. One technique for doing this is the well-proven method of fourdrinier "shake", or oscillating the breast roll along its axis at a controlled frequency and stroke. The goal of this action is to produce better formation in the sheet by preventing premature flocculation. A more recent development involving vibration for paper quality improvement is the "resonating wire" system, which induces vibration in the wire at a high frequency and thus produces micro-turbulence in the stock.

Another potential benefit derived from vibration in the paper machine is the use of vibration for "predictive" maintenance. Because vibration is always present in machinery, it can be readily measured at any number of points on the paper machine. Regular checking of vibration levels (or "trending") at these points can be used to determine when and where vibration levels are increasing. These increases can usually be related to a deterioration of some component in the system, possibly a bearing or a wet felt. It is then possible to plan for a part replacement or adjustment during a scheduled machine down, rather than being forced to make unexpected repairs during unscheduled shutdowns due to failure. Hence, the highest possible machine productivity is maintained. With experience, it is possible through vibration monitoring and the analysis of the obtained data to eliminate virtually all unscheduled downtime due to mechanical problems around the paper machine.

2. Measurement of vibration

Several levels of sophistication of vibration measurement exist, each particularly suited to one purpose or another. The simplest vibration meter, which yields only an overall level of vibration at a measurement point, can be a very useful tool for trending overall vibration levels and planning for the most efficient use of machine downtime. A more sophisticated instrument is the real-time spectrum analyzer, which decomposes the vibration signal into its various frequency components. The resulting vibration "spectrum" can be used to pinpoint the source of the high vibration and further reduce downtime through even more accurate planning. Finally, vibration measurement transducers can be mounted permanently at various locations on the machine and wired to a dedicated computer for data acquisition and analysis. This system provides for continuous monitoring of critical points and can offer the built-in diagnostic capabilities of the spectrum analyzer.

In order to best utilize mechanical vibrations and to avoid the possible pitfalls of misunderstanding the significance of vibrations on the paper machine, it is necessary to first obtain a fundamental awareness of vibration theory and terminology. This background understanding can then be applied in subsequent discussions of specific areas of concern about vibration around the paper machine.

B. FUNDAMENTALS OF VIBRATION THEORY

Vibration is the oscillatory motion which results from forces applied to mechanical systems. Such vibrational motion displays characteristics which can be predicted, measured and analyzed. Vibration analysis is the study of this vibrational motion, and has as one of its goals the understanding of the causes and consequences of

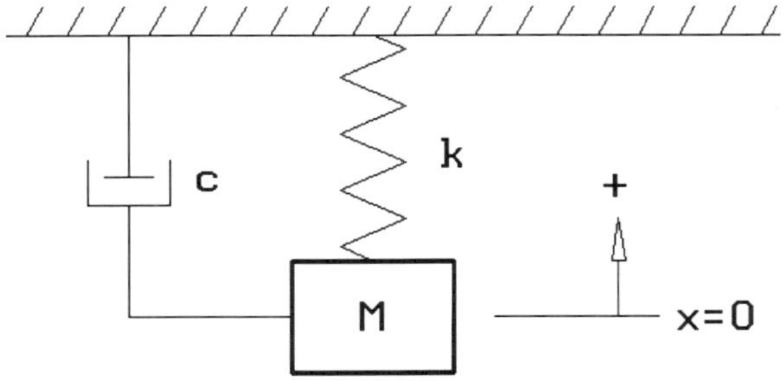

Fig. 776. Spring/mass/damper system schematic.

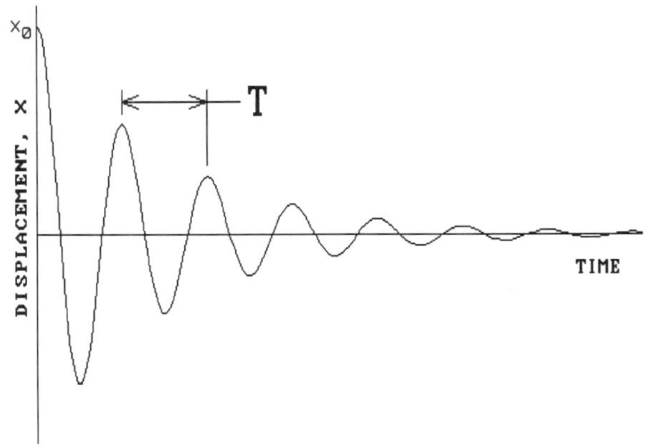

Fig. 777. Displacement of spring/mass/damper system.

vibrations. In the paper industry, vibration analysis has found application in efforts to understand machinery problems and maintain high levels of mill productivity. A basic understanding of vibration theory is useful if one is to effectively apply vibration analysis to mill problems. The theory which follows is intended to provide only a brief glimpse of this complicated but well-developed field. For a more detailed presentation, one of the many available vibration theory texts can be consulted [1–4].

One of the most fundamental systems one can study in vibration analysis is the single degree of freedom (SDOF) spring-mass-damper system shown in Fig. 776. Since the mass is constrained to move only in one direction, the "degree of freedom" referred to is the motion of the mass, M, in the positive and negative x-direction. The mass is restrained by a spring with a spring constant, k, and its motion is impeded by a damper with a damping coefficient, c. In the paper mill, the components of this system might be the mass of a roll, the spring stiffness of the supporting structure, and the damping supplied by shock absorbers or load bags.

When the mass is disturbed from its rest position by some initial displacement X_0 and released, the mass will oscillate about its rest position as shown in Fig. 777. Observe that the mass oscillates with a decaying sinusoidal motion about its rest position ($x = 0$), with a clearly defined period, T, and a decreasing peak-to-peak amplitude. Due to the effect of damping, the motion eventually dies out and the mass returns to its rest position. This type of motion, resulting from only an initial disturbance, is called "free vibration".

Referring to the system of Fig. 776, it is possible to write an equation governing the motion of the mass as follows:

$$m\,\ddot{x} + c\,\dot{x} + k\,x = F(t) \tag{1}$$

where
m = mass
c = damping coefficient
k = stiffness of the spring
x = displacement of the mass
F = vibration forcing function
t = time

The notation in Eq. 1 of $\ddot{x}$ implies the second derivative of x with respect to time, or acceleration. Velocity is indicated by $\dot{x}$, the first time derivative of x.

When the forcing function, F(t), is zero, the case of free vibration results. It can be shown [1] that the solution to this equation yields the displacement, x, of the mass as a function of time and the other parameters of the system. When graphed, this equation displays an oscillation about the rest position of the mass which decays exponentially in time, as shown in Fig. 777. The period, T, of the motion of the mass is defined as the time necessary for the mass to complete one full oscillation. Another way to describe the motion is to use the concept of "frequency", where frequency is the inverse of the period,

$$f = 1/T \tag{2}$$

From Eq. 2 it can be seen that frequency has units of 1/time, commonly referred to as "cycles". Common units of frequency are cycles/minute (CPM) and cycles/second (or Hertz). In the analysis of rotating equipment such as gears and rolls, "orders" are also used as a frequency measuring unit. The order is defined as the ratio of the vibration frequency to the rotating frequency of the element.

For the SDOF system shown in Fig. 776, there is one frequency of oscillation apparent regardless of the decaying amplitude. This frequency is referred to as the "natural frequency" of the system, f_n. This is the frequency at which the system naturally *prefers* to oscillate or vibrate. One good example of natural frequency vibration is the tuning fork. When struck (or given an initial disturbance), the tuning fork "resonates" at its natural frequency in a decaying fashion. Different tuning forks will resonate at different frequencies, depending on their geometry and material of construction and not on the initial impact or disturbance.

In a similar manner, any mechanical system will display one or more natural frequencies which are a function of their geometry and material composition. The concept of natural frequency is an important one in vibration of paper machine systems, since high vibration amplitudes can result in systems resonating at their natural frequency.

A different form of vibration results when the mass, m, is acted upon by a force, F(t), which itself varies with time in a cyclic manner. For example, the excitation force may be mathematically described as

$$F(t) = A \sin(wt) \tag{3}$$

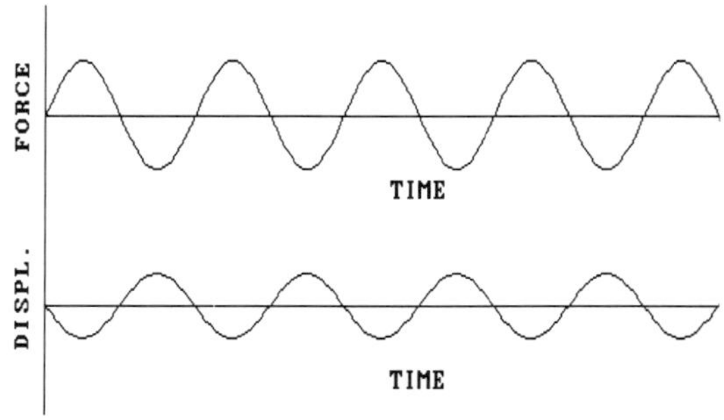

Fig. 778. Force and displacement in forced vibration of an SDOF system.

where
F = forcing function or exciting force
A = constant maximum amplitude
w = frequency of excitation force
t = time

This type of excitation is called "harmonic excitation". The force due to roll unbalance is an example of such an excitation. A harmonic excitation force and the resulting motion are shown graphically in Fig. 778. In this case, the mass is forced to vibrate at the frequency of the excitation, and the oscillatory motion of the mass will continue at some steady-state peak-to-peak amplitude as long as the force amplitude and frequency do not change. The frequency of the vibration will be determined by the frequency of the exciting force. This type of vibration is referred to as "forced vibration". Governing equations of motion and displacement solutions can be written for forced vibration, and can be found in most vibration theory references [1, 2].

As the excitation frequency in forced vibration changes, the steady-state peak-to-peak vibration amplitude of the motion can change significantly. First, consider a system with no damping, termed an "undamped system". While the forcing frequency increases, the amplitude of vibration increases until the natural frequency of the system is reached. At this point, the amplitude of the vibration theoretically becomes infinite. As the frequency of the excitation force continues to increase, the vibration amplitude decreases again. This phenomenon is illustrated by curve "A" of Fig. 779. In this figure, peak-to-peak vibration amplitude is graphed as a function of the ratio of the excitation frequency to the natural frequency of the system, f/f_n, called the "frequency ratio".

Most real vibrating systems possess some degree of damping. Engineering materials commonly have internal damping due to their

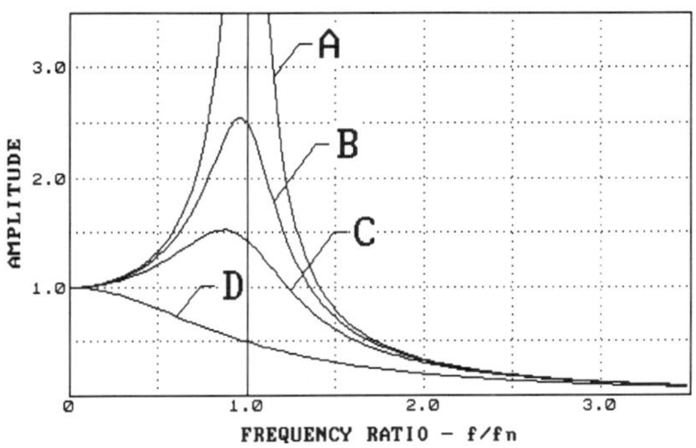

Fig. 779. Displacement in forced vibration as function of damping and forcing frequency.

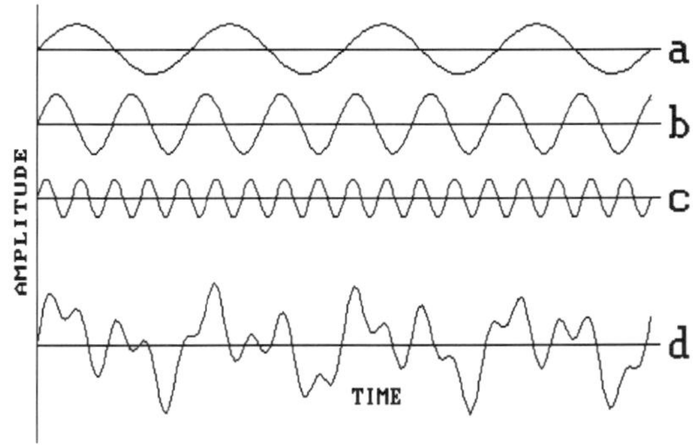

Fig. 780. Complex waveform "d" produced by superposition of waves "a", "b", and "c".

microstructure, and other physical dampers can be added to assist in vibration control. When the system being excited has damping, the motion does *not* become infinite as the frequency of the excitation force passes through the system's natural frequency. Instead, the amplitude increases markedly to a maximum when the forcing frequency equals the natural frequency of the system. After reaching this maximum, the amplitude drops again as in the undamped system, as shown in Fig. 779, curve "B". As increasing amounts of damping are added to the system, the amplitude of the motion at the system's natural frequency decreases as curve "C" illustrates. Systems which behave this way are referred to as "underdamped" systems.

As damping is increased, the system behavior eventually reaches a condition where oscillatory motion no longer results under forced vibration. Systems with this behavior are called "overdamped". For the simple system of Fig. 776, the limiting case between under- and overdamped vibration occurs when the damping, c, reaches a value called the "critical damping". This value can be calculated for the system to be

$$c_c = 2 * (k * m)^{1/2} \tag{4}$$

The case of critically damped vibration is shown in Fig. 779 as curve "D".

Both free and forced vibrations are generally present in paper machines under normal operation. Free vibration effects are seen in the passage of a rough felt seam through a press nip, for example. More common is the forced vibration resulting from roll unbalance. Either type of vibration can result in structural failures, roll degradation, or shortened bearing life. Both types of vibration can be highly detrimental to machine operation and productivity.

The motion of the SDOF system discussed above is perhaps the simplest in the field of vibration analysis. Unfortunately, such simple systems are rare in the industrial environment. Most machines have not one but many degrees of freedom in which to vibrate. The loading may actually consist of many forcing functions acting on the system simultaneously. Such systems also display not one but many natural frequencies. Moreover, there is a constant interaction among the many systems in a given vicinity, further complicating the situation. Although this presents a formidable challenge to the vibration analyst, it is possible to reduce most systems down into several simpler ones to assist in understanding and analyzing the motion.

Complicated motion often results from a superposition of simple motions. For example, consider the motion illustrated in Fig. 780. In this figure, three simple periodic motions are shown as waves "a", "b", and "c". Each is a simple sinusoidal function of time with different but fixed frequencies and amplitudes. If these motions are combined by adding them linearly using the principle of "superposition", the more complicated motion shown as curve "d" results. This combined motion is shown in the "time domain", where amplitude is graphed as a function of time. Analysis of the time-domain data can provide significant insight to vibration problems [5]. It is also possible to show the same waveform in the "frequency domain", where the amplitude of vibration is graphed against the frequency of the motion. In this case, since the waveform was composed of three sinusoidal waves, the resulting frequency-domain description of the combined motion is as shown in Fig. 781. The three components of the combined motion are labelled "a", "b", and "c" corresponding to the components in Fig. 780. Figure 781 is an example of a "frequency spectrum" for the motion of Fig. 780d. The description of complicated motion in the frequency domain greatly facilitates the analysis of complex motion

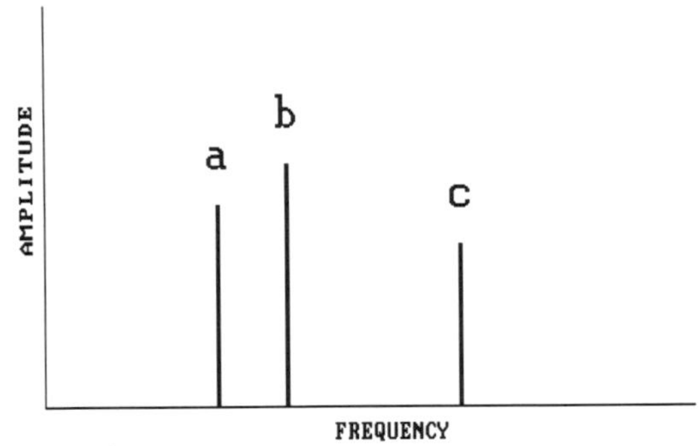

Fig. 781. Frequency spectrum for waveform shown in Fig. 780.

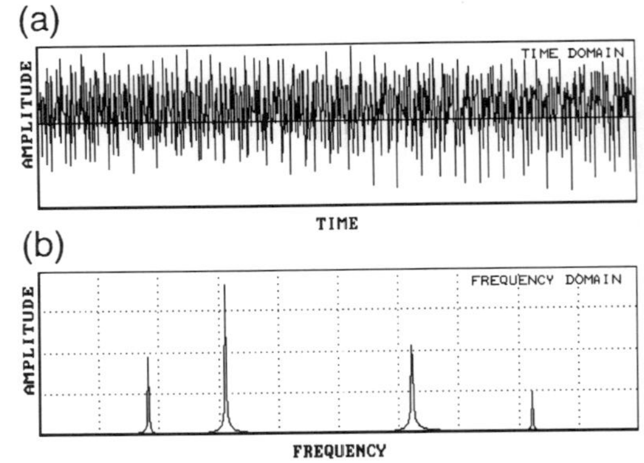

Fig. 782. Time-domain waveform as digitized by a spectrum analyzer and the frequency-domain representation of the waveform.

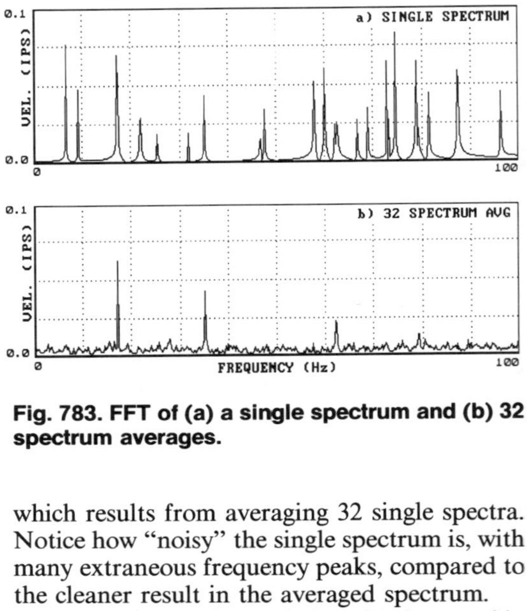

Fig. 783. FFT of (a) a single spectrum and (b) 32 spectrum averages.

since the individual effects of components of the motion are discernable.

It is theoretically possible to reduce any motion to a set of simple motions in the manner shown in Figs. 780 and 781. A mathematical technique, called the "Fourier transform", enables one to effectively "decompose" a complicated waveform into its frequency components and produce a frequency spectrum of the motion [1, 6]. This technique is widely used in dedicated spectrum analyzers [7] and machine monitoring systems to produce frequency spectra from measured motion waveforms. The spectrum analyzer collects data via a vibration transducer which usually measures displacement, velocity, or acceleration. The analyzer consists of a data acquisition unit which digitizes the signal from the transducer at a high rate, producing a time-domain representation of the waveform such as that shown in Fig. 782a. Then the Fourier transform is applied to the waveform through a microprocessor in the analyzer, producing the frequency spectrum shown in Fig. 782b. In this case, four periodic motions were found in the combined signal.

Sometimes, the motion being analyzed consists of both periodic and non-periodic or random components. The introduction of non-periodic motion into the analysis of a waveform can lead to results which may be misleading due to extraneous "noise" in the frequency-domain presentation of the data. Averaging of spectra or waveforms is often used to eliminate the effects of truly random disturbances in producing a frequency spectrum [8]. It is common to average as many spectra as practical to produce the final spectrum. The effect of any truly random events will be reduced to an insignificant level through averaging. Figure 783 shows two frequency spectra taken from the same source. In Fig. 783a, a single spectrum is shown, while Fig. 783b shows the spectrum which results from averaging 32 single spectra. Notice how "noisy" the single spectrum is, with many extraneous frequency peaks, compared to the cleaner result in the averaged spectrum.

A special averaging technique is often used in the paper industry to produce a modified spectrum for analysis of vibration problems related to rotating machinery. This technique is called "synchronous averaging", since the data acquisition for each spectrum used in the average is synchronized to the rotation of a specific rotating element. When sufficient averages are taken using this technique, the spectrum which is produced is indicative of the vibration caused by the synchronizing element only. Using synchronous averaging can thus eliminate many candidates as potential causes of vibration in the multi-roll environments often encountered around paper machinery. A similar technique involves the synchronous averaging of the time signal data before the Fourier transform is performed. This can also lead to a "cleaner" spectrum, with the vibration induced by the synchronizing element remaining after the average is calculated.

As has been seen, complicated vibration patterns can be decomposed into their frequency components to facilitate analysis, and each component may be treated as a simple motion much like that of the SDOF system of Fig. 776. It was shown that, for forced vibration, the damping in the system has a strong influence on the vibration amplitudes in the system. This is especially true when the frequency of the excitation force approaches the natural frequency of the system. The amplitude of the vibration is also a direct function of the amplitude of the excitation force. Hence, to understand the vibrational behavior of the system, one must also understand the sources of excitation on the system.

In the paper mill, there is no shortage of excitation forces for vibration of machinery! Indeed, the abundance of these forces makes the task of the vibration analyst in the mill quite difficult. Perhaps the primary source of vibration in paper machines is roll unbalance. It is difficult (if not impossible) to obtain "perfect" balance for a roll, and then for only one rotational speed. Rolls tend to excite vibrations through simple unbalance, whipping, misalignment, and bent journals. The construction of rolls, some with soft covers, aggravates their tendency to vibrate. The frequency of the excitation forces generated by rolls will usually be once or twice the roll rotating frequency. Since many rolls with different diameters can be present in a paper machine, each generating a different frequency, many forcing frequencies can exist in a running machine. Sometimes a forcing frequency will be approximately equal to the natural frequency of the machine's supporting structure and can lead to severe framing vibrations, as was illustrated in Fig. 779.

Other sources of vibration include irregularities such as faults in gearing and bearings. Often these produce a higher frequency excitation than roll unbalance and thus are more likely to affect bearing life and framing connections than to excite natural frequencies in the equipment. Other mechanical components may also serve as vibration sources through misadjustment or improper installation. Shaft couplings and clutches are two examples. Finally, during operation geometric irregularities may develop which can become significant sources of vibration, such as the roll corrugations (flat spots) sometimes developed in a press.

Vibration may also be excited by clothing or other auxiliary equipment on the machine. Felt seams are seen by presses as a sudden impulse when they pass through the nips, producing what is termed "press bounce". The resulting vibrations can both produce compacted bands (or "bars") on the felt and induce roll corrugations if roll sizes are not properly selected. Another source of vibration from auxiliary equipment is the "chatter" which can be present in doctor blades running on dryers, particularly yankee dryers. This chatter can cause deterioration of the dryer surface and lead to runnability problems if amplitudes are high enough.

In addition to mechanical sources of vibration, hydraulic phenomena can also produce vibration in the paper machine. The pulsing action of pumps can introduce vibrations in approach piping and the headbox. These vibrations can degrade paper quality as well as lead to mechanical difficulties with equipment. Another potential source of vibration comes from condensate loading in dryers. High levels of condensate can induce an "unbalance" in the dryers which can lead to serious vibrations. Temporarily high vibrations can be caused if dryers with spoiler bars are loaded with condensate during startup, causing a "paddle wheel" effect.

Finally, the paper itself can be an exciter of vibrations if it has a periodic basis weight variation. Such a variation, passing through the nips of presses or calender stacks, can cause a bouncing action much like that produced by felt seams. The bouncing can lead to roll corrugation or to cyclic variations in the paper caliper, both of which are unacceptable.

As can be seen from this brief description, *many* sources of vibration exist in the paper mill. Understanding the sources of vibration can help one understand the vibration itself and its consequences. Some level of vibration will *always* be present in paper mill machinery. Unless the amplitude of the vibration is sufficient to cause operating or maintenance difficulties, it is often not cost effective to attempt to reduce the levels of vibration. The deciding factor then must be: At what amplitude does vibration become a "problem", and at what level is it not? The definitive answer to this question depends on the specifics of each individual combination of equipment, operation, and excitation force. Accurate determination of "damaging" levels of vibration for a situation can only be made after much data gathering and analysis. There are, however, some guidelines which can be used to estimate the severity of vibration in common cases [3, 4, 9, 10]. A graphical representation of some of these

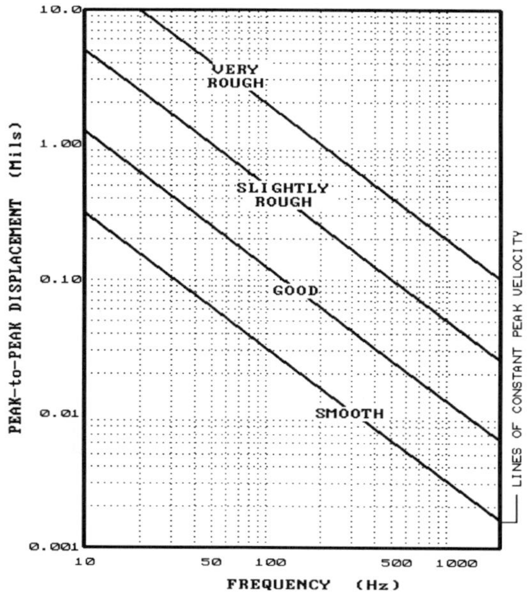

Fig. 784. Vibration severity chart.

guidelines is shown in Fig. 784. In this figure, called a "vibration severity chart", regions of vibration severity are described by their frequency and amplitude. It should be noted that many such charts exist, and each is valid only for limited types of measurements on certain classes of machines. Figure 784 is limited to vibration readings taken on bearing caps and machine structures.

C. VIBRATION MEASUREMENT TECHNIQUES AND EQUIPMENT

The science of vibration measurement has been well developed for many years, benefiting recently from the continuing miniaturization of electronic equipment and the widespread use of computers for data acquisition and reduction. However, long before such equipment was available, papermakers were using vibration analysis as a tool for diagnosing machinery problems and improving paper properties. Starting with the human senses of touch and hearing and using broomsticks to enhance their sensitivity, papermakers were able to "monitor" the condition of bearings and gears on their running equipment. Limitations in frequency resolution and the subjective nature of these tests led to the development of mechanical and electrical equipment designed specifically for vibration analysis.

Electronic instrumentation has been developed which quickly provides data to completely characterize the vibration at a measurement point. Several levels of sophistication in vibration analysis instrumentation are available. Each depends upon a "transducer" to generate an electrical signal from vibratory motion. There are different types of transducers available which measure either the displacement, velocity, or acceleration of the vibrating machine component. Each has its strengths and weaknesses from a theoretical viewpoint, but when properly used each is capable of providing the electrical signal needed for vibration analysis. Most often, transducers are attached to machinery temporarily with magnetic bases during measurements. Some instruments use transducers which are simply held in contact with the machine. However, to obtain the best vibration signature, a better method of attachment is to bolt the transducer to the structure when possible and practical. There have been a number of attempts in the literature [9, 11-14] to define which transducer is "best" for a given application, but the choice inevitably remains with the vibration analyst.

Displacement transducers measure the displacement of the body from a rest condition, and supply a voltage signal which is proportional to the displacement of the vibratory motion. Signal conditioning can usually be accomplished within the housing of displacement transducers, eliminating the requirement for power supplies and additional signal conditioning equipment. This type of transducer generally is limited in its frequency response, displaying a reduction, or "roll off", of signal response at high frequencies and very low frequencies. These transducers also tend to be rather large and heavy; it is possible to alter the vibration response of a light structure by attaching a heavy transducer to it.

A different type of displacement transducer is the "proximity probe", a device which returns a voltage signal proportional to the distance from the probe to a metallic surface. These probes are useful in determining relative displacements between two surfaces, and are reliable over a wide frequency range from DC (0 Hz) to several kilohertz. These probes find application in some on-line paper machine monitoring systems (to be discussed later; see Sec. F), and in bearing monitoring, where they are used to examine minute shaft motions in a bearing housing.

Velocity transducers are popular for many measurements in the paper mill. They are typically smaller than displacement transducers and also require no power supply or signal conditioning equipment. Although velocity transducers have good medium frequency range response, they also tend to roll off in response below about 10 Hz. While compensation for this can be made in the analysis, this limitation reduces the transducer sensitivity for some applications.

Accelerometers are perhaps the most popular transducer for vibration analysis in use today. They are available in a wide range of sizes and frequency ranges, but are typically the smallest transducers available. Most accelerometers work on the piezoelectric crystal principle [15], which states that some materials become electrically charged when placed under stress. A typical accelerometer may be constructed by placing a piezoelectric crystal under a small mass. When the transducer is vibrated, the mass alternatingly stresses and unstresses the crystal, producing a varying electrical charge. An amplifier in the transducer boosts the signal and transmits it to an external signal conditioner, where the signal is prepared for analysis. Accelerometers have very wide frequency responses, and exhibit roll off only at very low frequencies. They are limited in high frequency by the resonant frequency of the crystal/mass system, which can be designed to be well above the range of interest. One drawback to the use of accelerometers is the low acceleration of low-frequency vibration.

Just as there are different transducers available for vibration analysis, there are several types of vibration analysis instrumentation available also. Each receives the transducer output and "translates" it into a useable form. Depending on the instrument used, one or many vibration signals can be processed at once. Different instruments also supply different levels of detail in the analysis. The simplest instrument is a "vibration meter", which provides an overall level of vibration at the measurement point. This instrument can be very effective in isolating the area of concern when dealing with a vibration problem and quantifying the level of vibration [9, 16]. Some meters have the capability of filtering the vibration signal to include only certain frequency bands, a technique which is helpful in eliminating extraneous information when focusing on the cause of a problem. Vibration meters are small and portable, making them suitable instruments for quick checks of overall vibration levels around the paper machine. They are also well-suited to "trending" vibration levels on the machine. Routine checks of specific points (bearings, roll ends, etc.) can be made and the vibration levels recorded and/or graphed. Any significant increases in overall vibration level at a measurement point can be noted, and steps can be taken to further define and correct the problem before a failure occurs or it becomes necessary to stop production to make repairs. This technique forms the basis of many machinery monitoring systems, regardless of their complexity.

"Tunable filter" instruments can be used to determine the frequency content of the vibration signal. The signal from the transducer is passed through an adjustable, very narrow band filter, eliminating all but certain frequencies. The meter on the instrument is then used to determine which frequencies are prevalent in the signal. Some of these meters also provide for a strobe light at the tuned frequency, which can be used to "stop" the motion of the different machine components in question. This very simple and useful technique can greatly facilitate isolating some problems, particularly belt drives and pedestal movements. The "swept filter" vibration meter automates the tuning of a narrow-band filter through a frequency range and records the vibration amplitude as it does so. The result is a frequency spectrum showing the amplitudes of vibration at all frequencies covered. The swept filter instrument formed the basis of many vibration analysis programs for many years, and is still in use today due to its simplicity and portability.

Further refinements in vibration analysis instrumentation are available in the form of "real-time spectrum analyzers". These powerful instruments are dedicated microcomputers which receive vibration signals, rapidly process them to determine the frequency content, and display the results graphically on a built-in cathode ray tube in virtually real time. Spectrum analyzers use the Fast Fourier Transform (FFT) algorithm discussed previously to decompose complex time-domain signals into frequency-domain components, and so are often referred to as "FFT Spectrum Analyzers". Analyzers can be categorized according to the number of channels of data they can process simultaneously. Single-channel and dual-channel analyzers are common. With recent advances in microprocessor technology, analyzers have been developed which can simultaneously process eight or more channels of vibration input.

Multiple-channel analyzers provide not only frequency-domain information about vibrations, but can also perform mathematical correlations between channels to help determine how vibrations at different measurement points are related. This feature can be very useful in efforts to solve complex vibration problems on paper machines. An example of a dual-channel FFT analyzer display is shown in Fig. 785. The upper portion of the display shows the frequency spectrum of the vibration at one location, while the lower portion shows the phase relationship

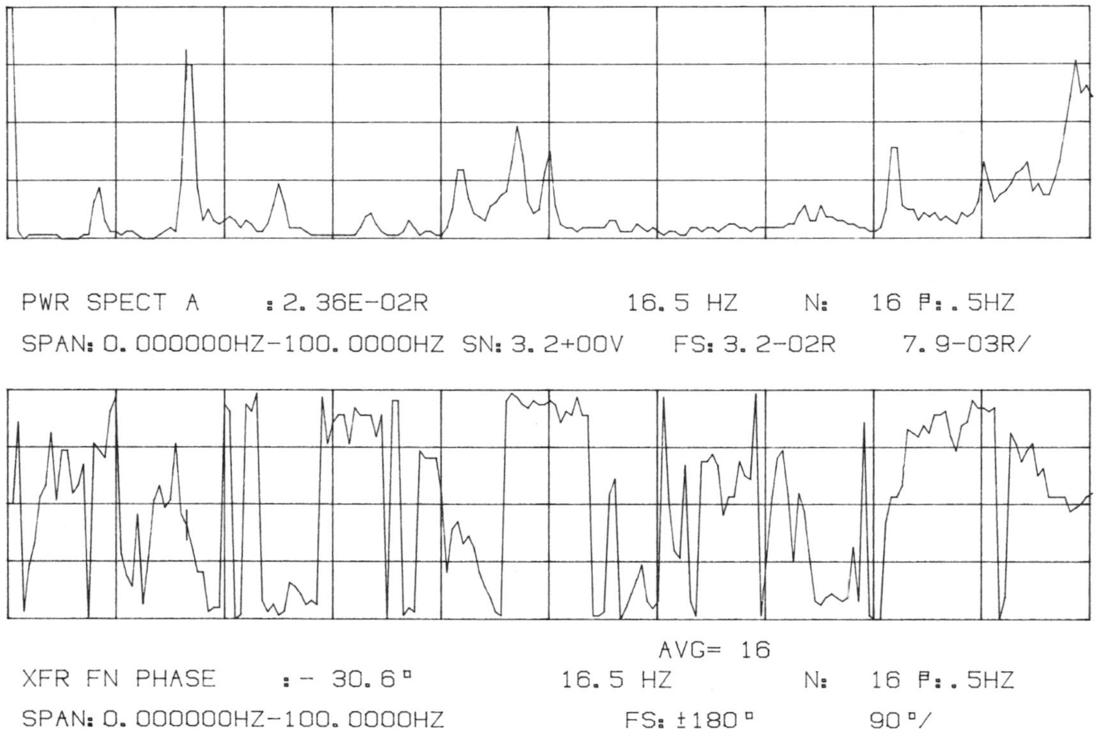

Fig. 785. Display from a two-channel spectrum analyzer.

of that signal and the vibration at another location.

The functions available in the real-time spectrum analyzer can also be programmed into computers, and with the addition of available data acquisition hardware the computer can become a powerful multi-channel vibration analyzer. This configuration is often found in dedicated machinery monitoring systems, with many vibration signals periodically scanned and processed continuously. Thus, as experience and understanding of vibration analysis grows, a vibration trending program in the paper mill which might have started with a hand-held vibration meter can evolve into such a state-of-the-art system with built-in data acquisition, diagnostics and reporting capabilities.

D. EXAMPLES OF PAPER MACHINE VIBRATION

Paper machines are very complex systems, combining structural, fluid dynamic, and heat transfer technologies to produce a variety of paper grades dependably and efficiently. In many cases, new machines are installed only to be modified and rebuilt several times in subsequent years in an effort to obtain higher production speeds or different paper grades. Throughout the life of the machine, economics dictate that it be able to run continuously while maintaining a high degree of uniformity in the paper it produces. Also, safety must not be compromised by increased speeds or structural modifications.

Vibration is always present in mechanical systems, even very simple ones. In the paper mill, imperfections in the manufacture of machine components and their installation and operation make a measurable degree of vibration a certainty. Efforts to eliminate the vibration of simple systems can certainly lead to reductions of vibration, but its total elimination is an elusive goal at best. When the system in question is as complex as a paper machine, with hundreds of rotating components, roll nips, and felts, the presence of vibration becomes a matter of course. The paper mill engineer, maintenance engineer, and papermaker must learn to live with and deal with vibration if they hope to obtain and maintain optimum production from a machine.

There are almost an infinite number of sources of vibration on a typical paper machine. The following paragraphs discuss some of the more common ones, categorized by the section of the paper machine where they are commonly found. The list is by no means complete; indeed, each paper machine will exhibit a unique set of vibration problems during its life and must be dealt with as such.

1. Wet end vibrations

The wet end of the paper machine is susceptible to vibrations from several sources [17, 18]. Vibration can be induced by fluid dynamic forces acting on the approach piping and headbox or from mechanical forces such as roll unbalance or bearing faults. In some cases, vibration may even be introduced into the wet end intentionally to enhance web formation.

Fluid-induced vibration at the wet end usually manifests itself as a motion in the headbox. Any such motion can be detrimental to proper papermaking as it can cyclically change the geometry around the slice and stock jet, where uniformity and steadiness are critical to consistent paper quality. A common source of fluid-induced vibration in the headbox is pressure pulsation created in the approach piping system due to the operation of the fan pump(s) and/or pressure screen(s) in the system. These pulses are produced in the fan pump, for example, whenever an impeller vane passes the pump outlet. A similar condition exists when the foils of a pressure screen pass the screen inlet or outlet. Knowing the pump or screen rotation speed and the number of vanes or foils present allows one to calculate the frequency of the pressure pulse expected from a device. For example, a fan pump with 6 vanes running 1200 rpm would produce pressure pulses at 120 Hz. If the pump condition were poor, it would also be likely to produce a stronger pulse from one vane than from the others, producing a 20 Hz pulse. This could also result from a foreign object lodged in a vane of the pump. An example of a pressure pulsation spectrum is shown in Fig. 786. The peak pressure pulse is seen to be almost 0.2 psig at a frequency of 21.38 Hz, which was the fan pump rotating speed. The strong peak indicates a problem with the pump. Pressure pulsation attenuators are available from machine suppliers to reduce these pulses before they reach the headbox, but not all machines have these devices installed.

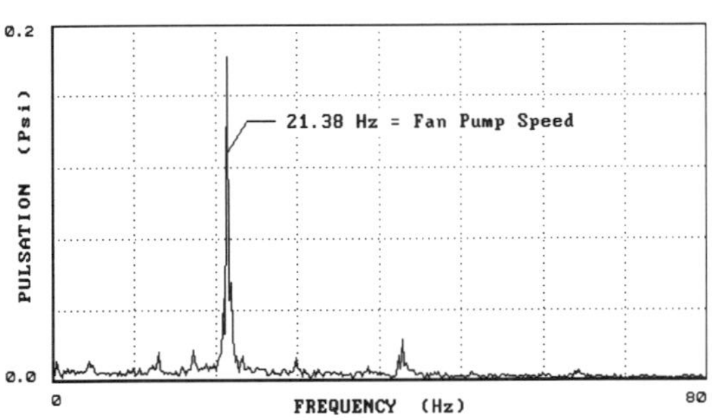

Fig. 786. Pressure pulsation spectrum from headbox header.

Usually, the structure on which the headbox is mounted is specifically designed to resist vibration. However, older or modified machines may be operating outside the design envelope of the original machine. Hence, when high-pressure pulses are transmitted through the stock to the headbox, it is possible to develop vibrational motion in the headbox. If the motion is in the machine direction (from the wet end to the dry end), there will be cyclic variations in the jet-to-wire speed ratio. If these are severe enough, cyclic variations in fiber orientation and paper properties can result. While headbox vibration is not a common problem faced by papermakers, it is generally detrimental to machine operation and should be reduced whenever present. The usual solution to the problem is to reduce the pressure pulsations by repairs or modifications to the approach flow system since high-pressure pulsations commonly cause cyclic machine-direction basis weight variations which are unacceptable. It is also possible to stiffen the headbox support structure to reduce its susceptibility to vibrations.

The headbox is also subject to vibrational excitation from other sources around the machine, including breast roll unbalance and forces transmitted from outside sources through the machine room floor and foundation. It is necessary to isolate and correct these sources of vibration through measurements whenever they appear to affect machine operation.

Twin-wire formers can also suffer from structural vibration around the wet end of the machine. In addition to vibration of the headbox, these formers are susceptible to vibrations of the breast rolls for the two wires. If relative motion is developed between the two rolls, such parameters as the wire approach angle and the jet impingement point can change cyclically, degrading the uniformity of the paper produced.

Other common causes of vibration in the wet end of the paper machine are related to the mechanical condition of the equipment itself. Indeed, this is a topic of concern for every area of the paper machine and the paper mill in general. Bearing condition, alignment, journal condition, and drive component condition are critical issues here. Looseness can result from corrosion of sole plates and fasteners in the wet environment.

Table 84. Characteristics of vibration problems.

Probable Source	Frequency	Direction	Amplitude
Mass Unbalance	1 × rpm	Radial	Steady
Bent Shaft	1 × rpm	Axial	Steady
Parallel Misalignment	1–2 × rpm	Radial	Steady
Angular Misalignment	1–2 × rpm	Axial	Steady
Mechanical Looseness	1–10 × rpm	Radial	Steady
Anti-Friction Bearing	30k–60k cpm	Radial	Increasing
Oil Whirl	<0.5 × rpm	Radial	Steady
Gear Transmission Error	No. Teeth × rpm	Radial/Axial	Steady
Mismatched Belt Tension	1–2 × rpm	Axial	Steady
Pump Vane Pass	No. Vanes × rpm	Radial	Varying
Cavitation	Broadband	All	Varying
Flat Spot on Roll	1 × rpm	Radial	Steady
Corrugated Roll	rpm × No. Flats	Radial	Steady
Felt Seam Passage	1 × Felt Pass	Radial	Steady
Electrical Ground Loops	1–3 × Elec Freq.	Radial	Drops w/o Power

Vibration in wet end mechanical components can sometimes affect paper quality. For example, unbalance or a bent journal in a dandy roll can result in cyclical variations in contact with the forming sheet, producing poor machine-direction uniformity in the paper. Water marking can be seriously degraded by this type of vibration.

Moreover, vibrations in equipment can lead to the overstressing of parts, and eventually serious failures can result. It is imperative for continued efficient operation that the mechanical condition of all components be monitored routinely to avoid unscheduled machine shutdowns for the repair of failed components. Routine inspection is also a must for reasons of safety.

The vibrational characteristics of mechanical faults in shafts, couplings, motors, gears, and bearings are well researched and understood [8–10]. It is usually possible to isolate the cause of high vibrations in these components using the FFT analyzer to produce a frequency spectrum of the vibration. Then, relating the peak frequencies to the rotating speed of the element, it is possible to determine the cause of the vibration. A list of some characteristic vibration frequencies for mechanical faults is given in Table 84. For example, a spectrum from a shaft (or roll) which is out of balance will exhibit a peak frequency at the rotating speed of the shaft in a direction perpendicular to the shaft axis. Such a spectrum from an unbalanced shaft is shown in Fig. 787. The rotating speed of the shaft was 510 rpm, or 8.50 Hz. The peak acceleration of 0.85 *g* at that frequency is well above the amplitudes at other frequencies recorded.

A great deal of study has surrounded the vibrational behavior of rolling element anti-friction bearings and their various faults [3, 19–21]. Knowing the geometry of the bearing (i.e., number and diameter of balls, ball contact angle, pitch diameter, etc.) and the running speed, specific frequencies can be determined which are characteristic of different flaws in the bearings. The five different types of bearing flaws are inner and outer race flaws, ball or roller flaws, cage defects, and looseness caused by wear of other components. For example, a bearing which has a flawed inner race will have a vibration peak at a frequency defined by the following equation:

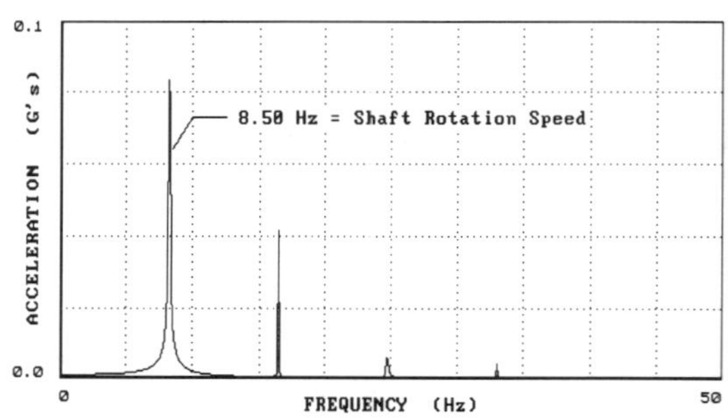

Fig. 787. Frequency spectrum from an unbalanced shaft.

$$f = N/2 * w * (1 + d/D * \cos \phi) \qquad (5)$$

where
N = number of balls or rollers
w = rotating speed of shaft (rpm)
d = ball or roller diameter
D = pitch diameter of bearing
ϕ = contact angle of bearing

Similarly, if the flaw is in the outer race the peak frequency will be

$$f = N/2 * w * (1 - d/D * \cos \phi) \qquad (6)$$

A flaw in a ball or roller will produce a frequency defined by

$$f = D/2d * w * [1 - (d/D)^2 * \cos^2\phi] \quad (7)$$

A cage defect will exhibit vibration at a frequency of

$$f = w/2 * (1 - d/D * \cos\phi) \quad (8)$$

Defects associated with looseness will produce high-frequency peaks which are integer multiples of the inner race speed of the bearing.

The formulas above for bearing vibration are among the most valuable in the field of vibration analysis since much of the vibration analysis effort in paper mills is concentrated on maintenance of the papermaking equipment and there are so many bearings in the mill. The use of vibration analysis and sophisticated instrumentation in maintenance has resulted in an evolution from "preventive" maintenance, where bearings are replaced on a time-in-service basis, to true "predictive" maintenance, where the failure of bearings is predicted scientifically. The former method leads to the unnecessary replacement of some bearings and unexpected premature failures of others. Predictive maintenance allows the maximum useful life to be achieved from the bearings, yet routine monitoring prevents unexpected failures from occurring [22, 23].

Before leaving the wet end of the paper machine, it should be noted that there are also intentional applications of vibration here which are beneficial to good paper formation. One previously mentioned is the use of "shake" on fourdriniers of relatively slow paper machines [24]. In this application, a lateral (cross-machine) motion is imposed on the breast roll for the purpose of introducing turbulence into the stock as it lands on the wire. The best performance of these systems is obtained with small amplitudes (0.25–0.50 cm; 0.1″–0.2″) and high frequencies (up to 250 cycles per minute). This will, of course, introduce mechanical vibrations into other components on the wet end and must be accounted for in the machine design and maintenance.

A more recent development is the application of the toothed roll, or "Sheraton roll", to the table [25]. This roll introduces small vertical displacements into the wire at a frequency determined by the roll speed and the "tooth" rotation. When operated at a surface speed equal to the wire speed, these displacements vibrate the wire at a moderate frequency, enhancing formation through the introduction of small-scale turbulence. It has been found, however, that a much more pronounced effect can be generated if the Sheraton roll is operated at a speed which produces resonance in the paper machine wire ahead of the roll. In this application, part of the wire acts as a two-dimensional membrane supported by the Sheraton roll and the table element which precedes the roll, and vibrates much as a string would. The precise roll speed needed to vibrate this section of wire at its natural frequency will be a function of the geometry of the installation, the wire weight and stock, and the wire tension. An advantage of this technique of formation enhancement by vibration is the low amplitude of the vibration and hence the reduction of detrimental effects on other wet end equipment.

2. Press section vibrations

Vibration in the press section of the paper machine has traditionally been an area of great concern to papermakers. As such, it has warranted considerable attention in the literature [26–29]. Vibrations in the press section are perhaps more noticeable than others due to the hard rolls and press nips found there. The presence of expensive press felts, which must be in good condition to assure maximum runnability, makes the control of press vibration essential. Excessive press vibration can shorten felt life by many days and lead to gross non-uniformity in paper quality. In extreme cases, it may not be possible to run the paper machine safely at all. Other mechanical components in the press section can be very expensive. For example, triple race bearings in presses can easily cost more than $25,000 each. This high component cost provides ample justification for paying careful attention to vibration levels around the press.

There are two prominent areas of concern in press vibration in addition to those of routine maintenance of mechanical components discussed above. These are roll vibrations and nip vibrations. Felt roll vibrations are important since these rolls are typically smaller in diameter than other press rolls and hence turn at a higher frequency. Unbalance in felt rolls can lead to vibrations in the press framing, especially if the natural frequency of the press frame is near the rotational frequency of the felt rolls. If felt roll vibrations are high, the best course of action is to remove and rebalance the vibrating rolls. If the press framing has a natural frequency near the roll rotation frequency, it would be better either to stiffen the framing to alter its natural frequency or change to larger diameter felt rolls to reduce their rotational frequency. The more cost effective course of action to take would depend on the specifics of the problem.

The second area of concern for press vibrations is in the area of press nips. Vibration in the nip is exacerbated by the nip geometry, with hard rolls highly loaded together and a relatively thin, somewhat compressible felt between them. The press loading, accomplished with hydraulics or air loading, has a spring effect also, further complicating the situation. Any unbalance or roll imperfection will lead to vibrations synchronized with the rotation of one or the other roll in the press nip. Other vibrations can be introduced into the press nip by faults in other mechanical elements such as drive components and bearings. When encountered, these problems must be identified through the use of careful measurements and frequency analysis.

The element in the press nip which makes press vibrations unique and difficult to deal with is the press felt. To some extent, press felts serve to dampen vibrations in the nips, especially when they are newly installed. However, as they wear, the felts display changing stiffness and as a result, "normal" vibration levels will increase in the nips. Furthermore, the construction of the felt itself introduces vibration into the press section. The seam of the press felt passes through the nip once for each complete felt revolution. Since the felt seam has a higher stiffness than the rest of the felt, when it goes through the nip it can produce an impact load in the system. For a felt which is 100 feet long on a paper machine running 914 m/min (3000 ft/min), the frequency of these impacts will be 30 per minute, or 0.5 Hz. With each seam passage through the nip, the press rolls may appear to "bounce" on each other. The extent of the press bounce will be a function of the stiffness of the seam and the press nip stiffness. In extreme cases, the press bounce can visibly affect the paper passing the nip, producing a "barring" effect with each seam passage. In time, the press bounce can also create a barring or corrugated pattern on the felt where repeated impacts will have produced narrow, cross-machine direction bands of compressed felt relative to the original felt condition. This felt non-uniformity can have a drastic effect on paper quality and uniformity in addition to further increasing the vibration levels in the press. The eventual result of this development is the removal of the felt, sometimes after only a few days in service. A temporary reduction in the vibration severity can be obtained by "cocking the seam" of the felt, or skewing the felt on the machine so the seam does not pass through the nip at once but in a more gradual manner.

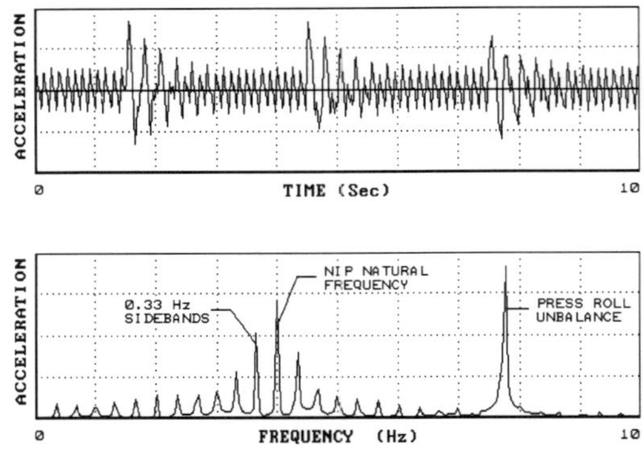

Fig. 788. Time- and frequency-domain data for a bouncing press.

Consider the example shown in Fig. 788. In the top portion of the figure the nip acceleration is graphed against time. The "ringing" of the nip is apparent in the time trace of acceleration, which is similar to that shown for the SDOF spring/mass/damper system in Fig. 777. From the time trace, one can determine that the felt seam passes the nip about every 3 s. The frequency domain graph for the signal is shown in Fig. 788b. As can be seen, the unbalance of the press roll is apparent at about 7.75 Hz. Also evident is a family of peaks composed of a relatively high 4 Hz peak surrounded by many lesser evenly spaced peaks called "sidebands". The 4 Hz peak can be related to the "natural frequency" of the nip, while the sideband spacing of 0.33 Hz is equal to the felt pass frequency. Thus, the frequency domain FFT contains a great deal of information about the press nip dynamics. Any other facet of the press felt construction, whether design or workmanship, which produces a variation in mass, stiffness, or air permeability in the felt in the machine direction can similarly cause escalating vibration problems leading to the premature removal of the felt from the machine.

Many press rolls are covered with a hard synthetic cover which serves to help minimize vibrations in the press nip. However, under the repeated action of felt seam passes or possibly mechanical faults, it is possible for the roll cover to develop flat areas around the circumference, called "corrugations" [30]. These areas can develop unusually rapidly when the felt length is an integral multiple of the roll circumference. In this case, the felt seam will impact the same point on the roll circumference with each passage. The cover will soon be flattened locally in the impact region. When the seam then passes, the combined effect of the seam and the flat in the roll will serve to produce even higher vibrations in the press. Similar results are seen when multiple corrugations develop around the circumference of the roll. In this case, both the felt and the rolls of the press must be replaced. To avoid this problem, the papermaker should take care to use felts which are not even multiples of the roll circumference in length.

3. Dryer section vibrations

As a factor in paper quality, vibration in the dryer section is less critical to papermakers than press vibration. However, dryer section vibration is as important as that in any other machine section for maintaining high productivity on the paper machine. The dryer section is typically composed of many steam-heated drying cylinders and many more smaller felt rolls. These rolls are supported on bearings and are often gear driven in enclosed ventilation hoods under high temperature. Each component in this system is a potential source of downtime and machine failures if their vibrations are ignored. Hence, mechanical condition monitoring of the components of the dryer section is critical to the efficient operation of the paper machine. The aspects of routine vibration analysis of machine components and problem diagnosis which were discussed in conjunction with wet end vibrations are equally applicable to the dryer section.

As was the case in the press section, felt rolls become important in the vibrational health of the dryer section. Dryer cylinders themselves are typically 91–183 cm (3–6 ft) in diameter, and as such rotate much slower than the felt rolls which are from 25–61 cm (10–24 in.) in diameter. This results in less bearing wear and less susceptibility to unbalance in the dryer cylinders. Certainly, these aspects of dryer vibration are not to be ignored; good maintenance practice mandates thoroughness. However, the felt rolls are likely to develop vibration problems in the dryer section first. Simple unbalance should be corrected quickly when it is detected during inspections, and bearings which show deterioration should be replaced at an early convenient routine down time.

As machines are pushed to higher speeds, often exceeding their design speed, it is sometimes the case that the dryer section exhibits gradually increasing vibration levels. It is possible that the dryer section framing has a natural frequency which is excited by the felt roll unbalance, as discussed in relation to press vibrations. In this case, vibration analysis techniques can be used to determine the frequency being excited in the framing and to isolate the source of the excitation. While balancing suspected unbalanced roll(s) will

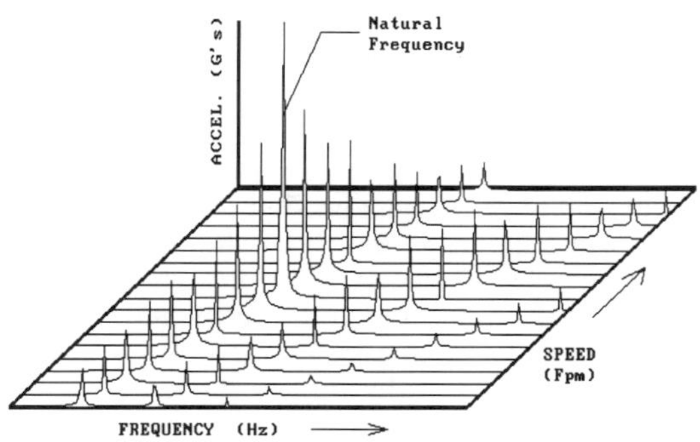

Fig. 789. Waterfall plot of dryer frame vibration.

reduce the vibration temporarily, it is desirable to make a more permanent correction to the problem to avoid future problems. New felt rolls of increased diameter can be installed to reduce the excitation frequency below the natural frequency again, or modifications to the structure can be made. Analytical techniques such as finite element analysis (see Sec. E1) can be used to determine the probable effect of proposed structural changes, and should always be employed before frame changes are actually made.

Figure 789 shows vibration measurements made on a typical dryer frame of an older paper machine after a wet end rebuild resulted in a speed increase. The mill was experiencing problems with felt roll journal and bearing failures. The spectra recorded while the machine was running 381 m/min (1250 ft/min), showed a high vibration peak at 10 Hz, which was the rolling frequency of the 20.3 cm (8 in.) diameter felt rolls in the dryer section. In order to determine whether a natural frequency was being approached in the machine, spectra were produced at several machine speeds at each measurement point. Figure 789 shows a "waterfall" plot of the spectrum obtained from one location during the test of the dryer section. The spectrum for a given speed, plotted on the horizontal (frequency) and vertical (acceleration) axes, yields a familiar picture. Several such spectra are stacked along a third speed axis to produce a 3-dimensional picture of the vibration. From this graph, it can be seen that the system vibration is much higher at one point than at any other. This defines a natural frequency of the structure, which in this case was about 14.25 Hz excited by felt rolls at 549 m/min (1800 ft/min). Thus, further speed increases for the machine will only serve to increase the vibration levels in the framing and aggravate the roll problems. The favorable course of action for this case is an increase in felt roll diameters while modifications to the dryer section framing are designed.

A different type of vibration can result from internal damage to the dryer condensate removal system. If condensate removal syphons are damaged or the system becomes clogged, one or more dryer cylinders may fill with condensate. In this case, high vibrations may be introduced into the dryer section framing due to the unbalanced nature of the heavy condensate load in the rotating dryer(s). This type of excitation will often produce a cross-machine direction vibration at a frequency equal to the dryer rotating frequency. While certainly also causing problems from a process standpoint due to loss of drying capacity, this high vibration can be very detrimental to the mechanical condition of the machine and should be corrected as soon as possible. Many mills use dryer temperature measurements to locate waterlogged dryers during operation. The dryers are then valved off until repairs can be made to the condensate removal system.

Tissue machines with Yankee dryers present other vibration problems to consider. The mass of these dryers and their high speed and temperature require special bearings to be fitted on the journals. Typical bearings for Yankee dryers can cost many thousands of dollars and take several hours to replace if damaged. Their proper installation is critical and usually done with the help of field engineers from the bearing manufacturer. These factors combine to make bearing condition monitoring essential to good Yankee dryer operation.

Tissue machines also are susceptible to higher frequency vibration or "chatter" in the doctor blades used for crepeing and cleaning the Yankee surface. Vibrations in these blades can cause uneven wear on the Yankee surface, eventually leading to corrugations in the Yankee. These "chatter marks", although only a few thousandths of an inch deep, can create serious difficulties in papermaking. When they grow too deep, it becomes necessary to grind the Yankee smooth again, an expensive process which consumes several days of production time. To minimize the probability of this happening, the doctor blade holders on tissue machines are made very stiff, and special clamps are used to hold the blades. During the course of normal operation, doctor blades wear down and must periodically be replaced to maintain good papermaking performance. Some monitoring systems use vibration in the doctor blade holder to determine when a blade should be replaced, a technique which maximizes the blade life and helps prevent the development of chatter marks on the Yankee.

4. Coater, size press, and calender stack vibrations

Other sections of the paper machine also suffer from vibration problems under some conditions. In all sections, the mechanical condition of bearings, gears, and other components can cause excessive vibration. Routine vibration monitoring can reduce the negative impact of normal wear of these components and help determine when and where maintenance is required.

Other vibration problems can directly affect the quality of the paper being produced. In the coater sections of coated paper machines, vibration can lead to poor uniformity of coating application. Size presses can exhibit some of the problems of press sections, where roll unbalance and geometric imperfections can lead to uneven size distribution or moisture and caliper variations in the sheet. Vibrations in the reel can make it difficult to build an even, well-tensioned reel of paper and can make converting the paper difficult. Winder vibrations can quickly turn good paper into waste if left uncorrected [31].

One of the most vibration sensitive pieces of equipment on the dry end of the machine is the calender stack. These stacks of hard steel rolls are very susceptible to problems of roll bounce, roll corrugation, and loading system vibration [32-34]. Calender stacks will sometimes respond to cyclic changes in the basis weight of the paper in a manner similar to the press nip's response to felt seam passage. This can be particularly bad if the calender stack has a natural frequency near the frequency of the basis weight variation. Calender stacks with several rolls and multiple nips are even more sensitive to vibration. Often, air loading systems are filled with water or hydraulic fluid in an effort to dampen vibrations in calender stacks and presses.

As an example of the effect of calender stack vibration, consider the graph of paper caliper along the sheet in the machine direction shown in Fig. 790. These caliper measurements were obtained from paper which was produced on a paper machine with two one-nip calender stacks. The curve marked "a" shows the caliper of the paper without any calendering. Curve "b" shows the caliper after passing through one calender stack, and curve "c" shows the caliper after being calendered in both stacks. Analysis of the graphs, knowing the machine speed and the measurement frequency, shows that the caliper variation of "b" has a frequency equal to the rotation frequency of one roll in the stack. The

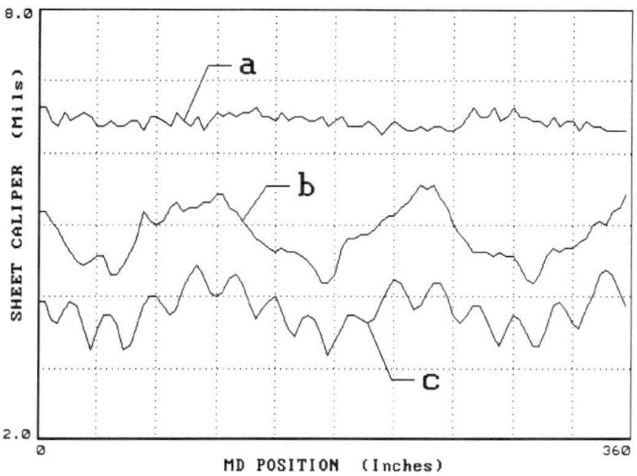

Fig. 790. Sheet caliper variations after calendering.

two-stack caliper curve in "c" appears to superimpose another smaller variation of higher frequency on the first variation. The added frequency was found to be 5 times the roll rotation of the second calender stack. Vibration spectra taken from the two calender stacks showed vibration peaks which matched the variations in the caliper data. It was subsequently found that one calender stack was suffering from a faulty drive coupling. The other stack had a roll which was in the early stages of roll corrugation. Both conditions were corrected with a resulting improvement in caliper uniformity.

Another source of vibration in calender stacks is the loading system itself when low nip pressures are used. In this case, the entire dynamic system may have inadequate stiffness to prevent vibration resulting from paper variations or roll unbalance. When the centrifugal (or shaking) force on the loaded roll in a nip exceeds the nip loading imposed by the stack, the nip can open and close at the frequency of the roll rotation. This will produce caliper variations as described above at first, and eventually lead to sheet breaks when the nips reclose. In the extreme, the vibration around the nip can become the limiting factor for paper machine speed. Often, hydraulic loading or counter loading arrangements are used to stiffen the system and ensure nip integrity.

Many other types of auxiliary equipment around the paper machine are susceptible to vibration and must be carefully maintained and monitored. Line shafts are often driven by steam turbines, which rotate at high speeds. When undetected, small vibrations in these components can quickly lead to serious failures. Other areas of concern should be sleeve bearings and governor controls.

As can be seen from the examples presented above, vibration is indeed present in all sections of the paper machine and in each case can lead to serious equipment failures, increased downtime, and degradation of paper quality. Other paper handling equipment, from the rewinder through the converting plant, is equally subject to the effects of vibration and should be treated accordingly.

E. CONTROL OF VIBRATION AND ITS IMPACT

With machine vibration a fact of life for the papermaker and mill engineer, the important question becomes "What can be done to avoid problems and control vibration?" The answer to this question involves a coordinated effort from three areas of the paper mill. Engineering staff must ensure that new equipment is designed and installed properly to minimize vibration. Paper machine operators must be aware of vibration and operate the machine in a manner which maintains low vibration levels. Finally, maintenance personnel must routinely monitor vibrations in the equipment and effect repairs before serious problems develop and lead to failures. With this kind of effort, the negative impact of vibration on the productivity of the entire paper mill can be minimized.

1. Design for vibration control

There are several steps which should be taken during the design of new equipment or modifications to existing machines which will serve to reduce the potential for excessive vibration in the machine. Among these are proper design of the rotating elements (rolls, bearings, gears) in the machine, careful attention to component mounting and vibration isolation, and recognition of all possible sources of vibration. It is particularly important that the rolls in the paper machine be chosen so that their operating rotational speed is not near their "critical speed" [3]. A roll's critical speed is similar to a natural frequency for a roll, or a rotating frequency at which the roll vibrates most severely. When operated at or near its critical speed, a roll produces high shaking forces which can damage the roll, its bearings, or surrounding equipment. All rolls possess several critical speeds, and should not be operated at speeds higher than the lowest critical speed for safe operation.

Another important design step is the careful analysis of the machine framing to ensure that the natural frequencies of the frame will be high enough to be outside the normal frequency range of the excitation forces in the system. Usually, the excitation forces come from unbalanced rolls in the system. Typically, the lowest natural frequency should be 25% higher than the

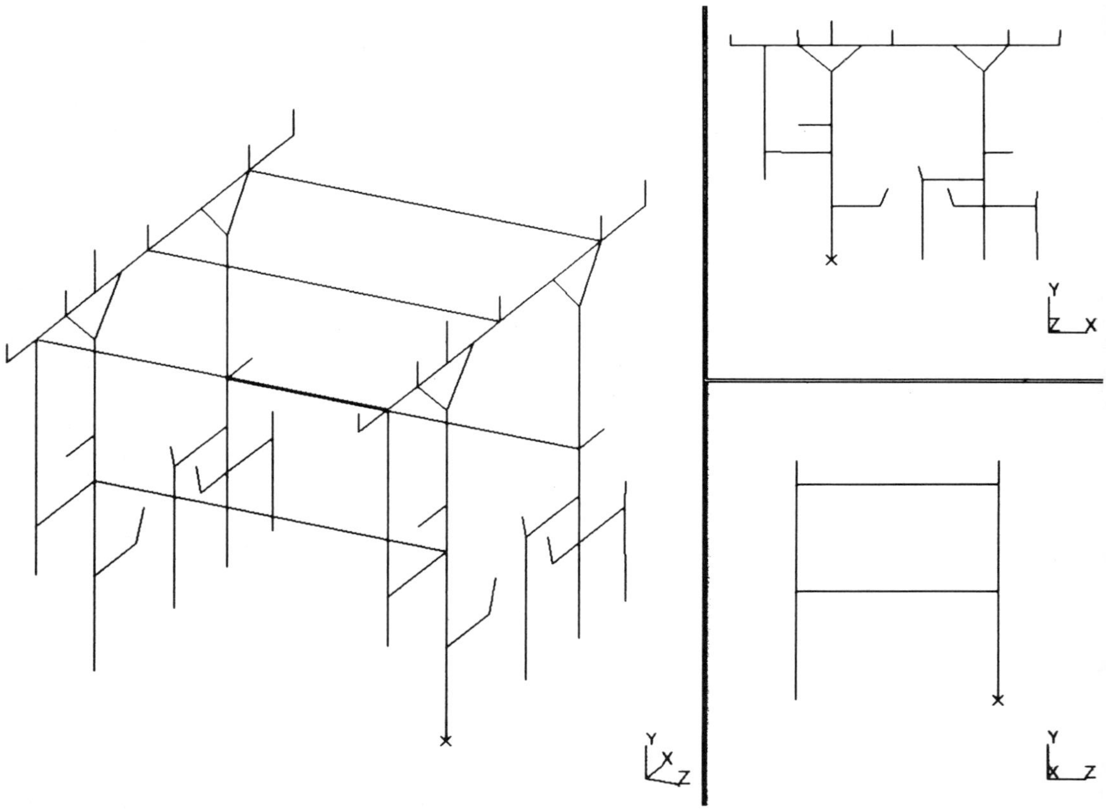

Fig. 791. Finite element model of press frame.

excitation force frequency. Since smaller diameter felt and wire return rolls turn with the highest frequency, they become the governing criterion for this frequency.

Methods to approximate the natural frequencies of simple structures have existed for quite some time. While providing good results for simple structures, however, they were inadequate for estimating the vibrational response of a structure as complicated as a press frame. Hence, it is unusual to find good estimates of the vibrational response in frames of older (pre-1970) machinery. However, early advances in digital computing led to the use of the finite element method of structural analysis for complex designs. This approximation technique, actually developed in the Thirties but too complex for "hand" solution, was found to be solvable using computers for the solution of the many simultaneous equations generated by the theory [27, 35].

In brief, the finite element method allows the analyst to disassemble a complex structure into many small pieces, or "elements", which are connected at points called "nodes". The basic equations governing the behavior of each piece can then be written. Compatibility conditions at element intersections and boundary conditions for the structure complete the set of equations. These simultaneous equations are then solved on the computer using one of several available approximation techniques. Interactive computer programs are available which allow the designer to "build" a "model" of the structure graphically on a computer terminal, supplying information about the material, geometry, and loading conditions for the model. The solution to the vibration problem is then accomplished and the results can be displayed graphically to assist the designer in understanding the behavior of the structure.

A simplified example of a finite element model of a press section frame is shown in Fig. 791. Each element is modeled as a straight beam with properties governed by the geometry of the actual design. Some approximation is necessary to reduce a complex design down to a model of this type; the skill of the designer is critical here if a valid result is to be obtained. Figure 792 shows the analysis result for the first (lowest) natural

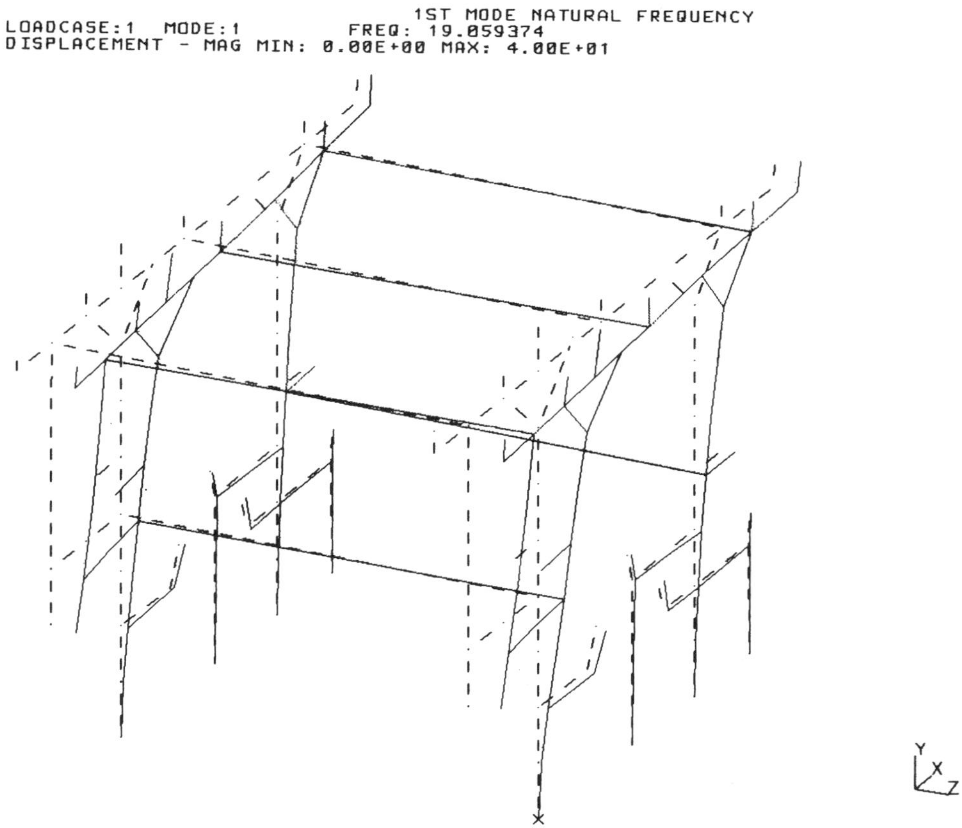

Fig. 792. First national frequency mode of press.

frequency of the structure, in this case about 19 Hz. The model is shown in its distorted shape (first mode shape) in solid lines, while the undistorted model is displayed in dashed lines. It can be seen that this mode is a cross-machine sway with some slight twisting. In order for this design to be acceptable, the rotating frequencies of all rolls should be below about 15 Hz. If this were not the case, the designer would make conceptual modifications to the structure and repeat the analysis of the model until the criterion is met.

While the finite element method should always be employed in the design of new equipment to avoid fundamental frame problems, there may be times when certain components develop high vibrations due to increases in speeds or loads. In these cases, there are some techniques available to the engineer for reducing the vibration to acceptable levels. These include the installation of vibration dampers and active vibration control systems, with the simpler approach being the installation of vibration dampers. A damper was shown schematically in Fig. 776 and the effect of damping on the response of a system was illustrated in Fig. 779. The installation of vibration dampers on the load system of a press, for example, will have the effect of increasing the system damping and reducing the vibration in the press nip. Isolation pads are also used extensively in pump and motor bases. These systems can be termed "passive" in operation.

In some cases, it is possible to install "active" vibration damping systems to control vibration in troublesome components. These systems depend on a transducer to sense the vibration in a system and feed this information to a control computer. The computer then determines an appropriate response to be applied to an actuator in the system. Hydraulic actuators are usually used to supply a load on the system which "cancels" the normal vibration being sensed. These systems are complex and expensive, however, and are usually the last resort of the engineer to control serious vibration problems.

2. Operation for vibration control

Once equipment is designed and installed, its vibrational behavior will certainly be influenced by how the equipment is operated. Papermakers are usually keenly aware of the "normal" behavior of their machines. Any overall changes in machine behavior are quickly detected and corrective steps are taken. For example, felt barring caused by press bounce is usually noticed as an increased vibration around the press section. The offending felt is typically replaced when the vibration grows to be excessive. However, there are other changes the papermaker can choose to make to help reduce the chance of similar patterns developing in the new felt. The practice of cocking the felt to reduce the impact of the seam passing the nip is commonly used on presses sensitive to this problem. In addition, the papermaker can often turn to suppliers of clothing for assistance in selecting the best felt for minimizing the vibration.

If a paper machine is susceptible to vibration at certain running speeds, the papermaker should avoid running the machine at those speeds if at all possible. Often, a modest reduction in speed will decrease the excitation frequencies of the machine and significantly reduce the shaking forces due to unbalance. While this may not seem desirable from a production standpoint, minimizing the vibration in the machine will reduce the likelihood of a machine failure; production lost to minor speed changes will be offset by avoiding unscheduled downtime due to a failure. This approach to vibration control is only temporary, of course, and eventually machine changes must be made to allow running at the higher production speeds.

The papermaker can also help reduce the impact of vibration on mill productivity by maintaining a constant awareness of the machine condition. By paying careful attention to the details of the operating machine, unusual operating conditions which can lead to high vibration can often be avoided or at least corrected before serious damage can occur. For example, it has been noted that dryer cylinders which become waterlogged can induce high vibrations in the dryer section. By discovering and correcting such a condition early, the papermaker can avoid subjecting the dryer section components to high vibrations which can lead to equipment failures. Cyclic basis weight variations in the paper should also be avoided since they can influence the vibration of presses and calender stacks. Maintaining good press nips is also beneficial in minimizing the press vibrations and avoiding non-uniform felt compression or barring.

3. Vibration control through maintenance

The major burden of vibration control in running equipment falls on the mill's maintenance department personnel. These individuals, responsible for maintaining the equipment in good, safe operating condition, can make the difference between a successful vibration control effort and an ineffective one. Each repair and change they make to the machine has the potential to either reduce or increase the vibration in the machine [36-38]. Care must always be taken when correcting machine faults not to introduce new ones through poor workmanship or improper procedures. The replacement of bearings is one example where the life of a new bearing may be greatly reduced by an improper installation. Loose bolts can be a major cause of serious vibration problems on paper machines. Improper torquing of bolts during installation or looseness developed during operation can result in major failures, and great care must be taken to avoid these conditions.

By maintaining a good level of general equipment condition, the maintenance engineer and crew can keep overall vibration levels low and contribute dramatically to a machine's productivity. Properly installed and lubricated couplings, clutches, gears, and bearings produce much less vibration than faulty ones. This in turn lengthens the life of the components and keeps machine downtime at a minimum. The same can be said for pumps and motors around the machine.

Rolls in the paper machine present special challenges. When rolls are removed from paper machines, they are usually reconditioned in the mill shops or sent to vendors for repairs. As with other components, journals must be in good condition and straight with no material cracks. Even slightly bent journals can result in very high vibrations at frequencies equal to the roll rotation speed. Bearings must be carefully installed on the journals to avoid premature failures in the new installation. Finally, during installation it is important to obtain good alignment of the roll and its drive or nip. Misalignment of roll can produce high vibrations at frequencies once and twice the rotating speed of the roll.

It is also critical that any roll installed in the paper machine be balanced properly [3]. This is often done in the mill as part of the reconditioning of the roll. If the mill does not have the facilities for balancing rolls, the reconditioned roll must be sent out to a shop for balancing. It is important that the roll balancing be done to proper specifications. Balancing specifications can be written in two broad classes, "static balance" and "dynamic balance".

As the name implies, static balancing is performed on a stationary roll. It is used for short rolls where the length is not more than twice the diameter and rolls running at low speeds. Static balancing requires that a roll, when freely turned, will not exhibit any "heavy" regions when stopping. Balancing weights (usually steel bar stock) are added to the end(s) of the roll, or weight is removed by drilling, until the roll appears to be balanced. No effort is made in static balancing to determine the axial location of the unbalance. Therefore, it is possible that the unbalance and the balancing weight will combine to produce a force couple when the roll is rotating. For short rolls or rolls at low speeds, this effect is not important. However, "low speeds" must be identified for each type of roll in the machine. For example, static balancing may be sufficient for breast rolls at surface speeds below 305 m/min (1000 ft/min), where 914 m/min (3000 ft/min) might be the speed limit for dryer cylinders. A good guideline for the maximum speed for static balancing of most smaller rolls in the machine is 244–305 m/min (800–1000 ft/min).

A more precise standard for balancing rolls is "dynamic balancing". In dynamic balancing, the roll is balanced while turning in bearings attached to very flexible mounts. The balancing standard for the roll specifies the speed at which the roll should be balanced, and the maximum radial runout allowable for the roll during balancing. Dynamic balancing is important for longer slender rolls and rolls turning at higher speeds.

Different standards exist for rolls which will be mounted rigidly in the machine and those which are mounted on flexible supports. Examples of the former are felt rolls, table rolls, reel spools, and dryers. Flexible standards apply to press rolls mounted on swing arms, or rolls on rockers and tensioning devices. In both cases, the balancing weight can be proportioned to both ends of the roll to prevent the development of force couples when running in the machine. The maximum radial runout for balance is generally expressed in "mils", or thousandths of an inch, measured at the front, center and back of the roll. Typical values would be 4 mils (0.10 mm; 0.004 in.) for breast rolls and press rolls, and 2 mils (0.05 mm; 0.002 in.) for felt rolls.

Care must be taken not to assume a roll is balanced at all operating speeds of the machine. Particularly when the machine speed is increased, rolls which displayed little unbalance at one speed may vibrate badly at the higher speed. Rolls should be dynamically balanced at their anticipated operating speed. Since unbalance is a major source of vibration in the paper mill, the proper specification of balancing method and standard is critical to the continued smooth operation of the machine.

In addition to routine repairs and replacements of rolls, the paper machine maintenance department is often called upon to perform regular inspections of the machinery for the purpose of identifying potential problems before they become serious. Sometimes this function is placed with a separate group of trained "inspectors", but the goal is the same. By isolating potential problems in bearings and rolls before failures occur, the mill's productivity is maintained. Vibration analysis provides the most widely used technique for machinery condition monitoring in use today. Its use as a preventive/predictive maintenance tool began with the broomstick placed against the bearing housing to "listen" for bearing defects. The vibration meters and analyzers in use currently provide an objective means to quantify the vibration at points on the machine and track the development of problems through increased vibration levels. Reports from many industries attest to the effectiveness of vibration monitoring in reducing unscheduled maintenance downtime for bearing and other component replacement [39–42].

The most basic machinery monitoring program consists of a vibration meter for measuring overall vibration levels and a means to record the measurements at specific intervals over a period of time. A "route" of measurement points on the machines is determined and a routine established for checking the vibration at each point regularly. The measurements for each point are typically graphed to visually portray the vibration levels at the point and accentuate changes which might occur. This is called "trending" the vibrations at the point; a typical trend graph for a bearing is shown in Fig. 793. Vibration severity levels are indicated on the plot, which covers a time period of almost one year. Recent increases in vibration into the "rough" region imply that the bearing should be changed out at the first convenient opportunity.

While it is possible to conduct a vibration monitoring program with a simple meter and hand recording the results, systems are now available which automate many of the most tedious functions of the program [43]. Mechanics or inspectors can make their rounds with a

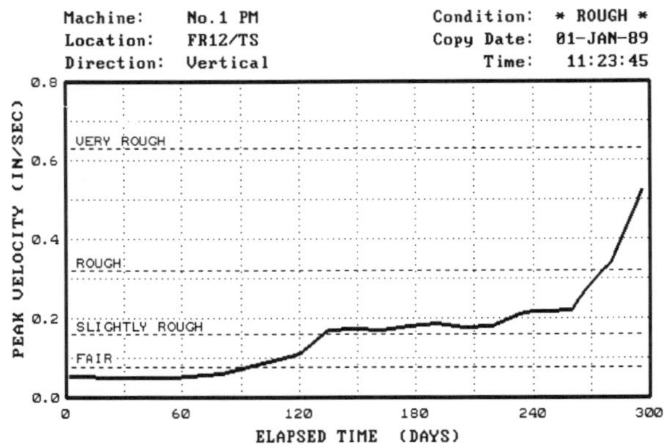

Fig. 793. Trend plot of bearing vibrations.

hand-held data logger, sequentially measuring the vibrations at each point on a route. The data logger will automatically save the overall vibration level while this is done. Some data loggers have advanced features, such as a display and built-in FFT analyzer, to allow the inspector to obtain more information from a point if desired. Upon completion of the route, the data logger is connected to a host computer and the measurements are transferred to a database of readings maintained in the computer. The computer can then be used to produce reports of where vibrations have increased, or draw graphs of vibration trends which are increasing. This type of system allows an inspector to cover a much larger area of the mill than would be possible with a simpler system, and eliminate error-prone activities such as the manual recording of vibration levels and data entry into the computer.

Regardless of the sophistication of the system employed, the use of regular vibration monitoring can produce a significant reduction of unscheduled maintenance downtime when properly used. "Proper use" in this case means maintaining a high level of effort to ensure all points are measured when due and that all critical equipment in the mill is included in an inspection route. Experience has shown that well executed monitoring programs can be one of the most effective vibration control methods available to the mill.

F. ON-LINE VIBRATION MONITORING SYSTEMS

The logical extension of routine vibration monitoring via hand-held data loggers and computerized databases is full-time monitoring of machine condition via dedicated computer systems and permanently installed transducers. Reductions in computer costs, coupled with higher costs for machine downtime, have made these systems a reality in the paper mill today. Different levels of sophistication and machine coverage are available, but regardless of which is chosen these systems are replacing manual inspection on new machines and being retrofitted on critical existing machines.

The primary goal of automated machinery monitoring systems is the reduction of unscheduled downtime due to unexpected or premature failures of equipment and clothing. On-line systems can also be used to maintain high quality levels in the paper produced on the machine. It has been found that these systems can help papermakers run their machines efficiently, often at higher average speeds than was previously common for the machine [44–46]. Some systems include built-in diagnostic capabilities which help isolate the cause of increased vibration as it occurs. Other systems combine vibration analysis capabilities with other functions such as wet end pulsation analysis, process variable monitoring, and lubrication system monitoring. One type of full-machine system is shown schematically in Fig. 794.

Vibration monitoring is accomplished through permanently installed transducers at critical points on the machine. A complete machine monitoring system might include monitoring of wet end parameters, such as pressure pulsations and breast roll vibrations, as well as press, coater, and calender vibrations. Several types of transducers can be used in a monitoring system. Typically, proximity probes and accelerometers are used to gather mechanical vibration data. Other devices, such as pressure, flow, and temperature transducers, are used to gather other process data. In addition, monitoring systems can also receive data from the machine's

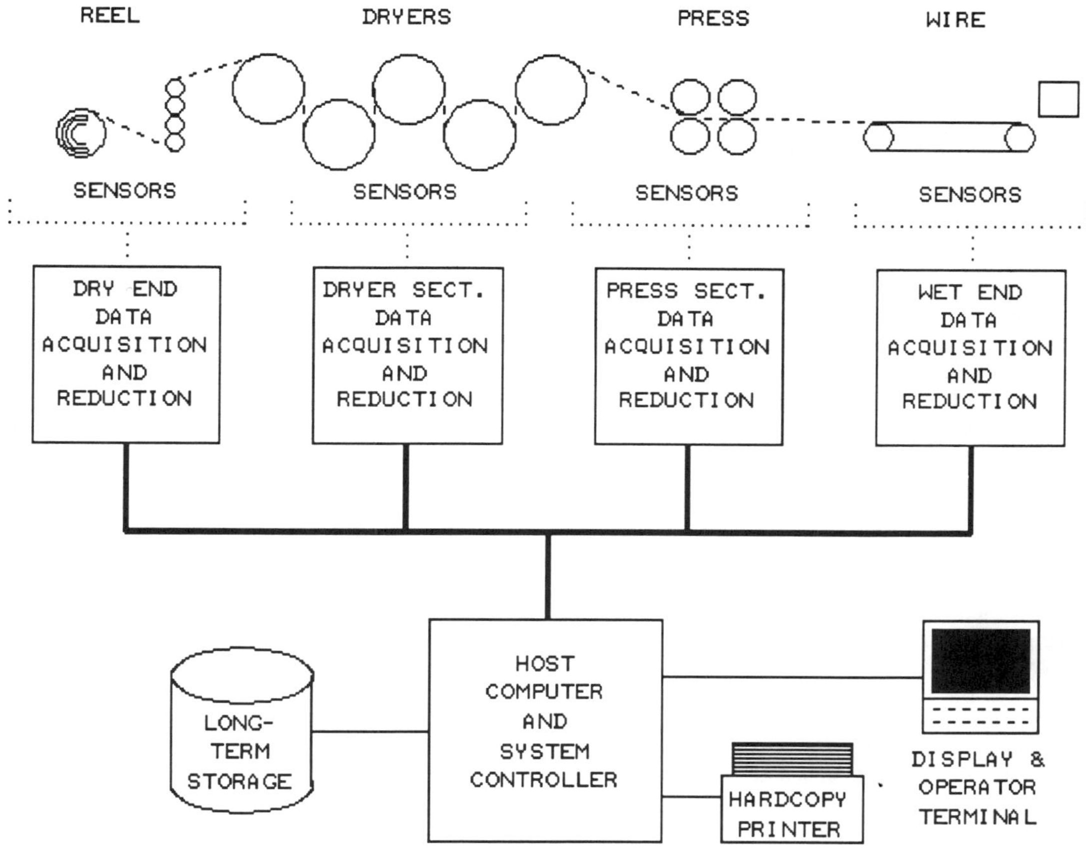

Fig. 794. Schematic diagram of paper machine monitoring system.

basis weight and moisture scanners for correlation with vibration data.

One critical set of measurement points is found in the press section, particularly at the press nips where felt life may be extended through careful monitoring. Transducers installed here must be manufactured to withstand the surrounding environment of heat and oil. Usually the transducers are mounted at the measurement points on both the tending and drive sides of the machine. Mounting can be accomplished by bolting the transducer to the housing, or using an epoxy-based glue. Cabling is run from the transducer to a remote signal conditioning/power supply cabinet, and from there to the machine control room. The entire installation can be accomplished in a routine down and can be compact and out of the way for normal operations.

The heart of an automated vibration monitoring system is the host computer data acquisition unit, with high-speed analog to digital (A/D) conversion capabilities. Depending on the extent of the installation and the number of monitoring points included, this computer can range from a desktop personal computer to a large minicomputer. It is often the case, however, that the processing power available in the system's computer(s) is the limiting factor in how many inputs can be handled, or how much time elapses between scans of a given input. Some systems use a "distributed control" philosophy in which data acquisition and reduction are handled by dedicated processors on each section of the machine to reduce the scan-to-scan time requirement. The results of the analysis then can be transmitted to a central host computer for display, archiving, and reporting functions. In other systems, all data acquisition and reduction are handled by the host, which also provides operator-interface functions. In either case, the data acquisition portion of the system must be capable of acquiring data at a controlled rate sufficient to cover the frequency range of interest on the machine. This does not usually present a problem to state-of-the-art equipment. The unit periodically scans each channel of vibration input and digitizes the vibration signal received.

It is possible to provide synchronization signals from roll rotation as triggers for the data acquisition. In this case, the computer can perform synchronous averaging or time averaging of the signal to help define the vibration at a point due to a given roll. Synchronization pulses can also be picked up from the press felts and machine wires by using an optical trigger to detect the passage of a light or dark spot on the fabric.

The analysis of the transducer data obtained can include overall vibration level determination as well as FFTs of the signal. Commonly, both overall level and frequency-domain information are used for comparison to previous readings and historical "baseline" spectra. Sudden changes in level at a given point may trigger an alarm in the system to warn the operators that a vibration problem may exist. Trend graphs and spectra are available to the operator on request. When unusual problems arise, it is often possible to use the vibration monitoring system to study the problem and help isolate the cause of the change without resorting to a more detailed and costly analysis.

Dedicated monitoring systems have made vibration analysis part of the routine operation of new paper machines. Powerful computers and increasing usage have eased the burden of the vibration inspectors on these machines. However, costs can be high for these systems, and for old or slow paper machines there may not be a clear return on the investment required for their installation. Existing machines without these systems still require routine vibration monitoring to avoid the pitfalls of unexpected high vibrations and their effect on paper machines and mills.

G. SUMMARY

Vibration is an ever-present aspect of papermaking, induced by mechanical and/or hydraulic phenomena. Regardless of the effort expended, it is virtually impossible to completely eliminate this vibration. Indeed, the amplitudes of vibration may be reduced and the frequency of vibration may be altered by various techniques, but the *total* elimination of vibration is an unrealistic goal. For these reasons, it is necessary that papermakers and paper mill engineers understand vibration and its effects on the papermaking process and equipment, and that they are prepared to deal with vibration problems appropriately.

Vibration analysis should be an integral part of any serious papermaking effort if one hopes to minimize machine downtime, both scheduled and unscheduled, and maintain high productivity and paper quality. Another benefit of an active vibration analysis program is the preservation of the highest possible paper quality from a given paper machine, free from the detrimental effects of vibration on quality.

For the papermaker and the engineer alike, understanding vibration and careful vibration monitoring of the paper machine can minimize the adverse impacts of vibrations on machine productivity and thus the paper mill as a whole.

REFERENCES

1. Thompson, W.T., Theory of Vibration with Applications, Prentice-Hall, Englewood Cliffs, NJ (1981).
2. Baumeister, T., Mark's Standard Handbook for Mechanical Engineers, McGraw-Hill, New York, Ch. 5, p. 67–75 (1978).
3. Jackson, C., The Practical Vibration Primer, Gulf Publishing Co., Houston, TX (1979).
4. Harris, C.M., ed., Shock and Vibration Handbook, 3rd Ed., McGraw-Hill, New York (1988).
5. Eschelman, R.L., "Studying Vibration with Time-Domain Analysis", Machine Design 55(1):128 (Jan. 6, 1983).
6. Barasch, M., "Modulation: A Problem in Vibration Troubleshooting", PIMA 66(3):51–55 (Mar. 1984).
7. Eshleman, R.L., "Machinery Diagnostics and Your FFT", Sound and Vibration 17(4):12–18 (Apr. 1983).
8. Fox, R.L., "Successful Vibration Analysis Techniques", Tappi 62(3):59–62 (Mar. 1979).
9. Thorp, B.A. and Barasch, M., "Maintaining the Paper Mill Through Vibration Analysis", PIMA 66(2):47–54 (Feb. 1984).

10. "Turbomachinery Problems and Their Correction", Sawyer's Turbomachinery Maintenance Handbook, Vol. 2, Ch. 7, Turbomachinery International Publications, Norwalk, CT (1980).
11. Chenco, G., "Review of Vibration Measurement and Analysis Applied to Machinery Maintenance", Appita 35(5):417–419 (Mar. 1982).
12. Gilstrap, M., "Acceleration vs. Velocity vs. Absolute Vibration Measurement Transducers", Sensors 2(8):16–20 (Aug. 1984).
13. Taylor, J.I., "Diagnosing Paper Machine Problems with Vibration Analysis", PIMA 67(6):61–62 (June 1985).
14. Mitchell, J.S., Machinery Analysis and Monitoring, PennWell Books, Tulsa, OK (1981).
15. Maines, R., "The Basics of Piezoelectric Sensors", Sensors 6(5):26–27 (May 1989).
16. Application Brief: "Improved Machine Maintenance using Simple Vibration Measurements", Bruel & Kjaer Instruments, Marlborough, MA (May 1986).
17. Wochinger, A.L. and Danforth, H., "Wet End Troubleshooting", Tappi 63(4):71–75 (Apr. 1980).
18. Perrault, J., "The Influence of Wet End Vibrations on MD Grammage Variations", Tappi J. 67(7):62–65 (July 1984).
19. Whyte, D.McG., "The Application of Predictive Maintenance to the Pulp and Paper Industry", Appita 36(1):47–51 (July 1982).
20. Taylor, R.R., "Predict Bearing Failures with Portable 'Checkers' ", Hydrocarbon Processing 61(1):88–90 (Jan. 1982).
21. Philips, G.J., "Measuring the Size of Rolling Element Bearing Flaws", Sound and Vibration 23(4):6–16 (Apr. 1989).
22. Hanna, A.J., "Predictive Maintenance Via Vibration Analysis", Tappi 57(5):153–157 (May 1974).
23. Yolton, J.R., "Applications of Predictive Maintenance Technology", TAPPI Focus on Predictive Maintenance, p. 17–20 (Aug. 1988).
24. Manson, D.W., "The Practical Aspects of Formation", TAPPI Wet End Operations Seminar Notes, p. 49–55 (1988).
25. Kallmes, O., et al., "Forming a Sheet on a Fabric at or Near Its Resonant Frequency", TAPPI Wet End Operations Seminar Notes, p. 463–470 (1989).
26. Kokkonen, O. and Karvinen, M., "Vibration Control and Analysis of Modern High Speed Press Sections", Proc. 1981 Engineering Conf., TAPPI PRESS, Atlanta, p. 161–168.
27. Anema, J. and Clarke, B., "The Cause and Cure of Press Section Vibrations", Tappi J. 63(3):58–61 (Mar. 1984).
28. Royo, M.A. and Bullerwell, A.C., "Press Vibrations", TAPPI Practical Aspects of Pressing and Drying Seminar Notes, p. 99–102 (1989).
29. Vinicki, J., "Vibration Analysis in the Press Section", Tappi J. 70(11):91–95 (Nov. 1987).
30. Taylor, J.I., "Troubleshooting with Vibration Analysis", PIMA 67(9):51 (June 1985).
31. Roisum, D.R., "Winder Vibration Can Reduce Operating Efficiency and Increase Maintenance Costs", Tappi J. 71(1):87–96 (Jan. 1988).
32. Chen, Y.N. and Boos, G., "Calender Barring on Paper Machines: Practical Conclusions and Recommendations", Tappi 58(7):93–97 (July 1975).
33. Chen, Y.N. and Boos, G., "Calender Barring on Paper Machines: Theoretical Model Development", Tappi 58(8):147–151 (Aug. 1975).
34. Emmanuel, A., "Some Experiences with Calender Barring on a Newsprint Machine and Diagnosis of Roll Corrugations", Appita 38(4):269–274 (July 1985).
35. Fenner, D.N., Engineering Stress Analysis: A Finite Element Approach, Wiley, New York (1987).
36. Fitch, J. and Wood, W.R., "The Fundamentals of Paper Mill Maintenance", Tappi 63(12):31–34 (Dec. 1980).
37. Brower, M.W., "Maintenance Productivity: The Hands-on Approach", Tappi 63(12): 11–12 (Dec. 1980).
38. Elphick, J.C., "Maintenance Programs Extend Useful Life of Equipment", American Papermaker 52(8):23–27 (Aug. 1989).
39. "Managing Maintenance with Computers", Chemical Week 131(4):36–42 (Oct. 1982).
40. Lafontaine, Y., "Preventive Maintenance Through Vibration Analysis", PIMA Mag. 64(11):21–23 (Nov. 1982).
41. Corcoran, T., "Fraser's Vibration Analysis Program", PIMA 64(11):13–16 (Nov. 1982).
42. Fisher, L., "The Payoff: On Stream Factors", Orbit 5(2):11–13 (July 1984).
43. Dodd, V.R., "Condition Monitoring of Major Turbomachinery Cuts Costs Over Four-year Period", Orbit 5(3):8–11 (Dec. 1984).
44. Teo, S., "Condition Monitoring of Slow Speed Rolling Element Bearings in a Mechanically Noisy Environment", Appita 42(3):206–208 (May 1989).
45. Yeager, R., "Runnability Monitoring System Measures More Than Vibrations", American Papermaker 52(3):26–30 (Apr. 1989).
46. Yeager, R., "Online Condition Monitoring Boosts Runnability at Lake Superior Mill", Pulp & Paper 63(2):61–64 (Feb. 1989).

XXV COVERED ROLL CARE AND GRINDING

J.A. ZWINAK

Joe Zwinak is Director of Technical Services for Beloit Manhattan, Division of Beloit Corporation, a Harnischfeger Industry Company. He studied Chemical Engineering at Rutgers University. During his 30 years with Beloit Manhattan he has been involved in Research and Development, Technical Sales, Application Engineering and Technical Services. He has written several papers for various trade publications, is a member of TAPPI and is an Instructor for the TAPPI Practical Aspects of Pressing and Drying Short Course.

REVIEWED BY

James P. McNamee
Consultant
Pawtucket, RI

A. INTRODUCTION

Today's paper industry demands dictate the necessity for a good maintenance schedule for the various rolls in a paper machine. This chapter deals with the proper roll care and grinding procedures to be followed to maximize paper machine efficiency and meet sheet quality demands. The paper mill rolls which are most frequently changed out and ground are covered press rolls and chilled iron calender rolls. The first item to be discussed is when and why it is necessary to grind rubber-covered rolls since these rolls are much more affected by the vagaries of the paper machine.

The grinding cycle is invariably the longest time in the machine before the surface contour of the roll (crown) deteriorates to an extent that will affect the sheet, the felt or the integrity of the cover itself. While it is possible to give some approximate times for grinding cycles, the actual times in any mill must be determined for that mill. In addition, each position on the machine will generally have a different cycle. Grinding cycles are established through the examination and interpretation of roll records.

Wear of the ordinary abrasion type is often considered to be the primary cause of surface contour degradation. Uneven wear is the real problem and there are many things which will cause uneven wear.

The first of these is an uneven load distribution across the face. The most obvious cause of this is an improper crown which will result in greater loads on one part of the cover than another effectively causing differences in wear due to speed differentials and temperature differences as well as actual load differentials.

Temperature differences across the roll face can be the result of things other than varying loads due to surface contours not matching the core deflections exactly. Among the many factors which have caused temperature variations are web temperature differences, temperature carryover from spray nozzles used for cleaning or chemical application, massive roll heads, truss construction of the core, improper design of water-cooling equipment and problems with oil circulation in crown compensating type rolls.

Chemical attack always acts to increase the rate of wear. Chemical attack also is very temperature dependent so that cross-machine temperature differences accelerate uneven wear.

The hardness history of the cover can reveal a great deal to the experienced eye. The normal behavior of a cover is to harden with age. This will vary from being almost imperceptible to quite rapid depending on the type of compound used to cover the roll. Once such a rate of hardening has been established for a given application, changes in that rate are most likely to be due to changes in the operating conditions. More rapid hardening is indicative of much more severe operating conditions while less rapid hardening, which may carry over into softening, indicates chemical attack. It is possible for chemical attack to result in hardening but for our purposes such effects are so rare as to be ignored. In today's environment, changes in hardness behavior are much more likely to be due to changes in operating conditions than to changes or variations in the covers.

Rolls operating in a paper machine under load will deflect and it is necessary to compensate for this deflection by grinding a convex shape or crown into the cover. The crown is measured on diameter and is equal to twice the roll deflection.

When comparing the crown of a newly installed roll (in-crowns) and the crown of a roll that was removed (out-crowns), it is important to have a benchmark, which is why an accurate roll centerline diameter is important. In this way, a plot of the in-crown and out-crown can be made to see if what we are dealing with is wear or whether growth due to chemical attack is the problem.

Figure 795 is an illustration of roll crown specifications and definitions. Roll crown specification and terminology are addressed in detail in TAPPI Technical Information Sheet (TIS) 0420-06 [1].

Since the questions of when and why to grind a covered roll have been answered, actual roll handling and operating and grinding procedures can be addressed. These will be covered in three separate sections:

1. Roll handling – receiving, shipping; inspection; storage.
2. In-service operation – installation; taking a nip impression; roll operation.
3. General grinding instructions – grinding techniques and equipment; surface finish; dimensional tolerances.

B. ROLL HANDLING

1. Receiving, shipping and inspection

All paper mill rolls must be handled with care to avoid damage to the roll surface, journals and bearings. Extreme care is to be exercised when handling covered rolls of any type (rubber, polyurethane, plastic) since they are much more susceptible to impact damage and environmental changes.

Rolls should always be supported on journals and not on the roll body or cover.

Shipping cases are to be constructed of either wood or steel and contain journal rests to support the roll and prevent shifting of the roll in transit. There should be a minimum of 2″ (5.08 cm) clearance between the cover surface and any part of the shipping case. Upon receipt of any covered roll, thoroughly inspect the roll and case to ensure that the cover was not touching the top,

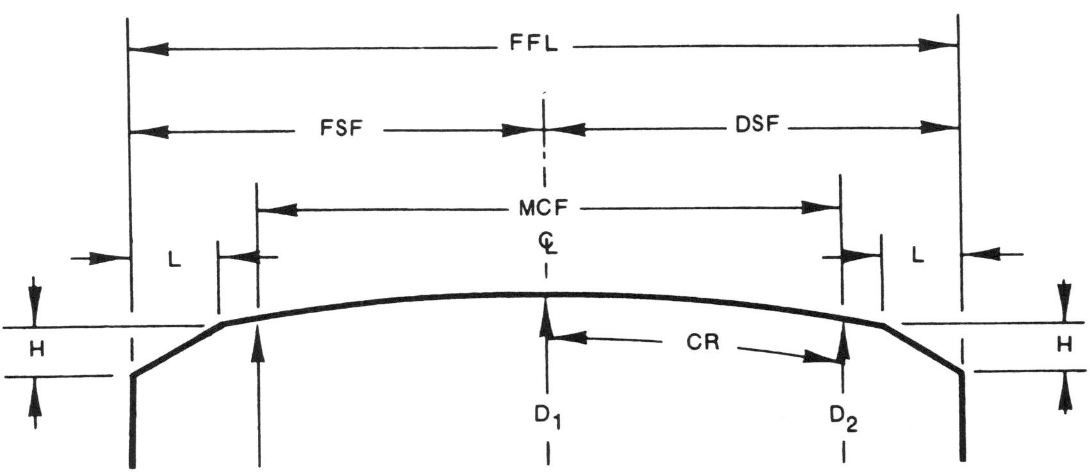

Fig. 795. Roll crown specifications and definitions [1]. FFL — full face length, inches; FSF — front side face, inches; DSF — drive side face, inches; MCF — measured crown face; D_1–D_2 — crown amount, thousandths of inches; CR — cam rotation, degrees; L — dub length, inches; H — dub depth, thousandths of inches; D_1 — basic roll diameter measured at centerline of measured crown face.

bottom or sides of the case. If the cover has been allowed to come in contact with any part of the case, the cover more than likely is severely damaged, and may have to be replaced or, at a minimum, reground. Contact the transportation company immediately and then your cover supplier if such damage has occurred in transit.

The roll should be carefully removed from the shipping case using the properly rated slings or cables. Using two hooks, pick the roll up by the journals, taking care not to allow the slings or cables to contact any bearing surface.

When journals cannot be used, use wide nylon slings to pick the roll up by its face and wrap the roll with multiple layers (3 to 4) of protective paper wrapper. Slippage will occur between the sheets of paper and prevent the cover from splitting from the sling stretching under load. Your roll cover supplier can provide you with information regarding the proper sling width to be used. Avoid using a single hook to handle covered rolls; but when it is absolutely necessary, use wood blocks at the contact point of the cables and the cover (see Fig. 796).

Once the roll has been unboxed, it should be supported on roll journal stands, blocks or on the bearing housings. Inspect the wrapping paper to see if it is torn in any area which could indicate damage to the cover. Remove the paper and inspect the cover surface for any damage. Check the hardness of the cover, centerline diameter, crown profile and end relief (when applicable). All these data should be recorded in the roll records. Figure 797 illustrates a typical roll record card.

2. Storage

The roll is now ready for installation in the paper machine or storage. Rolls to be placed in storage should be rewrapped with heavy kraft paper, and stored in a cool (15–21°C; 60–70°F), dark, damp area away from any type of electrical equipment. This will help prevent ozone attack which can cause rapid oxidation of the cover surface which manifests itself as thousands of fine cracks. Roll covers must also be protected from oil, chemical, and mechanical hazards while in storage (see Fig. 798). Since rolls naturally sag, it is advisable to rotate them 90° or 180° periodically (30–60 days). This is especially important for rolls of lightweight construction, such as, table rolls, wire rolls, lumpbreakers, coater backing rolls, etc. Rolls stored for long periods of time (over 6 months) should be reground before installation in the paper machine.

Roll covers should never be allowed to freeze [below 0°C (32°F)] during transit or in storage. Rolls received which have been subjected to freezing temperatures, must be moved with extreme care and slowly allowed to reach room temperature to prevent damage to the cover. Even though the cover may not have any visible evidence of cover surface cracking, irreparable damage to the cover substrate may have occurred as a result of exposure to freezing temperatures. Roll covers shipped during winter months must be wrapped in insulation and shipped in heated vans to prevent thermal shock to the cover.

C. IN-SERVICE OPERATION

1. Installation

When a covered roll is received from the roll cover supplier, it has been prepared in most cases to be put directly into the machine. If the roll being used is coming from storage, it theoretically should be lightly reground before being installed. This is to remove any possible surface oxidation or skin hardening which could lead to surface cracking. Unfortunately this is not

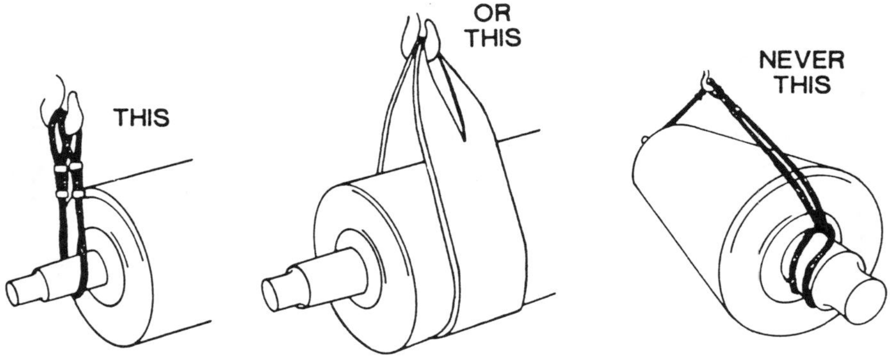

Fig. 796. Handling covered rolls.

Position .... Type of Roll .... Roll No. .... Ser. No. ....

Length of Face .... Diameter of Core .... Finished Diameter ....

Cover Material and Hardness .... Thickness of Cover ....

Weight of Roll .... Crown .... Drawing No. .... Bearings....................

Installed	Taken Out	No. Days Service	Date Reground	Crown	Hardness Reading	Finish O.D.	Covered by	REMARKS

Fig. 797. Roll covering service record.

always possible but, if the roll has been stored for six months or longer, it is indeed a must.

During installation, extreme care should be used during lifting and moving of the roll. Once in place, it is good practice to verify the alignment relative to its mating roll. (Obviously, this is when a mating roll is present, as in any press position.)

Having proper alignment is extremely important in achieving a uniform pressure across the nip. Non-uniform nip pressure due to misalignment will cause premature localized wear on the covers reducing the normal running time between regrinds. Misalignment, if severe enough, can cause a cover failure. Exact roll placement for proper alignment is usually marked or pinned but these can shift or wear over a period of time and should be rechecked at each roll installation. Roll alignment, in relation to its mating roll, can easily be checked by use of a roll alignment gauge, an example of which is shown in Fig. 799, or any level long enough to reach the tangent between the rolls.

Another important check which should be done when a roll is installed, is to ensure that the loading mechanism of the press is in top working condition. Consideration should be given to installing gauges at each load cylinder to verify the load readings set on the floor console. More often than not, the load readings on the floor console frequently get out of calibration causing the operator, unknowingly, to overload or underload the press. Having the gauges on the individual load cylinders takes away any inaccuracies in the air system between the cylinders and the floor console.

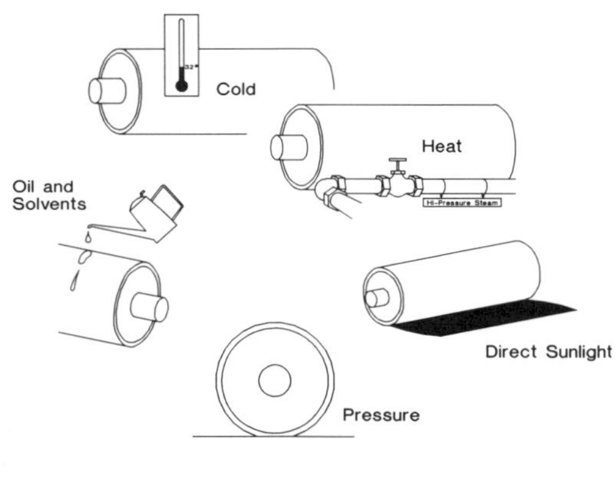

Fig. 798. Roll damaging agents.

Once it has been verified that the press-loading mechanism is in good working order, it is recommended that a nip impression be taken of the nip formed between the two rolls under the desired press loading. This nip impression, or footprint, of the nip will allow us to verify if we have an even load distribution across the nip by measuring the nip width. This nip width should be the same across the entire nip.

It is a good idea, at this point, before taking the impression, to check another extremely important detail. It should be verified that the desired operating nip pressure is the correct pressure for which the two rolls forming the nip have been crowned. Every press should have two important curves usually supplied by the press manufacturer relating the loading applied, and the required crown for the rolls. One curve shows the required crown for both rolls for any given operating load pli (pounds per linear inch) [kN/m (kilonewtons per meter)] up to the maximum design load. The other curve relates pli (kN/m) load to the required psi (kg/cm^2) loading applied to the press to achieve any desired pli (kN/m). The total crown of both rolls together need to match exactly the desired pli (kN/m) on the curve to achieve a uniform nip as long as the curves are up-to-date.

Figure 800 illustrates how too much or not enough crown for any given load will affect the nip profile. These illustrations also show the effect of too low or too high of applied load (kN/m) for which the rolls are crowned.

2. Taking a nip impression

Taking a nip impression is relatively quick and easy to do. A nip impression kit with instructions can be obtained from your roll cover supplier. The nip paper is usually some type of either carbon paper or embossed foil so that a footprint can be obtained of the nip between the two rolls. The procedure is basically either holding or taping the paper in the nip area, loading the press

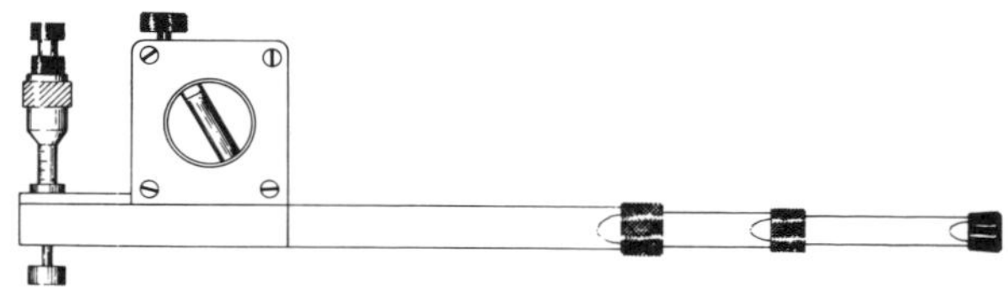

Fig. 799. Alignment gauge (Courtesy of Appleton Mills).

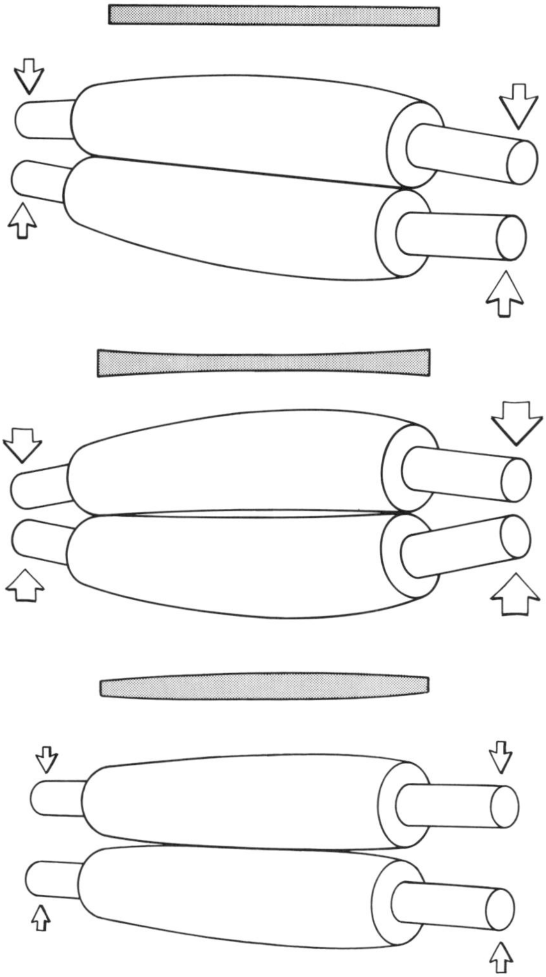

Fig. 800. Effect of crown.

to the desired load, releasing the press and removing the paper. One should then label the paper with the date, the press, the press loading, the roll's number, the crown in each roll, and which end is the front (or tending end) or back end.

The next step is to measure the impression. It is always good to measure the nip impression length and compare it to the crowned face length of the rolls. This is to verify that the entire crowned face was in the nip. The other important measurement is the nip impression width and how it might vary across the length of the impression. In many cases, people will eyeball the impression and say that it looks uniform. However, a uniform nip impression (nip width) should be uniform usually within 0.05 in. (0.127 cm) and this is too small to just eyeball. When measuring the nip width, one should take at least five measurements; one on each end of the impression, usually 1–2 in. (2.54–5.08 cm) inside to be sure you are out of any dub area, one each at the quarter points between the ends and center, and one, of course, at the centerline. Some prefer to take up to twenty measurements across the length matching the measurements with each crown station. They would then graph the measurements to take a closer look of the uniformity of the nip.

The next step is to evaluate these measurements. The first thing to look for is that both ends should be the same; i.e., the impression should be symmetrical about the centerline. If the impression width of the ends are significantly different, a number of causes could be present. The impression is sensitive to differences in the rate of loading between the front and back ends of the press. If one end should reach full load before the other, that end will also be applying a larger portion of roll weight causing a higher nip loading on this end until the other side comes to full load and balances out the roll weight. Even though the press nip is now evenly loaded, the impression will still show a heavy side on the end that loaded faster. This could be one cause for different end measurements. Another could be simply that the heavy end was indeed loaded heavier (this is where the gauges at the load cylinders help) or, possibly, the light end did not reach full load. Another cause could be misalignment. Whatever the cause, it is important that it be found and corrected.

If the impression does measure to be fairly symmetrical about the centerline, then one can compare the end measurements to the centerline measurement to verify the proper crowning for the applied load. If the ends and center measurements are equal, then the nip has a uniform pressure profile and the press is ready for operation. If the center measurement is different from the ends, then a change in the amount of crown or the amount of pli (kN/m) applied needs to be adjusted. Figure 801 shows the formula used for calculating the amount of crown correction needed to give a uniform nip profile under the present pli conditions. As can be seen, this equation relates the end and center nip width measurements to the roll diameters to come up with a value "C", which is the amount of additional or reduction of crown from the total crown of both rolls needed to make the nip uniform. If the center measurement is lower than the ends, then "C" will be positive and will call for the addition of "C" to the total crown. If, on the other hand, the center is larger than the ends, then "C" will be negative and this amount of crown needs to be reduced from the total crown.

C = additional crown required i.e., the difference in diameter between the center and 2" in from the ends of the roll

N_1 = nip width at center of roll

N_2 = nip width 2" in from the ends of the roll

D_1 = diameter of top roll

D_2 = diameter of bottom roll

then

$$C = \frac{(N_2^2 - N_1^2)\ (D_1 + D_2)}{2 D_1 D_2}$$

or if rolls have equal diameters

$$C = \frac{N_2^2 - N_1^2}{D}$$

Fig. 801. Crown correction formula. Nip width and diameters determined.

Since the rolls are already in the machine and it is not logical to take them out for regrinding at this point, the pli (kN/m) should then be adjusted up (if not at maximum) or down, depending on "C". If "C" is positive, the load should be decreased; if "C" is negative, the load should be increased. In most cases, pli (kN/m) and total crown are directly proportional so that, if "C" calls for a 5% increase in crown, then a 5% decrease in pli will do the same to achieve a uniform nip. In any event, it is important to make sure that the press rolls have a uniform nip profile when installed.

Nip impressions are very sensitive to how they are taken. If not done properly, it could be impossible to determine if any variances in the nip width were caused by improper crowning or just by carelessness when taken. Below are listed a number of specific concerns which will help to ensure that the nip impression taken will be helpful.

1. When using a nip impression to verify if the crowns are correct, both rolls in the nip must be freshly ground.
2. The nip needs to be clean and dry with all felts out of the way.
3. If a number of impressions are desired, always rotate the rolls some amount after each impression to prevent a possible set in the covers.
4. If a symmetrical nip impression exists, the crown correction formula can be used to find the proper load or crown.
5. When taking a nip impression with a suction roll involved, the vacuum area must be sealed and the impression taken with the vacuum on. If this is not achievable, the vacuum effect can be calculated and added to the pli (kN/m). The vacuum in the nip does contribute to the total deflection of the rolls and the machinery builder of the press has included the pli (kN/m) equivalent to compensate for the vacuum on the original crown curves for that press.
6. When taking an impression with a crown compensating-type roll, special procedures need to be taken. Each roll should have instructions to do this from the roll manufacturer.
7. When taking an impression with a multiple nip press, (such as a 3 roll/2 nip or 4 roll/3 nip configuration), all nips must be loaded to obtain a representative impression. The total deflection found in one nip at a given pli (kN/m) is dependent upon the pli (kN/m) applied in the other nips.
8. Every new crown calculation or crown change should be verified by a nip impression.

3. Roll operating

Internal water-cooling [2] is often necessary to prevent temperature buildup and avoid cover failure. Water-cooling should keep the metal/rubber interface below 71.1°C (160°F). Cover temperatures can exceed 93.3°C (200°F) without water cooling. A general guideline is that press rolls operating over 400 pli (70 kN/m) should be water-cooled. The rate of cooling water should be sufficient so that inlet and outlet water temperatures are within 2.8°–5.6°C (5–10°F). Inadequate cooling water flow can cause uneven roll temperature profiles. This, in turn, can cause uneven wear, premature loss of crown shape, poor runnability and possible cover failure.

When crown compensating-type rolls are in use, the outlet oil temperature should be below 62.8°C (145°F). Some roll manufacturers specify lower outlet oil temperatures.

As mentioned, for optimum running time in between necessary regrinding of the cover, one must make sure that the cover is seeing a uniform pressure profile upon start-up. Obviously, this is also true during its run cycle. However, this is not always obtainable just by ensuring that the degree of loading isn't changed. Many other machine factors can cause a cover to wear nonuniformly and thus cause an uneven pressure profile over time. Observing a roll cover's surface periodically during its run cycle can alert a machine operator to a potential problem usually before it becomes a major problem. When time allows, a quick visual inspection of the roll covers should be done during any interruption of running, looking for possible irregularities of the cover surface.

If surface cracking is noticed, this is a sure sign of a problem. If the cracks appear straight and parallel to the roll axis and forming a circumferential band of cracks, it is most likely due to overloading in that area and one needs to find its cause. Irregular cracks not parallel to the roll axis and not all the way around the roll are usually due to some impact damage by a paper wad or anything foreign going through the nip. Excessive heat in an area can also cause cracking. Whenever surface cracking is noticed, it is most important to remove the roll as soon as possible for regrinding to prevent any further cracking and hopefully save the cover. It is also important when grinding a cracked cover that you make sure to grind below the cracked area because any cracks still in the surface will grow when operated again. If any surface cracks appear to be significantly deep, there is the possibility that the cover adhesion may be affected and your roll cover supplier should be consulted.

One should look for any changes in the surface appearance, such as washout areas, abraded or rough areas, rounded land areas on grooved rolls, hole cupping of drilled rolls, etc. Any of these appearances is an indication that the cover should be reground. Corrugations in the cover parallel to the roll axis are the result of anything which imparts an impulse to the nip. Such things as out-of-balance rolls, felt seams, bad bearings, universal joints, etc., can cause bouncing which in turn will lead to corrugations. One other cause of corrugations is leaving the rolls under pressure without them rotating. If the cover takes a bad enough set from this, the roll will need to be reground below the affected area.

Chemical attack [3] on the roll surface is one more potential problem that early detection can minimize. This sort of attack can happen at any time, due to the ever-changing conditions on the machine. Chemical attack on roll covers is not likely to be sudden and catastrophic, and can usually be detected by noticing any irregularities on the surface, such as pits, roughness, orange peel, hole cupping, fine cracking, premature or increased wear, etc. A change in hardness is the most indicative sign of attack. It is almost always softening, but on rare occasion hardening can occur.

There is no way to predict, except in a general fashion, the action of a new chemical on existing covers even if the exact chemical composition is known. It should be noted that most chemicals are proprietary products. Their compositions are given only in a generalized form. In addition, combinations of chemicals may have a greater effect than the sum of the effects of each individually. This synergistic effect on the covering cannot always be predicted.

Obviously, under these conditions, and with the multitude of chemical additives, both the roll covering manufacturer and the roll covering user have a common interest. Processing chemicals, chemicals which become a part of the product, cleaning agents, defoamers and inhibitors are increasingly important and are growing in variety and in quantity. It is essential that the roll cover user provide complete information to the roll covering manufacturer. The problem is to determine the offending chemical or combination of chemicals. Once this is done, those chemicals may be eliminated or, if this is impossible because of process restrictions, the covering manufacturer may be able to supply a covering which will be more resistant to those chemicals.

Not only the chemical itself but also the temperature at which the chemical is introduced to the press as well as the operating temperature of the press itself are important. Higher temperatures invariably accelerate attack. Concentration is another factor which must be considered. All coverings have what can be called a threshold of concentration of some chemicals below which they will not be attacked. By contrast, other chemicals, even in trace amounts, will be selectively absorbed by the covering. Time and contact with the chemical will have a bearing on this problem. Very often, the change from a one-shift to a two-shift usage can result in irreversible chemical attack on the covering. Dynamic changes might, in themselves, influence the rate of chemical attack.

Roll covering manufacturers have libraries and they cross-reference the field results and laboratory tests available to estimate the behavior of a roll covering in a given environment. Laboratory tests are complicated, time consuming, and require experience in the interpretation of the results. It is, therefore, important to both the manufacturer and the user that complete and accurate information regarding the service conditions be given to the roll covering manufacturer if chemical immersion testing is to be done.

It is important that roll covering users advise roll covering manufacturers of contemplated changes in any of the materials and conditions. Only in this way will any potential problem be minimized.

When regrinding a roll cover that has been subjected to chemical attack, and has changed in hardness, one must grind down far enough to ensure that all of the affected cover is removed and the hardness has come back to or very close to the original value.

D. GENERAL GRINDING INSTRUCTIONS

1. Grinding techniques and equipment

Roll grinding is accomplished using a one- or two-wheel grinder which traverses the roll face as the roll rotates on its axis, supported at each end on journals or bearings. The quality of the grind depends on a number of factors, such as the condition of the roll-grinding equipment, selection of the proper type and grit grinding wheel or belt, roll and traverse speeds and, of course, operator proficiency. Dimensional control is very critical and temperature control of the grinding area is necessary to prevent deformation of the roll or grinder. Wide variations of temperature caused by drafts from open doors, sunlight, heaters, etc., particularly when concentrated on one part of the roll or grinder, will affect dimensional accuracy. Therefore, it is quite important to select an area in the mill with good climatic controls and minimal vibration levels as a site for a roll grinder installation.

Prior to setting up a roll in a grinder, inspect the roll surface for damage, particularly covered rolls. Severely damaged roll covers will more than likely have to be recovered and your roll cover supplier should be contacted to determine the extent of the damage. Also, the roll cover hardness, centerline diameter and crown profile are to be measured and recorded. Unusual cover hardness changes and crown profile wear should be addressed at this time, since the cover may have been subjected to a chemical which generally softens the cover and causes premature cover wear.

Due to the sheer weight of press rolls, they should be ground on bearings whenever possible. Journal rests may be used; however, they must be continuously lubricated to prevent heating up the journals which will result in roll distortion. As noted above, temperature control of the grinding area is critical to maintain dimensional control. Likewise, the temperature of the roll cover and metal core must be stabilized at room temperature before attempting to grind a cover. Since rubber has a coefficient of thermal expansion about ten times that of steel, a difference of 2.8 to 5.6°C (5–10°F) between the cover and the metal can result in unacceptable surface contour variations. Rolls, therefore, should not be ground immediately upon removal from the paper machine since there is substantial heat generated in the cover during operation. In general, the cover/metal temperature should stabilize in approximately 24 h.

Once the roll is set up on the grinding stands and rotating freely, the cover runout should be checked across the entire face of the cover to detect any flat spots in the cover. These flat spots can be caused by either leaving the roll loaded in the paper machine during shutdown or allowing the roll to rest on the floor. If a cover is ground round after having such a flat spot imposed on it, the flat spot will recover after flexing in the nip and the result will be a cross-machine bump or ridge. In any case, the grinding of the roll should be delayed until the affected area has had sufficient time to recover. Another type of set is that which is unavoidably caused by the metal core as a roll body. The result of this set is the assumption of a banana shape (sag) by the roll body, particularly on lightweight rolls. Rotating press rolls in the grinder at a slow speed for a few hours will remove this sag.

Removal of a minimum of 0.76 mm (0.030 in.) thickness of the cover will usually restore the cover to its original hardness and surface properties. Covers which have been subjected to chemical attack or are corrugated will have to have additional cover material removed to regain the original cover physical properties. Rolls can be ground either dry or wet, using the appropriate coolant. As a rule, when grinding dry, the depth of the cut should not exceed 0.13 mm (0.005 in.) for roughing passes and should be reduced to 0.05 mm (0.002 in.) for finishing passes. When regrinding a roll cover, a 32–36 grit silicon carbide grinding wheel or 80–180 grit silicon carbide belts are recommended. The average grinding wheel speed should be 1220–1372 m/min (4000–4500 surface ft/min) and the roll speed 30.5–38.1 m/min (100–125 surface ft/min). The traverse is to be set at approximately two thirds the wheel width during roughing passes and reduced to a minimum for finishing passes and the wheel should be allowed to travel off the end of the cover during each grinding pass. The ends of all press rolls are to be dubbed or tapered in order to relieve pressure on the edges of the felt and roll covers involved. This relief should be renewed to original specification at each regrind. Wet grinding, in general, produces a smoother finish on covered rolls and is a necessity when grinding chilled iron rolls. The selection of the appropriate grinding coolant, an effective filtering system and the application of a generous flow of coolant between the wheel and the roll all contribute to obtaining a finish free from surface imperfections.

2. Surface finish

As stated above (Sec. A), the paper mill rolls most frequently ground are covered press rolls and chilled iron calender rolls. The proper grinding wheel must be selected to guarantee

Table 85. Tolerances for covered rolls.

STANDARD TOLERANCES — PAPER INDUSTRY										
Class	Finished diameters		Tolerance on diameter		Tolerance on concentricity (TIR)		Tolerance on face length		Tolerance on taper (diameter)	
	mm	in.	mm	in.	mm	in.	mm	in.	mm	in.
Standard	**all**	**all**	**±1.5**	**±1/16**	**0.05**	**0.002**	**±6**	**±1/4**	**±0.05**	**±0.002**
Precision	up to 200	up to 8	±0.04	±1/64	0.03	0.001	±6	±1/4	±0.05	±0.002
Precision	200 to 750	8 to 30	±0.2%	±0.2%	0.03	0.001	±6	±1/4	±0.05	±0.002
Precision	over 750	over 30	±1.5	±1/16	0.03	0.001	±6	±1/4	±0.05	±0.002

Amount of Crown	Crown Tolerance
up to and including 0.25 mm (0.010″)	±0.03 mm (±0.001″)
0.26 mm — 1 mm (0.011″ — 0.040″)	±0.05 mm (±0.002″)
Over 1 mm (Over 0.040″)	±5% of the total specified crown

that the required finish is obtained and, therefore, a wheel change is recommended when switching from grinding covered rolls to metal rolls.

Factors to be considered when selecting grinding wheels are grit size, grade, density and bond. Refer to TAPPI TIS 0420-1 (Old No. TIS 405-2), specifically Table 405-2.1, "Recommended Practice for Roll Grinding", for details concerning grinding wheel selection and roll grinding operating parameters for the various types of covered rolls as well as metal rolls [4].

3. Dimensional tolerances

Table 85 gives the paper industry dimensional tolerances for covered rolls. The table is an excerpt from the roll covering handbook published by the Rubber Manufacturers Association [3]; this publication is an excellent reference material for roll cover care and maintenance. Consult the individual paper machine supplier for dimensional tolerances for metal rolls.

REFERENCES

1. TAPPI TIS 0420-06, Roll Crown Specifications (rev. 1985).
2. TAPPI TIS 0404-37, Press Roll Cover Application Guidelines, p. 2.
3. Roll Covering Handbook, Rubber Manufacturers Assoc., IP-5, p. 1.5, 3.4, 3.5 (1989).
4. TAPPI TIS 0420-1, Grinding Paper Mill Rolls.

XXVI PAPER MACHINE SAFETY

R.C. VONDERAU

Ray Vonderau recently retired in 1991 as Regional Vice President after more than 31 years of association with the Beloit Corporation. He received a B.S. degree in Mechanical Engineering from Purdue University in 1949. His work experience includes managerial assignments with Beloit Paper Machinery Division, Beloit Canada, Ltd., Beloit Jones Division and, most recently, with Beloit Nippon, Ltd. in Tokyo, Japan. He has written technical articles and presented technical papers and has been active with TAPPI as a Sustaining Member Committee Member and Session Chairman. He is currently a member of TAPPI, ASME, ASME — Japan Chapter, and is a Registered Professional Engineer in the State of Wisconsin.

REVIEWED BY

William S. Fuller
Director of Safety and Health
Weyerhaeuser Paper Company
Tacoma, WA

Jack E. Ugulini
Safety and Plant Protection Supervisor
James River Corp.
Camas, WA

A. INTRODUCTION

Most of the material contained in this chapter has been taken from the Beloit Corporation Manual "Partners – Paper Machine Safety". Since many paper mill safety organizations shared their safety regulations to help create this original manual, we are indeed grateful for their contribution of the material in this chapter which is dedicated to help make the paper mill a safe working environment.

Federal, State and local safety regulations and codes exist and are made into law to ensure compliance with practices deemed safe in the paper machine environment. If any statements or recommendations in this text are found not to be in full agreement with such regulations and codes, the Federal, State and local laws must be given preference for compliance.

B. PAPER MACHINE SAFETY

Today's paper mill management recognizes that its primary responsibility is to provide a safe working environment for its employees. The safety of mill employees is one of the prerequisites for successful and profitable operation. One papermaker's philosophy is as follows: "We believe that safety is an inseparable part of every job at every level. A job is done right only when it is done safely".

This is all very well and good; however, one of the major causes of occupational injuries and illness is the failure of employees to observe safety rules. One successful program initiated by a paper company was to have the employees write their own safety rules. This was accomplished by having them go through their jobs step by step and write down all the ways they could think of in which they could be injured. A list was then compiled with the actions that must be taken to keep from being injured. While developing these lists, familiarity with the safety rules increased.

The results of this program was a job safety analysis conducted by the most qualified persons – the operators themselves. Employee knowledge of the job increased as well as increased awareness of the hazards of the job.

Management is responsible for providing a safe working environment. A program like the one described above improves employee attitude because they have a voice in it. The safe practices come from them and the rules are generated from the employee peer group. New employees also accept rules more rapidly when the source is the workers.

Also, government legislation and regulations have created an entirely new environment in the paper mill, creating new operating conditions.

This government involvement has made everyone safety conscious.

The paper machine and supportive operations are by their very nature potentially hazardous to work around. Stock preparation with its grinders, shredders and refiners; the paper machine with its rolls rotating at tremendous speeds, high-pressure nips and great physical mass and momentum; paper finishing with its slitters, sheeters and high-speed winding equipment are all in this category.

However, with the proper mechanical and electrical protection devices, proper training of mill personnel, and good on-going safety programs, the hazards of the papermaking industry can be minimized. The result: an injury-free safety record for every mill that subscribes to the premise that a safe place to work encourages maximum efficiency and improves morale – the benefits are higher productivity and improved profits.

Machine builders and suppliers of machinery stand ready to assist the papermaker in finding solutions for safeguarding existing machinery and enhance safety through safety-engineered principles related to paper machinery development and design.

Following are some of the ways the machine builder may be of assistance to the papermaker in his safety programs.

- Review safe operating procedures throughout the papermaking process.
- Provide a source of guards and barriers.
- Review use of footwalks, steps and ladders for increased effective use.
- Provide a source of safety signs designed specifically for use on papermaking machinery.
- Recommend use of color for designated danger areas.
- Recommend and provide sources for warning alarm systems.

Although guards and barriers do protect and signs are excellent reminders of hazardous areas, there is no substitute for a good continuing safety training program. Safety, like machine operations, is a teachable skill. FOR THE PAPERMAKER, TRAINING IS THE GREATEST DETERRENT TO ACCIDENTS.

1. Mill safety program

Most mills have a safety program of some kind established and working. Some safety programs are quite sophisticated. The safety records of these mills substantiate this. Where a safety program is established, and followed, accidents can be virtually eliminated.

Good sources of information on how to establish safety programs and run safety meetings are the federal and state safety departments, American Paper Institute (API), Paper Industry Management Association (PIMA), The National Safety Council, Occupational Safety and Health Administration (OSHA), and other civic and industrial organizations.

C. GENERAL PRACTICES FOR OPERATION AND MAINTENANCE

1. Nip protection/threading devices

The paper machines consist almost entirely of rolls. Some of these rolls are used to press moisture from the sheets of paper or the felts; others are used to smooth the surfaces of the paper. These rolls contacting each other form one of the greatest hazards on the machine, that of ingoing nips. Nips are also formed where wire, felt, paper webs and threading ropes wrap rolls and pulleys. Accidents resulting from nip points not only occur in the highest frequency, but they are also the most costly injuries.

Hand feeding of the paper web into nips must be eliminated.

For other special threading problems such as splicing on winders and coaters, please refer to the manufacturer's operating manuals.

Equipment is continually being improved so machines can be threaded without operating personnel being physically involved close to an in-running nip.

2. Stairways, ladders and footwalks

Stairways, ladders, and footwalks are a vital part of the paper machine. Because of the great size of the machine and the need to have access, there are stairways, ladders, and footwalks extending to the uppermost reaches at strategic points and across machines.

Footwalks provide safe access to critical areas. Walking on rolls and other cross-machine members must be discouraged because of the danger to the operator and damage that may be done to the machine. The footwalks are designed to cross the machine at specific locations. It is unwise to cross the machine when it is operating except at points that are specifically designated. These areas would be where there is ample clearance for an operator to cross the machine without coming close to any moving component.

Stairways have a maximum pitch of 50° so the operator may ascend or descend walking forward. On a ladder with a pitch steeper than 50°, the operator must always face the ladder when ascending or descending.

Footwalks must be a minimum of 24 in. wide except where unique situations dictate; however, these arrangements must conform to OSHA regulations. Railings must be 42 in. high from the walk surface with a knee rail 18 in. down from the top. Railings should be 1.5 in. diameter pipe.

Walk surfaces should be non-skid open grating heavy enough to support 100 pounds per square foot. Grating pattern, though open, must be small enough to keep tools from falling through. A 4 in. toe board should be included on the footwalk to keep tools, etc., from being kicked over the side and endangering personnel working below. Use of removable sections should be minimized. When used, these sections must be secure during machine operation.

Ladders have a pitch of 75° or greater and the operator must always face the ladder when ascending or descending. A ladder with a 75° to 90° pitch rising more than twenty feet must be supplied with a cage. The cage will extend down to 7 ft. (min.) to 8 ft. (max.) above the base of the ladder.

3. Housekeeping

Poor housekeeping results in cluttered areas, tripping hazards, lack of room and the possibility of slipping and falling. Housekeeping has an important environmental influence. It has been found that when housekeeping is good, work attitudes are better, efficiency is up, and there are fewer injuries. Some ways to improve housekeeping are:

- Have a definite place for everything.
- Keep everything in its place and return it when removed.
- Put scrap and waste in proper containers.
- Put cigarette butts, cans, bottles, wrappers, etc., in receptacles.

Examples of poor housekeeping that cause accidents are:

- Loose objects on the floor – tripping.
- Slippery materials on the floor – slips and falls.
- Loose objects overhead – may fall on someone.
- Objects out of place – persons may bump against.
- Insecurely piled material – may cause cuts and bruises.
- Improperly stored tools – may cause cuts and bruises.
- Trash and disorder – disorderly work habits.
- Disorderly surroundings – bad attitudes.

REMEMBER, NO JOB IS FINISHED UNTIL IT IS CLEANED UP! Safety starts with good housekeeping. Experience has shown time and time again that good housekeeping is essential for a successful safety program.

4. Protective clothing and equipment

It is becoming common practice in most mills to wear safety glasses, safety-toed shoes, and hard hats. During downtime maintenance and mill construction, hard hats are to be worn as they are during any other major construction job involving overhead work [1].

Protective clothing must be worn while using the caustic solutions to dissolve slime buildups during "washdowns" on the wet end and when working on or near any of the acidic chemicals used in the stock preparation and felt cleaning. Eye protection must be worn when using high-pressure hoses.

Safety showers must be provided close to areas where caustic solutions or acids are in use.

If employees are assigned to use protective clothing and other equipment because of the hazardous environment in which they have to work, they must be trained in the proper use of the equipment.

Personal protective clothing and equipment worn when the extent of a potential hazard warrants their use include: foot and shin protection, hard hats, ear plugs or other sound attenuation devices, glasses, goggles, masks, rubber gloves, and rubber boots. In severe environments, respirators, protective masks, rubber coats or coveralls are recommended.

Individuals required to wear breathing respirators for extended periods must obtain a medical release and complete fitness testing [7].

When an employee's hair is long enough to obscure vision or become tangled in machinery or a nip, a hair net must be worn. Rings, neckties and other jewelry must not be worn where they create a hazard.

5. Hazardous materials

The standard in Ref. [11] is designed to make sure that chemical hazards are assessed and the information made known to the worker by the employer. By reading the Material Safety Data Sheets (MSDS) and container labels, the worker will gain increased awareness and knowledge about chemical hazards. It is the responsibility of the employer to train personnel how to interpret labels and MSDS. The employer must obtain or develop the MSDS for all hazardous chemicals used throughout the mill.

Hazards from materials are divided into two categories: physical hazards and health hazards.

Materials posing physical hazards are those that burn easily, explode or react in some other dangerous way.

Materials posing health hazards are those substances that are: irritating, corrosive, sensitizing or toxic to the skin, eyes, mucous membranes, lungs or other body organs, or the substance has been shown to cause cancer in humans or lab animals.

A list of hazardous chemicals should be posted in plain view at the work site. The worker should be trained in methods and observations to detect the presence of the hazardous chemical. The worker must then be trained in protective measures to take with specific procedures to implement the worker's protection. Training must include use of personal protective clothing and equipment, work methods, and ventilation procedures and control.

6. Emergency stops

Paper machines must be equipped with devices that will stop the machine in an emergency. OSHA recommends that the devices consist of pushbuttons for electric motive power (or electrically operated engine stops), pull cords connected directly to the prime mover, control clutches, or other devices interlocked with adequate braking action. These emergency stops must be located so any person working on the machine can stop the machine section, or the entire machine, in the quickest possible manner [2].

Emergency stop devices on hazardous machines should be red. Stop buttons or electrical switches with letters or other markings used for emergency stopping or machinery should be red.

Quick stop devices should be tested periodically by making use of them during normal operations, or during planned shutdown periods, to make certain they are operational at all times.

New employees must be made aware of the emergency stops in their section as part of their safety training and indoctrination.

7. Safety signs

Safety signs are an integral part of well-managed safety programs. They cannot take the place of an ongoing safety program or a physical guard. They can only warn personnel of a potential hazard or hazardous area.

OSHA directs that a "*Danger*" sign be used only where an immediate hazard exists.

All employees are to be instructed that danger signs indicate immediate danger and that special precautions are necessary.

"*Caution*" signs have the word caution on a black square, the message is in black letters on a yellow field. All employees shall be instructed that caution signs indicate a possible hazard against which proper precaution should be taken [3].

Generally accepted safety colors now being used to a great extent are red and green.

- *Red* is the designation for fire protection equipment and apparatus, danger and stop.
- *Green* designates the basis color for "*Safety*" and the location of first-aid equipment.
- *Orange*. This basic color designates hazardous points on machinery which might cause personal injury. Examples would be nip guards and barriers, roll ends, electrical components, prohibited walkways, rotary shafts and points on machinery to be totally avoided during operation.
- *Yellow*. This basic color designates caution and is used for marking physical hazards such as "striking against", stumbling, falling, tripping, and potential pinch points. Solid yellow, yellow and black stripes, or yellow and black checkers can be used interchangeably. Examples would be all guards, crossing walks, and ladders or steps.

8. Air hose safety

The use of high-pressure air delivered, directed, and controlled with hand-held hoses is a vital part of paper machinery operation for a myriad of functions including threading. It is important that air hose safety be discussed in detail:

1. Prior to use:
   a. Inspect air lines to ensure that a bulbous safety nozzle is securely attached to hose. The bulbous nozzle prevents the hose from being drawn into a closed nip.
   b. Always grip hose and nozzle in hands.
   c. Position hose between yourself and equipment.
   d. Never allow arms, legs or any part of the body in hose loops.
   e. When turning on air hose for someone else, be sure the person holding the nozzle knows when the air is to be turned on and has a secure grip on the nozzle.
2. During use:
   a. Always use eye protection when using the air hose.
   b. Hold air hose at the outgoing nip, never at the ingoing nip.
   c. When blowing broke from running dryers, do not extend hands beyond ropes.
   d. Do not extend nozzle beyond ropes.

   e. Never direct air stream at a person – never horseplay with air hoses.
   f. Air hoses are not designed for blowing clothes.
   g. Be sure of direction of the nozzle at all times. Never attempt to answer someone while using an air hose.
3. Following use:
   a. When not in use, the hoses on attending side of the machine should be coiled and placed as close to the machine as possible. All other hoses should also be coiled or hung on racks.
   b. Periodically inspect air hoses for:
      1) Cracks in hoses.
      2) Hose clamps.
      3) Nozzle threads.

9. Lock-out safety

One of the greatest potential hazards in the paper mill is failure to lock out power sources and drives. Because of the large size of the equipment, components may be started up while machine tenders are out of sight of the operator. A warning blast should be given and a final blast before starting large equipment. Safety with controls cannot be emphasized enough. Complete safety instructions and periodic refresher courses should be given to operating personnel so they are keenly aware of lock-out procedures.

The following lock-out procedures are meant to be suggestions only. Most mills have their own lock-out procedures. Lock-out procedures established by mills must comply with OSHA regulations and by the state in which the mill is located. Some states have very specific procedures regarding these important safety precautions [4].

Prevent personal injury to employees and/or damage to machinery by following these procedures.

1. Notify the operating supervisor or operator or both of the particular equipment or section that has to be shut down and locked out.
2. If the source of power is electrical, an electrician, usually the one responsible for the equipment in question, should be notified that a disconnect has to be opened.
3. When the electrician opens the disconnect, the circuit should be tested with a meter to make sure the current is broken and the disconnect switch locked out.
4. A group supervisor and/or individual worker must then apply a lock to the disconnect or switch box. Many mills have set up color codes for locks and provide them from a central source, or the workers may have their own locks identified by color, name, clock number, and/or combination thereof.

   EACH EMPLOYEE WORKING ON THE MACHINE IS REQUIRED TO PUT A PERSONALIZED LOCK ON THE SWITCH.

   There are a variety of devices on the market that permit a number of locks to be placed on a single disconnect or switch.
5. A worker applying a lock should have the electrician test the circuit to confirm the identification of the circuit that has been taken out of service.
6. Where electrical disconnects may be hazardous, an electrician may apply or remove the lock for a non-electrical worker . . . but only while the worker is present because each worker is responsible for his/her own lock.
7. When working on mechanically driven devices powered by hydraulics, gas, air or steam, the control levers, electrical power to the pump motors, and valves must be locked out to prevent starting of the equipment. Raised components must be mechanically blocked. Operating personnel are normally responsible for closing valves, draining lines, etc. Each employee working on this type of equipment is to attach his/her own lock to the device before initiating work on it.
8. When the work has finished, the last person to remove the lock must inform the operating supervisor or operator or both that the equipment is ready to be put back in service.

Provisions for lock removal in case a worker is relieved before the job is completed or leaves the mill premises for an extended time should be set up by the mill to conform with its own policy. A lock must never be removed in the absence of a worker unless it has been established that this worker will not return within a reasonable work period. Removal of this lock, if necessary, must be handled through proper mill management channels.

On major shutdowns, such as clothing changes, i.e., wires, felts or dry felts, it is recommended that a senior shift supervisor also be required to lock-out the equipment involved. The supervisor should follow lock-out procedures as established by the mill which, in turn, conform to OSHA and state-regulated safety acts.

10. Tanks, chests and pressure vessels

When a worker must enter a tank, chest, barking drum, dryer cylinder or other potentially hazardous area, it is extremely important that all controls be locked out and the employee follow an OSHA-approved confined space entry procedure. Valves for all inputs to these vessels must be secured and locked.

Note: Pressure vessels must be properly ventilated before the worker enters them.

In addition to keeping airborne contaminants in breathing atmospheres at or below recommended limits, ventilation in confined spaces must also (1) ensure adequate oxygen for life support but not an oxygen-enriched atmosphere, and (2) prevent accumulation of flammable mixtures. Asphyxiation can quickly result in unconsciousness and death without warning. A confined space must not be entered unless (1) it is well ventilated, or (2) the welder is properly trained to work in such spaces and is wearing an approved air-supplied breathing apparatus. A similarly equipped second person must be present. Confined spaces should be tested for (1) toxic or flammable gases and vapors, and (2) adequate or excess oxygen before entering, using instruments approved by the U.S. Bureau of Mines. Heavier-than-air gases, such as argon, MPS (methyl acetylene propadene stabilized; sometimes called MAPS gas), propane and carbon dioxide, may accumulate in pits, tank bottoms, low areas and near floors. Lighter-than-air gases, such as helium and hydrogen, may accumulate in tank tops, high areas and near ceilings. The precautions for confined spaces also apply to these areas. If practical, a continuous monitoring system with audible alarms should be used for confined space work.

Oxygen-enriched atmospheres can pose great danger to occupants of confined areas. They are especially hazardous at oxygen concentrations above 25%.

Protection in confined spaces must be provided not only for welders but also for other persons in the enclosure. Only clean, respirable air must be used for ventilation. Oxygen, other gases, or mixtures of gases must never be used for ventilation.

All personnel entering tanks, chests or other pressure vessels must wear a safety body harness with a lifeline to permit rapid rescue. At least one individual must be stationed outside to observe the person or persons working inside. In addition to following lock-out procedures as previously described, the worker must securely attach a safety lock to all access doors or openings through which tanks, chests, drums, or cylinders are entered. Upon leaving any of the above, the worker is to remove this lock. If other locks are at the entrance, they must not be disturbed.

11. Welding in confined spaces

Special consideration must be given to the safety and health of welders and other workers in confined spaces. Gas cylinders must be located outside of the confined space to avoid possible contamination of the space with leaking gases or volatiles. Welding power sources should also be located outside to avoid danger of engine exhaust and electric shock.

A means for removing persons quickly in case of emergency must be provided. Safety belts and lifelines, when used, should be attached to the worker's body in a manner that avoids the possibility of the person becoming jammed in the exit. A trained helper should be stationed outside the confined space with a preplanned rescue procedure to be put into effect in case of emergency.

The cautions previously stated, relating to possible hazards for airborne contaminants, are especially important during welding operations. When welding, cutting or related processes are performed in confined areas immediately dangerous to life and health, positive pressure self-contained breathing apparatus must be used. It must have an emergency air supply of at least 5 min duration in the event that the main source fails.

12. Corrosion inspection

A major common problem in all paper mills is corrosion.

Maintenance personnel and everyone else associated with the paper machine should be constantly on the alert for corroded parts in all areas of the machine and report them so they may be repaired.

Corrosion is the deterioration of a substance (usually a metal) or its properties due to a reaction with its environment. Properties, as well as the material itself, can and do deteriorate. In some forms of corrosion, there is no weight change or visible deterioration, yet the material may fail unexpectedly. Such changes may defy ordinary visual observation but may be detected using proper methods of examination with which maintenance engineers should become familiar. See, for example, the chapter on Corrosion in Volume 9 of this series.

(a) Visible forms of corrosion

- uniform
- galvanic
- pitting
- intergranular

(b) Forms of corrosion not readily visible

- crevice
- selective leaching stress
- erosion

(c) Mill environment

In general, white water closure affects corrosion by causing a substantial increase of dissolved solids, gases, temperature and biological activity. Both organic and inorganic constituents in the dissolved solids influence the severity of corrosion.

Most significant among these are chlorides, sulfates, lignin-related sulfur compounds, organic acids, and carbonates.

(d) Corrosion monitoring

1. Visual inspection – identifies rusted, eroded or corroded bolts, brackets, walkways, etc. so repairs may be implemented as soon as possible.
2. Electrical resistance – measures corrosion rate as a result of electrical resistance changes in a probe which corrodes in the environment.
3. Linear polarization – measures instantaneous corrosion rate.
4. Coupons (racks) – longer term used to monitor corrosion rates and compare numerous samples relative to each other.
5. Mill environment – analytical testing should be employed to monitor the effect of process changes and to identify corrosive chemicals. The most important values to monitor are pH, acidity, TDS, chlorides, sulfates, thiosulfates and conductivity.

All of the above monitoring methods are used readily in white waters and pipe lines.

13. Safe use of lifting devices

There are many types of lifting devices used in a paper mill to remove rolls and various equipment from the paper machine and to transport it to a storage location or maintenance area. The same lifting facilities are also used to re-install the rolls or equipment after the necessary maintenance work has been performed. It is essential that the mill maintenance people receive proper training and supervision to be sure these routine operations are done safely in accordance with the manufacturer's specifications and according to the appropriate ANSI/ASME and OSHA Standards [5].

ANSI/ASME Standard B30.9-1984 and OSHA Regulation 1910.184 provide useful safety standards for cableways, cranes, derricks, hoists, hooks, jacks and slings and should be studied by all mill personnel involved in the use of lifting devices. It is important to be aware of the safe capacity of each of the elements of the lift; ie. crane, hoist, hook, chain or sling at the angle of lift produced by the specific hitch for each lift.

It is also important to maintain an awareness of the condition of the lifting equipment to be sure the original rated capacity has not been diminished with age or misuse. Qualified personnel responsible for routine periodic inspection of the condition of these devices should be specifically assigned and trained to be sure safe lifting practices published by ANSI/ASME and OSHA are rigorously followed at all times.

14. Emergency preparedness

At many paper mill sites, hazardous substances and chemicals are present depending on the type of pulping/bleaching/processing systems in use. Even though these materials may not always be used or stored within the machine room, paper machine operators and personnel working around the paper machine area must also be familiar with the mill emergency response procedures should an accident occur at the mill site which involves these hazardous materials.

OSHA regulations require employee training, written procedures for emergency response and periodic inspections to ensure the emergency response procedures are known and continue to be adequate for any new chemical added which may result from a change to the existing mill facilities. The applicable regulations are very specific and mill management must be knowledgeable about the publications listed in the bibliography [11].

D. NOISE SAFETY

Dangerous noise levels may be capable of producing permanent damage to the hearing of paper mill employees. A noise control program should be an important part in a mill's safety program. With the establishment of health and safety guidelines through the noise exposure provisions of OSHA, the paper industry has moved to establish noise survey procedures and controls. The level of the noise and exposure time to that noise over a normal working period can determine how damaging it may be to human hearing. Warning signs should indicate noise areas.

Awareness in paper mills of the noise hazard to operating personnel has initiated safety policies and procedures regarding the use of protective devices and "*sound safety areas*". A wide variety of approved personal protective equipment is available to reduce sound levels. Sound and vibration-proofed rooms may also be

MACHINE NO:		MACHINE NO:	
TYPE:		TYPE:	
SPEED:		SPEED:	
PAPER GRADE:		PAPER GRADE:	

SECTION AREAS	READING LOCATION	REMARKS	NOISE LEVEL		OCTAVE BANDS						
			A	AP	63	125	250	500	1000	2000	4000
WINDERS	C 100	SPEED									
	3-B 100										
	4-B 100										
REEL	C 90										
	3-B 90										
	4-B 90										
CALENDER	C 80	NIPS –									
	3-B 80	ROLLS –									
	4-B 80										
DRYERS	C 70										
	3-B 70										
	4-B 70										
BREAKER CALENDER	C 60										
	3-B 60										

Fig. 802. Typical intermediate environmental noise survey form.

provided on or just off the machine room floors so the operators are protected but are still in close proximity to the paper machine. Signs should be in evidence to identify these "*quiet*" areas.

Safety procedures regarding noise must be strictly adhered to. Hearing protection should be a major part of any safety training program and proper instructions with refresher courses should be given to new employees and old hands alike. Noise is an irritant to workers and reduction of noise has been proven to increase production [6].

Injury to hearing may be attributed to the following unsafe practices:

- Being in an area identified as hazardous without hearing protection.
- Spending too much time in a hazardous area with inadequate or no hearing protection.

Adequate noise surveys of the workers' environment is a necessary part of noise control. Environmental surveys should be conducted periodically to determine existing noise levels. A good procedure to follow is the TAPPI recommended TIS 0420-03 "Noise Survey Techniques for Noise Mapping in Pulp and Paper Mills" shown in Fig. 802.

Table 86. OSHA occupational noise exposure regulations, Section 1910.95, Table G1-16.

Duration of Daily Exposure (h)	Allowable Level db(A)
8	90
6	92
4	95
3	97
2	100
1–1.5	102
1	105
0.5	110
0.25	115

1. Hearing conservation program

The employer shall administer a continuing, effective hearing conservation program, whenever employee noise exposures equal or exceed an 8 h time-weighted average sound level (TWA) of 85 decibels measured on the A scale (slow response) or, equivalently, a dose of 50%. For purposes of the hearing conservation program, employee noise exposures shall be computed in accordance with Table 86.

E. PAPER MACHINE LIGHTING

Proper lighting in and around the paper machine is crucial to maintain good machine operation and a safe working environment. All traffic areas must be well lighted. Also, major papermaking functions should be well lighted so they can be clearly observed. For example, the stock inlet and forming zone on a twin-wire former should be illuminated both at the front and back of the machine for ease of observation during operation. Critical paper runs between machine sections should be illuminated.

Lighting should be provided under footwalks where machine room lighting may be reduced or blocked out. When the overhead crane is not furnished with lighting, the lights on the paper machine must provide adequate illumination while the crane is positioned overhead.

On the wet end of the machine, properly sealed, waterproof lighting fixtures designed for application in wet environments must be used. The lenses of these light fixtures must be kept free of fiber and pulp buildup to ensure maximum effectiveness. All lighting fixtures must be maintained to ensure a continuously safe environment. Some areas may be illuminated by spot lights, while other areas will require flood lights.

It is the responsibility of the mill to provide proper and adequate lighting on and around the paper machine.

If requested, the machine builder may be able to assist by recommending the type of lighting and locations for maximum effect, determined by the specific needs of a given installation.

Lighting levels should conform to Federal and State regulations for industrial lighting. National Electrical Code, 1970 Standard (subject to current review) recommends lighting standards for papermaking (foot candle on tasks) as follows:

Beaters, grinding	30
Finishing, cutting, trimming, papermaking machines	50
Hand counting, wet end of paper machine	70
Paper machine, calender and reel paper inspection, laboratories	100
Rewinder	150

F. RECOMMENDED SAFETY REMINDERS

Do not operate the equipment without first reviewing the following instructions.

1. Do not remove warning signs. They are installed to warn personnel of possible danger. Observe all instructions given on the signs.
2. Observe all color coding.
3. Footwalks, handrails, barriers, and guards must be in place before starting the machine.
4. Do not over-reach, climb or stand on places other than properly designed and designated ladders, steps, or walkways, except during machine/mill outages when repairs are authorized.
5. Aisles must be clean and clear of obstructions. Wipe up spilled oil, grease, unnecessary water, and stock. Good housekeeping prevents injuries.
6. Keep clothing and all parts of the body away from ingoing nips, travelling belts, fabrics, wires, ropes, and rotating or pivoting loading mechanisms.
7. Beware of head-high obstacles in and around the machinery area. Wear proper head protection when indicated.
8. Exhaust blasts from air motors may blow dirt, scale, or other foreign materials into eyes causing eye injury. Wear proper eye protection when indicated.
9. Keep all parts of the body away from the belt drives.
10. Do not walk under operating equipment in the machine room basement.
11. A nip point on operating paper machinery may be dangerous. Keep clothing and all parts of the body away from these points while machine is operating. Do not wear loose clothing, long hair or jewelry that could become entangled in the roll nips.
12. Do not operate equipment until all personnel are accounted for and outside of safety lines.
13. Guards should be provided for all exposed head bolts on rolls. Rotating nuts or capscrews on roll heads may catch clothing or loose paper. Use caution in these areas while the machine is running. Never climb between guards and moving machinery.
14. Keep hands away from belt and chain drives. Make certain that all guards or drive components are in place.
15. Lock out all drives and controls before working on the machinery.
16. All non-operating personnel are to be out of the danger area before activating drives and operating controls. Mirrors should be used to provide the operator with a view of the drive side area, if possible.
17. Inspect slings and cables for worn or weak spots before using them. Keep all personnel away from under machine components when lifting. Use lifting points specified by manufacturer. Do not allow chains or other

lifting devices to hang in the aisles.

18. Do not walk under machinery, rolls, or other items being transported by overhead crane equipment.
19. Do not allow broke holes to become covered by paper or broke which may hide them from operating personnel.
20. Do not depend upon hydraulic or pneumatic devices to hold equipment in a raised position while performing maintenance; pin, chain, or block in a raised position.
21. Inspect chains and clevis pins at frequent intervals for wear and damage. Block under or around units raised by chains when performing maintenance to prevent injury to personnel.
22. Tie sling securely when attempting to lift machine components. Rotation of out-of-balance pieces could be hazardous.
23. Release pressure from oil and air lines before disassembly. Oil and air under pressure can be dangerous to personnel in the area.
24. Use proper stops when applying hydraulic movement equipment to bearings, heads, gears, etc. These items may travel at a high rate of speed once they have broken loose from the fit.
25. Be sure that all slings and cables are designed to lift the loads taking into consideration the angles of the hookup and the load to be lifted.
26. Use lifting points as specified by manufacturers. Where provisions have been made for lifting eyes to be screwed into a tapped hole, make certain that eyebolt is tightened to the shoulder and that eye is parallel to the lifting plane to prevent breakage.
27. Do not stand astraddle of the machine line shaft or motor couplings.
28. Personal articles are not to be stored in electrical switchboxes, panels, or in other potentially hazardous places.
29. All drain covers and barricades must be replaced over and around openings before returning function to operation.
30. When threading machinery, feet must be squarely and properly placed for adequate balance.
31. When lifting rolls, reel spools or reels of paper, make sure both hooks are securely in place on the journals before making the lift.

G. A CLOSER LOOK AT THE PAPER MACHINE

1. Wet-end safety

Floors, footwalks, handrails and designated walkways must be properly used in the wet environment of the headbox, fourdrinier and former. Poor footing may cause injuries from falls at this end of the paper machine. Personnel assigned to work at the wet end should be given complete safety instructions and periodic refresher courses so they are keenly aware of safety procedures and will constantly be on guard for potential dangers.

2. Headbox

Proper footwalks, non-skid floors, etc. must be used to ensure safe footing on and around the headbox. Workers who will enter the headbox during clean-up must use caution and adhere to the confined space procedures in Sec. C. The fan pump and all rectifier roll drives must be locked out before the headbox is entered. At shutdown the headbox contains far more water vapor than oxygen; the headbox must be properly vented before anyone enters it. And it must be continuously vented while the worker is inside. A helper should always be stationed outside the headbox while the worker is inside. Protection from cleaning chemicals used inside and outside the headbox must be available to wet end personnel. Mill personnel assigned to the wet end should be thoroughly drilled in the proper cleaning procedures and use of equipment and materials.

Pad all manholes in the stock surfaces before entering. These surfaces are accurately machined to prevent stock hang-up and are very sharp. They can cut clothing and unprotected flesh.

Injuries around the headbox may be attributed to:

- Slippery floor surfaces.
- Improper use of footwalks and handrails.
- Failure to lock out drives during shutdown.
- Improper protection from caustic cleaning solutions.
- Improper handling of manhole covers and other access doors.
- Unpadded entry points.
- Mishandling of slice lip components.
- Wash-up hoses not stored properly.

3. Fourdrinier

A continuously wet environment creates an extremely hazardous condition for footing on and around the fourdrinier. Operators and other personnel with work assignments on the fourdrinier should be continually warned about the hazardous footing in that area.

Modern fourdriniers are designed so removal of footwalks or handrails during wire stringing is minimal. This applies to cross-machine walks as well. Footwalks and hand rails may be designed to assist or be part of the wire stringing operations.

However, on some older fourdriniers, handrails are required to be removed for the wire stringing operation and may create a hazardous working area. Extreme caution must be used when working around all older equipment in the mill.

Injuries on the fourdrinier may be attributed to the following:

- Hazardous, slippery footing.
- Removing or making improper adjustments to table components on the run.
- Improper use of footwalks and crosswalks.
- Use of table rolls and other table components to cross the machine.
- Careless use of slings on table component removal and replacement.
- Coming in contact with the moving wire.
- Inadequate barriers, especially around the couch and wire pits.
- Start-up of remote controlled devices while all personnel are not clear.
- Inadequate barriers, especially around the couch, wire pits and drive shafts.
- Removal of safety interlocks, safety equipment and safety signs.
- Failure to keep working area clear of tripping hazards.
- Improper lighting.

4. Wire change

The fourdrinier and surrounding area should receive a thorough washing and rinsing to remove slime from surfaces that people may be required to work on before each wire change. Wire stringing apparatus furnished with the fourdrinier should always be used. It is designed to apply the wire to the table with mechanical helps and avoid short cuts which may result in serious injuries.

Removable fourdriniers and wire stringing equipment are operated by power to assist machine tenders.

5. Safety requirements

1. Heavy, slow-moving loads with high torque running against stops can be very hazardous to both operators and equipment.
2. Use safety lines when working near open floor areas. Areas around a couch pit may be slippery.
3. Do not walk across table components. They may rotate and cause a serious accident.
4. Keep clothing and all parts of the body away from remover wheels or gear rack teeth. These parts move slowly and quietly and may cause serious injury.
5. Remotely operated winches can be extremely dangerous if not operated properly:
   a. Avoid winching at angles which may place overloads on the cables.
   b. Avoid running loads into a sheave or snatch block which may break the cable.
   c. Inspect for frayed or damaged cables.
6. Careless handling of wire poles may result in pinched fingers or cuts. Lock poles in place.
7. Exercise caution when working near motors which may be started from a remote location by ensuring that lockout procedures have been followed.
8. Replace security lanyards attached to remover pins if they are broken. This will prevent pins from falling into the machine.
9. Wear protective industrial gloves when handling wires, worn doctor blades or fourdrinier table components.
10. Lock out all drives and controls before performing maintenance or during washups. Accidental startup may injure personnel.
11. When changing wires, fill the couch pit with water. Stop and lock out the agitator.
12. Be sure to securely retighten all swing bolts on blocks, showers, arms and stands.

6. Top wire former

Precautions observed around a twin-wire former and fourdrinier would apply to top-wire formers which convert a fourdrinier into a horizontal twin-wire unit. Because the top-wire former is accessible from the fourdrinier footwalk, extra caution should be taken when approaching the twin-wire nip. Contact with the fabrics or wires must be avoided while the unit is operating.

The wet environment of the forming area creates hazardous footing. A non-skid surface should be applied to the operating area around the fourdrinier. Use caution when raising a top wire former out of the fourdrinier for wire changing. De-energize drives at the power source and lock them out during wire change. Before working on the unit, spacer blocks must be placed around the jack posts. These blocks are removed long enough to permit wire stringing. The top-wire former is then supported by a porter bar. Proper equipment must be used and instructions closely followed during the fabric or wire stringing operation to minimize potentially unsafe situations.

When lifting the porter bar into position, practice safe overhead crane operation. Be sure lifting apparatus is in good condition.

When changing the fourdrinier wire, follow the instructions provided. The top-wire former must be supported by a porter bar or cantilever arrangement to allow for fabric stringing. Caution must be exercised during this phase of stringing as the hazard potential is greatest at this time.

Savealls are removed for wire stringing clearance. Use caution when handling savealls – the edges can be sharp. After savealls have been replaced, make sure all the fasteners are tightly secured before start-up.

Eliminate injuries on top wire formers by avoiding the following:

- Misjudgment of clearance between moving components.
- Improper use of footwalks and crosswalks.
- Failure to keep footwalks and work areas free of tools and other materials that create tripping hazards.
- Improperly guarded nip points.
- Disregard for hazardous footing due to wet end environment.
- Improper handling of support blocks.
- Failure to exercise caution when handling equipment.

7. Twin wire

Precautions must be practiced when working on or around the twin-wire former because of the continuously wet conditions. The primary danger is hazardous footing. A non-skid surface is recommended around the operating area (epoxy paint with hard sand imbedded in the surface). The operators must be aware of the sections of footwalk that pivot or are removed during wire changing. Safety chains and warning signs must be used to block off the areas where sections of footwalk have been temporarily removed.

To facilitate wire changing, the stretcher rolls and guide rolls are retracted by means of pivoting arms. When preparing for wire stringing, these rolls should be avoided until fully retracted. The controls to activate the retracting mechanisms must not be activated until the area is clear.

The main drive must be de-energized at the power source and locked out before wire changes. When wire changing has been completed, the controls should not be reactivated until all personnel have been accounted for and the area inspected and cleared of workers.

When overhead cranes are used for wire and roll changes, all precautions recommended for safe overhead crane operations must be exercised. Personnel working near or on top of the former must be especially cautious of the overhead crane travel. The crane operator must use prescribed audio signals when approaching the former to warn personnel who may be in the crane's path.

Safety requirements include:

- Do not enter saveall troughs without a light, air hose, safety rope and an observer at the entrance.
- Do not remove warning signs. They have been installed to warn personnel of potential hazards. Observe all instructions on the signs.
- Footwalks, handrails, barriers, guards, and shields must be in place before starting.
- Do not overreach; keep proper footing at all times. Do not stand on handrails.
- Keep area around the former clean and clear of obstructions.
- Wipe up spilled oil and grease.
- Keep clothing and all parts of the body away from the nip points, especially the convergence of the two wires.
- All non-operating personnel must stay outside of safety lines when the machine is running.
- Become familiar with machine operations. Avoid contact with intermittent moving parts.
- Do not operate machine with remover blocks out.
- Do not enter wire loop when machine is operating.
- Avoid contact with moving fabric or wire during operation of the machine.
- Do not cam a doctor away from a roll when machine is operating.
- Do not remove doctor blades when machine is operating.
- Doctor blades are extremely sharp. Wear gloves when handling them.
- Do not exceed recommended fabric or wire tension.

Eliminate injuries on the twin-wire former by avoiding the following:

- Improper use of footwalks, crosswalks, and ladders.
- Disregarding safety signs and other safety warning methods.
- Removal of warning signs and railings.
- Coming in contact with fabric or wire while machine is running.
- Failure to keep working areas and footwalks clear of tools and other tripping hazards.
- Disregard hazardous footing due to wet environment.
- Improper handling of doctor blades.
- Unguarded nip points and rotating machinery.
- Carelessness around the couch pit.

- Misjudgment of overhead clearances.
- Failure to completely lock out power sources during wire changing.
- Improper use of wire stringing equipment.
- Crossing the machine at undesignated areas.
- Wash-up hoses not stored properly.
- Unauthorized removal of shields and failure to replace shields after machine has started up.

8. Press safety

The press removes moisture from the sheet kith heavy rolls and high-pressure nips. A safety training program for personnel assigned to the press must alert them of its potential hazards. Foot boards, crosswalks, railings and ladders should be used as directed to prevent operating personnel from falling. Workers must never go into hazardous locations or dangerous positions during press operation.

When overhead cranes are used for clothing and roll changes, all precautions recommended for safe overhead crane operation must be exercised.

Personnel working near the top of press frames should be especially cautious of the overhead crane travel. The crane operator should similarly be aware of personnel in the path of operation and give ample warning for them to clear. Special care must be given during clothing and roll changing to see that slings, hooks and other lifting devices are in good condition for handling felts and rolls. Proper lift points must be used to pick up and transport equipment.

If unsafe conditions exist on or around the press section, the press should be shut down until the condition has been corrected.

Eliminate injuries in the press section by avoiding the following:

- Hand threading.
- Improper use of footwalks and crosswalks.
- Failure to keep working area and footwalks free of broke, tripping hazards, and foreign materials that result in hazardous footing.
- Improper handling of broke.
- Removal of warning signs and railings.
- Failure to completely lock out power sources during felt and roll changing.
- Use of improper hoisting equipment.
- Unguarded nip points.
- Improper protection of doctor blades when press section is down.
- Unguarded rotating machinery.
- Working in felt pits when suction rolls are being cleaned with chemicals or during boil-outs.
- Inadequate barriers.
- Misjudgment of clearance between moving members, swing arms and frame.
- Carelessness around couch and broke pits.
- Walking across slow moving felts.
- Unguarded nip points.
- Unguarded rotating machinery.
- Wash-up hoses not stored properly.

(a) Extended Nip Press®

The previously mentioned safety precautions apply, in general, to the Extended Nip Press (ENP). In addition, there are some areas that are unique to the ENP® as well as some areas that deserve special emphasis.

The components that are being handled during felt, blanket and roll change are significantly heavier than those in a conventional press section. This requires procedures for clothing and roll changes to be followed carefully. For example:

1. Use proper sequence of jacking for clothing and roll change; follow instructions from supplier.
2. Cross-direction stabilizer link must remain connected during blanket and felt change.
3. Load beam must be lowered in proper sequence for roll removal. Lower drive side of beam onto stabilizer legs before disconnecting front link pins and lowering tending side.
4. Lock-up pins must be used when load links are raised before working in this area.
5. Maintain proper lanyards on link pins to prevent complete removal of link pins. Link pins are heavy and lanyards are designed to prevent complete removal from load beam ears.
6. There is a potential for oil-spills associated with the ENP®. To prevent hazardous footing, the source of any leakage, mist or spill, must be corrected and the oil cleaned up as soon as possible.

(b) Safety requirements—press

Operation

1. Felt rolls are not to be tapped during operation.
2. Limit switches are provided to prevent injuries to personnel and damage to the machine. Be sure all limit switches are working properly.
3. Nuts or capscrews on rotating roll heads may catch clothing or broke. Use caution in these areas while machine is running. Roll ends must be guarded.
4. Rope sheaves must have a minimum clearance of 3 in. at the rope nips to prevent injury

to hands.

5. Pivot pins should be lubricated periodically so they will rotate when required.
6. Ingoing press nips should be guarded on the front and back sides of the machine to prevent an operator's hand from going into the ingoing nip.
7. Be sure footwalks, handrails and guards are securely in place before starting machine. They have been placed for operator's safety. Be sure pins are secured in removable sections.
8. Work at outgoing felt nips, never at an ingoing nip.
9. Felt pits should be periodically cleaned to prevent slime from building up and creating a potential slipping hazard.

Maintenance

1. Remover blocks and swing bolts can cause serious injury to fingers during felt change operation. Handle with care. Use hand holds on remover blocks to avoid injury to fingers and hands.
2. Do not walk on felt or press rolls. Use only the walkways provided.
3. Do not depend on hydraulic, pneumatic or screwjack devices to hold equipment in raised positions while performing maintenance. Pin, chain or block in raised positions as precautionary measures.
4. Use hydraulic or screwjacks (when provided) for felt change rather than an overhead crane.
5. Lock out all drives and controls before performing maintenance or during washups. Accidental startup may injure personnel.
6. Inspect chains and clevis pins for possible corrosion.
7. Release pressure from steam, water and air lines before disassembly. Hot steam, or high water and air pressure may be dangerous to personnel in the area.
8. Use proper stops when applying hydraulic removal equipment to bearings, head, gears, etc. These items may travel at a high rate of speed once they have broken loose from the fit.
9. Do not use excessively long porter bars (over 36 in.) on felt rolls during felt change. Long porter bars may overstress the roll journals.
10. When rotating suction rolls for cleaning, work on the opposite side to the guard.
11. When finished turning suction roll, make sure bar or wrench is removed and guard replaced.

(c) Felt change safety

1. Disconnect main drive from power source and lock out.
2. Put lock-up pins in press so the open nips will not close.
3. Shut off and block off water, steam and lubrication systems so they will not leak and create hazardous footing.
4. Use proper spacer blocks. Have appropriate wrenches nearby.
5. Watch fingers on blocks and wrenches.
6. Do not drop sockets off torque wrenches.
7. Do not carry loose articles in pockets or on belts.
8. Use safety chains with warning signs to block off footwalk where portions have been removed.
9. Close all doctor blades against their associated rolls.
10. Be aware of crane travel over the top of the press and watch clearances.
11. Crane operator must watch for personnel working in the area of the lift and give proper signal when approaching.
12. Inspect slings and other lifting devices.
13. Be aware of, and watch for, revolving cross shafts.
14. One person only must give directions to the crane operator.
15. Do not assume an awkward position when pulling on the felt.
16. Remove pins carefully. Do not permit rolls to slam together when closing the nips.
17. Make sure all components that are removed or loosened are replaced and properly tightened.
18. Be sure to have positive footing before trimming felt.
19. Replace all footwalks and secure them to the permanent footwalks.
20. Replace all guards.
21. Observe all warning signs on the press before, during and after felt change.
22. Before starting up the press, make sure all personnel are clear and accounted for.

Footwalks and railings on the press section must be approached with caution because of their height. Also many sections are removable or pivot for ease of roll removal and clothing changes. The latter must be clearly identified. Pivoting sections create safety gates where sections are removed.

9. Dryer section safety

Since dryer sections require use of high temperatures and pressures, they must be treated with caution. The nature of the dryers' function and required operating care make it important that they be given close attention in the mill safety program. The key to dryer safety is to keep operating personnel from getting into dangerous areas or unsafe positions, especially while the dryers are operating. If unsafe conditions exist, the dryers must be shut down until the condition is corrected.

Machine tenders and other mill personnel with work assignments in the dryer section must be given complete safety instructions and be on guard against potential hazards. They should be carefully instructed on the proper use of equipment and potential dangers of the steam systems, high pressures and high temperatures associated with dryer operation. Also, adequate barriers, footwalks and signs should be installed at appropriate locations to protect and warn personnel of the potential hazards.

Preventive inspection and subsequent maintenance of the following will avert potentially hazardous situations for machine tenders and protect the dryer section from the possibility of serious damage.

Felts and felt seams should be inspected at least once a shift on high-speed machines, and during all clothing changes. Should a felt seam part, due to fatigue or wear, serious damage could occur.

Steamfits can be hazardous if not protected by a guard due to exposed rotating capscrews or bolt heads. The guard must completely cover the rotating parts that provide the seal between the rotating dryer and the stationary piping that supplies the steam to the dryers and carries away the condensate from the dryers. It is recommended that the guard be constructed from solid sheet metal as opposed to expanded metal plate, to provide protection against a sudden escape of steam. Watch closely for wads and remove them as soon as possible. Wads cause excessive strain on the felt and felt rolls and may lead to the failure of both. Keep broke cleaned up. Paper broke is a serious fire hazard. Fires can be caused as a result of friction from broke against moving or rotating equipment.

Loose balance weights, siphons and other debris that may be inside the dryer cylinder can cause severe erosion that may result in shell failure. If detected during routine inspection, the loose items must be removed. If erosion is excessive, the dryer should be replaced or at least run at lower pressures.

Loose or broken siphons can cause locally severe erosion that may result in shell failure.

If the steam in the dryer or the outlet is turned off, the dryer can become half filled with water. The steam lines should be disconnected from the dryer. Dryers one half full of water add unnecessary load to the drive, increase dryer journal stress, and pose a potentially dangerous situation in the section. Also, condensate in flooded dryers has on occasion eroded dryer shells where it flowed around balance weights and manhole covers.

When opening steam valves, care must be taken by operators. Open valves very slowly. First crack them open slightly, then listen a moment to hear if water hammering is taking place. If no hammering occurs, open the valve a little more, by degrees, until one is sure it is safe to proceed. Be sure no one is working the steam line by ensuring that lockout procedures have been followed.

Loose plates on manholes and clean-out holes must be reported to arrange for immediate repair.

Insulation and steam pipe welds should be continually inspected to eliminate the hazard of burns.

Rope systems should be continually inspected. Ropes showing wear should be replaced. Dislodged ropes must be replaced.

When a machine tender has to enter an enclosed hood under operating conditions, an observer or fellow worker must accompany or keep the tender in sight at all times. Other confined space entry procedures must be followed.

(a) Dryer warm-up procedure

The warm-up procedure that follows must be adhered to, to avoid excessive stress concentrations in the dryers due to large variations in temperature. Failure to follow this procedure can be dangerous to life and machinery. The controlled warm-up should be tailored to your operation. The following steps in warming up the dryers from a room temperature condition are recommended.

1. Run dryer at crawl speed and insert 5 psi steam pressure and warm the dryers for 1.5 h. Speed may be increased after admitting steam.
2. Slowly increase the steam pressure to 30 psi and run 45 min. Speed may be increased to operating speed.
3. Slowly increase the steam pressure to 60 psi and run for 45 min.
4. After 3 h lapse time in warm-up cycle the steam pressure can be increased to operating pressure.

If the machine has been down for only 4 to 8 h, the length of time for step 1 may be reduced.

If the machine has been down for 4 h or less, omit step 1 and reduce the time of steps 2 and 3 to 30 min each, for a total warm up time of 1 h.

(b) Broke removal in dryers

When a person has been assigned to pull broke under the dryers, that person must notify the back tender that he or she is going under the dryers. An observer or assistant must constantly accompany the worker. In the event of a sheet break, upon hearing the break horn, the worker and assistant must get out from under the dryers immediately. As a back-up to the automatic break detectors, it is the responsibility of the backtender to locate the observer and the person under the dryers to see that they are clear.

- If one or more dryers are loaded with broke or wads, start removing wads nearest the wet end.
- Never reach into dryers with a hose.
- Keep away from nip side of felt or ropes.
- If it is necessary to remove a wad from the hitch roll or a guide roll, dryers must be locked and main power switch must be pulled and locked out.
- Never keep any length of hose behind you or between your legs.
- When removing small pieces of broke from the edge of a doctor blade, use safety gloves to eliminate contact with the sharp edge.
- When paper wraps around a dryer cylinder, wait until the machine is shut down, then use spears to remove the paper from the cylinder.
- Dryer spears must not be used to remove paper while the machine is running. The spears may be drawn into a nip. When not in use, spears are to be stored in a safe place.
- Operators must not climb on dryers or frames.

Eliminate injuries in the dryer section by avoiding the following unsafe practices;

- Failure to lock-out the source of drive power during maintenance.
- Improper disposal of broke in walk areas.
- Hand threading.
- Personnel going into hot enclosed hoods.
- Inadequate barriers.
- Removing wads and wraps with hand-held tools while dryers are rotating.
- Walking in, around and under operating equipment.
- Hand wiping moving rolls.
- Cutting tails with a hand-held knife.
- Deteriorating insulation on steam pipes.
- Ropes dislodged from sheaves and rope grooves.
- Working outside footwalks and standing on handrails.
- Steam and condensate systems not valved off tightly before breaking the line.
- Not venting cylinders properly before entering and during maintenance.
- Hosing down hot cylinders with water.
- Contact with hot steamfits.
- Unprotected counter-weights for felt stretchers.
- Improper maintenance by failing to remove loose parts in rotating dryers or not repairing the steam leaks.

(c) Paper dryer head-to-shell capscrews

The most probable cause of failure of capscrews is stress corrosion cracking.

Stress corrosion cracking is an intergranular mode of failure, a result of the combination of stress and corrosion. The stress in this case is due to the original torque applied at assembly. The corrosive agents needed to produce stress cracking varies for different metals. However, for the quenched and tempered steel used for these capscrews, the following are known to cause stress corrosion cracking:

1. Sodium hydroxide solutions.
2. Ammonium nitrate solutions.
3. Hydrogen cyanide solutions.
4. Carbonate and bicarbonate solutions.
5. Carbon dioxide plus carbon monoxide and water.

Many of these corrosive elements may result from steam leaks. The most likely corrosive materials are the sodium hydroxide and the carbonate or bicarbonate solutions which may be present in steam and condensate. Also, if the steam is treated with neutralizing amines, nitrates may form. All paper dryer head-to-shell capscrews should be tested every two years for fractures. Furthermore, all capscrews in a leak area of a dryer-to-shell joint should not be tightened or retorqued, but replaced immediately upon leak detection.

The preferred method of finding cracks in capscrews is by ultrasonic tests. There are many companies available that do ultrasonic testing. Tests can be carried out in the paper mill at the machine with the capscrews in place. Corrosion and excessive paint must be wire brushed from the top surface of the capscrew head to ensure good transducer contact.

When installing a new capscrew, use ASME Code designated material with specification SA 354 Grade BC. The threads and washer face of the capscrew must be lightly lubricated with a thread lubricant. Install the new capscrew and torque to the following amount of torque:

Capscrew Size (inches)	Torque (foot-pounds)
3/4	185
7/8	300
1	450

No retorque is necessary after warm up.

(d) Over-torqued dryer head capscrews

Paper mills will, at times, retorque dryer head capscrews in an effort to stop steam leaks at the dryer head to shell joint. This practice often leads to cracked or broken dryer head capscrews which can result in accidents that damage machinery and are a threat to personal safety.

(e) Dryer journal-shaft failure

During the life of a paper dryer cylinder, it may be subjected to severe paper wraps and/or felt failures and subsequent wrap-up. These accidents, if extreme, can overload the journal shaft and cause immediate failure or start a crack with eventual shaft failure resulting.

Recommendations

1. When an accident occurs, as described above, the dryer journals must be dye checked or magnetic particle examined for detection of cracks, particularly in the area of the fillet radius where the journal shaft blends into the head.
2. There are many early vintage dryers still in use in paper machines; some 60–80 years old. All old dryer journal shafts and handhole or manhole rims should be examined every two years for cracks or metal failure.
3. When a crack is detected the part must be taken out of service. There is no known way to safely repair cracked cast iron pressure vessel parts. DEFECTIVE PARTS MUST BE REPLACED.

It is essential that a written procedure, including maintaining detailed records, be established to ensure that journal shafts are examined as described in the above recommendations.

(f) Dryer protection during a fire

Fires in paper machine dryer sections are dangerous and difficult to fight. The use of water to put out or contain the fire can create a greater hazard to personnel and equipment than the fire.

To douse a HOT dryer shell or head with COLD water results in a sudden thermal shock that creates extremely high stress in the shell. The result may be shell failure or explosion.

A fine spray should be used, about the volume of a garden hose, NOT the full volume of the fire hose. The dryer cylinder must be rotating and the spray should be worked back and forth across the cylinder so the wetting is uniform.

Automatic sprinklers are also likely to be activated over hot dryers. A water temperature of 120°F would not be too hot to prevent fire fighters from entering the area.

In case of fire, the following rules MUST be followed:

1. Sound all proper alarms (alarms must be periodically tested to be sure they are in working order).
2. Turn off steam to all dryers, either from the control cabinet or manual shutoff valves.
3. If possible, never stop rotating the dryers. Rotating dryers are less likely to be damaged from water spray.
4. When a fire hose is used on a dryer, move the spray around; do not hold it in one place. Maintain a uniform coverage at all times.

(g) Trapped steam in dryer journal

Caution should be taken when removing the journal flange from the end of the dryer journal on some older designs. Steam pressure can be trapped in the cavity formed between the dryer journal and steamfit insulating sleeve. The journal flange should be loosened enough to let the trapped steam escape before removal of the fasteners.

Check for moisture on the flange. This will indicate the presence of the trapped steam. Remove the flange carefully, it can blow off or the escaping trapped steam can cause burns, or both.

The steam becomes trapped in this void due to deterioration of the "O" ring between the flange and sleeve. This pressure may exist although the steam pressure in the dryer is released.

To eliminate this problem, a ⅛ in. weep hole should be drilled in the journal flange. This can be easily accomplished during routine service to the steamfit.

(h) Care of dryer cylinders

Do not spray hot paper machine dryers or Yankee dryers with water because the thermal shock may be severe enough to cause immediate failure or initiate formation of small cracks. Repeated applications of water may extend small cracks, resulting in the weakening of the dryer cylinder and destruction of equipment.

Some typical instances where the use of water is hazardous:

– Removing paper wraps from running or stopped dryers.

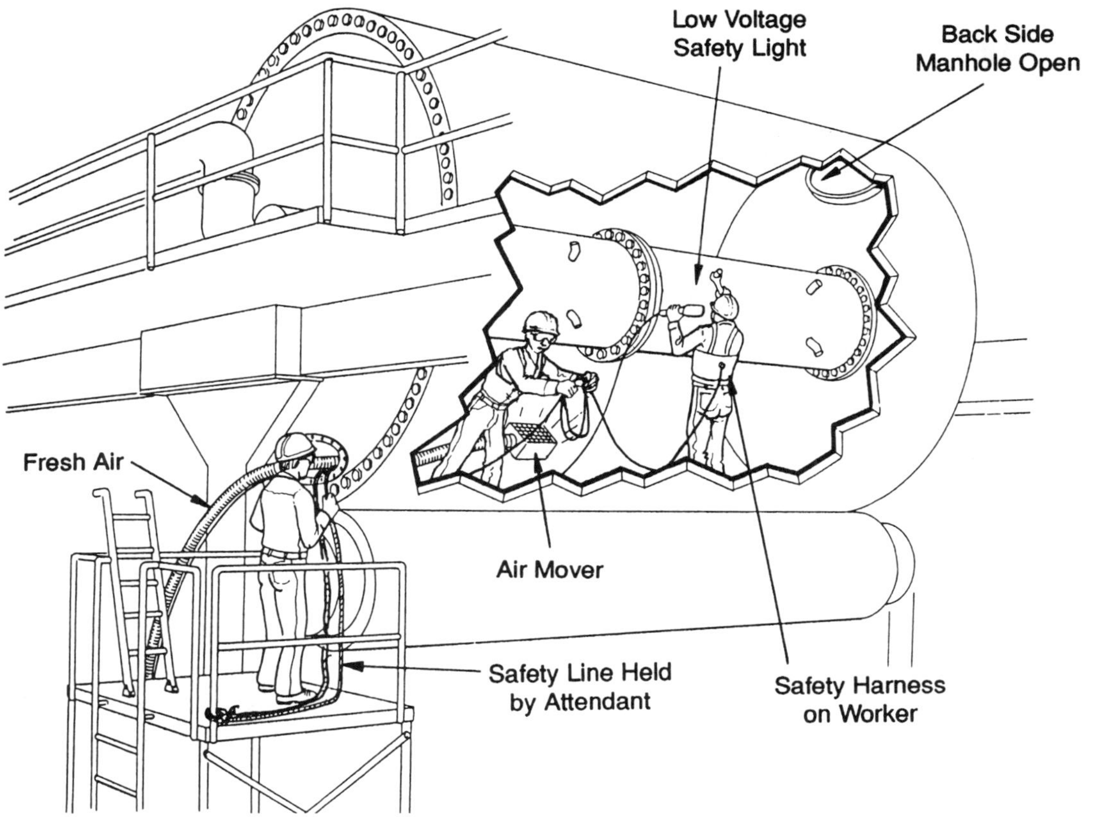

Fig. 803. Safe entry into a closed vessel.

- Cleaning dryers or felts when dryers are stopped.
- Cooling stopped dryers so they can be entered for quick maintenance.
- Cooling around manhole opening so it can be entered for quick maintenance.
- Cooling dryer head/journal for quick change of failed bearing.
- Spraying stopped dryers to control hood or felt fire.
- Any other conditions where sprayed water can cause localized cooled areas on dryers.

To ensure safe operation of dryers IT IS RECOMMENDED THAT A FORMAL PROCEDURE FOR PERIODIC INSPECTION OF ALL DRYERS BE IMPLEMENTED. The inspection procedure should take into account the following:

- Inside erosion due to loose condensate parts or balance weights (loose parts must be removed immediately to prevent damage from tumbling).
- Inside erosion under the siphon due to condensate flow.
- Inside erosion due to condensate in flooded dryers acting on balance weights and manhole reinforcement.
- Steam or condensate leaks at the head/shell gasket (this may cause contaminated fasteners to fail due to stress corrosion cracking).
- Decrease in dryer shell thickness due to outside diameter wear (severe wear would normally affect sheet runnability; however, shell thickness awareness must be part of an inspection program).

(i) Entrance into dryers

Dryer cylinders are classified as unfired pressure vessels, in the same category as closed storage tanks, tank cars, and process vessels that have limited access openings for personnel. Dryers have to be entered for inspection, cleaning, repair and other functions associated with the dryer's operation (Fig. 803).

Some hazards that may be encountered upon entering a dryer are:

- Lack of oxygen, causing asphyxiation.
- Accidental rotation of the dryer, because of poor lockout procedures.

- Electric shock from portable lights, tools or associated electrical equipment.
- Slipping and falling.
- Loose falling objects.
- Toxic vapors in fatal concentrations from welding.
- Heat stress.
- Burns from accidental opening of steam valves in a line not properly disconnected, blocked off, or locked out.
- Thermal burns from contact with hot water, hot shell, condensate or steam lines.
- Unprotected or improperly insulated hot water, condensate or steam lines.
- Poorly maintained manhole entrance with sharp edges and rust.

The following procedure is recommended for entering a dryer cylinder to lessen the possibility of one of the above happening, thereby causing an injury.

1. Before the dryer is entered, all the proper stops must be installed and secured. The dryer drive and inching drive (if applicable) must be disconnected and locked out with the lock or locks of the personnel who will be entering the dryer.
2. The inlet steam and outlet condensate lines must be physically disconnected and blanked off ahead of the dryer inlet and outlet.
3. The dryer must be allowed to cool.
4. When the manhole covers are removed, the dryer must be thoroughly ventilated by some method of mechanical exhaust, or blower system, which is arranged to avoid recirculating stale or contaminated air.

   Oxygen deficiency tests must verify that the oxygen concentration is stable and maintained at not less than 19.5% or more than 23% in the air by volume. Explosion meter tests, taken where there is any possibility of an explosion or flammable gas or vapor present, must not exceed 25% of the "lower explosive limit".
5. After the dryer is ventilated, the exhaust or blower system shall be kept operating to provide a continuous supply of fresh air to the dryer.
6. Lighting inside the dryer shall be of low voltage or battery operated and protected.
7. A worker shall never enter a dryer unless an attendant is outside the dryer. The attendant must not leave the dryer at any time or for any reason while the worker is inside. This holds true for entrance of the center shaft (Yankee dryers) where another attendant shall be inside the dryer as well as the one outside while the worker is in the center shaft.

   The attendant may pass tools but must have no other function that will take attention away from the person in the dryer.
8. Chemicals or material capable of giving off hazardous fumes must not be taken inside of the dryer without proper safety equipment.
9. A dryer lacking permanent access to enter through the manhole must have a ladder secured in place until the work is finished and the manhole replaced.

   It is the responsiblity of the individual or individuals entering the dryer and their supervisor to make sure the above procedure is followed.

(j) Emergency and rescue

Where rescue may be necessary due to an unusual exposure to a hazardous atmosphere in a dryer, the worker entering the dryer must be provided with a body harness and lifeline*. This permits standby personnel to move the worker close to the dryer entrance without having to enter the dryer. A rescue team can then enter the dryer and remove the worker. The free end of the life line must be secured to a fixed object outside the dryer and attended by the observer. The observer or attendant must know who is in the dryer and keep in voice contact and if possible in visual contact with the worker at all times.

In case of an emergency, the outside attendant must never enter the dryer until help arrives. The attendant is to summon aid by using an available alarm (whistle, siren, bell, etc.). Removal of the victim and whatever rescue functions are necessary must be done from outside the shell until help arrives.

Upon the arrival of help, should the attendant be unable to remove the victim, the attendant may enter the dryer to provide whatever rescue assistance is required. The attendant should be trained in basic first aid principles, especially rescue breathing techniques. The attendant must be equipped with the proper safety equipment before entering the dryer, as should any others who may assist [7].

10. Calender safety

The purpose of the calender is to apply a uniform caliper and finish to the paper web. The essence of calender safety is to never allow personnel to get into a dangerous area or unsafe position. Personnel must not be in an unsafe area while the calender is operating. If unsafe conditions exist, such as broke buildup around

---

* Caution should be used when wearing a body harness and lifeline inside a Yankee dryer to avoid becoming entangled with equipment, hoses, etc.

the calender, kerosene on the floor, etc., the calender must be shut down until the unsafe condition is corrected.

Calender operators and other mill personnel with work assignments in the area of the calender must be given complete safety instructions and periodic refresher courses so they know safe procedures. They must always be on guard against potential hazards. Operators should be carefully instructed on the proper use of the equipment and the potential dangers associated with in-running calender nips. In addition, adequate barriers, footwalks, and signs should be installed to protect and warn personnel of potential calender hazards.

Calender safety is improved when the following practices are eliminated.

- Hand threading.
- Cutting paper tail on the in-going calender nip.
- Becoming excited when paper adheres to or wraps a roll and grabbing it by hand.
- Scabbing from opposite sides at the same time.
- Removing scabs from moving rolls by means of hand-held tools.
- Hand wiping moving rolls.
- Failure to lock-out calender during maintenance.
- Improper disposal of broke or feeding broke through calender nip.
- Working outside footwalks and standing on hand rails.
- Reversing calender roll rotation without warning other personnel.
- Failure to keep calender area clear of broke and other tripping hazards.
- Static electricity buildups.

(a) Roll removal, open framing

The following methods for roll removal are recommended for open frame calender stocks. Other methods may be used depending upon style and size of the unit and the lifting equipment available at the mill.

Before attempting to remove a calender roll, disconnect all air, steam, water and lubrication hoses from the roll and outboard doctor and lockout the electric drive. Remove the outboard doctor and nip guard and, using an overhead crane, place them carefully on the floor supported at each end by blocks. The holder should be pointed up. The nip guards should be placed close to their respective doctors.

Inboard doctors are not removed; they are unloaded. To remove all the rolls of the calender, start at the top and remove them one at a time. To remove a single roll from within the stack, using the roll lift bar or offset hook, separate the rolls to maximum and block them open. This will provide clearance for removal of the single roll.

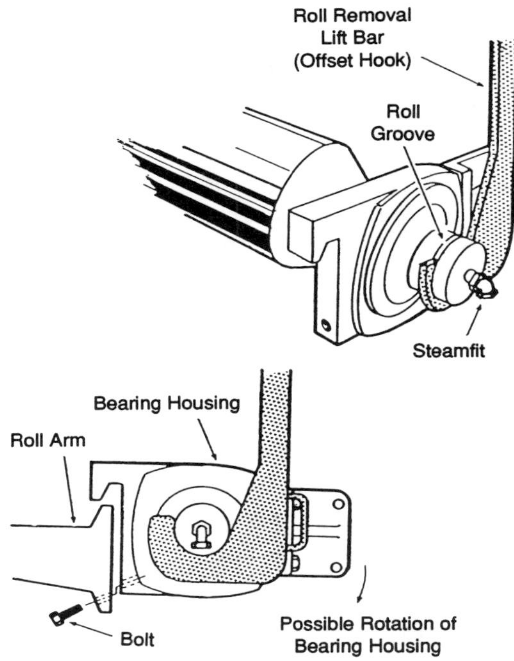

Fig. 804. Roll removal details.

Recommended equipment for removal should include an overhead crane or cranes with two or three hooks. As illustrated in Figs. 804–806, the overhead crane may utilize a "roll removal lift bar" or a sling around the journal or roll handling groove.

Do not remove the mounting bolt or bolts that hold the bearing housing to the roll arm until the roll is firmly held by the overheaad crane. Remove bolts only from the roll being held by the crane. Rolls unbolted and not supported by the crane could be bumped off the roll arm causing serious injury to personnel working on or near the calender.

(b) Roll lift bar

Roll lift bars shown in Fig. 805 simplify the roll removal procedures. With these bars, slings, "come-a-longs" and other devices are not necessary. Attach the lift bar to the "roll handling grooves" provided in the roll journal as shown and lift the roll slowly. Allow the bearing housing to rotate to a balanced position. The unbalanced weight of the calender bearing housing may rotate suddenly, so treat it with caution (Fig. 804).

Take up the weight of the roll and remove the bolts attaching the bearing housing to the lift arms. Protect the roll face during a removal . . . never sling the roll face unless it is properly protected or nylon slings are used.

To avoid rotation of the bearing housing, obtain a bar with a threaded end. Screw the threaded end into one of the doctor mounting holes. Use this bar to steady the housing during removal.

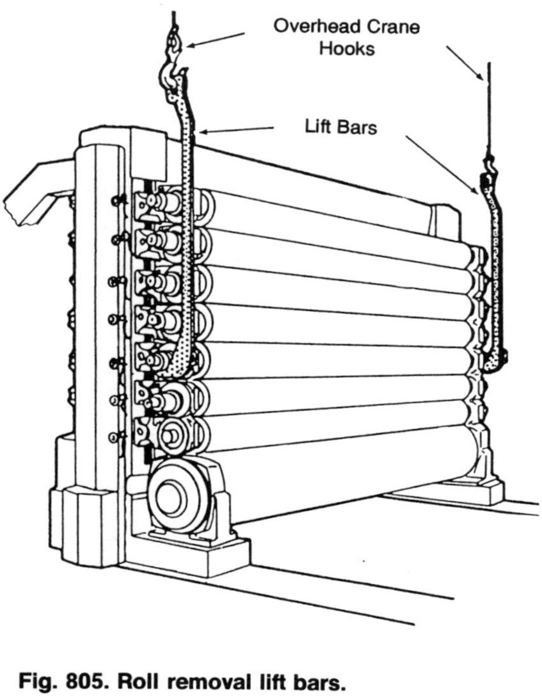

Fig. 805. Roll removal lift bars.

Remove the roll from the stack horizontally. When clear of the stack, transport it to prepared blocking.

Another method of roll removal is the use of slings and "come-a-longs" shown in Fig. 806. The wire rope slings should be placed in the roll handling grooves. Attach the "come-a-longs" from the slings to the calender frame. Take up the weight of the roll. Remove the bolts attaching bearing housing to the roll arms.

Move the overhead crane outward from the stack slowly while working the ratchet of the "come-a-long". This permits the roll to move horizontally from the stack. When the roll is free of the calender, remove the "come-a-longs" and move the roll away from the stack. Place the roll on prepared blocking.

(c) "A" frame calenders and supercalenders

Roll removal from an "A" frame calender or supercalender varies widely with mill practice. Multiple crane hooks are required because of the nature of the design.

For an "A" frame calender with built-in roll removal hoists use the following procedure:

- Attach the roll lift bar to the end of the roll to be removed. Attach a sling to the other end. The lift bar and overhead crane hook will be on the outside of the frame. The hook with the sling will be on the inside of the frame.
- Take up the weight of the roll and unbolt the bearing housings from the roll arms. Protect the roll face during removal. Move the calender (toward the lift bar). A "come-a-long" may be used on the sling side if needed.

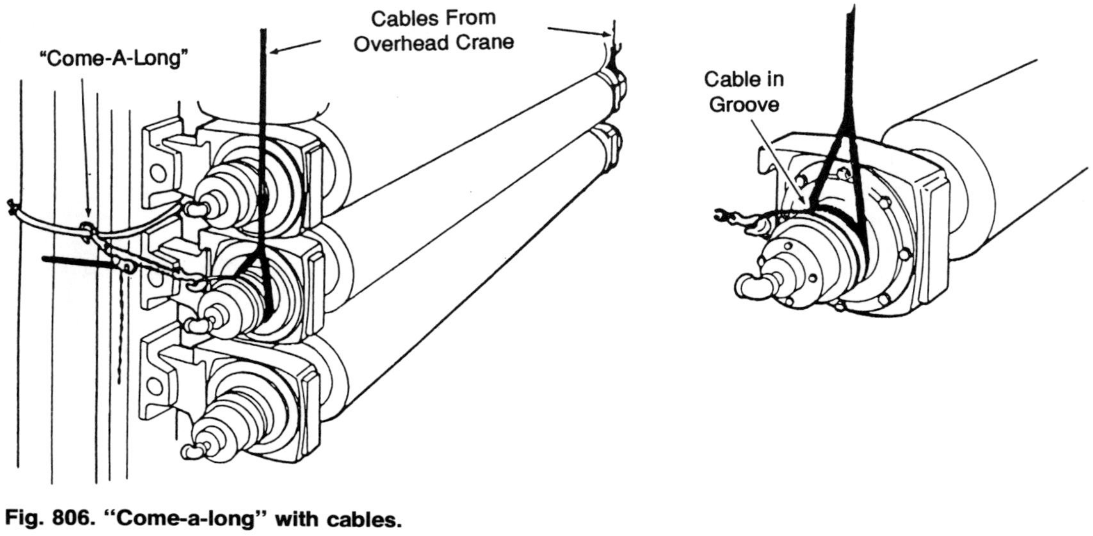

Fig. 806. "Come-a-long" with cables.

- When the crane hook and sling inside the framing reach the inside of the overhead framing, a third hook is required outside of the framing. This hook, using a nylon sling, takes the weight of the first sling. Remove the first sling, move the roll away from the calender to prepared blocks.
- Separate the rolls to and including the roll to be removed using the roll lift screw to its limit. Position hoists over the outboard and inboard ends of the roll. Attach a yoke to the journal of the roll outboard of the bearing housing. From the inboard hoist attach a yoke to the shoulder of the roll, inboard of the bearing housing. From the inboard hoist attach a yoke to the shoulder of the roll, inboard of the bearing. *The yokes must not be attached to the roll face.*
- Take up the weight of the roll and free the bearing housings from the frame. Pick up the roll with hoists and move it horizontally out through the frame to the awaiting house crane. The house crane should use nylon slings or a strong back with slings on each end. The roll should be lowered to roll blocks and transferred to the house crane. The roll may be lowered onto a dolly designed to support the calender roll during transportation.
- The calender rolls are transported to properly blocked roll storage by house crane or dolly.
- Some mills use a roll removal cradle to remove the calender roll horizontally from the stack using the roll below as a guide and support. A house crane picks up the roll as it clears the stack.

(d) Safety requirements — calender

Operation

1. Do not thread by hand. If an automated system is not available when threading paper into a calender nip, always use spears, sticks, or other accepted recommended devices. Do not open nips to thread.
2. Do not remove broke by passing it through the calender roll nips. Hands and/or entire body can become entangled in the broke and be pulled through the calender nip.
3. Never use a rubber hose near the calender stack without a hard safety nozzle end that cannot go into a roll nip.
4. All scrapers are to have long handles to keep hands and arms out of nip areas.
5. Do not scrape stack with operators working on opposite side. Loose particles may cause personal injury.
6. Keep area around calender stack free from obstructions of any kind.
7. Do not hand dress rolls.
8. Stop calender stack before removing broke from nip areas.
9. Do not lower calender rolls with stack running. Roll particles could dislodge, cause personal injury and damage rolls.
10. Do not run calender rolls bare for any extended period of time. When running without a sheet between the rolls, operating personnel must not enter areas between calender frames. Metal particles could become dislodged and cause personal injury.
11. Do not spray calender rolls with water or steam. Sudden temperature change will cause cracking or splitting on chilled iron calender rolls. The rate of temperature change should not exceed 50°F (28°C) surface temperature per 20 min.
12. Use static eliminators on the sheet and properly ground the calender journals with a brush and ground wire.
13. Wear goggles or face shield when threading calender stacks.

11. Reel safety

The reel is the transfer point from the machine to the finished product processing equipment. Special safety measures must be taken during reel spool transfer and transport of the extremely heavy weights represented by the finished reel logs of paper or board. Also, total automated reel operation is available.

Adequate barriers and signs should be installed to protect and remind personnel of potential reel operating hazards (Fig. 807).

(a) Safety requirements—reel

Operation

1. The reel crane should not require manual spool unhooking. The necessity for operating personnel to reach over equipment must be eliminated. Strongback should be cable supported at each end, not pivoted from a center support hook. Reel hooks are to pivot only in machine direction.
2. All non-operating personnel are to be out of the reel area before activating drives and operating controls. Injuries may result if this is not done. Mirrors should be used to provide operator with view of drive side area.
3. When placing reel spools in the primary arms, clamp the reel spool in position with primary arm "load" pressure before removing crane hooks.

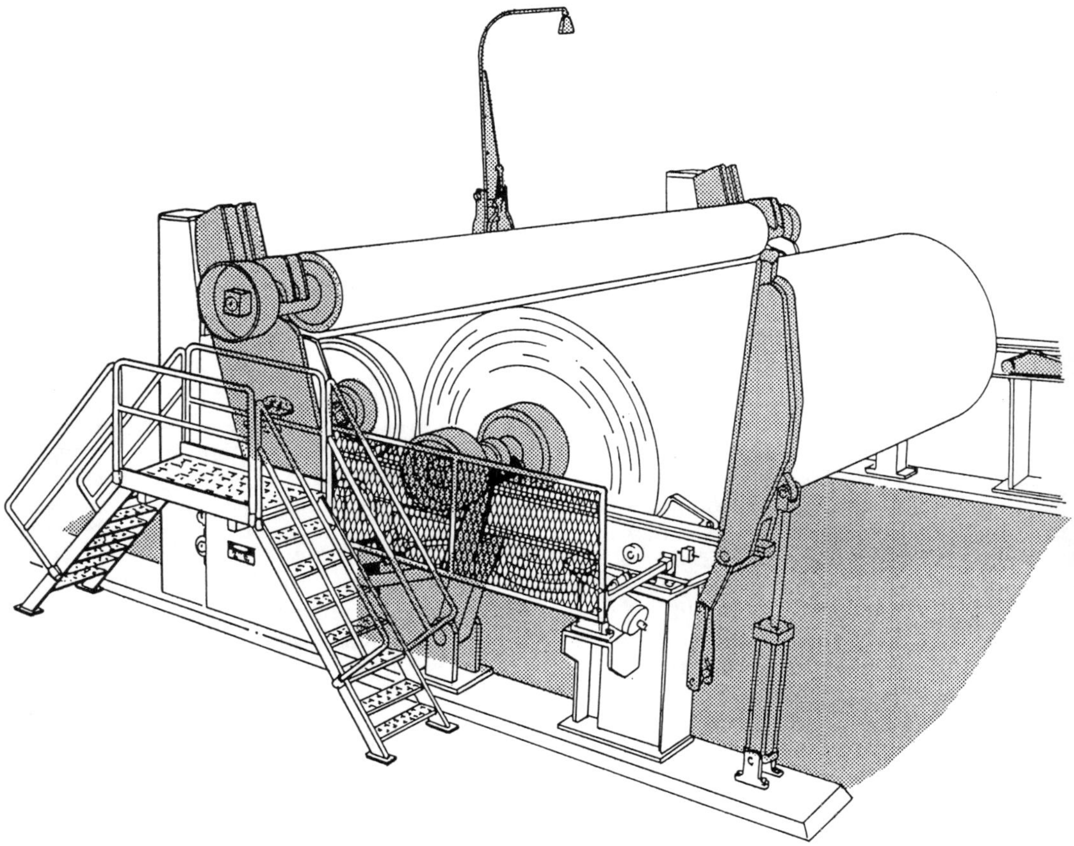

Fig. 807. Safety guards at the reel.

4. Do not use hand-held devices to turn up new reels. If an emergency air hose is required, hose must be equipped with a metal nozzle approximately 2 ft. in length. Hose length and hose connection placement must prohibit the possibility of an operator becoming encircled with the hose. The operator should wear a safety belt and rope when working around the broke hole area or above the reel.
5. Make certain that rope sheaves have a minimum clearance of 3 in. at the rope nips to prevent injury to hands.
6. Primary arm air cylinder hoses and limit switch wiring must be positioned and anchored to prohibit loops being caught in gear nips.
7. Do not mark paper defects manually by placing cards into the nip. A mechanical device should mark the edge of the roll.
8. Stop reel to remove broke between the calender and reel. Broke may become entangled with rotating equipment and cause injury to personnel.
9. Do not operate the reel with damaged parts or with reel spools that may have bent journals or be out-of-balance.
10. Do not stack empty reel spools.
11. Paper samples must not be taken when the reel is suspended or traveling on the hoist.
12. Slabs must not be taken from a finished reel while the reel is on a moving hoist or crane.
13. Empty spools must not be left suspended from a hoist. Reel hoist must not be left in the lower position.
14. Hoist chains or hooks must not be engaged around the reel spool until the spool has stopped revolving.
15. A minimum of 10 in. must be maintained between full reels of paper.
16. When lifting reel spools or reels of paper, be sure both hooks are securely in place on reel spool journals before making the final lift.

17. Do not set paper reels on the machine room floor. Place the paper reels on support stands or rails. Reels set on the floor cause flat spots that create vibration on the winder unwind stand that can cause accidents.
18. Do not replace worn or broken ropes while machine is in operation. Operators must not work on the rope system while reel drum and other rolls are revolving.

Reel safety can be enhanced by eliminating the following practices:

- Operating personnel between the reel and calender during machine operation.
- Failure to bring reel spools up to speed before sheet transfer.
- Making sheet transfers by hand.
- Cleaning reel drums during operation.
- On-the-run adjustments of reel drum doctors.
- Accidental contact with rope nips.
- Utilization of bent or unbalanced reel spools.
- Hand marking or hand insertion of tags in winding reel to mark defects or breaks.
- Failure to keep working area clear of broke and other tripping hazards.
- Improper handling of broke.
- Failure to completely lock out reel controls and drives during maintenance operations.
- Placing hands on surface of winding reel.
- Hoisting reel logs while they are still rotating.
- Use of improper hoisting equipment and hardware.
- Static electricity build-up.
- Improper use of handrails and footwalks.
- Inadequate barriers.
- Standing on guard rails around "broke" holes.

(b) Automatic turnup devices

When working around the reel, operational personnel must be cautious, especially as the paper roll builds toward its maximum diameter. Operators should not be working on or near the reel when it is operating. Lightweight grades utilize a "gooseneck" air knife that automatically comes down to the sheet, breaking the sheet with air.

The heavyweight sheet is automatically cut and wrapped across the face of the spool. Where cutting devices are in use, the operator must always be cautious.

(c) Reel spool limits

Reel spools carry more load for their size than any other roll in the paper mill. Because spool journals are more highly stressed than other rolls and are subjected to a fatigue stress, a hazardous situation may develop should they become bent. Bent journals can break, causing the log to fall from the reel or storage reels onto the floor and bump into equipment and personnel [8].

The following practices and precautions are recommended:

1. Bent reel spool journals must be replaced. Metal fatigue occurs in a bent reel spool journal due to residual stress. As the spool revolves, the fatigue becomes greater. When the reel spool bends, the metal becomes work hardened, making the journal brittle at the fatigue point, resulting in notch sensitivity. When this occurs, the journal can break.
2. Bent journals must never be straightened. Journals subjected to fatigue stresses will not tolerate yielding. Three problems are created in straightening the journal: First, it is impossible to straighten a journal without overloading the head. Second, in straightening the journal, the headfit may be lost. Third, if the journal could be straightened, it would probably creep back 20 to 30% due to residual stresses.
3. Care should be taken to prevent stress raisers. Due to notch sensitivity of metals in fatigue, a stress raiser can be detrimental. Some common causes of stress raisers are:
   a. Punch or chisel marks on fillet radii due to improper bearing removal.
   b. Scored crane pickup areas from picking up rotating spools.
   c. Dents or scrapes from mishandling.
4. Slowly lower jumbo logs into storage stands or unwind. Journal failures can easily be caused by lowering jumbo logs too quickly with the mill crane. An impact load may cause both bent journals and journal failures.
5. All replacement journals should be 4340 heat treated steel. Journal replacement steel should be ordered to ASM Spec. A434 class Bd. This specification gives the chemical composition, heat treating procedure and mechanical properties required for replacement journals.
6. Replacement journals should have polished fillet radii. When new journals are machined, care should be taken to obtain the proper fillet radii and each radius should be polished. Since fillet radii are areas of high stress concentration, tool marks can be detrimental.

In all cases, personnel must stay clear of logs being transferred by crane or logs in motion on storage rails. Warnings are to be given when a log is being transferred by crane from storage rails to the winding station. Because of the massive weight of paper rolls, they must be handled with caution.

Proper attention to spool journals can reduce potential hazards at the dry end of the paper machine.

12. Coater safety

On machine or off machine, the coater applies a slippery or sticky coating to the paper web. The presence of the coating creates hazardous footing for operating personnel in and around the coating station. Cleanliness and care in handling the coating is paramount. Continuous training programs should be instituted to instill positive, safety-conscious work habits around the coater. Standard mill operating procedures must be critically assessed to eliminate potentially unsafe operations.

Caution should be exercised at all times when working in the area of the coater head. Coater blades are extremely sharp. The blade guard should be in place at all times during clean-up. Controls must be locked out to prevent movement of the blade head during service.

Safe practices must be established for wash-up procedures. Use automated or mechanical devices employed during these operations.

Proper lighting of the sheet run before and after the coating stations is extremely important to observation of the coated sheet. Operations approaching the sheet run for inspection must be warned of the potential danger and work cautiously.

Adequate barriers and signs should be installed at appropriate points and they should remind personnel of potential coater operating hazards.

Eliminate injuries at the coater by avoiding the following unsafe practices:

- Unsafe handling of coater blades.
- Hand threading.
- Inadequate barriers or warning signs at tension rolls and nip points.
- Horizontal tension device swing at sheet break.
- Reaching in the coating pan to remove buildup while the machine is running.
- Improperly cleaned drain pit and footwalks.
- Improper use of guards.
- Inadequate lighting of the sheet run.
- Improper air cap clean-up procedures.
- Improper disposal of broke.
- Working outside footwalks and standing on handrails.
- Failure to lock out coater station and flying splice during maintenance.
- Static electricity build-up.
- Improper precautionary procedures at the flying splice.
- Improper procedure for adjustment of applicator edge wipe.
- Walking or working under gate roll size press, size press, or flying splice while they are in operation.

13. Paper machine drive safety

The standard in Ref. [9] applies to the back side or drive side of the paper machine and includes: lineshafts, inshafts, couplings, pulleys, and belts revolving at high speeds. With proper guarding and a safety training program, potential hazards can be minimized. Personnel assigned to work on the drive side of the paper machine must be given complete safety instructions and periodic refresher courses so they remain keenly aware of safety procedures and be constantly on guard for potential dangers (see Figs. 808, 809).

Eliminate injuries on the drive side of the paper machine by avoiding the following unsafe practices:

- Loose clothing (neckties, shirts) caught on unprotected rotating couplings and inshafts.
- Head injuries from low, unguarded shafts.
- Inattention to locations of shafts.
- Unprotected shaft stickouts.
- Inadequate guarding of pulleys and belt nips.
- Improper use of indicated walkways or stepping over shafting.
- Failure to keep work areas free of tools and other foreign materials.
- Removal of signs and barriers.
- Failure to disconnect and lock out power source.
- Contact with shafts at shutdown before they have stopped revolving.
- Making adjustments on the run.
- Hazardous footing from lubricant deposits.
- Inadequate lighting.

Here are some of the ways your safety programs can be improved:

- Review of safe drive maintenance and operating procedures.
- Consult a source for guards and barriers.
- Install warning alarm systems.
- Color designate danger areas.
- Install safety signs.
- Check out controls.

14. Winder safety

Many of today's winders run in the 6000–8000 ft/min range, making them the highest speed units in the papermaking process. Their high

Fig. 808. Guards on a belt drive.

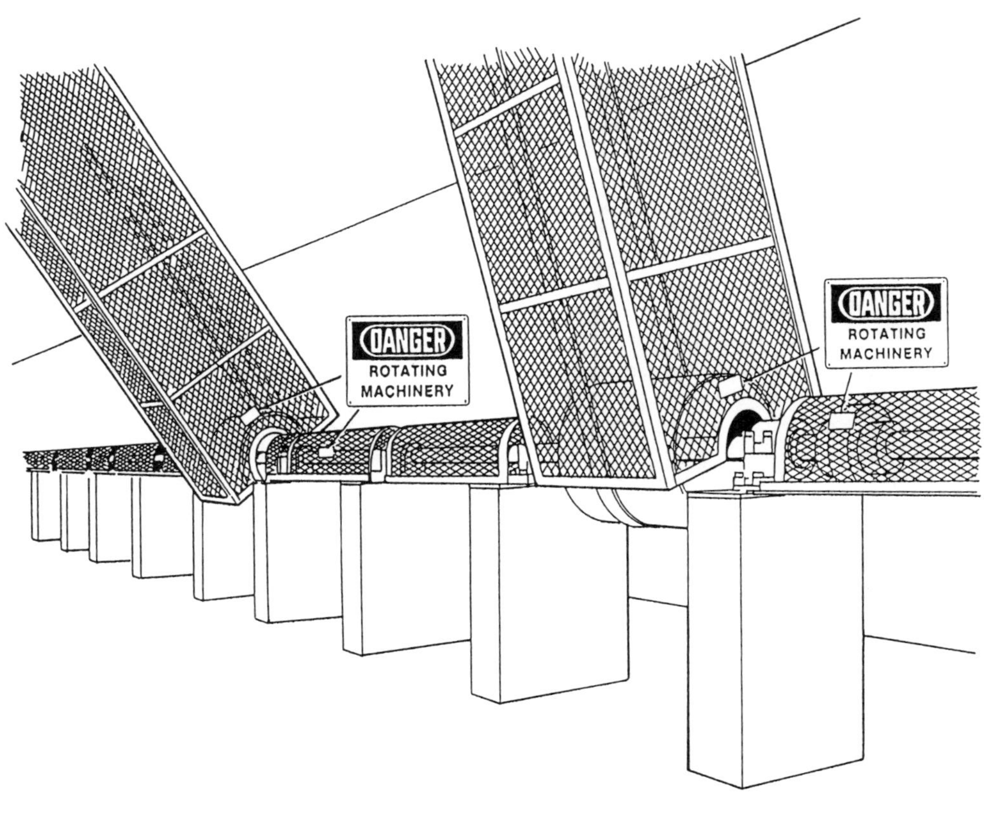

Fig. 809. Guards on the line shaft in the machine room basement for a belt drive.

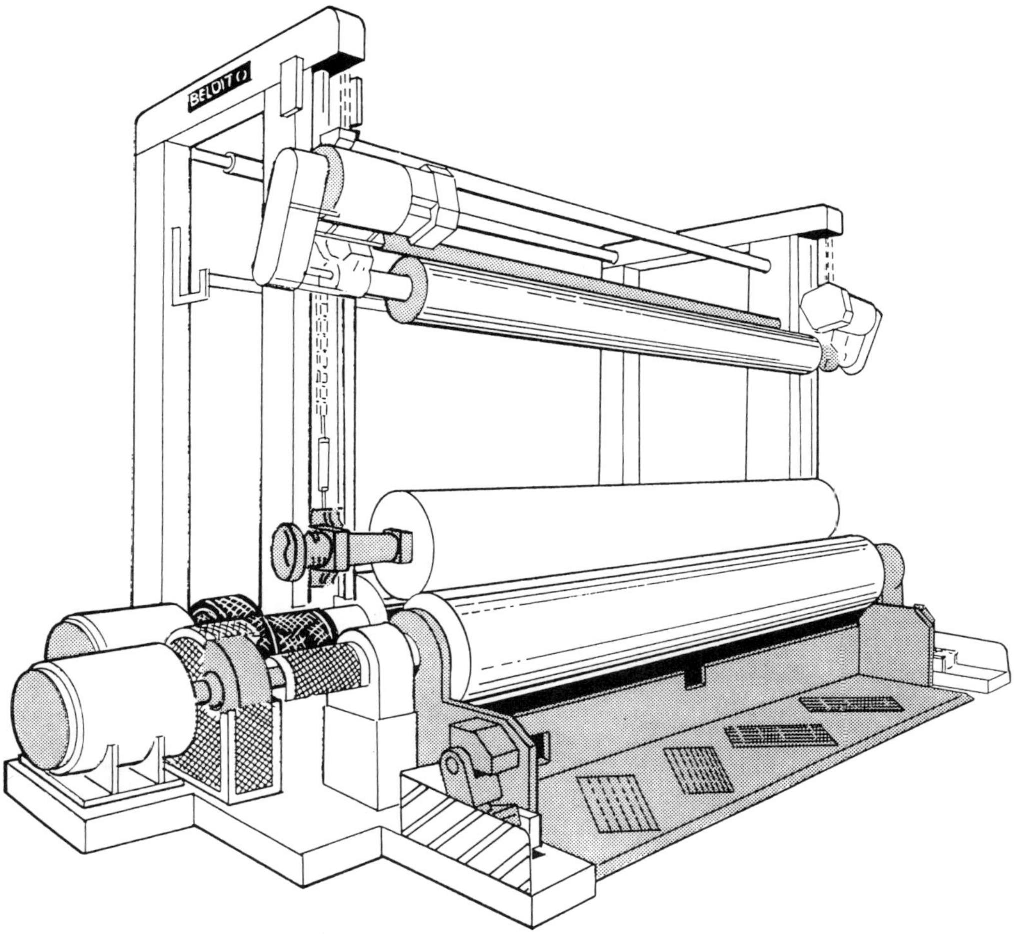

Fig. 810. Safety guards at the winder.

speed reflects the need to keep up with today's paper machines. To maintain high-speed operation and peak efficiency, many of the winder's functions are remotely controlled or automated. Although the winder operations are being automated, many are still labor intensive. In many cases, the operators must assist the threading device or thread by hand. Splicing, although it too is being automated, still remains pretty much a manual operation. Both of these functions can be hazardous unless the operator has been given the proper training so safety becomes an integral part of the operator's movements in and around the winder.

Guards and barriers located at the front of the winder should be built into the framework to prevent operator contact with the paper roll as it is being wound. When winding in the shaftless mode, the paper roll must be contained within the winding station by barriers to prevent ejection from the winder at high speeds. The roll lowering cradle is a barrier that is an integral part of the winding operation. Barriers should be designed to work in conjunction with the winder and the ejection sequence (Fig. 810).

The winder must be operated at low speeds for threading while guards are opened. Guards which have to be opened during threading should be interlocked with the winder drive. This prevents the winder from running when the guards are open. Interlocks must be constantly checked to ensure safe performance. Threading systems are available and can be applied to existing winders to eliminate operator hazards.

Guards must be provided for the ends of core chucks, rider roll belt drives, unwind spools, over inshafts, couplings, unwind clutch, brake drums and belt drives [10].

Cylinders which activate various lowering and raising functions should also be guarded if they create a hazardous pinch point.

Following are some recommended safety

requirements for safe winder operation.

1. Do not start the winder unless hydraulic unit, pneumatic and electrical power are on.
2. Interlocks must be working properly. Do not run winder above thread speed until:
   a. Nip barrier or cradle is in running position.
   b. Core chucks are engaged.
   c. Rider roll is lowered.
   d. Roll ejector is retraced.
3. Do not eject rewound roll until winder has stopped. If ejection is automatic, operators must stay clear of the lowering cradle or other ejection devices.
4. When winding shaftless, do not raise the rider roll off the rewound roll until winder has come to complete stop.
5. The barrier must not be moved to the eject position until the winder has come to complete stop.
6. The operator must not start the winder until absolutely sure everyone is outside the established safety lines (yellow lines). Mirrors should be used to increase the operator's range of vision.
7. Intermediate slitters and paper rolls are to be stopped before the paper is threaded over the winder.
8. Keep aisle around the winder clear of obstacles.
9. All nonoperating personnel must remain outside the safety lines while the winder is running.
10. A sheet threader should be used to eliminate the operator function of entering into the winder and passing the tail through the drum gap.
11. All operators must be out of the rider roll area before the rider roll is lowered or loaded. All hands and fingers must be out of the way of the rider roll and ingoing nips. Mirrors should be used to increase the operator's range of vision.
12. Do not disengage core chucks while winder is running.
13. Do not operate winder with bent core shafts or bent reel spool journals.
14. A knife must not be used to cut rolls apart until the winder has completely stopped.
15. When working in the slitter area, the following practice should be observed:
    a. Slitter blades are sharp and must be handled with caution.
    b. Do not adjust slitters when they are in motion.
    c. Be sure slitter guards are in place before starting up slitters.
16. An audiovisual warning signals that the automated slitter section is about to go into operation.
17. Automated, programmed slitter sections have an emergency stop switch on the control console; when depressed, the slitters will not operate. However, when performing major work on the slitter section, disconnect the power source to the drive bus.

Use of safety color codes and prominent display or safety signs is encouraged to warn operators of the potential hazards around the winder. These warnings and the use of barriers and guards are effective only if the operating personnel are familiar with their purpose. This is best achieved by an ongoing safety training program.

(a) Shaft-type winders

Barrier and guard designs permit full visibility during the winder cycle. Nip barriers can also be designed to serve as roll-lowering cradles/tables. Prominently displayed safety signs remind personnel of potential danger.

(b) Shaftless winding

Barriers on shaftless winders perform the function of containing the winding sets within the winder framework in case of roll ejection at high speeds. Roll-lowering cradles serve as part of the overall nip barrier. Prominent use of safety signs warn personnel of hazards.

(c) Drive side

Brake drums, couplings, drive inshafts, nip points and rider roll drives must be thoroughly guarded. Safety signs should also be prominently displayed on this area of the winding equipment. Personnel should stay out of this area during winder operation.

(d) Winder drum brakes

The purpose of winder drum brakes is to assist in "spotting" splices and to prevent the winder drums from rotating when the web is under a stalled tension condition while the operators are making up a splice.

Should the web break while under stalled tension or during the splicing procedure, there is a danger that the drums may rotate forward slightly due to sudden loss of tension. This movement may catch an operator's hand in the front drum nip. To eliminate this hazard, the addition of drum brakes will prevent the drums and rewinding set of rolls from rotating.

Winder drum brakes are not sized to stop the winder during high speed. They are designed to be used only as "parking" brakes and for stopping the drums at thread to jog speed.

If the drum brakes are to be furnished for an "emergency stop" capability from high speeds, this should be clearly specified so larger capacity brakes may be selected.

If winders do not incorporate drums, these brakes should be added to eliminate a potentially unsafe condition.

(e) Threading two-drum winders

Efficient sheet threading adds to the time-savings and safety of winder operation. Winders are equipped with efficient threading systems to reduce threading time. Using belt conveyors, air trays, air showers and nip rollers, the systems thread the tail from the unwind to the lead-in roll to an operator at the winder drum. Or, the tail may be threaded directly to the drum by an air shower. These systems ensure the safety of the operator, especially when threading the slitter section (horizontal or vertical systems).

(f) Two-drum sheet splicing

There are several methods for splicing the sheet on a two-drum winder. The winder should be stopped or in crawl condition during the splicing procedure. When the winder is moving slowly during the threading or splicing procedures, extreme caution must be exercised when nips are unguarded. An operator must be within reach of an emergency stop button at all times when the winder is in motion during one of the above modes.

(g) Winder roll lowering cradle

Winders equipped with roll-lowering cradles or similar type equipment pose a potential danger to personnel if proper safety precautions are not practiced.

Recommended safety practices include:

1. There must be one, and only one, pushbutton for lowering the cradle provided on the winder. This pushbutton must be located where the operator has a full view of the area to be occupied by the lowering cradle. IF YOUR WINDER HAS A CRADLE LOWERING CONTROL SWITCH AT THE BENCHBOARD, OR ANY OTHER LOCATION THAT DOES NOT PROVIDE FULL VISIBILITY OF THE FRONT OF THE CRADLE, THE CONTROL MUST BE DISCONNECTED AND RELOCATED TO AN AREA WHICH PROVIDES FULL VISIBILITY.
2. The lowering cradle must be painted on both sides with orange or yellow with black cross-hatch stripes to denote an area of potential danger.
3. The area of the floor which is occupied by the cradle in its lowered position must also be painted orange or yellow with black cross-hatch stripes to denote an area of potential danger.
4. A lock-out switch or a disconnect switch must be available for the operator to attach a tag. This lockout switch or disconnect must be operated to deactivate all winder controls and drive when an operator must enter the danger area under the winder. The mill should have lock-out procedures so that, when an operator tags out or locks out a switch, no one removes that tag except the operator.
5. The cradle lower control must be interlocked with the drive so the cradle cannot be lowered until the winder is stopped.
6. An additional mechanical lock can easily be made by the mill maintenance department. A 6 in. diameter pipe with 0.75 in. thick plate top and bottom with a handle attached will provide an adequate barrier. The proper height can be determined by measuring the distance from the floor to the bottom of the cradle and allowing a 0.5 in. clearance for ease of placement and removal. If desired, a power controlled lockup device can be supplied.
7. To prevent an operator from standing too close to the side of the cradle and getting toes caught and pinched, a lip of rubber can be attached to the outside edges of the cradle. The extended lip will brush against operators who are too close and alert them to move away from the cradle lowering area.

The roll-lowering cradle should promote safe operation of the winder by containing the winding roll so it cannot leave the winder during high speed winding.

(h) Winder hydraulic chucks

Winder personnel should never place their hands in the area between the handwheel hub and the end of the winder hydraulic chuck cylinder. To do so may result in serious injury if the chuck is activated. Winder operators must be advised of this danger. Protective boots may be purchased from the winder supplier and installed on the hydraulic chucks by the mill. The boot will prevent operating personnel from getting their fingers caught between the hand-wheel hub and the cylinder when the chuck is activated. A bearing housing guard has also been designed for the inboard chucking apparatus.

REFERENCES

1. Personal Protective Clothing and Equipment
29 CFR 1910.261 (b) Safe Practices (2).
2. Emergency Stops
29 CFR 1910.261 (k) Machine Room (1).
3. Specification for Accident Protection Signs, Z35.1 — 1968
29 CFR 1910.261 (a) General Requirements (3) (xxiv).
4. Lockouts
29 CFR 1910.261 (b) Safe Practices (4).
29 CFR 1910.147.
5. Slings ANSI/ASME B 30.9c — 1988.
ANSI/ASME B 30.9a — 1985.
ANSI/ASME B 30.9 — 1984.
Lifting devices
OSHA Regulation 1910.184.
6. Method of Measurement of Real-Ear Attenuation of Ear Protection at Threshold, Z24.22 — 1957.
29 CFR 1910.261 (a) General Requirements (3) (xxii).
29 CFR 1910.95.
7. Closed Vessel Entering
29 CFR 1910.261 (b) Safe Practices (5).
29 CFR 1910.134.
8. Lifting Reels
29 CFR 1910.261 (k) Machine Room (24)(i)(ii)(iii).
9. Drives
29 CFR 1910.261 (k) Machine Room (2).
10. Winder
29 CFR 1910.261 (k) Machine Room (26) through (31).
11. Emergency Response
29 CFR 1910.120.

BIBLIOGRAPHY

ANSI/ASME Standards
American Society of Mechanical Engineers
345 East 47th Street
New York, NY, USA
10017

ANSI Standards
ANSI Standards
1430 Broadway
New York, NY, USA
10018

OSHA Regulations
Superintendent of Documents
Government Printing Office
Washington, DC, USA
20402

TAPPI Methods
TAPPI PRESS
Technology Park/Atlanta
P.O. Box 105113
Atlanta, GA,USA
30348-5113

INDEX

A

Absorbance (paper), 14-15
  tests, 14-15
Additives
  dry strength agents, 164-168
    natural, 164-167
    synthetic, 167-168
  fillers, 150-153
  pigments, 152-153
  retention aids, 158-159
  wet strength agents, 168-169
Agitation (wire), 209-211
Air leak testers for smoothness
  Bekk, 4
  Bendtsen, 4, 5
  Porosity, 4
  Print-Surf, 4, 5
  Sheffield, 4, 5
Air resistance, 5
  tester, 5
Alkaline papermaking, 178-179
Alginate, in surface sizing, 311
American Gear Manufacturer's Association, 534
American Paper Institute, *see* API
American Society of Mechanical Engineers, 325, 679
Anionic trash (papermaking), 146, 174
Animal glue, in surface sizing, 311
Anisotropy, 6-7
  square sheet, 6
ANSI/ASME safety standards, 656
API
  Containers and Kraft Paper Group, 24
  mill safety programs, 651
Approach systems, *see* Flow approach systems
Apron size press, 315-316
Arcu former, 250
Ash, distribution in sheet, 223, 229-230
ASME safety standards, 656

B

Basis weight, 2-3, 220, 223, 239, 247, 262
  beta gauges, 483-484
  CD control, 476-477, 480-481, 487-490
  variation, 481-486
Bel Bond former, 251
Bending stiffness (paper), 12
Bi-Vent press, 276
Blade formers, *see* Gap blade formers
Box compressive strength, 76
Breast roll, 194
Brightness (paper), 16-17
  fluorescent component, 17
  test, 17
Bristow test, 14-15
Britt jar, 184-185
Broke storage chest, 468-469
Broke systems, 462-471
  coated broke, 470-471
  colored broke, 471
  control, 463-466
  design, 462-463
  dry broke repulping, 466-467
    deflakers, 467
    dry end pulper control, 468
    safety, 467-468, 665
    sheet handling, 467
  finishing broke, 470

  - wet broke
    - couch pit, 469
    - furnish control, 469-470
    - press pulper, 469
    - saveall, 469
  - wet strength broke, 471
- Bulk (paper), 2-3
  - effect of calendering on, 447-448
  - effect of calendering roll loading on, 449-450
- Bursting strength (paper), 11-12
  - test, 11

C

- Calendering, 436-461
  - CD variation reductions, 455-456
  - coated paper, 450-452
  - distortion-energy theory, 453-454
  - doctors, 521
    - effect on sheet bulk, 447-450
  - gloss calenders, 457-459
  - machinery variables, 447-455
    - fly rolls, 454
    - roll hardness, 454
    - roll temperature, 454
  - MD variations, 456
  - nip passes, 447-449
  - reel building, 455-456
  - replication, 450-454
  - process control, 455-456
  - sheet variables, 446-447
  - steam showers, 454-455
  - substrata thermal molding, 460
  - supercalendering, 439-440, 450-452
  - temperature-gradient, 458
- Calender rolls
  - anti-deflection rolls, 442-443
  - construction, 441
  - deflection control, 442-443
  - drives, 543
  - electric power requirements, 543
  - fly rolls, 454
  - hardness, 454
  - heating (metal), 441-442, 443-444
  - non-metallic covering, 444-445
  - resilient, 450
  - safety requirements, 668-671
  - soft nip calenders, 459-460
  - surface temperature of metal rolls, 460
- Calender sizing of paperboard, 320-321
- Calender stacks
  - operating variables, 445-446
  - vibration, 630-631
- Caliper
  - CD control, in calendering, 443, 455-456, 500-501, 630-631
  - effective caliper, 62
  - uniformity, 479-481, 485
- Canson, 193
- Cantilever beam tester, 10
- Carbon steel corrosion in white water, 601
- Carboxymethyl cellulose, in surface sizing, 311
- Cationic demand, 146, 174
- CD control, *see* Cross-direction control
- Centrifugal exhausters, 555-557
- Chemical pulps, 34
- Chemimechanical pulps, 37
- Chemistry, papermaking, 140-191
- Chemithermomechanical pulp, *see* CTMP
- Chests
  - for broke storage, 468-469
  - in white water systems, 94-95
- Chromium complexes, in surface sizing, 312
- CIE tristimulus functions, 18-19
  - STARLAB system, 19
- Coagulation (papermaking), 145-146
- Coated broke control, 466
- Coated paper calendering, 450-452
- Coater safety, 674
- Coater vibration, 630-631
- Coat weight CD control, 501-502
- Cobb test, 14
- Collaborative testing services, 24
- Colloid chemistry, in papermaking, 141-160,
  - association colloids, 174-177
  - emulsions, 176
- Colorants, 170-174
- Color measurement (paper), 18-19
  - CIE tristimulus functions, 18-19
    - STARLAB system, 19
  - Hunter Lab system, 18-19
- Combination pickup/suction press, 274
- Computer control
  - papermaking chemistry, 185-187
- Condensate and steam systems in drying, 287-293, 336
- Conductivity/resistivity (paper), 19
- Consistency
  - in headboxes, 207-209
- Consolidation, *see* Web consolidation
- Contra flow vat, 241-243
- Contra flow former, 253
- Convertibility
  - influence of sheet properties on, 75-78
- Core problems, 363-367
- Corrosion
  - closure of papermaking systems, 598-601
  - crevice and pitting, chloride, 601-603
    - thiosulfate, 603
  - headboxes, 595-596
  - mechanism, 601-604
  - microbiological, 603-604
  - monitoring, 655-656
  - paper machines, 594-604
  - pitting and crevice, chloride, 601-603
  - roll journals, 596-598
  - stock piping, 595
  - suction rolls, 604-614
  - vacuum pumps, 596
  - wet end structures, 596
  - wires, paper machine, 596
  - Yankee dryers, 334-335
- Corrugating medium
  - fluting, 76-78
- Couch roll, 201-202, 559-560
- Covered rolls
  - care and grinding, 640-649
  - dimensional tolerances, 649
  - grinding, 640-649, 648-649
  - inspection, 641-642
  - installation, 642-644
  - nip impressions, 644-646
    - crown effect on, 645-646
  - operation, 646
  - receiving, 641-642
  - shipping, 641-642
  - storage, 642
  - surface cracking, 646-647

surface finish, 648-649
Creping process, 338-340
Crompton, 194
Cross-direction control, 472-506
actuators, 486-487, 489-502,
benefits, 478-481
caliper uniformity, 443, 455-456, 479-481, 500-501
coat weight control, 501-502
gloss, 502
moisture profile, 490-500
production costs, 477-478
smoothness, 502
topographical maps, 472, 482
variability costs, 474-476
CTMP, 31-34, 39-41
aspen, 33
eucalyptus, 33
Cylinder machines, 237-259
Contra flow vats, 241-243
pressure former (short former), 247-248
Rotoformer, 246
suction former, 246-247
Uniflow vat, 243
web formation mechanism, 244-245

D

Dandy roll, 194, 211-213
machine speed to roll diameter ratio, 211
Deaerators, 557-558
Deflectors, 195-197
Density
Wood, 30, 49-50
Deposit control, 177-178
Dielectric strength, 19
Delamination (paper), 12
Density (paper), apparent, 3-4
Didot, Leger, 192-193
Dickinson, John, 193
Distributed control
broke systems, 465
Doctors, on paper machines, 507-524
applications, 519-521
bearings, 519
blades, 518
wear patterns, 523
doctor back, 518
journals, 518-519
installation, 521-524
loading and lifting, 519
maintenance, 524
oscillators, 519
Donkin, Bryan, 192-193
Double-felted linerboard press, 276
DPCV, differential pressure control valve, 327
DLVO theory, 144
Drainage
foils, 205-207
Sankey diagram, 223
split diagrams, 224-232
table rolls, 205-207
Drainage devices
deflectors, 195-197
foils, 197
Sheraton roll, 210
table rolls, 195-197
vacuum boxes, 198-201, 559
vacuum foil units, 197-199, 558
Drainage, pulps, 31, 159-160
dry line, 199
wet line, 199
Drainage and wire flow, 202-207
Sheraton roll, 210
Draws, influence on sheet structure, 70-73
Drives
ac drives, 526-527
cable and transformer ratings, 533-534
dc drives, 525-526
calender section, 543
dryer sections, 541-543
forming sections, 535-537
maintenance, 529-530
normal running load, 531-544
performance, 534-535
press sections, 538-541
recommended drive capacity, 533-544
reel section, 543-544
regulators, 527-528
torsional frequency calculations, 535-536
safety, 674
speed reducing gears, 534
transformer and cable ratings, 533-534
winders, 544-546
Drives, paper machines, 525-547
electric power requirements, 530-538
forming sections, 535-537
operator control function, 528-529
wire drive ability, 537-538
Drives, press sections, 538-543
Drives, winders, 544-546
Dryer fabrics, 585-593
air permeability, 586
application, 586-589
classifications, 586
cleaning, 516-517, 591-592
construction, 585-586
functions, 585
installation, 590-591
operation, 591
repairs, 592
seams, 589-590
Dryers
bars, 290
blowthrough flow control, 293
cascade system, 290-291
double-felted, 305
electric power requirements, 541-543
pistol grip section, 588
runnability, 301-305
safety requirements, 663-668
Serpentine, 303
sheet flutter, 302-302
single-felted (Unorun), 302-304, 588
single-tier arrangement, 304-305, 588-589
syphons, 288-290
thermocompressor system, 291-293
ventilation system, 293-297
vibration, 629-630
Yankee dryers, 323-342
safety, 330-331
vibration, 630
Drying, 58-60, 282-305
condensate system, 287-293
layer effect of, 287-288
rimming, 288, 326
cylinder drying, 282-283
doctors, 520

fabric cleaning, 516-517
efficiency, influence of variables on, 286
energy consumption, 287
evaporation, 285-287
heat transfer, 283-285
heat transfer coefficients, 336-338
mechanism of Yankee drying, 335-338
pocket ventilation, 285-286, 298-301, 587
rates for different paper grades, 286-287
rates in different dryer zones, 286
sheet dryness, performance guidelines, 260-261
sheet stabilization, 303-304
steam and condensate system, 287-293
steam pressure for different paper grades, 284
surface-sized sheet, 319-320
theory, 283-285
Yankee dryers, 323-342
Drying rates (TAPPI), 286-287
Drying zones, 286
Dry forming, 255
Dry line, 199
Dry strength additives, 164-168
Duoformer, 251
Dusting and linting, 12
Dyes, 171-174

E

ECT, 76
Edgewise compressive strength, *see* ECT
Electrical double layer, 144-145
Electrical properties (paper), 19-20
Electric drives, *see* Drives
Electric power requirements
calender section, 543
dryer sections, 541-543
forming sections, 535-538
paper machines, 535-538
press sections, 538-541
reel section, 543-544
winders, 544-546
Elmendorf tear tester, 11
Extended nip press, 70, 267-269, 277, 540, 662

F

Fabric presses, 266
Fatty acids, in surface sizing, 312
Felts
conditioning, 560-561
making felt (cylinder machines), 244
showers on press felts, 516
Fiber bonding, *see also* interfiber bonding
relative bonded area, 61
Fiber orientation (MD/CD), 6-7, 68-69
Fiber properties, 26-53, 44-48, 58-60, 147-150
pulping processes, influence on, 32-35, 148
Fibers
cellulosic, 34
cell wall, 57-60, 147
characteristics, 147-150
chemistry of, 147-148
coarseness, 30, 46
cohesiveness, 48
compactability, 47-48
earlywood, 50-51
features, 44-48
hardwoods and softwoods, 30
interfiber bonding, 28-29, 48, 58
latewood, 50-51
morphology, 43-44, 147
influence on strength, 30
nonwood, 43-44
properties, 44-48
raw materials for papermaking, 26-53
relative bonded area, 28, 61
softwoods and hardwoods, 30, 42-43
sources, 26-27, 41-44
strength, 29-30, 47
breaking length, 61
surface area, 31
water retention value, 31-32
wood species, 41-43
zero-span tensile strength, 30
Fibrils, 58
Fillers, in papermaking, 150-153
specific surface area of, 152
Film splitting, in sizing, 313-316
Fines, 6, 147-150
distribution in newsprint, 216
retention of, 155-158
Finishing
broke handling, 470
calendering, 436-461
productivity, 426-431
slitting, 406-423
supercalendering, 439-440, 450-452
web control, 347-349, 390-406
winding, 343-435
unwinding, 423-424
Finite element calculation in stress analysis, 330
Fire-retardant paper, 312
Flexonip press, 268
Flocculating agents, 153
Flocculation (papermaking), 145-146, 154-158, 207-209, 220
wire agitation influence on, 209-211
Flow approach systems
hydraulic headboxes, 118-119
manifold systems, 100-104
Multiplex system, 101
piping recommendations, 98-100
Flow rate measurement, 181
Fluting, 76-78
Fluorescence, 17
Fluorochemicals, in surface sizing, 312
Flocculation, 105
Foams, 176-177
Foils, 197
vacuum foils, 197-199
Formation, 6, 68, 110, 155, 158, 202-214, 220, 223
oriented hydraulic shear, influence on, 209-211
QNS formation curve, 212-213
STFI units, 202
two-sidedness, 220, 228
Formation aids, 211-214
dandy roll, influence on, 211-213
machine speed to dandy roll diam ratio, 211
shake, 213-214
Formers, 216-219
cylinder machines, 237-259
doctors, 519
electric power requirements, 535-538
fourdrinier, 192-215
fourdrinier/cylinder, 248-249
gap (blade and roll), 217-228, 233-234
hybrid, 218-219, 228-233
multiple-wire, 216-236

multi-ply, 239
"on top" formers, 248, 250, 252
preformers, 228-233
retrofit, 228-233
tissue, 223-226
twin-wire, 216-236
Forming board, 194-195
Forming fabrics, 564-571
air permeability, 570
application, 568-569
cleaning, 570
construction, 564, 566-567
functions, 564-567
maintenance, 570-571
operation, 569-570
plastic, 566-567
terminology, 568
material, 567-568
Fourdrinier brothers, 192-193
Fourdrinier machine, 194-202
multi-wire, 248-249
safety procedures, 659-660
secondary headbox on, 248
showers on, 515
Fourdrinier papermaking, 192-215, 229
Fourier's heat conduction law, 335-336
Fourier transform, in vibration analysis, 620, 623
Furnish components, 141-142, 174

G
Gamble, John, 192-193
Gap blade forming, 226-228, 233-234
Gap roll former, 222-223, 233-234
Sankey diagram for, 223
Gate roll size press, 353
Gears, speed reducing for drives, 534
American Gear Manufacturer's Association, 534
Gloss, 17-18
CD control, 502
test, 17-18
Gloss calenders, 457-459
Grade changes
in long circulation systems, 91
Gurley "densometer," 6
Gurley tester, 10

H
Headbox consistency, 207-209, 254
streaking, 108, 132, 230
Headboxes
corrosion, 595-596
high-consistency, 254-255
hydraulic headboxes, 117-139
basis weight uniformity, 130-132, 137-139
design, 119-129
distribution system, 120
evaluation of, 129-130
flow approach, 118-119
flow nozzles, 122
flow stabilization, 121-122, 123-125
header control, 136-139
high-consistency, 129
jet angle, 134-135
jet speed control, 136-139
jet/wire ratio, 137
lip opening (consistency), 135
manifolds, 120-121
multilayered, 128
profile problems, 132-133
skimmer operations, 135-136
slices, 126-128
step diffusors, 123
streaking, 132
multilayer headboxes, 238, 251-252
roll headboxes, 97-116
design of, 100-104
distributor rolls, 110-111
flow approach system, 97-100
liquid level control, 111-114
manifold systems, 100-104
no-wake formula, 105
silo design, 97-99
slices and nozzles, 104-110, 115
streaking, 108
safety procedures, 659
secondary headboxes, 248
Hercules size test, 14
High-consistency forming, 254-255
High-consistency headbox, 254
High-yield pulps, 32-34
Hoods and exhaust, 297-298
high humidity, 298
Hornbostle tube, 113
Hunter Lab system, 18-19
Hybrid formers, 218-219, 228-233
Hydrophilic systems, 142-143
Hydrophobic systems, 142-144
total potential energy curve, 145
Hydrogen bonds, 58
Hygroexpansivity, 5-6
Hygroexpansimeter, 5-6

I
IGT pick test, 13
IGT printability tester, 14
Impulse drying, 268-269
Infrared heaters, 494-497
Institute of Paper Chemistry, 23
Institute of Paper Science and Technology, 23
Intensa press, 268
Inverform machine, 238, 250-251
Interfiber bonding, 28-29, 48, 58
Intl. Standards Organization, *see* ISO
ISO, 23

J
Jet speed to wire speed ratio, 202-204
Jet setting, 221-222
Jones, J.F., 237

K
K&N ink smears, 14
Ketene dimers, in surface sizing, 312
Kraft pulps, 34

L
Light-weight coated paper, *see* LWC
Linerboard
double-felted press for, 276
Linting and dusting, 12
Load-elongation paper properties, 7-9
Lubrication, Yankee dryers, 327-328
LVDT, linear variable differential transformer, 489
L&W tester, 10
LWC paper, 38
Lyophilic systems, 142-143

Lyophobic systems, 142-143

M

Machine clothing, 563-593
- dryer fabrics, 585-593
- fabric life, as affected by drive power, 538
- felt conditioning, 560-561
- forming fabrics, 564-571
- forming fabric tension, 222
- monofilament fabric coefficient of friction, 537
- press clothing, 278, 571-584

Machine-direction control, *see* Cross-direction control
Maintenance
- safety requirements, 650-679
- test equipment, 22-23

Market pulps, 31
Marshall, John, 194
MD control, *see* Machine-direction control
Mechanical pulps, 32-34
- hardwoods, low density, 33
- Norscan spruce, 33

Metallurgy, of Yankee dryer shells, 331-332
Metering size presses
- blade/rod, 316-318
- wire-wound rods, 317

Methyl cellulose, in surface sizing, 311
Mill safety, 650-679
- air hoses, 653-654
- chests, 654-655
- calenders, 668-671
- coater, 674
- clothing, protective, 652
- corrosion monitoring, 655-656
- drives, 674
- dryers, 663-668
- emergency stop devices, 653
- equipment operation, 658-659
- extended nip press, 662
- footwalks, 651-652
- fourdrinier, 659-660
- hazardous materials, 652-653, 656
- headbox, 659
- housekeeping, 652
- ladders, 651-652
- lifting devices, 656
- lighting, 658
- Material Safety Data Sheets, 652
- nip protection, 651
- noise, 656-657
- OSHA regulations, 679, *see also* OSHA
- paper machine, 650-651, 659-668
- power sources lock-out, 654
- press operation, 662-663
- pressure vessels, 654-655
- programs
  - American Paper Institute (API), 651
  - National Safety Council, 651
  - Paper Industry Management Association (PIMA), 651
- reels, 671-674
- signs, 653
- stairways, 651-652
- tanks, 654-655
- threading devices, 651
- twin wire former, 660-662
- welding, 655
- wet-end, 659
- winder, 674-678
- wire changes, 660

Modulus of elasticity (paper), 15-19
Moisture profile, 477-481, 485, 490-500
- infrared heaters, 494-497
- steam showers, 497-499
- streak control, 492-494
- tissue and towel grades, 499-500

Multi-cylinder forming, 237-259
Multiple-wire forming, 216-236
Multi-ply forming, 237-259

N

National Electric Code, 533, 658
National Safety Council, 651
Newsprint, production on twin-wire formers, 228
Nipco roll, 265
Nips, 264-265
- double-felted, 267, 273
- double-vented, 267
- mechanics in winding, 353-356
- pressure profiles, 268
- triple nip, 273

Noise safety, 656-657
- survey techniques, 657

Non-wood fibers, 43-44

O

Occupational Safety and Health Administration, *see* OSHA
Optical properties, 30
- scattering coefficient, 30

OSHA, 653, 655, 656, 679

P

Packaging papers, 11
Paperboard, 40-41
- bending stiffness and tensile, 41
- boxboard, 40
- calender sizing, 320-321
- compressive strength, 76-78
- end-use performance, 78-82
- fluting, 76-78
- foodboard, 40
- multi-ply paperboard, 239-240, 255-258
- solid bleached board, 40

Paper grades
- bag, 39
- creped, 323
  - creping process, 338-340
- glassine, 34, 38, 312
- greaseproof, 34, 38
- industrial, 38-39
- light-weight sheets, 262
- light-weight coated, 38
- linerboard, 252
  - double-felted press for, 276
- machine-glazed, 323
- MG cylinders, 323
- multi-ply, 258
- napkin, 323
- newsprint, 34, 36
  - fines distribution in, 216
- packaging, 11, 38-39
- paperboard, 40-41
- parchment, 312
- printing, 36-38, 252
- sack, 39
- supercalendered, 38

tissue, 39-40, 323, 499-500
towel, 499-500
wadding, 323
waxing stock, 312
wrapping, 38-39
writing, 36-38
Paper Industry Management Association (PIMA), 651
Paper machine clothing, *see* Machine clothing
Paper machines
corrosion, 594-604
cylinder machines, 237-259
electric power requirements, 535-538
fourdrinier, 192-215
multi-cylinder board, 244
multiple wire, 216-236
multi-ply, 237-259
operating data, 181-182
twin-wire, 216-236
two-drum winder, 429-430
vacuum systems, 548-562
vibration, 615-639
wet end, 625-627
Papermaking chemistry, 140-191
computer control, 185-187
corrosion of closed system, 598-601
laboratory measurements, 179-182
on-line measurements, 180
pulp suitability for, 35-41
Paper properties, *see also* under Sheet
absorbance, 14-15
air resistance, 5
anisotropy, 6-7
basis weight, 2-3
bending stiffness, 10-11
blistering, 12
brightness, 16-17
bulk, 2-3
bursting strength, 11-12
caliper, effective, 61
conductivity/resistivity, 19
density, 34, 62, 275
density (apparent), 3-4
dielectric strength, 20
dimensional stability, 59
elastic properties, 64-66
electrical, 19-20
end-use properties, 30-32, 78-82
fiber properties, influence on, 31
formation, 6
gloss, 17-18
hygroexpansivity, 5-6
interfiber bonding, 28-29
light scattering, 62-63
load-elongation, 7-9
mechanical strength, 29-30, 61-63
dry strength agents, 164-168
wet strength agents, 168-170
modulus of elasticity, 8, 10
moisture influence on, 5
optical, 15-19
pulp properties, influence on, 27-32
relative bonded area, 29
relative humidity influence on, 20-22
rupture resistance, 12
smoothness, 4, 30
square sheet, 6
sizing, effect of, 318-319
stiffness, of multi-ply structure, 239
strain relaxation, 21-22
surface properties, 30
surface strength, 12-13
TEA, 8
tearing strength, 11
tensile strength, 8, 28-29
thickness, 3-4
two-sidedness, 220, 228
zero-span tensile, 9-10
Picking, 12
test, 13
Pigments, 152-153, 170-171
Piling, 12
Piping design for vacuum systems, 561
Pitch control, 177-178
Plasticizers, in surface sizing, 312
Pocket ventilation, 285-286, 298-301
Poisson ratio, 64-66
Polyelectrolyte adsorption, 153-154
Polyester, polyethylene terephthalate (q.v.)
Polyethylene terephthalate (PET), 567
Polymeric sizes, 312
Polyvinyl alcohol, in surface sizing, 311
Preformers, 228-233
Press arrangements, 272-277
transfer presses, 272-273
Twinver press, 273
Press drying, 268, *see also* Presses, Pressing
Press clothing, 278, *see also* Press fabrics
Press fabrics, 571-584
applications, 575-581
cylinder machines, 575-576
fine paper machines, 578-580
kraft machines, 580
multi-ply board machines, 575-576
newsprint machines, 580
pulp machines, 580-581
tissue machines, 576-578
conditioning, 581-582
construction, 573-575
functions, 572
installation and operation, 581
performance, 573
problems, 582-584
properties, 572-573
repairing, 582
storage, 581
Presses, 265-270, *see also* Nips
Bi-Vent, 276
blind-drilled rolls, 266
clothing, 278
combination pickup/suction, 274
double-felted, 277
double-felted, for linerboard, 276
electric power requirements, 538-541
extended nip press, 267-269, 277, 540
fabric presses, 267
felt change safety, 663
felt roll vibration, 627
grooved rolls, 265-267
maintenace safety, 663
no-draw, 273
safety, 662-663
showers on press sections, 516
smoothing, 275
steam showers, 277-278
suction rolls, 265-266, 273
Sym-press, 275

terminology, 578
three-nip no draw, 580
triple-nip, 273-274
Tri-Vent, 275
types, 272-277
vibration, 627-629
felt roll, 627
Pressing, 260-281
doctors, 519-520
fundamentals, 261-262
limitations, 263
performance, variables affecting, 278-280
sheet dryness, 260-261, 275
sheet marking, 273
sheet transfer from wire, 269-272
variables affecting water removal, 263
vibration, 584
Press rolls, 263-265
crowning, 264-265
nip pressure, 264-265, 268
Pressroom problems, 369-372
Pressure former (short former), 247-248
Printing
pressroom problems, 369-372
Printing inks, 14
opacity, 16
papers, 13-14
problems, 12
Printing properties, 33-34, 36-37
aspen CTMP, 33
ink penetration, 37
newsprint, 34, 36
Printing tests, 14
IGT printability tester, 14
K&N ink smears, 14
Process control
CD control, 472-506
papermaking chemistry, 185-187
Production costs, 477-478
Productivity
in winding, 426-431
Profiles, *see* Cross-direction control
Pulping processes, 32-35
influence on fiber properties, 32-35
pulps, 32-35
wood species for, 33
Pulp properties, 27-32
chemical, 29
drainage, 31
evaluating pulps, 28
mechanical, 29-30
optical properties, 30
strength relative to drainage, 31
suitability for papermaking, 35-41
tear-tensile relationship, 31
wood density, effect on, 50
Pulps
BIVIS, 33
chemical pulps, 34
chemimechanical, 37
extrusion mechanical, 33
high-yield mechanical, 32-34
kraft, 34
kraft, bleached, 37
mechanical, 32-34, 63
refiner mechanical, 32-34
secondary fibers, 34-35
STAKE, 33
sulfite, ultra-high-yield, 33
suitability for papermaking, 35-41
Pulpwood characteristics, 48-51
Pumps
centrifugal exhausters, 555-557
corrosion, 596
liquid ring, 550-555, 596

Q

QNS (Quebec North Shore) formation curve, 212

R

RBA, *see* relative bonded area
Recycled fibers, see Secondary fibers
Reel building, 455-456
Reel doctors, 521
Reel drives, 543
Reel safety, 671-674
Refiner mechanical pulp, see RMP
Refining, 66-68
Refining
high-consistency, 39
in strength development, 31-32
Relative bonded area, 28, 61
Relative humidity influence on paper properties, 20-22
Replication in calendering, 450-452
Retention, 155-158, 220-221
aids, effect of, on, 158-159
Britt jar, 184-185
dyes, 173-174
first pass, 182-185
Retention and Drainage Subcommittee (TAPPI), 180
water retention value, 31-32
Rider rolls in winding, 358
Robert, Louis, 192-193
Roll defects, 350-353, 394, 428-429
Roll density, 361-363
Roll formers, see Gap roll forming
Roll journals, corrosion of, 596-598
Roll quality measurement, 356-357
Rolls, *see also* Covered rolls
Covered, care and grinding, 640-649
grinding, 640-649
Roll structure, 349-353
Roll vibration, 615-639
Rotoformer, 246
Round robin testing, 23-24
RMP, 32-34
Rupture resistance (paper), 12

S

Safety, *see also* Mill safety
ANSI/ASME standards, 679
broke pulpers, 467-468
mill safety, *see* Mill safety
OSHA regulations, 679, *see also* OSHA
paper machine, 650-679
drives, 674
winders, 674-678
Yankee dryers, 330-331
Sampling
reference samples, 23-24
Sankey diagram, 223
Scattering coefficient (pulps), 30
Secondary fibers, 34-35, 224
Serpentine dryer, 303
SGW, 31-34

Shear modulus, 64-66
Sheet structure, 54-84
  draws, effect of on, 70-73
  drying, effect of on, 75
  refining, effect of on, 66-68
Sheet flutter, drying, 302
Sheet transfer from press to dryer, 301-302
Sheet transfer from wire to press section, 269-272
Sheffield units, 4
Showers, on paper machines, 507-524
  fixed orifice, 511-512
  fourdrinier, location of showers, 515
  forming section, 513-515
  maintenance, 517-518
  oscillating showers, 507-508
  single-jet traversing, 508-511
  stationery showers, 507
Silicone resins, in surface sizing, 312
Size presses, 312-315
  designs of, 315-318
  film-splitting, 314
  metering press, 316
  pickup variables, 313-315, 319
  solution viscosity, 313
  vibration, 630-631
Sizing, internal, 160-164
  cellulose-reactive sizes, 163-164
  dispersed sizes, 162-163
  fortified rosin sizes, 162-163
  rosin-alum sizing, 161-163
  Washburn equation, 160
Sizing, surface, 306-322
  alginate, 311
  animal glue, 311
  calender sizing of paperboard, 320-321
  carboxymethyl cellulose, 311
  chromium complexes, 312
  drying, 319-320
  fatty acids, 312
  film formers, 311
  fluorochemicals, 312
  ketene dimers, 312
  methyl cellulose, 311
  plasticizers, 312
  polymeric sizes, 312
  polyvinyl alcohol, 311
  silicone resins, 312
  specialty sizes, 312
  starch, 307-311
  wax emulsions, 311-312
Slice delivery, 202-205
Slice geometry, 204
Slices, 104-110, 126-128
  jet setting, 221-222
  jet velocity, 110, 202-204
  rush/drag ratio, 110, 222
Slitters, 418-421
Slitter table, 421-423
Slitting, 406-423
  problems, 406-417
  theory, 407-411
  trim systems, 417-418
Smoothness, 4, 30
  CD control, 502
  testers, *see* air leak testers
Soft nip calenders, 459-460
Spreading, web, in winding, 398-406
Square sheet, 6
Standardization of test methods, 22-24
Starch, as surface size, 307-311, 318-319
STFI units of formation, 202
Stickies, 224
Steam systems, drying, 287-293
Stock delivery, 68-69, 239-240
  slice lip adjustment, 487-490
    linear variable differential transformer, 489
Stone groundwood pulp, *see* SGW
Strain relaxation (paper), 21-22
Strength properties of paper, *see* Paper properties
Suction boxes, *see* Vacuum boxes
Suction former, 246-247
Suction rolls, 265-266, 273
  alloys, 605-606
  corrosion, 604-614
    cracking, 608-610
    fatigue testing, 607-608
  corrosion resistance measures, 611-613
  failures, 608-610
  inspection, 613
  rubber cover failures, 610
  shell manufacturing, 610-611
  showers on, 516
  vacuum application, 560
Supercalendering, 439-440, 450-452
  supercalenders, special features, 456-457
Surface area measurements (BET), 149, 152
Surface properties (paper), 30
Surface sizing, 306-322
Surface strength (paper), 12-13
Surfactants, 175-176
Sym-Press, 275
Syphons, dryers, 288-290

T

Taber tester, 10
Table rolls, 195-197
Tanks
    in white water systems, 94-95
TAPPI brightness, 17
TAPPI opacity, 16
TAPPI methods, 23
TAPPI noise survey techniques, 657
TAPPI Retention and Drainage Subcommittee, 180
Taylor vortices, 124, 132
TEA, 8
Tear tester, 11
Tearing strength (paper), 11
Temperature gradient calendering, 458
Tensile energy absorption, see TEA
Tensile strength of paper, 8
Tension control, in winding, 390-394
Testing
  procedures, 23
  round-robin testing, 23-24
  sampling, 23-24
  standardization and calibration of instruments, 22-24
Testing environments, 21
  standardization and calibration, 22-24
Testing paper properties, 1-25, 64-66
Thermomechanical pulp, *see* TMP
Thickness (paper), 4
Tissue formers, 223-226
Tissue moisture profile control, 499-500
TMP, 31-34, 36
Topographical maps, process control, 472, 482

Traction control, in winding, 394-398
Trim systems, in slitting, 417-418
Tri-nip press, 274
Tri-Vent press, 275
Twinver press, 273
Twin-wire forming, 216-236, 250
    safety procedures, 660-662
Two-sidedness, *see* Paper properties

U

Ultra former, 252, 253-254
Unorun, single-felted dryer, 302
Uniflow vat, 243
Unwinding
    braking, 424-426
    drives, 545
    shafted, 423-424
    shaftless, 424

V

Vacuum boxes, 198-201
Vacuum foils, 197-199
Vacuum systems, 548-562
    centrifugal exhausters, 555-557
    deaerators, 557-558
    drainage element requirements, 558
    felt conditioning, 560-561
    fundamentals, 548-550
    liquid ring pumps, 550-555
    paper machine requirements, 557-561
        couch rolls, 559-560
    piping design, 561
    suction rolls, 560
Valley headboxes, 101, 112
Van der Waals forces, 143-146, 149, 153
Ventilation
    dryers, 293-297
    hoods, 297-298
Vibration
    calender stack, 630-631
    coater, 630-631
    control, 631-636
    dryers, 629-630
    measurement of, 616, 622-624
        Fourier transform, 620, 623
    monitoring, on-line, 636-638
    paper machine, 615-639
        advantages of, 616
    presses, 584
    rolls, 615-639
    single degree of freedom system, 617-621
    size press, 630-631
    theory, 616-622
    winders (bouncing), 426-429

W

Water
    process water systems, 512-513
Water retention value, *see* WRV
Wax emulsions, in surface sizing, 311-312
Web breaks, 277
Web consolidation, 56-60, 261
Web shrinkage, 59
Webs, multilayered, 237-259
Webster, Daniel, 216
Web structure, 54-84
Web tension control, in winding, 390-399
Welding safety, 655
Wet end chemistry, 140-191
    process control, 185-187
Wet end structures corrosion, 596
Wet line, 199
Wet pressing, 69-70
    extended nip press, 70
Wet strength additives, 168-170
White water systems, 85-96, 199
    closing the system, 92-94, 598-601
    composition, 599
        chlorides, 600
        sulfates, 600
        sulfites, 600
        thiosulfates, 600
    conductivity, 600-601
    corrosion of closed system, 598-601
    long circulation, 89-92
        broke pulp operation, 89-91
        material losses, 92
        saveall, importance in, 91
    pH, 599-600
    short circulation, 86-88
        air entrainment, 89
        consistency variations, 87
        process control points, 95-96
    temperature, 599
Winders, *see also* Winding equipment
    drives, 397-398, 525-547
    duplex, 386-388, 430
        drives, 388
    carbide coated drums, 361, 380-381
        combining tissue winder, 388-390
        drives, 390
    core support/roll support station, 388
    electric power requirements, 544-546
    safety, 674-678
    three-drum, 358
    two drum station, 379-386
        drives, 385-386
    vibration, 426-429
    windup drives, 546
Winding, 343-435
    combining winding, 378
    converting winding, 378
    core problems, 363-367
    core shaft winding, 382-384
    definitions, 343-344
    drum diam vs paper grade, 357-359
    drum size vs nips, 359-360
    floating webs, 395
    glossary, 431-435
    machine winding, 377
    nip mechanics, 353-361, 367
        effect on web profile, 360-361
        J-line, 355-356, 364
    reclaim or salvage winding, 377, 430
    reeling, 377
    reprocessing winding, 377-378, 430
    rider rolls, 358, 384-385, 387-388
    roll characteristics, 368
    roll defects, 350-353, 394
    roll grooving, 396-397
    roll structure, 349-353
    roll structure control, 378-390
    shaftless winding, 382
    tension control, 390-394
    traction control, 394-398
    theory, 344-346

torque, 361-362, 367
unwind control, 346-349
unwinding functions, 423-426
variables, tension, nip and torque, 353
web control, 347-348, 390-406
web separation, 349
web spreading, 398-406
Winding equipment definitions, 372-376, *see also* Winders
center winder, 376
combining (tissue) winder, 376
converting winder, 376
duplex winder, 376
reel, 372-373
re-reeler, 373
two-drum winder, 373-376
Wire agitation, 209-211
Wire changes safety, 660
Wire corrosion, 596
Wire drive ability, power requirements, 537-538
WIT-WOT, in winding, 351-353, 355-357
WRV, in fibers, 31-32
Wood
aspen, 33
density, 49
Douglas-fir, 42
eucalyptus, 33
hardwoods, 34, 38, 42-43
hardwoods, low density, 33
hemlock, 33, 42
Norscan spruce, 33
pine, 33
Populus, 33
pulpwood, 41-43
softwoods, 33, 42
southern pine,

Y

Yankee dryers, 323-342
bearings, 327-328
condensate removal, 326-327, 336
corrosion wear, 334-335
deformation analysis, 332-334
design and construction, 325-330
design stresses, 328-330
lubrication, 327-328
heat transfer coefficients, 336-338
mechanism of drying, 335-338
rimming, 326
safety, 330-331
shell metallurgy, 331-332
steam systems, 326-327
stress equations, 329
thermal stress equations, 329
tissue moisture profile control, 499-500
vibration, 630
Young's Modulus, 64-66

Z

ZDT, edge compressive strength, 78
Zeta potential, 144, 160
on-line measurement of, 180
Zero-span tensile (paper), 9-10

SI CONVERSION UNITS

Quantity or Test	Value In Trade or Customary Unit	x	Conversion Factor	=	Value in SI Unit	Symbol
Area	square inches		6.45		square centimeters	cm^2
	square feet		0.0929		square meters	
	square yards		0.836		square meters	
	acres		0.405		hectares	ha
Basis Weight* or substance (500-sheet ream) or Grammage* when expressed in g/m^2	17 × 22		3.760		grams per square meter	g/m^2
	24 × 36		1.627		grams per square meter	g/m^2
	25 × 38		1.480		grams per square meter	g/m^2
	25 × 40		1.406		grams per square meter	g/m^2
	pounds per 1000 square feet (Paperboard)		4.882		grams per square meter	g/m^2
Breaking Length	meters		0.001		kilometers	km
Burst Index	$\frac{g/m^2}{g/m^2}$		0.0981		$\frac{g/m^2}{g/m^2}$	
Bursting Strength	pounds per square inch		6.89		kilopascals	kPa
Caliper	mils		0.0254		millimeters	mm
Concora Crush	pounds		4.45		newtons	N
Edge Crush	pounds per inch		0.175		kilonewtons per meter	kN/m
Energy	British thermal units (Btu)		1005		joules	J
Flat Crush	pounds per square inch		6.89		kilopascals	kPa
Force	kilograms		9.81		newtons	N
	pounds		4.45		newtons	N
Length	angstroms		0.1		nanometers	nm
	microns		1		micrometers	μm
	mils		0.0254		millimeters	mm
	feet		0.305		meters	m
Mass	tons (2000 lb)		0.907		metric tons	t
	pounds		0.454		kilograms	kg
	ounces (avd p)		28.3		grams	g
Mass per Unit Volume	ounces per gallon		7.49		kilograms per cubic meter	kg/m^3
	pounds per cubic food		1.60		kilograms per cubic meter	kg/m^3
Puncture Resistance	foot pounds		1.36		joules	J
Ring Crush	pounds (for a 6-in. length)		0.0292		kilonewtons per meter	kN/m
Stillness (Taber)	gram centimeters (Taber Units)		0.0981		millinewton meters	mN·m
Tear Strength	grams		9.81		millinewtons	mN
Tensile Breaking Load	pounds per inch		0.175		kilonewtons per meter	kN/m
	kilograms per 15 millimeters		0.654		kilonewtons per meter	kN/m
Volume, Fluid	ounces (US fluid)		29.6		milliliters	mL
	gallons		3.79		liters	L
Volume, Solid	cubic inches		16.4		cubic centimeters	cm^3
	cubic feet		0.0283		cubic meters	m^3
	cubic yards		0.765		cubic meters	m^3

*See TAPPI Technical Information Sheet 0800-01.

Production Editor: Catharine Findlay
Set in type by Southam Business Communications Inc.
Printed by Bowne de Montréal, Inc., Saint-Laurent, QC, Canada